Fundamental Physical Constants

Quantity	**Symbol**	**Value**	**Comments**
Speed of light	c	2.99792458×10^{8} m-s^{-1}	
Charge of an electron	q_e	$-1.602177 \times 10^{-19}$ C	
Gravitational constant	G	6.67259×10^{-11} N-m^{2}-kg^{-2}	
Gravitational acceleration	g	9.78049 m-s^{-2}	Sea level, at equator
Planck's constant	h	$6.6260755 \times 10^{-34}$ J-s	
Boltzmann's constant	k_{B}	1.38066×10^{-23} J-(K)$^{-1}$	
Avogadro's number	N_{A}	6.022×10^{23} (mole)$^{-1}$	Molecules per mole
Permittivity of free space	ϵ_0	8.854×10^{-12} F-m^{-1}	$\sim(36\pi)^{-1} \times 10^{-9}$ F-m^{-1}
Permeability of free space	μ_0	$4\pi \times 10^{-7}$ H-m^{-1}	
Rest mass of an electron	m_e	9.10939×10^{-31} kg	
Rest mass of a proton	m_p	1.67492×10^{-27} kg	$(m_p/m_e) \simeq 1838$
Rest mass of a neutron	m_n	1.67292×10^{-27} kg	
Bohr radius	a	0.529177×10^{-10} m	Radius of hydrogen atom
1 electron-volt	eV	1.60219×10^{-19} J	
Frequency of 1-eV photon	f	2.41796×10^{14} Hz	

Powers of Ten

Power	**Symbol**	**Prefix**
10^{18}	E	Exa
10^{15}	P	Peta
10^{12}	T	Tera
10^{9}	G	Giga
10^{5}	M	Mega
10^{3}	k	kilo
10^{-3}	m	milli
10^{-6}	μ	micro
10^{-9}	n	nano
10^{-12}	p	pico
10^{-15}	f	femto
10^{-18}	a	atto

ENGINEERING Electromagnetics

UMRAN S. INAN
Stanford University

AZIZ S. INAN
University of Portland

An imprint of Addison Wesley Longman, Inc.

Menlo Park, California • Reading, Massachusetts • Harlow, England
Berkeley, California • Don Mills, Ontario • Sydney • Bonn • Amsterdam • Tokyo • Mexico City

Acquisitions Editor: Paul Becker
Assistant Editor: Anna Eberhard Friedlander
Production Manager: Pattie Myers
Senior Production Coordinator: Rhonda Zachmeyer
Art and Design Supervisor: Kevin Berry
Composition: Publication Services, Inc.
Cover Design: Yvo Riezebos Design
Cover Photos: Main photo: Umran S. Inan, STAR Laboratory; Insets: (top) Lockheed Martin Missiles and Space Advanced Technology Center; (middle) Image #1264 © 1998 PhotoDisc; (bottom) Image #4107 © 1998 PhotoDisc
Text Design: Publication Services, Inc.
Text Printer and Binder: Courier Westford
Cover Printer: Phoenix Color Corporation

Library of Congress Cataloging-in-Publication Data

Inan, Umran S.
Engineering electromagnetics / by Umran S. Inan, Aziz S. Inan.
p. cm.
Includes bibliographical references and index.
ISBN (invalid) 0-8053-4423-3
1. Electromagnetic theory. I. Inan, Aziz S. II. Title.
QC670.I5 1998
530.14′1–dc21 98-19864
CIP

Instructional Material Disclaimer
The programs presented in this book have been included for their instructional value. They have been tested with care but are not guaranteed for any particular purpose. Neither the publisher or the authors offer any warranties or representations, nor do they accept any liabilities with respect to the programs.

The full complement of supplemental teaching materials is available to qualified instructors.

0–8053–4423–3

1 2 3 4 5 6 7 8 9 10—CRW—02 01 00 99 98

Addison Wesley Longman, Inc.
2725 Sand Hill Road
Menlo Park, California 94025

To our parents, wives, and children

Contents

Preface

This book provides engineering students with a solid grasp of electromagnetic fundamentals by emphasizing physical understanding and practical applications. The topical organization of the text starts with an initial exposure to transmission lines and transients on high-speed distributed circuits, naturally bridging electrical circuits and electromagnetics.

Engineering Electromagnetics is designed for upper-division (3rd and 4th year) college and university engineering students, for those who wish to learn the subject through self-study, and for practicing engineers who need an up-to-date reference text. The student using this text is assumed to have completed typical lower-division courses in physics and mathematics as well as a first course on electrical engineering circuits.

KEY FEATURES

The key features of this textbook are

- Modern chapter organization, covering transmission lines before the development of fundamental laws
- Emphasis on physical understanding
- Detailed examples, selected application examples, and abundant illustrations
- Numerous end-of-chapter problems, emphasizing selected practical applications
- Historical notes and abbreviated biographies of the great scientific pioneers
- Emphasis on clarity without sacrificing rigor and completeness
- Hundreds of footnotes providing physical insight, leads for further reading, and discussion of subtle and interesting concepts and applications

Modern Chapter Organization

We use a physical and intuitive approach so that this engineering textbook can be read by students with enthusiasm and with interest. We provide continuity with circuit theory by first covering transmission lines—an appropriate step, in view of the newly emerging importance of transmission line concepts, not only in microwave and millimeter-wave applications but also in high-speed digital electronics, microelectronics, integrated circuits, packaging, and interconnect applications. We then cover the fundamental subject material in a logical order, following the historical development of human understanding of electromagnetic phenomena. We base the fundamental laws on experimental observations and on physical grounds, including brief discussions of the precision of the fundamental experiments, so that the physical laws are easily understood and accepted. Once the complete set of fundamental laws is established, we then discuss their most important manifestation: the propagation, reflection, transmission, and guiding of electromagnetic waves.

Emphasis on Physical Understanding

Future engineers and scientists need a clear understanding and a firm grasp of the basic principles so that they can understand, formulate, and interpret the results of complex practical problems. Engineers and scientists nowadays do not and should not spend time working out formulas and obtaining numerical results by substitution. Most of the number crunching and formula manipulations are left to computers and packaged application and design programs, so a solid grasp of fundamentals is now more essential than ever before. In this text we maintain a constant link with established as well as new and emerging applications (so that the reader's interest remains perked up), while at the same time emphasizing fundamental physical insight and solid understanding of basic principles. We strive to empower the reader with more than just a working knowledge of a dry set of vector relations and formulas stated axiomatically. We supplement rigorous analyses with extensive discussions of the experimental bases of the laws, of the microscopic versus macroscopic concepts of electromagnetic fields and their behavior in material media, and of the physical nature of the electromagnetic fields and waves, often from alternative points of view. Description of the electrical and magnetic properties of material media at a sufficiently simple, yet accurate manner at the introductory electromagnetics level has always been a challenge, yet a solid understanding of this subject is now more essential than ever, especially in view of many applications that exploit these properties of materials. To this end we attempt to distill the essentials of physically-based treatments available in physics texts, providing quantitative physical insight into microscopic behavior of materials and the representation of this behavior in terms of macroscopic parameters. Difficult three-dimensional vector differential and integral concepts are discussed when they are encountered—again, with the emphasis being on physical insight.

Detailed Examples and Abundant Illustrations

We present the material in a clear and simple yet precise and accurate manner, with interesting examples illustrating each new concept. Many examples emphasize selected applications of electromagnetics. A total of 180 illustrative examples are detailed over eight chapters, with four of the chapters having more than 30 examples each. Each example is presented with an abbreviated topical title, a clear problem statement, and a detailed solution. In recognition of the importance of visualization in the reader's understanding, especially in view of the three-dimensional nature of electromagnetic fields, over 400 diagrams, graphs, and illustrations appear throughout the book.

Numerous End-of-Chapter Problems

Each chapter is concluded with a variety of homework problems to allow the students to test their understanding of the material covered in the chapter, with a total of over 300 exercise problems spread over seven chapters. The topical content of each problem is clearly identified in an abbreviated title (e.g., "Digital IC interconnects" or "Inductance of a toroid"). Many problems explore interesting applications, and most chapters include several practical "real-life" problems to motivate students.

Historical Notes and Abbreviated Biographies

The history of the development of electromagnetics is laden with outstanding examples of pioneering scientists and development of scientific thought. Throughout our text, we maintain a constant link with the pioneering giants and their work, to bring about a better appreciation of the complex physical concepts as well as to keep the reader interested. We provide abbreviated biographies of the pioneers, emphasizing their scientific work in electromagnetics as well as in other fields such as optics, heat, chemistry, and astronomy. We illustrate the apparatus used by discoverers such as Coulomb and Faraday so that the reader can have a feel for how one would carry out such an experiment.

Emphasis on Clarity without Sacrificing Rigor and Completeness

This textbook presents the material at a simple enough level to be readable by undergraduate students, but it is also rigorous in providing references and footnotes for in-depth analyses of selected concepts and applications. We provide the students with a taste of rigor and completeness at the level of classical reference texts—combined with a level of physical insight that was so well exemplified in some very old texts—while still maintaining the necessary level of organization and presentation clarity required for a modern textbook. We also provide not just a superficial but a rigorous and in-depth exposure to a diverse range of applications of electromagnetics, in the body of the text, in examples, and in end-of-chapter problems.

Hundreds of Footnotes

In view of its fundamental physical nature and its broad generality, electromagnetics lends itself particularly well to alternative ways of thinking about physical and engineering problems and also is particularly rich in terms of available scientific literature and many outstanding textbooks. Almost every new concept encountered can be thought of in different ways, and the interested reader can explore its implications further. We encourage such scholarly pursuit of enhanced knowledge and understanding by providing many footnotes in each chapter that provide further comments, qualifications of statements made in the text, and references for in-depth analyses of selected concepts and applications. A total of 450 footnotes are spread over eight chapters. These footnotes do not interrupt the flow of ideas and the development of the main topics, but they provide an unusual degree of completeness for a textbook at this level, with interesting and sometimes thought-provoking content to make the subject more appealing.

ELECTROMAGNETICS IN ENGINEERING

The particular organization of this textbook, as well as its experimentally and physically based philosophy, are motivated by our view of the current status of electromagnetics in engineering curricula. Understanding electromagnetics and appreciating its applications require a generally higher level of abstraction than most other topics encountered by electrical engineering students. Beginning electrical engineers learn to deal with voltages and currents, which appear across or flow through circuit elements or paths. The relationships between these voltages and currents are determined by the characteristics of the circuit elements and by Kirchhoff's current and voltage laws. Voltages and currents in lumped electrical circuits are scalar quantities that vary only as a function of time, and are readily measurable, and the students can relate to them via their previous experiences. The relationships between the quantities (i.e., Kirchhoff's laws) are relatively simple algebraic or ordinary differential equations. On the contrary, electric and magnetic fields are *three-dimensional* and *vector* quantities that in general *vary in both space and time* and are related to one another through relatively complicated vector *partial differential* or vector *integral equations.* Even if the physical nature of electric and magnetic fields were understood, visualization of the fields and their effects on one another and on matter requires a generally high level of abstract thinking.

Most students are exposed to electromagnetics first at the freshman physics level, where electricity and magnetism are discussed in terms of their experimental bases by citing physical laws (e.g., Coulomb's law) and applying them to relatively simple and symmetrical configurations where the field quantities behave as scalars, and the governing equations are reduced to either algebraic equations of first-order integral or differential relationships. Freshman physics provides the students with their first experiences with fields and waves as well as and some of their measurable manifestations, such as electric and magnetic forces, electromagnetic induction (Faraday's law), and refraction of light by prisms.

The first course in electromagnetics, which most students take after having had vector calculus, aims at the development and understanding of Maxwell's equations, requiring the utilization of the full three-dimensional vector form of the fields and their relationships. It is this very step that makes the subject of electromagnetics appear insurmountable to many students and turns off their interest, especially when coupled with a lack of presentation and discussion of important applications and the physical (and experimental) bases of the fundamental laws of physics. Many authors and teachers have attempted to overcome this difficulty by a variety of topical organizations, ranging from those that start with Maxwell's equations as axioms to those that first develop them from their experimental basis.

Since electromagnetics is a mature basic science, and the topics covered in introductory texts are well established, the various texts primarily differ in their organization as well as range and depth of coverage. Teaching electromagnetics was the subject of a special issue of *IEEE Transactions on Education* [vol. 33, February, 1990]. Many of the challenges and opportunities that lie ahead in this connection were summarized well in an invited article by J. R. Whinnery.[1] Challenges include (1) the need to return to fundamentals (rather than relying on derived concepts), especially in view of the many emerging new applications that exploit unusual properties of materials and that rely on unconventional device concepts,[2] submillimeter transmission lines,[3] and optoelectronic waveguides,[4] and (2) the need to maintain student interest in spite of the decreasing popularity of the subject of electromagnetics and its reputation as a difficult and abstract subject.[5] Opportunities are abundant, especially as engineers working in electronics and computer science discover that as devices get smaller and faster, circuit theory is insufficient in describing system performance or facilitating design. It is now clear, for example, that transmission line concepts are not only important in microwave and millimeter-wave applications but also necessary in high-speed digital electronics, microelectronics, integrated circuits, interconnects,[6] and packaging applications.[7] The need for a basic understanding of electromagnetic waves and their guided propagation is underscored by the explosive expansion of the use of optical fibers, the use of extremely high data rates, ranging to 10 Gbits/s,[8] and the emerging use of high-performance, high-density cables for communication within systems that will soon be required to carry digital signals at Gb/s rates over distances of a few meters.[9] In addition, issues of electromagnetic

---

[1] J. R. Whinnery, The teaching of electromagnetics, *IEEE Trans. on Education,* 33(1), pp. 3–7, February 1990.

[2] D. Goldhaber-Gordon, M. S. Montemerlo, J. C. Love, G. J. Opiteck, and J. C. Ellenbogen, Overview of nanoelectronic devices, *Proc. IEEE,* 85(4), pp. 521–540, April 1997.

[3] L. P. B. Katehi, Novel transmission lines for the submillimeter region, *Proc. IEEE,* 80(11), pp. 1771–1787, November 1992.

[4] R. A. Soref, Silicon-based optoelectronics, *Proc. IEEE,* 81(12), December, 1993.

[5] M. N. O. Sadiku, Problems faced by undergraduates studying electromagnetics, *IEEE Trans. Education,* 29(1), pp. 31–32, February, 1986.

[6] A. Deutsch, Electrical characteristics of interconnections for high-performance systems, *Proc. IEEE,* 86(2), pp. 315–355, February 1998.

[7] H. B. Bakoglu, *Circuits, Interconnections, and Packaging for VLSI,* Addison Wesley, 1990.

[8] R. Heidelmann, B. Wedding, and G. Veith, 10-Gb/s transmission and beyond, *Proc. IEEE,* 81(11), pp. 1558–1567, November 1993.

[9] H. Falk, Prolog to electrical characteristics of interconnections for high-performance systems, *Proc. IEEE,* 86(2), pp. 313–314, February 1998.

interference (EMI) and electromagnetic compatibility (EMC) are beginning to limit the performance of system-, board-, and chip-level designs, and electrostatic discharge phenomena have significant impacts on the design and performance of integrated circuits.[10] Other important applications that require better understanding of electromagnetic fields are emerging in biology[11] and medicine.[12]

In organizing the material for our text, we benefited greatly from a review of the electromagnetic curriculum at Stanford University that one of us conducted during the spring quarter of 1990. A detailed analysis was made of both undergraduate and graduate offerings, both at Stanford and selected other schools. Inquiries were also made with selected industry, especially in the Aerospace sector. Based on the responses we received from many of our colleagues, and based on our experience with the teaching of the two-quarter sequence at Stanford, it was decided that an emphasis on fundamentals and physical insight and a traditional order of topics would be most appropriate. It was also determined that transmission line theory and applications can naturally be studied before fields and waves, so as to provide a smooth transition from the previous circuits and systems experiences of the typical electrical engineering students and also to emphasize the newly emerging importance of these concepts in high-speed electronics and computer applications.

RECOMMENDED COURSE CONTENT

This book is specifically designed for a one-term first course in electromagnetics, nowadays typically the only required fields and waves course in most electrical engineering curricula. The recommended course content for a regular three-unit one-semester course (42 contact hours) is provided in Table 1. The sections marked under "Cover" are recommended for complete coverage, including illustrative examples, whereas those marked "Skim" are recommended to be covered lightly, although the material provided is complete in case individual students want to go into more detail. The sections marked with a superscript asterisk are intended to provide flexibility to the individual instructor. For example, one may want to cover magnetic materials (Sec. 6.8) and skim magnetic forces and torques (Sec 6.10), or vice versa. Similarly, one may want to cover guided waves (Sec. 8.3) but skim reflection from multiple or lossy interfaces (Sec. 8.2.3 and 8.2.4), instead of the other way around.

Table 1 also shows a recommended course content for a 4-unit one-quarter course (32 contact hours) identical to the course titled "Engineering Electromag-

[10] J. E. Vinson and J. J. Liou, Electrostatic discharge in semiconductor devices: an overview, *Proc. IEEE*, 86(2), pp. 399–418, February 1998.

[11] J. Raloff, Electromagnetic fields exert effects on and through hormones, *Science News,* 153, pp. 29–31, January 10, 1998; J. Raloff, Electromagnetic fields may trigger enzymes, *Science News,* 153, pp. 29–31, February 21, 1998.

[12] R. L. Magin, A. G. Webb, and T. L. Peck, Miniature magnetic resonace machines, *IEEE Spectrum,* pp. 51–61, October 1997.

TABLE 1 Suggested course content

	Quarter Course (32 Hours)			Semester Course (42 Hours)		
Chapter	**Cover**	**Skim**	**Skip**	**Cover**	**Skim**	**Skip**
1	All			All		
2	2.1–2.5.1 2.6.1	2.6.2, 2.7	2.5.2	2.1–2.5.1 2.6.1, 2.7	2.6.2 2.5.2*	
3	3.1–3.5	3.6, 3.7	3.8, 3.9	3.1–3.7	3.8*	3.9*
4	4.1–4.9 4.11	4.10, 4.12	4.13	4.1–4.9, 4.10* 4.11–4.12*	4.13*	
5	5.1–5.6	5.7,5.8		5.1–5.8		
6	6.1–6.7 6.9	6.8	6.10	6.1–6.7 6.8*–6.9	6.10*	
7	7.1–7.2 7.4–7.5	7.3		7.1–7.2, 7.3* 7.4–7.5		
8	8.1.1–8.1.2	8.1.3	8.1.4, 8.2–8.3	8.1, 8.2.1, 8.2.2 8.2.3*, 8.2.4*	8.3*	

netics" (required for BSEE) that one of us has been teaching at Stanford for the past seven years. This topical coverage provides the students with (1) a working knowledge of transmission lines, (2) a solid, physically based background and a firm understanding of Maxwell's equations and their experimental bases, and (3) a first exposure to the most important manifestations of Maxwell's equations: electromagnetic waves. At Stanford, this required course is followed by a course titled "Electromagnetic Waves," which serves as the entry course for students opting for the fields and waves specialization.

ACKNOWLEDGMENTS

We gratefully acknowledge those who have made significant contributions to the successful completion of this text. We thank Professor J. W. Goodman of Stanford, for his generous support of textbook writing by faculty throughout his term as deparment chair; Professor G. Kino and Dr. T. Bell of Stanford, for course-testing a preliminary version of the manuscript, and Professor R. N. Bracewell, who inspired our use of the abbreviated biographies of great scientists. We thank many students at both Stanford and the University of Portland who have identified errors and suggested clarifications, and Mrs. Jun-Hua Wang for typing parts of the manuscript and drawing some of the illustrations. We owe special thanks to our reviewers for their valuable comments and suggestions, including J. Bredow of University of Texas—Arlington; S. Castillo of New Mexico State University; R. J. Coleman of University of North Carolina—Charlotte; A. Dienes of University of California—Davis; J. Dunn of University of Colorado; D. S. Elliott of Purdue University; R. A. Kinney of Louisiana State University; L. Rosenthal of Fairleigh Dickinson University; E. Schamiloglu of

University of New Mexico; T. Shumpert of Auburn University; D. Stephenson of Iowa State University; E. Thomson of University of Florida; J. Volakis of University of Michigan; and A. Weisshaar of Oregon State University. We greatly appreciate the efforts of our developmental editor Judy Ziajka and the Addison Wesley Longman staff including Anna Eberhard Friedlander, Pattie Myers, Kevin Berry, and especially our editor Paul Becker, whose dedication and support was crucially important in completing this project.

As teachers with a good deal of experience, we firmly believe that practice is the key to learning, and that homeworks and exams are all instruments of teaching, although they may often not be regarded as such by the students at the time. In our own courses, we take pride in providing the students with detailed solutions of homework and exam problems, rather than cryptic and abbreviated answers. To aid the instructors who choose to use this text, we have thus taken it upon ourselves to prepare a thorough and well-laid-out solutions manual, describing the solution of *every* end-of-chapter problem, in the same step-by-step detailed manner as our illustrative examples withing the chapters. The solution for each end-of-chapter problem has been typeset by the authors themselves, with special attention to pedagogical detail. This solutions manual is available to instructors upon request from Addison Wesley Longman.

As authors of this book, we are looking forward to interacting with its users, both students and instructors, to collect ad respond to their comments, questions, and corrections. We can most easily be reached by email at inan@nova.stanford.edu (url:http://nova.stanford.edu/ṽlf) and at ainan@up.edu. Supplemental information about the book and errata will be available at http://www2.awl.com/cseng/titles/0-8053-4423-3.

We dedicate this book to our parents, Mustafa and Hayriye Inan, for their dedication to our education; to our wives, Elif and Belgin, for their persistent support and understanding as this project expanded well beyond our initial expectations and consumed most of our available time for too many years; and to our children, Ayse, Ali, Baris, and Cem, for the joy they bring to our lives.

Umran S. Inan
Aziz S. Inan

CHAPTER 1

Introduction

1.1 Lumped versus Distributed Electrical Circuits
1.2 Electromagnetic Components
1.3 Maxwell's Equations and Electromagnetic Waves
1.4 Summary

This book is an introduction to the fundamental principles and applications of *electromagnetics.* The subject of electromagnetics encompasses *electricity, magnetism,* and *electrodynamics,* including *all* electric and magnetic phenomena and their practical applications. A branch of electromagnetics, dealing with electric charges at rest (static electricity) named *electrostatics,* provides a framework within which we can understand the simple fact[1] that a piece of amber, when rubbed, attracts to itself other small objects. Another branch dealing with static magnetism, namely *magnetostatics,* is based on the facts that some mineral ores (e.g., lodestone) attract iron[2] and that current-carrying wires produce magnetic fields.[3] The branch of electromagnetics known as *electrodynamics* deals with the time variations of electricity and magnetism and is based on the fact that magnetic fields that change with time produce electric fields.[4]

Electromagnetic phenomena are governed by a compact set of principles known as Maxwell's equations,[5] the most fundamental consequence of which is that electromagnetic energy can propagate, or travel from one point to another, as *waves.* The propagation of electromagnetic waves results in the phenomenon of *delayed action at a distance;* in other words, electromagnetic fields can exert forces, and hence can do work, at distances far away from the places where they are generated and at later times. Electromagnetic radiation is thus a means of transporting energy and

[1]First discovered by the Greek mathematician, astronomer, and philosopher Thales of Miletus [640–548 B.C.].

[2]First noted by the Roman poet and philosopher Lucretius [99?–55? B.C.], in his philosophical and scientific poem titled *De rerum natura (On the Nature of Things).*

[3]First noted by Danish physicist H. C. Oersted in 1819.

[4]First noted by British scientist M. Faraday in 1831.

[5]J. C. Maxwell, *A Treatise in Electricity and Magnetism,* Clarendon Press, Oxford, 1892, Vol. 2, pp. 247–262.

momentum from one set of electric charges and currents (at the source end) to another (those at the receiving end). Since whatever can carry energy can also convey information, *electromagnetic waves* thus provide the means of transmitting energy and information at a distance.

The concept of waves is one of the great unifying ideas of physics and engineering.[6] Our physical environment is full of waves of all kinds: seismic waves, waves on oceans and ponds, sound waves, heat waves, and even traffic waves. The idea of delayed action as manifested in wave phenomena is familiar to us when we hear a sound and its echo or when we create a disturbance[7] in a pool of water and observe that waves reach the edge of the pool after a noticeable time. We also appreciate that it might take minutes or hours for heat to penetrate into objects; that the thunderclap is delayed with respect to the lightning flash by many seconds; and that when we are lined up in front of a traffic light, it often takes a long time for us to be able to move after the light turns green. Light, or electromagnetic waves, travel so fast that their delayed action is not perceptible to our senses in our everyday experiences. On the other hand, in astronomy and astrophysics we deal with vast distances; light waves from a supernova explosion may arrive at earth millions of years after the brightness that created them has been extinguished.

At a qualitative level, we recognize a wave as some pattern in space that appears to move in time. Wave motion does not necessarily involve repetitive undulations of a physical quantity (e.g., the height of the water surface for water waves in a lake). If a disturbance that occurs at a particular point in space at a particular time is related in a definite manner to what occurs at distant points in later times, then there is said to be wave motion. To express this mathematically, let z be distance, t be time, and v be a fixed positive parameter. Consider any arbitrary function $f(\zeta)$ of the argument $\zeta = (z - vt)$. Figure 1.1a shows a sketch of $f(\zeta)$, identifying a point P on the curve that corresponds to $f(\zeta_1)$. Note that this peak value of the function (i.e., point P) is obtained whenever its argument equals ζ_1. Shown in Figure 1.1b is $f(z - vt_1)$, where the point P is at $z_1 = \zeta_1 + vt_1$. A plot of the function at $t = t_2$ (i.e., $f(z - vt_2)$) is also shown in Figure 1.1 b, where we see that the point P has now moved to the right, to the new location $z_2 = \zeta_1 + vt_2$. It is clear that the entire curve $f(\zeta)$, which comprises the function $f(z - vt)$, moves in the z direction as time elapses. The velocity of this motion can be determined by observing a fixed point on the curve—for example, point P. Since this point is defined by the argument of the function being equal to ζ_1, we can set $(z - vt) = \zeta_1$, which upon differentiation gives $dz/dt = v$, since ζ_1 is a constant. It thus appears that the speed with which point P moves to the right is v, which is identified as the velocity of the wave motion. Note that the function

---

[6]For an excellent qualitative discussion, see J. R. Pierce, *Almost All About Waves,* MIT Press, Cambridge, Massachusetts, 1974. For more extensive treatment of waves of all kinds, see K. U. Ingard, *Fundamentals of Waves and Oscillations,* Cambridge University Press, 1990.

[7]On the scale of a pond, we can simply think of dropping a stone; on a larger scale, earthquakes in oceans produce giant *tsunami* waves. A 9-meter-high tsunami produced by the 1964 Alaskan earthquake hit the Hawaiian islands (at a distance of 2000 km) about 5 hours later, causing more than 25 million dollars of damage.

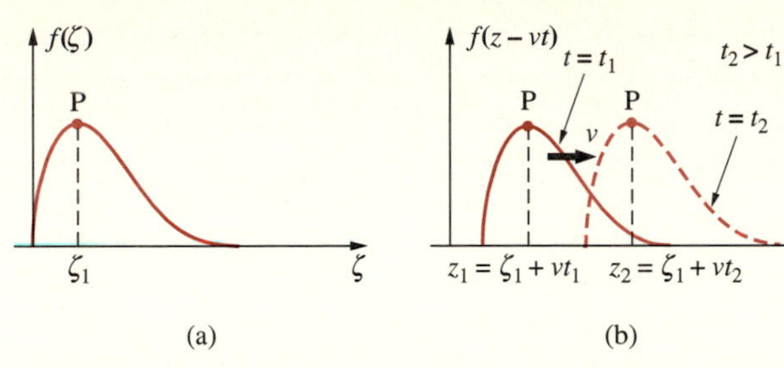

FIGURE 1.1. **Example of a wave.** (a) An arbitrary function $f(\zeta)$. (b) The function $f(z - vt)$, where v is a positive constant, plotted versus z at $t = t_1$ and $t = t_2$. The wave nature is evident as the pattern in space at time t_1 is shifted to other values of z at a later time t_2.

$f(\cdot)$ could represent any physically observable entity; it may be a scalar,[8] such as voltage, or it may be a vector, such as the velocity of an object in motion. If $f(\cdot)$ is a vector, each of its components must be a function of $(z - vt)$ for it to be a propagating wave. Quantities varying as functions of $(z - vt)$ constitute natural solutions of the fundamental equations of electromagnetics and distributed electrical circuits. Chapters 2, 3, and 8 of this textbook is devoted to the study of voltage, current, and electromagnetic waves that vary in space and time as functions of $(z - vt)$.

Most waves travel through substances, whether they be earth, water, air, steel, or quartz, without actually carrying the substance bodily with them.[9] Like moving objects, moving waves carry energy, albeit by different amounts depending on the nature of the waves and the medium they propagate in. Electromagnetic waves have the special property that they can also propagate in a vacuum, without any matter present. However, the propagation of electromagnetic waves is nevertheless affected by the presence of matter, and this property often allows us to confine or guide waves and utilize them more efficiently. Electromagnetic engineering problems generally involve the design and use of materials that can generate, transmit, guide, store, distribute, scatter, absorb, and detect electromagnetic waves and energy.

The 20th century has witnessed rapid advances in electrical engineering, which have largely come about by our ability to predict the performance of sophisticated electrical circuits accurately. Central to this tremendous progress is our ability to utilize the simple but powerful tool called *circuit theory.* Classical circuit theory

---

[8]A scalar is a quantity that is completely specified by its value, such as the number of coins in your pocket, the number of people or the density of air in a room, pressure, or temperature. Other physical quantities have direction: for example, velocity, momentum, force, or displacement. Specification of a vector quantity requires both a magnitude and direction. A brief review of basic principles of vector analysis is provided in Appendix A.

[9]Leonardo da Vinci [1452–1519] wrote of waves, "The impetus is much quicker than the water, for it often happens that the wave flees the place of its creation, while the water does not; like the waves made in a field of grain by the wind, where we see the waves running across the field while the grains remain in place" [J. R. Pierce, *Almost All About Waves,* MIT Press, 1974].

considers a voltage or current source applied to an electrical circuit consisting of series and/or parallel connection of simple *lumped* (see Section 1.1) circuit elements, such as resistances, capacitances, inductances and dependent sources, which may be idealized models of more complex physical devices. The behavior of circuits is described by ordinary differential equations that are derived on the basis of Kirchhoff's voltage and current laws. Circuit theory is a simplified approximation to the more exact electromagnetic theory.[10] The classical theory of electricity and magnetism relies on a set of physical laws known as Maxwell's equations, which are based on experimental facts, and which govern all electromagnetic phenomena. Electromagnetic theory is inherently more complicated than circuit theory, primarily because of the larger numbers of variables involved. In general electromagnetic problems, most of the physical quantities that we deal with are vectors, whose values may depend on all three coordinates of space (i.e., x, y, and z in rectangular coordinates) and time (t). In classical circuit theory, on the other hand, voltages and currents are scalar quantities and are typically functions of only one variable, namely time. The theory of *distributed* circuits (see Section 1.1), or transmission lines, represents an intermediate level of complexity where, in many cases, we can continue to deal with scalar quantities, such as voltages and currents, that are now functions of two variables, namely a single spatial dimension and time. In this regime, we can continue to benefit from the relative simplicity of circuit theory, while treating problems for which the lumped circuit theory is not applicable.

In this text, and in view of the preceding discussion, we choose to study distributed circuits or transmission lines using a natural extension of circuit theory before we formally introduce the physical laws of electricity and magnetism. This approach presents the general fundamental concepts of waves and oscillations at the outset, which are expanded upon later when we study propagation of electromagnetic waves. In this way, the reader is provided an unhindered initial exposure to properties of waves such as frequency, phase velocity, wavelength, and characteristic impedance; to energy relations in oscillating systems and waves; and to concepts such as reflection and bandwidth, as well as to fundamental mathematical techniques necessary to describe waves and oscillations. All of these concepts subsequently extend to more complicated problems and applications where a full electromagnetic treatment becomes necessary. An initial exposure to transmission line analysis also enables us to address a wide range of increasingly important engineering applications that require the use of wave techniques but for which a full vector electromagnetic analysis may not be necessary.[11]

---

[10]Kirchhoff's voltage and current laws, which provide the basis for classical circuit theory, can be derived from the more general electromagnetics equations; see Chapter 4 of S. Ramo, J. R. Whinnery, and T. Van Duzer, *Fields and Waves in Communication Electronics,* 3rd ed., John Wiley & Sons, Inc., New York, 1994.

[11]Examples are on-chip and chip-to-chip interconnections in digital integrated circuits and many other computer engineering applications. See A. Deutsch, et al., When are transmission-line effects important for on-chip interconnections, *IEEE Trans. Mic. Th. MTT,* 45(10), pp. 1836–1846, October 1997.

In our coverage of transmission lines or distributed circuits, we assume that the reader is familiar with the elementary physics of electricity and magnetism (at the level of freshman physics) and with electrical circuits at the level of understanding Kirchhoff's voltage and current laws and terminal behavior (i.e., voltage-current relationships) of circuit elements such as inductors, capacitors, and resistors. A more complete discussion of the concepts of inductance, capacitance, and resistance is provided in later chapters using the concepts and principles of electromagnetic fields as we introduce the fundamental laws of electromagnetics.

1.1 LUMPED VERSUS DISTRIBUTED ELECTRICAL CIRCUITS

A typical electrical engineering student is familiar with circuits, which are described as *lumped, linear,* and *time-invariant* systems and which can be modeled by *ordinary, linear,* and *constant-coefficient* differential equations. The concepts of linearity and time invariance refer to the relationships between the inputs and outputs of the system. The concept of a lumped circuit refers to the assumption that the entire circuit (or system) is at a single point (or in one "lump"), so that the dimensions of the system components (e.g., individual resistors or capacitors) are not important. In other words, current and voltage do not vary with space across or between circuit elements, so that when a voltage or current is applied at one point in the circuit, currents and voltages of all other points in the circuit react instantaneously. Lumped circuits consist of interconnections of lumped elements. A circuit element is said to be lumped if the instantaneous current entering one of its terminals is equal to the instantaneous current throughout the element and leaving the other terminal. Typical lumped circuit elements are resistors, capacitors, and inductors. In a lumped circuit, the individual lumped circuit elements are connected to each other and to sources and loads within or outside the circuit by conducting paths of negligible electrical length. Figure 1.2a illustrates a lumped electrical circuit to which an input voltage of $V_{\text{in}} = V_0$ is applied at $t = 0$. Since the entire circuit is considered in one lump, the effect of the input excitation is instantaneously felt at all points in the circuit, and all currents and voltages (such as I_1, I_2, V_1, and the load current I_{L}) either attain new values or respond by starting to change at $t = 0$, in accordance with the natural response of the circuit to a step excitation as determined by the solution of its corresponding differential equation. Many powerful techniques of analysis, design, and computer-aided optimization of lumped circuits are available and widely used.

The behavior of lumped circuits is analogous to rigid-body dynamics. In mechanics, a rigid body is postulated to have a definite shape and mass, and it is assumed that the distance between any two points on the body does not change, so that its shape is not deformed by applied forces. Thus, an external force applied to a rigid body is assumed to be felt by all parts of the body simultaneously, without accounting for the finite time it would take for the effect of the force to travel elastically from one end of the body to another.

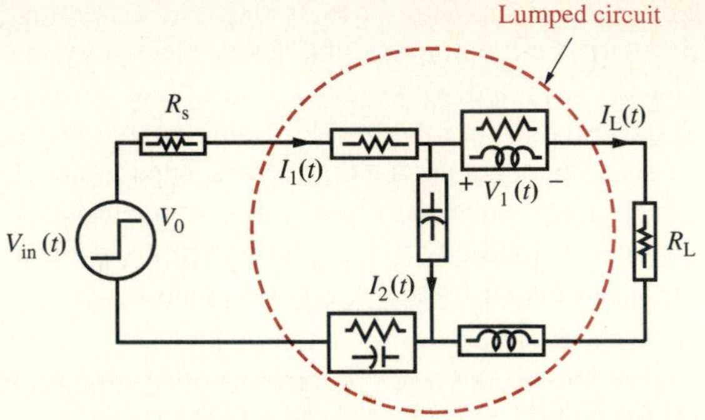

(a)

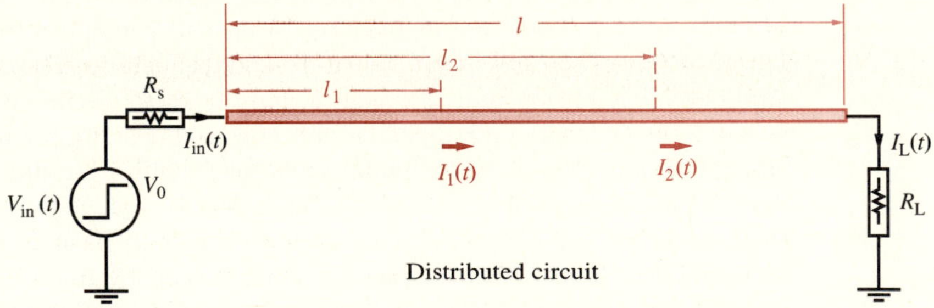

(b)

FIGURE 1.2. Lumped versus distributed electrical circuits. (a) When a step voltage $V_{\text{in}}(t)$ is applied to a lumped circuit, we assume that all currents and voltages start to change at $t = 0$, implicitly assuming that it takes zero travel time for the effect of the input to move from any point to any other point in the circuit. (b) In a distributed circuit, the nonzero travel time of the signal from one point to another cannot be neglected. For example, when a step voltage $V_{\text{in}}(t)$ is applied at one end of the circuit at $t = 0$, the load current at the other end does not start to change until $t = l/v$.

With the "lumped" assumption, one does not have to consider the travel time of the signal from one point to another. In reality, however, disturbances or signals caused by any applied energy travel from one point to another in a nonzero time. For electromagnetic signals, this travel time is determined by the speed of light,[12] $c \simeq 3 \times 10^8$ m-s^{-1} = 30 cm-(ns)$^{-1}$. In practical transmission systems, the speed of signal propagation is determined by the electrical and magnetic properties of the surrounding media and the geometrical configuration of the conductors and may in general be different from c, but it is nevertheless of the same order of magnitude as c. Circuits for which this nonzero travel time cannot be neglected are known as *distributed* circuits. An example of a distributed circuit is a long wire, as shown

[12]The more accurate empirical value of the speed of light is c = 299,792,458 m-s^{-1} [*CRC Handbook of Chemistry and Physics,* 76th ed., CRC Press, Inc., Boca Raton, Florida, 1995].

in Figure 1.2b. When an input voltage V_0 is applied at the input terminals of such a distributed circuit (i.e., between the wire and the electrical ground) at $t = 0$, the voltages and currents at all points of the wire cannot respond simultaneously to the applied excitation, because the energy corresponding to the applied voltage propagates down the wire with a finite velocity v. Thus, while the input current $I_{\text{in}}(t)$ may change from zero to I_0 at $t = 0$, the current $I_1(t)$ does not flow until $t = l_1/v$, $I_2(t)$ does not flow until $t = l_2/v$, and no load current $I_{\text{L}}(t)$ can flow until after $t = l/v$. Similarly, when a harmonically (i.e., sinusoidally) time-varying voltage is connected to such a line, the successive rises and falls of the source voltage propagate along the line with a finite velocity, so that the currents and voltages at other points on the line do not reach their maxima and minima at the same time as the input voltage.

In view of the fundamentally different behavior of lumped and distributed circuits as illustrated in Figure 1.2, it is important, in practice, to determine correctly whether a lumped treatment is sufficiently accurate or whether the circuit in hand has to be treated as a distributed circuit. In the following, we quantify on a heuristic basis the circumstances under which the travel time and/or the size of the circuit components or the length of the interconnects between them can be neglected. Note that in problems related to heat, diffusion, sound waves, water waves, traffic waves, and so on, the travel time is readily observable and almost always has to be accounted for. In the context of electromagnetics, on the other hand, we find a wide range of applications where lumped analysis is perfectly valid, is substantially simpler, and provides entirely satisfactory results. However, in an equally wide range of other applications we find that the lumped treatment is not sufficiently accurate and that one has to resort to field and wave techniques, which are generally more involved, both mathematically and conceptually. It is thus important, particularly in the context of electromagnetic applications, that we develop criteria by which we can determine the applicability of lumped circuit formulations. We provide below a heuristic discussion from three different but related points of view.

1.1.1 Rise Time versus Travel Time

It is apparent from the preceding discussion that we can consider the delay time (travel time) over the signal path as long or short, important or negligible only relative to or in comparison with some other quantity. In terms of positive step changes in an applied signal, we note that the input signal typically exhibits a nonzero *rise time* (usually measured as the time required for the signal to change from 10% to 90% of its final value), which may be denoted as t_r (Figure 1.3). We can then compare t_r to the one-way propagation time delay through the signal path (also called the one-way transit time or time of flight), $t_d = l/v$, where v is the velocity of propagation and l is the length of the signal path. For example, in practical design of interconnects between integrated circuit chips, one rule of thumb is that the signal path can be treated as a lumped element if $(t_r/t_d) > 6$, whereas lumped analysis is not appropriate for $(t_r/t_d) < 2.5$. Whether lumped analysis is appropriate for the in-between range of $2.5 < (t_r/t_d) < 6$ depends on the application in hand and the required accuracy.

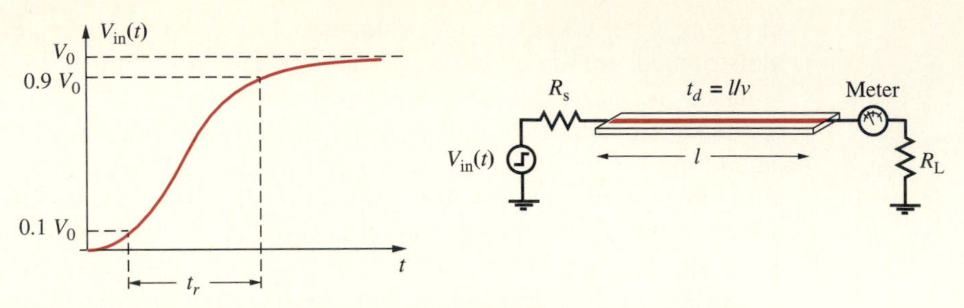

FIGURE 1.3. **Rise time versus one-way travel time.**

In terms of the particular application of high-speed integrated circuits, the on-chip rise times range from 0.5–2 ns for CMOS to 0.02–0.1 ns for GaAs technologies. The speed of signal propagation within the chips depends on the material properties; for example, it is $\sim 0.51c$ for SiO_2. Thus, for on-chip interconnections (typical $l \simeq 1$ cm), lumped circuit analysis breaks down (i.e., $t_r/t_d < 2.5$) for rise times less than ~ 0.165 ns. For printed circuit boards made of a commonly used glass epoxy material, the speed of propagation is $\sim 0.47c$, so that for a ~ 10 cm interconnect, lumped analysis is not appropriate for rise times less than ~ 1.8 ns. As clock speeds increase and rise times become accordingly shorter, distributed analyses will be required in a wider range of digital integrated circuit applications.

The importance of considering rise time versus travel time is underscored by advances in the generation of picosecond pulses.[13] For such extremely short pulse durations, with subpicosecond rise times, distributed circuit treatment becomes necessary for one-way propagation time delay of $t_d > 10^{-13}$ seconds. Assuming propagation at the speed of light, the corresponding distances are > 0.03 mm! In other words, for picosecond rise times, lumped analysis is not appropriate for circuits with physical dimensions longer than a few tens of microns (1 micron $= 1\ \mu$m).

1.1.2 Period versus Travel Time

For sinusoidal steady-state applications, the suitability of a lumped treatment can be assessed by comparing the one-way propagation delay t_d with the period T of the propagating sine wave involved. As an illustration, consider the telephone line shown in Figure 1.4. If a sinusoidal voltage at frequency f is applied at the input of the line so that $V_{AA'}(t) = V_0 \cos(2\pi f t)$, the voltage at a distance of l from the input is delayed by the travel time, $t_d = l/v$. In other words,

$$V_{BB'}(t) = V_0 \cos[2\pi f(t - t_d)] = V_0 \cos\left[2\pi f t - 2\pi \frac{t_d}{T}\right] \qquad [1.1]$$

where $T = 1/f$ is the period of the sinusoidal signal. It is apparent from [1.1] that if $t_d \ll T$, then the voltage $V_{BB'}(t)$ is very nearly the same as the input voltage $V_{AA'}(t)$ at all times, and the telephone line can be treated as a lumped system. If, on the other

[13] See D. W. Van der Weide, All-electronic generation of 0.88 picosecond, 3.5 V shockwaves and their application to a 3 Terahertz free-space signal generation system, *Appl. Phys. Letters,* 62(1), pp. 22–24, January 1993.

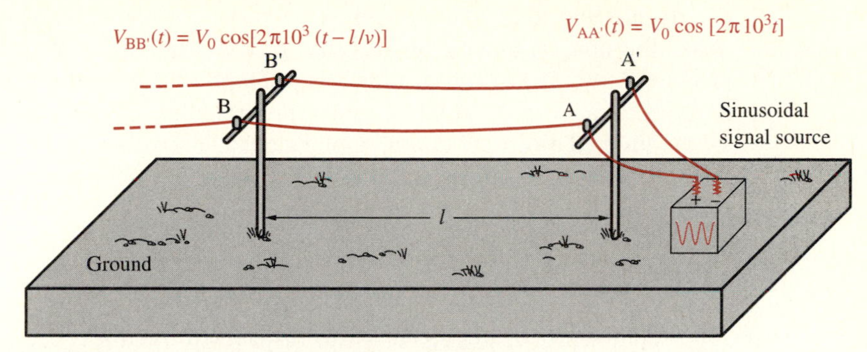

FIGURE 1.4. **Period versus one-way travel time.**

hand, t_d is comparable to T, the instantaneous voltage at a certain time t at points further down the line can be quite different from that at the input; for example, for $t_d = 0.5T$, $V_{BB'}(t) = -V_{AA'}(t)$ at any time t!

In practice, a useful rule of thumb is that lumped analyses can be used when $t_d < 0.1T$ or, under more stringent applications, when $t_d < 0.01T$. In the telephone line application, the signal frequency is of order $\sim$1 kHz, so that $T \simeq 1$ ms, and a lumped treatment would be appropriate for $t_d < 0.01$ ms. Assuming propagation at the speed of light,[14] $t_d = 0.01$ ms corresponds to a line length of $l = 3$ km, which is the maximum length for which the lumped approximation is valid. If the line shown in Figure 1.4 were instead a power transmission line, operating typically at 60 Hz, we have $T = (60)^{-1} \simeq 16.7$ ms, and $0.01T \simeq 0.167$ ms. Once again assuming propagation at the speed of light, 0.167 ms corresponds to a line length of $\sim$50 km, which then is the maximum distance for which lumped analyses can be used. Table 1.1 provides examples of other kinds of waves. We note that the speeds of propagation of different types of waves are in general also different, as shown in Table 1.1.

TABLE 1.1. Maximum lengths for lumped analysis in different applications

Application	**Frequency**	**Propagation speed m-s^{-1}**	**Maximum length (based on $t_d = 0.01T$)**
Power transmission	60 Hz	$c \simeq 3 \times 10^8$	50 km
Telephone line	1 kHz	c	3 km
AM radio broadcast	$\sim$1 MHz	c	3 m
TV broadcast	$\sim$150 MHz	c	2 cm
Radar/microwave	$\sim$10 GHz	c	0.3 mm
Sound waves in air	60 Hz	$\sim$340	5.8 cm
Sound waves in water	60 Hz	$\sim$1500	25 cm
Heat in a concrete dam	1 cycle/hr	$\sim$1.64	2.2 mm
Visible light	$\sim 5 \times 10^{14}$ Hz	c	6×10^{-9} m
X-rays	10^{18} Hz	c	3×10^{-12} m

[14]We shall see in later chapters that speed of propagation of voltage and current waves on a transmission line such as that shown in Figure 1.4 is indeed c.

It thus appears that for circuits excited by sinusoidal signals, whether one can make the lumped assumption depends on the relationship between period (or frequency $f = 1/T$), speed of travel, and length of the signal path. Note that in the frequency range $f < 10$ MHz, the lumped assumption is indeed very good for most of the lumped electrical circuit applications and models, as long as the lengths involved, or the sizes of the components and interconnecting wires or traces, are less than 30 cm!

1.1.3 Component Size versus Wavelength

The applicability of a lumped versus distributed treatment can also be determined in terms of *wavelength* λ. For a periodic wave, wavelength is the length of one complete wave pattern at any given instant of time. In other words, wavelength is the distance between any two points at corresponding positions on successive repetitions of the wave shape. In many applications, electromagnetic quantities (electric fields, magnetic fields, currents, voltages, charges, etc.) can be assumed to vary as

$$I(z, t) = A\cos\left[(2\pi f)t - \left(\frac{2\pi}{\lambda}\right)z\right] \qquad [1.2]$$

where the amplitude A, the frequency f, and the wavelength λ are constants. In later chapters, we shall see that for sinusoidal steady-state applications, quantities with variations as described by [1.2] are natural solutions of the fundamental equations describing wave motion, which are in turn derived from the fundamental physical laws of electromagnetics known as Maxwell's equations. Note also that the form of [1.2] is similar to the general form of a wave function, or $f(z - vt)$, discussed in connection with Figure 1.1.

Now consider the behavior of this quantity, or in the case of [1.2], the current $I(z, t)$, in both space and time. At fixed points in space (i.e., $z =$ const.; e.g., $z = 0$ and $z = \lambda/4$), we have

$$I(z = 0, t) = A\cos(2\pi f t - 0) \qquad \text{and} \qquad I(z = \lambda/4, t) = A\cos(2\pi f t - \pi/2)$$

The variation of current at these two fixed locations is shown in Figure 1.5. We see that at every fixed point the current varies sinusoidally with time, as expected for sinusoidal steady state. However, as is apparent from Figure 1.5, at different points along z, the current reaches its maxima and minima at different times.

At fixed points in time (i.e., $t =$ const.; e.g., $t = 0$ and $t = T/4$), we have

$$I(z, t = 0) = A\cos\left[0 - \left(\frac{2\pi}{\lambda}\right)z\right] \quad \text{and} \quad I(z, t = T/4) = A\cos\left[(\pi/2) - \left(\frac{2\pi}{\lambda}\right)z\right]$$

The variation of current in space at these two different times is shown in Figure 1.6. We see that the behavior of $I(z, t)$ at a given time is also sinusoidal, with its peak value (i.e., the point where the argument of the cosine is zero) occurring at different points in space at different times. The speed $v = dz/dt$ with which the crests (the

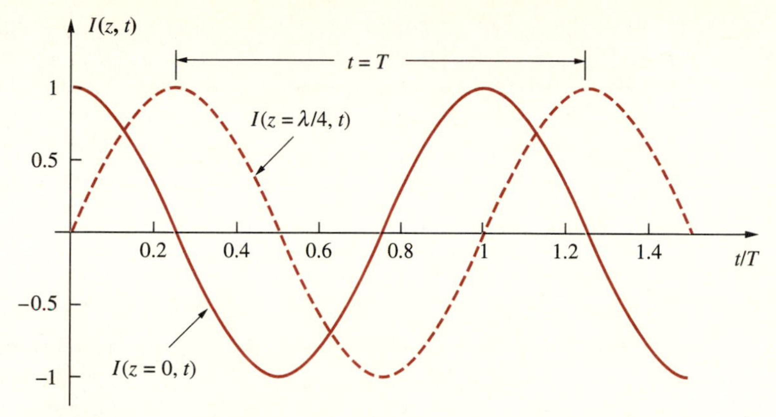

FIGURE 1.5. **Oscillations in time.** Plots versus time of $A\cos[(2\pi f)t - (2\pi/\lambda)z]$ at $z = 0$ and $z = \lambda/4$. Note that the abscissa is t/T, where $T = 1/f$, and the plot shown is for $A = 1$.

maxima) or the valleys (the minima) of the cosinusoid propagate (move from one point in space to another) can be determined by setting its argument equal to a constant (for example, zero, which corresponds to the maximum) and differentiating:

$$\left[(2\pi f)t - \left(\frac{2\pi}{\lambda}\right)z\right] = \text{const.} \quad \longrightarrow \quad v = \frac{dz}{dt} = \lambda f$$

For electromagnetic waves in free space, and for voltage and current waves on some simple transmission lines (e.g., those consisting of two wires separated by air or air-filled coaxial lines), the speed of propagation v is equal to the speed of light c.

Since the current at any given instant of time varies sinusoidally in space, whether we can assume a given circuit to be lumped (i.e., that the instantaneous

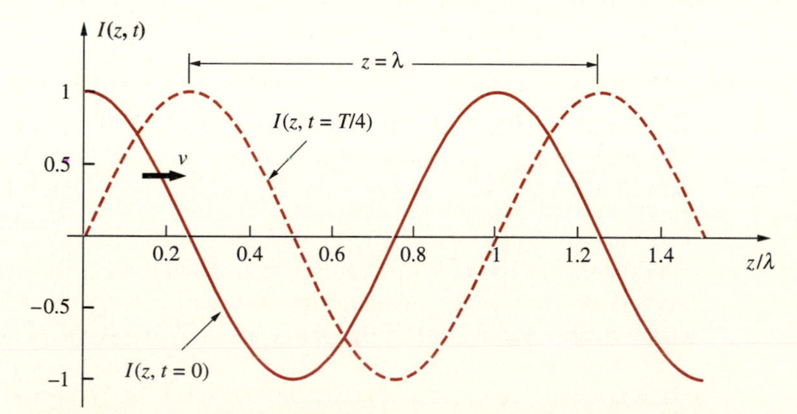

FIGURE 1.6. **Wave motion in space.** Plots versus z of $A\cos[(2\pi f)t - (2\pi/\lambda)z]$ at $t = 0$ and $t = T/4$. Note that the abscissa is z/λ and the plot shown is for $A = 1$.

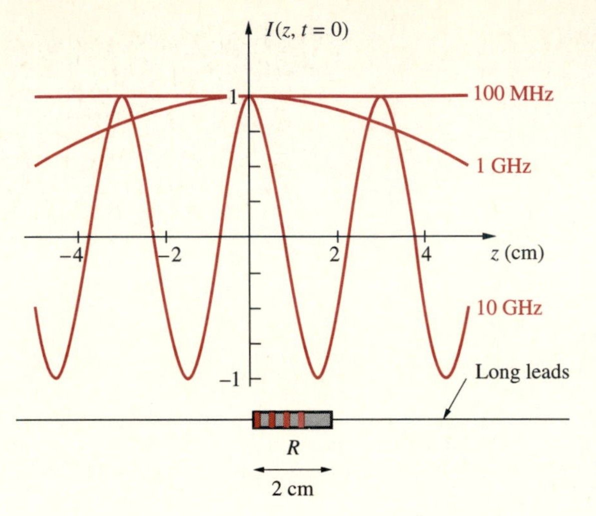

FIGURE 1.7. Current through 2-cm long resistor. Current $I(z, t) = A\cos[2\pi f t - (2\pi/\lambda)z]$ is shown as a function of z at $t = 0$ for frequencies of 100 MHz, 1 GHz, and 10 GHz. For the purposes of this plot, we have taken $A = 1$.

current at a given time is approximately the same at different points across the component carrying the current) depends on the size of the element compared with wavelength, λ. This concept is illustrated in Figure 1.7, which shows a 2-cm long resistor (including leads) in comparison with wavelength at three frequencies: 100 MHz, 1 GHz, and 10 GHz. For 10 GHz, it is clear that the notion of a current through the resistor is not useful, since, at any given time, one lead of the resistor has a current of different polarity than the other,[15] since according to the current expression [1.2] and Figure 1.6 the current at any fixed time switches polarity every $\lambda/2$ change in position, where $\lambda = c/f = 3$ cm for 10 GHz. Even for the 1 GHz case, the dimensions of the resistor are a significant fraction of a wavelength, so that the current at any time is significantly different at different points across the circuit component. Thus, of the cases shown, the lumped assumption is appropriate only for 100 MHz.

1.2 ELECTROMAGNETIC COMPONENTS

The electromagnetic circuit components used at high frequencies can differ conspicuously in appearance from the often more familiar lumped-element circuits used at low frequencies. The connecting wires of conventional circuits provide conductors for the electric currents to flow, and the resistors, capacitors, and inductors possess simple relationships between their terminal currents and voltages. Often overlooked is the fact that the wires and circuit components merely provide a

[15]Note that at 10 GHz, not only would we not be able to define the value of current but there would be substantial radiation losses, because the resistor leads, being comparable in size to λ, would serve as efficient antennas.

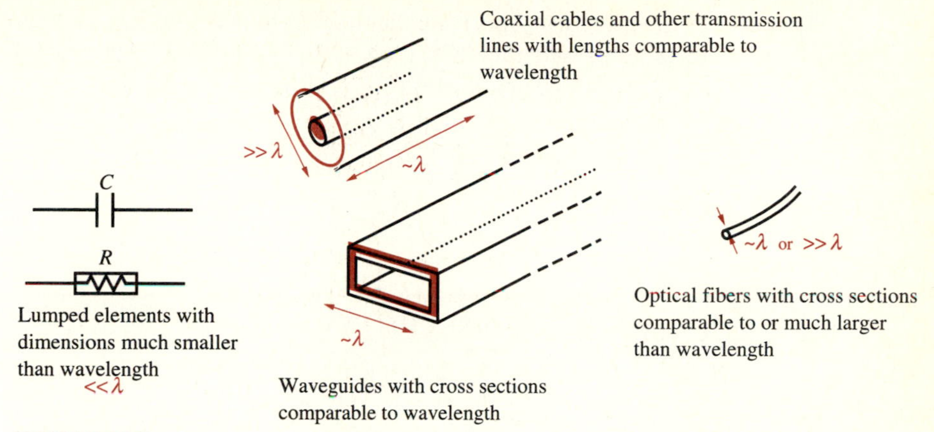

FIGURE 1.8. **Different categories of electromagnetic circuit components.**

framework over which charges move and disperse. These charges set up electric and magnetic fields that permeate the circuit, often having almost indescribably complicated configurations. It would, in principle, be possible to treat the behavior of circuits entirely in terms of these electromagnetic fields instead of the usual practice of working in terms of circuit voltages and currents. It can be argued, however, that much of the progress in modern electrical and electronic applications would not have come about if it were not for the simple but powerful circuit theory.

As the operating frequency of circuits increases, however, the effects of stray capacitances and inductances alter the effective circuit behavior radically from its low-frequency form. Radiation from the circuit also increases rapidly with frequency, and much power may be lost from the system in this way.[16] This radiative power loss may be prevented by confining the fields to the interior of metallic enclosures. For example, in hollow metallic tubes (referred to as waveguides) the charges move exclusively on the interior surfaces of conductors, and because of the simple geometry of the enclosures, the electromagnetic fields can have simple analytical forms. Since it is often not possible to define voltages and currents uniquely within waveguides, analysis of waveguides is usually carried out on the basis of the full electromagnetic theory. At even higher frequencies, metallic enclosures become too lossy and impractical. Efficient guiding of electromagnetic energy at optical frequencies occurs in optical fibers, consisting of hair-thin glass strands. The light wave in an optical fiber is guided along the fiber by means of multiple reflections from its walls. The principles underlying the guiding of electromagnetic energy in metallic enclosures is discussed in Chapter 8.

In summary, component sizes in electromagnetic applications can be categorized as shown in Figure 1.8:

---

[16]In antenna applications, we take advantage of such radiation "losses" when we design antennas to maximize the radiated power in selected directions.

- Component sizes much smaller than λ (lumped elements). Examples: Resistors, capacitors, inductors, ICs, transistors, and interconnects used for $< \sim 30$ MHz.
- Component sizes comparable to λ. Examples: Hollow waveguides, long coaxial cables, some optical fibers, and some antennas.
- Component sizes much greater than λ. Examples: Graded index optical fibers and some antennas.

1.3 MAXWELL'S EQUATIONS AND ELECTROMAGNETIC WAVES

The principles of guiding and propagation of electromagnetic energy in these widely different regimes differ in detail, but are all governed by one set of equations, known as *Maxwell's equations*,[17] which are based on experimental observations and which provide the foundations of *all* electromagnetic phenomena and their applications. The sequence of events in the late 19th century that led to the development of these fundamental physical laws is quite interesting in its own right. Many of the underlying concepts were developed by earlier scientists, especially Michael Faraday, who was a visual and physical thinker but not enough of a mathematician to express his ideas in a complete and self-consistent form to provide a theoretical framework. James Clerk Maxwell translated Faraday's ideas into strict mathematical form and thus established a theory that predicted the existence of electromagnetic waves. The experimental proof that electromagnetic waves do actually propagate through empty space was supplied by the experiments of Heinrich Hertz, carried out many years after Maxwell's brilliant theoretical work.

Maxwell's equations are based on experimentally established facts, namely Coulomb's law,[18] Ampère's law,[19] Faraday's law,[20] and the principle of conservation of electric charge. As we introduce and discuss the fundamental bases of Maxwell's equations in Chapters 4 through 7, we also emphasize the historical context by presenting abbreviated biographies of the key scientists whose work led to their establishment.

---

[17]J. C. Maxwell, *A Treatise in Electricity and Magnetism,* Clarendon Press, Oxford, 1892, Vol. 2, pp. 247–262.

[18]Coulomb's law states that electric charges attract or repel one another in a manner inversely proportional to square of the distance between them; C. A. de Coulomb, *Première mémoire sur l'électricité et magnétisme (First Memoir on Electricity and Magnetism), Histoire de l'Académie Royale des Sciences,* p. 569, 1785.

[19]Ampère's law states that current-carrying wires create magnetic fields and exert forces on one another, with the amplitude of the magnetic field (and thus force) depending on the inverse square of the distance; A. M. Ampère, *Recueil d'observations électrodynamiques,* Crochard, Paris, 1820–1833.

[20]Faraday's law states that magnetic fields that change with time induce electromotive force or electric field; M. Faraday, *Experimental Researches in Electricity,* Taylor, London, 1839, Vol. I, pp. 1–109.

Maxwell's equations embody all of the essential features of electromagnetics, including the ideas that light is electromagnetic in nature, that electric fields that change in time create magnetic fields in the same way as time-varying voltages induce electric currents in wires, and that the source of electric and magnetic energy resides not only on the body that is electrified or magnetized but also, and to a far greater extent, in the surrounding medium. However, arguably the most important and far-reaching implication of Maxwell's equations is the idea that electric and magnetic effects can be transmitted from one point to another through the intervening space, whether that be empty or filled with matter.

To appreciate the concept of propagation of electromagnetic waves in empty space, it is useful to think of other wave phenomena that we may observe in nature. When a pebble is dropped into a body of water, the water particles in the vicinity of the pebble are immediately displaced from their equilibrium positions. The motion of these particles disturbs adjacent particles, causing them to move, and the process continues, creating a wave. Because of the finite velocity of the wave, a finite time elapses before the disturbance causes the water particles at distant points to move. Thus the initial disturbance produces, at distant points, effects that are *retarded* in time. The water wave consists of ripples that move along the surface away from the initial disturbance. Although the motion of any particular water particle is essentially a small up-and-down movement, the cumulative effects of all the particles produce a wave that moves radially outward from the point at which the pebble is dropped. Another excellent example of wave propagation is the motion of sound through a medium. In air, this motion occurs through the to-and-fro movement of the air molecules, but these molecules do not actually move along with the wave.

Electromagnetic waves consist of time-varying electric and magnetic fields. Suppose an electrical disturbance, such as a change in the current through a conductor, occurs at a point in a region. The time-changing electric field resulting from the disturbance generates a time-changing magnetic field. The time-changing magnetic field, in turn, produces an electric field. These time-varying fields continue to generate one another in an ever-expanding region, and the resulting wave propagates away from the location of the initial disturbance. When electromagnetic waves propagate in vacuum, there is no vibration of physical particles as in the case of water and sound waves. Nevertheless, the velocity of wave propagation is limited by the speed of light, so that the fields produced at distant points are *retarded* in time with respect to those near the source. Propagation of electromagnetic waves in empty space or material media, their reflection from boundaries, and their guiding within metallic boundaries are subjects of Chapter 8.

1.4 SUMMARY

There are two kinds of electrical circuits: *lumped* and *distributed.* In classical lumped circuit theory, the entire circuit is assumed to be at a single point such that all parts of the circuit respond to an excitation at the same time. Distributed circuits or transmission lines can be treated using a natural extension of classical circuit theory by taking account of the nonzero travel time of signals from one point to another in a distributed circuit, the speed of travel being bounded by the speed of light. In any given application, a number of criteria can be used to determine the applicability of a lumped analysis. A given system must be treated as a distributed one if:

- The rise time t_r of the applied signal is less than 2.5 times the one-way travel time t_d across the circuit, i.e., $t_r < 2.5t_d$.
- The one-way travel time t_d across the circuit is greater than one-hundredth of the period T of the applied sinusoidal signal, i.e., $t_d > 0.01T$.
- The physical dimensions of the circuit are a significant fraction of the wavelength at the frequency of operation.

Electromagnetic components can be categorized in three basic groups with sizes much smaller than, comparable to, or much larger than the wavelength. Electromagnetic applications involving any one of these three classes of components are governed by a set of physical laws known as Maxwell's equations, which are based on experimental facts and describe the propagation of electromagnetic waves through empty space or material media and, in general, *all* other electrical and magnetic phenomena.

In the next two chapters, we will study a special class of electromagnetic waves called *transverse electromagnetic* (TEM) waves, which propagate on electrical transmission lines consisting of two metallic conductors. In the case of TEM waves, both the electric and magnetic fields are everywhere perpendicular to each other and to the direction of propagation. The behavior of TEM waves on two-conductor lines can be understood in the context of a voltage-current description by modeling the transmission line as a distributed circuit and utilizing many of our electrical circuits concepts. The fact that the voltage and current waves that propagate on such two-conductor transmission lines are indeed transverse electromagnetic waves will become clear in Chapter 8.

CHAPTER 2

Transient Response of Transmission Lines

Wave motion is said to occur when a disturbance of a physical quantity at a particular point in space at a particular time is related in a definite manner to what occurs at distant points in later times. *Transient* waves occur in response to sudden, usually brief disturbances at a source point, leading to temporary disturbances at distant points at later times. They are thus different from *steady-state* waves, which are sustained by disturbances involving periodic oscillations at the source point.

Transient waves are of importance in many different contexts. Consider, for example, a line of cars waiting at a red traffic light. When the light turns green, the cars do not all start moving at the same time; instead, the first car starts to move first, followed by the car behind it, and so on, as the act of starting to move travels backwards through the line. This wave travels at a speed that depends on the response properties of the cars and the reaction times of the drivers. As another example, when the end of a stretched rope is suddenly moved sideways, the action of moving sideways travels along the rope as a wave whose speed depends on the tension in the rope and its mass. If the rope is infinitely long, the disturbance simply continues to propagate away from its source. If, on the other hand, the distant end of the rope is held fixed, the wave can be reflected back toward the source.

Other examples of transient waves include the thunderclap, the sound wave emitted from an explosion, and seismic waves launched by an earthquake.

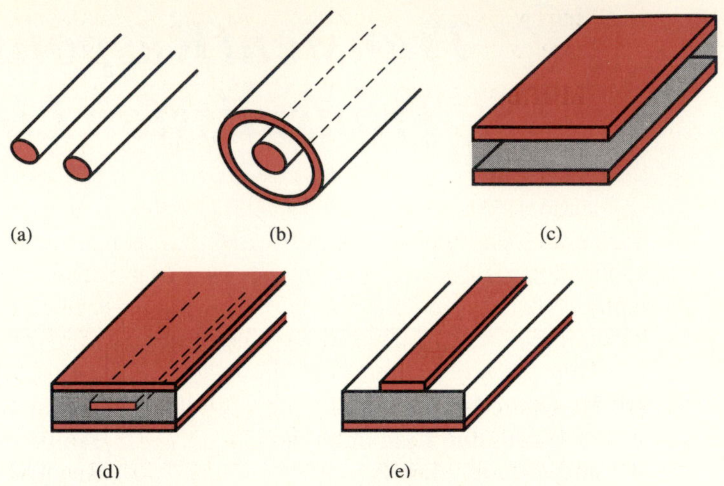

FIGURE 2.1. Different types of uniform transmission lines: (a) parallel two-wire; (b) coaxial; (c) parallel-plate; (d) stripline; (e) microstrip.

Transient waves are often used as tools to study the disturbances that create them. The sound wave from a blast can be used for detecting the source of the blast from a long distance; a seismograph measures the strength of a distant earthquake based on tiny transient motions of the earth. Transient waves can also have destructive effects far away from the sources of their initial disturbances. The great Alaskan earthquake of March 27, 1964, produced a giant tsunami, which radiated seaward from Prince William Sound, causing enormous damage when it hit the Hawaiian islands some 5 hours after the disturbance; its remnants produced numerous seiches[1] that sloshed back and forth for more than 24 hours in the various bays, inlets, and harbors along the western coast of North America. Seismic surface waves launched by the same earthquake, propagating by rippling the earth's crust, took 14 minutes to travel from Prince William Sound to the Gulf Coast region of Texas and Louisiana, where they triggered seiches in bays, harbors, rivers, and lakes. Cajun trappers and night fishermen in a Louisiana bayou were surprised to be violently rocked back and forth just after they heard of a large Alaskan earthquake on their radios.

The purpose of this chapter is to study voltage and current transients on electrical transmission lines. A transmission line may consist of two parallel wires (as it will often be illustrated in this book), of coaxial conductors, or of any two conductors separated by an insulating material or vacuum. Some types of two-conductor transmission lines are shown in Figure 2.1.

---

[1]*Seiche* is a French word that has become the internationally adopted scientific term for transient free (or unforced) surges and oscillations that develop primarily in closed and semiclosed bodies of water. For an excellent, easy-to-read article, see B. J. Korgen, Seiches, *American Scientist,* 83(4), pp. 330–341, July–August, 1995.

Many important electrical engineering applications involve *transients:* temporary variations of current and voltage that propagate down a transmission line. Transients are produced by steplike changes (e.g., sudden on or off) in input voltage or current. Digital signals consist of a sequence of pulses, which represent superposition of successive steplike changes; accordingly, the transient response of transmission lines is of interest in most digital integrated circuit and computer communication applications. Transient transmission line problems arise in many other contexts. The transient response of lines can be used to generate rectangular pulses; the earliest applications of transmission lines involved the use of rectangular pulses for telegraphy. When lightning strikes a power transmission line, a large surge voltage is locally induced and propagates to other parts of the line as a transient.

This chapter is unique; the following chapters are mostly concerned with applications that either involve static quantities, which do not vary in time, or involve steady signals that are sinusoids or modified sinusoids. However, with the rapid advent of digital integrated circuits, digital communications, and computer communication applications, transient responses of transmission lines are becoming increasingly important. Increasing clock speeds make signal integrity[2] analysis a must for the design of high-speed and high-performance boards. Managing signal integrity in today's high-speed printed circuit boards and multichip modules involves features such as interconnect lengths, vias, bends, terminations, and stubs and often requires close attention to transmission line or distributed circuit effects.[3] It is thus fitting that we start our discussion of engineering electromagnetics by studying the transient response of transmission lines. Also, analysis of transients on transmission lines requires relatively little mathematical complexity and brings about an intuitive understanding of concepts such as wave propagation and reflection, which facilitates a better understanding of the following chapters.

Our discussions in this chapter start in Section 2.1 with a heuristic analysis of transmission line behavior, in particular the response to a step (on or off) input, and a discussion of lumped circuit models. Section 2.2 introduces the fundamental circuit equations for a transmission line and their solution for lossless lines in terms of traveling waves. The reflection of the waves at the termination of a transmission line and the step response of lossless lines with open and short circuit terminations are presented in Section 2.3. Section 2.4 covers the step and pulse responses of lossless transmission lines terminated in resistive loads or in other transmission lines, while Section 2.5 treats the cases of reactive loads and loads with nonlinear current-voltage characteristics. Selected practical topics are presented in Section 2.6, followed by a brief discussion of the parameters of some commonly used practical transmission lines in Section 2.7.

---

[2]The term *signal integrity* refers to the issues of timing and quality of the signal. The timing analysis is performed to ensure that the signal arrives at its destination within a specified time interval and that the signal causes correct logic switching without false switching or transient oscillations, which can result in excessive time delays. See R. Kollipara and V. Tripathi, Printed wiring boards, Chapter 71 in J. C. Whitaker (ed.), *The Electronics Handbook,* CRC Press, 1996, pp. 1069–1083.

[3]See R. Goyal, Managing signal integrity, *IEEE Spectrum,* pp. 54–58, March 1994.

2.1 HEURISTIC DISCUSSION OF TRANSMISSION LINE BEHAVIOR AND CIRCUIT MODELS

Typically, we explain the electrical behavior of a two-conductor transmission line in terms of an equal and opposite current flowing in the two conductors, as measured at any given transverse plane. The flow of this current is accompanied by magnetic fields set up around the conductors (Ampère's law), and when these fields change with time, a voltage (electromotive force) is induced in the conductors (Faraday's law), which affects the current flow.[4] This behavior can be represented by a small inductance associated with each short-length segment of the conductors. Also, any two conductors separated by a distance (such as the short sections of two conductors facing one another) have nonzero capacitance, so that when equal and opposite charges appear on them, there exists a potential drop across them. Hence, each short section of a two-conductor line exhibits some series inductance and parallel capacitance.[5] The values of the inductance and capacitance depend on the physical configuration and material properties of the two-conductor line, including the surface areas, cross-sectional shape, spacing,[6] and layout of the two conductors as well as the electrical and magnetic properties[7] of the substance filling the space between and around the conductors.

2.1.1 Heuristic Discussion of Transmission Line Behavior

We can qualitatively understand the behavior of a two-conductor transmission line by considering a lossless circuit model of the line, consisting of a large number of series inductors and parallel capacitors connected together, representing the short sections Δz of the line, as illustrated in Figure 2.2.

To illustrate the behavior of a lossless transmission line, we now consider the simplest possible transient response: the step response, which occurs upon the sudden application of a constant voltage. At $t = 0$, a battery of voltage V_0 and source resistance R_{s1} is connected to the infinitely long two-conductor transmission line represented by the lossless circuit shown in Figure 2.2, where each pair of

---

[4]Detailed discussion of these experimentally based physical laws will be undertaken later; here we simply rely on their qualitative manifestations, drawing on the reader's exposure to these concepts at the freshman physics level.

[5]Neglecting losses for now.

[6]This can be seen at a qualitative level, from the reader's understanding of capacitance and inductance at the freshman physics level. For example, the closer the conductors are to each other, the larger the capacitance is. On the other hand, the inductance of a two-conductor line is smaller if the conductors are closer together, since the magnetic field produced by the current flow is linked by a smaller area, thus inducing less voltage.

[7]At the simplest level, the magnetic properties of a material represent the ability of a material medium to store magnetic energy. Similarly, by electrical properties we refer to the ability of a material to store electric energy or its response to an applied electric field. The microscopic behavior of the materials that determines these properties will be discussed in Chapters 4 and 6.

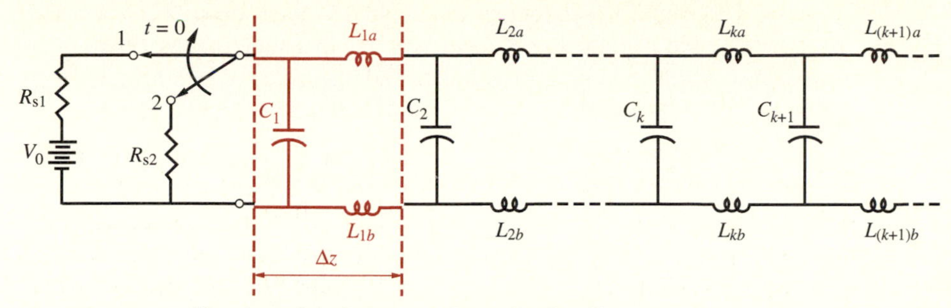

FIGURE 2.2. **Circuit model of a two-conductor lossless line.**

inductances corresponds to the inductance of a short section of the line of length Δz, and each capacitor corresponds to the capacitance of the same section of length Δz. Initially, the transmission line is completely discharged, so all the capacitances have zero charge (and thus zero voltage) and the inductances have zero current flowing through them. The switch in Figure 2.2 is moved to position 1 at $t = 0$, so that, starting at $t = 0^+$, the source voltage V_0 appears across the source resistance R_{s1} and the terminals of the capacitance C_1, which takes time to charge;[8] and until it charges, there is no voltage across it to drive currents through L_{1a} and L_{1b}. As the voltage across C_1 builds up, the currents in L_{1a} and L_{1b} also take time to increase.[9] When these currents increase enough to cause appreciable flow through C_2, this capacitance now takes some time to charge. As it charges, current starts to flow in inductors L_{2a} and L_{2b}, but this takes time. This same process continues all the way down the line, with the capacitor C_k not starting to charge until the preceding capacitors are charged, just as if it did not know yet that the voltage step had been applied at the input. In this way, the information about the change in the position of the switch travels down the line.

When the switch moves back to position 2 at $t = t_1$, the reverse happens: C_1 has to discharge through R_{s2}, which pulls current (not suddenly) from L_{1a} and L_{1b}, which in turn allows C_2 to discharge, and so on. All of this takes time, so C_k is not affected by the removal of the input signal until the preceding capacitors are discharged first. The rate of charging and discharging depends only on the circuit element values, so the charging and discharging disturbances both continue down the line at the same speed, since $L_{ia} = L_{jb}$ for all i, j and all C_i values are equal, assuming a uniform transmission line.

Note from the above discussion that if the inductance of the line segments is negligible, the line can be approximated as a lumped capacitor (equal to the parallel combination of all of its distributed shunt capacitances); all the points on the line are then at the same potential, and traveling-wave effects are not important. The line inductance becomes important if the line is relatively long or if the rise time of the applied signal (as defined in Figure 1.3) is so fast that the current through the inductor

[8]Voltage across a capacitance cannot change instantaneously.

[9]Current through an inductor cannot change instantaneously.

increases very rapidly, producing appreciable voltage drop ($\mathcal{V} = L\, d\mathcal{I}/dt$) across the inductor even if the value of L is small. By the same token, it is intuitively clear that, even if the line is long (and thus L is large), transmission line effects will be negligible for slow enough rise times, as was discussed in Chapter 1.

2.1.2 Circuit Models of Transmission Lines

It is often more useful to describe transmission line behavior in terms of inductance and capacitance *per unit length,* rather than viewing the line as an infinite number of discrete inductances and capacitances, as implied in Figure 2.2. We must also note that, in general, the conductors of a transmission line exhibit both inductance and resistance and that there can be leakage losses through the material surrounding the conductors. The inductance per unit length (L) of the line, in units of henrys per meter, depends on the physical configuration of the conductors (e.g., the separation between conductors and their cross-sectional shape and dimensions) and on the magnetic properties of the material surrounding the conductors. The series resistance per unit length R, in units of ohms per meter, depends on the cross-sectional shape, dimensions, and electrical conductivity of the conductors[10] and the frequency of operation. Between the conductors there is a capacitance (C), expressed as farads per meter; there is also a leakage conductance (G) of the material surrounding the conductors, in units of siemens per meter. The capacitance depends on the shape, surface area, and proximity of the conductors as well as the electrical properties of the insulating material surrounding the conductors. The conductance depends on the shape and dimensions of the outside surface of the conductors[11] and on the degree to which the insulating material is lossy. A *uniform* transmission line consists of two conductors of uniform cross section and spacing throughout their length, surrounded by a material that is also uniform throughout the length of the line. An equivalent circuit of a uniform transmission line can be drawn in terms of the per-unit-length parameters, which are the same throughout the line. One such circuit is shown in Figure 2.3, where each short section of length Δz of the line is modeled as a lumped circuit whose element values are given in terms of the per-unit parameters of the line. The electrical behavior of a uniform transmission line can be studied in terms of such a circuit model if the length of the line (Δz) represented by a single L-R-C-G section is very small compared to, for example, the wavelength of electromagnetic waves in the surrounding material at the frequency of operation. Four different circuit models are shown in Figure 2.4.

Expressions for L, R, C, and G for some of the commonly used uniform transmission lines shown in Figure 2.1 are provided in Section 2.7. The values of these quantities depend on the geometric shapes and the cross-sectional dimensions

[10]The resistance simply represents the ohmic losses due to the current flowing through the conductors; hence, it depends on the cross-sectional area and the conductivity (see Chapter 5) of the conductors.

[11]The leakage current flows from one conductor to the other, through the surrounding material, and in the direction transverse to the main current flowing along the conductors of the line; hence, it depends on the outer surface area of the conductors.

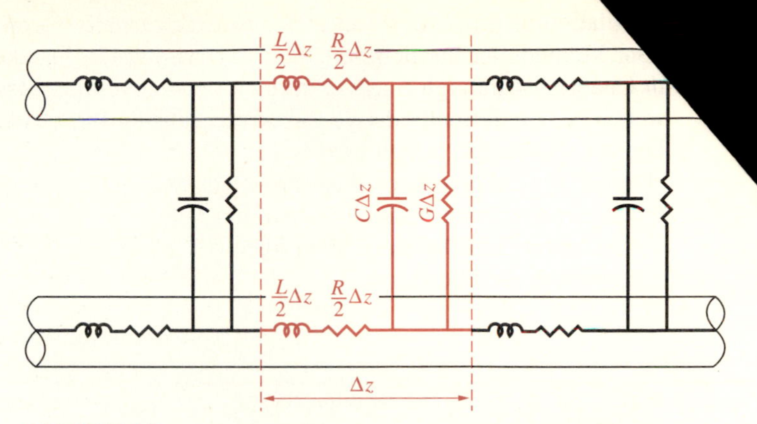

FIGURE 2.3. Distributed circuit of a uniform transmission line.

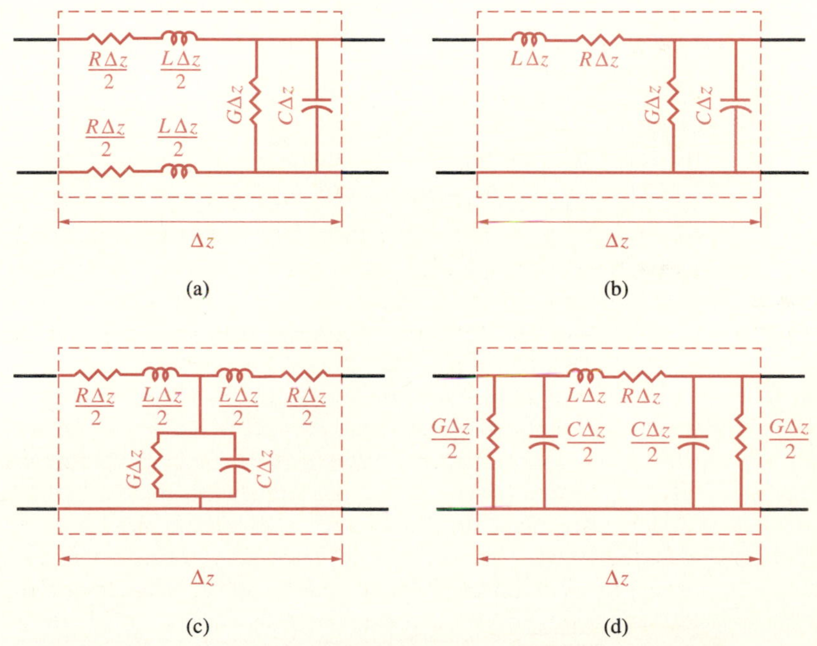

FIGURE 2.4. Lumped circuit models for a short segment of a uniform transmission line.

of the line, the electrical conductivity of the metallic conductors used, the electrical and magnetic properties of the surrounding medium, and the frequency of operation. The expressions for L, R, C, and G for the various lines can be obtained by means of electromagnetic field analysis of the particular geometries involved. For some cases (such as the parallel wire, coaxial line, and parallel plate lines shown in Figure 2.1), compact analytical expressions for R, L, C, and G can be found. For other, more complicated structures (e.g., the stripline and microstrip lines in Figure 2.1),

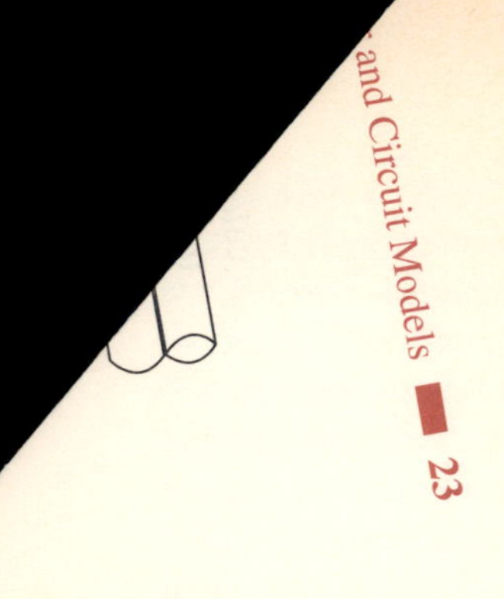

C, and G parameters usually requires numerical computa-
ermination of transmission line parameters are discussed
s we introduce the governing electromagnetic equations,
rive the expressions for the transmission line parameters.
analyses of transmission lines in this and the following
that the values of L, R, C, and G are directly calculable
n line configuration. We can thus proceed by using their
xpressions in Section 2.7, as given in handbooks, or as

TRANSMISSION LINE EQUATIONS AND WAVE SOLUTIONS

In this section we develop the fundamental equations that govern wave propagation along general two-conductor transmission lines. Various lumped circuit models of a single short segment of a transmission line are shown in Figure 2.4. In the limit of $\Delta z \rightarrow 0$, any one of the circuit models of Figure 2.4 can be used to derive the fundamental transmission line equations. In the following, we use the simplest of these models (Figure 2.4b), shown in further detail in Figure 2.5.

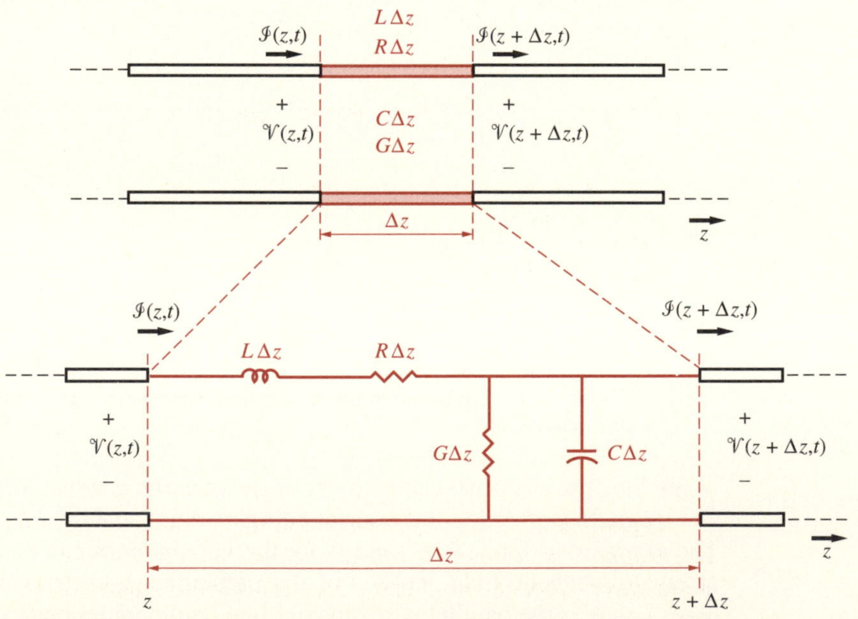

FIGURE 2.5. **Equivalent circuit of a short length of Δz of a two-conductor transmission line.**

2.2.1 Transmission Line Equations

The section of line of length Δz in Figure 2.5 is assumed to be located at a distance z from a selected point of reference along the transmission line. We consider the total voltage and current at the input and output terminals of this line section: that is, at points z and $(z + \Delta z)$. The input and output voltages and currents are denoted as $\mathcal{V}(z, t)$, $\mathcal{I}(z, t)$ and $\mathcal{V}(z + \Delta z, t)$, $\mathcal{I}(z + \Delta z, t)$, respectively. Using Kirchhoff's voltage law, we can see that the difference in voltage between the input and output terminals is due to the voltage drop across the series elements $R\Delta z$ and $L\Delta z$, so we have

$$\mathcal{V}(z + \Delta z, t) - \mathcal{V}(z, t) = -R\Delta z\ \mathcal{I}(z, t) - L\Delta z\frac{\partial \mathcal{I}(z, t)}{\partial t}$$

Note that we shall consider Δz to be as small as needed so that the lumped circuit model of the segment can accurately represent the actual distributed line. In the limit when $\Delta z \to 0$, we have

$$\lim_{\Delta z \to 0} \frac{\mathcal{V}(z + \Delta z, t) - \mathcal{V}(z, t)}{\Delta z} = \frac{\partial \mathcal{V}(z, t)}{\partial z} = -R\mathcal{I}(z, t) - L\frac{\partial \mathcal{I}(z, t)}{\partial t}$$

or

$$\boxed{\frac{\partial \mathcal{V}(z, t)}{\partial z} = -\left(R + L\frac{\partial}{\partial t}\right)\mathcal{I}(z, t)} \qquad [2.1]$$

Similarly, using Kirchhoff's current law, the difference between the current at the input and output terminals is equal to the total current through the parallel elements $G\Delta z$ and $C\Delta z$, so we have

$$\mathcal{I}(z + \Delta z, t) - \mathcal{I}(z, t) = -G\Delta z\ \mathcal{V}(z + \Delta z, t) - C\Delta z\frac{\partial \mathcal{V}(z + \Delta z, t)}{\partial t}$$

Upon dividing by Δz and expanding $\mathcal{V}(z + \Delta z, t)$ in a Taylor series,[12] and taking $\Delta z \to 0$, we have:

$$\lim_{\Delta z \to 0}\left\{\frac{\mathcal{I}(z + \Delta z, t) - \mathcal{I}(z, t)}{\Delta z}\right\} = -G\mathcal{V}(z, t) - C\frac{\partial \mathcal{V}(z, t)}{\partial t} - \lim_{\Delta z \to 0}\{\Delta z(\cdots)\}$$

or

$$\boxed{\frac{\partial \mathcal{I}(z, t)}{\partial z} = -\left(G + C\frac{\partial}{\partial t}\right)\mathcal{V}(z, t)} \qquad [2.2]$$

Equations [2.1] and [2.2] are known as the *transmission line equations* or *telegrapher's equations.* We shall see in Chapter 8 that uniform plane electromagnetic wave propagation is based on very similar equations, written in terms of the

[12] $\mathcal{V}(z + \Delta z, t) = \mathcal{V}(z, t) + [\partial \mathcal{V}(z, t)/\partial z]\Delta z + \cdots$

components of electric and magnetic fields instead of voltages and currents. Most other types of wave phenomena are governed by similar equations; for acoustic waves in fluids, for example, one replaces voltage with pressure and current with velocity.

2.2.2 Traveling-Wave Solutions for Lossless Lines

Solutions of [2.1] and [2.2] are in general quite difficult and require numerical treatments for the general transient case, when all of the transmission line parameters are nonzero. However, in a wide range of transmission line applications the series and parallel loss terms (R and G) can be neglected, in which case analytical solutions of [2.1] and [2.2] become possible. In fact, practical applications in which transmission lines can be treated as lossless lines are at least as common as those in which losses are important. Accordingly, our transmission line analyses in this chapter deal exclusively with lossless transmission lines. A brief discussion of transients on lossy lines is provided in Section 2.6.3, and steady-state response of lossy lines is discussed in Section 3.8.

We now apply [2.1] and [2.2] to the analysis of the transient response of lossless transmission lines. For a lossless line we have $R = 0$ and $G = 0$, so that [2.1] and [2.2] reduce to

$$\frac{\partial \mathcal{V}}{\partial z} = -L\frac{\partial \mathcal{I}}{\partial t} \qquad [2.3]$$

$$\frac{\partial \mathcal{I}}{\partial z} = -C\frac{\partial \mathcal{V}}{\partial t} \qquad [2.4]$$

By combining [2.3] and [2.4] we obtain the *wave equations* for voltage and current,

$$\boxed{\frac{\partial^2 \mathcal{V}}{\partial z^2} = LC\frac{\partial^2 \mathcal{V}}{\partial t^2}} \qquad [2.5]$$

or

$$\boxed{\frac{\partial^2 \mathcal{I}}{\partial z^2} = LC\frac{\partial^2 \mathcal{I}}{\partial t^2}} \qquad [2.6]$$

Either one of [2.5] or [2.6] can be solved for the voltage or current. We follow the usual practice and consider the solution of the voltage equation [2.5], which can be rewritten as

$$\frac{\partial^2 \mathcal{V}}{\partial z^2} = \frac{1}{v_p^2}\frac{\partial^2 \mathcal{V}}{\partial t^2}; \qquad v_p = \frac{1}{\sqrt{LC}} \qquad [2.7]$$

Note that once $\mathcal{V}(z,t)$ is determined, we can use [2.3] or [2.4] to find $\mathcal{I}(z,t)$. The quantity v_p represents the speed of propagation of a disturbance, as will be evident in the following discussion. For reasons that will become clear in Chapter 3, v_p is also referred to as the *phase velocity;* hence the subscript p.

Any function[13] $f(\cdot)$ of the variable $\xi = (t - z/v_p)$ is a solution of [2.7]. To see that

$$\mathcal{V}(z, t) = f\left(t - \frac{z}{v_p}\right) = f(\xi)$$

is a solution of [2.7], we can express the time and space derivatives of $\mathcal{V}(z, t)$ in terms of the derivatives of $f(\xi)$ with respect to ξ:

$$\frac{\partial \mathcal{V}}{\partial t} = \frac{\partial f}{\partial \xi}\frac{\partial \xi}{\partial t} = \frac{\partial f}{\partial \xi} \quad \text{and} \quad \frac{\partial^2 \mathcal{V}}{\partial t^2} = \frac{\partial^2 f}{\partial \xi^2}\frac{\partial \xi}{\partial t} = \frac{\partial^2 f}{\partial \xi^2}$$

since $\partial \xi/\partial t = 1$. Similarly, noting that $\partial \xi/\partial z = -(1/v_p)$, we have

$$\frac{\partial \mathcal{V}}{\partial z} = -\frac{1}{v_p}\frac{\partial f}{\partial \xi} \quad \text{and} \quad \frac{\partial^2 \mathcal{V}}{\partial z^2} = \frac{1}{v_p^2}\frac{\partial^2 f}{\partial \xi^2}$$

Substituting in [2.7] we find that the wave equation is indeed satisfied by any function $f(\cdot)$ of the variable $\xi = (t - z/v_p)$.

That an arbitrary function $f(t - z/v_p)$ represents a wave traveling in the $+z$ direction is illustrated in Figure 2.6 for $v_p = 1$ m-s^{-1}. By comparing $f(t - z/v_p)$ at two different times $t = 2$ and $t = 3$ s, we note that the function maintains its shape and moves in the $+z$ direction as time t advances, as seen in Figure 2.6a. Similarly, by comparing $f(t - z/v_p)$ at two different positions $z = 0$ and $z = 1$ m, we note that the function maintains its shape but appears at $z = 1$ m exactly 1 s after its appearance at $z = 0$, as seen in Figure 2.6b. Figure 2.6c shows a three-dimensional display of $f(t - z/v_p)$ as a function of time at different points z_1, z_2, and z_3. To determine the speed with which the function moves in the $+z$ direction, we can simply keep track of any point on the function by setting its argument to a constant. In other words, we have

$$t - \frac{z}{v_p} = \text{const} \quad \rightarrow \quad \frac{dz}{dt} = v_p$$

The speed of propagation of waves on a transmission line is one of its most important characteristics. It is evident from [2.7] that v_p depends on the line inductance L and C. In the case of most (all except the microstrip line) of the commonly used two-conductor transmission lines shown in Figure 2.1, the phase velocity v_p in the absence of losses is not a function of the particular geometry of the metallic conductors

[13]An important function of $(t - z/v_p)$ that is encountered often and that we shall study in later chapters is the sinusoidal traveling-wave function, $A\cos[\omega(t - z/v_p)]$. Depending on the location of the observation point z along the z axis, this function replicates the sinusoidal variation $A\cos(\omega t)$ observed at $z = 0$, except delayed by (z/v_p) seconds at the new z. Thus, (z/v_p) represents a time shift, or delay, which is a characteristic of the class of wave functions of the variable $(t - z/v_p)$.

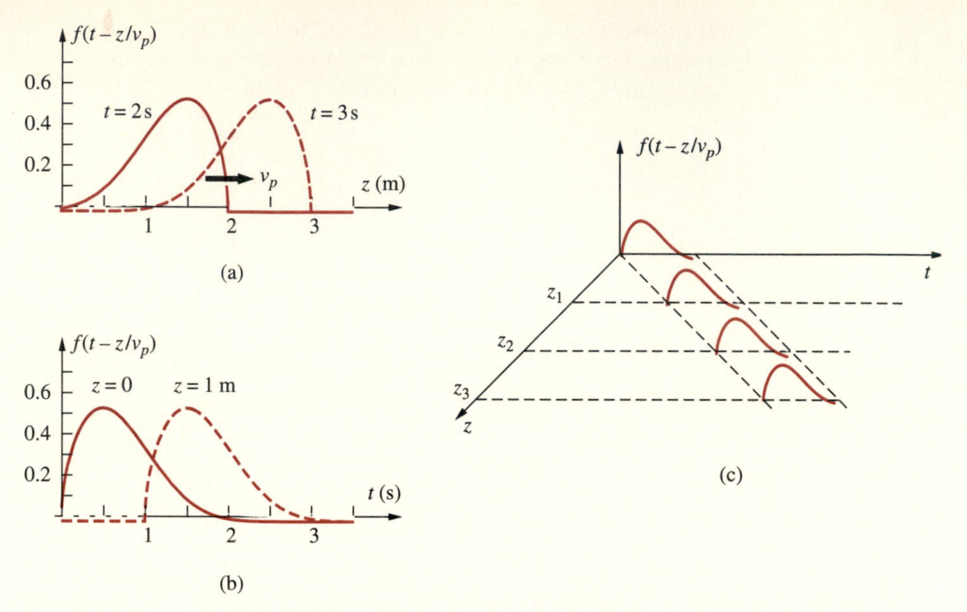

FIGURE 2.6. **Variation in space and time of an arbitrary function $f(t - z/v_p)$.** The phase velocity is taken to be equal to unity, i.e., $v_p = 1$ m/s. (a) $f(t - z/v_p)$ versus z at two different times. (b) $f(t - z/v_p)$ versus t at two different locations. (c) Three-dimensional display of $f(t - z/v_p)$ as a function of time at different points z_1, z_2, and z_3.

but is solely determined by the electrical and magnetic properties of the surrounding medium.[14] When the medium surrounding the metallic conductors is air, the phase velocity is equal to the speed of light in free space, namely $v_p = c$. The propagation speeds for other insulating materials are tabulated in Table 2.1.

It is clear from the above analysis that any function of the argument $(t + z/v_p)$ is an equally valid solution of [2.7]. Thus, the general solution for the voltage $\mathcal{V}(z, t)$ is

$$\mathcal{V}(z, t) = f^+\left(t - \frac{z}{v_p}\right) + f^-\left(t + \frac{z}{v_p}\right) \qquad [2.8]$$

where $f^+(t - z/v_p)$ and $f^-(t + z/v_p)$, respectively, represent waves traveling in the $+z$ and $-z$ directions. Note that $f^+(\cdot)$ and $f^-(\cdot)$ can in general be completely different functions.

[14]This result will become evident in Chapters 4 and 6 as we determine the capacitance and inductance of selected transmission lines from first principles. That $v_p = (LC)^{-1/2}$ does not depend on the geometrical arrangement of the conductors can also be seen by considering the inductance and capacitance expressions given in Table 2.2, Section 2.7. For transmission lines that do not exhibit symmetry in the cross-sectional plane, such as the microstrip line of Figure 2.1e, the phase velocity depends in a complicated manner on the properties of the surrounding material, the shape and dimensions of the conductors, and the operating frequency. See Section 8.6 of S. Ramo, J. R. Whinnery, and T. Van Duzer, *Fields and Waves in Communication Electronics,* 3rd ed., John Wiley & Sons, New York, 1994.

TABLE 2.1. Propagation speeds in some materials

Material	Propagation speed at ~20°C (cm/ns at 3 GHz)
Air	30
Glass	(3 to 15)*
Mica (ruby)	12.9
Porcelain	(10 to 13)*
Fused quartz (SiO_2)	15.4
Alumina (Al_2O_3)	10.1
Polystyrene	18.8
Polyethylene	20.0
Teflon	20.7
Vaseline	20.4
Amber (fossil resin)	18.6
Wood (balsa)	27.2
Water (distilled)	3.43
Ice (pure distilled)	16.8**
Soil (sandy, dry)	18.8

*Approximate range valid for most types of this material.
**At −12°C.

To find the general solution for the current $\mathcal{I}(z, t)$ associated with the voltage $\mathcal{V}(z, t)$, we can substitute [2.8] in [2.3] and [2.4], integrate with respect to time, and then take the derivative with respect to z to find

$$\mathcal{I}(z, t) = \frac{1}{Z_0}\left[f^+\left(t - \frac{z}{v_p}\right) - f^-\left(t + \frac{z}{v_p}\right)\right]; \qquad Z_0 \equiv \sqrt{\frac{L}{C}} \qquad [2.9]$$

where Z_0 is known as the *characteristic impedance* of the transmission line.[15] The characteristic impedance is the ratio of voltage to current for a single wave propagating in the $+z$ direction, as is evident from [2.8] and [2.9]. Note that the current associated with the wave traveling in the $-z$ direction (i.e., toward the left) has a negative sign—as expected, since the direction of positive current as defined in Figure 2.5 is to the right. In other words, since the polarity of voltage and the direction of current are defined so that the voltage and current have the same signs for forward (to the right)-traveling waves, the voltage and current for waves traveling to the left have opposite signs.

[15]To find $\mathcal{I}(z, t)$, we can also note that the wave equation [2.6] for the current is identical to equation [2.5] for voltage, so its solution should have the same general form. Thus, the general solution for the current should be

$$\mathcal{I}(z, t) = g^+\left(t - \frac{z}{v_p}\right) + g^-\left(t + \frac{z}{v_p}\right)$$

Substituting this expression for $\mathcal{I}(z, t)$ and [2.8] into [2.3] or [2.4] yields $g^+ = Z_0^{-1} f^+$ and $g^- = -Z_0^{-1} f^-$.

The characteristic impedance of a line is one of the most important parameters in the equations describing its behavior as a distributed circuit. For lossless lines, as considered here, Z_0 is a real number having units of ohms. Since Z_0 for a lossless line depends only on L and C, and since these quantities can be calculated from the geometric shape and physical dimensions of the line and the properties of the surrounding material, Z_0 can be expressed in terms of these physical dimensions for a given type of line. Formulas for Z_0 for some common lines are provided in Section 2.7. Characteristic impedances for other types of transmission lines are given elsewhere.[16]

The following example illustrates the meaning of the characteristic impedance of a line. An infinitely long and initially uncharged line (i.e., all capacitors and inductors in the distributed circuit have zero initial conditions) is considered, so there is no need for the $f(t + z/v_p)$ term in either [2.8] or [2.9], which would be produced only as a result of the reflections of the voltage disturbance at the end of the line.

Example 2-1: Step response of an infinitely long lossless line. As a simple example of the excitation of a transmission line by a source, consider an infinitely long lossless transmission line characterized by L and C and connected to an ideal step voltage source of amplitude V_0 and source resistance R_s, as shown in Figure 2.7a. Find the voltage, the current, and power propagating down the transmission line.

Solution: Before $t = 0$, the voltage and current on the line are identically zero, since the line is assumed to be initially uncharged. At $t = 0$, the step voltage source changes from 0 to V_0, launching a voltage $\mathcal{V}^+(z, t)$ propagating toward the right. In the absence of a reflected wave (infinitely long line), the accompanying current $\mathcal{I}^+(z, t) = (Z_0)^{-1}\mathcal{V}^+(z, t)$, as can be seen from equations [2.8] and [2.9]. In other words, the characteristic impedance $Z_0 = \sqrt{L/C}$ is the resistance that the transmission line *initially* presents to the source, as shown in the equivalent circuit of Figure 2.7b. Accordingly, the initial voltage and current established at the source end ($z = 0$) of the line are

$$\mathcal{V}_s(t) = \mathcal{V}^+(z = 0, t) = \frac{V_0 Z_0}{Z_0 + R_s}$$

and

$$\mathcal{I}_s(t) = \mathcal{I}^+(z = 0, t) = \frac{V_0}{Z_0 + R_s}$$

The propagation of the voltage $\mathcal{V}^+(z, t)$ and the current $\mathcal{I}^+(z, t)$ down the line are illustrated in Figures 2.7c and 2.7d at $t = l/v_p$ as a function of z.

---

[16] *Reference Data for Engineers,* 8th ed., Sams Prentice Hall Computer Publishing, Carmel, Indiana, 1993.

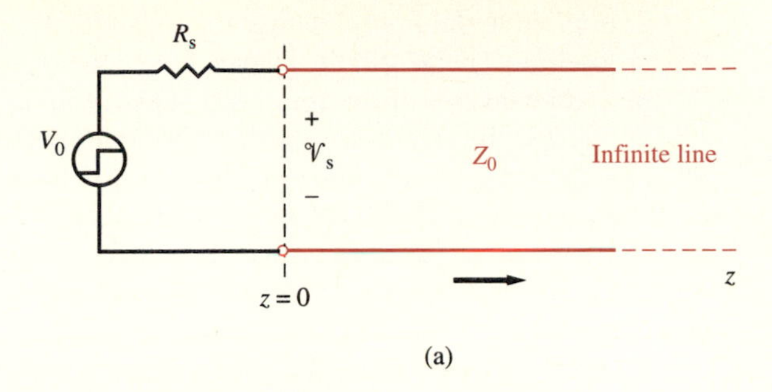

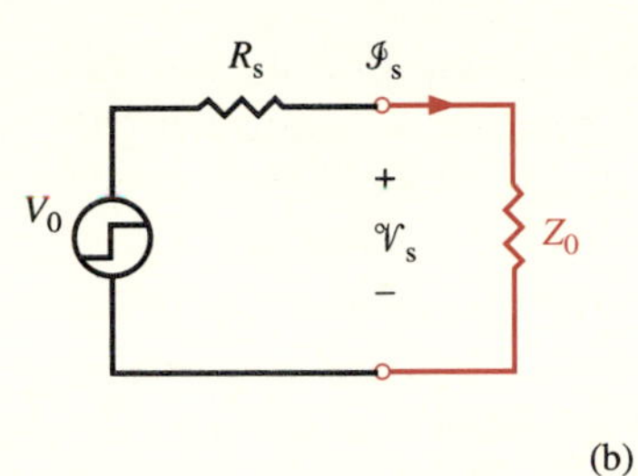

$\mathcal{V}^+(z, t = l/v_p)$

$(V_0Z_0)/(R_s + Z_0)$

v_p

l

z

(c)

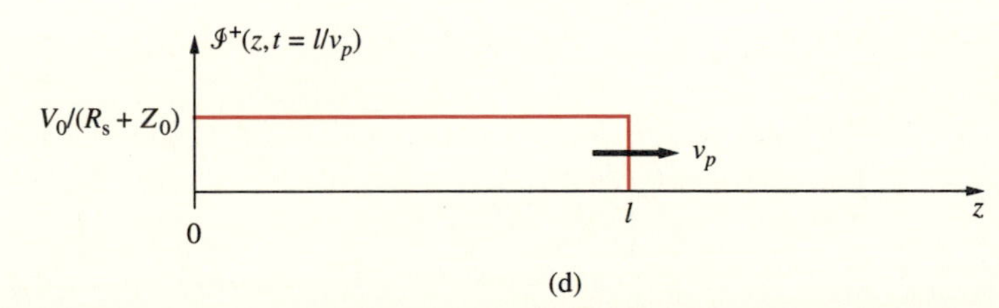

FIGURE 2.7. Step excitation of a lossless line. (a) Step voltage applied to an infinite lossless line. (b) Initial equivalent circuit seen from the source. (c) The voltage disturbance $\mathcal{V}^+(z, t)$. (d) The current disturbance $\mathcal{I}^+(z, t)$.

Note that the flow of a current $\mathcal{I}_s$ outward from a source producing a voltage V_0 represents a total power of $P_0 = \mathcal{I}_s V_0$ supplied by the source. Part of this power, given by $\mathcal{I}_s^2 R_s$, is dissipated in the source resistance. The remainder, given by

$$P_{\text{line}}^+ = \mathcal{I}^+(0, t)\mathcal{V}^+(0, t) = \mathcal{I}_s \mathcal{V}_s = \frac{V_0^2 Z_0}{(Z_0 + R_s)^2}$$

is supplied to the line. Because the line is lossless, there is no power dissipation on the line. Therefore, the power P^+_{line} goes into charging the capacitances and the inductances[17] of the line, as discussed in connection with Figure 2.2. Note that as P^+_{line} travels down the line the amounts of energy stored respectively in the capacitance and inductance of a fully charged portion of the line of length dl are given by $\frac{1}{2}(C\,dl)\mathcal{V}_s^2$ and $\frac{1}{2}(L\,dl)\mathcal{I}_s^2$.

2.3 REFLECTION AT DISCONTINUITIES

In most transmission line applications, lines are connected to resistive loads, other lines (with different characteristic impedances), reactive loads, combinations of resistive and reactive loads, or loads with nonlinear current-voltage characteristics. Such discontinuities impose boundary conditions, which cause reflection of the incident voltages and currents from the discontinuities, while new voltages and currents are launched in the opposite direction. In this section we consider the reflection process and also provide examples of step responses of transmission lines terminated with short- and open-circuited terminations.

Consider a transmission line terminated in a load resistance R_L located at $z = l$, as shown in Figure 2.8, on which a voltage of $\mathcal{V}_1^+(z, t)$ is initially ($t = 0$) launched by the source at $z = 0$. Note that for an ideal step voltage source of amplitude V_0 and a source resistance R_s, as shown in Figure 2.8, the amplitude of $\mathcal{V}_1^+(z, t)$ is given by

$$\mathcal{V}_1^+(0, 0) = \frac{Z_0 V_0}{R_s + Z_0}$$

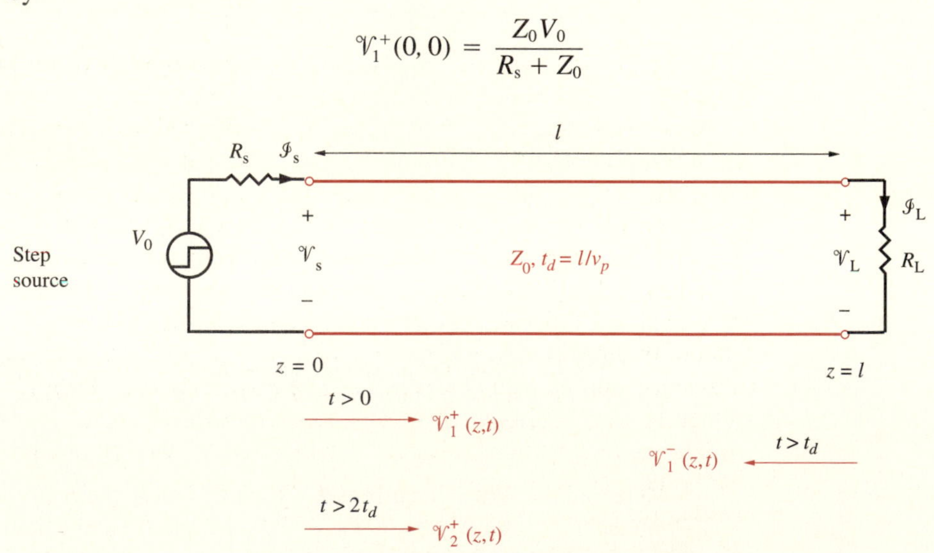

FIGURE 2.8. A terminated transmission line. The load R_L is located at $z = l$, while the source end is at $z = 0$.

[17]"Charging" an inductance can be thought of as establishing a current in it.

In general, a new reflected voltage $\mathcal{V}_1^-(z,t) = \mathcal{V}_1^-(l,t)$ is generated when the disturbance $\mathcal{V}_1^+(z,t)$ reaches the load position at time $t = t_d$, where t_d is the one-way travel time on the line, or $t_d = l/v_p$. In order to determine the amplitude of the reflected wave, we write the total voltage and current at the load position (i.e., $z = l$) at $t = t_d^+$ (i.e., immediately after the arrival of the incident wave) as

$$\mathcal{V}_L(t) = \mathcal{V}_1^+(l,t) + \mathcal{V}_1^-(l,t) \qquad [2.10a]$$

$$\mathcal{I}_L(t) = \frac{1}{Z_0}\mathcal{V}_1^+(l,t) - \frac{1}{Z_0}\mathcal{V}_1^-(l,t) \qquad [2.10b]$$

Using [2.10a] and [2.10b] and the *boundary condition* $\mathcal{V}_L(t) = \mathcal{I}_L(t)R_L$ imposed by the purely resistive termination R_L, we can write

$$\mathcal{I}_L(t) = \frac{\mathcal{V}_L(t)}{R_L} \quad \rightarrow \quad \frac{\mathcal{V}_1^+(l,t)}{Z_0} - \frac{\mathcal{V}_1^-(l,t)}{Z_0} = \frac{\mathcal{V}_1^+(l,t) + \mathcal{V}_1^-(l,t)}{R_L} \qquad [2.11]$$

From [2.11] we can find the ratio of the reflected voltage $\mathcal{V}_1^-(l,t)$ to the incident one $\mathcal{V}_1^+(l,t)$. This ratio is defined as the *load voltage reflection coefficient,* Γ_L,

$$\boxed{\Gamma_L \equiv \frac{\mathcal{V}_1^-(l,t)}{\mathcal{V}_1^+(l,t)} = \frac{R_L - Z_0}{R_L + Z_0}} \qquad [2.12]$$

and it follows that

$$\frac{R_L}{Z_0} = \frac{1 + \Gamma_L}{1 - \Gamma_L} \qquad [2.13]$$

The reflection coefficient is one of the most important parameters in transmission line analysis. Accordingly, the simple expression [2.12] for Γ_L should be memorized. For lines terminated in resistive loads, Γ_L can have values in the range $-1 \leq \Gamma_L \leq +1$, where the extreme values of -1 and $+1$ occur when the load is, respectively, a short circuit (i.e., $R_L = 0$) and an open circuit (i.e., $R_L = \infty$). The special case of $\Gamma_L = 0$ occurs when $R_L = Z_0$; meaning that the load resistance is the same as the characteristic impedance,[18] that there is no reflected voltage, and that all the power carried by the incident voltage is absorbed by the load. It is important to note that expression [2.12] for the reflection coefficient was arrived at in a completely general fashion. In other words, whenever a voltage $\mathcal{V}_k^+(z,t)$ is incident on a load R_L from a transmission line with characteristic impedance Z_0, the amplitude of the reflected voltage $\mathcal{V}_k^-(l,t)$ is $\Gamma_L\mathcal{V}_k^+(l,t)$, with Γ_L given by [2.12].

The generality of [2.12] can also be used to determine the reflection of the new voltage $\mathcal{V}_1^-(z,t)$ when it arrives at the source end of the line. Having originated at the load end at $t = t_d$, the reflected voltage $\mathcal{V}_1^-(z,t)$ arrives at the source end (terminated with the source resistance R_s) at $t = 2t_d$. At that point, it can be viewed as a new voltage disturbance propagating on a line with characteristic

[18]This condition is referred to as a *matched* load and is highly preferred in most applications.

impedance Z_0 that is incident on a resistance of R_s. Thus, its arrival at the source end results in the generation of a new reflected voltage traveling toward the load. We denote this new voltage $\mathcal{V}_2^+(z, t)$, where the subscript distinguishes it from the original voltage $\mathcal{V}_1^+(z, t)$ and the superscript underscores the fact that it is propagating in the $+z$ direction. The amplitude of the new reflected voltage $\mathcal{V}_2^+(z, t)$ is determined by the *source reflection coefficient* Γ_s, which applies at the source end of the line and is defined as

$$\Gamma_s \equiv \frac{\mathcal{V}_2^+(0, t)}{\mathcal{V}_1^-(0, t)} = \frac{R_s - Z_0}{R_s + Z_0} \qquad [2.14]$$

Thus we have $\mathcal{V}_2^+(z, t) = \mathcal{V}_2^+(0, t) = \Gamma_s \mathcal{V}_1^-(0, t)$.

Note that the voltage $\mathcal{V}_1^+(z, t)$ was created at $t = 0$ and still continues to exist. Thus the source-end voltage and current at $t = 2t_d^+$ are

$$\begin{aligned} \mathcal{V}_s(t) &= \mathcal{V}_1^+(0, t) + \mathcal{V}_1^-(0, t) + \mathcal{V}_2^+(0, t) \\ &= \mathcal{V}_1^+(0, t)(1 + \Gamma_L + \Gamma_s\Gamma_L) \end{aligned} \qquad [2.15a]$$

$$\begin{aligned} \mathcal{I}_s(t) &= \frac{1}{Z_0}\mathcal{V}_1^+(0, t) - \frac{1}{Z_0}\mathcal{V}_1^-(0, t) + \frac{1}{Z_0}\mathcal{V}_2^+(0, t) \\ &= \frac{1}{Z_0}\mathcal{V}_1^+(0, t)(1 - \Gamma_L + \Gamma_s\Gamma_L) \end{aligned} \qquad [2.15b]$$

The newly generated voltage $\mathcal{V}_2^+(z, t)$ will now arrive at the load end at $t = 3t_d$ and lead to the creation of a new reflected voltage $\mathcal{V}_2^-(z, t)$, and this process will continue indefinitely. In practice, the step-by-step calculation of the successively generated voltages becomes tedious, especially for arbitrary resistive terminations. In such cases, the graphical construction of a *bounce diagram* is very helpful. We introduce the bounce diagram in the following subsection.

2.3.1 Bounce Diagrams

A bounce diagram, illustrated in Figure 2.9, also called a reflection diagram or lattice diagram, is a distance-time plot used to illustrate successive reflections along a transmission line. The distance along the line is shown on the horizontal axis, and time is shown on the vertical axis. The bounce diagram is a plot of the time elapsed versus distance z from the source end, showing the voltages traveling in the $+z$ and $-z$ directions as straight lines sloping[19] downward from left to right and right to left, respectively. Each sloping line corresponds to an individual traveling voltage and is labeled with its amplitude. The amplitude of each reflected voltage is obtained by multiplying the amplitude of the preceding voltage by the reflection coefficient at the position where the reflection occurs.

[19]The slope of the lines on the bounce diagram (i.e., dt/dz) can be thought of as corresponding to $(\pm v_p)^{-1}$.

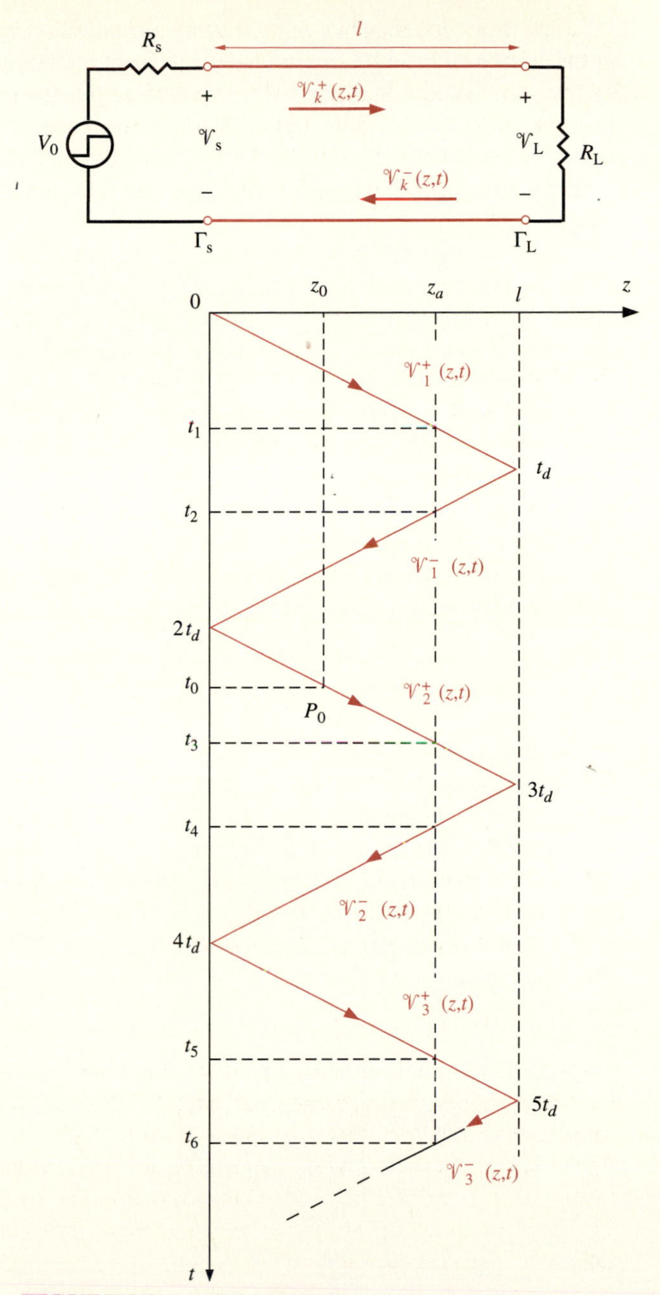

FIGURE 2.9. **Bounce diagram.**

The time sequence of events starting with the first application of the step voltage at the source end can be easily visualized from the bounce diagram. The application of the step voltage launches $\mathcal{V}_1^+(z, t)$ toward the load. This voltage arrives at the load end at $t = t_d$, and its arrival leads to the generation of $\mathcal{V}_1^-(z, t) = \Gamma_L \mathcal{V}_1^+(z, t)$ propagating toward the source. This new voltage $\mathcal{V}_1^-(z, t)$ arrives at the source end and leads to the generation of $\mathcal{V}_2^+(z, t)$, and this process continues back and forth indefinitely.

Once constructed, a bounce diagram can be used conveniently to determine the voltage distribution along the transmission line at any given time, as well as the variation of voltage with time at any given position. Suppose we wish to find the voltage distribution $\mathcal{V}(z, t_0)$ along the line at $t = t_0$, chosen arbitrarily to be $2t_d < t_0 < 3t_d$. To determine $\mathcal{V}(z, t_0)$, we mark t_0 on the time axis of the bounce diagram and draw a horizontal line from t_0, intersecting the sloping line marked $\mathcal{V}_2^+(z, t)$ at point P_0, as shown in Figure 2.9. Note that all sloping lines below the point P_0 are irrelevant for $\mathcal{V}(z, t_0)$, since they correspond to later times. If we now draw a vertical dashed line through P_0, we find that it intersects the z axis at z_0. At time $t = t_0$, all points along the line have received voltages $\mathcal{V}_1^+(z, t)$ and $\mathcal{V}_1^-(z, t)$. However, only points to the left of z_0 have yet received the voltage $\mathcal{V}_2^+(z, t)$, so a discontinuity exists in the voltage distribution at $z = z_0$. In other words, we have

$$\mathcal{V}(z, t_0) = \begin{cases} \mathcal{V}_1^+(z, t)(1 + \Gamma_L + \Gamma_s\Gamma_L) & z < z_0 \\ \mathcal{V}_1^+(z, t)(1 + \Gamma_L) & z > z_0 \end{cases}$$

Alternatively, we may wish to determine the variation of voltage as a function of time at a fixed position, say z_a. To determine $\mathcal{V}(z_a, t)$, we simply look at the intersection points with the sloping lines of the vertical line passing through z_a, as shown in Figure 2.9. Horizontal lines drawn from these intersection points, crossing the time axis at $t_1, t_2, t_3, t_4, \ldots$, are the time instants at which each of the new voltages $\mathcal{V}_1^+(z, t), \mathcal{V}_1^-(z, t), \mathcal{V}_2^+(z, t), \mathcal{V}_2^-(z, t), \ldots$, arrive at $z = z_a$ and abruptly change the total voltage at that point. Thus, the time variation of the voltage at $z = z_a$, namely $\mathcal{V}(z_a, t)$ is given as

$$\mathcal{V}(z_a, t) = \begin{cases} 0 & 0 < t < t_1 \\ \mathcal{V}_1^+(z_a, t) & t_1 < t < t_2 \\ \mathcal{V}_1^+(z_a, t)(1 + \Gamma_L) & t_2 < t < t_3 \\ \mathcal{V}_1^+(z_a, t)(1 + \Gamma_L + \Gamma_s\Gamma_L) & t_3 < t < t_4 \\ \mathcal{V}_1^+(z_a, t)(1 + \Gamma_L + \Gamma_s\Gamma_L + \Gamma_L\Gamma_s\Gamma_L) & t_4 < t < t_5 \\ \cdots & \cdots \end{cases}$$

where $\mathcal{V}_1^+(z_a, t) = \mathcal{V}_1^+(z, t)$.

A bounce diagram can also be constructed to keep track of the component current waves. However, it is also possible and less cumbersome to derive the component line current, $\mathcal{I}_k^{\pm}(z, t)$, simply from the corresponding voltage $\mathcal{V}_k^{\pm}(z, t)$. In this connection, all we need to remember is that the current associated with any voltage disturbance $\mathcal{V}_k^+(z, t)$ propagating in the $+z$ direction is simply $\mathcal{I}_k^+(z, t) = (Z_0^{-1})\mathcal{V}_k^+(z, t)$, whereas that associated with a voltage disturbance $\mathcal{V}_k^-(z, t)$ propagating in the $-z$ direction is $\mathcal{I}_k^-(z, t) = -(Z_0^{-1})\mathcal{V}_k^-(z, t)$.

The uses of the bounce diagram in specific cases are illustrated in Examples 2-2 through 2-9.

2.3.2 The Reflection Process

Before we proceed with specific examples, we now provide a heuristic discussion of the reflection process for the case of a transmission line terminated in an open circuit.[20] This discussion involves the same considerations as the heuristic discussion in connection with Figure 2.2 of the propagation of disturbances along a transmission line in terms of successive charging of capacitors and inductors. When the voltage disturbance reaches the open-circuited end of the line, its orderly progress of successively charging the distributed circuit elements is interrupted. Consider the approach of a disturbance to the end of an open-circuited transmission line, shown in Figure 2.10a. Figure 2.10b shows the voltage and current progressing together; L_y carries current but L_z does not, and C_y is charged to the source voltage V_0 but C_z

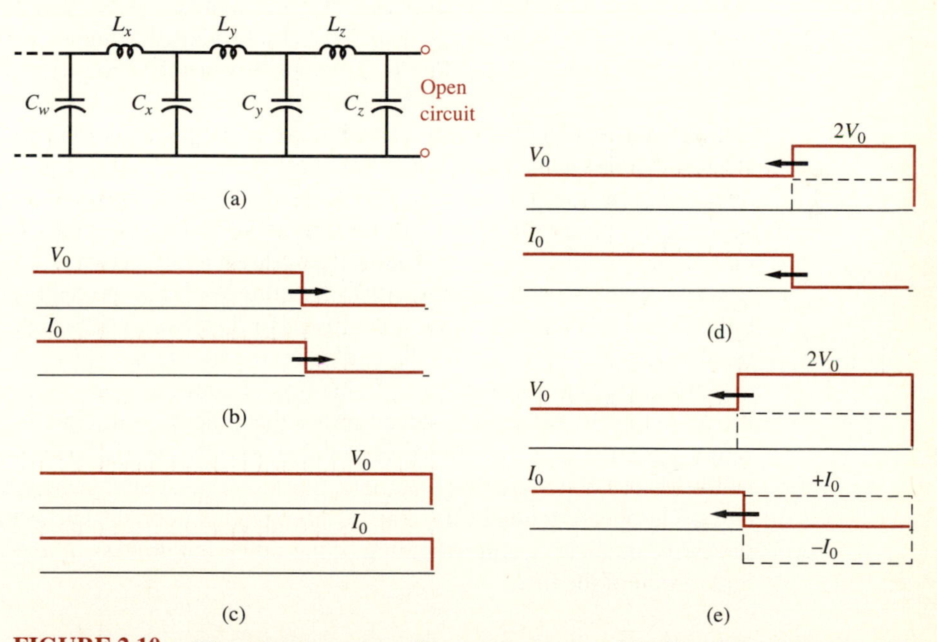

FIGURE 2.10. The reflection process. The orderly progress of the disturbances of current and voltage, propagating initially from left to right, is interrupted when they reach the open end of the line, leading to the reflection of the disturbance.

[20]The qualitative discussion presented herein is based on a similar discussion in Chapter 14 of H. H. Skilling, *Electric Transmission Lines,* McGraw-Hill, 1951.

is not. The voltage on C_y, however, causes current to flow through L_z, and thus through C_z, charging it to a voltage V_0 (Figure 2.10c).

At the time C_z is charged, all of the inductances (including L_z) carry the full current I_0. The progress of the disturbance along the line cannot continue any more, since there is no inductance beyond C_z to serve as an outlet for current as C_z is charged. As a result, C_z becomes overcharged, since the current through L_z cannot stop until the stored magnetic energy is exhausted. Thus, current continues to flow through C_z until it is charged[21] to twice its normal value ($2V_0$), at which time the current through L_z drops to zero (i.e., L_z now acts like an open circuit) (Figure 2.10d).

When L_z stops carrying current, the current carried by L_y is now driven solely into C_y, doubling its voltage and forcing the current in L_y to stop. At the same time, the voltages at the two ends of L_z are both $2V_0$, so that the current through L_z remains zero and the doubly charged capacitor C_z stays at $2V_0$. We now have the condition depicted in Figure 2.10e, where C_y and C_z are both at $2V_0$ and L_y and L_z both have zero current. This procedure now continues along the line from right to left as the voltage on the line is doubled and the current drops to zero.

The above described phenomenon can be viewed as a reflection, since the original disturbance, traveling from left to right, appears to be reflected from the end of the line and to begin to progress from right to left. In Figure 2.10e, the reflected voltage disturbance of amplitude V_0 travels toward the left, and adds to the previously existing line voltage, making the total voltage $2V_0$. It is accompanied by the reflected current of amplitude $-I_0$, which is added on top of the existing line current I_0, making the total current on the line zero. Although the front of the current disturbance is progressing toward the left, it should be noted that this does not imply any reversal of current flow. The current originally flowing from the source toward the end of the line (i.e., $\mathcal{I}_1^+(z, t)$) continues to do so even after reflection. This current flows from left to right to charge the capacitances when the disturbance progresses toward the end of the line, and it continues to flow from left to right as the reflected disturbance returns, doubling the charge on the line as the voltage is raised to $2V_0$.

All of these physical effects of charging of capacitors and current flows through the inductors are simply accounted for by the general solutions for voltage and current as given respectively by [2.8] and [2.9] and by the application of the boundary condition at the termination—namely, that there must always be zero current at the end of an open-circuited line. The purpose of this heuristic discussion is simply to provide a qualitative understanding of the reflection process in terms of the equivalent circuit of the line.

---

[21] It is not obvious why the capacitor would charge to precisely twice its normal value. The circuit model of Figure 2.10a, consisting of discrete elements, is not adequate for the determination of the precise value of the reflected voltage, which is unambiguously determined by the governing wave equations [2.5], [2.6] and their solutions [2.10] as applied to an open-circuited termination, as shown in the next section. Nevertheless, consider the fact that the amount that the capacitor voltage is charged to is determined by $\int I\,dt$; thus, with no other outlet for the inductor current, twice the normal current goes through the capacitance, charging it to twice its normal value.

2.3.3 Open- and Short-Circuited Transmission Lines

We now consider examples of step responses of transmission lines with the simplest type of terminations, namely an open or a short circuit. The circumstances treated in Examples 2-2 and 2-3 are commonly encountered in practice, especially in computer-communication problems; for example, when the voltage at one end of an interconnect switches to the HIGH state due to a change in the status of a logic gate. The resultant response is then similar to the short-circuited line case (Example 2-2) if the interconnect is a short-circuited matching stub or drives (i.e., is terminated in) a subsequent gate (or another interconnect) with low input impedance ($R_L \ll Z_0$). Alternatively, and more commonly in practice, the interconnect would be driving a gate with a very high input impedance ($R_L \gg Z_0$), corresponding to an open-circuited termination (Example 2-3).

Example 2-2: Step voltage applied to a short-circuited lossless line. Consider the transmission line of length l terminated with a short circuit at the end (i.e., $R_L = 0$), as shown in Figure 2.11a. Sketch the voltages $\mathcal{V}_s$ and $\mathcal{V}_{l/2}$ as a function of time.

Solution: Initially, the applied voltage is divided between the source resistance R_s and the line impedance Z_0 in the same manner as for the infinite line in Example 2-1, and at $t = 0^+$, a voltage $\mathcal{V}_1^+(z, t)$ of amplitude $\mathcal{V}_1^+(0, 0) = (V_0 Z_0)/(Z_0 + R_s)$ is launched at the source end of the line. The corresponding

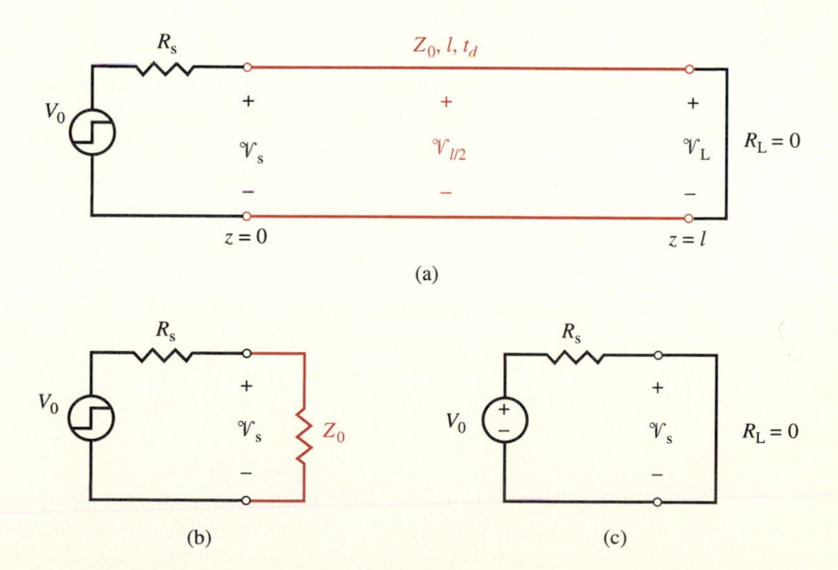

FIGURE 2.11. A short-circuited lossless line. (a) Step voltage applied to a short-circuited lossless line. (b) The initial equivalent circuit. (c) Steady-state (final) equivalent circuit.

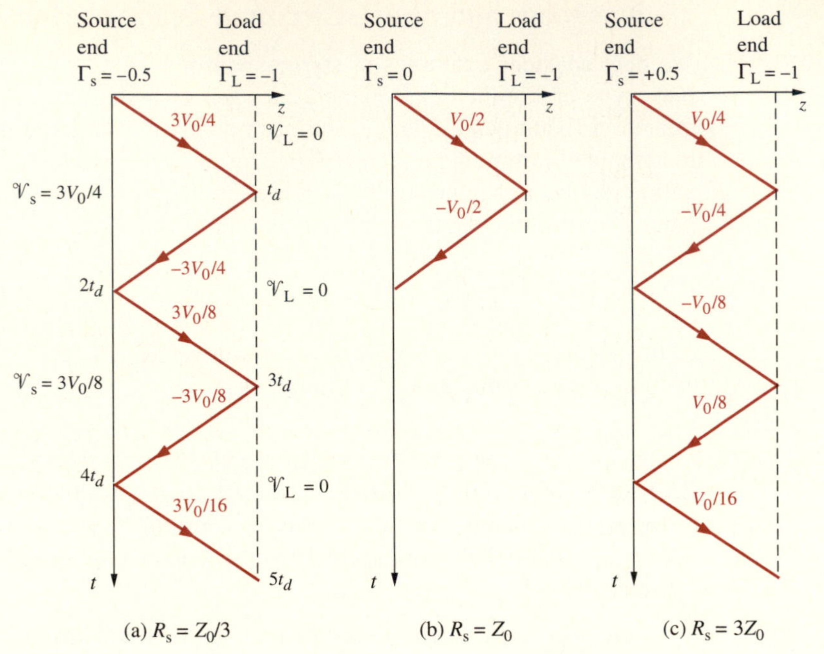

FIGURE 2.12. Bounce diagram for Example 2-2 (Figure 2.11). (a) $R_s = Z_0/3$, (b) $R_s = Z_0$, and (c) $R_s = 3Z_0$.

current is $\mathcal{I}_1^+(z, t) = \mathcal{V}_1^+(z, t)/Z_0$ and has an amplitude of $\mathcal{I}_1^+(0, 0) = V_0/(Z_0 + R_s) = \mathcal{V}_1^+(0, 0)/Z_0$. The equivalent circuit initially presented to the source by the line is thus simply a resistance of Z_0, as shown in Figure 2.11b. Eventually, when all transients die out, the equivalent circuit of the line is a short circuit (Figure 2.11c); thus, the voltage everywhere on the line (e.g., $\mathcal{V}_s$, $\mathcal{V}_{l/2}$, and $\mathcal{V}_L$) must eventually approach zero.

To analyze the behavior of the line voltage, we use a bounce diagram as shown in Figure 2.12. When the voltage disturbance $\mathcal{V}^+(z, t)$ reaches the end of the line (at $t = t_d = l/v_p$), a reflected voltage of $\mathcal{V}^-(z, t) = -\mathcal{V}^+(z, t)$ is generated,[22] since the total voltage at the short circuit ($R_L = 0$) has to be $\mathcal{V}_L(t) = \mathcal{V}^+(l, t) + \mathcal{V}^-(l, t) = 0$. In other words, the reflection coefficient at the load end is

$$\Gamma_L = \frac{\mathcal{V}_1^-(l, t)}{\mathcal{V}_1^+(l, t)} = \frac{0 - Z_0}{0 + Z_0} = -1$$

[22]The reflection process at the short circuit occurs rather differently than that from an open circuit as discussed in connection with Figure 2.10. As the next-to-the-last capacitance (C_y) is charged to V_0, L_z begins to carry current; however, C_z cannot take any charge, since it is short-circuited. Current flows freely from L_z through the short circuit and into the return conductor (just like the flow of water from the open end of a pipe), until the charge on C_y is exhausted. As a result, the current through L_z becomes twice as much as its normal value ($2I_0$), and the voltage across C_y drops to zero. In this way, "reflection" reduces the voltage from V_0 to zero and increases current from I_0 to $2I_0$.

However, the current $\mathcal{I}_1^-(z, t)$ associated with the voltage traveling in the $-z$ direction is $-\mathcal{V}_1^-(z, t)/Z_0$, resulting in a reflected current of $V_0/(Z_0 + R_s)$, which adds directly to the incident current $\mathcal{I}_1^+(z, t) = V_0/(Z_0 + R_s)$ traveling in the $+z$ direction, doubling the total current on the line.

As it travels toward the source during $t_d < t < 2t_d$, the reflected voltage makes the total voltage everywhere on the line zero and the total current on the line equal to $2V_0/(Z_0 + R_s)$. When this disturbance reaches the source end at $t = 2t_d$, the source presents an impedance of R_s, and the reflection coefficient at the source end (i.e., Γ_s) depends on the value of R_s relative to Z_0. For $R_s = Z_0$, we have $\Gamma_s = 0$, and no further reflections occur, so the voltage on the line remains zero; in other words, a steady state is reached. However, for $R_s \neq Z_0$, it takes further reflections to eventually reach steady state, as illustrated in Figure 2.13, where the time evolution of the voltages at the source end ($\mathcal{V}_s(t)$) and at the center ($\mathcal{V}_{l/2}(t)$) of the transmission line are shown. Note that the load voltage $\mathcal{V}_L(t)$ is identically zero at all times, as dictated by the short-circuit termination. Note also that the voltage everywhere on the line, including at the center and at the source end, eventually approaches zero; however, we see from Figure 2.13 that the particular way in which $\mathcal{V}_s(t)$ and $\mathcal{V}_{l/2}(t)$ approach their final value of zero depends critically on the ratio R_s/Z_0.

Example 2-3: Overshoot and ringing effects. The distributed nature of a high-speed digital logic board commonly leads to *ringing*, the fluctuations of the voltage and current about an asymptotic value. Ringing results from multiple reflections, especially when an unterminated (i.e., open-circuited)[23] transmission line is driven by a low-impedance buffer. Consider the circuit shown in Figure 2.14a, where a step voltage source of amplitude V_0 and a source resistance $R_s = Z_0/4$ drives a lossless transmission line of characteristic impedance Z_0 and a one-way propagation delay of t_d seconds. Sketch $\mathcal{V}_s$, $\mathcal{V}_L$, and $\mathcal{I}_s$ as a function of t.

Solution: At $t = 0$, the source voltage rises from 0 to V_0, and a voltage $\mathcal{V}_1^+(z, t)$ of amplitude $\mathcal{V}_1^+(0, 0) = Z_0V_0/(R_s + Z_0) = 0.8V_0$ is applied from the source end of the line. During $0 < t < t_d$, the line charges as the front of this voltage disturbance travels toward the load. At $t = t_d$, the disturbance front reaches the open end of the line, and a reflected voltage of amplitude $\mathcal{V}_1^-(l, t_d) = \Gamma_L\mathcal{V}_1^+(l, t_d) = 0.8V_0$ is launched back toward the source, since $\Gamma_L = 1$. (In other words, the total current at the open end of the line remains zero, so we have $\mathcal{I}_1^+(l, t) + \mathcal{I}_1^-(l, t) = 0$, and thus $\mathcal{V}_1^-(l, t) = \mathcal{V}_1^+(l, t) = 0.8V_0$.) Note that as long as the source voltage does not change, $\mathcal{V}_1^+(z, t)$ remains constant in time and also constant with z once it reaches the end of the line ($z = l$) at $t = t_d$. Once $\mathcal{V}_1^-(z, t)$ is launched (at $t = t_d^+$), it travels toward the source, reaching the source end at $t = 2t_d$, after which time it also remains constant and coexists along the

[23]Note that if the input impedance of a load device is very high compared to Z_0 (i.e., $R_L \gg Z_0$), R_L can be approximated as an open circuit.

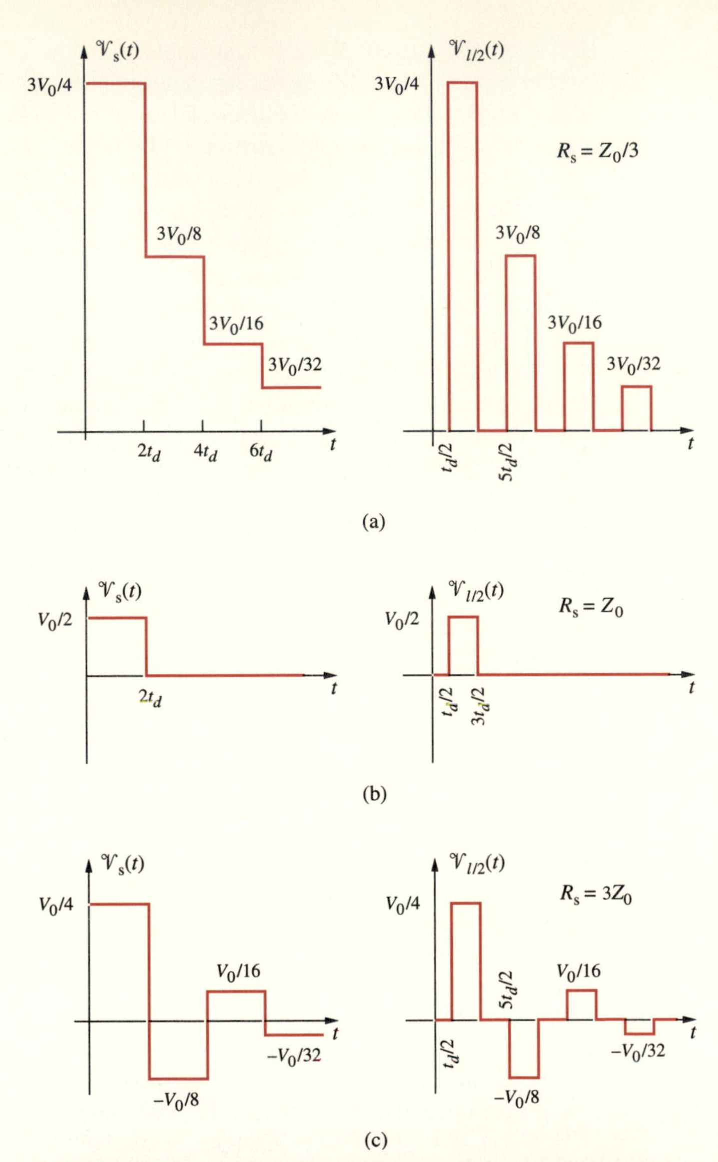

FIGURE 2.13. Step voltage applied to a short-circuited lossless line. Voltages at the source end ($\mathcal{V}_s$) and at the center ($\mathcal{V}_{l/2}$) of a short-circuited transmission line for source impedances of (a) $R_s = Z_0/3$, (b) $R_s = Z_0$, and (c) $R_s = 3Z_0$.

line with $\mathcal{V}_1^+(z, t)$. The arrival of the reflected voltage $\mathcal{V}_1^-(z, t)$ at the source end of the line at $t = 2t_d$ leads to the generation of a new reflected voltage disturbance of amplitude $\mathcal{V}_2^+(0, 2t_d) = \Gamma_s \mathcal{V}_1^-(0, 2t_d) = -0.48V_0$ [since $\Gamma_s = (R_s - Z_0)/(R_s + Z_0) = -0.6$] is launched toward the load. Note that $\mathcal{V}_2^+(z, t)$ is now superimposed on top of $\mathcal{V}_1^+(z, t)$ and $\mathcal{V}_1^-(z, t)$.

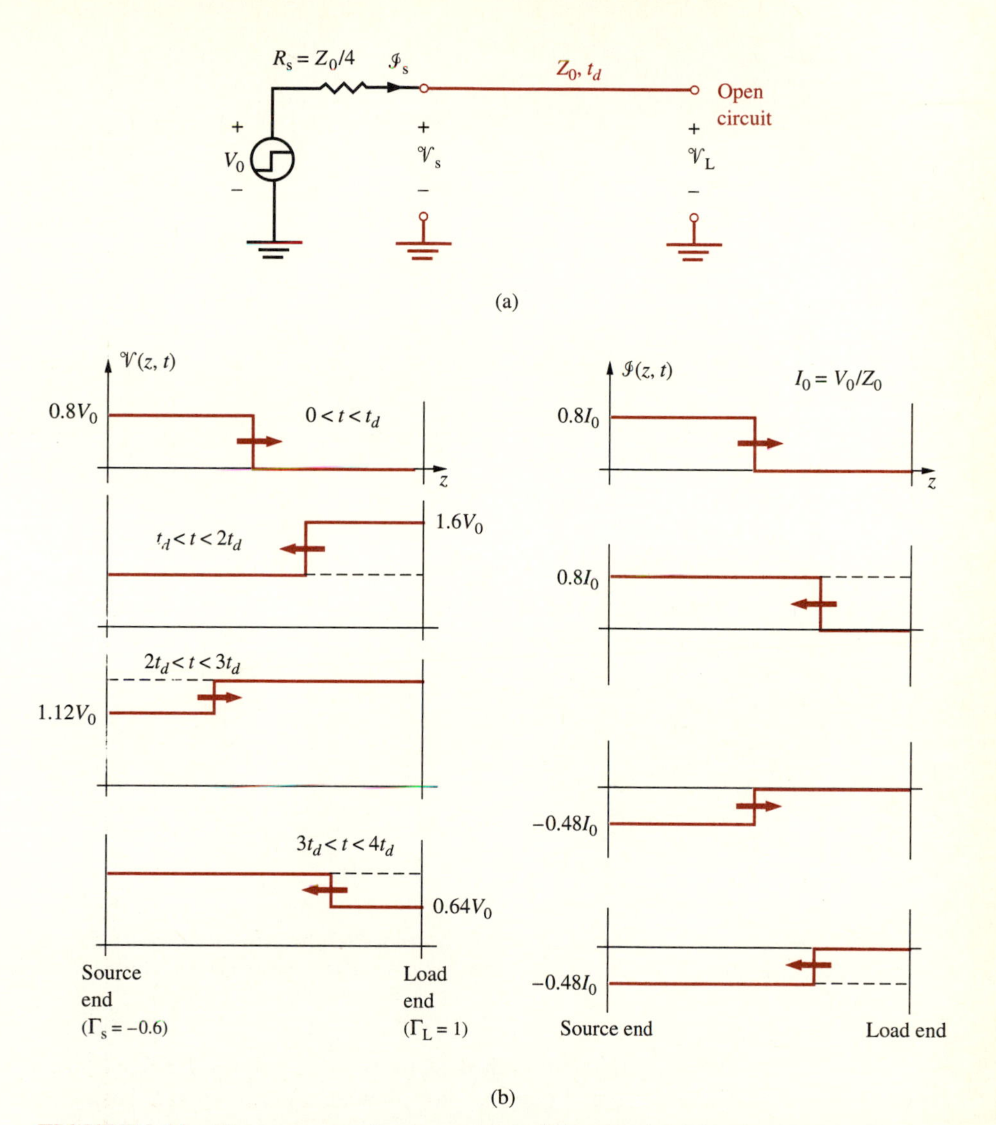

FIGURE 2.14. Step response of an open-circuited line. (a) Circuit for Example 2-3. (b) Voltage and current distributions along the line at different time intervals.

This process continues indefinitely, with the total voltage and current on the line gradually approaching their steady-state values. The voltage and current distribution along the line is shown in Figure 2.14b for various time intervals. The variation of the voltage (and thus the current) with time can be quantitatively determined by means of a bounce diagram, as shown in Figure 2.15b, which specifies the values of the source- and load-end voltages at any given time. The temporal variations of the source- and load-end voltages and the source-end current, as derived from the bounce diagram, are also shown in Figure 2.15c. Both voltage waveforms oscillate about and asymptotically approach their final

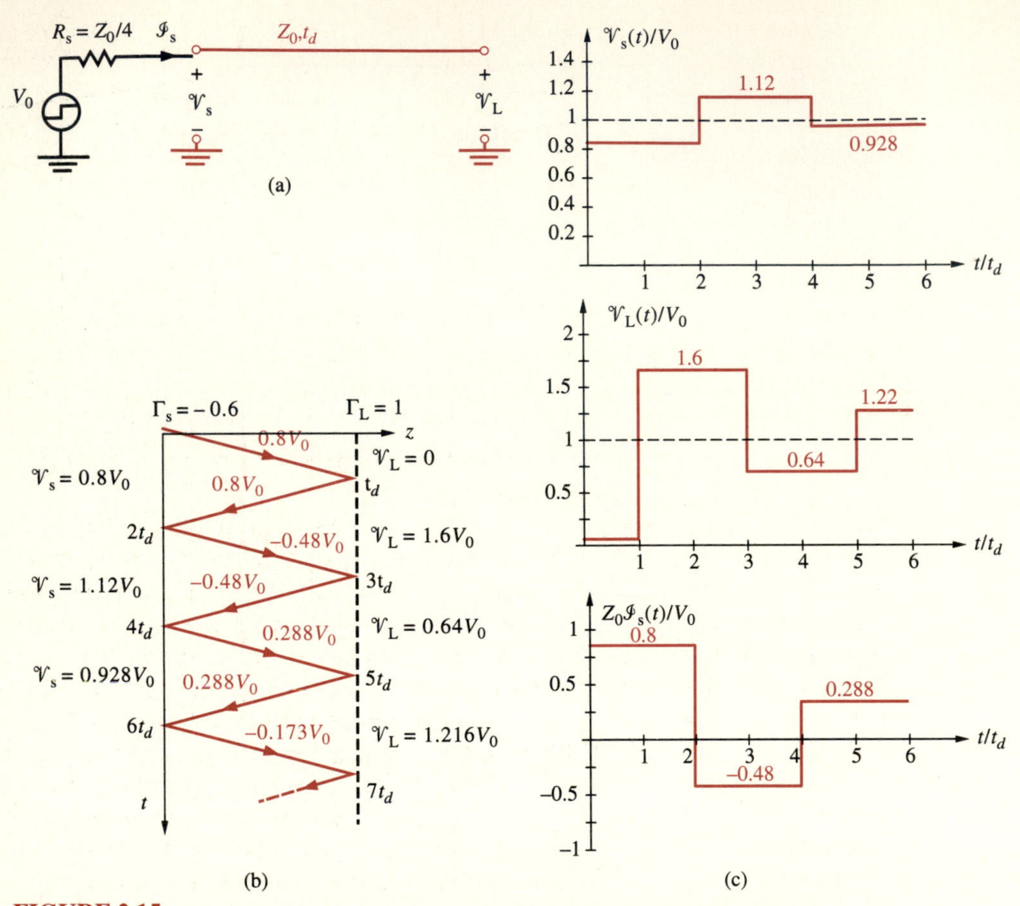

FIGURE 2.15. **Step response of an open-circuited lossless line.** (a) Circuit for Example 2-3. (b) Bounce diagram. (c) Normalized source- and load-end voltages and source-end current as a function of t/t_d.

value V_0 (shown as dashed lines)—the process referred to earlier as ringing. The source-end current waveform eventually approaches zero, as expected for an open-circuited termination. For the case shown, the percentage maximum overshoot, defined as the percentage difference between the maximum value and the asymptotic value, for $\mathcal{V}_L$ is $[(1.6V_0 - V_0)/V_0] \times 100 = 60\%$.

2.4 TRANSIENT RESPONSE OF TRANSMISSION LINES WITH RESISTIVE TERMINATIONS

Our discussions in the preceding section served to introduce the concepts of reflection at discontinuities in the context of the relatively simple open- and short-circuited terminations. In this section, we study the response of transmission lines terminated with an arbitrary resistance R_L to excitations in the form of a step change in voltage (e.g., an applied voltage changing from 0 to V_0 at $t = 0$) or a short pulse of a given duration. Step excitation represents such cases as the output voltage of a driver gate changing from LOW to HIGH or HIGH to LOW state at a specific time, while pulse excitations are relevant to a broad class of computer communication problems. We consider two cases of resistive terminations: (i) single lossless transmission lines terminated in resistive loads and (ii) lossless transmission lines terminated in other lossless transmission lines. Resistively terminated lines and transmission lines terminated in other lines are encountered very often in practice. In digital communication applications, for example, logic gates are often connected via an interconnect to other gates with specific input resistances, and interconnects often drive combinations of other interconnects.

2.4.1 Single Transmission Lines with Resistive Terminations

We start with a general discussion of the step response of transmission lines with resistive terminations. Consider the circuit shown in Figure 2.16a where a step voltage source (0 to V_0 at $t = 0$) with a source resistance R_s drives a lossless transmission line of characteristic impedance Z_0 and one-way time delay t_d, terminated in a load resistance R_L. At $t = 0$, when the source voltage jumps to V_0, a voltage disturbance of amplitude $\mathcal{V}_1^+(0, 0) = Z_0V_0/(R_s + Z_0)$ is launched at the source end of the line; it travels down the line (during $0 < t < t_d$) and arrives at the load end at $t = t_d$, when a reflected voltage of amplitude $\mathcal{V}_1^-(l, t_d) = \Gamma_L\mathcal{V}_1^+(l, t_d)$, where $\Gamma_L = (R_L - Z_0)/(R_L + Z_0)$, is launched back toward the source. Note that $\mathcal{V}_1^+(z, t)$ remains constant in time (and also with z once it reaches $z = l$ at $t = t_d$), with its value given by $Z_0V_0/(R_s + Z_0)$. The reflected voltage travels along the line (during $t_d < t < 2t_d$) and reaches the source end at $t = 2t_d$, when a new voltage is reflected toward the load. The amplitude of the new reflected voltage is $\mathcal{V}_2^+(0, 2t_d) = \Gamma_s\mathcal{V}_1^-(0, 2t_d)$, where $\Gamma_s = (R_s - Z_0)/(R_s + Z_0)$. As this process continues with successive reflections at both ends, the total voltage at any time and at any particular position along the line

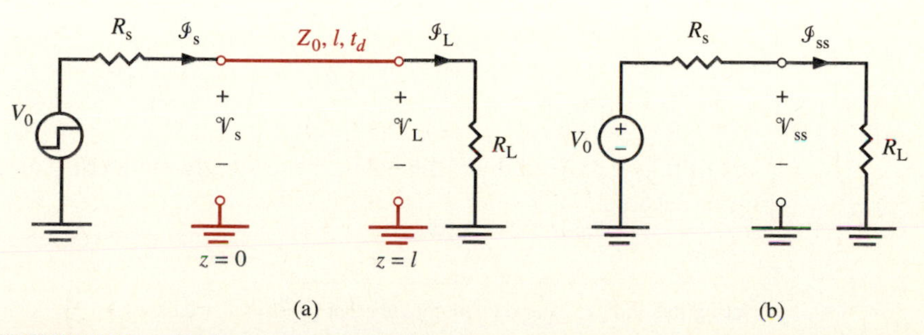

FIGURE 2.16. Resistively terminated line. (a) Step excitation of a lossless transmission line terminated with a resistive load. (b) Steady-state equivalent circuit seen by the source.

is given as an algebraic sum of all the voltages at that particular location and at that time. For example, at $t = 4t_d$, the total voltage at the center of the line is given by

$$\begin{aligned}\mathcal{V}_{l/2}(4t_d) &= \mathcal{V}_1^+(l/2, 4t_d) + \mathcal{V}_1^-(l/2, 4t_d) + \mathcal{V}_2^+(l/2, 4t_d) + \mathcal{V}_2^-(l/2, 4t_d)\\ &= \mathcal{V}_1^+(l/2, 4t_d)[1 + \Gamma_L + \Gamma_s\Gamma_L + \Gamma_s\Gamma_L^2]\\ &= \frac{Z_0V_0}{R_s + Z_0}[1 + \Gamma_L + \Gamma_s\Gamma_L + \Gamma_s\Gamma_L^2]\end{aligned}$$

where we have used the fact that $\mathcal{V}_1^+(l/2, 4t_d) = \mathcal{V}_1^+(0, 0) = Z_0V_0/(R_s + Z_0)$; in other words, as long as the source voltage does not change again, $\mathcal{V}_1^+(z, t)$ remains constant in time and also is the same everywhere (i.e., at all z) once it reaches the end of the line $z = l$ at $t = t_d$. Similarly, the total current at the center of the line at $t = 4t_d$ is

$$\begin{aligned}\mathcal{I}_{l/2}(4t_d) &= \mathcal{I}_1^+(l/2, 4t_d) + \mathcal{I}_1^-(l/2, 4t_d) + \mathcal{I}_2^+(l/2, 4t_d) + \mathcal{I}_2^-(l/2, 4t_d)\\ &= \frac{V_0}{R_s + Z_0}[1 - \Gamma_L + \Gamma_s\Gamma_L - \Gamma_s\Gamma_L^2]\end{aligned}$$

In general, since $|\Gamma_L| \le 1$ and $|\Gamma_s| \le 1$, we have $|\mathcal{V}_{i+1}^{\pm}| \le |\mathcal{V}_i^{\pm}|$, so the contribution of new individual reflected components to the total voltage or current at any position along the line diminishes as $t \to \infty$. The sum of the contributions of the voltage components traveling in both directions converges[24] to a finite steady-state value for the voltage at any position z, given as

$$\begin{aligned}\mathcal{V}(z, \infty) &= \mathcal{V}_1^+(z, \infty) + \mathcal{V}_1^-(z, \infty) + \mathcal{V}_2^+(z, \infty) + \mathcal{V}_2^-(z, \infty) + \mathcal{V}_3^+(z, \infty) + \mathcal{V}_3^-(z, \infty) + \cdots\\ &= \mathcal{V}_1^+(z, \infty)[1 + \Gamma_L + \Gamma_s\Gamma_L + \Gamma_s\Gamma_L^2 + \Gamma_s^2\Gamma_L^2 + \Gamma_s^2\Gamma_L^3 + \Gamma_s^3\Gamma_L^3 + \cdots]\\ &= \mathcal{V}_1^+(z, \infty)\{[1 + (\Gamma_s\Gamma_L) + (\Gamma_s^2\Gamma_L^2) + \cdots] + \Gamma_L[1 + (\Gamma_s\Gamma_L) + (\Gamma_s^2\Gamma_L^2) + \cdots]\}\\ &= \mathcal{V}_1^+(z, \infty)\left[\left(\frac{1}{1 - \Gamma_s\Gamma_L}\right) + \left(\frac{\Gamma_L}{1 - \Gamma_s\Gamma_L}\right)\right] = \mathcal{V}_1^+(z, \infty)\left(\frac{1 + \Gamma_L}{1 - \Gamma_s\Gamma_L}\right)\end{aligned}$$

This expression can be further simplified by substituting for $\mathcal{V}_1^+(z, \infty) = Z_0V_0/(R_s + Z_0)$, $\Gamma_L = (R_L - Z_0)/(R_L + Z_0)$, and $\Gamma_s = (R_s - Z_0)/(R_s + Z_0)$, yielding

$$\mathcal{V}(z, \infty) = V_{ss} = \left(\frac{R_L}{R_s + R_L}\right)V_0$$

a result that is expected, on the basis of the steady-state equivalent circuit shown in Figure 2.16b.

[24]Noting that $|\Gamma_s\Gamma_L| < 1$ and using the fact that for $|x| < 1$ we have

$$1 + x + x^2 + x^3 + \cdots = \frac{1}{1 - x}$$

At steady state it appears from Figure 2.16b as if the transmission line is simply not there and that the source is directly connected to the load. While this is essentially true, the transmission line is of course still present and is in fact fully charged, with all of its distributed capacitors charged to a voltage V_{ss} and all of its distributed inductors carrying a current V_{ss}/R_L. If the source were to be suddenly disconnected, the energy stored on the line would eventually be discharged through the load resistance, but only after a sequence of voltages propagating back and forth, reflecting at both ends and becoming smaller in time (see Example 2-5).

We now consider three specific examples. Example 2-4 illustrates the step response of a resistively terminated line, whereas Example 2-5 illustrates the process of discharging of a charged transmission line. Example 2-6 illustrates the pulse response of a resistively terminated line.

Example 2-4: Step response of a resistively terminated lossless line. Consider the circuit shown in Figure 2.17a for the specific case of $R_s = 3Z_0$ and $R_L = 9Z_0$. Sketch $\mathcal{V}_s$, $\mathcal{V}_L$, $\mathcal{I}_s$, and $\mathcal{I}_L$ as a function of t.

Solution: Based on the above discussion, an incident voltage $\mathcal{V}_1^+(z, t)$ of amplitude $\mathcal{V}_1^+(0, 0) = Z_0V_0/(3Z_0 + Z_0) = V_0/4$ is launched on the line at $t = 0$. When this disturbance reaches the load at $t = t_d$, a reflected voltage $\mathcal{V}_1^-(z, t)$ of amplitude $\mathcal{V}_1^-(l, t_d) = \Gamma_L\mathcal{V}_1^+(l, t_d) = V_0/5$, where $\Gamma_L = (9Z_0 - Z_0)/(9Z_0 + Z_0) = 4/5$, is launched toward the source. The reflected voltage arrives the source end at $t = 2t_d$, and a new voltage $\mathcal{V}_2^+(z, t)$ of amplitude $\mathcal{V}_2^+(0, 2t_d) = \Gamma_s\mathcal{V}_1^-(0, 2t_d) = V_0/10$, where $\Gamma_s = (3Z_0 - Z_0)/(3Z_0 + Z_0) = 1/2$, is produced, traveling toward the load. At $t = 3t_d$, a voltage $\mathcal{V}_2^-(z, t)$ of amplitude $\mathcal{V}_2^-(l, 3t_d) = \Gamma_L\mathcal{V}_2^+(l, 3t_d) = 2V_0/25$ is launched from the load end, traveling toward the source, and so on. The bounce diagram is shown in Figure 2.17b. The source- and load-end voltages and the source- and load-end currents are shown in Figure 2.17d. The steady-state circuit seen by the source is also shown in Figure 2.17c.

Example 2-5: A charged line connected to a resistor. Consider a transmission line that is initially charged to a constant voltage $\mathcal{V}(z, t) = V_0$ (such as the steady-state condition of the circuit in Example 2-3), as shown in Figure 2.18a. At $t = 0$, the switch is moved from position 1 to position 2. Analyze and sketch the variation of the source- and load-end voltages $\mathcal{V}_s$ and $\mathcal{V}_L$ as a function of t for three different cases: (a) $R_{s2} = Z_0/3$, (b) $R_{s2} = Z_0$, and (c) $R_{s2} = 3Z_0$.

Solution: Before $t = 0$, the steady-state condition holds, and $\mathcal{V}_s(0^-) = \mathcal{V}_L(0^-) = V_{ss} = V_0$ and $\mathcal{I}_s(0^-) = \mathcal{I}_L(0^-) = I_{ss} = 0$. At $t = 0$, the switch moves to position 2, which causes both $\mathcal{V}_s$ and $\mathcal{I}_s$ to change immediately. The change in the source-end voltage $\mathcal{V}_s$ (and the source-end current $\mathcal{I}_s$) can be interpreted as a new voltage disturbance $\mathcal{V}_1^+(z, t)$ (and a new current disturbance

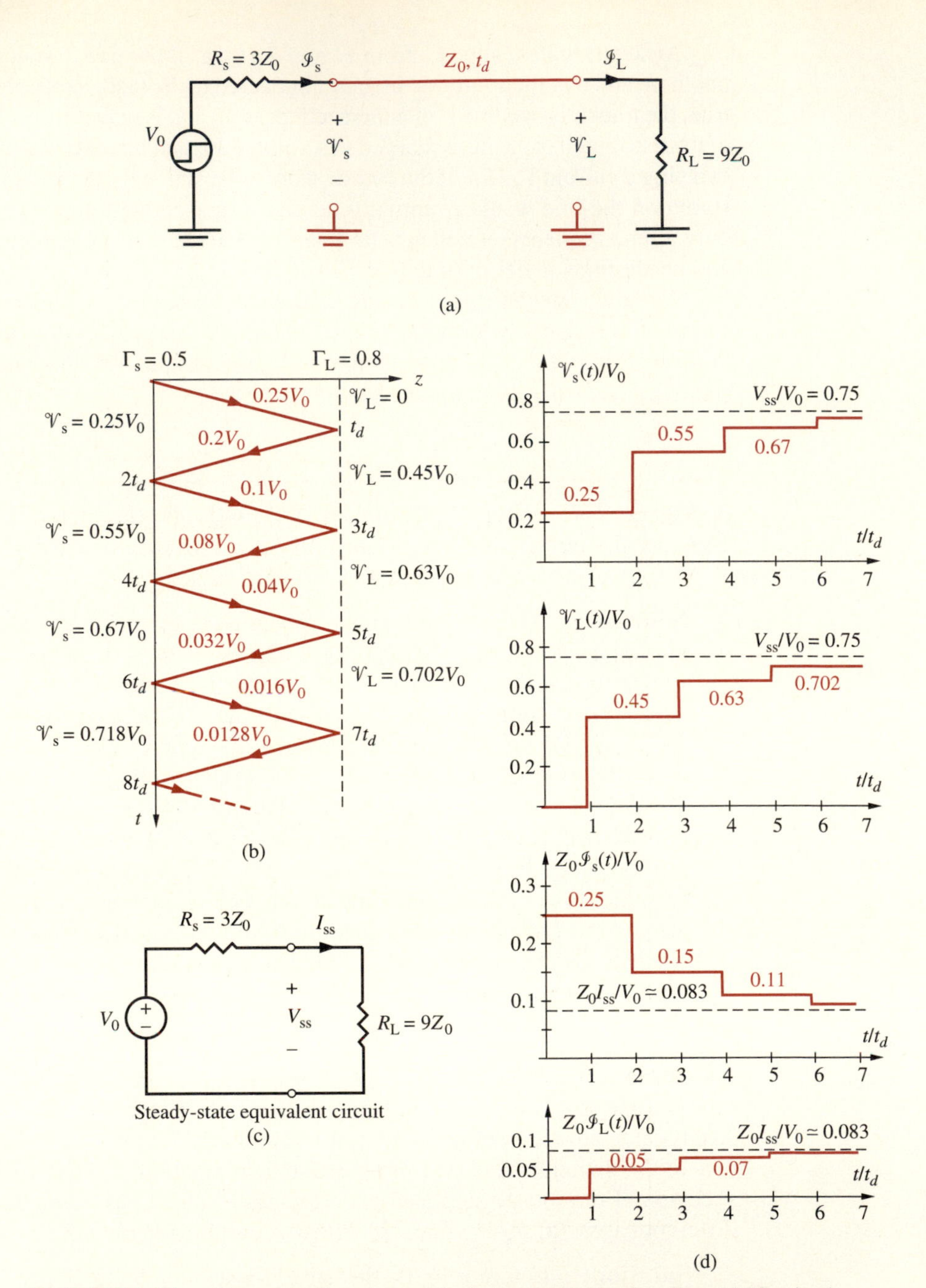

FIGURE 2.17. Step excitation of a resistively terminated lossless line. (a) The circuit configuration. (b) Bounce diagram. (c) Steady-state equivalent circuit seen by the source. (d) Normalized source- and load-end voltages and source- and load-end currents as a function of t/t_d.

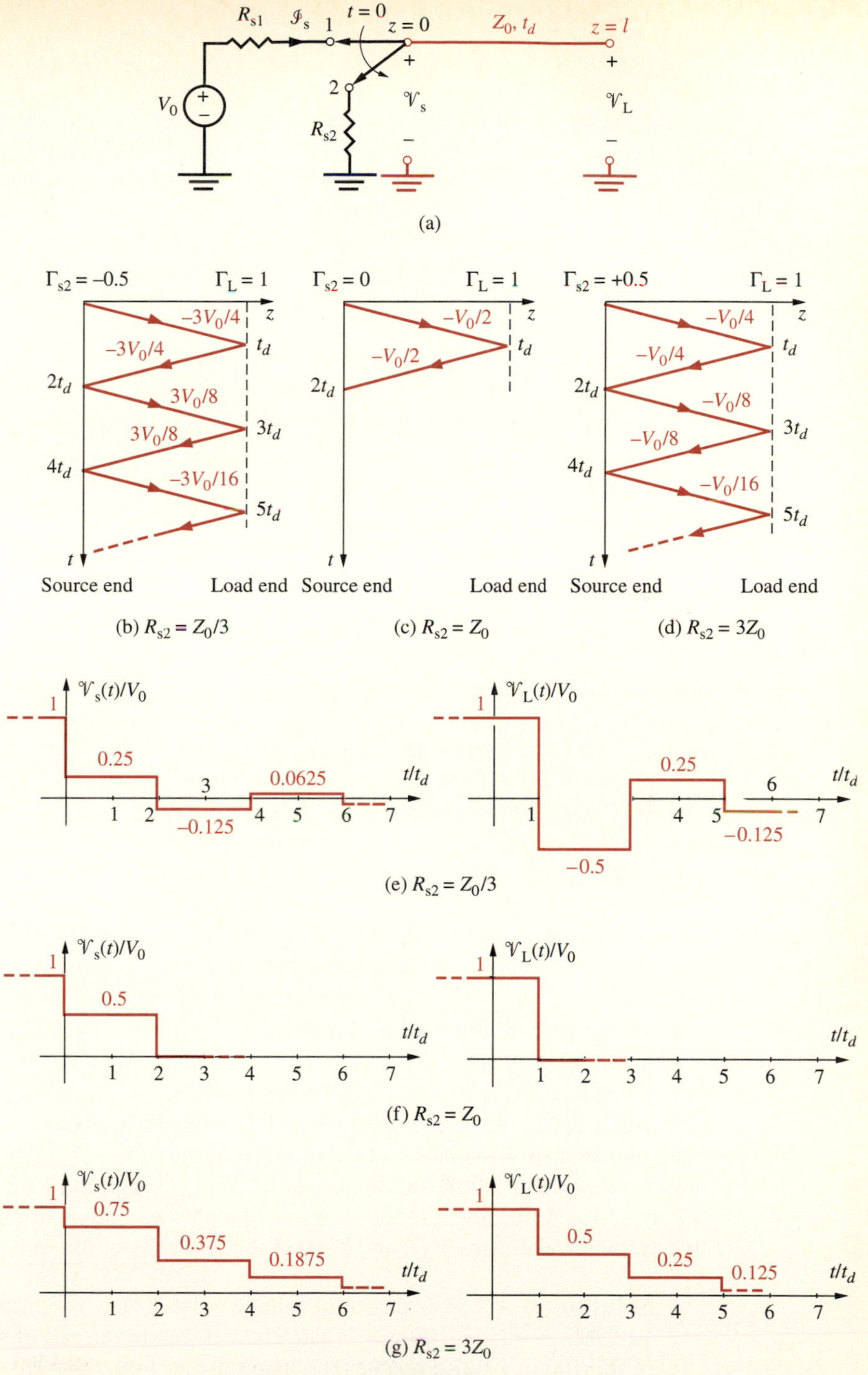

FIGURE 2.18. **Discharging of a charged line.** (a) A charged line connected to a resistor R_{s2} for Example 2-5. (b), (c), and (d) Bounce diagrams for $R_{s2} = Z_0/3$, $R_{s2} = Z_0$, and $R_{s2} = 3Z_0$. (e), (f), and (g) Normalized source- and load-end voltages as a function of t for the three different cases.

$\mathcal{I}_1^+(z,t)$) launched on the line from the source end. The amplitude of this new voltage $\mathcal{V}_1^+(z,t)$ (and the new current $\mathcal{I}_1^+(z,t)$) is determined by the change in $\mathcal{V}_s$ (or in $\mathcal{I}_s$) between $t = 0^-$ and $t = 0^+$, namely,

$$\mathcal{V}_1^+(0,0) = \mathcal{V}_s(0^+) - \mathcal{V}_s(0^-) = \mathcal{V}_s(0^+) - V_0$$

and

$$\mathcal{I}_1^+(0,0) = \mathcal{I}_s(0^+) - \mathcal{I}_s(0^-) = \mathcal{I}_s(0^+)$$

Using the new boundary condition at the source end imposed by R_{s2}, namely,

$$\mathcal{V}_s(0^+) = -R_{s2}\mathcal{I}_s(0^+) = -R_{s2}\mathcal{I}_1^+(0,0) = -R_{s2}\mathcal{V}_1^+(0,0)/Z_0$$

we can write

$$\mathcal{V}_1^+(0,0) = -R_{s2}\frac{\mathcal{V}_1^+(0,0)}{Z_0} - V_0 \rightarrow \mathcal{V}_1^+(0,0) = -\frac{Z_0V_0}{R_{s2}+Z_0}$$

Note that the negative sign in the source-end boundary condition $\mathcal{V}_s = -R_{s2}\mathcal{I}_s$ is due to the defined direction of $\mathcal{I}_s$, with positive current coming out of the terminal of positive voltage. At $t = t_d$, the new voltage disturbance reaches the open end of the line, where a reflected voltage $\mathcal{V}_1^-(z,t)$ with amplitude $\mathcal{V}_1^-(l,t_d) = \mathcal{V}_1^+(l,t_d)$ is produced, traveling toward the source end. At $t = 2t_d$, $\mathcal{V}_1^-(z,t)$ arrives at the source end, and a reflected voltage $\mathcal{V}_2^+(z,t)$ with amplitude $\mathcal{V}_2^+(0,2t_d) = \Gamma_s\mathcal{V}_1^-(0,2t_d) = (R_{s2}-Z_0)\mathcal{V}_1^-(0,2t_d)/(R_{s2}+Z_0)$ is launched back toward the load. This process continues until a new steady-state condition is reached, when the line voltage eventually becomes zero. Figures 2.18b, c, and d show the bounce diagrams for three different values of R_{s2}, namely $Z_0/3$, Z_0, and $3Z_0$. Figures 2.18e, f, and g show the variation of the source- and load-end voltages for all three cases as a function of time t.

Example 2-6: Pulse excitation of a transmission line. A high-speed logic gate represented by a pulse voltage source of amplitude 1 V, pulse width 200 ps, and output impedance 900Ω drives a load of 25Ω through a 100Ω line, as shown in Figure 2.19. Assuming a lossless line with a one-way delay of $t_d = 400$ ps, sketch the voltage waveforms $\mathcal{V}_s(t)$ and $\mathcal{V}_L(t)$ at the two ends of the line.

Solution: At $t = 0$, an initial voltage pulse $\mathcal{V}_1^+(z,t)$ of amplitude $\mathcal{V}_1^+(0,0) = V_0Z_0/(R_s+Z_0) = 100$ mV is launched at the driver end of the line. The front of the 100 mV pulse reaches the load (at $z = l$) at $t = 400$ ps, and a reflected pulse of amplitude $\mathcal{V}_1^-(l,t_d) = \Gamma_L\mathcal{V}_1^+(l,t_d) = -60$ mV starts traveling back toward the driver. Both pulses $\mathcal{V}_1^+(z,t)$ and $\mathcal{V}_1^-(z,t)$ exist at the load end for a period of only 200 ps, adding up to a total voltage of 40 mV. The reflected

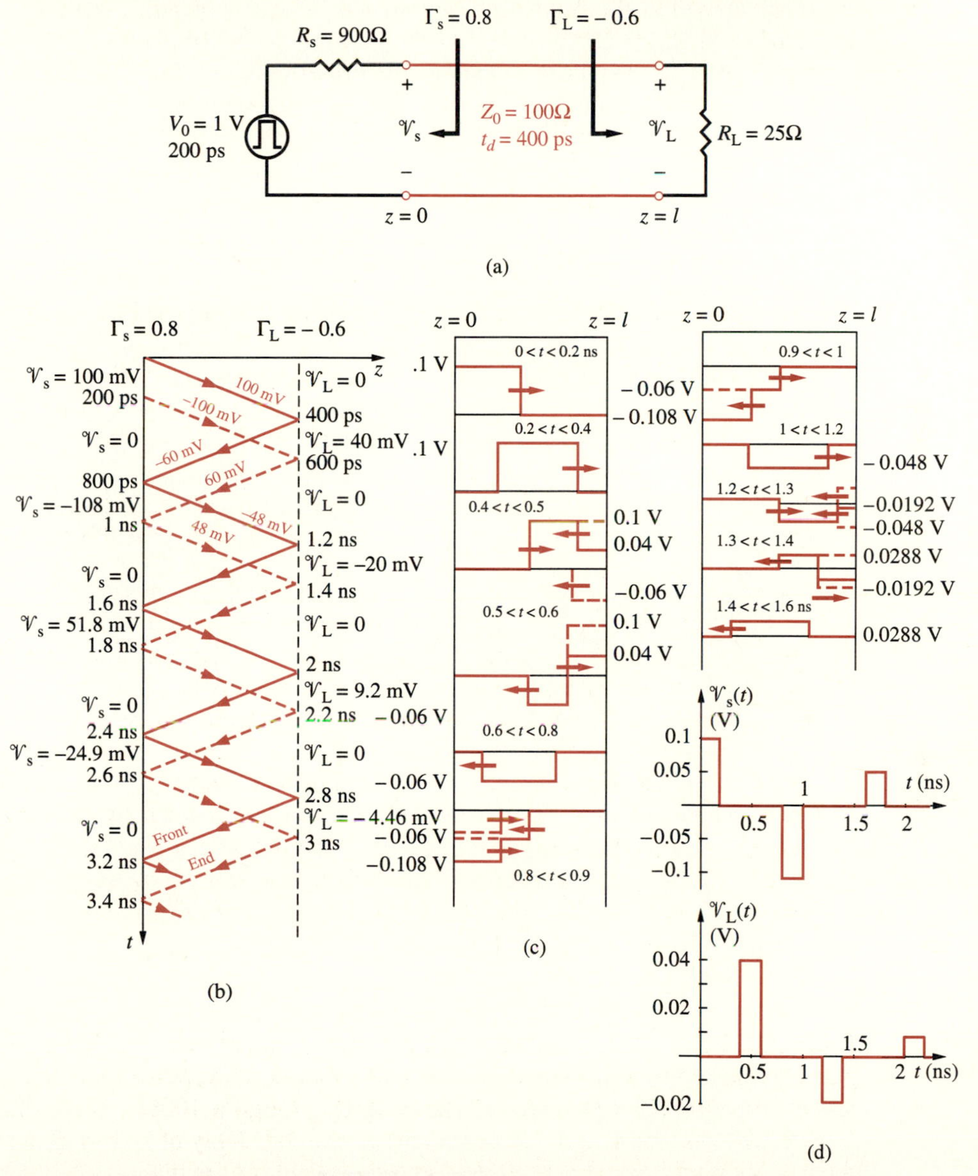

FIGURE 2.19. **Pulse excitation of a transmission line (Example 2-6).** (a) Circuit diagram. (b) Bounce diagram. (c) Distribution of voltage along the line at different times. (d) The source- and load-end voltages as a function of time.

pulse arrives at the driver end at $t = 800$ ps and launches a pulse of amplitude $\mathcal{V}_2^+(0, 2t_d) = \Gamma_s \mathcal{V}_1^-(0, 2t_d) = -48$ mV back toward the load end. The two pulses $\mathcal{V}_1^-(z, t)$ and $\mathcal{V}_2^+(z, t)$ exist at the driver end only for 200 ps, totaling to a voltage of -108 mV. This process continues on, with the pulse amplitude reduced by 40% and inverted at the load end and reduced by 20% at the driver end each time. Figure 2.19b shows the bounce diagram, and Figure 2.19c shows the snapshots of the voltages along the line at various time intervals. The variation with time of the driver- and load-end voltages are also shown in Figure 2.19d.

The Transmission Line as a Linear Time-Invariant System In some cases it is useful to think of the transmission line as a linear time-invariant system, with a defined input and output. For this purpose, the input $\mathcal{V}_{\text{in}}(t)$ can be defined as the voltage or current at the input of the line, while the output $\mathcal{V}_{\text{out}}(t)$ can be a voltage or current somewhere else on the line—for example, the load voltage $\mathcal{V}_{\text{L}}(t)$ as indicated in Figure 2.20.

Note that since the fundamental differential equations ([2.1] and [2.2]) that govern the transmission line voltage and current are linear, the relationship between $\mathcal{V}_{\text{in}}(t)$ and $\mathcal{V}_{\text{out}}(t)$ is linear. In other words, for two different input signals $\mathcal{V}_{\text{in}_1}$ and $\mathcal{V}_{\text{in}_2}$, which individually produce two different output signals $\mathcal{V}_{\text{out}_1}$ and $\mathcal{V}_{\text{out}_2}$, the response due to a linear superposition of the two inputs, $\mathcal{V}_{\text{in}_{1+2}} = a_1\mathcal{V}_{\text{in}_1} + a_2\mathcal{V}_{\text{in}_2}$, is

$$\mathcal{V}_{\text{out}_{1+2}} = a_1\mathcal{V}_{\text{out}_1} + a_2\mathcal{V}_{\text{out}_2}$$

Since the physical properties of the transmission line (L, C, t_d, Z_0) do not change with time, the relationship between $\mathcal{V}_{\text{in}}(t)$ and $\mathcal{V}_{\text{out}}(t)$ is also time-invariant. In

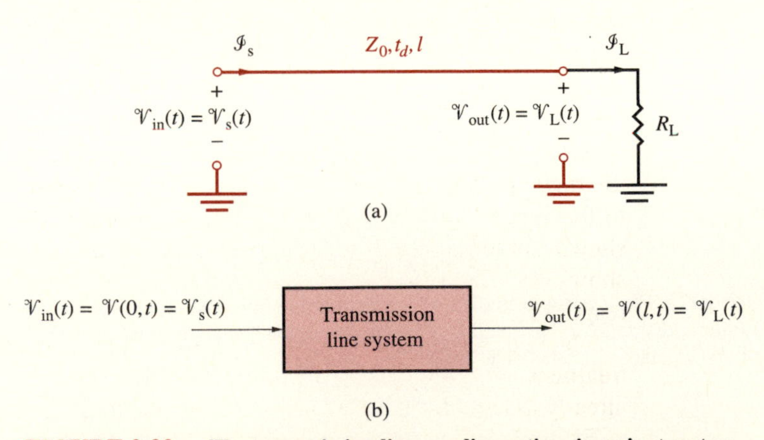

FIGURE 2.20. The transmission line as a linear time-invariant system. The input $\mathcal{V}_{\text{in}}(t)$ to the system can be defined as the line input voltage $\mathcal{V}_s(t)$, whereas the output $\mathcal{V}_{\text{out}}(t)$ could be any voltage or current of interest anywhere on the line, such as the load voltage $\mathcal{V}_{\text{L}}(t)$.

other words, if the output due to an input $\mathcal{V}_{\text{in}_1}(t)$ is $\mathcal{V}_{\text{out}_1}(t)$, then the output due to a time-shifted version of the input, namely $\mathcal{V}_{\text{in}_1}(t-\tau)$, is simply a similarly shifted version of the output, namely $\mathcal{V}_{\text{out}_1}(t-\tau)$.

As with any linear time-invariant system, the response of a transmission line to any arbitrary excitation signal can be determined from its response to an impulse excitation. In the transmission line context, an input pulse can be considered to be an impulse if its duration is much shorter than any other time constant in the system or the one-way travel time t_d in the case of lossless lines with resistive terminations. In most applications, however, it is necessary to determine the response of the line to step inputs, as were illustrated in Examples 2-3 through 2-5. For this purpose, it is certainly easier to determine the step response directly rather than to determine the pulse (or impulse) response first and then use it to determine the step response.

Treatment of a transmission line as a linear time-invariant system can sometimes be useful in determining its response to pulse inputs, as illustrated in Example 2-7.

Example 2-7: The transmission line as a linear time-invariant system. Consider the transmission line system of Figure 2.21a, the step response of which was determined in Example 2-4. Determine the load voltage $\mathcal{V}_{\text{L}}(t)$ for an input excitation in the form of a single pulse of amplitude V_0 and duration $0.5t_d$ (i.e., $\mathcal{V}_{\text{in}}(t) = V_0[u(t) - u(t - t_d/2)]$).

Solution: For the circuit of Figure 2.21a, it is convenient to define the input signal to be the excitation (source) voltage and the output as the load voltage, as indicated. As shown in Figure 2.21b, the input pulse of amplitude V_0 and duration $t_d/2$ can be viewed as a superposition of two different input signals: a step input starting at $t = 0$, namely $\mathcal{V}_{\text{in}_1}(t) = V_0 u(t)$, and a shifted negative step input, namely, $\mathcal{V}_{\text{in}_2}(t) = -V_0 u(t - t_d/2)$, where $u(\zeta)$ is the unit step function, $u(\zeta) = 1$ for $\zeta > 0$, and $u(\zeta) = 0$ for $\zeta < 0$. The output $\mathcal{V}_{\text{out}_1}(t)$ due to the input $\mathcal{V}_{\text{in}_1}(t)$ was determined in Example 2-4 and is plotted in the top panel of Figure 2.21c. Since the transmission line is a linear time-invariant system, the response $\mathcal{V}_{\text{out}_2}(t)$ due to the input $\mathcal{V}_{\text{in}_2}(t)$ is simply a flipped-over and shifted version of $\mathcal{V}_{\text{out}_1}(t)$, as shown in the middle panel of Figure 2.21c. The bottom panel of Figure 2.21c shows the superposition of the two responses, which is the desired pulse response.

Note that our solution of this problem using the linear time-invariant system treatment was simplified by the fact that the step response of the system was already in hand from Example 2-4. Note also that the pulse response in the case of Example 2-6 could also have been determined by a similar method. However, one then has to first determine the step response, and whether or not the system approach is easier in general depends on the particular problem.

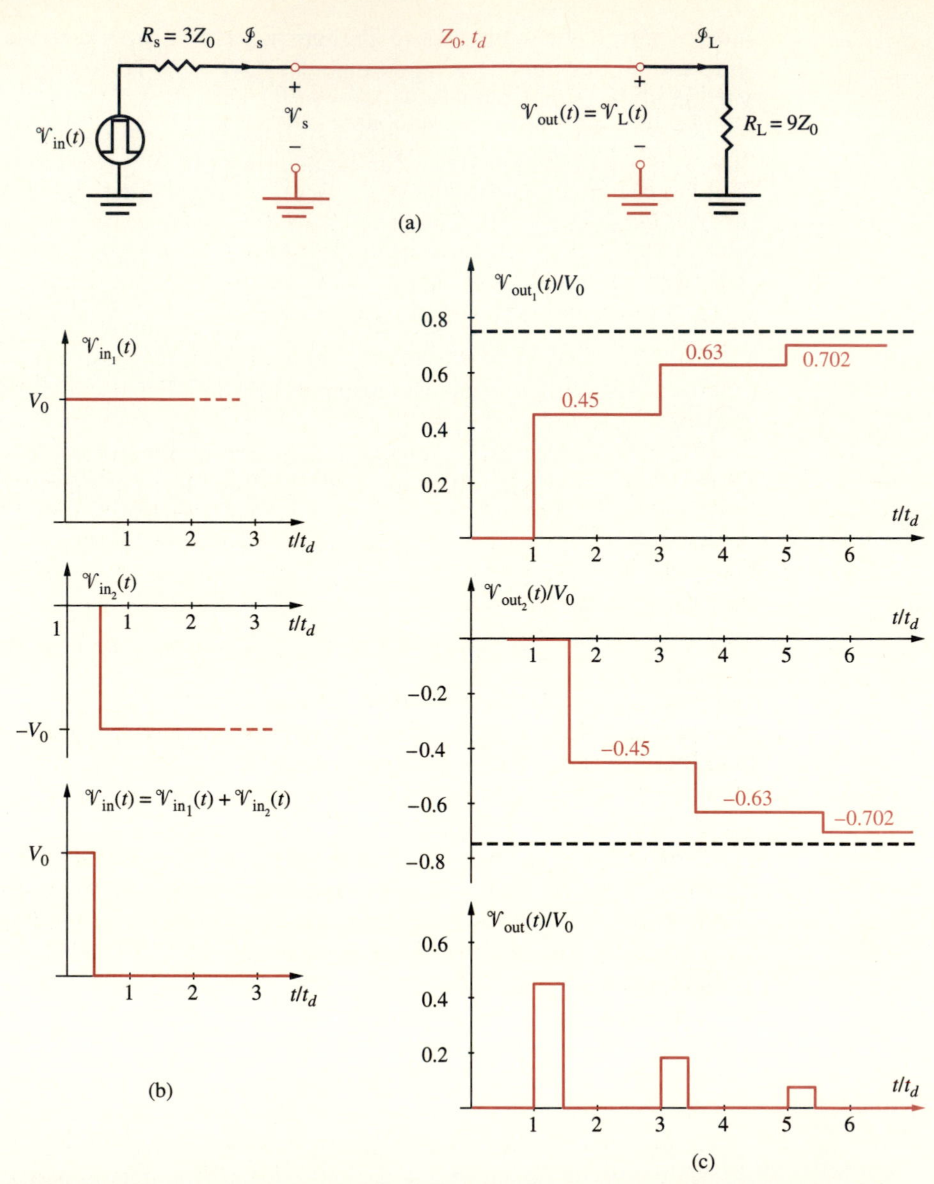

FIGURE 2.21. Pulse response of a transmission line. (a) The line configuration and source and load impedances. (b) The input pulse of amplitude V_0 and duration $t_d/2$ is represented as a superposition of a $\mathcal{V}_{in_1}$ and $\mathcal{V}_{in_2}$. (c) The response is computed as a superposition of the responses due to the individual step inputs. Note that $\mathcal{V}_{out_1}(t)$ was already computed in Example 2-4.

2.4.2 Junctions between Transmission Lines

We have seen that reflections from terminations at the source and load ends of transmission lines lead to ringing and other effects. Reflections also occur at discontinuities at the interfaces between transmission lines, connected either in cascade or in parallel, as shown, for example, in Figures 2.22a and 2.23a and as often encountered in practice. For example, consider the case of the two lossless transmission lines A and B (with characteristic impedances Z_{0A} and Z_{0B}) connected in tandem (i.e., in series) as shown in Figure 2.22a. Assume a voltage disturbance of amplitude $\mathcal{V}_{1A}^{+}(z,t)$ (with an associated current of $\mathcal{I}_{1A}^{+}(z,t) = \mathcal{V}_{1A}^{+}/Z_{0A}$) to arrive at the junction between lines A and B (located at $z = l_j$) from line A at $t = t_0$. A voltage $\mathcal{V}_{1A}^{-}(z,t)$ of amplitude $\mathcal{V}_{1A}^{-}(l_j,t_0) = \Gamma_{AB}\mathcal{V}_{1A}^{+}(l_j,t_0)$ reflects back to line A, where the reflection coefficient Γ_{AB} is given by $\Gamma_{AB} = (Z_{0B} - Z_{0A})/(Z_{0B} + Z_{0A})$, since line B presents a load impedance of Z_{0B} to line A. In addition, a voltage $\mathcal{V}_{1B}^{+}(z,t)$ of amplitude $\mathcal{V}_{1B}^{+}(l_j,t_0) = \mathcal{T}_{AB}\mathcal{V}_{1A}^{+}(l_j,t_0)$ is transmitted into line B, where $\mathcal{T}_{AB}$ is called the *transmission coefficient,* defined as the ratio of the transmitted voltage to the incident voltage, i.e., $\mathcal{T}_{AB} \equiv \mathcal{V}_{1B}^{+}(l_j,t_0)/\mathcal{V}_{1A}^{+}(l_j,t_0)$. To find $\mathcal{T}_{AB}$, we apply the boundary condition at the junction, which states that the total voltages on the left and right sides of the junction must be equal:

$$\mathcal{V}_{1A}^{+}(l_j,t_0) + \mathcal{V}_{1A}^{-}(l_j,t_0) = \mathcal{V}_{1B}^{+}(l_j,t_0)$$

yielding $\mathcal{T}_{AB} = 1 + \Gamma_{AB} = 2Z_{0B}/(Z_{0B} + Z_{0A})$. The transmission coefficient $\mathcal{T}_{AB}$ represents the fraction of the incident voltage that is transferred from line A to line B. Note that depending on the value of the reflection coefficient, the transmitted voltage can actually be larger in amplitude than the incident voltage, so that we may have $\mathcal{T}_{AB} > 1$ (in those cases when $\Gamma_{AB} > 0$).

Similarly, if a voltage disturbance $\mathcal{V}_{1B}^{-}(z,t)$ of amplitude $\mathcal{V}_{1B}^{-}(l_j,t_1)$ (produced by reflection when $\mathcal{V}_{1B}^{+}(z,t)$ reaches the end of line B) arrives at the same junction between A and B from line B at $t = t_1$, a voltage $\mathcal{V}_{2B}^{+}(z,t)$ of amplitude $\mathcal{V}_{2B}^{+}(l_j,t_1) = \Gamma_{BA}\mathcal{V}_{1B}^{-}(l_j,t_1)$ reflects back to line B and a voltage $\mathcal{V}_{2A}^{-}(l_j,t_1) = \mathcal{T}_{BA}\mathcal{V}_{1B}^{-}(l_j,t_1)$ is transmitted into line A, where Γ_{BA} and $\mathcal{T}_{BA}$ are given by

$$\Gamma_{BA} = \frac{Z_{0A} - Z_{0B}}{Z_{0A} + Z_{0B}} = -\Gamma_{AB}$$

$$\mathcal{T}_{BA} = 1 + \Gamma_{BA} = \frac{2Z_{0A}}{Z_{0A} + Z_{0B}} = \frac{Z_{0A}}{Z_{0B}}\mathcal{T}_{AB}$$

The following two examples are both associated with junctions between transmission lines.

Example 2-8: Cascaded transmission lines. Consider the transmission line system shown in Figure 2.22a, where a step voltage source of amplitude 1.5 V and source resistance 50Ω excites two cascaded lossless transmission lines (A and B) of characteristic impedances 50Ω and 25Ω and lengths 5 cm and 2 cm, respectively. The speed of

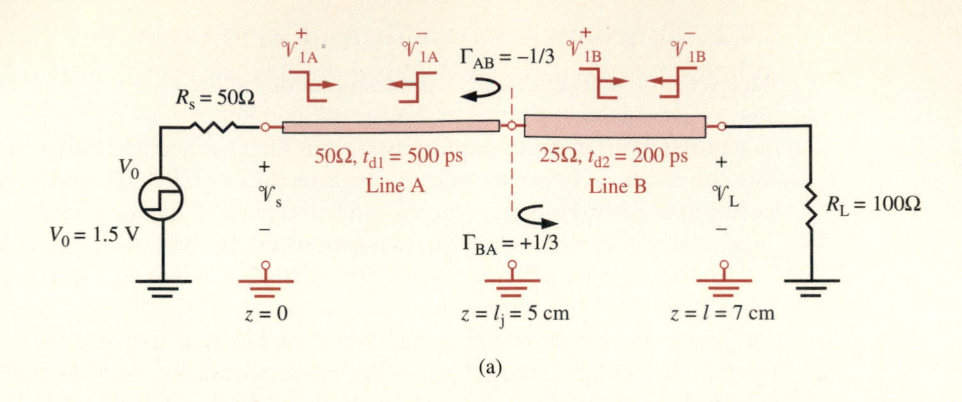

(a)

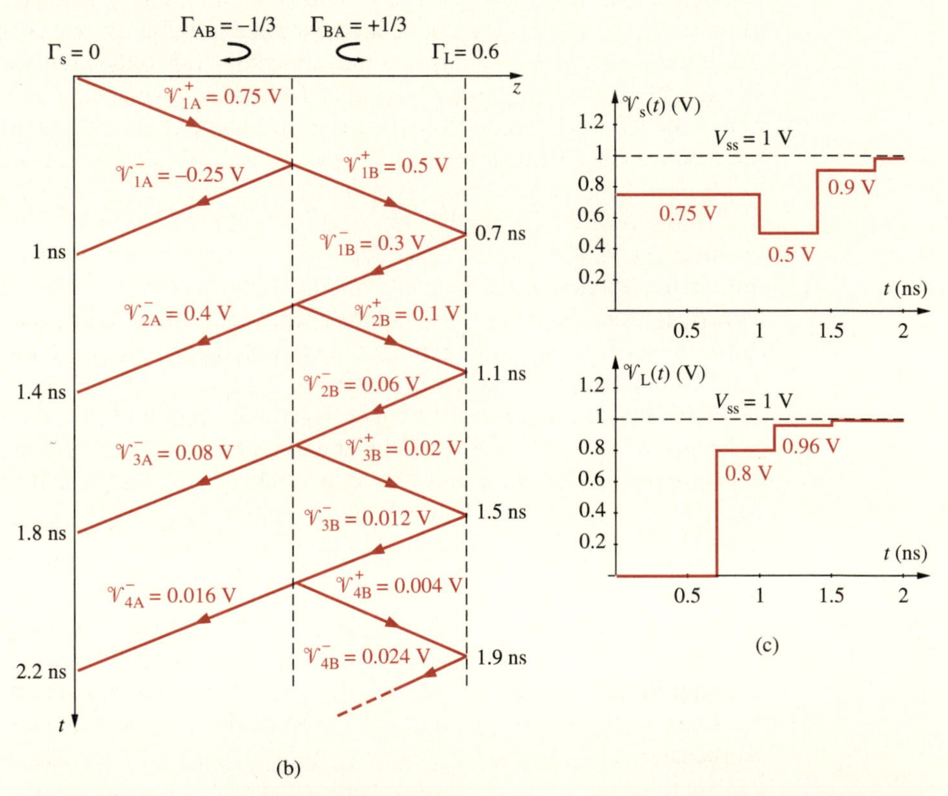

(b)

FIGURE 2.22. Cascaded transmission lines. (a) Circuit diagram for Example 2-8. (b) Bounce diagram. (c) Source-end voltage $\mathcal{V}_s$ and load-end voltage $\mathcal{V}_L$ as a function of t.

propagation in each line is 10 cm-ns^{-1}.[25] The second line (B) is terminated with a load impedance of 100Ω at the other end. Draw the bounce diagram and sketch the voltages $\mathcal{V}_s(t)$ and $\mathcal{V}_L(t)$ as functions of time.

Solution: With respect to Figure 2.22a, we note that $l_j = 5$ cm and $l = 7$ cm. At $t = 0^+$, voltage $\mathcal{V}_{1A}^+(z, t)$ of amplitude $\mathcal{V}_{1A}^+(0, 0) = 0.75$ V is launched on line A. This voltage reaches the junction between the two lines at $t = t_{d1} = 5$ cm/(10 cm/ns) $= 500$ ps, when a reflected voltage $\mathcal{V}_{1A}^-(z, t)$ of amplitude $\mathcal{V}_{1A}^-(l_j, t_{d1}) = \Gamma_{AB}\mathcal{V}_{1A}^+(l_j, t_{d1}) = (-1/3)(0.75) = -0.25$ V, and a transmitted voltage $\mathcal{V}_{1B}^+(z, t)$ of amplitude $\mathcal{V}_{1B}^+(l_j, t_{d1}) = \mathcal{T}_{AB}\mathcal{V}_{1A}^+(l_j, t_{d1}) = (2/3)(0.75) = 0.5$ V are created. The reflected disturbance arrives at the source end at $t = 2t_{d1} = 1$ ns and is absorbed completely since $\Gamma_s = 0$. The transmitted wave reaches the load at $t = t_{d1} + t_{d2} = 700$ ps, and a reflected voltage $\mathcal{V}_{1B}^-(z, t)$ of amplitude $\mathcal{V}_{1B}^-(l, 700 \text{ ps}) = \Gamma_L\mathcal{V}_{1B}^+(l, 700 \text{ ps}) = (0.6)(0.5)$ V $= 0.3$ V is launched toward the source. This reflected disturbance arrives at the junction from line B at $t = t_{d1} + 2t_{d2} = 900$ ps, and reflected and transmitted voltages $\mathcal{V}_{2B}^+(z, t)$ and $\mathcal{V}_{2A}^-(z, t)$ of amplitudes respectively $\mathcal{V}_{2B}^+(l_j, 900 \text{ ps}) = \Gamma_{BA}\mathcal{V}_{1B}^-(l_j, 900 \text{ ps}) = (1/3)(0.3) = 0.1$V and $\mathcal{V}_{2A}^-(l_j, 900 \text{ ps}) = \mathcal{T}_{BA}\mathcal{V}_{1B}^-(l_j, 900 \text{ ps}) = (4/3)(0.3) = 0.4$ V are launched respectively toward the load and the source. The continuation of this process can be followed by means of a bounce diagram, as shown in Figure 2.22b. The source- and load-end voltages are plotted as a function of t in Figure 2.22c.

Example 2-9: Three parallel transmission lines. Consider three identical lossless transmission lines, each with characteristic impedance Z_0 and one-way time delay t_d, connected in parallel at a common junction as shown in Figure 2.23a. The main line is excited at $t = 0$ by a step voltage source of amplitude V_0 and a source resistance of $R_s = Z_0$. Find and sketch the voltages $\mathcal{V}_s(t)$, $\mathcal{V}_{L1}(t)$, and $\mathcal{V}_{L2}(t)$ for the following two cases: (a) $R_{L1} = R_{L2} = Z_0$ and (b) $R_{L1} = Z_0$ and $R_{L2} = \infty$.

Solution: For any voltage disturbance $\mathcal{V}_i(z, t)$ of amplitude V_i arriving at the junction from any one of the three lines, the parallel combination of the characteristic impedances of the other two lines acts as an equivalent load impedance at the junction. Once a voltage is incident at the junction, a voltage of amplitude $V_r = \Gamma_j V_i = (Z_0/2 - Z_0)V_i/(Z_0/2 + Z_0) = -V_i/3$ is reflected and a voltage of amplitude $V_t = \mathcal{T}_j V_i = (1 + \Gamma_j)V_i = 2V_i/3$ is transmitted to both of the other lines. Note that the reflection coefficient at the junction is denoted simply as Γ_j, because all of the transmission lines are identical and the reflection coefficient at the junction is thus the same regardless of which line the wave is incident from. For the circuit shown in Figure 2.23a, the bounce diagram and the sketches of voltages $\mathcal{V}_s$, $\mathcal{V}_{L1}$, and $\mathcal{V}_{L2}$ for both cases are shown in Figure 2.23b and c. When $R_{L1} = R_{L2} = Z_0$, only one of lines 1 and 2 is shown in the bounce diagram, since

[25]Note from Table 2.1 that 10 cm-ns^{-1} is approximately the speed of propagation in alumina (Al_2O_3), a ceramic commonly used for electronics packaging.

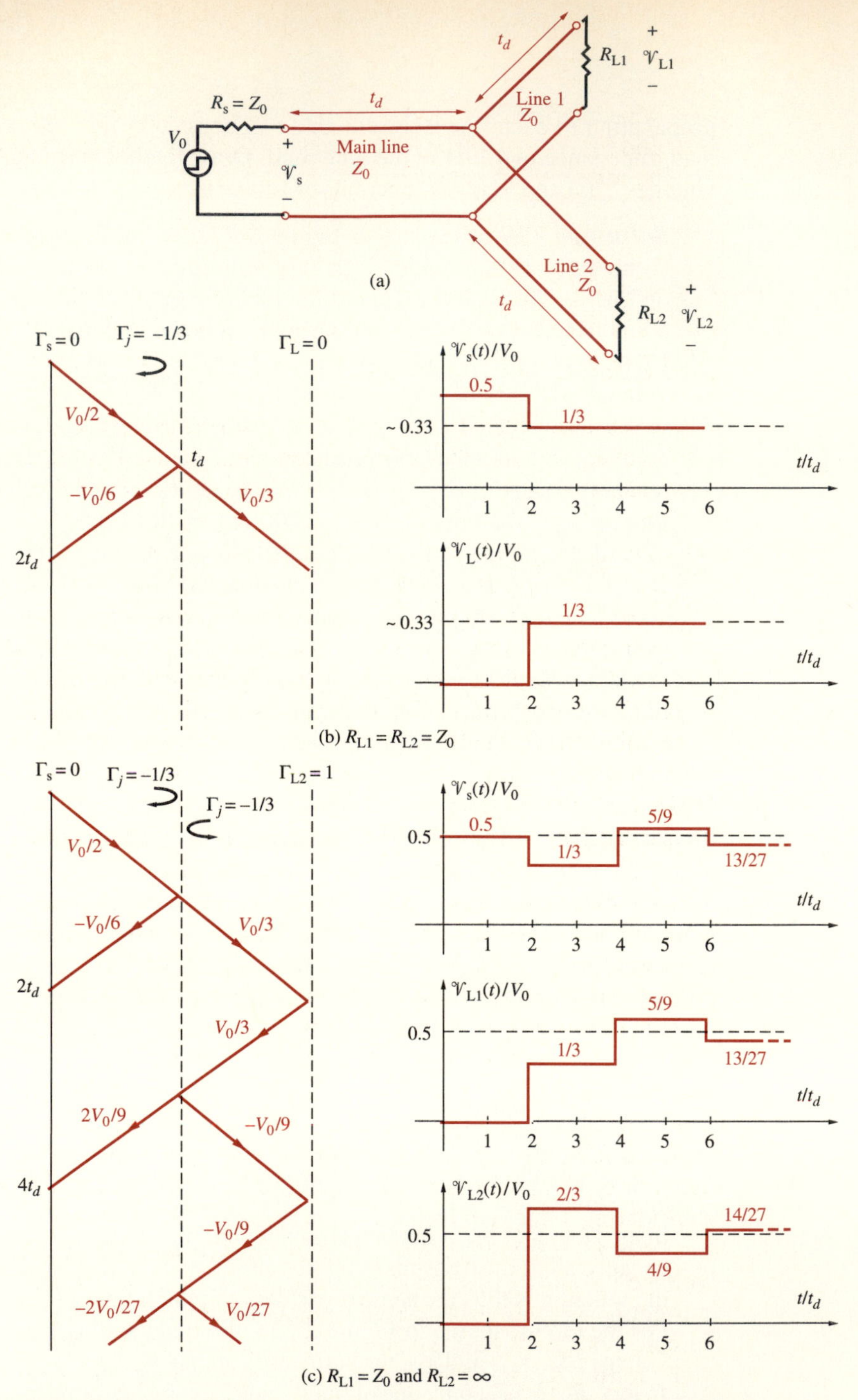

FIGURE 2.23. **Three transmission lines connected in parallel at a common junction.** (a) Circuit diagram for Example 2-9. (b) Bounce diagram and the variation of the $\mathcal{V}_s$ and $\mathcal{V}_{L1} = \mathcal{V}_{L2} = \mathcal{V}_L$ voltages as a function of t for $R_{L1} = R_{L2} = Z_0$. (c) Bounce diagram and the variation of the $\mathcal{V}_s$, $\mathcal{V}_{L1}$, and $\mathcal{V}_{L2}$ voltages as a function of t for $R_{L1} = Z_0$ and $R_{L2} = \infty$.

what happens on the other is identical, as seen in Figure 2.23b. When $R_{L1} = Z_0$ and $R_{L2} = \infty$, only line 2 is shown in the bounce diagram in Figure 2.23c, since no reflection occurs on line 1.

2.5 TRANSIENT RESPONSE OF TRANSMISSION LINES WITH REACTIVE OR NONLINEAR TERMINATIONS

Up to now, we have studied only transmission lines with resistive terminations. In this section, we consider reactive loads and loads with nonlinear current-voltage characteristics.

2.5.1 Reactive Terminations

Reactive loads are encountered quite often in practice; in high-speed bus designs, for example, capacitive loading by backplanes (consisting of plug-in cards having printed circuit board traces and connectors) often becomes the bottleneck when high-speed CPUs communicate with shared resources on the bus. Inductive loading due to bonding wire inductances is also important in many integrated-circuit packaging technologies. Packaging pins, vias between two wiring levels, and variations in line width can often be modeled as capacitive and inductive discontinuities. The capacitances and inductances of these various packaging components can range between 0.5 and 4 pF and between 0.1 and 35 nH, respectively.[26]

For transmission lines with resistive loads, the reflected and transmitted voltages and currents have the same temporal shape as the incident ones and do not change their shape as a function of time. For a step excitation, for example, the reflected voltage produced by a resistive termination remains constant in time, as discussed in preceding sections. However, in the case of capacitive or inductive terminations, the reflected and transmitted voltages and currents do not have the same temporal shape as the incident ones. The terminal boundary condition at the reactive termination must now be expressed as a differential equation whose general solution can be exceedingly complicated, whether the solutions are carried out in the time domain or by the use of Laplace transformation. We illustrate the basic principles by considering a line terminated at an inductance, as shown in Figure 2.24.

When the traveling disturbance $\mathcal{V}_1^+(z, t)$ (taken in Figure 2.24b, c as a constant voltage V_0) and its associated current $\mathcal{I}_1^+(z, t)$ (taken in Figure 2.24b, c as a constant current $I_0 = V_0/Z_0$) first reach the end of the line, the inductive load acts as an open circuit, since its current cannot change instantaneously. Thus the disturbance is initially reflected in the same manner as an open circuit, the terminal voltage jumping

[26]See Chapter 6 of H. B. Bakoglu, *Circuits, Interconnections, and Packaging for VLSI,* Addison Wesley, Reading, Massachusetts, 1990.

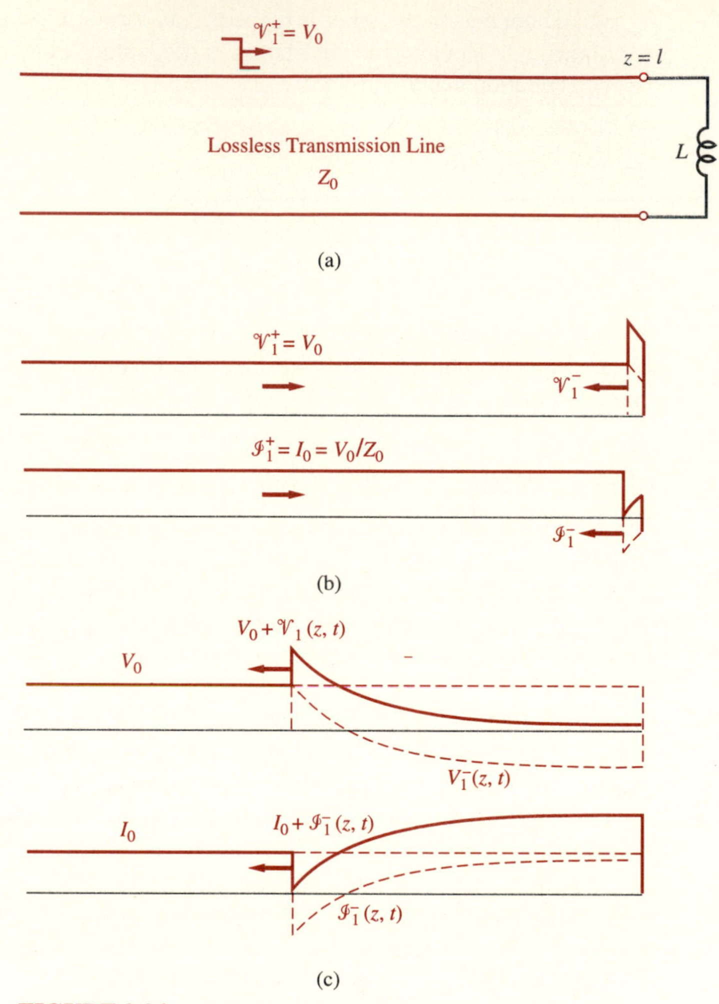

FIGURE 2.24. **Reflection from a purely inductive load.** The source is assumed to be matched (i.e., no reflections back from the source) and to supply a constant voltage V_0. The distributions of the voltage and current are shown at two different times: (b) immediately after reflection when the inductor behaves like an open circuit, and (c) later in time when the inductor behaves as a short circuit.

to $2V_0$ and the terminal current being zero, for that instant (Figure 2.24b). However, since a voltage of $2V_0$ now exits across the inductor, its current builds up, flowing more and more freely in time, until it is practically equivalent to a short circuit. At steady state, when the voltage across the inductance reduces to zero, the current through it becomes $2I_0$ (Figure 2.24c), similar to the case of a short-circuit termination (see Example 2-2).

Between an initial open circuit and an eventual short circuit, the voltage across the inductive load goes through all intermediate values, and the reflected voltage changes accordingly. To determine the analytical expression describing the variation of the voltage across the inductance, we need to simultaneously solve the transmission line equations (or the general solutions dictated by them, namely [2.8] and [2.9]) along with the differential equation describing the boundary condition imposed by the inductive load as

$$\mathcal{V}_{\mathrm{L}}(t) = L\frac{d\mathcal{I}_{\mathrm{L}}(t)}{dt}$$

For a general incident voltage $\mathcal{V}_1^+(z, t)$, the load voltage $\mathcal{V}_{\mathrm{L}}(t)$ and current $\mathcal{I}_{\mathrm{L}}(t)$ are given by

$$\mathcal{V}_{\mathrm{L}}(t) = \mathcal{V}_1^+(l, t) + \mathcal{V}_1^-(l, t)$$

$$\mathcal{I}_{\mathrm{L}}(t) = \mathcal{I}_1^+(l, t) + \mathcal{I}_1^-(l, t) = \frac{\mathcal{V}_1^+(l, t)}{Z_0} - \frac{\mathcal{V}_1^-(l, t)}{Z_0}$$

where the location of the load (i.e., the inductive termination) is assumed to be at $z = l$. We thus have

$$\mathcal{V}_1^+(l, t) + \mathcal{V}_1^-(l, t) = L\frac{d}{dt}\left(\frac{\mathcal{V}_1^+(l, t)}{Z_0} - \frac{\mathcal{V}_1^-(l, t)}{Z_0}\right)$$

or

$$\frac{d\mathcal{V}_1^-(l, t)}{dt} + \frac{Z_0}{L}\mathcal{V}_1^-(l, t) = \frac{d\mathcal{V}_1^+(l, t)}{dt} - \frac{Z_0}{L}\mathcal{V}_1^+(l, t)$$

which is a differential equation with $\mathcal{V}_1^-(l, t)$ as the dependent variable, since $\mathcal{V}_1^+(l, t)$ is presumably known, it being the incident voltage disturbance arriving from the source end of the line. The right-hand side is therefore a known function of time, and the equation is simply a first-order differential equation with constant coefficients. Note that we assume the source to be either far enough away or matched, so that the voltage $\mathcal{V}_1^-(z, t)$ does not reach the source end and generate a reflected voltage $\mathcal{V}_2^+(z, t)$ before $\mathcal{V}_1^-(l, t)$ reaches its "steady-state" value (nearly zero in the case when $\mathcal{V}_1^+(z, t) = V_0$ as shown in Figure 2.24c).

To study the simplest case, let us consider an incident voltage with a constant amplitude V_0 (i.e., $\mathcal{V}_1^+(l, t) = V_0$) launched by a step source reaches the inductor at $t = 0$. The above differential equation then simplifies to

$$\frac{d\mathcal{V}_1^-(l, t)}{dt} + \frac{Z_0}{L}\mathcal{V}_1^-(l, t) = -\frac{Z_0 V_0}{L}$$

The solution of this first-order differential equation is[27]

$$\mathcal{V}_1^-(l, t) = -V_0 + Ke^{-(Z_0/L)t}$$

[27]The solution can be found via Laplace transformation or as a superposition of the homogeneous solution and the particular solution; the validity of the solution can be shown by simply substituting it into the differential equation.

where the coefficient K needs to be determined by the known initial conditions as they relate to $\mathcal{V}_1^-(l, t)$. At the instant of the arrival of the voltage disturbance ($t = 0$), the current through the inductance $\mathcal{I}_L(t = 0) = 0$, and we thus have the incident voltage fully reflected, or $\mathcal{V}_1^-(l, t = 0) = V_0$. Thus, we must have $K = 2V_0$. The solution for the reflected voltage is then

$$\mathcal{V}_1^-(l, t) = -V_0 + 2V_0 e^{-(Z_0/L)t}$$

which varies from its initial value of V_0 to an eventual value of $-V_0$, as shown in Figure 2.24c.

In Examples 2-10 and 2-11, we illustrate two specific cases of reactive loads, in which relatively simple time-domain solutions are possible and provide useful insight.

Example 2-10: Lossy capacitive load. Consider the transmission line system shown in Figure 2.25a where a step voltage source of amplitude V_0 and source resistance $R_s = Z_0$ excites a lossless transmission line of characteristic impedance Z_0 and one-way time delay t_d connected to a load consisting of a parallel combination of R_L and C_L. Find and sketch the source- and load-end voltages as a function of time.

Solution: At $t = 0$, an incident voltage $\mathcal{V}_1^+(z, t)$ of amplitude $\mathcal{V}_1^+(0, 0) = V_0/2$ is launched at the source end of the line. This disturbance reaches the capacitive load at $t = t_d$, when a voltage $\mathcal{V}_1^-(z, t)$ of initial amplitude $\mathcal{V}_1^-(l, t_d)$ reflects toward the source. For $t \geq t_d$, the total load voltage $\mathcal{V}_L(t)$ and current $\mathcal{I}_L(t)$ are given by

$$\mathcal{V}_L(t) = \mathcal{V}_1^+(l, t) + \mathcal{V}_1^-(l, t)$$

$$\mathcal{I}_L(t) = \mathcal{I}_1^+(l, t) + \mathcal{I}_1^-(l, t) = \frac{\mathcal{V}_1^+(l, t)}{Z_0} - \frac{\mathcal{V}_1^-(l, t)}{Z_0}$$

where $\mathcal{V}_1^+(l, t) = \mathcal{V}_1^+(0, 0) = V_0/2$. These two equations are related by the boundary condition imposed by the load; that is,

$$\mathcal{I}_L(t) = \frac{\mathcal{V}_L(t)}{R_L} + C_L \frac{d\mathcal{V}_L(t)}{dt}$$

Substituting the first two equations into the third equation yields

$$\frac{d\mathcal{V}_1^-(l, t)}{dt} + \left[\frac{R_L + Z_0}{R_L Z_0 C_L}\right]\mathcal{V}_1^-(l, t) = \left[\frac{R_L - Z_0}{R_L Z_0 C_L}\right]\mathcal{V}_1^+(l, t)$$

which is a first-order differential equation for $\mathcal{V}_1^-(t)$. Note that in deriving it, we have used the fact that the incident voltage is constant in time, so that $d\mathcal{V}_1^+(l, t)/dt = 0$. The solution of this first-order differential equation can be found by noting that $\mathcal{V}_1^+(l, t) = V_0/2$ and by writing the general solution as

$$\mathcal{V}_1^-(l, t) = K_1 + K_2 e^{-[(R_L + Z_0)/(R_L Z_0 C_L)](t - t_d)}$$

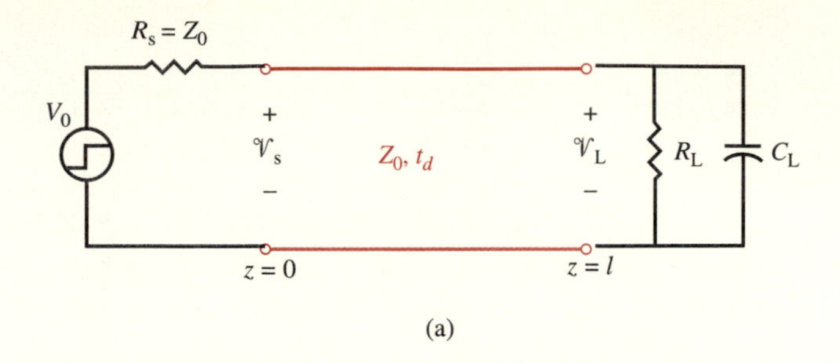

(a)

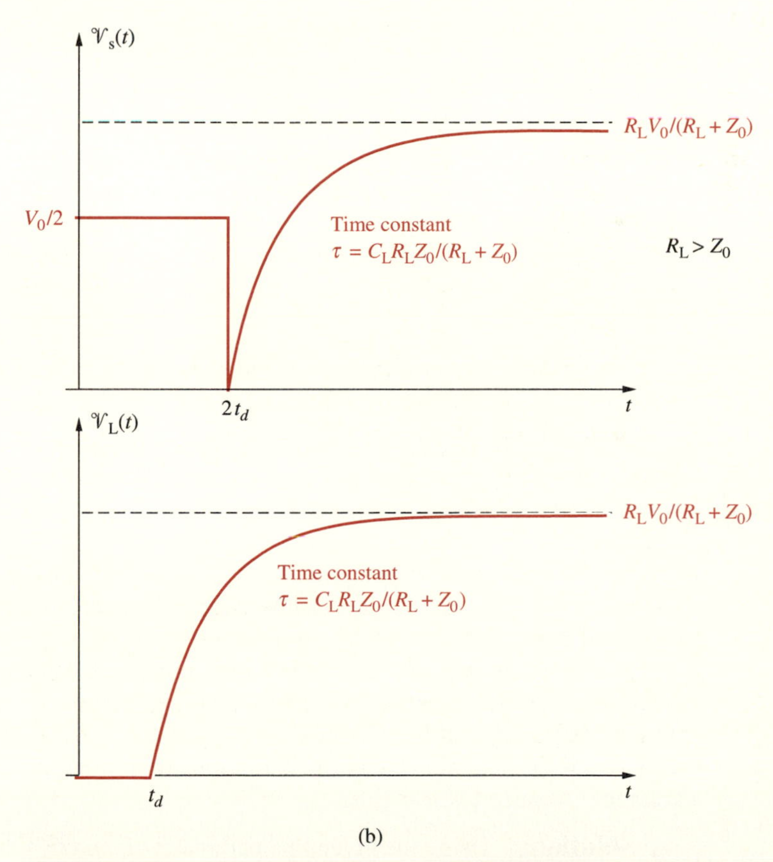

(b)

FIGURE 2.25. Step response of a capacitively loaded line. (a) Circuit diagram. (b) Time variation of the source- and load-end voltages.

in which case the constants K_1 and K_2 can be determined by using the initial and final conditions. Note that the reflected voltage must vary exponentially from $-\mathcal{V}_1^+(l, t)$ at $t = t_d$ to $(R_L - Z_0)\mathcal{V}_1^+(l, t)/(R_L + Z_0)$ for $t \to \infty$. We thus have

$$\mathcal{V}_1^-(l, t) = \mathcal{V}_1^+(l, t)\left[\frac{R_L - Z_0}{R_L + Z_0} - \frac{2R_L}{R_L + Z_0} e^{-[(R_L+Z_0)/(R_L Z_0 C_L)](t-t_d)}\right]$$

which is valid for $t \geq t_d$. This behavior can be understood as follows: When the incident voltage reaches the capacitive load at $t = t_d$, the capacitor C_L is initially uncharged and acts like a short circuit, resulting in $\mathcal{V}_1^-(l, t_d) = -\mathcal{V}_1^+(l, t_d) = -V_0/2$. However, at steady state the capacitor is fully charged and acts like an open circuit, resulting in $\mathcal{V}_1^-(l, \infty) = (R_L - Z_0)\mathcal{V}_1^+(l, \infty)/(R_L + Z_0) = (R_L - Z_0)V_0/[2(R_L + Z_0)]$, as expected. Note also that the time constant of the exponential variation is $\tau = R_L Z_0 C_L/(R_L + Z_0) = R_{Th} C_L$, where R_{Th} is the Thévenin equivalent resistance,[28] as seen from the terminals of the capacitor. Substituting $\mathcal{V}_1^-(l, t)$ into $\mathcal{V}_L(t)$ yields

$$\mathcal{V}_L(t) = \frac{R_L V_0}{R_L + Z_0}[1 - e^{-[(R_L+Z_0)/(R_L Z_0 C_L)](t-t_d)}]$$

valid for $t \geq t_d$. When the front of the reflected voltage reaches the source end at $t = 2t_d$, it is completely absorbed, since the source end of the line is matched (i.e., $R_s = Z_0$). The voltage at the source end is given by $\mathcal{V}_s(t) = \mathcal{V}_1^+(0, t) = V_0/2$ for $t < 2t_d$, and

$$\mathcal{V}_s(t) = \frac{R_L V_0}{R_L + Z_0}[1 - e^{-[(R_L+Z_0)/(R_L Z_0 C_L)](t-2t_d)}]$$

is valid for $t \geq 2t_d$. Sketches of $\mathcal{V}_s(t)$ and $\mathcal{V}_L(t)$ are shown in Figure 2.25b.

Example 2-11: A lumped series inductor between two transmission lines. Microstrip transmission lines on printed circuit boards are often connected together with bonding wires, which are inherently inductive. In this example, we consider a typical model for such a connection, namely a lumped series inductor between two different transmission lines. The measurement of the bonding-wire inductance is considered later in Example 2-16. Two microstrip transmission lines having equal line parameters of $L = 4$ nH-(cm)$^{-1}$ and $C = 1.6$ pF-(cm)$^{-1}$ and lengths 15 cm and 10 cm are connected by a wire represented by a series lumped inductance of $L_w = 5$ nH, shown in Figure 2.26a. The end of the shorter line is matched with a 50Ω load, and the circuit is excited at $t = 0$ by a unit step voltage source of $R_s = 50\Omega$. Find and sketch the variations with time of the source and load end voltages. Assume lossless lines.

Solution: Using the given line parameters L and C, the characteristic impedance and the phase velocity of the microstrip lines can be calculated as

$$Z_0 = \sqrt{\frac{L}{C}} = \sqrt{\frac{4 \times 10^{-9}}{1.6 \times 10^{-12}}} = 50\Omega$$

[28] In this case being simply equal to the parallel combination of the load resistance R_L and the characteristic impedance Z_0 of the line.

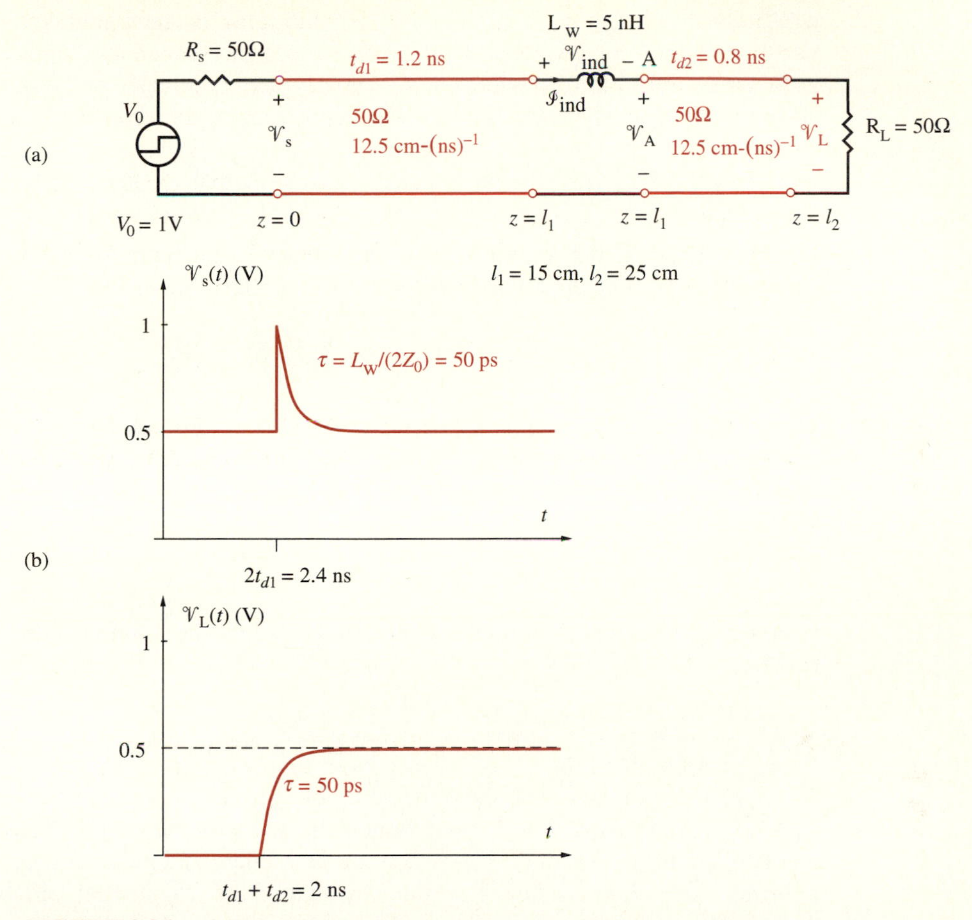

FIGURE 2.26. A lumped series inductor between two different transmission lines. (a) Circuit diagram. (b) Time variation of source- and load-end voltages for Example 2-11.

and

$$v_p = \frac{1}{\sqrt{LC}} = \frac{1}{\sqrt{4 \times 10^{-9} \times 1.6 \times 10^{-12}}} = 12.5 \times 10^9 \text{ cm-s}^{-1}$$

Therefore, the one-way delay times of the two lines are respectively $t_{d1} =$ 15 cm/(12.5 cm-(ns)$^{-1}$) = 1.2 ns and $t_{d2} =$ 10 cm/(12.5 cm-ns^{-1}) = 0.8 ns, as indicated in Figure 2.26a.

As in the previous example, at $t = 0$, an incident voltage $\mathcal{V}_1^+(z, t)$ of amplitude $\mathcal{V}_1^+(0, 0) = 0.5$ V is launched from the source end of the line. Note that the amplitude of this incident voltage remains constant in time, as long as it is supplied by the source. When this disturbance arrives at the junction at $t = t_{d1} = 1.2$ ns, the uncharged inductor initially acts like an open circuit (i.e., it resists the flow of current), producing a reflected voltage $\mathcal{V}_1^-(z, t)$ of initial amplitude $\mathcal{V}_1^-(l_1, t_{d1}) = \mathcal{V}_1^+(l_1, t_{d1}) = 0.5$ V, the same as one would have if the two lines were not connected. Current flows into the second line through the inductor as the inductor charges exponentially and eventually behaves like a short circuit at steady state. The following equations apply for $t \geq t_{d1}$:

$$\mathcal{V}_1^+(l_1, t) + \mathcal{V}_1^-(l_1, t) = \mathcal{V}_{\text{ind}}(t) + \mathcal{V}_{\text{A}}(t) = L_{\text{w}}\frac{d\mathcal{I}_{\text{ind}}(t)}{dt} + Z_0\mathcal{I}_{\text{ind}}(t)$$

where $\mathcal{V}_1^+(l_1, t) = 0.5$ V and $\mathcal{V}_{\text{A}}(t)$ is the voltage at position A, at which the impedance seen looking toward the load is $Z_0 = 50\Omega$. We also have

$$\mathcal{I}_{\text{ind}}(t) = \mathcal{I}_1^+(l_1, t) + \mathcal{I}_1^-(l_1, t) = \frac{\mathcal{V}_1^+(l_1, t)}{Z_0} - \frac{\mathcal{V}_1^-(l_1, t)}{Z_0}$$

where $\mathcal{I}_{\text{ind}}(t)$ and $\mathcal{V}_{\text{ind}}(t)$ are, respectively, the current through and the voltage across the inductor, as defined in Figure 2.26a. Substituting the second equation into the first yields

$$\frac{d\mathcal{V}_1^-(l_1, t)}{dt} + \frac{2Z_0}{L_{\text{w}}}\mathcal{V}_1^-(l_1, t) = 0$$

which is a first-order differential equation for $\mathcal{V}_1^-(l_1, t)$. The solution of this equation is

$$\mathcal{V}_1^-(l_1, t) = \mathcal{V}_1^+(l_1, t_{d1})e^{-(2Z_0/L_{\text{w}})(t-t_{d1})} = 0.5e^{-2\times10^{10}(t-1.2\times10^{-9})} \text{ V}$$

which is valid for $t \geq t_{d1} = 1.2$ ns. Note that we have used the fact that $\mathcal{V}_1^-(l_1, t_{d1}) = 0.5$ and that the reflected voltage varies exponentially from $\mathcal{V}_1^-(l_1, t_{d1}) = \mathcal{V}_1^+(l_1, t_{d1})$ at $t = t_{d1}$ (when the inductor initially behaves like an open circuit) to zero at $t \to \infty$ (when the fully energized inductor eventually behaves like a short circuit). The time constant of the exponential variation is $\tau = L_{\text{w}}/R_{\text{Th}} = 50$ ps, where $R_{\text{Th}} = 2Z_0 = 100\Omega$ is the Thévenin equivalent resistance as seen from the terminals of the inductor.

To satisfy the boundary condition at the junction, and noting that the inductor L_{w} is a lumped element, the current on both sides of the inductor must be the same. Thus, the inductor current $\mathcal{I}_{\text{ind}}(t)$ can be written as

$$\mathcal{I}_{\text{ind}}(t) = \mathcal{I}_1^+(l_1, t_{d1}) + \mathcal{I}_1^-(l_1, t) = \frac{\mathcal{V}_1^+(l_1, t_{d1})}{Z_0}[1 - e^{-(2Z_0/L_{\text{w}})(t-t_{d1})}]$$

which is valid for $t \geq t_{d1}$. At $t = t_{d1}$, a transmitted voltage $\mathcal{V}_{\text{A}}^+(z, t) = Z_0\mathcal{I}_{\text{A}}^+(z, t)$ is launched at position A on the second line, where $\mathcal{I}_{\text{A}}^+(z, t) = \mathcal{I}_{\text{ind}}(t)$. Therefore, the transmitted voltage $\mathcal{V}_{\text{A}}^+(z, t)$ can be written as

$$\mathcal{V}_{\text{A}}^+(z, t) = Z_0\mathcal{I}_{\text{ind}}(t) = \mathcal{V}_1^+(l_1, t_{d1})[1 - e^{-(2Z_0/L_{\text{w}})(t-t_{d1})}]$$

valid for $t \geq t_{d1}$.

The voltage $\mathcal{V}_{\text{A}}^+(z, t)$ arrives at the load end at $t = t_{d1} + t_{d2} = 2$ ns, where it is completely absorbed, since $R_{\text{L}} = Z_0 = 50\Omega$. The load voltage is given by

$$\mathcal{V}_{\rm L}(t) = \mathcal{V}_{\rm A}^{+}(l_2, t) = \mathcal{V}_1^{+}(l_1, t_{d1})[1 - e^{-(2Z_0/L_{\rm w})(t-(t_{d1}+t_{d2}))}]$$
$$= 0.5[1 - e^{-2\times10^{10}(t-2\times10^{-9})}] \text{ V}$$

valid for $t \geq (t_{d1} + t_{d2}) = 2$ ns. Note that $\mathcal{V}_1^{+}(l_1, t_{d1}) = 0.5$ V for $t \geq t_{d1}$, since the incident voltage remains constant unless the source changes. Similarly, the source-end voltage is given by $\mathcal{V}_{\rm s}(t) = \mathcal{V}_1^{+}(0, t) = 0.5$ V for $t < 2t_{d1} = 2.4$ ns and

$$\mathcal{V}_{\rm s}(t) = \mathcal{V}_1^{+}(0, t)[1 + e^{-(2Z_0/L_{\rm w})(t-2t_{d1})}] = 0.5[1 + e^{-2\times10^{10}(t-2.4\times10^{-9})}] \text{ V}$$

for $t \geq 2t_{d1} = 2.4$ ns. Sketches of the time variations of source- and load-end voltages are shown in Figure 2.26b.

In the preceding two examples we used time-domain methods to determine the step response of transmission lines with terminations or discontinuities involving reactive elements. The basis for our analyses was the simultaneous solution of [2.10], describing the transmission line voltage and current, together with the differential equations that describe the terminal voltage-current relationship of the reactive load. These time-domain solutions were tractable partly because the excitation voltage was a simple step function and also because the reactive discontinuities that we analyzed involved only one energy storage element (i.e., a single capacitor or inductor). When the input voltage function is more complicated, or when the reactive discontinuity involves more than one energy storage element, it is often easier to use Laplace transform methods to determine the response of the line. We demonstrate the use of the Laplace transform method in Example 2-12.

Example 2-12: Reflections due to inductance of resistor leads. Consider the transmission line system shown in Figure 2.27, where the source voltage amplitude increases linearly from zero to V_0 over a time[29] of t_r. The output

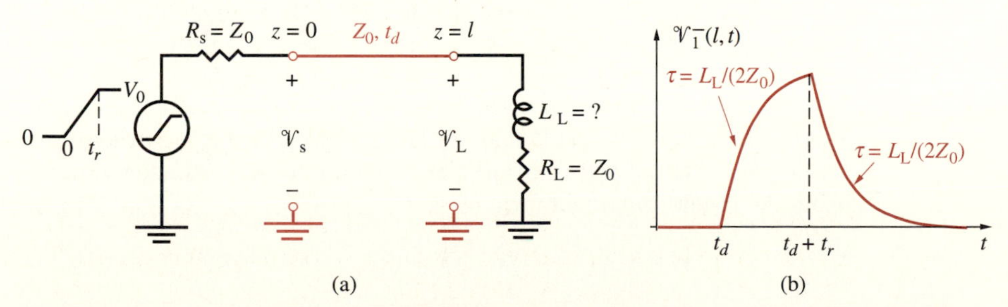

FIGURE 2.27. Reflections due to inductance of resistor leads. (a) Circuit diagram. (b) Time variation of the reflected voltage $\mathcal{V}_1^{-}(l, t)$ for the case $R_{\rm L} = Z_0$.

[29]Note that t_r is not exactly the rise time discussed in Section 1.1, which was defined as the time required for the signal to change from 10% to 90% of its final value.

resistance of the source is $R_s = Z_0$, while the transmission line having a characteristic impedance Z_0 and a one-way time delay t_d is terminated in a reactive load consisting of a series combination of R_L and L_L. Find an expression for the reflected voltage at the load [i.e., $\mathcal{V}_1^-(l, t)$], and determine its maximum value for $R_L = Z_0$.

Solution: From Figure 2.27 we note that, starting at $t = 0$, an incident voltage is launched at the source end of the line, given by

$$\mathcal{V}_1^+(0, t) = \frac{V_0}{2t_r}[tu(t) - (t - t_r)u(t - t_r)]$$

where $u(\cdot)$ is the unit step function. The Laplace transform of this voltage waveform is

$$\tilde{\mathcal{V}}_1^+(s) = \frac{V_0}{2t_r}\frac{1 - e^{-t_r s}}{s^2}$$

This incident voltage propagates to the load end, and no reflected voltage exists until it arrives there at $t = t_d$. For $t \geq t_d$, the total load voltage and current $\mathcal{V}_L(t)$ and $\mathcal{I}_L(t)$ are given by

$$\mathcal{V}_L(t) = \mathcal{V}_1^+(l, t) + \mathcal{V}_1^-(l, t)$$

$$\mathcal{I}_L(t) = \mathcal{I}_1^+(l, t) + \mathcal{I}_1^-(l, t) = \frac{\mathcal{V}_1^+(l, t)}{Z_0} - \frac{\mathcal{V}_1^-(l, t)}{Z_0}$$

and are related by the boundary condition imposed by the load,

$$\mathcal{V}_L(t) = \mathcal{I}_L(t)R_L + L_L\frac{d\mathcal{I}_L(t)}{dt}$$

Substituting the first two equations into the third equation yields

$$L_L\frac{d\mathcal{V}_1^-(l, t)}{dt} + (R_L + Z_0)\mathcal{V}_1^-(l, t) = L_L\frac{d\mathcal{V}_1^+(l, t)}{dt} + (R_L - Z_0)\mathcal{V}_1^+(l, t)$$

Note that unlike the case of Example 2-10, the derivative of the incident voltage is not zero, since $\mathcal{V}_1^+(l, t)$ is not constant in time during the time interval $t_d \leq t \leq (t_d + t_r)$. To solve this differential equation for the reflected voltage $\mathcal{V}_1^-(l, t)$ using the given functional form of $\mathcal{V}_1^+(l, t)$, we can take its Laplace transform,

$$(sL_L + R_L + Z_0)\tilde{\mathcal{V}}_1^-(s) = (sL_L + R_L - Z_0)\tilde{\mathcal{V}}_1^+(s)$$

where $\tilde{\mathcal{V}}_1^\pm(s)$ is the Laplace transform of $\mathcal{V}_1^\pm(l, t)$. Using the previously noted form of $\tilde{\mathcal{V}}_1^+(s)$, we then have

$$\tilde{\mathcal{V}}_1^-(s) = \left[\frac{sL_L + R_L - Z_0}{sL_L + R_L + Z_0}\right]\left[\frac{V_0}{2t_r}\frac{1 - e^{-t_r s}}{s^2}\right] = \left[\frac{s + (R_L - Z_0)/L_L}{s + (R_L + Z_0)/L_L}\right]\left[\frac{V_0}{2t_r}\frac{1 - e^{-t_r s}}{s^2}\right]$$

which can be expanded into its partial fractions as

$$\tilde{\mathcal{V}}_1^-(s) = \frac{K_1}{s + (R_L + Z_0)/L_L} + \frac{K_2}{s} + \frac{K_3}{s^2}$$

where the coefficients are

$$K_2 = -K_1 = \frac{V_0 Z_0 L_L}{t_r (R_L + Z_0)^2} \quad \text{and} \quad K_3 = \frac{V_0}{2t_r}\frac{R_L - Z_0}{R_L + Z_0}$$

Taking the inverse Laplace transform[30] yields

$$\mathcal{V}_1^-(l,t) = \frac{V_0}{t_r}\frac{Z_0 L_L}{(R_L + Z_0)^2}\{[1 - e^{-(R_L+Z_0)t'/L_L}]u(t') + [1 - e^{-(R_L+Z_0)(t'-t_r)/L_L}]u(t' - t_r)\}$$
$$+ \frac{V_0}{2t_r}\frac{R_L - Z_0}{R_L + Z_0}[t'u(t') - (t' - t_r)u(t' - t_r)]$$

where $t' = t - t_d$. Note that the solution for $\mathcal{V}_1^-(l,t)$ is valid only for $t \geq t_d$, or $t' \geq 0$.

A practical case of interest is that in which $R_L = Z_0$. When a microstrip is terminated at a matched load resistance to avoid reflections, the nonzero inductance of the resistor leads may nevertheless produce reflections. To determine the maximum reflection voltage due to the inductance of the resistor leads, we substitute $R_L = Z_0$ in the solution for $\mathcal{V}_1^-(l,t)$ to find

$$\mathcal{V}_1^-(l,t) = \frac{V_0}{t_r}\frac{L_L}{4Z_0}\{[1 - e^{-(2Z_0/L_L)(t-t_d)}]u(t - t_d) - [1 - e^{-(2Z_0/L_L)(t-t_d-t_r)}]u(t - t_d - t_r)\}$$

The time variation of $\mathcal{V}_1^-(l,t)$ is plotted in Figure 2.27b, showing that the reflected voltage rises and falls exponentially with a time constant of $L/(2Z_0)$. The maximum reflected voltage occurs at $t = t_d + t_r$ and is given by

$$[\mathcal{V}_1^-(l,t)]_{\max} = \frac{V_0}{t_r}\frac{L_L}{4Z_0}[1 - e^{-(2Z_0/L_L)t_r}]$$

Note that in practice the maximum reflected voltage can easily be measured—for example, by using a time-domain reflectometer (see Section 2.6.1), from which the value of the inductance L_L of the resistor leads can be calculated, since Z_0 and t_r are known in most cases.

---

[30] We use the following Laplace transform pairs:

$$e^{-a(t-b)}u(t-b) \Longleftrightarrow \frac{e^{-bs}}{s+a}$$

$$(t-b)u(t-b) \Longleftrightarrow \frac{e^{-bs}}{s^2}$$

where a and b are constants.

2.5.2 Nonlinear Terminations

Up to now, we have considered responses of transmission lines with linear sources and linear loads, and we have analyzed reflections using the simultaneous analytical solution of the transmission line voltage and current expressions together with the load characteristics. In some high-speed digital circuits, both the sources (driver gates) and loads (receiving gates) can have nonlinear current-voltage characteristics. This is particularly the case for transistor-transistor logic (TTL) and complementary metal oxide semiconductor (CMOS) logic gates. In cases where transmission lines are terminated in or driven by nonlinear loads, a graphical technique known as the Bergeron method[31] is quite useful.

The Bergeron graphical technique can be applied to transmission line circuits that involve linear or nonlinear devices. It provides the same basic information as the bounce diagram (voltage and current versus time) but with fewer calculations. It relies on a graphical means of describing the reflections on the transmission line. The graphs involved in using this approach can become quite complex, especially if reactive elements are present. Also, the graphical technique requires accurate knowledge of the current-voltage characteristics of the nonlinear devices.

To illustrate the graphical methodology before we apply it to a nonlinear load, we consider in Example 2-13 the step response of a simple transmission line terminated in a linear resistive load.

Example 2-13: Graphical solution of the step response of a resistively terminated lossless line. Consider a transmission line with a characteristic impedance of 100Ω, driven by a 1 V step source with source resistance of 25Ω, and terminated in a 300Ω load, as shown in Figure 2.28a. Use the graphical Bergeron method to analyze the effects of reflections on the source- and load-end voltages and currents.

Solution: We start by plotting the current-voltage characteristics of the source and load ends (i.e., $\mathcal{I}_s$ versus $\mathcal{V}_s$ and $\mathcal{I}_L$ versus $\mathcal{V}_L$), using the same set of axes. At the source end we have

$$\mathcal{V}_s = V_0 - R_s\mathcal{I}_s = 1 - 25\mathcal{I}_s$$

which is a straight line (referred to as the source-end line) with slope $-1/R_s = -(1/25)$ S, as shown in Figure 2.28b. At the load end we have

$$\mathcal{V}_L = R_L\mathcal{I}_L = 300\mathcal{I}_L$$

---

[31] L. Bergeron was a French hydraulic engineer and developed this graphical method in 1949 to study water hammer waves in pipes. For an English translation of Bergeron's original work see L. B. J. Bergeron, *Water Hammer in Hydraulics and Wave Surges in Electricity,* John Wiley & Sons, Inc., New York, 1961. The use of this method in high-speed digital switching circuits was initially suggested in 1968. See R. S. Singleton, No need to juggle equations to find reflection—just draw three lines, *Electronics,* pp. 93–99, October 28, 1968.

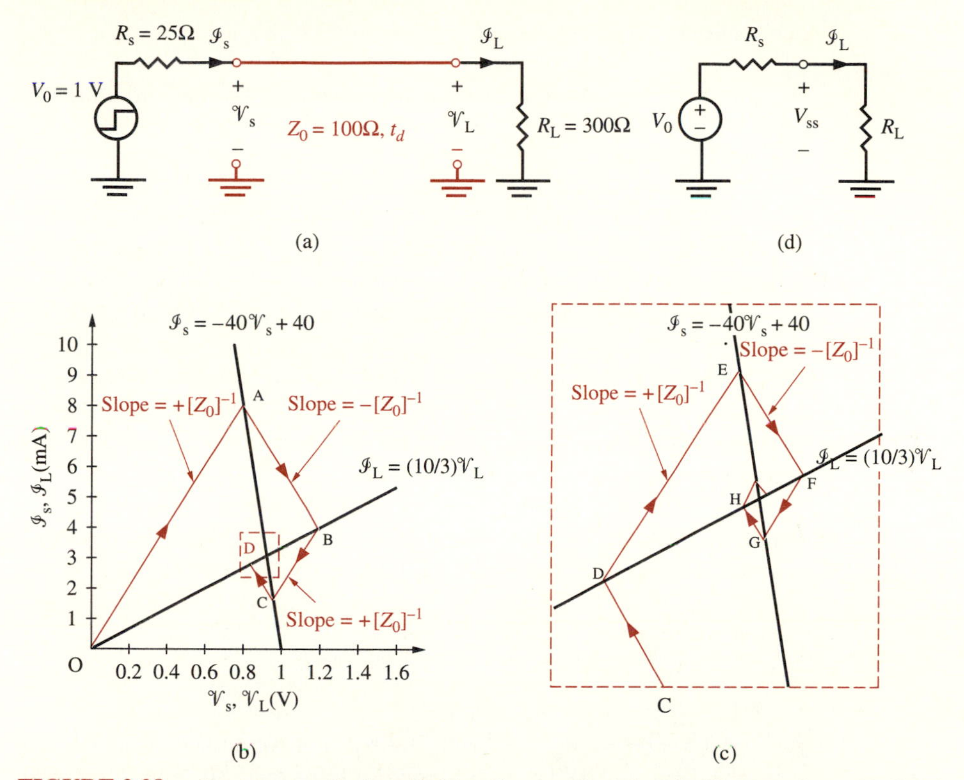

FIGURE 2.28. Bergeron method applied to a line terminated with a linear resistor. (a) Circuit diagram. (b) Current versus voltage characteristics at the source and load ends are shown as heavy lines. The lighter lines are lines with slopes of $+[Z_0]^{-1}$ and $-[Z_0]^{-1}$ in units of mA-(V)$^{-1}$, or mS. The intersection points of the lighter lines with the source- and load-end current-voltage characteristics represent respectively the values of the source- or load-end voltages at the corresponding times. (c) Enlarged view of the rectangular dashed-line region in (b). (d) Steady-state equivalent circuit.

which is also a straight line (the load-end line) with slope $1/R_L = (1/300)$ S, as shown in Figure 2.28b. These two straight lines, sometimes also referred to simply as the "load lines," define the relation between current and voltage at the source and load ends. At $t = 0$, a voltage $\mathcal{V}_1^+$ is launched from the source end. This voltage disturbance is accompanied by its associated current given by $\mathcal{I}_1^+ = [Z_0]^{-1}\mathcal{V}_1^+ = [100]^{-1}\mathcal{V}_1^+$. The relationship between $\mathcal{I}_1^+$ and $\mathcal{V}_1^+$ is another straight line with slope $[100]^{-1}$ S, shown as the line segment OA in Figure 2.28b. Noting that at $t = 0$, $\mathcal{V}_s = \mathcal{V}_1^+$ and $\mathcal{I}_s = \mathcal{I}_1^+$, the intersection of the line OA with the source-end line at point A determines the amplitude of the initial voltage and current launched at the source end of the transmission line. This intersection point essentially is the solution of the two linear equations describing the source-end voltage and current, with the resultant voltage $V_A = \mathcal{V}_s(t = 0) = \mathcal{V}_1^+$ being what we would have if we simply divided the applied voltage V_0 between the source resistance and the characteristic impedance of the line. The resultant voltage and current values can easily be read from point A in Figure 2.28b as $\mathcal{V}_1^+ = V_A = 0.8$ V and $\mathcal{I}_1^+ = I_A = 8$ mA, respectively.

At $t = t_d$, when the voltage disturbance reaches the load end of the line, a reflected voltage $\mathcal{V}_1^-$ is generated, which is accompanied by its associated current $\mathcal{I}_1^-$. Namely, we have

$$\mathcal{I}_1^- = -[Z_0]^{-1}\mathcal{V}_1^- \quad \rightarrow \quad \mathcal{I}_L - \mathcal{I}_1^+ = -[Z_0]^{-1}(\mathcal{V}_L - \mathcal{V}_1^+)$$

which represents a straight line passing through point A and having a slope $-[Z_0]^{-1} = -[100]^{-1}$, if we now view the axes in Figure 2.28b as $\mathcal{I}_L$ versus $\mathcal{V}_L$. The intersection of this straight line (shown in Figure 2.28b) with the load-end line (i.e., $\mathcal{V}_L = 300\mathcal{I}_L$) at point B determines the values of $\mathcal{V}_L$ and $\mathcal{I}_L$ at $t = t_d$, which can easily be read from the graph as $\mathcal{V}_L(t = t_d) = \mathcal{V}_1^+ + \mathcal{V}_1^- = V_B = 1.2$ V and $\mathcal{I}_L(t = t_d) = \mathcal{I}_1^+ + \mathcal{I}_1^- = I_B = 4$ mA, respectively. The value of $\mathcal{V}_L$ at $t = t_d$, for example, can in turn be used to determine the amplitude of the reflected voltage as $\mathcal{V}_1^- = (\mathcal{V}_L - \mathcal{V}_1^+) = (1.2 - 0.8) = 0.4$ V.

The reflected disturbance $\mathcal{V}_1^-(z, t)$ reaches the source end at $t = 2t_d$, at which time a new voltage $\mathcal{V}_2^+$ is created, accompanied by its corresponding current $\mathcal{I}_2^+ = [Z_0]^{-1}\mathcal{V}_2^+$. Noting that at $t = 2t_d$, we have

$$\mathcal{V}_s = \mathcal{V}_1^+ + \mathcal{V}_1^- + \mathcal{V}_2^+ = V_B + \mathcal{V}_2^+$$

$$\mathcal{I}_s = \mathcal{I}_1^+ + \mathcal{I}_1^- + \mathcal{I}_2^+ = I_B + \mathcal{I}_2^+$$

the relationship between $\mathcal{I}_2^+$ and $\mathcal{V}_2^+$ can be rewritten as $[Z_0]^{-1}(\mathcal{V}_s - V_B) = (\mathcal{I}_s - I_B)$, representing a straight line passing through the point B and having a slope $[Z_0]^{-1} = [100]^{-1}$ S, as shown in Figure 2.28b. The intersection of this straight line with the source-end line ($\mathcal{V}_s = 1 - 25\mathcal{I}_s$) at point C determines the values of $\mathcal{I}_s$ and $\mathcal{V}_s$ at $t = 2t_d$, which can be read from Figure 2.28b as $\mathcal{I}_s = \mathcal{I}_1^+ + \mathcal{I}_1^- + \mathcal{I}_2^+ = I_C = 1.6$ mA and $\mathcal{V}_s = \mathcal{V}_1^+ + \mathcal{V}_1^- + \mathcal{V}_2^+ = V_C = 0.96$ V, respectively. From the value of $\mathcal{V}_s$ at $t = 2t_d$, one can determine the amplitude of the reflected voltage $\mathcal{V}_2^+ = \mathcal{V}_s(t = 2t_d) - V_B = 0.96 - 1.2 = -0.24$ V.

We can continue this procedure by drawing a straight line with slope $-[Z_0]^{-1}$ passing through the point C and find its intersection with the load-end line at point D and from there read $\mathcal{I}_L$ and $\mathcal{V}_L$ values at $t = 3t_d$ to be 2.8 mA and 0.84 V, respectively. A straight line with slope $[Z_0]^{-1}$ passing through the point D can then be drawn to intersect the source-end line at point E, representing $\mathcal{I}_s$ and $\mathcal{V}_s$ values at $t = 4t_d$ of 3.52 mA and ~0.912 V, respectively. The process continues in this manner until both the source- and the load-end voltages and currents converge to the steady-state values as determined by the intersection of the source- and load-end lines. At that point, we have $\mathcal{V}_s = \mathcal{V}_L$ and $\mathcal{I}_s = \mathcal{I}_L$, the transmission line is charged to its final voltage $V_{ss} \simeq 0.923$ V, and the source and load ends are essentially connected together, as shown in the steady-state equivalent circuit shown in Figure 2.28d.

The graphical technique illustrated in Example 2-13 is not particularly needed when dealing with transmission lines with linear terminations. However, it becomes very useful for cases in which the source or load terminations are nonlinear. In Example 2-14, we consider a transmission line terminated by a diode, which in general has a nonlinear current-voltage characteristic.

Example 2-14: A nonlinear termination. Consider a transmission line with characteristic impedance of 50Ω driven by a 0.7 V step voltage source with an internal impedance of 25Ω and terminated by a diode, as shown in Figure 2.29a. The diode has a current-voltage characteristic given by

$$\mathcal{I}_{\mathrm{L}} = I_0(e^{\mathcal{V}_{\mathrm{L}}/V_{\mathrm{T}}} - 1)$$

where I_0 is called the saturation current, given as $I_0 = 10^{-15}$ A, and V_{T} is the thermal voltage, having the value $V_{\mathrm{T}} \simeq 26$ mV at room temperature. Use the graphical

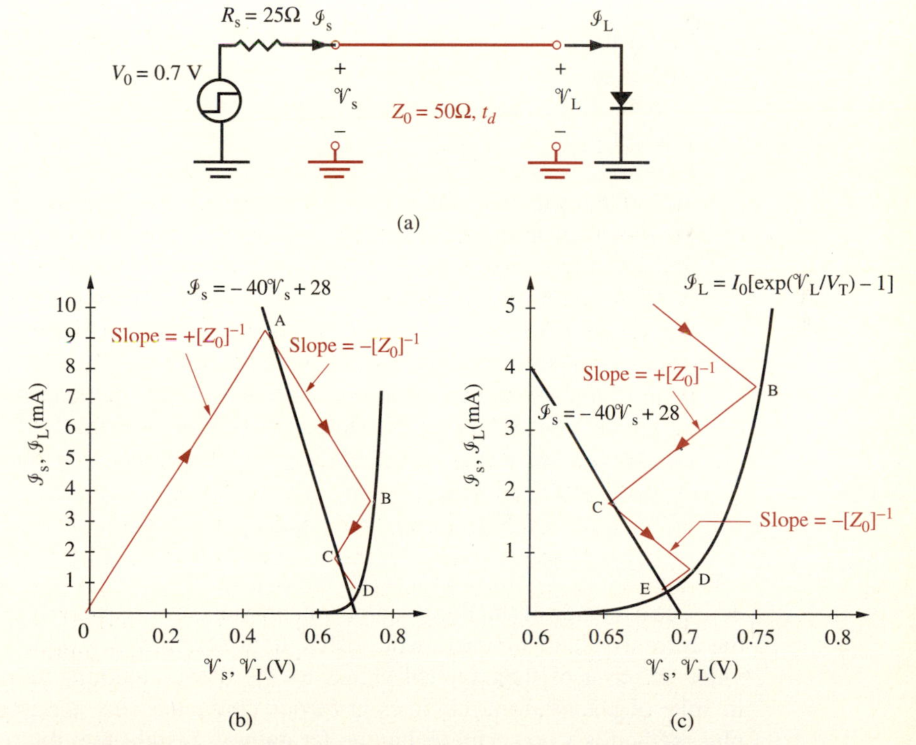

FIGURE 2.29. Step response of a line terminated in a nonlinear load. (a) Circuit diagram. (b) Current-voltage characteristics on the source- and load-ends are shown as heavy colored lines. The lighter lines are lines with slopes of $+[Z_0]^{-1}$ and $-[Z_0]^{-1}$. The intersection points of the lighter lines with the source- and load-end current-voltage characteristics represent the values of the source- and load-end voltages, respectively, at the corresponding times. (c) Enlarged view of the region around the steady-state point (i.e., the intersection of the source- and load-end current-voltage characteristics).

Bergeron technique to determine how long it will take for the circuit to reach steady state. Neglect the charging effects of the diffusion capacitance of the diode.

Solution: The solution of this problem can proceed in a manner quite similar to Example 2-13, except that the current-voltage characteristic of the load is nonlinear instead of a straight line. Following a procedure similar to that in Example 2-13, we first plot the source-end current-voltage characteristic, which is a straight line given by $\mathcal{V}_s = 0.7 - 25\mathcal{I}_s$, and then the load-end characteristic (i.e., the nonlinear diode equation) on the same graph, as shown in Figure 2.29b. At $t = 0$, the voltage $\mathcal{V}_1^+$ and its associated current $\mathcal{I}_1^+ = [Z_0]^{-1}\mathcal{V}_1^+$ are launched from the source end of the line. Since at $t = 0$ we have $\mathcal{V}_s = \mathcal{V}_1^+$ and $\mathcal{I}_s = \mathcal{I}_1^+$, the relationship between $\mathcal{I}_1^+$ and $\mathcal{V}_1^+$ is a straight line given as $\mathcal{V}_s = 50\mathcal{I}_s$ with a slope $[Z_0]^{-1} = [50]^{-1}$ S, which is plotted on the same graph. The intersection of this line with the source-end line is denoted by A, as shown in Figure 2.29b. The coordinates of point A can be read from Figure 2.29b as $I_A \simeq 9.33$ mA and $V_A \simeq 0.467$ V, which are also the values of the incident current and associated voltage disturbances $\mathcal{I}_1^+$ and $\mathcal{V}_1^+$, respectively. Next, we draw a straight line with slope $-[Z_0]^{-1} = -[50]^{-1}$ passing through point A and find its intersection with the diode characteristic, which is denoted as point B, as shown. The coordinates of point B are approximately given by $V_B = \mathcal{V}_L(t = t_d) \simeq 0.7519$ V and $I_B = \mathcal{I}_L(t = t_d) \simeq 3.63$ mA, respectively. Using the value of $\mathcal{V}_L(t_d)$, the amplitude of the reflected voltage can be determined as $\mathcal{V}_1^- = \mathcal{V}_L(t_d) - \mathcal{V}_1^+ \simeq 0.285$ V. We then draw a straight line with slope $+[Z_0]^{-1}$ and passing through point B and find its intersection with the source-end line at point C, where we have $V_C \simeq 0.657$ V and $I_C \simeq 1.73$ mA. This process continues until we reach the intersection point of the source-end line with the diode characteristic approximately at time $t = 4t_d$ and at point E, which is indistinguishably close to the intersection point of the source- and load-end current-voltage characteristics. Thus, in this particular case, it takes approximately $4t_d$ units of time for the circuit to reach steady state. At steady state (i.e., $t \to \infty$), we have $\mathcal{V}_s = \mathcal{V}_L \simeq 0.6912$ V and $\mathcal{I}_s = \mathcal{I}_L \simeq 0.3514$ mA.

The accuracy of the Bergeron plot technique depends heavily on the accuracy of the current-voltage characteristics of the nonlinear devices, which are usually provided in terms of their typical values by the device manufacturers. Nevertheless, in spite of potential inaccuracies in device characteristics, the graphical Bergeron plot method is a powerful technique for gaining insight into the response of a line terminated by a nonlinear device, for example to estimate the approximate duration of ringing effects. Finally, it should be noted that the graphical method is entirely suitable to circumstances where both the load- and the source-end current-voltage characteristics might be nonlinear. Such cases may arise in high-speed digital applications, since both the driver gates and the receiving gates are inherently nonlinear transistor devices, especially when used in on/off modes.

2.6 SELECTED PRACTICAL TOPICS

In this section, we discuss two selected practical topics, namely (a) time-domain reflectometers, and (b) the effects of source rise time. We also briefly comment on transients on lossy transmission lines. A brief discussion on time-domain reflectometry and two associated examples serve to introduce this simplest and most direct method of measuring characteristic impedance of a line, the nature of its termination, and the presence of discontinuities on the transmission line. A discussion of source rise time effects is necessary because up to now we have primarily (except in Example 2-12) considered responses to ideal step function excitations. In most practical applications, the sources used and the outputs of driver gates have finite rise and fall times.

2.6.1 Time-Domain Reflectometry

In practice, it is often necessary to make a number of measurements on a given transmission line system to characterize its transient response. The quantities that need to be measured include the nature (capacitive, inductive, or resistive) of the load termination, the characteristic impedance of the line, the maximum voltage level at which the line can be used, and others of a more specialized character. A time-domain reflectometer[32] (commonly abbreviated as TDR) is an instrument which is used to test, characterize, and model a system involving transmission lines and their accessories. In general, it consists of a very-fast-rise-time (typically less than 50 ps) step pulse source and a display oscilloscope in a system that operates like a closed-loop radar, as shown in Figure 2.30. The source produces an incident step voltage, which travels down the transmission line under investigation, and the incident and the reflected voltages at a particular point (typically the source end) on the line are monitored by the display oscilloscope using a high-impedance probe. The output impedance of the step source is typically well matched to the nominal characteristic impedance of the line to eliminate reflections from the source end.

The most common use of time-domain reflectometry involves the measurement of the characteristics of an unknown load termination or a discontinuity on the line. The former application is illustrated in Example 2-15 for resistive loads. A discontinuity on a transmission line could, for example, be a point of breakage on a buried coaxial line, an unwanted parasitic capacitance on an interconnect, or the inductance of a bonding wire between two interconnects. The latter case is illustrated in Example 2-16.

[32]B. M. Oliver, Time domain reflectometry, *Hewlett-Packard Journal,* 15 (6), February 1964.

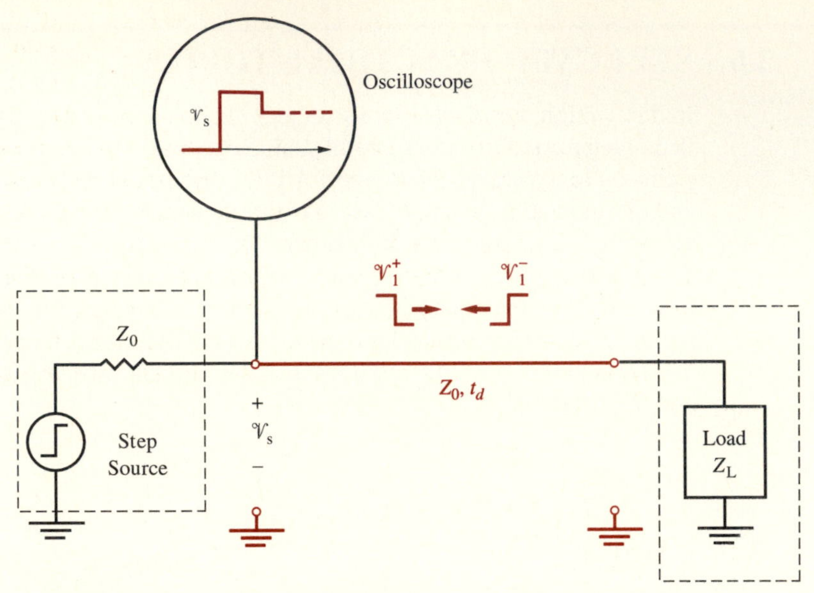

FIGURE 2.30. **Time-domain reflectometry.** Essential components of a typical TDR system.

Example 2-15: TDR displays for resistive loads. A TDR system (represented by a step pulse source of amplitude V_0 and output impedance $R_s = Z_0$) is connected to a transmission line of characteristic impedance Z_0 terminated with a resistive load R_L, as shown in Figure 2.31. Three TDR waveforms monitored at the source end are shown for three different values of R_L. Find the load resistance R_L for each case.

Solution: The initial value (immediately after the application of the step input) of the source-end voltage $\mathcal{V}_s(0)$ is equal to

$$\mathcal{V}_s(0) = \mathcal{V}_1^+(z = 0, 0) = \frac{V_0}{2} = 0.2 \text{ V}$$

from which the amplitude of the step voltage is found to be $V_0 = 0.4$ V. At $t = 1$ ns, the reflected voltage arrives at the source end and is completely absorbed. So

$$\mathcal{V}_s(1\text{ns}) = \mathcal{V}_1^+(0, 1 \text{ ns}) + \mathcal{V}_1^-(0, 1 \text{ ns}) = \mathcal{V}_1^+(0, 1 \text{ ns})(1 + \Gamma_L)$$

where $\Gamma_L = (R_L - Z_0)/(R_L + Z_0)$. For $R_L = R_{L1}$, we have

$$0.2(1 + \Gamma_{L1}) = 0.1$$

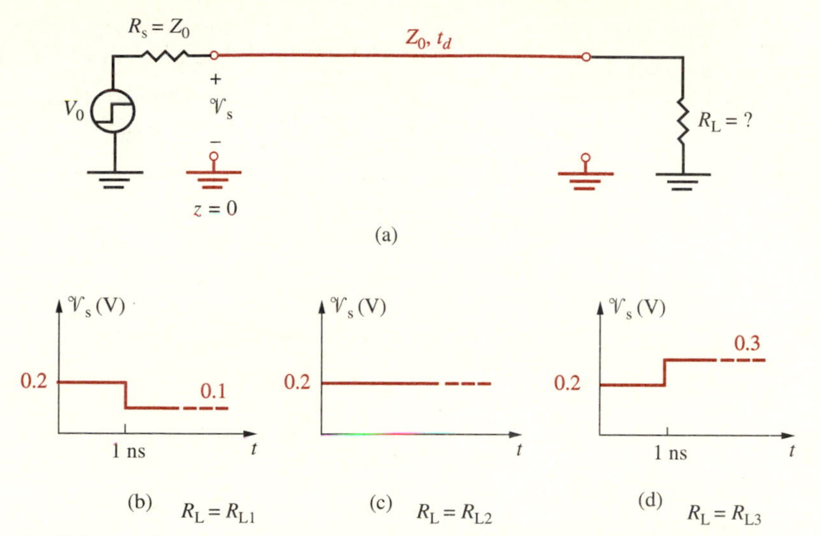

FIGURE 2.31. **TDR displays for resistive loads.** (a) A TDR system connected to a transmission line terminated with an unknown load resistor R_L. (b) $R_L = R_{L1}$. (c) $R_L = R_{L2}$. (d) $R_L = R_{L3}$.

from which $\Gamma_{L1} = -0.5$, yielding $R_{L1} = Z_0/3$. Similarly, we find $R_{L2} = Z_0$ and $R_{L3} = 3Z_0$.

A simple summary of the TDR waveforms observed at the source end for purely resistive, capacitive, and inductive terminations is provided in Figure 2.32. Note that the case of a resistive termination was discussed in the preceding example, while a simple inductive termination was discussed in Section 2.5.1 and in connection with Figure 2.24. The result for the capacitive termination case corresponds to that of Example 2-10 for $R_L = \infty$.

We now illustrate (Example 2-16) the use of the TDR technique for the measurement of the value of a reactive element connected between two transmission lines.

Example 2-16: TDR measurement of the inductance of a bonding wire connecting two transmission lines. Consider a bonding wire between two microstrip interconnects (each with characteristic impedance Z_0) on an integrated circuit board, as shown in Figure 2.33a. To measure the value of the bonding-wire inductance L_w, the circuit is terminated with a matched load (Z_0) on one side and is excited by a matched TDR system (i.e., $R_s = Z_0$) at the input side, as shown in Figure 2.33b. The TDR waveform $\mathcal{V}_s(t)$ measured is shown in Figure 2.33c, which is similar to Figure 2.26b (Example 2-11). Determine the value of the bonding-wire inductance in terms of the area under the "glitch" seen in the TDR waveform.

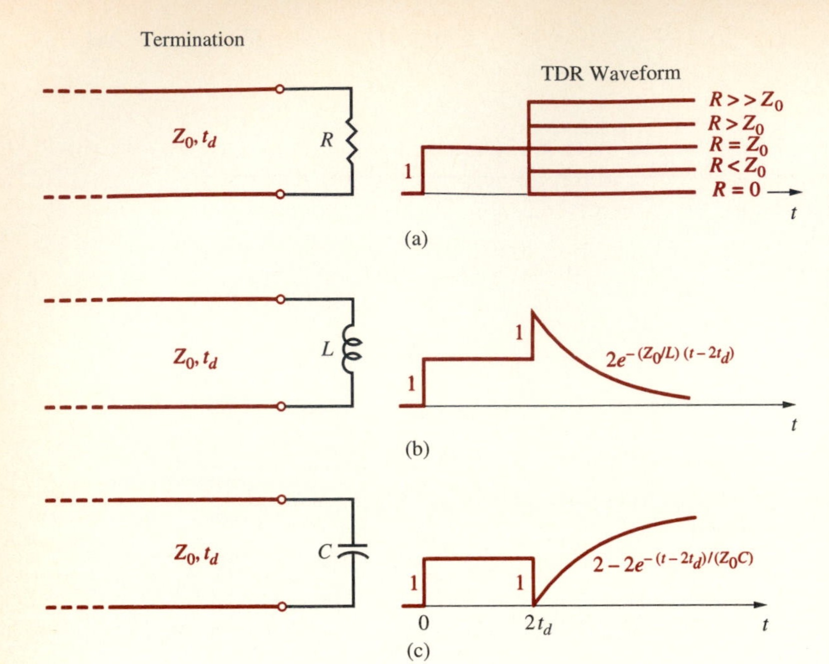

FIGURE 2.32. TDR signatures produced by simple terminations. Source-end TDR voltage signatures for purely (a) resistive, (b) inductive, and (c) capacitive terminations. In terms of excitation by a step voltage source of amplitude V_0 and source resistance R_s, the TDR traces shown are drawn for $V_0 = 2$ V and $R_s = Z_0$. (This figure was adapted from Figure 5 of B. M. Oliver, Time domain reflectometry, *Hewlett-Packard Journal,* 15 (6), pp. 14–9 to 14–16, February 1964. ©Hewlett-Packard Company 1964. Reproduced with permission.)

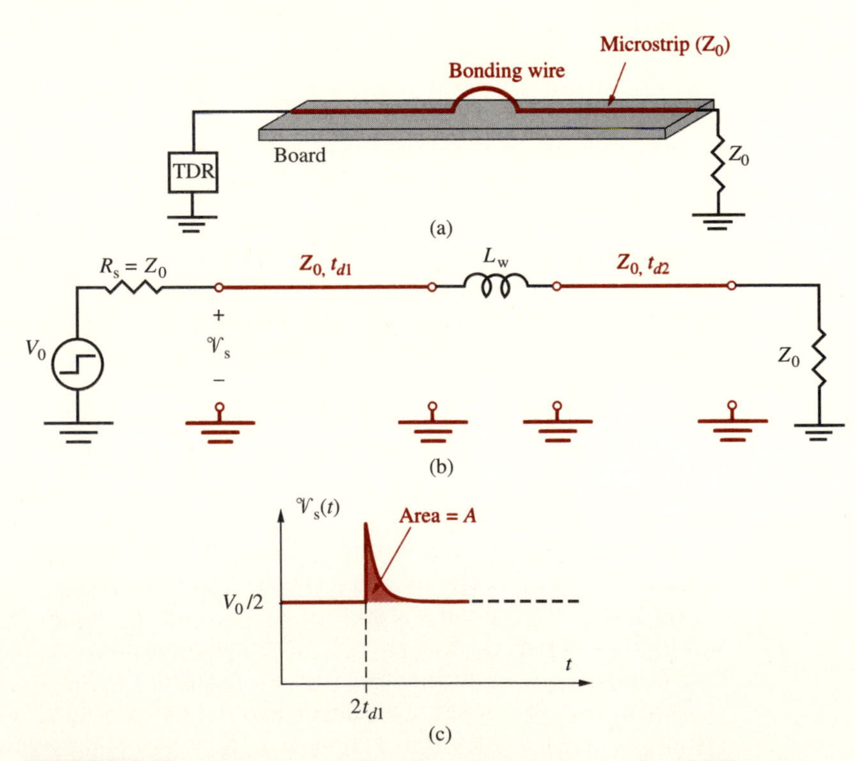

FIGURE 2.33. Measurement of bonding-wire inductance using a TDR system. (a) Actual circuit configuration. (b) Equivalent transmission line circuit model. (c) Measured TDR waveform.

Solution: In principle, the bonding-wire inductance L_w can be determined from the curvature of the glitch by accurately fitting an exponential function. However, a more accurate method is to determine L_w from the area under the curve, which can be measured more accurately. Following an approach similar to that used in Example 2-11, it can be shown that the source-end voltage is

$$\mathcal{V}_s(t) = \begin{cases} \dfrac{V_0}{2} & 0 < t < 2t_{d1} \\ \dfrac{V_0}{2}[1 + e^{-(2Z_0/L_w)(t-2t_{d1})}] & t \geq 2t_{d1} \end{cases}$$

To find the area A under the glitch, we integrate $(\mathcal{V}_s(t) - V_0/2)$ from $t = 2t_{d1}$ to $t = \infty$:

$$A = \int_{2t_{d1}}^{\infty} \frac{V_0}{2} e^{-(2Z_0/L_w)(t-2t_{d1})}\, dt = \frac{V_0}{2}\int_0^{\infty} e^{-(2Z_0/L_w)t'}\, dt' = -\frac{L_w V_0}{4Z_0} e^{-(2Z_0/L_w)t'}\Big|_0^{\infty} = \frac{L_w V_0}{4Z_0}$$

where $t' = t - 2t_{d1}$. Therefore, the bonding-wire inductance L_w is given in terms of the area A as

$$L_w = \frac{4Z_0 A}{V_0}$$

A simple summary of TDR signatures of purely resistive or purely reactive discontinuities on a transmission line is provided in Figure 2.34. The signatures of discontinuities involving combinations of reactive and resistive elements are dealt with in several problems at the end of this chapter.

2.6.2 Effects of Source Rise Time

Up to now, as we considered the consequences of propagation time delays that result from transmission line effects, we assumed that the excitations are ideal step sources that rise and fall instantaneously, with rise times and fall times (t_r, t_f) being identically zero. In practice, however, driver devices possess finite rise and fall times, which can be comparable to delays due to propagation effects. As discussed briefly in Chapter 1, the ratio of the source rise time and the one-way time delay (along a transmission line) can often be a useful determinant of whether or not lumped analyses are applicable. For example, a trace of length l on a printed circuit board behaves mostly in a lumped fashion as long as t_r and $t_f > 6t_d$, where $t_d = l/v_p$ is the one-way propagation delay of the signal on the trace. However, for high-speed drivers (small t_r and t_f) or longer trace lengths l, the trace behaves like a transmission line, or a distributed circuit. Example 2-17 illustrates the relationship between signal rise or fall time and the one-way time delay along a printed circuit board trace.

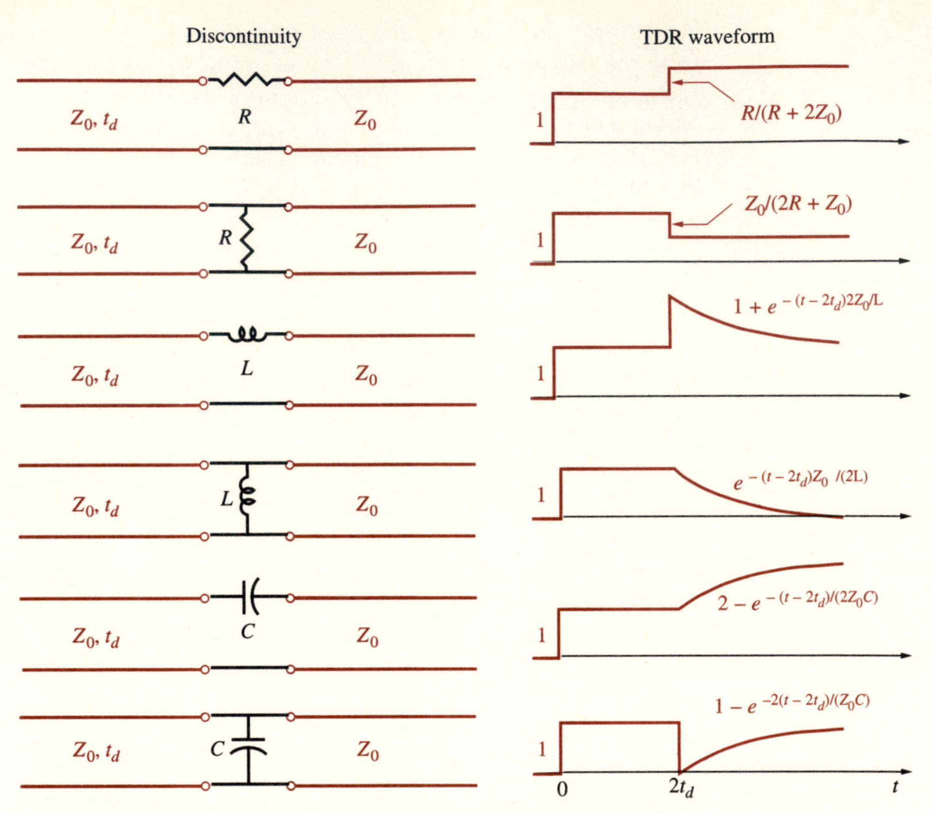

FIGURE 2.34. TDR signatures produced by simple discontinuities. Source-end TDR voltage signatures for shunt or series purely resistive, inductive, and capacitive discontinuities. In terms of excitation by a step voltage source of amplitude V_0 and source resistance R_s, the TDR voltage waveforms shown are drawn for $V_0 = 2$ V and $R_s = Z_0$, and one-way travel time t_d from the source to the disconuity. (This figure was adapted from Figure 6 of B. M. Oliver, Time domain reflectometry, *Hewlett-Packard Journal,* 15(6), pp. 14–9 to 14–16, February, 1964. ©Hewlett-Packard Company 1964. Reprinted with permission.)

Example 2-17: Rise time versus one-way time delay. A digital integrated circuit chip with 1 ns rise and fall times drives another chip with a very large input impedance (100kΩ) through a microstrip trace with characteristic impedance 60Ω, a phase velocity of 20 cm/ns, and length 6 cm on a printed circuit board. Does the line need to be terminated (at a matched impedance) to reduce transmission line effects (such as ringing)?

Solution: Comparing the rise and fall times of the driver with the one-way propagation delay along the trace (i.e., $t_d = 6/20 = 0.3$ ns), we find $t_r = t_f \simeq$

$3.33t_d$, indicating that transmission line effects may not be neglected.[33] Accordingly, it might be useful in this case to terminate the line with a 60Ω resistive load to eliminate reflections and possible ringing.

We can also examine the interplay between t_r and t_d from a more quantitative perspective. When the length of a transmission line is short enough so that t_r/t_d is large, the shape of the output waveform (i.e., the voltage at the end of the line) strongly depends on the finite rise time of the source signal. To see this, consider that it is possible for a new component voltage to arrive at the load or source positions before the previous voltage rises to its final value; for example, $\mathcal{V}_1^-(z, t)$ could arrive at $z = 0$ before the input voltage has risen to its full value. In such cases, the temporal variation of the total voltage or current at any position along the line can be found by summing all of the component voltages $\mathcal{V}_i^-(z, t)$ and $\mathcal{V}_i^+(z, t)$, $i = 1, 2, 3, \ldots$, each of which, as mentioned before, exists for all time after its generation. As an example, consider the circuit of Example 2-3, where the source was considered to be an ideal step voltage source with a rise time $t_r = 0$. We now assume that the same circuit, shown again in Figure 2.35a, is driven by a source having a finite rise time[34] $t_r \neq 0$, such that the source voltage changes linearly from 0 to V_0 in t_r seconds. Note that we have chosen $R_s = Z_0/4$. At $t = 0$, a voltage of $\mathcal{V}_1^+(z, t)$ given by

$$\mathcal{V}_1^+(0, t) = \begin{cases} \left[\dfrac{Z_0 V_0}{(R_s + Z_0)t_r}\right]t = \left(\dfrac{0.8V_0}{t_r}\right)t & t \leq t_r \\ \dfrac{Z_0 V_0}{(R_s + Z_0)} = 0.8V_0 & t \geq t_r \end{cases}$$

is applied from the source side of the line. This voltage reaches the open-circuited end of the line ($z = l$) at $t = t_d$, and a voltage $\mathcal{V}_1^-(z, t)$ of amplitude given by

$$\mathcal{V}_1^-(l, t) = \begin{cases} \Gamma_L \mathcal{V}_1^+(l, t) = \left(\dfrac{0.8V_0}{t_r}\right)(t - t_d) & t_d \leq t \leq t_d + t_r \\ 0.8V_0 & t \geq t_d + t_r \end{cases}$$

reflects toward the source, since $\Gamma_L = 1$. This reflected voltage arrives at the source end of the line at $t = 2t_d$, and a new voltage $\mathcal{V}_2^+(z, t)$ of amplitude given by

$$\mathcal{V}_2^+(0, t) = \begin{cases} \Gamma_s \mathcal{V}_1^-(0, t) = -\left(\dfrac{0.48V_0}{t_r}\right)(t - 2t_d) & 2t_d \leq t \leq 2t_d + t_r \\ -0.48V_0 & t \geq 2t_d + t_r \end{cases}$$

---

[33]Note that, as mentioned in Chapter 1, a rule-of-thumb criterion for interconnects between integrated circuit chips is that lumped analysis is appropriate only if $t_r/t_d > 6$, and inappropriate for $t_r/t_d < 2.5$, with the applicability and the required accuracy being the determining factors in the intervening range ($2.5 < t_r/t_d < 6$) depending on the particular application in hand.

[34]Note that t_r is not exactly the rise time discussed in Section 1.1, which was defined as the time required for the signal to change from 10% to 90% of its final value.

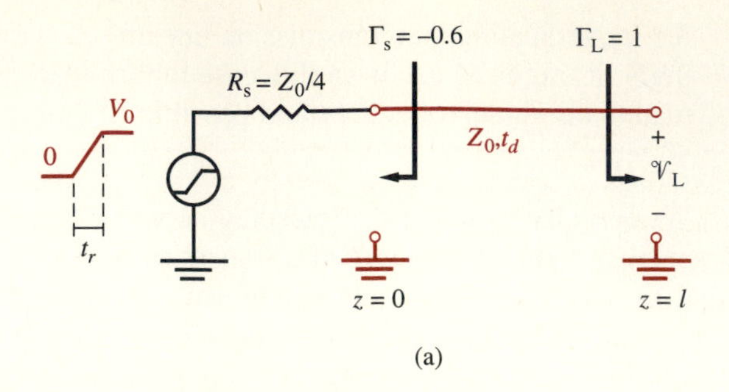

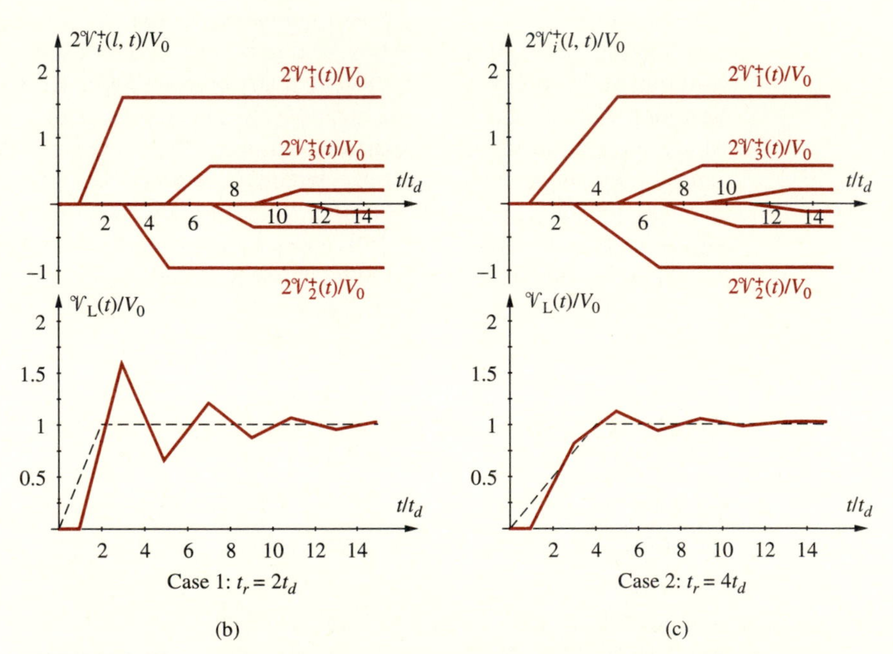

FIGURE 2.35. Effect of source rise time. (a) Circuit diagram for a transmission line excited with a step source of amplitude V_0 and rise time t_r and terminated with an open circuit at the load end. (b) The individual component voltages at the load end (top panel) and the load voltage $\mathcal{V}_L(t)$ (bottom panel) versus time for $t_r = 2t_d$. (c) The individual component voltages at the load end (top panel) and the load voltage $\mathcal{V}_L(t)$ (bottom panel) versus time for $t_r = 4t_d$.

reflects toward the load, since $\Gamma_s = -0.6$. At $t = 3t_d$, a new reflected voltage $\mathcal{V}_2^-(z, t)$ of amplitude given by

$$\mathcal{V}_2^-(l, t) = \begin{cases} \Gamma_L \mathcal{V}_2^+(l, t) = -\left(\dfrac{0.48V_0}{t_r}\right)(t - 3t_d) & 3t_d \le t \le 3t_d + t_r \\ -0.48V_0 & t \ge 3t_d + t_r \end{cases}$$

is launched toward the source. This process continues indefinitely, with the total voltages and currents gradually approaching their steady-state values. The top panels of Figure 2.35b and 2.35c show the individual contributions of the various component voltages $\mathcal{V}_i^+(z, t)$ and $\mathcal{V}_i^-(z, t)$, where $i = 1, 2, 3, \ldots$, that are generated as a result of reflections at both ends of the line, for two different rise times, $t_r = 2t_d$ and $t_r = 4t_d$. Note that since $\Gamma_L = 1$ and thus $\mathcal{V}_i^-(l, t) = \mathcal{V}_i^+(l, t)$, the quantity actually plotted in the top panels of Figure 2.35b and 2.35c is $2\mathcal{V}_i^+(l, t)$. The bottom panels show the time variation of the total load voltage, as determined by the summation of the component voltages. Shown in dashed lines in the lower panels are the load voltage waveforms predicted by a simple lumped analysis (i.e., neglecting transmission line effects). Note that $\mathcal{V}_L(t)$ is simply the superposition of all the component voltages at $z = l$, namely $\mathcal{V}_i^{\pm}(l, t)$, for $i = 1, 2, 3, \ldots$.

It is clear from Figure 2.35b that for $t_r = 2t_d$, the output voltage waveform is substantially different from that expected based on a lumped treatment (i.e., by neglecting transmission line effects or in effect assuming that $l = 0$ or $t_d = 0$), shown for comparison as a dashed line in the lower panel of Figure 2.35b. For the case of $t_r = 4t_d$, however, the load voltage variation deviates only slightly from the lumped case, as shown in Figure 2.35c. For larger values of t_r/t_d, $\mathcal{V}_L(t)$ are even more similar to that predicted based on a lumped assumption, and transmission line effects can be neglected for these applications.

2.6.3 Transients on Lossy Transmission Lines

At a qualitative level, losses on a transmission line lead to *distortion* of the information being transmitted. Distortion is defined as the change in the shape of the signal, as a function of distance, as it travels down the line. For example, a signal that is in the form of a rectangular pulse at the beginning of the line does not retain its rectangular shape as it propagates further; a steplike change in the input voltage is rounded off when observed at points further down the line. The distortion is a result of the fact that the general solutions of the transmission line equations for the lossy case ($R, G \neq 0$) are no longer in the form $f(t - z/v_p)$. In Chapter 3, we shall see that the phase velocity v_p for sinusoidal signals, which is independent of frequency on a lossless line, becomes a function of frequency for lossy lines. If we imagine a transient signal (e.g., a pulse) at some point $z = z_1$ to consist of a superposition of its Fourier components, each of these components travels to a new point $z = z_2$ at a different speed. In addition, each frequency component is in general attenuated by a different amount. Even if there were no reflected pulses (i.e., an infinitely long or a matched line), the differently attenuated and time-shifted sinusoidal components of the signal at $z = z_2$ do not add up to reconstruct the original shape of the signal at $z = z_1$.

The treatment of the propagation of transient signals on lossy lines is a difficult problem, generally requiring extensive analyses using Laplace transformation methods or numerical time domain solutions.[35] The special case of RC lines (i.e., lines with $L = 0$ and $G = 0$), which represents most on-chip interconnect structures and some thin-film package wires that exhibit small inductance but significant resistance, can be treated analytically.[36]

2.7 TRANSMISSION LINE PARAMETERS

We have seen in previous sections that the response of a lossless transmission line to a given excitation depends on its characteristic impedance Z_0 and the propagation speed v_p (or the one-way travel time $t_d = l/v_p$), which in turn depends on the line inductance L and capacitance C per unit line length. The response of lossy lines is additionally influenced by the values per unit line length of the series resistance R and shunt conductance G. In general, the values of these transmission line parameters depend on (i) the geometric shapes, physical dimensions, and proximity of the two conductors that form the line; (ii) the electromagnetic properties of the material surrounding the conductors; and (iii) the electrical conductivity of the conductors and the frequency of operation. In later chapters, after we have introduced the governing electromagnetic equations, we will discuss methods by which the line capacitance, inductance, resistance, and conductance per unit length can be defined and determined from basic principles. In the case of the common transmission lines shown in Figure 2.36, we will be able to find convenient analytical expressions for the line parameters. For other, more complicated, structures R, L, C, and G can be either

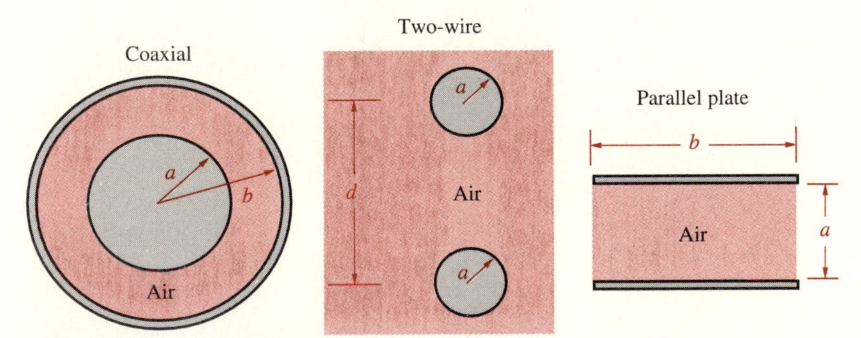

FIGURE 2.36. Cross-sectional view of three common uniform transmission lines. Expressions for the circuit parameters L, R, C, and G for these coaxial, two-wire, and parallel-plate lines are provided in Table 2.2.

---

[35] As an example, see F. Chang, Transient analysis of lossy transmission lines with arbitrary initial potential and current distributions, *IEEE Trans. Circuits Syst.—I: Fundamental Theory and Applications,* 39(3), pp.180–198, March 1992.

[36] See H. B. Bakoglu, *Circuits, Interconnections, and Packaging for VLSI,* Addison-Wesley, 1990; and A. Wilnai, Open-ended RC line model predicts MOSFET IC response, *EDN,* pp. 53–54, December 1971.

TABLE 2.2. Transmission line parameters for some uniform two-conductor transmission lines surrounded by air

	Coaxial	Two-wire	Parallel-plate*
L (μH-m^{-1})	$0.2\ln(b/a)$	$0.4\ln\left[\dfrac{d}{2a}+\sqrt{\left(\dfrac{d}{2a}\right)^2-1}\right]$	$\dfrac{1.26a}{b}$
C (pF-m^{-1})	$\dfrac{55.6}{\ln(b/a)}$	$\dfrac{27.8}{\ln\left[\dfrac{d}{2a}+\sqrt{\left(\dfrac{d}{2a}\right)^2-1}\right]}$	$\dfrac{8.85b}{a}$
R (Ω-m^{-1})	$\dfrac{4.15\times10^{-8}(a+b)\sqrt{f}}{ab}$	$\dfrac{8.3\times10^{-8}\sqrt{f}}{a}$	$\dfrac{5.22\times10^{-7}\sqrt{f}}{b}$
G** (S-m^{-1})	$\dfrac{7.35\times10^{-4}}{\ln(b/a)}$	$\dfrac{3.67\times10^{-4}}{\ln\left[\dfrac{d}{2a}+\sqrt{\left(\dfrac{d}{2a}\right)^2-1}\right]}$	$\dfrac{1.17\times10^{-4}b}{a}$
Z_0 (Ω)	$60\ln(b/a)$	$120\ln\left[\dfrac{d}{2a}+\sqrt{\left(\dfrac{d}{2a}\right)^2-1}\right]$	$\dfrac{377a}{b}$

*Valid for $b \gg a$.
**For polyethylene at 3 GHz.

evaluated using numerical techniques or measured. Parameters for many different transmission lines are also extensively available in handbooks.[37]

Expressions for L, R, C parameters and for Z_0 for the common uniform transmission lines shown in Figure 2.36 are given in Table 2.2. The characteristic impedances (Z_0) provided are for lossless lines (i.e., $Z_0 = \sqrt{L/C}$). In Table 2.2, we have assumed the transmission line conductors to be made of copper and the surrounding medium to be air. Note that the parameters depend on the geometric shapes and the physical dimensions of the lines (d, a, and b). The line capacitance C and characteristic impedance Z_0 for the case when the surrounding medium is a nonmagnetic[38] material other than air can be derived from those given in Table 2.2 by using the propagation speed v_p for these media as given in Table 2.1. Specifically we have

$$[C]_{\text{material}} = \frac{c^2}{v_p^2}[C]_{\text{air}} \quad \text{and} \quad [Z_0]_{\text{material}} = \frac{v_p}{c}[Z_0]_{\text{air}}$$

where c is the speed of light in free space, or $c \simeq 3\times10^8$ m-s^{-1}. The line inductance L remains the same, since it is governed by the magnetic properties of the surrounding material.

---

[37]*Reference Data for Engineers,* 8th ed., Sams Prentice Hall Computer Publishing, Carmel, Indiana, 1993.

[38]Magnetic properties of materials are discussed in Section 6.8. In the transmission line context, all materials can be considered nonmagnetic except for iron, nickel, cobalt, a few of their alloys, and some special compounds involving mixtures of magnetic materials with barium titanate.

TABLE 2.3. Relative conductivities of metals versus copper

Material	Relative conductivity	Material	Relative conductivity
Aluminum	0.658	Silver	1.06
Brass	0.442	Sodium	0.375
Copper	1.00	Stainless steel	0.0192
Gold	0.707	Tin	0.151
Lead	0.0787	Titanium	0.0361
Magnesium	0.387	Tungsten	0.315
Nickel	0.250	Zinc	0.287

The series resistance (R) is inversely proportional to the electrical *conductivity* of the particular metal that the conductors are made of, with the values given in Table 2.2 being relative to that of copper. The physical underpinnings of electrical conductivity are discussed in Chapter 5. For now, it suffices to know that it is a quantitative measure of the ability of a material to conduct electrical current and that the values of conductivity for different materials are tabulated extensively in various handbooks (see Table 5.1). A brief list of conductivities of some common metals relative to that of copper is provided in Table 2.3. The series resistance R is proportional to the square root of the frequency because of the so-called *skin effect,* which results from the nonuniform distribution of electrical current in a metal at higher frequencies, and which is discussed in Chapter 8.

With air as the surrounding medium, the shunt conductance $G = 0$, since air is an excellent insulator and leakage losses through it are generally negligible. In the case of other surrounding media for which leakage losses may not be negligible, the value of G depends on the geometrical layout of the conductors (as do the values of C, L, and R) but is more strongly determined by the loss properties of the insulating medium surrounding the conductors and is, in general, a rather complicated function of the frequency of operation. Table 2.2 provides a representative expression for G for polyethylene as the surrounding medium at an operating frequency of 3 GHz. High-frequency losses in insulating materials are discussed in Sections 7.4 and 8.1.

Examples 2-18, 2-19, and 2-20 illustrate the use of the formulas given in Table 2.2 for selected transmission lines.

Example 2-18: Television antenna lead-in wire. A student measures the dimensions of a television antenna lead-in wire made of two copper wires. The diameter of the wires is found to be $\sim$1 mm each, while the spacing between the centers of the conductors is $\sim$0.7 cm. Assume the conductors to be surrounded by air, although they might in fact be held together by some plastic material. Determine the values of the line parameters and the characteristic impedance at an operating frequency of 200 MHz.

Solution: We can directly use the formulas given in Table 2.2 for the two-wire line. Noting that $d = 0.7$ cm and $2a = 1$ mm, we have

$$L = 0.4 \ln\left[\frac{d}{2a} + \sqrt{\left(\frac{d}{2a}\right)^2 - 1}\right] \simeq 1.05 \ \mu\text{H-m}^{-1}$$

$$C = \frac{27.8}{\ln\left[\frac{d}{2a} + \sqrt{\left(\frac{d}{2a}\right)^2 - 1}\right]} \simeq 10.6 \ \text{pF-m}^{-1}$$

$$R = \frac{8.3 \times 10^{-8}\sqrt{200 \times 10^6}}{a} \simeq 2.35\Omega\text{-m}^{-1}$$

$$Z_0 = 120 \ln\left[\frac{d}{2a} + \sqrt{\left(\frac{d}{2a}\right)^2 - 1}\right] \simeq 316\Omega$$

Note that the characteristic impedance of this line is quite close to 300Ω, within the tolerances of the measurement. Indeed, the household television lead-in line is usually referred to as a 300-ohm line. Note also that $G \simeq 0$ for this wire, since the leakage losses are negligible.

Example 2-19: Coaxial line. A coaxial line consists of inner and outer conductors made of copper and having radii of $a = 0.65$ mm and $b = 2.75$ mm, the space between the conductors being filled with air. The line is to be used at 1 GHz. Find the values of the distributed parameters and the characteristic impedance of this line.

Solution: We can directly use the formulas given in Table 2.2 for the coaxial line:

$$L = 0.2 \ln\left(\frac{b}{a}\right) \simeq 0.289 \ \mu\text{H-m}^{-1}$$

$$C = \frac{55.6}{\ln\left(\frac{b}{a}\right)} \simeq 38.5 \ \text{pF-m}^{-1}$$

$$R = \frac{4.15 \times 10^{-8}(a + b)}{ab}\sqrt{10^9} \simeq 2.5\Omega\text{-m}^{-1}$$

$$Z_0 = 60 \ln\left(\frac{b}{a}\right) \simeq 86.5\Omega$$

Note once again that $G = 0$ for this air-filled coaxial line.

Example 2-20: RG58/U coaxial line. RG58/U is a commonly used coaxial line with an inner conductor of diameter 0.45 mm and an outer conductor of inside diameter 1.47 mm constructed using copper conductors and filled with polyethylene as its insulator. The line is to be used at 3 GHz. Find the line parameters (i.e., R, L, C, G, and Z_0). Note from Table 2.1 that the propagation speed for polyethylene at 3 GHz is $v_p = 20$ cm-(ns)$^{-1} = 2 \times 10^8$ m-s^{-1}.

Solution: We use the expressions provided in Table 2.2, except for the multipliers needed for C and Z_0 to correct for the fact that the filling is polyethylene rather than air. Note that we can also use the expression from Table 2.2 for G, since it was also given for polyethylene and for 3 GHz. We have

$$R = \frac{4.15 \times 10^{-8}(0.45 + 1.47) \times 10^{-3}\sqrt{3 \times 10^9}}{(0.45 \times 10^{-3})(1.47 \times 10^{-3})} \simeq 6.6\Omega/\text{m}$$

$$L = 0.2 \ln\left(\frac{1.47}{0.45}\right) \simeq 0.237\ \mu\text{H/m}$$

$$C = \left(\frac{c^2}{v_p^2}\right)\frac{55.6}{\ln\left(\dfrac{b}{a}\right)} = \left(\frac{3}{2}\right)^2 \frac{55.6}{\ln\left(\dfrac{1.47}{0.45}\right)} \simeq 106\ \text{pF/m}$$

$$G = \frac{7.35 \times 10^{-4}}{\ln\left(\dfrac{1.47}{0.45}\right)} \simeq 6.21 \times 10^{-4}\ \text{S/m}$$

$$Z_0 = \left(\frac{v_p}{c}\right) 60 \ln\left(\frac{b}{a}\right) \simeq 47.4\Omega$$

Note that the value of Z_0 is close to the nominal 50-ohm impedance of this coaxial line. Note also that the difference between 47.4Ω and 50Ω is within the range of accuracy (i.e., two digits after the decimal point) by which the physical quantities were specified (e.g., the radii of conductors, the value of v_p, etc.).

2.8 SUMMARY

This chapter discussed the following topics:

- **Transmission line parameters.** A transmission line is commonly characterized by its distributed parameters R (in Ω/m), L (in H/m), G (in S/m), and C (in F/m), whose values are determined by the line geometry, the conductivity of the metallic conductors, the electrical and magnetic properties of the surrounding insulating material, and the frequency of excitation. Formulas for calculating R, L, and C are provided in Table 2.2 for coaxial, two-wire, and parallel-plate transmission lines surrounded by air. Formulas for G are also provided in Table 2.2 for the same lines surrounded by polyethylene.

- **Transmission line equations.** The distributed parameters of the line are used in an equivalent lumped-circuit model to represent a differentially short segment of the line. Using this model as a basis, the transmission line equations are derived from Kirchhoff's laws in the limit where the length of the line segment approaches zero. The differential equations governing the behavior of voltage and current on a lossless ($R = 0$ and $G = 0$) transmission line, and the wave equation for voltage derived from them, are

$$\left.\begin{aligned} \frac{\partial \mathcal{V}}{\partial z} &= -L\frac{\partial \mathcal{I}}{\partial t} \\ \frac{\partial \mathcal{I}}{\partial z} &= -C\frac{\partial \mathcal{V}}{\partial t} \end{aligned}\right\} \longrightarrow \frac{\partial^2 \mathcal{V}}{\partial z^2} = LC\frac{\partial^2 \mathcal{V}}{\partial t^2}$$

- **Propagating-wave solutions, characteristic impedance, and phase velocity.** The general solution of the transmission line equations leads to mathematical expressions for voltage and current along the line that are wave equations in nature, depending on both distance and time. These are

$$\mathcal{V}(z,t) = f^+\left(t - \frac{z}{v_p}\right) + f^-\left(t + \frac{z}{v_p}\right)$$

$$\mathcal{I}(z,t) = \frac{1}{Z_0}\left[f^+\left(t - \frac{z}{v_p}\right) - f^-\left(t + \frac{z}{v_p}\right)\right]$$

where Z_0 is the characteristic impedance of a lossless line, which is defined as the voltage-to-current ratio of a single disturbance propagating in the $+z$ direction and is given by $Z_0 = \sqrt{L/C}$. The characteristic impedance is one of the most important quantities that determine the response of a transmission line, and its value is tabulated for most practical transmission lines. Formulas for calculating Z_0 for three different types of lines (coaxial, two-wire, and parallel-plate) are provided in Table 2.2 for lossless transmission lines surrounded by air. The velocity with which waves on a transmission line propagate is called the phase velocity and is given by $v_p = (LC)^{-1/2}$. For a lossless transmission line, the phase velocity is determined by the properties of the material surrounding the transmission line conductors and is equal to the speed of light in free space ($c = 3 \times 10^8$ m/s) if the conductors constituting the line are surrounded by free space or air. For uniform transmission lines, v_p is a constant, regardless of the shape of the voltage (or current) signal traveling down the line. Values of v_p for selected materials are tabulated in Table 2.1.

- **Transmission lines terminated in resistive loads, reflection coefficient.** Transient response of a lossless transmission line to step or pulse excitation involves reflections from discontinuities along the line or from loads at its termination. Reflection effects are described in terms of the reflection coefficient Γ, defined as the ratio of the reflected to the incident voltage at a given point. The reflection coefficients at the load and source ends of a transmission line are given by

$$\Gamma_L = \frac{R_L - Z_0}{R_L + Z_0} \qquad \text{Load end}$$

$$\Gamma_s = \frac{R_s - Z_0}{R_s + Z_0} \qquad \text{Source end}$$

where R_L is the resistance terminating the load end of the line, the line's characteristic impedance is Z_0, and R_s is the source resistance. In the special case of a matched termination, we have $R_L = Z_0$, so $\Gamma_L = 0$, and thus no reflection occurs from the termination. Similarly, when $R_s = Z_0$, $\Gamma_s = 0$, and no reflection occurs from the source end. In general, when a voltage disturbance is launched from the source end of a transmission line (e.g., due to a step change in input voltage), a sequence of reflections from both the load and source ends of the line occurs. The process of multiple reflections from the load and source ends of a transmission line can be described using a bounce diagram.

- **Transmission lines terminated in reactive or nonlinear loads.** To determine the transient behavior of lossless lines terminated in reactive or nonlinear elements, it is necessary to solve the differential equations that determine the voltage-current relationships of the terminations subject to the appropriate initial conditions. For reactive loads, the reflected voltage due to a step excitation is no longer a simple step function but, in general, varies continuously at a fixed position with respect to time depending on the nature of the reactive termination. For nonlinear terminations, a graphical approach known as the Bergeron plot can be used to find the voltage and current following each reflection, using the known current-voltage characteristics of the nonlinear device.

2.9 PROBLEMS

2-1. Open-circuited line. Consider the circuit shown in Figure 2.37, with an ideal unit step source connected to a lossless line of characteristic impedance $Z_0 = 50\Omega$ having an open-circuited termination at the other end. Assuming a one-way propagation delay

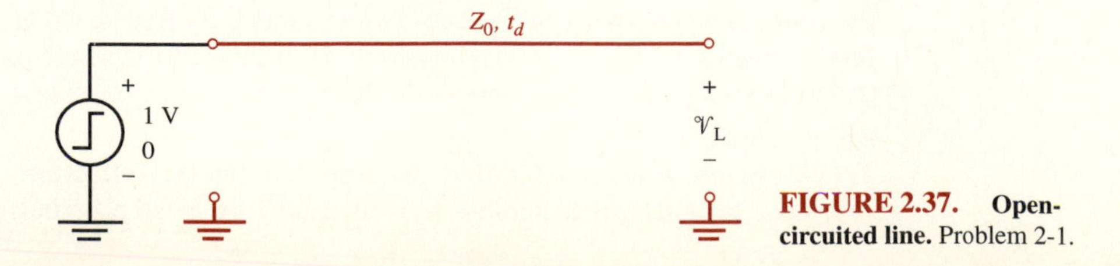

FIGURE 2.37. Open-circuited line. Problem 2-1.

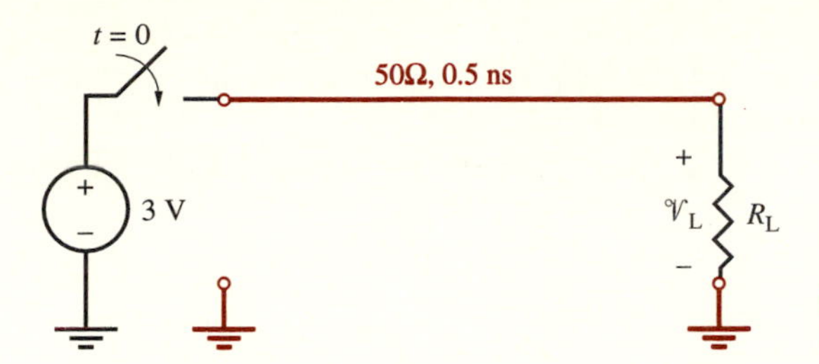

FIGURE 2.38. Resistive loads. Problems 2-2 and 2-4.

across the line of t_d, use a bounce diagram to sketch the voltage $\mathcal{V}_L(t)$ versus time for $0 \leq t \leq 10t_d$.

2-2. Resistive loads. The circuit shown in Figure 2.38 consists of an uncharged transmission line connected to a load resistance R_L. Assuming that the switch closes at $t = 0$, sketch the load voltage $\mathcal{V}_L(t)$ over the time interval $0 \leq t \leq 3$ ns for the following load resistances: (a) $R_L = 25\Omega$, (b) $R_L = 50\Omega$, (c) $R_L = 100\Omega$.

2-3. Ringing. The transmission line system shown in Figure 2.39 is excited by a step-voltage source of amplitude 3.6 V and source impedance 15Ω at one end, and is terminated with an open circuit at the other end. The line is characterized by the line parameters $L = 4.5$ nH-(cm)$^{-1}$, $C = 0.8$ pF-(cm)$^{-1}$, $R = 0$, $G = 0$, and has a length of $l = 30$ cm. Sketch the load voltage $\mathcal{V}_L(t)$ over $0 \leq t \leq 10$ ns with the steady-state value indicated.

2-4. Discharging of a charged line. For the circuit of Problem 2-2, assume that the switch has been closed for a long time before it opens at $t = 0$. Sketch the load voltage $\mathcal{V}_L(t)$ over $0 \leq t \leq 3$ ns for the same three cases.

2-5. Pulse excitation. The circuit shown in Figure 2.40 is excited by an ideal voltage pulse of 1 V amplitude starting at $t = 0$. Given the length of the line to be $l = 10$ cm and the propagation speed to be 20 cm-(ns)$^{-1}$, (a) sketch the voltage at the source end of

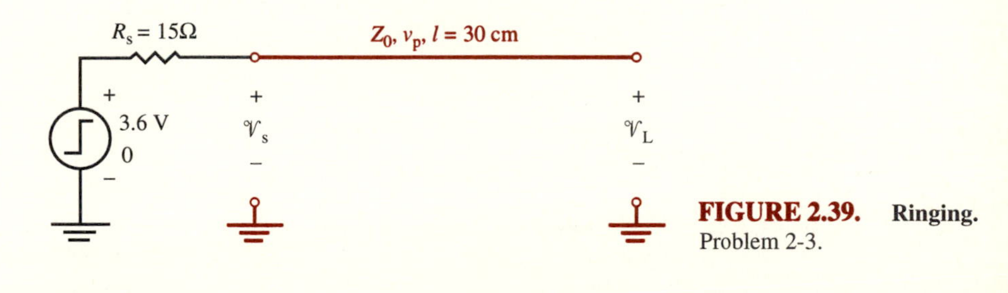

FIGURE 2.39. Ringing. Problem 2-3.

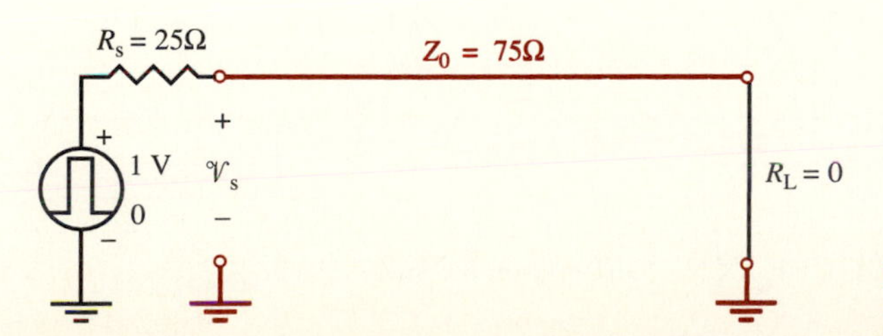

FIGURE 2.40. Pulse excitation. Problem 2-5.

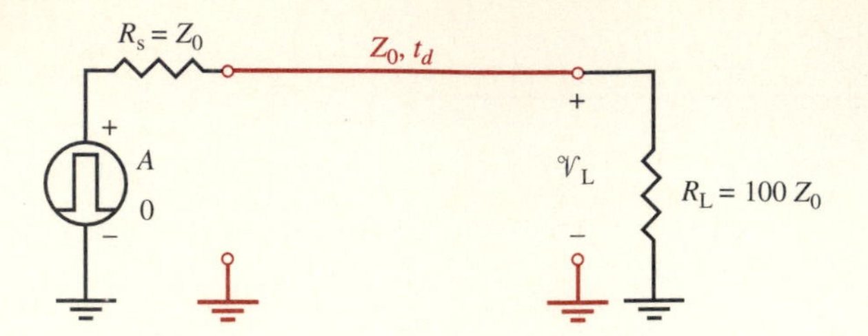

FIGURE 2.41. Pulse excitation. Problem 2-6.

the line, $\mathcal{V}_s(t)$, for an input pulse duration of 10 ns; (b) repeat part (a) for an input pulse duration of 1 ns.

2-6. Pulse excitation. The circuit shown in Figure 2.41 is excited with a voltage pulse of amplitude A and pulse width t_w. Assuming the propagation delay of the line to be t_d, sketch the load voltage $\mathcal{V}_L(t)$ versus t for $0 \leq t \leq 10t_d$ for (a) $t_w = 2t_d$, (b) $t_w = t_d$, and (c) $t_w = t_d/2$.

2-7. Observer on the line. A transmission line with an unknown characteristic impedance Z_0 terminated in an unknown load resistance R_L, as shown in Figure 2.42, is excited by a pulse source of amplitude 1 V and duration $t_w = 3t_d/4$, where t_d is the one-way flight time of the transmission line. An observer at the center of the line observes the voltage variation shown. (a) Determine Z_0 and R_L. (b) Using the values found in (a), sketch the voltage variation (up to $t = 4t_d$) that would be observed by the same observer if the pulse duration were $t_w = 1.5t_d$.

2-8. Cascaded transmission lines. For the transmission line circuit shown in Figure 2.43, sketch $\mathcal{V}_s(t)$ and $\mathcal{V}_L(t)$ over $0 \leq t \leq 5$ ns.

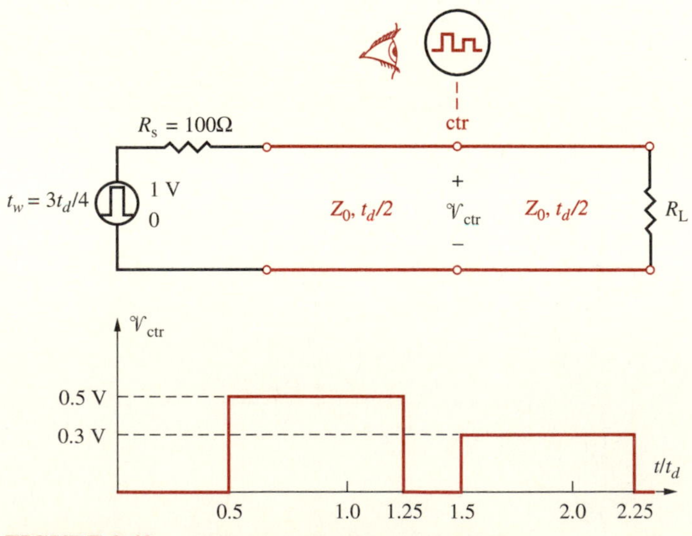

FIGURE 2.42. Observer on the line. Problem 2-7.

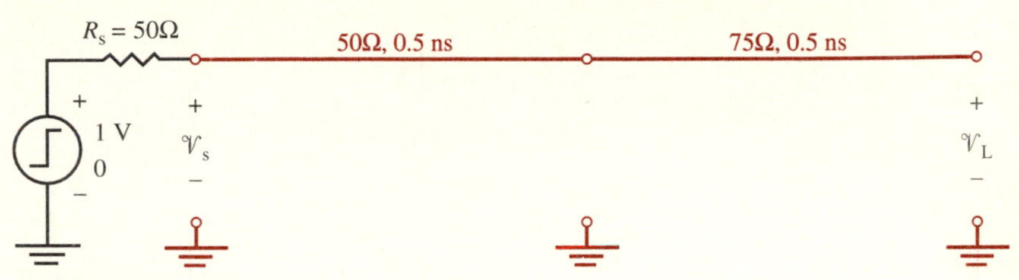

FIGURE 2.43. **Cascaded lines.** Problem 2-8.

2-9. Time-domain reflectometry (TDR). A TDR is used to test the transmission line system shown in Figure 2.44. Using the sketch of $\mathcal{V}_s(t)$ observed on the TDR scope as shown, determine the values of Z_{01}, l_1, and R_1. Assume the phase velocity of the waves to be 20 cm-(ns)$^{-1}$ on each line. Plot $\mathcal{V}_{R1}(t)$ versus t for $0 \leq t \leq 4$ ns.

2-10. Time-domain reflectometry (TDR). TDR measurements can also be used in cases with more than one discontinuity. Two transmission lines of different characteristic impedances and time delays terminated by a resistive load are being tested by a TDR, as shown in Figure 2.45. (a) Given the TDR display of the source-end voltage due to a 3-V, 100Ω step excitation starting at $t = 0$, find the characteristic impedances (Z_{01} and Z_{02}) and the time delays (t_{d1} and t_{d2}) of both lines, and the unknown load R_L. (b) Using the values found in part (a), find the time and magnitude of the next change in the source-end voltage $\mathcal{V}_s(t)$, and sketch it on the display.

2-11. Time-domain reflectometry (TDR). Consider the circuit shown in Figure 2.46. The two line segments are of equal length l. Assuming the propagation speeds on the two lines are equal to 15 cm-(ns)$^{-1}$ each, find Z_{01}, Z_{02}, R_L and l using the TDR display of the source voltage $\mathcal{V}_s(t)$, as shown.

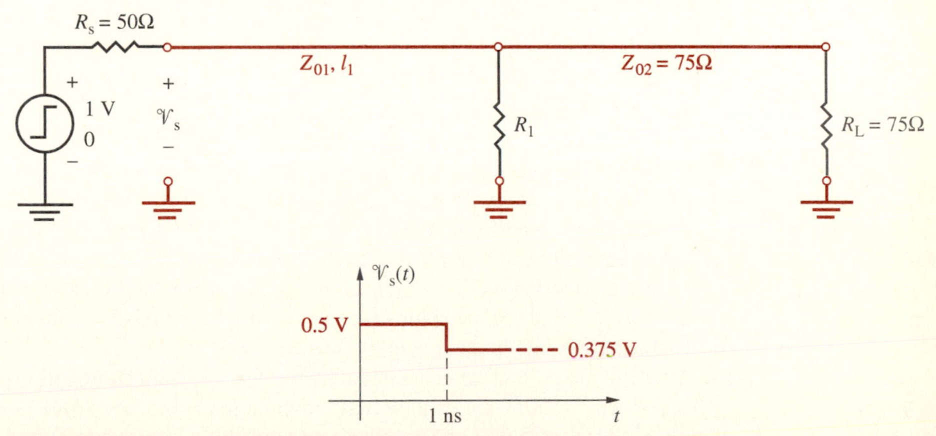

FIGURE 2.44. **Time-domain reflectometry.** Problem 2-9.

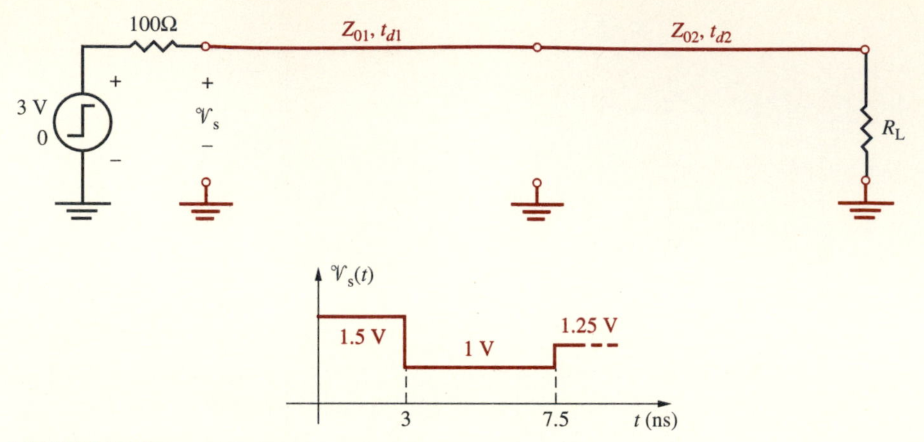

FIGURE 2.45. **Time-domain reflectometry.** Problem 2-10.

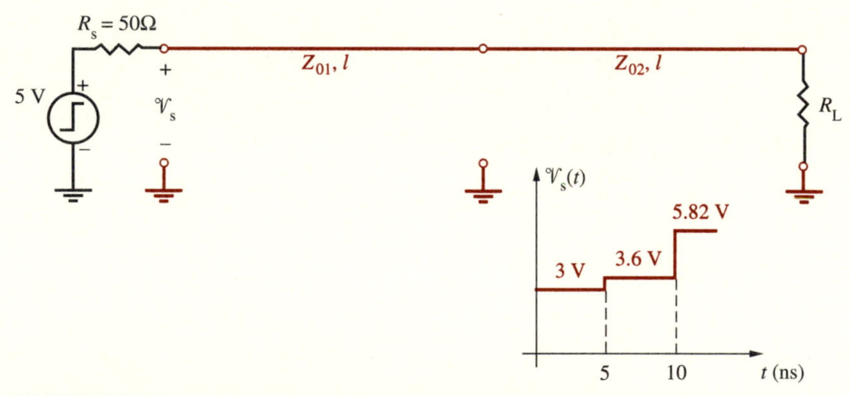

FIGURE 2.46. **Time-domain reflectometry.** Problem 2-11.

2-12. Multiple lines. For the distributed transmission line system shown in Figure 2.47, sketch the voltages $\mathcal{V}_1(t)$, $\mathcal{V}_2(t)$, and $\mathcal{V}_3(t)$ versus t for $0 \leq t \leq 7t_d$. Assume the one-way propagation delay to be t_d on each transmission line.

2-13. Digital IC chips. Two impedance-matched, in-package-terminated Integrated Circuit (IC) chips are driven from an impedance-matched IC chip, as shown in Figure 2.48. Assuming the lengths of the interconnects to be 15 cm each and the propagation velocity on each to be 10 cm-(ns)$^{-1}$, do the following: (a) Sketch the voltages $\mathcal{V}_{L1}$ and $\mathcal{V}_{L2}$ for a time interval of 10 ns. Indicate the steady-state values on your sketch. (b) Repeat part (a) if one of the load chips is removed from the end of the interconnect connected to it (i.e., the lead point A is left open-circuited).

2-14. Multiple lines. For the distributed interconnect system shown in Figure 2.49, and for $Z_{01} = Z_{02} = 50\Omega$, find and sketch the three load voltages $\mathcal{V}_1(t)$, $\mathcal{V}_2(t)$, and $\mathcal{V}_3(t)$ for a time interval of 5 ns. Assume each interconnect to have a one-way time delay of 1 ns. (b) Repeat part (a) for $Z_{01} = Z_{02} = 25\Omega$.

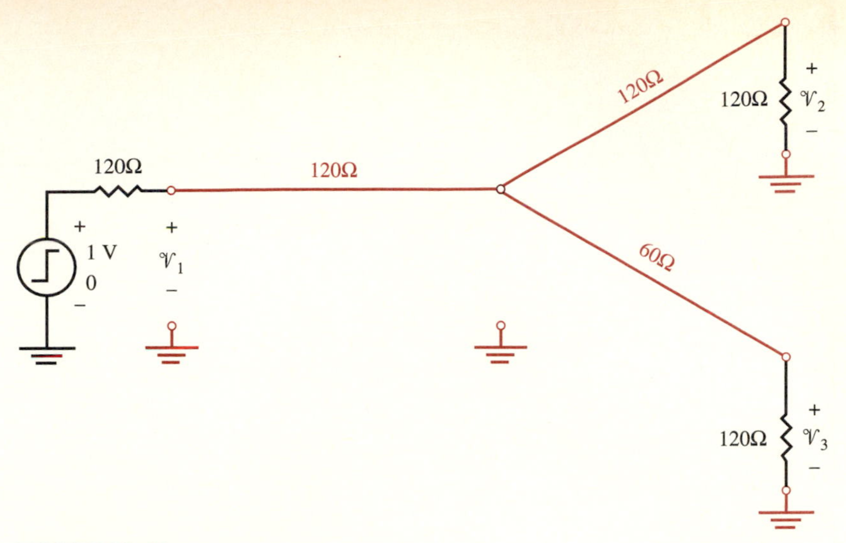

FIGURE 2.47. **Multiple lines.** Problem 2-12.

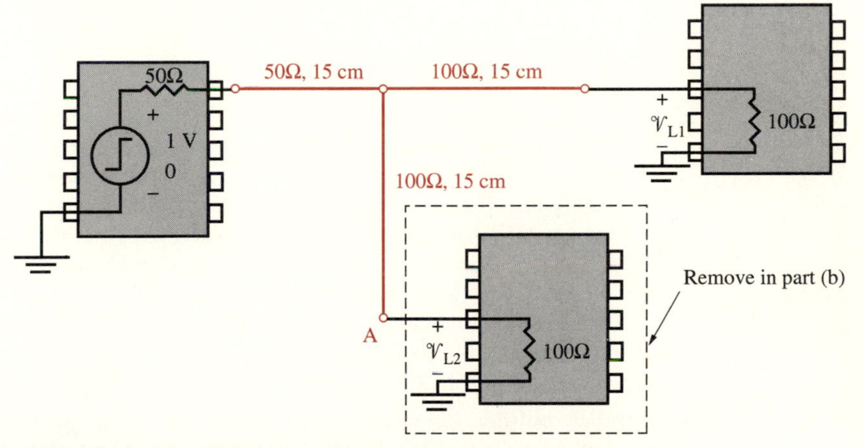

FIGURE 2.48. **Digital IC chips.** Problem 2-13.

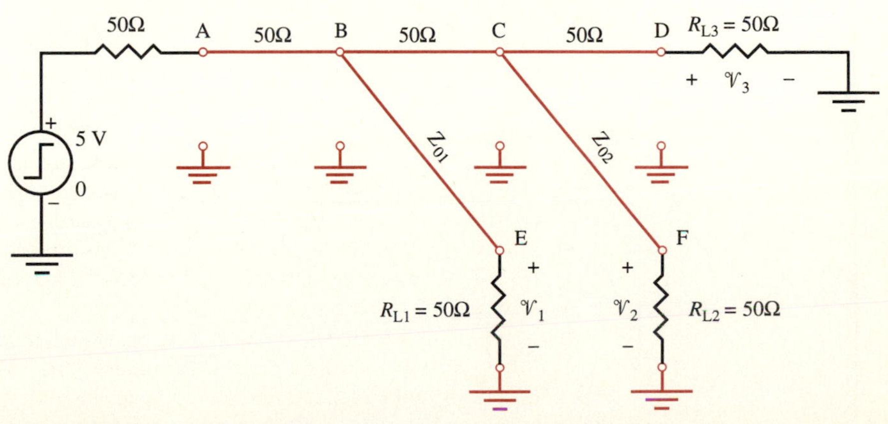

FIGURE 2.49. **Multiple lines.** Problem 2-14.

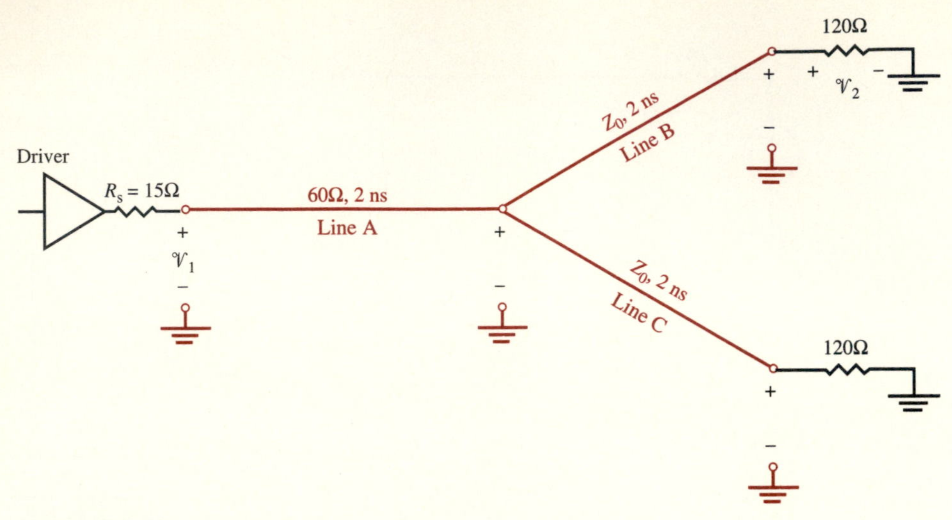

FIGURE 2.50. **Reflections due to mismatches.** Problem 2-15.

2-15. Reflections due to parasitic effects. The circuit shown in Figure 2.50 consists of a low-impedance driver driving a distributed interconnect system that was intended to be impedance-matched, with $Z_0 = 120\Omega$. An engineer performs some tests and measurements and observes reflections due to parasitic effects associated with the two interconnects terminated at the 120Ω loads. Assuming that the effective characteristic impedances of these interconnects (i.e., taking parasitic effects into account) is such that we have $Z_0 = 80\Omega$ instead of 120Ω, find and sketch the voltages $\mathcal{V}_1(t)$ and $\mathcal{V}_2(t)$ for $0 \leq t \leq 12$ ns, assuming the one-way time delay on each interconnect to be 2 ns. Comment on the effects of the mismatch caused by parasitic effects. Assume the initial incident wave launched at the driver end of the 60Ω line to be $\mathcal{V}_1^+ = 4$ V.

2-16. Parallel multiple lines. The transmission line system shown in Figure 2.51 consists of three lines, each having $Z_0 = 50\Omega$ and a one-way propagation delay of 1 ns. (a) Find and sketch the voltages $\mathcal{V}_s(t)$ and $\mathcal{V}_L(t)$ versus t for $0 \leq t \leq 10$ ns.

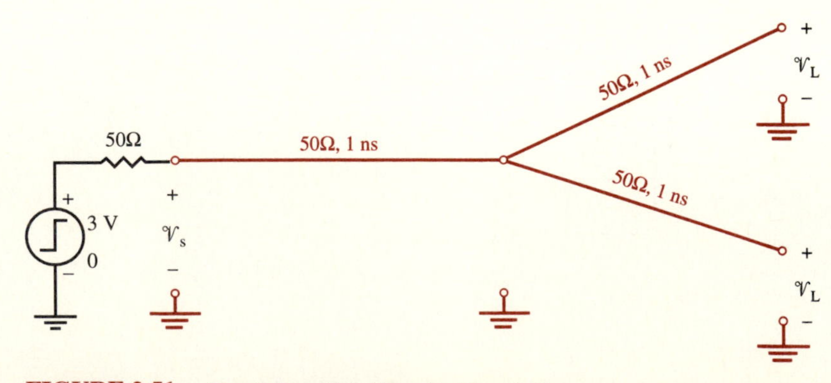

FIGURE 2.51. **Parallel multiple lines.** Problem 2-16.

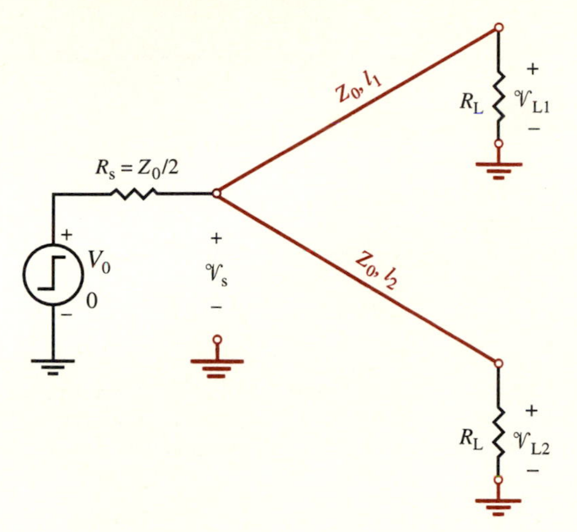

FIGURE 2.52. Optimized multiple lines. Problem 2-17.

(b) Repeat part (a) when the open-circuited ends are terminated with a load resistance of 50Ω each.

2-17. Optimized multiple lines. The transmission line system shown in Figure 2.52 consists of an ideal step source of amplitude V_0 and output impedance R_s connected to two identical parallel lossless transmission lines terminated by equal resistive loads R_L. The parameters of the lines are $L = 2.5$ nH/cm and $C = 1$ pF/cm. (a) Calculate Z_0 and v_p for the lines. (b) For $R_s = Z_0/2$, $R_L \gg Z_0$, and line lengths of $l_2 = 1.5 l_1 = 30$ cm, sketch the voltages $\mathcal{V}_s$, $\mathcal{V}_{L1}$, and $\mathcal{V}_{L2}$ versus t for the time interval $0 \leq t \leq 7.5$ ns. (c) Repeat part (b) if the line lengths are optimized to be $l_1 = l_2 = 25$ cm each to minimize ringing effects, and compare with the results of part (b).

2-18. Optimized multiple lines. A multisection transmission line consists of three lossless transmission lines used to connect an ideal step source of 5-V amplitude and 6Ω output impedance to two separate load resistances of 66Ω each, as shown in Figure 2.53.

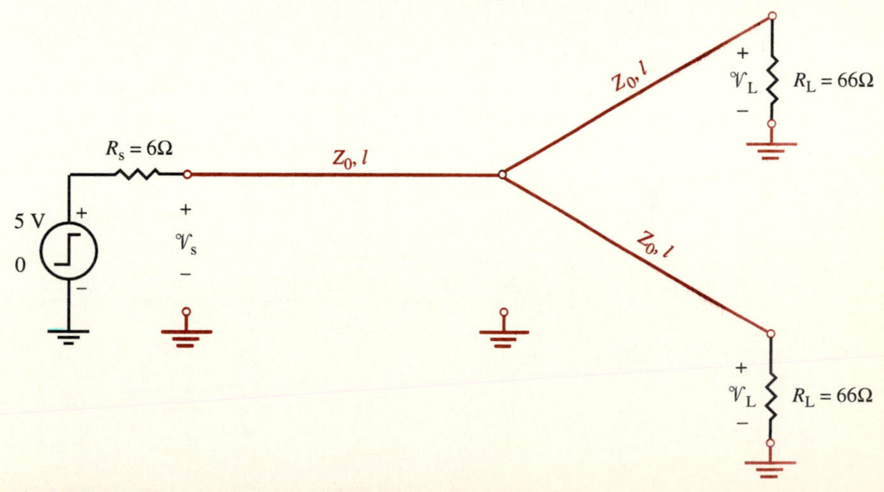

FIGURE 2.53. Optimized multiple lines. Problem 2-18.

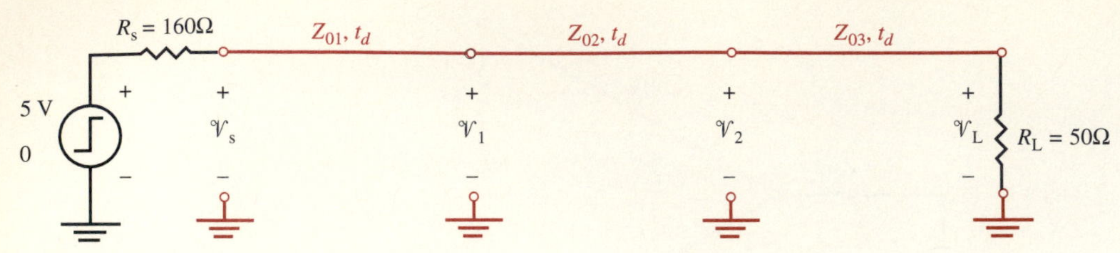

FIGURE 2.54. **Multiple lines.** Problem 2-19.

All three lines are characterized by line parameters $L = 364.5$ nH-m^{-1}, and $C = 125$ pF-m^{-1}. To minimize ringing effects, the line lengths are optimized to be of equal length. If each line length is $l = 40$ cm, sketch the voltages $\mathcal{V}_s$ and $\mathcal{V}_L$ versus t for $0 \leq t \leq 20$ ns, and comment on the performance of the circuit in minimizing ringing.

2-19. Minimizing ringing on multiple lines. The three-section equal-delay lossless transmission line system shown in Figure 2.54 is to be designed to minimize ringing effects at the load due to impedance discontinuities along the transmission path. For a step source of 5-V amplitude and 160Ω source impedance and a load impedance of 50Ω, one design consists of three lines with $Z_{01} = 148\Omega$, $Z_{02} = 200\Omega$, and $Z_{03} = 69\Omega$. Assuming a one-way time delay of $t_d = 250$ ps for each of the lines, sketch the voltages $\mathcal{V}_s$, $\mathcal{V}_1$, $\mathcal{V}_2$, and $\mathcal{V}_L$ versus t for $0 \leq t \leq 1.5$ ns, and comment on the performance of the design.

2-20. Charging and discharging of a line. For the transmission line system shown in Figure 2.55, the switch S_2 is closed at $t = 2t_d$ (where t_d is the propagation delay of each line) after the switch S_1 is closed at $t = 0$. Find and sketch the voltage $\mathcal{V}_L$ versus t for $0 \leq t \leq 6t_d$.

2-21. Digital IC interconnect. The circuit shown in Figure 2.56 consists of a logic gate driving another logic gate via a 50Ω interconnect. At $t = 0$, the driver output voltage switches from LOW to HIGH state. The Thévenin equivalent of the driver gate at its output at LOW and HIGH states can be approximated respectively as a -1.67 V or a -0.85 V voltage source, each in series with a 7Ω resistor. The input impedance of the load gate can be approximated by a very large impedance (say, 50 kΩ). Assume the signal delay and the length of the interconnect to be 200 ps-(in.)$^{-1}$ and 6 in., respectively. (a) Sketch the voltage at the load end and comment on the performance of the circuit. (b) Connect an additional terminating network at the load end as shown. Select $R_T = 50\Omega$ and $V_T = -2$ V and repeat part (a).

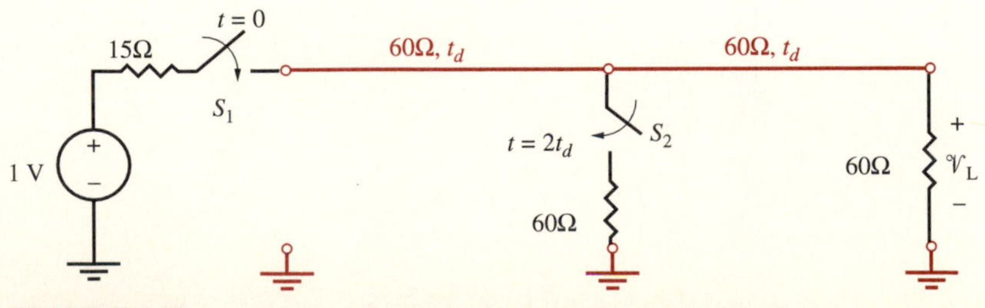

FIGURE 2.55. **Charging and discharging of a line.** Problem 2-20.

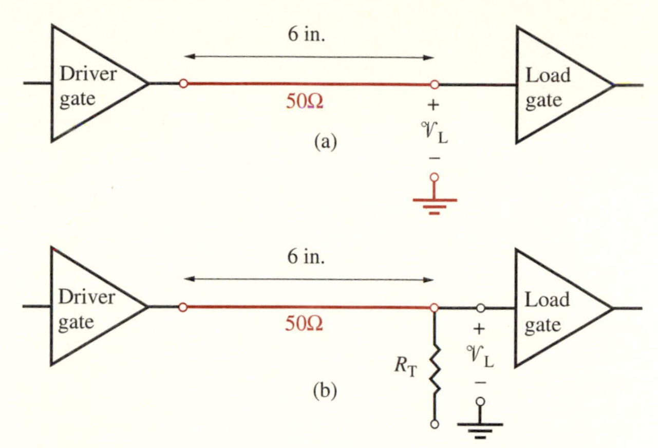

FIGURE 2.56. **Digital IC interconnect.** Problem 2-21.

2-22. Digital IC interconnect. A circuit consists of one logic gate driving an identical gate via a 1-ft-long, 50Ω interconnect. Before $t = 0$, the output of the driver gate is at LOW voltage state and can simply be approximated as a 14Ω resistor. At $t = 0$, the output of the driver gate goes from LOW to HIGH state and can be approximated with a 5-V voltage source in series with a 14Ω resistor. The input of the load gate can be approximated to be an open circuit (i.e., $R_L = \infty$). Assuming that a minimum load voltage of 3.75 V is required to turn and keep the load gate on, (a) find the time at which the load gate will turn on for the first time. (b) Find the time at which the load gate will turn on permanently. (Assume a signal time delay of 1.5 ns-(ft)$^{-1}$ along the interconnect for both parts.)

2-23. Terminated IC interconnects. The logic circuit of Problem 2-22 needs to be modified to eliminate ringing. Two possible solutions are to terminate the line in its characteristic impedance at either the source (series termination) or receiver (parallel termination) end. Both of these circuits are shown in Figure 2.57. (a) Select the value of the termi-

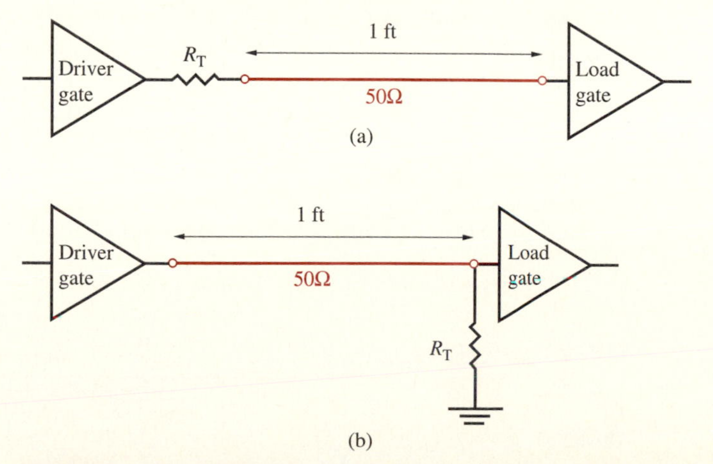

FIGURE 2.57. **Terminated IC interconnects.** (a) Series termination. (b) Parallel termination. Problem 2-23.

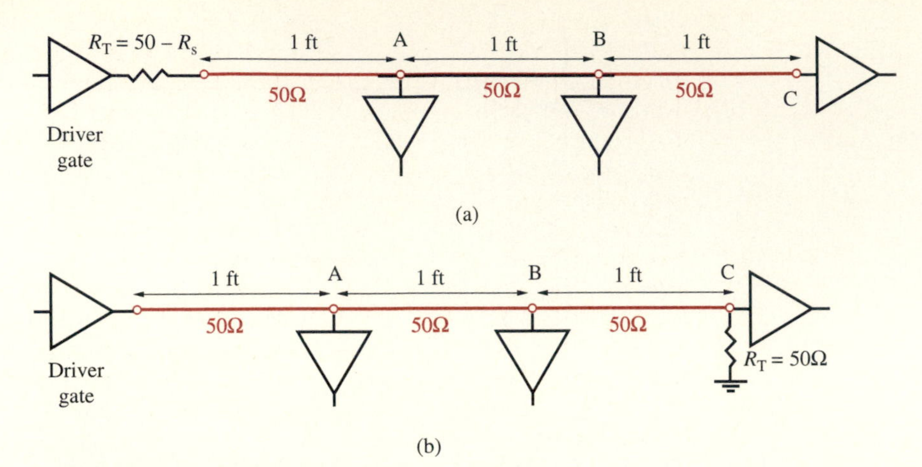

FIGURE 2.58. IC gate interconnects. (a) Series termination. (b) Parallel termination. Problem 2-24.

nation resistance R_T in both circuits to eliminate ringing. (b) Compare the performance of these two circuits in terms of their speed and dc power dissipation. Which technique is the natural choice for a design to achieve low-power dissipation at steady state?

2-24. Digital IC gate interconnects. A disadvantage of the series termination scheme in Problem 2-23 is that the receiver gate or gates must be near the end of the line to avoid receiving a two-step signal. This scheme is not recommended for terminating distributed loads. The two circuits shown in Figure 2.58 have three distributed loads equally positioned along a 3-ft-long 50Ω interconnect on a pc board constructed of FR4 material (take $v_p \simeq 14.3$ cm-(ns)$^{-1}$). Each circuit uses a different termination scheme. Assuming the driver and all the loads to be the same gates as in Problem 2-22, find the times at which each load gate changes its logic state after the output voltage of the driver gate switches to HIGH state at $t = 0$. Comment on the performance of both circuits and indicate which termination scheme provides faster speed. (Use some of the data provided in Problem 2-22.)

2-25. Open-ended stub. An electrical engineer is assigned the task of designing the circuit in Problem 2-24 that has the parallel termination scheme. After the design is complete, she performs some tests and measurements on the circuit. Noticing some peculiar effects in the test results, she decides to check the design. She realizes that she used a 4-ft-long interconnect, where the extra foot extends beyond the farthest element away from the driver and is not terminated at the end (i.e., an open-circuited stub). (See Figure 2.59.) Does this open stub affect the overall performance of the circuit? Explain.

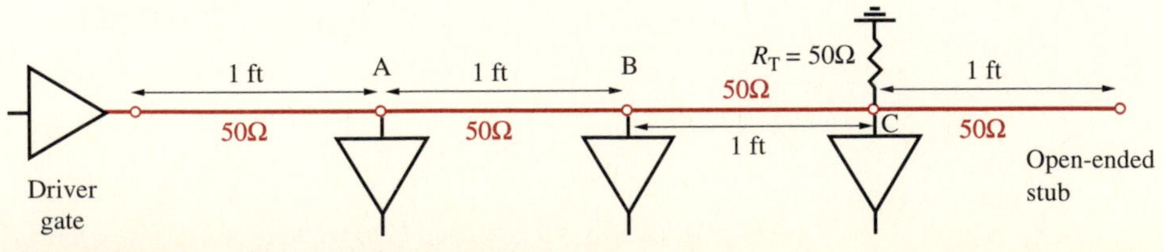

FIGURE 2.59. Open-ended stub. Problem 2-25.

2-26. Digital IC circuit. For the digital IC circuit shown in Figure 2.60, the driver gate goes from LOW to HIGH state at $t = 0$, and its Thévenin equivalent circuit (including the series termination resistance) can be approximated as a 5-V voltage source in series with a 50Ω resistor. If the time delay for all interconnects is given to be 2 ns-(ft)$^{-1}$, find the time(s) at which each receiver gate changes its state permanently. Assume each load gate to change state when its input voltage exceeds 4 V. Also assume each load gate to appear as an open circuit at its input. Support your solution with sketches of the two load voltages $\mathcal{V}_1$ and $\mathcal{V}_2$ as functions of time for a reasonable time interval.

2-27. Two driver gates. Two identical logic gates drive a third identical logic gate (load gate), as shown in Figure 2.61. All interconnects have the same one-way time delay t_d and characteristic impedance Z_0. When any one of these driver gates is at HIGH state, its Thévenin equivalent as seen from its output terminals consists of a voltage source with voltage V_0 in series with a resistance of value $R_s = Z_0$. At LOW state, its Thévenin equivalent is just a resistance of value $R_s = Z_0$. The input impedance of the load gate is very high compared to the characteristic impedance of the line (i.e., $Z_{in} \gg Z_0$). (a) Assuming steady-state conditions before both driver gates change to HIGH state at $t = 0$, sketch the load voltage $\mathcal{V}_L$ as a function of time for $0 \leq t \leq 7t_d$.

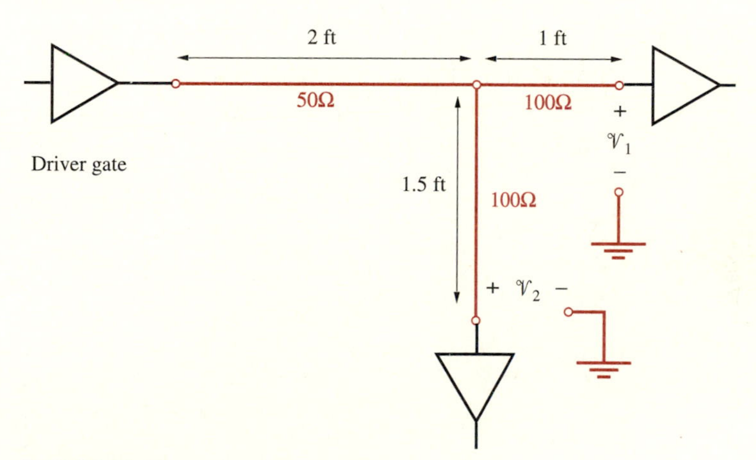

FIGURE 2.60. Digital IC circuit. Problem 2-26.

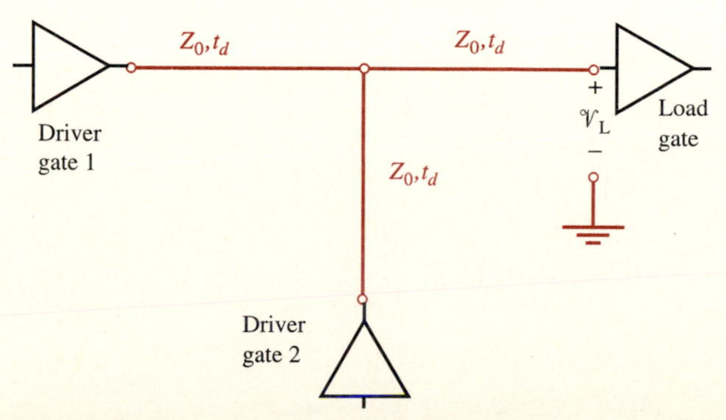

FIGURE 2.61. Two driver gates. Problem 2-27.

What is the eventual steady-state value of the load voltage? (b) Assume steady-state conditions before $t = 0$ to be such that driver gate 1 is at HIGH state and gate 2 is at LOW state. At $t = 0$, gate 1 and gate 2 switch states. Repeat part (a).

2-28. Capacitive load. For the transmission line system shown in Figure 2.62, the switch is closed at $t = 0$. Each of the two transmission lines has a one-way time delay of 2 ns. Assuming both transmission lines and the 5 pF capacitor to be initially uncharged, find and sketch the voltage $\mathcal{V}_1(t)$ across the resistor R_1.

2-29. Inductive load. The circuit shown in Figure 2.63 consists of a voltage source of amplitude 5 V and a source resistance of 50Ω driving a lossless 50Ω transmission line having a one-way time delay of 3 ns terminated with an ideal inductor of 25 nH. The circuit has been in steady state for a long time with the switch at position A. At $t = 0$, the switch is moved to position B. (a) Find the mathematical expressions and sketch the voltages at the source and load ends of the line. (b) Repeat part (a) for the case of the same line terminated with a 25 nH inductor in series with a 50Ω resistor.

2-30. Unknown lumped element. The transmission line circuit has an unknown lumped element, as shown in Figure 2.64. With the source-end voltage due to step excitation measured to be as plotted, determine the type of the unknown element, and find its value in terms of the shaded area A.

2-31. Unknown lumped element. The following circuit consists of two transmission lines of characteristic impedances Z_{01} and Z_{02} connected with an unknown lumped series element, as shown in Figure 2.65. The circuit is excited by a step source of amplitude V_0 and a source resistance $R_s = 50\Omega$, starting at $t = 0$. The source-end voltage is observed as a function of time, as shown. (a) Assuming the second line to be terminated

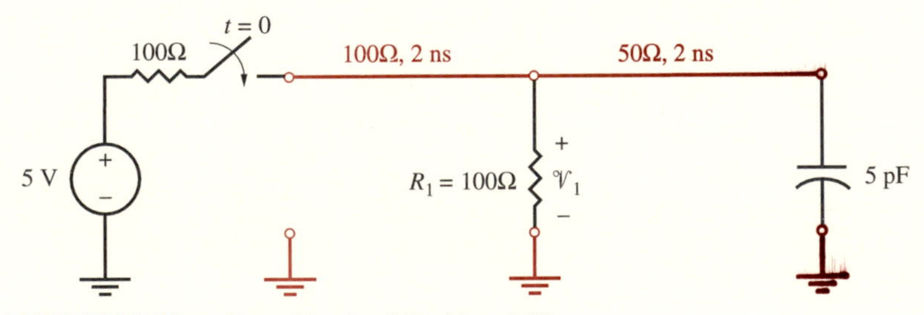

FIGURE 2.62. Capacitive load. Problem 2-28.

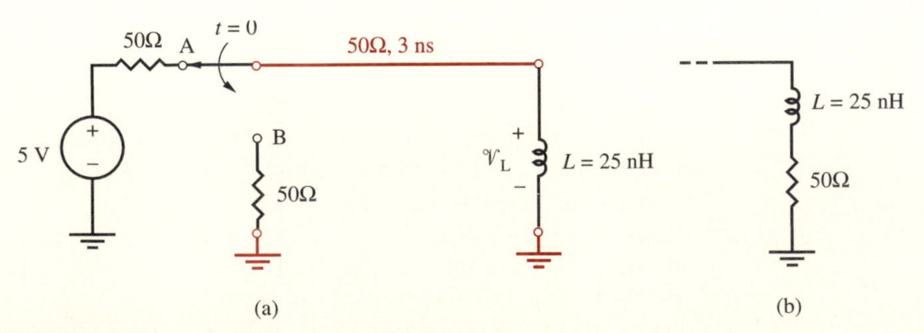

FIGURE 2.63. Inductive load. Problem 2-29.

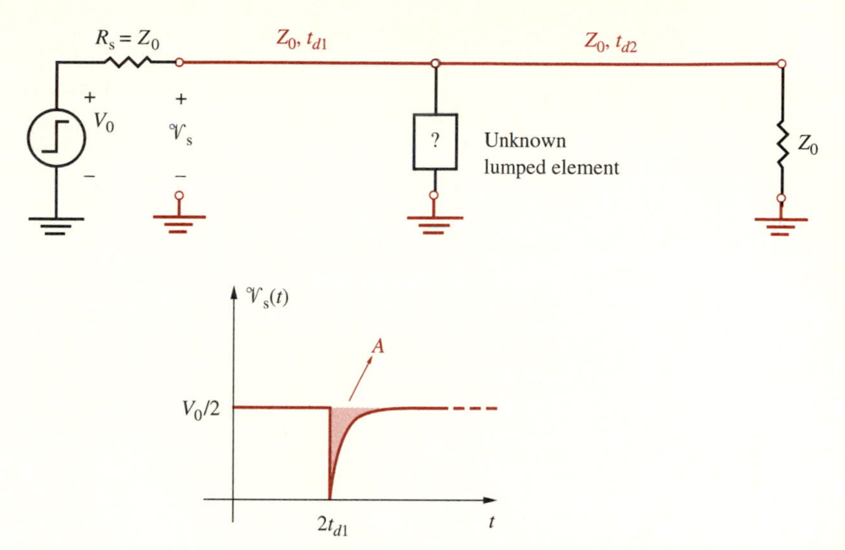

FIGURE 2.64. **Unknown lumped element.** Problem 2-30.

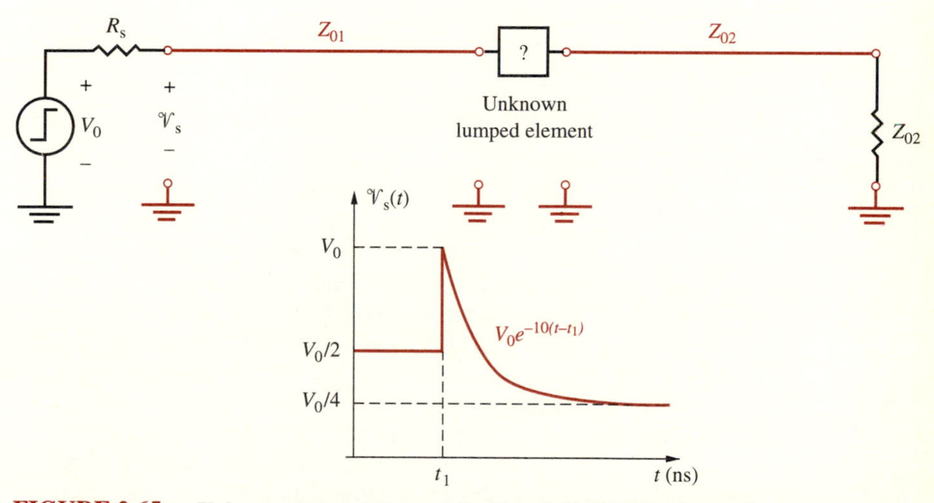

FIGURE 2.65. **Unknown lumped element.** Problem 2-31. Both t and t_1 are in ns.

with $R_L = Z_{02}$ at the far end, determine Z_{01}, Z_{02}, the type (e.g., inductance, capacitance, resistance) of the unknown circuit element, and its value (i.e., nH, pF, or Ω). (b) Find and sketch the voltage across this element.

2-32. Capacitive load. Two transmission lines of characteristic impedances 75Ω and 50Ω are joined by a connector that introduces a shunt resistance of 150Ω between the lines, as shown in Figure 2.66. The load end of the 50Ω line is terminated with a capacitive load with a 30 pF capacitor initially uncharged. The source end of the 75Ω line is excited by a step function of amplitude 3.6 V and a series resistance of 75Ω, starting

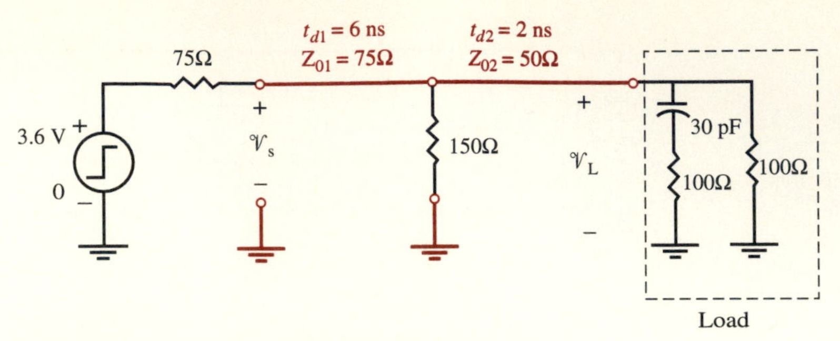

FIGURE 2.66. **Capacitive load.** Problem 2-32.

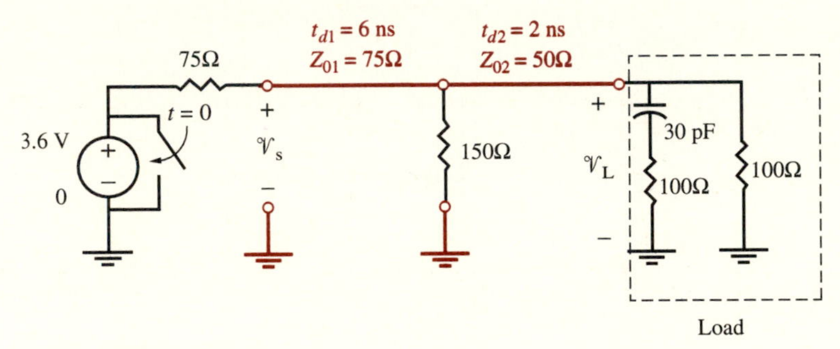

FIGURE 2.67. **Discharging with a capacitive load.** Problem 2-33.

at $t = 0$. Assuming the total time delay of each line to be $t_{d1} = 6$ ns and $t_{d2} = 2$ ns, respectively, find and sketch (a) the voltage $\mathcal{V}_L(t)$ at the load end of the 50Ω line and (b) the voltage $\mathcal{V}_s(t)$ at the source end of the 75Ω line.

2-33. Discharging with a capacitive load. In the circuit shown in Figure 2.67, the 3.6 V source voltage is shorted by a switch at $t = 0$, after being connected to the circuit for a long time. Find and sketch the source- and the load-end voltages $\mathcal{V}_s(t)$ and $\mathcal{V}_L(t)$.

2-34. Inductive load. Two transmission lines of characteristic impedances 50Ω and 75Ω are joined by a connector that introduces a series resistance of 25Ω between the lines, as shown in Figure 2.68. The load end of the 75Ω line is terminated with an inductive load. The inductor is initially uncharged. Find and sketch the voltage $\mathcal{V}_L$. First determine the initial and final values and accurately mark all points of your sketch.

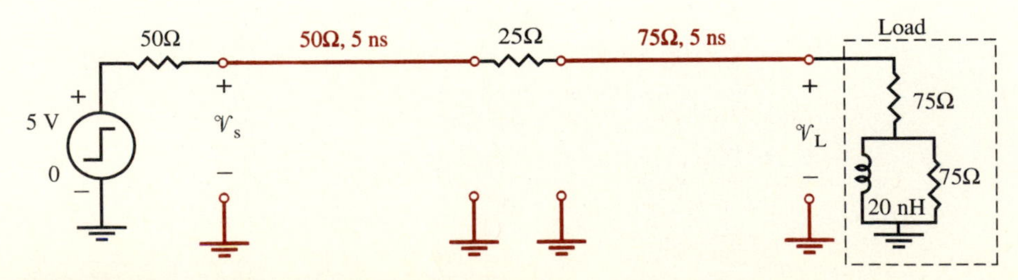

FIGURE 2.68. **Inductive load.** Problem 2-34.

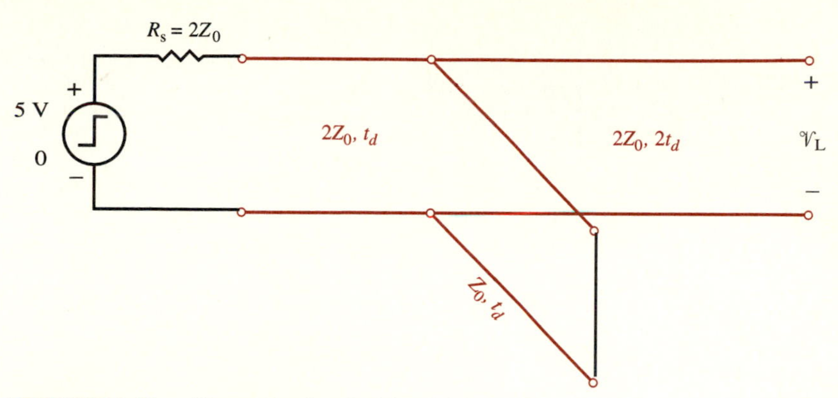

FIGURE 2.69. **Step excitation.** Problem 2-35.

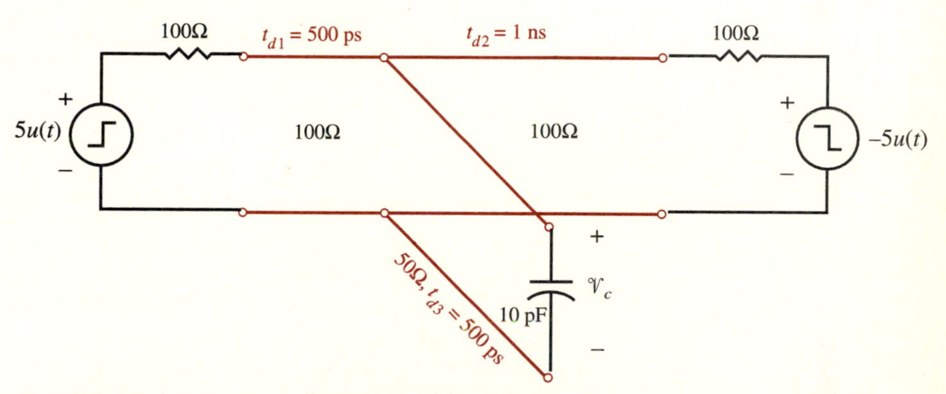

FIGURE 2.70. **Capacitive load excited by two sources.** Problem 2-36.

2-35. Step excitation. The circuit shown in Figure 2.69 is excited by a step-voltage source of amplitude 5 V and source resistance $R_s = 2Z_0$, starting at $t = 0$. Note that the characteristic impedance of the shorted stub is half that of the main line and that the second segment of the main line is twice as long, so its one-way time delay is $2t_d$. (a) Assuming the load to be an open circuit (i.e., a very large resistance), sketch the load voltage $\mathcal{V}_L(t)$ versus t for $0 \leq t \leq 11t_d$. (b) Repeat part (a) for the case when the input is a pulse of duration $t_w = 4t_d$.

2-36. Capacitive load excited by two sources. For the transmission line system shown in Figure 2.70, find the mathematical expression for the capacitor voltage $\mathcal{V}_c(t)$ and sketch it for $t > 0$. Assume the capacitor to be initially uncharged.

2-37. Two sources. For the circuit shown in Figure 2.71, sketch the voltages $\mathcal{V}_{s1}(t)$ and $\mathcal{V}_{s2}(t)$ for $0 \leq t \leq 7t_d$.

2-38. Nonlinear termination. Consider a 50Ω, 2-ns transmission line used to connect a driver logic gate to a load gate. At $t = 0$, the driver gate switches from LOW to HIGH state and can be modeled at HIGH state with an output voltage of 5 V in series with an output impedance of 10Ω. The load gate has a nonlinear voltage-current characteristic represented by

$$\mathcal{I}_L = 0.35(1 - e^{-\mathcal{V}_L/2})$$

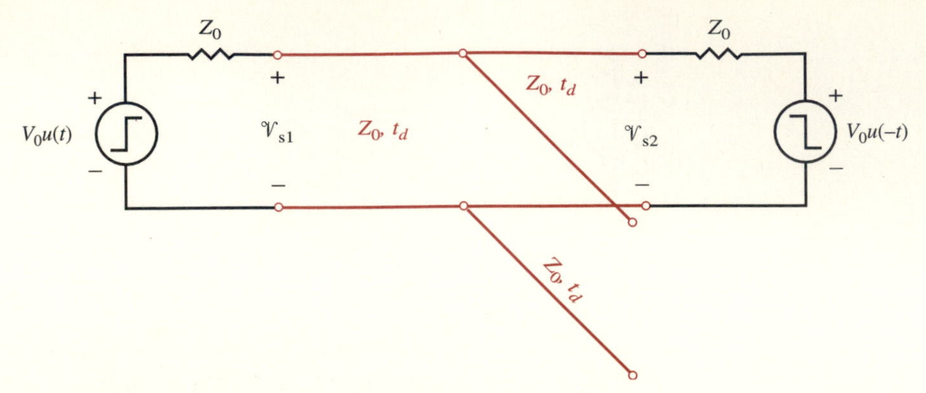

FIGURE 2.71. **Two sources.** Problem 2-37.

where $\mathcal{I}_L$ is in A and $\mathcal{V}_L$ is in V. Use the graphical Bergeron technique to determine approximately the time it would take for the circuit to reach a steady state.

2-39. Nonlinear source. A circuit consists of a driver gate connected to a load gate via a 50Ω, 2-ns transmission line. At $t = 0$, the driver gate switches from LOW state to HIGH state. The output of the driver at HIGH state has a nonlinear voltage-current characteristic represented by

$$\mathcal{I}_s = 8(5 - \mathcal{V}_s) - (5 - \mathcal{V}_s)^2$$

where $\mathcal{I}_s$ is in mA and $\mathcal{V}_s$ is in V. The load gate has a very large input impedance (~10 MΩ). Use the graphical Bergeron technique to (a) sketch the load voltage $\mathcal{V}_L$ versus t and (b) determine the approximate time it takes for the circuit to reach steady state. Assume the line to be uncharged at $t = 0$.

2-40. Effects of source rise time. Consider a lossless transmission line trace excited by a voltage source with output impedance 12.5Ω at one end and terminated by a short circuit at the other end. The characteristic impedance, the propagation delay, and the length of the trace are equal to 50Ω, 80 ps-(cm)$^{-1}$, and 25 cm, respectively. The source voltage increases linearly from zero at $t = 0$ to an amplitude of 5 V at $t = t_r$. (a) Find and sketch the source end voltage of the trace if the source rise time $t_r = 1$ ns. (b) Repeat part (a) for $t_r = 250$ ps.

2-41. Effects of source rise time. Consider a step voltage source of 3 V amplitude, 1 ns rise time, and 25Ω output impedance connected to a transmission line with $Z_0 = 50\Omega$ characteristic impedance and $C = 1$ pF-(cm)$^{-1}$ line capacitance terminated with a load resistance $R_L \gg Z_0$. Find and sketch the voltages at the two ends of the transmission line for a line length of (a) $l = 5$ cm, and (b) $l = 50$ cm. Compare the results and comment on the difference.

2-42. RG 8 coaxial line. A student buys an RG 8 low-loss coaxial cable from Radio Shack for a VHF antenna project. He looks up the specifications of the RG 8 coax in the Radio Shack product catalog and finds out that its characteristic impedance is $Z_0 = 50\ \Omega$, its velocity factor is $v_p/c = 0.66$, and its line capacitance is $C = 26.4$ pF-(ft)$^{-1}$. He then cuts a portion of this coax and measures the diameter of the inner conductor and the outer diameter of the dielectric to be approximately 2 mm and 7.5 mm, respectively. Using these values, find or verify the values of the unit length line parameters L, C, R, and G and the characteristic impedance Z_0 of this cable at 100 MHz. Note that the dielectric inside RG 8 coax is polyethylene and that the leakage conductance per unit length of a polyethylene-filled coaxial line at 100 MHz is approximately given by $G \simeq 1.58 \times 10^{-5}/\ln(b/a)$ S-m^{-1}.

2-43. Two-wire line. Calculate the per-unit-length line parameters L, C, R, and G and the characteristic impedance Z_0 of an air-insulated two-wire line made of copper wires with wire separation of 2.1 cm and wire diameter of 1.2 mm at a frequency of 200 MHz.

2-44. Distributed capacitive load. A transmission line system consists of a driver gate, a transmission line trace, and a load gate. The transmission line trace is $l = 25$ cm long and is characterized by the trace parameters $L = 2.46$ nH-(cm)$^{-1}$ and $C = 0.984$ pF-(cm)$^{-1}$. The driver output resistance is 20Ω for driving a HIGH-to-LOW signal and 25Ω for a LOW-to-HIGH one, and its driver output voltage is between 3.5 V (at HIGH state) and 0 V (at LOW state). The load gate has a very large input resistance of ~50 MΩ. Consider the case when the driver changes from HIGH to LOW state at $t = 0$. (a) Sketch the voltage at each end of the trace by using a bounce diagram. Neglect the fall time of the output voltage of the driver gate. (b) Repeat part (a) for the case in which the same circuit has an additional total load capacitance of $C_L =$ 15 pF that is uniformly distributed along the length of the trace, and comment on the difference. (Hint: Combine the load capacitance C_L with the line capacitance C as if the combination is the new "effective" line capacitance of the trace.)

CHAPTER 3

Steady-State Waves on Transmission Lines

In digital-integrated electronics, computer communication, and many other applications, it is important to understand the response of transmission lines to steplike changes in their inputs, as we studied in detail in Chapter 2. We have seen that waves travel down a transmission line by successively charging the distributed capacitors of the line and establishing current in the distributed inductors. In this context, a wave is a function of both time and space but does not necessarily involve periodic oscillations of a physical quantity (e.g., the height of water in ocean waves). If a disturbance that occurs at a certain point and time causes disturbances at other points in the surrounding region at later times, then wave motion is said to exist. Because of the nonzero travel time along a transmission line, disturbances initiated at one location induce effects that are retarded in time at other locations. The natural response of transmission lines to sudden changes in their inputs typically involves wave motion, with disturbances traveling down the line, producing reflections at terminations, or discontinuities, which in turn propagate back to the source end, and so on. However, except in special cases, the natural response eventually decays after some time interval and is primarily described by the intrinsic properties of the transmission line (characteristic impedance and one-way travel time, or length) and its termination (i.e., the load), rather than the input excitation.

In many engineering applications, however, it is also important to understand the steady-state response of transmission lines to sinusoidal excitations. Electrical power and communication signals are often transmitted as sinusoids or modified sinusoids. Other nonsinusoidal signals, such as pulses utilized in a digitally coded system, may be considered as a superposition (i.e., Fourier series) of sinusoids of different frequencies. In sinusoidal signal applications, the initial onset of a sinusoidal input produces a natural response. However, this initial transient typically decays rapidly in time, while the forced response supported by the sinusoidal excitation continues indefinitely. Once the steady state is reached, voltages and currents on the transmission line vary sinusoidally in time at each point along the line while also traveling down the line. The finite travel time of waves leads to phase differences between the voltages (or currents) at different points along the line. The steady-state solutions for voltage and current waves when reflections are present lead to *standing waves,* which also vary sinusoidally in time at each point on the line but do not travel along the line. The differences between traveling and standing waves will become clear in the following sections.

In discussing the steady-state response of transmission lines to sinusoidal excitation, we take full advantage of powerful tools[1] commonly used for analysis of alternating-current phenomena in lumped electrical circuits, including *phasors* and complex *impedance.* The phasor notation eliminates the need to keep track of the known sinusoidal time dependence of the various quantities and allows us to transform the transmission line equations from partial differential equations to ordinary differential equations. The magnitude of the complex impedance of a device or a load is a measure of the degree to which it opposes the excitation by a sinusoidal voltage source; a load with a higher impedance (in magnitude) at any given frequency requires a higher voltage in order to allow a given amount of current through it *at the given frequency.* The phase of the impedance represents the phase difference between the voltage across it and the current through it. In lumped circuits, the impedance of a load is determined only by its internal dynamics (i.e., its physical makeup as represented by its voltage-current characteristics, e.g., $V = L\, dI/dt$ for an inductor). In transmission line applications, the impedance of a load presented to a source via a transmission line depends on the characteristic impedance and the *electrical length*[2] (physical length per unit of wavelength) of the line connecting them. This additional length dependence makes the performance of transmission line systems dependent on frequency—to a greater degree than is typical in lumped circuit applications.

In this chapter, we exclusively consider excitations that are pure sinusoids. In most applications, transmission line systems have to accommodate signals made up of modified sinusoids, with the energy or information spread over a small bandwidth around a central frequency. The case of excitations involving waveforms that

---

[1]These tools were first introduced by a famous electrical engineer, C. P. Steinmetz, in his first book, *Theory and Calculation of AC Phenomena,* McGraw-Hill, 1897.

[2]The electrical length of a transmission line is its physical length divided by the wavelength at the frequency of operation. For example, the electrical length of a 1.5-m-long air-filled coaxial line operating at 100 MHz (wavelength $\lambda = 3$ m) is 0.5λ, while the electrical length of the same line operating at 1 MHz (wavelength $\lambda = 300$ m) is 0.005λ.

are other than sinusoidal can usually be handled by suitably decomposing the signal into its Fourier components, each of which can be analyzed as described in this chapter. In the first seven sections we consider lossless (i.e., $R = 0$ and $G = 0$) transmission lines. Lossy transmission lines (i.e., $R \neq 0$ and $G \neq 0$) and resonant transmission line elements are discussed in Sections 3.8 and 3.9, respectively. We shall see that the effects of small but nonzero losses can be accounted for by suitably modifying the lossless analysis. Note also that the loss terms are truly negligible in many applications, so that the lossless cases considered in detail are of practical interest in their own right.

Our topical presentation in this chapter starts in Section 3.1 with a discussion of the solutions of transmission line equations expressed in terms of voltage and current phasors. The important special cases of transmission lines with short- and open-circuited terminations are covered in Section 3.2, followed in Section 3.3 by the analysis of the response of lines terminated in an arbitrary impedance, and in Section 3.4 by a discussion of power and energy relations. The subject of impedance matching, crucially important in practice, is covered in Section 3.5, and the Smith chart, a graphical method useful for both quantitative analysis and qualitative visualization of transmission line behavior, is discussed in Section 3.6. Selected application examples involving the steady-state response of lossless transmission lines are presented in Section 3.7. Lossy and resonant transmission lines are discussed respectively in Sections 3.8 and 3.9.

3.1 WAVE SOLUTIONS USING PHASORS

The fundamental transmission line equations for the lossless case were developed in Chapter 2 (see equations [2.3] and [2.4]):

$$\frac{\partial \mathcal{V}}{\partial z} = -L\frac{\partial \mathcal{I}}{\partial t} \qquad [3.1]$$

$$\frac{\partial \mathcal{I}}{\partial z} = -C\frac{\partial \mathcal{V}}{\partial t} \qquad [3.2]$$

Equations [3.1] and [3.2] are written in terms of the space-time functions describing the instantaneous values of voltage and current at any point z on the line, denoted respectively as $\mathcal{V}(z, t)$ and $\mathcal{I}(z, t)$. When the excitation is sinusoidal, and under steady-state conditions, we can use the phasor concept to reduce the transmission line equations [3.1] and [3.2] to ordinary differential equations (instead of partial differential equations, as they are now) so that we can more easily obtain general solutions. As in circuit analysis, the relations between phasors and actual space-time functions are as follows:

$$\mathcal{V}(z, t) = \Re e\{V(z)e^{j\omega t}\} \qquad [3.3a]$$

$$\mathcal{I}(z, t) = \Re e\{I(z)e^{j\omega t}\} \qquad [3.3b]$$

Here, both phasors $V(z)$ and $I(z)$ are functions of z only and are in general complex.

We can now rewrite[3] equations [3.1] and [3.2] in terms of the phasor quantities by replacing $\partial/\partial t$ with $j\omega$. We have

$$\boxed{\frac{dV(z)}{dz} = -j\omega L I(z)} \qquad [3.4]$$

and

$$\boxed{\frac{dI(z)}{dz} = -j\omega C V(z)} \qquad [3.5]$$

Combining [3.4] and [3.5], we can write a single equation in terms of $V(z)$,

$$\frac{d^2V(z)}{dz^2} = -(\omega^2 LC)V(z) \quad \text{or} \quad \frac{d^2V(z)}{dz^2} = -\beta^2 V(z) \qquad [3.6]$$

where, $\beta = \omega\sqrt{LC}$ is called the *phase constant.* Equation [3.6] is referred to as the complex *wave equation* and is a second-order ordinary differential equation commonly encountered in analysis of physical systems. The general solution of [3.6] is of the form

$$V(z) = V^+ e^{-j\beta z} + V^- e^{+j\beta z} \qquad [3.7]$$

where, as we shall see below, $e^{-j\beta z}$ and $e^{j\beta z}$ represent, respectively, wave propagation in the $+z$ and $-z$ directions, and where V^+ and V^- are complex constants to be determined by the boundary conditions. The corresponding expression for the current $I(z)$ can be found by substituting [3.7] in [3.4]. We find

$$I(z) = I^+ e^{-j\beta z} + I^- e^{+j\beta z} = \frac{1}{Z_0}[V^+ e^{-j\beta z} - V^- e^{+j\beta z}] \qquad [3.8]$$

where $Z_0 = V^+/I^+ = -V^-/I^- = \sqrt{L/C}$ is the characteristic impedance of the transmission line.

Using [3.3], and the expressions [3.7] and [3.8] for the voltage and current phasors, we can find the corresponding space-time expressions for the instantaneous voltage and current. We have

$$\begin{aligned} \mathcal{V}(z,t) &= \Re e\{V(z)e^{j\omega t}\} = \Re e\{V^+ e^{-j\beta z} e^{j\omega t} + V^- e^{+j\beta z} e^{j\omega t}\} \\ &= V^+ \cos(\omega t - \beta z) + V^- \cos(\omega t + \beta z) \end{aligned} \qquad [3.9]$$

[3]The actual derivation of [3.4] from [3.1] is as follows:

$$\frac{\partial \mathcal{V}}{\partial z} = -L\frac{\partial \mathcal{I}}{\partial t} \longrightarrow \frac{\partial}{\partial z}\underbrace{[\Re e\{V(z)e^{j\omega t}\}]}_{\mathcal{V}(z,t)} = -L\frac{\partial}{\partial t}\underbrace{[\Re e\{I(z)e^{j\omega t}\}]}_{\mathcal{I}(z,t)}$$

$$\longrightarrow \Re e\left\{e^{j\omega t}\frac{dV(z)}{dz}\right\} = \Re e\{-L(j\omega)e^{j\omega t}I(z)\} \longrightarrow \frac{dV(z)}{dz} = -j\omega L I(z)$$

where we have assumed[4] V^+ and V^- to be real. Similarly, we have

$$\mathscr{I}(z, t) = \frac{1}{Z_0}[V^+ \cos(\omega t - \beta z) - V^- \cos(\omega t + \beta z)] \qquad [3.10]$$

The voltage and current solutions consist of a superposition of two waves, one propagating in the $+z$ direction (i.e., toward the load) and the other in the $-z$ direction (i.e., a reflected wave moving away from the load). To see the wave behavior, consider the case of an infinitely long transmission line; in this case no reflected wave is present, and thus $V^- = 0$. The voltage and current for an infinitely long line are

$$V(z) = V^+ e^{-j\beta z}; \qquad \mathscr{V}(z, t) = V^+ \cos(\omega t - \beta z) \qquad [3.11]$$

and

$$I(z) = \frac{V^+}{Z_0} e^{-j\beta z}; \qquad \mathscr{I}(z, t) = \frac{V^+}{Z_0} \cos(\omega t - \beta z) \qquad [3.12]$$

Note that, everywhere along the line, the ratio of the voltage to current phasors is Z_0; hence, Z_0 is called the characteristic impedance of the line. Note, however, that this is true not only for an infinitely long line but also for a line of finite length that is terminated at a load impedance $Z_L = Z_0$, as we discuss later.

The solutions [3.11] and [3.12] are in the form of [1.2], which was introduced in Chapter 1 by stating that electromagnetic quantities with such space-time dependencies are often encountered. Here we see that this form of solution is indeed a natural solution of the fundamental transmission line equations.[5] The space-time behavior of the voltage wave given by [3.11] is illustrated in Figure 3.1. We note from Figure 3.1a that the period of the sinusoidal oscillations (as observed at fixed points in space) is $T_p = 2\pi/\omega$. The voltage varies sinusoidally at all points in space, but it reaches its maxima at different times at different positions. Figure 3.1b indicates that the voltage distribution as a function of distance (observed at a fixed instant of time) is also sinusoidal. The distance between the crests of the voltage at a fixed instant of time is the *wavelength* $\lambda = 2\pi/\beta$. As time progresses, the wave propagates to the right ($+z$ direction), as can be seen by observing a particular point on the waveform (e.g., the crest or the minimum) at different instants of time. The speed of this motion is the *phase velocity,* defined as the velocity at which an observer must travel to observe a stationary (i.e., not varying with time) voltage. The voltage observed would be the same as long as the argument of the cosine in [3.11] is the same; thus we have

$$\omega t - \beta z = \text{const.} \quad \longrightarrow \quad v_p = \frac{dz}{dt} = \frac{\omega}{\beta} = \frac{\omega}{\omega\sqrt{LC}} = \frac{1}{\sqrt{LC}} \qquad [3.13]$$

---

[4]If instead V^+ and V^- were complex, with $V^+ = |V^+|e^{j\phi^+}$ and $V^- = |V^-|e^{j\phi^-}$, we would have

$$\mathscr{V}(z, t) = |V^+|\cos(\omega t - \beta z + \phi^+) + |V^-|\cos(\omega t + \beta z + \phi^-)$$

[5]We will encounter the same type of space-time variation once again in Chapter 8, as the natural solution of Maxwell's equations for time-harmonic (or sinusoidal steady-state) electric and magnetic fields.

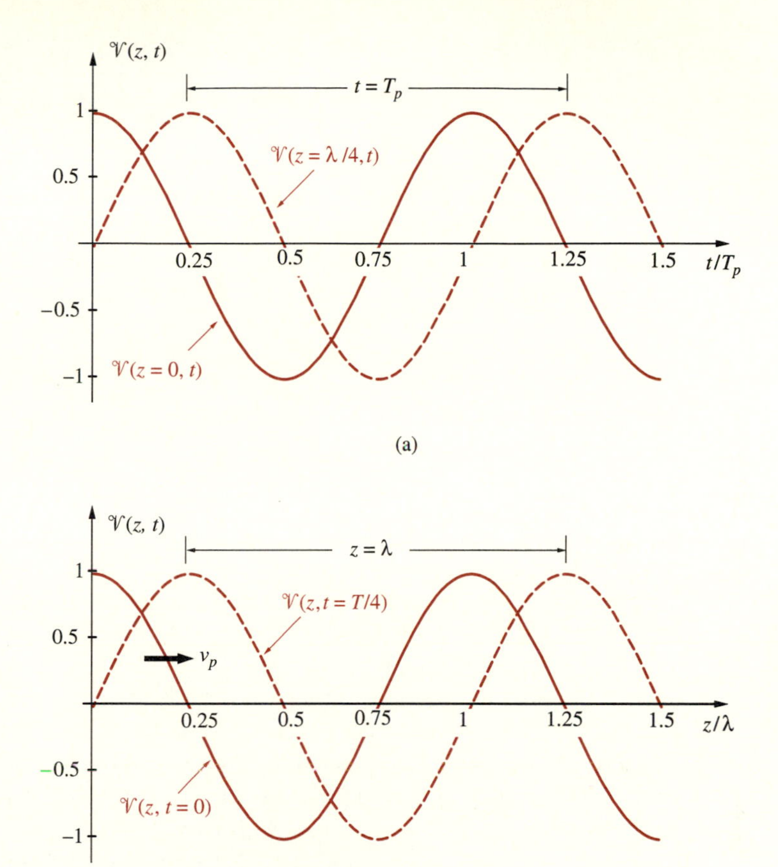

FIGURE 3.1. **Wave behavior in space and time.** (a) $\mathcal{V}(z, t) = V^+ \cos[2\pi(t/T_p) - 2\pi(z/\lambda)]$ versus t/T_p for $z = 0$ and $z = \lambda/4$. (b) $\mathcal{V}(z, t) = V^+ \cos[2\pi(t/T_p) - 2\pi(z/\lambda)]$ versus z/λ for $t = 0$ and $t = T_p/4$. In both panels we have taken $V^+ = 1$.

where v_p is the phase velocity, which was also introduced in Sections 1.1.3 and 2.2 (see equation [2.7]).

As discussed in Section 2.2, for most of the commonly used two-conductor transmission lines (Figure 2.1), the phase velocity v_p is not a function of the particular geometry of the metallic conductors but is instead solely determined by the electrical and magnetic properties of the surrounding insulating medium. When the surrounding medium is air, the phase velocity is the speed of light in free space, namely, $v_p = c \simeq 3 \times 10^8$ m-s^{-1} = 30 cm-(ns)$^{-1}$. The phase velocity v_p for some other insulating materials was tabulated in Table 2.1. Phase velocities for some additional materials are given in Table 3.1, together with the corresponding values of wavelength at a frequency of 300 MHz. Note that since $\lambda = 2\pi/\beta$, we have $v_p = \omega/\beta = \lambda f$, so that the phase constant β and wavelength λ depend on the electrical

TABLE 3.1. Phase velocity and wavelength in different materials

Material	**Wavelength (m at 300 MHz)**	**Phase velocity speed (cm-(ns)$^{-1}$ at 300 MHz)**
Air	1	30
Silicon	0.29	8.7
Polyethylene	0.67	20.0
Epoxy glass (PC board)	0.45	13.5
GaAs	0.30	9.1
Silicon carbide (SiC)	0.15	4.6
Glycerin	0.14	4.2

and magnetic properties of the material surrounding the transmission line conductors as well as on the frequency of operation.

3.2 VOLTAGE AND CURRENT ON LINES WITH SHORT- OR OPEN-CIRCUIT TERMINATIONS

Most sinusoidal steady-state applications involve transmission lines terminated at a load impedance Z_L. Often, voltages and currents near the load end are of greatest interest, since they determine the degree of matching between the line and the load and the amount of power delivered to (versus that reflected from) the load. A portion of a lossless transmission line terminated in an arbitrary load impedance Z_L is shown in Figure 3.2. We can use this setup to explore the concept of reflected waves on transmission lines, a fundamental feature of distributed circuits in general.

Assume that a forward-propagating ($+z$ direction) wave of the form $V^+e^{-j\beta z}$ produced by a source located at some position z ($z < 0$) is incident on load Z_L located at $z = 0$. Contrary to the case of an infinitely long transmission line, here the ratio of the total voltage $V(z)$ to the total current $I(z)$ at any position z along the line is not equal to Z_0. For example, at the load position ($z = 0$), we must satisfy the

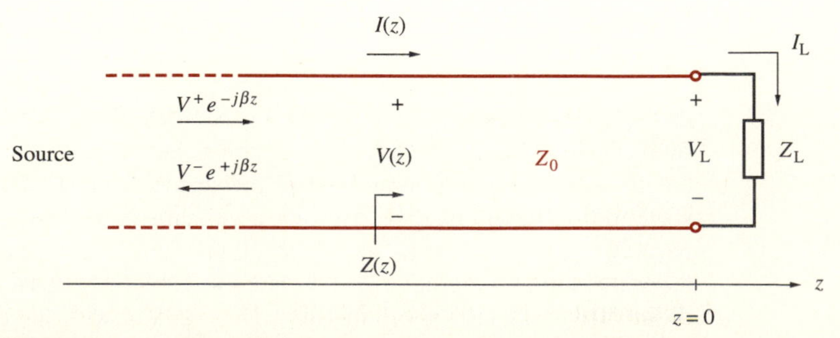

FIGURE 3.2. **A terminated lossless transmission line.** For convenience, the position of the load is taken to be $z = 0$.

boundary condition $[V(z)/I(z)]_{z=0} = Z_L$, where Z_L is in general not equal to Z_0. Thus, since $V^+/I^+ = Z_0$, and in general $Z_0 \neq Z_L$, a reverse-propagating ($-z$ direction) reflected wave of the form $V^- e^{+j\beta z}$ with the appropriate value for V^- must be present so that the load boundary condition is satisfied. The total voltage and current phasors, $V(z)$ and $I(z)$, at any position on the line consist of the sum of the forward and reverse waves as specified by [3.7] and [3.8], namely

$$V(z) = V^+ e^{-j\beta z} + V^- e^{+j\beta z} \qquad [3.14]$$

$$I(z) = \frac{1}{Z_0}[V^+ e^{-j\beta z} - V^- e^{+j\beta z}] \qquad [3.15]$$

where V^+ and V^- are in general complex constants to be determined by the boundary conditions.

When a transmission line has only a forward-traveling wave with no reflected wave (e.g., in the case of an infinitely long line), the ratio of the total voltage to the current is the characteristic impedance Z_0, as was discussed in Section 3.1. When the line is terminated so that, in general, a reflected wave exists, the ratio of the total line voltage $V(z)$ to the total line current $I(z)$ at any position z—the *line impedance*—is not equal to Z_0. The line impedance is of considerable practical interest; for example, the impedance that the line presents to the source at the source end of the line (called the input impedance of the line, denoted by Z_{in}) is the line impedance evaluated at that position. The source, or the generator, does not know anything about the characteristic impedance of the line or whether a reflected wave exists on the line; it merely sees that when it applies a voltage of V_s to the input terminals of the line, a certain current I_s flows, and thus the source interprets the ratio of V_s/I_s as an impedance of a particular magnitude and phase.

The line impedance as seen by looking toward the load Z_L at any position z along the line (see Figure 3.2) is defined as

$$\boxed{Z(z) \equiv \frac{V(z)}{I(z)} = Z_0 \frac{V^+ e^{-j\beta z} + V^- e^{j\beta z}}{V^+ e^{-j\beta z} - V^- e^{j\beta z}}} \qquad [3.16]$$

In general, the line impedance $Z(z)$ is complex and is a function of position z along the line. From electrical circuit analysis, we know that a complex impedance $Z(z)$ can be written as

$$Z(z) = R(z) + jX(z)$$

where the real quantities $R(z)$ and $X(z)$ are, respectively, the resistive and the reactive parts of the line impedance.

The following example considers the case of a *matched load,* defined as a load impedance equal to the characteristic impedance of the line, or $Z_L = Z_0$.

Example 3-1: Matched load. A lossless transmission line is terminated with a load $Z_L = Z_0$, as shown in Figure 3.3a. Find the magnitude of the reflected wave V^- and the line impedance $Z(z)$.

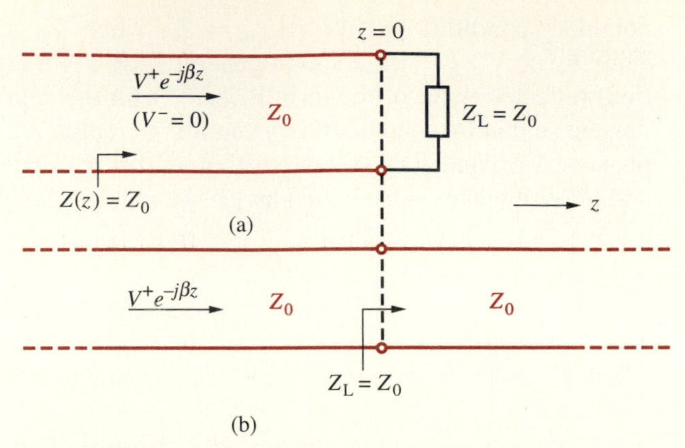

FIGURE 3.3. **Matched load.** (a) Circuit diagram. (b) Z_L is replaced by an infinitely long extension of the same line.

Solution: If $Z_L = Z_0$, the load boundary condition $V(z = 0)/I(z = 0) = Z_L = Z_0$ is satisfied without any reflected wave, so that $V^- = 0$. The line impedance at any position z along the line is $Z(z) = V(z)/I(z) = Z_0 = Z_L$ independent of z. In this case, the load impedance Z_L can be viewed as an infinite extension of the same transmission line, as shown in Figure 3.3b.

Transmission line segments terminated in short or open circuits are commonly used as tuning elements for impedance matching networks (see Sections 3.2.3 and 3.5) and also as resonant circuit elements (see Sections 3.2.3 and 3.9). In the next two subsections, we study these two special cases in detail before considering (in Section 3.3) the more general case of lines terminated in an arbitrary impedance Z_L.

3.2.1 Short-Circuited Line

Figure 3.4 shows a transmission line of length l terminated in a short circuit. Short-circuited termination forces the load voltage V_L to be zero, so we have

$$V_L = [V(z)]_{z=0} = [V^+e^{-j\beta z} + V^-e^{j\beta z}]_{z=0} = V^+ + V^- = 0$$

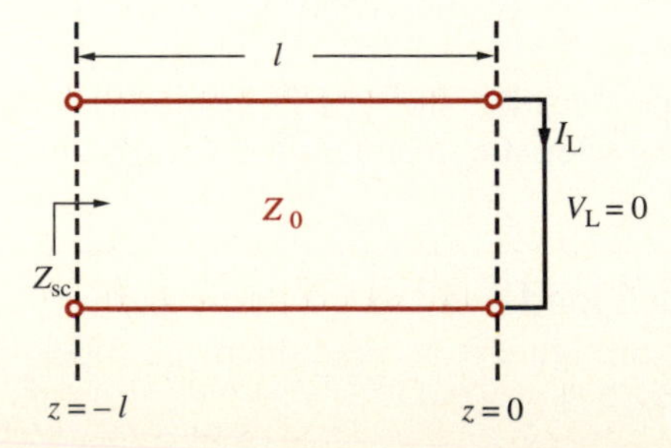

FIGURE 3.4. **Short-circuited line.** The input impedance Z_{sc} of a short-circuited line can be capacitive or inductive, depending on the length l of the line.

leading to

$$V^- = -V^+ \quad \text{or} \quad \frac{V^-}{V^+} = -1$$

Note that the load current flowing through the short circuit can be found from [3.15] using $V^- = -V^+$:

$$I_L = [I(z)]_{z=0} = \frac{1}{Z_0}(V^+ - V^-) = \frac{2V^+}{Z_0}$$

Anywhere else along the line we have

$$V(z) = V^+(e^{-j\beta z} - e^{j\beta z}) = -2V^+ j \sin(\beta z)$$

$$I(z) = \frac{V^+}{Z_0}(e^{-j\beta z} + e^{j\beta z}) = \frac{2V^+}{Z_0} \cos(\beta z)$$

The instantaneous space-time function for the voltage can be obtained from $V(z)$ using [3.3a]. We have

$$\begin{aligned}\mathcal{V}(z,t) \equiv \Re e\{V(z)e^{j\omega t}\} &= \Re e\{V^+(e^{-j\beta z} - e^{j\beta z})e^{j\omega t}\} \\ &= \Re e\{2|V^+|e^{j\phi^+} \sin(\beta z)e^{-j\pi/2}e^{j\omega t}\} \\ &= 2|V^+|\sin(\beta z)\cos\left(\omega t - \frac{\pi}{2} + \phi^+\right)\end{aligned}$$

where $V^+ = |V^+|e^{j\phi^+}$ is in general a complex constant, to be determined from the boundary condition at the source end of the transmission line. However, note that we can assume $\phi^+ = 0$ (i.e., V^+ is a real constant) without any loss of generality because V^+ is a constant multiplier that appears in front of all voltages and currents everywhere along the line. Accounting for a possible finite phase ϕ^+ simply amounts to shifting the time reference, with no effect on the relationships among the various quantities. Accordingly, we assume $\phi^+ = 0$ throughout the following discussion.

Note that the voltage $\mathcal{V}(z,t)$ along a short-circuited line is a cosinusoid (in time) whose amplitude varies as $2|V^+|\sin(\beta z)$ with position z along the line. It is not a traveling wave, since the peaks (or minima) of the $\cos(\omega t - \pi/2)$ always stay (i.e., stand) at the same positions (i.e., z/λ) along the line as the voltage varies in time (see Figure 3.5b). The voltage $\mathcal{V}(z,t)$ is thus said to represent a pure *standing wave*. Similarly, for current we have

$$\mathcal{I}(z,t) = \frac{2|V^+|}{Z_0}\cos(\beta z)\cos(\omega t)$$

which is also a standing wave (see Figure 3.5c). The absolute amplitudes of both the voltage and current phasors, namely $|V(z)|$ and $Z_0|I(z)|$, are shown as functions of z/λ in Figure 3.5d.

Note that the current and voltage are in time quadrature (i.e., 90° out of phase), such that when $\mathcal{I}(z,t)$ at a given point z is zero, the absolute amplitude of $\mathcal{V}(z,t)$ is

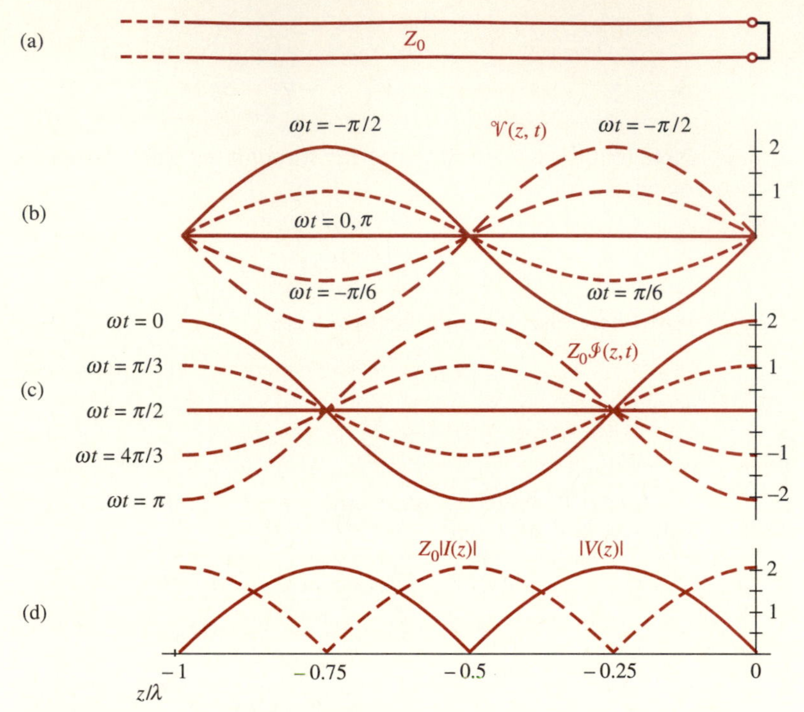

FIGURE 3.5. Voltage and current on a short-circuited line. (a) Schematic of a short-circuited line. (b) Instantaneous voltage $\mathcal{V}(z, t)$ versus z at different times. (c) Corresponding instantaneous current times the characteristic impedance, namely $Z_0\mathcal{I}(z, t)$. (d) Magnitudes of the voltage and current phasors, showing $|V(z)|$ (solid line) and $Z_0|I(z)|$ (dashed line) as functions of z/λ. All of the plots shown are for $V^+ = 1e^{j0°}$.

a maximum, and vice versa. Standing waves stand on the line but do not travel or carry any time-average power to the load;[6] they represent reactive power in a manner

---

[6]To see this, consider that the instantaneous power carried by the wave is given by

$$\mathcal{P}(z,t) = \mathcal{V}(z,t)\mathcal{I}(z,t) = \left[2|V^+|\sin(\beta z)\cos\left(\omega t - \frac{\pi}{2}\right)\right]\left[\frac{2|V^+|}{Z_0}\cos(\beta z)\cos(\omega t)\right]$$

$$= \frac{2|V^+|^2}{Z_0}\sin(2\beta z)\cos\left(\omega t - \frac{\pi}{2}\right)\cos(\omega t)$$

$$= \frac{|V^+|^2}{Z_0}\sin(2\beta z)\sin(2\omega t)$$

where we have used the trigonometric identities of $\cos(\zeta - \pi/2) = \sin\zeta$ and $2\sin\zeta\cos\zeta = \sin(2\zeta)$. Note that the instantaneous power carried by the standing wave oscillates in time at a rate twice that of the voltage and current and that its average over one period (i.e., $T = 2\pi/\omega$) is thus zero, as expected on the basis of the fact that the voltage and current are 90° out of phase.

analogous to the voltage and current, also in time quadrature, of a capacitor or inductor. The power relationships on a transmission line are discussed in more detail in Section 3.4.

The line impedance seen looking toward the short circuit at any position z along the short-circuited line is

$$Z(z) = \frac{V(z)}{I(z)} = Z_0 \frac{-2jV^+ \sin(\beta z)}{2V^+ \cos(\beta z)} = -jZ_0 \tan(\beta z)$$

where $z < 0$. The input impedance of a short-circuited line segment of length l can then be found by evaluating the preceding equation at the source end, where $z = -l$:

$$\boxed{Z_{sc} = jZ_0 \tan(\beta l) = jX_{sc}} \qquad [3.17]$$

We note from [3.17] that the input impedance of a short-circuited line of length l is purely reactive. As illustrated in Figure 3.6, the input impedance depends on line length l, or more generally, on the electrical length, which is defined as the ratio of the physical length of the line to the wavelength, i.e., l/λ, where $\lambda = 2\pi/\beta$. The input impedance can be varied by varying the length or the frequency, or both, and can be capacitive (negative X_{sc}) or inductive (positive X_{sc}). It makes physical sense that the input impedance is inductive for very short line lengths ($l < \lambda/4$); a shorted two-wire line of relatively short length resembles a small loop of wire. The fact that any reactive input impedance can be realized by simply varying the length of a short-circuited line (and the open-circuited line as will be discussed in the next subsection) is very useful in tuning- and impedance-matching applications at microwave frequencies, as discussed in Section 3.5.

Example 3-2 illustrates the use of [3.17] to determine the input impedance of a television antenna lead-in wire.

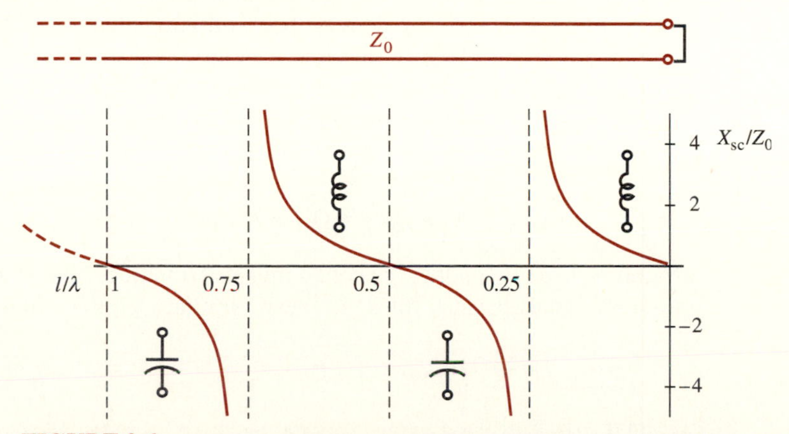

FIGURE 3.6. Input impedance of a short-circuited line. The normalized reactance X_{sc}/Z_0 of a short-circuited line segment of length l is shown as a function of electrical length l/λ.

Example 3-2: Inductance of a short television antenna lead-in wire. Consider a television antenna lead-in wire of length $l = 10$ cm having a characteristic impedance of $Z_0 = 300\Omega$, shorted at one end. Find the input impedance of this line if it is to be used at 300 MHz.

Solution: We can first determine the electrical length of the line. As in Example 2-18, we assume the conductors to be mostly surrounded by air (although they might in fact be held together by some plastic material) so that the phase velocity $v_p \simeq c$. Since the wavelength at 300 MHz is $\lambda = v_p/f \simeq 3 \times 10^8/(300 \times 10^6) = 1$ m, the electrical length is $(l/\lambda) = 0.1$. The input impedance of the shorted line can then be found directly from [3.17]:

$$Z_{sc} = jZ_0 \tan(\beta l) = j(300)\tan\left(\frac{2\pi}{\lambda}l\right) \simeq j218\Omega$$

This is an inductive input impedance. At a frequency of 300 MHz, an inductive reactance of $X_{sc} = \omega L_{sc} = 218\Omega$ corresponds to a lumped inductor having an inductance of $L_{sc} = 218/(2\pi \times 300 \times 10^6) \simeq 0.116$ μH.

According to [3.17] and Figure 3.6, the input impedance of a short-circuited line of length $l = \lambda/4$ is infinite; that is, the line appears as if it is an open circuit. In practice, however, the input impedance of such a line is limited by its distributed conductance. Note that considering the circuit model of the line (Figure 2.5) the shunt conductance per unit length G presents a resistance proportional to $(G)^{-1}$ across the input terminals of a line, even when the input impedance looking toward the load is infinite. Although G was assumed to be zero for our lossless analysis, it nevertheless is a nonzero value and thus limits the input impedance of the line to a finite value. Similarly, equation [3.17] and Figure 3.6 indicate that the input impedance of a short-circuited line of length $l = \lambda/2$ is zero, in other words, that the line appears as if it is a short circuit. In practice, however, the minimum value of input impedance of a short-circuited half-wavelength long line is determined by its series resistance R, which, although small, is nevertheless a nonzero value. The behavior of a short-circuited line of length $l = \lambda/4$ is similar to that of a lumped resonant circuit, as discussed in Section 3.9.

3.2.2 Open-Circuited Line

Figure 3.7 shows an open-circuited transmission line. The analysis of open-circuited lossless transmission lines is very similar to that of short-circuited lines.

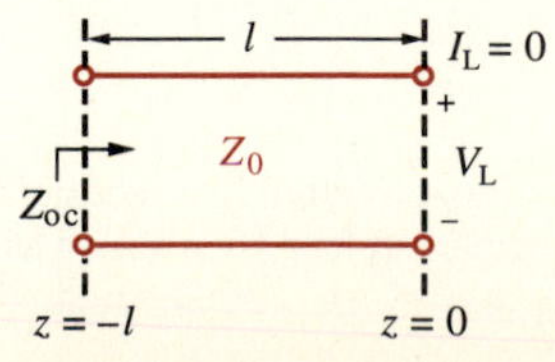

FIGURE 3.7. Open-circuited line. The input impedance Z_{oc} of an open-circuited line can be capacitive or inductive, depending on the length of the line.

Open-circuited termination forces the load current I_L to be zero, so that using [3.15] we have

$$I_L = [I(z)]_{z=0} = \left[\frac{V^+}{Z_0}e^{-j\beta z} - \frac{V^-}{Z_0}e^{j\beta z}\right]_{z=0} = \frac{V^+ - V^-}{Z_0} = 0$$

leading to

$$V^- = V^+ \quad \text{or} \quad \frac{V^-}{V^+} = +1$$

Note that the load voltage appearing across the open circuit can be found from [3.14] using $V^- = V^+$. We have

$$V_L = [V(z)]_{z=0} = [V^+e^{-j\beta z} + V^-e^{j\beta z}]_{z=0} = (V^+ + V^-) = 2V^+$$

Anywhere else along the line we have

$$V(z) = V^+(e^{-j\beta z} + e^{j\beta z}) = 2V^+\cos(\beta z)$$

$$I(z) = \frac{V^+}{Z_0}(e^{-j\beta z} - e^{j\beta z}) = -2\frac{V^+}{Z_0}j\sin(\beta z) = 2\frac{V^+}{Z_0}(e^{-j\pi/2})\sin(\beta z)$$

The instantaneous space-time expressions for the voltage and current[7] are

$$\mathcal{V}(z, t) = \Re e\{V(z)e^{j\omega t}\} = 2|V^+|\cos(\beta z)\cos(\omega t)$$

$$\mathcal{I}(z, t) = \Re e\{I(z)e^{j\omega t}\} = 2\frac{|V^+|}{Z_0}\sin(\beta z)\cos\left(\omega t - \frac{\pi}{2}\right)$$

where we have assumed $V^+ = |V^+|e^{j\phi^+}$, with $\phi^+ = 0$, without any loss of generality.

As for the short-circuited line, the current and voltage on an open-circuited line are in time quadrature (i.e., out of phase by 90°) so that the average power carried is again zero. Both $\mathcal{V}(z, t)$ and $\mathcal{I}(z, t)$ are purely standing waves. Their absolute amplitude patterns are shown in Figure 3.8b, in the same format as in Figure 3.5d.

The line impedance seen looking toward the open circuit at any position z along the line is

$$Z(z) = \frac{V(z)}{I(z)} = jZ_0\cot(\beta z)$$

---

[7]Note for the current that

$$\Re e\{I(z)e^{j\omega t}\} = \Re e\left\{\frac{V^+}{Z_0}[e^{-j\beta z} - e^{j\beta z}]e^{j\omega t}\right\} = \Re e\left\{2\frac{|V^+|}{Z_0}\sin(\beta z)e^{j\phi^+}e^{-j(\pi/2)}e^{j\omega t}\right\}$$

$$= \Re e\left\{2\frac{|V^+|}{Z_0}\sin(\beta z)\left[\cos\left(\omega t - \frac{\pi}{2} + \phi^+\right) + j\sin\left(\omega t - \frac{\pi}{2} + \phi^+\right)\right]\right\}$$

$$= \frac{2|V^+|}{Z_0}\sin(\beta z)\cos\left(\omega t - \frac{\pi}{2} + \phi^+\right)$$

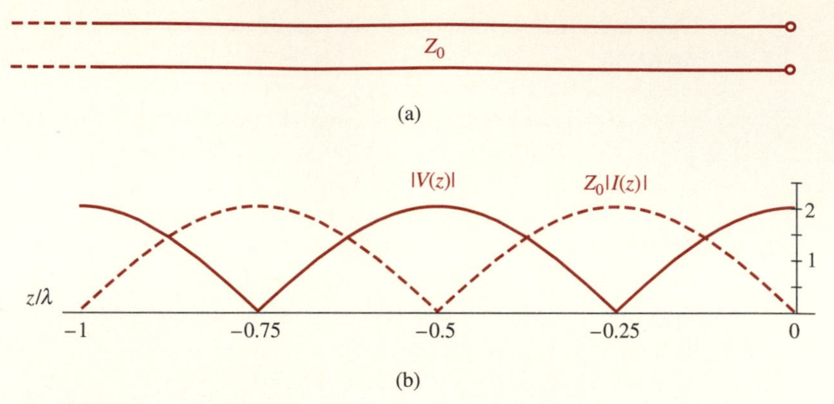

FIGURE 3.8. Voltage and current on an open-circuited line. (a) Schematic of an open-circuited line. (b) Magnitudes of the voltage and current phasors, showing $|V(z)|$ (solid line) and $Z_0|I(z)|$ (dashed line) as functions of z/λ, for $V^+ = 1e^{j0°}$.

where $z < 0$. The input impedance of an open-circuited line of length l (i.e, at $z = -l$) is then given by

$$\boxed{Z_{\text{oc}} = Z(z = -l) = -jZ_0 \cot(\beta l) = jX_{\text{oc}}} \qquad [3.18]$$

As for a short-circuited line, the input impedance of an open-circuited line of length l is purely reactive. The normalized reactance X_{oc}/Z_0 is plotted in Figure 3.9 as a function of electrical length l/λ. A capacitive or inductive reactance can

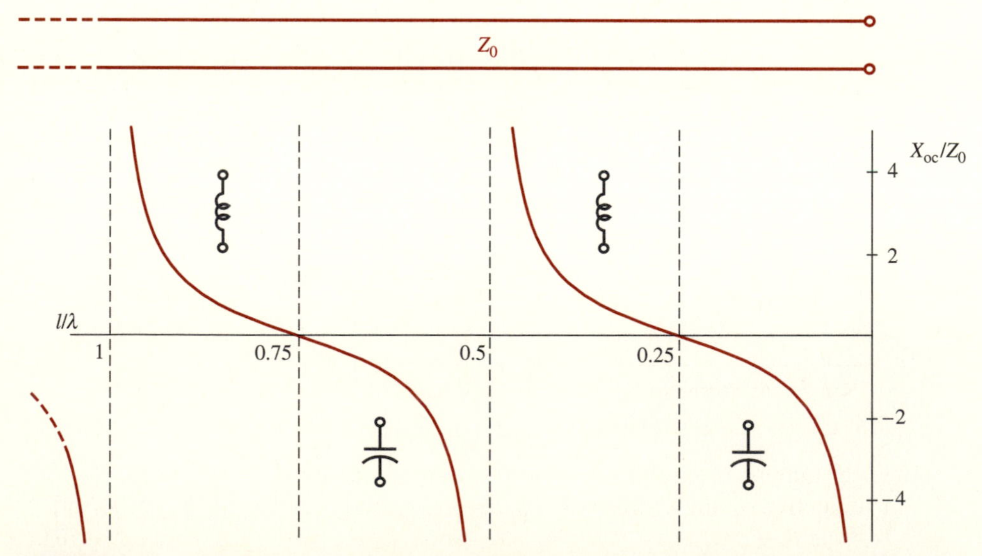

FIGURE 3.9. Input impedance of an open-circuited line. The normalized reactance X_{oc}/Z_0 of an open-circuited line segment of length l is shown as a function of electrical length l/λ.

be obtained simply by adjusting the line length l for a fixed wavelength λ (or frequency ω), or by adjusting the wavelength λ (or frequency ω) for fixed line length l. The fact that for $l < \lambda/4$ the impedance is capacitive makes physical sense, since a relatively short-length open-circuited line consists of two conductors with some separation between them, resembling an ordinary lumped capacitor.

Example 3-3 illustrates the use of [3.18] for a practical transmission line.

Example 3-3: Capacitance of a short television antenna lead-in wire. Consider a television antenna lead-in wire of length $l = 20$ cm having a characteristic impedance of $Z_0 = 300\Omega$ and nothing connected at its end (i.e., open-circuited). Find the input impedance of this line if it is to be used at 300 MHz.

Solution: Once again we assume the phase velocity to be $v_p \simeq c$ and first determine the electrical length of the line. Since the wavelength at 300 MHz is $\lambda = v_p/f \simeq 3 \times 10^8/(300 \times 10^6) = 1$ m, the electrical length is $(l/\lambda) = 0.2$. The input impedance of the open-circuited line can then be found directly from [3.18]:

$$Z_{oc} = -jZ_0 \cot(\beta l) = -j(300)\cot\left(\frac{2\pi}{\lambda} l\right) \simeq -j97.5\Omega$$

This is a capacitive input impedance. At a frequency of 300 MHz, a capacitive reactance of $X_{oc} = -(\omega C_{oc})^{-1} \simeq -97.5\Omega$ corresponds to a lumped capacitor of capacitance $C_{oc} = [(97.5)(2\pi \times 300 \times 10^6)]^{-1} \simeq 5.44$ pF.

According to [3.18] and Figure 3.9, the input impedance of an open-circuited line of length $l = \lambda/4$ is zero; that is, the line appears as if it is short-circuited. In practice, however, the minimum value of input impedance is determined by its series resistance R, which, although small, is nevertheless a nonzero value. Similarly, equation [3.18] and Figure 3.9 both indicate that the input impedance of an open-circuited line of length $l = \lambda/2$ is infinite, that is, that the line appears as if it is an open circuit. In practice, however, the input impedance of such a line is limited by its nonzero distributed conductance G. Open- or short-circuited lines of lengths equal to integer multiples of $\lambda/4$ or $\lambda/2$ are analogous to lumped resonant circuits, and such line segments constitute highly efficient resonators, as discussed in Section 3.9.

3.2.3 Open- and Short-Circuited Lines as Reactive Circuit Elements

An important application of transmission lines involves their use as capacitive or inductive tuning elements in microwave circuits at frequencies between a few gigahertz to a few tens of gigahertz. In this frequency range, lumped inductors and capacitors become exceedingly small and difficult to fabricate. Furthermore, the wavelength is small enough that the physical sizes and separation distances of ordinary circuit components are no longer negligible. On the other hand, transmission line sections of appropriate sizes can be constructed with relative ease. For frequencies higher than ~100 GHz, the physical size of transmission lines is too small, although novel transmission line implementations can operate[8] at frequencies as high as 500 GHz, corresponding to submillimeter wavelengths.

That transmission lines behave as reactive circuit elements is quite evident from Figures 3.6 and 3.9. Consider, for example, the input impedance as a function of frequency of a short-circuited line of length l such that $l = \lambda_0/4$. At a frequency of f_0, for which $\lambda = \lambda_0$, this line presents an infinite impedance (i.e., appears as an open circuit) at its input terminals. For frequencies slightly smaller than f_0, namely $f < f_0$ so that $\lambda > \lambda_0$, the length of the line is slightly shorter than $\lambda/4$, so it presents a very large inductive impedance (Figure 3.6). For frequencies slightly greater than f_0, the electrical length of the line is slightly larger than $\lambda/4$, so it has a very large capacitive impedance (Figure 3.6). Such behavior is similar to that of a lumped circuit consisting of a parallel combination of an inductor and a capacitor.

A similar analysis of a short-circuited line of length l such that $l = \lambda_0/2$ indicates that a half-wavelength line behaves as a lumped circuit consisting of a series combination of an inductor and a capacitor. As can be seen from Figure 3.6, the magnitude of the input impedance of a short-circuited line of length $l = \lambda_0/2$ is very small in the vicinity of its resonant frequency f_0, and the input impedance is inductive for $f > f_0$ and capacitive for $f < f_0$.

Corresponding observations can also be made for open-circuited line segments, for which the input impedance is given as a function of electrical length in Figure 3.9. Lumped circuit counterparts of various transmission line segments are summarized in Figure 3.10.

We see from the preceding discussion that short- or open-circuited transmission line elements act as resonant circuits. In the absence of losses, a short- or open-circuited line segment would store its electrical energy forever, even if the source were removed. When losses are taken into account, transmission line resonators consisting of short- or open-circuited lines of lengths equal to integer multiples of $\lambda/4$ or $\lambda/2$ are highly efficient energy storage devices and exhibit a high degree of frequency selectivity, as discussed in Section 3.9.

The use of a short-circuited transmission line segment as a microwave inductance is illustrated in Example 3-4.

Example 3-4: Transmission line inductor. A short-circuited coaxial line with $v_p = 2.07 \times 10^8$ m-s^{-1} is to be designed to provide a 15-nH inductance for a microwave filter operating at 3 GHz. (a) Find the shortest possible length l if the characteristic impedance of the line is $Z_0 = 50\Omega$, and (b) find the lumped element value of the short-circuited line designed in part (a) at 4 GHz.

[8] Linda P. B. Katehi, Novel transmission lines for the submillimeter-wave region, *Proceedings of the IEEE,* 80(11), p. 1771, November 1992.

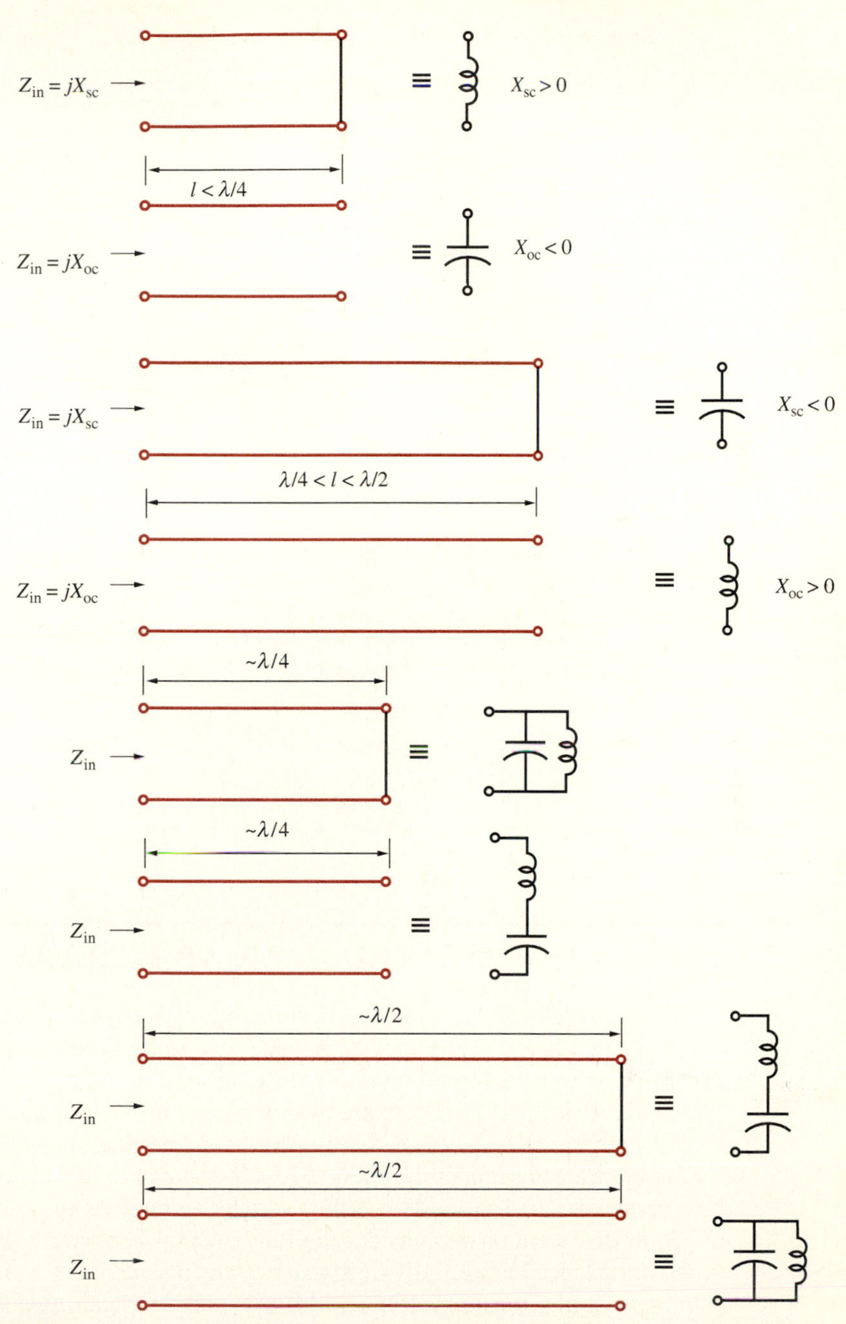

FIGURE 3.10. **Lumped circuit models of various open- and short-circuited line segments.**

Solution:

(a) Equating the input impedance of a short-circuited line of length l to the impedance of a lumped inductor, we have

$$Z_{sc} = jZ_0 \tan\left(\frac{2\pi}{\lambda}l\right) = j\omega L_{sc}$$

where $\lambda = v_p/f = (2.07 \times 10^{10})/(3 \times 10^9) = 6.9$ cm. For $Z_0 = 50\Omega$, we can write

$$l = \frac{6.9}{2\pi}\tan^{-1}\left(\frac{2\pi \times 3 \times 10^9 \times 15 \times 10^{-9}}{50}\right) \simeq 1.53 \text{ cm}$$

(b) At 4 GHz, $\lambda = 2.07 \times 10^{10}/(4 \times 10^9) = 5.175$ cm. Thus, the input impedance of the short-circuited 50Ω coaxial line of length $\sim$1.53 cm is

$$Z_{sc} \simeq j(50)\tan\left(\frac{2\pi \times 1.53}{5.175}\right) \simeq -j167.4\Omega$$

Therefore, at 4 GHz, the short-circuited 50Ω coaxial line designed in part (a) represents a lumped capacitor of element value given by

$$-\frac{j}{2\pi \times 4 \times 10^9 C} = -j167.4\Omega \quad \rightarrow \quad C_{sc} \simeq 0.238 \text{ pF}$$

The results are summarized in Figure 3.11.

3.3 LINES TERMINATED IN AN ARBITRARY IMPEDANCE

Most sinusoidal steady-state applications involve transmission lines terminated in arbitrary complex load impedances. The load to be driven may be an antenna, the feed-point impedance of which depends in a complicated manner on the antenna characteristics and operating frequency and is in general quite different from the characteristic impedance of the transmission line that connects it to a source. For efficient transmission of the energy from the source to the load, it is often necessary to match the load to the line, using various techniques to be discussed in Section 3.5. In this section, we consider the fundamental behavior of line voltage, current, and impedance for arbitrarily terminated transmission lines. Consider a transmission line of length l terminated in an arbitrary complex load impedance Z_L and excited by a sinusoidal voltage source, the phasor of which is represented by V_0, as shown in Figure 3.12. The line is uniform and lossless (i.e., $Z_0 =$ const., $R = 0$ and $G = 0$), so the voltage and current phasors at any position $z < 0$ along the line are in general given by

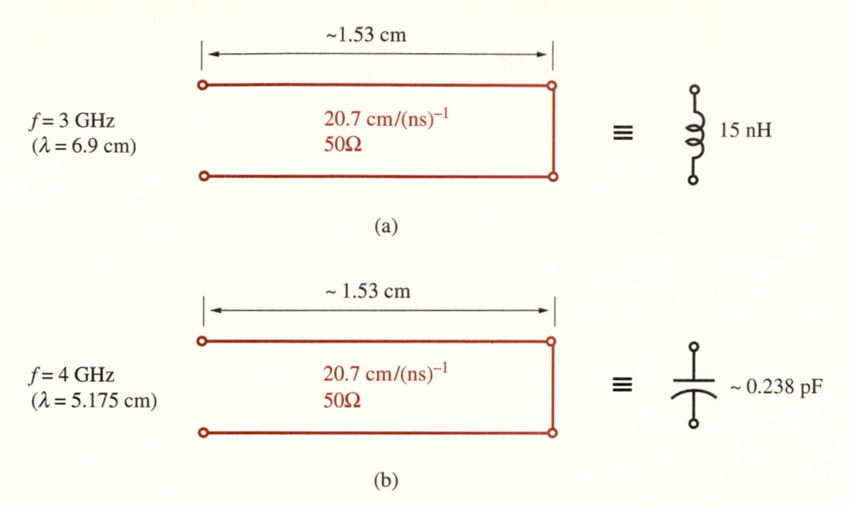

FIGURE 3.11. A short-circuited 50Ω line as an inductor or capacitor. (a) Designed to provide a 15 nH inductance at 3 GHz. (b) Equivalent to a ~ 0.238 pF capacitance at 4 GHz.

$$V(z) = V^+ e^{-j\beta z} + V^- e^{+j\beta z} \qquad [3.19]$$

$$I(z) = \frac{1}{Z_0}[V^+ e^{-j\beta z} - V^- e^{+j\beta z}] \qquad [3.20]$$

The boundary condition at the load end ($z = 0$) is simply

$$V_L = Z_L I_L \quad \rightarrow \quad V(z)|_{z=0} = Z_L I(z)|_{z=0}$$

or

$$Z_L = Z_0 \frac{V^+ + V^-}{V^+ - V^-}$$

The ratio of the phasors of the reverse and forward waves at the load position ($z = 0$) is the *load voltage reflection coefficient,* defined as $\Gamma_L \equiv V^-/V^+$ such that

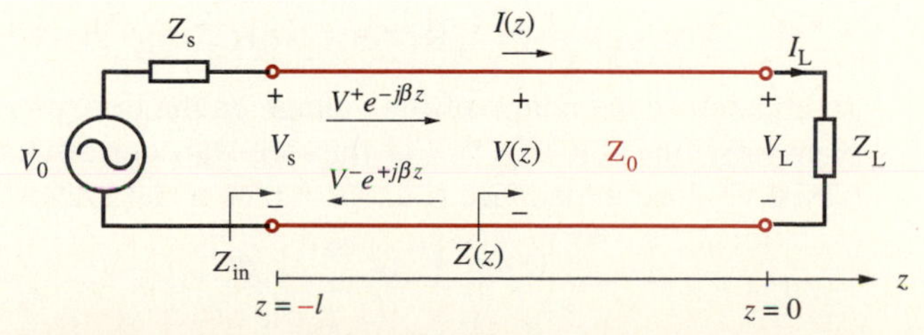

FIGURE 3.12. A terminated line. A lossless transmission line excited by a sinusoidal source terminated in a complex load impedance Z_L.

$$Z_L = Z_0 \frac{1+\Gamma_L}{1-\Gamma_L} \quad \rightarrow \quad \Gamma_L = \rho e^{j\psi} = \frac{Z_L - Z_0}{Z_L + Z_0} \qquad [3.21]$$

where we explicitly recognize that Γ_L is in general a complex number with a magnitude ρ and phase angle ψ, where $0 \leq \rho \leq 1$. Note that, as shown before, $\Gamma_L = -1$ (i.e., $V^- = -V^+$) for a short-circuited line, $\Gamma_L = +1$ (i.e., $V^- = V^+$) for an open-circuited line, and $\Gamma_L = 0$ (i.e., $V^- = 0$) for a matched load (i.e., $Z_L = Z_0$).

The voltage and current phasors given by equations [3.19] and [3.20] can now be written in terms of Γ_L as

$$V(z) = V^+(e^{-j\beta z} + \Gamma_L e^{j\beta z}) = V^+ e^{-j\beta z}[1 + \Gamma(z)] \qquad [3.22]$$

$$I(z) = \frac{V^+}{Z_0}(e^{-j\beta z} - \Gamma_L e^{j\beta z}) = \frac{V^+}{Z_0} e^{-j\beta z}[1 - \Gamma(z)] \qquad [3.23]$$

where

$$\Gamma(z) \equiv \frac{V^- e^{j\beta z}}{V^+ e^{-j\beta z}} = \Gamma_L e^{j2\beta z} = \rho e^{j(\psi + 2\beta z)}$$

is the *voltage reflection coefficient* at any position z along the line, defined as the ratio of the phasors of the reverse and forward propagating waves at that position. The voltage reflection coefficient is a complex number with a constant magnitude ρ (equal to the magnitude of Γ_L) and a phase angle varying with position z. We can view the quantity $\Gamma(z) \equiv \Gamma_L e^{j2\beta z}$ as a generalized reflection coefficient defined not only at the load but also at any point z along the line. Noting that the voltage along the line is given by

$$V(z) = \underbrace{V^+ e^{-j\beta z}}_{\text{forward wave}} + \underbrace{\Gamma_L V^+ e^{j\beta z}}_{\text{reflected wave}}$$

we can see that the quantity $\Gamma(z) = \Gamma_L e^{j2\beta z}$ is indeed the ratio of the reflected wave at point z to the forward wave at that same position.

Note that the line impedance seen looking toward the load at any position z along the line can be written in terms of $\Gamma(z)$ as

$$Z(z) = \frac{V(z)}{I(z)} = Z_0 \frac{1 + \Gamma(z)}{1 - \Gamma(z)} \qquad [3.24]$$

3.3.1 Voltage and Current Standing-Wave Patterns

To understand the nature of the voltage on the line, it is useful to examine the complete time function $\mathcal{V}(z, t)$. For this, we start our discussion with the case of a real (resistive) load impedance (i.e., $\psi = 0$ or π) and rewrite [3.22] as

$$V(z) = V^+[e^{-j\beta z} \pm \rho e^{j\beta z}]$$
$$= V^+[(1 \pm \rho)e^{-j\beta z} \pm \rho(-e^{-j\beta z} + e^{j\beta z})]$$

which, by using $\mathcal{V}(z, t) = \Re e\{V(z)e^{j\omega t}\}$, gives

$$\mathcal{V}(z,t) = \underbrace{|V^+|(1 \pm \rho)\cos(\omega t - \beta z + \phi^+)}_{\text{propagating wave}} \pm \underbrace{|V^+|(2\rho)\sin(\beta z)\cos\left(\omega t + \phi^+ + \frac{\pi}{2}\right)}_{\text{standing wave}}$$

where we have taken $V^+ = |V^+|e^{j\phi^+}$. In other words, the voltage on the line consists of a standing wave plus a propagating wave.

According to [3.22], the magnitude of the voltage phasor (i.e., $|V(z)|$) alternates between the maximum and minimum values of V_{max} and V_{min} given by

$$V_{\text{max}} = |V(z)|_{\text{max}} = |V^+|(1 + |\Gamma_{\text{L}}|) = |V^+|(1 + \rho)$$
$$V_{\text{min}} = |V(z)|_{\text{min}} = |V^+|(1 - |\Gamma_{\text{L}}|) = |V^+|(1 - \rho)$$

Similarly, from Equation [3.23], the magnitude of the current phasor (i.e., $|I(z)|$) alternates between the maximum and minimum values of

$$I_{\text{max}} = |I(z)|_{\text{max}} = \frac{|V^+|}{Z_0}(1 + \rho)$$

$$I_{\text{min}} = |I(z)|_{\text{min}} = \frac{|V^+|}{Z_0}(1 - \rho)$$

where I_{max} occurs at the same position as V_{min}, and I_{min} occurs at the same position as V_{max}. For example, Figure 3.13 shows the variations of both voltage and current magnitudes (represented by $|V(z)|$ and $Z_0|I(z)|$) as functions of position with respect to wavelength along the line, for the case of a purely resistive load, with $Z_{\text{L}} = R_{\text{L}} = 2Z_0$. As is apparent from Figure 3.13, the distance between successive voltage maxima (or minima) is $\lambda/2$. Note that for the case shown, with $Z_{\text{L}} = R_{\text{L}} > Z_0$, the reflection coefficient Γ_{L} is purely real with $\psi = 0$ and $0 \leq \rho \leq 1$.

The following example illustrates the concepts of reflection coefficient and standing-wave pattern.

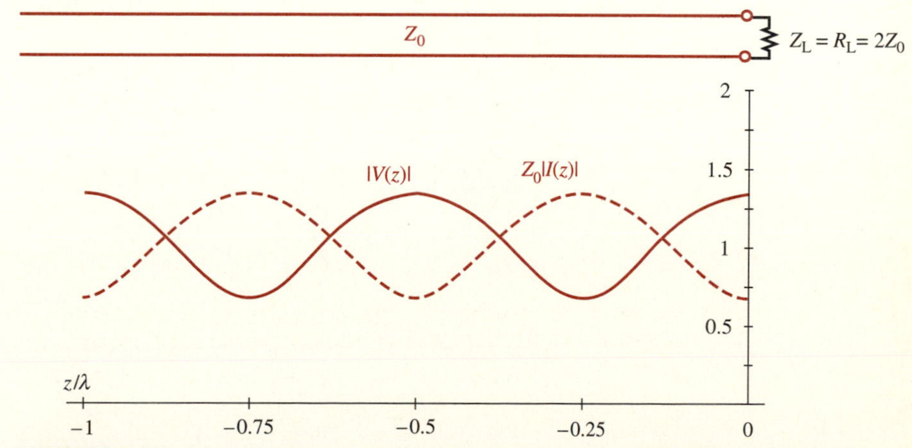

FIGURE 3.13. **Standing-wave patterns for $Z_{\text{L}} = R_{\text{L}} = 2Z_0$.** Magnitudes of the voltage and current phasors (i.e., $|V(z)|$ and $Z_0|I(z)|$) are shown as functions of electrical distance z/λ away from the load, for $V^+ = 1$.

Example 3-5: A Yagi antenna array driven by a coaxial line. To increase the geographic coverage area of a broadcast station, four Yagi antennas, each having a feed-point impedance of 50Ω, are stacked in parallel on a single antenna tower and connected to the transmitter by a 50Ω coaxial line, as shown in Figure 3.14a. (a) Calculate the load reflection coefficient Γ_L. (b) Calculate $V_{\max}$, $V_{\min}$, $I_{\max}$, and $I_{\min}$ along the line, assuming $V^+ = 1$ V. (c) Sketch $|V(z)|$ and $|I(z)|$ as functions of z/λ, taking the position of the antenna array terminals to be at $z = 0$.

Solution:

(a) The total load impedance seen by the coaxial line is a parallel combination of the four 50Ω impedances, resulting in

$$Z_L = \frac{50}{4} = 12.5\Omega$$

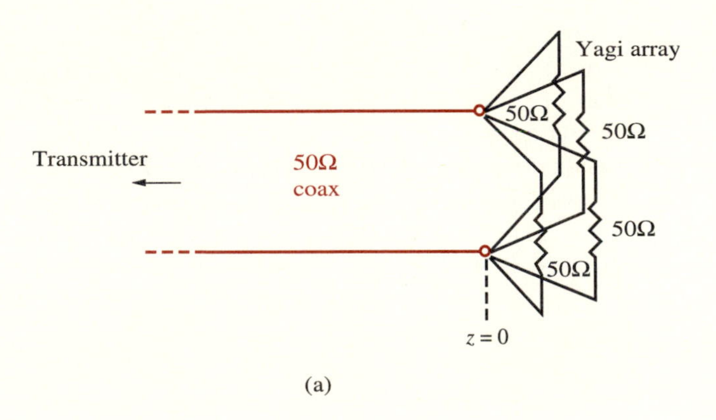

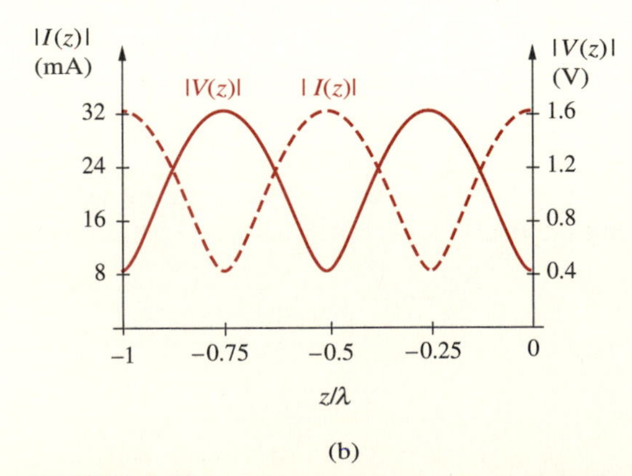

FIGURE 3.14. Yagi array driven by a coaxial line. (a) Array of a stack of four Yagi antennas fed by a coaxial line. (b) Voltage and current standing-wave patterns.

The load reflection coefficient is then given by

$$\Gamma_L = \frac{Z_L - Z_0}{Z_L + Z_0} = \frac{12.5 - 50}{12.5 + 50} = -0.6 = 0.6e^{j180°}$$

so that $\rho = 0.6$ and $\psi = 180°$.

(b) $V_{max} = |V^+|(1 + \rho) = 1.6$ V $\qquad V_{min} = |V^+|(1 - \rho) = 0.4$ V

$$I_{max} = \frac{|V^+|}{Z_0}(1 + \rho) = 32 \text{ mA} \qquad I_{min} = \frac{|V^+|}{Z_0}(1 - \rho) = 8 \text{ mA}$$

(c) The voltage reflection coefficient at any position z along the line is given by

$$\Gamma(z) = \rho e^{j(\psi + 2\beta z)} = 0.6e^{j\pi(1+4z/\lambda)}$$

Using [3.22] and [3.23] with $V^+ = 1$ V, the magnitudes of the line voltage and current (i.e., $|V(z)|$ and $|I(z)|$) are plotted in Figure 3.14b as functions of electrical distance z/λ.

Standing-wave patterns such as those in Figure 3.13 are important in practice because, although the rapid temporal variations of the line voltages and currents are not easily accessible, the locations of the voltage minima and maxima and the ratio of the voltage maxima to minima are often readily measurable. A key parameter that is commonly used to describe the termination of a transmission line is the standing-wave ratio (SWR), or S, defined as

$$S = \frac{V_{max}}{V_{min}} = \frac{I_{max}}{I_{min}} = \frac{1 + \rho}{1 - \rho} \longrightarrow \rho = \frac{S - 1}{S + 1} \qquad [3.25]$$

Note that S varies in the range $1 \leq S \leq \infty$.

The following example illustrates the concepts of reflection coefficient and standing-wave ratio for a UHF antenna.

Example 3-6: UHF blade antenna. A UHF blade antenna installed in the tail-cap of a small aircraft is used for communication over the frequency band 225 MHz to 400 MHz. The following table provides the measured values of the feed-point impedance of the antenna at various frequencies:[9]

f(MHz)	$Z_L(\Omega)$
225	$22.5 - j51$
300	$35 - j16$
400	$45 - j2.5$

[9]R. L. Thomas, *A Practical Introduction to Impedance Matching*, Artech House, Inc., 1976.

A 50Ω coaxial line is used to connect the communication unit to the antenna. Calculate the load reflection coefficient Γ_L and the standing-wave ratio S on the line at (a) 225 MHz, (b) 300 MHz, and (c) 400 MHz.

Solution:

(a) At 225 MHz, the load reflection coefficient is given by

$$\Gamma_L = \frac{Z_L - Z_0}{Z_L + Z_0} = \frac{22.5 - j51 - 50}{22.5 - j51 + 50} \simeq \frac{57.9e^{-j118°}}{88.6e^{-j35.1°}} \simeq 0.654e^{-j83.2°}$$

The standing-wave ratio S at 225 MHz can then be obtained as follows:

$$S = \frac{1+\rho}{1-\rho} \simeq \frac{1+0.654}{1-0.654} \simeq 4.78$$

(b) Similarly, at 300 MHz, we have

$$\Gamma_L = \frac{35 - j16 - 50}{35 - j16 + 50} \simeq \frac{21.9e^{-j133°}}{86.5e^{-j10.7°}} \simeq 0.254e^{-j122°}$$

and

$$S \simeq \frac{1+0.254}{1-0.254} \simeq 1.68$$

(c) Similarly, at 400 MHz, we have

$$\Gamma_L = \frac{45 - j2.5 - 50}{45 - j2.5 + 50} \simeq \frac{5.59e^{-j153°}}{95e^{-j1.51°}} \simeq 0.0588e^{-j152°}$$

and

$$S \simeq \frac{1+0.0588}{1-0.0588} \simeq 1.13$$

We see that the reflections on the coaxial line are quite significant near 225 MHz but are much reduced near 400 MHz.

Another quantity that can sometimes be measured in experimental settings is $z_{\min}$, or the distance from the load to the *first* minimum of the voltage standing-wave pattern.[10] From [3.22] we have

$$V(z) = V^+(e^{-j\beta z} + \Gamma_L e^{j\beta z}) = V^+ e^{-j\beta z}(1 + \rho e^{j\psi} e^{j2\beta z})$$

Since $|V(z)| = V_{\min}$ when $e^{j(\psi + 2\beta z)} = -1$, or

$$\psi + 2\beta z_{\min} = -(2m+1)\pi \qquad m = 0, 1, 2, 3, \ldots$$

---

[10]In practice, it may often be difficult to actually measure the first minimum; however, if the location of *any* of the minima can be measured, the location of the first minimum can be deduced by using the fact that successive minima are separated by λ/2.

where $-\pi \leq \psi < \pi$, and $z_{\min} \leq 0$. For any given frequency, measuring the wavelength (by measuring the distance between successive minima) provides a means to determine the phase velocity $v_p = f\lambda$.

At the location of the first minimum we have

$$\psi + 2\beta z_{\min} = -\pi \quad \longrightarrow \quad \psi = -\pi - 2\beta z_{\min} \qquad [3.26]$$

We see that $z_{\min}$ is directly related to the phase ψ of the reflection coefficient Γ_L, whereas S determines its magnitude through [3.25]. Once Γ_L is known, the load can be fully specified (assuming the characteristic impedance Z_0 is known), or Z_0 can be found (if Z_L is known). Thus, the two measurable quantities, S and $z_{\min}$, completely characterize the transmission line terminated in an arbitrary load impedance.

It is often useful to rewrite [3.22] and [3.23] in terms of the load voltage V_L and load current I_L. Using the fact that $V_L = V(z)|_{z=0}$ and $I_L = I(z)|_{z=0}$, and after some manipulation, we have

$$V(z) = V_L \cos(\beta z) - jI_L Z_0 \sin(\beta z) \qquad [3.27]$$

$$I(z) = I_L \cos(\beta z) - j\frac{V_L}{Z_0} \sin(\beta z) \qquad [3.28]$$

The voltages and currents at the source end ($z = -l$) can be found from [3.27] and [3.28] by substituting $z = -l$. Note that for a general complex load impedance Z_L, the load voltage V_L and current I_L are in general complex, so that equations [3.27] and [3.28] do not necessarily constitute a decomposition of $V(z)$ and $I(z)$ into their real and imaginary parts.

In general, the voltage and current standing-wave patterns on a terminated line depend on the nature of the load. Typically what is plotted is $|V(z)|$, as was shown in Figure 3.13 for a purely resistive load $R_L = 2Z_0$. In the general case, with a complex load Z_L, the reflection coefficient Γ_L is complex, with $\psi \neq 0$. From [3.22] we have

$$V(z) = V^+[\cos(\beta z) - j\sin(\beta z) + \rho\cos(\psi + \beta z) + j\rho\sin(\psi + \beta z)]$$

and

$$|V(z)| = |V^+|\sqrt{[\cos(\beta z) + \rho\cos(\psi + \beta z)]^2 + [-\sin(\beta z) + \rho\sin(\psi + \beta z)]^2} \qquad [3.29]$$

which is the quantity plotted in various figures as the voltage standing-wave pattern. For $\psi = 0$ or π (i.e., load is purely resistive) [3.29] reduces to

$$|V(z)| = |V^+|\sqrt{(1 \pm \rho)^2\cos^2(\beta z) + (-1 \pm \rho)^2\sin^2(\beta z)} \qquad [3.30]$$

where the lower signs correspond to the case for $\psi = \pi$. Similar expressions can also be written for $|I(z)|$. Voltage and current standing wave patterns for different types of load impedances are shown in Figures 3.15 and 3.16. The interpretation of some of these patterns will become clearer after the discussion of line impedance in the following subsection.

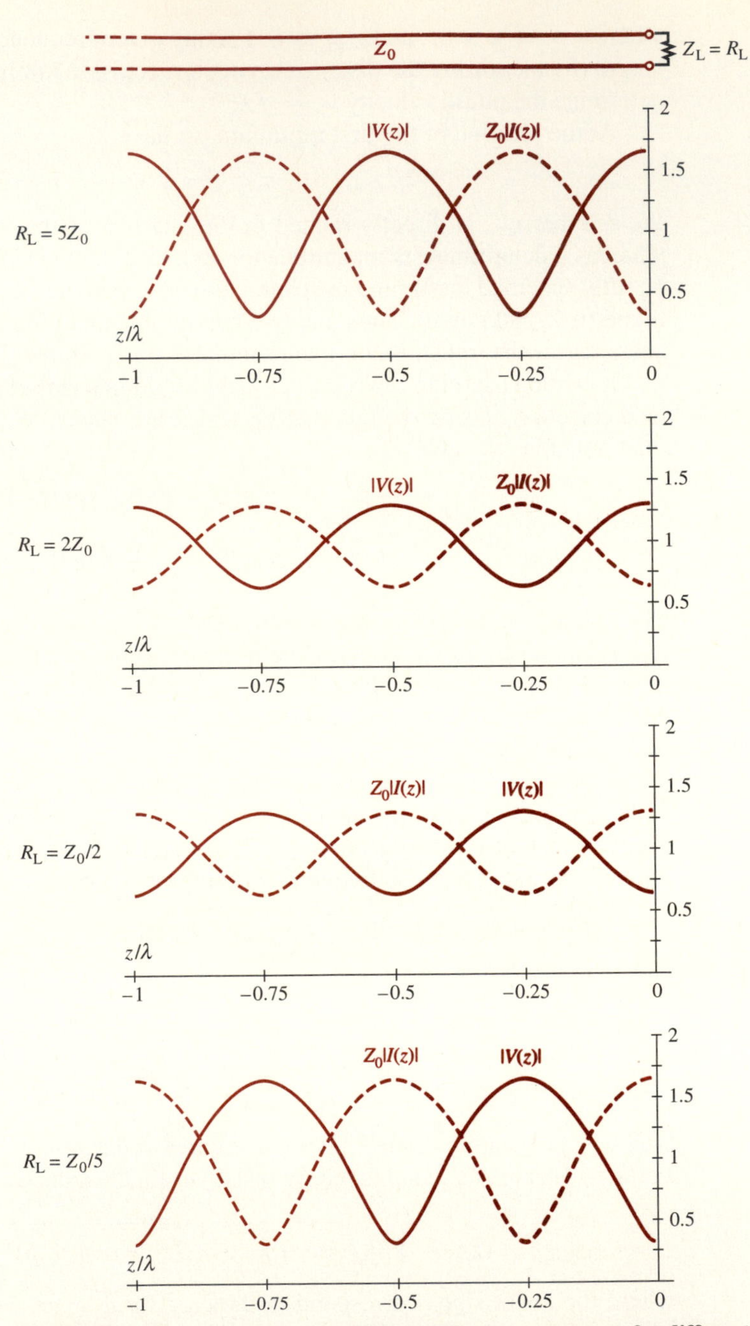

FIGURE 3.15. Voltage and current standing-wave patterns for different purely resistive loads. Magnitudes of the voltage and current phasors (i.e., $|V(z)|$ and $Z_0|I(z)|$) are shown as functions of electrical distance from the load position z/λ, for $V^+ = 1$ and for $R_L = 5Z_0, 2Z_0, Z_0/2$, and $Z_0/5$.

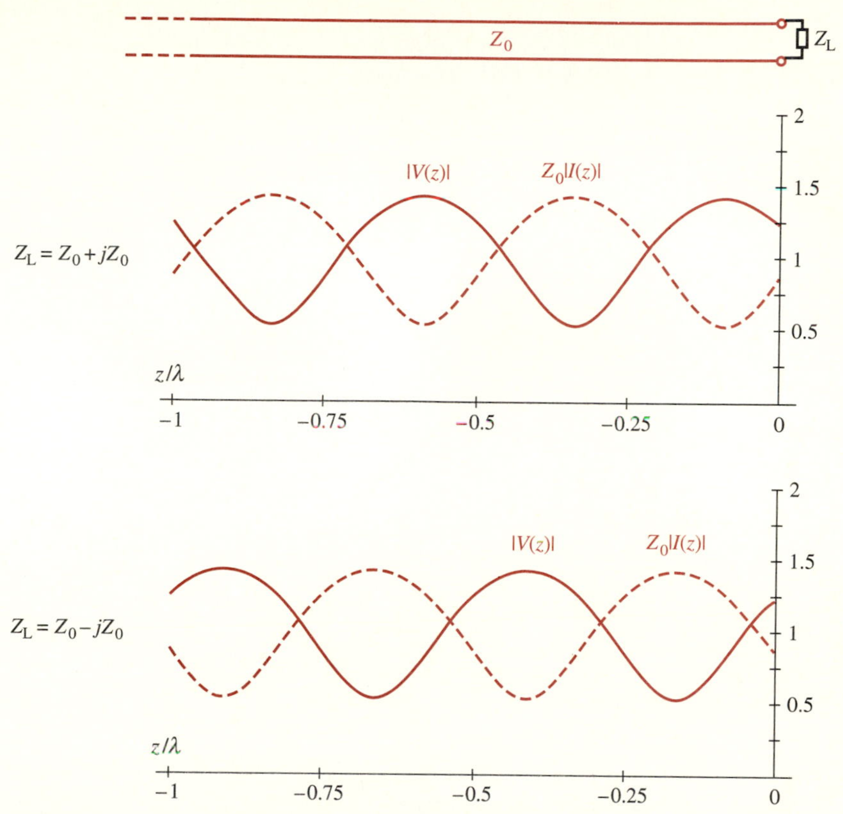

FIGURE 3.16. Voltage and current standing-wave patterns for complex loads. Magnitudes of the voltage and current phasors (i.e., $|V(z)|$ and $Z_0|I(z)|$) are shown as functions of electrical distance from the load z/λ for $V^+ = 1$ and for $Z_L = Z_0 + jZ_0$ (inductive load) and $Z_L = Z_0 - jZ_0$ (capacitive load).

In general, for purely resistive loads ($Z_L = R_L + j0$), the load position is a point of a voltage maximum or minimum, depending on whether $R_L > Z_0$ or $R_L < Z_0$, respectively. This behavior is apparent from Figure 3.15 and can also be seen by considering [3.22] and [3.23]. For $R_L > Z_0$, $0 < \Gamma_L \leq 1$ and $|V(z = 0)| = V_{max} = |V^+|(1 + \rho)$, whereas for $R_L < Z_0$, $-1 \leq \Gamma_L < 0$ and $|V(z = 0)| = V_{min} = |V^+|(1 - \rho)$.

The standing-wave patterns in Figure 3.16 for $Z_L = Z_0 \pm jZ_0$ illustrate specific cases of the general behavior for loads with a reactive (capacitive or inductive) component. In general, the sign of the reactance (positive or negative) can be determined by inspection of the voltage standing-wave pattern. For $Z_L = R_L + jX_L$, X_L is negative (i.e., the load is capacitive) when the first minimum is at a distance smaller than one quarter of wavelength from the load (i.e., $-z_{min} < \lambda/4$) and X_L is positive (inductive) when the first minimum is at a distance greater than one quarter of a wavelength from the load (i.e., $\lambda/4 < -z_{min} < \lambda/2$), as illustrated in Figures 3.17a,b.

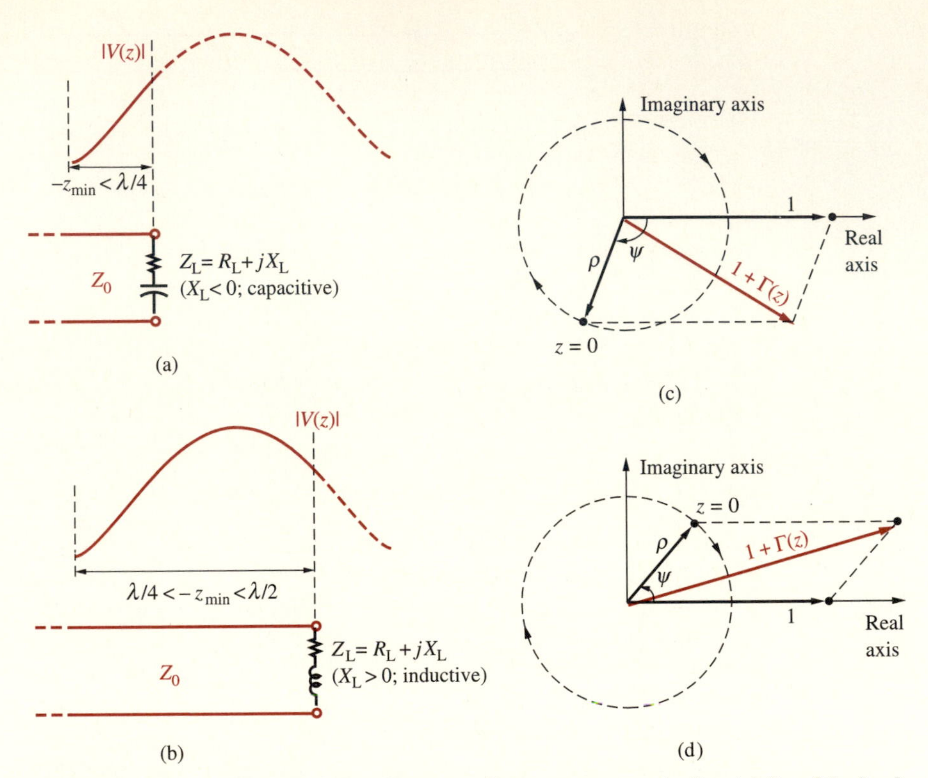

FIGURE 3.17. Variation of the voltage standing-wave pattern in the vicinity of the load for inductive or capacitive loads. In descriptive terms, starting from the load (i.e., $z = 0$), the standing-wave voltage at first increases (decreases) as one moves away from the load (i.e., clockwise in the diagrams shown) for an inductive (capacitive) load. (a) $|V(z)|$ for a capacitive load. (b) $|V(z)|$ for an inductive load. (c) $|V(z)| = |1 + \Gamma(z)|$ as the sum of two complex numbers, 1 and $\Gamma(z)$, for a capacitive load ($-\pi < \psi < 0$). (d) $|V(z)| = |1 + \Gamma(z)|$ as the sum of two complex numbers, 1 and $\Gamma(z)$, for an inductive load ($0 < \psi < \pi$). Note once again that $V^+ = 1$.

The behavior illustrated in Figures 3.17a,b can be understood upon careful examination of [3.21] and [3.22]. For a general complex load impedance, [3.21] can be rewritten as

$$\Gamma_L = \rho e^{j\psi} = \frac{R_L + jX_L - Z_0}{R_L + jX_L + Z_0} = \frac{(R_L - Z_0) + jX_L}{(R_L + Z_0) + jX_L}$$

The phase angle ψ of the reflection coefficient is $0 < \psi < \pi$ if the load impedance is inductive ($X_L > 0$) and $-\pi < \psi < 0$ if the load impedance is capacitive ($X_L < 0$). The magnitude of the voltage along the line can be written from [3.22] as

$$|V(z)| = |V^+||1 + \Gamma(z)| = |V^+||1 + \rho e^{j(\psi + 2\beta z)}|$$

where $z \leq 0$. Noting that $|V^+|$ is a constant, consider the second term and its variation with z (note that z decreases as one moves away from the load (at $z = 0$) along the transmission line). This term is the magnitude of the sum of two numbers, one being the real number 1 and the other being a complex number, $\Gamma(z) = \rho e^{j(\psi+2\beta z)}$, which has a constant magnitude ρ ($0 \leq \rho \leq 1$, as determined by Z_L and Z_0) and a phase angle $\psi + 2\beta z$ that decreases with decreasing z (corresponding to clockwise rotation of this complex number on a circle with radius ρ centered at the origin on the complex plane). The two different cases of capacitive and inductive load are shown respectively in Figures 3.17c and 3.17d. For an inductive load, we see from Figure 3.17d that as we move away from the load (i.e., starting at $z = 0$ and rotating clockwise), $|V(z)|$ (which is proportional to $|1 + \Gamma(z)|$) first increases, reaches a maximum (at $\psi + 2\beta z = 0$), and then decreases, consistent with the variation of $|V(z)|$ for the inductive load as shown in Figure 3.17b. Similarly, for a capacitive load, we see from Figure 3.17c that as z decreases (starting with $z = 0$), $|V(z)|$ first decreases, reaches a minimum (at $\psi + 2\beta z = -\pi$), and then increases, consistent with Figure 3.17a.

3.3.2 Transmission Line Impedance

An important property of a transmission line is its ability to transform impedances. In Section 3.2, we saw that the input impedance of a short- or open-circuited transmission line segment can be made equal to any arbitrary reactive impedance by simply adjusting its electrical length (l/λ). The input impedance of a transmission line terminated in an arbitrary load impedance Z_L is similarly dependent on the electrical length of the line, or the distance from the load at which the impedance is measured. For the transmission line shown in Figure 3.12, the impedance seen looking toward the load Z_L at any position z along the line ($-l \leq z \leq 0$) given by [3.24] can be rewritten as

$$Z(z) = \frac{V(z)}{I(z)} = Z_0 \frac{e^{-j\beta z} + \Gamma_L e^{j\beta z}}{e^{-j\beta z} - \Gamma_L e^{j\beta z}} = Z_0 \frac{Z_L - jZ_0 \tan(\beta z)}{Z_0 - jZ_L \tan(\beta z)} \quad [3.31]$$

using [3.21], [3.22], and [3.23]. Expression [3.31] for the line impedance is often written as

$$Z(z) = Z_0 \frac{Z_L \cos(\beta z) - jZ_0 \sin(\beta z)}{Z_0 \cos(\beta z) - jZ_L \sin(\beta z)}$$

Note that since $\beta = 2\pi/\lambda$, the impedance varies periodically with electrical distance (z/λ) along the line, with the same impedance value attained at intervals in z of $\pm\lambda/2$. At the load, where $z = 0$, we have

$$Z(z = 0) = Z_0 \frac{1 + \Gamma_L}{1 - \Gamma_L} = Z_L$$

as expected. In particular, the input impedance seen by the source at the source end, where $z = -l$, is

$$Z_{in} = [Z(z)]_{z=-l} = \frac{V_s}{I_s} = Z_0 \frac{Z_L + jZ_0 \tan(\beta l)}{Z_0 + jZ_L \tan(\beta l)} \quad [3.32]$$

For example, for a short-circuited line ($Z_L = 0$), the input impedance is

$$Z_{in} = jZ_0 \tan(\beta l)$$

and for an open-circuited line ($Z_L = \infty$), it is

$$Z_{in} = -jZ_0 \cot(\beta l)$$

as was shown in Sections 3.2.1 and 3.2.2.

The following example illustrates the dependence of the input impedance of a line on its electrical length.

Example 3-7: Input impedance of a line. Find the input impedance of a 75-cm long transmission line where $Z_0 = 70\Omega$, terminated with a $Z_L = 140\Omega$ load at 50, 100, 150, and 200 MHz. Assume the phase velocity v_p to be equal to the speed of light in free space.

Solution:

(a) At $f = 50$ MHz, we have $\lambda \simeq (3 \times 10^8)/(5 \times 10^7) = 6$ m, so the electrical length of the line is $l/\lambda \simeq 0.75/6 = 0.125$. Noting that $\beta l = 2\pi l/\lambda$, we then have, from [3.32],

$$Z_{in} \simeq 70\frac{140 + j70\tan(2\pi \times 0.125)}{70 + j140\tan(2\pi \times 0.125)} = 70\frac{(140 + j70)(70 - j140)}{(70)^2 + (140)^2}$$
$$= \frac{(2 + j)(70 - j140)}{5} = \frac{280 - j210}{5} = 56 - j42\Omega$$

since $\tan(2\pi \times 0.125) = 1$. Note that the input impedance of the line at 50 MHz is capacitive.

(b) At $f = 100$ MHz, we have $\lambda \simeq 3$ m and $l/\lambda \simeq 0.75/3 = 0.25$. From [3.32] we have

$$Z_{in} \simeq 70\frac{140 + j70\tan(2\pi \times 0.25)}{70 + j140\tan(2\pi \times 0.25)} = \frac{(70)^2}{140} = 35\Omega$$

since $\tan(2\pi \times 0.25) = \infty$. Note that the input impedance of the line at 100 MHz is purely resistive.

(c) At $f = 150$ MHz, $\lambda \simeq 2$ m and $l/\lambda \simeq 0.75/2 = 0.375$. We have from [3.32]

$$Z_{in} \simeq 70\frac{140 + j70\tan(2\pi \times 0.375)}{70 + j140\tan(2\pi \times 0.375)} = 70\frac{(140 - j70)(70 + j140)}{(70)^2 + (140)^2}$$
$$= \frac{(2 - j)(70 + j140)}{5} = 56 + j42\Omega$$

since $\tan(2\pi \times 0.375) = -1$. Note that the input impedance of the line at 150 MHz is inductive.

(d) At $f = 200$ MHz, we have $\lambda \simeq 1.5$ m and $l/\lambda \simeq 0.75/1.5 = 0.5$. We have from [3.32]

$$Z_{\text{in}} \simeq 70\frac{140 + j70\tan(2\pi \times 0.5)}{70 + j140\tan(2\pi \times 0.5)} = \frac{70 \times 140}{70} = 140\Omega$$

since $\tan(2\pi \times 0.5) = 0$. Note that the input impedance of the line at 200 MHz is purely resistive and is exactly equal to the load impedance.

Normalized Line Impedance In transmission line analysis, it is often convenient and common practice to normalize all impedances to the characteristic impedance Z_0 of the transmission line. Denoting the normalized version of any impedance by using a bar at the top, we can rewrite [3.31] to express the normalized line impedance $\bar{Z}(z)$ in terms of the normalized load impedance $\bar{Z}_{\text{L}}$

$$\bar{Z}(z) = \frac{\bar{Z}_{\text{L}} - j\tan(\beta z)}{1 - j\bar{Z}_{\text{L}}\tan(\beta z)} = \frac{\bar{Z}_{\text{L}}\cos(\beta z) - j\sin(\beta z)}{\cos(\beta z) - j\bar{Z}_{\text{L}}\sin(\beta z)} \qquad [3.33]$$

where

$$\bar{Z}_{\text{L}} = \frac{Z_{\text{L}}}{Z_0} = \frac{1 + \Gamma_{\text{L}}}{1 - \Gamma_{\text{L}}} = \frac{1 + \rho e^{j\psi}}{1 - \rho e^{j\psi}}$$

Using [3.25] and [3.26], we can further write

$$\bar{Z}_{\text{L}} = \frac{Z_{\text{L}}}{Z_0} = \frac{1 + jS\tan(\beta z_{\min})}{S + j\tan(\beta z_{\min})} = \frac{\cos(\beta z_{\min}) + jS\sin(\beta z_{\min})}{S\cos(\beta z_{\min}) + j\sin(\beta z_{\min})}$$

which expresses the normalized load impedance $\bar{Z}_{\text{L}}$ in terms of the measurable quantities S and $z_{\min}$.

Sometimes it is useful to express the real and imaginary parts of the load impedance $Z_{\text{L}} = R_{\text{L}} + jX_{\text{L}}$ explicitly in terms of $z_{\min}$ and S:

$$\bar{R}_{\text{L}} = \frac{R_{\text{L}}}{Z_0} = \frac{S}{S^2\cos^2(\beta z_{\min}) + \sin^2(\beta z_{\min})}$$

$$\bar{X}_{\text{L}} = \frac{X_{\text{L}}}{Z_0} = \frac{(S^2 - 1)\cos(\beta z_{\min})\sin(\beta z_{\min})}{S^2\cos^2(\beta z_{\min}) + \sin^2(\beta z_{\min})}$$

The relationship between the polarity of X_{L} (i.e., inductive versus capacitive) and the distance to the first minimum, as depicted in Figure 3.17, can also be deduced by careful consideration of the preceding equation for $\bar{X}_{\text{L}}$.

Example 3-8 illustrates the determination of an unknown load impedance from measurements of S and $z_{\min}$.

Example 3-8: Unknown load. Determine an unknown load Z_{L} from S and $z_{\min}$ measurements. The following measurements are carried out on a 100Ω

FIGURE 3.18. Transmission line terminated in an unknown impedance.

transmission line terminated with an unknown load Z_L, as shown in Figure 3.18. The voltage standing-wave ratio S is 5, the distance between successive voltage minima is 25 cm, and the distance from Z_L to the first voltage minimum is 8 cm. (a) Determine the load reflection coefficient Γ_L. (b) Determine the unknown load impedance Z_L. (c) Determine the location of the first voltage maximum with respect to the load.

Solution:

(a) Using [3.25] and [3.26], we have

$$\rho = \frac{S-1}{S+1} = \frac{5-1}{5+1} \simeq 0.667$$

$$\frac{\lambda}{2} = 25 \text{ cm} \quad \rightarrow \quad \lambda = 50 \text{ cm}$$

$$\psi = -\pi - 2\beta z_{\text{min}} = -\pi + 2\left(\frac{2\pi}{50}\right)(8) = -0.36\pi \text{ rad or } -64.8^\circ$$

$$\Gamma_L = \rho e^{j\psi} \simeq 0.667 e^{-j64.8^\circ}$$

(b) Using the expression derived previously for Z_L in terms of z_{min} and S, we have

$$(\beta z_{\text{min}}) = \frac{-\pi + 0.36\pi}{2} = -0.32\pi \text{ rad} \quad \rightarrow \quad \tan(\beta z_{\text{min}}) \simeq -1.58$$

and so

$$Z_L = Z_0 \frac{1 + jS\tan(\beta z_{\text{min}})}{S + j\tan(\beta z_{\text{min}})} \simeq 100\frac{1 - j5(1.58)}{(5 - j1.58)}$$

$$= 100\frac{(1 - j7.88)(5 + j1.58)}{(5 - j1.58)(5 + j1.58)} \simeq 63.4 - j137.6\Omega$$

(c) The location of the first voltage maximum is $\lambda/4$ away from the location of the voltage minimum. Thus, we have

$$z_{\text{max}} = z_{\text{min}} - \lambda/4 = -8 - 12.5 = -20.5 \text{ cm}$$

Transmission Line Admittance In Sections 3.5 and 3.6, when we discuss impedance matching and the Smith chart, it will be useful at times to work with the line *admittance* rather than the impedance. From [3.31], we can find the expression for line admittance as

$$Y(z) = \frac{1}{Z(z)} = Y_0 \frac{1 - \Gamma_L e^{+j2\beta z}}{1 + \Gamma_L e^{+j2\beta z}} = Y_0 \frac{Y_L - jY_0 \tan(\beta z)}{Y_0 - jY_L \tan(\beta z)} \qquad [3.34]$$

where $Y_0 = (Z_0)^{-1}$. Using [3.33], the normalized line admittance $\bar{Y}(z) = Y(z)/Y_0$ can be written as

$$\bar{Y}(z) = \frac{\bar{Y}_L - j\tan(\beta z)}{1 - j\bar{Y}_L \tan(\beta z)} = \frac{\bar{Y}_L \cos(\beta z) - j\sin(\beta z)}{\cos(\beta z) - j\bar{Y}_L \sin(\beta z)} \quad [3.35]$$

The load reflection coefficient Γ_L can also be written in terms of admittances as

$$\Gamma_L = \frac{Y_0 - Y_L}{Y_0 + Y_L}$$

Line Impedance for Resistive Loads The variation with z of the real and imaginary parts of the normalized line impedance is illustrated in Figure 3.19a for the case of a resistive load with $Z_L = R_L = 2Z_0$. The voltage and current standing-wave

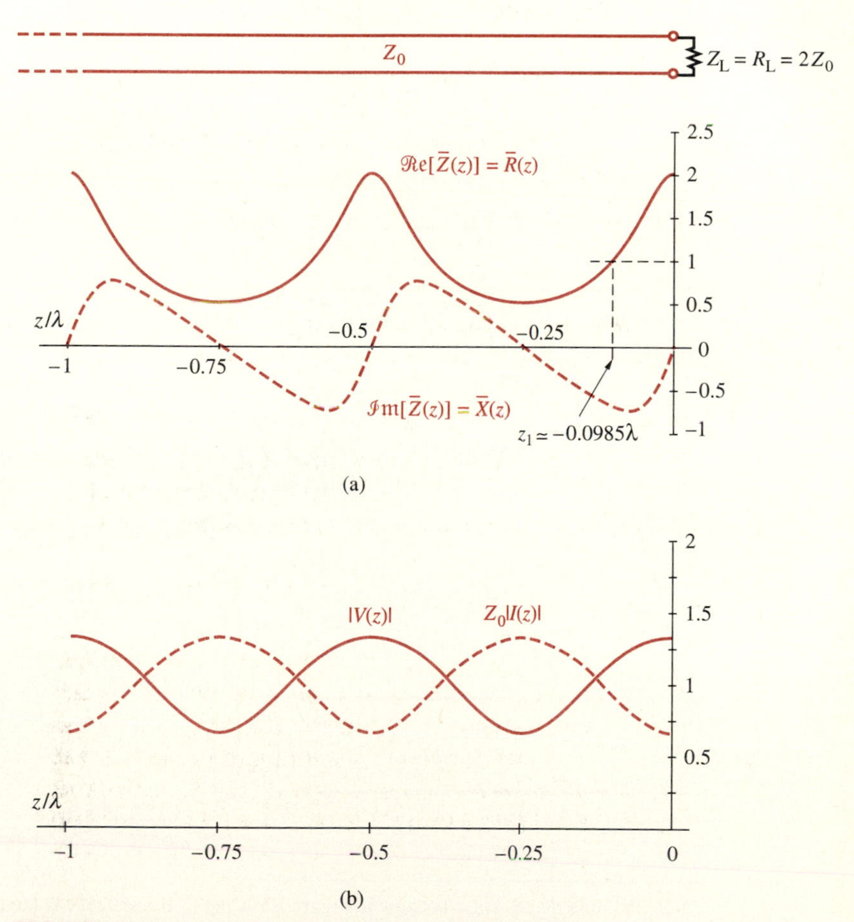

FIGURE 3.19. **Impedance along a line terminated with $Z_L = R_L = 2Z_0$.** (a) The real and imaginary parts of the normalized line impedance $\bar{Z}(z)$ are shown as functions of z/λ. (b) Magnitudes of the voltage and current phasors (i.e., $|V(z)|$ and $Z_0|I(z)|$) are shown as functions of electrical distance from the load z/λ, for $V^+ = 1$.

patterns (from Figure 3.13) are also shown for reference in Figure 3.19b. Note from Figure 3.19a that, as viewed from different positions at a distance z from the load, the real part of the normalized line impedance varies between $\bar{R}(z) = 2$ and $\bar{R}(z) = 0.5$, with the distance between successive maxima being $\lambda/2$. The line impedance is purely real at the load and at distances of integer multiples of $\lambda/4$ from the load. These positions also correspond to the positions of voltage maxima and minima along the line. For $-z < \lambda/4$, the line impedance $Z(z)$ is capacitive (i.e., its imaginary part is negative; $X(z) < 0$), reminiscent of the behavior of the open-circuited line[11] (see Figure 3.9). For $\lambda/4 < -z < \lambda/2$, the line impedance $Z(z)$ is inductive ($X(z) > 0$), switching thereafter back and forth between being inductive and capacitive at intervals of $\lambda/4$.

An interesting aspect of the result in Figure 3.19a is the fact that $\Re e\{\bar{Z}(z)\} = 1$ at $z_1 \simeq -0.0985\lambda$.[12] If the imaginary part of the line impedance at that position could somehow be canceled (as we shall see in Section 3.5), the line would appear (from all positions at locations $z < -0.0985\lambda$) as if it were matched (i.e., terminated with an impedance Z_0). For example, such cancellation can in principle be achieved by introducing a purely reactive series impedance that is opposite in sign to the reactive part of $Z(z)$ at that position, as will be discussed in Section 3.5. The following example illustrates the determination of the point at which $\Re e\{\bar{Z}(z)\} = 1$ for a specific complex load impedance.

Example 3-9: An inverted-V antenna. A 50Ω coaxial line filled with teflon ($v_p \simeq 21$ cm-(ns)$^{-1}$) is connected to an inverted-V antenna represented by Z_L, as shown in Figure 3.20. At $f = 29.6$ MHz, the feed-point impedance of the antenna is approximately measured to be $Z_L \simeq 75 + j25\Omega$.[13] Find the two closest positions to the antenna along the line where the real part of the line impedance is equal to the characteristic impedance of the line (i.e., Z_0).

Solution: The line impedance at any position z is given by

$$Z(z) = 50\frac{(75 + j25) - j50\zeta}{50 - j(75 + j25)\zeta} = 50\frac{3 + j(1 - 2\zeta)}{(2 + \zeta) - j3\zeta}$$

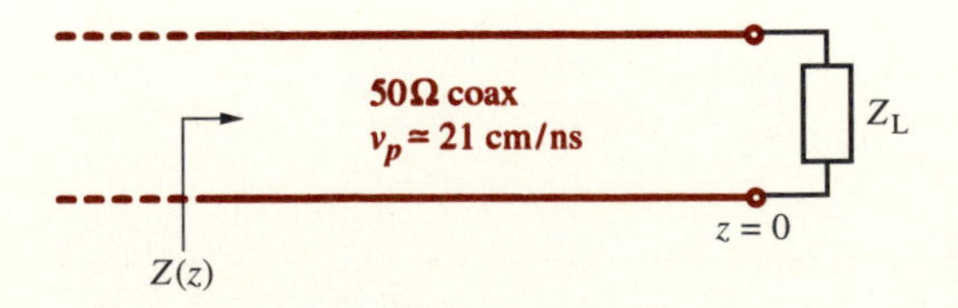

FIGURE 3.20. A coaxial line connected to an antenna.

[11]Note that this makes sense because in Figure 3.19a the load resistance is larger than the characteristic impedance ($R_L > Z_0$), which is also the case for the open circuit.

[12]The value of z_1 can be read roughly from Figure 3.19a or accurately evaluated from [3.33], by letting $\bar{Z}(z) = 1 + j\bar{X}(z)$.

[13]*The ARRL Antenna Book,* 17th ed., American Radio Relay League, pp. 27-28 and 27-29, 1994–1996.

where $\zeta = \tan(\beta z)$, $\beta = 2\pi/\lambda$, and $\lambda = v_p/f \simeq (2.1 \times 10^8)/(29.6 \times 10^6) \simeq 7.09$ m. Multiplying both the numerator and the denominator with the complex conjugate of the denominator, we can extract the real part of $Z(z)$ as

$$\mathcal{R}e\{Z(z)\} = \mathcal{R}e\left\{50\frac{3 + j(1 - 2\zeta)}{(2 + \zeta) - j3\zeta} \cdot \frac{(2 + \zeta) + j3\zeta}{(2 + \zeta) + j3\zeta}\right\} = 50\frac{3(2 + \zeta) - 3\zeta(1 - 2\zeta)}{(2 + \zeta)^2 + (3\zeta)^2}$$

The value of z for which we have $\mathcal{R}e\{Z(z)\} = Z_0 = 50\Omega$ can then be found as

$$\mathcal{R}e\{Z(z)\} = Z_0 \quad \rightarrow \quad 50\frac{3(2 + \zeta) - 3\zeta(1 - 2\zeta)}{(2 + \zeta)^2 + (3\zeta)^2} = 50$$

$$\rightarrow \quad 2\zeta^2 + 2\zeta - 1 = 0 \quad \rightarrow \quad \zeta_1, \zeta_2 \simeq -1.37, 0.366$$

Using

$$\tan(\beta z) = \tan\left(\frac{2\pi z}{\lambda}\right) \simeq \tan\left[\frac{2\pi z}{7.09}\right] = \zeta$$

and noting that $z < 0$, we find

$$z_1 \simeq -0.149\lambda \simeq -1.06 \text{ m} \quad \text{and} \quad z_2 \simeq -0.444\lambda \simeq -3.15 \text{ m}$$

as the locations at which the real part of the line impedance is equal to the characteristic impedance of the line.

Some aspects of the behavior of the real and imaginary parts of the line impedance shown in Figure 3.19 for $\bar{Z}_L = 2$ can be generalized. For example, the line impedance seen at the positions of voltage maxima (minima) is always purely real and has maximum (minimum) magnitude. To see this, consider the voltage along the line at the position of a voltage maximum (i.e., $z = z_{max}$) given by

$$V(z = z_{max}) = V^+ e^{-j\beta z_{max}}[1 + \rho e^{j(\psi + 2\beta z_{max})}] = V^+ e^{-j\beta z_{max}}(1 + \rho)$$

with a maximum magnitude of

$$|V(z = z_{max})|_{max} = V_{max} = |V^+|(1 + \rho)$$

occurring at

$$\psi + 2\beta z_{max} = -m2\pi \qquad m = 0, 1, 2, 3, \ldots$$

where $-\pi \leq \psi < \pi$, $z_{max} \leq 0$, and where $m = 0$ does not apply if $-\pi \leq \psi < 0$. At the same position, the current is equal to

$$I(z = z_{max}) = \frac{V^+}{Z_0} e^{-j\beta z_{max}}(1 - \rho)$$

with a minimum magnitude given by

$$|I(z = z_{max})|_{min} = I_{min} = \frac{|V^+|}{Z_0}(1 - \rho)$$

so that the line impedance $Z(z_{max}) = V(z_{max})/I(z_{max})$ is clearly purely resistive and has a maximum magnitude given by

$$|Z(z_{max})|_{max} = R_{max} = \frac{V_{max}}{I_{min}} = Z_0\frac{1+\rho}{1-\rho} = SZ_0 \quad [3.36]$$

whereas at the voltage minima ($z_{min} = z_{max} - \lambda/4$), the line impedance $Z(z_{min})$ is purely resistive with a minimum magnitude given by

$$|Z(z_{min})|_{min} = R_{min} = \frac{V_{min}}{I_{max}} = Z_0\frac{1-\rho}{1+\rho} = \frac{Z_0}{S} \quad [3.37]$$

Also, for purely resistive terminations ($Z_L = R_L$), the load is at a position of either minimum or maximum for the voltage and therefore, the load impedance R_L is either equal to the minimum ($R_L = Z_0/S$ when $R_L < Z_0$) or maximum ($R_L = SZ_0$ when $R_L > Z_0$) magnitude for the line impedance. Note that we have

$$S = \frac{1+\rho}{1-\rho} = \begin{cases} \dfrac{R_L}{Z_0} & \text{for} \quad R_L > Z_0 \\[2ex] \dfrac{Z_0}{R_L} & \text{for} \quad R_L < Z_0 \end{cases}$$

Line Impedance for Complex Load Impedances For general load impedances that are not purely resistive (i.e., $Z_L = R_L + jX_L$), the behavior of the line impedance $Z(z) = R(z) + jX(z)$ is similar to that for purely resistive loads, in that its real part $R(z)$ varies between a maximum value of SZ_0 and a minimum value of Z_0/S, and the imaginary part $X(z)$ alternates sign at intervals of $\lambda/4$. (SZ_0 occurs at the positions of the voltage maxima when the line impedance is purely resistive and therefore is also the maximum magnitude of the line impedance. Z_0/S occurs at the positions of the voltage minima when the line impedance is also purely resistive and therefore is also the minimum magnitude of the line impedance.) However, the maxima and minima of the magnitudes of either the voltage or the line impedance are not at the load position. Figures 3.21 and 3.22 show plots of the real and imaginary parts of the normalized line impedance $\bar{Z}(z)$ as functions of z/λ for selected load impedances.

Example 3-10 illustrates the numerical evaluation of the reflection coefficient, standing-wave ratio, and maximum and minimum line resistances for a complex load termination.

Example 3-10: Reflection coefficient, standing-wave ratio, and maximum and minimum resistances. A radio transmitter is connected to an antenna having a feed-point impedance of $Z_L = 70 + j100\Omega$ with a 50Ω coaxial line, as shown in Figure 3.23. Find (a) the load reflection coefficient, (b) the standing-

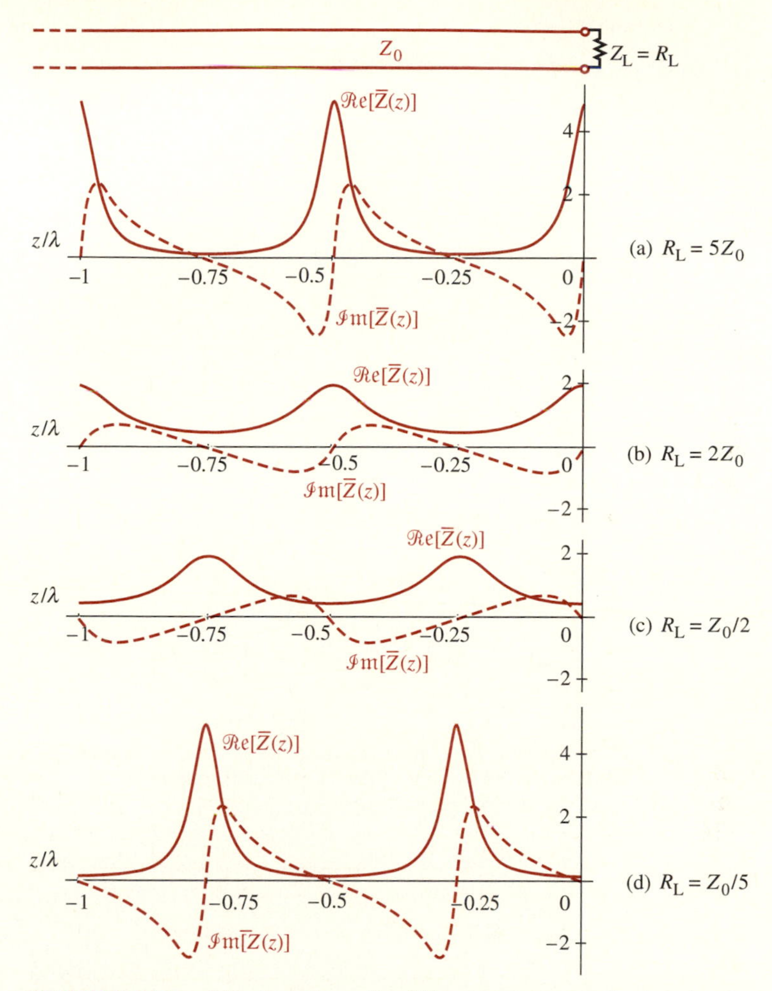

FIGURE 3.21. Line impedance for different purely resistive terminations. The real and imaginary parts of the normalized line impedance $\bar{Z}(z)$ are shown as functions of electrical distance z/λ along the line for Z_L equal to (a) $5Z_0$, (b) $2Z_0$, (c) $Z_0/2$, and (d) $Z_0/5$.

wave ratio, and (c) the two positions closest along the line to the load where the line impedance is purely real, and their corresponding line impedance values.

Solution:

(a) The load reflection coefficient is

$$\Gamma_L = \frac{Z_L - Z_0}{Z_L + Z_0} = \frac{70 + j100 - 50}{70 + j100 + 50} = \frac{1 + j5}{6 + j5}$$

$$= \frac{\sqrt{26}e^{j\tan^{-1}(5)}}{\sqrt{61}e^{j\tan^{-1}(5/6)}} \simeq 0.653e^{j38.9°}$$

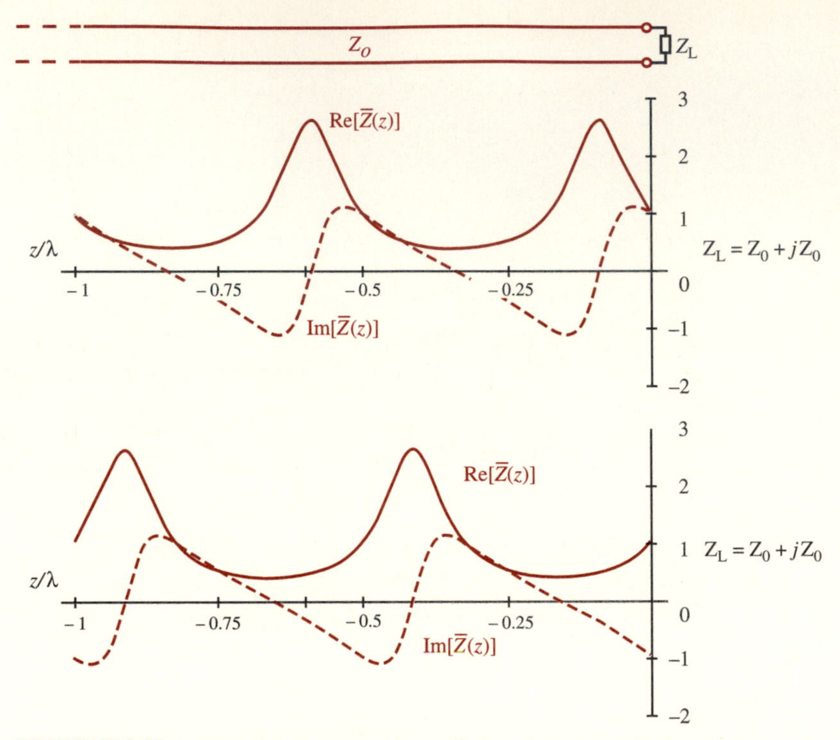

FIGURE 3.22. Line impedance for two different complex load impedances. The real and imaginary parts of the normalized line impedance $\bar{Z}(z)$ are shown as functions of electrical distance z/λ along the line for (a) $Z_L = Z_0 + jZ_0$ and (b) $Z_L = Z_0 - jZ_0$.

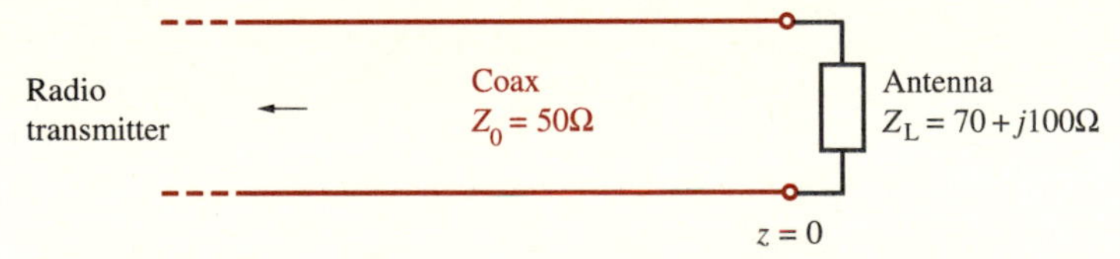

FIGURE 3.23. Transmission line terminated in an antenna. In general, the feed-point impedance of an antenna is complex.

(b) The standing-wave ratio is

$$S = \frac{1+\rho}{1-\rho} \simeq \frac{1+0.653}{1-0.653} \simeq 4.76$$

(c) The maximum voltage position is the position z_{max} at which the line impedance is purely real and the magnitude of the line impedance is a maximum; note that the position of the first voltage maximum is closer to the load than that of the first voltage minimum because the load impedance is inductive. To find the maximum voltage position, use $\psi + 2\beta z_{max} = 0 \quad \rightarrow \quad z_{max} \simeq -0.054\lambda$:

$$R_{max} = SZ_0 \simeq (4.76)(50) = 238\Omega$$

The minimum voltage position is the position $z_{\min}$ at which $|Z(z)|$ is minimum; note that this is the next closest position where the line impedance $Z(z)$ is real. To find the minimum voltage position, use

$$\psi + 2\beta z_{\min} = -\pi \quad \rightarrow \quad z_{\min} \simeq -0.304\lambda$$

Note that as expected, $z_{\min} = z_{\max} - \lambda/4$. We then have

$$R_{\min} = Z_0/S \simeq 50/(4.76) \simeq 10.5\Omega$$

3.3.3 Calculation of V^+

Up to now, we have primarily focused on the line impedance and the variation of the voltage and current along the line without particular attention to the source end of the line. The source that excites the transmission line shown in Figure 3.12 is a voltage source with an open-circuit phasor voltage V_0 and a source impedance Z_s. Using Equation [3.22], the voltage V_s at the source end of the line ($z = -l$) is

$$V_s = V(z = -l) = V^+ e^{j\beta l}(1 + \Gamma_L e^{-j2\beta l})$$

As seen from the source end, the transmission line can be represented by its input impedance, Z_{in}. We can thus also express the source-end voltage phasor V_s in terms of the source parameters V_0 and Z_s by noting the division of voltage between Z_{in} and Z_s, namely,

$$V_s = \frac{Z_{in}}{Z_{in} + Z_s} V_0$$

By equating the two preceding expressions for V_s, we can solve for the constant V^+:

$$V^+ = \frac{Z_{in} V_0}{(Z_{in} + Z_s) e^{j\beta l}(1 + \Gamma_L e^{-j2\beta l})}$$

Note that the knowledge of V^+ and the wavelength $\lambda = 2\pi/\beta$ completely specifies the transmission line voltage and current as given in [3.22] and [3.23], as for any given transmission line with characteristic impedance Z_0 and load Z_L (and hence Γ_L).

Example 3-11 illustrates the calculation of V^+ by considering the source end of a terminated transmission line circuit.

Example 3-11: Coaxial line feeding an antenna. A sinusoidal voltage source of $\mathcal{V}_0(t) = 10\cos(5\pi \times 10^7 t)$ V and $R_s = 20\Omega$ is connected to an antenna with feed-point impedance $Z_L = 100\Omega$ through a 3-m long, lossless coaxial transmission line filled with polyethylene ($v_p = 20$ cm-(ns)$^{-1}$) and with a characteristic impedance of $Z_0 = 50\Omega$, as shown in Figure 3.24a. Find (a) the voltage and current phasors, $V(z)$ and $I(z)$, at any location on the line and (b) the corresponding instantaneous expressions $\mathcal{V}(z, t)$ and $\mathcal{I}(z, t)$.

Solution:

(a) At $f = \omega/(2\pi) = 25$ MHz, the wavelength in a polyethylene-filled coaxial line is

$$\lambda = v_p/f = (20 \text{ cm-(ns)}^{-1})/(25 \text{ MHz}) = 8 \text{ m}$$

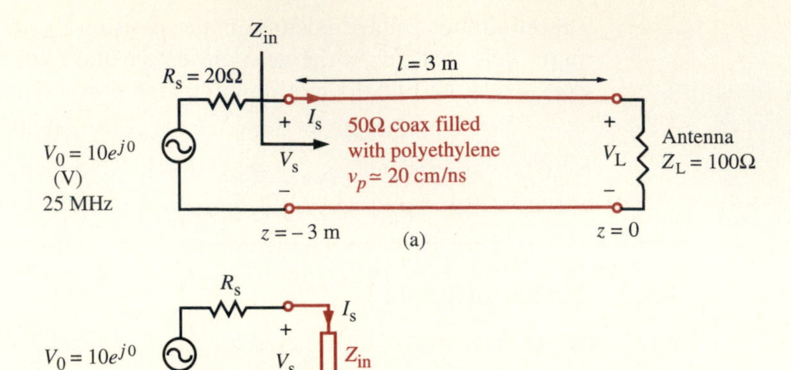

FIGURE 3.24. **Coaxial line feeding an antenna.** (a) Circuit configuration. (b) Thévenin equivalent circuit seen from the source end.

The electrical length of the 3-m line is then $l/\lambda = 3/8 = 0.375$, so we have $\beta l = 2\pi(0.375) = 3\pi/4$ and $\tan(\beta l) = -1$. The input impedance seen at the source end is

$$Z_{in} = Z(z = -l) = Z_0 \frac{Z_L + jZ_0 \tan(\beta l)}{Z_0 + jZ_L \tan(\beta l)} = 50 \frac{100 + j50(-1)}{50 + j100(-1)}$$

$$= \left(\frac{100 - j50}{1 - j2}\right)\left(\frac{1 + j2}{1 + j2}\right) = (20 - j10)(1 + j2) = 40 + j30\Omega$$

Using the equivalent circuit shown in Figure 3.24b, we have

$$V_s = \frac{Z_{in}}{R_s + Z_{in}} V_0 = \frac{40 + j30}{60 + j30}(10) \simeq \frac{5e^{j36.9°}}{3\sqrt{5}e^{j26.6°}}(10) \simeq 7.45e^{j10.3°} \text{ V}$$

We can also write an expression for V_s by evaluating $V(z)$ at $z = -3$ m as

$$V_s = V(z = -3 \text{ m}) = V^+ e^{j3\pi/4}(1 + \Gamma_L e^{-j3\pi/2})$$

where Γ_L is the load reflection coefficient given by

$$\Gamma_L = \frac{Z_L - Z_0}{Z_L + Z_0} = \frac{100 - 50}{100 + 50} = \frac{1}{3}$$

Equating the two expressions for V_s, we can determine the complex constant V^+

$$V_s = V^+ e^{j3\pi/4}\left(1 + \frac{j}{3}\right) \simeq 7.45e^{j10.3°} \quad \rightarrow \quad V^+ \simeq 7.07e^{-j143°} \text{ V}$$

so that the voltage phasor at any position z from the load is given as

$$V(z) \simeq 7.07e^{-j143^\circ}e^{-j\pi z/4}\left(1 + \tfrac{1}{3}e^{j\pi z/2}\right) \text{ V}$$

and the corresponding current phasor is

$$I(z) \simeq 0.141e^{-j143^\circ}e^{-j\pi z/4}\left(1 - \tfrac{1}{3}e^{j\pi z/2}\right) \text{ A}$$

(b) Using $\mathcal{V}(z,t) = \Re e\{V(z)e^{j\omega t}\}$ and $\mathcal{I}(z,t) = \Re e\{I(z)e^{j\omega t}\}$, we find

$$\mathcal{V}(z,t) \simeq 7.07\cos\left(5\pi 10^7 t - \frac{\pi z}{4} - 143^\circ\right) + 2.36\cos\left(5\pi 10^7 t + \frac{\pi z}{4} - 143^\circ\right) \text{ V}$$

and

$$\mathcal{I}(z,t) \simeq 0.141\cos\left(5\pi 10^7 t - \frac{\pi z}{4} - 143^\circ\right) - 0.0471\cos\left(5\pi 10^7 t + \frac{\pi z}{4} - 143^\circ\right) \text{ A}$$

3.4 POWER FLOW ON A TRANSMISSION LINE

In practice, the primary purpose of most steady-state sinusoidal transmission line applications is to maximize the time-average power delivered to a load. The power and energy flow on a transmission line can be determined from the line voltage and line current; the product of the instantaneous current $\mathcal{I}(z,t)$ and instantaneous voltage $\mathcal{V}(z,t)$ at any point z is by definition the power that flows into the line[14] at that point. In most applications, the quantity of interest is not the rapidly varying instantaneous power but its average over one sinusoidal period T_p, namely, the time-average power, which is given by

$$P_{\text{av}}(z) = \frac{1}{T_p}\int_0^{T_p} \mathcal{V}(z,t)\mathcal{I}(z,t)\,dt$$

where $T_p = 2\pi/\omega$. The time-average power can also be calculated directly from the voltage and current phasors:

$$P_{\text{av}}(z) = \tfrac{1}{2}\Re e\{V(z)[I(z)]^*\}$$

Consider the general expressions for the voltage and current phasors along a lossless uniform transmission line:

$$\begin{aligned} V(z) &= V^+e^{-j\beta z} + \Gamma_{\text{L}}V^+e^{j\beta z} \\ I(z) &= \underbrace{\frac{V^+}{Z_0}e^{-j\beta z}}_{\text{forward wave}} - \underbrace{\Gamma_{\text{L}}\frac{V^+}{Z_0}e^{j\beta z}}_{\text{reverse wave}} \end{aligned}$$

[14]Note that the product $\mathcal{V}(z,t)\mathcal{I}(z,t)$ represents power flow into the line, rather than out of the line, due to the defined polarity of the current $I(z)$ and voltage $V(z)$ in Figure 3.12.

We denote the time-average power carried by the forward and backward traveling waves as P^+ and P^-, respectively, and we evaluate them directly from the phasors. The time-average power carried by the forward wave is

$$P^+ = \frac{1}{2}\Re e\left\{\frac{(V^+e^{-j\beta z})(V^+e^{-j\beta z})^*}{Z_0}\right\} = \frac{V^+(V^+)^*}{2Z_0} = \frac{|V^+|^2}{2Z_0}$$

Note that although V^+ is in general complex, $V^+(V^+)^* = |V^+|^2$ is a real number.[15] The power carried by the reverse-propagating wave is

$$P^- = \frac{1}{2}\Re e\left\{\frac{(\Gamma_L V^+e^{j\beta z})(-\Gamma_L V^+e^{j\beta z})^*}{Z_0}\right\} = -\Gamma_L\Gamma_L^*\frac{V^+(V^+)^*}{2Z_0} = -\rho^2\frac{|V^+|^2}{2Z_0}$$

The fact that P^- is negative simply indicates that the backward wave carries power in the opposite direction with respect to the defined polarity of $I(z)$ and $V(z)$ in Figure 3.12. The net total power in the forward direction is then given by

$$P_{\text{av}} = P^+ + P^- = \frac{|V^+|^2}{2Z_0} - \rho^2\frac{|V^+|^2}{2Z_0} = \frac{|V^+|^2}{2Z_0}(1-\rho^2) \qquad [3.38]$$

Thus, the net time-average power flow on a transmission line is maximized when the load reflection coefficient Γ_L is zero, which, according to [3.21], occurs when $Z_L = Z_0$. Note that when $\Gamma_L = 0$ and thus $Z_L = Z_0$, the standing-wave ratio $S = 1$, as is evident from [3.25]. Since the transmission line is assumed to be lossless, all of the net power flowing in the $+z$ direction is eventually delivered to the load.

The same result can also be obtained by examining the power dissipation in the load, which is given by[16]

$$P_L = \tfrac{1}{2}\Re e\{V_L I_L^*\} = \tfrac{1}{2}|I_L|^2\Re e\{Z_L\} = \tfrac{1}{2}|I_L|^2 R_L$$

Noting that we have

$$V_L = V(z)|_{z=0} = V^+ + \Gamma_L V^+ = V^+(1+\Gamma_L)$$

$$I_L = I(z)|_{z=0} = \frac{V^+}{Z_0} - \Gamma_L\frac{V^+}{Z_0} = \frac{V^+}{Z_0}(1-\Gamma_L)$$

and substituting in the preceding expression for P_L, we find

---

[15] If $V^+ = A + jB$, then $V^+(V^+)^* = (A+jB)(A-jB) = A^2 + B^2 = |A+jB|^2 = |V^+|^2$.

[16] Note that P_L can also be written in terms of the load voltage V_L and load admittance Y_L as

$$P_L = \tfrac{1}{2}|V_L|^2\Re e\{Y_L\} = \tfrac{1}{2}|V_L|^2 G_L$$

where $Y_L = Z_L^{-1} = G_L + jB_L$.

$$P_L = \frac{1}{2}\Re e\{V_L I_L^*\} = \frac{1}{2}\Re e\left\{\frac{V^+(1+\Gamma_L)[V^+(1-\Gamma_L)]^*}{Z_0}\right\}$$
$$= \frac{1}{2}\Re e\left\{\frac{V^+(V^+)^*[1+(\Gamma_L-\Gamma_L^*)-\Gamma_L\Gamma_L^*]}{Z_0}\right\} = \frac{|V^+|^2}{2Z_0}(1-\rho^2) \qquad [3.39]$$

which is identical to the total net forward power P_{av} as derived in [3.38]. We see that $P_{av} = P_L$, as expected, since, for the case of a lossless line as assumed here, *all* of the net power traveling toward the load must be dissipated in the load.

The same result can further be obtained by evaluating $P(z)$ at any point along the line using the total voltage and current phasors (rather than separating them into forward and reverse traveling wave components). In other words,

$$P_{av}(z) = \tfrac{1}{2}\Re e\{V(z)[I(z)]^*\}$$

where $V(z)$ and $I(z)$ are given by [3.22] and [3.23], respectively. (This derivation is left as an exercise for the reader.)

In summary, the total net power propagating in the $+z$ direction is

$$\boxed{P_{av}(z) = \frac{|V^+|^2}{2Z_0}(1-\rho^2)} \qquad [3.40]$$

and the following observations concerning power flow on a lossless transmission line can be made:

- For a given V^+, maximum power is delivered to the load when $Z_L = Z_0$, $\Gamma_L = 0$ (i.e., $\rho = 0$), and $S = 1$. Noting that Z_0 is a real number, this condition is realized when the load is purely resistive, that is, when $Z_L = R_L = Z_0$. When $R_L = Z_0$, the load is said to be *matched* to the line and all of the power P^+ is delivered to the load. Detailed discussion of impedance matching is given in Section 3.5.
- To deliver a given amount of power (say, P_L) when the line is not matched (i.e., $S > 1$) requires higher wave power in the incident wave with correspondingly higher voltages ($P^+ = |V^+|^2/(2Z_0)$). The higher voltages are undesirable as they may cause breakdown[17] of the insulation between the two conductors of the line.
- The power efficiency achieved by matching can be assessed by considering the ratio of the power P_L that is dissipated in a given load to the forward wave power P^+ that would be delivered to the load if the line were matched:

$$\frac{P_L}{P^+} = 1 - \rho^2 = \frac{4S}{(1+S)^2}$$

[17]Electrical breakdown of insulating materials will be discussed in Section 4.10.

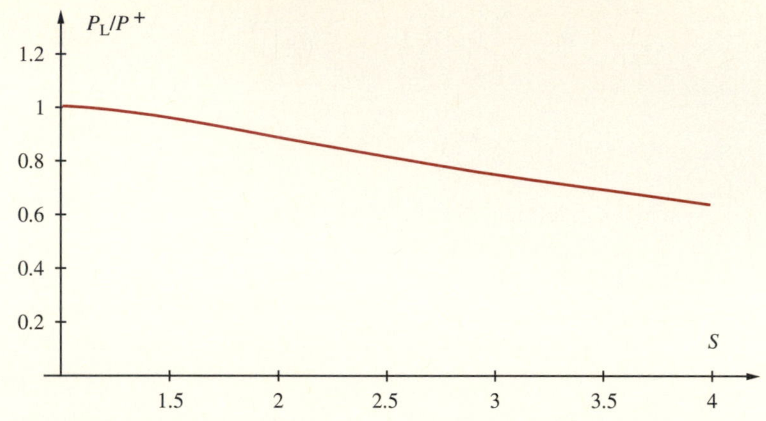

FIGURE 3.25. Power efficiency as a function of standing-wave ratio S.

The variation of P_L/P^+ with S is plotted in Figure 3.25. We see that $P_L/P^+ = 1$ for $S = 1$ and monotonically decreases to zero as S gets larger. Note that for voltage standing-wave ratios $S < 1.5$, which are relatively easy to achieve in practice, more than 90 percent of the power in the forward wave is delivered to the load. In other words, it is not necessary to strive for S very near unity to attain maximum power transfer to the load. Usually the more important issues are ensuring that the value of S is not so large as to make the line performance highly sensitive to frequency (see Section 3.5), and that the design of the line can accommodate large reactive voltages and currents that accompany a large value of S.

The degree of mismatch between the load and the line is sometimes described in terms of *return loss,* which is defined as the decibel value of the ratio of the power carried by the reverse wave to the power carried by the forward wave, given as

$$\text{Return loss} = -20 \log_{10} \rho = 20 \log_{10} \frac{S+1}{S-1}$$

If the load is perfectly matched to the line ($\rho = 0$), the return loss is infinite, which simply indicates that there is no reverse wave. If the load is such that $\rho = 1$ (i.e., a short-circuited or open-circuited line, or a purely reactive load), then the return loss is 0 dB. In practice, a well-matched system has a return loss of 15 dB or more, corresponding to a standing-wave ratio of $\sim$1.43 or less.

Examples 3-12, 3-13, and 3-14 illustrate the calculation of power flow and power delivery to the load for three different transmission line configurations.

Example 3-12: A 125-MHz VHF transmitter-antenna system. A VHF transmitter operating at 125 MHz and developing $V_0 = 100e^{j0°}$ V with a source resistance of $R_s = 50\Omega$ feeds an antenna with a feed-point impedance[18] of

[18] *The ARRL Antenna Compendium,* Vol. 4, p. 56, The American Radio Relay League, 1995–1996.

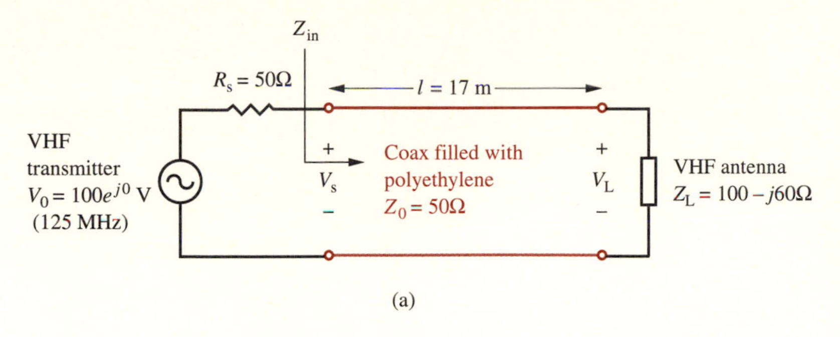

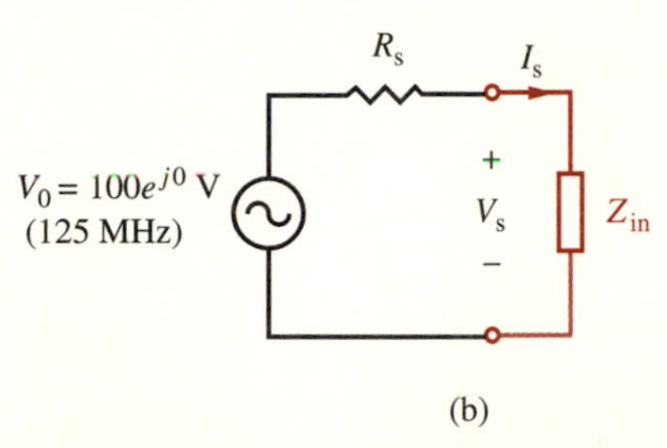

FIGURE 3.26. A 125-MHz VHF transmitter-antenna system. (a) Circuit diagram. (b) Equivalent circuit seen by the source.

$Z_L = 100 - j60$ through a 50Ω, polyethylene-filled coaxial line that is 17 m long. The setup is shown in Figure 3.26a. (a) Find the voltage $V(z)$ on the line. (b) Find the load voltage V_L. (c) Find the time-average power absorbed by the VHF antenna. (d) Find the power absorbed by the source impedance R_s.

Solution:

(a) First we note that for a polyethylene-filled coaxial line, the wavelength at 125 MHz is (using Table 3.1 and assuming v_p at 125 MHz is the same as that at 300 MHz) $\lambda = v_p/f = (2 \times 10^8 \text{ m/s})/(125 \text{ MHz}) = 1.6$ m. The length of the line is then

$$l = 17 \text{ m} = 10.625\lambda = 10.5\lambda + 0.125\lambda$$

Noting that $\tan(\beta l) = \tan[(2\pi/\lambda)(0.125\lambda)] = \tan(\pi/4) = 1$, and the input impedance of the line seen from the source end is then

$$Z_{in} = Z_0 \frac{Z_L + jZ_0 \tan(\beta l)}{Z_0 + jZ_L \tan(\beta l)} = 50\frac{(100 - j60) + j50\tan(\pi/4)}{50 + j(100 - j60)\tan(\pi/4)}$$

$$= 50\frac{100 - j10}{110 + j100} = \frac{50(100 - j10)(110 - j100)}{(110)^2 + (100)^2} \simeq 22.6 - j25.1\Omega$$

The equivalent circuit at the source end is as shown in Figure 3.26b. The source-end voltage V_s is then

$$V_s = \frac{Z_{in}}{R_s + Z_{in}} V_0 \simeq \frac{22.6 - j25.1}{50 + 22.6 - j25.1} 100e^{j0°}$$
$$\simeq \frac{33.8e^{-j48°}}{76.8e^{-j19.1°}} 100e^{j0°} \simeq 44.0e^{-j28.9°} \text{ V}$$

But we can also evaluate V_s from the expression for the line voltage $V(z)$ as

$$V_s = V(z = -17 \text{ m}) = V^+ e^{-j\beta z}(1 + \Gamma_L e^{j2\beta z})$$
$$\simeq V^+ e^{j5\pi/4}(1 + 0.483e^{-j28.4°} e^{-j\pi/2})$$
$$\simeq V^+ e^{-j3\pi/4}(0.770 - j0.425)$$

where we have used the facts that $e^{j21.25\pi} = e^{j1.25\pi} = e^{-j0.75\pi}$, $e^{-j42.5\pi} = e^{-j0.5\pi}$, and that the load reflection coefficient Γ_L is

$$\Gamma_L = \frac{100 - j60 - 50}{100 - j60 + 50} = \frac{50 - j60}{150 - j60} \simeq 0.483e^{-j28.4°}$$

Equating the two expressions for V_s, we can determine the unknown voltage V^+ as

$$V_s \simeq V^+ e^{-j3\pi/4}(0.770 - j0.425) \simeq 44.0e^{-j28.9°} \quad \rightarrow \quad V^+ = 50e^{j3\pi/4} \text{ V}$$

Thus the expression for the line voltage is

$$V(z) \simeq 50e^{j3\pi/4} e^{-j5\pi z/4}(1 + 0.483e^{-j(28.4° - 5\pi z/2)}) \text{ V}$$

(b) The voltage at the load end of the line is

$$V_L = V(z = 0) \simeq 50e^{j3\pi/4}(1 + 0.483e^{-j28.4°})$$
$$\simeq 50e^{j3\pi/4}(1.43 - j0.230) \simeq 72.2e^{-j126°} \text{ V}$$

(c) Using the value of V_L, the time-average power delivered to the VHF antenna can be calculated as

$$P_L = \frac{1}{2}|I_L|^2 R_L = \frac{1}{2}\left|\frac{V_L}{Z_L}\right|^2 R_L \simeq \frac{(72.2)^2(100)}{2(1.36 \times 10^4)} \simeq 19.2 \text{ W}$$

where $|Z_L|^2 = |100 - j60|^2 = (100)^2 + (60)^2 = 1.36 \times 10^4$. Note that P_L can also be found using the source-end equivalent circuit (Figure 3.26b). Since the line is lossless, all of the time-average power input to the line at the source end must be absorbed by the antenna. In other words,

$$P_L = \frac{1}{2}\left|\frac{V_s}{Z_{in}}\right|^2 R_{in} \simeq \frac{1}{2}\frac{(44.0)^2}{[(22.6)^2 + (25.1)^2]}(22.6) \simeq 19.2 \text{ W}$$

(d) Noting that $R_s = 50\Omega$ and $Z_{in} \simeq 22.6 - j25.1\Omega$, the current at the source end (again considering the lumped equivalent circuit shown in Figure 3.26b) is

$$I_s = \frac{V_0}{R_s + Z_{in}} \simeq \frac{100e^{j0°}}{50 + 22.6 - j25.1} \simeq 1.30e^{j19.1°} \text{ A}$$

Using the value of I_s, the time-average power dissipated in the source resistance R_s is then

$$P_{R_s} = \tfrac{1}{2}|I_s|^2 R_s \simeq \tfrac{1}{2}(1.30)^2(50) \simeq 42.3 \text{ W}$$

Therefore, the total power supplied by the VHF transmitter is $P_{\text{total}} = P_{R_s} + P_L \simeq 61.5$ W.

Example 3-13: Parallel transmission lines. Three lossless transmission lines are connected in parallel, as shown in Figure 3.27. Assuming sinusoidal steady-state excitation with a source to the left of the main line, find the reflection coefficient on the main line and the percentage of the total net forward power that is absorbed by the two loads Z_{L2} and Z_{L3} for the following cases: (a) $Z_{01} = Z_{02} = Z_{03} = Z_{L2} = Z_{L3} = 100\Omega$, (b) $Z_{01} = 50\Omega$, $Z_{02} = Z_{03} = Z_{L2} = Z_{L3} = 100\Omega$, and (c) $Z_{01} = Z_{02} = Z_{03} = 100\Omega$, $Z_{L2} = Z_{L3} = 50 + j50\Omega$.

Solution:

(a) $Z_{01} = Z_{02} = Z_{03} = Z_{L2} = Z_{L3} = 100\Omega$
Since the two parallel branches are both matched, the input impedance seen at the terminals of each branch is independent of its line length and is simply 100Ω. Thus, the line impedance Z_j seen from the main line is the parallel combination of two 100Ω impedances, or $Z_j = 50\Omega$. The reflection coefficient at the junction (as seen from the main line) is then

$$\Gamma_j = \frac{Z_j - Z_{01}}{Z_j + Z_{01}} = \frac{50 - 100}{50 + 100} = -\frac{1}{3}$$

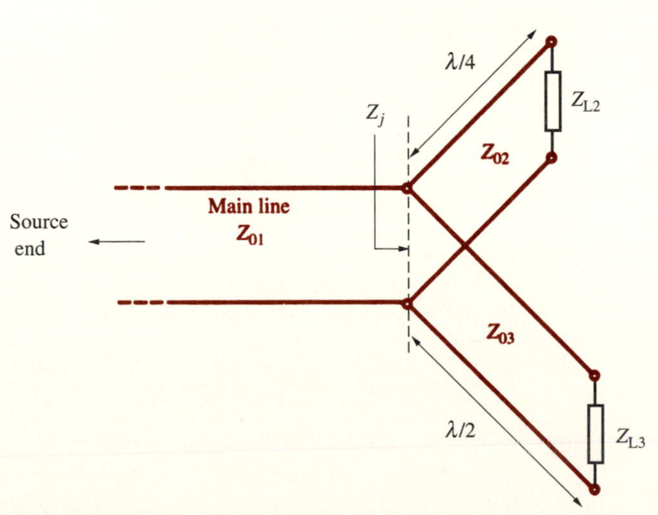

FIGURE 3.27. Parallel transmission lines. A main line with characteristic impedance Z_{01} drives two other lines of lengths $\lambda/4$ and $\lambda/2$, with characteristic impedances Z_{02} and Z_{03}, respectively.

In other words, the power efficiency, defined as the percentage of total power that is delivered to the loads versus that of the forward wave on the main line, is

$$\frac{P_L}{P^+} = (1 - |\Gamma_j|^2) \times 100 \simeq 88.9\%$$

Since each line presents the same impedance at the junction, each load absorbs half of the total power delivered, or approximately 44.4% of the total power of the incident wave.

(b) $Z_{01} = 50\Omega, Z_{02} = Z_{03} = Z_{L2} = Z_{L3} = 100\Omega$

The line impedance at the junction seen from the main line is once again $Z_j = 50\Omega$. However, the reflection coefficient is now zero since

$$\Gamma_j = \frac{Z_j - Z_{01}}{Z_j + Z_{01}} = \frac{50 - 50}{50 + 50} = 0$$

In other words, 100% of the power of the forward wave in this case is delivered, with each load absorbing 50%.

(c) $Z_{01} = Z_{02} = Z_{03} = 100\Omega, Z_{L2} = Z_{L3} = 50 + j50\Omega$

We now need to evaluate the input impedances of each of the two parallel sections at the junction as seen from the main line. Note that the length of line 2 is $\lambda/4$, so that using [3.32], its input impedance is

$$Z_{\text{in}_2} = Z_{02}\frac{Z_{L2} + jZ_{02}\tan(\frac{2\pi}{\lambda}\frac{\lambda}{4})}{Z_{02} + jZ_{L2}\tan(\frac{2\pi}{\lambda}\frac{\lambda}{4})}$$
$$= \frac{Z_{02}^2}{Z_{L2}} = \frac{(100)^2}{50 + j50} = \frac{200}{1 + j} \times \frac{1 - j}{1 - j} = 100 - j100\Omega$$

whereas line 3 has length $\lambda/2$ and thus presents Z_{L3} at the junction; in other words, $Z_{\text{in}_3} = Z_{L3} = 50 + j50\Omega$. The line impedance Z_j at the junction as seen from the main line is then the parallel combination of Z_{in_2} and Z_{in_3}, namely,

$$Z_j = \frac{(100 - j100)(50 + j50)}{100 - j100 + 50 + j50} = \frac{100(1 - j)(1 + j)}{3 - j}$$
$$= \frac{200}{10}(3 + j) = 60 + j20$$

and the reflection coefficient at the end of the main line is

$$\Gamma_j = \frac{Z_j - Z_{01}}{Z_j + Z_{01}} = \frac{60 + j20 - 100}{60 + j20 + 100} = \frac{-40 + j20}{160 + j20} = \left(\frac{-2 + j}{8 + j}\right)\left(\frac{8 - j}{8 - j}\right)$$
$$= \frac{-15 + j10}{65} = \frac{-3 + j2}{13} \simeq 0.277e^{j146^\circ}$$

The percentage of the incident power delivered to the two loads is thus

$$\frac{P_L}{P^+} = (1 - |\Gamma_j|^2) \times 100 \simeq 92.3\%$$

Since the two impedances Z_{in_2} and Z_{in_3} appear at the junction in parallel, they share the same voltage. In other words, we have

$$P_{L_2} = \frac{1}{2}\left|\frac{V_j}{Z_{\text{in}_2}}\right|^2 \Re e\{Z_{\text{in}_2}\} \quad \text{and} \quad P_{L_3} = \frac{1}{2}\left|\frac{V_j}{Z_{\text{in}_3}}\right|^2 \Re e\{Z_{\text{in}_3}\}$$

We can thus calculate the ratio of the powers delivered to the two loads as

$$\frac{P_{L_2}}{P_{L_3}} = \frac{|Z_{\text{in}_3}|^2 \Re e\{Z_{\text{in}_2}\}}{|Z_{\text{in}_2}|^2 \Re e\{Z_{\text{in}_3}\}} = \frac{[(50)^2 + (50)^2](100)}{[(100)^2 + (100)^2](50)} = 0.5$$

Therefore we have

$$P_{L_2} \simeq \tfrac{1}{3} \times 92.3\% \simeq 30.8\% \quad \text{and} \quad P_{L_3} \simeq \tfrac{2}{3} \times 92.3\% \simeq 61.5\%$$

Example 3-14: Cascaded transmission lines. An antenna of measured feed-point impedance of $72 + j36\Omega$ at 100 MHz is to be driven by a transmitter through two cascaded coaxial lines with the following characteristics:

$$Z_{01} = 120\Omega \quad l_1 = 3.75 \text{ m} \quad \text{air-filled} \quad v_{p1} = c \simeq 30 \text{ cm-(ns)}^{-1}$$
$$Z_{02} = 60\Omega \quad l_2 = 1.75 \text{ m} \quad \text{polyethylene-filled} \quad v_{p2} \simeq 20 \text{ cm-(ns)}^{-1}$$

where we have used Table 3.1 for the phase velocity v_{p2} for the polyethylene-filled coaxial line assuming it to be approximately the same at 100 MHz. (a) Assuming both lines to be lossless and assuming a source voltage of $V_0 = 100e^{j0}$ V and resistance of $R_s = 50\Omega$ for the transmitter, find the time-average power delivered to the load. (b) Repeat part (a) with $l_1 = 4.5$ m. The setup is shown in Figure 3.28a.

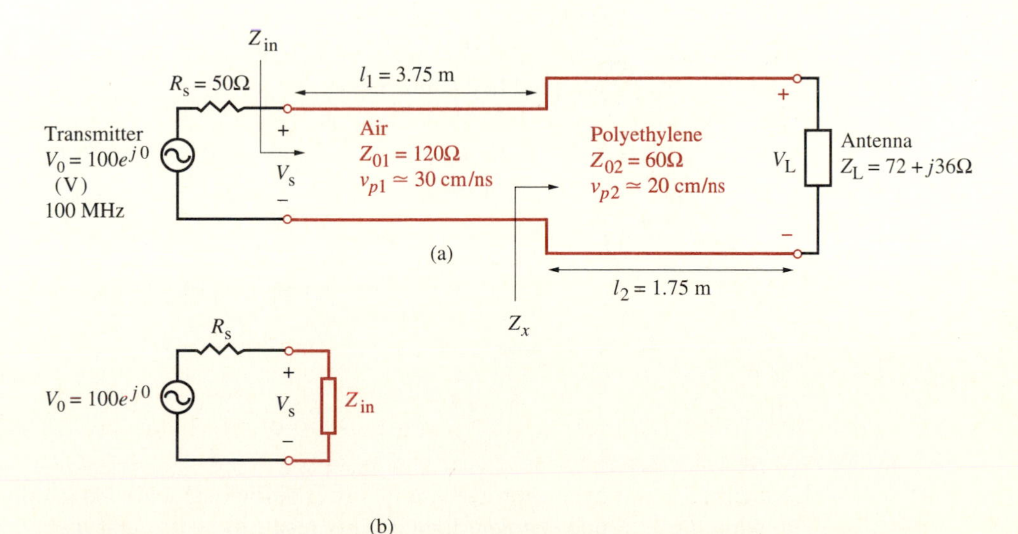

FIGURE 3.28. Cascaded transmission lines. (a) Circuit diagram showing an air-filled line of impedance 120Ω cascaded with a polyethylene-filled line of impedance 60Ω. (b) Equivalent circuit seen by the source.

Solution:

(a) At 100 MHz, the wavelengths for the two coaxial lines are respectively $\lambda_1 = v_{p1}/f \simeq 3 \times 10^8/10^8 = 3$ m, and $\lambda_2 = v_{p2}/f = 2 \times 10^8/10^8 = 2$ m. The lengths of the lines are then

$$l_1 = 3.75 \text{ m} \simeq 1.25\lambda_1 = \lambda_1 + 0.25\lambda_1$$
$$l_2 = 1.75 \text{ m} = 0.875\lambda_2 = 0.5\lambda_2 + 0.375\lambda_2$$

Note that the corresponding phase constants are $\beta_1 = 2\pi/\lambda_1$ and $\beta_2 = 2\pi/\lambda_2$. The impedance Z_x seen looking toward the load at the interface between the two coaxial lines is

$$Z_x = Z_{02}\frac{Z_L + jZ_{02}\tan(\beta_2 l_2)}{Z_{02} + jZ_L\tan(\beta_2 l_2)} = 60\frac{(72 + j36) + j60(-1)}{60 + j(72 + j36)(-1)}$$
$$= 60\frac{72 - j24}{96 + j72} = 60\frac{3 - j}{4 - j3} = 60\frac{(3 - j)(4 + j3)}{4^2 + 3^2} = 36 + j12\Omega$$

since $\tan[(2\pi/\lambda_2)l_2] = \tan(3\pi/4) = -1$. The input impedance is then

$$Z_{in} = Z_{01}\frac{Z_x + jZ_{01}\tan(\beta_0 l_1)}{Z_{01} + jZ_x\tan(\beta_0 l_1)} = \frac{Z_{01}^2}{Z_x} \simeq \frac{(120)^2}{36 + j12}$$
$$= \frac{(120)^2(36 - j12)}{36^2 + 12^2} = 360 - j120\Omega$$

With reference to the equivalent circuit in Figure 3.28b, we have

$$V_s = \frac{Z_{in}}{R_s + Z_{in}}V_0 = \frac{360 - j120}{410 - j120}(100) \simeq 88.8e^{-j2.12^\circ} \text{ V}$$

Thus the power delivered to the antenna is

$$P_L = P_{in} = \frac{1}{2}\left|\frac{V_s}{Z_{in}}\right|^2 \mathcal{Re}\{Z_{in}\} = \frac{1}{2}\frac{(88.8)^2}{(1.44 \times 10^5)}(360) \simeq 9.86 \text{ W}$$

(b) With $l_1 = 4.5$ m $= 1.5\lambda_1$, we have

$$Z_{in} = 36 + j12 \quad \rightarrow \quad V_s = \frac{36 + j12}{86 + j12}(100) \simeq 43.7e^{j10.5^\circ} \text{ V}$$

so that

$$P_L \simeq \frac{1}{2}\frac{(43.7)^2}{(1440)}(36) \simeq 23.9 \text{ W}$$

which is a significant improvement in power delivered, achieved simply by making the first line segment longer. This result indicates that the amount of power delivered to a load sensitively depends on the electrical lengths of the transmission lines.

3.5 IMPEDANCE MATCHING

We have already encountered the concept of impedance matching in previous sections, in connection with standing waves on transmission lines. It was shown that if the characteristic impedance Z_0 of the line is equal to the load impedance Z_L, the reflection coefficient $\Gamma_L = 0$, and the standing-wave ratio is unity. When this situation exists, the characteristic impedance of the line and the load impedance are said to be *matched,* that is, they are equal. In most transmission line applications, it is desirable to match the load impedance to the characteristic impedance of the line in order to reduce reflections and standing waves that jeopardize the power-handling capabilities of the line and also distort the information transmitted. Impedance matching is also desirable in order to drive a given load most efficiently (i.e., to deliver maximum power to the load), although maximum efficiency also requires matching the generator to the line at the source end. In the presence of sensitive components (low-noise amplifiers, etc.), impedance matching improves the signal-to-noise ratio of the system and in other cases generally reduces amplitude and phase errors. In this section, we examine different methods of achieving impedance matching.

3.5.1 Matching Using Lumped Reactive Elements

The simplest way to match a given transmission line to a load is to connect a lumped reactive element in parallel (series) at the point along the line where the real part of the line admittance (impedance) is equal to the line characteristic admittance (impedance).[19] This method is useful only at relatively low frequencies for which lumped elements can be used. The method is depicted in Figure 3.29, which shows a shunt (parallel) lumped reactive element connected to the line at a distance l from the load.

Since the matching element is connected in parallel, it is more convenient to work with line admittance rather than line impedance. The normalized admittance $\bar{Y}(z)$ seen on the line looking toward the load from any position z is given by [3.34]:

$$\bar{Y}(z) = \frac{Y(z)}{Y_0} = \frac{1 - \Gamma_L e^{j2\beta z}}{1 + \Gamma_L e^{j2\beta z}}$$

In Figure 3.29, matching requires that $\bar{Y}_2 = \bar{Y}_1 + \bar{Y}_s = 1$. Thus we first need to choose the position l along the line such that $\bar{Y}(z = -l) = \bar{Y}_1 = 1 - j\bar{B}$, that is, $\Re e\{\bar{Y}_1\} = 1$. Then we choose the lumped shunt element to be purely reactive with an appropriate value such that $\bar{Y}_s = j\bar{B}$, which results in $\bar{Y}_2 = \bar{Y}_1 + \bar{Y}_s = 1$. Substituting $z = -l$, we have

$$\bar{Y}_1 = \bar{Y}(z = -l) = \frac{1 - \Gamma_L e^{-j2\beta l}}{1 + \Gamma_L e^{-j2\beta l}} = 1 - j\bar{B}$$

[19]This possibility was noted earlier in Section 3.2.2 in connection with Figure 3.19, where $\Re e\{\bar{Z}(z_1)\} = 1$ at $z_1 \simeq -0.0985\lambda$. Note that, in general, the real part of the line admittance is equal to unity at some other point $z_2 \neq z_1$.

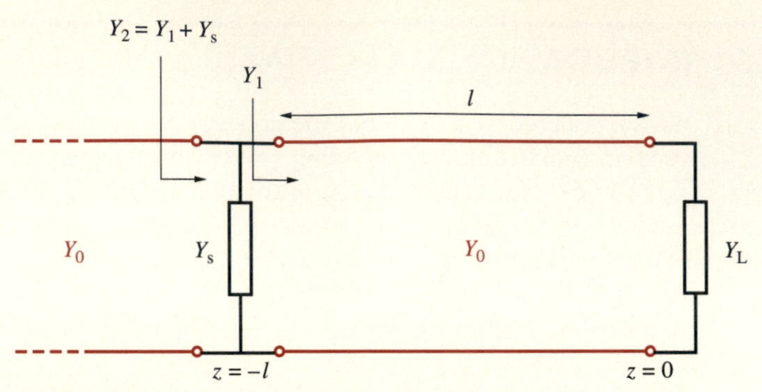

FIGURE 3.29. Matching by a lumped shunt element. The shunt element Y_s is connected at a distance l from the load such that the line admittance $Y_2(z = -l) = Y_1(z = -l) + Y_s = Y_0$.

from which we can solve for the position l of the lumped reactive element, its type (i.e., capacitor or inductor), and its normalized susceptance $-\bar{B}$. Since the load reflection coefficient Γ_L is, in the general case, a complex number given by

$$\Gamma_L = \rho e^{j\psi} = \frac{Z_L - Z_0}{Z_L + Z_0}$$

we have

$$\bar{Y}_1 = \frac{1 - \rho e^{j\psi} e^{-j2\beta l}}{1 + \rho e^{j\psi} e^{-j2\beta l}} = \frac{1 - \rho e^{j\theta}}{1 + \rho e^{j\theta}} \qquad [3.41]$$

where $\theta = \psi - 2\beta l$. By multiplying the numerator and the denominator by the complex conjugate of the denominator,[20] we obtain

$$\bar{Y}_1 = \underbrace{\frac{1 - \rho^2}{1 + 2\rho\cos\theta + \rho^2}}_{\mathcal{R}e\{\bar{Y}_1\} = 1} - j\underbrace{\frac{2\rho\sin\theta}{1 + 2\rho\cos\theta + \rho^2}}_{-\mathcal{I}m\{\bar{Y}_1\} = \bar{B}} \qquad [3.42]$$

Since $\mathcal{R}e\{\bar{Y}_1\} = 1$, we can write

$$\frac{1 - \rho^2}{1 + 2\rho\cos\theta + \rho^2} = 1$$

which yields

$$\theta = \psi - 2\beta l = \cos^{-1}(-\rho) \qquad [3.43]$$

[20] $[1 + \rho e^{j\theta}]^* = 1^* + (\rho e^{j\theta})^* = 1 + \rho e^{-j\theta}$

In other words, the distance l from the load at which $\bar{Y}(z = -l) = 1 - j\bar{B}$ is given by

$$l = \frac{\psi - \theta}{2\beta} = \frac{\psi - \cos^{-1}(-\rho)}{2\beta} = \frac{\lambda}{4\pi}[\psi - \cos^{-1}(-\rho)] \quad [3.44]$$

Note that, in general, $\theta = \cos^{-1}(-\rho)$ (with $\rho > 0$) has two solutions, one in the range $\pi/2 \leq \theta_1 \leq \pi$ and the other in the range $-\pi \leq \theta_2 \leq -\pi/2$. Also, if [3.44] results in negative values for l, then the corresponding physically meaningful solution can be found by simply adding $\lambda/2$.[21] To find $\bar{B}$, we substitute $\cos\theta = -\rho$ and $\sin\theta = \pm\sqrt{1-\rho^2}$ (where the plus sign corresponds to $\pi/2 \leq \theta_1 \leq \pi$ and the minus sign corresponds to $-\pi \leq \theta_2 \leq -\pi/2$) in the imaginary part of $\bar{Y}_1$ given by [3.42], resulting in

$$\bar{B} = -\mathcal{I}m\{\bar{Y}_1\}|_{\cos\theta=-\rho} = \pm\frac{2\rho\sqrt{1-\rho^2}}{1-2\rho^2+\rho^2} = \pm\frac{2\rho}{\sqrt{1-\rho^2}} \quad [3.45]$$

where the plus and minus signs correspond to a shunt capacitor ($\bar{B}_1 > 0$) and a shunt inductor ($\bar{B}_2 < 0$), respectively.[22] This susceptance also determines the value of the lumped reactive element $\bar{Y}_s$, which must be connected in parallel to the line in order to cancel out the reactive part of $\bar{Y}_1$. In particular, we should have $\bar{Y}_s = +j\bar{B}$ so that the total admittance $\bar{Y}_2$ seen from the left side of Y_s in Figure 3.29 is

$$\bar{Y}_2 = \bar{Y}_1 + \bar{Y}_s = 1 - j\bar{B} + j\bar{B} = 1$$

When matching with series lumped reactive elements, similar equations can be derived for the distance l away from the load at which $\bar{Z}_1 = \bar{Z}(z = -l) = 1 - j\bar{X}$,

$$l = \frac{\psi - \cos^{-1}\rho}{2\beta} = \frac{\lambda}{4\pi}(\psi - \cos^{-1}\rho) \quad [3.46]$$

and the normalized reactance of the series lumped element that would provide matching is given by

$$\bar{X} = \pm\frac{2\rho}{\sqrt{1-\rho^2}}$$

Example 3-15 illustrates impedance matching using a single shunt reactive element.

---

[21]For negative l, we have $-2\pi < 2\beta l < 0$. By adding $\lambda/2$, we have $2\beta(l+\lambda/2) = 2\beta l + 2(2\pi/\lambda)(\lambda/2) = 2\beta l + 2\pi > 0$, which lies between 0 and 2π.

[22]Note that $\theta = \cos^{-1}(-\rho)$ is an angle that is either in the second quadrant, $\pi/2 \leq \theta_1 \leq \pi$ (when $\sin\theta_1 > 0$, requiring a capacitive element based on the polarity of the imaginary part of $\bar{Y}_1$ as given in [3.42]) or the third quadrant, $-\pi \leq \theta_2 \leq -\pi/2$ (when $\sin\theta_2 < 0$, requiring an inductive element).

Example 3-15: Matching with a single reactive element. An antenna having a feed-point impedance of 110Ω is to be matched to a 50Ω coaxial line with $v_p = 2 \times 10^8$ m/s using a single shunt lumped reactive element, as shown in Figure 3.30. Find the position (nearest to the load) and the appropriate value of the reactive element for operation at 30 MHz using (a) a capacitor, and (b) an inductor.

Solution: The load reflection coefficient is

$$\Gamma_L = \frac{Z_L - Z_0}{Z_L + Z_0} = \frac{110 - 50}{110 + 50} = 0.375$$

Using [3.45], the reactive admittance (or susceptance) at the position of the shunt element is

$$\bar{B} = \pm \frac{2 \times 0.375}{\sqrt{1 - (0.375)^2}} \simeq \pm 0.809$$

(a) For $\bar{B}_1 \simeq +0.809$, the shunt element must be a capacitor. The nearest position of the capacitor with respect to the load can be found as

$$l_1 = -\frac{\theta_1}{2\beta} \simeq -\frac{\lambda}{4\pi} \underbrace{\cos^{-1}(-0.375)}_{\pi/2 \leq \theta_1 \leq \pi} \simeq -\frac{\lambda}{4\pi}(1.955) \simeq -0.156\lambda \simeq 0.344\lambda$$

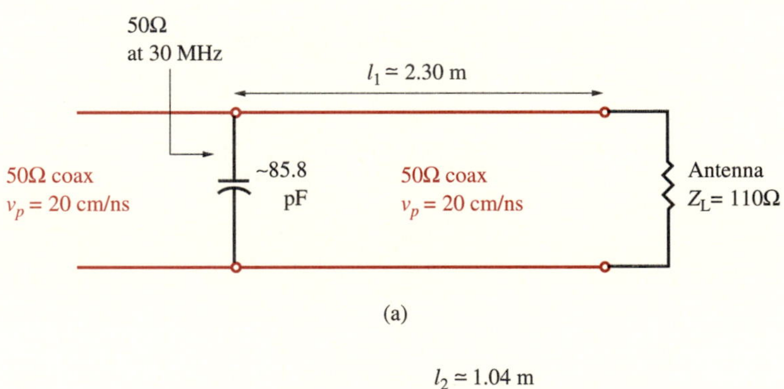

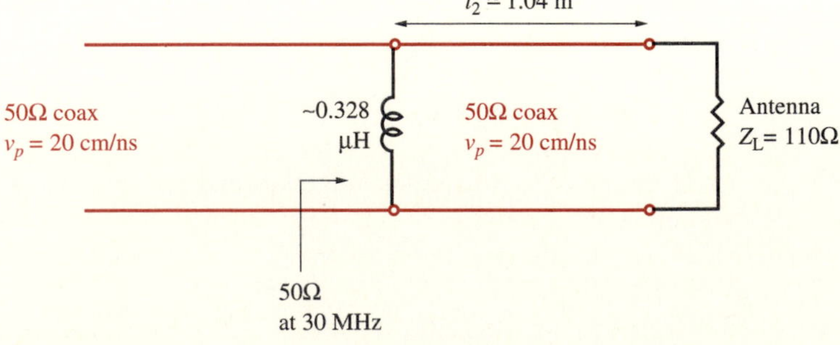

FIGURE 3.30. Matching with a single reactive element. The two solutions determined in Example 3-15. (a) Using a shunt capacitor. (b) Using a shunt inductor.

Since $\lambda = v_p/f = 2\times10^8/(30\times10^6) \simeq 6.67$ m, the actual position of the shunt capacitor is $l_1 \simeq 0.344\times6.67 \simeq 2.30$ m. To determine the capacitance C_s, we use

$$(j\omega C_s)(Z_0) = j\bar{B}_1 \quad\rightarrow\quad [j(2\pi\times30\times10^6 C_s)(50)] \simeq +j0.809$$
$$\rightarrow\quad C_s \simeq 85.8 \text{ pF}$$

(b) For $\bar{B}_2 \simeq -0.809$, the shunt element must be an inductor. Similarly, the nearest position of the inductor is

$$l_2 = -\frac{\theta_2}{2\beta} \simeq -\frac{\lambda}{4\pi}\underbrace{\cos^{-1}(-0.375)}_{-\pi\le\theta_2\le-\pi/2} \simeq -\frac{\lambda}{4\pi}(-1.955) \simeq 0.156\lambda$$

Using $\lambda \simeq 6.67$ m, the actual position of the shunt inductor is $l_2 \simeq 0.156\times 6.67 \simeq 1.04$ m. To determine the inductance L_s, we use

$$[-j/(\omega L_s)](Z_0) = j\bar{B}_2 \quad\rightarrow\quad [-j/(2\pi\times30\times10^6 L_s)](50) \simeq -j0.809$$
$$\rightarrow\quad L_s \simeq 0.328\ \mu\text{H}$$

3.5.2 Matching Using Series or Shunt Stubs

In Section 3.2, we saw that short- or open-circuited transmission lines can be used as reactive circuit elements. At microwave frequencies, it is often impractical or inconvenient to use lumped elements for impedance matching. Instead, we use a common matching technique that uses single open- or short-circuited stubs (i.e., transmission line segments) connected either in series or in parallel, as illustrated in Figure 3.31. In practice, the short-circuited stub is more commonly used for coaxial and waveguide applications because a short-circuited line is less sensitive to external influences (such as capacitive coupling and pick-up) and radiates less than an open-circuited line segment. However, for microstrips and striplines, open-circuited stubs are more common in practice because they are easier to fabricate. For similar practical reasons, the shunt (parallel) stub is more convenient than the series stub; the discontinuity created by breaking the line may disturb the voltage and current in the case of the series stub.

The principle of matching with stubs is identical to that discussed in Section 3.5.1 for matching using shunt lumped reactive elements. The only difference here is that the matching admittance Y_s is introduced by using open- or short-circuited line segments (or stubs) of appropriate length l_s, as shown in Figure 3.32. In the following, we exclusively consider the case of matching with a short-circuited stub, as illustrated in Figure 3.32. The corresponding analysis for open-circuited stubs is similar in all respects and is left as an exercise for the reader.

With the required location l and the normalized admittance $\bar{B}$ of the stub as determined from [3.44] and [3.45], we need only to find the length of the stub l_s necessary to present a normalized admittance of $\bar{Y}_s = +j\bar{B}$ at the junction. For this purpose, we can use expression [3.17] from Section 3.2 for the normalized

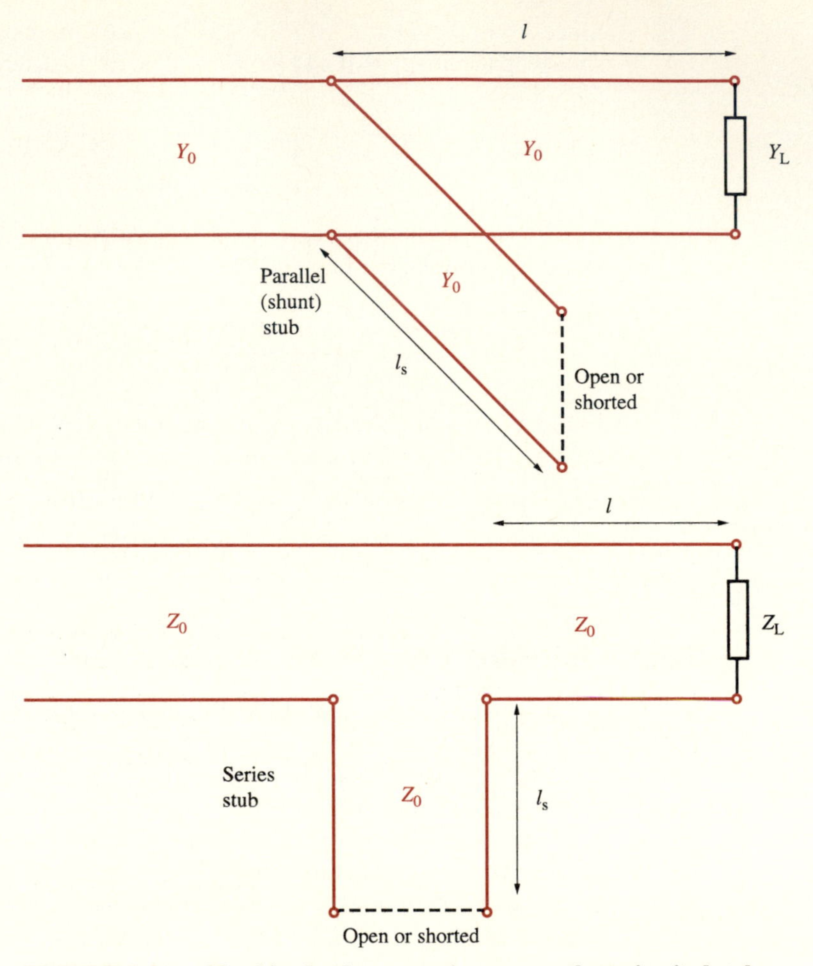

FIGURE 3.31. **Matching by shunt or series open- or short-circuited stubs.**

input impedance of a short-circuited line of length l_s and set the corresponding normalized admittance equal to $+j\bar{B}$. Recalling that for a short-circuited line $\bar{Z}_\mathrm{in} = j\tan(\beta l_\mathrm{s})$, we have

$$\bar{Y}_\mathrm{s} = \frac{1}{j\tan(\beta l_\mathrm{s})} = +j\bar{B}$$

or

$$\tan(\beta l_\mathrm{s}) = -\frac{1}{\bar{B}} \qquad [3.47]$$

The value of $\bar{B}$ determined from [3.45] can be used in [3.47] to find the length l_s of the short-circuited stub. Note that in [3.47], we have assumed the characteristic impedance of the short-circuited stub to be equal to that of the main line.

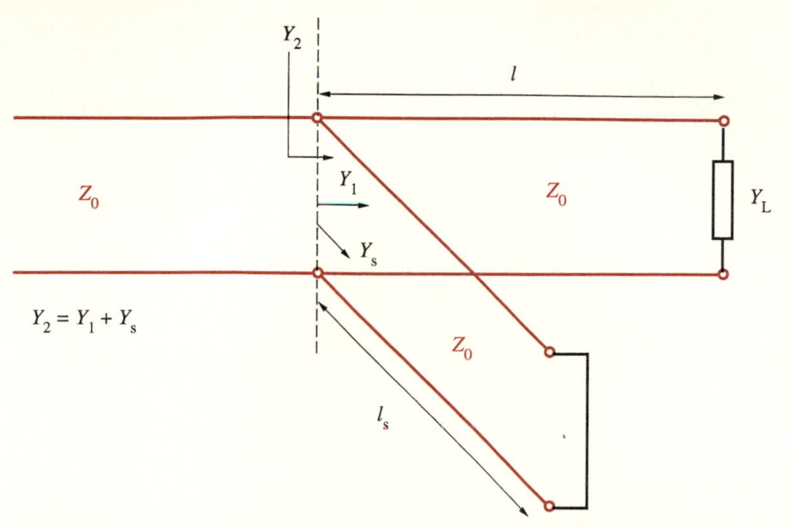

FIGURE 3.32. **Matching with a single parallel (shunt) short-circuited stub.**

In practice, single-stub matching can be achieved even if the load impedance Z_L is not explicitly known, by relying on measurements of S to determine ρ and measurements of the location of the voltage minimum or maximum to determine ψ. To see this, consider that the stub location l can be measured relative to the position z_{max} of the nearest voltage maximum toward the load end so that $l - \Delta l_{max} = |z_{max}| < \lambda/2$, as shown in Figure 3.33. Using [3.43], we can then write

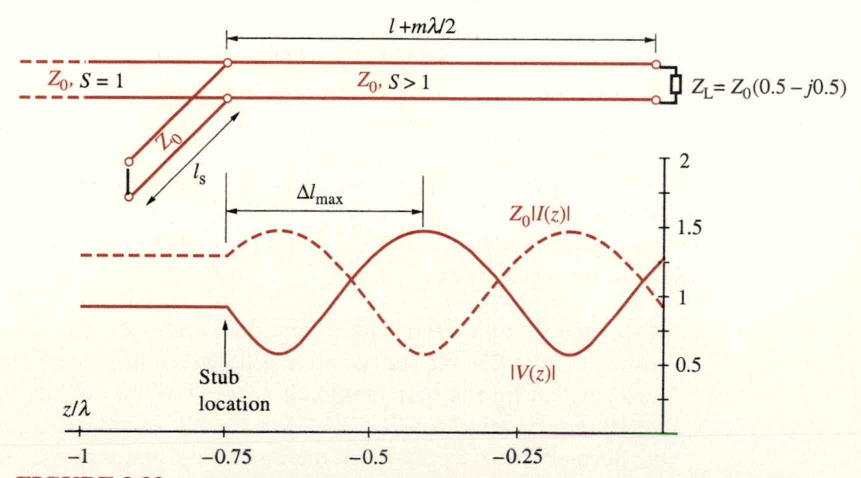

FIGURE 3.33. **Voltage standing-wave pattern on a transmission line with single-stub matching.** The standing-wave ratio S is unity to the left of the stub. The stub location is at a distance of Δl_{max} from the nearest voltage maximum toward the load end. The particular case shown is for $\bar{Z}_L = 0.5 - j0.5$. Note that, as usual, we have assumed $V^+ = 1$.

$$\theta = \psi - 2\beta l = \psi + 2\beta z_{\max} - 2\beta \Delta l_{\max}$$
$$= -m2\pi - 2\beta\Delta l_{\max} \qquad m = 0, 1, 2, \ldots$$

noting that $m = 0$ does not apply if $-\pi \leq \psi < 0$. Using the preceding expression for θ, we have

$$\Delta l_{\max} = -\frac{1}{2\beta}[\theta + m2\pi] = -\frac{1}{2\beta}[\cos^{-1}(-\rho) + m2\pi], \quad m = 0, 1, 2, \ldots \quad [3.48]$$

Thus, ρ can be directly determined from the measured standing-wave ratio, and [3.48] determines the stub location with respect to the measured location of the voltage maximum.

Figure 3.33 also illustrates the fact that, although the proper choice of the stub location l and its length l_s achieves matching so that the standing-wave ratio on the source side of the stub is unity, a standing wave does exist on the segment of line between the stub and the load.

Impedance matching using a single short-circuited transmission line stub is illustrated in Example 3-16.

Example 3-16: Single-stub matching. Design a single-stub system to match a load consisting of a resistance $R_L = 200\Omega$ in parallel with an inductance $L_L = 200/\pi$ nH to a transmission line with characteristic impedance $Z_0 = 100\Omega$ and operating at 500 MHz. Connect the stub in parallel with the line.

Solution: At 500 MHz, the load admittance is given by

$$Y_L = \frac{1}{R_L} - j\frac{1}{\omega L_L} = \frac{1}{200} - j\frac{1}{[2\pi(500 \times 10^6)(200/\pi)(10^{-9})]} = 0.005 - j0.005 \text{ S}$$

The reflection coefficient at the load is

$$\Gamma_L = \frac{Y_0 - Y_L}{Y_0 + Y_L} = \frac{0.01 - (0.005 - j0.005)}{0.01 + (0.005 - j0.005)} = \frac{1 + j}{3 - j} \simeq 0.447e^{j63.4°}$$

so that $\rho \simeq 0.447$ and $\psi \simeq 63.4°$. Using [3.44], we have

$$l \simeq \frac{1.11 - \cos^{-1}(-0.447)}{2(2\pi/\lambda)} \simeq \left(\frac{1.11 \mp 2.034}{4\pi}\right)\lambda \quad \rightarrow \quad \begin{matrix} l_1 \simeq -0.073\lambda \\ l_2 = 0.25\lambda \end{matrix}$$

where the first solution with a negative value of l can be realized by simply adding 0.5λ so that the stub position is between the load and the source. Thus, the stub position for the first solution is $l_1 \simeq -0.073\lambda + 0.5\lambda = 0.426\lambda$. Note that we have used $\theta = \cos^{-1}(-0.447) \simeq \pm 117° \simeq \pm 2.034$ radians. Using [3.45], the normalized susceptance of the input admittance of the short-circuited shunt stub needed is

$$\bar{B} \simeq \pm\frac{2(0.447)}{\sqrt{1 - (0.447)^2}} \simeq \pm 1$$

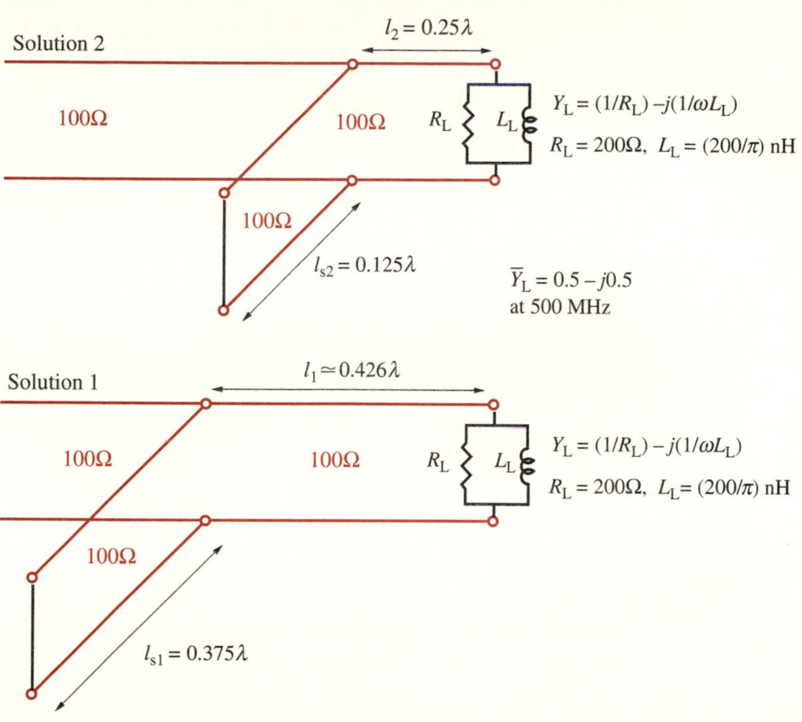

FIGURE 3.34. **Two alternative single-stub matching solutions.** Solution 2 would in general be preferred since it has shorter segments of line (and a shorter stub) over which the standing-wave ratio differs from unity.

where the plus sign corresponds to the stub position l_1 and the minus sign corresponds to l_2. From [3.47], the length l_{s1} of the short-circuited stub at a distance l_1 from the load is

$$l_{s1} = \frac{\lambda}{2\pi}\tan^{-1}\left(-\frac{1}{\overline{B}}\right) \simeq \frac{\lambda}{2\pi}\tan^{-1}(-1) = -0.125\lambda \quad \rightarrow \quad 0.375\lambda$$

and similarly, the stub length l_{s2} needed at position l_2 is

$$l_{s2} = \frac{\lambda}{2\pi}\tan^{-1}(1) = 0.125\lambda$$

Both of the alternative solutions are shown in Figure 3.34. Since Solution 2 gives a stub position closer to the load and a shorter stub length, it would usually be preferred over Solution 1. In general, standing waves jeopardize power-handling capabilities of a line and also lead to signal distortion. Thus, it is desirable to minimize the lengths of line over which the standing-wave ratio is large. In the case shown in Figure 3.34, more of the line operates under matched ($S = 1$) conditions for Solution 2.

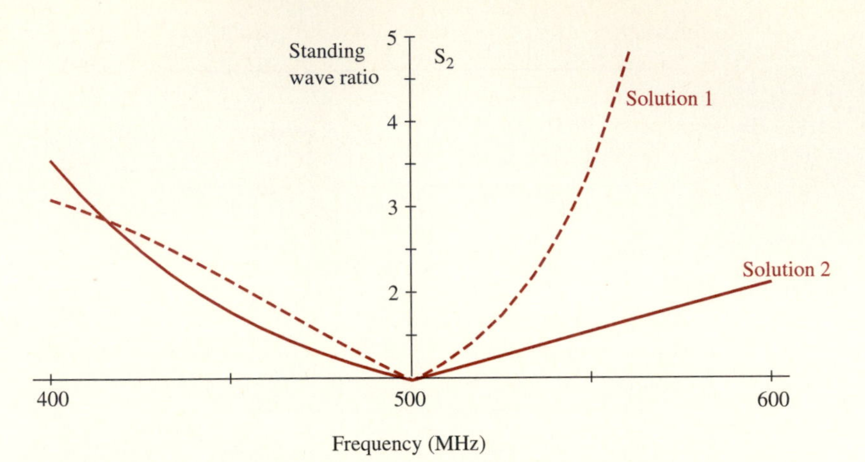

FIGURE 3.35. Frequency sensitivity of single-stub matching. The standing-wave ratio S_2 versus frequency for the two alternative solutions given in Figure 3.34.

The frequency dependence of the various designs can also be important in practice. A comparison of the frequency responses of the two alternative solutions for the previous example is given in Figure 3.35. Note that the load admittance as a function of frequency $f = \omega/(2\pi)$ is given by

$$Y_L(f) = \frac{1}{200} - j\frac{10^9}{400f}$$

Assuming that the phase velocity along the line is equal to the speed of light c, we have $\beta l = 2\pi f l/c$. The line admittance seen just to the right of the short-circuited stub is given as a function of frequency:

$$Y_1(f) = Y_0 \frac{Y_L + jY_0 \tan(2\pi f l/c)}{Y_0 + jY_L \tan(2\pi f l/c)}$$

If we also assume $\beta l_s = 2\pi f l_s/c$, the total line admittance seen from the source side of the short-circuited stub (see Figure 3.32) is

$$Y_2(f) = Y_s + Y_1 = \frac{-jY_0}{\tan(2\pi f l_s/c)} + Y_0 \frac{Y_L + jY_0 \tan(2\pi f l/c)}{Y_0 + jY_L \tan(2\pi f l/c)}$$

The reflection coefficient Γ_2 and the standing-wave ratio S_2 are then given as

$$\Gamma_2(f) = \frac{Y_0 - Y_2(f)}{Y_0 + Y_2(f)}; \quad S_2(f) = \frac{1 + |\Gamma_2(f)|}{1 - |\Gamma_2(f)|}$$

The quantity $S_2(f)$ is plotted in Figure 3.35 as a function of frequency between 400 and 600 MHz. Note that the bandwidths[23] of the two designs are dramatically

[23]Defined as the frequency range over which the standing-wave ratio S_2 is lower than a given amount, for example, $S_2 < 2$. The particular value of S_2 used depends on the application in hand.

different. Which solution to choose depends on the particular application in hand, although in most cases minimizing reflections ($S_2 < 2$, for example) over a wider frequency range is desirable. Note that, for the case shown, Solution 2 provides matching over a substantially broader range of frequencies than does Solution 1.

3.5.3 Quarter-Wave Transformer Matching

A powerful method for matching a given load impedance to a transmission line that is used to drive it is the so-called quarter-wave transformer matching. This method takes advantage of the impedance inverting property of a transmission line of length $l = \lambda/4$, namely, the fact that the input impedance of a line of length $l = \lambda/4$ is given by

$$Z_{\text{in}}\Big|_{l=\lambda/4} = Z_0 \frac{Z_{\text{L}} + jZ_0 \tan(\beta l)}{Z_0 + jZ_{\text{L}} \tan(\beta l)}\Bigg|_{l=\lambda/4} = \frac{Z_0^2}{Z_{\text{L}}}$$

or in terms of normalized impedances, we have

$$\bar{Z}_{\text{in}}\Big|_{l=\lambda/4} = \frac{Z_{\text{in}}}{Z_0}\Bigg|_{l=\lambda/4} = \frac{Z_0}{Z_{\text{L}}} = \frac{1}{\bar{Z}_{\text{L}}} \qquad [3.49]$$

hence the term "impedance inverter." Thus, a quarter-wave section transforms impedance in such a way that a kind of inverse of the terminating impedance appears at its input.

Consider a quarter-wavelength transmission line segment of characteristic impedance Z_{Q}, as shown in Figure 3.36. In the general case of a complex load impedance $Z_{\text{L}} = R_{\text{L}} + jX_{\text{L}}$, we have

$$Z_{\text{in}} = \frac{Z_{\text{Q}}^2}{R_{\text{L}} + jX_{\text{L}}} \quad \longrightarrow \quad Y_{\text{in}} = \frac{R_{\text{L}}}{Z_{\text{Q}}^2} + j\frac{X_{\text{L}}}{Z_{\text{Q}}^2}$$

so that a load consisting of a series resistance (R_{L}) and an inductive reactance ($X_{\text{L}} > 0$) appears at the input of the quarter-wave section as an admittance consisting of a conductance $R_{\text{L}}/Z_{\text{Q}}^2$ (or a resistance $Z_{\text{Q}}^2/R_{\text{L}}$) *in parallel* with a capacitive susceptance $X_{\text{L}}/Z_{\text{Q}}^2 > 0$. Similarly, if the load were capacitive ($X_{\text{L}} < 0$), it would appear as a conductance $R_{\text{L}}/Z_{\text{Q}}^2$ in parallel with an inductive susceptance $X_{\text{L}}/Z_{\text{Q}}^2 < 0$.

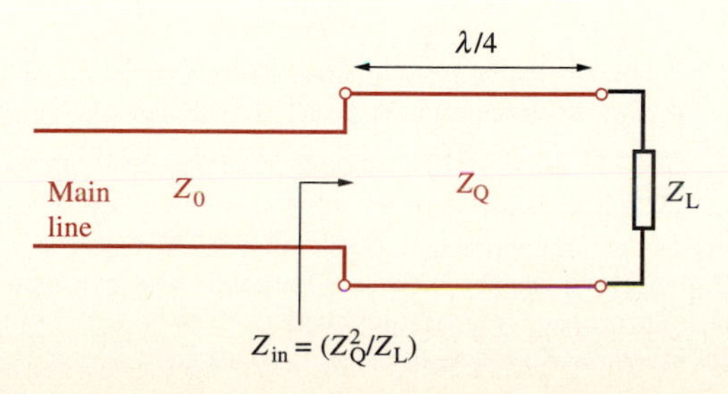

FIGURE 3.36. Quarter-wave transformer. A load Z_{L} to be driven by a transmission line of characteristic impedance Z_0 connected to the line via a quarter-wavelength-long line of characteristic impedance Z_{Q}.

Purely Resistive Loads The practical utility of the impedance-inverting property of a quarter-wavelength transmission line becomes apparent when we consider a purely resistive load. Any arbitrary purely resistive load impedance $Z_L = R_L$ is transformed into a purely resistive input impedance of Z_Q^2/R_L. Thus, by appropriately choosing the value of the characteristic impedance Z_Q of the quarter-wavelength line, its input impedance can be made equal to the characteristic impedance Z_0 (a real value for a lossless line) of the main line that is to be used to drive the load. This property of the quarter-wave line can be used to match two transmission lines of different characteristic impedances or to match a load impedance to the characteristic impedance of a transmission line. Note that the matching section must have a characteristic impedance of

$$Z_Q = \sqrt{R_1 R_2} \qquad [3.50]$$

where R_1 and R_2 are the two resistive impedances to be matched. Note that in the case shown in Figure 3.36, $R_1 = Z_0$ and $R_2 = R_L$. Alternatively, R_2 could be the characteristic impedance of another transmission line that may need to be matched to the main line (Z_0) using the quarter-wave section.

The following example illustrates quarter-wave transformer matching of a purely resistive load.

Example 3-17: Quarter-wave transformer for a monopole antenna. Design a quarter-wavelength section to match a thin monopole antenna of length 0.24λ[24] having a purely resistive feed-point impedance of $R_L \simeq 30\Omega$ to a transmission line having a characteristic impedance of $Z_0 = 100\Omega$.

Solution: With reference to Figure 3.37 and according to [3.50], the $\lambda/4$ section must match two impedances—$R_2 = R_L$ and $R_1 = Z_0 = 100\Omega$—and thus

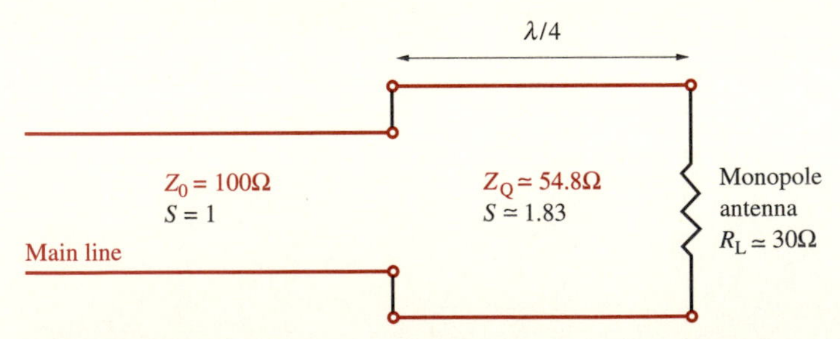

FIGURE 3.37. Quarter-wave transformer. The load is a monopole antenna of length 0.24λ, which has a purely resistive impedance of 30Ω.

---

[24]The reactive part of the impedance of such monopole antennas with length just shorter than a quarter wavelength is nearly zero. Monopole or dipole antennas with purely resistive input impedances are referred to as resonant antennas and are used for many applications. (See Section 14.06 of E. Jordan and K. Balmain, *Electromagnetic Waves and Radiating Systems*, Prentice Hall, 1968.)

must have a characteristic impedance Z_Q of

$$Z_Q = \sqrt{R_1 R_2} = \sqrt{Z_0 R_L} = \sqrt{(100)(30)} \simeq 54.8\Omega$$

The standing-wave ratio is unity beyond the quarter-wave section. However, note that $S \simeq 1.83$ within the $\lambda/4$ section.

Complex Load Impedances In using quarter-wave transformers to match a complex load impedance to a lossless transmission line (i.e., where Z_0 is real), it is necessary to insert the quarter-wave segment at the point along the line where the line impedance $\bar{Z}(z)$ is purely resistive. As discussed in previous sections, this point can be the position of either the voltage maximum or minimum. In most cases, it is desirable to choose the point closest to the load in order to minimize the length of the transmission line segment on which $S \neq 1$ because the presence of standing waves jeopardizes power-handling capabilities of the line, tends to reduce signal-to-noise ratio, and may lead to distortion of the signal transmitted. Example 3-18 illustrates quarter-wave matching of a complex load.

Example 3-18: Thin-wire half-wave dipole antenna. A thin-wire half-wave dipole antenna[25] has an input impedance of $Z_L = 73 + j42.50\Omega$. Design a quarter-wave transformer to match this antenna to a transmission line with characteristic impedance $Z_0 = 100\Omega$.

Solution: We start by evaluating the reflection coefficient at the load

$$\Gamma_L = \rho e^{j\psi} = \frac{Z_L - Z_0}{Z_L + Z_0} = \frac{73 + j42.5 - 100}{73 + j42.5 + 100} \simeq 0.283 e^{j109°}$$

so that $\psi \simeq 109° \simeq 1.896$ radians. Note that the standing-wave ratio is

$$S = \frac{1+\rho}{1-\rho} \simeq \frac{1+0.283}{1-0.283} \simeq 1.79$$

From previous sections, we know that, for an inductive load, the first voltage maximum is closer to the load than the first voltage minimum (see Figure 3.17). The first voltage maximum is at

$$\psi + 2\beta z_{max} = 0 \quad \longrightarrow \quad z_{max} = -\frac{\psi}{2\beta} \simeq -\frac{1.896\lambda}{4\pi} \simeq -0.151\lambda$$

Thus the quarter-wave section should be inserted at $z \simeq -0.151\lambda$, as shown in Figure 3.38.

---

[25] See, for example, Section 14.06 of E. C. Jordan and K. Balmain, *Electromagnetic Waves and Radiating Systems,* Prentice Hall, 1968.

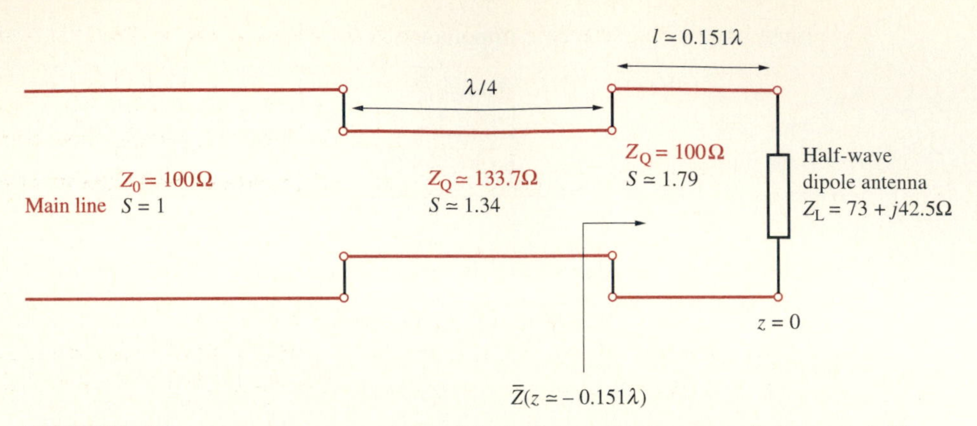

FIGURE 3.38. Quarter-wave matching of the half-wave dipole antenna in Example 3-18.

Noting that the normalized load impedance is $\bar{Z}_L = 0.73 + j0.425$ and that we have $\tan[\beta(0.151\lambda)] \simeq 1.392$, the normalized line impedance $\bar{Z}(z)$ seen toward the load at $z \simeq -0.151\lambda$ is

$$\bar{Z}(z)\big|_{z\simeq-0.151\lambda} = \left.\frac{\bar{Z}_L - j\tan(\beta z)}{1 - j\bar{Z}_L\tan(\beta z)}\right|_{z\simeq-0.151\lambda} \simeq \frac{0.73 + j0.425 + j1.392}{1 + j(0.73 + j0.425)(1.392)} \simeq 1.79$$

Note that $\bar{Z}(z \simeq -0.151\lambda) \simeq 1.79 = S$, as expected on the basis of the discussion in Section 3.3.2 (i.e., $\bar{R}_{\max} = S \simeq 1.79$). The characteristic impedance Z_Q of the quarter-wave section should thus be

$$Z_Q = \sqrt{Z(z \simeq -0.151\lambda)Z_0} \simeq \sqrt{(100)(1.79)(100)} \simeq 133.7\Omega$$

Note that in general, as in this specific example, we have $\bar{Z}(z = z_{\max}) = S$, and thus $Z_Q = Z_0\sqrt{S}$. Note also that the standing-wave ratio in the quarter-wave section is $S \simeq 1.34$, as can be calculated by using $Z(z \simeq -0.151\lambda)$ and Z_Q.

Frequency Sensitivity of Quarter-Wave Matching The frequency sensitivity of a quarter-wave transformer is a serious limitation since the design is perfect (i.e., provides $S = 1$) only at the frequency for which the length of the transformer segment is exactly $\lambda/4$. The bandwidth of the transformer can be assessed by plotting the standing-wave ratio S versus frequency, as was done in Section 3.5.2 for the single-stub tuning example and as is shown in the following example.

Example 3-19: Multiple-stage quarter-wave transformers. A resistive load of $R_L = 75\Omega$ is to be matched to a transmission line with characteristic impedance $Z_0 = 300\Omega$. The frequency of operation is $f_0 = 300$ MHz (which

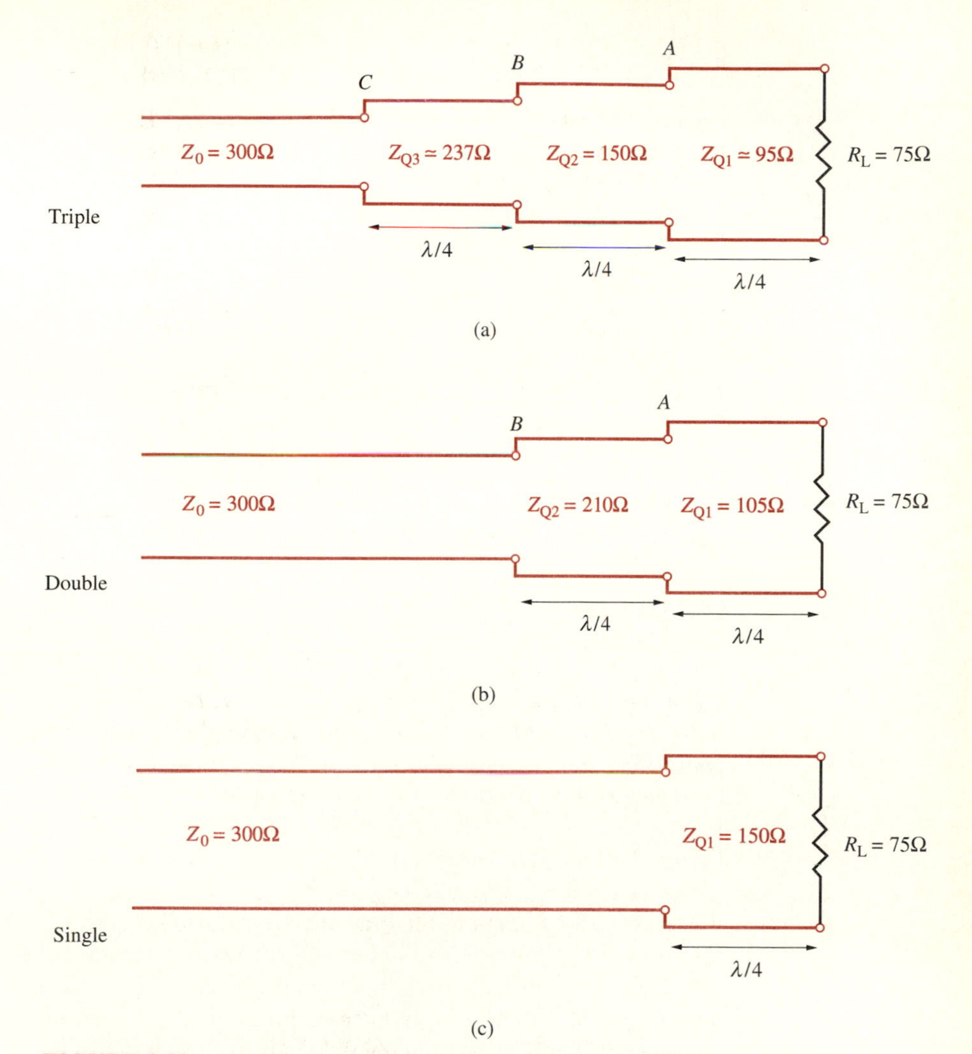

FIGURE 3.39. Multiple-stage quarter-wave matching. Matching with triple, double, and single $\lambda/4$ sections are illustrated respectively in panels (a), (b), and (c).

corresponds to $\lambda_0 = v_p/f_0 \simeq 3 \times 10^8/(300 \times 10^6) = 1$ m, assuming an air-filled coaxial line). Design multiple cascaded quarter-wave matching transformers and at 300 MHz compare their frequency responses between 200 and 400 MHz.

Solution: Three different designs are shown in Figure 3.39. Note that the choice of the impedance Z_{Q1} for the single-transformer case is straightforward. For the double-transformer case, the condition for exact quarter-wave matching is $Z_{Q2} = Z_{Q1}\sqrt{Z_0/R_L}$, allowing for different choices of Z_{Q2} and Z_{Q1} as long as the condition is satisfied. The design shown in Figure 3.39b is one that provides a standing-wave ratio in the first and second quarter-wave sections of

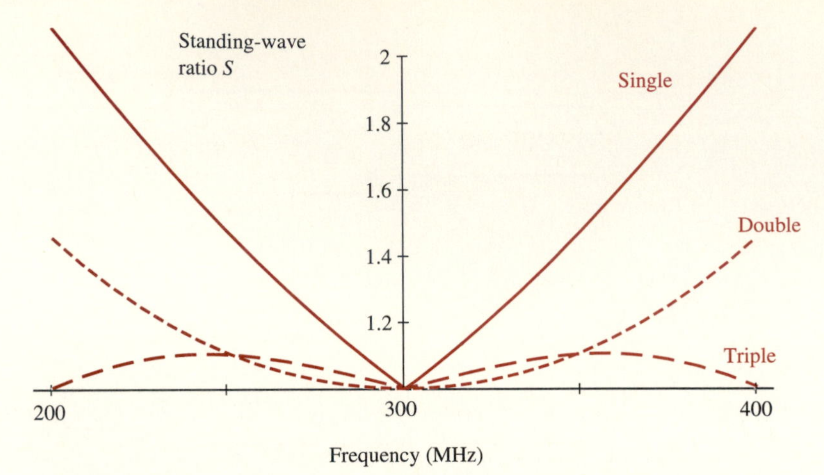

FIGURE 3.40. Quarter-wave transformer bandwidth. Standing-wave ratio S versus frequency for the three different transformers shown in Figure 3.39.

respectively $S \simeq 1.4$ and $S \simeq 1.6$. For the triple transformer case the condition for exact quarter-wave matching can be shown to be $Z_{Q1}Z_{Q3} = Z_{Q2}\sqrt{Z_0 R_L} = 150Z_{Q2}$. Once again, many different combinations of Z_{Q1}, Z_{Q2}, Z_{Q3} satisfy this condition, and other performance criteria (such as minimizing S in the quarter-wave sections) must be used to make particular design choices. One design approach[26] is to require the characteristic impedance of the second quarter-wave segment to be the geometric mean of the two impedances to be matched, namely $Z_{Q2} = \sqrt{(300)(75)} = 150\Omega$. The choices of Z_{Q1} and Z_{Q3} shown in Figure 3.39 is one that provides a relatively low value of $S <\sim 1.26$ in all the transmission line segments.

Figure 3.40 compares the three different designs in terms of their frequency response. As in Figure 3.35, we base this comparison on the behavior of the total standing-wave ratio $S(f)$ as a function of frequency. First we note that since the wavelength at 300 MHz is 1 m (assuming a transmission line with air as the material surrounding the conductors), all of the quarter-wave segments have physical lengths of 0.25 m each. To evaluate $S(f)$, we can start at the load end and transform impedances as we move toward the source end in accordance with

$$Z_i(f) = Z_{0i}\frac{Z_{i-1} + jZ_{0i}\tan[2\pi f(0.25\lambda_0/v_p)]}{Z_{0i} + jZ_{i-1}\tan[2\pi f(0.25\lambda_0/v_p)]}$$

where $\lambda_0 = v_p/f_0 = 1$ m, $v_p = c$, and Z_i and Z_{0i} are, respectively, the input impedance and the characteristic impedance of the ith quarter-wave transformer over which the impedance is being transformed, and Z_{i-1} is the input impedance of the $(i-1)$th quarter-wave transformer seen looking toward the load.

[26]There are various established methods for the design of multisection quarter-wavelength transformers. The approach described here is an ad-hoc one.

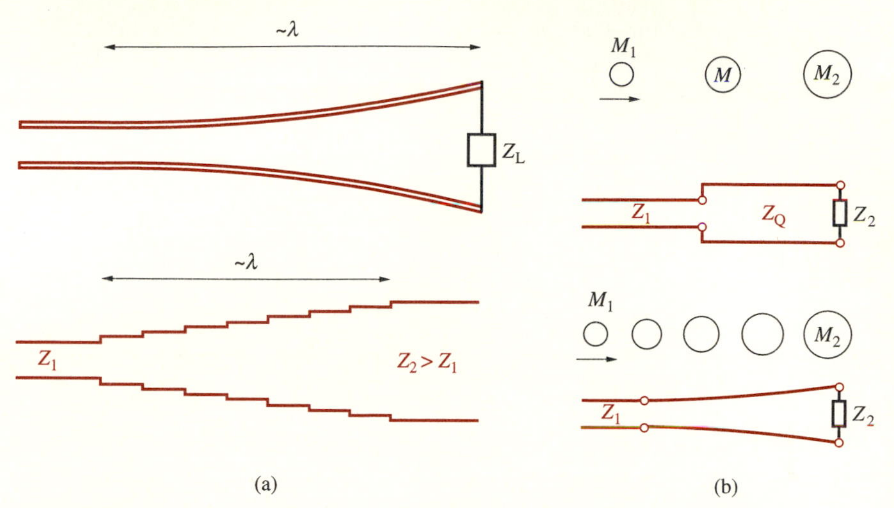

FIGURE 3.41. Tapered impedance transformer and a mechanical analogy. (a) A gradual taper provides wide bandwidth; many small reflections from a series of small incremental steps add with different phases to produce very small net total reflection. (b) Mechanical analog of impedance matching to provide better power absorption at a termination.

It is clear from Figure 3.40 that the bandwidth performance improves with the use of multiple segments. The improvement between a single and double quarter-wave section is very significant, with a tolerable standing-wave ratio of $S <\sim 1.2$ being achieved over a much larger range in a double quarter-wave section. The triple transformer provides for $S <\sim 1.1$ over the entire range of 200 to 400 MHz. In practice, little improvement is obtained by cascading more than four sections.[27]

In the limit of adding more and more sections, we would approach an infinitely long, smooth, gradually tapered transmission line with virtually no reflections. This is illustrated in Figure 3.41a. In practice, it is usually sufficient to make the tapered section of length $\sim\lambda$ or more.

A Mechanical Analogy Impedance matching to achieve maximum energy transfer is an essential aspect of not just electrical but also other types of physical systems. One analogy is the transfer of energy in an elastic head-on collision between a mass M_1 moving with a speed v_M and a stationary mass M_2. In the absence of losses, and based on the conservation of momentum, and given the elastic nature of the collision, we know that if $M_2 = M_1$, then all the energy resident in M_1 is transferred to M_2 (i.e., M_1 stops and M_2 moves away at velocity v_M). If the $M_2 = M_1$

[27]S. Guccione, Nomograms for speed design of λ/4 transformers, *Microwaves,* August 1975.

condition is not met, only a fraction of the total energy is transferred to M_2, and M_1 either reflects and moves in the reverse direction (if $M_1 < M_2$) or continues its motion at reduced speed (if $M_1 > M_2$).

The transfer of energy between M_1 and M_2 can be improved by the insertion of a third mass between them, as shown in Figure 3.41b. The energy transfer is optimum if the third mass M is the geometric mean of the other two, that is, if $M = \sqrt{M_1 M_2}$. Further improvement in the energy transfer can be achieved by using several bodies with masses varying monotonically between M_1 and M_2.

3.6 THE SMITH CHART

Many transmission-line problems can be solved easily with graphical procedures, using the so-called *Smith chart.*[28] The Smith chart is also a useful tool for visualizing transmission-line matching and design problems. Many aspects of the voltage, current, and impedance patterns discussed in previous sections can also be interpreted and visualized by similar means using the Smith chart. One might think that graphical techniques are not as useful in this age of powerful computers and calculators, but it is interesting to note that some commonly used pieces of laboratory test equipment have displays that imitate the Smith chart, with the line impedance and standing-wave ratio results presented on such displays. In this section, we describe the Smith chart and provide examples of its use in understanding transmission-line problems.

3.6.1 Mapping of Complex Impedance to Complex Γ

The Smith chart is essentially a conveniently parameterized plot of the normalized line impedance $\bar{Z}(z)$ of a transmission line and the generalized voltage reflection coefficient $\Gamma(z)$ as a function of distance from the load. To understand the utility of the Smith chart, we need to understand the relationship between $\bar{Z}(z)$ and $\Gamma(z)$.

From [3.24], we can write the normalized line impedance as

$$\bar{Z}(z) = \frac{1 + \Gamma(z)}{1 - \Gamma(z)} \qquad [3.51]$$

where we note that $\Gamma(z) = \rho e^{j(\psi + 2\beta z)}$ is the voltage reflection coefficient at any position z along the line. Denoting $\Gamma(z)$ simply as Γ, while keeping in mind that it is a function of z, we can rewrite [3.51] as

$$\boxed{\bar{Z} = \frac{1 + \Gamma}{1 - \Gamma}} \qquad [3.52]$$

[28] P. H. Smith, *Electronics,* January 1939. Also see J. E. Brittain, The Smith chart, *IEEE Spectrum,* 29(8), p. 65, August 1992.

where $\bar{Z} = \bar{R} + j\bar{X}$ and $\Gamma = u + jv$ are both complex numbers, so [3.52] represents a mapping between two complex numbers. Note that if the load is given, then we know Γ_L, and therefore Γ (and thus $\bar{Z} = \bar{R} + j\bar{X}$, from [3.52]), at any position at a distance z from the load. The Smith chart conveniently displays values of $\bar{Z}$ (or $\bar{R}$, $\bar{X}$) on the Γ (or u, v) plane for graphical calculation and visualization. From [3.52] we have

$$\bar{Z} = \bar{R} + j\bar{X} = \frac{1 + (u + jv)}{1 - (u + jv)} = \frac{[1 + (u + jv)][1 - (u - jv)]}{(1 - u)^2 + v^2} \qquad [3.53]$$

Equating real parts in [3.53] and rearranging, we have

$$\left(u - \frac{\bar{R}}{1 + \bar{R}}\right)^2 + v^2 = \left(\frac{1}{1 + \bar{R}}\right)^2 \qquad [3.54]$$

which is the equation of a circle in the uv plane centered at $u = \bar{R}/(1 + \bar{R})$, $v = 0$ and having a radius of $1/(1 + \bar{R})$. Examples of such circles are shown in Figure 3.42a. Note that $\bar{R} = 1$ corresponds to a circle centered at $u = \frac{1}{2}$, $v = 0$, having a radius $\frac{1}{2}$, and passing through the origin in the uv plane.

Similarly, by equating the imaginary parts in [3.53] and rearranging, we find

$$(u - 1)^2 + \left(v - \frac{1}{\bar{X}}\right)^2 = \frac{1}{\bar{X}^2} \qquad [3.55]$$

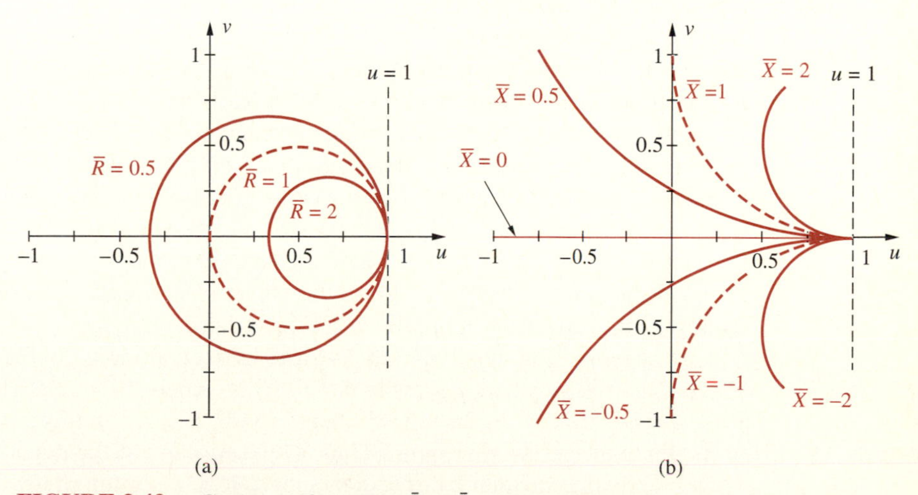

FIGURE 3.42. **Contours of constant $\bar{R}$ or $\bar{X}$.** (a)The circles in the uv plane are centered at $[\bar{R}(1 + \bar{R})^{-1}, 0]$, with radius $(1 + \bar{R})^{-1}$. Note the $\bar{R} = 1$ circle (dashed lines) passes through the origin (i.e., $u, v = 0$); this circle is centered at $(\frac{1}{2}, 0)$ with radius $\frac{1}{2}$. (b) The circles in the uv plane are centered at $(1, \bar{X}^{-1})$, with radius $\bar{X}^{-1}$. Note that for $\bar{X} = \pm 1$ we have two circles (dashed lines) with unity radii and centered at $(1, \pm 1)$.

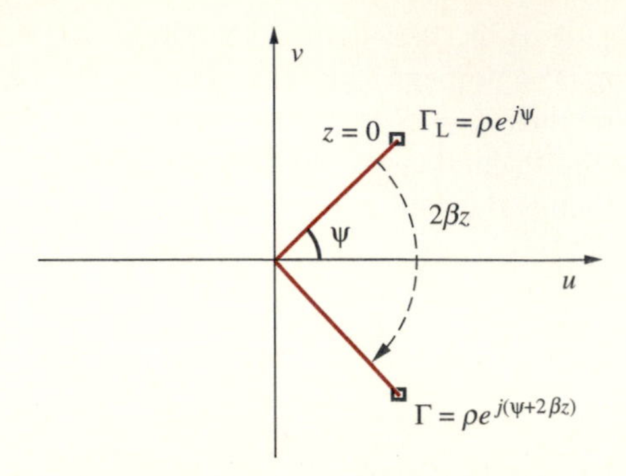

FIGURE 3.43. Complex reflection coefficient Γ. The complex number Γ is shown in the uv plane, together with its variation with z.

which is the equation for a circle in the uv plane centered at $u = 1$, $v = 1/\bar{X}$ and having a radius of $1/\bar{X}$. Examples of such circular segments are shown in Figure 3.42b. Note that $\bar{X} = \pm 1$ corresponds to a circle centered at $u = 1$, $v = \pm 1$; having a radius of 1; and tangent to the v axis at $v = \pm 1$.

The voltage reflection coefficient $\Gamma = u + jv$ is defined on the complex uv plane so that the locus of points of constant $|\Gamma| = |\Gamma_L| = \rho$ are circles centered at the origin. Once ρ is known (the value of which is set by Z_L and Z_0), motion along the line (i.e., variation of z) corresponds to motion around this Γ circle of fixed radius ρ. To see this, consider

$$\Gamma = \Gamma_L e^{j2\beta z} = \rho e^{j\psi} e^{j2\beta z} = \rho e^{j(\psi + 2\beta z)} \qquad [3.56]$$

As illustrated in Figure 3.43, motion away from the load (i.e., decreasing z) corresponds to clockwise rotation of Γ around a circle in the uv plane. Since $\beta = 2\pi/\lambda$, a complete rotation of $2\beta z = -2\pi$ occurs when z decreases by $-\lambda/2$, which is why a complete cycle of line impedance (or admittance) is repeated every $\lambda/2$ length along the line. A circle of constant radius ρ (corresponding to a given load) also corresponds to a fixed-voltage standing-wave ratio S, since $S = (1 + \rho)(1 - \rho)^{-1}$.

Although all Γ values along the line terminated with Z_L lie on the circle of radius ρ, each value of Γ corresponds (through [3.52]) to a different value of $\bar{Z} = \bar{R} + j\bar{X}$, which is the normalized line impedance seen at that position. On a typical Smith chart, shown in Figure 3.44, the contours of constant $\bar{R}$ or $\bar{X}$ are plotted and labeled on the uv plane so that the line impedance at any position along the constant ρ (or S) circle can be easily read from the chart. A summary of various Smith chart contours and key points is provided in Figure 3.45.

As shown in Figures 3.45 and 3.46, the horizontal radius to the left of the chart center (i.e., the negative u axis) is the direction where $\Gamma = u + jv = \rho e^{(\psi + 2\beta z)} = \rho e^{-j\pi}$, or $\psi + 2\beta z = -(2m + 1)\pi$ where $m = 0, 1, 2, \ldots$; in this case the magnitude of the line voltage is a minimum. Thus, every crossing of the negative u axis as one moves along the constant ρ circle corresponds to a minimum in the line voltage (and thus a maximum in the line current). The distance from the load to the first voltage minimum, namely $z_{\min}$, can thus be found simply by equating $2\beta z$ to the negative of the angle from the load point (i.e., $\Gamma = \Gamma_L$) to this axis measured in the clockwise direction (i.e., $2\beta z_{\min} = -(\pi + \psi)$, as was discussed in Section 3.3.1). Similarly,

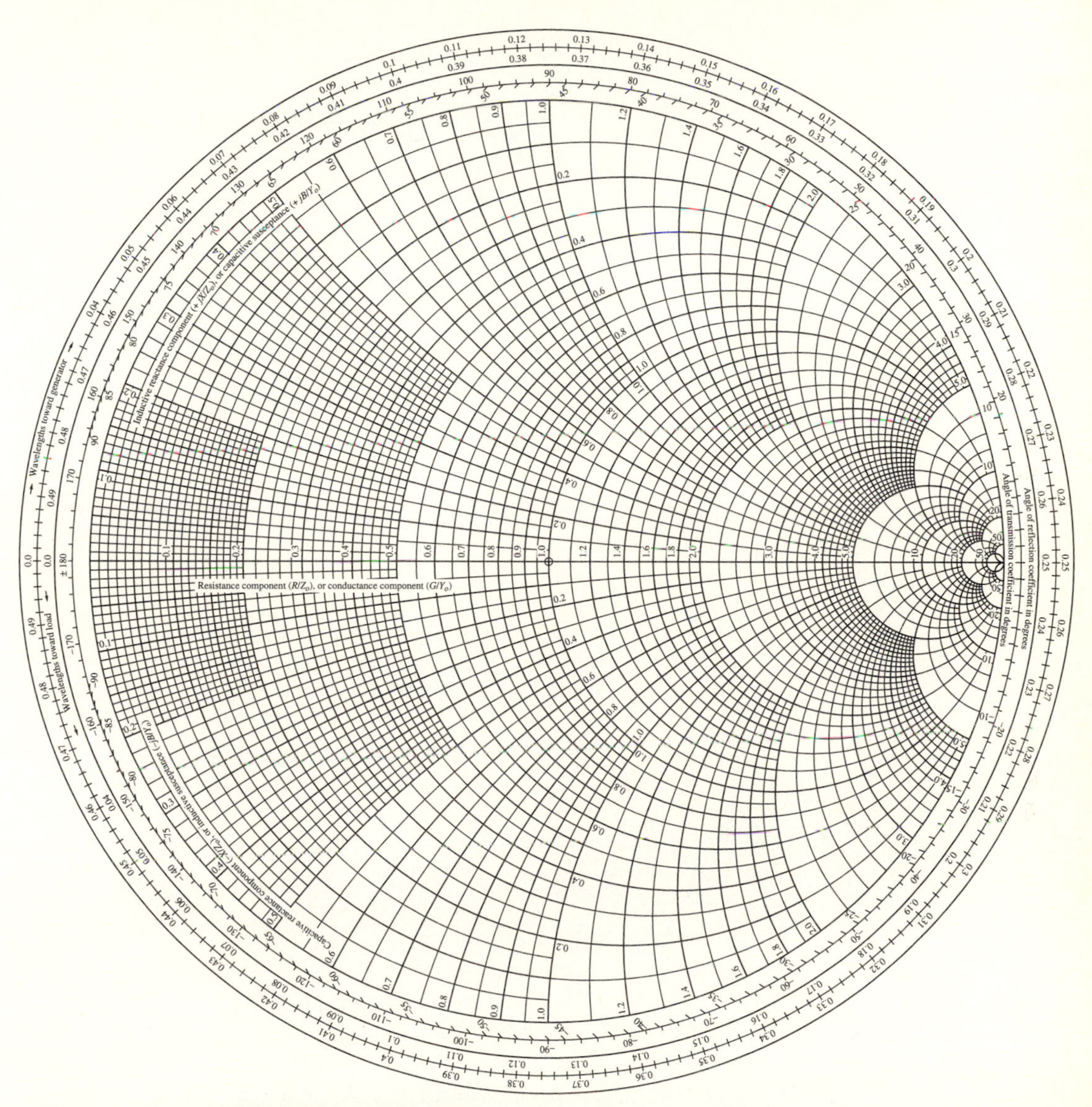

FIGURE 3.44. **Smith transmission line chart.**

the crossings of the horizontal radius to the right (i.e., the positive u axis) represent voltage maxima. Note that when $\psi + 2\beta z = -m2\pi$ ($m = 0, 1, 2, \ldots$) where $|V(z)| = V_{\max}$, we have

$$\bar{Z} = \frac{1+\rho}{1-\rho} = \bar{R}_{\max} = S$$

Hence if S is known (instead of ρ), the S circle (which is the same as the ρ circle) can be constructed with its center at the chart center and passing through the same

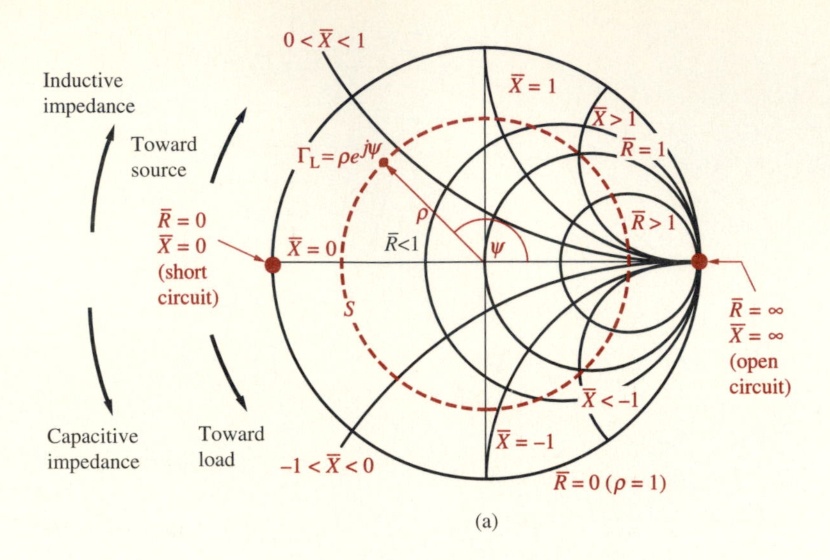

(a)

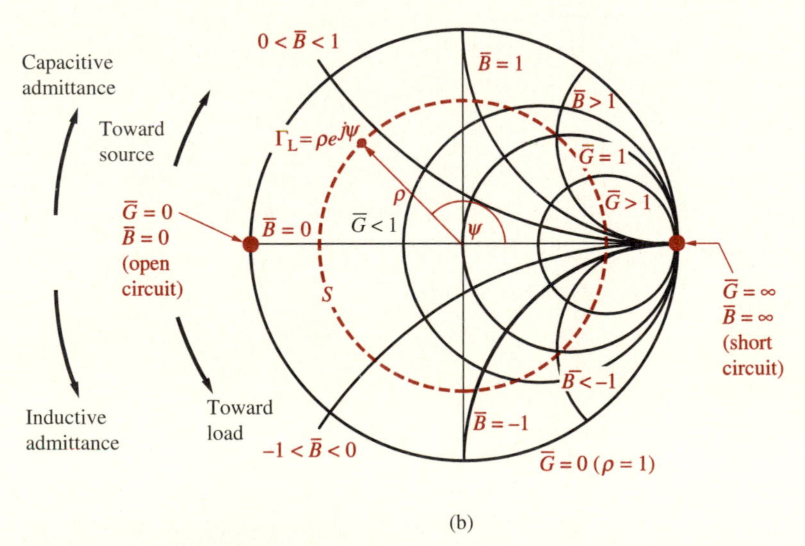

(b)

FIGURE 3.45. Summary of various Smith chart contours and locations. (a) For use as an impedance chart. (b) For use as an admittance chart.

point on the positive u axis as the $\bar{R} = S$ circle. This circle is then the locus of all impedances appearing at various positions along the transmission line, normalized to the characteristic impedance Z_0 of the line.

Once we realize that the upper (lower) half of the impedance Smith chart shown in Figure 3.45a corresponds to inductive (capacitive) reactances, that the negative u axis corresponds to a voltage minimum, and that moving away from the load corresponds to moving clockwise along a constant S (or constant ρ) circle around the chart, the interpretation of voltage standing-wave patterns for inductive versus capacitive

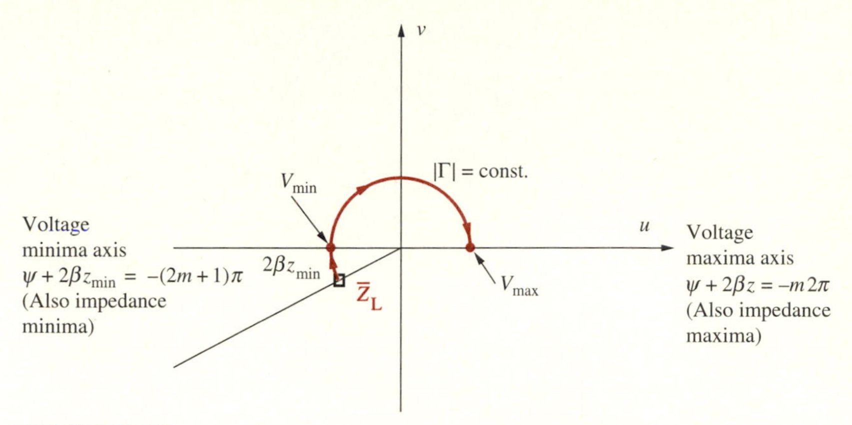

FIGURE 3.46. Location of voltage minima and maxima on the Smith chart.

loads (as depicted in Figure 3.17) becomes very clear. When we start anywhere in the upper half of the chart (i.e., inductive load) and move toward the source, we would encounter the voltage maxima (i.e., positive u axis) before the voltage minima so that the voltage magnitude would always first increase as we move away from an inductive load. The reverse would be true for a capacitive load. Many other aspects of the voltage, current, and impedance patterns discussed in previous sections can also be interpreted and visualized similarly using the Smith chart.

In cases for which it is more convenient to work with admittances than impedances, the Smith chart can be effectively used as an admittance chart. For this purpose, we note that since

$$\Gamma_L = \frac{Y_0 - Y_L}{Y_0 + Y_L} = -\frac{Y_L - Y_0}{Y_L + Y_0}$$

we have

$$\bar{Y}(z) = \bar{G} + j\bar{B} = \frac{1 - \Gamma(z)}{1 + \Gamma(z)} \qquad [3.57]$$

instead of [3.52]. In this case, the $\bar{R}$ and $\bar{X}$ circles can be treated, respectively, as $\bar{G}$ and $\bar{B}$ circles. However, note that the upper (lower) half of the chart now corresponds to capacitive (inductive) susceptances, which are represented by positive (negative) values of $\bar{B}$. A summary of various Smith chart contours and key points for its use as an admittance chart is provided in Figure 3.45b.

3.6.2 Examples of the Use of the Smith Chart

We now consider some applications of the Smith chart. The examples selected illustrate the relatively easy determination of line impedance for given resistive, reactive, and complex loads; determination of unknown load impedance based on measurements of standing-wave ratio and location of voltage minimum; single-stub impedance matching; and quarter-wave transformer matching.

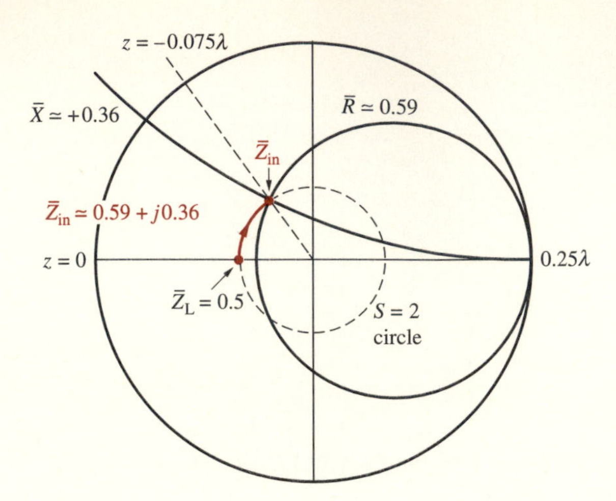

FIGURE 3.47. Graphical solution for Example 3-20.

Example 3-20: Input impedance with pure resistive load. Find the input impedance of a lossless transmission line with the following parameters: $Z_0 = 100\Omega$, $Z_L = 50 + j0\Omega$, line length $l = 86.25$ cm, wavelength $\lambda = 1.5$ m.

Solution: We first note that the electrical length of the line is 0.575λ. Since impedance goes through a full cycle every 0.5λ, the input impedance of this line would be identical to one with length $0.075\lambda = (0.575 - 0.5)\lambda$. The normalized load impedance is $\bar{Z}_L = Z_L/Z_0 = 0.5 + j0$. We enter the Smith chart at the point where the $\bar{R} = 0.5$ circle crosses the horizontal axis (note that the imaginary part of the load impedance is zero). We draw a circle passing through this point and centered at the origin; this is the constant ρ circle. We move along this circle by 0.075λ (from 0 mark to -0.075λ mark) away from the load (i.e., clockwise) and read the impedance as $\bar{Z}_{in}(z = -0.075\lambda) \simeq 0.59 + j0.36$. Since $\bar{Z}$ is the normalized impedance, the actual line impedance is $Z_{in} \simeq 59 + j36\Omega$. The details of the graphical solution are shown in Figure 3.47.

Example 3-21: Input impedance with a pure reactive load. Find the input impedance of a lossless transmission line given the following parameters: $Z_0 = 50\Omega$, $Z_L = 0 - j75\Omega$, line length $l = 1.202\lambda$ (i.e., $\lambda + 0.202\lambda$).

Solution: The normalized load impedance is $\bar{Z}_L = -j1.5$. For a purely reactive load, $R_L = 0$, so that $\rho = 1$ and $S = \infty$. We enter the chart at the point on the outermost circle (which corresponds to $\bar{R} = 0$), which is intersected by the $\bar{X} = -1.5$ circle. The length scale at that point reads $\sim 0.344\lambda$. The angle of Γ_L, or ψ, may be read to be $\sim -67°$. We now move along the outer circle (which in this case is our constant ρ circle) a distance of 0.202λ to the point $\sim 0.046\lambda$. The impedance at that point is $\bar{Z}_{in} \simeq j0.3$, or $Z_{in} \simeq j15\Omega$. The details of the graphical solution are given in Figure 3.48.

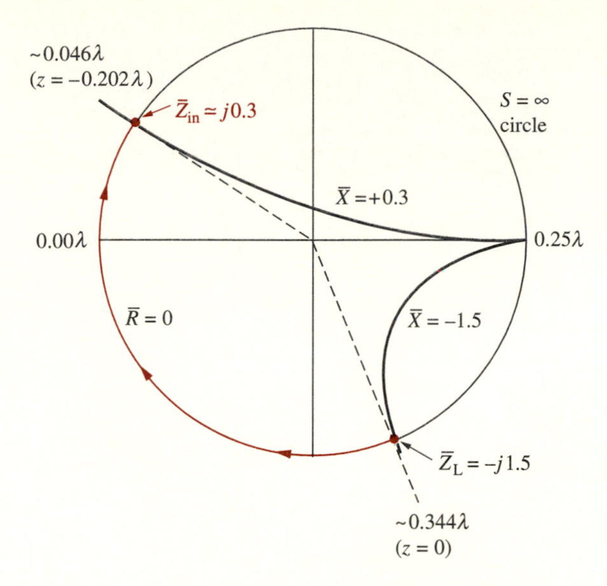

FIGURE 3.48. Graphical solution for Example 3-21.

Example 3-22: Input impedance with a complex load. Find the input impedance of a lossless transmission line given the following parameters: $Z_0 = 100\Omega$, $Z_L = 100 + j100\Omega$, line length $l = 0.676\lambda$ (i.e., $0.5\lambda + 0.176\lambda$).

Solution: The normalized load impedance is $\bar{Z}_L = 1.0 + j1.0$. We find the point on the chart corresponding to $\bar{R} = 1.0$ and $\bar{X} = 1.0$ (i.e., the intersection point of the $\bar{R} = 1.0$ and $\bar{X} = 1.0$ circles) and draw a circle passing through this point and centered at the origin. The intersection of this constant ρ circle with the right horizontal axis is at $\bar{R} \simeq 2.62$, which is also the value of S. To find the input impedance, we simply move along this circle (clockwise from the load position) a distance of 0.176λ and read $\bar{Z}_{in} = 1.0 - j1.0$. The input impedance of the line is then $Z_{in} = 100 - j100\Omega$. The details of the graphical solution are given in Figure 3.49.

Example 3-23: Unknown load impedance. Find the normalized load impedance on a transmission line with the following measured parameters: standing-wave ratio $S = 3.6$ and first voltage minimum $z_{min} = -0.166\lambda$.

Solution: We draw the constant ρ circle corresponding to $S = 3.6$ (i.e., intersecting the positive u axis at $\bar{R} = 3.6$). The point corresponding to z_{min} is that at which this circle crosses the negative u axis. We start at this point of intersection of the constant S circle with the left horizontal axis and move toward the load (i.e., counterclockwise) a distance of 0.166λ to find the normalized load impedance. This gives $\bar{Z}_L \simeq 0.89 - j1.3$. The details of the graphical solution are shown in Figure 3.50.

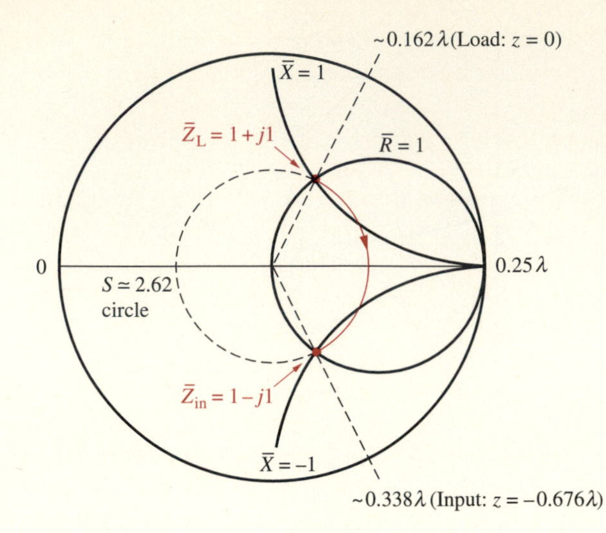

FIGURE 3.49. **Graphical solution for Example 3-22.**

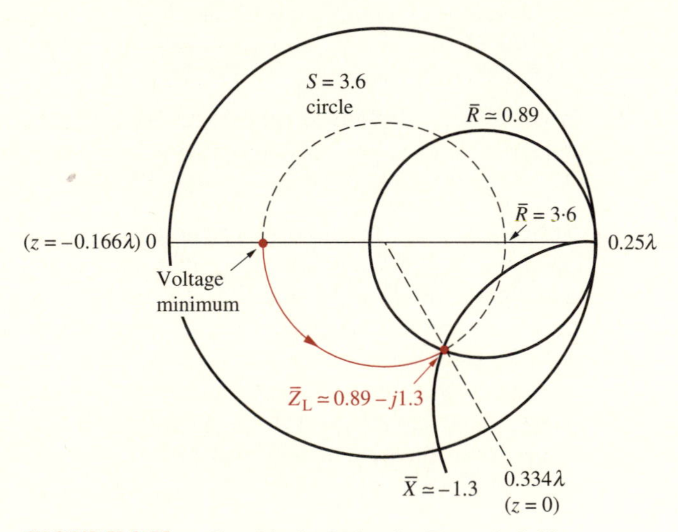

FIGURE 3.50. **Graphical solution for Example 3-23.**

Example 3-24: Single-stub impedance matching. Given a characteristic impedance $Z_0 = 80\Omega$ and a load impedance $Z_L = 160 - j80\Omega$, match the line to the given load by using a short-circuited shunt stub, as shown in Figure 3.32.

Solution: Refer to Figure 3.51 and to the discussion in Section 3.5 on impedance matching. Note that in view of the shunt connection of the stub, it is more convenient to deal with admittances. For this purpose, we use the Smith chart as an admittance chart. We require

$$\bar{Y}_1 = 1 - j\bar{B} \quad \text{and} \quad \bar{Y}_s = +j\bar{B}$$

where $\bar{Y}_1$ is the admittance seen looking toward the load at the position l where the stub is to be connected, and $\bar{Y}_s$ is the input admittance of the short-circuited stub of length l_s.

The normalized load impedance is $\bar{Z}_L = 2.0 - j1.0$. We enter the Smith chart at the point marked $\bar{Z}_L$ corresponding to the intersection of the resistance $\bar{R} = 2$ circle with the reactance $\bar{X} = -1.0$ circle, noting that negative reactances are in the lower half of the chart. The circle centered at the origin and passing through this point is our constant S (or constant ρ) circle along which the complex reflection coefficient Γ (or the line impedance) varies as we move away from the load. Noting that the Smith chart can be used equally for impedances and admittances, we choose to work with admittances in order to easily handle a parallel connected stub. The normalized load admittance can be found either directly (i.e., $\bar{Y}_L = (\bar{Z}_L)^{-1} = (2 - j)^{-1} = 0.4 + j0.2$) or by moving around the constant S circle by 180°, as shown in Figure 3.51. The normalized load admittance is thus $\bar{Y}_L = 0.4 + j0.2$. Note that when we change an impedance to an admittance on the Smith chart and work from there, all of the $\bar{R}$ and $\bar{X}$ circles can now be used as $\bar{G}$ and $\bar{B}$ circles.

We now move along the constant S circle up to its point of intersection with the conductance $\bar{G} = 1$ circle (P_1). The amount that we need to move determines the stub position at a distance l from the load. For this example, we find $l \simeq 0.126\lambda$. At the intersection point, the line impedance is $\bar{Y}_1 \simeq 1.0 + j1.0$, so that for matching we must have $\bar{Y}_s = -j1.0$.

To determine the length of a short-circuited stub that would present an admittance of $-j1.0$, we start from the point on the chart corresponding to a short circuit (i.e., $\bar{Y} = \infty$ on the right horizontal axis, or the positive u axis). We move clockwise until we intersect the circle corresponding to a susceptance $\bar{B} = -1.0$. This determines the length of the stub to be $l_s = 0.125\lambda$.

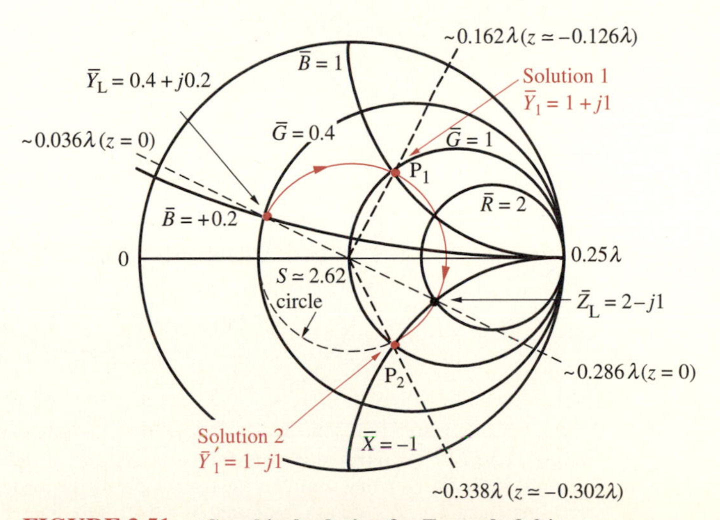

FIGURE 3.51. Graphical solution for Example 3-24.

Note that we might have taken the second intersection of the constant S circle with the $\bar{G} = 1$ circle (P_2), shown in Figure 3.51 as $\bar{Y}_1'$. This would have given $l \simeq 0.302\lambda$ and $\bar{Y}_1' \simeq 1.0 - j1.0$, requiring a stub impedance of $\bar{Y}_s = +j1.0$, which would be presented by a stub of length $l_s = 0.375\lambda$.

Example 3-25: Quarter-wave transformer matching. Given a transmission line with a characteristic impedance $Z_0 = 120\Omega$ and load impedance $Z_L = 72 + j96\Omega$, match the line to the given load using a quarter-wave transformer.

Solution: Refer to Figure 3.52, and the discussions in Section 3.5. We first move along the line a distance of l_1 such that the impedance Z_1 seen looking toward the line is purely resistive. Noting that the normalized load impedance is $\bar{Z}_L = (72 + j96)/120 = 0.6 + j0.8$, we enter the Smith chart at the point where the $\bar{R} = 0.6$ and $\bar{X} = 0.8$ circles intersect. We then move clockwise (away from load) along the constant S circle to its intersection with the horizontal axis, corresponding to the reactive part of the line impedance being zero. As shown in Figure 3.52, we need to move by 0.125λ, which means that the quarter-wave

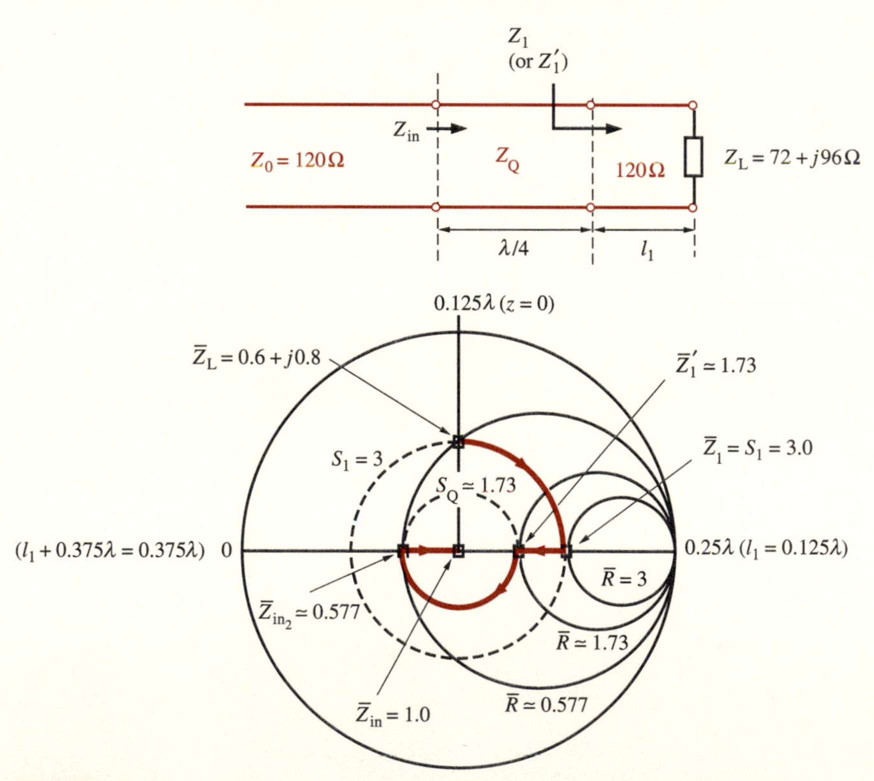

FIGURE 3.52. Quarter-wave transformer matching; graphical solution for Example 3-25. The path that we follow on the chart from the load impedance point to the origin (matching, i.e., $Z_{in} = Z_0$) is indicated by a thick dark line.

transformer can be placed at $l_1 = 0.125\lambda$. At that point, the line impedance normalized to a characteristic impedance of 120Ω is $\bar{Z}_1 = S_1 = 3.0$.

Noting that the quarter-wave transformation will occur on a line with characteristic impedance Z_Q, we now have to normalize the impedance to $Z_Q = \sqrt{Z_0 S_1 Z_0} \simeq 208\Omega$. The line impedance at $z = -0.125\lambda$, normalized to Z_Q, is then $\bar{Z}_1' = \bar{Z}_1(120/Z_Q) \simeq 1.73$. Following along the path on the Smith chart as shown in Figure 3.52, we thus move from $\bar{Z}_1$ to $\bar{Z}_1'$. The transformation along the quarter-wave segment is equivalent to a clockwise (away from load) rotation of 180°, which brings us to $\bar{Z}_{\text{in}_2} = 1/S \simeq 0.577$. Note that this rotation is along the circle of $S_Q \simeq 1.73$, which is the standing-wave pattern within the transformer. We now note that $\bar{Z}_{\text{in}_2} \simeq 0.577$ is an impedance normalized to Z_Q, whereas the characteristic impedance of the line to be matched is 120Ω. Re-normalizing back to 120Ω, we find the input impedance of the line looking into the quarter-wave segment to be $\bar{Z}_{\text{in}} \simeq \bar{Z}_1'(208/120) \simeq 1.0$; this brings us to the center of the chart, which represents a matched line.

3.6.3 Voltage and Current Magnitudes from the Smith Chart

Note that for a lossless transmission line, we have

$$V(z) = V^+ e^{-j\beta z}(1 + \underbrace{\rho e^{j\psi} e^{j2\beta z}}_{\Gamma})$$

so that the magnitude of the line voltage at any position z is given by

$$|V(z)| = |V^+||1 + \Gamma|$$

Since $|V^+|$ is just a scaling constant, the relative value of the line voltage can be obtained from the Smith chart by measuring the amplitude of the complex number $(1 + \Gamma)$. Note that at each position z along the line Γ is a new complex number as determined by a point on the uv plane (i.e., on the Smith chart). The length $|1 + \Gamma|$ can be determined graphically as shown in Figure 3.53. Note that as we move

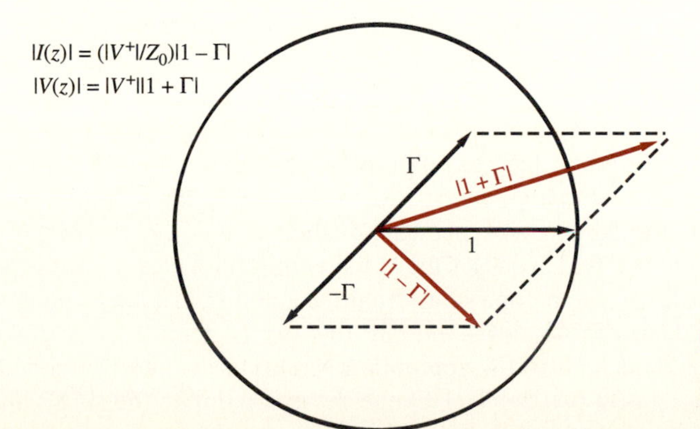

FIGURE 3.53. Line voltage and current from the Smith chart. We have assumed $V^+ = 1$.

along the line away from the load, the generalized reflection coefficient vector Γ rotates clockwise, and the voltage vector $(1 + |\Gamma|)$ rotates clockwise like a crank. By inspection of Figure 3.53, we can see that the maximum and minimum values of the voltage vector (i.e., maximum and minimum values of its length) are, respectively, $(1 + |\Gamma|)$ and $(1 - |\Gamma|)$, so that the standing-wave ratio is $S = (1 + |\Gamma|)/(1 - |\Gamma|) = (1 + \rho)/(1 - \rho)$, as was previously established in Section 3.3.1.

3.7 SELECTED APPLICATION EXAMPLES

In this section, we discuss two selected practical application topics, namely, (a) equivalent circuits for antennas or other loads with complex input impedance, and (b) matching networks.

3.7.1 Lumped Equivalent Circuits for Antennas and Other Loads with Complex Input Impedances

The determination of an unknown impedance from measurements of the standing-wave ratio and location of the voltage minimum, as discussed in Section 3.3, is a practical microwave method for measurement of unknown impedances that are difficult to calculate, such as the feed-point impedance of an antenna. Once the feed-point impedance is determined, an equivalent circuit model of the antenna can be constructed to determine the behavior of the antenna in various transmission line circuits. We illustrate the measurement of the unknown load impedance in Example 3-26 and the use of an equivalent circuit of a dipole antenna in Example 3-27.

Example 3-26: A meteor-damaged spaceship antenna. A spaceship has a microwave transmitter connected to an external antenna via a 50Ω coaxial line that is used to transmit radio waves at 1.5 GHz, as shown in Figure 3.54a. A small meteor strikes the external antenna and causes damage that results in a mismatch between the transmitter and the antenna. One of the crew members, an electrical engineer, decides to use a single short-circuited stub to correct the mismatch. However, the exact length of the coaxial line is unknown since a large portion of it goes through the hull of the spaceship. At first, the engineer measures the first voltage minimum position relative to the point where the coax exits on the inner surface of the hull to be at 8 cm and the next minimum position to be at 18 cm. She also measures the voltage standing-wave ratio on the line to be 4. Subsequently, a second crew member undertakes a spacewalk to short-circuit the external terminals of the damaged antenna, which causes the voltage minimum closest to the exit point of the cable on the hull (which, for a short-circuited antenna, is a deep voltage null) to move to 9.84 cm. Using these measurements, find (a) the feed-point impedance Z_L of the damaged antenna and (b) the appropriate length of the short-circuited 50Ω coaxial stub and the closest position (relative to the exit point on the ship's hull) at which it needs to be connected in parallel with the line to achieve matching.

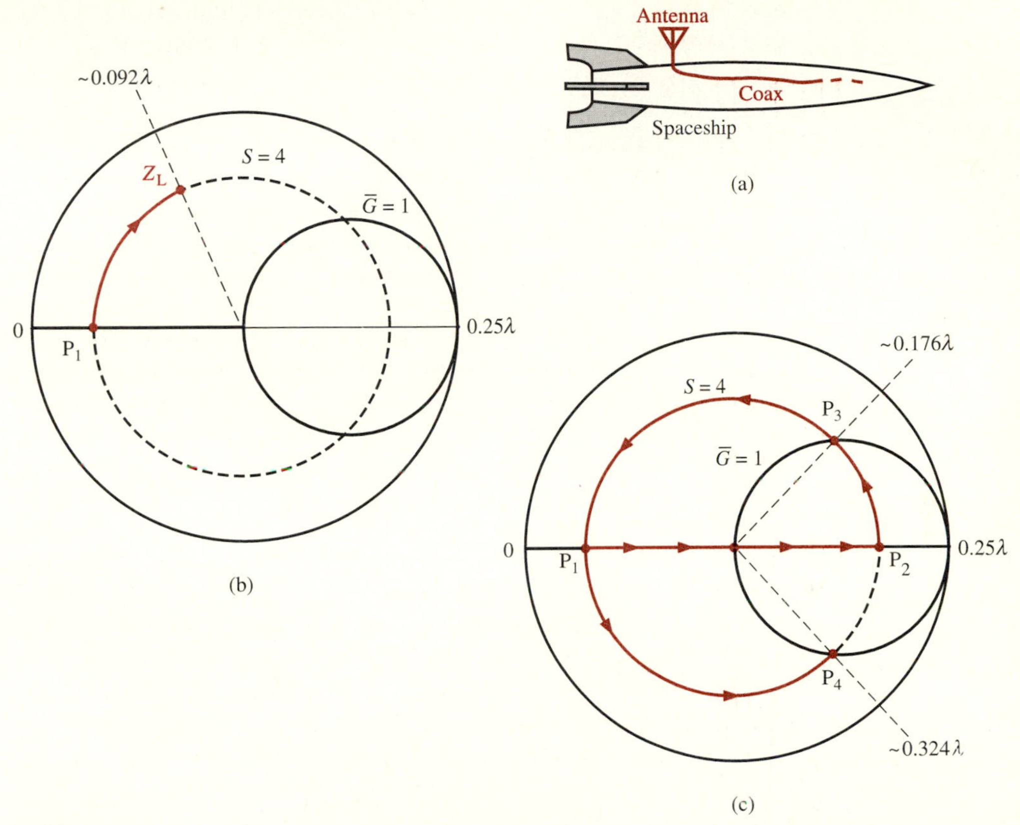

FIGURE 3.54. A meteor-damaged antenna on a spaceship. (a) A spaceship uses an externally mounted antenna at the end of a coaxial line. (b) Determination of the unknown terminal impedance of the damaged antenna from S and $z_{\min}$. (c) Determination of the length of a short-circuited stub.

Solution:

(a) Using the Smith chart, we draw the constant voltage standing-wave ratio ($S = 4$) circle, as shown in Figure 3.54b. The normalized line admittance seen looking toward the antenna from the appropriate location of the shunt stub, excluding the stub's admittance, lies on the $S = 4$ circle. Next, we determine the wavelength λ from the distance between successive voltage minima to be $\lambda/2 = 18 - 8 = 10$ cm $\rightarrow \lambda = 20$ cm. When the antenna is shorted all voltage minima along the coax shift toward the antenna by Δl, which can vary in the range $0 \leq \Delta l < \lambda/2$, depending on the feed-point impedance of the antenna Z_L (which lies on the $S = 4$ circle). If Z_L is capacitive, the shift is between 0 and $\lambda/4$, whereas if it is inductive, the shift is between $\lambda/4$ and $\lambda/2$. Since the shift is $18 - 9.84 = 8.16$ cm or 0.408λ in our case, we conclude that Z_L is inductive. Furthermore, if we move on the $S = 4$ circle, starting from point P_1 (i.e., minimum voltage point), a distance of $0.5\lambda - 0.408\lambda = 0.092\lambda$ in the clockwise direction, we reach the point that corresponds to $\bar{Z}_L$, which is read from the Smith chart as $\bar{Z}_L \simeq 0.35 + j0.6$. Thus, the unknown feed-point impedance of the damaged antenna is determined to be $Z_L = Z_0\bar{Z}_L \simeq 17.5 + j30\Omega$, since $Z_0 = 50\Omega$.

(b) We start with the normalized line impedance at 8 cm (point P_1 on the Smith chart) and convert it to line admittance at the same position (i.e., moving to point P_2) so that we can use the Smith chart as an admittance chart, since the matching network consists of a shunt stub (see Figure 3.54c). Next, to find the stub position closest to the exit point along the coaxial line, we move in the counterclockwise direction on the $S = 4$ circle (i.e., going toward the hull of the spaceship) until we find the intersection points with the $\bar{G} = 1$ circle corresponding to points on the inner side of the hull where the normalized line admittance is $\bar{Y}_1 = 1 - j\bar{B}$. As seen in Figure 3.54c, there are two such points, marked P_3 and P_4. Since P_4 is a point outside the ship (i.e., the length between P_2 and P_4 is $\sim 0.426\lambda \simeq 8.52 > 8$ cm), P_3 corresponds to the point on the coaxial line closest to the exit point on the ship's hull. Therefore, at point P_3, we read $\bar{Y}_1$ from the chart as $\bar{Y}_1 \simeq 1 + j1.5$ and its location relative to the exit point as $l \simeq 8 - 0.074\lambda = 8 - 1.48 = 6.52$ cm. The length of the shorted stub connected at $l \simeq 6.52$ cm can be found from $\bar{Y}_{sc} = -j\cot(\beta l_s) \simeq -j1.5$. From this we have $\beta l_s \simeq 0.588 \rightarrow l_s \simeq 0.0936\lambda \simeq 1.87$ cm.

In Example 3-27, we use a four-element equivalent lumped circuit to represent the feed-point impedance of a dipole antenna, as determined from measurements.[29] The circuit consists of a resistance, an inductance, and two capacitors, as shown in Figure 3.55. The values of these elements depend only on the physical dimensions of the antenna, not on the operation frequency. The empirical equations for these four elements are as follows:

$$C_1 = \frac{6.0337l}{\log(l/a) - 0.7245}\ \text{pF}$$

$$C_2 = l\left\{\frac{0.89075}{[\log(l/a)]^{0.8006} - 0.861} - 0.02541\right\}\ \text{pF}$$

$$L_a = 0.1l\{[1.4813\log(l/a)]^{1.012} - 0.6188\}\ \mu\text{H}$$

$$R_a = 0.41288[\log(l/a)]^2 + 7.40754(l/a)^{-0.02389} - 7.27408\ \text{k}\Omega$$

where l is the total length and a is the radius of the dipole, with both expressed in meters. These equations adequately represent the impedance of the dipole with length up to approximately 0.6 wavelength (i.e., $l \leq 0.6\lambda$).

[29] T. G. Tang, Q. M. Tieng, and M. W. Gunn, Equivalent circuit of a dipole antenna using frequency-independent lumped elements, *IEEE Transactions on Antennas and Propagation,* 41(1), pp. 100–103, 1993.

FIGURE 3.55. **A dipole antenna and its equivalent lumped circuit.**

Example 3-27: Dipole antenna fed by a coaxial line. Consider a dipole antenna of length 1.8 m and radius 2.64 mm connected at the end of an RG-213 (50Ω) coaxial line, as shown in Figure 3.56a. Calculate the standing-wave ratio S on the line at (a) 83.33 MHz and (b) 20.83 MHz.

Solution: For l = 1.8 m and a = 2.64 mm, we can use the empirical equations to calculate the values of the lumped elements in the equivalent circuit as

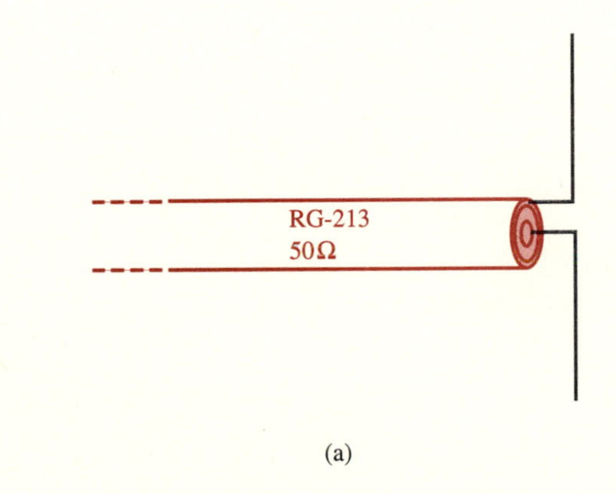

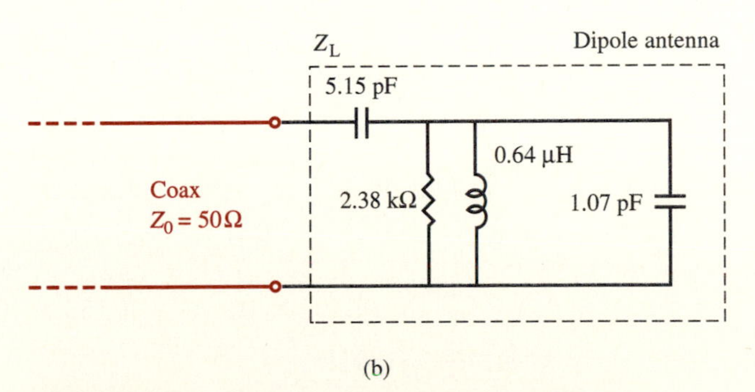

FIGURE 3.56. **Dipole antenna fed by a coaxial line.** (a) A coaxial line and the antenna wires connected to its inner and outer conductors. (b) Corresponding two-wire equivalent circuit.

$R_a \simeq 2.38$ kΩ, $L_a \simeq 0.6537$ μH, $C_1 \simeq 5.149$ pF, and $C_2 \simeq 1.067$ pF respectively. The corresponding two-wire transmission line circuit is shown in Figure 3.56b.

(a) At $f = 83.33$ MHz, the feed-point impedance of the dipole antenna can be calculated using the expression for the load impedance Z_L

$$Z_L = Z_{C_1} + \frac{1}{Y_{R_a} + Y_{L_a} + Y_{C_2}}$$

where $Z_{C_1} = 1/j\omega C_1 \simeq -j370.9\Omega$, $Y_{R_a} = G = 1/R_a \simeq 4.202 \times 10^{-4}$ S, $Y_{L_a} = 1/j\omega L_a \simeq -j2.906 \times 10^{-3}$ S, and $Y_{C_2} = j\omega C_2 \simeq j5.587 \times 10^{-4}$ S, respectively. Substituting these values in the Z_L expression, we find

$$Z_L \simeq -j370.9 + \frac{1}{4.202 \times 10^{-4} - j2.363 \times 10^{-3}}$$

$$\simeq -j370.9 + \frac{4.202 \times 10^{-4} + j2.363 \times 10^{-3}}{(4.202 \times 10^{-4})^2 + (2.363 \times 10^{-3})^2}$$

$$\simeq -j370.9 + 72.94 + j410.2 \simeq 72.94 + j39.28\Omega$$

The load reflection coefficient Γ_L is

$$\Gamma_L = \frac{Z_L - Z_0}{Z_L + Z_0} \simeq \frac{22.94 + j39.28}{122.9 + j39.28}$$

$$\frac{45.48e^{j59.71^\circ}}{129.1e^{j17.72^\circ}} \simeq 0.3524e^{j42^\circ}$$

and the standing-wave ratio can be found as

$$S = \frac{1+\rho}{1-\rho} \simeq \frac{1 + 0.3524}{1 - 0.3524} \simeq 2.088$$

At 83.33 MHz, the 1.8-m length of the antenna is equal to one-half the wavelength ($\lambda = c/f = 3.6$ m). Such an antenna is an efficient radiator,[30] and we see here that it can be fed by a standard 50Ω coaxial line at a reasonable standing-wave ratio of $S \simeq 2$.

(b) Similarly, at $f = 20.83$ MHz, we have

$$Z_L \simeq -j1484 + \frac{1}{4.202 \times 10^{-4} - j1.169 \times 10^{-2} + j1.396 \times 10^{-4}}$$

$$\simeq -j1484 + \frac{4.202 \times 10^{-4} + j1.155 \times 10^{-2}}{(4.202 \times 10^{-4})^2 + (1.155 \times 10^{-2})^2}$$

$$\simeq 3.146 - j1397\Omega$$

[30]See, for example, Section 14.06 of E. C. Jordan and K. Balmain, *Electromagnetic Waves and Radiating Systems,* Prentice Hall, 1968.

The load-reflection coefficient is then given by

$$\Gamma_L \simeq \frac{3.146 - j1397.4 - 50}{3.146 - j1397.4 + 50} \simeq 0.999839e^{-j4.098°}$$

and the standing-wave ratio is

$$S = \frac{1+\rho}{1-\rho} \simeq \frac{1+0.999839}{1-0.999839} \simeq 12430 \ !$$

The high value of the capacitive reactance of the feed-point impedance at 20.83 MHz ($\lambda \simeq 14.4$ m) is partly due to the fact that the dipole antenna is electrically short ($l/\lambda \simeq 0.125$) and is therefore not an efficient radiator.

3.7.2 Transmission Line Matching Networks

As discussed in Section 3.5, transmission lines are often used in matching networks. In this section, we provide two specific examples involving the use of microstrip lines to realize matching networks for a microwave amplifier (Example 3-28) and for a cellular phone base station (Example 3-29). Microstrip lines, easily fabricated using printed-circuit techniques, are widely used to match impedances in microwave transistor amplifiers. Most microwave transistor amplifiers can be classified as either low-noise amplifiers or power amplifiers. In both cases, the circuit design involves the selection of the appropriate transistor and the optimum design of the matching networks around it to satisfy the design considerations such as power gain, low noise, and bandwidth. Microstrip transmission line segments can be used as open- or short-circuited stubs. In fact, a microstrip line together with a short- or open-circuited shunt stub can transform a 50Ω resistor into any value of impedance.[31] Two such matching networks are illustrated in Example 3-28. In most applications, it is desirable to achieve matching over a broad range of frequencies. As discussed in Section 3.5, most of the simple matching techniques (e.g., quarter-wave transformation, single-stub matching, etc.) do not in general have good frequency response. In Example 3-29, we illustrate a simple and commonly used method called shunt compensation, which can greatly improve the frequency response of a matching network.

Example 3-28: Input matching network of a low-noise microwave amplifier. Design two separate microstrip-line matching networks, as shown in Figure 3.57b,c, for the input stage of a low-noise microwave transistor amplifier to transform a 50Ω load impedance to an input admittance $Y_{in} = G_{in} + jB_{in} =$

[31]See Section 2.5 of G. Gonzalez, *Microwave Transistor Amplifiers, Analysis and Design,* 2nd ed., Prentice Hall, 1997.

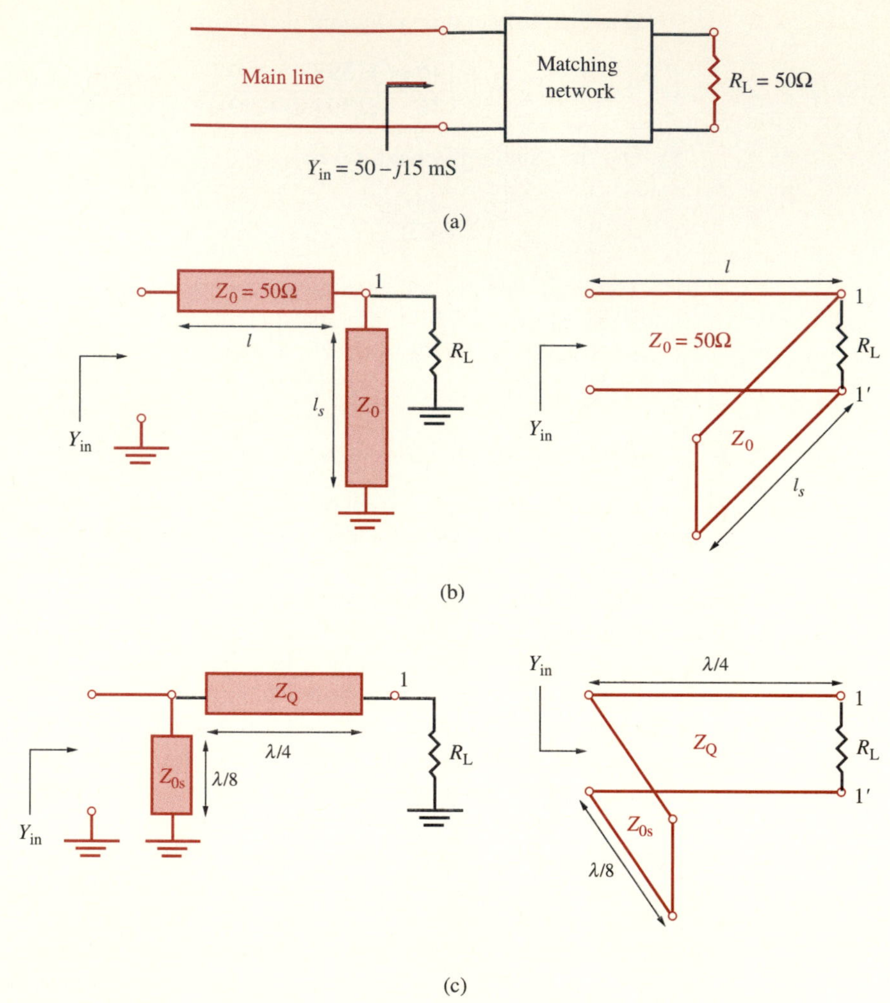

FIGURE 3.57. **Two different matching networks for a microwave amplifier.** (a) The purpose of the matching network is to transform the $R_L = 50\Omega$ load resistance to an input admittance of $Y_{in} = 50 - j15$ mS. (b) Matching network using a shunt short-circuited stub. (c) Matching network using quarter-wave transformation. The right hand panels in (b) and (c) show the two-wire equivalents of the microstrip-line circuits.

$50 - j15$ mS as required to achieve minimum noise performance.[32] (a) The first matching network consists of a short-circuited shunt microstrip stub connected in parallel with R_L followed by a microstrip line of length l, as shown in Figure 3.57b. The characteristic impedance of each of the two microstrip lines is 50Ω. Find the length of each line in terms of wavelength. (b) The second matching network consists

[32]See pp. 316–321 of T. Edwards, *Foundations for Microstrip Circuit Design,* 2nd ed., John Wiley and Sons, 1992.

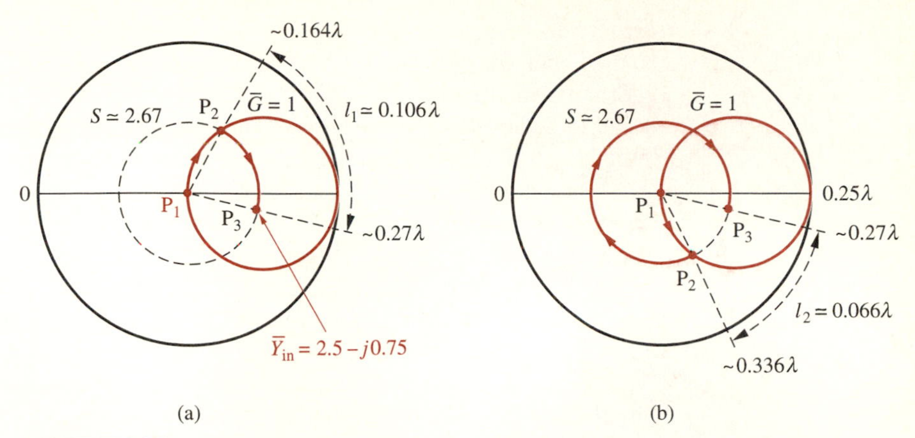

FIGURE 3.58. Smith chart solutions for Example 3-28, part (a). (a) First of two alternative solutions: $l_1 \simeq 0.106\lambda$, $l_{s1} \simeq 0.38\lambda$. (b) Second alternative solution: $l_2 \simeq 0.434\lambda$, $l_{s2} = 0.12\lambda$.

of a quarter-wavelength-long microstrip line of characteristic impedance Z_Q terminated by R_L at one end and having an eighth-wavelength-long short-circuited microstrip stub with characteristic impedance Z_{0s} connected in parallel with it at the other end, as shown in Figure 3.57c. Find Z_Q and Z_{0s}.

Solution:

(a) The matching network shown in Figure 3.57b is somewhat similar to the single-stub matching networks discussed in Section 3.5, except that the short-circuited shunt stub is located at the position of the load. In view of the shunt connections, it is more convenient to use the Smith chart as an admittance chart. Our desired goal is to achieve an input admittance of $Y_{in} = 50 - j15$ mS, which corresponds to a normalized admittance of $\bar{Y}_{in} = Y_{in}/Y_0 = 2.5 - j0.75$, since $Y_0 = (Z_0)^{-1} = (50)^{-1} = 0.02$ S. This normalized admittance point is marked as point P_3 in Figure 3.58a,b. The constant S circle passing through point P_3 is also shown in Figure 3.58a,b; this circle corresponds to $S \simeq 2.76$, as can be determined by reading off the S value from the chart. The standing-wave ratio of $S \simeq 2.76$ can also be calculated by noting that the reflection coefficient on the line with Y_0 terminated in Y_{in} is $\Gamma_{in} = \rho e^{j\psi} = (Y_0 - Y_{in})/(Y_0 + Y_{in})$ and that $S = (1+\rho)/(1-\rho)$. However, in the context of the graphical solution using the Smith chart, we do not need to know the numerical value of S explicitly.

The goal of the design is to determine the line length l and the stub length l_s so that we depart from point P_1 and arrive at point P_3 on the Smith chart. We enter the chart at the center point P_1 (i.e., $\bar{Z}_L = Z_L/Z_0 = 1$ or $\bar{Y}_L = Y_L/Y_0 = 1$), as shown in Figure 3.58a. Next, we determine the normalized admittance of the short-circuited shunt stub $\bar{Y}_{sc}$ such that the addition of $\bar{Y}_{sc} = j\bar{B}_{sc}$ to $\bar{Y}_L = 1$ brings us to the point P_2 on the constant S circle, where the normalized admittance is $\bar{Y}_{P_2} = 1 + j\bar{B}_{sc}$. There are two different ways in which this can be done, as illustrated in Figures 3.58a and 3.58b, respectively. In Figure 3.58a, the length of the shunt stub is designed such that it provides a capacitive admittance (i.e., $\bar{B}_{sc} > 0$), so that we move in the clockwise direction on the constant $\bar{G} = 1$ circle, from P_1 to P_2, where $\bar{Y}_{P_2} \simeq 1 + j1.06$. The minimum 50Ω short-circuited stub length that yields a normalized admittance of $\bar{Y}_{sc} = j1.06$ is determined as $\bar{Y}_{sc} = -j\cot(\beta l_s) \simeq j1.06$, so that we have $l_{s1} \simeq 0.38\lambda$. To find the line length l, we move from P_2 to P_3 around the constant S circle in the clockwise direction, yielding a length $l_1 \simeq 0.106\lambda$, as shown in Figure 3.58a. Similarly, in Figure 3.58b, the length of the shunt stub is chosen such that it provides an inductive admittance (i.e., $\bar{B}_{sc} < 0$), so that we move from P_1 to P_2 in the counter clockwise direction along the $\bar{G} = 1$ circle, where $\bar{Y}_{P_2} \simeq 1 - j1.06$. Once again, $\bar{Y}_{P_2}$ lies on the same constant S circle passing through $\bar{Y}_{in}$. The corresponding minimum short-circuited stub length is $l_{s2} \simeq 0.12\lambda$. Furthermore, to move from P_2 to P_3 along the S circle in the clockwise direction requires a minimum line length $l_2 \simeq 0.434\lambda$, as shown in Figure 3.58b.

For both matching circuits, if an open-circuited shunt stub were used instead of the short-circuited one, the only change in the design would have been the length of the open-circuited stub, the minimum value of which can be obtained by adding $\pm\lambda/4$ to the minimum length of the short-circuited stub, depending on whether the minimum length of the short-circuited stub is less than or greater than $\lambda/4$. In addition, although the two designs just discussed both involve a load impedance of 50Ω, this technique can also be applied to an arbitrary complex load impedance, since all of the calculations on the Smith chart were carried out using normalized admittances.

(b) For the second matching network, we can rely on quarter-wave transformer techniques and do not need to use the Smith chart. Noting that the real part (i.e., conductance) of the required Y_{in} is $G_{in} = 50$ mS, the input admittance of the quarter-wave transformer, not including the stub, is given by

$$Y_1 = \frac{Y_Q^2}{G_L} = G_{in} \quad \rightarrow \quad Y_Q \simeq 31.6 \text{ mS}$$

or $Z_Q \simeq 31.6\Omega$. The imaginary part (i.e., susceptance) of the required Y_{in}, that is, $B_{in} = -15$ mS, can be provided by the short-circuited shunt stub of length $\lambda/8$ using

$$Y_{sc} = -jY_{0s}\cot(\beta l_s) = jB_{in} = -j15 \text{ mS}$$

Since $\beta l_s = (2\pi/\lambda)(\lambda/8) = \pi/4$ and $\cot(\beta l_s) = 1$, we have $Y_{0s} = 15$ mS or $Z_{0s} \simeq 66.7\Omega$. Note that both of the characteristic impedances obtained (i.e., $Z_Q \simeq 31.6\Omega$ and $Z_{0s} \simeq 66.7\Omega$) are realizable using microstrips, since microstrip lines with characteristic impedance values ranging between 10Ω and 110Ω are easy to manufacture in practice.[33]

[33]T. Edwards, *Foundations for Microstrip Circuit Design,* 2nd ed., John Wiley and Sons, 1992.

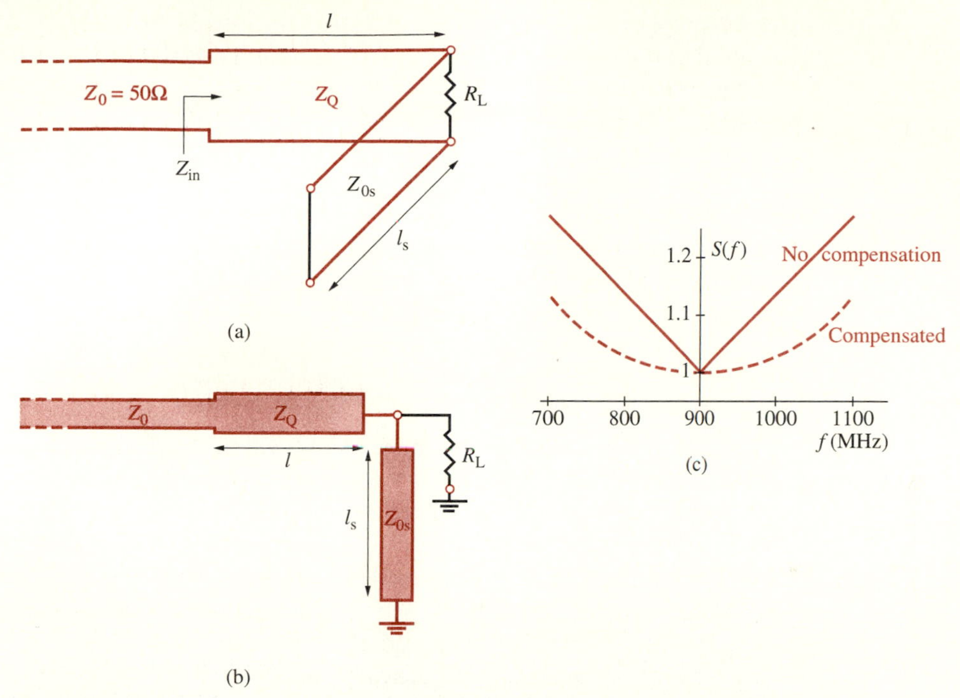

FIGURE 3.59. **Quarter-wave matching with shunt compensation.** (a) Transmission line circuit diagram. (b) Practical microstrip implementation. (c) Standing-wave ratio versus frequency for a simple quarter-wave transformer (solid line) and a quarter-wave transformer with shunt compensation (dashed line).

Example 3-29: Quarter-wave matching with shunt compensation. A quarter-wave matching network is to be designed for a cellular phone base station power amplifier operating at 900 MHz. The matching network is to match a resistive load of $R_L = 25\Omega$ to a transmission line with characteristic impedance $Z_0 = 50\Omega$, providing a standing-wave ratio $S < 1.1$ over the frequency range 800 to 1000 MHz. (a) Design a quarter-wave transformer to operate at 900 MHz and assess its standing-wave ratio across the specified frequency range. (b) Consider possible improvement of the bandwidth performance by the addition of a short-circuited quarter-wavelength-long (at 900 MHz) stub in shunt with the load, as shown in Figure 3.59a.

Solution:

(a) First consider the design of a quarter-wave transformer without any short-circuited compensation shunt stub. In order to match the 25Ω load to the 50Ω line, the quarter-wave transformer should have a characteristic impedance of $Z_Q = \sqrt{Z_0 R_L} = \sqrt{(50)(25)} \simeq 35.4\Omega$. Assuming an air-filled transmission line, the phase velocity is equal to the speed of light in free space, and the wavelength at the design frequency of 900 MHz is $\lambda_0 = v_p/f_0 = c/f_0 \simeq 3 \times 10^8/(900 \times 10^6) = \frac{1}{3}$ m. Thus, the length of the quarter-wave section should be $l = (\lambda_0/4) = \frac{1}{12}$ m. At the design frequency, the input impedance Z_{in} of the matching network designed is Z_0. At other frequencies, the quarter-wave transformer of length $l = \frac{1}{12}$ m and characteristic impedance Z_Q transforms the load R_L to its input as

$$Z_{\text{in}} = Z_Q \frac{R_L + jZ_Q \tan[2\pi(f/c)(\frac{1}{12})]}{Z_Q + jR_L \tan[2\pi(f/c)(\frac{1}{12})]}$$

The reflection coefficient, as observed at the junction between the main line and the matching network, and the standing-wave ratio on the main line are then

$$\Gamma = \frac{Z_{\text{in}} - Z_0}{Z_{\text{in}} + Z_0} = \rho e^{j\psi} \quad \rightarrow \quad S = \frac{1+\rho}{1-\rho}$$

The frequency dependence of the standing-wave ratio S is shown as a solid line in Figure 3.59c. We see that S varies sensitively with frequency and in fact exceeds the design criteria (i.e., $S > 1.1$) near the edges of the 800–1000 MHz band of interest.

(b) We now consider the use of a short-circuited quarter-wave line in shunt with the load, as shown in Figure 3.59a, to achieve better frequency response. If the length l_s of the short-circuited stub is chosen to be equal to a quarter wavelength at the design frequency of 900 MHz, that is, if $l_s = \frac{1}{12}$ m, the input impedance of the stub as viewed from the load terminals is an open circuit, so that the presence of the stub has no effect on the system performance at 900 MHz. However, at other frequencies, the input impedance of the short-circuited stub is

$$Z_{\text{sc}} = jZ_{0s} \tan[2\pi(f/c)(\tfrac{1}{12})]$$

where Z_{0s} is the characteristic impedance of the short-circuited stub. The stub impedance Z_{sc} appears in parallel with the load resistance R_L, so that the load impedance is

$$Z_L = \frac{Z_{\text{sc}} R_L}{Z_{\text{sc}} + R_L} = \frac{(jZ_{0s} \tan[2\pi(f/c)(\frac{1}{12})])R_L}{jZ_{0s} \tan[2\pi(f/c)(\frac{1}{12})] + R_L}$$

which is transformed to the input of the quarter-wave transformer as

$$Z_{\text{in}}^c = Z_Q \frac{Z_L + jZ_Q \tan[2\pi(f/c)(\frac{1}{12})]}{Z_Q + jZ_L \tan[2\pi(f/c)(\frac{1}{12})]}$$

where the superscript "c" indicates that this input impedance is for the compensated case, and thus it differs from the uncompensated Z_{in} found in part (a). The reflection coefficient and the standing-wave ratio on the main line for the compensated case are then

$$\Gamma^c = \frac{Z_{\text{in}}^c - Z_0}{Z_{\text{in}}^c + Z_0} = \rho^c e^{j\psi^c} \quad \rightarrow \quad S^c = \frac{1+\rho^c}{1-\rho^c}$$

The frequency dependence of the compensated standing-wave ratio S^c is shown as the dashed line in Figure 3.59c, for the case when $Z_{0s} = Z_0 = 50\Omega$. We see that S^c is substantially lower than S over the entire frequency range of interest, so that the compensation has significantly improved the frequency response. The design criteria of $S^c < 1.1$ is easily met over the range 800 to 1000 MHz. In general, both the length of the short-circuited stub and its characteristic impedance Z_{0s} can be optimally chosen to achieve a desired frequency response.

In practice, similar improvement in frequency response can be achieved by using a half-wavelength-long open-circuited stub or by using simple parallel lumped LC networks.

3.8 SINUSOIDAL STEADY-STATE BEHAVIOR OF LOSSY LINES

Our analyses so far have been based on the assumption that there is no power loss in the transmission line itself. A consequence of this assumption was the rather nonphysical result that the line current at a distance of a quarter wavelength from a short circuit is zero, and that the input impedance of a quarter-wavelength short-circuited line is infinite. In reality, every line consumes some power, partly because of the resistive losses (R) in the conductors and partly because of leakage losses (G) through the insulating medium surrounding the conductors. For lines with small losses, the effects of the losses on characteristic impedance, line voltage, line current, and input impedance are usually negligible, so that the lossless analysis is valid. However, in other cases, the losses and the resultant attenuation of signals cannot be ignored. The typical conditions under which losses cannot be neglected are (1) transmission of signals over long distances, (2) high-frequency applications, since resistive losses increase with frequency, and (3) use of quarter-wavelength or half-wavelength-long transmission line segments as circuit elements, when neglecting losses leads to nonphysical results such as zero current and/or infinite input impedance. In the third case, losses become the determining factor on resonant lines when the electrical quantities of interest tend toward zero or infinity. Thus, input impedance of a quarter-wavelength-long open-circuited transmission line is in fact not zero, but is a small nonzero value as determined by the losses. Similarly, the input impedance of a quarter-wavelength-long short-circuited transmission line is not infinite, but a large finite value determined by the losses.

The sinusoidal steady-state behavior of lossy lines can be formulated in a manner quite similar to that of lossless lines. We can start with the most general form of the transmission line equations that were obtained in Section 2.2:

$$\frac{\partial \mathcal{V}(z,t)}{\partial z} = -\left(R + L\frac{\partial}{\partial t}\right)\mathcal{I}(z,t) \qquad [3.58a]$$

$$\frac{\partial \mathcal{I}(z,t)}{\partial z} = -\left(G + C\frac{\partial}{\partial t}\right)\mathcal{V}(z,t) \qquad [3.58b]$$

Under sinusoidal steady-state conditions, it is more convenient to work with the voltage and current equations written in terms of the phasor quantities $V(z)$ and $I(z)$, such that $\mathcal{V}(z,t) = \Re e\{V(z)e^{j\omega t}\}$ and $\mathcal{I}(z,t) = \Re e\{I(z)e^{j\omega t}\}$. We have

$$-\frac{dV(z)}{dz} = (R + j\omega L)I(z) \qquad [3.59a]$$

$$-\frac{dI(z)}{dz} = (G + j\omega C)V(z) \qquad [3.59b]$$

Taking the derivative of [3.59*a*] and substituting from [3.59*b*], we find

$$\frac{d^2V(z)}{dz^2} = (RG)V(z) + (LG + RC)(j\omega)V(z) + (j\omega)^2(LC)V(z)$$
$$= (R + j\omega L)(G + j\omega C)V(z)$$

$$\frac{d^2V(z)}{dz^2} = \gamma^2 V(z) \qquad [3.60]$$

where

$$\gamma = \sqrt{(R + j\omega L)(G + j\omega C)} = \alpha + j\beta \qquad [3.61]$$

is the propagation constant. Note that the propagation constant γ is in general a complex number, and its real and imaginary parts, α and β, are known, respectively, as the *attenuation constant* and the *phase constant.*[34] For any given values of R, L, G, and C and the frequency f, the values of α and β can be directly calculated from [3.61].

Equation [3.60] is a second-order differential equation similar to the one we encountered for the lossless case. Its general solution is

$$V(z) = V^+e^{-\gamma z} + V^-e^{+\gamma z} \qquad [3.62]$$

where V^+ and V^- are complex constants to be determined by the boundary conditions.

The current phasor $I(z)$ can be determined by simply substituting [3.62] into [3.59*a*]. Thus, we have

---

[34]Note that β for the lossy case is not equal to that for the lossless case, which was defined earlier as $\beta = \omega\sqrt{LC}$. In the general lossy case, β is a function of R, L, G, and C and has a more complex dependence on the frequency f.

$$(R + j\omega L)I(z) = -\frac{dV(z)}{dz}$$

$$I(z) = \frac{-1}{R + j\omega L}\frac{dV(z)}{dz} = +\frac{\gamma}{R + j\omega L}(V^+e^{-\gamma z} - V^-e^{+\gamma z})$$

$$I(z) = \sqrt{\frac{G + j\omega C}{R + j\omega L}}(V^+e^{-\gamma z} - V^-e^{+\gamma z}) \quad [3.63]$$

$$= \frac{1}{Z_0}(V^+e^{-\gamma z} - V^-e^{+\gamma z})$$

where we have defined Z_0 as the *characteristic impedance,* namely,

$$Z_0 \equiv \sqrt{\frac{R + j\omega L}{G + j\omega C}} = |Z_0|e^{j\phi_z}$$

Compared to Z_0 for the lossless case, we see that for the lossy case Z_0 is in general a complex number. Note once again that Z_0 depends on the physical line constants R, L, G, and C (which in turn depend on the physical makeup and dimensions of the line as well as the properties of the surrounding media) but now also on the frequency of operation $\omega = 2\pi f$. For future reference, the general solutions of the transmission line equations for the voltage and current phasors are

$$V(z) = V^+e^{-\gamma z} + V^-e^{+\gamma z} \quad [3.64a]$$

$$I(z) = \frac{1}{Z_0}[V^+e^{-\gamma z} - V^-e^{+\gamma z}] \quad [3.64b]$$

3.8.1 Infinitely Long or Matched Line

To understand the behavior of the time harmonic solutions for a lossy transmission line, we first consider the case of an infinitely long or a matched-terminated line. By analogy with the lossless case, we can see that the second terms in [3.64*a*] and [3.64*b*], those multiplied by the constant V^-, are zero in these cases, since no reflected wave exists. Accordingly, the voltage and current phasors are

$$V(z) = V^+e^{-\gamma z}; \quad I(z) = \frac{1}{Z_0}V^+e^{-\gamma z} \quad \rightarrow \quad \frac{V(z)}{I(z)} = Z_0$$

Note that everywhere on the line the ratio of the voltage to current phasors is the characteristic impedance Z_0, once again underscoring the physical meaning of the characteristic impedance.

It is instructive to write the voltage and current phasors explicitly in terms of the real and imaginary parts of γ. In other words, we have

$$V(z) = V^+e^{-\alpha z}e^{-j\beta z} \quad [3.65a]$$

$$I(z) = \frac{V^+}{Z_0}e^{-\alpha z}e^{-j\beta z} \quad [3.65b]$$

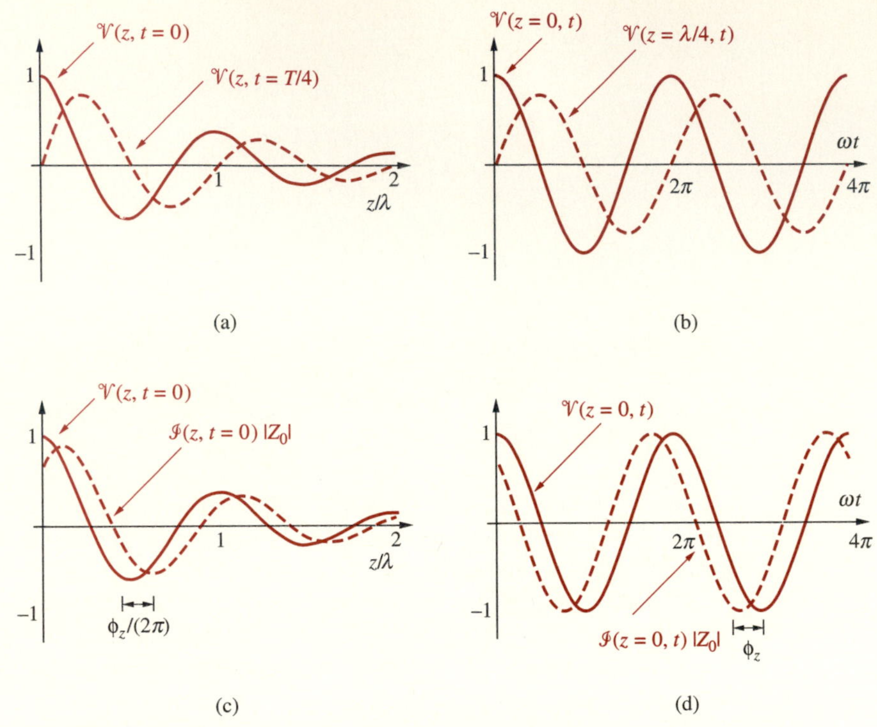

FIGURE 3.60. **Voltage and current on a matched lossy line.** (a) $\mathcal{V}(z, t) = V^+e^{-\alpha z}\cos(\omega t - \beta z)$ versus z/λ for $t = 0$ and for $t = T_p/4$, where $\beta = 2\pi/\lambda$ and $T_p = 2\pi/\omega$. (b) $\mathcal{V}(z, t)$ vs. ωt for $z = 0$ and $z = \lambda/4$. The attenuation constant was taken to be $\alpha = 1$ neper/λ. A comparison of voltage $\mathcal{V}(z, t)$ and current $\mathcal{I}(z, t)$ (c) at time $t = 0$ as a function of space and (d) at $z = 0$ as a function of time, for an assumed case where the R,L, G, and C values are such that $\phi_z = -\pi/4$. Note that we have assumed $V^+ = 1$.

Using [3.65], we can in turn obtain the space-time voltage and current functions as

$$\mathcal{V}(z, t) = \Re e\{V^+e^{-\alpha z}e^{-j\beta z}\} = V^+e^{-\alpha z}\cos(\omega t - \beta z) \qquad [3.66a]$$

$$\mathcal{I}(z, t) = \Re e\left\{\frac{V^+}{Z_0}e^{-\alpha z}e^{-j\beta z}\right\} = \frac{V^+}{|Z_0|}e^{-\alpha z}\cos(\omega t - \beta z - \phi_z) \qquad [3.66b]$$

where we have assumed V^+ to be real. The solutions for a lossy line are propagating waves with amplitudes exponentially decaying with increased distance. For physically realizable solutions, we must have $\alpha > 0$. Thus, in evaluating the propagation constant γ, the sign of the square root in [3.61] must be taken to be that which gives a value of $\alpha > 0$. To better visualize the behavior of the solutions, we show $\mathcal{V}(z, t)$ in Figure 3.60 as a function of position at fixed times and time at fixed positions. Also shown is the comparison between $\mathcal{V}(z, t)$ and $\mathcal{I}(z, t)$ as a function of space and time, clearly illustrating the phase difference ϕ_z between the two waveforms.

It can be shown from [3.65] that, for a matched or infinitely long line, the magnitude of the ratio of voltages or currents corresponding to two different positions separated by a length l is a constant. In other words, the magnitude of the ratio of the voltage at position $(z + l)$ to the voltage at position z is

$$\left|\frac{V(z)}{V(z+l)}\right| = \frac{e^{-\alpha z}}{e^{-\alpha(z+l)}} = e^{\alpha l} \qquad [3.67]$$

Taking the natural logarithm of both sides, we have

$$\alpha l = \ln\left[\left|\frac{V(z)}{V(z+l)}\right|\right] \qquad [3.68]$$

Note that αl is a dimensionless number, since the units of α are in m^{-1} and l is in meters. However, to underscore the fact that αl expresses the attenuation on the line in terms of the natural (Naperian) logarithm of the magnitude of the ratio of voltages (or currents) at different positions, it is common convention to express αl in units of *nepers* (np). Thus, in conventional usage, the unit of the attenuation constant α is nepers-m^{-1}.

In most engineering applications, a more commonly used unit for attenuation is the *decibel* (dB). The decibel is a unit derived from the *bel,* which in turn was named after Alexander Graham Bell and was used in early work on telephone systems. Specifically, the decibel is defined as

$$\text{Attenuation in decibels} \equiv 20\log_{10}\left[\left|\frac{V(z)}{V(z+l)}\right|\right] \qquad [3.69]$$

It is clear from [3.67] and [3.69] that a relation exists between attenuation expressed in decibels and that expressed in nepers. We have

$$\begin{aligned}\text{Attenuation in dB} &= 20\log_{10} e^{\alpha l} = (\alpha l)20\log_{10} e \simeq 8.686(\alpha l) \\ &\simeq 8.686\,(\text{attenuation in np})\end{aligned} \qquad [3.70]$$

The advantage of a logarithmic unit such as the decibel or neper is that the total loss of several cascaded transmission lines (and other networks connected to them) can simply be found by adding the losses in the individual units. As an example, if sections of a fiber-optic line have attenuations of 10 dB, 20 dB, and 5 dB, then the total attenuation of the signal in its passage through all three of the lines would be $10 + 20 + 5 = 35$ dB. Similarly, total gains of any number of amplifier stages in a system can also be easily calculated with the use of logarithmic units.

Examples 3-30 and 3-31 illustrate the numerical values of the line parameters, respectively, for an open-wire telephone line and a high-speed coplanar strip interconnect.

Example 3-30: Open-wire telephone line. An open-wire telephone line consists of two parallel lines made of copper with diameters ~0.264 mm and spaced ~20 cm apart on the crossarm of the wooden poles. Determine the propagation constant γ, its real and imaginary parts α and β, and the characteristic impedance Z_0. Assume it operates at 1.5 kHz.

Solution: Using the two-wire transmission line formulas given in Table 2.2, we find the transmission line parameters to be $R \simeq 24.4\Omega\text{-(km)}^{-1}$, $L \simeq$ 2.93 mH/km, and $C \simeq 3.80$ nF-(km)$^{-1}$. The value of G is assumed to be negligible. We have

$$\begin{aligned}
\gamma &= \sqrt{(R + j\omega L)(G + j\omega C)} \\
&\simeq \sqrt{(24.4 + j2\pi \times 1.5 \times 10^3 \times 2.93 \times 10^{-3})(0 + j2\pi \times 10^3 \times 3.80 \times 10^{-9})} \\
&\simeq \sqrt{(36.8e^{j48.6^\circ})(3.58 \times 10^{-5} e^{j90^\circ})} \simeq 0.0363e^{j69.3^\circ} \text{ (km)}^{-1} \\
&\simeq 0.0128 + j0.0339 \text{ (km)}^{-1} \\
&\rightarrow \alpha \simeq 1.28 \times 10^{-2} \text{ np-(km)}^{-1}; \quad \beta \simeq 3.39 \times 10^{-2} \text{ rad-(km)}^{-1}
\end{aligned}$$

The phase velocity and wavelength are

$$v_p = \frac{\omega}{\beta} \simeq \frac{2\pi \times 1.5 \times 10^3}{3.39 \times 10^{-2}} \simeq 2.78 \times 10^5 \text{ km-s}^{-1}; \qquad \lambda = \frac{2\pi}{\beta} \simeq 185 \text{ km}$$

We find that the waves on an open-wire telephone cable propagate at a speed somewhat smaller than the speed of light in free space, namely, $c = 3 \times 10^8$ m-s^{-1}. The characteristic impedance is given by

$$\begin{aligned}
Z_0 &= \sqrt{\frac{R + j\omega L}{G + j\omega C}} \\
&\simeq \sqrt{\frac{24.4 + j2\pi \times 1.5 \times 10^3 \times 2.93 \times 10^{-3}}{j2\pi \times 1.5 \times 10^3 \times 3.80 \times 10^{-9}}} \simeq 1014.5e^{-j20.7^\circ} \simeq 949 - j359\Omega
\end{aligned}$$

Example 3-31: High-speed GaAs digital circuit coplanar strip interconnects. Transmission line properties of typical high-speed interconnects are experimentally investigated by fabricating and characterizing coplanar strip interconnects on semi-insulating GaAs substrates.[35] Measurements are carried out up to 18 GHz, from which the pertinent per-unit line parameters can be extracted. In one case, the values of the propagation constant γ and characteristic impedance Z_0 at 10 GHz are determined from the measurements to be $\gamma \simeq$ 1.2 (np-(cm)$^{-1}$) + j6(rad-(cm)$^{-1}$) and $Z_0 \simeq 105 - j25\Omega$, respectively. Using these values, calculate the per-unit length parameters (R, L, G, and C) of the coplanar strip transmission line at 10 GHz.

---

[35]K. Kiziloglu, N. Dagli, G. L. Matthaei, and S. I. Long, Experimental analysis of transmission line parameters in high-speed GaAs digital circuit interconnects, *IEEE Transactions on Microwave Theory and Techniques,* 39(8), pp. 1361–1367, August 1991.

Solution: The per-unit length parameters of the transmission line can readily be computed from γ and Z_0 using the relations

$$R + j\omega L = \gamma Z_0$$

$$G + j\omega C = \frac{\gamma}{Z_0}$$

Using the measured values of γ and Z_0 at 10 GHz, we have

$$R + j\omega L = (1.2 + j6)(105 - j25) \simeq 276 + j600$$

from which $R \simeq 276\Omega\text{-(cm)}^{-1}$ and $L \simeq 600/(2\pi \times 10^{10}) \simeq 9.55\ \text{nH-(cm)}^{-1}$, respectively. Similarly, we have

$$G + j\omega C = \frac{1.2 + j6}{105 - j25} \simeq \frac{6.12e^{j78.9^\circ}}{107.9e^{-j13.4^\circ}} \simeq 0.0567e^{j92.1^\circ} \simeq -0.0021 + j0.0567$$

from which $G \simeq -0.0021\ \text{S-(cm)}^{-1}$ and $C \simeq 0.0567/(2\pi \times 10^{10}) \simeq 0.902\ \text{pF-(cm)}^{-1}$, respectively. The negative value of parameter G is nonphysical and is likely a result of measurement error.

The average power delivered into the line at any given point z can be found using the phasor expressions [3.65] for voltage and current:

$$\begin{aligned} P_{\text{av}}(z) &= \frac{1}{2}\Re e\{V(z)[I(z)]^*\} \\ &= \frac{1}{2}\Re e\left\{V^+e^{-\alpha z}e^{-j\beta z}\frac{(V^+)^*}{Z_0^*}e^{-\alpha z}e^{+j\beta z}\right\} \qquad [3.71] \\ &= \frac{|V^+|^2}{2|Z_0|}e^{-2\alpha z}\cos(\phi_z) \end{aligned}$$

We find that the time-average power decreases with distance as $e^{-2\alpha z}$, with an effective attenuation constant that is twice that of the voltage and current. The difference in time-average powers evaluated at any two points z_1 and z_2 is the amount of power dissipated in the segment of the line between z_1 and z_2.

Low-Loss Lines An important practical case is that in which the losses along the line are small but not negligible. If the line is low loss, we can assume that $R \ll \omega L$ and $G \ll \omega C$, which means that the resistive losses and leakage losses in the surrounding medium are both small. In such cases, useful approximate expressions can be derived for the characteristic impedance Z_0 and the propagation constant γ.

We first consider Z_0:

$$Z_0 = \sqrt{\frac{R + j\omega L}{G + j\omega C}}$$

$$= \sqrt{\frac{L}{C}}\sqrt{\frac{1 + R/(j\omega L)}{1 + G/(j\omega C)}} = \sqrt{\frac{L}{C}}\left[\frac{1 + R/(j2\omega L)}{1 + G/(j2\omega C)}\right]$$

where we have used the fact that for $\zeta \ll 1$ $(1+\zeta)^{1/2} = 1+(\zeta/2)+\cdots \simeq 1+(\zeta/2)$. By neglecting the higher-order terms in the numerator and the denominator, and using $(1+\zeta)^{-1} \simeq 1-\zeta$ for $\zeta \ll 1$, we can write the characteristic impedance as

$$Z_0 \simeq \sqrt{\frac{L}{C}}\left(1 + \frac{R}{j2\omega L}\right)\left(1 - \frac{G}{j2\omega C}\right)$$

$$= \sqrt{\frac{L}{C}}\left[\left(1 + \frac{RG}{4\omega^2 LC}\right) + j\frac{1}{2\omega}\left(\frac{G}{C} - \frac{R}{L}\right)\right] \quad [3.72]$$

In general, the second term in the real part of [3.72] is negligible since it involves the product of two small terms, namely, $R/(\omega L)$ and $G/(\omega C)$. Thus, the important effect of the losses on the transmission line is to introduce a small imaginary component to the characteristic impedance. In many cases, the imaginary part of Z_0 can be neglected, so that the characteristic impedance is, to the first order, equal to that for the lossless line.

A similar simplification can also be obtained for the propagation constant γ, again using $(1+\zeta)^{1/2} = 1 + (\zeta/2) + \cdots \simeq 1 + (\zeta/2)$ for $\zeta \ll 1$. We have

$$\gamma = [(R + j\omega L)(G + j\omega C)]^{1/2}$$

$$= \left[(j\omega L)(j\omega C)\left(1 + \frac{R}{j\omega L}\right)\left(1 + \frac{G}{j\omega C}\right)\right]^{1/2}$$

$$\simeq j\omega\sqrt{LC}\left[1 - j\frac{1}{2\omega}\left(\frac{R}{L} + \frac{G}{C}\right)\right]$$

The real and imaginary parts of γ for the low-loss line are thus

$$\alpha \simeq \frac{1}{2}\left[R\sqrt{\frac{C}{L}} + G\sqrt{\frac{L}{C}}\right] \quad [3.73a]$$

$$\beta \simeq \omega\sqrt{LC} \quad [3.73b]$$

As an example, coaxial lines used at high radio frequencies can be quite accurately represented by the above low-loss formulas of [3.73]. Note that the phase constant β is the same as that in the lossless case, so the phase velocity $v_p = \omega/\beta =$

$1/\sqrt{LC}$, independent of frequency. The loss constant α also does not depend on frequency; it simply accounts for a decrease in the overall signal intensity as the wave propagates along the line. Thus the distortion of an information-carrying signal (consisting of a finite band of frequencies), due to the different speed and attenuation of its frequency components, is minimized for a low-loss line.

Parameter values for a typical low-loss line are illustrated in Example 3-32.

Example 3-32: Low-loss coaxial line. RG17A/U is a low-loss radio frequency coaxial line. The following data for the nominal parameters of this line are available: characteristic impedance $Z_0 = 50\Omega$, line capacitance $C \simeq 96.8$ pF-m^{-1}, and line attenuation ~3 dB/100 m at 100 MHz. Determine the inductance L and resistance R per unit length of this line, assuming that G is negligibly small. Determine the velocity of propagation.

Solution: Using [3.70], we can express the attenuation in np-m^{-1}. We have

$$3 \text{ dB-}(100 \text{ m})^{-1} = 0.03 \text{ dB-m}^{-1} \simeq \frac{0.03}{8.686} \simeq 3.45 \times 10^{-3} \text{ np-m}^{-1} = \alpha$$

Using the low-loss formulas, we have

$$\alpha \simeq \frac{1}{2}\left[\frac{R}{Z_0} + GZ_0\right] = \frac{R}{2Z_0} \simeq 3.45 \times 10^{-3} \text{ np-m}^{-1}$$

which gives us $R \simeq 0.345\Omega\text{-m}^{-1}$ since $G \simeq 0$ and $Z_0 = 50\Omega$. The inductance can be determined from

$$Z_0 \simeq \sqrt{\frac{L}{C}} = 50\Omega \quad \rightarrow \quad L = Z_0^2 C \simeq (50)^2(96.8 \times 10^{-12}) = 0.242 \ \mu\text{H-m}^{-1}$$

We can check to see that the quantity $|R/(\omega L)| \simeq 2.27 \times 10^{-3}$, or is much smaller than 1, which is apparently why the characteristic impedance for this low-loss line is real. The phase velocity is given by

$$v_p = \frac{1}{\sqrt{LC}} = \frac{1}{\sqrt{Z_0^2 C^2}} = \frac{1}{Z_0 C} = \frac{1}{50 \times 96.8 \times 10^{-12}} \simeq 2.07 \times 10^8 \text{ m-s}^{-1}$$

3.8.2 Terminated Lossy Lines

An important result of losses is that both the forward wave traveling toward the load and the reflected wave traveling away from the load are attenuated exponentially with distance. As an observer moves away from the load on a terminated lossless line, the standing-wave pattern remains the same. However, on a lossy line, the same observer finds that the attenuation of the reflected wave causes this wave to be less important as he or she moves farther from the load. In addition, since the magnitude of the forward wave becomes larger as the observer moves away from the load, the

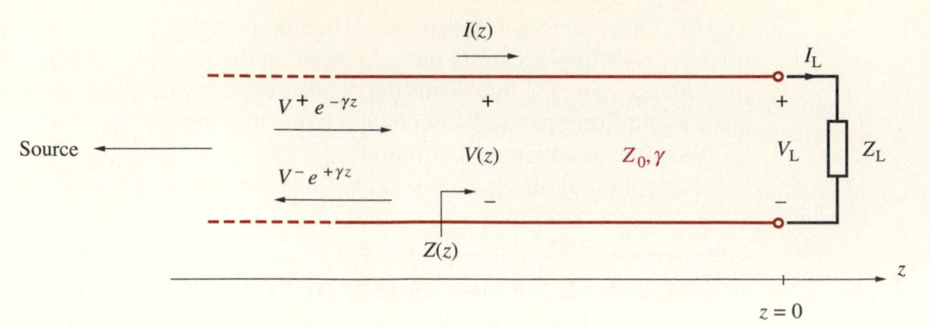

FIGURE 3.61. A terminated lossy transmission line.

relative size of the reflected wave is doubly reduced in moving toward the source. The net result of this effect is that, regardless of its termination, the transmission line begins to appear more and more like an infinite (or matched) line when viewed farther and farther from the load.

We consider a terminated lossy transmission line as shown in Figure 3.61, with $z = 0$ taken to be the position of the load as in the case of lossless lines. In general, the expressions for voltage and current on a terminated lossy transmission line are

$$\begin{aligned} V(z) &= V^+e^{-\gamma z} + V^-e^{\gamma z} = V^+(e^{-\alpha z}e^{-j\beta z} + \Gamma_L e^{\alpha z}e^{j\beta z}) \\ &= V^+e^{-\alpha z}e^{-j\beta z}[1 + \Gamma(z)] \end{aligned} \quad [3.74a]$$

$$I(z) = \frac{1}{Z_0}V^+e^{-\alpha z}e^{-j\beta z}[1 - \Gamma(z)] \quad [3.74b]$$

where $\Gamma(z)$ is the complex voltage reflection coefficient at any position z along the line defined as

$$\Gamma(z) \equiv \frac{V^-e^{\gamma z}}{V^+e^{-\gamma z}} = \Gamma_L e^{2\alpha z}e^{j2\beta z}$$

and where Γ_L is the complex load reflection coefficient given as

$$\Gamma_L \equiv \frac{V^-}{V^+} = \rho e^{j\psi} = \frac{Z_L - Z_0}{Z_L + Z_0} \quad [3.75]$$

The line impedance $Z(z)$ at any point z on the line is given by the ratio of the voltage and the current:

$$Z(z) = \frac{V(z)}{I(z)} = Z_0\frac{e^{-\alpha z}e^{-j\beta z} + \Gamma_L e^{\alpha z}e^{j\beta z}}{e^{-\alpha z}e^{-j\beta z} - \Gamma_L e^{\alpha z}e^{j\beta z}} \quad [3.76]$$

It is sometimes useful to rewrite the impedance as follows:

$$Z(z) = Z_0\frac{1 + \Gamma_L e^{2\alpha z}e^{j2\beta z}}{1 - \Gamma_L e^{2\alpha z}e^{j2\beta z}} = Z_0\frac{1 + \Gamma(z)}{1 - \Gamma(z)} \quad [3.77]$$

We compare [3.77] for $Z(z)$ to equation [3.31] for the lossless line. We see that Γ_L in [3.31] is replaced with $\Gamma_L e^{2\alpha z}$, so that the magnitude of the reflection coefficient for the lossy case is effectively reduced exponentially between the observation point and the end of the line (i.e., the load position). As viewed from larger and larger distances from the load (i.e., as $z \to -\infty$), the effect of reflections becomes negligible, and the line impedance approaches Z_0, as if the line were an infinitely long or matched line. To understand this effect, consider the general voltage reflection coefficient at any point z, namely

$$\Gamma(z) = \Gamma_{\rm L} e^{2\alpha z} e^{j2\beta z} = \rho e^{2\alpha z} e^{j(\psi + 2\beta z)} \qquad [3.78]$$

In Section 3.6, we noted that motion along the line away from the load corresponded to clockwise rotation of $\Gamma = u + jv$ in the uv plane (or on the Smith chart), while its magnitude $\Gamma = \rho$ remained constant. On lossy lines, [3.78] indicates that the same type of rotation occurs as determined by the $e^{j2\beta z}$ term, but that, in addition, the magnitude of $\Gamma(z)$, namely $|\Gamma| = \rho e^{2\alpha z}$, decreases as we move away from the load (i.e., as z decreases). Eventually, at some point, $|\Gamma(z)| \to 0$, and looking toward the load from the source side beyond this position, the line is indistinguishable from an infinitely long or matched line.

To examine the behavior of the line voltage, current, and impedance, we first consider a short-circuited line of length l, so that $Z_{\rm L} = 0$ in Figure 3.61. In this case, the load reflection coefficient is $\Gamma_{\rm L} = -1$, so the line voltage and current are

$$V(z) = V^+(e^{-\gamma z} - e^{\gamma z}) = -2V^+ \sinh(\gamma z) \qquad [3.79a]$$

$$I(z) = \frac{V^+}{Z_0}(e^{-\gamma z} + e^{\gamma z}) = \frac{2V^+}{Z_0} \cosh(\gamma z) \qquad [3.79b]$$

where we have used the defining expressions for the hyperbolic sine and cosine functions:

$$\sinh\zeta = \frac{e^{\zeta} - e^{-\zeta}}{2} \qquad \cosh\zeta = \frac{e^{\zeta} + e^{-\zeta}}{2}$$

Although the compact form of $V(z)$ in [3.79*a*] appears very similar to that for the lossless line (with sin replaced by sinh), the evaluation of $\sinh(\gamma z)$ is not trivial,[36] since γ is a complex quantity. Note that when $\alpha \to 0$, equations [3.79] for $V(z)$ and $I(z)$ reduce to their lossless equivalents, since $\sinh(\alpha z) \to 0$ and $\cosh(\alpha z) \to 1$.

Using [3.79*a*] and [3.79*b*], we can compactly write the input impedance of a short-circuited line of length l as

$$[Z_{\rm in}]_{\rm sc} = Z_0 \tanh(\gamma l)$$

[36]The hyperbolic sine of the complex number $\gamma = \alpha + j\beta$ can be expressed as

$$\begin{aligned}\sinh(\gamma z) &= \sinh[(\alpha + j\beta)z] \\ &= \sinh(\alpha z)\cosh(j\beta z) + \cosh(\alpha z)\sinh(j\beta z) \\ &= \cos(\beta z)\sinh(\alpha z) + j\cosh(\alpha z)\sin(\beta z)\end{aligned}$$

In practice, the evaluation of $\sinh[(\alpha + j\beta)z]$ would be straightforward using any reasonably sophisticated numerical evaluation tool (e.g., a software package or a scientific calculator); however, it is useful to note for insight the nature of the actual evaluation.

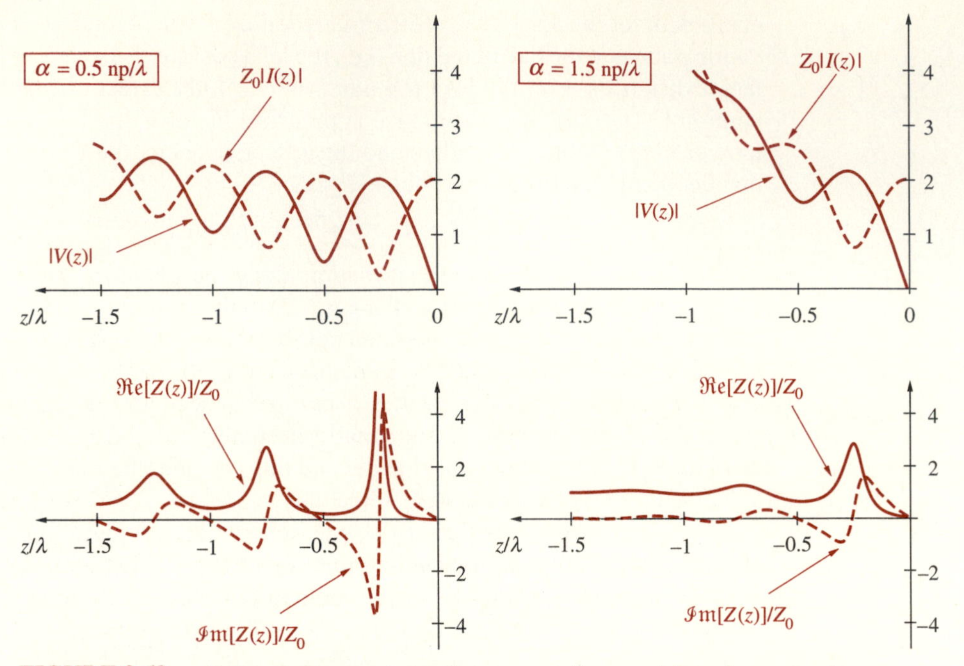

FIGURE 3.62. Voltage and current standing-wave patterns and impedance on a lossy short-circuited line. Results are shown for two different values of the attenuation constant, namely $\alpha = 0.5$ np/λ and 1.5 np/λ, where $\lambda = 2\pi/\beta$. For simplicity, we have assumed the characteristic impedance to be real, i.e., $\phi_z = 0$, and $V^+ = 1$.

Plots of magnitudes of the line voltage and current and the real and imaginary parts of the line impedance for a short-circuited line are provided in Figure 3.62, for two different values of the attenuation constant α, namely $\alpha = 0.5$ np/λ and 1.5 np/λ. For simplicity, we have assumed the phase of the characteristic impedance $\phi_z = 0$ in Figure 3.62. In general this phase angle is small, and leads to a phase difference between the voltage and current, as indicated in [3.66].

The resultant effects of the losses shown in Figure 3.62 become clear upon comparative examination of the lossless equivalents given in Figures 3.5 and 3.6. In the lossless case (Figures 3.5 and 3.6), the line voltage is zero at the load, and every half-wavelength thereafter, while the line current is a maximum at the same positions. The line impedance of the lossless line is zero at the load ($Z_L = 0$), inductive (i.e., $\mathcal{I}m\{Z(z)\} > 0$) in the range $-\lambda/4 < z < 0$, infinite (i.e., an open circuit) at $z = -\lambda/4$, capacitive in the range $-\lambda/2 < z < -\lambda/4$ back to zero at $z = -\lambda/2$, and repeating the same cycle thereafter.

For the lossy case, looking first at the relatively low-loss case of $\alpha = 0.5$ np/λ, we see that although the voltage and current exhibit generally similar cyclic behavior, the maximum and minimum values of both the line voltage and current increase with distance from the load. The line voltage is no longer zero at $z = -\lambda/2$. The differences between the values of the maxima and the minima also become smaller as z becomes increasingly negative, as is clearly evident from the relatively high-loss case of $\alpha = 1.5$ np/λ. At a sufficient distance away from the load, e.g., for $z < -1.5\lambda$ in the case of $\alpha = 1.5$ np/λ, the magnitude of line voltage and current do not vary significantly over a distance of half-wavelength (i.e., the standing-wave ratio is unity), as if the line were infinitely long or matched.

The line impedance for the relatively low-loss ($\alpha = 0.5$ np/λ) case exhibits similar behavior to the lossless case. The impedance is inductive (i.e., $\mathcal{I}m\{Z(z)\} > 0$) in the approximate range $-\lambda/4 < z < 0$, attains a large real value (but not quite an open circuit) at $z = -\lambda/4$, is capacitive in the approximate range $-\lambda/2 < z < -\lambda/4$, but does not quite return to zero at $z = -\lambda/2$. The peak in the real part of the impedance at $z = -3\lambda/4$ is considerably smaller than that at $z = -\lambda/4$. In general, the maxima and minima of the imaginary part of $Z(z)$ both approach zero as z attains larger and larger negative values, while the maxima and minima of the real part of $Z(z)$ both approach Z_0. At sufficient distances from the load, for example, for $z < -1.5\lambda$ in the case of $\alpha = 1.5$ np/λ, the line impedance $Z(z) \simeq Z_0$, just as if the line were infinitely long or matched.

For a lossy line terminated in an open circuit ($Z_L = \infty$), expressions for $V(z)$, $I(z)$, and Z_{in} can be obtained in a manner analogous to the preceding discussion for a short-circuited line. This straightforward procedure is left as an exercise.

The general behavior of the line voltage, current, and impedance for other terminations is quite similar, as illustrated in Figures 3.63 and 3.64 for a resistive load impedance of $Z_L = 5Z_0$. Results are shown for four different values of the

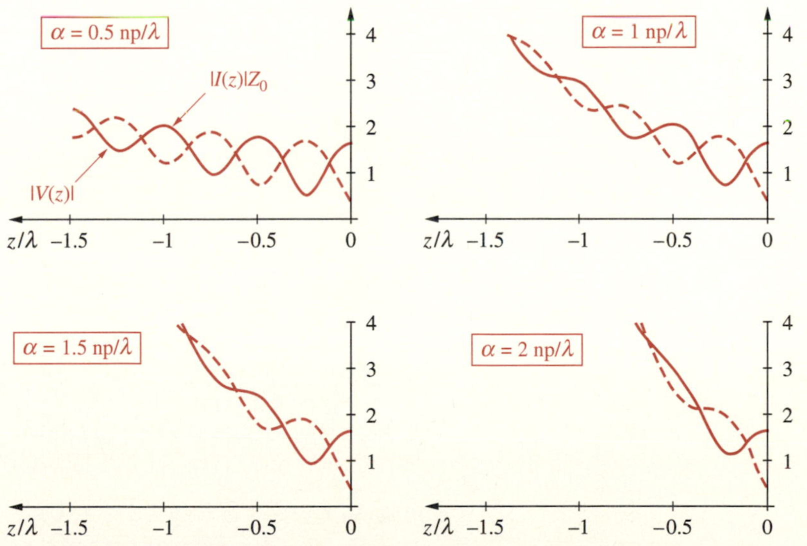

FIGURE 3.63. Voltage and current standing-wave patterns on a terminated lossy transmission line. The magnitudes of the voltage and current phasors (for current, the quantity plotted is $|I(z)|Z_0$) of a lossy line terminated in $Z_L = 5Z_0$ are shown for values of the attenuation constant $\alpha = 0.5$, 1, 1.5, and 2 np/λ, where $\lambda = 2\pi/\beta$.

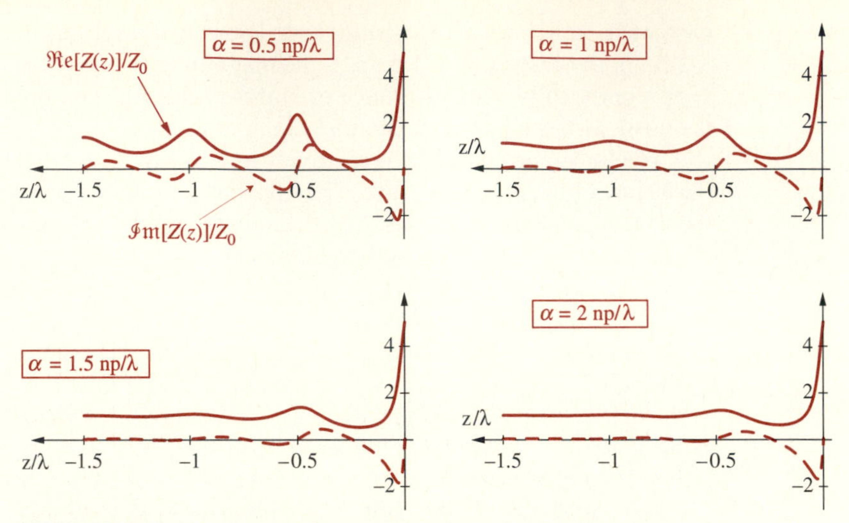

FIGURE 3.64. Line impedance standing on a terminated lossy transmission line. The real and imaginary parts of the line impedance (normalized to Z_0) of a lossy line terminated in $Z_L = 5Z_0$ are shown for values of the attenuation constant $\alpha = 0.5, 1, 1.5,$ and 2 np/λ, where $\lambda = 2\pi/\beta$.

attenuation constant. For simplicity, we have once again assumed the phase of the characteristic impedance $\phi_z = 0$.

The time-average power at any point z along the line can be evaluated using the expressions [3.74] for $V(z)$ and $I(z)$. We have

$$P_{\text{av}}(z) = \frac{1}{2}\Re e\{V(z)[I(z)]^*\}$$

$$= \frac{|V^+|^2 e^{-2\alpha z}}{2}\Re e\left\{\frac{[1+\Gamma_L e^{2\alpha z}e^{j2\beta z}][1-\Gamma_L e^{2\alpha z}e^{j2\beta z}]^*}{Z_0^*}\right\}$$

$$= \frac{|V^+|^2 e^{-2\alpha z}}{2}\Re e\left\{\frac{1-|\Gamma_L|^2 e^{4\alpha z}-\Gamma_L^* e^{2\alpha z}e^{-j2\beta z}+\Gamma_L e^{2\alpha z}e^{+j2\beta z}}{Z_0^*}\right\}$$

Consider a terminated transmission line of length l. The time-average power at its input, namely at $z = -l$, is given by $P_{\text{av}}(z = -l)$, whereas that at the load is given by $P_{\text{av}}(z = 0)$. The difference between these quantities is the average power dissipated in the lossy transmission line. Thus, power lost in the line is given by

$$P_{\text{lost}} = P_{\text{av}}(z = -l) - P_{\text{av}}(z = 0)$$

Reflections in lossy lines can lead to substantially increased losses since each time a wave travels down the line it is further attenuated. If the load reflects part of the incident power, more power is dissipated in the lossy line than would have been dissipated if the line were matched (i.e., $\Gamma_L = 0$). If the power dissipated in a lossy line under matched conditions is $P^{\rm m}_{\rm lost}$, it can be shown[37] that the extra power dissipated as a result of reflections is

$$\frac{P_{\rm lost} - P^{\rm m}_{\rm lost}}{P^{\rm m}_{\rm lost}} \simeq |\Gamma_{\rm L}|^2(e^{2\alpha l} - 1)$$

The extra power dissipated due to mismatch can be substantial, especially when $|\Gamma_{\rm L}| > 0.5$ and when the line is long.

Example 3-33 illustrates the concepts of power dissipation in lossy lines in the context of a high-speed microstrip interconnect.

Example 3-33: A high-speed microstrip interconnect. Consider a high-speed microstrip transmission line of length 20 cm used to connect a 1-V amplitude, 1-GHz, 50Ω sinusoidal voltage source to a digital logic gate having an input impedance of 1 kΩ, as shown in Figure 3.65. Based on measurements, the transmission line parameters of this interconnect at 1 GHz are approximately given by R = 5Ω-cm^{-1}, L = 5 nH-cm^{-1}, C = 0.4 pF-cm^{-1}, and G = 0 respectively.

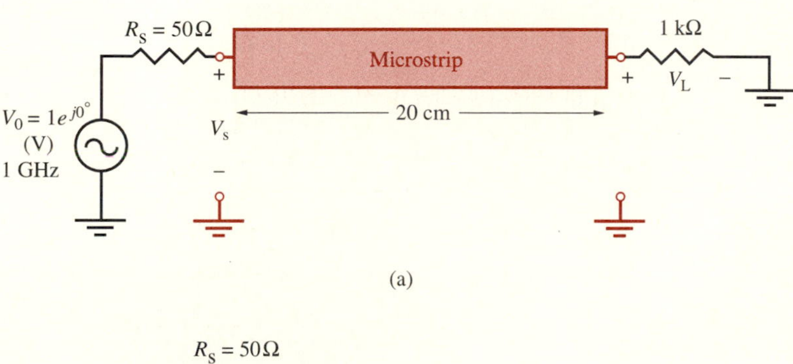

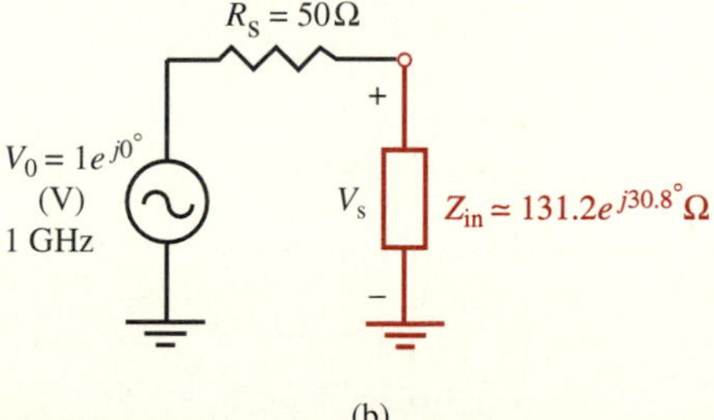

FIGURE 3.65. A lossy high-speed microstrip interconnect. (a) The microstrip transmission line connected to a 1-kΩ load and driven by a 1-GHz source. (b) Thévenin equivalent circuit as seen by the source, where $Z_{\rm in}$ is the input impedance at the source end of the microstrip.

[37] See Section 6-3 of R. K. Moore, *Traveling Wave Engineering*, McGraw-Hill, New York, 1960.

(a) Find the propagation constant γ and characteristic impedance Z_0 of the line. (b) Find the voltages at the source and the load ends of the line. (c) Find the time-average power delivered to the line by the source and the time-average power delivered to the load. What is the power dissipated along the line?

Solution:

(a) The propagation constant is given by

$$\gamma = \alpha + j\beta = \sqrt{(R + j\omega L)(G + j\omega C)}$$

$$= \sqrt{(500 + j2\pi \times 10^9 \times 500 \times 10^{-9})(j2\pi \times 10^9 \times 40 \times 10^{-12})}$$

$$\simeq \sqrt{3181e^{j80.96^\circ} \times 0.251e^{j90^\circ}}$$

$$\simeq 28.3e^{j85.5^\circ} \simeq 2.23 + j28.2$$

where $\alpha \simeq 2.23$ np-m^{-1} and $\beta \simeq 28.2$ rad-m^{-1}. The characteristic impedance is given by

$$Z_0 = \sqrt{\frac{R + j\omega L}{G + j\omega C}}$$

$$\simeq \sqrt{\frac{500 + j3142}{j0.251}} \simeq \sqrt{\frac{3181e^{j80.96^\circ}}{0.251e^{j90^\circ}}}$$

$$\simeq 112.5e^{-j4.522^\circ}\,\Omega$$

(b) The reflection coefficient at the load end can be found as

$$\Gamma_L = \frac{Z_L - Z_0}{Z_L + Z_0} \simeq \frac{1000 - 112.5e^{-j4.522^\circ}}{1000 + 112.5e^{-j4.522^\circ}}$$

$$\simeq \frac{887.8 + j8.869}{1112 - j8.869} \simeq 0.798e^{j1.03^\circ}$$

The reflection coefficient at any position along the line is given by

$$\Gamma(z) = \Gamma_L e^{2\gamma z} \simeq 0.798e^{j1.029^\circ}e^{4.458z}e^{j56.37z}$$

$$\simeq 0.798e^{4.458z}e^{j(56.37z+0.018)}$$

The input impedance of the line is given by

$$Z_{in} = Z(z)|_{z=-0.2\text{ m}} = Z_0\frac{1 + \Gamma(-0.2)}{1 - \Gamma(-0.2)}$$

We first find $\Gamma(-0.2)$ as

$$\Gamma(-0.2) \simeq 0.798e^{4.458(-0.2)}e^{j[56.37(-0.2)+0.018]} \simeq 0.327e^{j75.05^\circ}$$

We now use $\Gamma(-0.2)$ to find Z_{in}. We have

$$Z_{\text{in}} \simeq 112.5e^{-j4.522^\circ}\frac{1+0.327e^{j75.05^\circ}}{1-0.327e^{j75.05^\circ}}$$

$$\simeq 112.5e^{-j4.522^\circ}\frac{1.084+j0.316}{0.915-j0.316}$$

$$\simeq (112.5e^{-j4.522^\circ})(1.166e^{j35.31^\circ}) \simeq 131.2e^{j30.79^\circ}\Omega$$

Based on voltage division in the equivalent circuit of Figure 3.65b, the source-end voltage V_s is

$$V_s = \frac{Z_{\text{in}}}{Z_{\text{in}}+Z_s}V_0 \simeq \frac{131.2e^{j30.79^\circ}}{112.7+j67.16+50}(1)$$

$$\simeq \frac{131.2e^{j30.79^\circ}}{176e^{j22.43^\circ}} \simeq 0.745e^{j8.361^\circ}\text{ V}$$

The voltage at any position z along the line is given by

$$V(z) = V^+e^{-\gamma z}[1+\Gamma(z)]$$

from which V^+ can be written as

$$V^+ = \frac{V(z)}{e^{-\gamma z}[1+\Gamma(z)]}$$

At $z = -0.2$ m, we have

$$V(z=-0.2) = V_s \simeq 0.745e^{j8.361^\circ}\text{ V}$$

and

$$e^{-\gamma(-0.2)} \simeq e^{0.446}e^{-j37.01^\circ} \quad \text{and} \quad \Gamma(-0.2) \simeq 0.327e^{j75.05^\circ}$$

Using these values, the value of V^+ can be found as

$$V^+ \simeq \frac{0.745e^{j8.361^\circ}}{e^{0.446}e^{-j37.01^\circ}(1+0.327e^{j75.05^\circ})} \simeq 0.423e^{j29.12^\circ}\text{ V}$$

The voltage at the load end of the line is given as

$$V_L = V(z=0) = V^+[1+\Gamma_L] \simeq (0.423e^{j29.12^\circ})(1+0.798e^{j1.029^\circ}) \simeq 0.760e^{j29.57^\circ}\text{ V}$$

(c) The time-average power delivered to the line is given by

$$P_s = \frac{1}{2}\left|\frac{V_s}{Z_{\text{in}}}\right|^2 R_{\text{in}} \simeq \frac{1}{2}\left|\frac{0.745}{131.2}\right|^2(112.7)\text{ W} \simeq 1.82\text{ mW}$$

Similarly, the time-average power delivered to the load can be found as

$$P_L = \frac{1}{2}\frac{|V_L|^2}{R_L} \simeq \frac{1}{2}\frac{(0.760)^2}{1000}\text{ W} \simeq 0.289\text{ mW}$$

Thus, based on conservation of energy, the power dissipated in the lossy line is

$$P_{\text{lost}} = P_{\text{s}} - P_{\text{L}} \simeq 1.82 - 0.289 \simeq 1.53 \text{ mW}$$

3.9 TRANSMISSION LINES AS RESONANT CIRCUIT ELEMENTS

One of the basic elements in a wide variety of dynamical systems is a resonator. In electronic applications, resonant circuits are found in the design of systems that selectively amplify or transmit a single frequency or a narrow band of frequencies. Starting at frequencies of hundreds of MHz, transmission lines and other distributed devices are commonly used as resonant circuit elements in filters, oscillators, tuned amplifiers, phase equalizers, or frequency measuring devices. Since important aspects of the behavior of resonant circuit elements are largely determined by the degree to which the system is lossy, it is appropriate to discuss transmission-line resonators after the general discussion of lossy transmission lines.

Below microwave frequencies (<300 MHz), resonant circuits typically consist of lumped capacitances and inductances. Although microwave integrated circuit elements that behave as capacitances and inductances can be constructed for operation at microwave frequencies (>300 MHz), such elements usually have too high losses to be effective as resonant elements and also are physically too small to handle useful power levels. Accordingly, distributed circuit elements with dimensions comparable to a wavelength are used as resonant elements. Such elements typically consist of sections of transmission line elements (coaxial, two-wire line, parallel-plate line, microstrip, etc.) having lengths of a quarter wavelength or half wavelength.

3.9.1 Lumped Resonant Circuits

Since most concepts underlying lumped resonant circuits carry over to distributed resonators, we first provide a brief review of lumped resonant circuits. This discussion is also useful because transmission line resonators can often be analyzed and represented in terms of lumped equivalent circuits. We consider the series RLC circuit as the simplest example of an electrical resonator, while noting that a nearly identical analysis also applies to the parallel RLC circuit.

The input impedance of the series RLC circuit shown in Figure 3.66a is the ratio of the phasor of the applied voltage V to the phasor of the resultant current I, namely

$$Z_{\text{in}} = \frac{V}{I} = R + j\omega L - \frac{j}{\omega C} = R + j\left(\omega L - \frac{1}{\omega C}\right)$$

If we view the voltage $\mathcal{V}(t)$ as the input and the current $\mathcal{I}(t)$ as the output, it is clear that the magnitude of Z_{in} determines the magnitude of sinusoidal current fluctuations

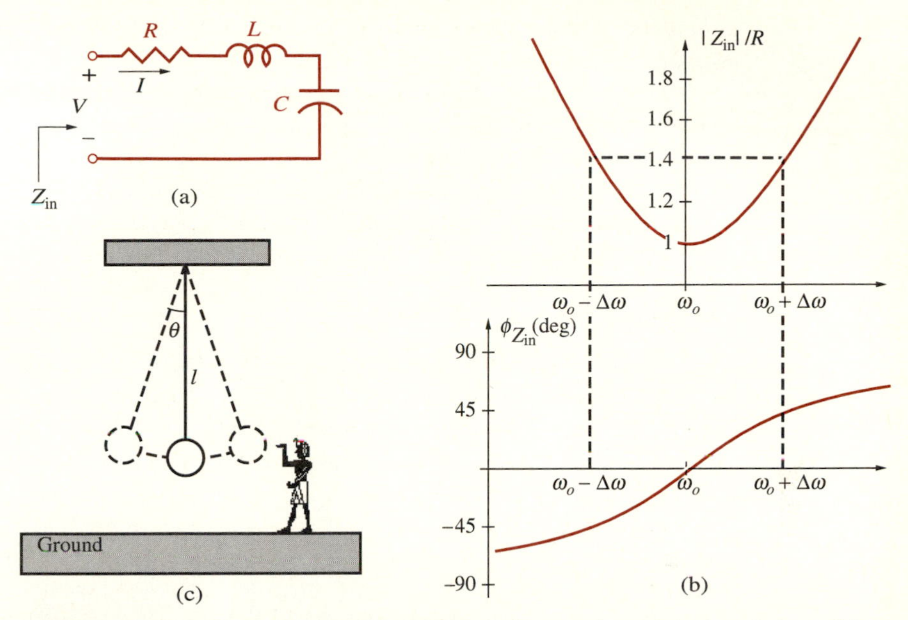

FIGURE 3.66. A series RLC circuit. (a) Circuit diagram. (b) Magnitude and phase of the input impedance as a function of frequency around the resonance frequency $\omega_0 = (LC)^{-1/2}$. (c) An analogous simple pendulum of large physical size. The angular resonant frequency for small oscillations ($\theta \ll \pi/2$) of a simple pendulum is $\omega_0 = \sqrt{g/l}$, where g is the gravitational acceleration and l is the length of the pendulum.

for a given magnitude of applied sinusoidal voltage; in terms of phasor quantities V and I, we have $|I| = |V|/|Z_{\text{in}}|$. In resonant circuit applications, the important parameter is the variation of Z_{in} with frequency, which is shown in Figure 3.66b. We note that the magnitude of the impedance $|Z_{\text{in}}|$ is a minimum at the *resonant* frequency, $\omega_0 = (LC)^{-1/2}$, for which $Z_{\text{in}} = R$. This in turn means that maximum sinusoidal current is established in this circuit when the sinusoidal input voltage is at a frequency $\omega = \omega_0$. Establishing an oscillatory current of the same amplitude at frequencies below or above ω_0 requires larger input voltages.

To better understand the principle of resonance, note that the resonant RLC circuit is analogous to a simple pendulum, which exhibits a natural frequency for small oscillations with an angular frequency of $\omega_0 = \sqrt{g/l}$, where l is the length of the pendulum and g is the gravitational acceleration. Consider a pendulum consisting of a very large and heavy ball hung with a long cable, as shown in Figure 3.66c. If we were to make the pendulum swing back and forth at different frequencies, we would find that it requires a trivially small force to set it into oscillations at ω_0; even for a rather heavy ball, a person could set the pendulum into oscillation (i.e., swinging back and forth repeatedly) with the periodic tap of a finger at time intervals of approximately $2\pi/\omega_0$. However, if we wanted to make the pendulum swing back and forth at a faster or slower rate than its natural frequency of oscillation, we would have to exert an enormous amount of force to carry the large weight of the ball across to make it go faster than its natural frequency or to hold it back in order to make it oscillate slower than its natural frequency.

The magnitude of the oscillatory current that can be established in an RLC circuit at ω_0 is determined by the losses, since $Z_{\text{in}} = R$, and thus $I(\omega_0) = V(\omega_0)/R$. In the absence of losses, that is, if $R = 0$, we can establish an oscillatory current at ω_0 with zero input voltage; in other words, if there were any initial stored energy in the circuit, oscillatory current would flow indefinitely even if we short-circuited the input terminals (i.e., if $V = 0$). In practice, one strives to make R as small as possible, but its value is necessarily nonzero and determines the sharpness, or quality, of resonance.

A useful measure of the sharpness of resonance is the quality factor Q, defined as

$$Q \equiv \omega_0 \frac{\text{time-average energy stored}}{\text{energy lost per second}}$$

with all quantities evaluated at the resonant frequency $\omega = \omega_0$. The energy lost per second (Joules-s^{-1}) is the power loss, given by $P_{\text{loss}} = \frac{1}{2}|I|^2 R$, in watts. The time-average energy stored in the inductance is

$$\bar{W}_{\text{L}} = \frac{1}{T_p}\int_0^{T_p} \frac{1}{2}|\mathcal{I}(t)|^2 L\, dt = \frac{1}{T_p}\int_0^{T_p} \frac{1}{2}[|I|\cos(\omega t + \phi_I)]^2 L\, dt = \frac{1}{4}|I|^2 L$$

where we have recognized that the current phasor is usually a complex number $I = |I|e^{\phi_I}$, so that $\mathcal{I}(t) = \mathcal{Re}\{Ie^{j\phi_I}e^{j\omega t}\} = |I|\cos(\omega t + \phi_I)$. Similarly, the time-average energy stored in the capacitance is

$$\bar{W}_C = \frac{1}{4}|V_c|^2 C = \frac{1}{4}\frac{|I|^2}{(\omega^2 C)}$$

where V_c is the phasor of the voltage across the series capacitance, the magnitude of which is given by $|V_c| = |I|/(|j\omega C|) = |I|/wC$. Note that at the resonance frequency $\omega = \omega_0$, we have $\bar{W}_{\text{L}} = \bar{W}_C$ since $\omega_0 = (LC)^{-1/2}$. The total stored energy in the RLC circuit at the resonance frequency is $\bar{W} = \bar{W}_{\text{L}} + \bar{W}_C = 2\bar{W}_{\text{L}}$. The quality factor Q is then given by

$$Q = \omega_0 \frac{\bar{W}}{P_{\text{loss}}} = \frac{\omega_0 2(\frac{1}{4}|I|^2 L)}{\frac{1}{2}R|I|^2} = \frac{\omega_0 L}{R} = \frac{1}{\omega_0 RC}$$

Let us now examine the behavior of the input impedance as a function of frequency in the vicinity of resonance. At $\omega = \omega_0 + \Delta\omega$, with $\Delta\omega \ll \omega_0$, we have

$$\begin{aligned}
Z_{\text{in}} &= R + j(\omega_0 + \Delta\omega)L + \frac{1}{j(\omega_0 + \Delta\omega)C} \\
&= R + j\omega_0 L + j\Delta\omega L + \left[\frac{1}{j\omega_0 C} - \left(\frac{1}{j\omega_0 C}\right)^2 j\Delta\omega C + \cdots\right] \\
&\simeq R + j\Delta\omega L + j\frac{\Delta\omega}{\omega_0^2 C} = R + j\frac{2\Delta\omega}{\omega_0^2 C} = R + j2L\Delta\omega
\end{aligned}$$

where we have used $\omega_0^2 = 1/LC$. Since $Q = 1/(\omega_0 RC)$, we can express Z_{in} in terms of Q as

$$Z_{\text{in}} \simeq R\left(1 + j2Q\frac{\Delta\omega}{\omega_0}\right)$$

The bandwidth of the series RLC circuit can be determined from the variation with frequency of its input impedance, as shown in Figure 3.66b. The so-called 3-dB bandwidth, defined to be the frequency range over which the magnitude of the impedance is within a factor of $\sqrt{2}$ of that at resonance,[38] is marked in Figure 3.66b by the points at which the real and imaginary parts of the input impedance are equal. Namely,

$$2Q\frac{\Delta\omega}{\omega_0} = 1 \quad \rightarrow \quad \frac{2\Delta\omega}{\omega_0} = \frac{1}{Q}$$

Note that the bandwidth is $2\Delta\omega$, since the behavior around resonance is approximately symmetrical (for $Q \gg 1$) on both sides of ω_0. Thus, we have

$$\text{Bandwidth} = \frac{\omega_0}{Q}$$

The higher the Q of a resonant circuit, the narrower is its bandwidth. All of the preceding concepts apply to different kinds of resonant systems, although they were specifically derived for the series RLC circuit.

3.9.2 Transmission Line Resonators

The simplest examples of distributed resonant circuit elements are short- or open-circuited transmission lines. Note that, in practice, these may be implemented in terms of any of the different two-conductor transmission line configurations shown in Figure 2.1, that is, coaxial, two-wire line, stripline, or others. Consider first the short-circuited half-wavelength-long line shown in Figure 3.67, which we will show

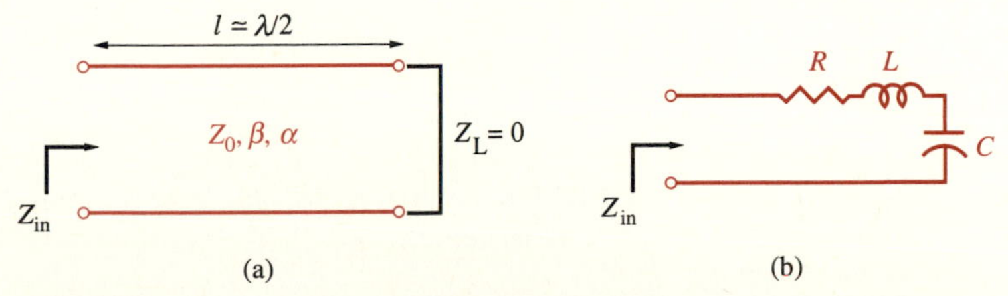

FIGURE 3.67. Short-circuited transmission line resonator. A shorted line shown in (a) exhibits behavior similar to a series RLC circuit (shown in (b)) for $l \simeq \lambda/2$.

[38]Note that, for a given applied voltage V, a reduction in impedance Z_{in} by a factor of $\sqrt{2}$ corresponds to a decrease in the current $I = V/Z_{\text{in}}$ by a factor of $\sqrt{2}$, or a decrease in power by a factor of 2, corresponding to 3 dB.

to be equivalent (in terms of the frequency variation of its input impedance) to a series RLC circuit.

The input impedance of such a lossy line of length l is

$$Z_{\text{in}}|_{z=-l} = Z_0 \frac{e^{\alpha l} e^{j\beta l} - e^{-\alpha l} e^{-j\beta l}}{e^{\alpha l} e^{j\beta l} + e^{-\alpha l} e^{-j\beta l}}$$

where we have taken into account the fact that $\Gamma_{\text{L}} = -1$ for a short-circuited termination. We consider a line of length $l = \lambda_0/2$ at $\omega = \omega_0$, where $\lambda_0 = 2\pi c/\omega_0$, with c being the speed of light in free space. At any other frequency ω for which $\lambda = 2\pi c/\omega$, we then have

$$\beta l = \frac{2\pi}{\lambda}\frac{\lambda_0}{2} = \frac{\pi\omega}{\omega_0} = \frac{\pi(\omega_0 + \Delta\omega)}{\omega_0} = \pi + \frac{\pi\Delta\omega}{\omega_0} \quad \to \quad e^{\pm j\beta l} = -e^{\pm j\pi\Delta\omega/\omega_0}$$

For a low-loss line as is considered here, we must have $\alpha l \ll 1$, so that the $e^{\pm\alpha l}$ terms in the preceding equation can be approximated by $1 \pm \alpha l$. Making the necessary substitutions, we have

$$\begin{aligned}
Z_{\text{in}} &\simeq Z_0 \frac{(1 + \alpha l)(-e^{j\pi\Delta\omega/\omega_0}) - (1 - \alpha l)(-e^{-j\pi\Delta\omega/\omega_0})}{(1 + \alpha l)(-e^{j\pi\Delta\omega/\omega_0}) + (1 - \alpha l)(-e^{-j\pi\Delta\omega/\omega_0})} \\
&= Z_0 \frac{-(\alpha l)\cos(\pi\Delta\omega/\omega_0) - j2\sin(\pi\Delta\omega/\omega_0)}{-j2(\alpha l)\sin(\pi\Delta\omega/\omega_0) - 2\cos(\pi\Delta\omega/\omega_0)} \\
&\simeq Z_0 \frac{-(\alpha l) - j(\pi\Delta\omega/\omega_0)}{-j2(\alpha l)(\pi\Delta\omega/\omega_0) - 2} \\
Z_{\text{in}} &\simeq Z_0 \left[\alpha l + j\frac{\pi\Delta\omega}{\omega_0}\right]
\end{aligned}$$

where we have assumed that $\sin(\pi\Delta\omega/\omega_0) \simeq \pi\Delta\omega/\omega_0$ and $\cos(\pi\Delta\omega/\omega_0) \simeq 1$ (since $\Delta\omega \ll \omega_0$).

The preceding expression can now be compared to the input impedance of a series RLC circuit. We have

$$[Z_{\text{in}}]_{\text{RLC}} \simeq R\left[1 + j2Q\frac{\Delta\omega}{\omega_0}\right] = R + j2L\Delta\omega; \qquad [Z_{\text{in}}]_{\lambda/2\ \text{line}} \simeq Z_0\left[\alpha l + j\frac{\pi\Delta\omega}{\omega_0}\right]$$

Thus, the short-circuited line of length $\lambda/2$ can be represented by an equivalent series RLC circuit, with element values

$$R_{\text{eq}} = Z_0\alpha l = Z_0\frac{\alpha\lambda}{2}; \qquad L_{\text{eq}} = \frac{\pi Z_0}{2\omega_0}; \qquad C_{\text{eq}} = \frac{2}{Z_0\pi\omega_0}$$

where $\omega_0 = 1/\sqrt{L_{\text{eq}}C_{\text{eq}}}$. By analogy, we can then deduce the expression for the Q of the short-circuited half-wavelength line to be

$$Q = \frac{1}{\omega_0 R_{\text{eq}} C_{\text{eq}}} = \frac{\pi}{\alpha\lambda_0} = \frac{\beta_0}{2\alpha}$$

where $\beta_0 = 2\pi/\lambda_0$ is the phase constant at $\omega = \omega_0$ and α has to be evaluated at $\omega = \omega_0$. Typical values of Q for short-circuited transmission lines range from several hundred to tens of thousands, much higher than is possible for low-frequency lumped circuits.

Transmission line resonators with high Q values are necessarily low-loss lines so that the simplified expressions [3.73*a*] and [3.73*b*] are valid, respectively, for α and β. Thus, the Q of a transmission line resonator can be written in terms of the transmission line parameters as

$$Q = \frac{\beta_0}{2\alpha} \simeq \frac{\omega_0 C Z_0}{G Z_0 + (R/Z_0)}$$

Example 3-34 illustrates the calculation of the Q of an air-filled coaxial line.

Example 3-34: *Q* of an air-filled coaxial line. Determine the Q of an air-filled coaxial line, shorted at one end, made of copper with dimensions $a = 1$ cm and $b = 3$ cm. The operating frequency is 300 MHz.

Solution: We have $\lambda = c/f \simeq 1$ m $= 100$ cm. Thus the resonant line length is $l = \lambda/2 \simeq 50$ cm. Assuming that the losses are low and neglecting shunt losses ($G = 0$), the attenuation constant from [3.73a] is $\alpha \simeq R/(2Z_0)$. From Table 2.2, the series resistance R for a coaxial line made of copper is $R = 4.15 \times 10^{-8}\sqrt{f}(a^{-1} + b^{-1})$, and $Z_0 = 60 \ln(b/a)$. Thus, we have

$$\alpha = \frac{4.15 \times 10^{-8}}{2 \times 60}\sqrt{f}\frac{(a^{-1} + b^{-1})}{\ln[b/a]} \simeq 7.27 \times 10^{-6} \quad \text{np-(cm)}^{-1}$$

and

$$Q = \frac{\pi}{\alpha\lambda} \simeq \frac{\pi}{7.27 \times 10^{-6} \times 100} \simeq 4321$$

Note that we have neglected the losses in the imperfect short circuit. Also, if the space between the conductors of the coaxial line were filled with an insulator other than air, additional high-frequency losses in the insulator would generally tend to reduce Q.

A transmission line resonator that behaves like a series RLC circuit can also be implemented using an open-circuited transmission-line section of length $\lambda/4$. Open-circuited line resonators are easier to implement for microstrip or striplines because short circuits cannot be easily placed on these structures. An analysis of the input impedance similar to that just given shows that for $l = \lambda/4$, we have

$$Z_{\text{in}} \simeq Z_0\left[\alpha l + j\frac{\Delta\omega}{\omega_0}\frac{\pi}{2}\right]$$

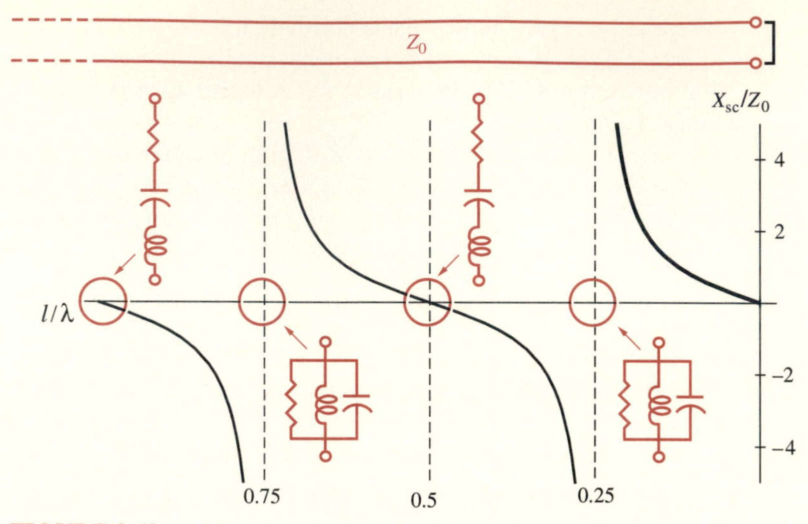

FIGURE 3.68. Resonant behavior of short-circuited transmission line segments. For the purposes of this diagram, the imaginary part of the line impedance for a lossless line of length l is shown, namely $X_{sc} = Z_0 \tan(\beta l)$. Note that in applications of transmission lines as resonators, the losses are generally quite small, so that the behavior of the impedance close to the load is only negligibly different from the lossless case.

so that the equivalent series RLC circuit parameters are

$$R_{eq} = Z_0(\alpha l)^{-1}; \qquad L_{eq} = \frac{\pi Z_0}{4\omega_0}; \qquad C_{eq} = \frac{4}{Z_0 \pi \omega_0}$$

Thus the Q is

$$Q = \frac{1}{\omega_0 R_{eq} C_{eq}} = \frac{\pi}{\alpha \lambda_0} = \frac{\beta_0}{2\alpha}$$

Although we have only considered the series RLC circuit and its transmission line analog, similar results can be obtained for the parallel RLC circuit. A summary of the behaviors of various lengths of shorted transmission lines as parallel or series RLC circuits is given in Figure 3.68. The reader is encouraged to construct an analogous diagram for an open-circuited transmission line.

3.10 SUMMARY

This chapter discussed the following topics:

- **Transmission line equations.** When a transmission line is excited by a sinusoidal source of angular frequency ω at steady state, the variations of the line voltage and current can be analyzed using the phasor form of the transmission line equations, which for a lossless line are

$$\frac{dV(z)}{dz} = -j\omega LI(z)$$

$$\frac{dI(z)}{dz} = -j\omega CV(z)$$

where L and C are the per-unit length distributed parameters of the line, and $V(z)$ and $I(z)$ are, respectively, the voltage and current phasors, which are related to the actual space-time voltage and current expressions as follows:

$$\mathcal{V}(z,t) = \Re e\{V(z)e^{j\omega t}\}; \qquad \mathcal{I}(z,t) = \Re e\{I(z)e^{j\omega t}\}$$

- **Propagating-wave solutions, characteristic impedance, phase velocity, and wavelength.** The solutions of the lossless transmission line equations consist of a superposition of waves traveling in the $+z$ and $-z$ directions. The voltage and current phasors and the corresponding space-time functions have the form

$$V(z) = V^+e^{-j\beta z} + V^-e^{+j\beta z}; \qquad \mathcal{V}(z,t) = V^+\cos(\omega t - \beta z) + V^-\cos(\omega t + \beta z)$$

$$I(z) = \frac{V^+}{Z_0}e^{-j\beta z} - \frac{V^-}{Z_0}e^{+j\beta z}; \qquad \mathcal{I}(z,t) = \frac{V^+}{Z_0}\cos(\omega t - \beta z) - \frac{V^-}{Z_0}\cos(\omega t + \beta z)$$

The characteristic impedance Z_0 of the line is the ratio of the voltage to the current phasor of the wave propagating in the $+z$ direction (or the negative of the ratio of the voltage to the current phasor of the wave traveling in the $-z$ direction) and, for a lossless line, is given by $Z_0 = \sqrt{L/C}$. The phase velocity and the wavelength for a lossless line are given as

$$v_p = 1/\sqrt{LC}; \qquad \lambda = 2\pi/\beta = v_p/f$$

Note that the phase velocity of a lossless line is independent of frequency.

- **Input impedance of short- and open-circuited lines.** The line impedance of a transmission line seen looking toward the load at any position along the line is defined as

$$Z(z) = \frac{V(z)}{I(z)} = Z_0\frac{V^+e^{-j\beta z} + V^-e^{+j\beta z}}{V^+e^{-j\beta z} - V^-e^{+j\beta z}}$$

The input impedances of short- or open-circuited transmission lines of length l are purely imaginary and are given by

$$Z_{sc} = jZ_0\tan(\beta l) \qquad \text{short-circuited line}$$

$$Z_{oc} = -jZ_0\cot(\beta l) \qquad \text{open-circuited line}$$

Since any arbitrary reactive impedance can be realized by simply adjusting the length l of open- or short-circuited stubs, these stubs are commonly used as reactive circuit elements for impedance matching and other applications.

- **Reflection coefficient.** It is common practice to treat steady-state transmission line problems by considering the wave traveling in the $+z$ direction (toward the load) as the incident wave and the wave traveling in the $-z$

direction (away from the load and toward the source) as the reflected wave. The ratio of the reflected to the incident voltage phasor at any position z along the line is defined as the reflection coefficient, represented by $\Gamma(z)$. The reflection coefficient at the load end of the line (where $z = 0$) is given by

$$\Gamma_L = \frac{V^-}{V^+} = (Z_L - Z_0)/(Z_L + Z_0) = \rho e^{j\psi}$$

The case of $Z_L = Z_0$ is referred to as a matched load, for which there is no reflected wave, since $\Gamma_L = 0$. The reflection coefficient $\Gamma(z)$ at any other location z (where $z < 0$) on a lossless line is given by

$$\Gamma(z) = \frac{V^- e^{j\beta z}}{V^+ e^{-j\beta z}} = \Gamma_L e^{j2\beta z} = \rho e^{j(\psi + 2\beta z)}$$

- **Standing-wave pattern.** The superposition of the incident and reflected waves constitutes a standing-wave pattern that repeats every $\lambda/2$ over the length of the line. The standing-wave ratio S is defined as the ratio of the maximum to minimum voltage (current) magnitude over the line and is given by

$$S = \frac{1 + \rho}{1 - \rho}$$

where $\rho = |\Gamma_L|$. The standing-wave ratio S has practical significance because it is easily measurable. The value of S varies in the range $1 \le S \le \infty$, where $S = 1$ corresponds to $\rho = 0$ (i.e., no reflection case) and $S = \infty$ corresponds to $\rho = 1$ (i.e., the load is either open or short circuit).

- **Transmission line as an impedance transformer.** The line impedance of a lossless transmission line terminated in an arbitrary load impedance, defined as the ratio of the total voltage to current phasor at position z, is in general complex and is a periodic function of z, with period of $\lambda/2$. The line impedance is purely real at locations along the line where the voltage is a maximum or minimum.
- **Power flow.** The net time-average power propagating toward the load on a lossless transmission line is given by

$$P(z) = \frac{|V^+|^2}{2Z_0}(1 - \rho^2)$$

and is equal to the power P_L delivered to the load. For a given value of $|V^+|$, the power delivered to the load is maximized under matched conditions, or $\rho = 0$. The degree of mismatch between the load and the line can be described in terms of return loss, given as

$$\text{Return loss} = 20 \log_{10} \frac{S + 1}{S - 1}$$

- **Impedance matching.** In most applications it is desirable to match the load impedance to the line in order to reduce reflections and standing waves. In single-stub matching, a short- or open-circuited stub is placed in shunt or series at a location $z = -l$ along the line at which the normalized line admittance or the impedance is given as

$$\bar{Y}_1(z)|_{z=-l} = 1 - j\bar{B}; \qquad \bar{Z}_1(z)|_{z=-l} = 1 - j\bar{X}$$

The matching is then completed by choosing the length l_s of a short- or open-circuited stub so that it presents an admittance or impedance at $z = -l$ of $\bar{Y}_s = j\bar{B}$ or $\bar{Z}_s = j\bar{X}$. In quarter-wave matching, it is first necessary to determine the location l along the line at which the line impedance is purely real, that is, where

$$Z(z)|_{z=-l} = R + j0$$

Matching to a line of impedance Z_0 is then completed by using a quarter-wavelength-long line of characteristic impedance $Z_Q = \sqrt{Z_0 R}$.

- **Smith chart.** The fact that the impedance $Z(z)$ and the reflection coefficient $\Gamma(z)$ on a lossless line are both periodic functions of position z along the line makes it possible to analyze and visualize the behavior of the line using a graphical display of $\Gamma(z)$, S, and $Z(z)$ known as the Smith chart. The Smith chart provides a convenient means of analyzing transmission line problems to determine values of impedance and reflection coefficient (or standing-wave ratio). The Smith chart is also a useful tool for matching network design.
- **Lossy transmission lines.** The solutions for voltage and current propagating in the z direction on a lossy transmission line have the form

$$\mathcal{V}(z, t) = V^+ e^{-\alpha z} \cos(\omega t - \beta z)$$

$$\mathcal{I}(z, t) = \frac{V^+}{|Z_0|} e^{-\alpha z} \cos(\omega t - \beta z - \phi_z)$$

where α and β are the real and imaginary parts of the propagation constant $\gamma = \alpha + j\beta = \sqrt{(R + j\omega L)(G + j\omega C)}$, R, L, G, and C are the per-unit distributed parameters of the line, and ω is the angular frequency of the excitation. The characteristic impedance for a lossy line is in general complex and is given by

$$Z_0 = |Z_0| e^{j\phi_z} = \sqrt{\frac{R + j\omega L}{G + j\omega C}}$$

For terminated lines, the general expressions for line voltage and current are

$$V(z) = V^+ e^{-\alpha z} e^{-j\beta z}[1 + \Gamma(z)]$$

$$I(z) = \frac{1}{Z_0} V^+ e^{-\alpha z} e^{-j\beta z}[1 - \Gamma(z)]$$

where $\Gamma(z) = \Gamma_L e^{2\alpha z} e^{j2\beta z}$, with $\Gamma_L = (Z_L - Z_0)/(Z_L + Z_0)$ being the complex load voltage reflection coefficient. The line voltage and current exhibit a standing-wave pattern near the load, but the differences between the maxima and minima become smaller as distance from the load increases. At sufficient distances from the load, the magnitudes of the line voltage and current do not vary significantly with distance, as if the line were matched. The impedance of a lossy transmission line is given by

$$Z(z) = Z_0 \frac{1 + \Gamma(z)}{1 - \Gamma(z)} = Z_0 \frac{1 + \Gamma_L e^{2\alpha z} e^{j2\beta z}}{1 - \Gamma_L e^{2\alpha z} e^{j2\beta z}}$$

The real and imaginary parts of $Z(z)$ exhibit maxima and minima near the load, similar to that of a lossless line. However, at sufficient distances from the load, $Z(z)$ approaches Z_0, as if the line were matched.

- **Transmission line resonators.** Short- or open-circuited transmission lines of lengths that are integer multiples of $\lambda/4$ behave as highly efficient resonators. The Q of a low-loss short-circuited half-wavelength-long line is

$$Q = \frac{\omega_0 C Z_0}{GZ_0 + (R/Z_0)}$$

where C, G, and R are the distributed constants of the line, and $\omega_0 = (LC)^{-1/2}$ is the resonant frequency.

3.11 PROBLEMS

3-1. Transmission line capacitor. An open-circuited 50Ω microstrip transmission line is used in a microwave amplifier circuit to provide a capacitance of 3.2 pF at 2.3 GHz. (a) Find the appropriate electrical length of the line. (b) Find the lumped element values of the open-circuited line designed in part (a) at 2 and 2.6 GHz.

3-2. Resistive load. A 50Ω transmission line is terminated with an antenna having a feed-point impedance of 150Ω. (a) Calculate V_{max}, V_{min}, I_{max}, and I_{min} along the line, assuming $V^+ = 1$ V. (b) Sketch $|V(z)|$ and $|I(z)|$ as functions of z, taking the antenna position to be $z = 0$. Assume $\lambda = 20$ cm.

3-3. Microwave filter. An air-filled coaxial line with $Z_0 = 75\Omega$ is designed to provide an inductive impedance of $j231\Omega$ for a microwave filter to operate at 2.5 GHz. Find the length of the coaxial line if (a) it is short-circuit terminated and (b) it is open-circuit terminated.

3-4. Capacitive termination. A lossless transmission line with $Z_0 = 100\Omega$ is terminated with a capacitive load of $40 - j50\Omega$. (a) Calculate the standing-wave ratio S. (b) Find the position of the first voltage minimum and maximum with respect to the load. (c) Sketch $|V(z)|$ as a function of z/λ. Assume $V^+ = 1$ V.

3-5. Input impedance. A 10-cm-long air transmission line segment with $Z_0 = 100\Omega$ is terminated at $z = 0$ with a resistive load of 200Ω and is operated at 1.5 GHz. Calculate the input impedance of the line if (a) a shunt capacitance of ~ 2.12 pF is connected at a point halfway ($z = -5$ cm) on the line, (b) a series capacitance of ~ 2.12 pF is connected at $z = -5$ cm.

3-6. Resistive load. A transmission line segment with $Z_0 = 50\Omega$ and of length l is terminated at a load resistance of R_L that can be varied . Sketch the input impedance Z_{in} as a function of R_L if (a) $l = \lambda/4$ and (b) $l = \lambda/2$. At what value of R_L do the two curves intersect?

3-7. Inductive termination. An air-filled coaxial line with $Z_0 = 50\Omega$ is terminated with a load of $100 + j50\sqrt{3}\Omega$. If the line is operated at $\lambda = 10$ cm, calculate (a) the standing-wave ratio S on the line, (b) the distance from the load to the first voltage maximum, and (c) the distance from the load to the first current maximum.

3-8. A wireless communication antenna. The following table provides the approximate values at various frequencies of the feed-point impedance of a circularly polarized patch antenna used in the wireless industry for making cellular phone calls in difficult environments, such as sport arenas and office buildings:

f (MHz)	$Z_L(\Omega)$
800	$21.5 - j15.4$
850	$38.5 + j2.24$
900	$43.8 + j9.74$
950	$55.2 - j10.2$
1000	$28.8 - j7.40$

If this antenna is directly fed by a 50Ω transmission line, find and sketch the standing-wave ratio S as a function of frequency.

3-9. Resistive line impedance. A 50Ω coaxial line is terminated with a load impedance of $40 + j80\Omega$ at $z = 0$. Find the minimum electrical length l/λ of the line at which the line impedance (i.e., $Z(z = -l)$) is purely resistive. What is the value of the resistive line impedance?

3-10. Resistive line impedance. A transmission line with $Z_0 = 100\Omega$ is terminated with a load impedance of $120 - j200\Omega$. Find the minimum length l of the line at which the line impedance (i.e., $Z(z = -l)$) is purely resistive. What is the value of the resistive line impedance?

3-11. Resistive load. A lossless line is terminated with a resistive load of 120Ω. If the line presents an impedance of $48 + j36\Omega$ at a position $3\lambda/8$ away from the load, what is the characteristic impedance Z_0 of the line?

3-12. Input impedance. For the lossless transmission-line system shown in Figure 3.69, find Z_{in} for the following load impedances: (a) $Z_L = \infty$ (open circuit), (b) $Z_L = 0$ (short circuit), and (c) $Z_L = Z_0/2$.

3-13. Input impedance. Repeat Problem 3-12 for the circuit shown in Figure 3.70.

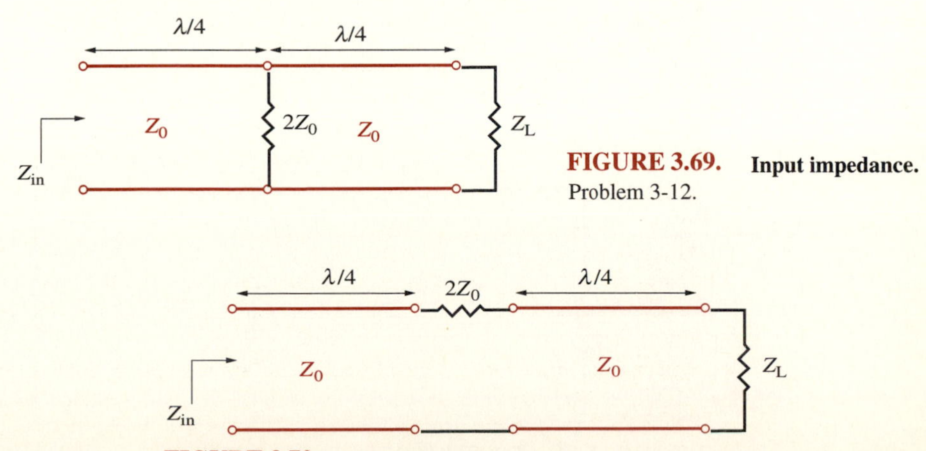

FIGURE 3.69. **Input impedance.** Problem 3-12.

FIGURE 3.70. **Input impedance.** Problem 3-13.

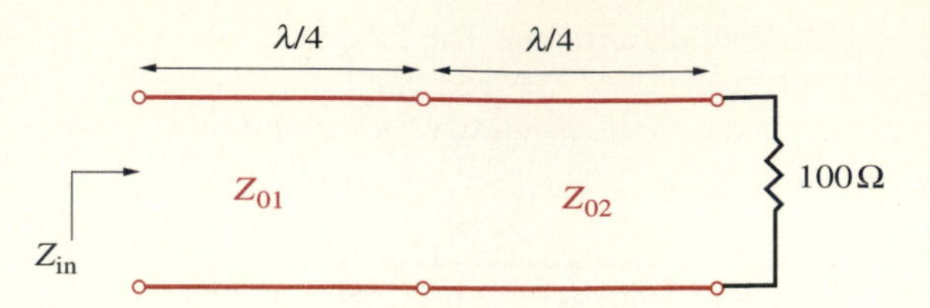

FIGURE 3.71. **Input impedance.** Problem 3-14.

3-14. Input impedance. For the lossless transmission-line system shown in Figure 3.71, what is the ratio Z_{01}/Z_{02} if $Z_{in} = 225\Omega$?

3-15. Unknown termination. Consider a transmission line with $Z_0 = 50\Omega$ terminated with an unknown load impedance Z_L. (a) Show that

$$Z_L = Z_0 \frac{1 - jS\tan(\beta l_{min})}{S - j\tan(\beta l_{min})}$$

where l_{min} is the length from the load to the first voltage minimum and S is the standing-wave ratio. (b) Measurements on a line with $Z_0 = 50\Omega$ having an unknown termination Z_L show that $S = \sqrt{3}$, $l_{min} = 25$ mm, and that the distance between successive minima is 10 cm. Find the load reflection coefficient Γ_L and the unknown termination Z_L.

3-16. Distance to the first maximum. Derive a formula similar to that in Problem 3-15 in terms of l_{max}, where l_{max} is the distance from the load to the first voltage maximum.

3-17. Power dissipation. For the lossless transmission line system shown in Figure 3.72, with $Z_0 = 100\Omega$, (a) calculate the time-average power dissipated in each load. (b) Switch the values of the load resistors (i.e., $R_{L1} = 200\Omega$, $R_{L2} = 50\Omega$), and repeat part (a).

3-18. Power dissipation. Consider the transmission line system shown in Figure 3.73. (a) Find the time-average power dissipated in the load R_L with the switch S open. (b) Repeat part (a) for the switch S closed. Assume steady state in each case.

3-19. Power dissipation. Repeat Problem 3-18 if the characteristic impedance of the transmission lines on the source side is changed from 50Ω to $25\sqrt{2}\Omega$.

3-20. Two antennas. Two antennas having feed-point impedances of $Z_{L1} = 40 - j30\Omega$ and $Z_{L2} = 100 + j50\Omega$ are fed with a transmission line system, as shown in Figure 3.74.

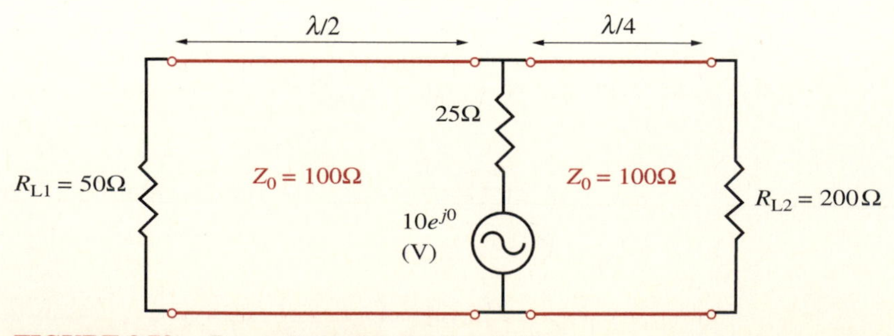

FIGURE 3.72. **Power dissipation.** Problem 3-17.

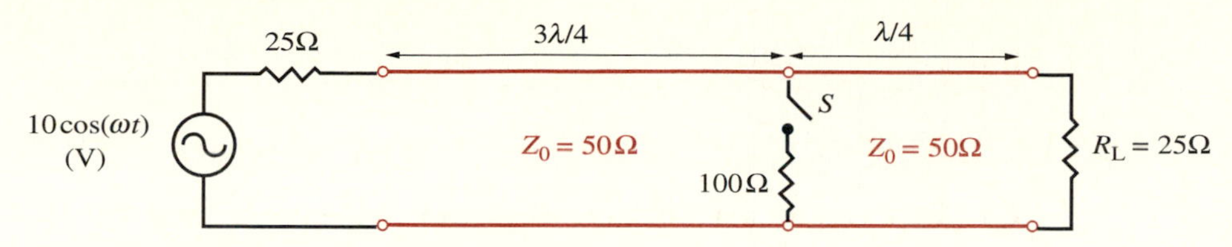

FIGURE 3.73. **Power dissipation.** Problem 3-18.

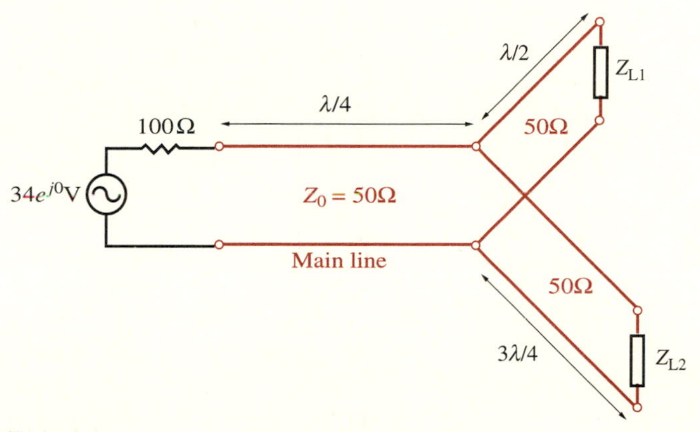

FIGURE 3.74. **Two antennas.** Problem 3-20.

(a) Find S on the main line. (b) Find the time-average power supplied by the sinusoidal source.

(c) Find the time-average power delivered to each antenna. Assume lossless lines.

3-21. Power dissipation. For the transmission line network shown in Figure 3.75, calculate the time-average power dissipated in the load resistor R_L.

3-22. Three identical antennas. Three identical antennas A1, A2, and A3 are fed by a transmission line system, as shown in Figure 3.76. If the feed-point impedance of each antenna is $Z_L = 50 + j50\Omega$, find the time-average power delivered to each antenna.

3-23. Power delivery. For the transmission system shown in Figure 3.77, calculate the percentage of time-average power delivered to R_{L1} and R_{L2} at (a) $f = f_1$, (b) $f = f_2 = 2f_1$, and (c) $f = f_3 = 1.5f_1$.

3-24. Matching with a single lumped element. The transmission line matching networks shown in Figure 3.78 are designed to match a 10Ω load impedance to a 50Ω line. (a) For the network with a shunt element, find the minimum distance l from the load where the unknown shunt element is to be connected such that the input admittance

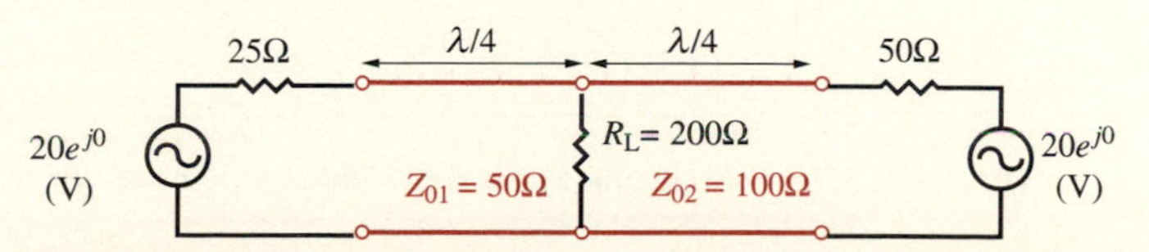

FIGURE 3.75. **Power dissipation.** Problem 3-21.

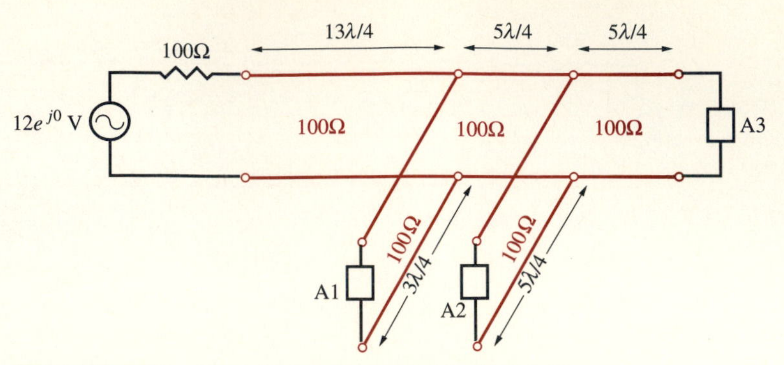

FIGURE 3.76. Three identical antennas. Problem 3-22.

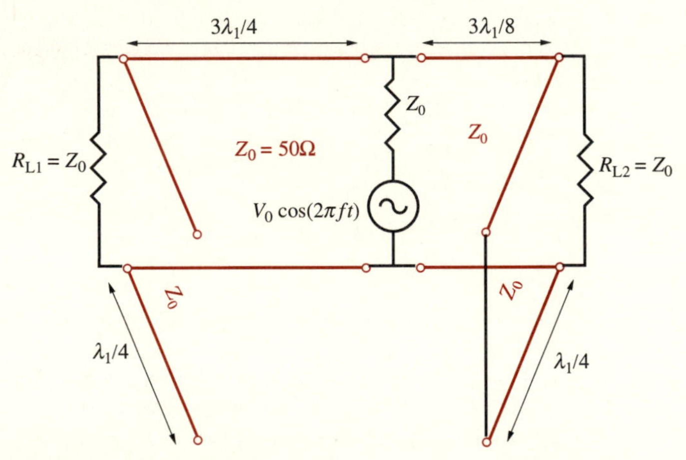

FIGURE 3.77. Power delivery. The normalized line lengths are given at $f = f_1$. Problem 3-23.

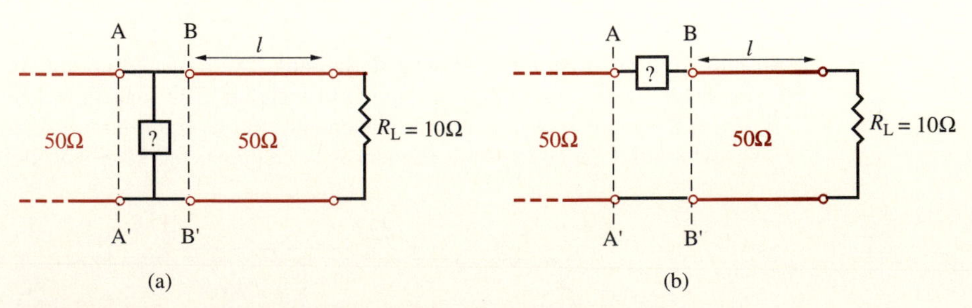

FIGURE 3.78. Matching with a single lumped element. Problem 3-24.

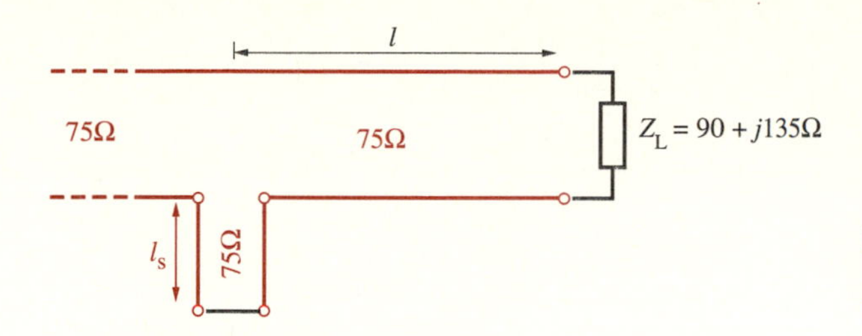

FIGURE 3.79. Matching with series shorted stub. Problem 3-25.

seen at B-B′ has a conductance part equal to 0.02 *S*. (b) Determine the unknown shunt element and its element value such that the input impedance seen at A-A′ is matched to the line (i.e., $Z_{A-A'} = 50\Omega$) at 1 GHz. (c) For the matching network with a lumped series matching element, find the minimum distance l and the unknown element and its value such that a perfect match is achieved at 1 GHz. Assume $v_p = 30$ cm-(ns)$^{-1}$.

3-25. Matching with series stub. A load impedance of $90 + j135\Omega$ is to be matched to a 75Ω lossless transmission line system, as shown in Figure 3.79. If $\lambda = 20$ cm, what minimum length of transmission line l will yield a minimum length l_s for the series stub?

3-26. Open-ended extension. A transmission line with $Z_0 = 50\Omega$ is terminated with a 100Ω load resistance shunted by an open-circuited line having $Z_0 = 50\Omega$ and length 7.4 mm as shown in Figure 3.80. If $\lambda = 10$ cm on both lines, find the length l_s and the position l (measured from the 100Ω load resistance) of the single short-circuited stub to match this load to the line.

3-27. Series stub matching. A series-shorted-stub matching network is designed to match a capacitive load of $R_L = 50\Omega$ and $C_L = 10/(3\pi)$ pF to a 100Ω line at 3 GHz, as shown in Figure 3.81. (a) The stub is positioned at a distance of $\lambda/4$ away from the load. Verify the choice of this position and find the corresponding electrical length of the stub to achieve a perfect match at the design frequency. (b) Calculate the standing-wave ratio S on the main line at 2 GHz. (c) Calculate S on the main line at 4 GHz.

3-28. Quarter-wave transformer. (a) Design a single-section quarter-wave matching transformer to match an $R_L = 20\Omega$ load to a line with $Z_0 = 80\Omega$ operating at 1.5 GHz. (b) Calculate the standing-wave ratio S of the designed circuit at 1.2 and 1.8 GHz.

3-29. Helical antenna. The feed-point impedance of an axial-mode helical antenna with a circumference C on the order of one wavelength is nearly purely resistive and is

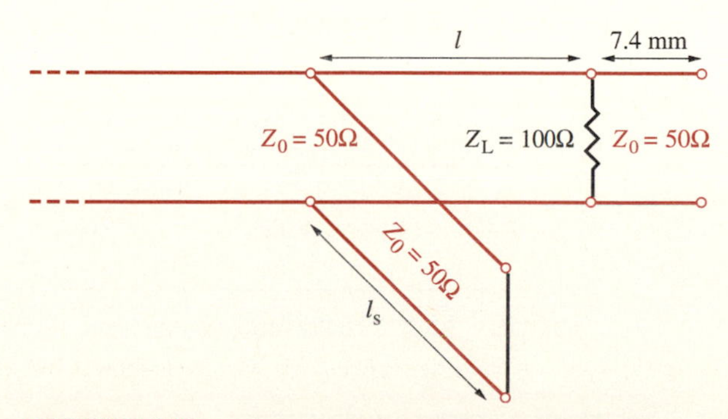

FIGURE 3.80. Open-ended extension. Problem 3-26.

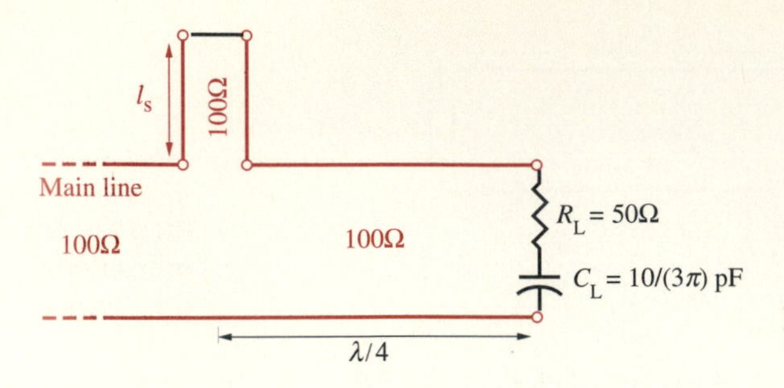

FIGURE 3.81. Series stub matching. Problem 3-27.

approximately given[39] by $R_L \simeq 140(C/\lambda)$, with the restriction that $0.8\lambda \leq C \leq 1.2\lambda$. Consider a helical antenna designed with a circumference of $C = \lambda_0$ for operation at a frequency f_0 and corresponding wavelength λ_0. The antenna must be matched for use with a 50Ω transmission line at f_0. (a) Design a single-stage quarter-wave transformer to realize the design objective. (b) Using the circuit designed in part (a), calculate the standing-wave ratio S on the 50Ω line at a frequency 15% above the design frequency. (c) Repeat part (b) at a frequency 15% below the design frequency.

3-30. Helical antenna. A helical antenna designed with a feed-point impedance of 125Ω is matched to a 52Ω line by inserting a coaxial transmission line section of characteristic impedance 95Ω and length 0.125λ at a distance of 0.0556λ from the antenna feed point. (See Figure 3.82.) (a) Verify the design by calculating the standing-wave ratio S on the line. (b) Using the same circuit as in part (a), calculate S on the main line at a frequency 20% above the design frequency. Note: Use the approximate expression given in Problem 3-29 to recalculate the feed-point impedance of the helical antenna.

3-31. Quarter-wave matching. Many microwave applications require very low values of S over a broad band of frequencies. The two circuits shown in Figure 3.83 are designed to match a load of $Z_L = R_L = 400\Omega$ to a line with $Z_0 = 50\Omega$, at 900 MHz. The first circuit is an air-filled coaxial quarter-wave transformer, and the second circuit consists of two air-filled coaxial quarter-wave transformers cascaded together. (a) Design both circuits. Assume $Z_{Q1}Z_{Q2} = Z_0 Z_L$ for the second circuit. (b) Compare the bandwidth of the two circuits designed by calculating S on each line at frequencies 15% above and below the design frequency.

3-32. Is a match possible? A 75Ω coaxial line is connected directly to an antenna with a feed-point impedance of $Z_L = 156\Omega$. (a) Find the load-reflection coefficient and the standing-wave ratio. (b) An engineer is assigned the task of designing a matching

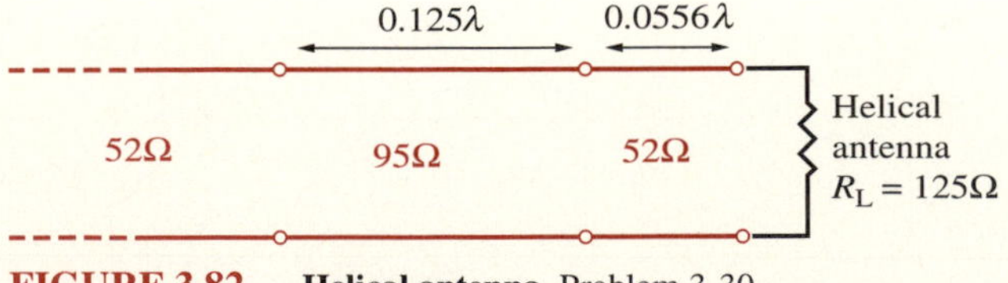

FIGURE 3.82. Helical antenna. Problem 3-30.

[39] See Chapter 7 of J. D. Kraus, *Antennas,* 2nd. ed., McGraw-Hill, 1988.

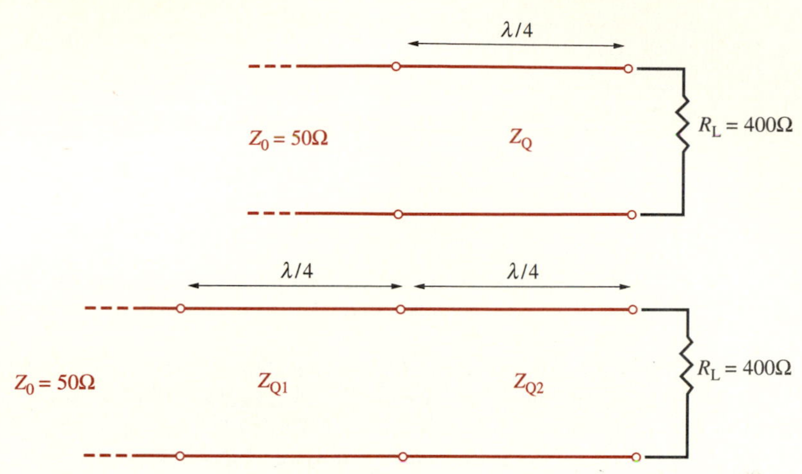

FIGURE 3.83. Quarter-wave matching. Problem 3-31.

network to match the feed-point impedance of the antenna to the 75Ω coaxial line. However, all he has available to use for this design is another coaxial line of characteristic impedance 52Ω. Is a match possible?

3-33. L-section matching networks. A simple and practical matching technique is to use the lossless L-section matching network that consists of two reactive elements. (a) Two L-section matching networks marked A1 and A2, each consisting of a lumped inductor and a capacitor, as shown in Figure 3.84, are used to match a load impedance of $Z_L = 60 - j80\Omega$ to a 100Ω line. Determine the L section(s) that make(s) it possible to achieve the design goal, and calculate the appropriate values of the reactive elements at 800 MHz. (b) Repeat part (a) for the two L-section networks marked B1 and B2, consisting of two inductances and two capacitors, respectively.

3-34. Variable capacitor. A shunt stub filter consisting of an air-filled coaxial line terminated in a variable capacitor is designed to eliminate the FM radio frequencies (i.e., 88–108 MHz) on a transmission line with $Z_0 = 100\Omega$, as shown in Figure 3.85. If the

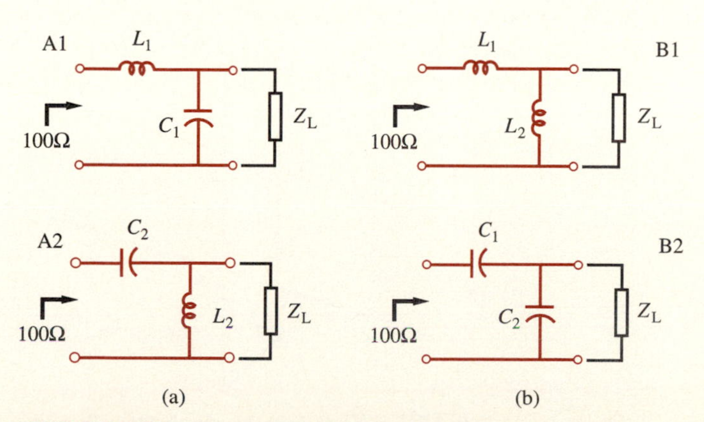

FIGURE 3.84. L-section matching networks. Problem 3-33.

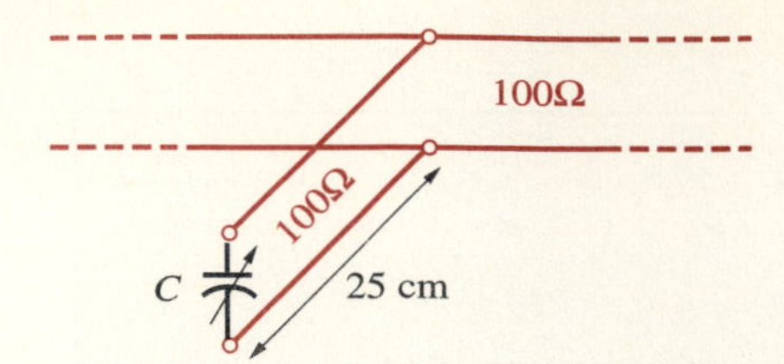

FIGURE 3.85. Variable capacitor. Problem 3-34.

stub length is chosen to be 25 cm, find the range of the variable capacitor needed to eliminate any frequency in the FM band. Assume the characteristic impedance of the stub to be also equal to 100Ω.

3-35. Matching with lumped reactive elements. Two variable reactive elements are positioned on a transmission line to match an antenna having a feed-point impedance of $100 + j100\Omega$ to a $Z_0 = 100\Omega$ air-filled line at 5 GHz, as shown in Figure 3.86. (a) Determine the values of the two reactive elements to achieve matching. (b) If the reactive elements are to be replaced by shorted 50Ω air-filled stubs, determine the corresponding stub lengths.

3-36. Fifth-harmonic filter. The circuit shown in Figure 3.87 has two shunt stubs (one open and one short) that are connected at the same position on a line with $Z_0 = 50\Omega$. The normalized lengths of the two stubs are given at a frequency f_0 (or wavelength λ_0). Assume each stub to have a characteristic impedance of 50Ω. What is the standing-

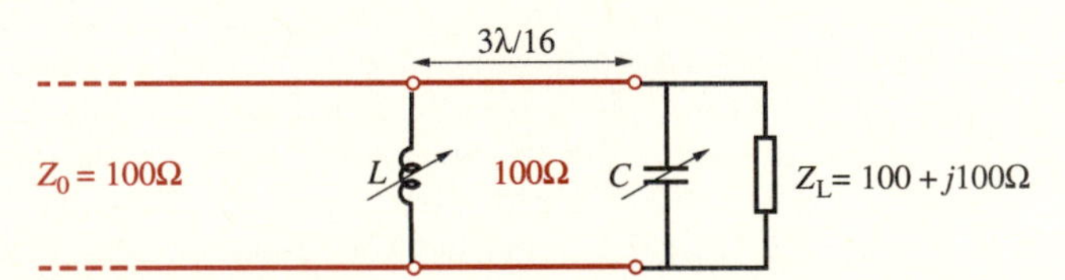

FIGURE 3.86. Matching with lumped reactive elements. Problem 3-35.

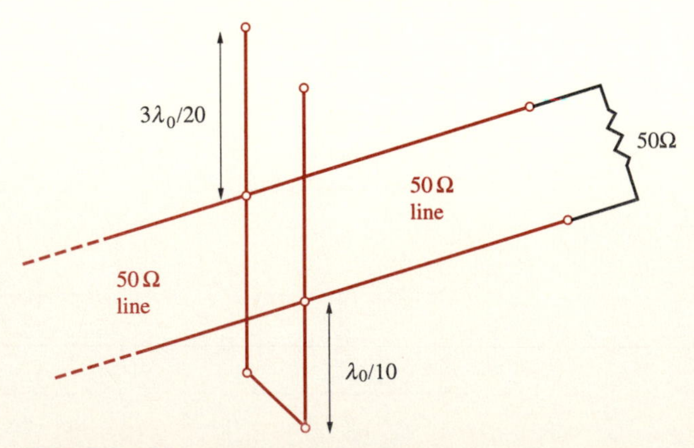

FIGURE 3.87. Fifth-harmonic filter. Problem 3-36.

wave ratio S on the line at f_0? At $3f_0$? At $5f_0$? (Note that this circuit is a fifth-harmonic filter.)

3-37. Standing-wave ratio. For the transmission line shown in Figure 3.88, calculate S on the main line at (a) 800 MHz, (b) 880 MHz, and (c) 960 MHz.

3-38. Quarter-wave matching. (a) For the transmission line system shown in Figure 3.89, determine the value of the characteristic impedance of a quarter-wave transformer (i.e., Z_Q) and its location l with respect to the load needed to achieve matching between Z_L and Z_0. (b) Repeat part (a) for $Z_L = 80 - j60\Omega$.

3-39. Single-stub matching. For the transmission line system shown in Figure 3.90, (a) design a single shorted stub to be as close as possible to the load such that the load is matched to the air line at 3 GHz ($\lambda = 10$ cm). (b) After the matching circuit is built, an engineer experiencing reflections on the main line discovers that the line has an open-circuited extension of 2.5 cm beyond the load position. With the design values of l_s and l found in part (a), what is the actual standing-wave ratio S on the main line caused by the open-circuit extension?

3-40. Quarter-wave transformer. Consider the double quarter-wave transformer system shown in Figure 3.91. (a) Find l, Z_{Q1}, and Z_{Q2} such that the load is matched to the 200Ω line at $\lambda_0 = 12$ cm. Assume $\sqrt{Z_{Q1}Z_{Q2}} = 60\Omega$. (b) Using the values of l, Z_{Q1} found in part (a), find the standing-wave ratio S on the main line at twice the operating frequency.

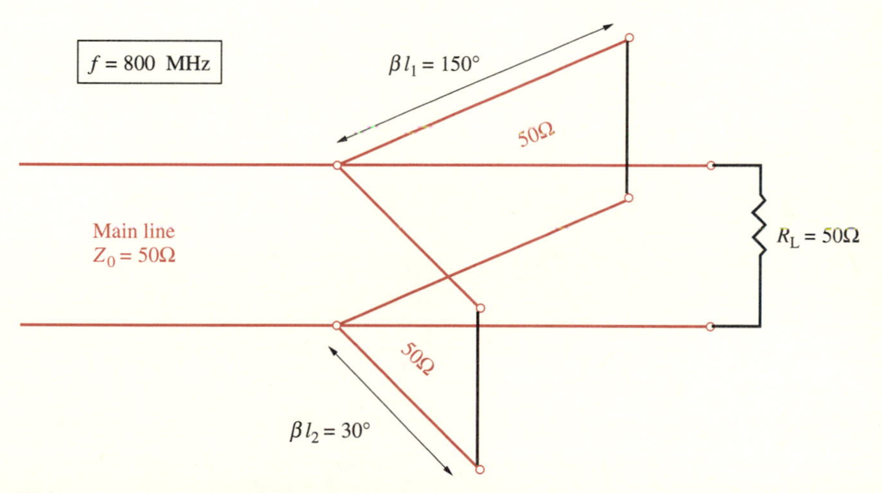

FIGURE 3.88. Standing-wave ratio. Problem 3-37.

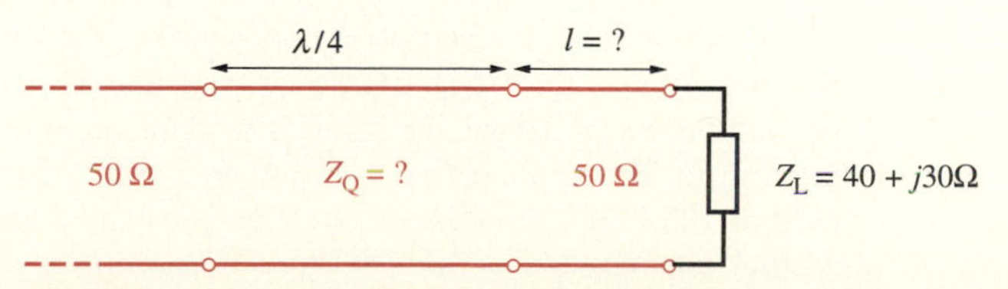

FIGURE 3.89. Quarter-wave matching. Problem 3-38.

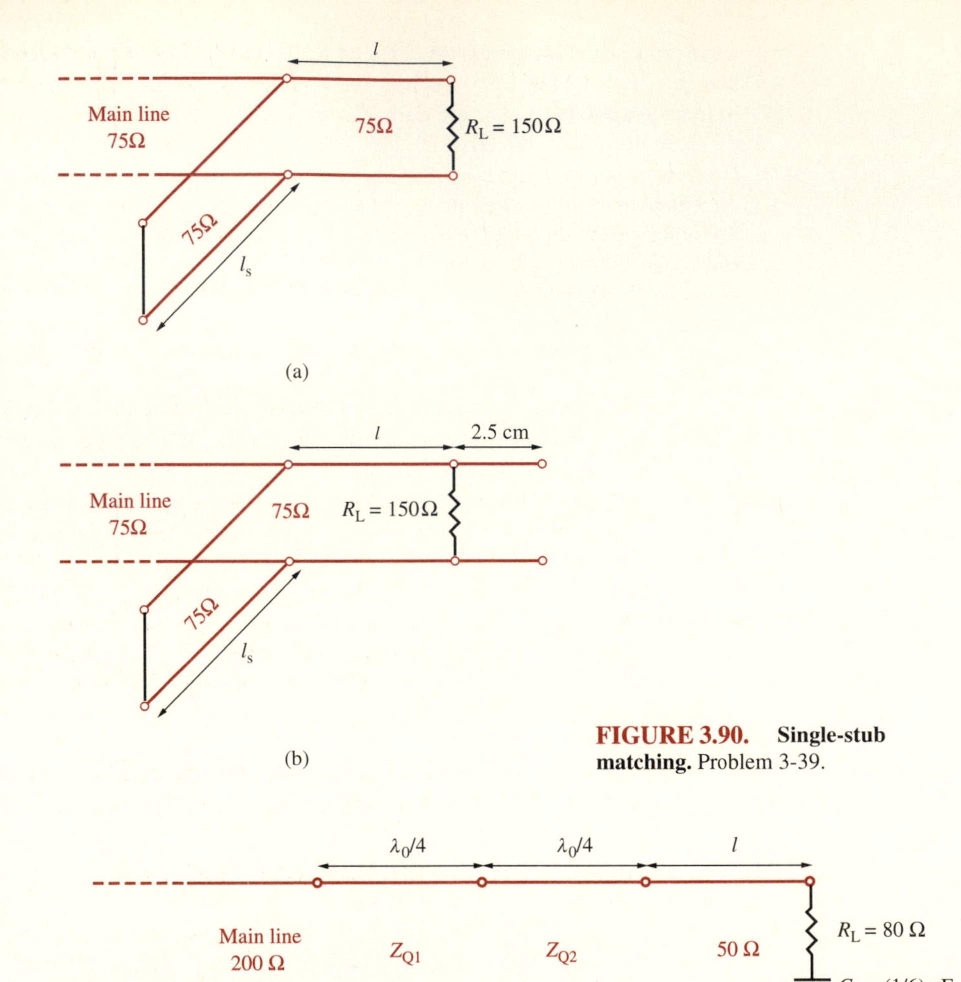

FIGURE 3.90. Single-stub matching. Problem 3-39.

FIGURE 3.91. Quarter-wave transformer. Problem 3-40.

3-41. Unknown feed-point impedance. A 50Ω transmission line is terminated with an antenna that has an unknown feed-point impedance. An engineer runs tests on the line and measures the standing-wave ratio, wavelength, and a voltage minimum location away from the antenna feed point to be, respectively, 3.2, 20 cm, and 74 cm. Use the Smith chart to find the feed-point impedance of the antenna.

3-42. RG218 coaxial line. The RG218 coaxial line is made of copper conductors with polyethylene as the insulator filling. The diameter of the inner conductor and the outer diameter of the insulator are 4.95 mm and 17.27 mm, respectively. The line is to be used at 100 MHz. Find the propagation constant γ and the characteristic impedance Z_0. For polyethylene at 100 MHz, $v_p \simeq 20$ cm-(ns)$^{-1}$, and for a polyethylene-filled coaxial line at 100 MHz, assume $G \simeq 1.58 \times 10^5/[\ln(b/a)]$.

3-43. Two-wire air line. An air-insulated two-wire line made of copper conductors has a characteristic impedance of 500Ω when it is assumed to be lossless. (a) Find the L, C, and R parameters of this line at 144 MHz. Assume a wire diameter of 1.024 mm. (b) Find the propagation constant γ and the characteristic impedance Z_0 with losses included.

3-44. Two-wire matching section. An air-insulated two-wire quarter-wave transmission line section is constructed using a copper wire with diameter 2.54 mm to match a 588Ω load impedance to a 75Ω line at 300 MHz. Assuming the lossless case, find the length and the spacing of the wires of the matching section. (b) Find γ and Z_0 of the matching section with the losses included.

3-45. A parallel-plate line. A certain parallel-plate line is to be made of two copper strips each 5 cm wide and separated by 0.5 cm. The dielectric is air and the frequency of operation is 1 GHz. Find the line parameters L, C, and R, the characteristic impedance Z_0, and the attenuation constant α of the line.

3-46. A lossy high-speed interconnect. The per-unit line parameters of an IC interconnect at 5 GHz are extracted using a high-frequency measurement technique resulting in $R = 143.5\Omega\text{-(cm)}^{-1}$, $L = 10.1\text{ nH-(cm)}^{-1}$, $C = 1.1\text{ pF-(cm)}^{-1}$, and $G = 0.014\text{ S-(cm)}^{-1}$, respectively.[40] Find the propagation constant γ and the characteristic impedance Z_0 of the interconnect at 5 GHz.

3-47. Characterization of a high-speed GaAs interconnect. The propagation constant γ and the characteristic impedance Z_0 at 5 GHz of the GaAs coplanar strip interconnects considered in Example 3-31 are determined from the measurements to be $\gamma \simeq 1.1\text{ np-(cm)}^{-1} + j3\text{ rad-(cm)}^{-1}$ and $Z_0 \simeq 110 - j40\Omega$, respectively. Using these values, calculate the per-unit length parameters (R, L, G, and C) of the coplanar strip line at 5 GHz.

3-48. A lossy high-speed interconnect. Consider a high-speed microstrip transmission line of length 10 cm used to connect a 1-V amplitude, 2-GHz, 50Ω sinusoidal voltage source to an integrated circuit chip having an input impedance of 50Ω. The per-unit parameters of the microstrip line at 2 GHz are measured to be approximately given by $R = 7.5\Omega\text{-(cm)}^{-1}$, $L = 4.6\text{ nH-(cm)}^{-1}$, $C = 0.84\text{ pF-(cm)}^{-1}$, and $G = 0$, respectively. (a) Find the propagation constant γ and the characteristic impedance Z_0 of the line. (b) Find the voltages at the source and the load ends of the line. (c) Find the time-average power delivered to the line by the source and the time-average power delivered to the load. What is the power dissipated along the line?

3-49. Half-wave coaxial line resonator. A $\lambda/2$ resonator is constructed using a piece of copper coaxial line, with an inner conductor of 2-mm diameter and an outer conductor of 8-mm diameter. If the resonant frequency is 3 GHz, find the Q of (a) the air-filled coaxial line and (b) the teflon-filled coaxial line, and compare the results. (c) For an air-filled line, determine the equivalent series RLC circuit parameters, namely R_{eq}, L_{eq}, and C_{eq}. For teflon at 3 GHz, $v_p \simeq 20.7\text{ cm-(ns)}^{-1}$. For a teflon-filled coaxial line at 3 GHz, the per-unit conductance G of a coaxial line is approximately given by $G \simeq 3.3 \times 10^{-4}/[\ln(b/a)]\text{ S-m}^{-1}$.

---

[40] W. R. Eisenstadt and Y. Eo, S parameter-based IC interconnect transmission line characterization, *IEEE Trans. on Components, Hybrids, and Manufacturing Technology,* 15(4) pp. 483–489, August 1992.

CHAPTER 4

The Static Electric Field

Following our discussion of transmission line behavior in Chapters 2 and 3, we begin in this chapter a more traditional study of electric and magnetic fields, following the historical development of ideas that led to the general laws of electromagnetics known as Maxwell's equations. The voltage and current waves on two-conductor transmission lines are but special cases of electromagnetic waves, which can propagate in empty space and in material media as well as being guided by a variety of conducting or insulating structures. Kirchhoff's voltage and current laws, which were used in Chapter 2 to derive the basic transmission line voltage and current equations, are special cases of Maxwell's equations, which in general are expressed in terms of electric and magnetic fields. This chapter is the first of four chapters in which we undertake the development of these fundamental laws of electromagnetics, culminating in the establishment of the full set of interconnected laws in Chapter 7. Our bases for the development of Maxwell's equations are experimentally established facts of nature, as verified by physical observations.

The interaction between electric and magnetic fields is at the root of electromagnetic wave propagation—the basis for all telecommunications, from the telegraph to today's satellite, wireless, and optical fiber networks. The interaction between electric and magnetic fields is also responsible for the behavior of electric circuit elements and networks as well as the workhorses of our industrial society: electrical motors and machinery. Our interest in electromagnetics stems from a need to understand the behavior of practical devices and systems, to describe such devices and systems mathematically, to predict their performance, and to design systems for particular applications. To achieve these goals requires an understanding of the physical bases of the fundamental laws of electromagnetics, which are developed in Chapters 4 through 7 and stated compactly as Maxwell's equations in Section 7.4. The branch of electromagnetics dealing with electric charges at rest, namely static electricity or *electrostatics,* involves the study of the first, and one of the most important, of these fundamental laws, known as Coulomb's law.[1] Coulomb's law is based upon physical observation and is not logically or mathematically derivable from any other concept. The experimental basis of this law, the mathematical formulations that it leads to, and its broad range of applications and implications are the subjects of this chapter.

The study of electrostatics constitutes a relatively simple first step in our quest to understand the laws of electromagnetics. In electrostatics we deal only with electric charges that are at rest and that do not vary with time. Yet, mastering the behavior of static electric fields and the techniques for the solution of electrostatic problems is essential to the understanding of more complicated electromagnetic phenomena. Furthermore, many natural phenomena and the principles of some important industrial and technological applications are electrostatic in nature.

Lightning, corona discharge, and arcing are natural phenomena that involve very strong electrostatic fields that cause ionization in the surrounding medium. Lightning discharges involve currents of tens of kiloamperes that flow for a few hundred microseconds and represent release of energies of up to 10^{10} joules. [2] Applications of electrostatics in industry encompass diverse areas—cathode ray tubes (CRTs) and flat panel displays, which are widely used as display devices for computers and oscilloscopes; ink jet printers, which can produce good quality printing at very fast speeds; and photocopy machines are all based on electrostatic fields. Electrostatic technologies are extensively used for sorting charged or polarized granular materials;[3] applications of this technology are reflected in hundreds of patents, extending from mineral beneficiation and seed conditioning to recycling of metals and plastics from industrial wastes. Among other applications, electrostatic spraying and painting, electrostatic precipitators and filters, electrostatic transducers, and electrostatic recording are utilized in numerous industrial and household applications.

Our discussion of electrostatics in this chapter also serves to bring us closer to a full understanding of the underlying physical basis of the transmission line

---

[1]That electric charges attract or repel one another in a manner inversely proportional to the square of the distance between them. C. A. de Coulomb, Première mémoire sur l'électricité et magnétisme (First memoir on electricity and magnetism), *Histoire de l'Académie Royale des Sciences,* p. 569, 1785.

[2]See Sections 1.3 and 1.8 of M. A. Uman, *The Lightning Discharge,* Academic Press, 1987.

[3]A. D. Moore (ed.), *Electrostatics and Its Applications,* John Wiley and Sons, 1973; F. S. Knoll, J. E. Lawver, and J. B. Taylor, Electrostatic separation, in *Ullman's Encyclopedia of Industrial Chemistry,* 5th ed., VCH, Weinheim, 1988, vol. B2, pp. 20-1–20-11.

behavior discussed in Chapters 2 and 3. One of the important physical properties of a transmission line is its distributed capacitance, which comes about as a result of the separation of charge induced on the two conductors that constitute a transmission line. In Chapters 2 and 3, we took it for granted that a two-conductor system exhibits capacitance, and we presented formulas (in Table 2.2) for the distributed capacitances of a few common transmission line structures. In this chapter we define the concept of capacitance and discuss how the capacitance of different conductor configurations can be determined using Coulomb's law.

In addition to starting our discussion of the fundamental underpinnings of electromagnetic theory, we also present in this chapter the important concepts of vector algebra. Although some of the most basic aspects of vector algebra are provided in Appendix A, we introduce important concepts such as gradient and divergence along with the relevant physical laws. This approach ensures that the physical significance of the vector operations can be best understood in the electromagnetic context. Throughout this text, vector quantities are often written using boldface symbols (e.g., **G**), although a bar above the symbol is also used (e.g., $\overline{\mathcal{G}}$).[4] In either case, the vector in question is understood in general to have three components, G_x, G_y, and G_z, and it is often written as $\mathbf{G} = \hat{\mathbf{x}}G_x + \hat{\mathbf{y}}G_y + \hat{\mathbf{z}}G_z$, where $\hat{\mathbf{x}}$, $\hat{\mathbf{y}}$, and $\hat{\mathbf{z}}$ are the unit vectors in the x, y, and z directions, respectively. The "hat" notation is always used to represent a unit vector, so we can also write the vector **G** as $\mathbf{G} = \hat{\mathbf{G}}G$, where $\hat{\mathbf{G}}$ is a unit vector in the direction of **G** and G is the magnitude of **G**, $G = |\mathbf{G}|$. Points in three-dimensional space are identified by means of their position vectors, which point from the origin of the coordinate system to the point in question. For example, the position vector for point P located at (x_p, y_p, z_p) is $\mathbf{r}_p = \hat{\mathbf{x}}x_p + \hat{\mathbf{y}}y_p + \hat{\mathbf{z}}z_p$.

We now proceed with our study of electrostatics with a brief discussion in Section 4.1 of our everyday experiences with electricity and the nature of electric charge. We then introduce and discuss Coulomb's law in Section 4.2 and the concept of an electric field in Section 4.3. The scalar electric potential and the concept of the gradient of potential are discussed in Section 4.4, followed by the introduction of the notion of electric flux and Gauss's law in Section 4.5. Section 4.6 introduces the important concept of the divergence of a vector field and the divergence theorem, followed by a discussion of metallic conductors in Section 4.7. Laplace's equation and its solution to determine electric potential and field distributions are presented in Section 4.8, leading to a discussion of the concept of the capacitance of a conductor configuration in Section 4.9. Section 4.10 presents the properties of dielectric materials, followed by a discussion of electrostatic boundary conditions in Section 4.11. The chapter concludes with the discussion of electrostatic energy and forces, respectively, in Sections 4.12 and 4.13.

---

[4]This alternative notation is used starting in Chapter 7 in order to distinguish between real physical quantities and their corresponding complex *phasors,* necessary for sinusoidal (or time-harmonic) applications.

4.1 ELECTRIC CHARGE

Our experiences with electricity date back to ancient times[5] and have their roots in the observation that, for example, a piece of glass and a piece of resin (or rubber) attract one another if they are first rubbed together and then separated. Also, if a second piece of glass is rubbed with another piece of resin, the two pieces of glass or resin repel one another, while each glass piece attracts each piece of resin. Various manifestations of electrification by friction are encountered in our daily experiences.[6]

4.1.1 Electrification by Friction, Induction, and Conduction

These electrical phenomena of attraction and repulsion that come about due to *friction* and that are part of our daily experiences[7] are understood in terms of *electric charge*. Electric charge is said to be acquired by the material as a result of rubbing. In actual fact, when glass and resin, for example, are rubbed together, a small amount of charge is transferred from one to the other, causing each material to become non-neutral, that is, charged. The glass becomes positively charged, while the resin acquires negative charge. The different behavior of glass and resin indicates that there must be two different types of charge. Materials that behave upon electrification like glass are said to be *positively charged,* and those that behave like resin are said to be *negatively charged.* Other materials also acquire charge as a result of being rubbed, although this property may be less apparent for some than others.

Metallic materials, such as brass, do not retain electricity for a sufficient time (after they are rubbed to another object) for us to observe it. However, a brass rod with a glass handle becomes electrified to a marked degree on rubbing. Once it is electrified, such a brass rod with a glass handle loses all of its electricity if

---

[5]For example, Thales of Miletus [640–540 B.C.] wrote that a piece of amber rubbed in silk attracts pieces of straw.

[6]In his text, *Electricity and Magnetism* (Cambridge Press, 1907), Sir James Jeans gives the following amusing account of Robert Symmer's [1759] observations: He was in the habit of wearing two pairs of stockings simultaneously: a worsted wool one for comfort and a silk pair for appearance. In pulling off his stockings, he noticed that they gave a crackling noise, and sometimes they even emitted sparks when taken off in the dark. On taking two stockings off together from the foot and then drawing the one from inside the other, he found that both became inflated to reproduce the shape of the foot, and exhibited attractions and repulsions at a distance as much as a foot and a half. "When this experiment is performed with two black stockings in one hand, and two white in the other, it exhibits a very curious spectacle; the repulsion of those in the same colour, and attraction of those of different colours, throws them into an agitation that is not unentertaining, and makes them catch each at that of its opposite colour, and at a greater distance than one would expect. When allowed to come together they all unite in one mass. When separate, they resume their former appearance, and admit of the repetition of the experiment as often as you please, till their electricity, gradually wasting, stands in need of being recruited."

[7]In our everyday lives, we experience frictional electrification every time we rub our feet across a wool carpet, pull clothes out of a dryer, or comb our hair with a plastic comb. On a dry summer day, it is not uncommon to receive a shock on touching a doorknob, hear an accompanying crackling noise, and sometimes see a spark. We see sparks in such cases because large amounts of charge create electric fields that cause local breakdown of air.

it comes in contact with water, flame, or the human body. Conversely, the same rod retains its power if it comes in contact with hard rubber, a piece of silk, or wood. We understand such behavior in terms of the classification of materials into *conductors* of electricity and *insulators.* Metals are excellent conductors of electricity. Solutions of salts, acids, and other electrolytes are also conductors. Examples of good insulators are oils, waxes, silk, glass, rubber, and plastics. Gases are ordinarily good insulators, but flames or other ionized gases are good conductors. Distilled water is almost a perfect insulator, but any other type of water contains impurities, which in general cause it to conduct reasonably well. Being made

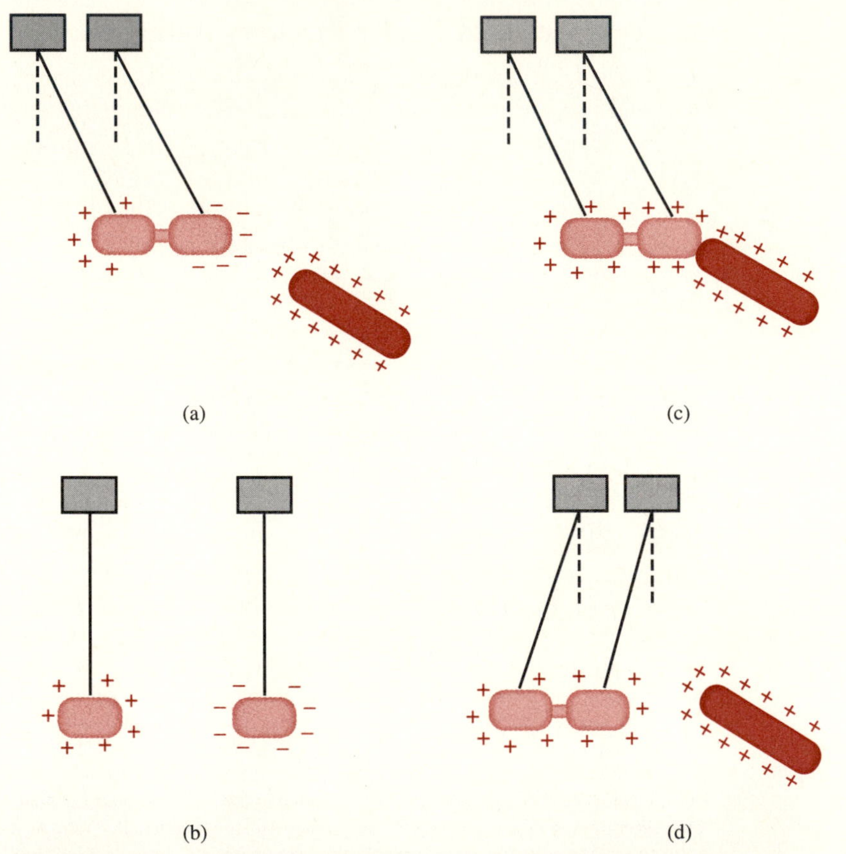

FIGURE 4.1. **Electrification by induction.** (a) A charged rod near a suspended neutral object induces opposite charge on the surface closer to it. Since the suspended object is initially neutral, an equal amount of positive charge remains on its far surface. Since the negative charge is closer to the charged rod, the attractive force is stronger than the repelling force, and the suspended object deflects toward the charged rod. (b) If we cut the suspended object in the middle while the charged rod is near it, each of its two parts remains electrified, one positive, one negative. This is an example of electrification by *induction.* (c) If the charged rod is brought in physical contact with the neutral body, the excess positive charge on the rod is shared between the two objects and the body is now positively charged. This is an example of electrification by *conduction.* (d) The two positively charged objects now repel one another.

largely of water, the human body conducts reasonably well and is generally a bad insulator.

Although electrification by *friction* is part of our daily experiences, material bodies can also acquire electric charge through *induction.* If we suspend an uncharged metallic (e.g., brass) object by silk threads, as shown in Figure 4.1a, and bring near one end of it a positively (or negatively) charged rod, the nearby end of the suspended object becomes negatively (or positively) charged, while the other end becomes primarily positive (or negative). If the charged rod is removed, the suspended object loses its electrification. However, if the suspended object is cut in the middle before the inducing charge is removed, each of its parts remains electrified, one positive and the other negative, as shown in Figure 4.1b.

When a charged conductor is connected to another conductor (charged or not), for example by means of a metal wire, the total charge on both bodies is shared between the two. The second body is then said to have become electrified via *conduction* through the metal wire. Such conduction would clearly not occur if the two bodies were connected with a silk thread. Once again we see a basis for classifying materials as conductors (e.g., metals) and insulators (e.g., silk). Electrification by conduction is essentially what occurs when we place the charged rod in contact with the neutral body as in Figure 4.1c; the charge on the rod is now shared between the two objects, which now repel one another, as shown in Figure 4.1d.

4.1.2 Faraday's Gold-Leaf Electroscope

Long series of experiments carried out by M. Faraday,[8] many years after the establishment of Coulomb's law, were instrumental in bringing about a physical understanding of electrical phenomena and demonstrated the principle of conservation of electric charge. In assessing the quantity of electrification or the quantity of charge associated with any body it is useful[9] to think in terms of a gold-leaf electroscope, as shown in Figure 4.2.

Under normal conditions the gold leaves in Figure 4.2a hang flat side by side. When an electrified body (e.g., a charged rod) touches (conduction) or is brought near the brass rod of the gold-leaf electroscope (induction), the two gold leaves separate because of electrostatic repulsion (Figure 4.2b), so that the electroscope can be used to examine the degree to which a body is charged.

Now consider a metal vessel placed on top of the brass rod (Figure 4.2c), closed but having a lid attached to a silk thread so that it can be opened or closed without touching it. When a charged glass ball is inserted into the vessel and the lid is closed, opposite charges are induced on the inner surface of the vessel. Since the metal vessel

---

[8]M. Faraday, *Experimental Researches in Electricity,* vol. 3, art. 3249, Bernard Quatrich, London, 1855.

[9]We do not pretend here that Faraday's experiments can be carried out under modern conditions to verify electrical laws accurately, or that a gold-leaf electroscope would be used today to measure charge. The discussion in this section should simply be viewed as a set of thought experiments that illustrate the properties and constitution of electricity remarkably well.

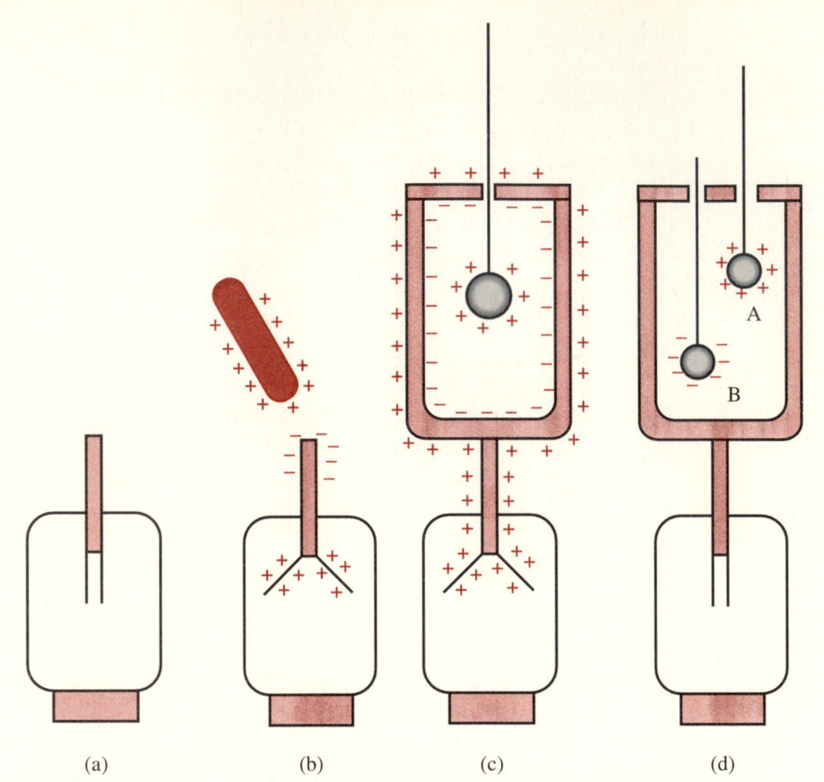

FIGURE 4.2. Faraday's gold-leaf electroscope. (a) An actual unit might consist of a glass vessel, through the top of which a metal (brass) rod is passed, to which are attached two gold leaves. (b) A charged object is brought near the brass rod; the gold leaves repel one another and diverge. (c) A closed metal vessel with a lid is placed on top of the brass rod, so that a charged glass ball can be inserted within it. The gold leaves deflect by the same amount, regardless of the position of the charged ball within the vessel. (d) If two equally but oppositely charged bodies are inserted into the vessel, the gold leaves stay flat.

and the gold leaves were originally neutral, unbalanced positive charge appears on the outside of the vessel, some of which is on the gold leaves, which repel one another and diverge. The separation of the gold leaves, which indicates the degree of electrification (i.e., amount of charge), remains exactly the same if we do the experiment by placing the electrified ball *at different positions* inside the vessel, as long as it does not come into contact with the vessel or other conductors. The separation of the gold leaves is also independent of any *changes in shape* of the glass piece (e.g., a thin rod versus a round ball or a straight rod versus a bent one) or any *changes in its state* (e.g., it might be heated[10] or cooled). It thus appears that the separation of

[10]Not with a flame, however; as was mentioned before, flames are good conductors because they consist of ionized matter.

the gold leaves is due only to a *quantity* of electricity, or *electric charge,* associated with the glass ball.

Now imagine two balls, one glass (A) and one rubber (B), electrified by rubbing against one another (Figure 4.2d). If we introduce A and B separately into the vessel, the gold leaves diverge by the same exact amount either way. If we introduce both A and B together into the vessel, we find that the leaves stay flat; that is, no electrification occurs outside the vessel. From this we conclude that the process of electrification by friction is *not* one that *creates* charge; rather, rubbing merely *transfers* charge[11] from one object to another, slightly disturbing the neutrality of each. If we inserted into the vessel two charged glass balls A and A′, the induced charge on the vessel would be the algebraic sum of what it would be when each ball was introduced separately. If, while A and A′ were inside the vessel, we connected them with a conducting wire, the induced charge on the vessel would not change, indicating once again that the *quantity of total charge remained constant.*

4.1.3 Electric Charge and the Atomic Structure

Our goal in this book is to develop and apply the fundamental laws of electrical and magnetic phenomena that are valid in the macroscopic, nonatomic realm. The physical experiments performed to verify Coulomb's law, and most applications of electrostatics, involve the use of ponderable macroscopic objects. The same is true for the other laws of electromagnetics that we shall study in later chapters. A brief review of modern notions concerning the atomic nature of matter is helpful in clarifying these macroscopic principles. Matter is composed of atoms with dimensions of order of one angstrom, or 10^{-8} cm, each containing a *nucleus* of dimensions $\sim 10^{-12}$ cm and electrons that move about the nucleus. It is the electrons' electrical influence on those of other atoms that effectively defines the atomic dimensions. Regardless of where they reside, electrons are all alike, each bearing a charge $q_e \simeq -1.602 \times 10^{-19}$ coulombs.[12] Most of the mass of an atom is contained in its nucleus,[13] which bears a positive charge of $N_a|q_e|$, where N_a is the atomic number[14] of the particular element. Thus, there essentially are two kinds of electrical charge in matter: positive and negative. Each atom or molecule in its normal state has as many electrons as it has units of positive charge in its nucleus or nuclei, and thus is electrically neutral.

Nearly all of the macroscopic effects of electricity arise from the fact that electrons may be separated from their atoms in some circumstances, leading to separation of positive and negative charges by distances appreciably greater than atomic dimensions. At present, the universe is believed to contain equal amounts of

---

[11]In actual fact, *electrons* are transferred, in this case from glass to rubber.

[12]The most precise value of the charge of an electron is $q_e = -1.60217733(49) \times 10^{-19}$ C, as specified in *The CRC Handbook of Physics & Chemistry,* 76th ed., CRC Press, Boca Raton, Florida, 1995–96.

[13]An electron is physically quite light, having a mass of $m_e \simeq 9.11 \times 10^{-31}$ kg, much smaller than the mass of a proton ($\sim 1.66 \times 10^{-27}$ kg) or a neutron, the two types of particles that together form the nucleus.

[14]The atomic number of an element is defined as the number of protons present in its nucleus.

positive and negative electricity, with the result that an excess of one polarity of charge at one place implies the presence elsewhere of an equal but opposite charge. Furthermore, net amounts of electric charge can neither be created nor destroyed (the law of conservation of electric charge).

The atomic nature of matter is inherently quantized in that electric charges are made up of integral multiples of the electron charge q_e. However, any charged object of macroscopic size (i.e., much larger than atomic dimensions) typically holds so many electrons (or, if positively charged, would have a deficiency of so many electrons) that it is considered to have a "continuous" distribution of charge. Since we restrict our attention here to the macroscopic realm, all quantities that are dealt with, whether they be charges, fields, or potentials, should be understood to be macroscopic in nature. For example, the discussion of Coulomb's law in the next section describes the electrostatic force between two *point* charges residing at rest at two different locations in vacuum. In the context of our discussion, we implicitly understand that these point charges may actually occupy physical space of many atomic dimensions in extent, but that the size of these regions is nevertheless negligible on a macroscopic scale.

4.2 COULOMB'S LAW

Although the preceding discussion introduces us to fundamentals of electrostatic principles, the law of action between electrified bodies needs to be specified for mathematical formulation of these concepts. This law of action, known as *Coulomb's law,* is the quantitative expression of the experimental observations discussed above. Coulomb's law states that the electric force between two point[15] charges Q_1 and Q_2 is proportional to the product of the two charges Q_1Q_2 and inversely proportional to the square of the distance between them. In vector form, the forces $\mathbf{F}_{12}$ and $\mathbf{F}_{21}$ felt by charges Q_1 and Q_2 are given as

$$\boxed{\mathbf{F}_{12} = -\mathbf{F}_{21} = \hat{\mathbf{R}}\frac{kQ_1Q_2}{R^2}} \qquad [4.1]$$

where $\mathbf{R} = \mathbf{r}_2 - \mathbf{r}_1$, with $\mathbf{r}_1$ and $\mathbf{r}_2$ being the position vectors[16] for points P_1 and P_2 (where charges Q_1 and Q_2 reside), respectively, shown in Figure 4.3. As for any other vector, we can write $\mathbf{R}$ as $\mathbf{R} = \hat{\mathbf{R}}R$, where $\hat{\mathbf{R}}$ is the unit vector in the direction of $\mathbf{R}$ and R is the magnitude of $\mathbf{R}$. Note that $\mathbf{R}$ is the vector pointing *from* point P_1 *to* P_2 and that $\hat{\mathbf{R}} = \mathbf{R}/R = (\mathbf{r}_2 - \mathbf{r}_1)/(|\mathbf{r}_2 - \mathbf{r}_1|)$, and $R = |\mathbf{r}_2 - \mathbf{r}_1|$. The force as given in [4.1] is repulsive (as shown in Figure 4.3) if the charges are alike in sign and

---

[15]By "point charges" we mean charges that occupy a macroscopic region of space much smaller in extent than the distance between the charges.

[16]As defined earlier, the position vector for any point P located in three-dimensional space is the vector pointing from the origin to the point P.

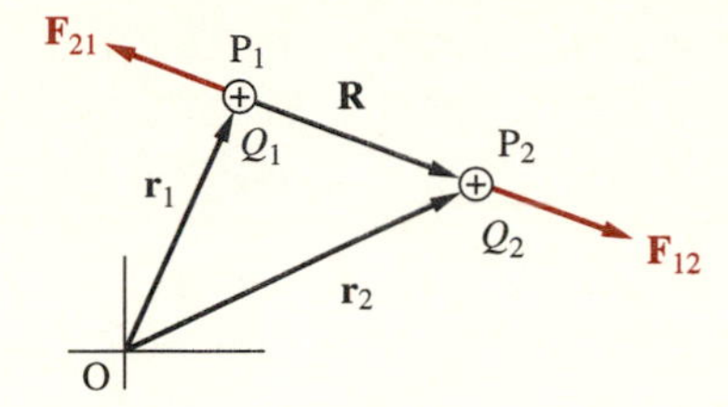

FIGURE 4.3. Coulomb's law. Forces on point charges Q_1 and Q_2 when the two charges are alike in sign.

is attractive if they are of opposite sign (i.e., $Q_1Q_2 < 0$, so that the directions of both $\mathbf{F}_{12}$ and $\mathbf{F}_{21}$ in Figure 4.3 are reversed). The proportionality constant k is equal to $(4\pi\epsilon_0)^{-1}$, with the 4π "rationalization" factor included so that a 4π factor does not appear in Maxwell's equations, which are more commonly used than [4.1].

Using the SI (*Le Système International d'Unités,* International System of Units) units of force, charge, and distance—respectively, newtons (N), coulombs (C), and meters (m)—the value of the quantity[17] ϵ_0, referred to as the permittivity of free space,[18] is $\epsilon_0 \simeq 8.854 \times 10^{-12}$ farads per meter (or F-m^{-1}). Since the proportionality constant k is needed in numerical evaluations, it is helpful to note that in SI units, $k = (4\pi\epsilon_0)^{-1} \simeq 9 \times 10^9$ m-F^{-1}. A coulomb can be defined as the quantity of charge that flows through a given cross section of a wire in one second when there is a steady current of one ampere (A) flowing in the wire.[19]

The following example introduces the application of Coulomb's law.

Example 4-1: Two point charges. Two point charges of $Q_1 = +37$ nC and $Q_2 = +70$ nC are located at points (1, 3, 0) m and (0, 0, 2) m, respectively, as shown in Figure 4.4. Find the force exerted on Q_2 by Q_1.

Solution: The repulsive force $\mathbf{F}_{12}$ on charge Q_2 exerted by Q_1 is

$$\mathbf{F}_{12} = \hat{\mathbf{R}}\frac{1}{4\pi\epsilon_0}\frac{Q_1Q_2}{R^2}$$

where

$$R = |\mathbf{r}_2 - \mathbf{r}_1| = |2\hat{\mathbf{z}} - \hat{\mathbf{x}} - 3\hat{\mathbf{y}}| = [(-1)^2 + (-3)^2 + 2^2]^{1/2} = \sqrt{14}\text{ m}$$

since $\mathbf{r}_2 = 2\hat{\mathbf{z}}$ and $\mathbf{r}_1 = \hat{\mathbf{x}} + 3\hat{\mathbf{y}}$. The unit vector $\hat{\mathbf{R}}$ in direction of $\mathbf{F}_{12}$ is

$$\hat{\mathbf{R}} = \frac{\mathbf{R}}{R} = \frac{\mathbf{r}_2 - \mathbf{r}_1}{|\mathbf{r}_2 - \mathbf{r}_1|} = \frac{2\hat{\mathbf{z}} - \hat{\mathbf{x}} - 3\hat{\mathbf{y}}}{\sqrt{14}}$$

---

[17] It will later become evident that the precise value of this quantity is $\epsilon_0 = (4\pi \times 10^{-7}c^2)^{-1}$, where c is the speed of light in vacuum.

[18] The physical meaning of the dimensions of ϵ_0 will become clearer in later sections, when we discuss capacitance.

[19] The definition of an ampere will be discussed in Chapter 6 in connection with Ampère's law.

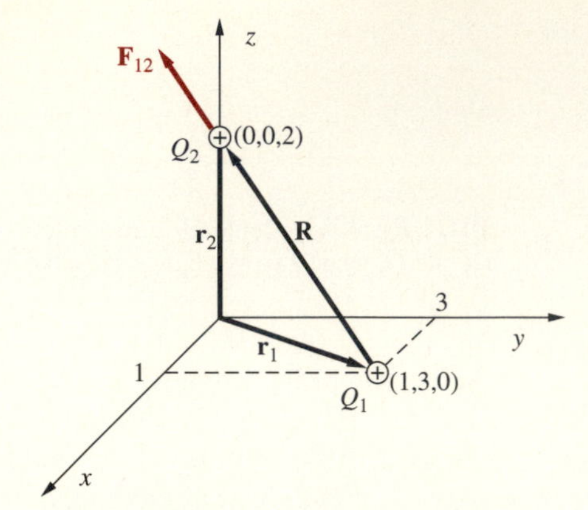

FIGURE 4.4. Two point charges. Configuration of two point charges for Example 4-1.

Therefore, we have

$$\mathbf{F}_{12} = \frac{9 \times 10^9 \text{ m-F}^{-1} \times 37 \times 10^{-9} \text{ C} \times 70 \times 10^{-9} \text{ C}}{14 \text{ m}^2} \underbrace{\left(\frac{-\hat{\mathbf{x}} - 3\hat{\mathbf{y}} + 2\hat{\mathbf{z}}}{\sqrt{14}} \right)}_{\hat{\mathbf{R}}}$$

$$\simeq 1.66 \times 10^{-6} \left(\frac{-\hat{\mathbf{x}} - 3\hat{\mathbf{y}} + 2\hat{\mathbf{z}}}{\sqrt{14}} \right) \simeq 445(-\hat{\mathbf{x}} - 3\hat{\mathbf{y}} + 2\hat{\mathbf{z}}) \text{ nN}$$

It is obvious that the repulsive force $\mathbf{F}_{21}$ on charge Q_1 exerted by Q_2 is $\mathbf{F}_{21} = -\mathbf{F}_{12}$.

To acquire a feel for the magnitude of the electric force as represented by [4.1], (for example, in comparison with the gravitational force), consider the force between two protons in the nucleus of a helium atom, which are $\sim 10^{-15}$ m apart.[20] The gravitational force between the two protons is $\sim 1.84 \times 10^{-34}$ N, but the electrical force from [4.1] is ~ 230 N! The electrical force, in other words, is $\sim 10^{36}$ times larger. Were it not for the nuclear force that keeps the nucleus together, the initial acceleration that each of the protons would acquire due to the electrical force would be $\sim 10^{28} g$!

To appreciate the enormity of one coulomb of charge, note that two 1 C charges 1 m apart would exert an electrical force upon one another of $\sim 9 \times 10^9$ newtons! In normal air, a single 1 C charge would cause electrical breakdown[21] at distances of ~ 50 m.

If there are more than two charges present, [4.1] holds for every pair of charges, and the principle of superposition can be used to determine the net force on any

---

[20] Approximate radius of helium nuclei.

[21] Breakdown of air occurs when the electric field intensity in air exceeds 3×10^6 V-m^{-1}, at which time atmospheric electrons and ions are accelerated to high energies as a result of the Coulomb force. As these energized particles have inevitable collisions with neutral air molecules, many other electrons and ions are knocked out of the neutral air molecules, resulting in arcing and corona discharges.

one of the charges due to all others. For this purpose, the force due to each charge is determined as if it alone were present; the vector sum of these forces is then calculated to give the resultant force, as illustrated in Example 4-2.

Example 4-2: Three point charges. Consider two point charges of $Q_1 = +1\ \mu\text{C}$ and $Q_2 = +2\ \mu\text{C}$ located respectively at (1, 0) m and (−1, 0) m, as shown in Figure 4.5. (a) What is the magnitude and direction of the electrical force felt by a third charge $Q_3 = +1$ nC when placed at (0, 1) m? (b) At what point(s) must the third charge $Q_3 = +1$ nC be placed in order to experience no net force?

Solution:

(a) Note from Figure 4.5a that $\mathbf{r}_3 = \hat{\mathbf{y}}$, $\mathbf{r}_1 = \hat{\mathbf{x}}$, and $\mathbf{r}_2 = -\hat{\mathbf{x}}$. Using Coulomb's law along with the superposition principle, the repulsive forces exerted on charge Q_3 by charges Q_1 and Q_2 are, respectively,

$$\mathbf{F}_{13} = \hat{\mathbf{R}}_{13}\frac{kQ_1Q_3}{R_{13}^2}; \qquad \mathbf{F}_{23} = \hat{\mathbf{R}}_{23}\frac{kQ_2Q_3}{R_{23}^2}$$

where

$$R_{13}^2 = |\mathbf{r}_3 - \mathbf{r}_1|^2 = |\hat{\mathbf{y}} - \hat{\mathbf{x}}|^2 = 2\ \text{m}^2; \qquad R_{23}^2 = |\mathbf{r}_3 - \mathbf{r}_2|^2 = |\hat{\mathbf{y}} + \hat{\mathbf{x}}|^2 = 2\ \text{m}^2$$

and the unit vectors $\hat{\mathbf{R}}_{13}$ and $\hat{\mathbf{R}}_{23}$ are

$$\hat{\mathbf{R}}_{13} = \frac{\mathbf{r}_3 - \mathbf{r}_1}{|\mathbf{r}_3 - \mathbf{r}_1|} = \frac{\hat{\mathbf{y}} - \hat{\mathbf{x}}}{\sqrt{2}}; \qquad \hat{\mathbf{R}}_{23} = \frac{\mathbf{r}_3 - \mathbf{r}_2}{|\mathbf{r}_3 - \mathbf{r}_2|} = \frac{\hat{\mathbf{y}} + \hat{\mathbf{x}}}{\sqrt{2}}$$

Therefore we have

$$\mathbf{F}_{13} = \left(\frac{\hat{\mathbf{y}} - \hat{\mathbf{x}}}{\sqrt{2}}\right)\frac{9 \times 10^9\ \text{m-F}^{-1} \times 10^{-6}\ \text{C} \times 10^{-9}\ \text{C}}{2\ \text{m}^2} = \frac{4.5}{\sqrt{2}}(-\hat{\mathbf{x}} + \hat{\mathbf{y}}) \quad \mu\text{N}$$

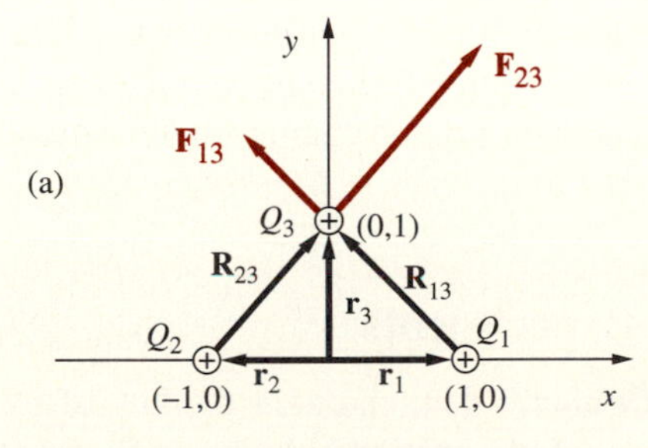

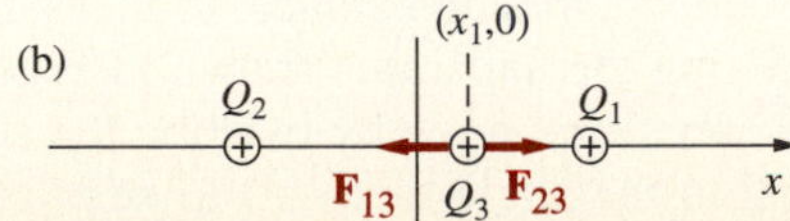

FIGURE 4.5. Three point charges. Configurations of the three point charges Q_1, Q_2, and Q_3 for Example 4-2.

and similarly

$$\mathbf{F}_{23} = \frac{9}{\sqrt{2}}(\hat{\mathbf{x}} + \hat{\mathbf{y}}) \quad \mu\text{N}$$

The total electrical force on charge Q_3 is thus

$$\mathbf{F}_3 = \mathbf{F}_{13} + \mathbf{F}_{23} = \frac{4.5}{\sqrt{2}}(\hat{\mathbf{x}} + 3\hat{\mathbf{y}}) \quad \mu\text{N}$$

(b) Since the forces exerted by the two charges on Q_3 must be equal and opposite in order to have no net force, Q_3 must be placed somewhere along the straight line between Q_1 and Q_2 (i.e., on the x axis), as shown in Figure 4.5b. Let us assume that Q_3 is located at $(x_1, 0)$ and find x_1. Referring to Figure 4.5b, we have

$$\mathbf{F}_3 = \mathbf{F}_{13} + \mathbf{F}_{23} = -\frac{9}{\underbrace{(1 - x_1)^2}_{R_{13}^2}}\hat{\mathbf{x}} + \frac{18}{\underbrace{(1 + x_1)^2}_{R_{23}^2}}\hat{\mathbf{x}} = 0 \ \mu\text{N}$$

Solving for x_1, we find $x_1 \simeq 0.172$ m.

It is important to remember that Coulomb's law as stated above is defined for "point" charges. To see this, consider two spheres A and B at a distance R from one another; sphere A is charged by an amount $+Q_A$ while sphere B is uncharged, $Q_B = 0$. Direct application of Coulomb's law specifies the force between the two bodies to be

$$|\mathbf{F}| = \frac{1}{4\pi\epsilon_0}\frac{(+Q_A)(0)}{R^2} = 0$$

However, based on the discussion in connection with Figure 4.1, we know that the force is not zero, because sphere A will induce negative charge on the side of sphere B closer to it and will be attracted toward it. For this induction effect to be negligible, we must have $b \ll a$, where a is the radius of the charged sphere A and b is the radius of the smaller sphere B.

4.2.1 Coulomb and His Experiments

In 1785, Charles Augustin de Coulomb demonstrated the law of electric force by employing one of his inventions, called a torsion balance, to measure the repulsive force between two like charges. Using the apparatus shown in Figure 4.6, he was able to compensate for the electric repulsion force by twisting the suspension head (thus exerting a force proportional to the angle of twist) to keep the two charged

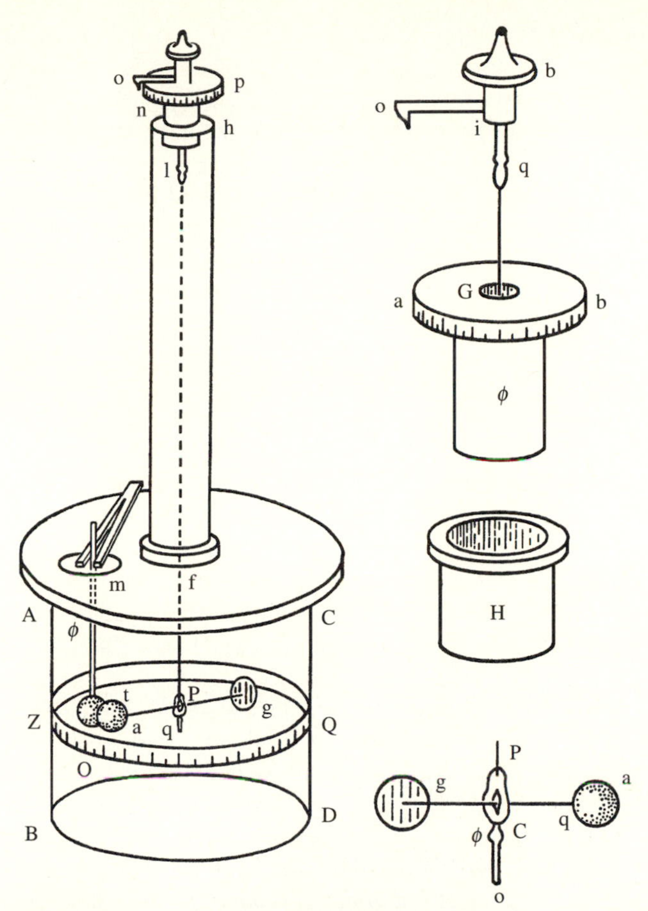

FIGURE 4.6. Coulomb's torsion balance apparatus. The glass cylinder ABCD, 12 inches in diameter by 12 inches high, was covered by a glass plate having two holes approximately one-twelfth of an inch in diameter, one at the center (at f) and the other at m. Cemented in the center was a vertical tube 24 inches in length, with a torsion micrometer at its top, consisting of a circumferential scale divided into degrees, a knob with a scale pointer, and a chuck or pincer to hold the torsion wire. The torsion wire was either silver, copper, or silk, and attached at its lower end was a pincer, which held the torsion wire taut and linear and also provided support for a thin horizontal suspended straw or silk thread coated with sealing wax. At one end of the straw was a gilded elderwood pith-ball (one-sixth of inch in diameter), while at the other end was a small vertical paper disk, which served as a counterweight and to damp out oscillations. A second scale, ZOQ, marked in degrees, was attached outside the glass cylinder. Coulomb turned the micrometer at the top until the horizontal thread ag lined up with the zero marking on scale ZOQ. He then introduced (through the hole m) a thin insulated rod with a second identical pith-ball t, and made it touch the pith-ball a. Coulomb then charged an insulated pin and touched the pin to the two pith-balls, which, having acquired equal charge, separated by a certain distance, allowing Coulomb to measure the repulsive force between them. Figure and description adapted with permission from C. S. Gillmor, *Coulomb and the Evolution of Physics and Engineering in Eighteenth Century France,* Princeton University Press, Princeton, NJ, 1971.

spheres apart at different distances and thus measure the force as a function of distance between them.[22,23]

In the context of current scientific standards and practices, Coulomb's experimental data as presented in his Memoir would have left a lot to be desired as a demonstration of the inverse square law. Only three data points were presented to support Coulomb's hypothesis, and the values deviated from the inverse square law by as much as a few percent, probably due to the limited accuracy of the torsion balance measurements. However, scientists at the time expected an inverse square law for electrostatic forces (similar to Newton's gravitational law of attraction, then known for over 100 years), and Coulomb's results received wide acclaim, with none of his contemporaries expressing doubts. It is interesting to note that some recent attempts to reproduce Coulomb's experimental results have run into difficulties, raising questions as to whether Coulomb may have manipulated his data to obtain a result he intuitively expected.[24,25]

It is also interesting that B. Franklin in 1755, J. Priestley in 1767, J. Robison in 1769, and H. Cavendish in 1773 conducted experiments of one kind or another that pointed to the inverse square law—in the case of Cavendish, with substantially better precision than Coulomb. However, these results were either not published or not well publicized and thus were unknown to the scientific community at the time of Coulomb's work.[26]

Coulomb, Charles Augustin *(b. Dec. 11, 1736, Angoulême, Charente; d. August 23, 1806, Paris) Coulomb was a military engineer in his younger days, serving in the West Indies for nine years beginning in 1764. There, he supervised the building of fortifications in Martinique. He returned to Paris in 1776 with his health impaired, and his search for a quieter life drew him toward scientific experimentation. When the French Revolution began, he combined discretion with inclination and retired to the provincial town of Blois to work in peace. His membership in the Academy of Sciences was later restored by Napoleon.*

By then he had made his name. In 1777 he invented a torsion balance (see Figure 4.6) that measured the quantity of a force by the amount of twist it produced in a thin, stiff fiber. Weight is a measure of the force of gravity upon an object, so a torsion balance can be used to measure weight. A similar instrument

---

[22]See Coulomb's Première mémoire sur l'électricité et magnétisme (First memoir on electricity and magnetism), *Histoire de l'Académie Royale des Sciences,* p. 569, 1785. For an English translation of excerpts, see W. F. Magie, *A Source Book in Physics,* McGraw-Hill Book Company, New York, 1935.

[23]For an excellent summary of Coulomb's experiments and related previous work of other researchers, see R. S. Elliott, *Electromagnetics,* IEEE Press, Piscataway, New Jersey, 1993, Chapter 3.

[24]S. Dickman, Could Coulomb's experiment result in Coulomb's law? *Science,* p. 500, 22 October 1993.

[25]P. Heering, On Coulomb's inverse square law, *Am. J. Phys.* 60 (11), p. 988, 1992.

[26]For an excellent account of these early works, see R. S. Elliott, *Electromagnetics,* IEEE Press, Piscataway, New Jersey, 1993, Chapter 3.

had been invented earlier by English geologist J. Michell, but Coulomb's discovery was independent and in 1781 he was elected to the French Academy.

Coulomb put the delicacy of his instrument at the service of electrical experiments. In 1785, in a course of experimentation that began out of a desire to improve the mariner's compass, he used his torsion balance to measure the electrical force of attraction or repulsion and put forth the inverse square law which is known as Coulomb's law. In his honor, the unit of electric charge is the coulomb. [Partly adapted with permission from I. Asimov, Biographical Encyclopedia of Science and Technology, *Doubleday, 1982].*

1700 *1736* *1806* *2000*

4.2.2 The Accuracy and Validity of Coulomb's Law

Since Coulomb's law is based solely upon physical observations (i.e., experimental fact), and in view of the fact that it provides one of the foundations of electromagnetics, it is important to assess the accuracy of the measurements by which it is established. It was mentioned above that Coulomb's measurements of the electric force were quite approximate, deviating by as much as a few percent from the r^{-2} law. Even with modern techniques, it is difficult to measure the *force* between two charged objects directly to a high degree of accuracy.[27] However, the accuracy of Coulomb's law can be assessed by verifying a fundamental consequence of the inverse square law, namely that the electric field inside a charged closed metallic surface (shell) is precisely zero. As we shall see in Section 4.7, this result is based on Gauss's law (Section 4.5), which in turn is derived from Coulomb's law. Original experiments conducted by H. Cavendish in 1772, and later repeated[28] by J. C. Maxwell in 1879, indicated that the electrostatic force depends on $r^{-2+\zeta}$, with $|\zeta| < 10^{-5}$. This basic experiment, which involves the placement of an electrometer (an instrument that measures charge) inside a large sphere and the observation of whether any deflections occur when the sphere is charged to a high voltage, has been carried out carefully over the years. A null result is always obtained, which allows one to bound $|\zeta|$ using the known configuration and geometry of the apparatus and the sensitivity of the electrometer. Such an experiment essentially constitutes a *comparison* of the force law to the inverse square law and thus provides a high degree of precision. Plimpton and Lawton[29] found $|\zeta| < 2 \times 10^{-9}$. A more recent experiment by Williams, Faller, and Hill[30] places a limit of $|\zeta| < (2.7 \pm 3.1) \times 10^{-16}$.

---

[27] See A. N. Cleland and M. L. Roukes, A nanometre-scale mechanical electrometer, *Nature,* 392, p. 160, 1998.

[28] H. Cavendish, *Electrical Researches,* J. C Maxwell (ed.), Cambridge University Press, 1879, pp. 104–113.

[29] S. J. Plimpton and W. E. Lawton, A very accurate test of Coulomb's law of force between charges, *Phys. Rev.,* 50, p. 1066, 1936.

[30] E. R. Williams, J. E. Faller, and H. A. Hill, New experimental test of Coulomb's law: A laboratory upper limit on photon rest mass, *Phys. Rev. Lett.,* 26, p. 721, 1971.

In terms of the range of circumstances under which the inverse square law is valid, the various experiments described above show that Coulomb's law holds at distances of a few tens of centimeters. What about at smaller distances? E. Rutherford's analysis[31] of experiments involving the scattering of alpha particles[32] by atomic nuclei shows that the inverse square law holds at distances of order 10^{-11} cm. The accuracy of the inverse square law at such small distances is difficult to ascertain; however, measurements of Lamb and Retherford[33] of the relative positions of the energy levels of hydrogen indicate that $|\zeta| < 10^{-9}$ even at distances as small as 10^{-8} cm. At distances smaller than 10^{-11} cm, nuclear forces and strong interaction effects come into play and partially mask the effects of the inverse square law. Nevertheless, measurements in nuclear physics indicate that electrostatic forces still vary approximately as r^{-2} at distances of 10^{-13} cm. At even smaller distances, the question of the validity of Coulomb's law is still an open question.

A natural question that comes to mind here is the validity of Coulomb's law at large length scales. While electrostatic experiments at distance scales of more than few meters are difficult to carry out, magnetic field measurements can be used to assess the applicability of Coulomb's law at large distances.[34] Geophysical measurements of the earth's magnetic field thus provide a highly accurate assessment of the inverse square law. According to such measurements,[35] the inverse square law holds at distances of $\sim 10^{9}$ cm (10^{4} km) to a degree of accuracy of $|\zeta| < 10^{-16}$, similar to that implied by the Williams, Faller, and Hill experiment.

In summary, Coulomb's law is known to be valid to a high degree of accuracy at distance scales ranging from 10^{9} to 10^{-11} cm, or 20 orders of magnitude! In thinking about the inverse square law, it is helpful to realize the similarities between it and the law of gravitational attraction, which also varies[36] as r^{-2}, whose validity is established to a very high degree of accuracy because it explains the motions of heavenly bodies so well. Laplace was the first to show[37] that no function of distance except the inverse square law satisfies the condition that a uniform spherical mass shell exerts no gravitational force on a particle within it. Similarly, it can be shown on

[31] E. Rutherford, Scattering of α- and β-particles by matter, and the structure of the atom, *Phil. Mag.*, 21, p. 669, May, 1911.

[32] α-particles are helium nuclei which move at speeds of $\sim 0.3c$ when they are emitted spontaneously by radioactive materials such as radium.

[33] W. E. Lamb, Jr., and R. C. Retherford, Fine structure of the hydrogen atom by a microwave method, *Phys. Rev.*, 72, p. 241, 1947; —, Fine structure of the hydrogen atom. Part IV, *Phys. Rev.*, 86, p. 1014, 1952, and references therein.

[34] In the context of special relativity theory, the fundamental laws of magnetostatics are related to Coulomb's law, so that a test of predictions of magnetostatics actually constitutes a test of the electrostatic inverse square law.

[35] For an excellent review, see A. S. Goldbaher and M. M. Nieto, Terrestrial and extraterrestrial limits on the photon mass, *Rev. Mod. Phys.*, 43, p. 277, 1971.

[36] For an interesting discussion of inverse square law forces, see Chapter 2 of N. Feather, *Electricity and Matter,* Edinburgh University Press, Edinburgh, 1968.

[37] P. S. Laplace, *Mécanique céleste,* vol. I., p. 163, Paris, 1799.

a purely mathematical basis[38] that no function of distance except the inverse square law is consistent with the fact that a spherical shell of charge exerts no electrostatic force on a charged particle within it.

In terms of the validity of aspects of Coulomb's law other than the r^{-2} dependence [for example, the assumptions that the electric force acts along the line connecting two point charges, that the force varies as the algebraic product of the two charges, and that the constant of proportionality is always $(4\pi\epsilon_0)^{-1}$], we rely on the accuracy of predictions based on these assumptions. Furthermore, since the charges used in various experiments were not really point charges but instead were distributed in space, the principle of superposition of forces was inherently assumed to hold. Once again, the validity of this particular assumption is verified only by virtue of the fact that predictions based on it are consistent with experiments.

4.3 THE ELECTRIC FIELD

Consider the region around a single point charge Q, as shown in Figure 4.7. Other charges (e.g., a test charge q) brought near this charge would experience a force whose magnitude depends on the distance between Q and q. The existence of this force can be described by saying that the charge Q produces an *electric field* in the region surrounding it. It is convenient to think of this region as being permeated by an electric field **E**, defined as the force per unit charge,

$$\mathbf{E} = \lim_{q\to 0} \frac{\mathbf{F}}{q} \qquad [4.2]$$

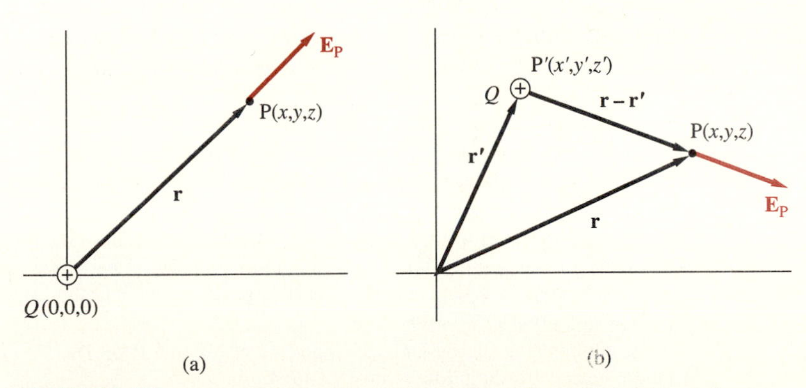

FIGURE 4.7. Electric field of single point charge. (a) Point charge at the origin. (b) Point charge at point P′(**r**′).

[38] For a discussion of Maxwell's original proof of this result, see R. S. Elliott, *Electromagnetics,* IEEE Press, Piscataway, New Jersey, 1993, Chapter 3.

The SI unit of electric field is the newton per coulomb ($N\text{-}C^{-1}$). A more commonly used equivalent unit for the electric field is the volt per meter ($V\text{-}m^{-1}$). In order for [4.2] to be valid, the test charge q must be negligibly small (both in terms of its amount of charge and physical size) so that its force on Q does not cause Q to move and so that induction effects due to the presence of q are negligible.[39] In short, the test charge must both be small (in terms of quantity of charge) and must be a point charge[40] so that it does not alter the charge distribution from what it would be in the absence of the test charge.

The notion of a *field*, however abstract, is nevertheless the most useful way of thinking and working with electromagnetic problems and of describing the effects of charges on other charges at a distance.[41] We can simply proceed by defining a field to be the set of values assumed by a physical quantity at various points in a region of space at various instants of time. However, a few more words of discussion are in order in view of the abstract nature of this concept. The field concept is used in other areas of physics governed by "action-at-a-distance" laws similar to Coulomb's law. For example, Newtonian mechanics is based on the effects of gravitational "forces" on objects having mass, without reference to the specific nature of the sources of these forces. Similarly, it is convenient to think that the source charge produces "something" at surrounding points, and that this something then interacts with other charges brought to these points. Since we can use [4.2] to determine the magnitude of the electric field *whether or not* there is a charge there to be subject to a force, it is tempting to regard the field as an actual physical entity in its own right. Most of these ideas originated with M. Faraday, who believed that the presence of charges actually changed the physical properties of space and that the electric field was a manifestation of this altered state. For Faraday, the electric field was a very real physical quantity.[42] If we view the electric field as a physical quantity, it becomes important that we are able to measure it. At first thought, it should be straightforward to simply bring a test charge q to point P and deduce the field from the electrical force that it experiences. However, the presence of this test charge q will in general subject the source charges to new forces and would thus change the field distribution that

---

[39]If the electric field is set up by voltage differences between conductors, as it usually is, the presence of the test charge should not affect the distribution of charges on the conductors that set up the electric field.

[40]A useful analogy is in using a small float or fishing line to deduce the direction of current in a lake, and implicitly assuming that the float is infinitesimal in size. Clearly, we could not infer the direction of the current from the orientation of a float as large as an anchored ship, since the presence of the ship would actually disturb the current flow.

[41]For an illuminating discussion, see Chapter 1 of Volume II of *The Feynman Lectures in Physics,* Addison Wesley, Reading, Massachusetts, 1964.

[42]"Faraday, in his mind's eye, saw lines of force traversing all space where the mathematicians saw centres of force attracting at a distance: Faraday saw a medium where they saw nothing but distance: Faraday sought the seat of the phenomena in real actions going on in the medium, they (i.e., the mathematicians) were satisfied that they had found it (i.e., the seat of the phenomena) in a power of action at a distance impressed on electric fluids." From J. C. Maxwell, *A Treatise on Electricity and Magnetism,* Clarendon Press, Oxford, 1892.

we set out to measure. It is in this context that the source charge q in [4.2] is required to be infinitesimally small.

For a positive charge Q located at the origin of our coordinate system and at rest in boundless free space as shown in Figure 4.7a, the electric field at any point P, identified by the position vector $\mathbf{r} = \hat{\mathbf{x}}x + \hat{\mathbf{y}}y + \hat{\mathbf{z}}z$, is

$$\mathbf{E}_P = \hat{\mathbf{r}}\frac{Q}{4\pi\epsilon_0 r^2} \qquad [4.3]$$

where $\hat{\mathbf{r}} = \mathbf{r}/|\mathbf{r}|$ and $r = |\mathbf{r}| = [x^2+y^2+z^2]^{1/2}$. We note that the electric field at point P is a vector that points along the direction of the position vector $\mathbf{r}$ (i.e., outward from the positive charge Q), as shown in Figure 4.7a. Conversely, if the source charge were negative (i.e., $Q < 0$), the electric field would point in the $-\mathbf{r}$ direction (i.e., inward toward the negative charge).

In general, the source charge Q may not be located at the origin but at a point P′ described by the position vector $\mathbf{r}' = \hat{\mathbf{x}}x' + \hat{\mathbf{y}}y' + \hat{\mathbf{z}}z'$ marked in Figure 4.7b. The electric field at point P produced by this charge is then in the $\hat{\mathbf{R}}$ direction, where $\mathbf{R} = (\mathbf{r} - \mathbf{r}')$, and is given by

$$\mathbf{E}_P = \hat{\mathbf{R}}\frac{Q}{4\pi\epsilon_0 R^2} = \hat{\mathbf{R}}\frac{Q}{4\pi\epsilon_0|\mathbf{r}-\mathbf{r}'|^2} \qquad [4.4]$$

Noting that $\hat{\mathbf{R}} \equiv \mathbf{R}/R = (\mathbf{r}-\mathbf{r}')/(|\mathbf{r}-\mathbf{r}'|)$, [4.4] can be written as

$$\mathbf{E}_P = \frac{Q}{4\pi\epsilon_0|\mathbf{r}-\mathbf{r}'|^2}\frac{\mathbf{r}-\mathbf{r}'}{|\mathbf{r}-\mathbf{r}'|} = \frac{Q}{4\pi\epsilon_0}\frac{\hat{\mathbf{x}}(x-x') + \hat{\mathbf{y}}(y-y') + \hat{\mathbf{z}}(z-z')}{[(x-x')^2+(y-y')^2+(z-z')^2]^{3/2}} \qquad [4.5]$$

where the expanded form is written out in Cartesian coordinates. In [4.5], as well as throughout the rest of this text, we use primed (e.g., $\mathbf{r}'$) quantities to represent the position vector or coordinates of a source point, while unprimed (e.g., $\mathbf{r}$) quantities are used to represent the position vector or coordinates for a point at which the electric field (or another quantity) is determined.

Equation [4.4] implies that the electric field becomes arbitrarily large as the observation point P is brought closer to the point charge (i.e., as $\mathbf{r} \rightarrow \mathbf{r}'$ or $R \rightarrow 0$). The resolution of this apparently unphysical result lies in our original premise (Section 4.1.3) of restricting our attention to effects on macroscopic scales, thus ensuring that the distance $R = |\mathbf{r} - \mathbf{r}'|$ between source and observation points is much larger than atomic dimensions.

Example 4-3 illustrates the evaluation of the direction and magnitude of the electric field in the vicinity of a point charge.

Example 4-3: Electric field. Find the electric field at P(3, 1, 0) due to a point charge $Q = +80$ nC located at (2,0,2) m as shown in Figure 4.8. Assume all distances to be in meters.

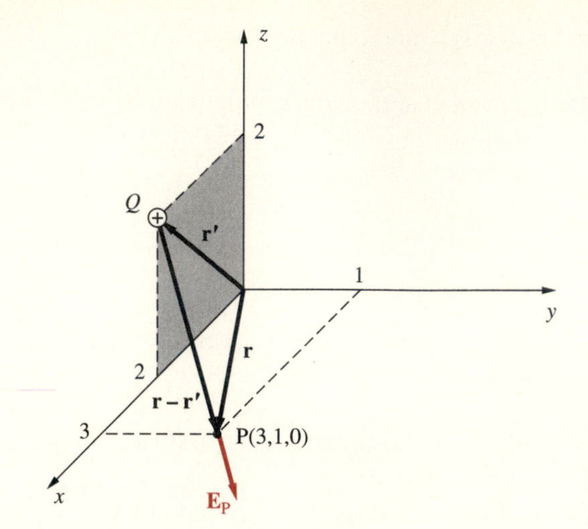

FIGURE 4.8. Electric field. Electric field $\mathbf{E}_P$ at point P(3,1,0) due to a positive point charge Q at (2,0,2) m.

Solution: Note from the geometry that $\mathbf{r} = 3\hat{\mathbf{x}} + \hat{\mathbf{y}}$ and $\mathbf{r}' = 2\hat{\mathbf{x}} + 2\hat{\mathbf{z}}$. Using [4.5] we have

$$\mathbf{E}_P = \frac{Q}{4\pi\epsilon_0}\frac{\mathbf{r}-\mathbf{r}'}{|\mathbf{r}-\mathbf{r}'|^3}$$

$$= \frac{80\text{ nC}}{(9\times 10^9)^{-1}}\frac{\hat{\mathbf{x}}+\hat{\mathbf{y}}-2\hat{\mathbf{z}}}{[(3-2)^2+1^2+(-2)^2]^{3/2}} \simeq 49(\hat{\mathbf{x}}+\hat{\mathbf{y}}-2\hat{\mathbf{z}})\text{ V-m}^{-1}$$

Note from Figure 4.8 that $\mathbf{E}_P$ points predominantly downward, consistent with the fact that the magnitude of the z component is twice as large as that for the other two components.

4.3.1 Electric Field Resulting from Multiple Charges

The electric field concept is most useful when large numbers of charges are present, since each charge exerts a force on all the others. Assuming that for a particular test charge the forces resulting from the different charges can be linearly superposed,[43] we can evaluate the electric field at a given point simply as a vector sum of the field contributions resulting from all charges.

[43]We are able to assume linearity simply because, at the macroscopic level, its use produces accurate results in many different experiments and applications involving groups of charges and currents. We note here that some material media can behave in a highly nonlinear manner and that such behavior often leads to interesting applications or consequences, such as in magnetic materials, nonlinear optics, crystals, and electrostatic breakdown phenomena (e.g., lightning discharges). As far as fields in vacuum are concerned, linear superposition is remarkably valid even at the atomic level, where field strengths of order 10^{11} to 10^{17} V-m^{-1} exist at the orbits of electrons. For further discussion, see Section I.3 of J. D. Jackson, *Classical Electrodynamics,* Wiley, New York, 1975.

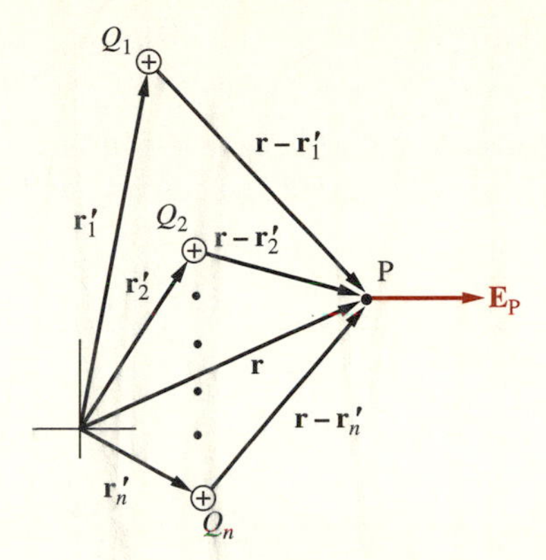

FIGURE 4.9. Electric field due to multiple charges. The electric field at the point P at position $\mathbf{r}$ is the sum of the electric fields due to the point charges $Q_1, Q_2, \ldots, Q_n$, located respectively at source points $\mathbf{r}_1', \mathbf{r}_2', \ldots, \mathbf{r}_n'$.

For a set of discrete point charges $Q_1, Q_2, \ldots, Q_n$ located at points with position vectors $\mathbf{r}_1', \mathbf{r}_2', \ldots, \mathbf{r}_n'$, as shown in Figure 4.9, the electric field at a point P with position vector $\mathbf{r}$ is given as

$$\mathbf{E}_P = \frac{1}{4\pi\epsilon_0}\sum_{k=1}^{n}\hat{\mathbf{R}}_k\frac{Q_k}{|\mathbf{r}-\mathbf{r}_k'|^2} \qquad [4.6]$$

where $\hat{\mathbf{R}}_k$ is the unit vector pointing in the direction from the source point $P_k'(\mathbf{r}_k')$ at which charge Q_k resides to the observation point $P(\mathbf{r})$. Noting that $\hat{\mathbf{R}}_k = (\mathbf{r}-\mathbf{r}_k')/(|\mathbf{r}-\mathbf{r}_k'|)$, [4.6] can be written as

$$\mathbf{E}_P = \frac{1}{4\pi\epsilon_0}\sum_{k=1}^{n}\frac{Q_k(\mathbf{r}-\mathbf{r}_k')}{|\mathbf{r}-\mathbf{r}_k'|^3} \qquad [4.7]$$

Example 4-4 illustrates the evaluation of the electric field resulting from multiple charges.

Example 4-4: Two point charges. Consider two point charges $Q_1 = +5$ nC and $Q_2 = -5$ nC located at points (1,0,0) m and (−1,0,0) m as shown in Figure 4.10.

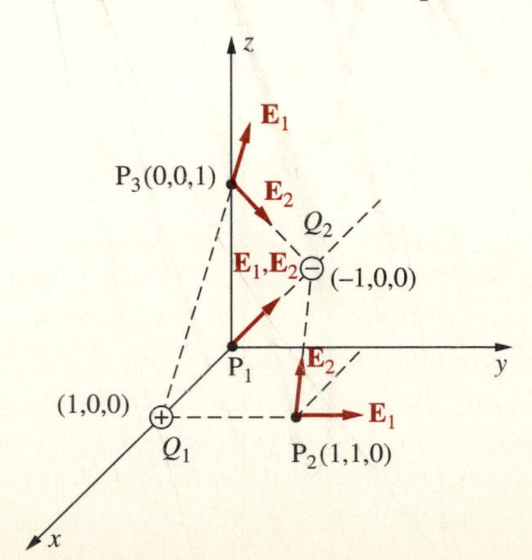

FIGURE 4.10. Two point charges. The electric field vectors $\mathbf{E}_1$ and $\mathbf{E}_2$, due respectively to charges $Q_1 = +5$ nC and $Q_2 = -5$ nC, are shown at points P_1, P_2, and P_3.

(a) Calculate the electric field at points $P_1(0,0,0)$, $P_2(1,1,0)$, and $P_3(0,0,1)$, respectively. (b) Repeat for $Q_2 = +5$ nC. Assume all coordinates to be in meters.

Solution: Using [4.7], the electric field at point P with a position vector $\mathbf{r}$ resulting from point charges Q_1 and Q_2 located respectively at positions $\mathbf{r}_1' = \hat{\mathbf{x}}$ and $\mathbf{r}_2' = -\hat{\mathbf{x}}$ is given as

$$\mathbf{E}_\mathrm{P} = \mathbf{E}_1 + \mathbf{E}_2 = \frac{Q_1(\mathbf{r} - \mathbf{r}_1')}{4\pi\epsilon_0|\mathbf{r} - \mathbf{r}_1'|^3} + \frac{Q_2(\mathbf{r} - \mathbf{r}_2')}{4\pi\epsilon_0|\mathbf{r} - \mathbf{r}_2'|^3}$$

(a) For point $P_1(0,0,0)$ we have $\mathbf{r} = 0$. Therefore,

$$\mathbf{E}_{\mathrm{P}_1} = \frac{5\times 10^{-9}}{(9\times 10^9)^{-1}}\frac{(-\hat{\mathbf{x}})}{1^3} - \frac{5\times 10^{-9}}{(9\times 10^9)^{-1}}\frac{(+\hat{\mathbf{x}})}{1^3} = -90\hat{\mathbf{x}} \text{ V-m}^{-1}$$

For point $P_2(1,1,0)$, we have $\mathbf{r} = \hat{\mathbf{x}} + \hat{\mathbf{y}}$ so that

$$\mathbf{E}_{\mathrm{P}_2} = \frac{5\times 10^{-9}}{(9\times 10^9)^{-1}}\frac{\hat{\mathbf{x}} + \hat{\mathbf{y}} - \hat{\mathbf{x}}}{1^3} - \frac{5\times 10^{-9}}{(9\times 10^9)^{-1}}\frac{\hat{\mathbf{x}} + \hat{\mathbf{y}} + \hat{\mathbf{x}}}{[2^2 + 1^2]^{3/2}}$$

$$= 45\hat{\mathbf{y}} - \frac{9}{\sqrt{5}}(2\hat{\mathbf{x}} + \hat{\mathbf{y}}) = -\frac{18}{\sqrt{5}}\hat{\mathbf{x}} + \frac{9(5\sqrt{5} - 1)}{\sqrt{5}}\hat{\mathbf{y}} \text{ V-m}^{-1}$$

For point $P_3(0,0,1)$ we have $\mathbf{r} = \hat{\mathbf{z}}$. Thus,

$$\mathbf{E}_{\mathrm{P}_3} = \frac{5\times 10^{-9}}{(9\times 10^9)^{-1}}\frac{\hat{\mathbf{z}} - \hat{\mathbf{x}}}{[1^2 + 1^2]^{3/2}} - \frac{5\times 10^{-9}}{(9\times 10^9)^{-1}}\frac{\hat{\mathbf{z}} + \hat{\mathbf{x}}}{[1^2 + 1^2]^{3/2}} = -\frac{45}{\sqrt{2}}\hat{\mathbf{x}} \text{ V-m}^{-1}$$

(b) When Q_2 changes sign, all $\mathbf{E}_2$ vectors reverse direction. As a result, we have

At P_1 $\quad \mathbf{E}_{\mathrm{P}_1} = 0$

At P_2 $\quad \mathbf{E}_{\mathrm{P}_2} = \dfrac{18}{\sqrt{5}}\hat{\mathbf{x}} + \dfrac{9(5\sqrt{5} + 1)}{\sqrt{5}}\hat{\mathbf{y}} \text{ V-m}^{-1}$

At P_3 $\quad \mathbf{E}_{\mathrm{P}_3} = \dfrac{45}{\sqrt{2}}\hat{\mathbf{z}} \text{ V-m}^{-1}$

4.3.2 Electric Field Resulting from Continuous Charge Distributions

The atomic theory of matter has shown us that matter is quantized and that electric charges are made up of integral multiples of a certain minimum electric charge, which is known as the charge of an electron, namely $q_e \simeq -1.602 \times 10^{-19}$ C. Although we know that charge is so quantized, it is often useful to ignore that charges come in packages of electrons and protons and to instead think of continuous distributions of charge. Such a view is valid because the "graininess" of electrical charge is not significant over the macroscopic scale sizes that are considered here. We can

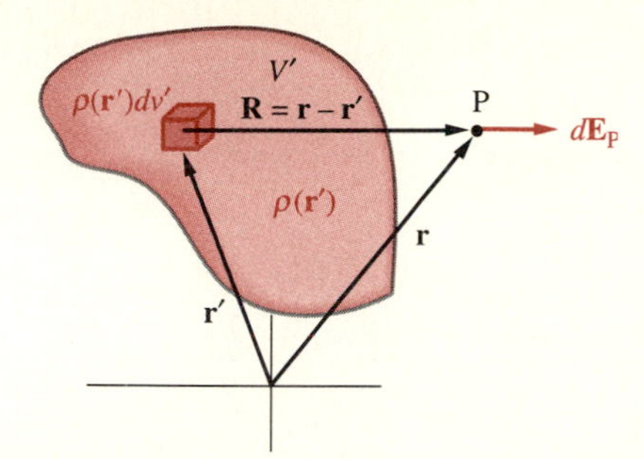

FIGURE 4.11. Electric field due to a continuous charge distribution. The field generated at point P (position vector **r**), due to the continuous charge distribution $\rho(\mathbf{r}')$ in the source volume V', is found by integrating the contributions of infinitesimal volume elements such as that shown.

consider charge as if it were "continuous" because we deal with very large numbers of charged particles that are extremely close (typically, confined to material bodies of macroscopic size). As an example, the current that flows in a wire connected to an ordinary 110-volt, 100-watt light bulb is about 1 ampere (or coulomb per second), so that $\sim 6 \times 10^{18}$ elementary charges flow per second through the cross-sectional area of the wire. We can describe such "continuous" distributions of charge in terms of density of charge, much like representing matter in terms of its mass density rather than counting the number of molecules. The *volume charge density,* $\rho(x, y, z)$ can in general be a function of all three coordinates of space and has units of C-m^{-3}. In other cases, it might be more convenient to represent the charge distribution in terms of a *surface charge density* ρ_s or a *line charge density* ρ_l.

Consider electric charge to be distributed over a volume V' with a volume density $\rho(\mathbf{r}')$ as shown in Figure 4.11. Since a differential element of charge behaves like a point charge, the contribution of a charge element $dQ = \rho(\mathbf{r}')\,dv'$ located in a differential volume element[44] dv' to the electric field at point P in Figure 4.11 is

$$d\mathbf{E}_{\mathrm{P}} = \hat{\mathbf{R}}\frac{\rho(\mathbf{r}')\,dv'}{4\pi\epsilon_0|\mathbf{r}-\mathbf{r}'|^2} \qquad [4.8]$$

The total electric field at point P due to all the charge in V' is then given as

$$\mathbf{E}_{\mathrm{P}} = \frac{1}{4\pi\epsilon_0}\int_{V'}\hat{\mathbf{R}}\frac{\rho(\mathbf{r}')\,dv'}{|\mathbf{r}-\mathbf{r}'|^2} \qquad [4.9]$$

Note once again that [4.9] can alternatively be written as

$$\mathbf{E}_{\mathrm{P}} = \frac{1}{4\pi\epsilon_0}\int_{V'}\frac{(\mathbf{r}-\mathbf{r}')\rho(\mathbf{r}')\,dv'}{|\mathbf{r}-\mathbf{r}'|^3} \qquad [4.10]$$

If the electric charge were instead distributed on a surface S' with a surface charge density ρ_s (having units of C-m^{-2}), then the integration would need to be carried out only over the surface. Thus,

[44]Note that although dv' can be made very small compared to the macroscopic dimensions of the problem in hand, it is nevertheless taken to be large enough compared to atomic scales so that it still does contain many discrete charges.

$$\mathbf{E}_{\mathrm{P}} = \frac{1}{4\pi\epsilon_0}\int_{S'} \hat{\mathbf{R}}\frac{\rho_s(\mathbf{r}')\,ds'}{|\mathbf{r}-\mathbf{r}'|^2} \qquad [4.11]$$

or, alternatively,

$$\mathbf{E}_{\mathrm{P}} = \frac{1}{4\pi\epsilon_0}\int_{S'} \frac{(\mathbf{r}-\mathbf{r}')\rho_s(\mathbf{r}')\,ds'}{|\mathbf{r}-\mathbf{r}'|^3} \qquad [4.12]$$

Surface charge distributions arise commonly in electromagnetics, especially when metallic conductors are placed in electric fields. As we shall see later, freely mobile electrons in a metallic conductor redistribute quickly (within $\sim 10^{-19}$ s) under the influence of an applied electric field by establishing distributions of charge confined to infinitesimally (on macroscopic scales) thin regions at the conductor surfaces.

For a line charge distribution ρ_l (having units of C-m^{-1}) we have

$$\mathbf{E}_{\mathrm{P}} = \frac{1}{4\pi\epsilon_0}\int_{L'} \hat{\mathbf{R}}\frac{\rho_l(\mathbf{r}')\,dl'}{|\mathbf{r}-\mathbf{r}'|^2} \qquad [4.13]$$

or, alternatively,

$$\mathbf{E}_{\mathrm{P}} = \frac{1}{4\pi\epsilon_0}\int_{L'} \frac{(\mathbf{r}-\mathbf{r}')\rho_l(\mathbf{r}')\,dl'}{|\mathbf{r}-\mathbf{r}'|^3} \qquad [4.14]$$

where L' is the line along which the charge is distributed. Note that L' need not be a straight line, just as S' need not be a flat surface. The concept of a line charge distribution becomes convenient when the objects on which the charges reside have significant extent only in one dimension. A good example is a thin wire with negligible cross-sectional dimensions.

Examples 4-5 and 4-6 illustrate the determination of the electric field in the vicinity of line and surface charge distributions.

Example 4-5: Line charge of finite length. Consider a cylindrical rod of charge of length $2a$ centered at the origin and aligned with the z axis, as shown in Figure 4.12, where the diameter d of the rod is much smaller than any other dimension in the system, so we can represent[45] the charge distribution in terms of a line charge density ρ_l. Determine the electric field at a point P$(r, \phi, 0)$ equidistant from the end points of the line charge.

Solution: In view of the symmetry in the azimuthal (i.e., ϕ) direction, it is appropriate to use the cylindrical coordinate system, with r, ϕ, z. Using [4.13], the electric field at point P with $\mathbf{r} = \hat{\mathbf{r}}r$ due to a line element of charge $dQ =$

---

[45]Note that the total charge on the cylindrical rod is given by $(2a)\rho_l$ or, in terms of volume charge density ρ, by $\rho(2a)\pi d^2/4$, so that $\rho_l = \rho\pi d^2/4$.

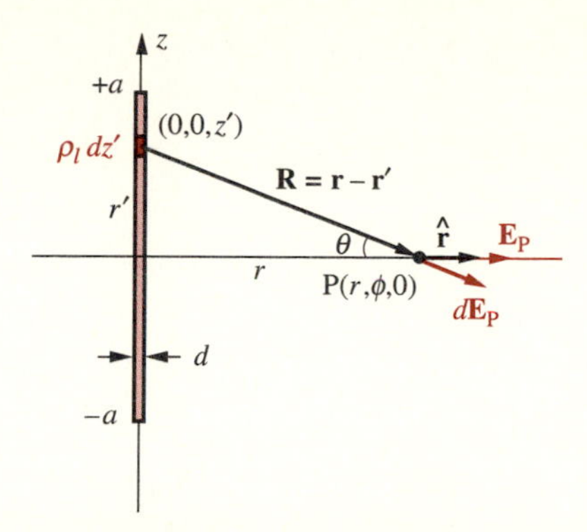

FIGURE 4.12. **Line charge of finite length.** A cylindrical rod of charge with linear charge density ρ_l C-m^{-1} produces the electric field $\mathbf{E}_\text{P}$ at point P equidistant from its endpoints.

$\rho_l\, dz'$ located at point $\mathbf{r}' = \hat{\mathbf{z}}z'$ and at a distance $|\mathbf{r} - \mathbf{r}'| = [r^2 + (z')^2]^{1/2}$ from point P is

$$d\mathbf{E}_\text{P} = \frac{\rho_l\, dz'}{4\pi\epsilon_0} \frac{(\mathbf{r} - \mathbf{r}')}{[r^2 + (z')^2]^{3/2}} = \frac{\rho_l\, dz'}{4\pi\epsilon_0} \frac{1}{[r^2 + (z')^2]} \underbrace{\frac{\mathbf{r} - \mathbf{r}'}{|\mathbf{r} - \mathbf{r}'|}}_{\hat{\mathbf{R}}}$$

This equation shows the field expressed in two slightly different ways; the expression on the right separates the magnitude of the vector from its direction (as expressed by the unit vector $\hat{\mathbf{R}}$, pointing from point $\mathbf{r}'$ to point $\mathbf{r}$). Note that the vector $d\mathbf{E}_\text{P}$ has components both in the radial (r) and z directions. However, in view of the symmetrical location of P, the contributions to the z component of charge elements located at positive z' locations cancel out those of charge elements at negative z' locations. Thus, the net total electric field at point P is in the radial direction, and we need only to sum (integrate) the radial components of $d\mathbf{E}_\text{P}$. The magnitude of the radial component is

$$dE_r = |d\mathbf{E}_\text{P}|\cos\theta = \frac{\rho_l\, dz'}{4\pi\epsilon_0} \frac{1}{[r^2 + (z')^2]} \frac{r}{\sqrt{r^2 + (z')^2}}$$

Integrating[46] over z' we find

$$\mathbf{E}_\text{P} = \hat{\mathbf{r}} \frac{\rho_l r}{4\pi\epsilon_0} \int_{-a}^{+a} \frac{dz'}{[r^2 + (z')^2]^{3/2}} = \hat{\mathbf{r}} \frac{\rho_l a}{2\pi\epsilon_0 r\sqrt{r^2 + a^2}}$$

For a very long rod, such that $a \gg r$, we have

$$\mathbf{E}_\text{P} \simeq \hat{\mathbf{r}} \frac{\rho_l}{2\pi\epsilon_0 r}$$

[46]Using either integral tables or change of variables (e.g., $\zeta = r\cot\theta$) it can be shown that

$$\int \frac{d\zeta}{\sqrt{(\zeta^2 + b^2)^3}} = \frac{\zeta}{b^2\sqrt{\zeta^2 + b^2}} + \text{const.}$$

where b is a constant.

which is a useful expression to remember. Note that the field of an infinitely long line charge falls off with distance as r^{-1}, although that for a point charge falls off with distance as r^{-2}.

Example 4-6: Ring of charge. Next we consider a ring of charge of radius a containing a total charge of amount Q, as shown in Figure 4.13. Assuming that the charge is uniformly distributed along the ring, and ignoring the thickness of the ring, the line charge density is $\rho_l = Q(2\pi a)^{-1}$. Evaluate the electric field at a point P$(0, \phi, z)$ along the axis of the ring at a distance z from its center.

Solution: Once again it is convenient to use cylindrical coordinates, and we take advantage of the symmetry of the problem to observe that the net total electric field at P is in the z direction. Using [4.13], a given line element $dl' = a\,d\phi'$ along the ring, containing a total charge of $dQ = \rho_l(a\,d\phi') = Q(a\,d\phi')/(2\pi a)$, produces a differential field at P of

$$d\mathbf{E}_{\mathrm{P}} = \frac{1}{4\pi\epsilon_0}\frac{Qa\,d\phi'}{2\pi a}\frac{(\mathbf{r}-\mathbf{r}')}{(a^2+z^2)^{3/2}} = \frac{1}{4\pi\epsilon_0}\frac{Qa\,d\phi'}{2\pi a}\frac{1}{(a^2+z^2)}\frac{\mathbf{r}-\mathbf{r}'}{|\mathbf{r}-\mathbf{r}'|}$$

where $\mathbf{r}-\mathbf{r}' = \hat{\mathbf{z}}z - \hat{\mathbf{r}}a$. Once again, we note that the electric field contributions of all elemental charges along the ring cancel out due to symmetry, except for the z components, so that the net total electric field at P is only in the z direction. The magnitude of the z component of the differential electric field is given by

$$dE_z = |d\mathbf{E}_{\mathrm{P}}|\cos\theta = \frac{1}{4\pi\epsilon_0}\frac{Qa\,d\phi'}{2\pi a}\frac{1}{(a^2+z^2)}\frac{z}{\sqrt{a^2+z^2}}$$

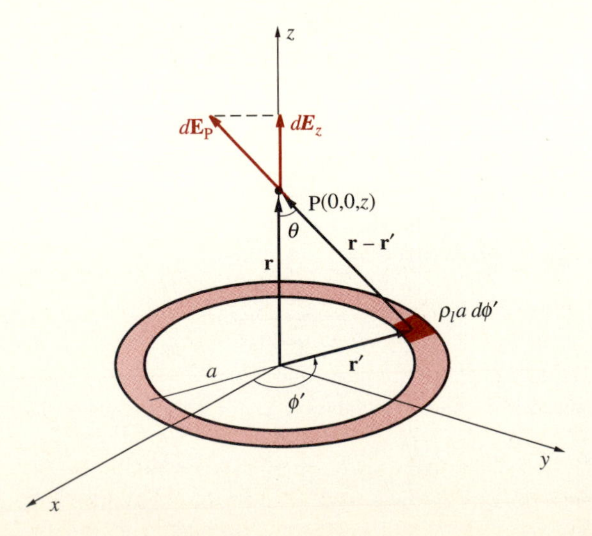

FIGURE 4.13. A ring of charge. Each elemental charge $\rho_l a\,d\phi'$ produces a differential electric field $d\mathbf{E}_{\mathrm{P}}$ at point P.

so that the total electric field at point P is

$$\mathbf{E}_{\mathrm{P}} = \hat{\mathbf{z}} \int_0^{2\pi} \frac{1}{4\pi\epsilon_0} \frac{Q\,d\phi'}{2\pi} \frac{1}{a^2 + z^2} \frac{z}{\sqrt{a^2 + z^2}}$$

$$= \hat{\mathbf{z}} \frac{1}{4\pi\epsilon_0} \frac{Qz}{(2\pi)(a^2 + z^2)^{3/2}} \int_0^{2\pi} d\phi'$$

$$\mathbf{E}_{\mathrm{P}} = \hat{\mathbf{z}} \frac{1}{4\pi\epsilon_0} \frac{Qz}{(a^2 + z^2)^{3/2}}$$

Note that for $z \gg a$, we have

$$\mathbf{E}_{\mathrm{P}} \simeq \hat{\mathbf{z}} \frac{1}{4\pi\epsilon_0} \frac{Q}{z^2}$$

In other words, at large enough distances the ring produces the same electric field as a point charge located at the origin.

4.3.3 Is This All There Is to Electrostatics?

It is clear from the above that *if the positions of all the charges* (either discrete or continuous) *are specified,* finding the electric field simply involves carrying out the appropriate summations and integrals. In other words, all of electrostatics appears to boil down to Coulomb's law and integration operations. These integrals are often complicated and may involve all three dimensions; however, they are nevertheless straightforward operations that can be evaluated numerically, if not solved or approximated analytically. For any given charge configuration, whether it be point, line, surface, or volume charge distribution or any combinations thereof, all we need to do is to evaluate the following integral(s):

$$\mathbf{E} = \frac{1}{4\pi\epsilon_0} \left[\sum_{k=1}^{n} \frac{Q_k(\mathbf{r} - \mathbf{r}')}{|\mathbf{r} - \mathbf{r}'|^3} + \int_{L'} \frac{(\mathbf{r} - \mathbf{r}')\rho_l(\mathbf{r}')\,dl'}{|\mathbf{r} - \mathbf{r}'|^3} + \int_{S'} \frac{(\mathbf{r} - \mathbf{r}')\rho_s(\mathbf{r}')\,ds'}{|\mathbf{r} - \mathbf{r}'|^3} + \int_{V'} \frac{(\mathbf{r} - \mathbf{r}')\rho(\mathbf{r}')\,dv'}{|\mathbf{r} - \mathbf{r}'|^3} \right]$$

Although evaluating the integrals may be a straightforward numerical problem, *understanding* the electric field and its relationship to charge configurations is enhanced by introducing additional concepts, such as field lines, electric potential, and divergence, and by expressing Coulomb's law in other ways, such as the so-called Gauss's law. Familiarity with such concepts may help us find easier ways of evaluating these integrals or, if they were to be evaluated for us by a number-crunching machine, of interpreting the numerical results.

It should also be realized that in many electrostatic problems—in fact, in most of the ones that are of practical interest—the distribution of at least some of the charges *is not known.* We know only the general rules (i.e., boundary conditions) by which charges reorganize themselves under the influence of electric fields. This is especially the case when metallic conductors are present in the vicinity of charges: When a charged body is brought near a metallic conductor, the charge carriers within the conductor redistribute themselves (see Section 4.7) in a manner that depends on the electric field, which in turn depends on the eventual charge configuration. In other words, although we may know the locations of some of the charges in our system, we do not know the actual distribution of charges on the conductors and thus cannot evaluate the total electric field simply by direct integration.

4.3.4 Visualization of the Electric Field: Field Lines

Michael Faraday was probably the greatest scientist in history who was completely innocent of mathematics. He made up for this "deficiency" through his intuitive ability to pictorialize, an ability perhaps unequaled in scientific history. Through his pictorial and nonmathematical imagination, Faraday understood magnetic and electric fields in terms of lines of force. We owe our present concept of field lines, introduced and discussed below, largely to Faraday. Faraday's ideas and points of view had a significant influence on J. C. Maxwell, who couched Faraday's qualitative thinking in a mathematical framework. Since we separately discuss the related concept of *electric flux* in a later section, we limit our attention here to simply the graphical representation of the electric field distribution using field lines.

The field line concept is useful for visualization and graphical display of any vector quantity, which in general may vary as a function of space. The need for such a concept becomes apparent if we think of how to "plot" even the simplest electrostatic field: the field generated by a point charge at the origin. This simple field is in the spherically radial direction and varies with distance as r^{-2}. One way to "plot" the field so as to display both the direction and the magnitude information is as shown in Figure 4.14a, from which it is apparent that this method provides for rather coarse spatial resolution. Note that we can easily display the r^{-2} dependence of the magnitude of the electric field by means of a three-dimensional plot in Figure 4.14b; however, such a display does not convey any information about the direction of the electric field. Graphical representation of the field becomes more important for more complicated field distributions.

Another way to plot the electric field distribution graphically is by drawing lines that are everywhere tangent to the electric field; such lines are called *field lines* or *lines of force*.[47] The magnitude information can be conveyed by the *density* of lines (i.e., number of lines per unit area through a surface perpendicular to the lines). Such displays are shown in Figures 4.15, 4.16, 4.17a, and 4.17b respectively for a single

[47] They are called lines of force because they literally represent the direction in which test charges initially placed at a point would move, or the direction of the *force* that the test charge would experience. In other words, a small charged test particle of negligible mass set free in the electric field would simply trace out a line of force, very much like a small float set free on a lake would trace out the line of water current.

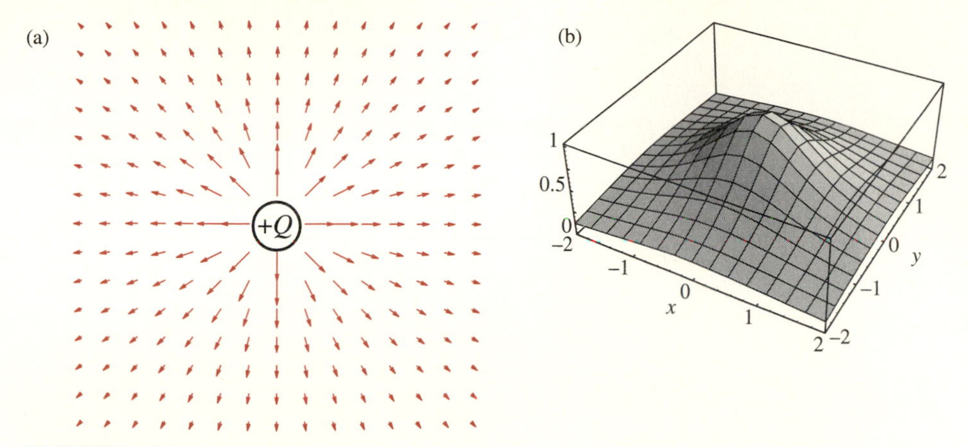

FIGURE 4.14. Electric field of a point charge. (a) A discretized vector plot of the electric field around a point charge. The sizes of the arrows at different positions represent the relative magnitude of the field at those positions. (b) The normalized magnitude $|\mathbf{E}(x, y, 1)|/|\mathbf{E}(0, 0, 1)|$ of the electric field due to a point charge $+Q$ at the origin is shown as a function of x, y at the $z = 1$ plane.

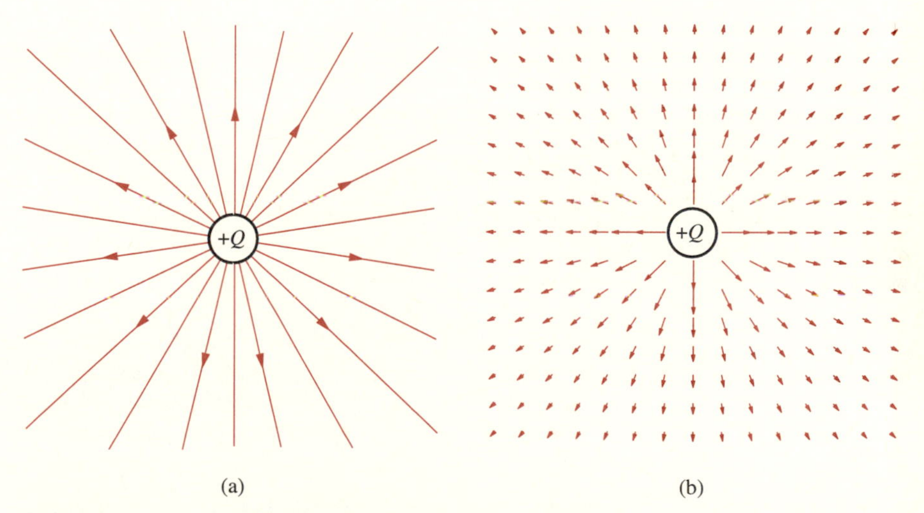

FIGURE 4.15. Electric field of a point charge. Graphical representation of the electric field of a point charge in terms of (a) field lines, (b) discrete vectors.

point charge, two equal point charges of opposite polarity, two equal point charges of the same polarity, and two charges of different magnitudes and polarity. Note that the electric field lines emanate from positive charges and terminate at negative charges, some of which (e.g., in Figure 4.17b) may be located at infinity.

Note that in order for a field line plot to convey useful information, a sensible number of lines has to be utilized—enough to indicate any symmetry in the system and the *relative* magnitude of the field at different points. It would not be possible to display the magnitude of the field on any absolute scale (e.g., a given number

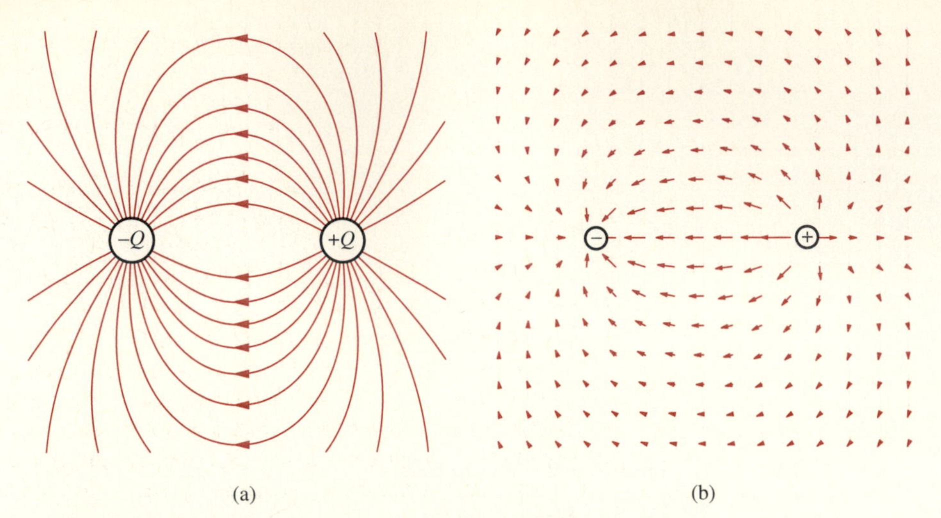

FIGURE 4.16. **Electric field of two equal but opposite point charges.** (a) Field line representation. (b) Representation using discrete vectors.

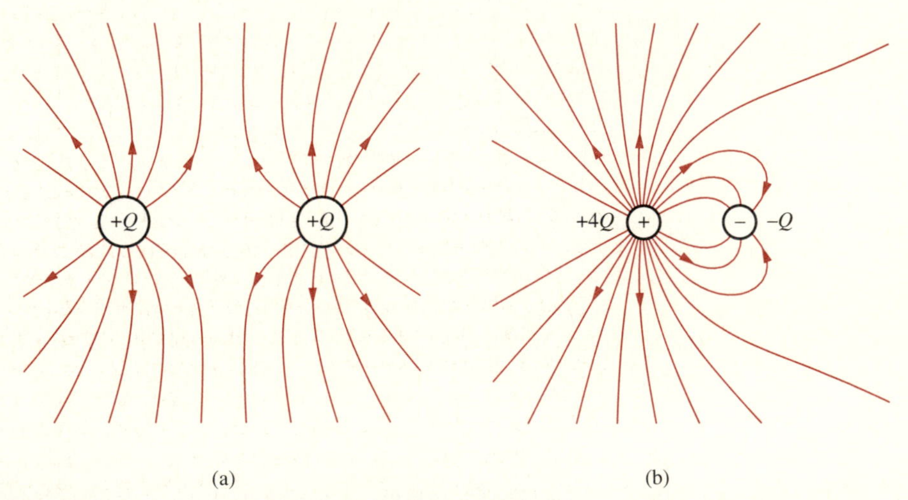

FIGURE 4.17. **Field maps for two charges Q_1 and Q_2.** (a) $Q_2 = Q_1 = Q$. (b) $Q_1 = -4Q_2 = 4Q$.

of lines per unit charge), since the relative magnitude of the charge might lead to unreasonably sparse or too dense field line distributions.

With respect to Figure 4.15a, one may at first think that the total number of lines, for example crossing a spherical surface of radius r, should decrease with distance as r^{-2} to represent relative magnitude of the field accurately. However, the area of such a spherical surface perpendicular to the lines increases as r^2, so the *density* of lines in Figure 4.15a is proportional to the magnitude of the field at any distance from the source charge, even though the total number of lines stays the same.

Since the electric field at any point has a definite direction, it follows that electric field lines cannot cross one another, because that would imply two possible directions at the crossing point. Realizing this constraint also highlights an important deficiency of the field line concept. Although the electric field in the vicinity of two point charges is calculated (see Example 4-4) by linear superposition of the fields due to the individual charges, the field line plot of Figure 4.16 is not a simple superposition of two plots of the kind in Figure 4.15.[48] The principle of superposition, an important principle concerning electric fields, does not have an easy interpretation in the context of field lines. Although we can in principle visualize a vectorial superposition, we cannot easily deduce the field line distribution in Figure 4.16 from that in Figure 4.15.

4.4 THE ELECTRIC POTENTIAL

An electric field is essentially a field of force. If a body being acted upon by a force is moved from one point to another, work will be done on the body. In the absence of losses (or dissipation of energy), the energy (i.e., work) thus put in must be stored as either kinetic or potential energy. The electrostatic field is an example of such a *conservative* field, since if a charge is moved in a static field in the absence of friction, no energy is dissipated. In moving a small test charge about in a field, either we have to do work against electric forces or we find that these forces do work for us. If we have two charges of opposite sign, work must be done to separate them in opposition to the attractive electrostatic force between them. This energy is "stored" in the separated configuration of the two charges and can be recovered if the charges are allowed to return to their original relative positions. The stored energy is called *potential* (rather than kinetic) energy, because it depends on the position (rather than motion) of the charges in a force field. The concept of *scalar electric potential* provides a quantitative measure of the work or energy required to move charges from one point to another under the influence of an electrostatic field.

4.4.1 Work Required to Move a Test Charge: Electrostatic Potential

We assume that there is some distribution of charges that sustains an electric field. The work W that we need to do *against* the electrical force ($\mathbf{F}$) in carrying a test charge q along some path from point a to point b is

$$W_{a\to b} = -\int_a^b \mathbf{F}\cdot d\mathbf{l} \qquad [4.15]$$

[48]Note that such a graphical superposition would create crossing field lines, which are not possible, as we just argued.

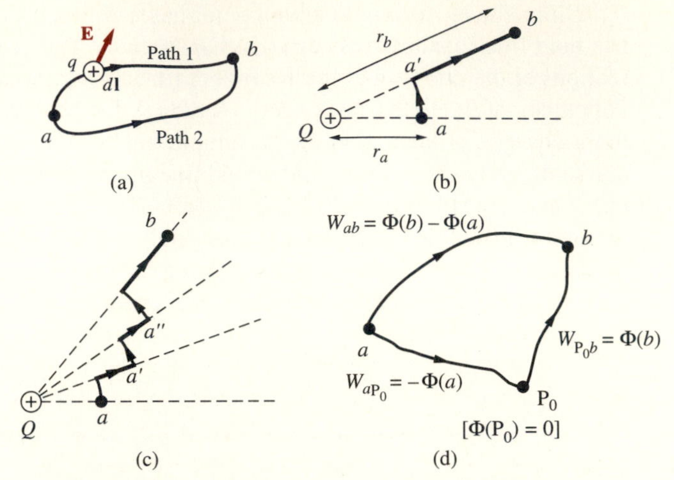

FIGURE 4.18. Work and voltage. The work done (by an external force against the electric field) in carrying a $+1$C charge from point a to b depends only on the end points, not on the path taken.

where $d\mathbf{l}$ is the differential vector displacement along the path, as shown in Figure 4.18a. Note that the units of W are joules (J). The minus sign in front of the integral is necessary because the work is done *against* the field. The dot product[49] accounts for the fact that it takes no work to move the test charge perpendicular to the field (no opposing force). Often we are more interested in work done (by an external force against the electrical force) in carrying a unit positive test charge (i.e., q = 1 C), in which case, using $\mathbf{F} = q\mathbf{E}$, we have

$$\left[\frac{W}{q}\right]_{a\to b} = -\int_a^b \mathbf{E}\cdot d\mathbf{l} \qquad [4.16]$$

where the units of the quantity $[W/q]$ are joules per coulomb (J-C^{-1}) or volts (V), and the subscript $a \to b$ indicates that the work is done in moving from a to b. In evaluating line integrals such as [4.15], it is customary to take $d\mathbf{l}$ in the direction of the increasing coordinate value (e.g., the $+x$ direction if the integration is along the

[49]The dot product of two vectors $\mathbf{A}$ and $\mathbf{B}$ is a scalar denoted by $\mathbf{A}\cdot\mathbf{B}$ and is equal to the product of the magnitudes $|\mathbf{A}|$ and $|\mathbf{B}|$ of $\mathbf{A}$ and $\mathbf{B}$ and the cosine of the angle θ_{AB} between them. Namely,

$$\mathbf{A}\cdot\mathbf{B} \equiv |\mathbf{A}||\mathbf{B}|\cos\theta_{AB}$$

Noting that in rectangular coordinates we have $\mathbf{A} = \hat{\mathbf{x}}A_x + \hat{\mathbf{y}}A_y + \hat{\mathbf{z}}A_z$ and $\mathbf{B} = \hat{\mathbf{x}}B_x + \hat{\mathbf{y}}B_y + \hat{\mathbf{z}}B_z$, an alternative expression for the dot product is

$$\mathbf{A}\cdot\mathbf{B} = (\hat{\mathbf{x}}A_x + \hat{\mathbf{y}}A_y + \hat{\mathbf{z}}A_z)\cdot(\hat{\mathbf{x}}B_x + \hat{\mathbf{y}}B_y + \hat{\mathbf{z}}B_z)$$

$$\mathbf{A}\cdot\mathbf{B} = A_xB_x + A_yB_y + A_zB_z$$

The dot product is sometimes referred to as the scalar product, since the result is a scalar quantity.

x axis) so that the manner in which the path of integration is traversed (i.e., $a \to b$ versus $b \to a$) is unambiguously determined by the limits of integration.

In general, a line integral of the type above depends on the path of integration. However, in a conservative field (i.e., one in which there is no mechanism for energy dissipation), the work done in moving from one point to another is independent of the path. If the integral depended on the path from a to b, we could extract energy out of the field by allowing the charge to move from a to b along the path for which $[W/q]$ is smaller and take it back from b to a along another. We could indefinitely extract work from the field[50] simply by repeating the process.

Now consider the field produced by a single point charge Q and two points a and b at distances of r_a and r_b, respectively, as shown in Figure 4.18b. Since the field is radial, it is clear that no work is done in moving a unit test charge along circular arc segments such as a–a' that are perpendicular to the field (i.e., $\mathbf{E} \cdot d\mathbf{l} = 0$ along the path). On the other hand, along path a'–b the field is in the direction of motion of the unit test charge, so $\mathbf{E} \cdot d\mathbf{l} = E\,dr$. Thus the work done by an external force against $\mathbf{E}$ in moving the unit test charge from point a to point b is

$$\left[\frac{W}{q}\right]_{a\to b} = \left[\frac{W}{q}\right]_{a\to a'} + \left[\frac{W}{q}\right]_{a'\to b} = -\int_a^{a'} \mathbf{E} \cdot d\mathbf{l} - \int_{a'}^{b} \mathbf{E} \cdot d\mathbf{l}$$

$$= 0 - \frac{Q}{4\pi\epsilon_0} \int_{r_{a'}}^{r_b} \frac{dr}{r^2} = -\frac{Q}{4\pi\epsilon_0} \left[-\frac{1}{r}\right]_{r_{a'}}^{r_b}$$

$$= \frac{Q}{4\pi\epsilon_0} \left(\frac{1}{r_b} - \frac{1}{r_a}\right)$$

since $r_{a'} = r_a$. It is also clear that if we took some other path, for example one that has smaller steps in the radial direction interconnected with smaller circular arc segments (Figure 4.18c), the result would be the same. Since any other path between a and b, however smooth it may be, can also be similarly divided into arc/radial segments, we conclude that the net work depends only on the end points of the path.

Since $[W/q]_{a\to b}$ depends only on the end points, it can be represented as a difference between two numbers. To see this, we can consider Figure 4.18d, which introduces a reference point P_0 and, for any point P, a function $\Phi(\mathrm{P})$ that is equal to the work done against the field in moving from P_0 to P. Since the work done in moving the test charge from P_0 to P depends only on the two end points, $\Phi(\mathrm{P})$ has only one value. Let us evaluate our line integral using a path that goes through P_0:

$$\left[\frac{W}{q}\right]_{a\to b} = -\int_a^b \mathbf{E} \cdot d\mathbf{l} = -\int_a^{P_0} \mathbf{E} \cdot d\mathbf{l} - \int_{P_0}^{b} \mathbf{E} \cdot d\mathbf{l} = \Phi(b) - \Phi(a)$$

[50]There is nothing inherently wrong about extracting work from a field. Energy could be extracted from the field if the motion of the test charge produced forces that could influence the field, for example, by moving the charges that produce the field. However, in electrostatics, we assume that the charges producing the fields are fixed in their locations, so that no work can be done on them.

since the work done in moving from a to P_0 is equal in magnitude but opposite in sign to that done in moving from P_0 to a.

The values of the function Φ constitute a relative *electrostatic potential* that can be assigned to every point in the field. This function Φ is a *scalar field* and is, in general, a function of x, y, and z. For convenience, the reference point P_0 is usually taken to be at infinity, so the *electrostatic potential* at any point P is

$$\Phi(P) \equiv \left[\frac{W}{q}\right]_{\text{at P}} = -\int_{\infty}^{P} \mathbf{E}\cdot d\mathbf{l} \qquad [4.17]$$

Note that $\Phi(P)$ represents the amount of energy *potentially* available when a unit positive test charge is at point P; this energy can be extracted from the system by allowing the test charge to move to a distant point (i.e., to infinity). It is thus appropriate to view $\Phi(P)$ as *potential* energy per coulomb, with units of joules per coulomb (J-C^{-1}) or volts (V). Note that for a single point charge Q at the origin ($\mathbf{r}' = 0$) as the source of the electric field, the electrostatic potential (i.e., the work done to move a unit positive test charge from infinity to point P) is

$$\Phi(P) = \Phi(x, y, z) = -\int_{\infty}^{r}\left(\hat{\mathbf{r}}\frac{Q}{4\pi\epsilon_0|\mathbf{r}|^2}\right)\cdot(\hat{\mathbf{r}}\,dr) = \frac{Q}{4\pi\epsilon_0}\frac{1}{r}$$

where, for simplicity, the path of integration is chosen to be in the radial direction.

The work done in moving a unit test charge from point a to b is the *electrostatic potential difference* between the two points and is denoted by Φ_{ab}. Based on the above, this quantity is given by

$$\left[\frac{W}{q}\right]_{a\to b} = \Phi_{ab} = -\int_{a}^{b} \mathbf{E}\cdot d\mathbf{l} = \Phi(b) - \Phi(a) \qquad [4.18]$$

where $\Phi(b)$ and $\Phi(a)$ are the electrostatic potentials respectively at points b and a.

One important corollary to the above discussion is that the line integral of the electrostatic field around any closed path is identically zero. This is readily seen from Figure 4.18d, if one imagines going around the closed path $a \to P_0 \to b \to a$. We thus have

$$\oint_{\text{any closed path}} \mathbf{E}\cdot d\mathbf{l} = 0 \qquad [4.19]$$

This equation turns out to describe a general property of any *conservative* vector field. We will refer back to this property of the electrostatic field as we discuss the concept of *curl* of a vector field in Section 6.4.

An important point to note here is that we arrived at this property simply through symmetry arguments (e.g., we did use the fact that the force from a single charge was spherically symmetric and in the radial direction), and without using the fact that the electrostatic force depends on distance as r^{-2}. Thus, the above integral equation contains only part of the known laws of electrostatics. The additional property, known as Gauss's law, will be described in a later section. Together, these two integral properties represent electrostatics as contained in Coulomb's law.

Examples 4-7 and 4-8, respectively, illustrate the evaluation of work done in carrying charges between two points and the electrostatic potential in the vicinity of a given charge distribution.

Example 4-7: Work done in carrying charges. Determine the work done in carrying a $q = +2\ \mu\text{C}$ point charge from point A(0, −3, 0) to point B(1, 0, 0) in a nonuniform electric field $\mathbf{E} = (3x + y)\hat{\mathbf{x}} + x\hat{\mathbf{y}} - 3\hat{\mathbf{z}}$ along (a) the shortest path (straight line), (b) an L-shaped path via the origin, and (c) along the parabola $y = 6x^2 - 3x - 3$, as shown in Figure 4.19.

Solution: In all three cases, we need to evaluate the line integral

$$W_{\text{A}\to\text{B}} = -q\int_{\text{A}}^{\text{B}} \mathbf{E}\cdot d\mathbf{l}$$

where

$$\begin{aligned}\mathbf{E}\cdot d\mathbf{l} &= [(3x + y)\hat{\mathbf{x}} + x\hat{\mathbf{y}} - 3\hat{\mathbf{z}}]\cdot[\hat{\mathbf{x}}\,dx + \hat{\mathbf{y}}\,dy + \hat{\mathbf{z}}\,dz]\\ &= (3x + y)dx + x\,dy - 3dz\end{aligned}$$

following the usual convention of taking $d\mathbf{l}$ to be equal to $+\hat{\mathbf{r}}$ so that the way we integrate along the path is determined by the limits of integration. Therefore,

$$W_{\text{A}\to\text{B}} = (-2\times 10^{-6})\left[\int_0^1 (3x + y)dx + \int_{-3}^{0} x\,dy - \int_0^0 3dz\right]$$

where the limits of each integral are provided with respect to its own variable of integration.

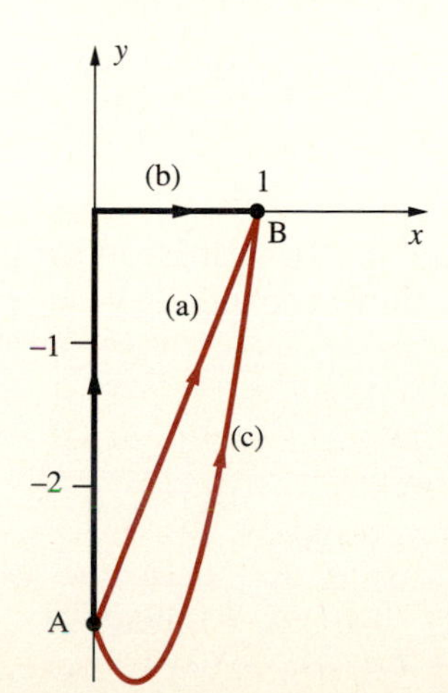

FIGURE 4.19. Example 4-7. The work done in moving a test charge from point A to point B is independent of the path.

(a) Along the shortest path, we have $y = 3x - 3$. Using this equation, we can write

$$W_{\text{A}\to\text{B}} = (-2\times10^{-6})\left[\int_0^1 (3x+3x-3)\,dx + \int_{-3}^0 \left[\left(\frac{y}{3}\right)+1\right]dy\right]$$

$$= (-2\times10^{-6})\left\{\left[3x^2-3x\right]_0^1 + \left[\frac{y^2}{6}+y\right]_{-3}^0\right\} = -3\ \mu\text{J}$$

(b) Along the L-shaped path, we have $x = 0$ from A to the origin and $y = 0$ from the origin to B. Thus, we can write

$$W_{\text{A}\to\text{B}} = (-2\times10^{-6})\left[\int_0^1 3x\,dx\right] = (-2\times10^{-6})\left[\frac{3x^2}{2}\right]_0^1 = -3\ \mu\text{J}$$

(c) Along the parabolic path, we have $y = 6x^2 - 3x - 3$, so that

$$W_{\text{A}\to\text{B}} = (-2\times10^{-6})\left[\int_0^1 (3x+6x^2-3x-3)\,dx + \int_{-3}^0 x\,dy\right]$$

Since along the parabola we have $dy = (12x - 3)\,dx$, and $x = 0$ when $y = -3$ and $x = 1$ when $y = 0$, we can rewrite the above as

$$W_{\text{A}\to\text{B}} = (-2\times10^{-6})\left[\int_0^1 (6x^2-3)\,dx + \int_0^1 x(12x-3)\,dx\right]$$

$$= (-2\times10^{-6})\left[\int_0^1 (18x^2-3x-3)\,dx\right]$$

$$= (-2\times10^{-6})\left[6x^3-\tfrac{3}{2}x^2-3x\right]_0^1 = -3\ \mu\text{J}$$

As expected, the work done is the same regardless of the path we follow in moving the charge.

Example 4-8: Electrostatic potential in the vicinity of an infinitely long line charge. The electric field in the vicinity of an infinitely long line charge was derived in Example 4-5:

$$\mathbf{E} = \hat{\mathbf{r}}\frac{\rho_l}{2\pi\epsilon_0 r}$$

Determine the electrostatic potential Φ in its vicinity.

Solution: Since the field is in the r direction, for simplicity, we take a line integral along a radial path, with $d\mathbf{l} = \hat{\mathbf{r}}\,dr$, so we have

$$\Phi(r) = -\int_{\infty}^{r} \mathbf{E} \cdot d\mathbf{l} = -\frac{\rho_l}{2\pi\epsilon_0}\int_{\infty}^{r} \frac{\hat{\mathbf{r}}}{r} \cdot (\hat{\mathbf{r}}\, dr)$$

$$= -\frac{\rho_l}{2\pi\epsilon_0}\left[\ln(r)\right]_{\infty}^{r} = \infty\ !$$

We can understand this seemingly unphysical result by recalling the meaning of electrostatic potential at any point as the work that needs to be done to bring a unit positive test charge from infinity to that point. With an infinitely long line charge, the electric field apparently does not fall rapidly enough with distance to allow moving a test charge from ∞ to any point r without requiring infinite energy.

Note, however, that the difference in electrostatic potential between any two points $r = a$ and $r = b$, such that $0 < b < a$, is straightforward to evaluate; we find

$$\Phi_{a \to b} = \Phi(b) - \Phi(a) = -\frac{\rho_l}{2\pi\epsilon_0}\int_a^b \frac{dr}{r} = \frac{\rho_l}{2\pi\epsilon_0} \ln\frac{a}{b}$$

which is the amount of work that we need to do to move a unit positive test charge closer to the line charge from $r = a$ to $r = b$. Note that for $a < b$, we have $\ln(a/b) < 0$, so $\Phi_{a \to b} < 0$, indicating that the field does work (i.e., work is extracted from the field) when the unit positive test charge moves away from the positive line charge.

4.4.2 The Electrostatic Potential and the Electric Field

Since forces on charges are determined by the electric field $\mathbf{E}$, the introduction of the additional concept of electrostatic potential may at first appear unnecessary. One advantage of the potential concept is that it is a scalar, which is relatively easier to evaluate in the vicinity of complicated charge distributions than the vector electric field, and from which the electric field can be readily calculated by simply taking derivatives, as described in the following paragraphs. Another advantage of the potential concept is that it allows us to link field concepts with the more familiar circuit concepts, which are couched in terms of potential (or voltage) differences between different points in a circuit and voltage drops across circuit elements.

We can now establish a differential relationship between the electrostatic potential Φ and the electric field $\mathbf{E}$. Consider two points at x and $(x + \Delta x)$ with the same y and z coordinates. The differential amount of work done (dW) in moving a test charge q between these points is directly proportional to the potential difference between them. Hence,

$$(dW)_x = q\Phi(x + \Delta x, y, z) - q\Phi(x, y, z) = q\frac{\partial \Phi}{\partial x}\Delta x$$

If the two points also had slightly different y, z coordinates, we then have

$$dW = q\left(\frac{\partial\Phi}{\partial x}\Delta x + \frac{\partial\Phi}{\partial y}\Delta y + \frac{\partial\Phi}{\partial z}\Delta z\right)$$

But the work done by an external force against the electric field is also given by $dW = -q\mathbf{E}\cdot\Delta\mathbf{l}$, where $\Delta\mathbf{l}$ is the differential length element given by $\Delta\mathbf{l} = (\hat{\mathbf{x}}\Delta x + \hat{\mathbf{y}}\Delta y + \hat{\mathbf{z}}\Delta z)$. Thus we have

$$\frac{\partial\Phi}{\partial x}\Delta x + \frac{\partial\Phi}{\partial y}\Delta y + \frac{\partial\Phi}{\partial z}\Delta z = -\mathbf{E}\cdot\Delta\mathbf{l}$$

$$\left(\hat{\mathbf{x}}\frac{\partial\Phi}{\partial x} + \hat{\mathbf{y}}\frac{\partial\Phi}{\partial y} + \hat{\mathbf{z}}\frac{\partial\Phi}{\partial z}\right)\cdot(\hat{\mathbf{x}}\Delta x + \hat{\mathbf{y}}\Delta y + \hat{\mathbf{z}}\Delta z) = -\mathbf{E}\cdot\Delta\mathbf{l}$$

$$\left(\hat{\mathbf{x}}\frac{\partial\Phi}{\partial x} + \hat{\mathbf{y}}\frac{\partial\Phi}{\partial y} + \hat{\mathbf{z}}\frac{\partial\Phi}{\partial z}\right)\cdot\Delta\mathbf{l} = -\mathbf{E}\cdot\Delta\mathbf{l}$$

from which it follows that

$$\boxed{\mathbf{E} = -\left(\hat{\mathbf{x}}\frac{\partial\Phi}{\partial x} + \hat{\mathbf{y}}\frac{\partial\Phi}{\partial y} + \hat{\mathbf{z}}\frac{\partial\Phi}{\partial z}\right)} \qquad [4.20]$$

Equation [4.20] indicates that the electric field $\mathbf{E}$ at any given point will have its largest component in the direction opposite to that along which the spatial rate of change of electrostatic potential is the largest. In other words, the electric field at any given point is the negative *gradient*[51] of the electrostatic potential at that point.

It is often customary to represent the gradient of a scalar by using the *del* operator, sometimes also called the *grad* or *nabla* operator, represented by the symbol ∇. For this purpose, the del operator is defined in rectangular coordinates as

$$\boxed{\nabla \equiv \hat{\mathbf{x}}\frac{\partial}{\partial x} + \hat{\mathbf{y}}\frac{\partial}{\partial y} + \hat{\mathbf{z}}\frac{\partial}{\partial z}} \qquad [4.21]$$

so [4.20] can be compactly written as

$$\boxed{\mathbf{E} = -\nabla\Phi} \qquad [4.22]$$

[51] The gradient of a scalar function at any given point is the maximum spatial rate of change (i.e., the steepest slope) of that function at that point. Gradient is thus naturally a vector quantity, because the maximum rate of change with distance must occur in a given direction. The direction of the gradient vector is that in which the scalar function (e.g., temperature) changes most rapidly with distance. A good analogy here is the gravitational potential. A marble placed on the slope of a mountain rolls down (i.e., its velocity vector is oriented) in the direction opposite to the gradient of the gravitational potential, and its speed (i.e., the magnitude of its velocity) is determined by the spatial derivative in that direction.

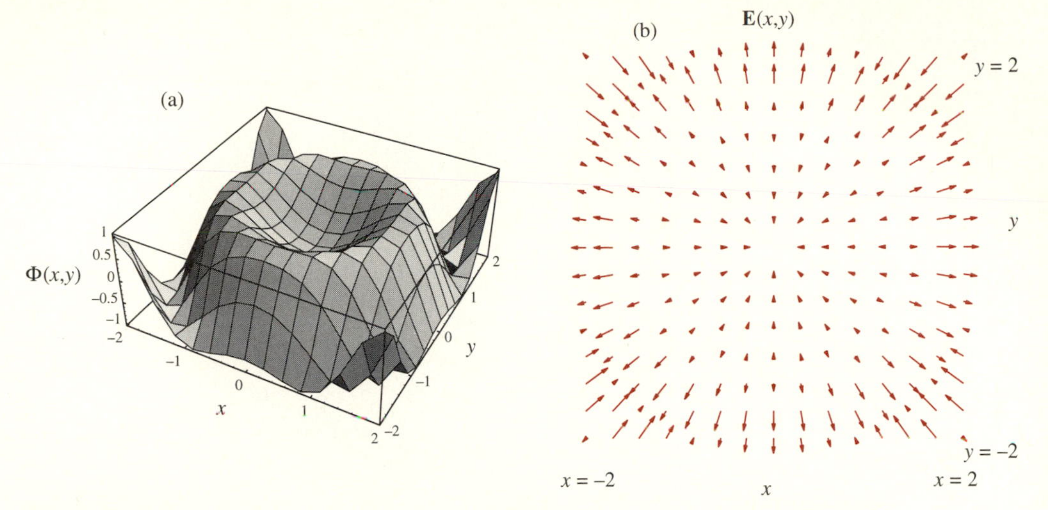

FIGURE 4.20. An example of a two-dimensional potential function and its gradient field. (a) The function shown is $\Phi(x, y) = \sin(x^2 + y^2)$. (b) The corresponding electric field is $\mathbf{E} = -\nabla\Phi = -\hat{\mathbf{x}}2x\cos(x^2 + y^2) - \hat{\mathbf{y}}2y\cos(x^2 + y^2)$. The units for the x and y axes are in radians for both panels.

As an example of gradients of potential functions that vary in two dimensions, consider the function shown in Figure 4.20. The function itself is illustrated using a three-dimensional surface plot, whereas the corresponding negative gradient (i.e., the electric field) is shown by means of a vector field plot. The plots illustrate that the electric field magnitudes (i.e., size of the arrows) are highest in regions of highest spatial slope of the potential functions and that the electric field has no components in directions along which there is no potential variation.

In problems for which a system of charges is specified and where it is necessary to determine the resultant electric field due to the charges, it is often simpler first to find the electrostatic potential $\Phi(x, y, z)$ and then determine $\mathbf{E}(x, y, z)$ by finding the negative gradient of the potential. This roundabout way of finding $\mathbf{E}$ is simpler because the electric field is a vector quantity, whereas Φ is a scalar and can be found as an algebraic (rather than vector) sum of the potentials due to each system of charge. In simple problems, there may be little advantage in using the potential method; however, in more complicated problems, the use of the scalar potential results in real simplification, as in the case of the electric dipole discussed in Section 4.4.3.

Gradient in Other Coordinate Systems The preceding definition of the concept of the gradient of scalar potential in terms of the motion of a test charge in the presence of an electric field was presented in a rectangular coordinate system. We now consider gradient in other coordinate systems (see Appendix A).

If we consider the motion of the test charge in a cylindrical (r, ϕ, z) coordinate system, we realize that in moving by a differential amount $d\phi$ in the ϕ direction, we span a distance of $r\,d\phi$ (see Figure 4.21a), so the E_ϕ is $r^{-1}\,\partial\Phi/\partial\phi$, rather than

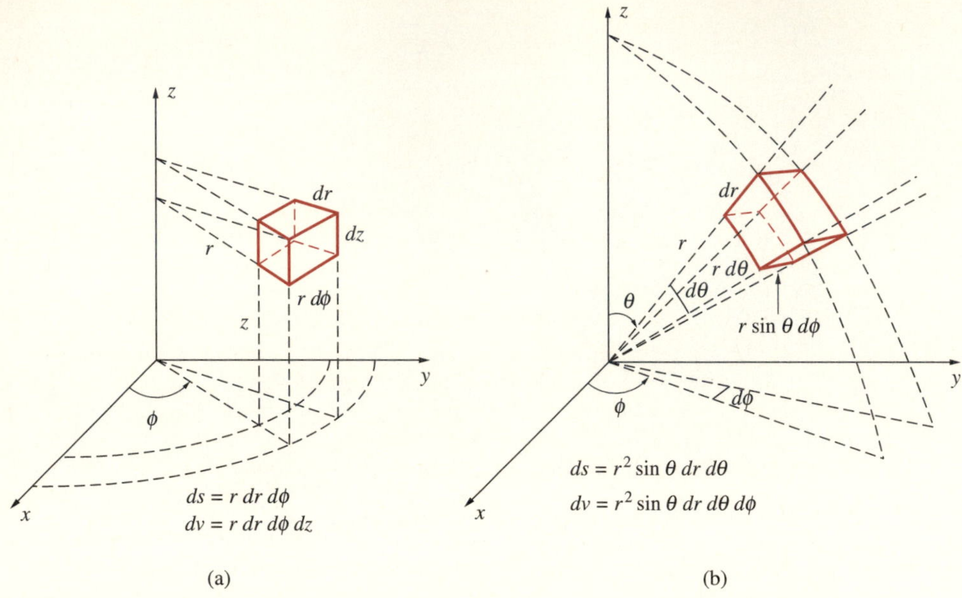

FIGURE 4.21. Gradient in other coordinate systems. (a) Cylindrical coordinate system. (b) Spherical coordinate system.

just $\partial\Phi/\partial\phi$. Motion in the r and z directions spans differential distances of dr and dz, so that we have

$$\mathbf{E} = -\nabla\Phi = -\left[\hat{\mathbf{r}}\frac{\partial\Phi}{\partial r} + \hat{\boldsymbol{\phi}}\frac{1}{r}\frac{\partial\Phi}{\partial\phi} + \hat{\mathbf{z}}\frac{\partial\Phi}{\partial z}\right] \quad [4.23]$$

Similar considerations for a spherical coordinate system (r, θ, ϕ) indicate that in moving by an amount $d\theta$ in latitude we span a distance of $r\,d\theta$, whereas the distance we span in moving by an amount $d\phi$ in azimuth (longitude) depends on the latitude coordinate θ and is $r \sin\theta\,d\phi$, as shown in Figure 4.21b. Accordingly we have

$$\mathbf{E} = -\nabla\Phi = -\left[\hat{\mathbf{r}}\frac{\partial\Phi}{\partial r} + \hat{\boldsymbol{\theta}}\frac{1}{r}\frac{\partial\Phi}{\partial\theta} + \hat{\boldsymbol{\phi}}\frac{1}{r\sin\theta}\frac{\partial\Phi}{\partial\phi}\right] \quad [4.24]$$

Equations [4.23] and [4.24] also indicate that the definition of the del operator ∇ as given in [4.21] is valid only for a rectangular coordinate system. Example 4-9 illustrates the use of [4.24] for the determination of gradient in a spherical coordinate system.

Example 4-9: Gradient in spherical coordinates. The potential distribution in the vicinity of a metal sphere placed in a uniform electric field is

$$\Phi(r, \theta) = E_0\left[1 - \left(\frac{a}{r}\right)^3\right] r\cos\theta$$

Find the corresponding electric field.

Solution: Using [4.24] we have

$$\mathbf{E} = -\nabla\Phi = -\frac{\partial\Phi}{\partial r}\hat{\mathbf{r}} - \frac{1}{r}\frac{\partial\Phi}{\partial\theta}\hat{\boldsymbol{\theta}} = -\hat{\mathbf{r}}E_0\left[1 + 2\left(\frac{a}{r}\right)^3\right]\cos\theta + \hat{\boldsymbol{\theta}}E_0\left[1 - \left(\frac{a}{r}\right)^3\right]\sin\theta$$

Note that the component of the electric field in the θ direction is generated as a result of the θ dependence of $\Phi(r, \theta)$, which has the simple $\cos\theta$ form, thus leading to a $\sin\theta$ dependence of the θ component of $\mathbf{E}$.

Graphical Representation of Potential: Equipotential Lines Just as it is convenient to graph and think of electric fields in terms of field lines, it is useful to visualize and graph scalar electric potential fields. The easiest way to represent potential is to draw surfaces on which Φ is a constant. Such surfaces are called *equipotentials.* Equipotentials for single and two point charges are shown in Figures 4.22 and 4.23, together with the electric field lines. We note from these figures that the equipotentials are at right angles to the field lines. This is a general property of equipotentials and field lines that follows from the fact that the electric field is the negative gradient of the potential. The gradient vector is in the direction of the most rapid change of potential and is therefore perpendicular to the equipotential surface. Another way of looking at it is to consider the fact that the potential is by definition constant over an equipotential surface, so any movement of charge over such a surface requires no work. Thus, equipotential surfaces must always be orthogonal to the electric field. Equipotential lines are analogous to contour elevation lines on maps, which connect the points of equal elevation.

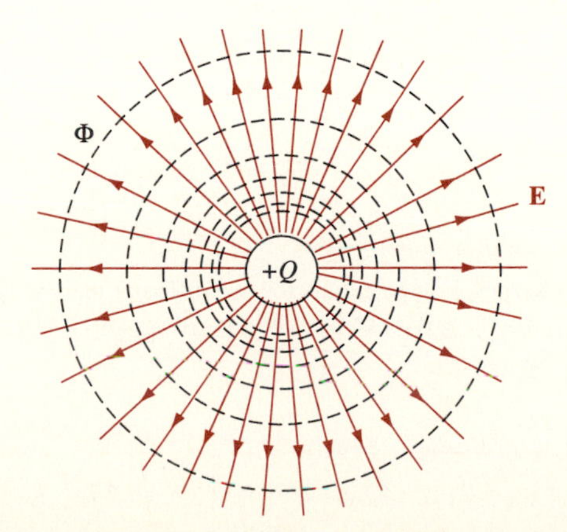

FIGURE 4.22. Equipotential lines. The dashed lines represent equipotential lines for a single positive point charge at the origin.

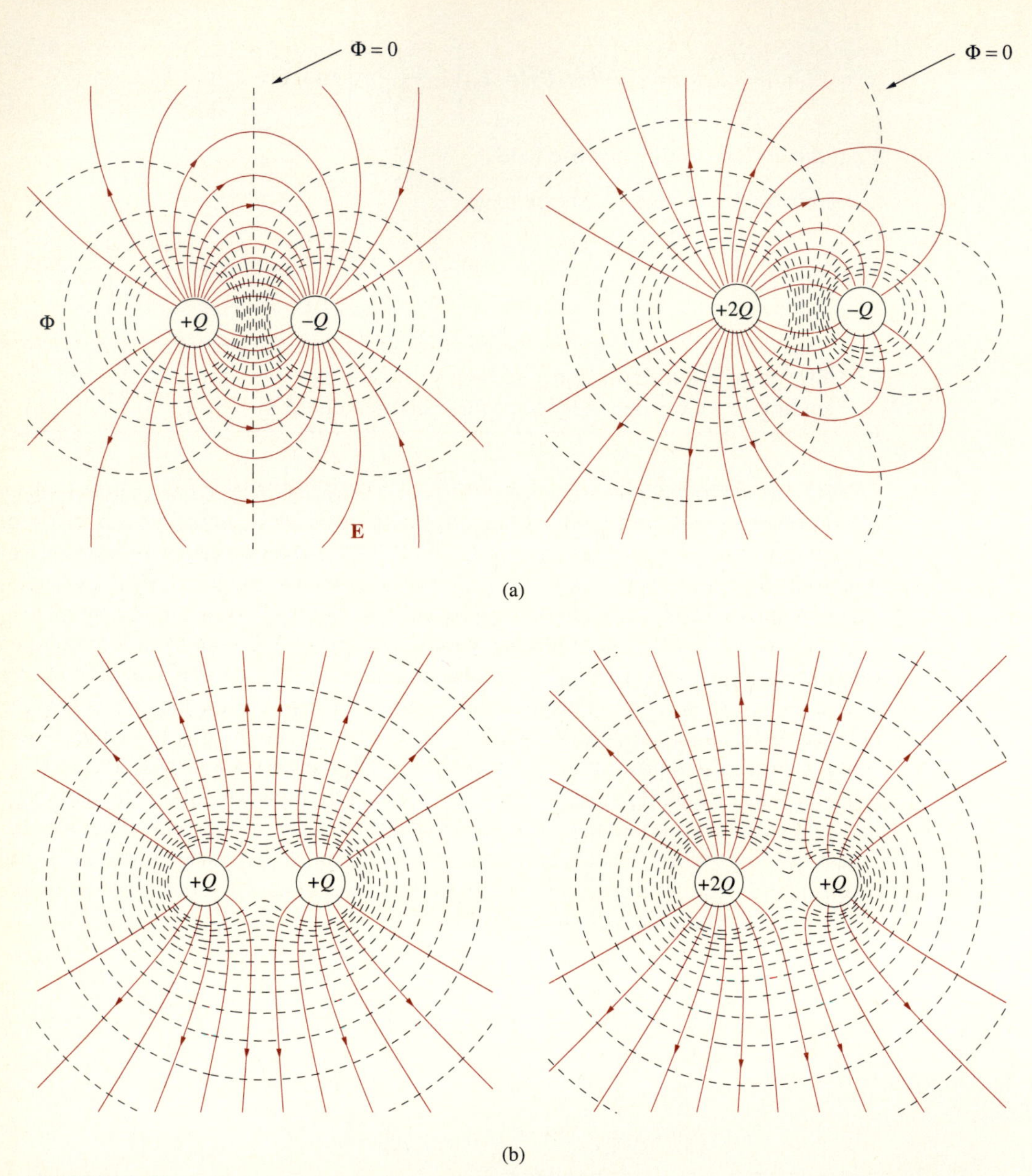

FIGURE 4.23. **Equipotentials for two point charges.** Equipotentials for two different pairs of charges are shown, together with the electric field lines for comparison. (a) Two charges of opposite polarity. (b) Two charges of the same polarity.

Equations that describe the equipotential surfaces can be derived simply by setting the total electrostatic potential to a constant value. For a single point charge Q located at the origin, the equation for the equipotential surfaces is

$$\frac{Q}{4\pi\epsilon_0 r} = \frac{Q}{4\pi\epsilon_0\sqrt{x^2 + y^2 + z^2}} = C_1 \quad \rightarrow \quad x^2 + y^2 + z^2 = C_2$$

where C_1 and C_2 are constants. At the plane defined by $z = 0$, the equipotential surfaces are described by $x^2 + y^2 = C_2$ and are circles, as shown in Figure 4.22.

For the case of two point charges Q_1 and Q_2 located respectively at points $(a, 0)$ and $(-a, 0)$, the superposition principle applies, so the equipotential surfaces are the set of curves corresponding to

$$\frac{Q_2}{r_2} + \frac{Q_1}{r_1} = \text{const.}$$

where r_1 and r_2 are respectively the distances to the point P located on the equipotential surface from the charges Q_1 and Q_2. The corresponding plots for $Q_2 = -Q_1 = Q$ and $Q_2 = -2Q_1 = 2Q$ are shown in Figure 4.23a. The orthogonality of the electric field lines and the equipotentials is clearly evident. For the case of charges of opposite sign (Figure 4.23a), note the clustering of the equipotentials at a point between the two charges where the polarity of the potential changes sign. The separating equipotential line corresponding to zero potential is indicated.

For the case of two equal positive charges of $+Q$ each, we note that the outermost equipotentials begin to look nearly circular. This is to be expected, since at far enough distances from the charges the field should be the same as that for a single point charge of $+2Q$ (left-hand panel of Figure 4.23b). The same is true for the $+2Q$ and $+Q$ pair shown in the right-hand panel of Figure 4.23b.

4.4.3 Electrostatic Potential Resulting from Multiple Point Charges

The electric potential at point $\mathbf{r}$ in a system of n discrete point charges $Q_1, Q_2, \ldots, Q_n$ located at points $\mathbf{r}'_1, \mathbf{r}'_2, \ldots, \mathbf{r}'_n$ is, by superposition, the sum of potentials resulting from the individual charges:

$$\boxed{\Phi = \frac{1}{4\pi\epsilon_0}\sum_{k=1}^{n}\frac{Q_k}{|\mathbf{r} - \mathbf{r}'_k|}} \qquad [4.25]$$

The most important example of a multiple charge distribution is the electrostatic dipole, considered next.

The Electric Dipole As an example, we consider the electrostatic dipole shown in Figure 4.24a, consisting of a pair of equal and opposite point charges $+Q$ and $-Q$ with a small separation d centered at the origin of the coordinate system. The distances from the charges to a point P are designated as r_+ and r_-, respectively.

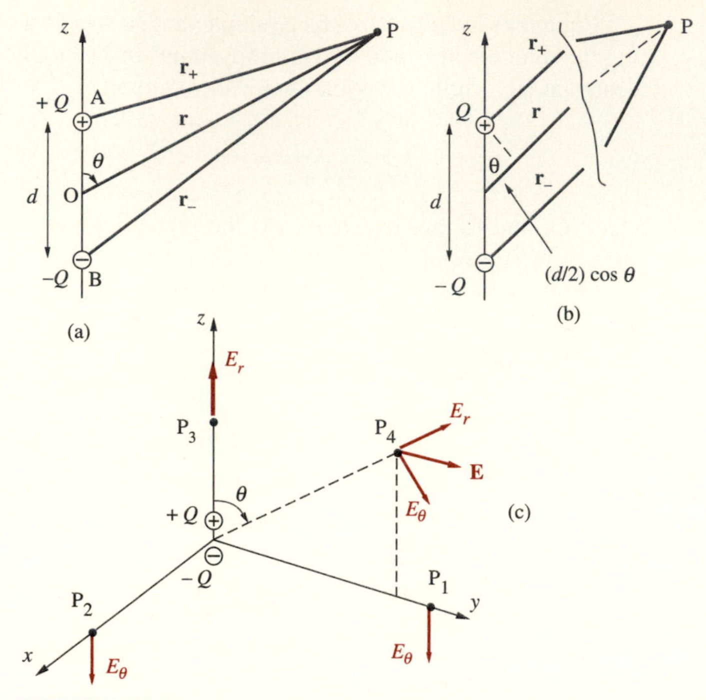

FIGURE 4.24. **Electric dipole.** The geometry of the problem of an electric dipole where $r \gg d$. (a) The general case of a point P at a distance r. (b) Approximate geometry when $r \gg d$. (c) The orientation of the electric field at different points P_i, all at large distances (i.e., $r \gg d$).

The potential at point P is obtained by superposing the potentials at P due to the individual charges, namely,

$$\Phi = \frac{Q}{4\pi\epsilon_0 r_+} - \frac{Q}{4\pi\epsilon_0 r_-} = \frac{Q}{4\pi\epsilon_0}\left(\frac{1}{r_+} - \frac{1}{r_-}\right)$$

where the distances r_+ and r_- follow from applying the law of cosines to the triangles AOP and BOP and are

$$r_\pm = \sqrt{r^2 + (d/2)^2 \mp rd\cos\theta} = r\sqrt{1 + (d/2r)^2 \mp (d/r)\cos\theta}$$

We are interested in the special case in which the charges are very close to one another so that $d \ll r_+, r_-$; in other words, the potential and electric field are observed at distances far away from the charges. Such closely spaced pairs of "dipole" charges are encountered often in physics and engineering. The fields produced by an electric dipole antenna can often be approximated as two charges separated by a small distance. More importantly, behavior of materials under the influence of electric fields can often be understood in terms of atomic dipoles established due to the relative displacement of electrons with respect to their nuclei, as discussed in Section 4.10.

The equation for the potential due to an electric dipole can be simplified further for the case $r \gg d$ (i.e., at a faraway point P). To obtain the simplified equation, we expand r_+^{-1} and r_-^{-1} expressions in a Taylor series in terms of d/r and neglect the terms of order $(d/r)^2$ and higher.[52] The Taylor series approximation is

$$\frac{1}{\sqrt{1+u}} = 1 - \frac{1}{2}u + \frac{3}{8}u^2 \cdots \simeq 1 - \frac{1}{2}u$$

as $u = \mp d/r \rightarrow 0$. Consequently, the r_+^{-1} and r_-^{-1} expressions are

$$\frac{1}{r_\pm} \simeq \frac{1}{r}\left(1 \pm \frac{d\cos\theta}{2r}\right)$$

Thus, the potential is

$$\Phi(r,\theta) \simeq \frac{Q}{4\pi\epsilon_0 r}\left[\left(1 + \frac{d\cos\theta}{2r}\right) - \left(1 - \frac{d\cos\theta}{2r}\right)\right] = \frac{Qd\cos\theta}{4\pi\epsilon_0 r^2}$$

It is useful to write the potential in terms of an electric dipole moment $\mathbf{p} \equiv (Qd)\hat{\mathbf{z}}$, and noting that $\hat{\mathbf{z}} \cdot \hat{\mathbf{r}} = \cos\theta$,

$$\boxed{\Phi = \frac{1}{4\pi\epsilon_0}\frac{\mathbf{p}\cdot\hat{\mathbf{r}}}{r^2} = \frac{Qd\cos\theta}{4\pi\epsilon_0 r^2}} \quad [4.26]$$

To find the electric field, we note that Φ does not have any ϕ dependence, so we have

$$\begin{aligned}\mathbf{E} &= -\nabla\Phi \\ &= -\left[\hat{\mathbf{r}}\frac{\partial\Phi}{\partial r} + \hat{\boldsymbol{\theta}}\frac{1}{r}\frac{\partial\Phi}{\partial\theta}\right] \\ &= -\left[-\hat{\mathbf{r}}\frac{p\cos\theta}{2\pi\epsilon_0 r^3} - \hat{\boldsymbol{\theta}}\frac{p\sin\theta}{4\pi\epsilon_0 r^3}\right]\end{aligned}$$

where $p = |p| = Qd$. This yields

$$\boxed{\mathbf{E} = \frac{p}{4\pi\epsilon_0 r^3}[\hat{\mathbf{r}}2\cos\theta + \hat{\boldsymbol{\theta}}\sin\theta]} \quad [4.27]$$

as the electric field of an electric dipole at a distance r. We note that the above result is valid only at distances far from the dipole, since the approximation $r \gg d$ was

---

[52]Note that the above result can be obtained in an even simpler manner by using the approximation illustrated in Figure 4.24b. For $r \gg d$, we have from Figure 4.24b that $r_+r_- \simeq r^2$ and $r_+ - r_- \simeq -d\cos\theta$, so that

$$\Phi = \frac{Q}{4\pi\epsilon_0 r_+} - \frac{Q}{4\pi\epsilon_0 r_-} = \frac{Q}{4\pi\epsilon_0}\frac{r_- - r_+}{r_+r_-} \simeq \frac{Q}{4\pi\epsilon_0}\frac{d\cos\theta}{r^2}$$

However, by keeping higher-order terms in d/r, it is possible to determine the value of the potential nearer to the dipole.

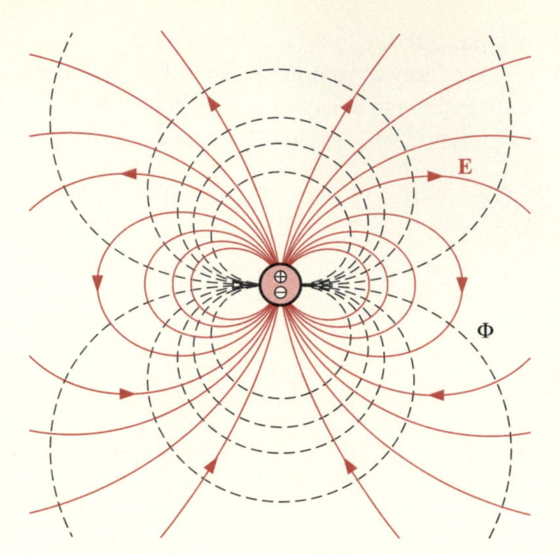

FIGURE 4.25. Electric field lines and equipotentials for a dipole. The dipole expressions [4.26] and [4.27] are not valid close to the charges, i.e., in the shaded region at the center. In that region, the electric field configuration is as shown in Figure 4.23a.

used in deriving the potential. Far from the dipole the electric field varies as r^{-3}, in contrast with the r^{-2} dependence for the field of a point charge and the r^{-1} variation of the field of a line charge (Example 4-5).

A two-dimensional sketch of the electric field lines and equipotential lines for an electric dipole is shown in Figure 4.25. Note that close to the charges the dipole approximations [4.26] and [4.27] are not valid and the field line configuration is equal to that for two point charges, shown in Figure 4.23a.

4.4.4 Electrostatic Potential Resulting from Continuous Charge Distributions

Since potential is a scalar quantity, its calculation for a given charge configuration simply involves an algebraic sum of the contributions of *all* charges in the system, whether they are point, line, surface, or volume charges. In general, we have

$$\Phi = \frac{1}{4\pi\epsilon_0}\left[\sum_{k=1}^{n}\frac{Q_k}{|\mathbf{r}-\mathbf{r}'_k|} + \int_{L'}\frac{\rho_l(\mathbf{r}')\,dl'}{|\mathbf{r}-\mathbf{r}'|} + \int_{S'}\frac{\rho_s(\mathbf{r}')\,ds'}{|\mathbf{r}-\mathbf{r}'|} + \int_{V'}\frac{\rho(\mathbf{r}')\,dv'}{|\mathbf{r}-\mathbf{r}'|}\right] \qquad [4.28]$$

Examples 4-10 and 4-11 illustrate the application of [4.28] to determine electrostatic potential in the vicinity of given charge distributions.

Example 4-10: Disk of charge. For a uniformly charged disk of radius a, and surface charge density ρ_s, as illustrated in Figure 4.26, determine the potential Φ and the electric field $\mathbf{E}$ at a point P along its axis.

Solution: We work in the cylindrical coordinate system (with the disk on the xy plane and centered at the origin). Consider an annular ring of radius r' and

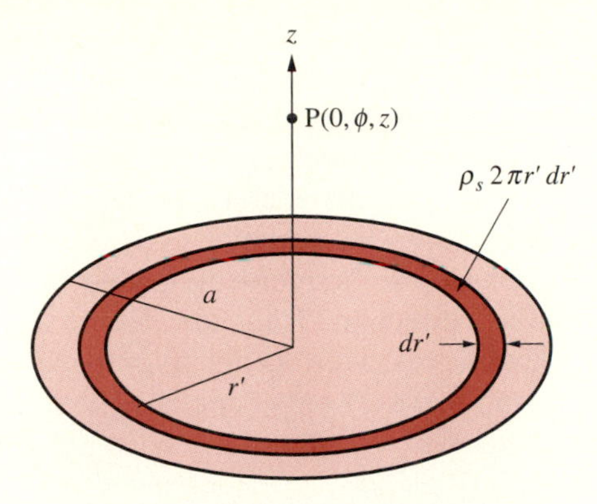

FIGURE 4.26. Disk of charge. Figure for Example 4-10.

differential width dr' as shown in Figure 4.26. The contribution of this ring to the electrostatic potential at point P along the z axis is

$$d\Phi_{\text{P}}(r = 0) = \frac{\rho_s 2\pi r'\,dr'}{4\pi\epsilon_0\sqrt{z^2 + (r')^2}}$$

where the 2π term comes from an integration along the ring (i.e., ϕ varying from 0 to 2π). If we now integrate the foregoing expression over the surface of the entire disk, we find

$$\Phi(r = 0) = \frac{\rho_s}{2\epsilon_0}\int_0^a \frac{r'\,dr'}{\sqrt{z^2 + (r')^2}} + C$$

$$= \frac{\rho_s}{2\epsilon_0}[(a^2 + z^2)^{1/2} - |z|] + C$$

where C is a constant and we retain $|z| = \sqrt{z^2}$ in order to ensure that Φ decreases[53] as $z \to \pm\infty$ (or $|z| \to \infty$) and to ensure that $\Phi(z) = \Phi(-z)$, as is necessary in view of the symmetry in z. Since we view electrostatic potential at a point in reference to infinity, we need to have $C = 0$ so that $\Phi \to 0$ as $z \to \infty$. The electrostatic potential at point P is thus

$$\Phi(r = 0) = \frac{\rho_s}{2\epsilon_0}[\sqrt{a^2 + z^2} - |z|]$$

Note that for $z \gg a$ we can make the approximation $\sqrt{a^2 + z^2} \simeq z + a^2/(2z)$, so that we have

$$\Phi(r = 0) \simeq \frac{\rho_s}{2\epsilon_0}\left(z + \frac{a^2}{2z} - z\right) = \frac{(\rho_s\pi a^2)}{4\pi\epsilon_0 z} = \frac{1}{4\pi\epsilon_0}\frac{Q}{z} \qquad z > 0$$

where $Q = \rho_s\pi a^2$ is the total charge on the disk. For $z < 0$ we would have $\Phi \simeq -Q/(4\pi\epsilon_0 z)$, since for the positively charged disk (i.e., $\rho_s > 0$) the potential Φ

[53] At large distances from the disk the electrostatic potential should vary like that of a point charge.

must be positive. To see this, we need to remember that electrostatic potential at a given point is the energy required to bring a unit positive test charge from infinity to that point. We note that, as expected, at large distances the potential due to the disk charge Q behaves like that of a point charge.

To find the electric field, we must in general take the negative gradient of Φ, which means that we need to know the variation of Φ as a function of r, ϕ, z. However, because of the symmetry of the problem, the electric field at a point on the axis can only be in the z direction. Thus, the electric field at point P is related only to $\partial\Phi/\partial z$. In other words, even though Φ in the vicinity of point P may in general depend on r and ϕ, we are only concerned with its z dependence as determined above. Thus we have

$$[E_z]_{\rm P} = -\frac{d\Phi}{dz} = \begin{cases} \dfrac{\rho_s}{2\epsilon_0}[1 - z(a^2 + z^2)^{-1/2}] & \text{for } z > 0 \\ -\dfrac{\rho_s}{2\epsilon_0}[1 + z(a^2 + z^2)^{-1/2}] & \text{for } z < 0 \end{cases}$$

Note that $E_z(z) = -E_z(-z)$, as is expected on physical grounds.

Example 4-11: Finite-length line charge. In Example 4-5, we found the electric field at a point equidistant from the end points of a thin cylindrical rod charge, which had a component only in the r direction due to the symmetry involved. For points off the central axis, such as point P(r, ϕ, z) as shown in Figure 4.27, it is clear

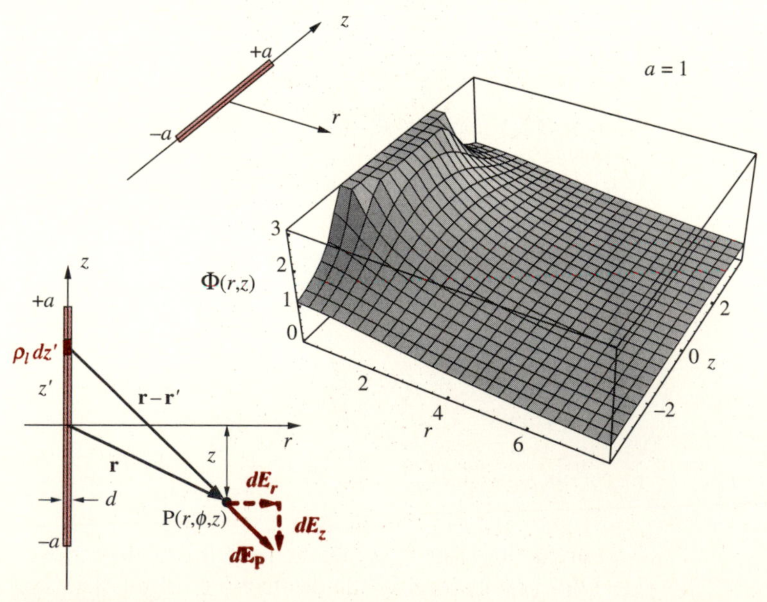

FIGURE 4.27. Electrostatic potential in the vicinity of a straight rod of charge. The amplitude range of the display is limited to 3 units, since the potential becomes arbitrarily large as $r \to 0$.

that the field would have both E_r and E_z components. This example illustrates the usefulness of the electrostatic potential concept by determining Φ at P via direct scalar integration and subsequently determining $\mathbf{E}$ from $\mathbf{E} = -\nabla\Phi$. Assume that $a \gg d$, so that the rod of charge can be represented by a line charge density ρ_l.

Solution: As shown in Figure 4.27, we consider an elemental charge $dQ = \rho_l\, dz'$ located at $(0, 0, z')$. The distance between this source element and point P is $|\mathbf{r} - \mathbf{r}'| = \sqrt{r^2 + (z - z')^2}$. The total potential Φ at P is then given by

$$\Phi = \int_{-a}^{+a} \frac{\rho_l\, dz'}{4\pi\epsilon_0[r^2 + (z - z')^2]^{1/2}}$$

$$= -\frac{\rho_l}{4\pi\epsilon_0} \ln\left(\frac{z - a + [r^2 + (z - a)^2]^{1/2}}{z + a + [r^2 + (z + a)^2]^{1/2}}\right)$$

A three-dimensional plot of $\Phi(r, z)$ is shown in Figure 4.27. For the purpose of this display, the normalized value of the electrostatic potential has been limited to 3 units, since for $-a \leq z \leq a$, the potential increases without limit as $r \to 0$.

The electric field components can now be found from $\mathbf{E} = -\nabla\Phi$; since Φ depends only on z and r, only E_r and E_z components exist and are given by

$$E_z = -\frac{\partial\Phi}{\partial z} = \frac{\rho_l}{4\pi\epsilon_0}\left(\frac{1}{[r^2 + (z - a)^2]^{1/2}} - \frac{1}{[r^2 + (z + a)^2]^{1/2}}\right)$$

$$E_r = -\frac{\partial\Phi}{\partial r} = -\frac{\rho_l}{4\pi\epsilon_0}\left(\frac{z - a}{[r^2 + (z - a)^2]^{1/2}} - \frac{z + a}{[r^2 + (z + a)^2]^{1/2}}\right)$$

It is useful to check the behavior of the field and potential in limiting cases. For $r \gg a$ or $z \gg a$, the potential approaches that of a point charge of $Q = \rho_l(2a)$, namely,

$$\lim_{r\to\infty} \Phi = \frac{\rho_l(2a)}{4\pi\epsilon_0 r} \quad \text{and} \quad \lim_{z\to\infty} \Phi = \frac{\rho_l(2a)}{4\pi\epsilon_0 z}$$

For $z = 0$, the solution boils down to that worked out in Example 4-5. For $r = 0$, and at points $|z| > a$, we have $E_r = 0$ and

$$E_z = \frac{\rho_l}{4\pi\epsilon_0}\left(\frac{1}{|z - a|} - \frac{1}{|z + a|}\right) = \frac{\pm\rho_l a}{2\pi\epsilon_0(z^2 - a^2)}$$

where the "+" and "−" respectively correspond to the cases of $z > a$ and $z < -a$.

The result in Figure 4.27 serves to give a better understanding of the relationship between the electric field and the electrostatic potential. Since $\mathbf{E} = -\nabla\Phi$, the direction of the electric field can be easily deduced from the three-dimensional surface plots of the potential. An analogy with gravitational potential is particularly useful here; if the quantity plotted, say, in Figure 4.27, were the gravitational potential, its gradient at any point is in the direction opposite to that in which a marble placed at that point rolls down (note that the top of the surface in Figure 4.27 looks flat only because of the limited amplitude range of the display). The electric field at any point specified by its r and z coordinates thus points in the direction of the most rapid change of slope of the potential distribution. The magnitude of the electric field is higher at points where the slope is higher—that is, closer to the disk charge.

4.5 ELECTRIC FLUX AND GAUSS'S LAW

The electric field line plots that we have seen suggest some sort of "flow," or *flux,* of electric energy emanating from positive charges and terminating at negative charges. If the field lines represented the velocity of fluid flow, we could indeed think of flux of fluid along the field lines. Although, in fact, electric field lines do not represent the flow of anything material, it is helpful to think of them as describing the flux of *something* that, like a fluid, is conserved and that emanates from the charges into the surrounding space.[54] In such a picture, each field line represents a unit of flux of that *something,* from each unit of positive charge to a unit of negative charge.

4.5.1 Electric Displacement and Flux Density D

Michael Faraday's experiments, carried out in 1837, well after Coulomb, and many years after his own work on induced electromotive force,[55] have established our present concepts of electric displacement and flux density. A schematic description of Faraday's experiments is shown in Figure 4.28.

Basically, Faraday used two concentric spheres. First, the inside sphere was charged by a known quantity of electricity (say, Q); then a larger outer sphere (uncharged) was placed around it[56] with a dielectric (insulating material) in between. The outer sphere was then grounded[57] momentarily; then the two spheres were separated, and the charge remaining on the outer sphere was measured. Faraday found that this charge was equal in magnitude and opposite in sign to that on the inner sphere, *regardless of the size of the sphere* and *for all types of dielectric materials*

[54]One possibility is to think of small "bullets" shot out of charges and to require that none of the bullets could disappear once they were produced; however, one needs to be very careful in carrying such a model any further than simply a thought exercise [R. P. Feynman, R. B. Leighton, and M. Sands, *The Feynman Lectures on Physics,* Addison Wesley, Reading, Massachusetts, 1964].

[55]See Chapter 7 for discussion of Faraday's law.

[56]Actually the outer sphere consisted of two hemispheres that could be firmly clamped together. Faraday prepared shells of dielectric material to occupy the entire volume between the spheres.

[57]In other words, connected to the ground, so that it can access as much charge as it required. In this context, 'ground" is an uncharged object at zero potential that has essentially an infinite number of electrons (and positive nuclei) and that can accept or provide any amount of excess charge from or to any object connected to it.

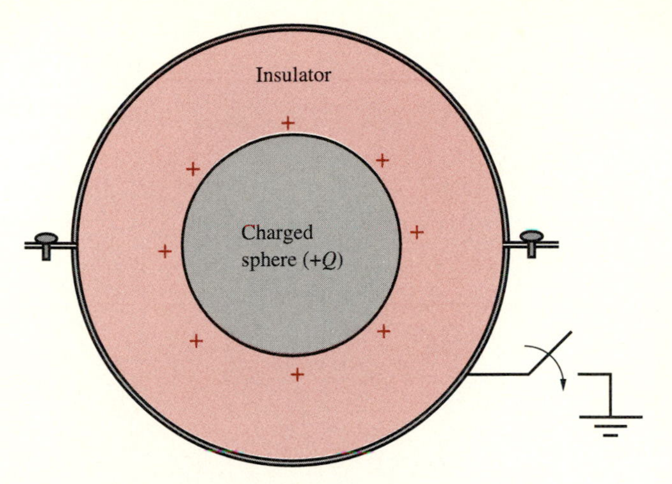

FIGURE 4.28. Faraday's experiments on electric displacement. The apparatus used in Faraday's experiments consisted of two concentric spheres separated by an insulating material.

filling the space between the spheres. He thus concluded that there was a sort of "displacement" from the charge on the inner sphere through the insulator to the outer sphere, the amount of this displacement depending only on the magnitude of the charge Q. This displacement is the closest we can come to identifying the *something* that must flow along the field lines, and we adopt this notion as our definition of the *electric flux.* In SI units, the electric displacement (or electric flux) is equal in magnitude to the charge that produces it; namely, it has units of coulombs (C).

Consider now the case of an isolated point charge as shown in Figure 4.29a; all other bodies, such as the outer sphere of Figure 4.28, can be considered to be at infinity. The *density* of electric displacement, or *flux density* vector **D**, at any point on a spherical surface S of radius r centered at the isolated point charge Q is defined as

$$\mathbf{D} \equiv \hat{\mathbf{r}}\frac{Q}{4\pi r^2} \quad \text{C-m}^{-2} \qquad [4.29]$$

Note that the electric displacement per unit area, **D**, depends on the orientation of the area; hence, it is a vector quantity. In simple materials, we can write

$$\boxed{\mathbf{D} = \epsilon\mathbf{E}} \qquad [4.30]$$

with the dielectric constant ϵ being a simple constant[58] so that the vector **D** is in the same direction as the electric field **E**. Note, however, that **D** is in units of C-m^{-2},

[58]We discuss dielectric materials and their behavior under applied electric fields in Section 4.10. At this point, it suffices to note that, in general, the relationship between **D** and **E** can be quite complex. In addition to the possibility of the value of ϵ not being a constant and depending instead on the magnitude of the electric field ($|\mathbf{E}|$), x, y, z, and t, in so-called *anisotropic* materials ϵ might also depend on the *direction* of the applied field; for example, an electric field in the x direction (i.e, E_x) may produce an electric flux density in the y direction (i.e., D_y).

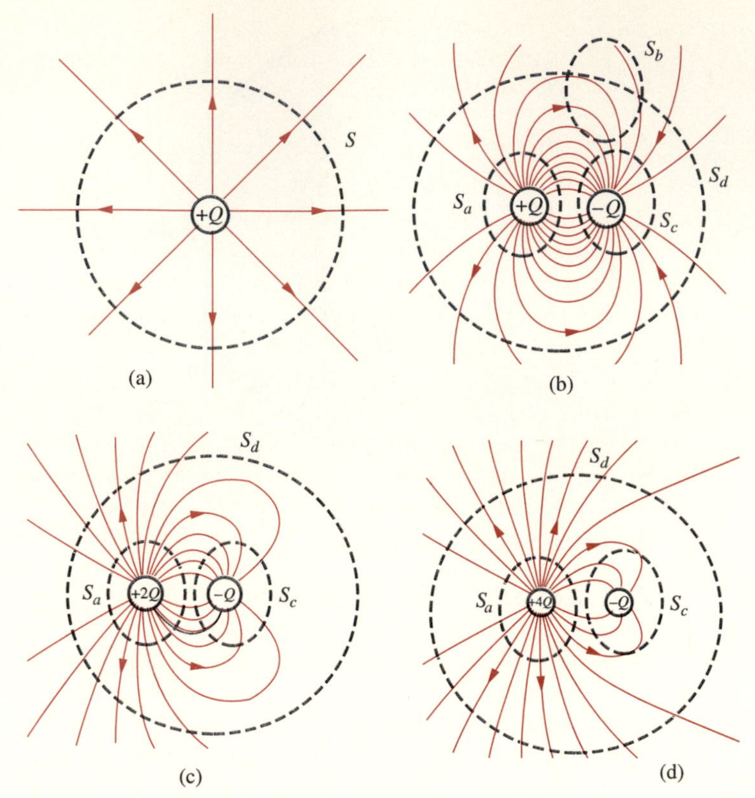

FIGURE 4.29. **Electric field lines around point charges.** (a) A positive point charge, (b) two opposite equal point charges, (c) two point charges $+2Q$ and $-Q$, and (d) two point charges $+4Q$ and $-Q$.

while **E** is in V-m^{-1}. In view of the simple proportionality of **D** and **E**, our previous plots showing electric field lines also represent electric flux lines. In the following, we confine ourselves to the consideration of simple materials with constant ϵ unless otherwise stated. In fact, until dielectric materials are discussed in a later section, we exclusively consider free space, in which **D** and **E** differ through the constant ϵ_0, (i.e., $\mathbf{D} = \epsilon_0 \mathbf{E}$).

Thinking of each unit of flux as a unit of charge enables us to have a deeper understanding of the distribution of electric field lines around charge configurations. Consider, for example, the flux lines around two opposite charges shown again in Figure 4.29b, and the closed surfaces S_a, S_b, S_c, and S_d. If a closed surface encloses a charge of $+Q$ (as is the case for S_a), all flux lines emanating from charge Q must go through it. Thus, by counting all the flux lines passing through this closed surface we should be able to know the amount of charges enclosed, independent of the properties of the material surrounding the charges. For this purpose, we would adopt the convention that lines outward (inward) represent positive (negative) enclosed charge.

Since equal numbers of lines enter and leave from surface S_b, the charge enclosed by it is zero, whereas surface S_c encloses negative charge, equal in magnitude to that enclosed by surface S_a, since *all* lines of flux coming out of surface S_a eventually enter surface S_c. Similarly, the large surface S_d encloses both charges $+Q$ and $-Q$, so equal numbers of field lines enter and leave, and the net total flux through it is zero, just as the net amount of charge enclosed in it is zero.

Consider the other two cases shown in Figure 4.29c,d. In the case of Figure 4.29c the positive charge is twice as large as the negative charge. Accordingly, twice as many (i.e., 24, versus 12 for the flux line density shown in Figure 4.29c) lines cross through surface S_a as do through surface S_c. Also, the large surface S_d now encloses a net charge of $2Q - Q = +Q$, so that it is crossed by the flux lines emanating from the $+2Q$ charge that do not connect back to the $-Q$ charge (i.e., 12 out of the total 24 lines[59] in Figure 4.29c), but instead terminate on charges located at infinity. We thus see that, indeed, the number of flux lines is an accurate measure of the amount of charge. Similar observations can be made in connection with Figure 4.29d, in which case the positive charge is four times as large as the negative charge; accordingly, four times as many flux lines (i.e., 24 in Figure 4.29d) cross through surface S_a as do through surface S_c (i.e., 6 in Figure 4.29d). The net number of outward flux lines (i.e., 18, Figure 4.29d) through S_d is determined by the net enclosed charge $(+4Q - Q = +3Q)$.

4.5.2 Gauss's Law

Consider now the region of space around a single positive point charge and an arbitrary closed surface S as shown in Figure 4.30a. If the electric field ($\mathbf{E}$) is indeed like a flow, the net flow out of this truncated conical box should be zero. Let us find the surface integral of the component of the electric flux density vector $\mathbf{D}$ normal to the surface. Using [4.29], we have $\mathbf{D} \cdot d\mathbf{s} = 0$ on the side faces, whereas on the radial faces we have $\mathbf{D} \cdot d\mathbf{s} = \pm|\mathbf{D}|ds$, plus on the outer surface S_b and minus on the inner surface S_a. Since the magnitude of the electric flux density $\mathbf{D}$ decreases as r^{-2}, and the surface areas of the radial faces increase as r^2, the total flow (or flux) inward through surface S_a completely cancels the flow outward though surface S_b. Since the total flow through the entire closed surface S is the integral of $\mathbf{D} \cdot d\mathbf{s}$ over S, we have $\oint_S \mathbf{D} \cdot d\mathbf{s} = 0$. Note that we would not have arrived at this conclusion if the electrostatic force law depended on distance in any other way than r^{-2}.

Considering Figure 4.30b, we conclude that the result we have just obtained is true even if the two end surfaces S_a and S_b were tilted with respect to the

[59] We note that the particular number of lines (e.g., 12) is determined entirely by the resolution of the field line plot. In the case of Figure 4.29 we have chosen a particular density of field lines that conveys the essential aspects of the field distributions. Too high a density of lines (e.g., 100 lines emanating from $+Q$ instead of 12) would clutter the figure, while too few lines (e.g., 2 instead of 12) would not be sufficient to convey the underlying symmetries and to illustrate the field intensity quantitatively in the regions surrounding the charges.

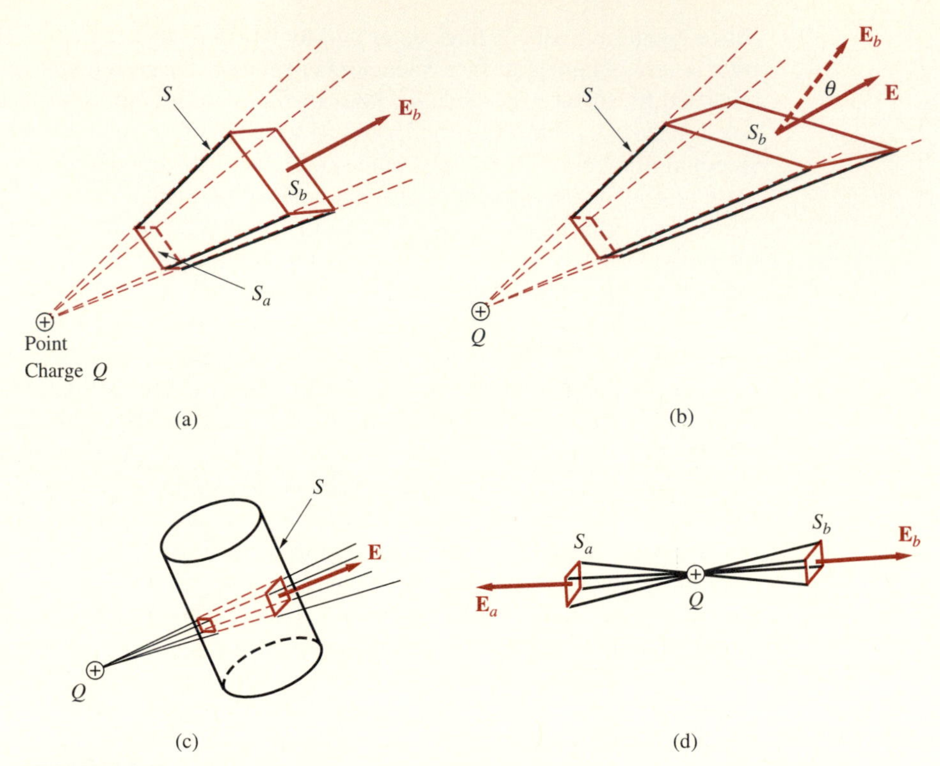

FIGURE 4.30. **The flux of E out of a surface is zero.** This figure is adapted with permission from R. P. Feynman, R. B. Leighton, and M. Sands, *The Feynman Lectures on Physics,* Addison Wesley, 1964. Copyright 1964 by California Institute of Technology.

radial distance. Note that when the surface is tilted at an angle θ, the area is increased by a factor $(\cos\theta)^{-1}$, but the component of $\mathbf{D}$ normal to the surface is decreased by the factor $\cos\theta$, so the product $\mathbf{D}\cdot d\mathbf{s}$ is unchanged. Thus, the net total electric flux out of the closed surface is still zero.

It is thus clear that the net flux out of *any* closed surface (that does not enclose any charge) must be zero, because, as shown in Figure 4.30c, any such surface can be constructed out of infinitesimal truncated cones such as those in Figure 4.30a,b. Now consider the surface shown in Figure 4.30d. We see that the magnitudes of the electric flux through the two surfaces S_a and S_b are once again equal (because the flux density decreases as r^{-2} but the area increases as r^2); however, they now have the same sign, so the net flux out of this surface is not zero.

To determine the net outward (or inward) flux from a surface enclosing a point charge, consider the general surface S shown in Figure 4.31, shown for simplicity as a two-dimensional cross section. Emanating from the point charge Q are cones of flux, which might cross the surface one or more times as shown. Also shown is another small spherical surface S' around the point charge Q. Note that the volume enclosed between S and S' has no enclosed charge, so that the total flux emanating

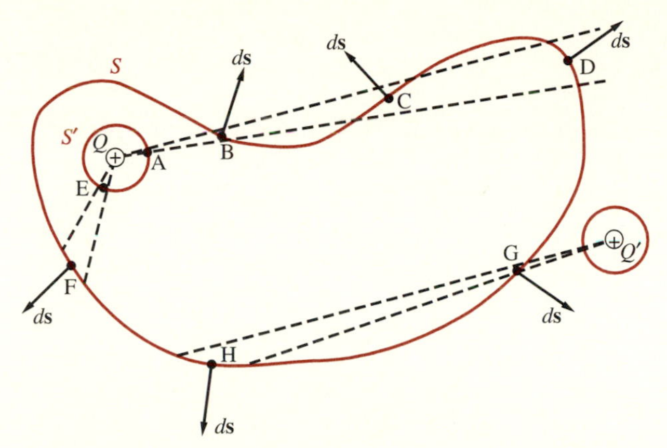

FIGURE 4.31. **Gauss's law.** The total flux emanating from the closed surface S' surrounding the charge Q is equal to that emanating from any other closed surface of arbitrary shape (i.e., S), as long as it encloses no other charges. This figure was inspired by Figures 8 and 9 of J. H. Jeans *The Mathematical Theory of Electricity and Magnetism,* 4th ed., Cambridge Press, 1920.

from this volume is zero, by the arguments given above.[60] Thus, the net total flux through the larger surface S must be identical to that through S'. In other words, the flux emanating from the end (near F) of the truncated cone between points E and F is precisely that which leaves the surface S' in the same solid angular range. Note that the above argument holds for a surface S' of any shape; we choose S' to be a sphere centered at the point charge Q for convenience so that we can easily calculate the flux through it.

Taking the radius of the sphere S' to be r, the electric flux density everywhere on its surface is given by [4.29]. Since S' is a spherical surface with surface area $4\pi r^2$, we have

$$\oint_{S'} \mathbf{D} \cdot d\mathbf{s} = \left(\frac{Q}{4\pi r^2}\right)(4\pi r^2) = Q$$

As expected, the net flux out of the spherical surface S' is independent of the radius of the sphere. Since *all* of the flux out of S' must go out of the closed surface S, and since we imposed no restrictions whatsoever on the shape of S, we have arrived at a rather general result:

$$\int_{\text{any closed surface enclosing } Q} \mathbf{D} \cdot d\mathbf{s} = Q$$

To illustrate the point further, a second point charge Q' that lies outside the closed surface S is shown in Figure 4.31. It is clear from the foregoing discussion in connection with Figure 4.30 that any solid angular range (i.e., conical tube of flux)

[60]Basically, the net flux emanating from any of the truncated cones, such as those between points A and B, B and C, or C and D, is zero, since these surfaces are just like those in Figure 4.30a, b and do not enclose any charge.

emanating from Q' cuts the closed surface S an even number of times (twice, four times, etc.), so the net flux from any of the pairs of truncated conical surfaces (such as G and H) is zero.

Although the result we have obtained is for a single point charge, it is obvious that it can be generalized to any charge distribution, since any continuous or discrete distribution of charges can be subdivided into a smaller number of elements, each of which can each be treated as a point charge. Using the superposition principle, the total flux out of any closed surface S is the sum of that from each individual enclosed charge element. We are thus ready to write down the most general form of Gauss's law:

$$\oint_S \mathbf{D} \cdot d\mathbf{s} = \int_V \rho\, dv = Q_{\text{enc}} \qquad [4.31]$$

where V is an arbitrary volume enclosed by the closed surface S; $\rho(x, y, z)$ is the volume density of electric charge inside the volume V; and Q_{enc} is the total charge in volume V enclosed by the surface S.

This general result is the most convenient method at our disposal for expressing the electrostatic inverse square action law of Coulomb. *Gauss's law is a direct consequence of Coulomb's law:* that electrostatic force between point charges varies with distance as r^{-2}. It does not provide us with any additional information about the way static charges interact; it simply allows us to state Coulomb's law in a way that may be more useful in the solution of some electrostatic problems, especially those that exhibit some sort of symmetry. Gauss's contribution was actually not in originating the law but in providing the mathematical framework for this statement.

Gauss, Johann Karl Friedrich *(German mathematician, b. April 30, 1777, Braunschweig; d. February 23, 1855, Göttingen, Hannover) Gauss, the son of a gardener and a servant girl, had no relative of more than normal intelligence, apparently, but he was an infant prodigy in mathematics who remained a prodigy all his life. He was capable of great feats of memory and of mental calculation. At the age of three, he was already correcting his father's sums.*

Some people consider him to be one of the three great mathematicians of all time, the others being Archimedes [287 B.C.–212 B.C.] and Newton [1642–1727]. His unusual mind was recognized early, and he was educated at the expense of Duke Ferdinand of Brunswick (Braunschweig). In 1795 Gauss entered the University of Göttingen, and in 1799 he received his doctor's degree.

While still in his teens he made a number of remarkable discoveries, including the method of least squares, advancing the earlier work of Legendre [1752–1833] in this area. While still in his early twenties, he was able to calculate an orbit for Ceres from Piazzi's [1746–1826] few observations so that the first asteroid might be located once more after it had been lost. Gauss also worked out theories of perturbations that were eventually used by Leverrier [1811–1877] and John C. Adams [1819–1892] in their discovery of the planet Neptune.

While still in the university, he also demonstrated a method for constructing an equilateral polygon of seventeen sides (a 17-gon) using only a straightedge and compass. Here was a construction all the ancient Greeks had missed. Gauss went further: he showed that only polygons of certain numbers of sides could be constructed with straightedge and compass alone. (These two tools were the only ones thought suitable for geometric constructions by Plato [427 B.C.–347 B.C.].) A polygon with seven sides (a heptagon) could not *be constructed in this fashion. This was the first case of a geometric construction being proved impossible. From that point on, the proof of the impossible in mathematics grew in importance, reaching a climax with Gödel [1906–1978] nearly a century and a half later.*

Gauss did important work on the theory of numbers, the branch of mathematics that P. Fermat [1601–1665] had founded, and on every other branch of mathematics. In 1799 Gauss proved the fundamental theorem of algebra, that every algebraic equation has a complex root of the form $a + jb$, *where a and b are real numbers and* $j = \sqrt{-1}$. *Gauss showed that these complex numbers can be represented as analogous to points on a plane. In 1801 he went on to prove the fundamental theorem of arithmetic: that every natural number can be represented as the product of primes in one and only one way.*

All this great work was not without a price, for his intense concentration on his work withdrew him from contact with humanity. There is a story that when he was told, in 1807, that his wife was dying, he looked up from the problem that engaged him and muttered, "Tell her to wait a moment till I'm through."

Outside the realm of pure mathematics it was his work on Ceres that gained Gauss his fame. In 1806 Gauss's sponsor, Ferdinand of Brunswick, was dead, fighting against Napoleon, and so Gauss had to have some other way of making a living. Through the influence of his friends, he was appointed director of the Göttingen Observatory in 1807. At Göttingen, Gauss devised a heliotrope: an instrument that reflected sunlight over long distances so that light rays could be put to work as straight lines marking the face of the earth, allowing more precise trigonometric determinations of the planet's shape. Gauss also worked on terrestrial magnetism and instituted the first geomagnetic observatory. He calculated the location of the magnetic poles, and, based on current geomagnetic observations, his calculations proved remarkably accurate. In 1832 he devised a logical set of units to measure magnetic phenomena; the unit of magnetic flux density was eventually named gauss.

Gauss remained on the faculty at Göttingen all his working life but hated teaching and had very few students. Each of his two wives died young, and only one of his six children survived him. His life was filled with personal tragedy, and though he died wealthy, he also died embittered. After his death, a medal was struck in his honor by the king of Hannover. A statue of him, raised by the city of his birth, stands on a pedestal in the shape of a 17-pointed star, celebrating his discovery of the 17-gon. [Adapted with permission from Biographical Encyclopedia of Science and Technology, *I. Asimov, Doubleday, New York, 1982.]*

1700 1777 1855 2000

4.5.3 Applications of Gauss's Law

We now consider examples of the use of Gauss's law, as expressed in [4.31], to determine the electric flux density $\mathbf{D}$ (and hence $\mathbf{E} = \mathbf{D}/\epsilon$) in cases where the charge distribution is known. Easy solutions of this integral equation (where the unknown $\mathbf{D}$ is under the integral) are possible only if we can take advantage of the inherent symmetries of the problem to identify a so-called *Gaussian surface* S such that $\mathbf{D}$ is everywhere either tangential or normal to the closed surface, so that $\mathbf{D} \cdot d\mathbf{s}$ is either zero or simply $D_n\, ds$, with D_n = const. on the surface. Gauss's law is thus useful only for symmetric charge distributions, which lead to symmetric distributions of $\mathbf{E}$ and $\mathbf{D}$. The following examples illustrate the use of Gauss's law for a few different charge distributions.

Example 4-12: Line charge. Consider an infinitely long line charge in free space with a line charge density ρ_l coulombs per meter, as shown in Figure 4.32. Find the electric field at a radius r from the line charge.

Solution: To find the field at a radius r, we can use Gauss's law and integrate the flux density $\mathbf{D}$ over a coaxial cylindrical surface S of length l and radius r, as shown in Figure 4.32. Due to symmetry, the electric field has a component only in the r direction, so there is no contribution to the integral over the closed surface from the end surfaces of the cylinder ($E_z = 0$). Thus, we can write

$$\oint_S \mathbf{D} \cdot d\mathbf{s} = \oint \epsilon_0 \mathbf{E} \cdot d\mathbf{s} = \epsilon_0 \int_0^{2\pi} E_r l r\, d\phi = Q_{\text{enc}} = \rho_l l \quad \rightarrow \quad E_r = \frac{\rho_l}{2\pi\epsilon_0 r}$$

When there is appropriate symmetry, the use of Gauss's law is clearly much simpler than direct use of Coulomb's law to find the electric field, as in Example 4-5.

Example 4-13: Spherical cloud of charge. Consider a spherical cloud of charge with radius a, as shown in Figure 4.33a. The electric charge is uniformly distributed over the spherical volume, with volume charge density ρ, so that the total charge in the cloud is $Q = \frac{4}{3}\pi a^3 \rho$. Find the electric field $\mathbf{E}(r)$, for both $r < a$ and $r > a$.

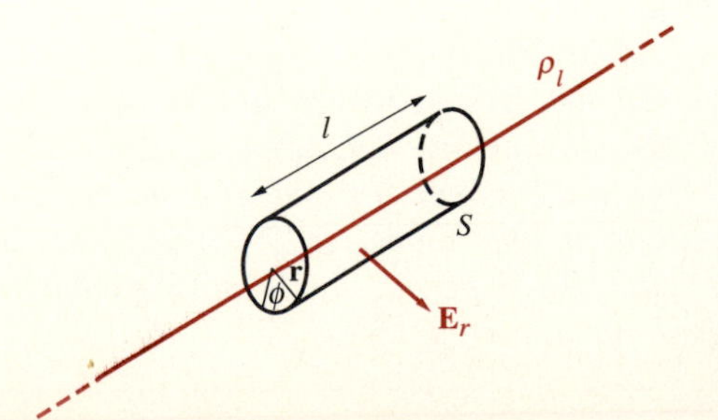

FIGURE 4.32. Line charge. A cylindrical Gaussian surface S is appropriate, since by symmetry E_r is constant on the side surface of the cylinder.

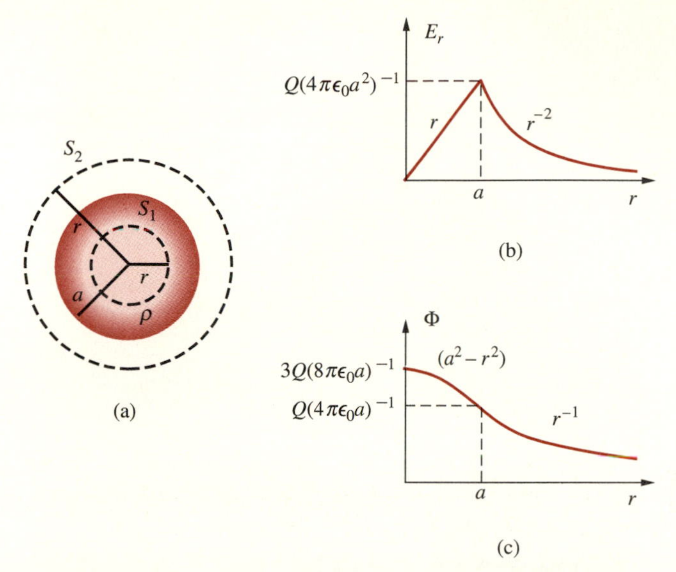

FIGURE 4.33. Electric field and potential due to a spherical cloud of uniform charge. (a) The spherical cloud of charge and Gaussian surfaces S_1 and S_2. (b) Electric field. (c) Potential.

Solution: We first observe that, due to the spherical symmetry, the electric flux density $\mathbf{D}$, and hence $\mathbf{E}$, are in the radial direction (i.e., $\mathbf{E} = \hat{\mathbf{r}}E_r$) and depend only on r. In other words, $\mathbf{D} =$ const. on any spherical surface that is concentric with the spherical charge cloud. We consider two spherical Gaussian surfaces S_1 and S_2, respectively, inside and outside the cloud of charge, and apply Gauss's law to both surfaces:

$$\oint_{S_1 \text{ or } S_2} \epsilon_0 \mathbf{E} \cdot d\mathbf{s} = \epsilon_0 E_r 4\pi r^2 = \begin{cases} \rho \dfrac{4}{3}\pi r^3 = Q\left(\dfrac{r}{a}\right)^3 & r < a \quad \text{on } S_1 \\[2ex] \rho \dfrac{4}{3}\pi a^3 = Q & r \geq a \quad \text{on } S_2 \end{cases}$$

Thus, we have

$$\mathbf{E} = \begin{cases} \hat{\mathbf{r}} \dfrac{Q}{4\pi\epsilon_0 r^2} & r \geq a \\[2ex] \hat{\mathbf{r}} \dfrac{Qr}{4\pi\epsilon_0 a^3} & r < a \end{cases}$$

The variation of E_r is plotted as a function of r in Figure 4.33b.

It is interesting to derive the expression for the electric potential $\Phi(r)$ from $E_r(r)$. Using the definition of electric potential as given in [4.17], and taking the path of integration along the radial direction, we have

$$\Phi = -\int_{\infty}^{r} (\hat{\mathbf{r}}E_r)\cdot(\hat{\mathbf{r}}\,dr) = \begin{cases} -\displaystyle\int_{\infty}^{r} \frac{Q\,dr}{4\pi\epsilon_0 r^2} & r > a \\ -\displaystyle\int_{\infty}^{a} \frac{Q\,dr}{4\pi\epsilon_0 r^2} - \int_{a}^{r} \frac{Qr\,dr}{4\pi\epsilon_0 a^3} & r \le a \end{cases}$$

Straightforward evaluation of these integrals yields

$$\Phi = \begin{cases} \dfrac{Q}{4\pi\epsilon_0 r} & r > a \\ \dfrac{Q}{4\pi\epsilon_0 a} + \dfrac{Q}{4\pi\epsilon_0 a^3}\dfrac{a^2 - r^2}{2} & r \le a \end{cases}$$

The variation of the potential Φ is plotted in Figure 4.33c. The relationship between $\mathbf{E}$ and Φ, namely, $\mathbf{E} = -\nabla\Phi$ (in this case, $E_r = -d\Phi/dr$), is apparent.

Example 4-14: Infinite sheet of charge. Consider the planar sheet of charge of infinite extent[61] located in the xy plane (i.e., $z = 0$), as shown in Figure 4.34. The surface charge density is given as ρ_s. Determine the electric field $\mathbf{E}$ and the electrostatic potential Φ.

Solution: We observe from the symmetry of the problem that the electric field can only be in the z direction: $\mathbf{E} = \hat{\mathbf{z}}E_z$ for $z > 0$ and $\mathbf{E} = -\hat{\mathbf{z}}E_z$ for $z < 0$. We choose to work with cylindrical (r, ϕ, z) coordinates using a cylindrical pillbox-

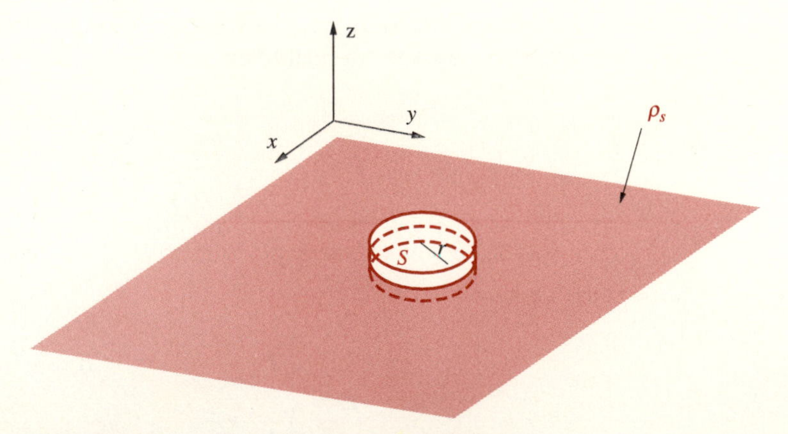

FIGURE 4.34. **An infinite planar sheet of charge.**

[61]This appears as a perfect example of an academic problem of no practical interest; after all, an infinite sheet of charge certainly cannot exist. However, it turns out that several practical results (e.g., evaluating the capacitance of planar conductor configurations) can actually be based on the result derived here, as on other "symmetrical" charge configurations we study in this section.

and since the electric field must be constant over the side surfaces (due to symmetry), from Gauss's law we have $E_r = 0$. We now consider the surface S_2 and apply Gauss's law, noting that the surface elements on the top, bottom, and side surfaces are, respectively, $d\mathbf{s} = \hat{\mathbf{z}} r\,dr\,d\phi$, $d\mathbf{s} = -\hat{\mathbf{z}} r\,dr\,d\phi$, and $d\mathbf{s} = \hat{\mathbf{r}} r\,d\phi\,dz$, and that the charge enclosed by this surface is $Q_{\text{enc}} = \rho_s(2\pi a h)$. We have

$$\oint_S \epsilon_0 \mathbf{E} \cdot d\mathbf{s} = Q_{\text{enc}}$$

$$\underbrace{\int_{\text{top}} (\hat{\mathbf{r}}E_r) \cdot (\hat{\mathbf{z}} r\,dr\,d\phi)}_{=0} + \underbrace{\int_{\text{bottom}} (\hat{\mathbf{r}}E_r) \cdot (-\hat{\mathbf{z}} r\,dr\,d\phi)}_{=0} + \int_{\text{side}} (\hat{\mathbf{r}}E_r) \cdot (\hat{\mathbf{r}} r\,d\phi\,dz) = \frac{Q_{\text{enc}}}{\epsilon_0}$$

$$E_r \int_{\text{side}} r\,d\phi\,dz = \frac{Q_{\text{enc}}}{\epsilon_0}$$

$$E_r(2\pi r h) = \frac{\rho_s 2\pi a h}{\epsilon_0}$$

$$E_r = \frac{\rho_s a}{\epsilon_0 r}$$

where E_r was taken outside the integral since it is constant on the side surface of S_2. In other words, the electric field is

$$\mathbf{E} = \begin{cases} 0 & r < a \\ \hat{\mathbf{r}}\dfrac{\rho_s a}{\epsilon_0 r} & a < r < b \\ 0 & r > b \end{cases}$$

The electrostatic potential in the region between the cylinders (i.e., $a \leq r \leq b$) can be found as usual by integrating the electric field

$$\Phi(r) = -\int_{\infty}^{r} (\hat{\mathbf{r}}E_r) \cdot (\hat{\mathbf{r}}\,dr) = -\int_{b}^{r} \frac{\rho_s a\,dr}{\epsilon_0 r} = \frac{\rho_s a}{\epsilon_0}\Big[-\ln r\Big]_b^r = -\frac{\rho_s a}{\epsilon_0} \ln\frac{r}{b}$$

Often it is more convenient to write $\Phi(r)$ in terms of the potential difference Φ_{ba} between the conductors. In the notation used in Section 4.4, we have $\Phi_{ba} = \Phi(a) - \Phi(b)$, and

$$\Phi(r) = \Phi_{ba}\frac{\ln(r/b)}{\ln(a/b)} \qquad a \leq r \leq b$$

a result that will be quite useful in later sections. Note that the quantity Φ_{ba} is positive in this case, since the inner shell (at $r = a$) is the one that has the positive charge. In other words, we would need to do work in order to move a positive test charge from $r = b$ to $r = a$.

4.6 DIVERGENCE: DIFFERENTIAL FORM OF GAUSS'S LAW

The concept of the divergence of a vector field is important for understanding the sources of fields. In the electrostatic context, divergence of the electric flux density **D** is directly related to the source charge density ρ. In this section, we discuss the definition and meaning of divergence and develop the relationship between **D** and ρ by applying Gauss's law to a very small (infinitesimal) volume.

If one envisions a closed surface S enclosing a source of any vector field **A**, then the strength or magnitude of the source is given by the net outward flow or flux of **A** through the closed surface S, or $\oint_S \mathbf{A} \cdot d\mathbf{s}$. However, a more suitable measure of the *concentration* or *density* of the source is the outward flux per unit volume, or

$$\frac{\oint_S \mathbf{A} \cdot d\mathbf{s}}{V}$$

where V is the volume enclosed by the surface S. By reducing V to a differential volume, we obtain a point relation that gives the source density per unit volume at that point. Since the source density in general varies from point to point, this measure of the source density is a scalar field and is called the *divergence* of the vector field **A**. The fundamental definition of divergence of a vector field **A** at any point is then

$$\text{div}\,\mathbf{A} \equiv \lim_{V \to 0} \frac{\oint_S \mathbf{A} \cdot d\mathbf{s}}{V} \qquad [4.32]$$

where S is the closed surface enclosing the volume V.

For the electrostatic field, we know from Gauss's law that the total outward flux through a surface S is equal to the total charge enclosed. For a differential volume element V, we can assume the volume charge density to be constant at all points within the infinitesimal volume, so the total charge enclosed is ρV. Accordingly, we have

$$\text{div}\,\mathbf{D} \equiv \lim_{V \to 0} \frac{\oint_S \mathbf{D} \cdot d\mathbf{s}}{V} = \rho \qquad [4.33]$$

so that the divergence of the electric flux density at any point is equal to the volume charge density at that point.

We now derive the proper differential expression for div **D**. Although the divergence of a vector is clearly independent of any coordinate system, for simplicity we use a rectangular coordinate system and consider a small differential cubical volume element Δv placed in an electric flux density field **D** as shown in Figure 4.36. To determine the net outward flux from this volume, we can separately consider each pair of parallel faces: left-right, front-back, and bottom-top. We assume that the flux density **D** at the center of the cube is given by $\mathbf{D}(x_0, y_0, z_0) = \hat{\mathbf{x}}D_{x0} + \hat{\mathbf{y}}D_{y0} + \hat{\mathbf{z}}D_{z0}$. We first consider the net flux through the left and right faces due to the y component (i.e., D_y)

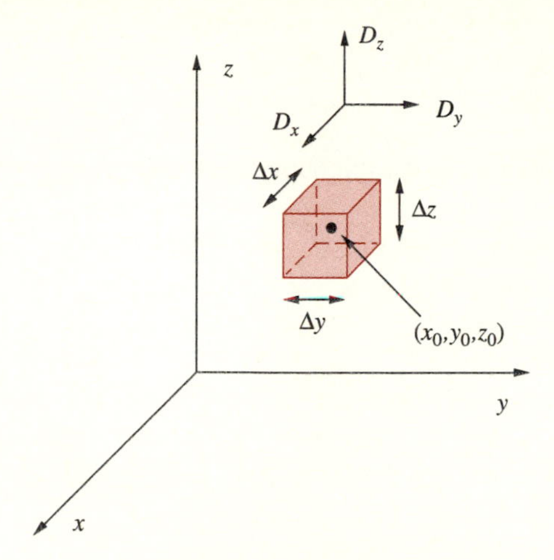

FIGURE 4.36. Divergence. A cubical volume element $\Delta v = \Delta x \Delta y \Delta z$ used to derive the differential expression for the divergence of **D**.

of **D**. Note that on the left face we have $y = y_0 - \Delta y/2$, whereas on the right face $y = y_0 + \Delta y/2$. On the left face of the cubical volume element shown in Figure 4.36, $d\mathbf{s} = -\hat{\mathbf{y}}\, dx\, dz$, so that the flux leaving from the left face is, to the first order,[63]

$$\left[\hat{\mathbf{y}} D_y\left(x_0, y_0 - \frac{\Delta y}{2}, z_0\right)\right] \cdot (-\hat{\mathbf{y}} \Delta z \Delta x) = -(\Delta z\, \Delta x) D_y\left(x_0, y_0 - \frac{\Delta y}{2}, z_0\right)$$

$$\simeq -(\Delta z\, \Delta x)\left(D_{y0} - \frac{\Delta y}{2}\frac{\partial D_y}{\partial y}\right)$$

That from the right face[64] is similarly

$$+(\Delta z\, \Delta x)\left(D_{y0} + \frac{\Delta y}{2}\frac{\partial D_y}{\partial y}\right)$$

Thus, the net outward flux between these two faces is

$$\Delta z\, \Delta x\, \Delta y \frac{\partial D_y}{\partial y}$$

The net outward flux between the other two pairs of faces (front-back and bottom-top) can be similarly evaluated to find the net total outward flux from the cubical surface:

---

[63] At a more formal level, we can expand D_y in a Taylor's series around $y = y_0$ so that

$$D_y(x_0, y, z_0) = D_y(x_0, y_0, z_0) + (y - y_0)\frac{\partial D_y}{\partial y} + (y - y_0)^2 \frac{\partial^2 D_y}{2\partial y^2} + \text{higher-order terms}$$

Note that for a differential volume element, with $\Delta y = (y - y_0) \to 0$, the higher-order terms involving $(\Delta y)^k$ are negligible compared to the first-order term multiplied by Δy.

[64] Note that we have $d\mathbf{s} = +\hat{\mathbf{y}}\, dx\, dz$.

$$\underbrace{\Delta z\, \Delta x\, \Delta y}_{\Delta v}\left(\frac{\partial D_x}{\partial x} + \frac{\partial D_y}{\partial y} + \frac{\partial D_z}{\partial z}\right)$$

Since divergence was defined as the total outward flux per unit volume, we have

$$\boxed{\text{div } \mathbf{D} = \frac{\partial D_x}{\partial x} + \frac{\partial D_y}{\partial y} + \frac{\partial D_z}{\partial z}} \qquad [4.34]$$

Using the definition of the del operator ∇ as a vector given in [4.21], the divergence of a vector $\mathbf{D}$ is typically written in a rectangular coordinate system as

$$\nabla \cdot \mathbf{D} = \left(\hat{\mathbf{x}}\frac{\partial}{\partial x} + \hat{\mathbf{y}}\frac{\partial}{\partial y} + \hat{\mathbf{z}}\frac{\partial}{\partial z}\right) \cdot (\hat{\mathbf{x}}D_x + \hat{\mathbf{y}}D_y + \hat{\mathbf{z}}D_z)$$

$$= \frac{\partial D_x}{\partial x} + \frac{\partial D_y}{\partial y} + \frac{\partial D_z}{\partial z}$$

An important result of our discussion of the concept of divergence of a vector field can now be stated. This result is the differential version of Gauss's law, namely

$$\boxed{\nabla \cdot \mathbf{D} = \rho} \qquad [4.35]$$

Equation [4.35] (or its integral form [4.31]) is one of the four fundamental equations of electromagnetics, the collection of which are referred to as Maxwell's equations. Note that this equation is simply a re-statement of Gauss's law, which in turn is a convenient expression of the fundamental experimentally established inverse square law known as Coulomb's law.

Examples 4-16 and 4-17 illustrate respectively vector fields with zero and nonzero divergence.

Example 4-16: Divergence-free (zero-divergence) fields. Show that the following vector fields are divergence-free:

$$\mathbf{A}_1 = -\hat{\mathbf{x}}x + \hat{\mathbf{y}}y$$
$$\mathbf{A}_2 = \hat{\mathbf{x}}\tfrac{1}{2}x^2 - \hat{\mathbf{y}}xy$$
$$\mathbf{A}_3 = \hat{\mathbf{x}}\sin y - \hat{\mathbf{y}}\cos x$$
$$\mathbf{A}_4 = \hat{\mathbf{x}}\sin x \cos y - \hat{\mathbf{y}}\cos x \sin y$$

Solution:

$$\nabla \cdot \mathbf{A}_1 = \frac{\partial A_x}{\partial x} + \frac{\partial A_y}{\partial y} = -1 + 1 = 0$$
$$\nabla \cdot \mathbf{A}_2 = x - x = 0$$
$$\nabla \cdot \mathbf{A}_3 = 0 - 0 = 0$$
$$\nabla \cdot \mathbf{A}_4 = \cos x \cos y - \cos x \cos y = 0$$

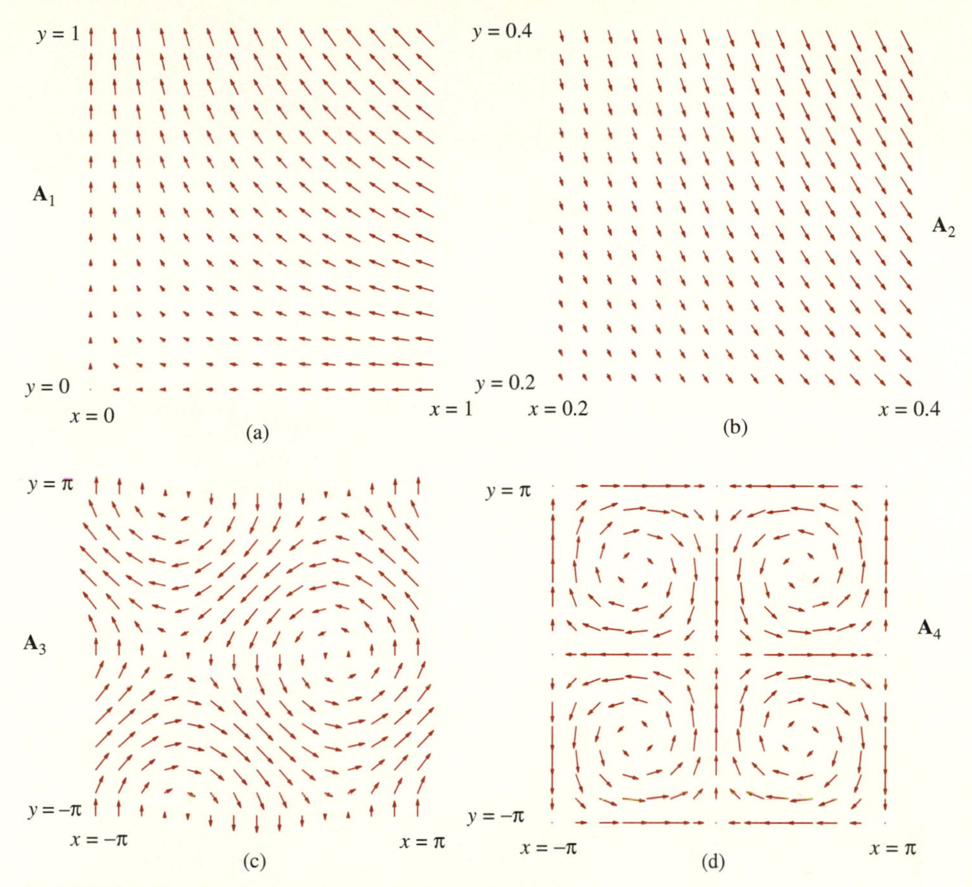

FIGURE 4.37 **Examples of divergence-free fields.** (a) $\mathbf{A}_1 = -\hat{\mathbf{x}}x + \hat{\mathbf{y}}y$. (b) $\mathbf{A}_2 = \hat{\mathbf{x}}\frac{1}{2}x^2 - \hat{\mathbf{y}}xy$. (c) $\mathbf{A}_3 = \hat{\mathbf{x}}\sin y - \hat{\mathbf{y}}\cos x$. (d) $\mathbf{A}_4 = \hat{\mathbf{x}}\sin x\cos y - \hat{\mathbf{y}}\cos x\sin y$.

These vector fields are shown in Figure 4.37. One common feature in the plots of all four vector fields is the fact that field vectors do not seem to converge toward or emerge from any "source" points.

Example 4-17: Vector fields with nonzero divergence. Find the divergence of the following vector fields:

$$\mathbf{A}_5 = \hat{\mathbf{x}}\frac{x^2}{2} + \hat{\mathbf{y}}xy$$

$$\mathbf{A}_6 = \hat{\mathbf{x}}\sin x - \hat{\mathbf{y}}\cos y$$

$$\mathbf{A}_7 = \hat{\mathbf{x}}\sin x\cos y + \hat{\mathbf{y}}\cos x\sin y$$

$$\mathbf{A}_8 = \hat{\mathbf{x}}\cos(xy) + \hat{\mathbf{y}}\sin(xy)$$

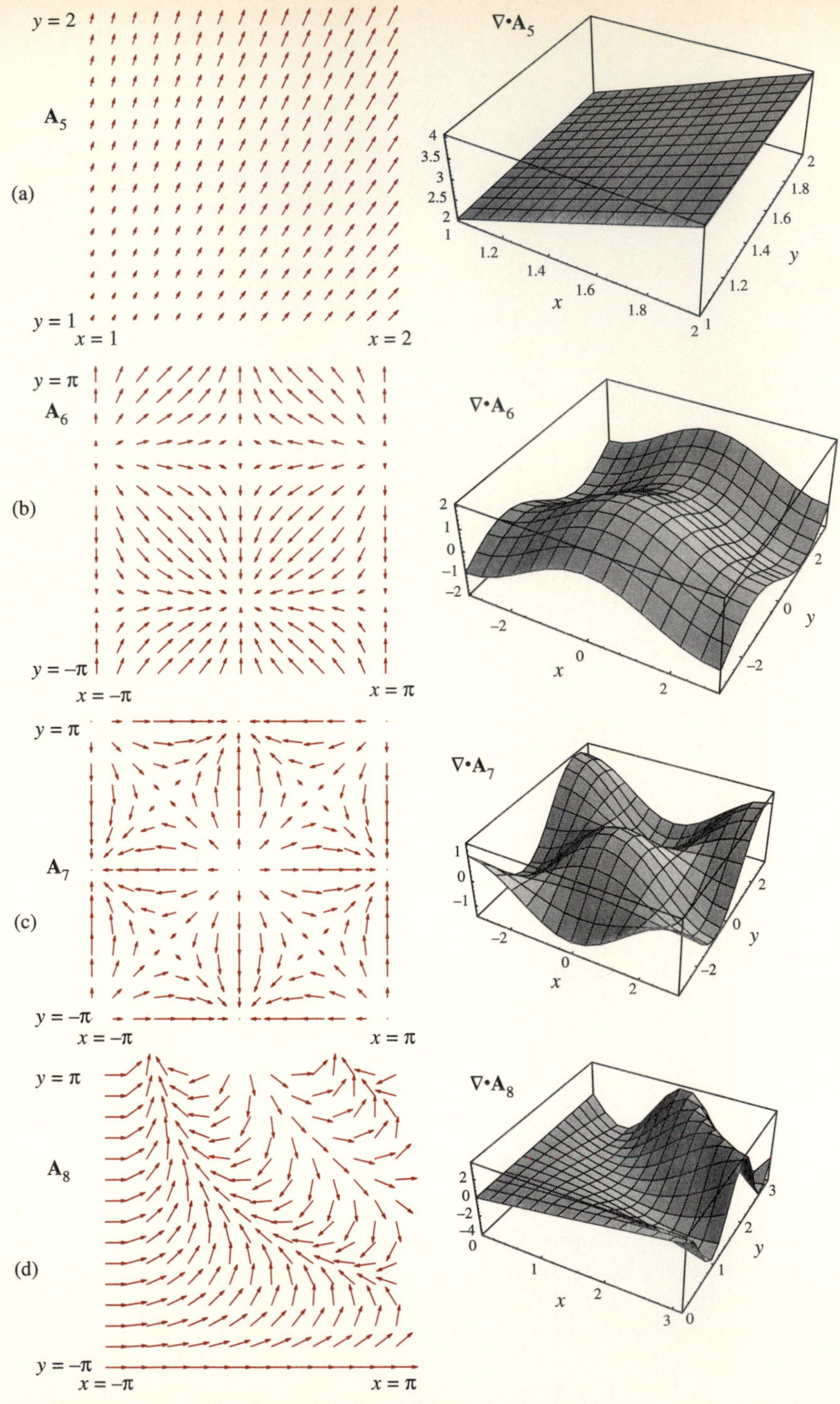

FIGURE 4.38 Examples of vector fields with nonzero divergence. (a) $\mathbf{A}_5 = \hat{\mathbf{x}}(x^2/2) + \hat{\mathbf{y}}xy$. (b) $\mathbf{A}_6 = \hat{\mathbf{x}}\sin x - \hat{\mathbf{y}}\cos y$. (c) $\mathbf{A}_7 = \hat{\mathbf{x}}\sin x \cos y + \hat{\mathbf{y}}\cos x \sin y$. (d) $\mathbf{A}_8 = \hat{\mathbf{x}}\cos(xy) + \hat{\mathbf{y}}\sin(xy)$. The units for the x and y axes are in radians for all panels.

Solution:

$$\nabla \cdot \mathbf{A}_5 = 2x$$

$$\nabla \cdot \mathbf{A}_6 = \cos x + \sin y$$

$$\nabla \cdot \mathbf{A}_7 = 2 \cos x \cos y$$

$$\nabla \cdot \mathbf{A}_8 = -y \sin(xy) + x \cos(xy)$$

Both the fields and their divergences are plotted in Figure 4.38. The emergence (or convergence) of the field vectors from source points is a common characteristic for $\mathbf{A}_5$, $\mathbf{A}_6$, $\mathbf{A}_7$, and $\mathbf{A}_8$. Note that the locations from which the field vectors appear to diverge are those at which the divergences are positive maxima, whereas those points at which the fields converge are negative minima. For $\mathbf{A}_5$, the divergence is steadily increasing with increasing x, representing a distributed source. In the case of $\mathbf{A}_8$, the arrows all have the same length, since the magnitude of the vector is a constant: $|\mathbf{A}_8| = \sqrt{\cos^2(xy) + \sin^2(xy)} = 1$.

Divergence in Other Coordinate Systems Although we use the notation $\nabla \cdot \mathbf{D}$ to indicate the divergence of a vector $\mathbf{D}$, it should be noted that a vector del operator as defined in [4.21] is useful only in a rectangular coordinate system. In other coordinate systems we still denote the divergence of $\mathbf{D}$ by $\nabla \cdot \mathbf{D}$, but note that the specific scalar derivative expressions to be used will need to be derived from the physical definition of divergence using a differential volume element appropriate for that particular coordinate system.

To derive the differential expression for divergence in the cylindrical coordinate system, we consider the cylindrical cuboid[65] volume element shown in Figure 4.39b. Note that the volume of this element is $\Delta v = r\,\Delta r\,\Delta\phi\,\Delta z$. The flux of the vector field $\mathbf{F}$ through the face marked as S_1 is

$$\int_{S_1} \mathbf{F} \cdot d\mathbf{s} = \int_{S_1} F_r\, ds \simeq F_r\left(r + \frac{\Delta r}{2}, \phi, z\right)\left(r + \frac{\Delta r}{2}\right)\Delta\phi\,\Delta z$$

while that through face 2 is

$$\int_{S_2} \mathbf{F} \cdot (-\hat{\mathbf{r}}\, ds) = -\int_{S_2} F_r\, ds \simeq -F_r\left(r - \frac{\Delta r}{2}, \phi, z\right)\left(r - \frac{\Delta r}{2}\right)\Delta\phi\,\Delta z$$

Adding these two and dividing by the volume $\Delta v = r\,\Delta r\,\Delta\phi\,\Delta z$ gives the net flux per unit volume out of the cube due to the r component of the vector field, namely

$$\frac{1}{\Delta v}\int_{S_1+S_2} \mathbf{F} \cdot d\mathbf{s} \simeq \frac{1}{r\Delta r}\left[\left(r + \frac{\Delta r}{2}\right)F_r\left(r + \frac{\Delta r}{2}, \phi, z\right) - \left(r - \frac{\Delta r}{2}\right)F_r\left(r - \frac{\Delta r}{2}, \phi, z\right)\right]$$

[65]This terminology and discussion are adapted from H. M. Schey, *div grad curl and All That, an Informal Text on Vector Calculus,* W. W. Norton, New York, 1992.

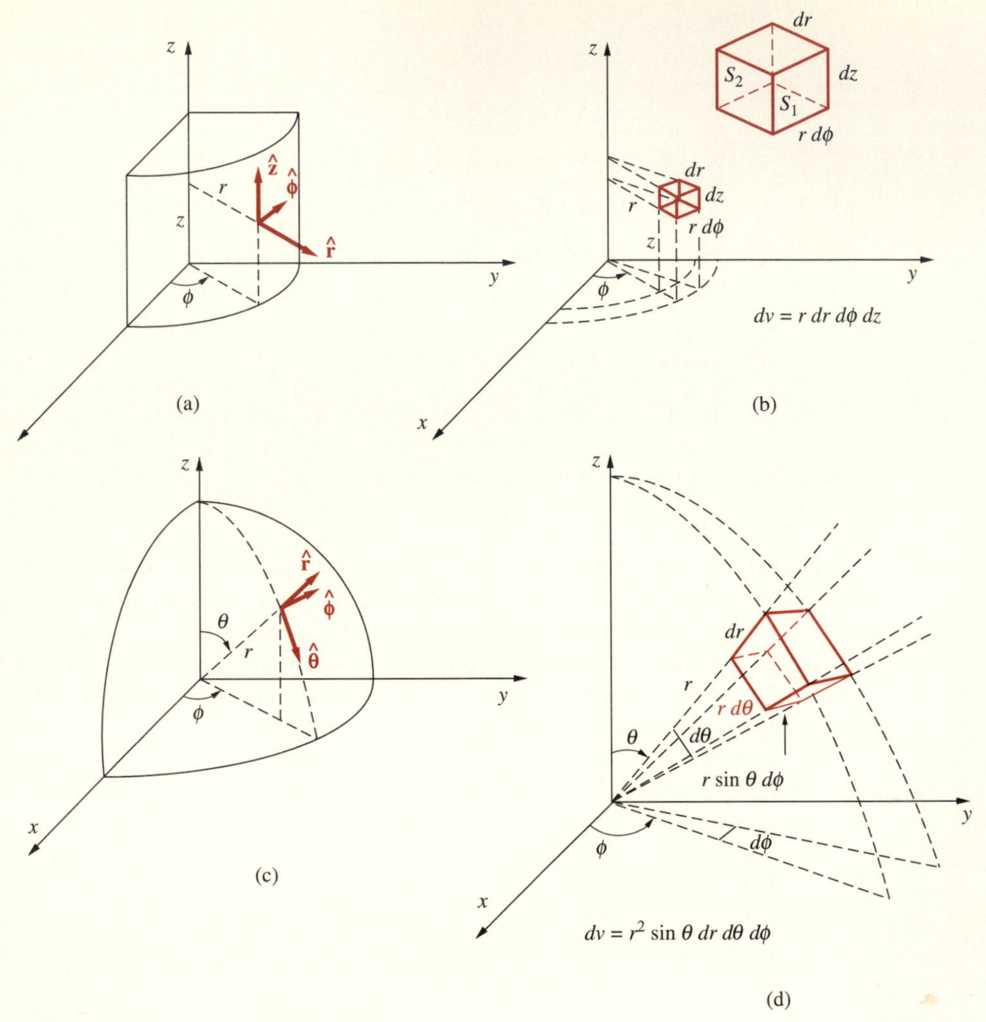

FIGURE 4.39. **Other coordinate systems.** (a, b) Unit vectors and volume element for cylindrical coordinates; (c, d) same for spherical coordinates.

which in the limit as $\Delta r \to 0$ (and thus $\Delta v \to 0$) becomes

$$\frac{1}{r}\frac{\partial}{\partial r}(rF_r)$$

Evaluating the contributions from the other two pairs of faces in a similar fashion, we find the expression for divergence in cylindrical coordinates:

$$\boxed{\operatorname{div}\mathbf{F} = \nabla\cdot\mathbf{F} = \frac{1}{r}\frac{\partial}{\partial r}(rF_r) + \frac{1}{r}\frac{\partial F_\phi}{\partial \phi} + \frac{\partial F_z}{\partial z}} \qquad [4.36]$$

Consideration of the flux in and out of the different opposing faces of the spherical cuboid shown in Figure 4.39d leads to the divergence expression in spherical coordinates:

$$\text{div}\,\mathbf{F} = \nabla\cdot\mathbf{F} = \frac{1}{r^2}\frac{\partial}{\partial r}(r^2 F_r) + \frac{1}{r\sin\theta}\frac{\partial}{\partial\theta}(\sin\theta F_\theta) + \frac{1}{r\sin\theta}\frac{\partial F_\phi}{\partial\phi} \qquad [4.37]$$

Examples 4-18 and 4-19 illustrate the use of expressions [4.36] and [4.37].

Example 4-18: Electrostatic field between two coaxial and concentric cylindrical shells of charge. In Example 4-15 we derived the electric flux density in the region between two concentric cylindrical shells of charge to be

$$\mathbf{D} = \epsilon_0\mathbf{E} = \hat{\mathbf{r}}\frac{\rho_s a}{r} \qquad a < r < b$$

Find the divergence of the electric flux density.

Solution: Since there are no sources (charges) in the region between the shells ($a < r < b$), the divergence of $\mathbf{D}$ must be identically zero. This is expected on physical grounds; the flux density varies with radial distance as r^{-1}, but the volume of an element subtended by an angular range $\Delta\phi$, a height range Δz, and a radial range Δr is $r\,\Delta r\,\Delta\phi\,\Delta z$ and hence varies as r, exactly compensating the r^{-1} variation of $\mathbf{D}$. Analytically, we have

$$\nabla\cdot\mathbf{D} = \frac{1}{r}\frac{\partial}{\partial r}\left(r\frac{\rho_s a}{r}\right) = 0$$

Example 4-19: Divergence of the dipole field. As an example of a field that has both r and θ components, consider the electric field of the electric dipole as derived in Section 4.4.3. For a dipole consisting of two opposite charges $\pm Q$ separated by a distance d, the electric flux density ($\mathbf{D} = \epsilon\mathbf{E}$) at faraway points ($r \gg d$) was shown to be

$$\mathbf{D} = \frac{Qd}{4\pi r^3}[\hat{\mathbf{r}}2\cos\theta + \hat{\boldsymbol{\theta}}\sin\theta]$$

Find the divergence of the electric flux density at faraway points.

Solution: The divergence of this flux density is

$$\text{div}\,\mathbf{D} = \frac{Qd}{4\pi}\left[\frac{1}{r^2}\frac{\partial}{\partial r}\left(r^2\frac{2}{r^3}\cos\theta\right) + \frac{1}{r\sin\theta}\frac{\partial}{\partial\theta}\left(\sin\theta\frac{\sin\theta}{r^3}\right)\right]$$

$$= \frac{Qd}{4\pi}\left[\frac{2\cos\theta}{r^2}\frac{\partial}{\partial r}\left(\frac{1}{r}\right) + \frac{1}{r^4\sin\theta}\frac{\partial}{\partial\theta}(\sin^2\theta)\right]$$

$$= \frac{Qd}{4\pi}\left[\frac{2\cos\theta}{r^2}\left(-\frac{1}{r^2}\right) + \frac{1}{r^4\sin\theta}2\sin\theta\cos\theta\right] = 0$$

Once again the divergence is zero as expected, since there are no charges at points far away from the source dipole.

4.6.1 The Divergence Theorem

For the electrostatic field, we can combine the integral and differential forms of Gauss's law to find

$$\left.\begin{aligned}\oint_S \mathbf{D}\cdot d\mathbf{s} &= \int_V \rho\, dv \\ \nabla\cdot\mathbf{D} &= \rho\end{aligned}\right\} \longrightarrow \boxed{\oint_S \mathbf{D}\cdot d\mathbf{s} = \int_V \nabla\cdot\mathbf{D}\, dv} \qquad [4.38]$$

This integral equation is entirely consistent with the interpretation of the divergence of a vector field as the outflow of flux per unit volume at a given point.

Consider a volume V, subdivided into smaller cells and surrounded by a surface S. The outward flux from any of the subcells is given by $(\nabla\cdot\mathbf{D})\Delta v$, where $(\nabla\cdot\mathbf{D})$ is the value of the divergence (i.e., source density) at that point. All of this outward flux enters the adjoining cells, unless the cell contains a portion of the outer surface. The divergence integrated over the volume V would give the total outward flux from that volume, which should thus be equal to the outward flux through the surface S enclosing the volume. Accordingly, we expect the above result to hold true for any arbitrary vector field **G**, rather than just the electrostatic flux density **D**. This general result, restated mathematically below for an arbitrary vector **G**, is known as the *divergence theorem* or *Gauss's theorem* or *Green's theorem* and is fundamentally important and useful in many branches of engineering and physics.

$$\boxed{\oint_S \mathbf{G}\cdot d\mathbf{s} = \int_V \nabla\cdot\mathbf{G}\, dv} \qquad [4.39]$$

4.7 METALLIC CONDUCTORS

Until now, we have considered charges to be situated in vacuum, without any matter nearby. We have assumed the charges to be stationary and have found that the calculation of electric fields and potentials due to these charges in surrounding regions can basically be reduced to carrying out (albeit often complicated) integration operations. However, when material media are in the presence of electric fields, complications arise, because the way in which charges are distributed is not initially known and is determined by the properties of matter. The most important property of matter in this context is *conduction,* which is its ability to allow motion of charges over macroscopic distances. Materials greatly vary in this property, all the way from good conductors (metals) to very poor conductors (insulators, or dielectrics).

In this section, we briefly discuss metallic conductors. Most electrostatic problems consist of configurations of metallic bodies that are connected to primary sources (e.g., a battery) so that potential differences exist between conductors. To be able to determine the resulting charge distributions, we need to have a basic understanding of the electrical properties of metallic conductors.

Metallic materials are in general good conductors. A good conductor is a material within which charges can move freely. Since we are dealing with electrostatics, we assume *a priori* that the electric charges have reached[66] their equilibrium positions and are fixed in space. Under such static conditions, there must be zero electric field inside the metallic conductor, since otherwise charges would continue to flow, contrary to our original premise. The electrostatic potential Φ must be the same throughout the metallic conductor for the same reason. It thus follows that if such a conductor is charged, for example by connecting it to ground (see Figure 4.28) or by bringing it near another charged metallic conductor, the charges rearrange themselves such that the net electric field due to all the charges becomes zero inside the conductor. If a metallic conductor is placed in an external electric field $\mathbf{E}_0$, charge once again flows temporarily within it to produce a second field $\mathbf{E}_1$, which precisely cancels out the existing electric field inside the conductor, so that $\mathbf{E}_0 + \mathbf{E}_1 = 0$. Regardless of the shape of the conductor, the field everywhere within it becomes zero once the charges come to rest. In most good conductors (metals) the establishment of this equilibrium occurs in $\sim 10^{-19}$ s, as will be shown in Section 5.5.[67]

There are two important consequences of the fact that the electrostatic field inside a metallic conductor is identically zero. Firstly, the entire space occupied by the conductor must be an equipotential volume. Secondly, all the charge, if any, on a conductor must reside entirely on its surface,[68] since if any charge did exist at any point within the body, then by Gauss's law a nonzero field would have to exist in the vicinity of a small Gaussian surface surrounding that point. At equilibrium, the surface charge is distributed in such a way that the total electric field inside the conductor and tangential to its surface is zero.[69] Note that the requirement for the tangential

---

[66]We note that if a charge distribution is suddenly disturbed, it would take a finite time (see Section 5.5) for the charges to redistribute themselves.

[67]The value $\sim 10^{-19}$ s for the rearrangement time of excess charge is particular to copper. It is given by the ratio ϵ/σ for the material, where σ is the electrical conductivity (to be introduced in Chapter 5) and ϵ is the electrical permittivity (to be discussed in Section 4.10). For other metallic conductors, ϵ/σ is similarly very small (of order 10^{-18} to 10^{-19} s).

[68]We consider the surface as having strictly infinitesimal thickness. Actually, the excess or deficiency of electrons will move around in a layer of the order of atomic dimensions, or an angstrom (10^{-8} cm) in thickness, producing on the average a volume density of charge in a layer too thin to be of macroscopic importance, but nevertheless containing many electrons.

[69]The charged metallic conductor can be thought of as a body that is internally electrically neutral but has a charge distribution over its exterior surface. The interior of the conductor is equivalent to vacuum (see footnote 23 in Section 5.5) once equilibrium is established, since the net effect of all the microphysics of the material has been to distribute the surface charge properly. This distribution is such that if the conducting body were somehow removed, leaving the surface charge distribution intact in vacuum, the field $\mathbf{E}$ at all points previously occupied by the conductor would still remain zero.

field to be zero also follows from the fact that the conductor is an equipotential, to which electric field lines must be orthogonal.

4.7.1 Macroscopic versus Microscopic Fields

The fundamental equations of electrostatics (Coulomb's law, Gauss's law), or for that matter the more general Maxwell's equations, which describe all electromagnetic phenomena, are written in terms of *macroscopic* quantities. For example, when we state that [4.35] is Gauss's law applied to a "point," we refer to a point that is very small compared to the physical dimensions of our system but very large compared to atomic dimensions, so that it actually contains many electrons, the volume charge density of which is ρ. A "point" in a macroscopic sense might have dimensions of 1 micron (or 1 μm $= 10^{-6}$ m $= 10^{-4}$ cm), whereas a microscopic point would have atomic dimensions of the order of one angstrom or 10^{-8} cm. Thus, a macroscopic "point" in the form of a cube of size 1 μm $\times$ 1 μm $\times$ 1 μm may well contain 10^{12} atoms, assuming that each atom has a radius of about one angstrom.

Matter consists of primarily empty space sparsely filled with atoms and molecules. An atom is itself mostly empty space, with the nucleus and electrons occupying a tiny fraction of its volume. Quantitatively, and in purely classical (nonquantum) terms, we understand an atom as consisting of a heavy nucleus of approximate radius $\sim 10^{-13}$ cm. Negatively charged electrons, with physical dimensions also of order $\sim 10^{-13}$ cm, travel around the nucleus in orbits of approximate radii $\sim 10^{-8}$ cm. At room temperature and pressure, the spacing between the centers of atoms varies from $\sim 10^{-8}$ cm in solids to $\sim 10^{-6}$ cm in gases. To determine the range of dimensions over which a macroscopic description is appropriate, we can rely on the fact that reflection and refraction of visible light (wavelength $\sim 10^{-5}$ cm) are properly described by fundamental equations using macroscopic parameters, whereas diffraction of X rays (wavelength $\sim 10^{-8}$ cm) reveals the atomic nature of matter.[70] Based on this, it is possible to take a dimension of 10^{-6} cm $= 10^{-2}$ μm $=$ 100 angstroms to be approximately the lower limit of the macroscopic domain.

Metals are conductors because electrons in them are free to move when electric fields of macroscopic extent are applied. In a classical picture, metallic structure can be thought of as an array of positive ions, which can vibrate about their lattice sites, and electrons as small ($<10^{-12}$ cm) points or spheres that are free to wander throughout the body under the influence of thermal agitation. The vibrations of the ions and the movement of the electrons are both random thermal effects, and local electric fields exist at any atom site, varying with the motions of nearby ions and electrons. These microscopic fields can have very large intensities but vary

---

[70]For further discussion of macroscopic versus microscopic fields and charges, see the following references, listed in increasing level of complexity: J. D. Jackson, *Classical Electrodynamics,* Wiley, New York, 1975, Sec. 6.7; W. B. Cheston, *Elementary Theory of Electric and Magnetic Fields,* Wiley, New York, 1964, Sec. 2.7; H. A. Lorenz, *The Theory of Electrons and Its Applications to the Phenomena of Light and Radiant Heat,* Dover, New York, 1952, Chap. 1.

extremely rapidly in both space and time. Instantaneous field strengths of order $\sim 10^9$ to $\sim 10^{15}$ V-cm^{-1} may exist at the orbits of electrons in atoms, and at the edge of a heavy nucleus the field intensities may be of order $\sim 10^{19}$ V-cm^{-1}. The spatial variations occur over distances of order $\sim 10^{-8}$ cm or less, and the temporal fluctuations occur with periods ranging from $\sim 10^{-13}$ to $\sim 10^{-17}$ seconds. Macroscopic measuring devices generally average over intervals in space and time that are much larger than the scales over which these microscopic fluctuations occur. Consequently, these variations are *averaged* out,[71] resulting in smooth (in space) and slowly varying (in time) macroscopic quantities. In a conductor, the *average* value of the electric field is zero unless there is a drift of the electron cloud (current). No such macroscopic drifts can occur in electrostatic equilibrium in the absence of an externally applied electric field. The influence of a macroscopic applied electric field $\mathbf{E}$ is to produce a net electron drift velocity $\mathbf{v}_d$ proportional to $-\mathbf{E}$, which leads to conduction currents, to be studied in Chapter 5.

In terms of quantum theory, the field $\mathbf{E}$ can be thought to set up a continuous process of electrons jumping into empty conduction band levels,[72] while at the same time collisions with thermally agitated atomic nuclei, lattice boundaries, and imperfections produce backward jumps that counteract the effect of $\mathbf{E}$, so a steady state of electron flow is established, with $\mathbf{v}_d$ proportional[73] to $-\mathbf{E}$. We also note that charges that freely move in a conductor are strictly electrons; when a conductor surface is positively charged, the conductor has a net deficiency of electrons, versus an excess of electrons in the case of a negatively charged conductor. It is important to note, however, that this excess or deficiency of electrons is extremely small compared to the total number of electrons in the metallic material, as will be illustrated in Example 4-20.

4.7.2 Electric Field at the Surface of a Metallic Conductor

We can formally show that the electric field tangential to a conductor surface should be zero by using the conservative property of the electrostatic field. Figure 4.40a shows an interface between a conductor and free space and a rectangular contour abcda having width Δw and height Δh, with sides bc and da parallel to the interface. Noting that the field inside the conductor is zero, letting $\Delta h \to 0$, and also noting

---

[71] Only a spatial averaging of the fields is necessary, since in any macroscopically significant region there are so many nuclei and electrons that temporal fluctuations are completely washed out by averaging over space. To appreciate this, note that any macroscopic amount of ordinary matter (with a lowest one-dimensional extent of say $\sim 10^{-6}$ cm, or volume of $\sim 10^{-18}$ cm^{-3}) has of order $\sim 10^6$ nuclei and electrons. For further discussion, see Section 6.7 of J. D. Jackson, *Classical Electrodynamics,* Wiley, New York, 1975.

[72] In metals, the lowest energy bands are only partially filled, with many vacant levels. For a brief elementary discussion, see Section 9.6 of R. L. Sproull, *Modern Physics, The Quantum Physics of Atoms, Solids, and Nuclei,* Wiley, New York, 1963. For more detail, see V. F. Weisskopf, *Am. J. Phys.,* 11:1943.

[73] This proportionality of $\mathbf{v}_d$ and $-\mathbf{E}$ is the basis for Ohm's law, which we will discuss in Chapter 5.

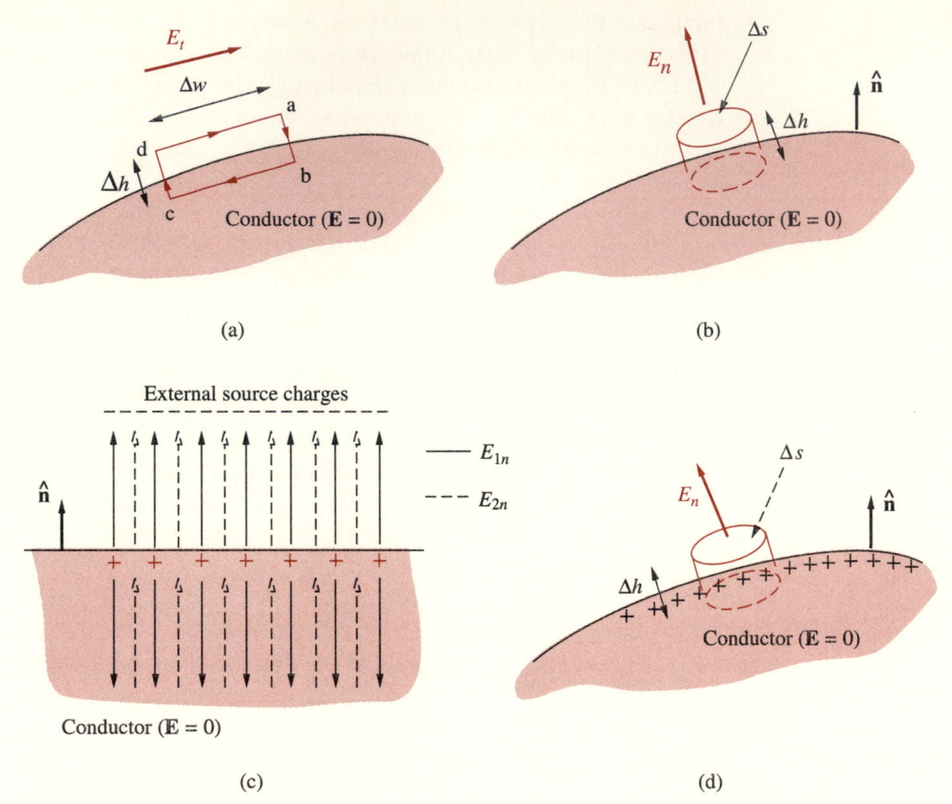

FIGURE 4.40. Field at the surface of a conductor. (a) Contour abcda used to show that $E_t = 0$. (b) Application of Gauss's law to relate normal electric field to surface charge. (c) Electric fields at the conductor surface; E_{1n} is the field due to the surface charge ρ_s, whereas E_{2n} is the field due to the external "source" charges of opposite polarity. (d) The fields at the surface of the conductor arise from two systems of charges.

that the line integral of the electrostatic field around any closed path is identically zero [4.19], we have

$$\oint_{\text{abcda}} \mathbf{E} \cdot d\mathbf{l} = \int_{\text{da}} E_t \, dl = E_t \, \Delta w = 0 \quad \rightarrow \quad \boxed{E_t = 0} \qquad [4.40]$$

Thus, the tangential component of the electric field at the surface of a conductor is identically zero. Electric field lines terminate perpendicularly on conductor surfaces. In Chapter 7, when we examine Faraday's law, we shall see that, for time-varying fields, the line integral of the electric field around a closed loop is not zero (i.e., $\oint \mathbf{E} \cdot d\mathbf{l} \neq 0$) but instead is given by the negative time rate of change of the magnetic flux passing through the surface enclosed by the closed loop. Nevertheless, the boundary condition on the conductor surface (i.e., the fact that $E_t = 0$) will be shown to be still valid in general.

Using Gauss's law it is possible to determine a relationship between the surface charge density on a conductor at equilibrium and the electrostatic field at the surface.

Consider a very small portion of any charged conducting surface, as shown in Figure 4.40b. An infinitesimal pillbox-shaped Gaussian surface is visualized as shown, with half of it above and the rest below the surface. In applying Gauss's law, we note that no flux crosses the lower surface, since the field inside the conductor is zero. No flux leaves through the sides either, since this would require that $\mathbf{E}$ have a component tangential to the surface. Furthermore, we lose no generality by considering the limit $\Delta h \rightarrow 0$, so that the total surface area of the closed surface is primarily due to the top and bottom faces. On the top surface, a normal component of $\mathbf{E}$ exists.[74] The net outward flux from this top surface is $\epsilon_0 \mathbf{E} \cdot \Delta \mathbf{s}$ and must equal the charge within it. Denoting the surface charge density on the conductor surface as ρ_s, and the electric field normal to the surface as E_n, we have

$$E_n \, \Delta s = \frac{\rho_s}{\epsilon_0} \Delta s \quad \longrightarrow \quad E_n = \hat{\mathbf{n}} \cdot \mathbf{E} = \frac{\rho_s}{\epsilon_0}$$

where $\hat{\mathbf{n}}$ is the unit vector outwardly perpendicular to the conductor surface, as shown. Note that since $\mathbf{D} = \epsilon_0 \mathbf{E}$, this condition can also be written as

$$\boxed{\hat{\mathbf{n}} \cdot \mathbf{D} = D_n = \epsilon_0 E_n = \rho_s} \qquad [4.41]$$

This more general relationship is also valid if the conductor is surrounded by an insulating material other than vacuum, except that in [4.41], ϵ_0 would have to be replaced with a different constant to account for the electrical behavior of the particular material (see Section 4.10).

If we think of the conductor surface as an infinite plane surface with charge density ρ_s, the result just obtained (i.e., $E_n = \rho_s/\epsilon_0$) seems to contradict the electric field that we found for an infinite sheet of charge in Example 4-14, which was $E_n = E_z = \rho_s/(2\epsilon_0)$. The reason for this difference is that the field at the surface of a charged conductor actually arises from two systems of charges: the local surface charge and the "source" charges, which are remote from the conductor and that are physically required by the "charged" state of the conductor.[75] In the simplest geometry, we could think of these distant charges as lying uniformly on a distant parallel planar surface, as shown in Figure 4.40c. By examination of Figure 4.40c, we can see that the field solved in Example 4-14 represented only the partial field caused by the surface charge (i.e., E_{1n}) and that the total field can be found only if the external source charges of opposite polarity are included. In other words, the conducting surface can have net induced positive charge only if these negative external source charges are present, typically because a potential difference (e.g., with a battery) is applied between the conducting surface and some other conductor (on which the external

[74]Presumably this field is the reason the conductor is charged; that is, the electric field emanates from (or converges on) the positive (negative) surface charges and connects to (emanates from) other charges of negative (positive) sign, which constitute the source of the "charging" of the conductor.

[75]Considering the common situation in which the conductor is charged by applying a potential difference between it and another conductor, these distant charges are the charges of opposite sign induced on the other conductor.

negative charges reside). The field E_{2n} due to the external charges cancels out E_{1n} inside the conductor and doubles it outside the conductor, so the net field normal to the surface of the conductor is $E_n = E_{1n} + E_{2n} = \rho_s/\epsilon_0$.

We can also understand the difference between the field near a sheet of charge and that at the surface of the conductor by considering Gauss's law with a cylindrical pillbox surface, as shown in Figure 4.40d. Since the field inside the conductor is zero, and since we take $\Delta h \to 0$, the only nonzero contribution to the surface integral is from the top surface. Thus, we have $\epsilon_0 E_n \Delta s = \rho_s \Delta s$, which gives $E_n = \rho_s/\epsilon_0$. However, in the case of the sheet of charge (Example 4-14), the field is nonzero on both the top and bottom sides of a similar pillbox surface, so we have $2\epsilon_0 E_n \Delta s = \rho_s \Delta s$, which gives $E_n = \rho_s/(2\epsilon_0)$. Both of these points of view (i.e., as discussed in relation to Figures 4.40c and d) are equally valid.

Example 4-20 quantitatively illustrates the relationship between surface electric field and charge density on a metallic conductor.

Example 4-20: Surface charge on a conductor. It is interesting to consider how small an excess or deficiency of electrons can be expected to occur on the surface of a metallic conductor in practice. Normal air breakdown occurs (see Section 4.10.3) when the magnitude of the electric field in air reaches approximately $E_{\max} = 3\times 10^6$ V-m^{-1}. (a) Find the charge density ρ_s on the surface of a metallic conductor that would produce such a field. (b) How many excess electrons would such a surface density represent in a square pillbox volume of surface area 1 μm $\times$ 1 μm and thickness one angstrom? Note that a surface density at which the field immediately outside the conductor would be at the breakdown level for air represents essentially the maximum surface charge that can be placed on a conductor in air before one sees corona breakdown.[76]

Solution:

(a) The maximum surface charge density can be found as

$$\rho_s = \epsilon_0 E_{\max} = 8.854 \times 10^{-12} \text{ F-m}^{-1} \times 3 \times 10^6 \text{ V-m}^{-1} \simeq 2.66 \times 10^{-5} \text{ C-m}^2$$

(b) Now consider the surface region of area $10^{-6} \times 10^{-6}$ m^2 and a thickness of 10^{-10} m. Such a region has a volume $10^{-6} \times 10^{-6} \times 10^{-10} = 10^{-22}$ m^3, with on average about $\sim 10^8$ free electrons and nuclei, assuming one free electron per atom and atomic dimensions of $\sim$1 angstrom^3 $= 10^{-30}$ m^3. The

---

[76]The process of corona breakdown can be understood as follows: The intense electric field accelerates the electrons and ions in air to high velocities. These high-speed particles collide with neutral air molecules and knock electrons out of the molecules. As a result, vast numbers of additional ions and electrons are produced, leading to *avalanche breakdown.* The air in the vicinity of the conductor becomes much more conducting, with the result that the charged conductor quickly loses most of its charge. The air might even glow (exhibit corona effect) because of light emitted from the air molecules during these collisions. Further discussion of breakdown effects is given in Section 4.10.3.

number of excess electrons represented by the charge density ρ_s determined above, in the same volume just described, can be found as

$$\frac{\rho_s A}{q_e} \simeq \frac{2.66 \times 10^{-5} \text{ C-m}^{-2}}{1.6 \times 10^{-19} \text{ C-e}^{-2}}(10^{-6} \times 10^{-6} \text{ m}^2) \simeq 1.66 \times 10^2 \quad \ll 10^8 \text{ electrons!}$$

Thus, even the maximum excess charge that can be placed on a metallic conductor represents a truly minute change in the total number of free electrons available in the conductor.

4.7.3 Induced Charges on Conductors

The boundary conditions $E_t = 0$ and $D_n = \rho_s$ at the surface of a conductor imply that if a conductor is placed in an externally applied electric field, then (1) the field distribution will be distorted so that the electric field lines are normal to the conductor surface, and (2) a surface charge will be induced on the conductor to support the electric field. We shall study examples of field distributions in the vicinity of conductors in Section 4.8. In this section, we consider some simple geometries to illustrate the behavior of the induced charge and its relationship to the electric field at a conceptual level.

Consider a horizontally oriented applied electric field E_{app}, into which we introduce a metallic slab, perpendicular to the electric field as shown in Figures 4.41a and b. Note that since E_t is already zero, the introduction of the conducting slab does not seem to distort the existing field configuration;[77] it appears as if the field enters the slab from the left and then re-emerges on the right side (Figure 4.41b). What actually happens is that the field pointing into the conductor on the left side induces negative charges in accordance with $D_n = \rho_s$. This negative surface charge consists of a very small fraction of the abundant supply of free electrons available in the metal. Since the metallic slab was initially neutral, movement of some negative charges toward the left side leaves an equal number of positive charges (a deficiency

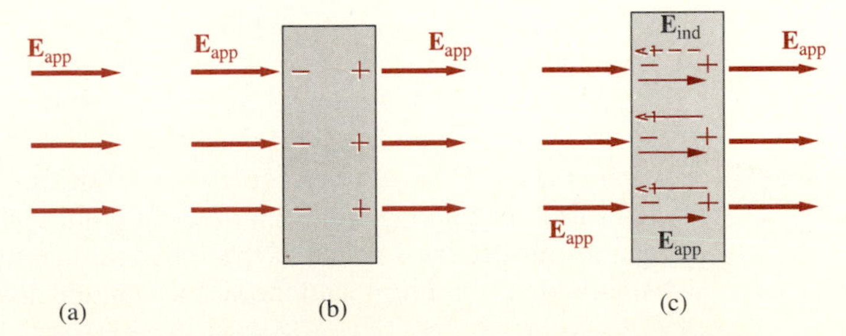

FIGURE 4.41. A conducting slab in an applied electric field.

[77]In contrast, if the conducting body were spherical, the field lines would have to bend in the vicinity of the sphere so as to be orthogonal on it. See Section 4.7 of D. K. Cheng, *Field and Wave Electromagnetics*, 2nd ed., Addison-Wesley, 1989.

of electrons). During the charge rearrangement time (typically $\sim 10^{-19}$ sec), electrons move around in such a way as to make the electric field inside the conductor identically zero. As a result, the positive charges (that are in effect produced by the fact that some electrons move to the left side) remain on the right-hand surface of the conductor, where they are supported by the electric field that emerges from the slab in accordance with $D_n = \rho_s$.

Another way of thinking about the behavior of the conductor is illustrated in Figure 4.41c. The induced negative and positive surface charge layers establish an induced electric field E_{ind} within the conductor, which balances the applied electric field E_{app} such that the total electric field in the metal is zero. In other words, the electric field induced in the metal is $E_{\text{ind}} = -E_{\text{app}}$. When the conductor is placed in the electric field, the free charges quickly establish a charge distribution to set up this internal electric field in order to cancel the applied electric field.

Example 4-21 illustrates the effect of a conducting shell on the electric field and potential distributions in the vicinity of a point charge.

Example 4-21: Spherical metallic shell. Consider a positive point charge Q at the center of a spherical metallic shell[78] of an inner radius a and outer radius b as shown in Figure 4.42a. Find the (a) electric field and (b) electrostatic potential everywhere as a function of radial distance r.

Solution:

(a) In view of the spherical symmetry, we apply Gauss's law to find $\mathbf{E}$ for three different spherical Gaussian surfaces, namely S_{o} with radius $r \geq b$, S_{m} with radius $a \leq r < b$, and S_{i} with radius $r < a$, as shown in Figure 4.42a with dashed lines. Note that due to spherical symmetry, in all three regions, we have $\mathbf{E} = \hat{\mathbf{r}}E_r$. Applying Gauss's law on the innermost surface S_{i}, we have

$$\oint_{S_{\text{i}}} \mathbf{E} \cdot d\mathbf{s} = E_r(4\pi r^2) = \frac{Q}{\epsilon_0} \quad \rightarrow \quad E_r = \frac{Q}{4\pi\epsilon_0 r^2} \qquad r < a$$

On the Gaussian surface S_{m}, which lies inside the metallic shell, the electrostatic field is identically zero as has been discussed. As a result of Gauss's law, the total charge enclosed inside S_{m} must be zero, so a negative surface charge of total amount $-Q$ (so that $Q_{\text{enclosed}} = Q - Q = 0$) is induced on the inner surface ($r = a$) of the spherical conducting shell where the electric field lines emanating from the positive point charge located at the center can terminate. As a result of the spherical symmetry, the induced charge distribution is uniform, and the surface charge density can be found from

$$\rho_s(4\pi a^2) = -Q \quad \rightarrow \quad \rho_s = -\frac{Q}{4\pi a^2}$$

[78]Note that the charge may have been placed through a small hole in the shell, or the shell may have consisted of two separate hemispheres enclosed on the charge and secured together, like that in Faraday's experimental setup as shown in Figure 4.28.

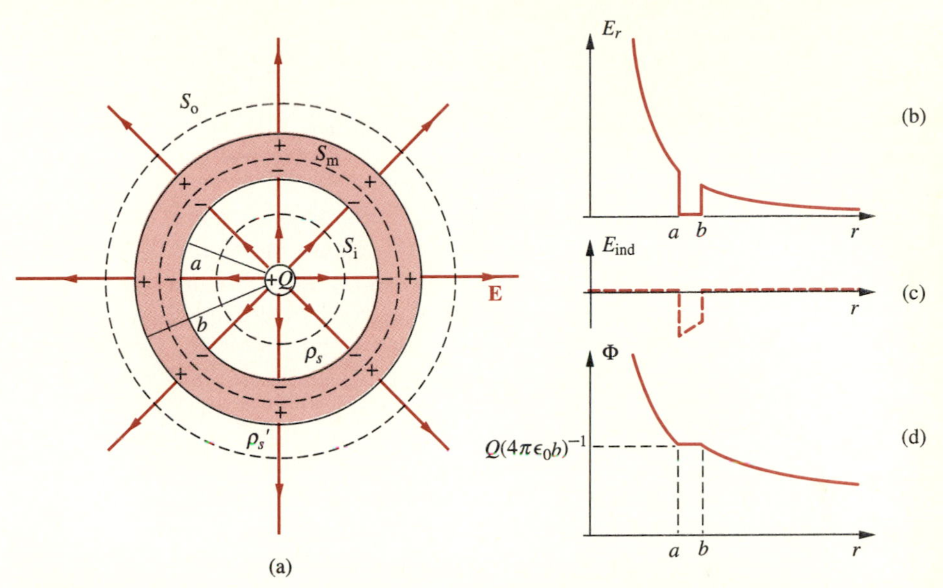

FIGURE 4.42 **Point charge at the center of a spherical conducting shell.** (a) The geometry. (b) The total electric field distribution. (c) The induced electric field distribution. (d) The electric potential distribution.

Note that since the conducting shell is initially neutral, a total amount of charge $+Q$ is also induced on the outer surface of the conductor, with a different surface charge density, given by $\rho_s' = Q/(4\pi b^2)$.

Application of Gauss's law to the outermost surface S_o yields

$$\epsilon_0 \oint_{S_o} \mathbf{E} \cdot d\mathbf{s} = E_r(4\pi r^2) = \underbrace{Q + (-Q) + Q}_{Q_{\text{enclosed}} \text{ in } S_o} = Q \rightarrow E_r = \frac{Q}{4\pi\epsilon_0 r^2} \qquad r \geq b$$

(b) The electrostatic potential Φ with respect to a point at infinity can be found by taking the line integral of $\mathbf{E}$. For points outside the spherical shell, we have

$$\Phi = -\int_{\infty}^{r} E_r\, dr = -\int_{\infty}^{r} \frac{Q}{4\pi\epsilon_0 r^2}\, dr = \frac{Q}{4\pi\epsilon_0 r} \qquad r \geq b$$

Since the conductor is an equipotential, the electrostatic potential Φ remains constant in the region $a \leq r < b$, namely

$$\Phi = -\int_{\infty}^{b} E_r\, dr - \underbrace{\int_{b}^{r} E_r\, dr}_{=0 \text{ since } E_r = 0} = \frac{Q}{4\pi\epsilon_0 b} \qquad a \leq r < b$$

For points in the region enclosed by the spherical metallic shell (i.e., $r < a$), we have

$$\Phi = -\int_{\infty}^{r} E_r\, dr = -\int_{\infty}^{b} \frac{Q}{4\pi\epsilon_0 r^2}\, dr - \int_{a}^{r} \frac{Q}{4\pi\epsilon_0 r^2}\, dr$$

$$= \frac{Q}{4\pi\epsilon_0}\left[\frac{1}{b} + \frac{1}{r} - \frac{1}{a}\right] \qquad r < a$$

The variations of E_r and Φ with radial distance are shown in Figure 4.42b and 4.42d. Also shown, in Figure 4.42c, is the variation of the induced field E_{ind}, which is set up within the conducting shell and cancels out the external field produced by the point charge Q at the center.

Further insight into the behavior of induced charges and metallic conductors in an electric field can be obtained by considering a variation of the geometry of Example 4-21 as shown in Figure 4.43, where the point charge Q is moved to a point off from the center of the spherical conducting shell. Although an analytical solution of this problem is much too involved to be considered here, we can qualitatively describe the resulting configuration. The electric field distribution in the region inside the shell is now distorted to ensure the termination of electric field lines perpendicularly to the inner surface of the shell. The negative charge induced in the inner surface of the shell is now distributed nonuniformly, in accordance with $D_n = \rho_s$,

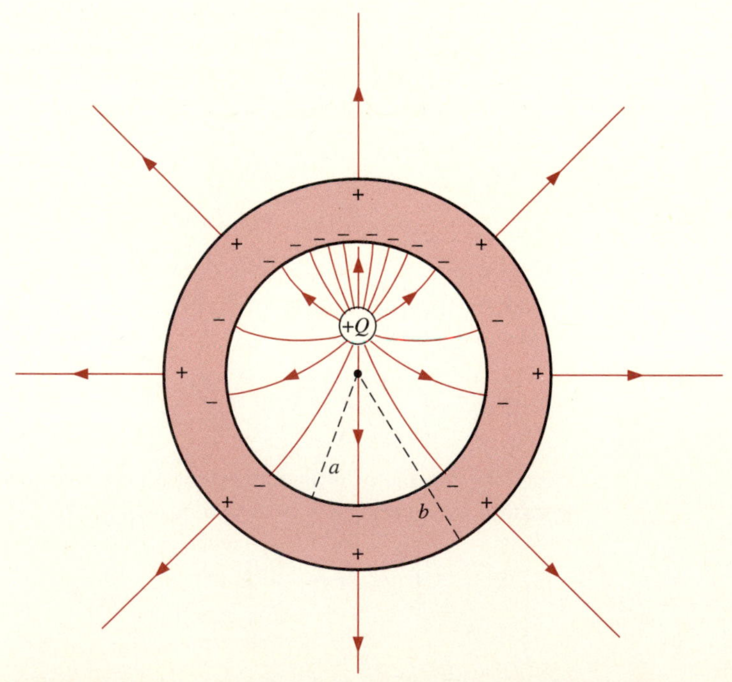

FIGURE 4.43. Off-centered point charge enclosed by a spherical conducting shell.

being denser at points of the surface closer to the charge Q, where the electric field magnitude is higher. The positive charge layer on the outer surface of the shell remains uniformly distributed, so the field at all points external to the shell remains the same, as if the exterior field originated from a point charge located at the center of the shell. In other words, the electric field produced by the charge induced in the inner surface of the conducting shell exactly cancels out the effect of the point charge for all radii larger than a.

We now can also better understand the charge displacement experiments carried out by Faraday, as depicted in Figure 4.28. The outer sphere in Faraday's experiments behaves exactly like the spherical shell considered here, with negative (positive) charge induced on its inner (outer) surface. When the outer sphere in Figure 4.28 is grounded, electrons move from the ground to the outer surface to neutralize it so that the outer sphere now has excess charge of amount exactly equal to $-Q$. With the ground connection open, and with the inner sphere removed, this $-Q$ charge stays on the conductor and is what Faraday measured.

4.7.4 Electrostatic Shielding and Tests of the Inverse Square Law

We now discuss an important consequence of Gauss's law: that the electric field inside an empty cavity completely enclosed by a conductor is identically zero, regardless of the shape of the cavity and the electric field distribution outside the cavity shell. Consider an arbitrarily shaped cavity such as that shown in Figure 4.44. For any Gaussian surface S lying entirely in the conductor as shown, the field is zero at all points on it, so there is no net electric flux through it. In other words, the total charge inside S is zero. Unless the cavity is spherical, as in the

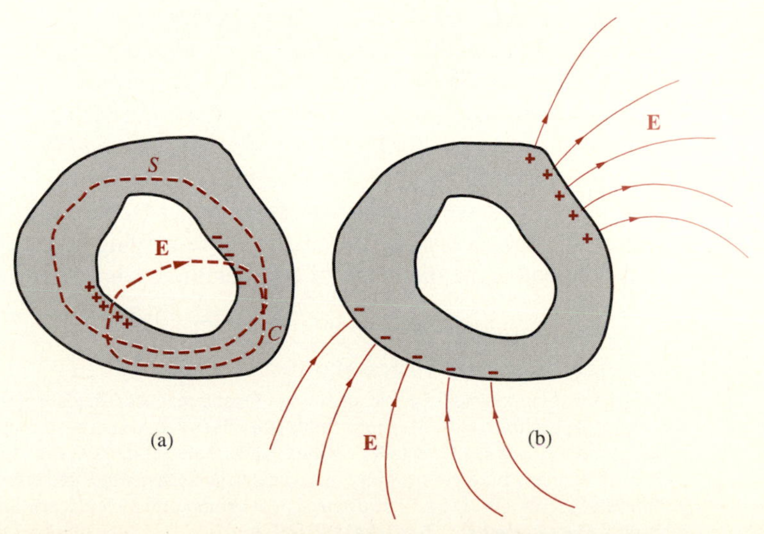

FIGURE 4.44. Electrostatic field inside an arbitrarily shaped cavity.

case of Figure 4.42, the fact that the net total charge on the inner conductor surface is zero does not necessarily imply a uniform distribution of charge. Thus, we cannot rule out the possibility of an uneven distribution of charge (positive in one region and negative in another) on this surface.

In fact, there can be absolutely no charge on the inner surface, as can be seen by the following arguments. On a qualitative basis, we observe that if equal and opposite charges existed in different parts of the inner surface, electrons would quickly move along the surface to neutralize the positive charges. We can also use the conservative property of the electrostatic field (i.e., the fact that the line integral of $\mathbf{E}$ along any closed contour is identically zero) to show that there can be no charge separation along the inner conductor surface. Assume that there is a clustering of positive charges on some part of the surface. Based on the use of Gauss's law in connection with surface S in Figure 4.44, we must then have an equal number of negative charges somewhere else. Between the positive and negative charges, there must exist an electric field $\mathbf{E}$, starting on the positive charges and ending on the negative charges (Figure 4.44a). Now consider a contour C that crosses the cavity along an electric field line and closes on itself via a path through the conductor. Since the line integral through the metal part is zero ($\mathbf{E} = 0$), we would then have

$$\oint_C \mathbf{E} \cdot d\mathbf{l} \neq 0$$

which contradicts the fact that the line integral of the electrostatic field along any closed contour is identically zero. Thus, we conclude that *there can be no fields inside the metallic cavity, nor any charges on its inner surface.* Note that our proof of this was not in any way dependent on any induced charges on the outer surface of the conductor. Thus, our conclusion is valid even in those cases when the cavity enclosure is charged or immersed in an external electric field, as shown in Figure 4.44b. In the latter case, the induced charges simply redistribute to support the external electric field, much like in the case described in connection with Figure 4.41.

The above result shows that if a cavity is completely enclosed by a metallic conductor, no static distribution of charges or fields outside can produce any fields inside. This result is the basis for the common practice of *shielding* electrical equipment by placing it in metal enclosures. Note that even extremely thin[79] conducting shells would give effective electrostatic shielding.[80] However, as we shall see in Chapter 8, this is not necessarily the case for time-varying fields. For such fields, the cavity is shielded if the thickness of the metallic enclosure is larger than the so-called

[79] Assuming macroscopic thickness of $> 10^{-6}$ m, or 1 micron.

[80] It is interesting to note here that shielding of static magnetic fields is quite a different problem from shielding of electric fields. Magnetic fields can effectively penetrate through conducting materials and thus cannot be confined to metallic cavities. Nevertheless, effective shielding of magnetic fields can be obtained by using high-permeability materials, which can confine and orient magnetic field lines away from selected regions. For a brief discussion at an appropriate level, see Section 10.2 of M. A. Plonus, *Applied Electromagnetics,* McGraw-Hill, 1978.

"skin depth." Nevertheless, it should be noted that at, for example, 3 MHz, the skin depth for copper is ~0.0382 mm, which means that, macroscopically speaking, the cavity shell can still be quite thin.

That the electrostatic field inside a metallic enclosure is zero is a direct consequence of Gauss's law, which in turn is based on Coulomb's inverse square law. Thus, this general conclusion was deduced from the single fact of the law of inverse square. This "prediction" of the inverse square law is valid for any shape of the conductor and allows for highly accurate verification of Coulomb's law. Even before Coulomb, the experiments carried out by H. Cavendish in 1772, and later repeated by J. C. Maxwell in 1879, indicated[81] that the instruments used would have shown charge on the inner sphere if the inverse square law were instead of the form $r^{-2+\zeta}$. In this way, Maxwell was able to show that $|\zeta| < 10^{-5}$. Later experimental verifications of Coulomb's law were also based on the same principle, with a more recent one by Williams, Faller, and Hill[82] establishing a bound of $|\zeta| < (2.7 \pm 3.1) \times 10^{-16}$.

It might occur to the reader that the assumption that the electrostatic force law is of the form $r^{-2+\zeta}$ might not be sufficiently general. What if the functional form of the force law as a function of distance were some other function $f(r)$? The experimental evidence that the force law depends on some power of inverse distance is the fact that conductors with similar geometric shapes have similar electrical properties when their dimensions are scaled. For example, the electric field line configurations between two spheres remain similar when we scale all physical dimensions. Also, the inverse square law can be mathematically derived from the single fact that the electric field inside a spherical conducting shell is zero.[83]

4.7.5 Forces on Metallic Conductors

A differential element of charge $dQ = \rho_s\, ds$ on the surface of a metallic conductor experiences the electrostatic field of all the other charges in the system (i.e., external charges as in Figure 4.40c). Since the charge $dQ = \rho_s\, ds$ is bound to the conductor and is prevented from leaving it by atomic forces,[84] the force acting on $dQ = \rho_s\, ds$ is transmitted to the solid conductor itself. To determine the force on the conducting body, we need only to evaluate the forces on the surface charges. For this purpose, any small area on the surface of a conductor can be treated as approximately

[81] H. Cavendish, *Electrical Researches,* J. C. Maxwell (ed.), Cambridge Univ. Press, 1879, pp. 104–113. For an excellent description of these and other earlier experiments, see R. S. Elliott, *Electromagnetics.* IEEE Press, Piscataway, New Jersey, Chapter 3, 1993.

[82] E. R. Williams, J. E. Faller, and H. A. Hill, A new experimental test of Coulomb's law: a laboratory upper limit on photon rest mass, *Phys. Rev. Lett.,* 26, p. 721, 1971.

[83] See Chapter 3 of R. S. Elliott, *Electromagnetics,* IEEE Press, Piscataway, New Jersey, 1993.

[84] For a discussion of why electrons cannot easily escape from a metal surface, see Sec. 2.3 of J. D. Jackson, *Classical Electrodynamics,* Wiley, 1975, and Sec. 2.12 of M. A. Plonus, *Applied Electromagnetics,* McGraw-Hill, New York, 1978.

planar, and we can separately consider the electric field E_{2n} due to the external sources and E_{1n} due to the surface charges (see Figure 4.40c). Since the latter cannot exert a force on the charge producing itself, the force on the surface charge element $dQ = \rho_s\,ds$ is entirely due to the external field $E_{2n} = \rho_s/(2\epsilon_0)$. Since the electrostatic force is $F = qE$, the differential force on this charge element is

$$dF = E_{2n}\,dQ = \frac{\rho_s}{2\epsilon_0}\rho_s\,ds$$

so that the differential force per unit area is

$$\frac{dF}{ds} = \frac{\rho_s^2}{2\epsilon_0} = \frac{\epsilon_0 E_n^2}{2}$$

where $E_n = E_{1n} + E_{2n} = \rho_s/\epsilon_0$ is the total electric field normal to the surface of the conductor.

Note that the above result was derived by assuming the surface charge to reside in an infinitely thin plane sheet, with the external charges also planar, as implied in Figure 4.40c. In practice, the surface charge resides in a layer of finite thickness, and the surface may have roughness on a microscopic scale. Nevertheless, the force per unit area for conductor surfaces of any other type of geometrical shape and in cases where the surface charge resides in a region of finite thickness, is given by $\epsilon_0 E_n^2/2$.

The above result implies that any charged conductor surface is under the influence of a force that pulls it in the direction normal to the surface. Note that since the force is proportional to ρ_s^2, an induced surface charge of any polarity, or equivalently incident electric field on the surface of any direction, results in a pull on the surface. The magnitude of the pull on different parts of the surface varies in accordance with the surface charge density as ρ_s^2 or the normal electric field as E_n^2. If the conductor surface is deformable, one can use the electrostatic force density together with the mechanical properties to determine the equilibrium shape that the conductor surface attains. For rigid conductor surfaces, the total force F_{total} on the conducting body can be found by integrating the force per unit area over the surface:

$$F_{\text{total}} = \int \frac{dF}{ds}\,ds = \int \frac{\epsilon_0 E_n^2}{2}\,ds$$

To appreciate the magnitude of the electrostatic force on a conductor surface, consider an aluminum surface region of size $10^{-2} \times 10^{-2}$ m^2 with a thickness of 10^{-3} m (1 mm). At an air breakdown electric field of 3×10^6 V-m^{-1}, the electrical force on this conductor per unit area would be

$$\frac{dF}{ds} = \frac{\epsilon_0 E_n^2}{2} = \frac{8.854 \times 10^{-12}\text{ F-m}^{-1} \times (3 \times 10^6\text{ V-m}^{-1})^2}{2} \simeq 39.8\text{ N-m}^{-2}$$

Thus, the total electrical force experienced by this thin 1 cm $\times$ 1 cm square shaped piece of aluminum would be $39.8 \times 10^{-2} \times 10^{-2} \simeq 4 \times 10^{-3}$ newtons. For comparison, we can consider the weight (or gravitational force) of such a piece of aluminum. Noting that the mass density of aluminum is 2.7×10^3 kg-m^{-3}, and the gravitational constant is $g \simeq 9.8$ m-s^{-2}, the piece of aluminum with a volume of $10^{-2} \times 10^{-2} \times 10^{-3} = 10^{-7}$ m^3 weighs $(2.7 \times 10^3)(10^{-7})(9.8) \simeq 2.65 \times 10^{-3}$ newtons, which is almost equal to the electrostatic force. In other words, the electrostatic force is barely able to lift such a piece of aluminum. Note, however, that the electrostatic force does not depend on the thickness of the material and is the same for any 1 cm $\times$ 1 cm piece of material having a macroscopic thickness of larger than, say 10^{-6} m. Thus, the electrostatic force produced at the surface of aluminum by an electric field at the air breakdown level is larger than the gravitational force for conductor thicknesses of less than or equal to $\sim$1 mm.

Example 4-22: Can electrostatic forces break aluminum? The tensile strength for commercial hard-drawn aluminum is 2.9×10^8 N-m^{-2}. Find (a) the charge density ρ_s required to apply enough force to break aluminum, (b) the corresponding number of excess electrons in a surface region of area 1 μm $\times$ 1 μm and thickness one angstrom, (c) the corresponding amount of total surface charge for an aluminum sphere of radius 10 cm, and (d) the corresponding value of the electric field E_n just outside the surface.

Solution:

(a) Since the force per unit area is $\rho_s^2/(2\epsilon_0)$, the surface charge density corresponding to 2.9×10^8 N-m^{-2} is

$$\rho_s = \sqrt{2 \times 8.854 \times 10^{-12} \times 2.9 \times 10^8} \simeq 7.17 \times 10^{-2} \quad \text{C-m}^{-2}$$

(b) Now consider the surface region of area A $= 10^{-6} \times 10^{-6}$ m^2. The total number of excess electrons in such an area is

$$\frac{\rho_s A}{q_e} = \frac{(7.2 \times 10^{-2}\ \text{C-m}^{-2})(10^{-6} \times 10^{-6}\ \text{m}^2)}{1.6 \times 10^{-19}\ \text{C}} \simeq 4.48 \times 10^5 \text{ electrons}$$

which is a small number compared with the total number of available free electrons in such a volume, on the order of $\sim 10^8$, as determined in Example 4-20. It thus appears that a relatively small number of excess electrons is required for the electrostatic force to break aluminum.

(c) On a sphere of radius $a = 10$ cm, the charge density of 7.2×10^{-2} C-m^{-2} corresponds to a total charge of $(7.2 \times 10^{-2}\ \text{C-m}^{-2})(4\pi a^2) \simeq 9.0 \times 10^{-3}$ coulombs. While this amount of charge may not seem to be large, we shall see later[85] that placing such a charge on an aluminum sphere of 10 cm radius requires the application of a potential difference (between the sphere and the ground) of $\sim 8.1 \times 10^8$ volts, or $\sim$800 megavolts!

---

[85] Any configuration of conductors has a finite *capacity* to hold charge per unit applied voltage, as we shall study in a later section on capacitance.

(d) The electric field E_n just outside the surface corresponding to a $\rho_s \simeq 7.2 \times 10^{-2}$ C-m^{-2} is

$$E_n = \frac{\rho_s}{\epsilon_0} \simeq \frac{7.2 \times 10^{-2}}{8.854 \times 10^{-12}} \simeq 8.1 \times 10^9 \text{ V-m}^{-1}$$

which is well above the breakdown field for air of 3×10^6 V-m^{-1}. Thus, it is completely impractical to attain charge densities large enough to break metallic conductors such as aluminum electrostatically.[86]

4.8 POISSON'S AND LAPLACE'S EQUATIONS

We noted in Section 4.3 that *if the positions of all the charges* (discrete or continuous) *are known,* the solution of electrostatic problems boils down to carrying out appropriate integration operations. We also noted that, often the locations of the charges is *not known,* and that we know only the general rules (i.e., boundary conditions) by which charges organize themselves under the influence of fields. We can appreciate this point better now that we have seen (Example 4-20) that slight rearrangements of free charge in conducting bodies can create very large electric fields. In practical problems, often the constraints on the field distributions are not expressed in terms of the charge distributions but rather are given in terms of the electrostatic potentials at which the various conducting bodies are held. In such cases, it is convenient to determine the electric field distributions by solving the equations written directly in terms of the electrostatic potential Φ. These equations, easily derived from Gauss's law as will be shown, are Poisson's and Laplace's equations, and are the subjects of this section.

Poisson, Simeon Denis *(b. June 21, 1781, Pithiviers, Loiret; d. April 25, 1840, Paris) French mathematician. Poisson studied at École Polytechnique under Laplace [1749–1827] and Lagrange [1736–1813], impressing both of them with his abilities. He was offered a teaching position upon his graduation in 1800 and in 1806 replaced Fourier [1768–1830] in an important professorial position. Poisson worked hard to refine the earlier work of Laplace and Lagrange and the work of Fourier on heat and applied mathematics to electricity and magnetism. He is best known for his work on probability; the so-called Poisson distribution deals with events that are in themselves improbable but occur because of the large*

[86]However, with a similar result for magnetic fields, this type of consideration becomes extremely important in the design of high-field magnetic coils.

number of chances for them to occur (e.g., airplane accidents). [Adapted with permission from Biographical Encyclopedia of Science and Technology, *I. Asimov, Doubleday, 1982].*

In Section 4.6, we derived the differential equation [4.35], Gauss's law at a point, which implies in turn that

$$\boxed{\nabla \cdot \mathbf{D} = \rho \quad \rightarrow \quad \nabla \cdot \mathbf{E} = \frac{\rho}{\epsilon_0}} \qquad [4.42]$$

which is valid for free space, for which ϵ_0 is a simple constant, so that it can be safely taken out of the spatial derivative. We also know from Section 4.4 that the electrostatic field is the negative gradient of the electrostatic potential Φ. Thus, we have

$$\nabla \cdot (-\nabla\Phi) = \frac{\rho}{\epsilon_0}$$

$$\nabla \cdot (\nabla\Phi) = -\frac{\rho}{\epsilon_0}$$

$$\left(\hat{\mathbf{x}}\frac{\partial}{\partial x} + \hat{\mathbf{y}}\frac{\partial}{\partial y} + \hat{\mathbf{z}}\frac{\partial}{\partial z}\right) \cdot \left(\hat{\mathbf{x}}\frac{\partial}{\partial x} + \hat{\mathbf{y}}\frac{\partial}{\partial y} + \hat{\mathbf{z}}\frac{\partial}{\partial z}\right)\Phi = -\frac{\rho}{\epsilon_0}$$

$$\left(\frac{\partial^2}{\partial x^2} + \frac{\partial^2}{\partial y^2} + \frac{\partial^2}{\partial z^2}\right)\Phi = -\frac{\rho}{\epsilon_0}$$

$$\nabla^2\Phi = -\frac{\rho}{\epsilon_0} \qquad [4.43]$$

Equation [4.43] is known as *Poisson's equation.* For a charge-free region of space, where $\rho = 0$, it reduces to *Laplace's equation,* namely

$$\boxed{\nabla^2\Phi = 0} \qquad [4.44]$$

The notation ∇^2 is referred to as the *Laplacian* and represents the operation

$$\nabla^2(\cdot) \equiv \nabla \cdot \nabla(\cdot) = \nabla \cdot [\nabla(\cdot)] \qquad [4.45]$$

Laplace, Pierre Simon *(b. March 28, 1749, Beaumont-en-Auge, Calvados; d. March 5, 1827, Paris) French astronomer and mathematician. Laplace became a professor of mathematics at an early age and initially worked on* *specific heats of substances. He then turned his attention to the application of perturbation theory to study the orbits of solar system planets and the stability of the solar system. He worked separately but cooperatively with Lagrange [1736–1813] on various anomalies in the orbits of the moon, Jupiter, and Saturn and showed that the solar system would remain stable indefinitely if it remained effectively isolated. The results of his studies were summed up in a monumental five-volume work called* Celestial Mechanics, *published between 1799 and 1825, notorious for stating that "it is obvious" that one equation follows from another, when in fact students must often spend hours or days to determine why it is so obvious. Laplace was reluctant to give credit to others when it was due and did less than justice to Lagrange's contributions to their joint work on celestial mechanics. Laplace was able to get along with political leaders who respected his reputation and his ability to apply his mathematics to problems involving artillery fire. Napoleon made Laplace minister of interior but, when he proved to be rather incompetent in that post, promoted him to the decorative position of senator. When Louis XVIII came to the throne after Napoleon's fall, Laplace was not penalized for working with Napoleon but was instead made a marquis. He became president of the French Academy in 1817. Between 1812 and 1820 Laplace wrote a treatise on the theory of probability, giving this branch of mathematics its modern form. [Adapted with permission from* Biographical Encyclopedia of Science and Technology, *I. Asimov, Doubleday, 1982].*

If the electric field is known everywhere in a region where metallic conductors are present, the charges on the surfaces of the conductors can be determined using [4.41]. Conversely, if the surface charge densities on the metallic conductors is given, the electric field **E** can be easily calculated by using [4.11] or [4.42]. However, in typical electrostatic problems, neither $\rho_s(x, y, z)$ nor $\mathbf{E}(x, y, z)$ is likely to be known *a priori.* The much more common electrostatic problem involves a given geometrical configuration of charged conductors[87] in space otherwise free of charge. The objective is usually to solve for the electrostatic potential $\Phi(x, y, z)$ that satisfies Laplace's equation along with the boundary conditions on charged conductors. The boundary conditions are usually in the form of $\Phi_i(x, y, z) = \text{const.}$, on the different metallic conductor surfaces $S_i(x, y, z)$. In addition, the interior of the metallic conducting bodies must also be at the same potential as their surfaces. In other words, the problem specification consists of the geometry of the conductor configuration and the potentials of the different conductors, and the goal is to obtain the scalar potential $\Phi(x, y, z)$ by solving the partial differential equation [4.44].

Relatively simple analytical solutions of [4.44] can be found only in a few cases involving symmetries of one form or another. A variety of powerful methods of

[87]Conductors are usually charged by applying a potential difference between two or more conductors or between a conductor and ground.

mathematical physics have been used to find solutions for more complicated geometries in terms of infinite sums of cylindrical and spherical harmonics.[88] A host of powerful numerical methods has also been employed and specifically developed[89] for the solution of [4.44]. As digital computer speed and memory capacity continue to increase, numerical solutions are becoming increasingly attractive, utilizing a variety of powerful techniques for the solution of partial differential equations subject to boundary conditions.[90] It can be shown[91] in general that the electrostatic potentials that satisfy the Laplace or Poisson equation in regions with prescribed boundary conditions are unique. In this section we confine our attention to examples of cases that lend themselves to a direct analytical solution.

Examples 4-23, 4-24, and 4-25 illustrate the use of Laplace's equation to determine electrostatic potential and electric field distributions for simple geometries.

Example 4-23: Two parallel plates. Consider two parallel conducting planes of infinite extent in the x and y directions and separated by a distance d in the z direction, as shown in Figure 4.45. A potential difference is applied between the plates, such that the bottom plate, located in the xy plane, is at zero potential while the top plate, located at $z = d$, is at potential V_0 (i.e., $\Phi(x, y, 0) = 0$ and $\Phi(x, y, d) = V_0$). The region between the plates is free space and contains no free charges. (a) Find $\Phi(x, y, z)$ and $\mathbf{E}(x, y, z)$ in the region between the plates. (b) Find the surface charge densities induced on the two plates.

Solution:

(a) Laplace's equation [4.44] in rectangular coordinates is

$$\nabla^2\Phi = \frac{\partial^2\Phi}{\partial x^2} + \frac{\partial^2\Phi}{\partial y^2} + \frac{\partial^2\Phi}{\partial z^2} = 0$$

---

[88] W. R. Smythe, *Static and Dynamic Electricity,* revised printing, Hemisphere Publishing Corp., Washington, DC, 1989; J. Van Bladel, *Electromagnetic Fields,* revised printing, Hemisphere Publishing Corp., Washington, DC, 1985; R. E. Collin, *Field Theory of Guided Waves,* revised printing, IEEE Press, Piscataway, New Jersey, 1991.

[89] R. Mittra, ed., *Computer Techniques for Electromagnetics,* Pergamon, Oxford, p. 7, 1973; K. Umashankar and A. Taflove, *Computational Electromagnetics,* Artech House, 1993.

[90] These techniques include the so-called finite-element methods, finite-difference methods, Fourier transformations, and the method of moments. For a description of these techniques and their applications see P. P. Silvester and R. L. Ferrari, *Finite Elements for Electrical Engineers,* Cambridge Univ. Press, 1983, and R. F. Harrington, *Field Computation by Moment Methods,* IEEE Press, Piscataway, New Jersey, 1993.

[91] See Section 1.17 of S. Ramo, J. R. Whinnery, and T. Van Duzer, *Fields and Waves in Communication Electronics,* 3rd ed., Wiley, New York, 1994. Many other proofs, basically all along the same lines, can be found in other texts.

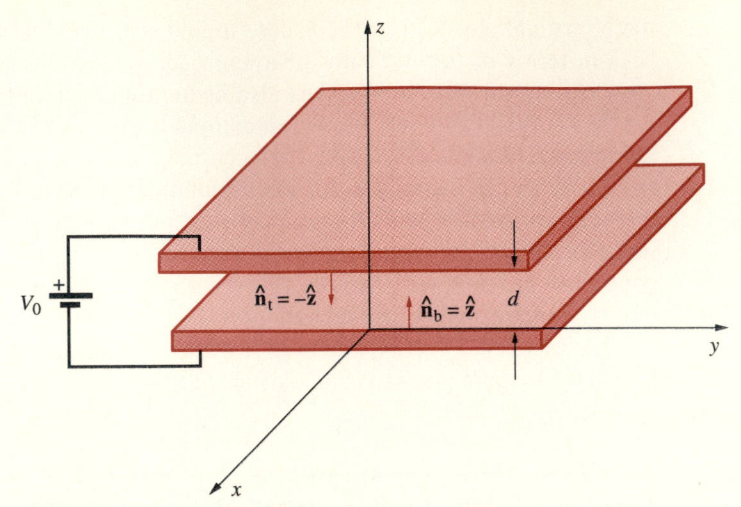

FIGURE 4.45. Two parallel plates. Two infinite-extent parallel plates of potential difference V_0 and separation distance d (Example 4-23).

It is clear from the symmetry of the problem that the potential does not vary with x or y but instead depends only on z. Hence Laplace's equation reduces to

$$\nabla^2\Phi = \frac{\partial^2\Phi}{\partial z^2} = 0$$

which can be directly solved by double integration,

$$\Phi(z) = C_1 z + C_2$$

where C_1 and C_2 are the integration constants to be determined by the boundary conditions given in the problem specification. Namely,

$$\left.\begin{aligned}\Phi(z = 0) &= 0\\ \Phi(z = d) &= V_0\end{aligned}\right\} \quad \rightarrow \quad \begin{aligned}C_2 &= 0\\ C_1 &= \frac{V_0}{d}\end{aligned}$$

Therefore, the potential $\Phi(z)$ is

$$\Phi(z) = \frac{V_0}{d}z$$

which varies linearly from 0 to V_0 in the region between the plates. The corresponding electric field can be found by taking the negative gradient of $\Phi(z)$, namely,

$$\mathbf{E} = -\nabla\Phi(z) = -\hat{\mathbf{z}}\frac{V_0}{d}$$

which indicates that the electric field is constant in the region between the plates and is directed toward the plate having the lower potential.

(b) The surface charge densities induced on the bottom ($\rho_{s,\mathrm{b}}$) and top ($\rho_{s,\mathrm{t}}$) conducting plates can be found as

$$\rho_{s,\mathrm{b}} = \epsilon_0(\hat{\mathbf{n}}_\mathrm{b} \cdot \mathbf{E}) = \epsilon_0 \left[\hat{\mathbf{z}} \cdot \left(-\hat{\mathbf{z}}\frac{V_0}{d}\right)\right] = -\epsilon_0 \frac{V_0}{d}$$

$$\rho_{s,\mathrm{t}} = \epsilon_0(\hat{\mathbf{n}}_\mathrm{t} \cdot \mathbf{E}) = \epsilon_0 \left[-\hat{\mathbf{z}} \cdot \left(-\hat{\mathbf{z}}\frac{V_0}{d}\right)\right] = +\epsilon_0 \frac{V_0}{d}$$

Note that the unit vectors $\hat{\mathbf{n}}_\mathrm{b}$ and $\hat{\mathbf{n}}_\mathrm{t}$ are defined to be the *outward* normals of the two plates as shown in Figure 4.45.

Example 4-24: Coaxial cylinders. The geometry of this problem is very similar to that in Example 4-15, where the surface charge densities on each of two coaxial cylindrical surfaces were specified and the electric field was directly evaluated using Gauss's law. In this case, illustrated in Figure 4.46, the problem specification is that the surfaces are conducting cylinders with a potential difference applied between the two cylinders such that

$$\Phi(r = a) = V_0 \qquad \text{and} \qquad \Phi(r = b) = 0$$

Determine the potential $\Phi(r, \phi, z)$ by direct solution of [4.44] and subsequently the electric field from $\mathbf{E}(r, \phi, z) = -\nabla\Phi(r, \phi, z)$. We are interested only in the electric field and potential in the region between the cylinders, which is taken to be free space.

Solution: Note that cylindrical coordinates are most appropriate in view of the cylindrical conductor boundaries. Equation [4.44] written in cylindrical

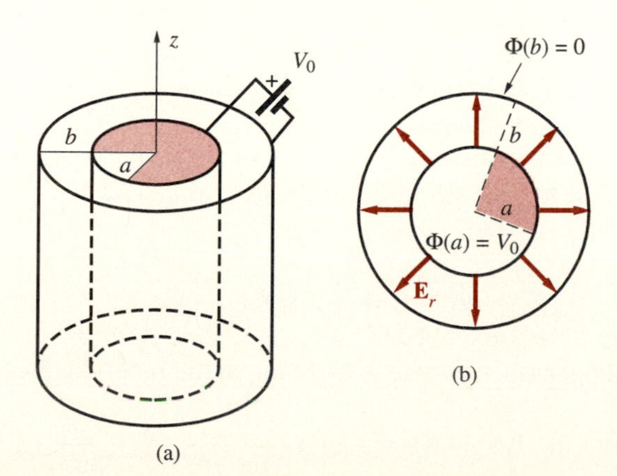

FIGURE 4.46. Two concentric cylinders. (a) Two concentric cylindrical conductors of radii a and b ($a < b$). (b) Cross-sectional view with the electric field lines shown.

coordinates (as can be found using the definition of the nabla operator as given in [4.45]) is

$$\nabla^2\Phi = \frac{1}{r}\frac{\partial}{\partial r}\left(r\frac{\partial \Phi}{\partial r}\right) + \frac{1}{r^2}\frac{\partial^2\Phi}{\partial \phi^2} + \frac{\partial^2\Phi}{\partial z^2} = 0$$

Assuming that the conductors are sufficiently long, we do not expect any dependence of the potential on z, except near the ends, which we shall neglect for our purposes here. Also, there cannot be any dependence of Φ on ϕ, because of the cylindrical symmetry. Thus, we have

$$\frac{1}{r}\frac{\partial}{\partial r}\left(r\frac{\partial \Phi}{\partial r}\right) = 0$$

This equation can be solved by two successive direct integrations, yielding

$$\Phi(r) = C_1 \ln r + C_2$$

where C_1 and C_2 are the integration constants, which may be determined from the boundary conditions given in the problem specification. Namely, we have

$$\Phi(r = a) = V_0 = C_1 \ln a + C_2$$

$$\Phi(r = b) = 0 = C_1 \ln b + C_2$$

From these two equations, we can solve for C_1 and C_2 to find

$$C_1 = \frac{V_0}{\ln(a/b)} \qquad \text{and} \qquad C_2 = -\frac{V_0 \ln b}{\ln(a/b)}$$

which fully specifies the potential $\Phi(r)$ as

$$\Phi(r) = \frac{V_0}{\ln(a/b)}(\ln r - \ln b) = \frac{V_0}{\ln(a/b)} \ln\frac{r}{b}$$

which is identical to the electrostatic potential $\Phi(r)$ found in Example 4-15, when we note that $\Phi_{ba} = V_0$. Note that the two solutions are identical only because the cylindrical surface charge densities in the specification of Example 4-15 were appropriately chosen.

The electric field in the region between the conducting cylinders can be found by taking the negative gradient of the potential. Since $\Phi(r)$ is only a function of r, we have

$$\mathbf{E} = \hat{\mathbf{r}}E_r = -\hat{\mathbf{r}}\frac{\partial \Phi}{\partial r} = \hat{\mathbf{r}}\frac{V_0}{r\ln(b/a)}$$

The corresponding surface charge density ρ_{si} on the inner conductor is

$$\rho_{si} = \epsilon_0 \hat{\mathbf{n}} \cdot \mathbf{E}(r = a) = \epsilon_0 \hat{\mathbf{r}} \cdot \hat{\mathbf{r}}\frac{V_0}{a\ln(b/a)} = \frac{V_0\epsilon_0}{a\ln(b/a)}$$

where $\hat{\mathbf{n}}$ is the outward normal. On the outer conductor we have

$$\rho_{so} = \epsilon_0 \hat{\mathbf{n}} \cdot \mathbf{E}(r = b) = \epsilon_0(-\hat{\mathbf{r}}) \cdot \hat{\mathbf{r}} \frac{V_0}{b \ln(b/a)} = \frac{-V_0 \epsilon_0}{b \ln(b/a)} = -\rho_{si}\left(\frac{a}{b}\right)$$

just as was specified in the statement of Example 4-15.

Example 4-25: Potential around a conducting sphere. Consider a spherical conductor that is maintained at a potential of $\Phi = V_0$ by means of a battery connection as shown in Figure 4.47. Noting that spherical coordinates are appropriate, find $\Phi(r, \theta, \phi)$, and $\mathbf{E}(r, \theta, \phi)$. Assume that the region around the sphere is free space and that no other charges or conductors are nearby, so that we have $\Phi \to 0$ as $r \to \infty$.

Solution: Laplace's equation in spherical coordinates is

$$\nabla^2 \Phi = \frac{1}{r^2} \frac{\partial}{\partial r}\left(r^2 \frac{\partial \Phi}{\partial r}\right) + \frac{1}{r^2 \sin\theta} \frac{\partial}{\partial \theta}\left(\sin\theta \frac{\partial \Phi}{\partial \theta}\right) + \frac{1}{r^2 \sin^2\theta} \frac{\partial^2 \Phi}{\partial \phi^2} = 0$$

In view of the spherical symmetry, Φ cannot be a function of θ or ϕ, so we have

$$\frac{1}{r^2} \frac{\partial}{\partial r}\left(r^2 \frac{\partial \Phi}{\partial r}\right) = 0 \quad \to \quad \frac{\partial}{\partial r}\left(r^2 \frac{\partial \Phi}{\partial r}\right) = 0$$

Integrating twice with respect to r, we find

$$r^2 \frac{\partial \Phi}{\partial r} = C_1 \quad \to \quad \Phi(r) = -\frac{C_1}{r} + C_2$$

where C_1 and C_2 are integration constants to be determined by the boundary conditions. One boundary condition is that $\Phi(r = a) = V_0$, which gives

$$-\frac{C_1}{a} + C_2 = V_0 \quad \to \quad \Phi(r) = \frac{V_0 a}{r} + C_2\left(1 - \frac{a}{r}\right)$$

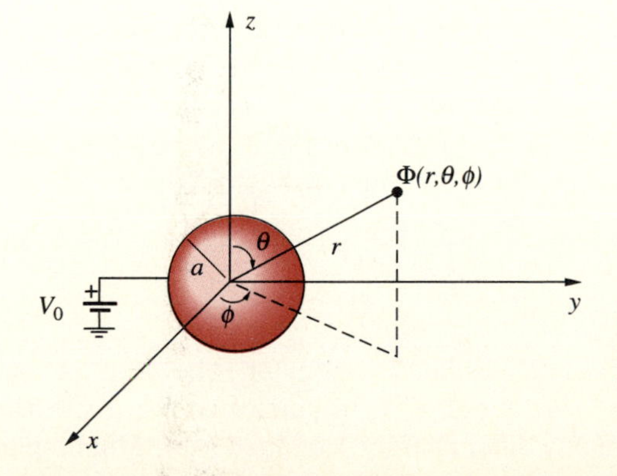

FIGURE 4.47. Potential around a conducting sphere. The sphere of radius a is held at a potential (with respect to ground) of V_0. Also shown are the spherical coordinates r, θ, ϕ.

To determine C_2, we need a second boundary condition. For a sphere of finite size, we would expect the potential $\Phi(r)$ to vanish as $r \to \infty$, which in turn implies that $C_2 = 0$. Our solution is then

$$\Phi(r) = \frac{V_0 a}{r} \quad \to \quad \mathbf{E} = -\hat{\mathbf{r}}\frac{\partial \Phi}{\partial r} = \hat{\mathbf{r}}\frac{V_0 a}{r^2}$$

This solution can be compared with what can be found using Gauss's law, if we were to assume that raising the sphere to a potential V_0 deposits a charge Q on it. In such a case, the electrostatic potential for $r > a$ would be

$$\Phi(r) = \frac{Q}{4\pi\epsilon_0 r}$$

For the two solutions to be identical, we must have

$$Q = V_0(4\pi\epsilon_0 a)$$

We shall see in the next section that the ratio of Q/V_0 represents the capacity of the sphere to hold charge per unit applied voltage.

4.9 CAPACITANCE

The relation between induced charges on a set of conductors and the resulting potentials in their vicinity depends only on the geometric arrangement of the conductors. If we double the surface charge densities at every point on the conductors (presumably by doubling the applied potential difference between the conductors), the configuration (i.e., spatial variation) of the electric field remains unchanged, but the magnitude of the field is doubled, and hence the work required to take a unit positive test charge from one conductor to another (i.e., potential difference between them) is also doubled.

Consider a single isolated conductor of any shape. If we place a charge Q on it, the charge is distributed in some equilibrium pattern on the surface of the conductor. If we then increase the surface charge density ρ_s everywhere by the same factor, we still have an equilibrium arrangement, since the electrostatic potential is everywhere greater by the same factor. This constant charge-to-potential ratio of an isolated conductor is its *capacitance*. In other words, we define

$$\boxed{C \equiv \frac{Q}{\Phi}} \qquad [4.46]$$

where Φ is the potential of the isolated conductor. The units of capacitance are coulombs per volt (C-V^{-1}), or farads (F). In particular, for a spherical conductor of radius a, far away from any other charges, conductors, or ground planes, we have $\Phi(r = a) = \Phi = Q/(4\pi\epsilon_0 a)$ (see Example 4-25), so that we have

$$C = 4\pi\epsilon_0 a \qquad [4.47]$$

For example, a sphere of radius $a = 10$ cm would have a capacitance of

$$C = 4\pi \underbrace{(8.854 \times 10^{-12})}_{\text{farads-m}^{-1}} \underbrace{(10 \times 10^{-2})}_{\text{m}} \simeq 11.1 \text{ pF}$$

In Example 4-20 it was found that a surface charge density of $\rho_s \simeq 2.66 \times 10^{-5}$ C-m^{-2} on a metallic conductor would cause electrical breakdown of the surrounding air. Such a charge density uniformly distributed over the surface of a sphere of radius $a = 10$ cm corresponds to a total charge of $Q = \rho_s 4\pi a^2 \simeq 3.34 \times 10^{-6}$ C. Using the capacitance value of $C \simeq 11.1$ pF, we find that placement of that much charge on a 10-cm radius sphere requires $\Phi = Q/C \simeq (3.34\times10^{-6})/(11.1\times10^{-12}) \simeq 3 \times 10^5$ V, or ~300 kilovolts!

In the case of an isolated conductor, the electric flux that leaves the conductor can be thought to terminate at infinity. If the amount of charge on the conductor is increased, the flux pattern remains the same, but the flux density **D** proportionally increases. If we have a pair of conductors and if these conductors are given equal and opposite charges (by applying a potential difference $\Phi_{12} = \Phi_2 - \Phi_1$ between them[92]), then all of the flux emanating from the positively charged conductor terminates on the negatively charged one. If the potential difference is increased, the magnitude of charge on both conductors, as well as the flux density **D**, also increases, without a change in the distribution of flux. Once again, we see that the amount of charge and the potential difference are proportional to one another, and we define their ratio to be the *capacitance* of the two-conductor configuration:

$$C \equiv \frac{Q}{\Phi_2 - \Phi_1} = \frac{Q}{\Phi_{12}} \qquad [4.48]$$

In [4.48], Q is the amount of charge on one of the conductors, usually taken to be the positive one. Note that the case of the single isolated conductor can be considered as a special case with the second conductor being at "infinity," so that $\Phi_2 = 0$. In summary, capacitance is the measure of the ability of a conductor configuration to hold charge per unit applied voltage between the conductors (i.e., their capacity to store charge).

In the following examples, we evaluate the capacitance of some simple two-conductor configurations, including the parallel-plate and coaxial transmission line configurations that were considered in Section 2.7, the capacitance expressions for which were given in Table 2.2. Although the capacitance expressions for these

[92]When a battery is connected between the two conductors, each conductor becomes an extended battery terminal. Within $\sim 10^{-19}$ s following the connection, electrons flow from the conductor connected to the positive terminal through the battery and connecting wires and to the conductor connected to the negative terminal, and equilibrium is established. As mentioned before, the time involved here ($\sim10^{-19}$ s) is the time in which charge is rearranged in a material, which is in the range of 10^{-18} to 10^{-19} s for most metallic conductors. The subject of redistribution of charge is discussed in Section 5.5.

configurations are derived on the basis of electrostatic formulations, they are nevertheless quite valid up to optical frequencies.[93]

Example 4-26: Parallel-plate capacitor. Consider two parallel conducting plates separated by free space as shown in Figure 4.48. Determine the capacitance C.

Solution: We start by assuming that the plates are charged so that the upper plate has a total charge of $+Q$ while the lower plate has $-Q$. The charge on both plates is uniformly distributed on the inner surfaces, supporting an electric field pattern as shown. Note that when the smallest linear dimension of the plates is sufficiently larger than the plate separation d, we can neglect "fringing" fields on the sides and assume the electric field between the plates to be identical to that in Example 4-23. If the area of the plates is A, the surface charge density is $\rho_s = Q/A$, positive on the upper plate and negative on the lower one.

Based on the results of Section 4.7.2 and Example 4-23, the electric field between the plates is

$$\mathbf{E} = -\hat{\mathbf{z}}\frac{\rho_s}{\epsilon_0}$$

The potential difference between the plates can be found from

$$\Phi_{12} = -\int_0^d \left(-\hat{\mathbf{z}}\frac{\rho_s}{\epsilon_0}\right)\cdot(\hat{\mathbf{z}}\,dz) = -\left[-\frac{\rho_s}{\epsilon_0}z\right]_0^d = \frac{\rho_s d}{\epsilon_0} = \frac{Qd}{\epsilon_0 A}$$

The capacitance is then

$$C = \frac{Q}{\Phi_{12}} = \frac{Q}{Qd/(\epsilon_0 A)} = \frac{\epsilon_0 A}{d}$$

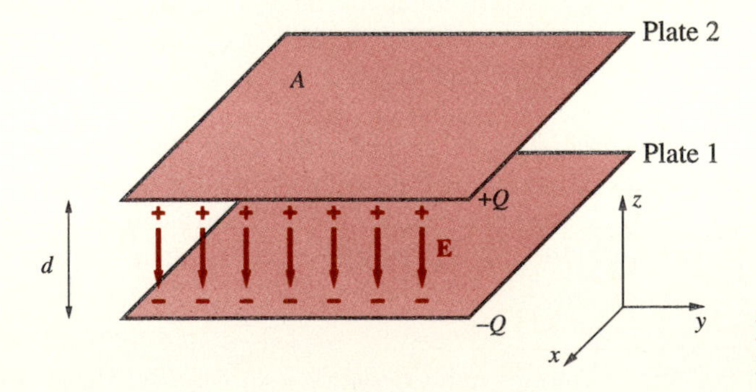

FIGURE 4.48. Parallel-plate capacitance.

[93]The basic reason for this is the extremely small ($\sim 10^{-19}$ s) time within which charges can redistribute in metallic conductors, as mentioned in Section 4.7 and the previous footnote and further discussed in Section 5.5.

We note that the capacitance is a function only of the geometry of the configuration (i.e., A and d). If the region between the conductors is an insulating material other than air, then the constant ϵ_0 is replaced by another constant characteristic to the material (Section 4.10). Thus, capacitance is also a function of the material medium around the conductors.

Example 4-27: Coaxial capacitor. Consider the coaxial capacitor, consisting of two concentric cylinders separated by free space, as shown in Figure 4.35 in connection with Example 4-15. Find the capacitance per unit length.

Solution: We can find the capacitance of this configuration by the usual method: (1) Assume a charge density ρ_s on the inner conductor (i.e., a total charge of $Q = \rho_s(2\pi ah)$, where h is the height of the cylinder); (2) find $\mathbf{E}$ in terms of ρ_s; (3) integrate $\mathbf{E}$ to find the potential difference between the inner conductor ($r = a$) and the outer conductor ($r = b$), denoted Φ_{ba}, in terms of ρ_s; and (4) take the ratio of Q and Φ_{ba}. However, in this case we already have the solution of the problem worked out in Example 4-15, where two different expressions were found for $\Phi(r)$:

$$\Phi(r) = -\frac{\rho_s a}{\epsilon_0}\ln\frac{r}{b} \quad \text{and} \quad \Phi(r) = \Phi_{ba}\frac{\ln(r/b)}{\ln(a/b)}$$

Equating the two expressions for $\Phi(r)$ we find

$$-\frac{\rho_s a}{\epsilon_0}\ln\frac{r}{b} = \Phi_{ba}\frac{\ln(r/b)}{\ln(a/b)}$$

$$-\frac{Q}{2\pi\epsilon_0 h} = -\frac{\Phi_{ba}}{\ln(b/a)}$$

$$\frac{Q}{\Phi_{12}} = \frac{2\pi\epsilon_0 h}{\ln(b/a)}$$

$$C = \frac{2\pi\epsilon_0 h}{\ln(b/a)}\ \text{F}$$

$$\frac{C}{h} = \frac{2\pi\epsilon_0}{\ln(b/a)}\ \text{F-m}^{-1}$$

where the last expression is the capacitance per unit length of a coaxial line. A useful expression for free space is

$$\frac{C}{h} = \frac{2\pi\epsilon_0}{\ln(b/a)}\ \text{F-m}^{-1} \simeq \frac{55.6}{\ln(b/a)}\ \text{pF-m}^{-1}$$

where C is the capacitance per unit length, identical to the expression listed in Table 2.2 as the per-unit-length capacitance of a coaxial line.

As a numerical example, consider a practical coaxial line such as RG-11, which has an inner-conductor diameter of $\sim$1.21 mm and the inner diameter of its outer conductor of $\sim$7.24 mm. Assuming the conductors to be separated by free space, the capacitance per meter of this line is $C \simeq 55.6/[\ln(7.24/1.21)] \simeq 31$ pF-m^{-1}.

Example 4-28: Capacitance of concentric spheres. Consider a spherical capacitor consisting[94] of two concentric conducting spheres as shown in Figure 4.49a. Find the capacitance.

Solution: If we assume a charge $+Q$ on the inner sphere and $-Q$ on the outer one, the electric field between the two spheres is

$$\mathbf{E} = \hat{\mathbf{r}}E_r = \hat{\mathbf{r}}\frac{Q}{4\pi\epsilon_0 r^2} \qquad a \leq r \leq b$$

consistent with Gauss's law. The potential difference between the spheres, denoted Φ_{ba} (note that Φ_{ba} is positive, since the inner sphere is positively charged) is given by

$$\Phi_{ba} = -\int_b^a \frac{Q}{4\pi\epsilon_0 r^2}\,dr = -\frac{Q}{4\pi\epsilon_0}\int_b^a \frac{dr}{r^2} = \frac{Q}{4\pi\epsilon_0}\left(\frac{1}{a} - \frac{1}{b}\right)$$

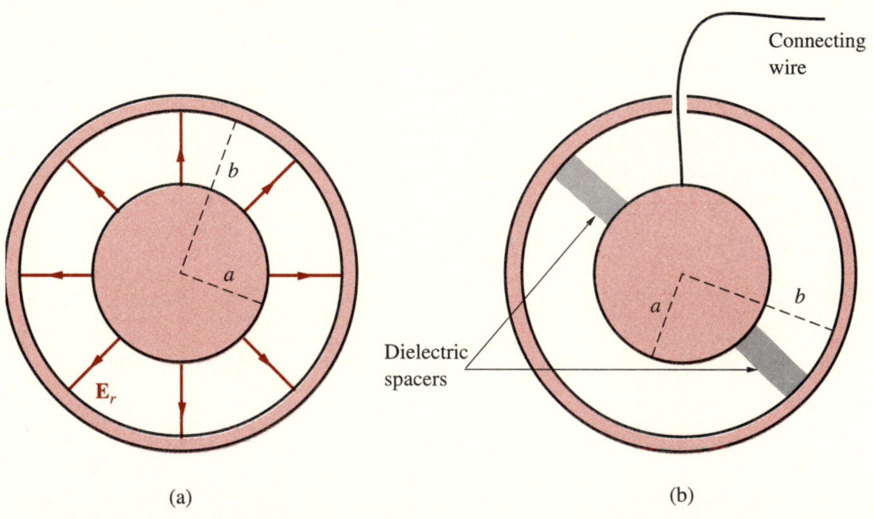

FIGURE 4.49. A spherical capacitor. (a) The geometry of a spherical capacitor consisting of two concentric conductor spheres of radii a and b. (b) In practice, the inner conductor would need to be supported by insulating spacers and can be accessed by means of a wire through a small hole as shown. The surface charges on the inner and outer spheres are not shown to avoid cluttering the figure.

[94]Note that, in practice, the inner sphere would need to be held in place by some insulating spacers and also a small opening through the outer sphere is needed so that a wire can be connected to the inner sphere, by means of which a potential difference can be applied between the two spheres, as shown in Figure 4.49b.

so that the capacitance is

$$C = \frac{Q}{\Phi_{ba}} = 4\pi\epsilon_0 \frac{ab}{b - a}$$

Note that if we let $b \to \infty$, we get the capacitance of an isolated sphere discussed earlier. In such a case, the flux lines originating on the positively charged inner conductor terminate at "infinity."

In general, regardless of the conductor configuration, and with reference to Figure 4.50, we can define capacitance as

$$\boxed{C \equiv \frac{Q}{\Phi_{12}} = \frac{\oint_S \mathbf{D} \cdot d\mathbf{s}}{-\int_L \mathbf{E} \cdot d\mathbf{l}}} \qquad [4.49]$$

where S is any surface enclosing the positively charged conductor and L is any path from the negative (lower potential) to the positive (higher potential) conductor.

The general procedure for determining the capacitance of any two-conductor configuration can be summarized as follows: (1) Choose a coordinate system appropriate to the geometrical layout and shapes of the two conductors. (2) Assume charges $+Q$ and $-Q$ on the conductors. (3) Find $\mathbf{E}$ from Gauss's law, by direct integration, or by other methods; the result will be proportional to Q. (4) Integrate $\mathbf{E}$ along any path between the two conductors to determine Φ_{12}; result will be proportional to the assumed charge Q. (5) Find C by taking the ratio Q/Φ_{12}; note that Q will cancel out and the result will depend only on the geometry of the conductors.

We now apply this general expression [4.49] to determine the capacitance of a two-wire transmission line, which was discussed in Section 2.7 and the capacitance expression for which was tabulated in Table 2.2.

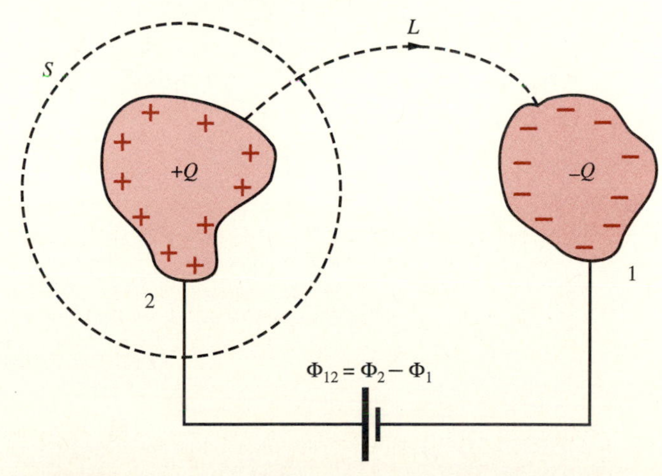

FIGURE 4.50 **Two arbitrary charged conductors forming a capacitor.**

Example 4-29: Capacitance of the two-wire line. A rather practical transmission line configuration used in telephony, radio engineering, and electric power distribution is the two-wire line, consisting of two parallel cylindrical conductors, each of radius a, separated by a distance d, as shown in Figure 4.51a. General solution of the capacitance for this configuration is more involved than that for the coaxial cable, since, in general, the charge distribution on the conductor surfaces is not uniform, because of the Coulomb attraction between the induced surface charges on the two conductors. This so-called "proximity effect" causes the charge density to be larger (smaller) on the sides of each conductor facing toward (away from) the other. The resultant distribution of the electric field lines and equipotentials is shown in Figure 4.51b. Closed-form analytical expressions for the electric field and potential can be found using various methods. Find a simplified expression of the capacitance of the two-wire line, valid for cases where $d \gg a$. Note that this approximation is indeed quite valid in most cases, including the well-known examples of the power transmission lines (for which a is a few cm, while d is many meters) and the TV antenna wire ($a < 1$ mm and $d \sim 1$ cm).

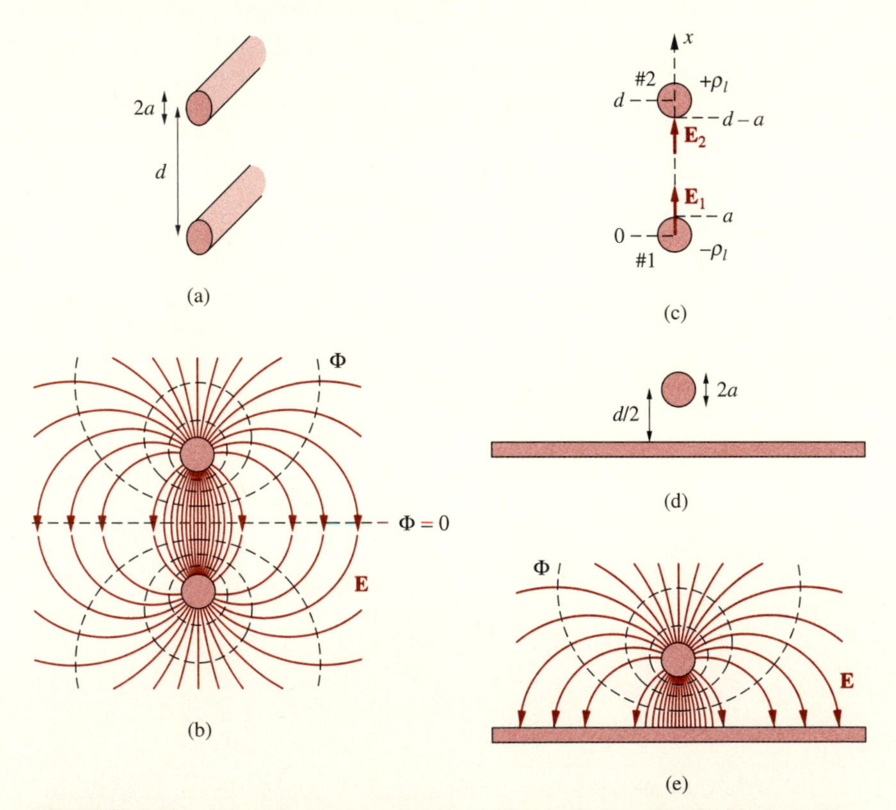

FIGURE 4.51. Two-wire transmission line and a single wire above ground. (a) The two-wire line. (b) Electric field configuration. (c) Representation in terms of two line charges. (d) Single wire above ground. (e) Field configuration for single wire above ground.

Solution: For $d \gg a$, the nonuniformity of the charge distribution on the conductor surfaces can be neglected and thus the electric field outside the conductors is the same as what would be produced by two line charges of opposite polarity (ρ_l and $-\rho_l$). From Example 4-11 we know that the electric field at a distance r from a line charge of linear charge density ρ_l is given as $E_r = \rho_l/(2\pi\epsilon_0 r)$. Taking the $+\rho_l$ and $-\rho_l$ line charges to be located respectively at $r = 0$ and $r = d$, the electric field as a function of distance x along the x axis (i.e., $y = 0$ and $z = 0$) is given as

$$E_x(x, 0, 0) = \frac{-\rho_l}{2\pi\epsilon_0 x} + \frac{\rho_l}{2\pi\epsilon_0(x - d)}$$

To find the capacitance using its general definition [4.49], we need to integrate the electric field *along any path* from one conductor to the other. It is obvious from Figure 4.51c that one convenient integration path[95] is along the x axis, from $x = a$ to $x = d - a$. We find

$$\begin{aligned}
\Phi_{12} &= -\frac{1}{2\pi\epsilon_0}\int_a^{d-a}\left[\frac{-\rho_l}{x} + \frac{+\rho_l}{x-d}\right]dx = \frac{\rho_l}{2\pi\epsilon_0}\int_a^{d-a}\left[\frac{1}{x} - \frac{1}{x-d}\right]dx \\
&= +\frac{\rho_l}{2\pi\epsilon_0}\Big[\ln x - \ln(x - d)\Big]_a^{d-a} \\
&= \frac{\rho_l}{2\pi\epsilon_0}\left[\ln\left(\frac{x}{x-d}\right)\right]_a^{d-a} \\
&= \frac{\rho_l}{2\pi\epsilon_0}\left[\ln\left(\frac{d-a}{d-a-d}\right) - \ln\left(\frac{a}{a-d}\right)\right] = \frac{\rho_l}{2\pi\epsilon_0}2\ln\left(\frac{d-a}{a}\right) \\
&\simeq \frac{\rho_l}{\pi\epsilon_0}\ln\frac{d}{a}
\end{aligned}$$

since $d \gg a$. Noting that the total charge on one of the conductors of unit length is $Q = \rho_l$, the capacitance *per unit length* of the two-wire line is

$$C = \frac{\rho_l}{|\Phi_{12}|} \simeq \frac{\pi\epsilon_0}{\ln(d/a)}\ \text{F-m}^{-1} \simeq \frac{27.8}{\ln(d/a)}\ \text{pF-m}^{-1}$$

As an example, consider the capacitance per kilometer between two Drake-type steel-reinforced aluminum conductors (ACSR) of a 115-kV power

[95]Note that the electric field at points other than along the x axis in general has components in both the x and y directions, as can be determined by a vector superposition of $\mathbf{E}_1$ and $\mathbf{E}_2$. However, since Φ_{12} can be found by integrating the total electric field along *any* path between the two conductors, it is convenient to choose this path along the x axis, where the electric field has only one nonzero component (i.e., E_x).

transmission line, separated by $d = 3$ m. For Drake ACSR, we have $a \simeq 1.407$ cm, so that the approximation $d \gg a$ is valid. Using the above, we find $C \simeq 5.19$ nF-(km)$^{-1}$.

The above expression is valid for $d \gg a$. A more accurate analysis for the general case yields

$$C = \frac{\pi\epsilon_0}{\ln[d/2a + \sqrt{(d/2a)^2 - 1}]} \text{ F-m}^{-1} \simeq \frac{27.8}{\ln[d/2a + \sqrt{(d/2a)^2 - 1}]} \text{ pF-m}^{-1}$$

which is the expression listed in Table 2.2 for the per-unit-length capacitance of a two-wire line.

It is clear from [4.49] that the capacitance of a two-conductor configuration is primarily dependent on the electric field distribution. This realization allows us to simply infer the capacitance of some conductor configurations from others for which we have derived expressions. A good example is the case of a single cylindrical conductor above a ground plane, as shown in Figure 4.51d. We can simply obtain this configuration by placing an infinitely large conducting sheet (i.e., ground) coincident with the flat equipotential plane of the two-wire line configuration of Figure 4.51b. Such a conducting plane does not influence the existing field distribution, since the electric field lines are already perpendicular to it at all points. For a given value of ρ_l, the electric field everywhere is the same for the single-wire above-ground case as that for the two-wire line. However, if the charge density of ρ_l corresponded to a potential difference of V_0 between the two wires, the potential difference between a single conductor and the ground plane (for the same ρ_l) is $V_0/2$. In other words, the capacitance of the configuration shown in Figure 4.51d is

$$C_{\text{single wire above ground}} = 2C_{\text{two-wire line}} \simeq \frac{2\pi\epsilon_0}{\ln(d/a)} \text{ F-m}^{-1} \qquad [4.50]$$

Another way to analyze the problem of the single wire above a ground plane is the so-called method of images, which involves the replacement of boundary surfaces by appropriately placed "image" charges instead of a formal solution of Laplace's equation. In the case in hand, the conducting plane can be replaced by an image line charge of line charge density $-\rho_l$ at a distance $d/2$ below the plane so that $\Phi = 0$ on the ground plane. The electric field at any given point is then simply a superposition of the field due to the original line charge (single wire) and the image line charge, which of course is entirely equivalent to the case of the two-wire line. The method of images is a powerful method for finding the fields produced by charges in the presence of dielectric and conducting boundaries with certain symmetries.[96]

---

[96]See, for example, Section 4.4 of D. K. Cheng, *Field and Wave Electromagnetics,* 2nd ed., Addison Wesley, Reading, Massachusetts, 1989.

4.10 DIELECTRIC MATERIALS

An insulator is a substance whose electrons and ions cannot move over macroscopic distances under the influence of an applied field. Although no material is a perfect insulator, in this section we deal with materials within which the motion of charges over macroscopic distances is negligible. Insulators are said to be *polarized* when the presence of an applied field displaces the electrons within a molecule away from their average positions. When we consider the polarizability of insulators, we refer to them as *dielectrics*.[97] Although all substances are polarizable, the effects of polarizability can be readily observed only when the material does not conduct electricity—that is, when it is an insulator.

4.10.1 Polarizability

In Section 4.7 we saw (e.g., Example 4-20) that redistribution of charges in a metallic conductor involves the transfer of a very small percentage of the amount of free charge in a conductor over macroscopic distances. Polarization of dielectrics, on the other hand, involves the displacement of one or more electrons per atom over subatomic or microscopic distances. The atoms of dielectric materials have their outermost electron shells almost completely filled. As a consequence, the electrons are tightly bound, and only a negligible number of electrons are available for conduction of electric current. A measure of the conducting ability of a material is the charge rearrangement time, which as we shall see in Chapter 5 is $\sim 10^{-19}$ s for metallic conductors. For a good dielectric, such as fused quartz, the charge rearrangement time is $\sim$50 days. Thus, free charge deposited inside quartz stays in place for all practical purposes and can be considered as bound charge. The difference in the electrical behavior of dielectrics versus conductors is essentially due to the different numbers of free versus bound charges, with semiconductors exhibiting intermediate behavior.

When a dielectric material is placed in an electric field, the electrons respond by shifting with respect to the nuclei in the direction opposite to the applied field, essentially establishing many small electric dipoles, as shown in Figure 4.52. Note that electric dipoles aligned with the field are produced in both polar and nonpolar materials. *Nonpolar* materials consist of molecules that do not possess a permanent dipole moment; the external field both induces the dipoles and orients them as shown in Figure 4.52a. The molecules of *polar* materials (e.g., NaCl) have a permanent dipole moment, but the individual molecular dipoles are usually randomly oriented

---

[97] The word *dielectric* was coined by Faraday, who extensively studied the behavior of different insulators placed between electrodes and found that the capacitances of spherical capacitors filled with dielectrics were higher. His results and insights were documented in *Experimental Researches in Electricity,* vol.1, Sec. 1252–1270, B. Quatrich, London, 1839. For a historical overview and for an excellent and thorough discussion of dielectric materials see R. S. Elliott, *Electromagnetics,* IEEE Press, Piscataway, New Jersey, 1993. Another excellent reference on the topic of dielectrics is A. von Hippel (ed.), *Dielectric Materials and Applications,* Artech House, Boston, 1995.

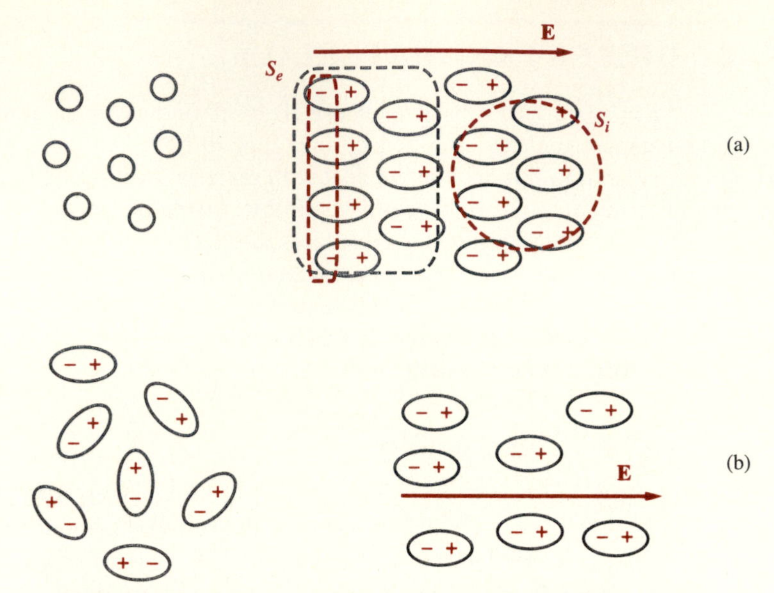

FIGURE 4.52. **Dipoles oriented in an applied field.** (a) Nonpolar molecules with no permanent dipole moment; the external field induces the dipoles and orients them along the field. (b) Polar molecules have permanent dipole moments randomized by thermal agitation; the applied field aligns the dipoles.

due to thermal agitation, as shown in Figure 4.52b; when an applied field is present, the individual dipoles tend to align in the direction of the field.

It is important to note that polarization does not produce net charge inside the dielectric. Any interior volume of macroscopic dimensions[98] (e.g., that enclosed by the surface S_i in Figure 4.52a) contains equal amounts of positive and negative charge. However, a net amount of surface polarization charge does appear on the surface of the polarized dielectric, as shown in Figure 4.52a for the volume enclosed by the surface S_e. We see that a layer of charge adjacent to the boundary remains unneutralized and appears as polarization charge with surface charge density ρ_{sp}. We shall see that the amount of surface charge is a direct indication of the degree of polarization of a material and that ρ_{sp} is proportional to the magnitude of the electric field.

Electronic Polarizability Figure 4.53a shows a very simplified model of the atom, consisting of a positively charged nucleus surrounded by a negatively charged

[98]It is important to understand that S_i has *macroscopic* dimensions; otherwise, Figure 4.52 may give the impression that there is net surface charge on the outer edges of S_i just as in the case of S_e. However, since S_i has macroscopic dimensions, it contains a very large number of individual molecules and is thus neutral. Since the precise location of the boundary of S_i is defined only on a macroscopic scale, it does not make sense to think of it as cutting through a molecule of microscopic dimensions.

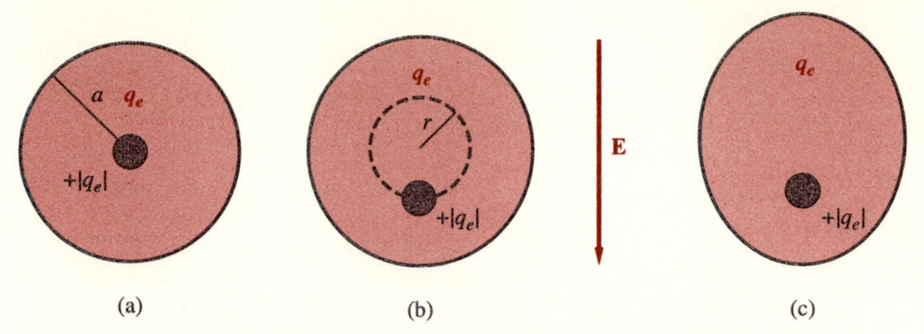

FIGURE 4.53. Simple model of an atom. (a) The simplified model of an atom with no applied field. (b) Displacement of the nucleus by an applied electric field. (c) The spherical electron cloud may in fact be somewhat elongated as a result of the applied field.

spherical electron cloud.[99] With an applied steady field the nucleus and the electron cloud are displaced (as illustrated in Figure 4.53b) until their mutual attractive force is just equal to the force due to the applied field. We can thus calculate the amount of displacement r by equating the two forces.

We can treat the electron cloud with radius a as a spherical cloud of charge much like that in Example 4-13, with uniform volume charge density $\rho = 3q_e/(4\pi a^3)$, as shown in Figure 4.53a. Note that a is the effective atomic radius. The electric field due to such a spherical cloud of total charge Q at a distance r from its center was derived in Example 4-13 to be $E_r = Qr/(4\pi\epsilon_0 a^3)$ for $r \leq a$, which for the case in hand (replacing Q by $|q_e|$) gives

$$E_r = \frac{|q_e|r}{4\pi\epsilon_0 a^3} \qquad r \leq a$$

When the atom is placed in an external electric field, the electron cloud is displaced with respect to the nucleus as shown in Figure 4.53b. Although the shape of the electron cloud may also be elongated as shown in Figure 4.53c, we assume that this effect is quite small and consider the displaced electron cloud to be spherical as shown in Figure 4.53b. Since the charge on the nucleus is $+|q_e|$, the Coulomb restoring force F_r of the spherical charge cloud acting on the nucleus when it is displaced by an amount r (where $r \leq a$) is[100]

$$F_r = E_r(+|q_e|) = \frac{q_e^2 r}{4\pi\epsilon_0 a^3}$$

[99]Note that the nucleus has a diameter of order $\sim 10^{-15}$ m, while that of the electron cloud is $\sim 10^{-10}$ m, so that the nucleus can be considered a point source. Nevertheless, it should be realized that the atomic model shown in Figure 4.53 is highly simplified and corresponds to the early classical model of the atom. In the Bohr model of the atom, the electron orbits at a fixed discrete radius, whereas the modern quantum mechanical model specifies the location of the electron on the basis of a wave function that describes the probability of the electron being located at different radii.

[100]We note that this is a highly simplified argument. Not only is the classical model of the atom not precisely valid, but the displaced electron cloud is in general not circular in shape (Figure 4.53c), and the charge density within the sphere is not uniform.

Equating this force on the nucleus to that due to the external electric field E (which is equal in magnitude but opposite to the direction of E_r), we find the electric dipole moment per atom as

$$p \equiv |q_e|r = 4\pi\epsilon_0 a^3 E$$

or, in vector form,

$$\mathbf{p} = 4\pi\epsilon_0 a^3 \mathbf{E} \quad \text{C-m-atom}^{-1} \qquad [4.51]$$

The proportionality between the dipole moment per atom $\mathbf{p}$ and the electric field $\mathbf{E}$ holds only for small applied fields, and thus for small displacements. If the material has N atoms per unit volume, then the dipole moment per unit volume is $\mathbf{P} = N\mathbf{p}$, referred to as the *polarization vector.* We often write $\mathbf{p} = \alpha_e \mathbf{E}$, where $\alpha_e = 4\pi\epsilon_0 a^3$ is called the *electronic polarizability* of the material. Although the atomic model shown in Figure 4.53 is extremely primitive, the result (α_e) is accurate to within a factor of four or so for many simple atoms. With atomic radii a for typical materials being of order 1 angstrom, the numerical value of α_e is of order $\sim 10^{-40}$ F-m^2.

Ionic Polarizability In some materials, two different atoms may join together as a molecule by forming a chemical bond. We can think of such molecules as consisting of positively and negatively charged ions, with the Coulomb forces between them serving as the binding force. Examples of such materials are hydrochloric acid (HCl), carbon dioxide (CO_2), and water (H_2O). Depending on whether or not the electrons are transferred or shared, the bond can be *ionic* or *covalent.* In both cases, the material may possess a permanent dipole moment, and the application of an electric field to any such molecule also displaces the positive ions with respect to the negative ones and thereby induces a dipole moment. A similar analysis as given above[101] shows that this process leads to the creation of an average dipole moment per molecule given by

$$\mathbf{p} = \alpha_i \mathbf{E} \quad \text{where} \quad \alpha_i = \frac{64\epsilon_0 d_a}{7(n+1) - 16}$$

where d_a is the average interatomic spacing between the centers of the two ions constituting the molecule; n is a number between 5 and 12, depending on the material[102]; and α_i is referred to as the *ionic polarizability.* For typical values of d_a, the ionic polarizability has values of $\sim 10^{-40}$ F-m^2, comparable to α_e.

Orientational Polarizability Some polyatomic molecules, such as H_2O, may be at least partially ionic and may consist of *polar* molecules, which carry a permanent

---

[101]For a discussion at an appropriate level, see R. S. Elliott, *Electromagnetics,* IEEE Press, Piscataway, New Jersey, 1993, Sec. 6.7.

[102]L. Pauling, *The Nature of the Chemical Bond*, Cornell University Press, Ithaca, New York, 1948, p. 339.

dipole moment. With no electric field, the individual dipole moments point in random directions, so the net dipole moment is zero, as shown in the left-hand panel of Figure 4.52b. When an electric field is applied, such materials exhibit the electronic and ionic polarization effects just discussed, but in addition, the electric field tends to line up the individual dipoles to produce an additional net dipole moment per unit volume. If all the dipoles in a material were to line up, the polarization would be very large; however, at ordinary temperatures and relatively small electric field levels, the collisions of the molecules in their thermal motion allow only a small fraction to line up with the field. The resulting effective polarization can be shown[103] to be

$$\alpha_o = \frac{p^2}{3k_{\mathrm{B}}T}$$

where T is the absolute temperature in kelvins (K), $k_{\mathrm{B}} \simeq 1.38 \times 10^{-23}$ J-K^{-1} is the Boltzmann's constant, and $p = |q_e|d$ is the molecular dipole moment. For $q_e \simeq -1.602 \times 10^{-19}$ C and $d = 1$ angstrom (10^{-10} m), we have $p = |q_e|d \simeq 1.6 \times 10^{-29}$ C-m, so that at room temperature $\alpha_o \simeq 2 \times 10^{-38}$ m^3, comparable with α_e and α_i. The quantity α_o is referred to as the *orientational polarizability* of a material.

Polarization per Unit Volume The total polarization of a material may arise as a result of electronic, ionic, and orientational polarizability, leading in general to a dipole moment per unit volume of

$$\mathbf{P} = N\underbrace{\left(\alpha_e + \alpha_i + \frac{p^2}{3k_{\mathrm{B}}T}\right)}_{\alpha_{\mathrm{T}}}\mathbf{E} \qquad \text{C-m}^{-2} \qquad [4.52]$$

where α_o is explicitly written to emphasize its temperature dependence. Note that while α_e and α_i depend only on atomic and molecular configurations, α_o is inversely proportional to temperature, as well as dependent on the atomic properties.

In general, the electric field $\mathbf{E}$ in [4.52] should not be taken as the external field that exists in the absence of the dielectric, because the presence of a large number density N of neighboring dipoles also contributes to the polarization of a given atom or molecule. The electric field $\mathbf{E}$ in [4.52] should thus be taken to be the total local electric field $\mathbf{E}_{\mathrm{loc}}$, consisting of the external field $\mathbf{E}$ plus the molecular field $\mathbf{E}_{\mathrm{mol}}$. The local field $\mathbf{E}_{\mathrm{loc}}$ is the field that actually exists at the molecule. For gases, the separation of the molecules is sufficiently large that $\mathbf{E}_{\mathrm{loc}}$ is closely approximated by the applied field $\mathbf{E}$. For solid dielectrics, the effect of the fields of adjacent molecules cannot be neglected. The molecular field $\mathbf{E}_{\mathrm{mol}}$ and thus the local field $\mathbf{E}_{\mathrm{loc}}$ can be calculated by removing the molecule in question, maintaining all the other molecules in their time-average polarized positions, and calculating the space-averaged electrostatic

[103] R. S. Elliott, *Electromagnetics,* IEEE Press, Piscataway, New Jersey, 1993, Sec. 6.7.

field in the cavity previously occupied by the molecule. Such an analysis[104] shows that the effective local field acting to polarize the molecule is

$$\mathbf{E}_{\text{loc}} = \mathbf{E} + \frac{\mathbf{P}}{3\epsilon_0}$$

so the total polarization per unit volume $\mathbf{P}$ from [4.52] is given as

$$\mathbf{P} = N\alpha_{\text{T}}\mathbf{E}_{\text{loc}} = N\alpha_{\text{T}}\left(\mathbf{E} + \frac{\mathbf{P}}{3\epsilon_0}\right)$$

where α_{T} is the total polarizability. The above equation can be solved for $\mathbf{P}$ as

$$\mathbf{P} = \frac{N\alpha_{\text{T}}}{1 - N\alpha_{\text{T}}/(3\epsilon_0)}\mathbf{E} \qquad [4.53]$$

The concepts of polarizability and the local electric field were introduced only so that we can relate the macroscopic behavior of the fields in a dielectric to the underlying microscopic causes. In practice, it is useful to lump together the various factors contributing to the polarization in the following expression:

$$\boxed{\mathbf{P} = \epsilon_0\chi_{\text{e}}\mathbf{E}} \qquad [4.54]$$

where

$$\chi_{\text{e}} = \frac{N\alpha_{\text{T}}/\epsilon_0}{1 - N\alpha_{\text{T}}/(3\epsilon_0)}$$

is a dimensionless quantity called the *electric susceptibility* and $\mathbf{E}$ is the macroscopic field present everywhere in the dielectric.

A brief discussion of macroscopic versus microscopic fields, along the lines of Section 4.7.1, is in order here. The quantity $\mathbf{P}$, polarization per unit volume, was defined to be $N\mathbf{p}$, where $\mathbf{p}$ is the atomic or molecular dipole moment. In this sense, $\mathbf{P}$ may appear to be a microscopic quantity. If that were the case, its use in [4.54] in connection with the macroscopic electric field $\mathbf{E}$ may be inappropriate. However, we should remember here that matter consists primarily of empty space interspersed with atoms and molecules, as discussed in Section 4.7.1. We note once again that the radii of nuclei are of order $\sim 10^{-15}$ m, whereas the spacing between adjacent atoms is $\sim 10^{-10}$ m. If we looked at the variation of $\mathbf{P}$ on a microscopic spatial scale, we would see it change drastically in magnitude and direction from one atomic nucleus to another, with zero values in the intervening empty space. Fortunately, a description

[104]This calculation was first carried out in 1906 by H. A. Lorentz, *Theory of Electrons,* Dover, New York, 1952. For reference, see Section 6.5 and Appendix K of R. S. Elliott, *Electromagnetics,* IEEE Press, Piscataway, New Jersey, 1993; Section 3.9 of R. Plonsey and R. E. Collin, *Principles and Applications of Electromagnetic Fields,* McGraw-Hill, New York, 1961; Chapter 13 of C. Kittel, *Introduction to Solid State Physics,* Wiley, New York, 1971.

of **P** or the electric field **E** on such a scale is not of interest to us. When we consider macroscopic dimensions of, say, 1 micron (10^{-6} m), the net polarization **P** and the external field **E** represent the spatial averages over many ($\sim 10^{12}$ atoms for a cube of size 1 μm) atoms, so these quantities vary smoothly over macroscopic distances and can properly be used as macroscopic quantities.

4.10.2 The Permittivity Concept

We have seen in the previous section that the effect of an applied electric field on the atoms or molecules of a dielectric material is to create a dipole moment distribution **P**. This distribution sets up induced secondary fields, so the net field in the presence of the dielectric is modified from its free-space value. We now proceed to the consideration of fields generated by the induced dipole moments in a dielectric material. For this purpose, we use the methods studied earlier in Section 4.4 to find electrostatic potentials and fields from given distributions of charge. The pertinent expression for the potential is

$$\Phi(\mathbf{r}) = \frac{1}{4\pi\epsilon_0}\int_{V'}\frac{\rho(\mathbf{r}')\,dv'}{|\mathbf{r}-\mathbf{r}'|} \qquad [4.55]$$

and the electric field can be found from the potential using $\mathbf{E} = -\nabla\Phi$.

When a dielectric is introduced in an external electric field set up by an arbitrary charge distribution,[105] a dipole moment distribution **P** is induced in the volume occupied by the dielectric. This distribution constitutes a secondary electric field source and must be included to determine the complete electric field in the presence of the dielectric.

The electrostatic potential at a faraway point of a z-directed dipole with dipole moment $\mathbf{p} = p\hat{\mathbf{z}}$ located at the origin was found in Section 4.4.3 to be[106]

$$\Phi(\mathbf{r}) = \frac{p\cos\theta}{4\pi\epsilon_0 r^2} = \frac{\mathbf{p}\cdot\hat{\mathbf{r}}}{4\pi\epsilon_0 r^2} = \frac{1}{4\pi\epsilon_0}\left[-\mathbf{p}\cdot\nabla\left(\frac{1}{r}\right)\right]$$

The above is obviously valid for any orientation of the dipole moment and can be adapted to an arbitrary position $\mathbf{r}'$ of the dipole. Thus, if $|\mathbf{r}-\mathbf{r}'|$ is the distance from the source point (location of the dipole) $\mathbf{r}'$ to the observation point $\mathbf{r}$, the electrostatic potential at a faraway point is

$$\Phi(\mathbf{r}) = \frac{1}{4\pi\epsilon_0}\left[-\mathbf{p}\cdot\nabla\left(\frac{1}{|\mathbf{r}-\mathbf{r}'|}\right)\right] = \frac{1}{4\pi\epsilon_0}\left[\mathbf{p}\cdot\nabla'\left(\frac{1}{|\mathbf{r}-\mathbf{r}'|}\right)\right]$$

---

[105]Typically the external field is set up by applying a potential difference between two conductors, such as two parallel plates, and we are considering the introduction of the dielectric material between the plates.

[106]Note that in spherical coordinates we have

$$\nabla\left(\frac{1}{r}\right) = \nabla r^{-1} = \hat{\mathbf{r}}\frac{\partial r^{-1}}{\partial r} + \hat{\boldsymbol{\theta}}\frac{1}{r}\frac{\partial r^{-1}}{\partial\theta} + \hat{\boldsymbol{\phi}}\frac{1}{r\sin\theta}\frac{\partial r^{-1}}{\partial\phi} = -\frac{1}{r^2}\hat{\mathbf{r}} + 0 + 0 = -\frac{1}{r^2}\hat{\mathbf{r}}$$

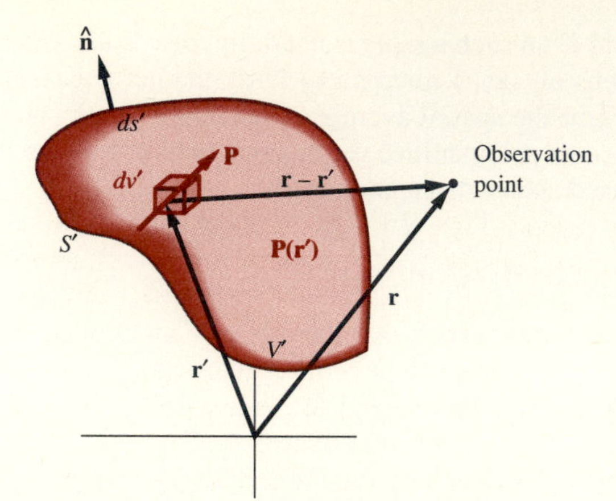

FIGURE 4.54. A volume distribution of polarization. Note that $\mathbf{P}(\mathbf{r}')$ is the distribution of volume polarization, and $\mathbf{P}$ is the polarization vector.

where ∇' represents the del operator applied to the source coordinates ($\mathbf{r}'$). Note that since $R = |\mathbf{r} - \mathbf{r}'|$, we have $\nabla R^{-1} = -\nabla' R^{-1}$.

Now consider a large number of infinitesimal dipoles distributed throughout a given volume V' as shown in Figure 4.54. Since $\mathbf{P}$ is the dipole moment per unit volume, the dipole moment in a volume element dv' is $\mathbf{P}\,dv'$ and the potential at the observation point is given by

$$\Phi(\mathbf{r}) = \frac{1}{4\pi\epsilon_0}\left[\int_{V'} \mathbf{P}\cdot\nabla'\left(\frac{1}{|\mathbf{r}-\mathbf{r}'|}\right)dv'\right]$$

We can now use a vector identity,

$$\nabla\cdot(a\mathbf{A}) = a\nabla\cdot\mathbf{A} + \mathbf{A}\cdot\nabla a$$

where $\mathbf{A}$ is any vector and a is any scalar quantity, to rewrite the potential as

$$\Phi(\mathbf{r}) = \frac{1}{4\pi\epsilon_0}\left[\int_{V'} \nabla'\cdot\left(\frac{\mathbf{P}}{|\mathbf{r}-\mathbf{r}'|}\right)dv' - \int_{V'}\frac{\nabla'\cdot\mathbf{P}}{|\mathbf{r}-\mathbf{r}'|}dv'\right]$$

We can further use the divergence theorem (see Section 4.6) to express $\Phi(\mathbf{r})$ as

$$\Phi(\mathbf{r}) = \frac{1}{4\pi\epsilon_0}\left[\oint_{S'}\frac{\mathbf{P}\cdot\hat{\mathbf{n}}}{|\mathbf{r}-\mathbf{r}'|}ds' - \int_{V'}\frac{\nabla'\cdot\mathbf{P}}{|\mathbf{r}-\mathbf{r}'|}dv'\right] \qquad [4.56]$$

where $\hat{\mathbf{n}}$ is the outward normal to the surface S' that encloses the volume V' as shown in Figure 4.54. Equation [4.56] allows for a simple and physically based interpretation of electrostatic potential due to a volume distribution of electric dipoles. By comparison with [4.55], we see that a volume distribution of dipoles may be represented as an equivalent volume (ρ_{p}) and surface (ρ_{sp}) distribution of charges given by

$$\rho_{\text{sp}} = \mathbf{P}\cdot\hat{\mathbf{n}} \qquad \text{on } S'$$

$$\rho_{\text{p}} = -\nabla'\cdot\mathbf{P} \qquad \text{in } V'$$

Such a result is not surprising on physical grounds; at points in space where $\nabla' \cdot \mathbf{P} \neq 0$, we have creation of a net dipole moment per unit volume, which means that charge density from adjacent dipoles does not completely cancel. The surface charge density occurs because the dipoles with one end on the surface cannot be neutralized since they do not have an adjacent dipole layer on that end (see Figure 4.52a).

If in addition there are also free charges within the dielectric medium as represented by volume charge density ρ, the total electrostatic potential is due to the superposition of the effects of polarization charges ρ_p and ρ_{sp} and the free charge ρ. In other words, we have

$$\Phi(\mathbf{r}) = \frac{1}{4\pi\epsilon_0}\left[\oint_{S'} \frac{\rho_{sp}}{|\mathbf{r}-\mathbf{r}'|}ds' + \int_{V'} \frac{\rho+\rho_p}{|\mathbf{r}-\mathbf{r}'|}dv'\right] \qquad [4.57]$$

In general, the polarization charges may react back on the free charges and modify the distribution ρ. In many cases, however, such effects are small.

Similar to [4.57], the polarization charge density ρ_p must also be added to the free charge density ρ in applying Gauss's law in the dielectric as

$$\epsilon_0 \nabla \cdot \mathbf{E} = \rho + \rho_p$$

Noting that $\rho_p = -\nabla \cdot \mathbf{P}$ we have

$$\nabla \cdot (\epsilon_0 \mathbf{E}) + \nabla \cdot \mathbf{P} = \rho \quad \rightarrow \quad \nabla \cdot (\epsilon_0 \mathbf{E} + \mathbf{P}) = \rho$$

This suggests that we can use a displacement density or flux density vector defined as

$$\boxed{\mathbf{D} = \epsilon_0 \mathbf{E} + \mathbf{P}} \qquad [4.58]$$

which permits us to write the familiar form of the differential form of Gauss's law,

$$\boxed{\nabla \cdot \mathbf{D} = \rho} \qquad [4.59]$$

Note that in materials for which [4.54] holds we have

$$\boxed{\mathbf{D} = \epsilon_0 \mathbf{E} + \epsilon_0 \chi_e \mathbf{E} = \epsilon_0 (1 + \chi_e)\mathbf{E} = \epsilon \mathbf{E}} \qquad [4.60]$$

where the parameter $\epsilon = \epsilon_0(1 + \chi_e)$ is called the *electrical permittivity* or the *dielectric constant* of the material. Typically we write $\epsilon = \epsilon_r \epsilon_0$, where ϵ_r is a dimensionless quantity called the *relative permittivity* or the *relative dielectric constant.*

The introduction of the permittivity concept greatly simplifies our analysis and understanding of electrostatic phenomena in the presence of dielectric materials. The polarizability and local molecular electric field concepts were introduced in Section 4.10.1 so that we can relate microscopic physical behavior to macroscopic fields. For most future problems we shall describe materials by means of their permittivity ϵ, since this constant is easily measured. If ϵ is known, we can find the polarization from the relation

$$\boxed{\mathbf{P} = \epsilon_0 \chi_e \mathbf{E} = \mathbf{D} - \epsilon_0 \mathbf{E} = (\epsilon - \epsilon_0)\mathbf{E}} \qquad [4.61]$$

The total polarizability α_T is thus

$$\epsilon - \epsilon_0 = \epsilon_0 \chi_e = \frac{N\alpha_T}{1 - N\alpha_T/(3\epsilon_0)} \quad \rightarrow \quad \alpha_T = \frac{3\epsilon_0(\epsilon - \epsilon_0)}{N(\epsilon + 2\epsilon_0)}$$

Note that [4.60] holds only for a class of dielectrics under certain conditions. Since [4.60] originates in [4.54], it depends primarily on the *linear* relationship between the polarization **P** and the electric field **E**. An important example of nonlinear behavior occurs when the applied electric field is intense enough to pull electrons completely out of their bound locations, causing dielectric breakdown, discussed in Section 4.10. Equation [4.60] also requires that the material be *isotropic;* that is, χ_e must be independent of the direction of **E**. In *anisotropic* materials (e.g., crystals), electric field in one direction can produce polarization (and thus **D**) in another direction. Accordingly, the relation between **D** and **E** must be expressed as a matrix, more commonly called a tensor.[107]

Permittivities of Selected Materials The relative permittivities of some selected materials are listed in Table 4.1. The values given are at room temperature and for static or low-frequency applied electric fields; the electrical permittivity of most materials depends on frequency and can in general be quite different from the low-frequency values, especially in the vicinity of resonances (see Section 7.4.4). The value of ϵ_r for most materials listed lies in the range 1 to 25, with the exception of distilled water, titanium dioxide, and barium titanate. The high permittivity of distilled water arises from the partial orientation of permanent dipole moments of its polar molecules. However, other polar materials such as mica and quartz do not have unusually large values of ϵ_r, so the substantially higher permittivity of distilled water is largely a result of the specific microscopic molecular structure of this ubiquitous material. Titanium dioxide is also a polar material but is also an anisotropic crystalline substance whose permittivity depends on the direction of the applied electric field. The value of ϵ_r for TiO_2 is 89 when the applied field is in the direction of one of its crystal axes and 173 when the applied field is perpendicular to this axis. Other materials such as quartz also exhibit anisotropic behavior, but with a narrower range of permittivities (e.g., 4.7 to 5.1 for quartz).

Titanates (combinations of TiO_2 with other oxides) also belong to a class of materials known as *ferroelectric* materials, which exhibit spontaneous polarization just as ferromagnetic materials (e.g., iron) exhibit permanent magnetization (see Chapter 6). Some of these materials have relative permittivities at room temperature of 500 to 6000; a good example is barium titanate ($BaO{\cdot}TiO_2$ or $BaTiO_3$),[108] with $\epsilon_r \simeq 1200$.

---

[107]For a brief discussion, see Section 3.3 of D. H. Staelin, A. W. Morgenthaler, and J. A. Kong, *Electromagnetic Waves,* Prentice Hall, 1994. For further discussion of electromagnetics of anisotropic media, see Section 13.8 of S. Ramo, J. R. Whinnery and T. Van Duzer, *Fields and Waves in Communication Electronics,* 3rd ed., Wiley, 1994.

[108]$BaTiO_3$ is also anisotropic, with $\epsilon_r \simeq 160$ along its principal ferroelectric axis and $\epsilon_r \simeq 5000$ along the perpendicular axis at room temperature. For an aggregation of randomly oriented $BaTiO_3$ crystals, the permittivity can be taken to be $\epsilon_r \simeq 1200$.

TABLE 4.1. Relative permittivity and dielectric strength of selected materials

Material	Relative permittivity (ϵ_r) (at room temperature)	Dielectric strength (MV-m^{-1}) (at room temp. and 1 atm)
Air	1	~3
Alumina (Al_2O_3)	~8.8	
Amber	2.7	
Bakelite	~4.8	25
Barium titanate ($BaTiO_3$)	1200	7.5
Freon	1	~8
Fused quartz (SiO_2)	3.9	~1000
Gallium arsenide (GaAs)	13.1	~40
Germanium (Ge)	16	~10
Glass	~4–9	~30
Glycerin	50	
Ice	3.2	
Mica (ruby)	5.4	200
Nylon	~3.6–4.5	
Oil	2.3	15
Paper	1.5–4	15
Paraffin wax	2.1	30
Plexiglass	3.4	
Polyethylene	2.26	
Polystyrene	2.56	20
Porcelain	~5–9	11
Rubber	~2.4–3.0	25
Rutile (TiO_2)	100	
Silicon (Si)	11.9	~30
Silicon nitride (Si_3N_4)	7.2	~1000
Sodium chloride (NaCl)	5.9	
Styrofoam	1.03	
Sulphur	4	
Tantalum pentoxide (Ta_2O_5)	~25	
Teflon (PTFE)	2.1	
Vaseline	2.16	
Water (distilled)	81	
Wood (balsa)	1.4	

Such materials can be used to construct compact general-purpose capacitors with very small dimensions but relatively high capacitances, in the range of 5 pF to 0.1 μF. The permittivity of the titanates depends sensitively on temperature and also on the applied electric field strength. The latter property can be very useful in nonlinear circuit applications. Other crystals that have similar structure to $BaTiO_3$ and show similar behavior include $SrTiO_3$, $NaTaO_3$, and $LaFeO_3$. Other examples of ferroelectric materials include lithium selenite $LiH_3(SeO_3)_2$ and related salts, ammonium cadmium sulfate $(NH_4)_2Cd_2(SO_4)_3$, and lead metaniobate $Pb(NbO_3)_2$.

Further discussion of electrical properties of different dielectric materials can be found elsewhere.[109]

[109] See A. von Hippel (ed.), *Dielectric Materials and Applications,* Artech House, 1995.

4.10.3 Dielectric Breakdown

The linear relationship [4.60] between **E** and **D** holds only for relatively small values of the applied electric field **E**. If the applied field is sufficiently high, a dielectric material is suddenly transformed from a good insulator into an extremely good conductor, causing substantial current to flow. *Dielectric breakdown* is the term used for this sudden loss of the insulating property of a dielectric under the influence of an electric field. The electric field strength at which dielectric breakdown occurs is referred to as the *dielectric strength* and is denoted by E_{BR}. The values of E_{BR} for selected materials are given in Table 4.1.

The dielectric strength of a given material may vary by several orders of magnitude depending on the conditions under which it is used. For gases, E_{BR} is proportional to pressure. For solids, microstructural defects, impurities, the shape of the dielectric, the manner in which it was prepared, and its environment are all factors that may strongly affect E_{BR}. Under normal conditions, dielectric breakdown does not permanently affect gaseous or liquid dielectrics. In solids, on the other hand, the breakdown leads to the formation of highly conductive channels, leaving behind a characteristic damage to the texture of the material, typically in the form of a channel of molten material, jagged holes, or a tree-like decomposition pattern of carbonized matter.

We now comment briefly on the nature of dielectric breakdown in gaseous and solid dielectrics.

Breakdown in Gases Consider the parallel-plate capacitor of Figure 4.48 to be filled with a gas and connected to a constant voltage applied across its plates, and arrangements made so that the current flowing through the dielectric region between the plates can be measured. As the applied voltage is slowly increased, the current initially rises to a small value of a small fraction of a microampere and levels off to a saturation level. As the voltage is increased further, the current remains nearly constant until a certain critical voltage, $V_{\text{BR}} = E_{\text{BR}}d$, where V_{BR} is called the breakdown voltage and d is the separation of the plates, is reached. When a voltage greater than V_{BR} is applied, the current through the dielectric quickly rises (typically in a time interval of the order of $\sim 10^{-8}$ s) to the maximum value allowed by the power supply.

This behavior was first analyzed by Townsend,[110] who introduced the concept of *electron avalanche breakdown.* Avalanche breakdown occurs when the applied field is large enough to accelerate a free electron (a few of which are always present in any gas due to cosmic ray ionization) to sufficiently large energies so that it can liberate a new electron by impact ionization of a neutral gas molecule, leaving behind a positively charged gas ion. Both the initial electron and the newly produced electron can then be subsequently accelerated by the field again and produce more electrons by impact ionization. This process thus develops into an avalanche of impact ionization, rapidly creating many free electrons and positive ions in the gas, and accounts for the highly conducting behavior of the gas. The acceleration of the initial

[110] J. S. Townsend, *Electricity in Gases,* Oxford University Press, Oxford, 1914.

electron to energies at which it can impact-ionize the neutrals must occur between two successive collisions of the electron with gas molecules. Since the density of gas molecules (i.e., the average distance between them) is less at higher pressures, the dielectric strength for gases is proportional to pressure[111]; in other words, pressurized gases can, in general, withstand stronger electric fields. This property of gases, and the fact that a gas recovers its insulating properties if the applied field is removed after breakdown, accounts for the frequent use of pressurized gases as dielectrics, especially in high-voltage applications.

In most cases, the electric field distribution in a gas is not uniform, as was implicitly assumed to be the case for the gas-filled parallel-plate capacitor just described. For example, in a single-wire-above-ground configuration as shown in Figure 4.51e, the electric field is strongest on the surface of the conductor facing the ground. Similarly, the electric field in a coaxial line is strongest on the inner conductor, decreasing with distance as r^{-1}. In general, on any charged conductor the electric field is higher in the vicinity of sharp points or conductor surfaces with the lowest radius of curvature. These high fields can sometimes locally exceed the dielectric strength of air and produce a breakdown in the vicinity of the high-field region although, further away from the conductor, the field is not sufficiently strong to sustain the breakdown. This type of a local discharge is referred to as a *corona discharge*. Consider, for example, a single-wire power transmission line above ground as shown in Figure 4.51e, with a typical height above ground of tens of meters. As the air breaks down in the vicinity of the conductor, it is ionized and behaves like a conductor. At night, the ionized air is apparent as a reddish (or sometimes blue-yellow) glow that crowns the conductor, hence the term "corona." To maximize power transmission, power lines typically operate very close to their highest voltage ratings. On days of bad weather, increased humidity and other effects may lower the dielectric strength of air, leading to corona discharges. Corona also comes about because of accumulation of rust and dirt on the lines, which create local inhomogeneities, resulting in large local electric fields in their vicinity. Corona discharge on power transmission lines must be avoided, because it substantially increases losses and emits electromagnetic waves that can interfere with certain types of nearby communication systems. The reader may have experienced the effects of corona discharge while driving under a transmission line with the radio on.

It was mentioned above that corona discharge occurs when the field is above the breakdown level in the vicinity of a conductor but not away from it. If the applied voltage is increased even further, a continuously ionized path in the form of a bright luminous arc is formed, extending to the nearest point of opposite polarity. An intense current (typically hundreds to thousands of amperes) flows through the gas,

---

[111]This important result was first demonstrated by Paschen in 1889 and is valid at moderate pressures. At higher pressure levels the breakdown voltage rises with increased pressure, since too many collisions take place and an excessive amount of energy is wasted in various excitation processes. For a discussion and references, see A. von Hippel (ed.), *Dielectric Materials and Applications,* Artech House, 1995; A. H. Beck, *Handbook of Vacuum Physics,* Macmillan, Pergamon Press, New York, 1965.

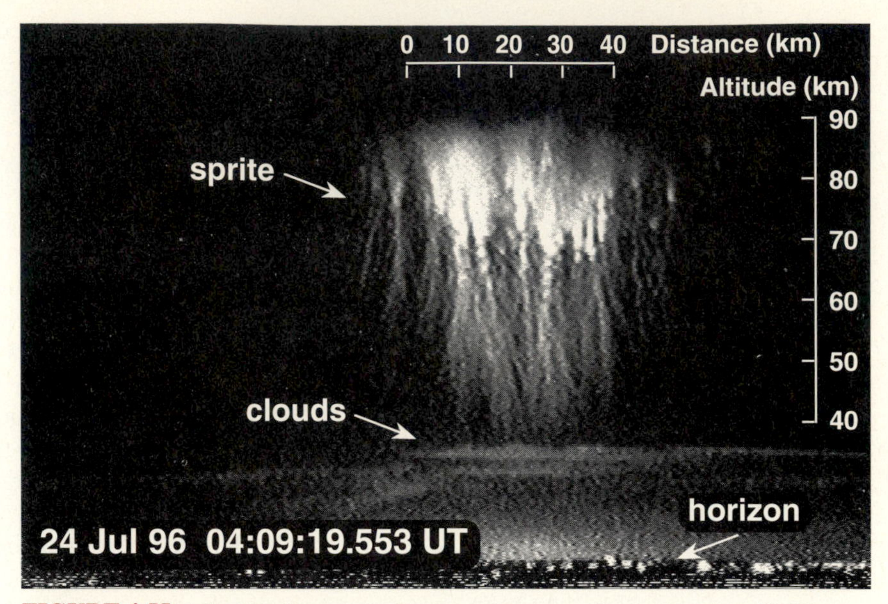

FIGURE 4.55. **Sprites.** Spectacular luminous glows occurring in clear air at 50–90-km altitudes above large active thunderstorms have been recently discovered and are called sprites. This image was taken from Fort Collins, Colorado, using a low-light-level television camera pointed toward the east. The event occurred above a thunderstorm in northern Kansas.

and an *arc discharge* is said to occur. The giant arcs of lightning are the best-known examples of an arc discharge.[112]

Rather spectacular examples of dielectric breakdown phenomena in air are the recently discovered luminous glows that occur at high altitudes above thunderstorms, referred to as *sprites.*[113] An example of sprites observed in the midwestern United States is shown in Figure 4.55. Sprites are believed to occur as a result of dielectric breakdown of air under the influence of quasi-static electric fields that appear at high altitudes following intense cloud-to-ground lightning discharges.

Breakdown in Solids The processes by which electrical breakdown occurs in liquids and solids are quite complex and depend on the particular periodic lattice structure of the material, the number of electrons in its conduction band, as well as imperfections due to the presence of foreign atoms. The externally measured voltage-current characteristics are quite similar to those observed for gases. When a pure, homogeneous solid dielectric is placed between suitable electrodes across which an

---

[112]M. A. Uman, *The Lightning Discharge,* Academic Press, Orlando, 1987.

[113]D. D. Sentman, E. M. Wescott , D. L. Osborne, D. L. Hampton, and M. J. Heavner, Preliminary results from the Sprites94 aircraft campaign, 1: red sprites, *Geophysical Research Letters,* 22 (10), pp. 1205–1208, 1995. For two descriptive articles, see Heaven's new fires, *Discover,* pp. 100–107, July 1997, and Lightning between earth and space, *Scientific American,* pp. 56–59, August 1997.

increasing voltage is applied, a small (microampere) current flows and levels off to a saturation value until the applied voltage exceeds a certain critical voltage. At this point, the current through the material suddenly (within $\sim 10^{-8}$ s) increases to the maximum value allowed by the supply. The dielectric strength of solids may vary by several orders of magnitude depending on the purity of the material, the shape of the material, the manner in which it was manufactured, the ambient temperature, and the duration of the applied field. In general, it is not possible to determine E_{BR} for solids on purely theoretical grounds; practical experimental tests under precisely defined conditions are generally required to specify the dielectric strength of the material reliably in the context of a particular application.

Three basic mechanisms of breakdown can be identified. So-called *intrinsic breakdown* is electronic in nature and depends on the presence of conduction-band electrons capable of migration through the lattice. The underlying process is an avalanche process similar to what occurs in gases; however, the microphysics can be understood only on the basis of the band theory of electronic structure and by including electron–lattice interactions—topics well beyond the scope of this book. The electric field levels at which intrinsic breakdown occurs typically represent the theoretical upper limits for E_{BR} for a solid dielectric, reached only by extremely pure and homogeneous materials. More typically, impurities and structural defects cause the material to break down at lower E_{BR} values. The second type of breakdown is called *thermal breakdown* and arises from the fact that small leakage currents flowing through a dielectric lead to resistive losses, which are dissipated as heat in the material. If the local heat is generated faster than it can be dissipated, the temperature of the material rises and breakdown occurs, typically by melting or decomposition. The dielectric strength E_{BR} due to this breakdown mechanism depends on the duration of the application of the field as well as the ambient temperature. The third type of breakdown mechanism is often referred to as *discharge breakdown* and depends on the presence of voids or cracks within the material. The presence of air in such cracks or voids produces a reduction in the dielectric strength. Partial discharges in the vicinity of such imperfections can in time lead to arcing, localized melting, and chemical transformations, creating conducting channels. Even in relatively pure and homogeneous materials, microscopic cracks can initially occur because of temperature or mechanical stresses, oxidation, or other aging (progressive degradation) effects.

Example 4-30 illustrates the use of dielectric strength to determine the voltage rating of a parallel-plate capacitor with two dielectric layers.

Example 4-30: Parallel-plate capacitor with sandwiched dielectrics. A parallel-plate capacitor is constructed using two separate dielectric materials, stacked between two conductor plates, as shown in Figure 4.56. (a) Find the capacitance of this configuration. (b) If glass ($\epsilon_{r1} = 6$ and $(E_{\text{BR}})_1 = 30$ MV-m^{-1}) and mica ($\epsilon_{r2} = 5$ and $(E_{\text{BR}})_2 = 200$ MV-m^{-1}) are chosen to be used with 2 cm^2 cross-sectional areas and equal thicknesses of 5 mm each (i.e., $d_1 = d_2 = 5$ mm), find the capacitance and voltage rating of the capacitor, using a safety factor of 10.

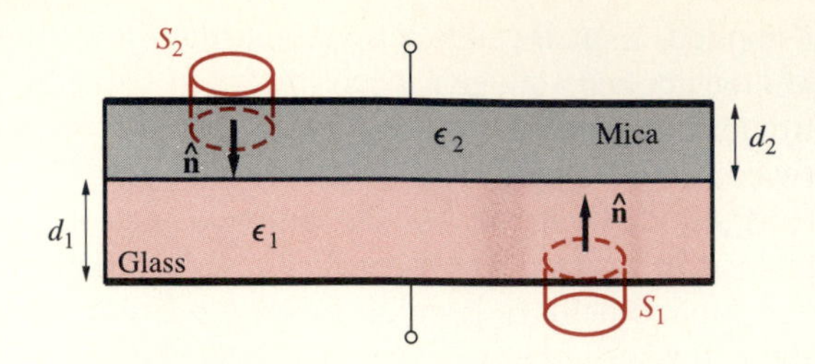

FIGURE 4.56. Dielectric strength. Parallel-plate capacitor with two sandwiched dielectric layers.

Solution:

(a) As usual in capacitance problems, we start by assuming that a potential difference is applied between the plates and that charges $+Q$ and $-Q$ are induced respectively on the lower and upper plates. The surface charge densities on the plates are, respectively, $\rho_{s1} = Q/A$ and $\rho_{s2} = -Q/A$, where A is the area of the plates. Taking a Gaussian surface S_1 as shown, we have

$$[\epsilon_1 E_1]A = \rho_{s1}A \quad \rightarrow \quad E_1 = \frac{\rho_{s1}}{\epsilon_1}$$

where we arbitrarily assume the positive direction for the electric field to be upward. Similarly, using the Gaussian surface S_2, we can show that $E_2 = -\rho_{s2}/\epsilon_2 = \epsilon_1 E_1/\epsilon_2$. The magnitude of the total potential drop between the plates is

$$|\Phi| = E_1 d_1 + E_2 d_2 = \frac{Q}{A}\left(\frac{d_1}{\epsilon_1} + \frac{d_2}{\epsilon_2}\right)$$

and the capacitance is then

$$C = \frac{Q}{|\Phi|} = \frac{A\epsilon_0}{(d_1/\epsilon_{r1}) + (d_2/\epsilon_{r2})}$$

Using the permittivity values given with $d_1 = d_2 = 5$ mm, we find $C \simeq 0.966$ pF.

(b) The voltage rating of the capacitor is determined by the dielectric layer that breaks down first. We have $(E_{\text{BR}})_1 = 30$ MV-m^{-1} and $(E_{\text{BR}})_2 = 200$ MV-m^{-1}. For a safety factor of 10, we must have

$$(E_1)_{\max} = \frac{(E_{\text{BR}})_1}{10} = 3 \text{ MV-m}^{-1} \quad \text{and} \quad (E_2)_{\max} = \frac{(E_{\text{BR}})_2}{10} = 20 \text{ MV-m}^{-1}$$

But since $E_{2y} = \epsilon_1 E_{1y}/\epsilon_2 = 6E_{1y}/5$, the voltage rating of this capacitor is determined by the glass layer, since the glass layer breaks down first. Thus, the voltage rating is

$$\begin{aligned}
|\Phi|_{\max} &= (E_1)_{\max} d_1 + \epsilon_{r1}(E_1)_{\max}\frac{d_2}{\epsilon_{r2}} \\
&= (E_1)_{\max}\left[d_1 + \frac{\epsilon_{r1}d_2}{\epsilon_{r2}}\right] \\
&= (30 \text{ MV-m}^{-1})(5 \times 10^{-3} \text{ m})\left[1 + \frac{6}{5}\right] = 330 \text{ kV}
\end{aligned}$$

Note that if the dielectric were filled uniformly with mica, the breakdown voltage with a safety factor of 10 would be $|\Phi|_{\max} = (E_2)_{\max}(d_1 + d_2) = (200 \text{ MV-m}^{-1})(0.01 \text{ m}) = 2 \text{ MV}$. Thus, the presence of the glass layer lowers the voltage rating of the capacitor.

4.11 ELECTROSTATIC BOUNDARY CONDITIONS

In dealing with electrostatic problems, it is often necessary to relate the polarization charges induced in dielectric media to electric fields produced by external charges (e.g., on conducting bodies). This is facilitated by relations such as [4.60], which are called *constitutive relations*. In most electrostatic applications, we also deal with interfaces between two or more different dielectric media. The manner in which **D** and **E** behave across such interfaces is described by the boundary conditions. The boundary conditions are derived from fundamental laws of electrostatics, which we reiterate below.

Starting with the experimental facts as expressed in Coulomb's law, namely that the electric force between charges is a function of the distance between them[114] and that its magnitude is proportional to the inverse square of the distance, we have derived the fundamental laws of electrostatics[115] as

$$\nabla \times \mathbf{E} = 0 \qquad \oint_C \mathbf{E} \cdot d\mathbf{l} = 0 \qquad \text{Conservative field}$$

$$\nabla \cdot \mathbf{D} = \rho \qquad \oint_S \mathbf{D} \cdot d\mathbf{s} = \int_V \rho \, dv \qquad \text{Coulomb's law}$$

The electrical properties of different physical media are accounted for by the *permittivity* ϵ of the material. For linear (polarization **P** linearly proportional to **E**) and isotropic (polarization **P** not dependent on the *direction* of **E**) media, **D** and **E** are related via [4.60], namely $\mathbf{D} = \epsilon\mathbf{E}$, which is valid even for inhomogeneous materials as long as we allow for ϵ to be a function of position $\epsilon(x, y, z)$.

A very common type of inhomogeneity occurs in practice at the interface between two electrically different materials. To establish a basis for solving such problems, we now study the behavior of **D** and **E** in crossing the boundary between two different materials and derive the conditions that **E** and **D** must satisfy at such

[114]It is important to remember that our arguments concerning the conservative nature of the electrostatic force—that $\oint_C \mathbf{E} \cdot d\mathbf{l} = 0$, or that $\int_C \mathbf{E} \cdot d\mathbf{l}$ is path-independent (Section 4.4)—were based only on the fact that the force from a single charge was radial and spherically symmetric. The fact that the force was proportional to r^{-2} was not important in this regard; any other r dependence would give the same result. Gauss's law, on the other hand, is based entirely on the fact that the electric force is proportional to r^{-2}.

[115]We do not discuss the definition and physical meaning of the notation $\nabla \times \mathbf{E}$ until Chapter 6, but it follows from $\oint_C \mathbf{E} \cdot d\mathbf{l}$ by Stokes's theorem, also to be introduced and discussed in Chapter 6.

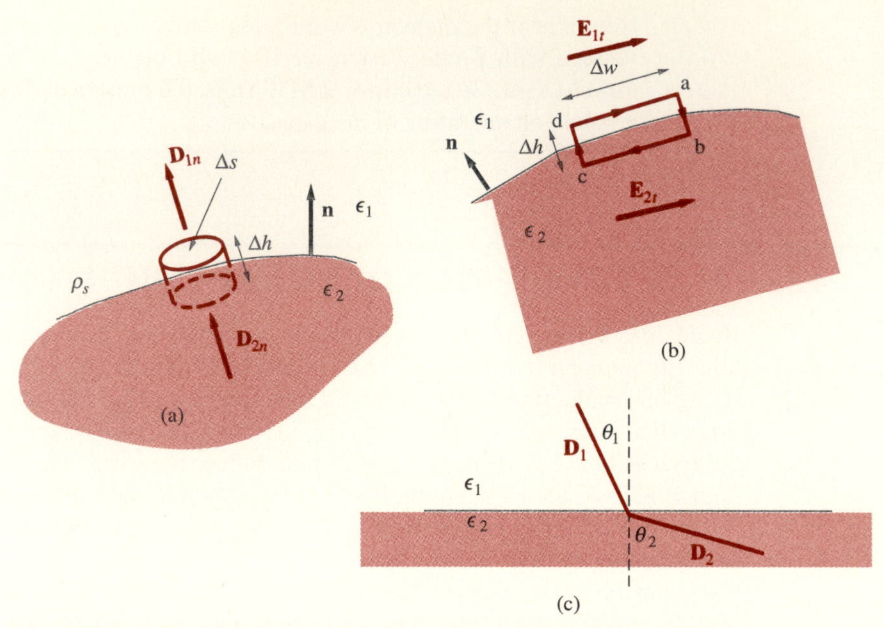

FIGURE 4.57. The boundary between two electrically different media. (a) A differential pillbox-shaped closed surface. (b) A differential closed contour. (c) **D** vectors at the boundary for the case $\epsilon_2 > \epsilon_1$.

interfaces. These conditions are referred to as the boundary conditions. The boundary conditions must be derived using the integral forms of the fundamental electrostatic laws, because the differential forms apply only at a point.

For the normal component of the electric flux density, we consider a pillbox-shaped closed differential surface as shown in Figure 4.57a. Since the contributions from the sides can be made infinitesimally small by taking $\Delta h \to 0$, we have

$$\oint_S \mathbf{D} \cdot d\mathbf{s} = (\hat{\mathbf{n}} \cdot \mathbf{D}_1)\Delta s - (\hat{\mathbf{n}} \cdot \mathbf{D}_2)\Delta s = \rho_s \Delta s \quad \rightarrow \quad \boxed{D_{1n} - D_{2n} = \rho_s} \qquad [4.62]$$

where the minus sign in front of the second term on the left-hand side is due to the unit vector normal to the surface of the pillbox being $-\hat{\mathbf{n}}$, and ρ_s is the surface charge density at the interface. Hence, the normal component of the electric flux density **D** is discontinuous through a surface that has free surface-charge density present; however, at interfaces between two dielectrics, the surface charge density ρ_s is ordinarily zero unless a surface charge density is actually placed at the interface. For $\rho_s = 0$, [4.62] becomes

$$\boxed{\hat{\mathbf{n}} \cdot \mathbf{D}_1 = \hat{\mathbf{n}} \cdot \mathbf{D}_2 \quad \text{or} \quad D_{1n} = D_{2n}} \qquad [4.63]$$

In other words, the normal component of **D** is continuous across a charge-free dielectric boundary. The normal component of **E** is then not continuous, since we have

$$\epsilon_1(\hat{\mathbf{n}} \cdot \mathbf{E}_1) = \epsilon_2(\hat{\mathbf{n}} \cdot \mathbf{E}_2) \quad \rightarrow \quad \epsilon_1 E_{1n} = \epsilon_2 E_{2n} \qquad [4.64]$$

The discontinuity in $\hat{\mathbf{n}} \cdot \mathbf{E}$ or E_n can be understood physically as follows. The normal component of the polarization vector $\mathbf{P}$ can be related to the surface polarization charge by integrating $\mathbf{P}$ over the surface of the differential pillbox shown in Figure 4.57a as $\Delta h \rightarrow 0$, similar to [4.62], as

$$\oint_S \mathbf{P} \cdot d\mathbf{s} = (\hat{\mathbf{n}} \cdot \mathbf{P}_1)\Delta s - (\hat{\mathbf{n}} \cdot \mathbf{P}_2)\Delta s = \rho_{sp}\Delta s$$

since $\mathbf{P} \cdot d\mathbf{s} = \mathbf{P} \cdot \hat{\mathbf{n}}\, ds = \rho_{sp}\, ds$. Simplifying, we can write

$$P_{1n} - P_{2n} = \rho_{sp} \qquad [4.65]$$

But, using [4.58], we have

$$P_{1n} = D_{1n} - \epsilon_0 E_{1n} = (\epsilon_1 - \epsilon_0)E_{1n}$$

$$P_{2n} = D_{2n} - \epsilon_0 E_{2n} = (\epsilon_2 - \epsilon_0)E_{2n}$$

so that

$$P_{1n} - P_{2n} = \underbrace{\epsilon_1 E_{1n} - \epsilon_2 E_{2n}}_{=0 \text{ from [4.64]}} + \epsilon_0(E_{2n} - E_{1n})$$

Using [4.65], we find the discontinuity in the normal component of the electric field at the interface in terms of the polarization surface charge as

$$E_{2n} - E_{1n} = \frac{\rho_{sp}}{\epsilon_0}$$

Thus, at the surface of the dielectric the normal component of the electric field $\mathbf{E}$ is in general discontinuous by an amount ρ_{sp}/ϵ_0, just as if ρ_{sp} is a surface layer of free charge at that location. The surface polarization charge represents the amount of charge on the ends of dipoles in medium 2 that are not canceled by the charge of opposite polarity on the ends of the dipoles in medium 1 (see Figure 4.52a).

In practice, surface polarization charge does not need to be taken into account explicitly, since the usual methodology involves finding suitable solutions for $\mathbf{E}$ and $\mathbf{D}$ in the two dielectric regions and then adjusting the magnitudes of the solutions to satisfy the boundary conditions. Using the different permittivities in the two regions fully accounts for all the requirements placed on the fields due to the different microphysical electrical behavior of the materials.

For the tangential components of the electric field, we consider the line integral of $\mathbf{E}$ around a closed differential contour such as that shown in Figure 4.57b. Since the contributions from sides ab and cd can be made infinitesimally small by taking $\Delta h \rightarrow 0$, we have

$$\int_{\mathrm{d}}^{\mathrm{a}} \mathbf{E}_1 \cdot d\mathbf{l} + \int_{\mathrm{b}}^{\mathrm{c}} \mathbf{E}_2 \cdot d\mathbf{l} = (E_{1t} - E_{2t})\Delta w = 0 \quad \rightarrow \quad \boxed{E_{1t} = E_{2t}} \qquad [4.66]$$

Thus the tangential component of the electric field is continuous across the interface between the two materials. Since $\mathbf{D} = \epsilon\mathbf{E}$, the tangential component of $\mathbf{D}$ is not continuous across a boundary. Instead we have

$$\frac{D_{1t}}{\epsilon_1} = \frac{D_{2t}}{\epsilon_2}$$

A consequence of [4.62] and [4.66] is the change in orientation of electric flux lines across material interfaces, as illustrated in Figure 4.57c. Consider an electric field $\mathbf{E}_1$ that is oriented at an angle θ_1 from the normal in medium 1. From [4.62] and [4.66] we have

$$E_1 \sin\theta_1 = E_2 \sin\theta_2$$

$$\epsilon_1 E_1 \cos\theta_1 = \epsilon_2 E_2 \cos\theta_2$$

where E_1 and E_2 are the magnitudes of $\mathbf{E}_1$ and $\mathbf{E}_2$ and where we have assumed $\rho_S = 0$. Therefore, we can find

$$\tan\theta_2 = \frac{\epsilon_2}{\epsilon_1}\tan\theta_1$$

and

$$E_2 = E_1\sqrt{\left(\frac{\epsilon_1}{\epsilon_2}\cos\theta_1\right)^2 + \sin^2\theta_1}$$

The above relationships indicate that electric field lines are *bent further away from the normal* to the boundary in the medium with the *higher* permittivity.

Example 4-31 illustrates the application of the electrostatic boundary conditions.

Example 4-31: Spherical dielectric shell. A positive point charge Q is at the center of a spherical dielectric shell of an inner radius a and an outer radius b. The relative dielectric constant of the shell is ϵ_r. Determine $\mathbf{E}$, Φ, $\mathbf{D}$, $\mathbf{P}$, and the polarization charge densities ρ_p and ρ_{sp}.

Solution: This example is very similar to Example 4-21 except that the conducting shell has now been replaced by a dielectric shell (see Figure 4.58). In view of the spherical symmetry, we apply Gauss's law to find $\mathbf{E}$ and $\mathbf{D}$ in three regions: (a) $r \geq b$; (b) $a \leq r < b$; and (c) $r < a$. The electric potential Φ is then determined from the negative line integral of $\mathbf{E}$, and polarization $\mathbf{P}$ follows from the relation

$$\mathbf{P} = \mathbf{D} - \epsilon_0\mathbf{E} = \epsilon_0(\epsilon_r - 1)\mathbf{E}$$

Note that the $\mathbf{E}$, $\mathbf{D}$, and $\mathbf{P}$ vectors have only radial components.

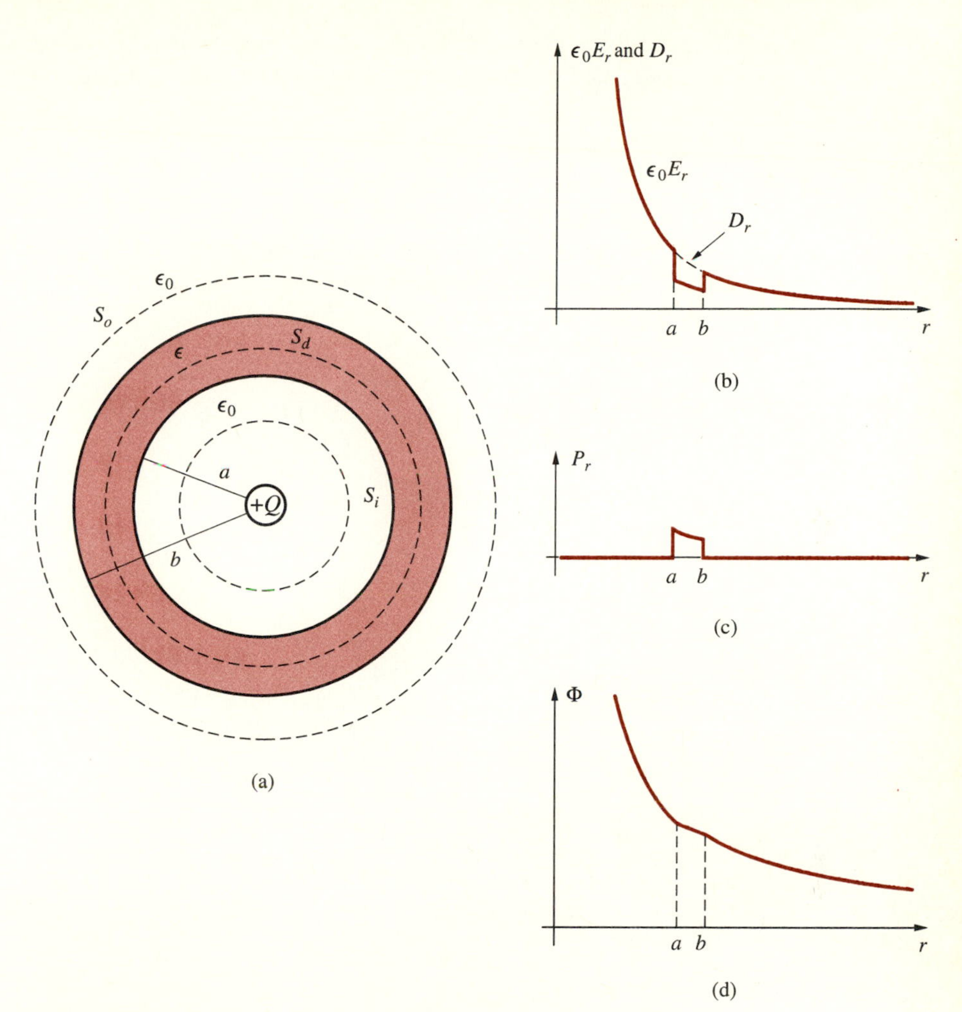

FIGURE 4.58. Point charge at the center of a spherical dielectric shell. (a) The geometry. (b) D_r and $\epsilon_0 E_r$ versus r. (c) P_r versus r. (d) Φ versus r. Note that $D_r = P_r + \epsilon_0 E_r$, as can be seen from the plots (b) and (c).

(a) For the region $r \geq b$, **E** and Φ are exactly the same as in Example 4-21. We have

$$E_r = \frac{Q}{4\pi\epsilon_0 r^2} \qquad r \geq b \qquad \Phi = \frac{Q}{4\pi\epsilon_0 r} \qquad r \geq b$$

From equations [4.58] and [4.60] we obtain

$$P_r = 0 \qquad r \geq b$$

$$D_r = \epsilon_0 E_r = \frac{Q}{4\pi r^2} \qquad r \geq b$$

(b) For the region $a \leq r < b$ (i.e., inside the dielectric shell), **E** can be found using Gauss's law using the spherical Gaussian surface S_d as

$$E_r = \frac{Q}{4\pi\epsilon_0\epsilon_r r^2} = \frac{Q}{4\pi\epsilon r^2} \qquad a \leq r < b$$

and **D** follows as

$$D_r = \epsilon E_r = \frac{Q}{4\pi r^2} \qquad a \leq r < b$$

The polarization vector **P** is

$$P_r = D_r - \epsilon_0 E_r = \frac{Q}{4\pi r^2}\left(1 - \frac{1}{\epsilon_r}\right) \qquad a \leq r < b$$

The electrostatic potential is

$$\begin{aligned}\Phi &= -\int_\infty^b \frac{Q}{4\pi\epsilon_0 r^2}dr - \int_b^r \frac{Q}{4\pi\epsilon r^2}dr \\ &= \frac{Q}{4\pi\epsilon_0 b} + \left[\frac{Q}{4\pi\epsilon r}\right]_b^r \\ &= \frac{Q}{4\pi\epsilon_0 r}\left[\frac{r}{b}\left(1 - \frac{1}{\epsilon_r}\right) + \frac{1}{\epsilon_r}\right] \qquad a \leq r < b\end{aligned}$$

(c) For the region $r < a$, the application of Gauss's law on surface S_i yields the same expressions for E_r, D_r, and P_r as in region $r \geq b$:

$$E_r = \frac{Q}{4\pi\epsilon_0 r^2} \qquad r < a$$

$$D_r = \frac{Q}{4\pi r^2} \qquad r < a$$

$$P_r = 0 \qquad r < a$$

To find Φ, we must add to Φ at $r = a$ the negative line integral of E_r in the region $r < a$:

$$\begin{aligned}\Phi &= \Phi(a) - \int_a^r \frac{Q}{4\pi\epsilon_0 r^2}dr \\ &= \frac{Q}{4\pi\epsilon_0 r}\left[1 + \left(\frac{r}{b} - \frac{r}{a}\right)\left(1 - \frac{1}{\epsilon_r}\right)\right] \qquad r < a\end{aligned}$$

The variations of $\epsilon_0 E_r$ and D_r versus r are plotted in Figure 4.58b. The difference $(D_r - \epsilon_0 E_r)$ is P_r and is also shown in Figure 4.58c. We note that D_r is a continuous curve exhibiting no sudden changes in going from one medium to another and that P_r exists only in the dielectric region. The plot for Φ is also shown in Figure 4.58d. The polarization charge densities associated with the dielectric shell are

$$\rho_{\rm p} = -\nabla \cdot \mathbf{P} = -\frac{1}{r^2}\frac{\partial}{\partial r}\left[r^2 \frac{Q}{4\pi r^2}\left(1 - \frac{1}{\epsilon_r}\right)\right] = 0$$

inside the dielectric shell,

$$\rho_{sp}\big|_{r=a} = \mathbf{P} \cdot (-\hat{\mathbf{r}})\big|_{r=a} = -\frac{Q}{4\pi a^2}\left(1 - \frac{1}{\epsilon_r}\right)$$

on the inner surface of the shell, and

$$\rho_{sp}\big|_{r=b} = \mathbf{P} \cdot \hat{\mathbf{r}}\big|_{r=b} = \frac{Q}{4\pi b^2}\left(1 - \frac{1}{\epsilon_r}\right)$$

on the outer surface of the shell, respectively.

These results indicate that there is no net polarization volume charge inside the dielectric shell. However, polarization surface charges of negative and positive polarity exist on the inner and the outer surfaces, respectively. These surface charges produce an electric field inside the dielectric shell that is directed radially inward, thus reducing the **E** field inside the dielectric shell produced by the point charge $+Q$ located at its center.

4.12 ELECTROSTATIC ENERGY

When we lift a flowerpot and place it on a windowsill, the work we do (i.e., the energy we expend) against gravity is stored in the form of potential energy. Any object with nonzero mass acquires gravitational potential energy when raised through a certain height. When we compress a spring, the work we expend to do so is stored in the form of elastic potential energy. The gravitational potential energy can be recovered by lowering the raised object, and the elastic potential energy can be recovered by releasing the spring. Bringing two like charges together from infinite separation against their electrostatic repulsion also requires work; *electrostatic energy* is thus stored in such a configuration of charges, and this energy may be recovered by allowing the charges to recede to an infinite separation. In Section 4.4, we defined the concept of electrostatic potential of a given point as the work that needs to be done to move a unit positive test charge from infinity to that point. In this section we further consider the relationship between electrostatic energy, electric potential, charges, and electric fields.

The absolute potential due to a single charge Q in a simple dielectric medium is given as

$$\Phi = \frac{Q}{4\pi\epsilon R}$$

where R is the distance between the point at which we measure the potential and the position of Q. In the presence of this potential distribution (or the corresponding electric field), the energy required to bring another charge q from infinity to a distance R from Q is

$$W_e = q\Phi = \frac{Qq}{4\pi\epsilon R} \qquad [4.67]$$

where $W_e = q\Phi$ follows from the definition of potential as the work per unit charge. This energy is what is required to hold these two charges at a distance R from one another, and is referred to as the *potential energy* of the charge configuration. This electric energy associated with an assembly of charges is only a function of the final position of the charges and can be determined by calculating the work required to gather the charges together. That any assemblage of point or continuous charge distributions "stores" energy can be seen by considering the motion of the charges if they are to be released from their fixed (i.e., externally imposed) points. The large Coulomb forces that exist between the charges cause the charge assembly to literally fly apart.[116] The net kinetic energy gained by the charges after they have been separated equals the initial potential energy stored in the assemblage.

4.12.1 Electrostatic Energy in Terms of Charges and Potential

We begin by considering a series of separated point charges at infinity and then calculate the energy required to bring these charges, one by one, from infinity to form the assemblage of charges as illustrated in Figure 4.59. The energy required to bring the first charge q_1 from infinity to the region of interest is $W_1 = 0$, since no other charges are present to exert any force on q_1. The energy required to bring the second charge q_2 to a distance R_{12} from q_1 is

$$W_2 = q_2\Phi_{12}$$

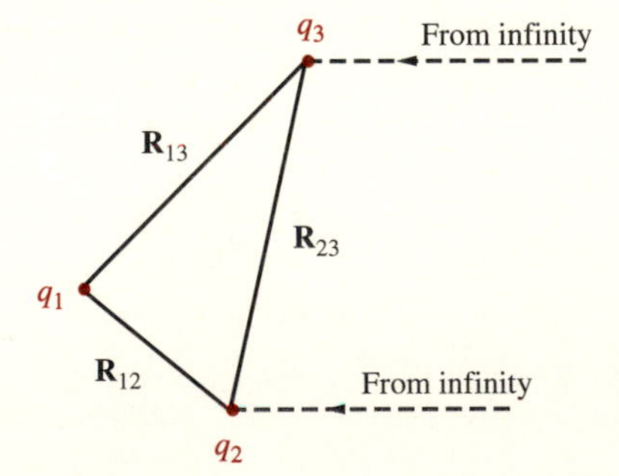

FIGURE 4.59. An assemblage of charges. Point charges q_i gathered together by bringing them one by one from infinity.

[116] Assuming that no other forces act on the particles—that is what we mean by having them "released" from their fixed locations.

where Φ_{12} is defined as the electric potential at the location of q_2 due to the presence of charge q_1; that is,

$$\Phi_{12} = \frac{q_1}{4\pi\epsilon R_{12}}$$

In Φ_{12}, the second subscript denotes the location where the potential is measured, whereas the first subscript denotes the source charge q_1 creating the potential at the position of q_2. If a third charge q_3 is brought from infinity to a distance R_{13} from q_1 and R_{23} from q_2, the energy required is

$$W_3 = q_3(\Phi_{13} + \Phi_{23})$$

The energy required to bring a fourth charge is

$$W_4 = q_4(\Phi_{14} + \Phi_{24} + \Phi_{34})$$

and thus the total energy required to gather together a series of charges is

$$\begin{aligned} W_e = \underbrace{q_2\Phi_{12}}_{W_2} + \underbrace{q_3(\Phi_{13} + \Phi_{23})}_{W_3} + \underbrace{q_4(\Phi_{14} + \Phi_{24} + \Phi_{34})}_{W_4} \\ + \underbrace{q_5(\Phi_{15} + \Phi_{25} + \Phi_{35} + \Phi_{45})}_{W_5} + \cdots \end{aligned} \qquad [4.68]$$

where the subscript "e" indicates that W_e is electrostatic energy, to be distinguished from magnetostatic energy W_m, discussed in Chapter 6. We can develop a compact series expression for W_e by recognizing that W_2 may be written alternatively as

$$W_2 = q_2\Phi_{12} = \frac{q_2 q_1}{4\pi\epsilon R_{12}} = q_1\frac{q_2}{4\pi\epsilon R_{12}} = q_1\Phi_{21}$$

which indicates that we have

$$q_i\Phi_{ki} = q_k\Phi_{ik}$$

The above is true simply because no matter what order the charges are assembled in, the partial energy associated with any two charges, say, the ith and the kth, equals $q_i q_k/(4\pi\epsilon R_{ki})$, and this term occurs only once during the assembly process.

It is then possible to rewrite [4.68] in its alternative form as

$$\begin{aligned} W_e = q_1\Phi_{21} + q_1\Phi_{31} + q_2\Phi_{32} + q_1\Phi_{41} + q_2\Phi_{42} + q_3\Phi_{43} \\ + q_1\Phi_{51} + q_2\Phi_{52} + q_3\Phi_{53} + q_4\Phi_{54} + \cdots \end{aligned} \qquad [4.69]$$

corresponding to assembling the charges in reverse order. We can now add [4.68] and [4.69] to find

$$\begin{aligned} 2W_e = q_1(\Phi_{21} + \Phi_{31} + \Phi_{41} + \Phi_{51} + \cdots) \\ + q_2(\Phi_{12} + \Phi_{32} + \Phi_{42} + \Phi_{52} + \cdots) \\ + q_3(\Phi_{13} + \Phi_{23} + \Phi_{43} + \Phi_{53} + \cdots) + \cdots \end{aligned}$$

which may be conveniently written as

$$W_e = \frac{1}{2}\sum_{i=1}^{N} q_i\Phi_i \qquad [4.70]$$

where

$$\Phi_i = \Phi_{1i} + \Phi_{2i} + \Phi_{3i} + \cdots + \Phi_{(i-1)i} + \Phi_{(i+1)i} + \cdots + \Phi_{Ni} \qquad [4.71]$$

is the total potential at the ith location due to all the other $(N-1)$ charges present, except for the ith charge. In other words, Φ_i includes the contributions to the total potential at the ith point of all charges other than the one that is at that point. Equation [4.70] represents the total energy of an assemblage of N charges in terms of a simple summation over the charges.

We can write [4.70] alternatively as

$$W_e = \frac{1}{2}\sum_{i=1}^{N}\sum_{j=1}^{N} q_i\Phi_{ji} \qquad (i \neq j) \qquad [4.72]$$

where the quantity inside the double summation is the electrostatic potential at the location of the ith charge produced by the jth charge multiplied by the value of the ith charge.

It might have been tempting to write [4.72] by inspection, without carrying out the indicated steps as was done above. However, any such intuitive argument that leads from the definition of the electrostatic potential to the electrostatic potential energy is more than likely to ignore the $\frac{1}{2}$ factor that stands outside the summation in [4.72]. This factor is a consequence of the fact that the potential energy associated with any two charges is a property of both charges but is not a property that can be assigned to each charge separately.

Example 4-32 provides a quantitative example of a configuration of three charges.

Example 4-32: Energy required to assemble three point charges. Determine the energy required to bring three charges from infinity to three points on the x axis: $(-1, 0, 0)$, $(0, 0, 0)$, and $(+1, 0, 0)$. Assume free space.

Solution: The total energy of the configuration can be calculated using [4.68]. Assume that charge q_1 is brought to $(0, 0, 0)$, charge q_2 to $(-1, 0, 0)$, and q_3 to $(+1, 0, 0)$. We have

$$W_e = q_2\Phi_{12} + q_3\Phi_{13} + q_3\Phi_{23} = q_2\frac{q_1}{4\pi\epsilon_0 R_{12}} + q_3\frac{q_1}{4\pi\epsilon_0 R_{13}} + q_3\frac{q_2}{4\pi\epsilon_0 R_{23}}$$

where we have $R_{12} = 1$, $R_{13} = 1$, and $R_{23} = 2$. If $q_1 = q_2 = q_3 = q$, we then have

$$W_e = \frac{5q^2}{8\pi\epsilon_0}$$

Note that the same result could have been obtained using [4.70], which is simply another way of writing [4.68]:

$$W_e = \frac{1}{2}q_1\left(\frac{q_2}{4\pi\epsilon_0 R_{12}} + \frac{q_3}{4\pi\epsilon_0 R_{13}}\right) + \frac{1}{2}q_2\left(\frac{q_1}{4\pi\epsilon_0 R_{12}} + \frac{q_3}{4\pi\epsilon_0 R_{23}}\right)$$
$$+ \frac{1}{2}q_3\left(\frac{q_1}{4\pi\epsilon_0 R_{13}} + \frac{q_2}{4\pi\epsilon_0 R_{23}}\right)$$

which upon adding terms with values substituted also gives $W_e = 5q^2/(8\pi\epsilon_0)$.

4.12.2 Electrostatic Energy in Terms of Fields

It is interesting to consider *where* the energy associated with the charge distribution is stored. This question is analogous to one we might ask in mechanics when we have two masses attached to the ends of a compressed spring. The stored energy might be considered to reside in the masses or in the stressed state of the spring. The first viewpoint is the same as that expressed by [4.70], where the stored energy is linked to the charges and their positions. This view ascribes physical reality only to the charges and their spatial distributions. On the other hand, if the field concept is to be complete, it should be capable of expressing the stored energy without recourse to a description of the charges producing it. The energy should be describable in terms of the "elastic" quality of the electric field, just as the spring stores the energy in the preceding mechanical example.

Another mechanical example is the flowerpot on the windowsill, which possesses potential energy, because if it is pushed over the edge, it gains kinetic energy. Is the potential energy stored in the windowsill, in the flowerpot, or in the gravitational field?

It turns out that the energy associated with a charge distribution can indeed be expressed solely in terms of the fields $\mathbf{E}$ and $\mathbf{D}$. We can generalize [4.70] to the case of a continuous distribution of charges in a volume V by making the following transformations and taking the limit as $N \to \infty$,

$$q_i \to \rho(r)dv, \qquad \Phi_i \to \Phi(r) \quad \text{and} \quad \sum_{i=1}^{N} \to \int_V$$

which gives

$$\boxed{W_e = \tfrac{1}{2}\int_V \rho\Phi\, dv} \qquad [4.73]$$

where Φ is now a continuous function of position and is the potential distribution due to all charges. It is no longer necessary to exclude the contribution due to $\rho\,\delta v$ at the point where Φ is evaluated, because in the limit $\delta v \to 0$, this contribution is zero anyway. Equation [4.73] can be interpreted as an integral over the volume V of an electric energy density $w_e = \frac{1}{2}\rho\Phi$, which exists in regions where $\rho \neq 0$.

To convert [4.73] into an expression involving **D** and **E**, note that **D** associated with the Φ in [4.73] is that generated by ρ, given by [4.59] or $\nabla \cdot \mathbf{D} = \rho$. Thus, we can write

$$W_e = \frac{1}{2} \int_V (\nabla \cdot \mathbf{D}) \Phi \, dv$$

We can now use the vector identity:

$$\nabla \cdot \Phi \mathbf{D} \equiv \Phi(\nabla \cdot \mathbf{D}) + \mathbf{D} \cdot \nabla \Phi$$

to find

$$W_e = \frac{1}{2} \int_V (\nabla \cdot \Phi \mathbf{D} - \mathbf{D} \cdot \nabla \Phi) \, dv$$

Since for electrostatic fields $\mathbf{E} = -\nabla \Phi$, we have

$$W_e = \frac{1}{2} \oint_S \Phi \mathbf{D} \cdot \hat{\mathbf{n}} \, ds + \frac{1}{2} \int_V \mathbf{D} \cdot \mathbf{E} \, dv$$

where S is the surface that encloses volume V, and $\hat{\mathbf{n}}$ is the outward normal to the differential surface element ds. The surface S, and thus the volume V, must be large enough to enclose all the charges in the region of interest. Since there are no other restrictions to the choice of the surface S, we can take it to extend to infinity. In that case the surface integral reduces to zero, since the dependencies of the various quantities at large distances from the charges (at which the entire charge distribution just looks like a point charge) are

$$\Phi \sim \frac{1}{r}, \qquad D \sim \frac{1}{r^2} \qquad \text{and} \qquad ds \sim r^2$$

We thus have

$$\lim_{r \to \infty} \Phi \mathbf{D} \cdot \hat{\mathbf{n}} \, ds \quad \rightarrow \quad \lim_{r \to \infty} \frac{1}{r} \frac{1}{r^2} r^2 \rightarrow 0$$

so the electrostatic energy W_e reduces to simply the volume integral term,

$$\boxed{W_e = \frac{1}{2} \int_V \mathbf{D} \cdot \mathbf{E} \, dv} \qquad [4.74]$$

Since we let S extend to infinity, we must have $V \rightarrow \infty$, meaning that the integral in [4.74] has to be carried out over all space in which the electric field is nonzero. Based on the above, we can define the *volume energy density* for the electrostatic field as

$$\boxed{w_e = \tfrac{1}{2} \mathbf{D} \cdot \mathbf{E} = \tfrac{1}{2} \epsilon E^2} \qquad [4.75]$$

We have thus found a way of expressing the stored energy of a charge distribution solely in terms of the electric field.

Note that the two alternative expressions we have found for electrostatic energy density, namely, $\frac{1}{2}\rho\Phi$ from [4.73] and $\frac{1}{2}\epsilon E^2$ from [4.75] are quite different. The former implies that energy exists where charges exist and is zero where $\rho = 0$. However, the latter indicates that electrostatic energy exists wherever the fields exist. Both points of view have merit, and it is neither necessary[117] nor possible to determine which one is more "correct." The dilemma here is the same as that in the mechanical examples discussed earlier: the two masses at the ends of a compressed spring and the flowerpot on the windowsill. It is not possible to "localize" energy or to decide whether it is associated with charge or the field. Thus, the quantities $\frac{1}{2}\rho\Phi$ and $\frac{1}{2}\epsilon E^2$ represent energy density only to the extent that their volume integral over space is the total potential energy.

We can also write W_{e} in terms of the polarization $\mathbf{P} = \mathbf{D} - \epsilon_0\mathbf{E}$ as

$$W_{\mathrm{e}} = \frac{1}{2}\int_V \epsilon_0 E^2 dv + \frac{1}{2}\int_V \mathbf{P}\cdot\mathbf{E}\,dv$$

In other words, to establish a given electric field $\mathbf{E}$ in a dielectric medium rather than in free space, additional energy needs to be supplied (by the source that sustains the field). This additional energy is given by the second integral in the last equation above and is the energy required to polarize the dielectric.

Energy Stored in a Capacitor Equation [4.73] or [4.75] can be used to derive the expression for energy stored in a capacitor, which we know from circuit theory to be $W_{\mathrm{e}} = \frac{1}{2}CV^2$. Consider any configuration of two conductors constituting a general capacitor as shown in Figure 4.50. For simplicity, assume one of the conductors to be at zero potential, while the second one is at potential V. Since the charge density is zero everywhere inside the conductors and is nonzero only on their surfaces, Equation [4.73] reduces to

$$W_{\mathrm{e}} = \frac{1}{2}\left[\int_{S_{c1}} \rho_{s1}\Phi_1\,ds + \int_{S_{c2}} \rho_{s2}\Phi_2\,ds\right]$$

where S_{c1} and S_{c2} are the surfaces of the conductors. The first integral reduces to zero, since the potential is identically zero (i.e., $\Phi_1 = 0$) on the first conductor. On the second conductor, the electrostatic potential everywhere is equal to V, so we have

---

[117]For an interesting discussion, see Vol II/Section 8 of *The Feynman Lectures on Physics,* Addison Wesley, Reading, Massachusetts, 1964. A more advanced discussion is given in Chapter 4 of W. B. Cheston, *Elementary Theory of Electric and Magnetic Fields,* Wiley, 1964. It is possible to formulate the subject of electrostatics such that the electrostatic energy and the fields produced by a system of charges are fundamental quantities and the force between charged particles is a derived concept—an approach more fruitful for treatment of quantum phenomena. In the context of the experimentally based approach adopted here, the electric force, as defined in Coulomb's law, is taken to be the fundamental quantity.

$$W_e = \tfrac{1}{2}\int_{S_{c2}} \rho_{s2}\Phi_2\,ds = \tfrac{1}{2}V\int_{S_{c2}} \rho_{s2}\,ds = \tfrac{1}{2}VQ = \tfrac{1}{2}CV^2 \qquad [4.76]$$

where Q is the total charge on each of the conductors, and where we have used $C = Q/V$.

Examples 4-33 and 4-34 illustrate the determination of the electrostatic energy of various charge and field configurations.

Example 4-33: Energy of a sphere of charge. Consider a sphere of radius b having uniform charge of volume density ρ in free space. Determine the electrostatic energy of this charge configuration using (a) [4.73] and (b) [4.75].

Solution: The potential $\Phi(r)$ and the electric field $E_r(r)$ were determined in Example 4-13:

$$E_r = \frac{1}{3}\frac{r\rho}{\epsilon_0} \qquad r < b \qquad \Phi(r) = -\frac{1}{6}\frac{r^2\rho}{\epsilon_0} + \frac{1}{2}\frac{\rho b^2}{\epsilon_0}$$

$$E_r = \frac{1}{3}\frac{\rho b^3}{\epsilon_0 r^2} \qquad r \geq b \qquad \Phi(r) = \frac{1}{3}\frac{\rho b^3}{\epsilon_0 r}$$

where we have made the appropriate substitutions of $a \to b$ and $\rho \to Q/[(4\pi/3)a^3]$.

(a) Using [4.73], the stored energy is

$$W_e = \frac{1}{2}\int_V \rho\Phi(r)\,dv = \frac{1}{2}4\pi\int_0^b \rho\Phi(r)r^2\,dr = \frac{2\pi\rho^2\left[\dfrac{b^5}{6} - \dfrac{b^5}{30}\right]}{\epsilon_0} = \frac{4\pi}{15}\frac{\rho^2 b^5}{\epsilon_0}$$

Note that ρ is finite only where $r < b$, so that we only need to evaluate Φ for $r < b$.

(b) Since $D_r = \epsilon_0 E_r$, the stored energy from [4.75] is

$$W_e = \frac{1}{2}\int_0^\infty \mathbf{D}\cdot\mathbf{E}\,dv = \frac{1}{2}4\pi\int_0^\infty D_r E_r r^2\,dr = \frac{2\pi\rho^2}{\epsilon_0}\frac{1}{9}\left[\int_0^b r^2 r^2 dr + \int_b^\infty \frac{b^6}{r^4}r^2 dr\right]$$

$$= \frac{2\pi\rho^2}{\epsilon_0}\frac{1}{9}\left[\frac{b^5}{5} + b^5\right] = \frac{4\pi\rho^2 b^5}{15\epsilon_0}$$

As expected, this result is identical to the stored energy as calculated from [4.73].

Example 4-34: Electrostatic energy around a charged conducting sphere. Consider a conducting sphere of radius a carrying a total charge of Q in free space. Determine the electrostatic energy stored around this sphere using (a) [4.76] and (b) [4.75].

Solution: The electrostatic potential for such a conducting sphere was determined in Example 4-25. The variations with r of the electric field $\mathbf{E}(r)$ and the potential $\Phi(r)$ are

$$\Phi(r) = \begin{cases} \dfrac{Q}{4\pi\epsilon_0 r} & r \geq a \\ \dfrac{Q}{4\pi\epsilon_0 a} & r < a \end{cases} \qquad \mathbf{E}(r) = \begin{cases} \hat{\mathbf{r}}\dfrac{Q}{4\pi\epsilon_0 r^2} & r \geq a \\ 0 & r < a \end{cases}$$

The capacitance of a charged conducting sphere is given by [4.47] as $C = 4\pi\epsilon_0 a$.

(a) We first calculate the stored energy by considering the sphere as a capacitor. Noting that the sphere is at a potential $V_0 = \Phi(r = a)$, the energy stored is given by

$$W_e = \frac{1}{2}QV_0 = \frac{1}{2}Q\left(\frac{Q}{4\pi\epsilon_0 a}\right) = \frac{Q^2}{8\pi\epsilon_0 a}$$

(b) We can also determine the stored energy by integrating the electric field over the entire space within which it exists (i.e., it is nonzero), by using [4.75]. We have

$$W_e = \frac{1}{2}\int_V \epsilon_0 E^2\, dv = \frac{1}{2}\int_0^{2\pi}\int_0^{\pi}\int_a^{\infty} \epsilon_0\left(\frac{Q}{4\pi\epsilon_0 r^2}\right)^2 r^2 \sin\theta\, dr\, d\theta\, d\phi$$

$$= \frac{1}{2}(4\pi)\int_a^{\infty} \epsilon_0\left(\frac{Q}{4\pi\epsilon_0}\right)^2 \frac{1}{r^2}\, dr = \frac{Q^2}{8\pi\epsilon_0}\left[-\frac{1}{r}\right]_a^{\infty} = \frac{Q^2}{8\pi\epsilon_0 a}$$

which is the same as that found in (a).

4.13 ELECTROSTATIC FORCES

The force experienced by a conducting body placed in an electric field was discussed in Section 4.7.5. The magnitude of this force per unit area at a point on the conductor on which an electric field E_n is incident is given by $\epsilon_0 E_n^2/2 = \rho_s^2/(2\epsilon_0)$, where ρ_s is the surface charge density at the same point. The fact that there is a force of attraction between the plates of a parallel-plate capacitor is thus to be expected, because of the presence on the plates of charges of opposite sign. In calculating the electrostatic force on an object, it is in principle possible simply to integrate the force per unit area [i.e., $\rho_s^2/(2\epsilon_0)$] over the entire surface of the conducting body. In fact, electrostatic forces on and between charged conductors are special cases of the subject of electrostatic force on matter in general, which can be formulated entirely in terms of electric fields that produce stress in the media which they permeate.[118]

In this book we confine our attention to selected special cases involving simple field geometries. To calculate the electrostatic forces, we consider how the energy of the system changes for a small virtual change in geometry. This method, referred to as the *principle of virtual work,* provides an entirely satisfactory means of determining the electrostatic forces for the cases considered. Any component of a force on an object at rest has a definite meaning only when we have defined the way in which the object might move in the direction indicated, at least infinitesimally—that is, if we can consider the force in question to perform at least an infinitesimal amount of work. The object may of course be in equilibrium under a balance of electrical and mechanical forces; by *motion* we mean motion against the mechanical restraints. If, for example, the possible motion is in the x direction, the work done by the electrical force for a displacement dx would be $F_x\, dx$. The calculation of this *virtual* work thus provides a means to calculate F_x.

When calculating electrostatic forces using the principle of virtual displacement, the conductors can be kept isolated (i.e., constant charge) or be maintained at a constant potential (i.e., by batteries) during their virtual displacement. The electrostatic force should not depend on this choice, since it could in principle be calculated from $\rho_s^2/(2\epsilon_0)$ as mentioned above. However, the energy change accompanying a displacement dx is different according to whether or not energy is available from the batteries. It is thus necessary to consider both possibilities separately. In the next two sections we illustrate the application of the principle of virtual work to determine the electrostatic force on a capacitor plate using a constant-charge constraint and on a dielectric slab with a constant-voltage constraint.

4.13.1 Electrostatic Forces on Charged Conductors

Consider a parallel-plate capacitor in free space with a total charge of $+Q$ in one plate and $-Q$ in the other, as shown in Figure 4.60a. We can visualize the upper plate as being displaced upward by an amount Δy. As a consequence, work must be performed, by an amount $-F_y\, \Delta y$, where F_y is the electrostatic force that must be overcome by the external force. From conservation of energy, this mechanical work must reappear as energy elsewhere, and in this case the stored energy of the field is the only other energy term that could possibly be involved. It must therefore have increased, and the amount of increase can be determined from

[118]Faraday was the first to speak of lines of force as elastic bands that transmit tension and compression. Maxwell also spent considerable time on this concept and placed Faraday's notions into clear mathematical focus. For brief discussions of the concept of electromagnetic stress see C. C. Johnson, *Field and Wave Electrodynamics,* Section 1.19, McGraw-Hill, New York, 1965; W. B. Cheston, *Elementary Theory of Electric and Magnetic Fields,* Sections 10.7–10.8, Wiley, New York, 1964. For more complete coverage, see L. M. Landau and E. M. Lifshitz, *Electrodynamics of Continuous Media,* Addison Wesley, Reading, Massachusetts, 1960; J. A. Stratton, *Electromagnetic Theory,* McGraw-Hill, New York, 1941.

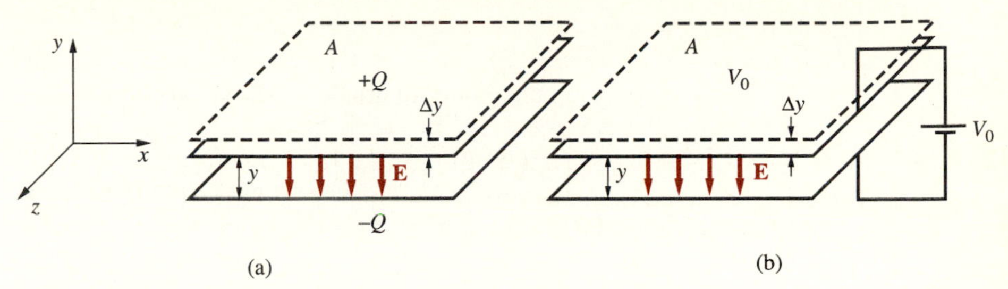

FIGURE 4.60. Virtual displacement of one plate in a parallel-plate capacitor. (a) Charge kept constant. (b) Potential difference between plates kept constant.

$$\Delta W_{\mathrm{e}} = W_{\mathrm{e}}\big|_{\mathrm{after}} - W_{\mathrm{e}}\big|_{\mathrm{before}} = \frac{1}{2}\frac{Q^2}{C_{\mathrm{after}}} - \frac{1}{2}\frac{Q^2}{C_{\mathrm{before}}}$$

$$= \frac{1}{2}\frac{Q^2(y+\Delta y)}{\epsilon_0 A} - \frac{1}{2}\frac{Q^2 y}{\epsilon_0 A} = \frac{Q^2\,\Delta y}{2\epsilon_0 A} = \frac{\epsilon_0}{2}E_y^2 A\,\Delta y$$

where we use the fact that $\epsilon_0 EA = Q$, as required by Gauss's law, and $C = \epsilon_0 A/y$, for the capacitance of the parallel plates separated by an amount y and of area A (Example 4-26), and also neglected fringing fields near the edge of the plates. Consequently, we have

$$-F_y\,\Delta y = \frac{\epsilon_0}{2}E^2 A\,\Delta y \quad\rightarrow\quad \boxed{F_y = -\frac{\epsilon_0 E^2 A}{2}} \qquad [4.77]$$

Note that the magnitude of the force per unit area, as calculated using the principle of virtual work, is the same as the magnitude that was deduced in Section 4.7.5 from field considerations, namely $\epsilon_0 E_n^2/2 = \rho_s^2/(2\epsilon_0)$. For the case of this simple geometry, we could have calculated the total force F_y on the capacitor plate by simply integrating $\epsilon_0 E_n^2/2$ over the area of the plates.

4.13.2 Electrostatic Forces in the Presence of Dielectrics

If metallic conductors are in the presence of a dielectric of finite extent (i.e., instead of being immersed in a dielectric permeating all space), then any virtual displacement of the dielectric may be expected to change the stored energy, depending on the geometries involved. Since change in stored energy under virtual displacement indicates the presence of mechanical force, it follows that dielectric bodies in electrostatic fields will experience a net force. The magnitude of this force can be calculated using the principle of virtual work, in a manner similar to the case of conducting bodies. We illustrate this technique in the following example.

Example 4-35: Force on a dielectric slab. Consider a uniform dielectric slab of permittivity ϵ partially inserted between the plates of a parallel-plate capac-

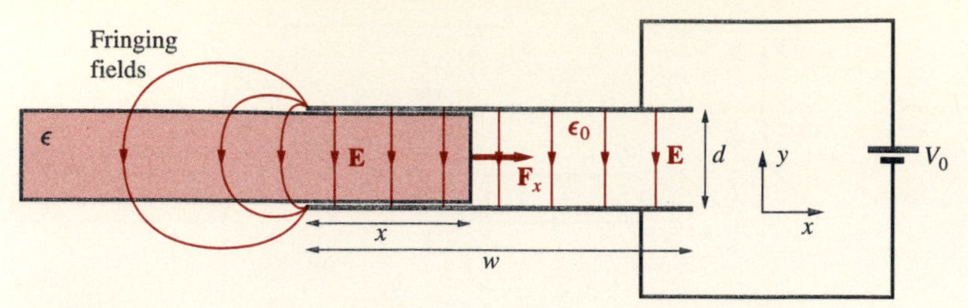

FIGURE 4.61. **Force on a dielectric slab between capacitor plates.**

itor of width w, and separation d, as shown in Figure 4.61 and having depth L (not shown). Determine the electrostatic force acting on the dielectric slab.

Solution: Suppose that the portion of the dielectric between the plates is of length x. Using [4.74], stored energy within the system is

$$W_e = \frac{1}{2}\epsilon_0\left(\frac{V_0}{d}\right)^2(w-x)Ld + \frac{1}{2}\epsilon\left(\frac{V_0}{d}\right)^2 xLd + W_{\text{fringing}}$$

where the $\mathbf{E}$ field in the region between the plates is assumed to be constant ($|\mathbf{E}| = V_0/d$) and the W_{fringing} term takes account of the fringing fields at both sides, which also store some of the energy. We now consider a virtual displacement Δx of the dielectric such that more of it is now between the plates. From the above expression for W_e, and noting that $\epsilon > \epsilon_0$, this displacement would increase the stored energy. We consider (arbitrarily[119]) the virtual displacement to occur under conditions of constant potential, so that the work done by the field (or the energy provided by the battery, which keeps V_0 constant) is given by $F_x\,\Delta x$ and equals the increase in the stored energy. The force on the dielectric can be found from

$$F_x\,\Delta x = \Delta W_e = -\frac{1}{2}\epsilon_0\left(\frac{V_0}{d}\right)^2\Delta x Ld + \frac{1}{2}\epsilon\left(\frac{V_0}{d}\right)^2\Delta x Ld + \frac{\partial W_{\text{fringing}}}{\partial x}\Delta x$$

$$= \frac{1}{2}E_y^2(\epsilon-\epsilon_0)Ld\,\Delta x \qquad \rightarrow \qquad F_x = \frac{1}{2}E_y^2(\epsilon-\epsilon_0)Ld$$

where we note that the fringing term does not change with the virtual displacement, as long as the ends of the slab are not too close to the plate edges. The direction of the force is such as to draw the dielectric slab further into the capacitor.

[119]If the charge on the capacitor plates were kept constant, the capacitance of the configuration and thus the potential difference Φ between the plates (and the electric field $E = \Phi/d$) changes with position x of the slab. For any given potential Φ (and thus E), the force on the dielectric is still given by $F_x = \frac{1}{2}E^2(\epsilon-\epsilon_0)Ld$. Since $Q = C\Phi$, the constant-charge case can be obtained by simply replacing Φ with Q/C.

It is interesting to calculate the work W required (i.e., the energy that needs to be supplied by the batteries) to place the dielectric slab completely inside the capacitor. We have

$$W = \int_0^w F_x\,dx = \tfrac{1}{2}E_y^2(\epsilon - \epsilon_0)Lwd = \tfrac{1}{2}\int_V \epsilon\mathbf{E}\cdot\mathbf{E}\,dv - \tfrac{1}{2}\int_V \epsilon_0\mathbf{E}\cdot\mathbf{E}\,dv$$

$$= \tfrac{1}{2}\int_V (\epsilon_0\mathbf{E}\cdot\mathbf{E} + \mathbf{P}\cdot\mathbf{E})\,dv - \tfrac{1}{2}\int_V \epsilon_0\mathbf{E}\cdot\mathbf{E}\,dv = \tfrac{1}{2}\int_V \mathbf{P}\cdot\mathbf{E}\,dv$$

which is the additional energy needed to polarize the inserted dielectric, as was discussed in Section 4.12.

The Physical Origin of Electrostatic Forces on Dielectrics Although we were able to calculate the force on the dielectric slab using the principle of virtual work, the physical origin of the force on a dielectric is interesting in its own right. This question relates to our most basic everyday experiences with electrostatics, which were discussed at the start of this chapter. Why does a charged (e.g., by rubbing) object pick up small dielectric objects? Why did Thales of Miletus observe that pieces of amber rubbed in silk attracted small pieces of straw? At first thought, one might conclude that the amber had one kind of charge and that the straws had the opposite charge. But the straws were not rubbed onto anything and thus were electrically neutral. Although they do not have any net charge, they were nevertheless attracted to the charged amber. What was the physical reason for this attraction?

The answer lies in the fact that the electric field produced by the charged amber (or any charged object of finite size) is nonuniform. As shown in Figure 4.62, a dielectric placed in an electric field is polarized. There are polarization charges of both signs, which experience attraction and repulsion forces due to the electric field. However, there is net attraction, because the electric field nearer to the source (at the top in Figure 4.62) is stronger than that farther away. If the electric field were uniform (constant intensity everywhere), there would be no net attraction. Indeed, if pieces of straw are placed between the plates of a large parallel-plate capacitor, they are not attracted to either of the plates, regardless of the intensity of the electric field. The spatial variation (or the nonuniformity) of the electric field is the fundamental reason why dielectrics experience a force in the presence of electric fields. A dielectric always tends to move towards regions of stronger fields, as illustrated in Figure 4.62.

As we have seen above, the force experienced by the dielectric depends on the square of the field. In fact, it can be shown that for small objects the force on a dielectric is proportional to the gradient of the square of the electric field, $-\nabla E^2$. This is because the induced polarization charges are proportional to the electric fields, and for given distributions of charges the electrostatic force is proportional to the electric field. As a result, the object experiences a net force only if the square of the field varies in space. The constant of proportionality that determines the magnitude of the force depends on the dielectric constant of the object as well as its size and shape.

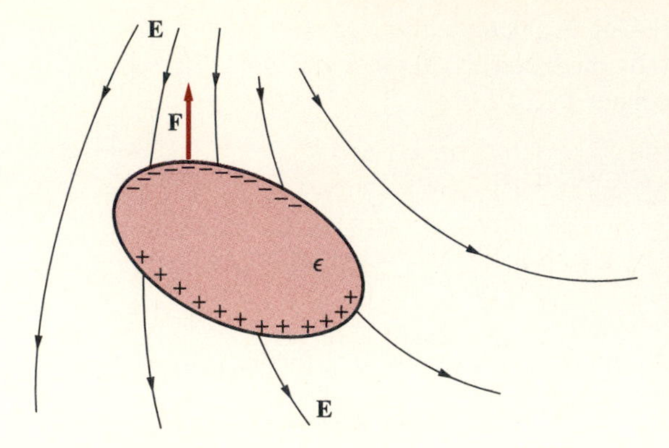

FIGURE 4.62. A dielectric object in an electric field. The force experienced by the object is proportional to the gradient of the square of the magnitude of the electric field. The dielectric tends to move toward regions of higher electric field.

The calculation of the electrostatic force on a dielectric object can in general be quite a difficult problem to solve.

The case of the dielectric slab between parallel plates considered in Example 4-35 is thus a very special case, where the calculation of the force is facilitated by the principle of virtual work. In actual fact, the force on the dielectric slab is produced by the nonuniform fringing fields (see Figure 4.61), whose intensity decreases with distance away from the plates. The direct calculation of the force would require the accurate determination of these fringing fields, evaluation of the gradient of the square of magnitude of the electric field, and its integration over the entire body of the dielectric. It is clear that the principle of virtual work comes in rather handy in this case.

4.14 SUMMARY

This chapter discussed the following topics:

- **Coulomb's law.** The electric force between two point charges Q_1 and Q_2 is given by

$$\mathbf{F} = \hat{\mathbf{R}}\frac{Q_1 Q_2}{4\pi\epsilon_0 R^2}$$

where $\hat{\mathbf{R}}$ is the unit vector pointing from one charge to the other and R is the distance between them. The force is repulsive between like charges and attractive between charges of opposite polarity. Coulomb's law is experimentally verified to a high degree of accuracy and is known to be valid over distance scales ranging from $\sim 10^{-11}$ to $\sim 10^{9}$ cm.

- **Electric field.** Any physical region of space in the vicinity of point charges Q_k and/or continuous distributions of charges $\rho(\mathbf{r})$ is said to be permeated by an electric field. The electric field is defined as the force per unit positive charge and is given by

$$\mathbf{E} = \frac{1}{4\pi\epsilon_0}\left[\sum_{k=1}^{n}\hat{\mathbf{R}}_k\frac{Q_k}{|\mathbf{r}-\mathbf{r}'_k|^2} + \int_{V'}\hat{\mathbf{R}}\frac{\rho(\mathbf{r}')dv'}{|\mathbf{r}-\mathbf{r}'|^2}\right]$$

where Q_k are discrete point charges and V' is the volume over which the continuous charge $\rho(\mathbf{r}')$ is distributed. Depending on the geometry of the continuous source charge distribution, it may be useful to describe the continuous distribution of charge in terms of surface charge density ρ_s or line charge density ρ_l.

- **Electrostatic potential.** Electrostatic potential at any point P is defined as the work required to move a unit positive test charge from infinity to the point P in the presence of an electrostatic field. The electrostatic potential at any point P($\mathbf{r}$) is related to the electric field as

$$\Phi(\mathbf{r}) = -\int_{\infty}^{r}\mathbf{E}\cdot d\mathbf{l}$$

so that the electric field is the negative gradient of potential:

$$\mathbf{E} = -\nabla\Phi$$

Electrostatic potential can also be directly evaluated from the known distributions of charge:

$$\Phi = \frac{1}{4\pi\epsilon_0}\left[\sum_{k=1}^{n}\frac{Q_k}{|\mathbf{r}-\mathbf{r}'_k|} + \int_{V'}\frac{\rho(\mathbf{r}')dv'}{|\mathbf{r}-\mathbf{r}'|}\right]$$

where Q_k are discrete point charges and V' is the volume over which the continuous charge $\rho(\mathbf{r}')$ is distributed.

- **Electric flux and Gauss's law.** The electric flux density vector $\mathbf{D}$ is a measure of electric displacement and is a function only of the electric charge producing it, independent of the type of medium surrounding the charge. In free space, $\mathbf{D} = \epsilon_0\mathbf{E}$, so for a point charge located at the origin we have

$$\mathbf{D} = \hat{\mathbf{r}}\frac{Q}{4\pi r^2}$$

Gauss's law states that the total electric flux out of any closed surface S is a constant and is equal to the total charge in the volume V enclosed by S:

$$\oint_S \mathbf{D}\cdot d\mathbf{s} = \int_V \rho\, dv = Q_{\text{enc}}$$

where ρ is the volume charge density in volume V, and Q_{enc} is the total charge in volume V. Gauss's law is a direct consequence of Coulomb's law, namely that the electrostatic force varies with distance as r^{-2}. It is particularly useful for the analysis of symmetric charge distributions.

- **Divergence.** Divergence is a measure of the source density per unit volume of a vector field. Qualitatively, divergence of a vector field is a scalar quantity and is nonzero only at those points where new flux lines emerge or terminate. For the electrostatic field, we have

$$\nabla\cdot\mathbf{D} = \rho$$

which is the differential version of Gauss's law. Thus, the electrostatic field has nonzero divergence only at points where charges exist. The divergence theorem is valid in general for any vector field **G**:

$$\oint_S \mathbf{G} \cdot d\mathbf{s} = \int_V \nabla \cdot \mathbf{G}\, dv$$

where V is the volume enclosed by the surface S.

- **Metallic conductors.** Metallic conductors have ample free electrons, which can move rapidly over macroscopic distances. When a conductor is charged or placed in an external electric field, electrons rapidly rearrange so that the net electric field inside the conductor is identically zero. Thus, the entire body of a charged metallic conductor is an equipotential volume, to which electric field lines are always orthogonal. The tangential electric field at the surface of a metallic conductor is identically zero, while the normal component of the electric flux density is equal to the surface charge density, namely,

$$\hat{\mathbf{n}} \cdot \mathbf{D} = \epsilon_0 \hat{\mathbf{n}} \cdot \mathbf{E} = \rho_s$$

In general, slight rearrangement of free charge in metallic conductors can create very large electric fields.

- **Laplace's equation.** When electrostatic fields are set up by applying potential differences between metallic conductors, the electric potential and field distributions at points free of charges can be found uniquely by solving Laplace's equation, namely,

$$\nabla^2 \Phi = 0$$

to find Φ, and $\mathbf{E} = -\nabla\Phi$ to find **E**. Laplace's equation can be analytically solved only for simple geometries; however, a host of numerical methods are available for its solution in more general cases. The more general form of Laplace's equation, which applies in the presence of charges, is Poisson's equation, namely $\nabla^2\Phi = -\rho/\epsilon_0$.

- **Capacitance.** The capacitance of a configuration of two conductors is a measure of its ability to hold charge per unit applied voltage between them. The capacitance is defined as

$$C \equiv \frac{Q}{\Phi_{12}} = \frac{\oint_S \mathbf{D} \cdot d\mathbf{s}}{-\int_L \mathbf{E} \cdot d\mathbf{l}}$$

where S is any surface enclosing the positively charged conductor, L is any path from the negative to the positive conductor, and Φ_{12} is the potential difference between them. To find the capacitance of a configuration, we typically assume charges $+Q$ and $-Q$ on the two conductors, find the electric field **E** from Gauss's law or Coulomb's law and integrate it along a path L to determine Φ_{12}.

- **Dielectric materials.** When a dielectric material is placed in an external electric field **E**, a dipole moment distribution **P** is created, which is called polarization and is given by $\mathbf{P} = \epsilon_0\chi_e\mathbf{E}$, where χ_e is the electric susceptibility, dependent on the particular microscopic atomic, molecular, and orientational properties of the material. The net effect of this induced dipole moment distribution within the dielectric is that the total electric flux density is now given by $\mathbf{D} = \epsilon_0\mathbf{E}+\epsilon_0\chi_e\mathbf{E}$, a fact that is typically accounted for by assigning to each material an electric permittivity $\epsilon = \epsilon_0(1 + \chi_e)$ such that $\mathbf{D} = \epsilon\mathbf{E}$.

- **Electrostatic boundary conditions.** Experimentally established laws of electrostatics dictate that the normal component of electric flux density is continuous across the interface between two materials, except when there is free surface charge present, which typically occurs at the surface of metallic conductors. The tangential component of the electric field is always continuous across any interface. In summary, we have

$$D_{1n} - D_{2n} = \rho_s \qquad \text{and} \qquad E_{1t} = E_{2t}$$

- **Electrostatic energy.** Any configuration of charges stores electrostatic energy of an amount equal to the work that was required to bring the charges together. The total energy stored in a distribution of charge is given by

$$W_e = \tfrac{1}{2}\int_V \rho\Phi\, dv$$

where V is the volume over which the charge is distributed. Alternatively, the electrostatic energy can be viewed as residing in the fields and can be found from

$$W_e = \tfrac{1}{2}\int_V \epsilon E^2\, dv$$

where V is the entire volume over which the electric field is nonzero. The volume energy density of the electrostatic field is $w_e = \frac{1}{2}\epsilon E^2$.

- **Electrostatic forces.** Both conducting and dielectric materials experience forces when in the presence of electric fields. A relatively easy means of calculating the electrostatic force on such objects is to use the principle of virtual work. The physical origin of electrostatic forces on metallic conductors is the attraction or repulsion between induced charges, whereas dielectrics experience a force only when placed in a nonuniform field.

4.15 PROBLEMS

4-1. Two point charges. Two identical point charges 1 m apart from each other in free space are experiencing a repulsion force of 1 N each. What is the magnitude of each charge?

4-2. Two point charges. Two small identical spheres have charges of +20 nC and −5 nC, respectively. (a) What is the force between them if they are apart by 10 cm? (b) The two spheres are brought into contact and then separated again by 10 cm. What is the force between them now?

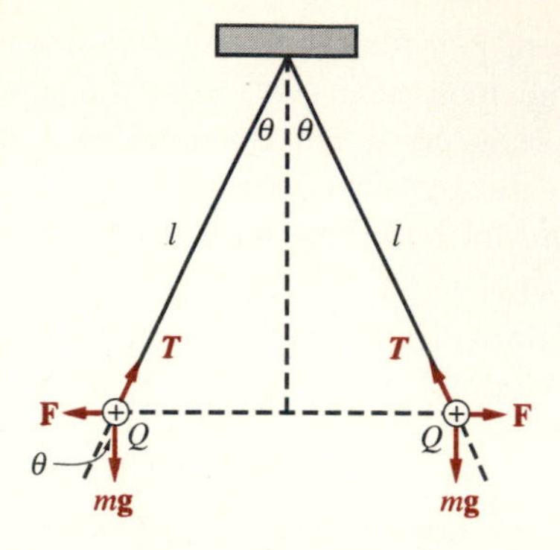

FIGURE 4.63 **Two suspended charges.** Problem 4-3.

4-3. Two suspended charges. Two small, identical, electrically charged conducting spheres of mass 2.5 g and charge +150 nC each are suspended by weightless strings of length 12 cm each, as shown in Figure 4.63. Calculate the deflection angle θ.

4-4. Zero force. (a) Three point charges of $Q_1 = +40$ nC, $Q_2 = -20$ nC, and $Q_3 = +10$ nC are all situated on the x axis in such a way that the net force on charge Q_2 due to Q_1 and Q_3 is equal to zero. If Q_1 and Q_2 are located at points $(-2,0,0)$ and $(0,0,0)$ respectively, what is the location of charge Q_3? (b) Q_3 is moved to a different position on the x axis such that the force on itself due to Q_1 and Q_2 is equal to zero. What is the new position of Q_3?

4-5. Three charges. Three identical charges of charge Q are located at the vertices of an equilateral triangle of side length a. Determine the force on one of the charges due to the other two.

4-6. Three point charges. Three point charges of values +150 nC, +100 nC, and +200 nC are located at points (0,0,0), (1,0,0), and (0,1,0), respectively. (a) Find the force on each charge due to the other charges. Which charge experiences the largest force? (b) Repeat part (a) for −100 nC as the second charge.

4-7. Two point charges. Two point charges of $+Q$ and $-Q$ are located at (0,0,0) and at $(0.4a,0)$ respectively. Find the electric field at $P_1(0,0,3a)$ and at $P_2(0,4a,3a)$. Sketch the orientations of the fields.

4-8. Two point charges. Two point charges, $+Q$ and $-3Q$, are located at points (0,2,0) and $(0,-1,0)$, respectively. (a) Find the electric field at the origin. (b) Find the coordinates of the point(s) where $\mathbf{E} = 0$. (c) Repeat parts (a) and (b) for a second charge of $+3Q$.

4-9. Zero field from three charges. Two point charges of $+Q$ each are located at (10, 0) cm, and $(-10,0)$ cm, while a third one of charge $-2Q$ is at $(0,-10)$ cm. Find the coordinates of the point where the electric field is zero.

4-10. Three point charges. Two point charges of +10 nC each are located at points (0.46,0,0) and $(-0.46,0,0)$, respectively. (a) Where should a third point charge of +15 nC be placed such that $\mathbf{E} = 0$ at point (0,1,0)? (b) Repeat part (a) for a third charge of −15 nC. (c) With the −15 nC charge located as you determined in part (b), is there another point at which the electric field $\mathbf{E} = 0$? If so, specify this point.

4-11. Four point charges on a square. Four point charges of +50 nC each are located at the corners of a square of side length 10 cm located on the xy plane and centered at the origin. (a) Find and sketch the electric potential on the z axis. (b) Find and sketch the electric field on the z axis.

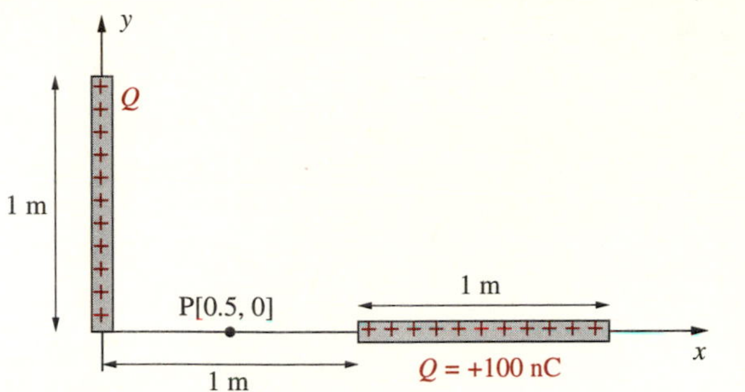

FIGURE 4.64 Two straight-line charges. Problem 4-13.

4-12. Six point charges on a hexagon. Six identical point charges of +25 nC each are situated in space at the corners of a regular hexagon whose sides are each of length 6 cm. (a) Find the electric potential at the center of the hexagon. (b) Determine the energy required to move a point charge of −25 nC from infinity to the center of the hexagon.

4-13. Two straight-line charges. Consider two uniformly charged wires, each of length 1 m and a total charge +100 nC, with their ends separated by 1 m, as shown in Figure 4.64. (a) Find the electric potential Φ at the point P, midway between the two wires. (b) Find the electric field $\mathbf{E}$ at point P.

4-14. Seven point charges on a cube. Seven identical point charges of +10 nC each occupy seven of the eight corners of a cube 3 cm on each side. Find the electric potential at the unoccupied corner.

4-15. Two line charges. A uniform line charge of $\rho_{l1} = -4\pi \times 8.85$ pC-m^{-1} is located between the points (−5,0) and (−2,0) m and another such line of positive charge (i.e., $\rho_{l2} = 4\pi \times 8.85$ pC-m^{-1}) between (5,0) and (2,0) m, as shown in Figure 4.65. (a) Find the electric potential Φ at (1,0) m. (b) Find the electric field $\mathbf{E}$ at the same point. At which points, if any, is the electric field $\mathbf{E}$ zero? At which points, if any, is the electric potential Φ zero?

4-16. Circular ring of charge. A total charge of Q_1 is distributed uniformly along a half-circular ring as shown in Figure 4.66. Two point charges, each of magnitude Q_2, are situated as shown. The surrounding medium is free space. (a) Find Q_2 in terms of Q_1 so that the potential Φ at the center of the ring is zero. (b) Find Q_2 in terms of Q_1 so that the electric field $\mathbf{E}$ at the center of the ring is zero.

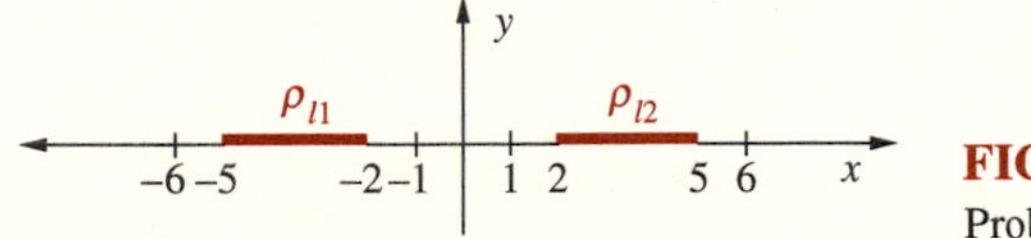

FIGURE 4.65 Two line charges. Problem 4-15.

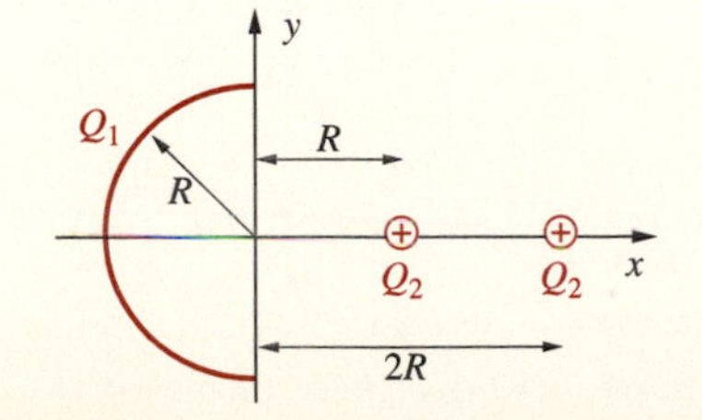

FIGURE 4.66 Circular ring of charge. Problem 4-16.

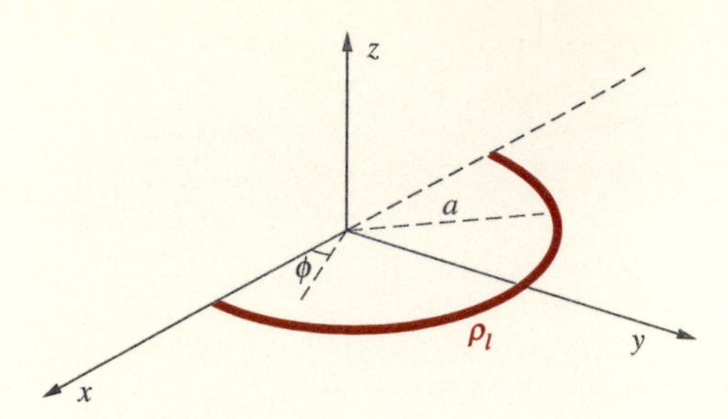

FIGURE 4.67 **Semicircular line charge.** Problem 4-17.

4-17. Semicircular line charge. A thin line charge of density ρ_l is in the form of a semicircle of radius a lying on the xy plane with its center located at the origin, as shown in Figure 4.67. Find the electric field at the origin for the cases in which (a) the line charge density $\rho_l = \rho_0$ is a constant, and (b) the line charge density varies along the semicircular ring as $\rho_l = \rho_0 \sin\phi$.

4-18. Charge on a hemisphere. The curved surface of a hemisphere of radius a centered at the origin carries a total charge of Q uniformly distributed over its curved surface, as shown in Figure 4.68. (a) Find the electric potential on the z axis. (b) Find the electric field on the z axis. (c) Repeat parts (a) and (b) if the charge Q is uniformly distributed throughout the volume of the hemisphere.

4-19. Sheet of charge with hole. An infinite sheet of uniform charge density ρ_s is situated coincident with the xy plane at $z = 0$. The sheet has a hole of radius a centered at the origin. Find (a) the electric potential Φ and (b) the electric field $\mathbf{E}$ at points along the z axis.

4-20. Spherical charge distribution. A charge density of

$$\rho(r) = Ke^{-br}$$

where K and b are constants, exists in a spherical region of space defined by $0 < r < a$. (a) Find the total charge in the spherical region. (b) Find the electric field at all points in space. (c) Find the electric potential at all points in space. (d) Show that the potential found in part (c) satisfies the equation $\nabla^2\Phi = -\rho(r)/\epsilon_0$ for both $r < a$ and $r > a$.

4-21. The electron charge density in a hydrogen atom. According to quantum mechanics, the electron charge of a hydrogen atom in its ground state is distributed like a cloud surrounding its nucleus, extending in all directions with steadily decreasing density such that the total charge in this cloud is equal to q_e (i.e., the charge of an electron). This electron charge distribution is given by

$$\rho(r) = \frac{q_e}{\pi a^3} e^{-2r/a}$$

where a is the *Bohr radius,* $a \simeq 0.529 \times 10^{-10}$ m. (a) Find the electric potential and the electric field due to the electron cloud only. (b) Find the total electric potential and

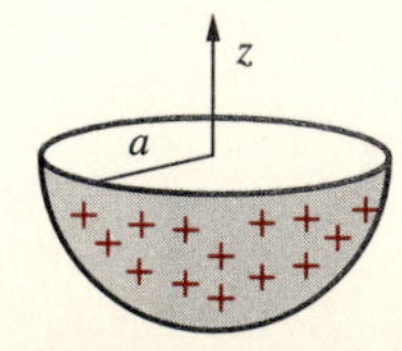

FIGURE 4.68 **Charge on a hemisphere.** Problem 4-18.

the electric field in the atom, assuming that the nucleus (proton) is localized at the origin.

4-22. Spherical shell of charge. The space between two concentric spheres of radii a and b ($a < b$) in free space is charged to a volume charge density given by

$$\rho(r) = \frac{K}{r^2}; \quad a \leq r \leq b$$

where K is a constant. (a) Find the total charge in the shell. (b) Find the electric field at all points in space. (c) Find the electric potential at all points in space. (d) What happens if $b \to a$?

4-23. Spherical charge distribution. A spherical charge distribution exists in free space in the region $0 < r < a$ given by

$$\rho(r) = \rho_0\left(1 - \frac{r^2}{a^2}\right)$$

(a) Find the total charge. (b) Determine $\mathbf{E}$ everywhere. (c) Determine Φ everywhere. (d) Sketch both $|\mathbf{E}|$ and Φ as a function of r.

4-24. Spherical charge with a cavity. A spherical region of radius b in free space is uniformly charged with a charge density of $\rho = K$, where K is a constant. The sphere contains an uncharged spherical cavity of radius a. The centers of the two spheres are separated by a distance d such that $d + a < b$. Find the electric field inside the cavity.

4-25. Charge on a hollow metal sphere. A hollow metal sphere of 20 cm diameter is given a total charge of 1 μC. Find the electric field and the electric potential at the center of the sphere.

4-26. A 1-farad capacitor. To get an idea about the physical size of a 1 F capacitor, consider a parallel-plate capacitor with the two metal plates separated by 1 mm thickness of air. Calculate the area of the metal plates needed so that the capacitance is 1 F.

4-27. Gate oxide capacitance of a MOS transistor. A basic MOS transistor consists of a gate conductor and a semiconductor (which is the other conductor), separated by a gate dielectric. Consider a MOS transistor using silicon dioxide (SiO_2) ($\epsilon_r = 3.9$) as the gate oxide. The gate oxide capacitance can be approximated as a parallel-plate capacitor. The gate oxide capacitance per unit area is given by

$$C_{ox} = \frac{\epsilon_{ox}}{t_{ox}}$$

where ϵ_{ox} and t_{ox} are the permittivity and the thickness of the gate dielectric. (a) If the thickness of the SiO_2 layer is 2×10^{-6} cm, find the gate oxide capacitance per unit area. (b) If the length and the width of the gate region are $L = 5 \times 10^{-4}$ cm and $W = 2 \times 10^{-3}$ cm respectively, find the total gate capacitance.

4-28. RG 6 coaxial cable. A coaxial cable (RG 6) designed for interior use, such as connecting a TV set to a VCR, has a per-unit-length capacitance listed as 17.5 pF/ft. If the relative dielectric constant of the insulator material in the cable is $\epsilon_r \simeq 1.64$, find the ratio of the inner and outer radii of the insulator.

4-29. Radius of a high-voltage conductor sphere. Consider an isolated charged metallic conductor sphere in a dielectric medium at an electric potential of 500 kV. Calculate the minimum radius of the sphere such that dielectric breakdown will not occur if the surrounding dielectric is (a) air (E_{BR} = 3 MV-m^{-1}); (b) a gaseous dielectric such as sulfur hexafluoride (SF_6) ($\epsilon_r \simeq 1$ and E_{BR} = 7.5 MV-m^{-1}); (c) a liquid dielectric such as oil ($\epsilon_r = 2.3$ and E_{BR} = 15 MV-m^{-1}); and (d) a solid dielectric such as mica (ϵ_r = 5.4 and E_{BR} = 200 MV-m^{-1}).

4-30. Parallel-plate capacitor. A parallel-plate capacitor is constructed from two aluminum foils of 1 cm^2 area each placed on both sides of rubber ($\epsilon_r = 2.5$ and E_{BR} = 25 MV-m^{-1}) of thickness 2.5 mm. Find the voltage rating of the capacitor using a safety factor of 10.

4-31. A parallel-plate capacitor with variable ϵ_r. A parallel-plate capacitor of cross-sectional area A and thickness d is filled with a dielectric material whose relative permittivity varies linearly from $\epsilon_r = 1$ at one plate to $\epsilon_r = 10$ at the other plate. Find the capacitance. (b) Compare the result to the case of the same capacitor filled with air instead of the dielectric.

4-32. Coaxial capacitor. Consider a coaxial capacitor as shown in Figure 4.69. Given a = 5 mm, l = 3 cm, and the voltage rating of the capacitor to be 2 kV with a safety factor of 10, what is the maximum capacitance that can be designed using (a) oil (ϵ_r = 2.3 and E_{BR} = 15 MV-m^{-1}) and (b) mica (ϵ_r = 5.4 and E_{BR} = 200 MV-m^{-1}).

4-33. Coaxial capacitor with two dielectrics. A coaxial capacitor consists of two conducting coaxial surfaces of radii a and b ($a < b$). The space between is filled with two different dielectric materials with relative dielectric constants ϵ_{1r} and ϵ_{2r}, as shown in Figure 4.70. (a) Find the capacitance of this configuration. (b) Assuming that l = 5 cm, $b = 3a$ = 1.5 cm, and oil and mica are used, calculate the capacitance. (c) Redo part (b) assuming that only oil is used throughout.

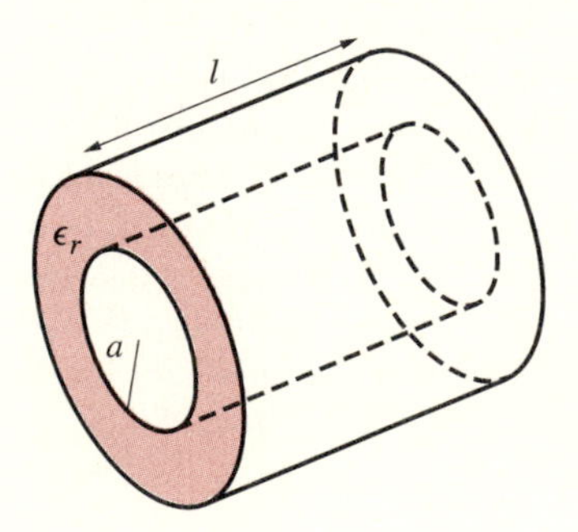

FIGURE 4.69 Coaxial capacitor. Problem 4-32.

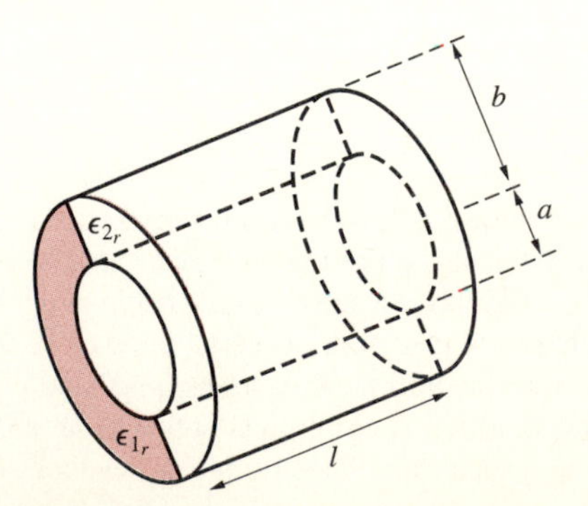

FIGURE 4.70 Coaxial capacitor with two dielectrics. Problem 4-33.

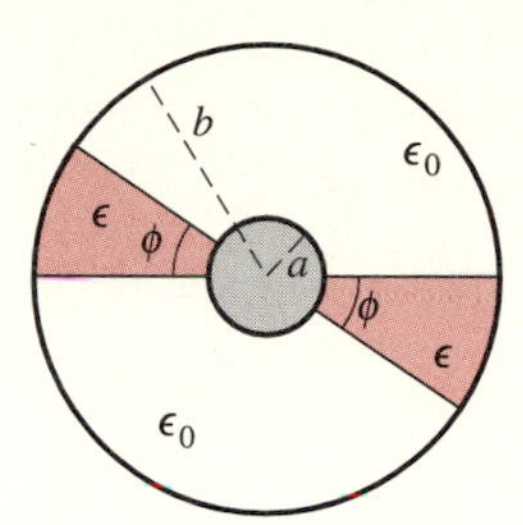

FIGURE 4.71 **Coaxial capacitor with spacers.** Problem 4-34.

4-34. Capacitor with spacers. The cross-sectional view of an air-filled coaxial capacitor with spacers made out of material with permittivity ϵ is shown in Figure 4.71. (a) Find the capacitance of this coaxial line in terms of ϵ, a, b, and ϕ. (b) If the spacers are to be made out of mica ($\epsilon = 6\epsilon_0$), determine the angle ϕ such that only 10% of the total energy stored by the capacitor is stored in the spacers. (c) Consider the capacitor without the spacers (i.e., $\phi = 0$). For a given potential difference V_0 between the inner and outer conductors and for a given fixed value of b, determine the inner radius a for which the largest value of the electric field is a minimum.

4-35. Earth capacitor. Consider the earth to be a large conducting sphere. (a) Find its capacitance (the earth's radius is $\sim 6.371 \times 10^6$ m). (b) Find the total charge and energy stored on the earth (take the electric field on the surface of the earth to be 100 V-m^{-1}). (c) Find the maximum charge and energy that can be stored on the earth.

4-36. A coaxial capacitor with variable permittivity. Consider a coaxial capacitor with two concentric metal cylinders of radii a and b ($a < b$), filled with a dielectric material whose permittivity ϵ varies linearly from ϵ_a at $r = a$ to ϵ_b at $r = b$. (a) Find the capacitance per unit length. (b) Find the numerical value of the capacitance if the radii are 2 mm and 6 mm and the relative dielectric constant varies from 2.25 to 8.5, respectively.

4-37. Planar charge. A surface charge distribution $\rho_s(x, z)$ exists on the xz plane, with no charge anywhere else (i.e., $\rho = 0$ for $|y| > 0$). Which of the following potential functions are valid solutions for the electrostatic potential in the half-space $y > 0$, and what is the corresponding charge distribution $\rho_s(x, z)$ on the xz plane?

$$\Phi_1 = e^{-y} \cosh x$$

$$\Phi_2 = e^{-y} \cos x$$

$$\Phi_3 = e^{-\sqrt{2}y} \cos x \sin x$$

$$\Phi_4 = \sin x \sin y \sin z$$

4-38. Parallel power lines. An isolated pair of parallel power lines a distance of d_1 apart have a potential difference of V_{AB} and are located a distance h above a pair of telephone wires, as shown in Figure 4.72. The parameter values are $d_1 = 1$ m, $a = 2$ cm, $V_{\text{AB}} = 440$ V, $h = 60$ cm, and $d_2 = 15$ cm. (a) Find the direction and magnitude of the electric field at points 1 and 2. Take the midpoint between the power lines as the origin of your coordinate system. (b) Determine the potential difference Φ_{12} between points 1 and 2.

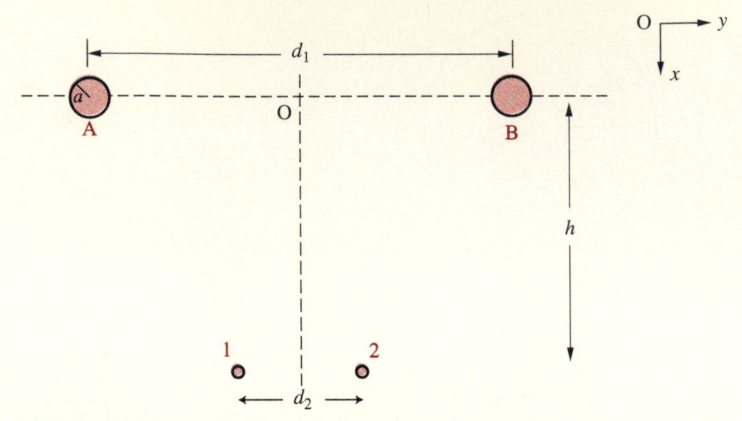

FIGURE 4.72 **Parallel power lines above telephone lines.** Problem 4-38.

4-39. Field under high-voltage line. Many 60 Hz high-voltage transmission lines operate at an rms alternating voltage of 765 kV. (a) What is the peak electric field at ground level under such a line if the wire is 12 m above the ground? (b) What is the peak potential difference between the head and feet of a 6-ft tall person? (c) Is the field sufficient to ignite a standard (110 V) fluorescent lamp of 2 ft length?

4-40. High-voltage dc transmission line. A high-voltage dc transmission line consists of a single 25-mm-diameter cylindrical conductor supported at an average height of 15 m above the ground. What is the electric field **E** at the ground level (a) directly under the conductor, (b) at a ground distance of 25 m, and (c) at a ground distance 50 m perpendicular to the line? Assume the line to be operating at 500 kV, and the ground to be flat and perfectly conducting.

4-41. Thundercloud fields. A typical thundercloud can be modeled as a capacitor with horizontal plates with 10 km^2 area separated by a vertical distance of 5 km. Just before a large lightning discharge, the upper plate may have a total positive charge of up to 300 C, with the lower plate having an equal amount of negative charge. (a) Find the electrostatic energy stored in the cloud just before a discharge. (b) What is the potential difference between the top and bottom plates? (c) What is the average electric field within the cloud? How does this value compare to the dielectric breakdown field of dry air (3 MV-m^{-1})?

4-42. Two conducting spheres. Consider a pair of small conducting spheres with radii a, b, small compared to the separation distance d between their centers (i.e., $a, b \ll d$). (a) Determine the electrostatic energy stored by this configuration, assuming that the spheres with radii a and b carry charges of Q and $-Q$, respectively. Your answer should depend on d. State all assumptions. (b) Repeat part (a) assuming that the spheres with radii a and b carry charges of $+Q$ and $+2Q$, respectively.

CHAPTER 5

Steady Electric Currents

The special properties of metallic conductors were noted in Chapter 4, namely that such materials hold a large number of free electronic charges that are readily able to move over macroscopic distances under the influence of applied electric fields. Although metals are particularly good conductors, many other materials also contain free charges that can move over macroscopic distances and are considered to be conductors. In this chapter, we introduce a measure of the ability of materials to conduct electricity, namely *conductivity,* and study the conditions that lead to steady flow of electric charge in conducting materials. Note that the steady flow of electric charge we discuss in this chapter is quite different from the transient redistribution of charge that occurs when a conductor is brought into a region of electric field or when charge is placed on a conductor. For steady currents to exist, we need not only a steady electric field to impart a velocity to the free charges, but also a mechanism that prevents the charge from piling up, which would tend to reduce the field. The subject of steady electric currents links field theory to several important circuit concepts, namely Ohm's law, Kirchhoff's voltage and current laws, and Joule's law. In addition, using the concept of steady current flow and the principle of conservation of charge, we can derive the fundamental equation of continuity, which relates current and charge density.

Our discussion of steady electric currents brings us closer to understanding the underlying physical bases of the transmission line behavior discussed in Chapters 2 and 3. The performance of transmission lines with losses (Section 3.8) is largely determined by the line parameters R and G, representing, respectively, resistive losses in the conductors and leakage losses in the dielectric media between them. In this chapter, we define the concept of resistance and discuss how the resistance of different configurations can be evaluated using fundamental laws of electrostatics.

We start our coverage of steady electric currents by defining current density and briefly discussing the microscopic view of conduction in Section 5.1. We then introduce Ohm's law and the concept of resistance in Section 5.2, followed by a discussion of electromotive force and Kirchhoff's voltage law in Section 5.3. Conservation of charge, the continuity equation, and Kirchhoff's current law are covered in Section 5.4, while redistribution of free charge in a conductor is discussed in Section 5.5. The boundary conditions for steady current flow are described in Section 5.6, followed by a discussion of the duality between current density and electric flux density in Section 5.7. We complete the chapter with a discussion of Joule's law in Section 5.8.

5.1 CURRENT DENSITY AND THE MICROSCOPIC VIEW OF CONDUCTION

We describe the flow of electric charge by introducing the concepts of current and current density. *Current* is a flow of charge, measured by the rate at which charge passes through any specified surface area, for example, the cross section of a wire. Current is a scalar quantity, and its unit, coulombs per second, is called the *ampere.* A current I is said to flow when the charge that passes through the surface area in time Δt is given by $\Delta Q = I\Delta t$. Since different amounts of charge may flow through different parts of a given cross-sectional area, it is often more appropriate to describe current in terms of *current density,* denoted as $\mathbf{J}$ and having units of amperes per square meter, or coulombs-s^{-1}-m^{-2}. The current density $\mathbf{J}(x, y, z)$ at any given point (x, y, z) in a conductor is a vector quantity having the direction of the flow of positive charge (which is opposite to the direction of electron flow) and a magnitude equal to the current per unit area that is normal to the direction of flow. The current through any small area Δs represented by a vector $\Delta\mathbf{s} = \hat{\mathbf{n}}\Delta s$ can be calculated as $\Delta I = \mathbf{J} \cdot \Delta\mathbf{s}$. For example, the current through the area element Δs in Figure 5.1 is the same as that through the area $\Delta s \cos\theta$ perpendicular to $\mathbf{J}$, or $J\Delta s \cos\theta$. To determine the current through a larger area S, we need to integrate $\mathbf{J}$ over S, namely

$$I = \int_S \mathbf{J} \cdot d\mathbf{s}$$

Although electric current is a macroscopic quantity, it results from the motion of many microscopic charges. At a simple level entirely sufficient for our purposes, we can think of the conducting material as composed of a lattice of fixed positive ions and an electron gas that is free to move about. The free electrons in a metallic conductor are in fact the conduction-band electrons, which are very loosely bound to their atoms and are essentially free to wander through the lattice. For example,

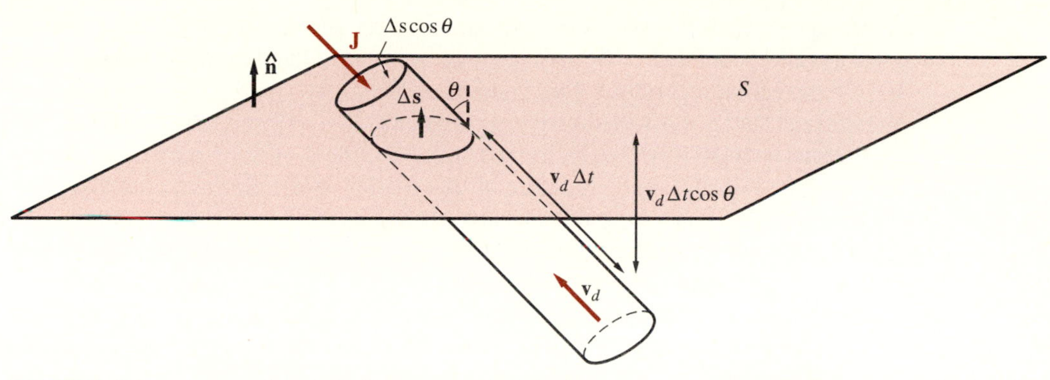

FIGURE 5.1. Current density J. Since current is defined as the rate of charge flow through a specified surface Δs, it is given by the projection of the selected surface perpendicular to the direction of the flow of conduction electrons.

atoms of copper have 29 electrons, 28 of which are in tightly bound shells. The remaining outermost electron is essentially free to move about the crystalline structure of copper. Ordinarily, these free electrons (each only loosely bound to an atom) are in a state of random motion because of their thermal energy. Without any external electric field applied, they move about randomly colliding with the atoms of the lattice and with one another, as illustrated in Figure 5.2a. Between collisions, however, the electrons in a metal at room temperature move at rather high thermal speeds of v_{th} of $\sim 10^5$ to $\sim 10^6$ m-s^{-1}. This rapid random motion is the source of the fluctuating

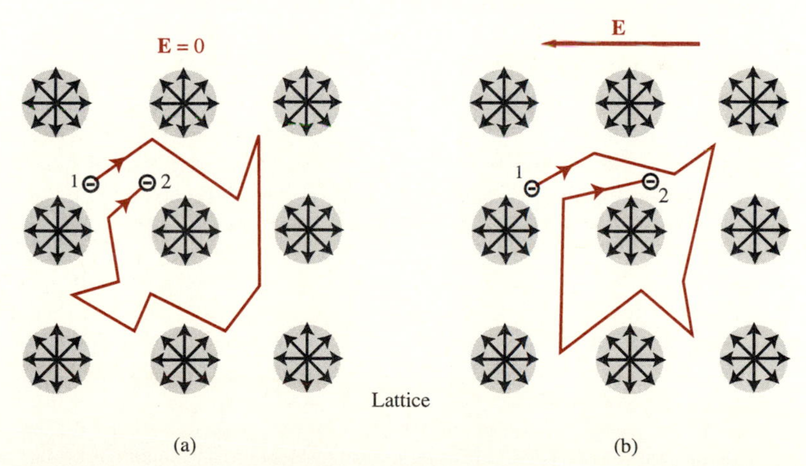

FIGURE 5.2. Representation of electron motion through the vibrating lattice structure of a solid. The shaded areas represent lattice sites, at which individual atoms or molecules are located; the atoms vibrate randomly in all directions as indicated by arrows. (a) Without an externally applied electric field, and (b) with an externally applied electric field. The field produces a small *drift* component toward the right (electrons move opposite to the field). The drift shown here is greatly exaggerated. In fact, for typical conduction current levels of a few amperes or less, the drift velocity in metals is $<10^{-4}$ m-s^{-1}, whereas the electron velocity between collisions due to thermal motion is $\sim 10^5$ m-s^{-1}.

currents known as *Johnson noise*,[1] which can be measured across any resistor. The mean free time t_c, defined as the average time between collisions, is typically of order $\sim 10^{-14}$ s, at room temperature. This value corresponds to a *mean free path* $\bar{l}$, defined as the average distance traveled between collisions, of $\sim 10^{-8}$ m. After each collision, the electrons acquire a new velocity nearly independent of their previous velocity. Because of the entirely random nature of this motion in the absence of any applied electric field, there is no net motion in any given direction, and the net charge density averaged over macroscopic spatial dimensions (i.e., $\gg 10^{-8}$ m) and time scales (i.e., $\gg 10^{-14}$ s) remains approximately zero.

Conduction, or net flow of free charge, occurs due to the relatively slow drift of electrons (compared to the rapid thermal motions of individual electrons) caused by an applied electric field. When an electric field E is established within a conductor (typically by applying a potential difference between the two ends of the conductor), the free electrons are accelerated opposite to the direction of the field. At first, one might think that the electrons would experience a steady acceleration[2] given by $(|q_e|E/m_e)$; however, before they can acquire any appreciable speed, the electrons collide with the lattice and acquire a new random velocity. Since the applied electric field essentially has to start accelerating all over again every $\sim 10^{-14}$ s, it can alter the random thermal velocities of the electrons only slightly, but in a systematic manner. This relatively slight systematic drift (shown on a highly exaggerated scale in Figure 5.2b) of the free electrons is the basis of conduction. Following a brief initial transient, the electrons acquire a steady-state average *drift velocity* $\mathbf{v}_d$, determined by the balance between the accelerating force of the applied field and the scattering effect of the collisions with the lattice. These collisions are also the mechanism by which some of the energy of the electrons, and thus of the electric field, is dissipated as heat.

If there are n_c conduction-band electrons[3] per cubic meter traveling with an average drift velocity $\mathbf{v}_d$, each travels a distance $v_d\Delta t$ in a time Δt, and all those in a cylinder of slant height $v_d\Delta t$ and base Δs (see Figure 5.1) pass through Δs in a time Δt. The volume of this cylinder is $v_d\Delta t\Delta s\cos\theta$, so that $n_c v_d\Delta t\Delta s\cos\theta$ electrons cross Δs, involving a total charge of $|q_e|n_c v_d\Delta t\Delta s\cos\theta$. Thus, the magnitude of the current density $\mathbf{J}$, equal to the total charge per unit time per unit area, is given as

$$|\mathbf{J}| = \frac{|q_e|n_c v_d\Delta t\Delta s\cos\theta}{\Delta t\Delta s\cos\theta} = |q_e|n_c v_d \quad\rightarrow\quad \mathbf{J} = n_c q_e \mathbf{v}_\mathbf{d}$$

Note that since q_e is negative, $\mathbf{J}$ is in the direction opposite to the electron drift velocity $\mathbf{v}_d$.

---

[1]The average size of random fluctuations in the voltage across a resistor depends on the resistor and the temperature. For a 400-kΩ resistor at room temperature, the amplitude of fluctuations with frequency < 20 kHz is ~ 10 μV. For more information, see A. L. King, *Thermophysics*, Freeman, San Francisco, 1962, pp. 212–214.

[2]A steady acceleration would mean that the current in the conductor would increase indefinitely with time, in contradiction with Ohm's law, which predicts a constant current for a constant applied potential (i.e., a constant electric field).

[3]For copper, the density of conduction electrons is $n_c = 8.45 \times 10^{28}$ el-m^{-3}, which is also the number of copper atoms per unit volume, since copper has one conduction-band electron per atom. Note that the number of atoms per unit volume is approximately the same for all solids.

We can crudely estimate the average drift velocity by assuming that each free electron has the same thermal speed v_{th}, the same mean free time t_c between collisions, and thus the same mean free path between collisions of $\bar{l} = v_{\text{th}} t_c$. The electric field gives each electron an acceleration of $q_e \mathbf{E}/m_e$, so that the change in the velocity of a free electron before the next collision is $\Delta v_{\text{th}} \simeq q_e \mathbf{E} t_c / m_e$. The drift velocity averaged over the entire population of electrons should be of the same order as Δv_{th}, although the actual value will depend on the lattice and band structure and other material properties, such as the energy distribution of free electrons. For metals, the average drift velocity is given as[4]

$$\mathbf{v}_d \simeq \frac{q_e \mathbf{E} t_c}{m_e} \quad \text{or} \quad \mathbf{v}_d \simeq \frac{q_e \mathbf{E} \bar{l}}{m_e v_{\text{th}}}$$

The proportionality constant between $\mathbf{v}_d$ and $\mathbf{E}$, namely $\mu_e = (|q_e| t_c / m_e)$, is referred to as the *mobility* of the conduction electrons in the particular material. In general, the drift velocity v_d is very small compared to v_{th}, since the electric field makes only a slight change in the velocity distribution that existed before the field was applied. Therefore, the average values of t_c and $\bar{l}$, which depend on the thermal velocity of the electrons, do not change appreciably due to the presence of the applied field. This means that the asymptotic value of the drift velocity is constant for a given applied electric field, and thus the current is constant. In most cases, the magnitude of drift velocity is quite small—less than $\sim 10^{-4}$ m-s^{-1} in metals (see Example 5-1)—for reasonable conduction currents.

The slowly[5] drifting electrons constitute a current of density

$$\mathbf{J} = n_c q_e \mathbf{v}_d = \frac{n_c q_e^2 \bar{l}}{m_e v_{\text{th}}} \mathbf{E} \quad \rightarrow \quad \boxed{\mathbf{J} = \sigma \mathbf{E}} \qquad [5.1]$$

where $\sigma = n_c q_e^2 \bar{l} / (m_e v_{\text{th}})$ is identified as the *conductivity* of the material. The conductivity is the macroscopic quantity that is hereafter used to account for the

---

[4]The *drift velocity* is the solution of an equation of motion with a damping term, namely

$$m_e \frac{d\mathbf{v}_d}{dt} + m_e \frac{\mathbf{v}_d}{t_c} = q_e \mathbf{E}$$

where the term t_c behaves essentially as a coefficient of friction. Note that the solution of this equation is

$$\mathbf{v}_d = \frac{q_e t_c}{m_e} \mathbf{E}[1 - e^{-(t/t_c)}]$$

so that for a time period t longer than t_c, the drift velocity asymptotically approaches $\mathbf{v}_d = (q_e t_c / m_e)\mathbf{E}$ as t increases (e.g., this asymptotic value is practically accurate when $t \geq 5t_c$). For further discussion, see R. L. Sproull, *Modern Physics: The Quantum Physics of Atoms, Solids, and Nuclei,* 3rd ed., John Wiley, New York, 1980, Sec. 8.5. For an excellent elementary review of conduction in metals, see V. F. Weisskopf, On the theory of electric resistance of metals, *Am. J. Phys.,* 11 (1), pp. 1–12, 1943.

[5]Note that the drift velocity is indeed quite small ($\sim 10^{-5}$ to $\sim 10^{-4}$ m-s^{-1}) considering that an applied signal on a transmission line propagates at the speed of light. It is clear that the propagation of a voltage/current disturbance along a wire does not depend on how much time it would take for a particular tagged electron to travel the distance through the wire. In fact, such an applied signal propagates as an electromagnetic wave outside the wire.

microscopic behavior of conductors. The unit of conductivity is Siemens-m^{-1} or S-m^{-1}, sometimes also referred to as mhos-m^{-1}. Values of σ for common materials vary over ~24 orders of magnitude, ranging from excellent conductors (e.g., 6.17×10^7 S-m^{-1} for silver) to excellent insulators (e.g., $\sim 10^{-17}$ S-m^{-1} for fused quartz). Equation [5.1] holds in cases when the electric field is constant or variable in time, as long as the field remains approximately constant for times much longer than the mean free time between collisions (i.e., $t_c \simeq 10^{-14}$ s). In some contexts, it might be more useful to work with *resistivity,* which is simply the inverse of conductivity (i.e., σ^{-1}), and has units of Ω-m.

The manner in which the conductivity depends on temperature differs drastically between metals and semiconductors and dielectrics. In the case of metals, *all* of the conduction-band electrons are already free to move at room temperature, and the bound electrons are not released from their shells even for very large temperatures or electric fields. The dominant effect of increased temperature for metals is the fact that the lattice atoms undergo larger vibrations (larger excursions around their central locations), thus increasing the probability of collisions and impeding the motion of the carriers. As a result, the conductivity σ of metallic conductors decreases with increasing temperature. Semiconductors and dielectrics, on the other hand, have relatively fewer free electrons available for conduction at room temperature.[6] As the temperature increases, the lattice vibrations become stronger, and, depending on the type of material, some bound electrons are knocked loose and become available for conduction. Although the increased lattice vibrations still tend to impede the flow of current carriers, the net dominant effect in semiconductors and dielectrics is increasing conductivity with increasing temperature, due to the larger number of conduction electrons.[7]

For pure metallic elements, the conductivity depends rather strongly on temperature, decreasing by ~0.1–0.5% per degree Celsius, depending on the type of material. Specifically, if σ_{20° is the conductivity of a metal at 20°C, its conductivity at a Celsius temperature T (in °C) is given by[8]

---

[6]For semiconductors, the conductivity is limited more by the small number of current carriers than by the size of t_c. By properly accounting for the energy distribution of conduction electrons, it can be shown that a more accurate expression for the conductivity of a semiconductor is

$$\sigma = \frac{4n_c q_e^2 \bar{l}}{3\sqrt{2\pi m_e k_B T}}$$

where k_B is Boltzmann's constant ($k_B \simeq 1.38 \times 10^{-23}$ J-K^{-1}) and T is the absolute temperature in °K. A more common expression for the conductivity of a semiconductor is

$$\sigma = |q_e|(\mu_e N_e + \mu_p N_p)$$

where N_e and N_p are respectively the densities of electrons and holes, and μ_e and μ_p are respectively the mobilities of the electrons and holes; see Problem 5-9.

[7]For further discussion, see Chapter 8 of R. L. Sproull, *Modern Physics: The Quantum Physics of Atoms, Solids, and Nuclei,* 3rd ed., John Wiley, New York, 1980.

[8]See Chapter 5 of W. T. Scott, *The Physics of Electricity and Magnetism*, 2nd ed., John Wiley, New York, 1966, and Section 8.5 of R. L. Sproull, ibid.

$$\frac{1}{\sigma_T} \simeq \frac{1}{\sigma_{20^\circ}}[1 + \alpha_\sigma(T - 20)]$$

where the coefficient α_σ is called the temperature coefficient and has the value 0.001 to 0.005 for most metals. Table 5.1 provides values of conductivity σ and temperature coefficient α_σ for selected materials. Note that the temperature coefficients for nonmetals are typically negative, representing increasing conductivity with increasing temperature. The preceding relation assumes that $(\sigma_T)^{-1}$ varies linearly with T; in fact, quadratic terms can be used for greater accuracy.

TABLE 5.1. Conductivities and temperature coefficients of selected materials

Material	Conductivity σ (S-m^{-1}) (at 20°C)	Temperature coefficient α_σ [(°C)$^{-1}$]
Aluminum	3.82×10^7	0.0039
Bismuth	8.70×10^5	0.004
Brass (66 Cu, 34 Zn)	2.56×10^7	0.002
Carbon (graphite)	7.14×10^4	−0.0005
Constantan (55 Cu, 45 Ni)	2.26×10^6	0.0002
Copper (annealed)	5.80×10^7	0.0039
Dry, sandy soil	$\sim 10^{-3}$	
Distilled water	$\sim 10^{-4}$	
Fresh water	$\sim 10^{-2}$	
Germanium (intrinsic)	~ 2.13	−0.048
Glass	$\sim 10^{-12}$	−0.07
Gold	4.10×10^7	0.0034
Iron	1.03×10^7	0.0052–0.0062
Lead	4.57×10^6	0.004
Marshy soil	$\sim 10^{-2}$	
Mercury (liquid)	1.04×10^6	0.00089
Mica	$\sim 10^{-15}$	−0.07
Nichrome (65 Ni, 12 Cr, 23 Fe)	1.00×10^6	0.00017
Nickel	1.45×10^7	0.0047
Niobium	8.06×10^6	
Platinum	9.52×10^6	0.003
Polystyrene	$\sim 10^{-16}$	
Porcelain	$\sim 10^{-14}$	
Quartz (fused)	$\sim 10^{-17}$	
Rubber (hard)	$\sim 10^{-15}$	
Seawater	~ 4	
Silicon (intrinsic)	$\sim 4.35 \times 10^{-4}$	
Silver	6.17×10^7	0.0038
Sodium	2.17×10^7	
Stainless steel	1.11×10^6	
Sulfur	$\sim 10^{-15}$	
Tin	8.77×10^6	0.0042
Titanium	2.09×10^6	
Tungsten	1.82×10^7	0.0045
Y $Ba_2Cu_3O_7$ (at $<$ 80K)	$\sim 10^{20}$	
Wood	10^{-11}–10^{-8}	
Zinc	1.67×10^7	0.0037

Superconductivity A dramatic deviation from the nearly linear dependence of resistance on temperature occurs in some materials that exhibit a phenomenon called *superconductivity,* discovered[9] by Kamerlingh Onnes in 1911. Below a critical temperature T_c, the conductivity of these materials becomes nearly infinite ($>10^{20}$ S-m^{-1} have been measured), although at slightly higher temperatures, the normal temperature-dependent resistance is observed. Most superconductors are metallic elements, compounds, or alloys and exhibit transitions into superconducting states at critical temperatures approaching absolute zero (0 K or -273°C). Some examples of metallic elements that exhibit superconductivity are aluminum ($T_c = 1.2$ K), lead ($T_c = 7.2$ K), and niobium ($T_c = 9.2$ K). Interestingly, some metals do not become superconducting; for example, copper is not superconducting even at 0.05 K.

Until recently, the Nb_3Al-Nb_3Ge ($T_c = 21$ K) alloy was believed to have the highest critical temperature, and alloys and compounds of niobium (e.g., niobium with a tin coating) are widely used as superconducting wire and tape in magnets operating at liquid-helium temperatures. A practical use for superconductivity has been in generating intense magnetic fields for research projects.[10] Niobium-tin compounds can support current densities of over 10^9 A-m^{-2}. In 1986, it was discovered that some oxides become superconducting at temperatures above the boiling temperature of liquid nitrogen (77 K). For example, yttrium-barium-copper oxide, or $YBa_2Cu_3O_7$, has $T_c = 80$ K, so its superconductivity can be utilized by cooling with liquid nitrogen. Since then, materials with T_c values as high as 125 K have been discovered.[11]

5.2 CURRENT FLOW, OHM'S LAW, AND RESISTANCE

Consider a piece of current carrying material such as the conductor shown in Figure 5.3. If at any time current is directed toward a boundary between the conductor and the surrounding dielectric, charge accumulates at the boundary and produces a steadily increasing electric field. To maintain a constant electric field and a steady current flow, we must thus have **J** and **E** parallel to the conductor boundaries. Consider any two surfaces A and A', both of which are perpendicular to **J**, with their circumferences defined by the conductor boundaries. The currents passing through these surfaces, namely $I = \int_A \mathbf{J} \cdot d\mathbf{s}$ and $I' = \int_{A'} \mathbf{J} \cdot d\mathbf{s}$, respectively, must be equal, since otherwise charge accumulates indefinitely in the region between the two surfaces. Because the two surfaces were chosen arbitrarily, the current must be the same through any other surface that cuts across the wire.

---

[9]See R. de Bruyn Outober, Heike Kamerlingh Onnes's discovery of superconductivity, *Scientific American,* pp. 98–103, March 1997.

[10]For more information on the practical uses of superconductivity, see the August 1989 issue of the *Proceedings of the IEEE.*

[11]V. Z. Kresin and S. A. Wolf, *Fundamentals of Superconductivity,* Plenum, New York, 1990.

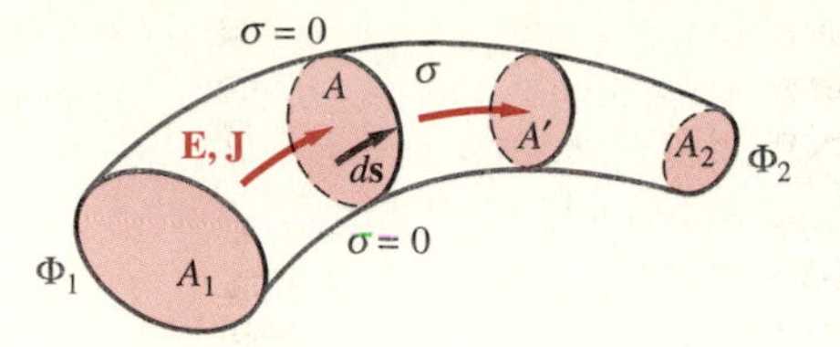

FIGURE 5.3. An arbitrary-shaped conductor. The surfaces A_1 and A_2 are respectively maintained at potentials Φ_1 and Φ_2 and are assumed to be equipotential surfaces (i.e., they are coated by material with conductivity much larger than that of the conductor).

For homogeneous media (i.e., where σ is not a function of position), the distribution of current flow lines is the same as that of the electric field, and the charge density ρ inside the conductor is zero[12] so that the current lines do not begin or end anywhere in the conductor. The electric field within the conductor is presumed to be set up and maintained by means of a potential difference applied between the cross-sectional surfaces at the ends of the conductor, namely A_1 and A_2 in Figure 5.3. For simplicity, we take these end surfaces to be equipotential surfaces.[13] Note that no current can flow outside the conductor ($\sigma = 0$) and that charge cannot indefinitely accumulate at the boundary between the conductor and the surrounding dielectric. Thus, the electric field lines, and thus the lines of current flow, *are always parallel to the side edges of the wire,* regardless of how curved or twisted the conductor may be. Since the conservative property of the electric field requires (see Section 4.11) that the tangential component of **E** is continuous across any interface, the electric field just outside the wire must have the same tangential component as that just inside. In general, the electric field outside the wire also has a component perpendicular to the side surface of the wires, which is supported by surface charge induced on the surface of the wire.[14] Further discussion of the structure of the electric field at the boundary between a conductor and an insulator is found in Section 5.6.

The flow of current (i.e., **J**) at each point in the material is in the direction of the electric field **E**. The potential difference Φ_{12} between the two equipotential surfaces A_1 and A_2 at potentials Φ_1 and Φ_2, respectively, is

$$\Phi_{12} = \Phi_2 - \Phi_1 = -\int_L \mathbf{E} \cdot d\mathbf{l} = -\int_L \frac{1}{\sigma}\mathbf{J} \cdot d\mathbf{l}$$

[12]It will be shown later in Section 5.4 that, based on the principle of conservation of charge, the divergence of steady current must be zero, so that current lines cannot originate or terminate anywhere in the conductor. For homogeneous media, $\nabla \cdot \mathbf{J} = 0$ implies that $\nabla \cdot \mathbf{E} = 0$, which in turn means that volume charge density is zero. However, if σ varies with position, we may then have volume charge density $\rho \neq 0$, with the electric field terminating on volume charges (see footnote 22 in Section 5.4).

[13]This could be achieved by coating the end surfaces with a material whose conductivity is much greater than that of the conductor. For example, if the conductor itself is made of graphite, the end surfaces may be copper- or silver-coated.

[14]This perpendicular field typically exists because of an applied potential difference between the conductor and other conductors in a circuit. For example, if the conductor in question is the inner conductor of a coaxial line, the perpendicular field is radially outward between the inner and outer conductors such that the negative line integral of this field from the inner conductor (of radius a) to the outer one (of radius b) is equal to the voltage difference Φ_{ab}. Also see Section 5.6 and Figure 5.12.

where the path of integration L is any path starting at any point on A_1 and ending at any point on A_2. Note that in order for Φ_{12} to be positive, A_1 must be at a lower potential than A_2 (i.e., $\Phi_2 > \Phi_1$). Note also that, as argued in the previous paragraph, the same total current must pass through all of the surfaces A_1, A_2, A, and A' in Figure 5.3. This current is given by

$$I = \int_A \mathbf{J} \cdot d\mathbf{s} = \int_A \sigma \mathbf{E} \cdot d\mathbf{s}$$

If the potential difference Φ_{12} between A_1 and A_2 is increased, the electric flux lines do not change shape, but $\mathbf{E}$ proportionally increases everywhere within the conductor. Since the current is linearly related to the electric field, assuming that the increased field and the resultant increased current are not sufficient to lead to a substantial temperature increase that might modify the conductivity σ, $\mathbf{J}$ increases everywhere in the same proportion as $\mathbf{E}$, as does the total current I. The ratio of the potential difference to the total current is thus a constant, defined as the *resistance,* and is given by

$$\boxed{R \equiv \frac{\Phi_{12}}{I} = \frac{-\int_L \mathbf{E} \cdot d\mathbf{l}}{\int_A \sigma \mathbf{E} \cdot d\mathbf{s}}} \qquad [5.2]$$

where L is any path from the end surface A_1 to A_2 (assuming $\Phi_2 > \Phi_1$), and where R remains constant for different Φ_{12} and I, but depends on the flow pattern between the end surfaces A_1 and A_2, and on the value of the conductivity σ. The relation between Φ_{12}, I, and R is the well known[15] *Ohm's law:*

$$\boxed{\Phi_{12} = IR} \qquad [5.3]$$

Equation [5.2] constitutes a general definition for resistance akin to that for capacitance given in [4.49]. The duality between $\mathbf{J} = \sigma\mathbf{E}$ and $\mathbf{D} = \epsilon\mathbf{E}$ is further discussed in Section 5.7.

Ohm, Georg Simon *(b. March 16, 1789, Erlangen, Bavaria; d. July 6, 1854, Munich, Bavaria) Ohm was the son of a self-taught master mechanic who was interested in science and who went to some pains to see that the youngster received a scientific education. Ohm entered the University of Erlangen and obtained his Ph.D. there in 1811. Science was not, however, to deal kindly with Ohm.*

He taught in high schools, but his ambition was to achieve a university appointment. To do this he had to produce some important research work, so he tackled the new field of current electricity that had been opened by Volta [1745–1827]. But he was poor and equipment was hard to get, so he made his own. In particular, he drew his own wires, and the influence of his mechanic father stood him in good stead.

---

[15]For an excellent discussion of Ohm's law and contributions of G. S. Ohm, see M. S. Gupta, Georg Simon Ohm and Ohm's Law, *IEEE Transactions on Education,* 23(3), pp. 156–162, August 1980.

Ohm decided to apply to the flow of electricity some of the discoveries made by Fourier [1768–1830] concerning the flow of heat. Just as the rate at which heat flowed from point A to point B depended in part on the temperature difference between those two points and in part on the ease with which heat was conducted by the material between the points, so the rate of flow of electric current should depend on the difference in electrical potential between points A and B and on the electrical conductivity of the material between the points.

By working with wires of different thicknesses and lengths, he found that the quantity of current transmitted was inversely proportional to the length and directly proportional to the cross-sectional area of the wire. He was in this way able to define the resistance of the wire and, in 1827, to show that a simple relation existed among the resistance, the electric potential, and the amount of current carried. This came to be called Ohm's law, which can be expressed as follows: The flow of current through a conductor is directly proportional to the potential difference and inversely proportional to the resistance. (Nearly half a century earlier, Cavendish had discovered this relationship, but he never published his results).

This was Ohm's only first-class contribution to science, but one first-class contribution is quite enough, and he deserved his university appointment. He did not get it, however. His work stirred up a good deal of opposition and resentment, apparently because Ohm tried to base his results on theory and some of his audience did not understand that good, thorough experimental work was also involved. In any case, Ohm met with so much criticism that he was forced to resign even his high school position.

For six years he lived in poverty and bitter disappointment, while very slowly his work became known and appreciated outside Germany. He found himself, probably to his own surprise, coming to be held in honor. The Royal Society gave him its Copley medal in 1841 and made him a member in 1842. Finally, prophet Ohm, with some help from Ludwig I of Bavaria, came to be honored even in his own country, and he was appointed to a professorship at the University of Munich in 1849. The last five years of his life were spent in the sun of ambition finally realized. What's more, a statue was raised to him in Munich after his death, and a street was named in his honor (for dead men, as always, are easy to appreciate).

His name is further immortalized in the unit of resistance, the ohm, denoted Ω. *Thus, when a current of one ampere passes through a substance under a potential difference of one volt, that substance has a resistance of one ohm. Furthermore, the unit of conductance (which is the reciprocal of resistance) was originally chosen to be the mho—Ohm's name spelled backward—a whimsical device introduced by Kelvin [1824–1907]. More recently, the unit mho has fallen out of favor, and the modern MKS unit for conductivity is now adopted to be the siemens, denoted simply as S. [Adapted with permission from* Biographical Encyclopedia of Science and Technology, *I. Asimov, Doubleday, 1982.]*

1700 *1789* *1854* *2000*

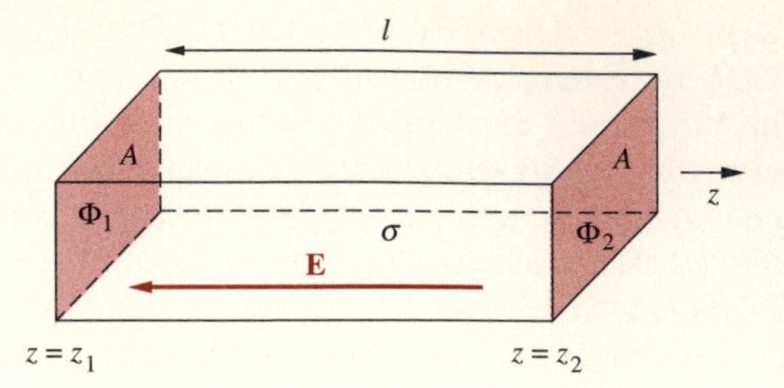

FIGURE 5.4. A rectangular conductor of uniform cross section. The end surfaces are coated with highly conducting material and are thus equipotential surfaces.

As a simple example of the application of [5.2], we consider a rectangular conductor of conductivity σ, cross-sectional area A, and length l, as shown in Figure 5.4. Assuming the two end surfaces (A_1 and A_2) to be coated with highly conducting material and maintained at potentials of Φ_1 and Φ_2, respectively, the electric field within the material is given[16] by $\mathbf{E} = -\hat{\mathbf{z}}E_0$, where E_0 can be found as

$$\Phi_{12} = \Phi_2 - \Phi_1 = -\int_{z_1}^{z_2} \mathbf{E}\cdot d\mathbf{z} = +\int_{z_1}^{z_2} E_0\hat{\mathbf{z}}\cdot\hat{\mathbf{z}}dz \quad \rightarrow \quad E_0 = \frac{\Phi_2 - \Phi_1}{z_2 - z_1} = \frac{\Phi_{12}}{l}$$

so that the current is given by

$$I = \int_A \sigma\mathbf{E}\cdot d\mathbf{s} = \frac{\sigma\Phi_{12}A}{l} \quad \rightarrow \quad R = \frac{\Phi_{12}}{I} = \frac{\Phi_2 - \Phi_1}{I} = \frac{l}{\sigma A}$$

The preceding expression is not specific to a rectangular block and is in fact valid for the resistance of a conductor of length l having a uniform cross-sectional area A of any shape.

Examples 5-1 through 5-5 illustrate the use of [5.2] to determine the resistance of different conductor configurations.

Example 5-1: Copper wire. Consider a copper wire 1 km long and 1 mm in radius. (a) Find the resistance R. (b) If the wire carries a current of 1 A, determine the duration of time in which the electrons drift across the length of the wire.

Solution:

(a) For copper, we have $\sigma = 5.8 \times 10^7$ S-m^{-1}. For $l = 10^3$ m and a 1-mm radius, which gives $A = \pi \times 10^{-6}$ m^2, we have

$$R = \frac{10^3 \text{ m}}{5.8 \times 10^7 \text{ S-m}^{-1} \times \pi \times 10^{-6} \text{ m}^2} \simeq 5.49\Omega$$

(b) A current of 1 A and a cross-sectional area $A = \pi \times 10^{-6}$ m^2 implies a current density of $\mathbf{J} \simeq 3.18 \times 10^5$ A-m^{-1}. Since $\mathbf{J} = n_c q_e \mathbf{v}_d$ and

[16]Note that we have assumed (on the basis of symmetry) the electric field to be constant everywhere inside the rectangular conductor.

$n_c = 8.45 \times 10^{28}$ el-m^{-3}, we have $|\mathbf{v}_d| \simeq 2.35 \times 10^{-5}$ m/s. In other words, the electrons can drift across a distance of 1 m in about ~12 hours. A particular tagged electron can traverse the entire 1-km length of this copper wire in approximately ~492 days! On the other hand, we know from transmission-line analysis that an applied signal (e.g., a change in current through or voltage across a line) travels to the other end of the line at nearly the speed of light in air (i.e., approximately ~3.33 μs for a 1-km wire). The transmission-line signal propagates outside the conductors (or wires) as an electromagnetic wave, not via the drifting motion of electrons within the conductors that constitute the line.

Example 5-2: A microstrip trace. Consider a printed circuit board microstrip trace, as shown in Figure 5.5. If the metal trace is made of copper with thickness $t = 34.3$ μm, find the trace resistance per centimeter for a trace width w of (a) 0.25 mm and (b) 0.5 mm.

Solution: We can use $R = l/(\sigma A)$ where $l = 1$ cm = 0.01 m, $\sigma = 5.8 \times 10^7$ S-m^{-1}, and $A = wt = 3.43 \times 10^{-5} w$ m^2:

(a) For $w_1 = 2.5 \times 10^{-4}$ m, we have $R_1 \simeq 20.1$ mΩ-cm^{-1}, and

(b) for $w_2 = 2w_1 = 5 \times 10^{-4}$ m, we have $R_2 = R_1/2 \simeq 10.05$ mΩ-cm^{-1}.

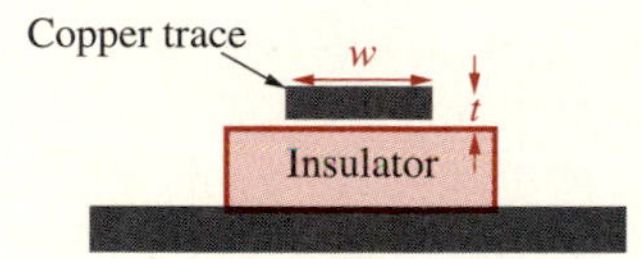

FIGURE 5.5. A microstrip trace. The cross-sectional view of a microstrip trace.

Example 5-3: Curved bar. The resistance of a rectangular block of metal was discussed in this section and found to be given by $R = l/(\sigma A)$, where σ is the conductivity, l is the length, and A is the cross-sectional area. As a more complicated geometry, consider the resistor in the form of a curved bar, which is made of the rectangular block bent to form the arc of a circle, with the two end surfaces remaining flat, as shown in Figure 5.6a. The edges are coated with perfectly conducting material so that they constitute electrodes at uniform potential. Find the resistance R between the electrodes.

Solution: This problem has a cylindrical symmetry with respect to the O-O′ axis, as shown. The equipotential surfaces are the intersections of the curved rectangular bar and the planes passing through the O-O′ axis. The current flow lines $\mathbf{J}$ are circular arcs centered on the O-O′ axis (i.e., $\mathbf{J}$ is in the $\hat{\boldsymbol{\phi}}$ direction). We may suppose from symmetry that the $\mathbf{E}$ vector is the same at all points inside the bar that are equidistant from the O-O′ axis. However, the electric field must vary with radial distance from the axis (r) across the curved bar, since the inward portions of the end faces are closer to each other than the outward portions are (i.e., $\mathbf{E}$ must be a function of r, or $\mathbf{E} = \mathbf{J}/\sigma = \hat{\boldsymbol{\phi}} E_\phi(r)$). To determine how the

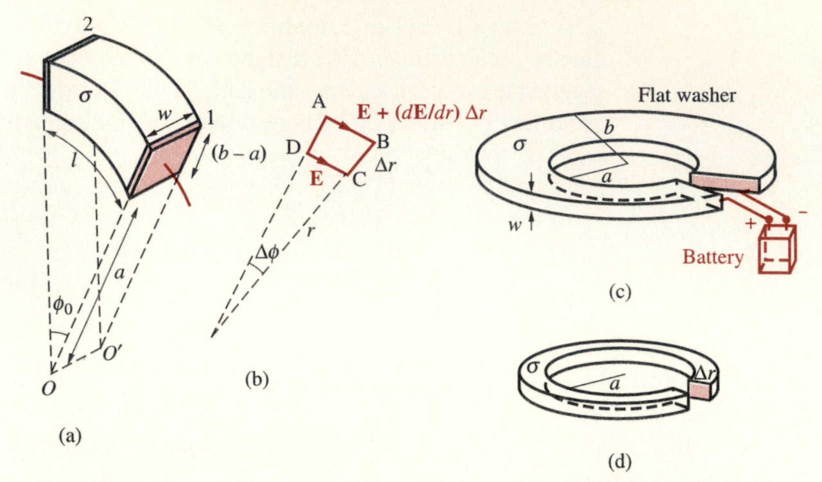

FIGURE 5.6. **A curved-bar resistor.** (a) A rectangular bar bent into a circular arc of rectangular cross section. (b) A differential angular element along the arc. (c) A flat washer is equivalent to a curved bar of arc length $\phi_0 \simeq 2\pi$. (d) The flat washer can be viewed as a parallel combination of elemental resistances of radial width Δr, as shown.

electric field varies with radial distance, we can use the fact that $\oint_C \mathbf{E} \cdot d\mathbf{l} = 0$. Consider a closed contour ABCD, as shown in Figure 5.6b. We have

$$\oint_C \mathbf{E} \cdot d\mathbf{l} = \left(E_\phi + \frac{dE_\phi}{dr}\Delta r\right)(r + \Delta r)\Delta\phi - E_\phi r \Delta\phi = 0$$

where the first term is the contribution from path segment AB and the second term is that from path segment CD. We thus have

$$E_\phi \Delta r + r\frac{dE_\phi}{dr}\Delta r + \frac{dE_\phi}{dr}(\Delta r)^2 = 0 \qquad \rightarrow \qquad \frac{dE_\phi}{E_\phi} = -\frac{dr}{r}$$

where the last term drops out as $\Delta r \rightarrow 0$. Therefore

$$\ln E_\phi = -\ln r + K_1 \qquad \rightarrow \qquad E_\phi = \frac{K}{r}$$

where K_1 and K are constants. The functional form of E_ϕ is thus determined. Note that we could alternatively have found $\Phi(r)$ using Laplace's equation, namely $\nabla^2\Phi = 0$, and subsequently determined $\mathbf{E}$ from $\mathbf{E} = -\nabla\Phi$. The voltage between the end faces is thus

$$\Phi_{12} = -\int_1^2 \mathbf{E} \cdot d\mathbf{l} = -\int_{\phi_0}^0 \frac{K}{r}\hat{\boldsymbol{\phi}} \cdot \hat{\boldsymbol{\phi}} r\, d\phi = \int_0^{\phi_0} \frac{K}{r} r\, d\phi = K\phi_0 = \frac{Kl}{a}$$

where we have implicitly assumed the end face at the lower potential to be at $\phi = \phi_0$, so that $\mathbf{E} = \hat{\boldsymbol{\phi}}(K/r)$. The total current between the end faces is given by

$$I = \int_S \mathbf{J} \cdot d\mathbf{s} = \int_S \sigma \mathbf{E} \cdot d\mathbf{s} = \sigma w K \ln \frac{b}{a}$$

Hence the resistance is given by

$$R = \frac{\Phi_{12}}{I} = \frac{l}{\sigma w a \ln(b/a)}$$

Note that the preceding expression for R applies for any arc length l of the curved bar. For example, consider the flat washer shown in Figure 5.6c, with resistance between the edges of a saw-cut through the radius of the washer (i.e., the shaded surface and the one facing it). In this case, and assuming the arc length of the cut region is very small ($\ll 2\pi a$), we simply replace the length l in the previous expression with $l \simeq 2\pi a$. Note from Figure 5.6d that the resistance of the flat washer between the indicated faces can also be thought of as being the parallel combination of elemental flat washer resistances, each of radial thickness Δr.

Example 5-4: Hemispherical electrode. A ground connection is made by burying a perfectly conducting hemispherical electrode of radius a in the earth, as shown in Figure 5.7. Assuming the earth's conductivity to be σ, find the resistance of the conductor to distant points in the ground (i.e., between the electrode and a concentric, perfectly conducting hemisphere of infinite radius).

Solution: We can set up a spherical coordinate system as shown in Figure 5.7, where the azimuth angle ϕ is not shown. Since the electrode is a perfect conductor, the potential will be constant at every point within and on the surface of this electrode. We can arbitrarily take the reference point for potential to be at infinity. Due to symmetry, all field quantities (and thus the current) are independent of the azimuth angle ϕ. The electric field is also independent of θ, although current flows only in the lower hemisphere. Accordingly, $\Phi(r) = K/r$, where K is a constant, is an admissible general solution of Laplace's equation for this configuration. Assuming that the hemispherical electrode is held at a certain potential V_0, we must have $\Phi(r) = V_0$ at $r = a$, so that $K = aV_0$. Thus, $\Phi(r) = aV_0/r$.

The corresponding electric field can be found from $\mathbf{E} = -\nabla\Phi$, which gives $\mathbf{E}(r) = \hat{\mathbf{r}}aV_0/r^2$. Thus, the current density $\mathbf{J}$ is also in the radial direction and is given by

$$\mathbf{J} = \sigma\mathbf{E} = \hat{\mathbf{r}}\frac{\sigma a V_0}{r^2}$$

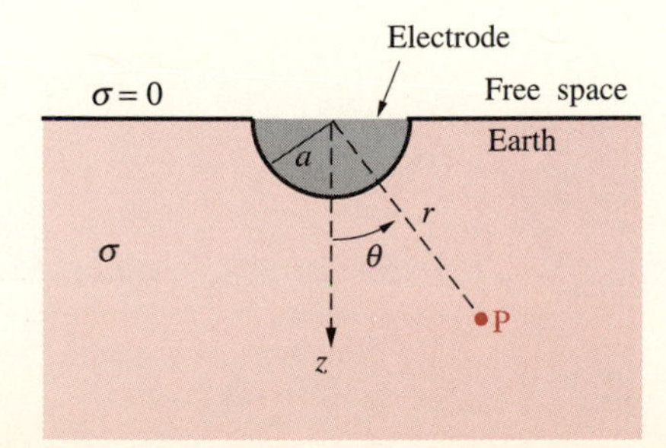

FIGURE 5.7. Hemispherical buried electrode. The electrode is made of perfectly conducting material and is thus an equipotential surface.

Note that this current satisfies all the boundary conditions since it is entirely tangential to the earth-free space interface at $\theta = \pi/2$. The total current I that crosses any concentric hemisphere of radius r is

$$I = \int_S \mathbf{J} \cdot d\mathbf{s} = \frac{\sigma a \Phi_0}{r^2} \frac{4\pi r^2}{2} = 2\pi\sigma a V_0$$

Therefore the total resistance between the electrode and distant points is

$$R = \frac{V_0}{I} = \frac{V_0}{2\pi\sigma V_0 a} = \frac{1}{2\pi\sigma a}$$

Note that because we took the reference point for potential to be at infinity, the voltage drop V_0 is that between the electrode and the hemisphere at infinity.

Example 5-5: A thin semiconductor film contact. Consider a thin semiconductor film of thickness h with concentric perfectly conducting electrodes (or contacts) of radii a and b, respectively ($h \leq a, h \leq b$). (See Figure 5.8.) The thin film of semiconductor material is deposited on an insulator, and the conductivity of this material is σ S-m^{-1}. Find the resistance between the electrodes.

Solution: We consider a cylindrical coordinate system with the z axis pointing up and passing through the center of the inner electrode. Due to symmetry, the electric field in the region between the electrodes is independent of ϕ and only has a component in the r direction. Considering circular electrodes of radii a and b, where $a \geq h, b \geq h$, and assuming the electrodes to be perfect conductors, the total radial current I flowing between the electrodes is given by

$$I = \int_S \sigma \mathbf{E} \cdot d\mathbf{s} = \int_0^h \int_0^{2\pi} \sigma E_r \hat{\mathbf{r}} \cdot \hat{\mathbf{r}} r \, d\phi \, dz$$

$$I = \sigma E_r(2\pi r h) \quad \rightarrow \quad E_r = \frac{I}{2\pi\sigma h r}$$

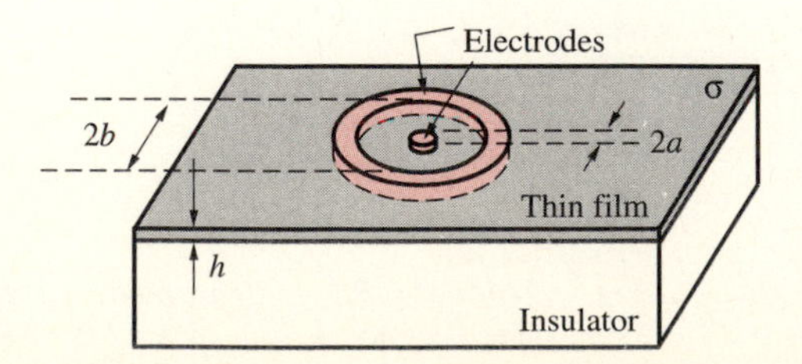

FIGURE 5.8. Thin-film contact resistance. Two concentric circular electrodes on a thin film of conducting material ($\sigma \neq 0$) lying on an insulator.

Taking the electrode at $r = b$ to be at zero potential and the electrode at $r = a$ to be at a potential V_0, we find that the potential at a radius r (where $a \le r \le b$) is

$$\Phi(r) = -\int_b^r \mathbf{E} \cdot d\mathbf{l} = -\int_b^r \frac{I}{2\pi\sigma h r} dr = \frac{I}{2\pi\sigma h} \ln \frac{b}{r}$$

Substituting $r = a$, we have the potential difference between the electrodes

$$\Phi_{ba} = V_0 = \frac{I}{2\pi\sigma h} \ln \frac{b}{a}$$

Therefore, the total resistance between the electrodes is

$$R = \frac{\Phi_{ba}}{I} = \frac{1}{2\pi\sigma h} \ln \frac{b}{a}.$$

5.3 ELECTROMOTIVE FORCE AND KIRCHHOFF'S VOLTAGE LAW

Until now, we have not discussed the requirements for the maintenance of steady currents, although it is clear that we must start with an electric field (i.e., $\mathbf{J} = \sigma\mathbf{E}$). It turns out that electrostatic fields produced by stationary charges cannot by themselves maintain a steady current, since they are conservative in nature (i.e., $\oint_C \mathbf{E} \cdot d\mathbf{l} = 0$). Thus electromotive force produced by another energy source, such as a chemical battery, is necessary to sustain steady currents. In this section, we discuss the maintenance of steady currents and electromotive force.

Suppose we wish to establish current in a rectangular conducting bar such as that shown in Figure 5.4. Since current needs to be driven by an electric field, we might think of inserting the conductor in an electrostatic field, such as that between the plates of a parallel-plate capacitor (see Figure 5.9a). At the instant the bar is inserted, the conductor is in a uniform electric field, and thus a current $I = \sigma E A$ flows. But this current cannot continue since it is essentially a redistribution of charge inside the conductor. Within a very short time scale (e.g., $\sim 10^{-19}$ s for copper) negative charge accumulates at the top of the bar, leaving behind positive charge at the bottom, and these separated charges set up a field within the conductor that completely cancels the external field and reduces the total field within the conductor to zero,[17] so that the current simply stops. It is thus apparent that an electrostatic field cannot sustain steady current flow.

[17]This internal field is the induced field E_{ind} that was discussed in Section 4.7.3 in connection with Figure 4.41.

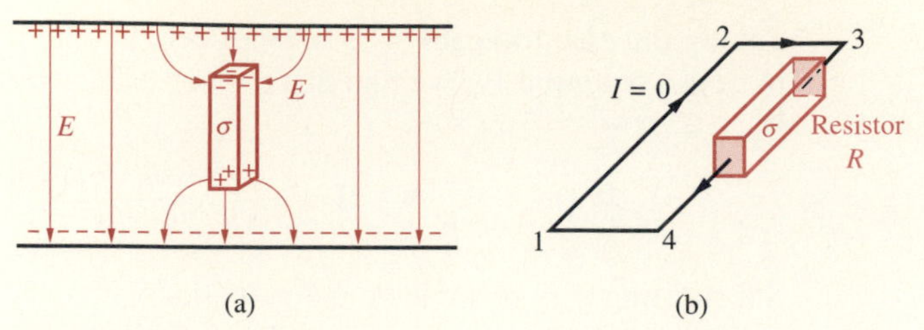

FIGURE 5.9. Steady currents cannot be maintained without an external energy source. (a) The placement of a rectangular conducting bar in a uniform field leads to rapid ($\sim 10^{-19}$ s) redistribution of the charge in the bar, but the field does not support steady current flow in the bar. (b) Current in a closed circuit cannot be maintained since the current carriers lose their energy in traveling through the resistor, and they cannot gain this energy back from the conservative electrostatic field.

We observed in connection with Figure 5.3 that steady currents cannot originate or terminate within the conductor, since this would imply nonstop removal or accumulation of charge. Accordingly, steady current flow requires a closed circuit. There may be branches in the circuit where three or more wires are connected, but any given flow line must close on itself without branching. Let us now consider one such closed circuit, as shown in Figure 5.9b, and the motion of the current carriers. For the sake of this argument, we neglect the wire resistances and assume that current can flow through the wires without any dissipation of energy. For steady-state current to exist, the current carriers (e.g., electrons) must make a complete circuit (closed loop), starting from point 1 and ending at point 1, repeating their motion ad infinitum. However, since current through the resistor (rectangular bar) produces energy loss in the form of heat (due to collisions), the electrons lose energy as they travel through this part of the circuit; they cannot gain this energy back from the conservative electrostatic field because the round-trip motion leads to zero net energy (i.e., $\oint_C \mathbf{E} \cdot d\mathbf{l} = 0$). Thus, another source of energy is necessary to maintain steady electric current flow through the circuit by continuously supplying the energy dissipated in the resistor as heat. In other words, as we go around the circuit starting at a given point, we observe a continual decrease in electrostatic potential. Thus, there *must* be some portion of the circuit that does not behave like a conductor (i.e., does not obey Ohm's law), but where potential *rises* along the direction of the current.

The sources of the external energy required to maintain steady currents and to create rises in potential can be chemical batteries, electric generators, thermocouples, photovoltaic cells, or other devices. These electrical energy sources, when connected in a circuit, provide the driving force for current carriers. This force manifests itself in the form of an external electric field, which we designate as $\mathbf{E}_{\text{emf}}$. (The subscript "emf" will be defined shortly.) Consider the simple electric circuit shown in Figure 5.10a. The circuit consists of a rectangular conducting bar that acts as a resistor R and a battery. A steady current I is assumed to be present at all points in the circuit. We interpret the internal action of the battery from the standpoint of producing a nonconservative electric field $\mathbf{E}_{\text{emf}}$. In general, an electrostatic field (designated

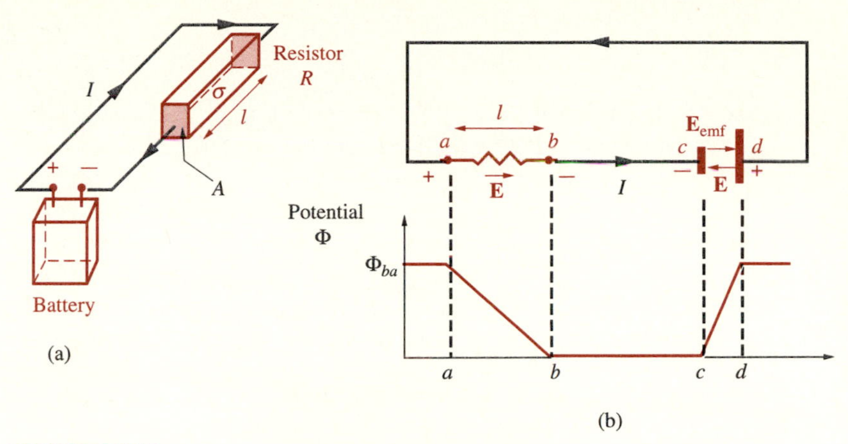

FIGURE 5.10. **A series circuit with a battery and resistor.** (a) Battery driving a circuit with a resistor. (b) Variation of potential around the circuit.

by **E**) is also created by the battery as a result of accumulation, at the battery terminals or elsewhere in the circuit, of stationary charge (e.g., such as that on a capacitor). The total electric field is then

$$\mathbf{E}_{\text{total}} = \mathbf{E}_{\text{emf}} + \mathbf{E}$$

Note also that although **E** is conservative and can be derived from the gradient of a scalar potential due to stationary charges (i.e., $\mathbf{E} = -\nabla\Phi$), $\mathbf{E}_{\text{emf}}$ is not related to any stationary charges. At all points in the circuit we have

$$\frac{\mathbf{J}}{\sigma} = \mathbf{E}_{\text{emf}} + \mathbf{E}$$

(where $|\mathbf{J}| = I/A$), although $\mathbf{E}_{\text{emf}}$ is actually zero at all points outside the battery. If we now integrate around a complete circuit (closed loop), we find

$$\oint_C \mathbf{E} \cdot d\mathbf{l} + \oint_C \mathbf{E}_{\text{emf}} \cdot d\mathbf{l} = \oint_C \frac{1}{\sigma}\mathbf{J} \cdot d\mathbf{l} = \int_a^b \frac{I}{\sigma A} dl$$

The first term is zero, since the line integral around a closed loop of an electrostatic field produced by stationary charges is always zero. However, the second term, involving the line integral of $\mathbf{E}_{\text{emf}}$, is not zero but is instead equal to a voltage called the *electromotive force,* or *emf,* of the circuit, typically denoted by $\mathcal{V}_{\text{emf}}$. The right-hand side of the preceding equation equals the total voltage drop around the circuit, which is simply equal to IR, assuming that the internal resistance of the battery is zero[18] and that the wire resistances are also zero (i.e., no voltage drops occur across the wires). Thus we have

[18]If the battery has an internal resistance R_s, the total voltage drop around the circuit is simply $IR + IR_s$.

$$\mathcal{V}_{\text{emf}} = I\int_a^b \frac{dl}{\sigma A} = IR \qquad \text{or} \qquad R = \int_a^b \frac{dl}{\sigma A}$$

If current were not allowed to flow (e.g., by disconnecting part of the circuit so that it is no longer a closed circuit), or if we consider a battery under open circuit conditions, an electrostatic field exists everywhere because of the accumulation of charge at the battery terminals. Within the battery, the electrostatic field is neutralized by the nonconservative E_{emf} field. From the point of view of a positive test charge, it is possible to acquire energy in moving from the positive to the negative terminal (i.e., in the direction of decreasing potential) external to the battery, but by completing the circuit through the battery, where there is zero field, the acquired energy is not returned to the field. The test charge can thus make a complete circuit with a net accumulation of energy.[19] In an actual circuit such as that in Figure 5.10a, the acquired energy is dissipated in the resistor as heat. Nevertheless, the test particle is capable of making repeated round trips, hence constituting a steady electric current.

In general, for a closed circuit containing many resistors and emf sources, we have

$$\boxed{\sum \mathcal{V}_{\text{emf}} = I\sum R} \tag{5.4}$$

This relation is known as *Kirchhoff's voltage law.* It states that the algebraic sum of the emfs around a closed circuit equals the algebraic sum of the voltage drops (i.e., IR) over the resistances around the circuit. Kirchhoff's voltage law applies not only for an isolated closed network but also for any single mesh (closed path) of a larger network. Although the preceding discussion relates to stationary electric fields and currents, Kirchhoff's voltage law also applies to time-varying situations, as long as the dimensions of the circuit are much smaller than a wavelength. In such cases, the algebraic sum of the *instantaneous* emfs equals the algebraic sum of the *instantaneous* IR drops around the circuit.

Further insight into the nature of the electromotive force and Kirchhoff's voltage law can be gained by examining the variation of potential around the circuit, as shown in Figure 5.10b. Taking the points b, c to be at zero potential, a positively charged current carrier starting at point a and traveling in the direction of the current undergoes a decrease in potential as it travels through the resistor and then an equivalent increase as it travels through the battery. Note that the potential difference Φ_{ab} is related to the electrostatic field $\mathbf{E}$, so that

$$\Phi_{ab} = \Phi_b - \Phi_a = -\int_a^b \mathbf{E}\cdot d\mathbf{l}$$

whereas the potential difference Φ_{cd} is simply equal to the electromotive force $\mathcal{V}_{\text{emf}}$ of the battery.

---

[19] Accumulation of kinetic energy in many such round trips through a nonconservative field in the absence of resistive losses is the basis for particle accelerators such as the cyclotron.

Kirchhoff, Gustav Robert *(b. March 12, 1824, Königsberg, Prussia (now Kaliningrad, Russia); d. October 17, 1887, Berlin) Kirchhoff, the son of a law councilor, studied at the University of Königsberg and graduated in 1847. He was the first to show that electrical impulses moved at the velocity of light, and he extended and generalized the work of Ohm.*

In 1854 he was appointed professor of physics at Heidelberg and began to deliver meticulous but very dull lectures. There he teamed up with Bunsen [1811–1899], whom he had worked with briefly four years earlier at Breslau. Bunsen was interested in photochemistry (the chemical reactions that absorb or produce light), and he studied the light produced through colored filters. Kirchhoff, with mathematical interests and a strong background in Newton's [1642–1727] laws, suggested using a prism. Once this was done, the two developed the first spectroscope by allowing the light to pass through a narrow slit before reaching the prism. The different wavelengths of light were refracted differently so that numerous images of the slit were thrown on a scale in different positions and, of course, with different colors.

The use of a Bunsen burner, first developed by Bunsen in 1857, was helpful. The burner produced so little light of its own that there was no luminous background to drown out and confuse the wavelengths of light produced by the reactions studied or by the minerals heated to incandescence. Previous workers, without Bunsen burners, had been misled by the background of luminous lines and bands produced by heated carbon compounds.

Through the use of a spectroscope it quickly became apparent to Kirchhoff that each chemical element, when heated to incandescence, produced its own characteristic pattern of colored lines. For example, incandescent sodium vapor produced a double yellow line. In a sense, the elements were producing their fingerprints, and thus the elementary composition of any mineral could be determined by spectroscopy.

By 1859 this new analytic method was moving along smoothly and was first publicly reported on October 27 of that year. As was inevitable, a mineral was found displaying spectral lines that had not been recorded for any of the known elements. The conclusion was that a hitherto unknown element was involved; the discovery of cesium was announced on May 10, 1860. The name of the element (from the Latin word for "sky-blue") was derived from the color of the most prominent line in its spectrum. Within a year a second element, rubidium, was discovered and that name (from the Latin term for "red") again marked the color of the line that had led to its discovery.

Kirchhoff went even further with spectroscopy. He noticed that the bright double line of the sodium spectrum was in just the position of the dark line in the solar spectrum that Fraunhofer [1787–1826] had labeled D. He allowed sunlight and sodium light to shine through the same slit, expecting that the dark line of the first and the bright line of the second might neutralize each other. Instead, the line was darker than ever.

From this and other experiments, he concluded that when light passes through a gas, those wavelengths that the gas would emit when incandescent were absorbed. This is sometimes called Kirchhoff's law, although it was discovered by others at about the same time.

If sunlight possessed the D line, then that sunlight must pass through sodium vapor on its way to the earth. The only place where the sodium vapor could exist would be in the sun's own atmosphere. Consequently, it was possible to say that sodium existed on the sun. In this way, Kirchhoff identified half a dozen elements in the sun, and others such as Angström [1814–1874], Donati [1826–1873], and Huggins [1824–1910] joined in these spectroscopic endeavors. Thus was shattered the French philosopher Auguste Comte's categorical statement in 1835 that the constitution of the stars was the kind of information that scientists would be eternally incapable of attaining. Comte died (insane) two years too soon to see spectroscopy developed.

Kirchhoff's banker, unimpressed by this ability to find elements in the sun, asked, "Of what use is gold in the sun if I cannot bring it down to earth?" When Kirchhoff was awarded a medal and a prize in golden sovereigns from Great Britain for his work, he handed it to his banker with the comment, "Here is gold from the sun."

But the gold of the discovery was greater still. Eventually the spectral lines proved to be a guide not only to the macro world of the outer cosmos, but to the micro world within the atom. Balmer [1825–1898] made the first steps in this direction.

Kirchhoff also pointed out that a perfect black body—one that absorbed all radiation falling on it, of whatever wavelength—would, if heated to incandescence, emit all wavelengths. This conclusion had been arrived at independently by Stewart [1828–1887]. Although no perfect black body actually existed, one could be constructed by the use of a trick, as Kirchhoff pointed out. A closed container with blackened inner walls and a tiny hole would serve the purpose. Any radiation, of whatever wavelength, that entered the hole would have only an infinitesimal chance of emerging again through the hole and could therefore be considered absorbed. Thus, if the box were heated to incandescence, all wavelengths of light should emerge from the hole.

The study of this "black-body radiation" was to prove of the utmost importance a generation later, when it led to Planck's [1858–1947] quantum theory. [Adapted with permission from Biographical Encyclopedia of Science and Technology, *I. Asimov, Doubleday, 1982.]*

1700 — *1824* — *1887* — *2000*

5.4 THE CONTINUITY EQUATION AND KIRCHHOFF'S CURRENT LAW

We now consider one of the fundamental principles of electromagnetics, namely the continuity equation. This important equation is a mathematical statement of the

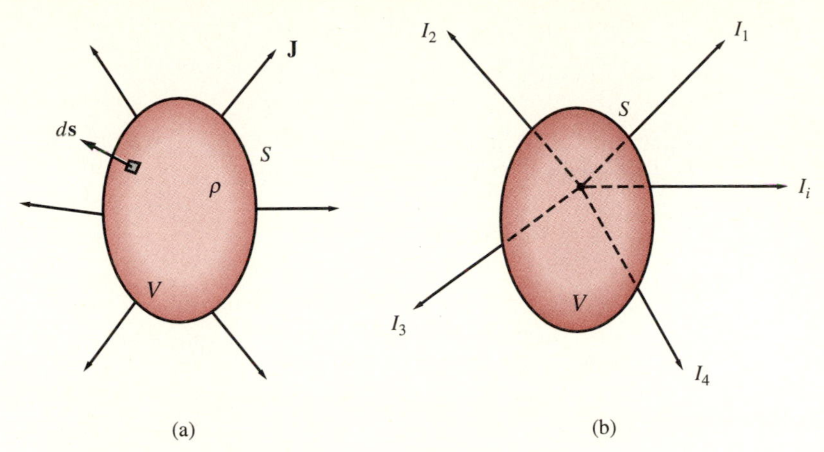

FIGURE 5.11. Current flow out of a volume V.

conservation of charge, and is also the basis of Kirchhoff's current law. Since current consists of the flow of charge and since charge is conserved,[20] we must have a build-up of charge within a region if there is net current flow into it. If we consider an arbitrary volume V bounded by a surface S, as shown in Figure 5.11a, we have

$$-\oint_S \mathbf{J} \cdot d\mathbf{s} = \frac{\partial}{\partial t} \int_V \rho \, dv$$

where the left-hand term gives the net inflow of current (hence the negative sign, since $d\mathbf{s}$ is by convention defined as being outward) and the right-hand side represents the net rate of increase of total charge enclosed by the surface S. Note from Figure 5.11a that $\mathbf{J}$ is defined as outward from the volume.

Using the divergence theorem (Section 4.6.1), we have $\oint_S \mathbf{J} \cdot d\mathbf{s} = \int_V \nabla \cdot \mathbf{J} \, dv$, and thus the preceding equation can be rewritten as

$$\int_V \left(\nabla \cdot \mathbf{J} + \frac{\partial \rho}{\partial t} \right) dv = 0$$

Since this relation must be true regardless of the choice of the volume V, the integrand itself must be zero, or

[20]Conservation of charge is one of the basic laws of physics. Charge is indestructible; it cannot be lost or created. Electric charge can move from place to place but can never appear from nowhere. The principle of conservation of charge is as important to physics as the constancy of the speed of light. The validity of this principle is verified by every experiment to date and is a direct consequence of Maxwell's equations. Like the speed of light, charge is the same for every frame of reference. Although other quantities such as mass, energy, and electric and magnetic fields all change with frames of reference, charge does not (see Sec. 11-8 of A. M. Portis, *Electromagnetic Fields,* John Wiley, 1978). The conservation of electric charge, based on experiments with electrical bodies and the transfer of electrification, was put forth in 1746 by William Watson and in 1747 by Benjamin Franklin. The first satisfactory experimental proof was provided by M. Faraday in *Phil. Mag.* vol. xxii, 1843, p. 200. For an excellent discussion of the history, see Chapters II and VI of E. Whittaker, *A History of the Theories of Aether and Electricity,* Thomas Nelson and Sons Ltd., London, 1951.

$$\boxed{\nabla \cdot \mathbf{J} + \frac{\partial \rho}{\partial t} = 0} \qquad [5.5]$$

Equation [5.5] is commonly referred to as the *continuity equation* and is essentially a differential form of the law of conservation of charge. For steady currents, we must have $\partial \rho / \partial t = 0$ so that

$$\nabla \cdot \mathbf{J} = 0$$

In other words, for stationary currents, the current density $\mathbf{J}$ is *solenoidal.*[21] We noted earlier that steady currents must flow in closed loops in order to avoid continuous accumulation of charge.

Since Ohm's law must hold in a conducting medium, we have $\mathbf{J} = \sigma \mathbf{E}$. If σ does not vary with position (i.e., if the medium is homogeneous), we must have $\nabla \cdot \mathbf{J} = \sigma \nabla \cdot \mathbf{E}$ for both steady and time-varying cases. In general, the continuity equation can then be written in terms of the electric field

$$\nabla \cdot \mathbf{E} + \frac{1}{\sigma} \frac{\partial \rho}{\partial t} = 0$$

For the static case ($\partial \rho / \partial t = 0$), the electric field in a homogeneous conducting medium must have zero divergence; that is, $\nabla \cdot \mathbf{E} = 0$. Since from Gauss's law we have $\nabla \cdot \mathbf{E} = \rho / \epsilon$, we conclude that the volume density of free charge ρ must be zero in a homogeneous conducting medium.[22]

Using the divergence theorem, the solenoidal nature of steady currents can be expressed in integral form as

$$\nabla \cdot \mathbf{J} = 0 \quad \rightarrow \quad \oint_S \mathbf{J} \cdot d\mathbf{s} = 0$$

[21] A vector field is said to be a solenoidal field if its divergence is zero everywhere. The zero divergence indicates that there are no sources or sinks in the field for the lines of flux to originate from or terminate on. Accordingly, the flux lines of a solenoidal vector field always close on themselves.

[22] In an inhomogeneous medium, with σ and ϵ both functions of position, we have

$$\nabla \cdot \mathbf{D} = \epsilon \nabla \cdot \mathbf{E} + \mathbf{E} \cdot \nabla \epsilon = \rho \quad \rightarrow \quad \nabla \cdot \mathbf{E} = \frac{\rho - \mathbf{E} \cdot \nabla \epsilon}{\epsilon}$$

and

$$\nabla \cdot \mathbf{J} = \sigma \nabla \cdot \mathbf{E} + \mathbf{E} \cdot \nabla \sigma = \frac{\sigma \rho}{\epsilon} - \frac{\sigma}{\epsilon} \mathbf{E} \cdot \nabla \epsilon + \mathbf{E} \cdot \nabla \sigma$$

In other words, $\nabla \cdot \mathbf{J} = 0$ does not imply that $\rho = 0$. Volume charge density can exist even under conditions of steady current flow (i.e., where $\nabla \cdot \mathbf{J} = 0$) in regions of variable conductivity, and its magnitude is proportional to the gradient of the conductivity, or $\rho = -(\epsilon / \sigma) \mathbf{E} \cdot \nabla \sigma + \mathbf{E} \cdot \nabla \epsilon$. Such regions of static free charge may exist in thin layers at the boundary between two materials of different conductivity. A good example is a parallel-plate capacitor with two different lossy materials sandwiched between its plates (see Problem 5-27).

The preceding relation is for steady currents and applies to any closed surface. The volume enclosed by this surface may be entirely inside a conducting medium, or it may only be partially filled with conductors. The conductors may form a network inside the volume, or they may all meet at a point. If the steady current is carried into the volume by different wires meeting at a node, as shown in Figure 5.11b, then the preceding relation implies that the *algebraic sum of all the currents at the junction is zero,* since a junction of wires can be neither a sink nor a source for charges. This condition is *Kirchhoff's current law,* which can be expressed mathematically as

$$\boxed{\sum I = 0} \quad [5.6]$$

Like Kirchhoff's voltage law, Kirchhoff's current law, as just derived for stationary fields and currents, also applies to time-varying situations, as long as the dimensions of the circuit (i.e., the length of the current-carrying wires) are much smaller than a wavelength (see Chapter 1).

5.5 REDISTRIBUTION OF FREE CHARGE

We have noted on several occasions that charge placed at a point inside a conducting body moves to the surface and redistributes itself in such a way that zero field exists within and tangent to the conductor surface. With the continuity equation now in hand, we can quantitatively evaluate the length of time required for this process. We shall show that there can be no permanent distribution of free charge within a homogeneous region of nonzero conductivity, and that the rearrangement or relaxation time is determined by the permittivity-to-conductivity ratio of the material (i.e., ϵ/σ).

Consider a homogeneous, isotropic, time-invariant, and linear conducting region where conductivity σ and permittivity ϵ are simple constants, and are not functions of position, direction, time, or the applied electric field. Using $\nabla \cdot \mathbf{D} = \rho$ and the continuity equation, and assuming that the relations between $\mathbf{D}$ and $\mathbf{E}$, and between $\mathbf{J}$ and $\mathbf{E}$ are linear, we have

$$\nabla \cdot \mathbf{J} + \frac{\partial \rho}{\partial t} = \nabla \cdot (\sigma \mathbf{E}) + \frac{\partial \rho}{\partial t} = \sigma \nabla \cdot \mathbf{E} + \frac{\partial \rho}{\partial t} = 0$$

On the other hand, we have

$$\nabla \cdot \mathbf{D} = \nabla \cdot (\epsilon \mathbf{E}) = \epsilon \nabla \cdot \mathbf{E} = \rho \quad \rightarrow \quad \nabla \cdot \mathbf{E} = \frac{\rho}{\epsilon}$$

Combining the two preceding equations we find

$$\frac{\partial \rho}{\partial t} + \frac{\sigma}{\epsilon}\rho = 0$$

Rewriting and integrating, we have

$$\frac{\sigma}{\epsilon}\,\partial t = -\frac{\partial\rho}{\rho}$$

$$\frac{\sigma}{\epsilon}\int_0^t \partial t = -\int_{\rho_0}^{\rho}\frac{\partial\rho}{\rho}$$

$$\frac{\sigma}{\epsilon}t = -\ln\left(\frac{\rho}{\rho_0}\right)$$

$$\rho(x, y, z, t) = \rho_0(x, y, z)e^{-(\sigma/\epsilon)t}$$

where $\rho_0(x, y, z)$ is the initial value of the charge density at $t = 0$. The initial charge distribution throughout the conductor decays exponentially with time at every point, completely independent of any applied electric fields. If the charge density is initially zero, it remains zero at all times thereafter.

The time $\tau_r = \epsilon/\sigma$ is is referred to as the *relaxation time* and is the time required for the charge at any point to decay to $1/e$ of its original value. The relaxation time is extremely short for good conductors and relatively large for insulators or dielectrics. In fact, whether a material is considered a conductor or an insulator is decided on the basis of the relaxation time. When τ_r is extremely short compared to measurable times or times of interest in a given application, the material is considered to be a conductor; whereas when τ_r is very long, the material behaves like an insulator. For most metals, τ_r is indeed far too short to measure or observe; for example, for copper we have $\sigma = 5.8\times10^7$ S-m^{-1} and $\epsilon = \epsilon_0$,[23] so that $\tau_r^{\text{copper}} \simeq 10^{-19}$ s. In fact, τ_r is very small for all but the poorest of conductors. Even for distilled water, $\tau_r^{H_2O} \simeq 10^{-5}$ s. On the other hand, for a good dielectric τ_r is very large; for example, $\tau_r^{\text{amber}} \simeq 4\times10^3$ s, $\tau_r^{\text{mica}} \simeq 10$ to 20 hours, and $\tau_r^{\text{quartz}} \simeq 50$ days.

Consider a thought experiment in which we suppose that at $t = 0$, charge is concentrated within a small spherical region located near the center of a very large conducting sphere. In every other region of the conductor, the charge density is initially zero. Starting at $t = 0$, the charge within the small spherical region begins to fade away exponentially, but since charge anywhere in the conductor can only *decrease* in time (because of the expression $\rho = \rho_0 e^{-(\sigma/\epsilon)t}$ derived previously), no charge can appear anywhere *within* the conductor. Where then does the charge in the small spherical region go? Because charge is conserved, the exponentially vanishing charge near the center must begin to appear at the surface of the conducting sphere, no matter how great the radius of the conducting sphere may be. However, the

---

[23]The dielectric constant ϵ for a metallic conductor is not easily measurable, since any polarization effect is completely overshadowed by conduction. Nevertheless, based on measurements of reflectivity of metals and the fact that atomic resonances for metals lie in the ultraviolet and x-ray ranges, metallic conductors can be treated as if their dielectric constant is ϵ_0 at frequencies up to and including the visible range (i.e., $\sim 10^{15}$ Hz).

surface charge must make its appearance at the exact instant that the interior charge begins to decay, because the total charge in the system is constant.[24]

Although we are concerned in this chapter with electrostatic fields, the concept of relaxation time is also used to determine the electrical nature (i.e., conductor versus insulator) of materials for time-varying fields. At any given frequency of operation f, a material is considered a good conductor if τ_r is much shorter than the period $T = 1/f$, that is, if $\tau_r \ll T$. Conversely, the material is considered an insulator if $\tau_r \gg T$. We can now see that some materials that are considered to be good conductors at certain frequencies tend to become insulators at sufficiently higher frequencies. For example, seawater ($\sigma = 4$ S-m^{-1}) is considered to be a good conductor at frequencies up to ~100 MHz, but is an insulator for frequencies above ~10 GHz. Further discussion of conducting or dielectric properties of materials at different frequencies is provided in Sections 7.4.4 and 8.1.3.

5.6 BOUNDARY CONDITIONS FOR STEADY CURRENT FLOW

Most applications of steady current flow involve considerations of the interfaces between current-carrying conductors and dielectrics (or insulators), or between two conducting materials of different conductivity. The manner in which the current density **J** and the electric field **E** behave across such interfaces is governed by the *boundary conditions,* which are formulated and discussed in this section. We separately consider conductor–dielectric and conductor–conductor interfaces.

5.6.1 Current and Electric Field at Conductor–Dielectric Interfaces

In Section 4.7, we showed that the surfaces of metallic conductors were equipotentials and that, therefore, the tangential component of electric field on a metallic surface must necessarily be zero (i.e., $E_t = 0$). On the other hand, when a conductor with finite conductivity carries a current, an electric field given by $\mathbf{E} = \mathbf{J}/\sigma$ exists within the medium. At the boundary of any conductor–dielectric interface, the

[24]This interesting observation at first appears to indicate that this phenomenon may be used to transmit signals with infinite velocity. However, in order to convey such a signal at a given moment ($t = 0$), we must, until this moment, prevent charge on the small sphere from being dispersed. This can be done by means of an insulating envelope (e.g., a thin membrane) that might be withdrawn at the given moment to initiate the signal. Before this could happen, however, an induced charge equal and opposite to the charge on the small sphere would appear on the external surface of the envelope, and at the same time an induced charge equal to the original charge in the small sphere would be produced on the surface of the large sphere. The withdrawal of the insulating envelope would merely cause the charges on its two sides to unite (M. Abraham and R. Becker, *Electricity and Magnetism,* Blackie & Son Limited, Glasgow, pp. 260, 272, 1944).

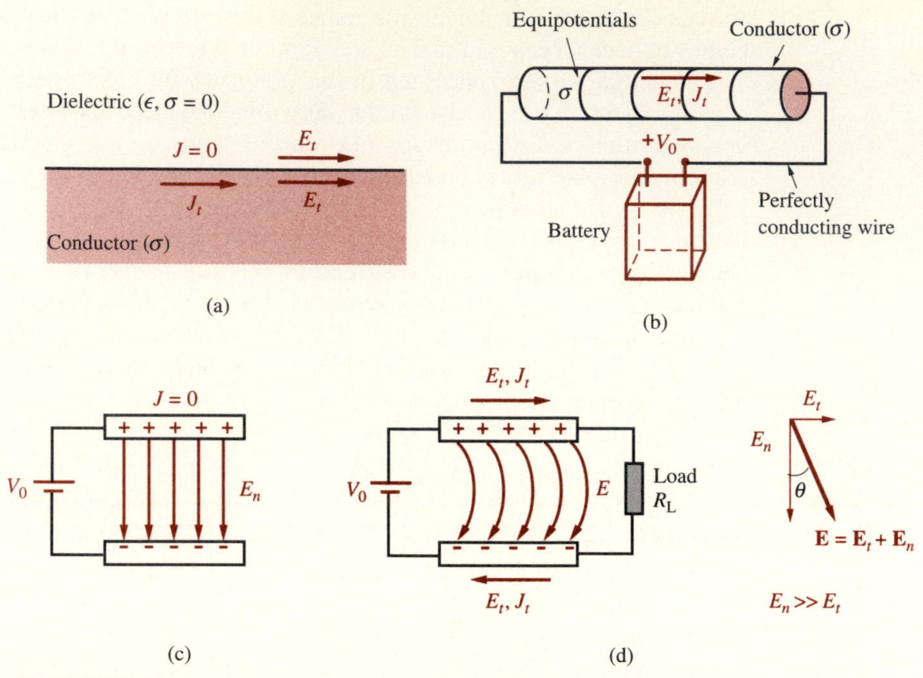

FIGURE 5.12. Conductor–dielectric interface. (a) Current must be tangential to the surface since it cannot flow into the dielectric region. (b) Battery connected to the two ends of a conducting bar. (c) Two metal strips separated by air or another dielectric are maintained at a potential difference of V_0. (d) The connection of a load across the metal strips leads to current flow, which in turn means that the electric fields on the surfaces of the metal strips are not entirely vertical.

current must flow tangentially to the boundary surface, as shown in Figure 5.12a, since current cannot flow across the boundary into the dielectric region with zero conductivity. Thus, on the conductor side of the interface, we have $E_t = J_t/\sigma$. By the continuity of the tangential electric field at a boundary, the tangential field on the dielectric side must also be E_t.

If a potential difference is applied between the two ends of a good (but not perfect) conductor,[25] current flows, and the body of the conductor is no longer an equipotential. Assuming that the two ends of the conducting bar are coated with perfectly conducting material so that they are equipotentials, the potential varies uniformly along such a current-carrying wire, as shown in Figure 5.12b. Assuming that the electric field E is uniform, the potential difference $d\Phi$ over any differential length dl of the wire is $E\,dl$. The integral $\int E\,dl$ carried out over the full length l of the wire is equal to the applied potential difference V_0 and is also given by $V_0 = IR$, where R is the resistance of the wire and I is the total current. The field is

[25]In practice this can be achieved by connecting the wire across a battery, which is rather unwise in most cases, since the resistance of a piece of conducting wire is typically extremely small (much smaller than the internal resistance of the battery), and the battery is shorted.

the same just inside and just outside of the wire and is entirely tangential (i.e., parallel to the axis of the wire).

Note, however, that in practice a conducting wire is very rarely used in the configuration implied in Figure 5.12b, since for typical applied potentials (e.g., $V_0 =$ 1 V from a battery) the current is rather large, and the wire shorts out the battery. As an example, consider a silver wire of radius 1 mm and length 10 cm. The resistance of such a wire is approximately $\sim 5 \times 10^{-4} \Omega$, which means that a current of $I \simeq$ 2000 amperes flows when this single wire is connected across a 1-V battery! In reality, however, any battery has an internal resistance much larger than the resistance of the wire, so that most of the electromotive force of the battery would appear across its internal resistance, resulting in all of the battery power being dissipated internally. This would lead to rapid generation of internal heat, which could cause the battery housing to melt down or burst.

In practice, a common use of conducting materials involves two or more conducting objects that are separated by dielectrics (or free space) and that are at different potentials.[26] In such cases, a surface charge distribution ρ_s exists on the conductor surfaces, and a component of electric field normal to the conductor–dielectric boundary exists, given by $E_n = \rho_s/\epsilon$. The amount of charge on any small portion of the surface of a conductor is equal to the average potential of that portion multiplied by its capacitance to ground or to the appropriate nearby conductor. An example of such a situation is shown in Figure 5.12c, where two metal strips are maintained at potential difference of V_0 by a battery connection. Just inside the conductors, we must have $E_n = 0$ since the the current density $\mathbf{J}$ (and hence $\mathbf{E}$) must be tangential to the boundary, as argued in Section 5.2. If no path is available between the metal strips for current flow (e.g., Figure 5.12c), then the current $\mathbf{J} = 0$, and hence we also have $E_t = 0$, so that the total field inside both conductors is zero.

If a load is now connected between the two conducting strips so that current can flow, the electric field lines no longer terminate on the conductors at right angles but are tilted, as shown in Figure 5.12d. Typically, the tangential electric field E_t inside the conductors is much smaller than the normal electric field E_n due to the external circuit connections. This circumstance is implied by the size of the arrows representing E_t and E_n in Figure 5.12d; however, the size of E_t and thus the slant of the electric field lines are nevertheless greatly exaggerated. In reality, the tilt angle θ of the field lines is much smaller than a degree, since $E_t \ll E_n$. Note that the small value of E_t is consistent with the fact that the strips are relatively good metallic conductors, in which very small fields can establish sizable currents. As an example, consider the strips to be copper ($\sigma = 5.8 \times 10^7$ S-m^{-1}) wires of 2-mm diameter and 1-cm length, the separation of the conductors to be 1 cm, $V_0 = 1$ V, and the load to be an $R_{\text{L}} = 1\Omega$ resistor. Since the wire resistance is entirely negligible compared to the 1Ω load, the current is determined by R_{L} alone and is $I = 1$ A. For a cylindrical

[26]The difference in potentials is due either to intentional application of a potential difference, such as in the case of the outer and inner conductors of a coaxial line, or to the unavoidable proximity of other conductors at different potentials.

wire of 1-mm radius, the current density is then $J = I/(\pi 10^{-6}) \simeq 3.2 \times 10^5$ A-m^{-2}, implying a tangential electric field of $E_t = J/\sigma \simeq 5.5 \times 10^{-3}$ V-m^{-1}. To determine E_n, we can use the expression derived in Section 4.9 for the capacitance per unit length of the two-wire line, namely, $C = (\pi\epsilon_0)/\ln(d/a)$, where in our case $d = 1$ cm and $a = 1$ mm. Substituting values, we find $C \simeq 12$ pF-m^{-1}. Noting that capacitance per unit length is $C = \rho_l/\Phi_{12}$ and with $\Phi_{12} = V_0 = 1$ V, in this case the equivalent line charge density on the conductors is $\rho_l \simeq 1.2 \times 10^{-11}$ C-m^{-1}. Again from Chapter 4, we know that the electric field at a distance r from a line charge density of ρ_l is $E_r = \rho_l/(2\pi\epsilon_0 r)$. Using the value of ρ_l just found, we find that the electric field midway between the wires (i.e., at $r = 0.5$ cm) due to one of the conductors is of order $E_n \simeq 43$ V-m^{-1}. It is thus obvious that $E_n \gg E_t$. In this case, we have $\theta = \tan^{-1}(E_t/E_n) \simeq 0.007°$,[27] so that the tilt of the electric field lines due to the current flow is indeed extremely small.

The charges that reside on the surfaces of wires carrying steady currents, which are induced via the potential differences between the wires and other conductors, thus play a fundamental role in the production of steady currents. The distributed capacitances (also called stray capacitances) by which these charges are induced are the sources of most of the field that supports the electron flow that constitutes the current. The surface charges must of course be in motion as are those inside the wire; however, the amount of charge (i.e., charge density ρ) at any place must remain constant, being maintained by steady currents.

5.6.2 Bending of Current Flow at Interfaces between Two Conductors

The boundary conditions for current flow across an interface between two conducting materials can be derived in a manner entirely analogous to the derivation of the electrostatic boundary conditions that was undertaken in Section 4.11. Figure 5.13 shows the interface between two media of conductivities σ_1 and σ_2. To determine the conditions for the component of the steady currents normal to the interface, we consider a cylindrical Gaussian surface, as shown in Figure 5.13a, and use the fact that

$$\int_V (\nabla \cdot \mathbf{J})\, dv = \oint_S \mathbf{J} \cdot d\mathbf{s} = 0$$

Assuming the height Δh of the cylinder to approach zero ($\Delta h \to 0$), the only nonzero contribution to the surface integral comes from the top and bottom surfaces. In other words,

$$(J_{1n} - J_{2n})\,\Delta s = 0 \quad \rightarrow \quad \boxed{J_{2n} = J_{1n}} \qquad [5.7]$$

[27]Note that the electric field due to the other conductor is in the same direction and essentially doubles the electric field. Also note that for the case of the two-wire configuration, the electric field varies with distance between the conductors, being much larger near the wires. Thus, the tilt of the lines is in fact substantially smaller than the ~0.007° calculated here.

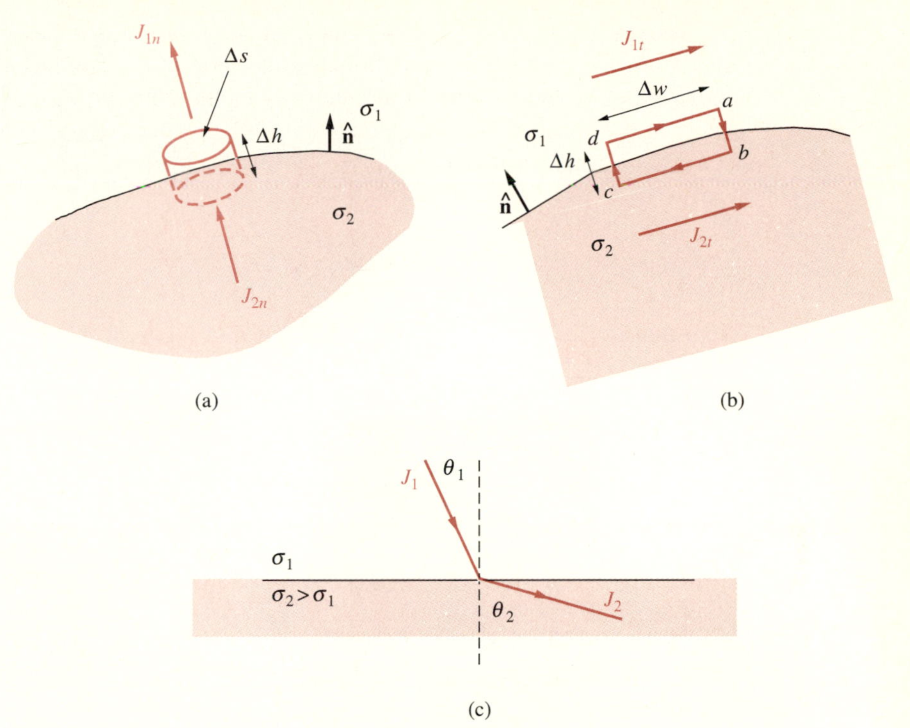

FIGURE 5.13. **The boundary between two different conductors.** (a) A differential pillbox-shaped closed surface. (b) A differential closed contour. (c) The bending of current flow at the boundary for $\sigma_2 > \sigma_1$.

which implies that

$$\sigma_1 E_{1n} = \sigma_2 E_{2n}$$

If medium 2 is a perfect dielectric ($\sigma_2 = 0$), we must have $J_{2n} = 0$, meaning that no current can flow perpendicular to the surface; that is,

$$J_{1n} = 0 \qquad \rightarrow \qquad E_{1n} = 0$$

Thus, the current flow at the surface of a conductor with a perfect insulator must be parallel to the surface, as mentioned earlier.

For the tangential components of the current, we consider the tangential component of the electric field around a closed differential contour, as shown in Figure 5.13b. The fundamental equation we use here is the same as that used in Section 4.11, namely, that the electrostatic field is conservative, or in other words, $\oint_C \mathbf{E} \cdot d\mathbf{l} = 0$. Naturally, the condition we find when we let $\Delta h \rightarrow 0$ is then also the same as in Section 4.11, namely, that $E_{1t} = E_{2t}$. In terms of the tangential components of the current, we then have

$$\boxed{\frac{J_{1t}}{\sigma_1} = \frac{J_{2t}}{\sigma_2}} \qquad [5.8]$$

Together, [5.7] and [5.8] indicate that upon crossing a boundary, the current line bends by an amount proportional to the ratio of the conductivities. Noting the definition of θ_1 and θ_2 from Figure 5.13c, we can write

$$\tan\theta_1 = \frac{J_{1t}}{J_{1n}} \qquad \tan\theta_2 = \frac{J_{2t}}{J_{2n}} = \frac{\sigma_2 J_{1t}}{\sigma_1 J_{1n}}$$

or

$$\tan\theta_2 = \frac{\sigma_2}{\sigma_1}\tan\theta_1$$

Note that if medium 1 is a good conductor and medium 2 is a low-loss dielectric (i.e., $\sigma_1 \gg \sigma_2$), the current enters medium 2 at a right angle to the boundary for practically all angles of incidence from medium 1 (i.e., $J_2 \simeq J_{2n}$). This corresponds to the requirement (noted in Section 4.7) that the electric field is normally incident to the surface of a good conductor.

The condition [5.7] requires in general that the normal component of the electric field be discontinuous across the interface. Using the boundary condition [4.62], we also have

$$\epsilon_1 E_{1n} - \epsilon_2 E_{2n} = \rho_s$$

indicating that the discontinuity of the normal component of electric flux density necessitates a surface charge layer ρ_s at the boundary. Using $\sigma_1 E_{1n} = \sigma_2 E_{2n}$, we can write

$$\rho_s = \left(\epsilon_1\frac{\sigma_2}{\sigma_1} - \epsilon_2\right)E_{2n} = \left(\epsilon_1 - \epsilon_2\frac{\sigma_1}{\sigma_2}\right)E_{1n}$$

Note that the surface charge ρ_s will vanish only for the special case of $\epsilon_2/\epsilon_1 = \sigma_2/\sigma_1$. Alternatively, we can write ρ_s as

$$\rho_s = J_n\left(\frac{\epsilon_1}{\sigma_1} - \frac{\epsilon_2}{\sigma_2}\right) = J_n(\tau_{r1} - \tau_{r2})$$

where τ_{r1} and τ_{r2} are, respectively, the charge rearrangement (or relaxation) times in media 1 and 2. If both media are metallic conductors, we have $\epsilon_1 \simeq \epsilon_2 \simeq \epsilon_0$, so that we have

$$\rho_s = \epsilon_0 J_n\left(\frac{1}{\sigma_1} - \frac{1}{\sigma_2}\right)$$

5.7 DUALITY OF J AND D: THE RESISTANCE–CAPACITANCE ANALOGY

In most materials, both the current density **J** and the electric flux density **D** are linearly proportional to the electric field. As a result, there exists a dual relationship between **J** and **D** in regions where nonconservative fields are not present, that is, in regions outside the interior of the batteries and in the absence of time-varying fields. In this section, we analyze some of these dual relationships.

If current enters and leaves a conducting medium via two "perfect" conductors (also referred to as electrodes), the equivalent resistance is given in terms of the potential difference $\Phi_{12} = \Phi_2 - \Phi_1$ between the perfect conductors as $R = \Phi_{12}/I$, where I is the total current leaving the positively charged electrode. Taking an arbitrary surface completely enclosing the positive electrode, this current is given by

$$I = \oint_S \mathbf{J} \cdot d\mathbf{s} = \sigma \oint_S \mathbf{E} \cdot d\mathbf{s} = \frac{\sigma}{\epsilon} \oint_S \mathbf{D} \cdot d\mathbf{s} = \frac{\sigma Q}{\epsilon}$$

where Q is the total charge induced on the positive electrode due to the capacitance between it and the other (negative) electrode. Noting that this capacitance is given by $C = Q/\Phi_{12}$, we thus have

$$R = \frac{\Phi_{12}}{I} = \frac{\epsilon \Phi_{12}}{\sigma Q} = \frac{\epsilon}{\sigma C}$$

Note that the definition of C depends on the existence of static charge on the electrodes. This charge is proportional to Φ_{12} and is independent of whether or not a current also exists. Note, however, that *all* of the electric flux, and thus current, that leaves one electrode enters the other.

Consider now a pair of electrodes that can be placed either in a dielectric (or free space) or in a conducting medium in which there are no boundaries that would disturb the field and the current patterns. The electric field pattern must be the same in both cases because charge is distributed in the same way on the conductors in each case, since $\nabla \cdot \mathbf{E} = 0$ holds in the intervening space. Thus, if we can find the capacitance for the case when the electrodes are immersed in a dielectric, we can then determine the resistance when the conducting medium is present simply by using the formula $R = \epsilon/(\sigma C)$. Note that since C is proportional to the permittivity ϵ (see, for example, the general definition of capacitance given by [4.49]), it is evident that the resistance R does not depend on the value of ϵ. Similarly, since R is inversely proportional to σ (see [5.2]), the capacitance C does not depend on σ.

This duality relationship can be easily verified by considering the case of a slab of low-conductivity material of thickness d sandwiched between and in contact with parallel planar electrodes of area A, as shown in Figure 5.14b. The resistance of this arrangement is $R = d/(\sigma A)$, and the capacitance of the same configuration when the conducting slab is replaced by a dielectric (Figure 5.14a) is $C = \epsilon A/d$, consistent with $RC = \epsilon/\sigma$.

The duality relationships between linear homogeneous dielectric and conducting media are summarized in Table 5.2 and in Figure 5.14.

In homogeneous, linear, isotropic, and time-invariant materials, an important consequence of the duality is

$$\boxed{RC = \frac{\epsilon}{\sigma} = \tau_r} \qquad [5.9]$$

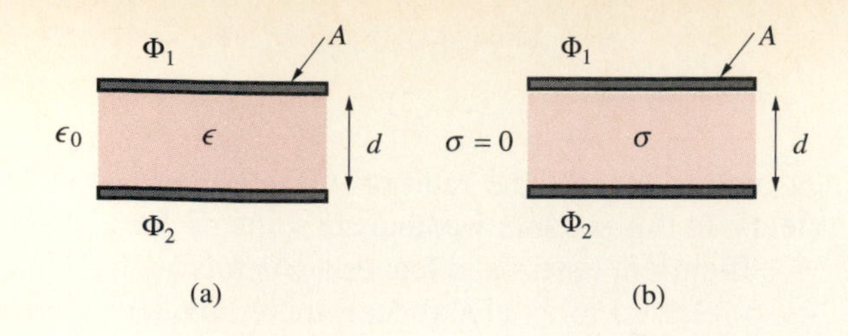

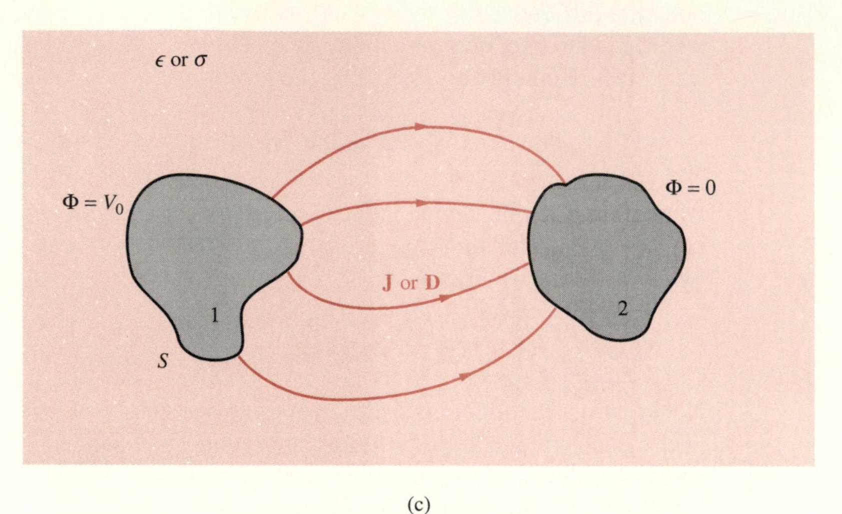

(c)

FIGURE 5.14. Duality between J and D. The fact that the configuration of the electric field lines is determined only by $\oint_C \mathbf{E} \cdot d\mathbf{l} = 0$ brings about the duality between resistance and capacitance between two arbitrary, perfectly conducting bodies (electrodes). (a) A dielectric medium sandwiched between two electrodes. (b) A conducting medium between two electrodes. (c) Two electrodes of arbitrary shape surrounded by a homogeneous medium, either a dielectric or a conductor.

TABLE 5.2 Duality between dielectric and conducting media

Conducting media	Dielectric media
$\oint_C \mathbf{E} \cdot d\mathbf{l} = 0$	$\oint_C \mathbf{E} \cdot d\mathbf{l} = 0$
$\mathbf{J} = \sigma \mathbf{E}$	$\mathbf{D} = \epsilon \mathbf{E}$
$\nabla \cdot \mathbf{J} = 0$	$\nabla \cdot \mathbf{D} = 0$
$\mathbf{J} = \sigma \mathbf{E} = -\sigma \nabla \Phi$	$\mathbf{D} = \epsilon \mathbf{E} = -\epsilon \nabla \Phi$
$J_{1n} = J_{2n}$	$D_{1n} = D_{2n}$
$\sigma_1^{-1} J_{1t} = \sigma_2^{-1} J_{2t}$	$\epsilon_1^{-1} D_{1t} = \epsilon_2^{-1} D_{2t}$
$R = \dfrac{-\int_L \mathbf{E} \cdot d\mathbf{l}}{\oint_S \sigma \mathbf{E} \cdot d\mathbf{s}}$	$C = \dfrac{\oint_S \epsilon \mathbf{E} \cdot d\mathbf{s}}{-\int_L \mathbf{E} \cdot d\mathbf{l}}$

which is consistent with our previous discussions of τ_r as the relaxation time constant of a material.[28] The relation [5.9] can be very useful in deriving expressions for resistance of electrode configurations for which we already have the capacitance or vice versa. For example, we had derived in Example 4.28 that the capacitance of two concentric spheres is

$$C = \frac{4\pi\epsilon ab}{b-a}$$

Using [5.9], we can immediately determine that, if the region between two concentric spheres is filled with a homogeneous conducting material, the resistance between the two spheres is

$$R = \frac{\epsilon}{\sigma C} = \frac{b-a}{4\pi\sigma ab}$$

Note that for $b \to \infty$, this resistance becomes one half of the resistance of the buried hemispherical electrode found in Example 5-4. This result is to be expected, since the geometry of the electric field (and thus that of the current) is exactly the same. But a full spherical electrode at the same potential V_0 has twice as much current flow. In other words, since half of the space in Example 5-4 is free space, half as much current flows for the same potential difference V_0, thus resulting in twice as much resistance.

Examples 5-6 and 5-7 illustrate the use of the resistance–capacitance duality to determine leakage resistance in a coaxial line and in measuring soil conductivity.

Example 5-6: Resistance of a coaxial shell. A cross section of a coaxial line consists of an inner conductor of radius a and an outer shell of radius b separated with a dielectric with ϵ, as shown in Figure 5.15. When the coaxial line is used as a transmission line, the current flows along the inner conductor and returns in the

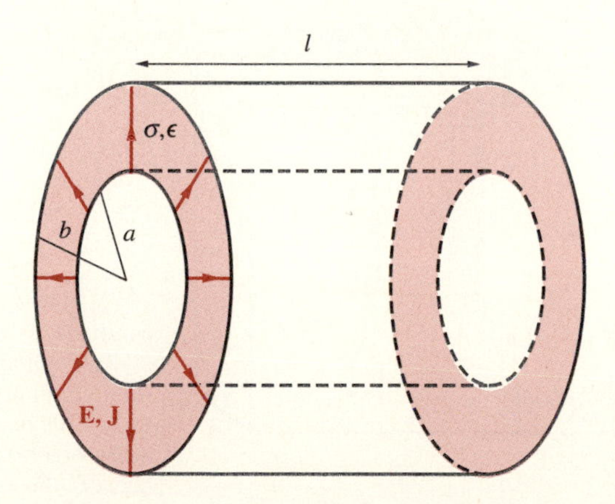

FIGURE 5.15. Resistance of a coaxial shell. When the space between the inner and outer conductors of a coaxial line is filled with lossy material, leakage current **J** flows between the two conductors, as the arrows show.

[28]Note that $\tau_r = \epsilon/\sigma$ is akin to the RC time constant of a first-order capacitive circuit, indicating that the behavior of a lossy dielectric medium under different applied voltage/current conditions can be modeled by such an RC equivalent circuit.

outer shell (or vice versa). The capacitance (per unit length) of this configuration was found in Example 4-27 to be $C = (2\pi\epsilon)/\ln(b/a)$. When the space between radii a and b is filled with imperfect dielectric with conductivity σ, a radial leakage current tends to flow between the inner conductor and the outer shell. Find the resistance that determines this leakage current.

Solution: Since the electric field line configuration is radial in both cases, we can use the duality of **D** and **J** to find the resistance per unit length R. We must have $RC = \epsilon/\sigma$, which yields

$$R = \frac{1}{C}\frac{\epsilon}{\sigma} = \frac{1}{2\pi\sigma}\ln\frac{b}{a} \quad \Omega\text{-m}$$

Note that the leakage resistance of the coaxial shell is identical to that of the thin circular semiconductor contact discussed in Example 5-5 (with $h = 1$ m). This result is to be expected, since the contact in Example 5-5 is simply a coaxial shell that has very small height $h \ll a, b$, instead of having a length l comparable to or larger than the inner and outer radii a and b. Also note that the unit of the resistance per unit length is Ω-m, rather than $\Omega\text{-m}^{-1}$. The reason for this is that longer coaxial shells would have less resistance (i.e., more leakage current flow per unit of applied voltage). In other words, the leakage resistances of two 1-m long segments of coaxial line connected together are connected in parallel, since they share the same voltage.

Example 5-7: Measuring soil conductivity. Conductivity of soil is an important parameter in a number of applications such as the design of AM broadcast systems, grounding electrical equipment, and locating buried objects in the ground. A simple experiment is proposed[29] to determine the conductivity of soil by using two parallel cylindrical metal electrodes, as shown in Figure 5.16a, that are buried

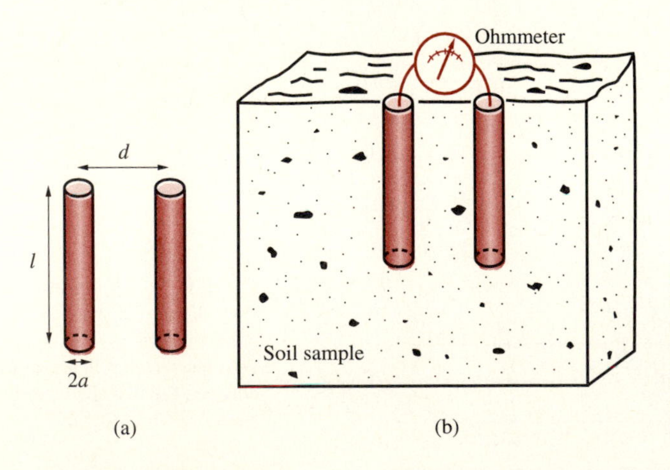

FIGURE 5.16. Soil conductivity measurement. (a) Two cylindrical electrodes. (b) Soil sample and electrodes in a plastic bucket several times larger than the separation between the electrodes.

[29] A. D. Wunsch, A backyard experiment for electromagnetics I, *IEEE Transactions on Education*, 34(1), pp. 142–144, February 1991.

in the soil. The soil and the electrodes are placed in a nonconducting container with diameter several times greater than the spacing between the two electrodes, as shown in Figure 5.16b. The electrodes are connected to an ohmmeter, which is used to measure the ohmic resistance R between the ends of the electrodes. The corresponding conductivity of the soil can then be calculated from the value of R using the geometric dimensions of the electrodes. A student is given a pair of copper rods, each of which is 5 mm in diameter and 15 cm in length, and an ordinary ohmmeter to conduct this experiment as an assignment in her backyard at home. The student conducts three sets of measurements at three different electrode spacings, with results summarized as follows:

d (cm)	R (kΩ)
2	1.47
4	2.10
8	3.04

Find the average conductivity of the soil.

Solution: In Chapter 4, Example 4-29, the per-unit-length capacitance of a two-wire line with diameter $2a$ and separation d was shown to be approximately given by

$$C = \frac{\pi\epsilon}{\ln(d/a)} \quad \text{F-m}^{-1}$$

for the case when $d \gg a$. Using the duality relationship $RC = \epsilon/\sigma$, we can write the per-unit-length resistance of the two electrodes embedded in the soil as

$$R = \frac{\epsilon}{C\sigma} = \frac{\ln(d/a)}{\pi\sigma} \quad \Omega\text{-m}^{-1}$$

From this, we can write the soil conductivity in terms of the measured resistance R between the two electrodes, each of length l, as

$$\sigma = \frac{\ln(d/a)}{\pi R l} \quad \text{S-m}^{-1}$$

where, in our case, $a = 0.25$ cm and $l = 0.15$ m. Thus, for the first measurement, with $d_1 = 2$ cm, we have

$$\sigma_1 = \frac{\ln(2 \text{ cm}/0.25 \text{ cm})}{\pi(1.47 \times 10^{-3}\Omega)(0.15 \text{ m})} \simeq 3.00 \times 10^{-3} \text{ S-m}^{-1}$$

Similarly, for $d_2 = 4$ cm, we have $\sigma_2 \simeq 2.80 \times 10^{-3}$ S-m^{-1}, and for $d_3 = 8$ cm, we have $\sigma_3 \simeq 2.42 \times 10^{-3}$ S-m^{-1}. Therefore, the average conductivity based on the three sets of measurements is

$$\sigma = \frac{\sigma_1 + \sigma_2 + \sigma_3}{3} \simeq 2.74 \times 10^{-3} \text{ S-m}^{-1}$$

5.8 JOULE'S LAW

Based on our understanding of the microscopic picture of conduction discussed in Section 5.1, steady current flow in a material must necessarily lead to dissipation of power, as individual current carriers (electrons) are accelerated by the applied field and transfer their energy to the material medium via collisions. In this section, we discuss the power dissipation resulting from steady current flow and express it in terms of such macroscopic parameters as the electric field **E** and current density **J**.

The energy required to maintain steady current flow can be determined by considering the definition of electrostatic potential. As current flows along a piece of wire of any shape, positive charge moves to lower values of potential, so that work is continually being done by the electric field. For each coulomb of charge that moves through one volt drop in potential, one *joule* of work is performed. The work done to move a charge Q through a potential difference $\Delta\Phi$ is given by $W = Q\Delta\Phi$. The rate at which the work is done is the *power* P expended, which is given by

$$\frac{dW}{dt} = P = \Delta\Phi\frac{dQ}{dt} = \Delta\Phi I$$

or, since $\Delta\Phi = IR$ by Ohm's law, we have

$$\boxed{P = I^2 R} \qquad [5.10]$$

in joules/second, or *watts*. Relation [5.10] is the well-known *Joule's law.*[30] Since in steady flow no kinetic energy is gained (i.e., the electron drift velocity $\mathbf{v}_d$ is constant in time) and no charges are redistributed, this power must all become heat; such dissipation of electrical energy is referred to as *Joule heating*. Note that even though we might reverse the polarity of voltage or the direction of current, P remains positive; thus, Joule heating is irreversible.

The expression [5.10] cannot be applied at a particular point. However, power dissipated per unit volume can be defined *at* a point by formulating a differential version of Joule's law. Consider a differential volume element, as shown in Figure 5.17a. The potential difference between the two ends of such an element is given by

[30] In 1841, J. P. Joule conducted a series of experiments by coiling wires of different lengths, cross sections, and composition into thin glass tubes, and then immersing the assemblies in separate containers filled with measured quantities of water. When the same intensity of steady current was passed through the different coils, the water was found to heat up to an equilibrium temperature that differed among the several containers, but in such a way that the change in temperature was proportional to the resistances of the coils. This led Joule to conclude that "...when a given quantity of voltaic electricity is passed through a metallic conductor for a given length of time, the quantity of heat evolved by it is always proportional to the resistance which it presents, whatever may be the length, thickness, shape, or kind of that metallic conductor" (J. P. Joule, On the heat evolved by metallic conductors of electricity, *Phil. Mag.*, August 1841, 19, pp. 260–265.) For further discussion, see Chapter 8 of R. S. Elliott, *Electromagnetics*, IEEE Press, Piscataway, NJ, 1993.

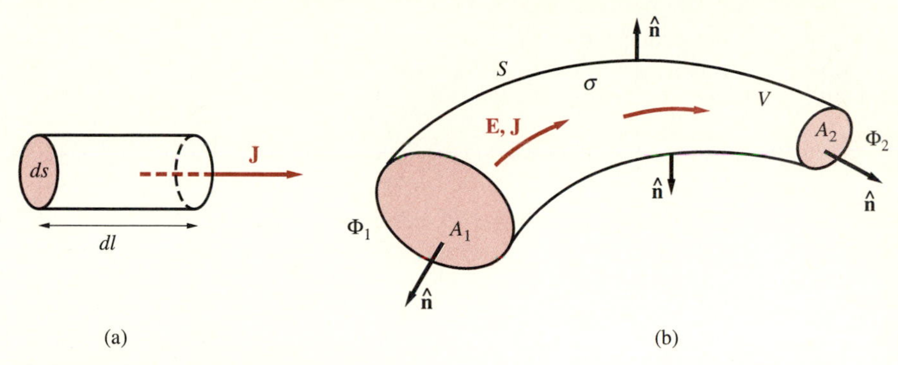

FIGURE 5.17. **Joule's law.** (a) A differential resistor element. (b) A resistor of arbitrary shape.

$$d\Phi = \mathbf{E} \cdot d\mathbf{l} = \frac{\mathbf{E} \cdot \mathbf{J}}{J} dl$$

where $\mathbf{E}$ is the electric field within the element and $d\mathbf{l} = (\mathbf{J}/J)\, dl$. The total current through the differential resistor element $dI = J\, ds$. Thus, according to Joule's law, the power dissipated in this differential element is

$$dP = dI\, d\Phi = \underbrace{\mathbf{E} \cdot \mathbf{J}}_{(\text{V-m}^{-1})(\text{A-m}^{-2})} \overbrace{dl\, ds}^{\text{m}^3} \quad \rightarrow \quad (\text{volts})(\text{amperes}) = \text{watts}$$

Thus, the differential form of Joule's law is

$$\frac{dP}{dv} = \mathbf{E} \cdot \mathbf{J} \quad \text{watts-m}^{-3}$$

so that $\mathbf{E} \cdot \mathbf{J}$ represents the power dissipated per unit volume. In other words, the electric field gives up $(\mathbf{E} \cdot \mathbf{J})$ watts-m^{-3} to the steady current flow of $\mathbf{J}$. This energy is converted into heat in the conducting material.

By considering an arbitrarily shaped resistor, as shown in Figure 5.17b, we can now derive a general expression for the power dissipated. With surfaces A_1, A_2 kept, respectively, at potentials Φ_1, Φ_2 (presumably because they are coated with material of conductivity much greater than σ), and noting that the current $\mathbf{J}$ flows normal to these end surfaces and parallel to the sides of the conductor, we have

$$P = I^2 R = \int_V \mathbf{E} \cdot \mathbf{J}\, dv = \int_V \frac{\mathbf{J} \cdot \mathbf{J}}{\sigma} dv = \int_V \sigma \mathbf{E} \cdot \mathbf{E}\, dv$$

and thus

$$R = \frac{P}{I^2} = \frac{\int_V \mathbf{E} \cdot \mathbf{J}\, dv}{I^2}$$

which gives the total resistance between the terminal faces of the resistor in terms of the total power dissipated. This expression for resistance is to be compared with equation [5.2], which expresses R in terms only of the electric field.

Joule, James Prescott *(b. December 24, 1818, Salford, Lancashire; d. October 11, 1889, Sale, Cheshire) Joule was the second son of a wealthy brewer, so he had the means to devote himself to a life of research. He also suffered poor health as a youngster, having a spinal injury, which meant he could withdraw to his books and studies. His father encouraged him and supplied him with a home laboratory. Like Faraday [1791–1867], Joule lacked in the area of mathematical rigor, but was almost a fanatic on the subject of measurement. In his teens, he was already publishing papers on measurements of heat in electric motors. By 1840 he had worked out Joule's law: The heat produced by an electric current is proportional to the square of the current intensity multiplied by the resistance of the circuit.*

He went on to devote a decade to measuring the heat produced by every process he could think of. He churned water and mercury with paddles. He passed water through small holes to heat it by friction. He expanded and contracted gases. In all cases he calculated the amount of work entering the system and the amount of heat exiting the system and found, as Rumford [1753–1814] had maintained fully half a century before, that a particular quantity of work always produced a particular quantity of heat. Joule's first full description of his experiments and his conclusion appeared in 1847 but were not well received by most scientists at the time. This may have been partly due to Joule's being a brewer and not an academician and partly to the fact that his results were based on small temperature differences in many cases (he used thermometers that could be read to 0.02°F).

Lord Kelvin [1824–1907] was the first to recognize the importance of Joule's work, and later Stokes [1819–1903] also supported Joule's work with enthusiasm. Full recognition came in 1849 when Joule read a paper on his work before the Royal Society, with Faraday himself as his sponsor. The concept of heat as a form of energy also forms the basis of the law of conservation of energy. Although Joule recognized the principle of conservation of energy, as did Mayer [1814–1878] before him, the first to present it to the world as an explicit generalization was Helmholtz [1821–1894], and it is usually Helmholtz who is given credit for its discovery.

Joule was elected to the Royal Society in 1850, received its Copley medal in 1866, and was president of the British Association for the Advancement of Science in 1872 and in 1887. Toward the end of his life he suffered economic reverses, but Queen Victoria granted him a pension in 1878. He was a modest and unassuming man, sincerely religious, and he regretted the increasing application of scientific discoveries to the art of warfare. [Adapted with permission from Biographical Encyclopedia of Science and Technology, *I. Asimov, Doubleday, 1982].*

1700 *1818* *1889* *2000*

5.9 SUMMARY

This chapter discussed the following topics:

- **Current density and conduction.** A constant electric field **E** in a conducting medium leads to steady current flow, with the current density **J** given by

$$\mathbf{J} = n_c q_e \mathbf{v}_d = \frac{n_c q_e^2 \bar{l}}{m_e v_{\text{th}}}\mathbf{E} \qquad \rightarrow \qquad \mathbf{J} = \sigma \mathbf{E}$$

  where parameters such as the number of conduction band electrons n_c, the mean free path $\bar{l}$, and thermal velocity v_{th} depend on the material properties; q_e and m_e are the charge and mass of the electron; and σ is the conductivity of the material. Microscopically, conduction occurs as the electrons, which are in a state of random agitation, slowly drift in the direction of the applied electric field with velocity $\mathbf{v}_d$, where $|\mathbf{v}_d| \ll v_{\text{th}}$.

- **Resistance and Ohm's law.** When the space between two perfectly conducting electrodes is filled with a conducting material, the resistance between the electrodes is given as

$$R = \frac{\Phi_{12}}{I} = \frac{-\int_L \mathbf{E} \cdot d\mathbf{l}}{\int_A \sigma \mathbf{E} \cdot d\mathbf{s}}$$

  where Φ_{12} is the potential difference between the two electrodes, I is the total current through them, A is the cross-sectional area of any of the electrodes, and L is any path from the electrode 1 at a lower potential to electrode 2 at a higher potential. This relation between Φ_{12} and I is also known as Ohm's law.

- **Kirchhoff's voltage law.** Steady current flow can be maintained only by an external energy source (e.g., a battery) supplying electromotive force. In general, for any closed circuit containing many resistors and emf sources, we have

$$\sum \mathcal{V}_{\text{emf}} = I \sum R$$

- **Continuity equation and Kirchhoff's current law.** The fundamental principle of conservation of electric charge indicates that the net inflow of current through a closed surface must be equal to the net rate of charge increase in the enclosed volume. The differential form of this fundamental principle is known as the continuity equation:

$$\nabla \cdot \mathbf{J} + \frac{\partial \rho}{\partial t} = 0$$

  For steady currents, we must thus have

$$\nabla \cdot \mathbf{J} = 0 \quad \text{or} \quad \oint_S \mathbf{J} \cdot d\mathbf{s} = 0$$

indicating that steady current must flow in closed loops. A more commonly known version of this result is Kirchhoff's current law, which states that the algebraic sum of all currents at a junction is zero.

- **Redistribution of free charge.** The ratio of permittivity of a material to its conductivity is known as the relaxation time $\tau_r = \epsilon/\sigma$, which defines the time period over which free charge is redistributed in the material. For metals, this time is extremely small, being of the order $\sim 10^{-19}$ s; for good dielectrics τ_r can be on the order of hours or days.
- **Boundary conditions for steady currents.** The component of current density normal to the boundary between two different conducting materials is continuous across the interface. The continuity of the tangential component of the electric field across any such interface indicates that the tangential component of current density is not continuous. In summary, we have

$$J_{1n} = J_{2n} \qquad \text{and} \qquad \frac{J_{1t}}{\sigma_1} = \frac{J_{2t}}{\sigma_2}$$

- **The resistance–capacitance analogy.** The resistance R and capacitance C between two perfectly conducting electrodes in a linear, homogeneous, and isotropic medium is related by

$$RC = \frac{\epsilon}{\sigma} = \tau_r$$

Many of the fundamental relationships for current density $\mathbf{J}$ are dual to those for the electric flux density $\mathbf{D}$, as summarized in Table 5.2.

- **Power dissipation and Joule's law.** The flow of a steady current in a material leads to dissipation of power in the form of heat. The rate at which work is done and power expended is $P = I^2R$, where I is the total current flowing through a resistance R. At a differential level, the volume density of power dissipation, in units of watts-m^{-3}, is represented by $\mathbf{E} \cdot \mathbf{J}$.

5.10 PROBLEMS

5-1. Current in a wire. If a current of 1 A is flowing in a conductor wire, find the number of electrons that pass through its cross section each second.

5-2. Current in a copper wire. If a copper wire of cross-sectional area 1 mm^2 is carrying a current of 1 A, find the average drift velocity of the conduction electrons in the wire. The density of conduction electrons in copper is $n_c = 8.45 \times 10^{28}$ el-m^{-3}.

5-3. Electric field in an aluminum wire. An aluminum wire of conductivity $\sigma = 3.82 \times 10^7$ S-m^{-1} is carrying a uniform current of density 100 A-cm^{-2}. (a) Find the electric field in the wire. (b) Find the electric potential difference per meter of wire.

5-4. Silver versus porcelain. The conductivity of silver is $\sigma = 6.17 \times 10^7$ S-m^{-1}, and that of porcelain is about $\sigma \simeq 10^{-14}$ S-m^{-1}. (a) If a voltage of 1 V is applied across a silver plate of 1 cm thickness, find the current density through it. Assume a uniform electric field. (b) Repeat part (a) for a porcelain plate of same thickness, and compare your results.

5-5. Electron mobility in copper. The average drift velocity of free electrons in a metal is proportional to the applied electric field approximately given by

$$v_d \simeq \frac{q_e t_c}{m_e} E = \mu_e E$$

where μ_e is a proportionality factor called the mobility of the electron. Since $v_d = \mu_e E$, the mobility describes how strongly the motion of an electron is influenced by an applied electric field. The conductivity of copper is $\sigma = 5.8 \times 10^7$ S-m^{-1} at room temperature and is due to the mobility of electrons that are free (one per atom) to move under the influence of an electric field. (a) Find the electron mobility in copper at room temperature and compare it with the mobility values of pure silicon and germanium (see Problem 5-9). (b) Find the average drift velocity of the electrons in the direction of the current flow in a copper wire of 1-mm diameter carrying a current of 1 A.

5-6. Resistance of a short, small-diameter conductor. A technique has been developed to measure the dc resistance at room temperature of conductors that are short in length and small in diameter.[31] To demonstrate this system, the resistance of commercial bare copper wire ($\sigma = 5.85 \times 10^7$ S-m^{-1} at 20°C) was measured in four different diameters. The manufacturer's reported diameters for these wires were 25.4 μm, 20.3 μm, 12.7 μm, and 7.6 μm. The resistances per unit length of these wires were measured to be 0.340, 0.527, 1.22, and 2.56, all in Ω-cm^{-1}. (a) Calculate the actual diameter of each wire using the measured resistance values. (b) Find the percentage of error in the reported diameter values by comparing them to the calculated values. (Note that the uniformity and the surface condition of each wire were examined with a scanning electron microscope. All of the wires had uniform diameters, and only the 7.6-μm diameter wire showed significant surface cracking and roughness. So a larger error for the 7.6-μm diameter wire is most likely due to its degraded surface condition.)

5-7. Resistance of a copper wire. Consider a copper wire of 1-mm diameter. Find its resistance per unit length if the wire temperature is (a) −20°C; (b) 20°C; and (c) 60°C. For copper, the conductivity is $\sigma = 5.8 \times 10^7$ S-m^{-1} at 20°C.

5-8. Resistance of a copper wire. Consider a copper wire at 20°C. To what temperature must it be heated in order to double its resistance?

5-9. Intrinsic semiconductor. A semiconductor is said to be intrinsic (pure) when it is free of any dopant impurity atoms. The only mobile carriers are those caused by thermal excitation, a process that creates an equal number of holes and electrons. The intrinsic charge carrier concentration N_i is a strong function of temperature. Under most conditions, it is given by

$$N_i^2 = N_c N_v e^{-E_g/k_B T}$$

where N_c and N_v are related to the density of allowed states near the edges of the conduction and valence bands, E_g is the energy gap between the conduction and the valence bands, k_B is Boltzmann's constant $k_B \simeq 1.38 \times 10^{-23}$ J-°K^{-1}, and T is the temperature in K. The conductivity of a semiconductor sample is given by

$$\sigma = |q_e|(\mu_e N_e + \mu_p N_p)$$

---

[31] C. A. Thompson, Apparatus for resistance measurement of short, small-diameter conductors, *IEEE Trans. Instrumentation and Measurement,* 43(4), pp. 675–677, August 1994.

where $|q_e| \simeq 1.6 \times 10^{-19}$ C; μ_e and μ_p are the electron and the hole mobilities, which are both functions of temperature; and N_e and N_p are the electron and hole concentrations, which, in the case of a pure semiconductor, are both equal to the intrinsic charge carrier concentration; that is, $N_e = N_p = N_i$. (a) For a bulk sample of pure silicon (Si) at 300K, $N_c = 2.8 \times 10^{19}$ cm^{-3}, $N_v = 1.04 \times 10^{19}$ cm^{-3}, $E_g = 1.124$ eV (where 1 eV $\simeq 1.602 \times 10^{-19}$ J), $\mu_e = 1450$ cm^2-(V-s)$^{-1}$, and $\mu_p = 450$ cm^2-(V-s)$^{-1}$. Calculate the intrinsic carrier concentration (in cm^{-3}) and the conductivity (in S-m^{-1}) of the Si sample at 300K. (b) Repeat part (a) for a bulk sample of pure germanium (Ge). For Ge at 300K, $N_c = 1.04 \times 10^{19}$ cm^{-3}, $N_v = 6.0 \times 10^{18}$ cm^{-3}, $E_g = 0.66$ eV, $\mu_e = 3900$ cm^2-(V-s)$^{-1}$, and $\mu_p = 1900$ cm^2-(V-s)$^{-1}$.

5-10. Extrinsic semiconductor. Semiconductors in which conduction results primarily from carriers contributed by impurity atoms are said to be extrinsic (impure). The impurity atoms, which are intentionally introduced to change the charge carrier concentration, are called dopant atoms. In doped semiconductors, only one of the components of the conductivity expression in Problem 5-9 is generally significant because of the very large ratio between the two carrier densities N_e and N_p, the product of which is always equal to the square of the intrinsic charge carrier concentration, that is, $N_e N_p = N_i^2$. In addition, the mobility values of the charge carriers vary with different doping levels. (a) Phosphorus donor atoms with a concentration of $N_d = 10^{16}$ cm^{-3} are added uniformly to a pure sample of Si (which is called n-type Si since the electrons of the phosphorus atoms are the majority carriers, that is, $N_e \simeq N_d \gg N_p$). Find the conductivity of the n-type Si sample at 300K. For the mobility of majority charge carriers, use $\mu_e = 1194$ cm^2-(V-s)$^{-1}$. (b) Repeat part (a) if boron acceptor atoms with a concentration of $N_a = 10^{16}$ cm^{-3} are added to a pure sample of Si (which is called p-type Si since the holes are now the majority carriers, that is, $N_p \simeq N_a \gg N_e$). For the mobility of majority charge carriers, use $\mu_p = 444$ cm^2-(V-s)$^{-1}$. (c) Compare your results in parts (a) and (b) with the conductivity of the intrinsic Si and comment on the differences.

5-11. A silicon resistor. A silicon bar 1 mm long and 0.01 mm^2 in cross-sectional area is doped with $N_d = 10^{17}$ cm^{-3} arsenic (As) atoms. Find the resistance of the bar and compare with the resistance of the same bar made of pure silicon (see Problem 5-9). For the mobility of the free arsenic electrons, use $\mu_e = 731$ cm^2-(V-s)$^{-1}$.

5-12. A silicon resistor. A sample of p-type silicon is 8 μm long and has a cross-sectional area of 2.5 μm×2.5 μm. If the measurements show that the average hole concentration and the resistance of the sample are $N_d = 10^{16}$ cm^{-3} and $R = 18$ kΩ respectively, find the hole mobility μ_p.

5-13. Sheet resistance R_{sq}. Consider the resistance of a uniformly doped n-type Si layer of length l, width w, and thickness t, as shown in Figure 5.18. This resistor can be divided

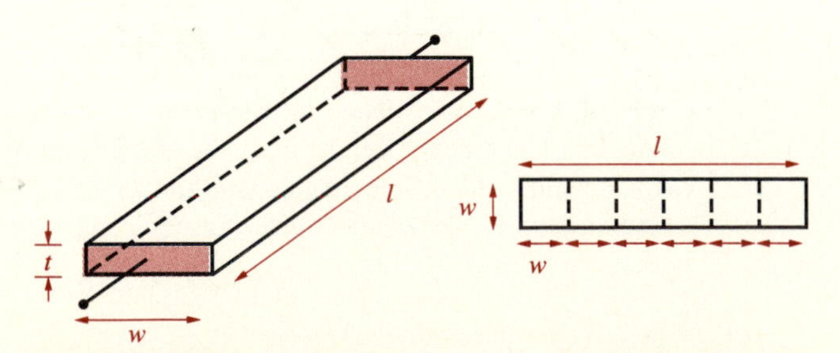

FIGURE 5.18. **Sheet resistance.** Problem 5-13.

into square sheets of dimension w on each side, as shown. The resistance of any one of these square sheets is called the sheet resistance, denoted by the symbol R_{sq}, in units of Ω-(sq)$^{-1}$. (a) Show that the sheet resistance is given by

$$R_{sq} = \frac{1}{|q_e|\mu_e N_e t}$$

(b) Show that the total resistance of the Si layer is given by

$$R = \frac{l}{w} R_{sq}$$

(c) Calculate the total resistance of a Si layer that has a length of 50 μm, width of 5 μm and a sheet resistance of 150Ω-(sq)$^{-1}$.

5-14. Integrated-circuit (IC) resistor. Consider an integrated-circuit (IC) resistor of length l, width w, and thickness t that is made of Si doped with 10^{16} cm^{-3} phosphorus atoms (i.e., $N_e \simeq 2.5 \times 10^{16}$ cm^{-3}). Given $t = 2$ μm, find the aspect ratio, w/l, such that the resistance of the IC resistor is 10 kΩ at 300K. Take $\mu_e = 1000$ cm^2-(V-s)$^{-1}$.

5-15. An ion-implanted IC resistor. An ion-implanted n-type resistor layer with an average doping concentration $N_d = 4 \times 10^{17}$ cm^{-3} is designed with 1 μm thickness and 2 μm width to provide a resistance of 3 kΩ for an IC chip. Find the sheet resistance and the required length of the resistor. Assume the electron mobility as $\mu_e = 450$ cm^2-(V-s)$^{-1}$.

5-16. Diffused IC resistor. An IC resistor is frequently fabricated by diffusing a thin layer of p-type impurity into an n-type isolation island, as shown in Figure 5.19. If contacts are made near the two ends of the p-type region and a voltage is applied, a current will flow parallel to the surface in this region. It is not possible to use $R = l/(\sigma A)$, however, to calculate the resistance of this region because the impurity concentration in it is not uniform. The impurity concentration resulting from the diffusion process is maximum near the surface ($x = 0$) and decreases as one moves in the x direction. (a) Show that the sheet resistance of the p-type layer is given by

$$R_{sq} = \left[\int_0^t \sigma(x)\, dx\right]^{-1}$$

(b) Assuming the conductivity of the p-type layer decreases linearly from σ_0 at the surface ($x = 0$) to $\sigma_1 \ll \sigma_0$ at the interface ($x = t$) with the n-type wafer, find the sheet resistance,

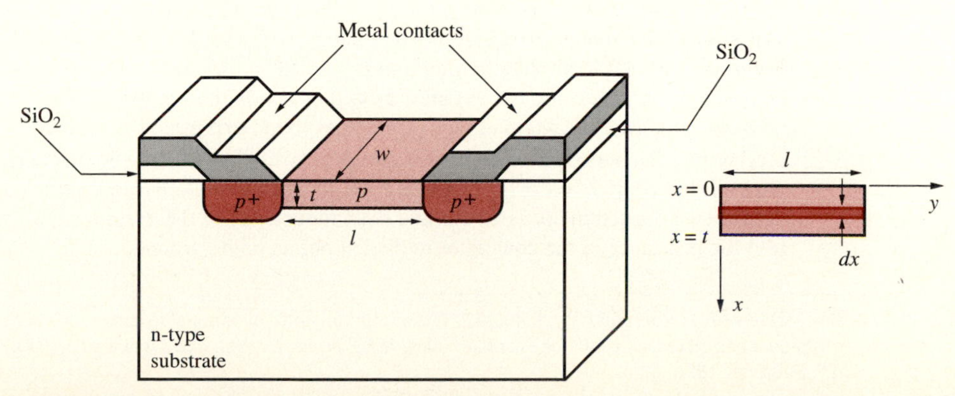

FIGURE 5.19. Diffused IC resistor. Problem 5-16.

R_{sq}. (c) Repeat part (b) if the conductivity decreases exponentially from σ_0 at $x = 0$ to $\sigma_1 \ll \sigma_0$ at $x = t$.

5-17. Resistance of a copper-coated steel wire. A stainless steel wire ($\sigma = 1.11 \times 10^6$ S-m^{-1}) 3 mm in diameter is to be coated with copper ($\sigma = 5.8 \times 10^7$ S-m^{-1}) in order to reduce its resistance per unit length by 50%. Find the thickness of the copper coating needed to achieve this goal.

5-18. Resistance of an aluminum conductor, steel-reinforced (ACSR) wire. A power utility company uses a steel-reinforced aluminum conductor (ACSR) wire of 3-cm diameter as an extra-high-voltage (EHV) transmission line. It is made of aluminum ($\sigma = 3.82 \times 10^7$ S-m^{-1}) with an inner core of stainless steel ($\sigma = 1.11 \times 10^6$ S-m^{-1}) along the center axis such that the steel content of the ACSR wire is only 10%. (a) Find the resistance per kilometer of the wire at 20°C. (b) If a current of 1000 A flows in the wire, find the current that flows in each metal. (Neglect the changes in the conductivity values due to changes in temperature because of the current flow.) (c) If the ACSR wire between two EHV towers is 300 m long, find the voltage drop between the two towers. (d) Repeat parts (a), (b), and (c) if the steel content of the ACSR wire is increased to 25%.

5-19. Conductivity of lunar soil. A simulated version of the lunar soil obtained from the Apollo 11 site on the moon is investigated as a possible material to be used for electrical insulation in high-voltage power systems in space.[32] A sample of this soil is placed as an insulator between the electrodes of a parallel-plate capacitor with 1 mm separation and 10 cm^2 area, and dc voltages are applied across the electrodes. Leakage currents of 10 nA, 20 nA, 30 nA, 40 nA, and 50 nA are recorded for electric field strengths of 2.5 kV/mm, 3 kV/mm, 3.6 kV/mm, 4 kV/mm, and 4.2 kV/mm respectively. (a) Using Ohm's law, find the conductivity of the lunar soil in each case. Note that the conductivity of this soil varies with the applied field. (b) Using a curve-fitting technique, the conductivity of the soil is approximated as

$$\sigma \simeq \frac{4^{(2E-5)/3}}{10^{14}E}$$

where σ is in S/mm and E is in kV/mm. Using this expression, find the value of σ and its percentage difference from its corresponding value found in part (a) for each case.

5-20. Resistance of a semicircular ring. The ends of a semicircular conductor ring of rectangular cross section are connected to a dc battery, as shown in Figure 5.20. (a) Write the total resistance R of the conductor using the result of Example 5-3. (b) Find R by assuming the conductor to be a straight conductor with a length $l = \pi(a+b)/2$. (c) Using $\sigma = 7.4 \times 10^4$ S-m^{-1} (graphite) and $b = 1.5a = 10t = 3$ cm, find R using both expressions obtained in parts (a) and (b) and compare the results. (d) Show that for $a \gg (b-a)$, the expression of part (a) reduces to the expression of part (b).

5-21 Resistance of a hemispherical conductor. A perfectly conducting hemispherical conductor is buried in the earth to achieve a good ground connection (see Figure 5.7). If the diameter of the conductor is 25 cm and the conductivity of the ground is 10^{-4} S-m^{-1}, find the resistance of the conductor to distant points in the ground.

[32] H. Kirkici, M. F. Rose, and T. Chaloupka, Experimental study on simulated lunar soil: high voltage breakdown and electrical insulation characteristics, *IEEE Trans. Dielect. and Elect. Insul.*, 3(1), pp. 119–125, February 1996.

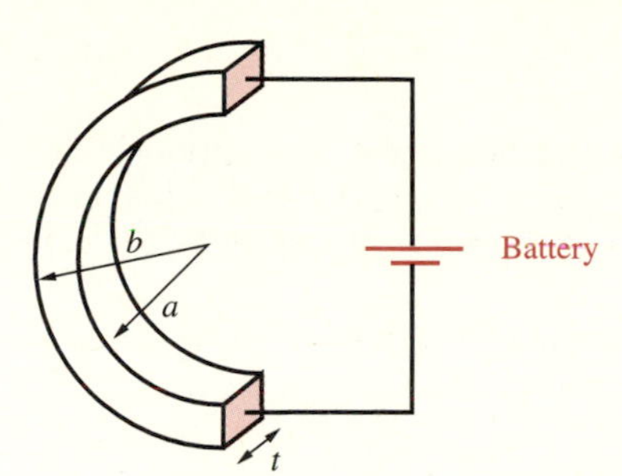

FIGURE 5.20. Semicircular ring. Problem 5-20.

5-22. Resistance of a toroidal conductor. Consider a segment of a toroidal (doughnut-shaped) resistor with a horizontal cross section, as shown in Figure 5.21. Show that the resistance between the flat ends having a circular cross section is given by

$$R = \frac{\phi_0}{\sigma\pi(\sqrt{b} - \sqrt{a})^2}$$

5-23. Leakage resistance. A 2-mm-diameter copper wire is enclosed in an insulating sheath of 4-mm outside diameter. If the wire is buried in a highly conducting ground, what is the leakage resistance per kilometer of the sheath to the ground? The sheath can be assumed to have a conductivity of 10^{-8} S-m^{-1}.

5-24. Pipeline resistance. Two parallel steel pipelines have centers spaced 10 m apart. The pipes are half buried in the ground, as shown in Figure 5.22. The diameter of the pipes is 1 m. The ground in which the pipes lie is marshy soil ($\sigma \simeq 10^{-2}$ S-m^{-1}). Find the resistance per kilometer between the two pipes.

5-25. Ground current. A vertical lightning rod discharges 10^5 amperes into the ground. Find the voltage produced by the discharge between two points 1 m apart on a radial line if the point nearest the rod is at a distance of (a) 3 m, and (b) 10 m from the rod. Assume the ground to consist of a 20-cm layer of conducting soil with $\sigma \simeq 10^{-2}$ S-m^{-1}.

5-26. Inhomogeneous medium. Consider an inhomogeneous medium in which both ϵ and σ are functions of position. Show that a steady current **J** flowing though such a medium would necessarily establish a charge distribution given by $\rho = -[\nabla\epsilon -$

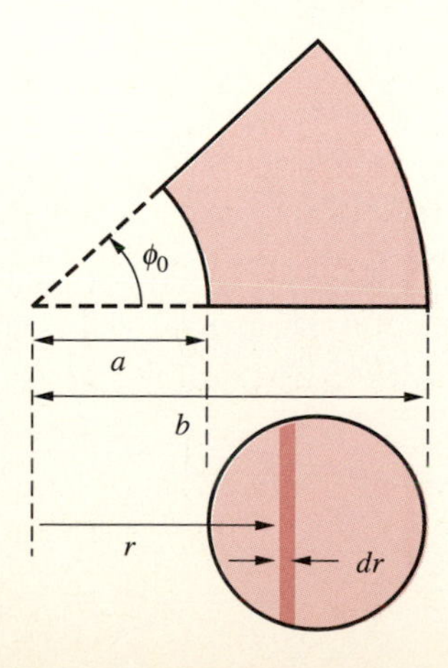

FIGURE 5.21. Toroidal conductor. Problem 5-22.

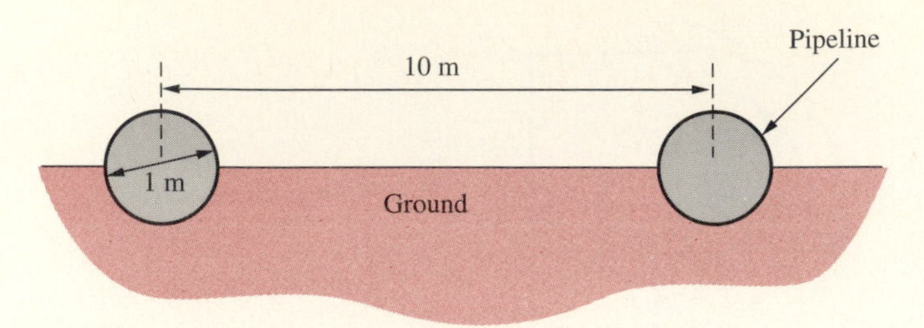

FIGURE 5.22. Pipeline resistance. Problem 5-24.

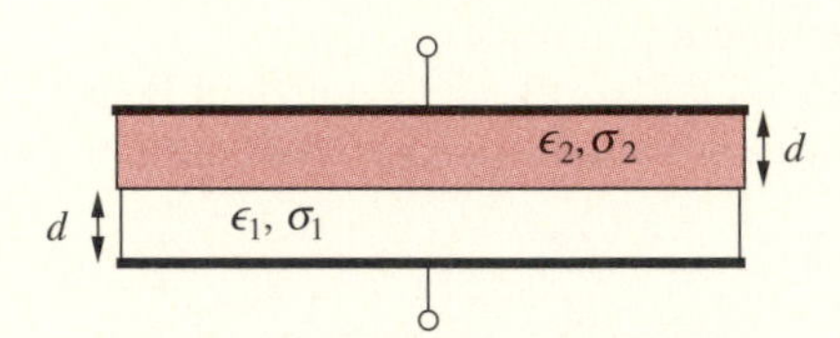

FIGURE 5.23. Leaky capacitor. Problem 5-27.

$(\epsilon/\sigma)\nabla\sigma] \cdot \nabla\Phi$. Note that the field arising from the flow of current can still be derived from a scalar potential, but the potential no longer satisfies Laplace's equation, since $\rho \neq 0$.

5-27. Leaky capacitor. Consider the parallel-plate capacitor with spacing $2d$ and plate area A, as shown in Figure 5.23. The region between the plates is filled with two lossy dielectric slabs, each of thickness d and with parameters ϵ_1, σ_1, and ϵ_2, σ_2, respectively. A potential V_0 is applied across the plates. (a) Find the steady-state value of the electric field in each of the two materials. (b) Find an expression for the surface charge density at the interface between the two materials.

5-28. Inhomogeneous medium. A spherical electrode of radius a is surrounded by a concentric spherical shell of radius b. The space between the two conductors is filled with material whose conductivity varies linearly with distance from the common center of the spheres (i.e., $\sigma = Kr$, where K is a proportionality constant). If a potential difference of V volts is maintained between the spheres, with the inner electrode grounded (i.e., at zero potential), what is the electric potential at a distance r ($a < r < b$) from the center?

5-29. Leakage resistance. A very long wire of radius a is suspended horizontally near the bottom of a deep lake. Assume the lake to have a plane bottom that is a very good conductor. The wire is parallel to the bottom and is at a height $h \gg a$ above it. If the conductivity of the water is σ, find the resistance (per unit length) between the wire and the planar bottom of the lake.

5-30. Two conducting spheres. Two metallic conducting spheres of radii a_1 and a_2 are buried deep in poorly conducting ground of conductivity σ and permittivity ϵ. The distance b between the spheres is much larger than both a_1 and a_2. Determine the resistance between the two spheres.

CHAPTER 6

The Static Magnetic Field

In the preceding chapters we discussed various aspects of electrostatics, or the configurations of static electric fields and their relationship to stationary charges and conducting or dielectric boundaries. Our discussions in Chapter 4 involved charges that were assumed to be at rest, and the various effects studied were consequences of Coulomb's law—the only experimental fact that we needed to introduce. In Chapter 5 we considered motion of charges under the influence of the Coulomb force resulting from an applied static electric field, constituting steady electric current. We now focus our attention on a series of new phenomena caused by the steady motion of charges. These phenomena constitute the subject of *magnetostatics* and can be understood as the consequences of a new experimental fact: namely, that current-carrying wires exert forces on one another.

In a certain sense, *magnetostatics* is an unfortunate name for the class of phenomena discussed in this chapter. First, the phenomena are caused by charges in motion, making the "static" part of the name a misnomer. Secondly, the "magnet" part of the name immediately suggests phenomena involving actual magnets. In reality, the physical phenomena involving natural magnets are highly complex, and it is very difficult, if not impossible, to construct a theory of magnetostatics based on experiments with magnets. It is much easier to formulate the theory of magnetostatics based on experiments involving moving charges or electric currents, a task which we undertake in this chapter.

As we noted in Chapter 4, our earliest experiences with electricity date back to ancient times and involve the attraction of objects by other objects. Interestingly enough, the subject that we now call *magnetism* also began with the observation that certain natural minerals, readily found near the ancient city of Magnesia in Anatolia, which is now part of Turkey, could attract other materials.[1] For centuries, magnetism was thought to be independent of electricity. Phenomena involving magnets were studied extensively because lodestone was readily available, and performing experiments with magnets was relatively simple.[2] However, construction of a theory of magnetostatics based upon results of such experiments requires that we introduce highly artificial concepts and fictions such as a "magnetic pole." After many years of experimentation, and the accidental discovery in 1820 (see the next paragraph) of the experimental fact that an electric current can exert a force on a compass needle, it was realized that the simple understanding of magnetostatic phenomena lay in the relatively difficult experiments involving moving charges rather than in the relatively simple experiments involving magnets. We adopt this modern viewpoint of magnetostatics, founded on Ampère's experimentally based law of force between current-carrying wires.

That wires carrying electric currents produce magnetic fields was first discovered[3] apparently by accident by H. C. Oersted in 1820. Within a few weeks after hearing of Oersted's findings, André-Marie Ampère announced[4] that current-carrying wires exert forces on one another, and J.-B. Biot and F. Savart repeated Oersted's experiments and put forth[5] a compact law of static magnetic fields generated by current elements in a circuit, which is known as the Biot–Savart law. The most thorough treatment of the subject was undertaken by Ampère in the following

[1]That lodestone attracts iron was first noted in print by Roman poet and and philosopher Lucretius [99?–55? B.C.] in his philosophical and scientific poem titled *De rerum natura (On the Nature of Things).* Lodestone is a form of iron oxide called magnetite, found in the shape of elongated fragments. The name *lodestone* comes from Middle English *lode,* "course," because a compass needle made of this material can be used for navigation. In the same manner, the lodestar is the pole star, which marks north in the sky. Also see G. L. Verschuur, *Hidden Attraction, The Mystery and History of Magnetism,* Oxford University Press, New York, 1993.

[2]An example is the work of Pierre de Maricourt, a native of Picardy [his *Epistola Petri Peregrini de Maricourt de magnete* was written in 1629], who experimented with lodestone and needles and introduced the concept of magnetic poles. Also notable was the work of William Gilbert [Gulielmi Gilberti *de Magnete, Magneticisque corporibus, et de magno magnete tellure*, London, 1600], who was the personal physician of Queen Elizabeth. Gilbert studied the properties of magnets and realized that magnets set themselves in definite orientations with respect to the earth because the earth is itself a giant spherical magnet. He was thus the first to discover the earth's magnetic field. Gilbert was rather carried away with this new discovery, however, and conjectured that magnetic forces also accounted for the earth's gravity and the motions of the planets.

[3]H. C. Oersted, *Experiments on the Effect of Current of Electricity on the Magnetic Needle,* privately distributed pamphlet dated July 21, 1820. English translation in *Ann. Philos.,* 16, p. 273, 1820.

[4]A.-M. Ampère, Memoir on the mutual action of two electric currents, *Ann. Chimie et Phys.,* 15, p. 59, 1820.

[5]J.-B. Biot and F. Savart, *Ann. Chimie et Phys.,* 15, p. 222, 1820.

three years, culminating in an extensive memoir[6] in 1825. This work established Ampère's law, which describes the law of force between two current elements and is analogous to Coulomb's force law in electrostatics. Ampère's studies also led him to postulate that magnetism itself was due to circulating currents on an atomic scale, thus closing the gap between the magnetic fields produced by currents and those produced by natural magnets.

This exciting period of development of the experimentally based underpinnings of magnetostatics was one of the most interesting in the history of science and was not without controversy.[7] After the development of the special theory of relativity, it was realized that magnetostatic theory may be derived from Coulomb's law.[8] Nevertheless, in this text we adopt an experimentally based formulation of Maxwell's equations, so Ampère's law of magnetic force and the Biot–Savart law of magnetic induction are considered quantitative statements of experimental facts.

Oersted, Hans Christian *(Danish physicist, b. August 14, 1777, Rudkøbing, Langeland; d. March 9, 1851, Copenhagen) The young Hans worked in his father's apothecary*[9] *shop, but his early training, which in most cases would have led straight to a career in chemistry, led to physics instead. He studied at the University of Copenhagen, where he obtained his Ph.D. in 1799 for a dissertation on Immanuel Kant's [1724–1804] philosophy. He then traveled through Europe, and in 1806 was appointed professor of physics and chemistry at his alma mater. He became an ardent adherent of the school of "nature philosophy," of which Oken [1779–1851] was an outstanding member. He accepted, with great gullibility, foolish theories and faked experimental work by men he admired, and for a while his scientific reputation lay under a cloud.*

His brother Anders, younger by a year and half, took to law, became attorney general of Denmark, and eventually the prime minister. He was a very unpopular prime minister and underwent impeachment proceedings after a forced resignation. It would seem, then, thanks to a single experiment, that Hans Oersted had taken the better road to fame.

It was in 1819 that Hans Oersted's great day came. He too was experimenting with the electric current, as half of Europe's scholars were doing. As part of a

---

[6]A.-M. Ampère, On the mathematical theory of electrodynamic phenomena uniquely deduced from experiment, *Mem. Acad.*, pp. 175–388, 1825. About 50 years later, J. C. Maxwell described Ampère's work as being "one of the most brilliant achievements in science." Also see L. P. Williams, André-Marie Ampère, *Scientific American,* pp. 90–97, January 1989.

[7]For an excellent account of the history and a thorough treatment of underlying fundamentals see R. S. Elliott, *Electromagnetics,* IEEE Press, Piscataway, New Jersey, 1993.

[8]L. Page, A derivation of the fundamental relations of electrodynamics from those of electrostatics, *Am. J. Sci.*, 34, p. 57, 1912; see Section 4.2 of R. S. Elliott, *Electromagnetics,* IEEE Press, Piscataway, New Jersey, 1993, for a clear treatment. It should be realized, however, that such derivations implicitly make other assumptions in addition to using special relativity; see Section 12.2 of J. D. Jackson, *Classical Electrodynamics,* 2nd ed., Wiley, 1975.

[9]A kind of pharmacy, where drugs were prepared and sold for medicinal purposes.

classroom demonstration, he brought a compass needle near a wire through which a current was passing. Scientists had long suspected there might be some connection between electricity and magnetism, and Oersted may have felt that the current in the wire might have some effect on the needle.

It did indeed. The compass needle twitched and pointed neither with the current nor against it, but in a direction at right angles to it. When he reversed the direction of the current, the compass needle veered and pointed in the opposite direction, but still at right angles. The astounded Oersted remained after class to repeat and continue his experiments.

This was the first demonstration of a connection between electricity and magnetism, and Oersted's experiment may be considered the foundation of the new study of electromagnetism.

Oersted's discovery (published in Latin, in the old-fashioned way) was announced in 1820, and it set off an explosion of activity. Coulomb [1736–1806] had developed views indicating that electricity and magnetism could not interact, and he had been very persuasive; but now it was clearly seen that he had been wrong. Arago [1786–1853] and Henry [1797–1878], especially, realized that electromagnetism was to grow into an entity that was eventually to change the world as drastically as the steam engine had changed the world a century before and as the internal combustion engine was to change it half a century later.

Oersted did not keep up with the whirlwind of activity his experiment had stirred up. He did show that the force of the current on the needle made itself felt through glass, metals, and other nonmagnetic substances, but except for that, he did nothing further to follow up his own momentous discovery. Nevertheless, the unit of magnetic field strength was officially named the oersted *in his honor in 1934.*

Outside electromagnetics, Oersted was the first to isolate the organic compound piperidine (1820) and the first to prepare metallic aluminum (1825). [Adapted with permission from I. Asimov, Biographical Encyclopedia of Science and Technology, *Doubleday, 1982.]*

The study of magnetostatics constitutes the second major step in our quest to understand the foundations of the laws of electromagnetics. In magnetostatics we deal with magnetic fields produced by steady currents, which are themselves constant in time and therefore do not allow inductive coupling between circuits or the coupling between electric and magnetic fields. Yet, mastering the behavior of static magnetic fields and the techniques of solution of magnetostatic problems is essential to the understanding of the more complicated electromagnetic phenomena. Furthermore, many natural phenomena and the principles of some important industrial and technological applications are based in magnetostatics. In this connection, it suffices to say that magnetic recording was a more than $20 billion industry[10] in 1984 and is believed to be more than $100 billion today.

Our coverage of magnetostatics in this chapter also brings us one more step closer to a full understanding of the underlying physical basis of the transmission line behavior discussed in Chapters 2 and 3. An important physical property of a transmission line is its distributed inductance, which comes about because of the magnetic fields generated by the transmission line currents. In Chapters 2 and 3, we took it for granted that any two-conductor system has some inductance and we relied on formulas (in Table 2.2) to determine the distributed inductances of a few common transmission line structures. In this chapter, we define the physical basis of the concept of inductance and discuss how the inductance of different types of current-carrying conductors can be determined using fundamental laws of magnetostatics.

We now proceed with our study of magnetostatics with Ampère's law of force in Section 6.1 and the concept of a magnetic field produced by steady electric currents, as expressed in the Biot–Savart law in Section 6.2. Ampère's circuital law is covered in Section 6.3, followed by a discussion of its differential form and the concept of the curl of the magnetic field in Section 6.4. Sections 6.5 and 6.6 respectively cover the important concept of vector magnetic potential and the magnetic dipole, while Section 6.7 discusses magnetic flux, divergence of the magnetic field, and inductance. Magnetic fields in material media is the topic of Section 6.8, followed by the derivation of magnetostatic boundary conditions in Section 6.9 and a discussion of magnetic forces and torques in Section 6.10.

6.1 AMPÈRE'S LAW OF FORCE

Before we formally write down the magnetic force law, it is useful to review the new experimental facts in simple terms. Figure 6.1 illustrates the direction of the force between current-carrying wires as first experienced by Oersted, and its dependence on the orientation of the wires, and the direction of the current flowing in them. The experimental facts indicate that two parallel wires carrying like (i.e., in the same direction) current attract one another while those carrying opposite current repel, and that when a small[11] current-carrying wire element is oriented perpendicular to another current-carrying wire (see Figure 6.1c) it feels no magnetic force. This set of

---

[10]See, for example, R. M. White (ed.), *Introduction to Magnetic Recording,* IEEE Press, Piscataway, New Jersey, 1984.

[11]The reason we consider a small element in Figure 6.1c is that a long wire oriented in the z direction and carrying current I_2 would in fact experience a torque (see Section 6.10 on magnetic forces and torques). More precisely, the portion of the long wire above the xy plane would feel a force in the $+x$ direction, while that below the xy plane would experience a force in the $-x$ direction.

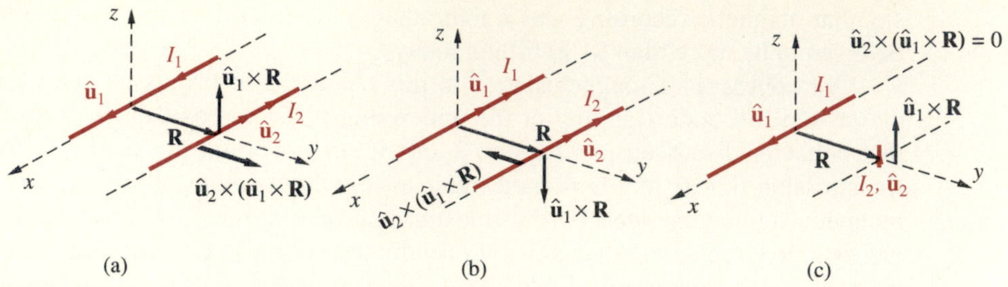

FIGURE 6.1. **Magnetic force between straight current-carrying wires.** (a) Two infinitely long straight parallel wires carrying current in opposite directions repel one another. (b) Two straight parallel wires carrying current in the same direction attract. (c) When a small wire element carrying a current I_2 is oriented perpendicular to the wire carrying current I_1, it feels no magnetic force, regardless of the direction of its current.

experimental facts is represented by expressing the force between two current-carrying wires as a double cross product,[12] namely

$$\mathbf{F}_{12} = kI_2\hat{\mathbf{u}}_2 \times (I_1\hat{\mathbf{u}}_1 \times \mathbf{R})$$

where $\mathbf{F}_{12}$ is the force exerted on wire 2 by wire 1, $\mathbf{R}$ is the vector from wire 1 to wire 2, $\hat{\mathbf{u}}_1$ and $\hat{\mathbf{u}}_2$ are unit vectors along wires 1 and 2 in the direction of currents I_1 and I_2, respectively, and k is a proportionality constant. Note from Figure 6.1a that the vector $(I_1\hat{\mathbf{u}}_1 \times \mathbf{R})$ is in the z direction, but the cross product of $\hat{\mathbf{u}}_2 \times (\hat{\mathbf{u}}_1 \times \mathbf{R})$ is in the y direction, indicating repulsion of the wires carrying oppositely directed current. Note that we could have just as well written an expression for the force $\mathbf{F}_{21}$ due to wire 2 at the location of wire 1, which points in the $-y$ direction. The magnitude of the magnetic force is experimentally determined to be inversely proportional to the square of the distance between wires.

The simplest form of Coulomb's law (Section 4.2) dealt with forces between individual point charges. By analogy, we might expect that we should similarly consider forces between elemental-length current-carrying wires. In practice, however, steady currents must necessarily flow in complete circuits, so our fundamental experimental law must describe the total force between two complete circuits. Consider

---

[12]The cross product of two vectors $\mathbf{A}$ and $\mathbf{B}$ is a vector, denoted by $(\mathbf{A} \times \mathbf{B})$, with its magnitude equal to the product of the magnitudes of the two vectors times the sine of the angle ψ_{AB} between them and its direction following that of the thumb of the right hand, when the fingers rotate from $\mathbf{A}$ to $\mathbf{B}$ through the angle ψ_{AB}. Namely,

$$\mathbf{A} \times \mathbf{B} \equiv \hat{\mathbf{n}}|\mathbf{A}||\mathbf{B}| \sin \psi_{\text{AB}}$$

where $\hat{\mathbf{n}}$ is normal to both $\mathbf{A}$ and $\mathbf{B}$ and its direction is in the direction of advance of a right-handed screw as $\mathbf{A}$ is turned toward $\mathbf{B}$. In rectangular coordinates, and noting that $\hat{\mathbf{x}} \times \hat{\mathbf{y}} = \hat{\mathbf{z}}$, $\hat{\mathbf{y}} \times \hat{\mathbf{z}} = \hat{\mathbf{x}}$, and $\hat{\mathbf{z}} \times \hat{\mathbf{x}} = \hat{\mathbf{y}}$, we can use the distributive property of the cross product to write

$$\mathbf{A} \times \mathbf{B} = (\hat{\mathbf{x}}A_x + \hat{\mathbf{y}}A_y + \hat{\mathbf{z}}A_z) \times (\hat{\mathbf{x}}B_x + \hat{\mathbf{y}}B_y + \hat{\mathbf{z}}B_z)$$

$$\mathbf{A} \times \mathbf{B} = \hat{\mathbf{x}}(A_yB_z - A_zB_y) + \hat{\mathbf{y}}(A_zB_x - A_xB_z) + \hat{\mathbf{z}}(A_xB_y - A_yB_x)$$

The cross product is sometimes referred to as the vector product, since the result is a vector quantity.

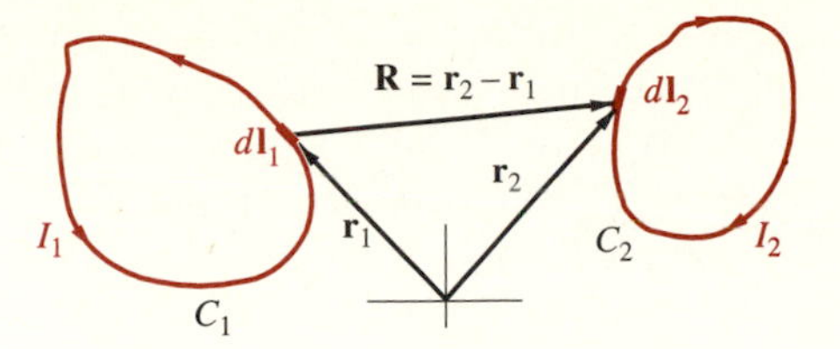

FIGURE 6.2. Ampère's law of force. Two separate circuits C_1 and C_2, carrying currents of I_1 and I_2, respectively, exert a force on one another.

two idealized[13] complete circuits C_1 and C_2, consisting of two very thin conducting loops (wires) carrying filamentary currents of I_1 and I_2, respectively. With respect to an arbitrary origin, as shown in Figure 6.2, the position vectors describing points on the two loops are taken to be $\mathbf{r}_1$ and $\mathbf{r}_2$ as indicated. We examine the force exerted on the circuit C_2 by the circuit C_1, neglecting for the purposes of this discussion the forces between current elements within the same loop. The vector distance from an elemental length $d\mathbf{l}_1$ along C_1 to another $d\mathbf{l}_2$ on C_2 is thus $(\mathbf{r}_2 - \mathbf{r}_1) = \mathbf{R} = R\hat{\mathbf{R}}$, $\hat{\mathbf{R}}$ is the unit vector directed *from*[14] $d\mathbf{l}_1$ *to* $d\mathbf{l}_2$, and $R = |\mathbf{R}| = |\mathbf{r}_2 - \mathbf{r}_1|$ is the distance between the two current elements. With $d\mathbf{l}_1$ at $\mathbf{r}_1 = \hat{\mathbf{x}}x_1 + \hat{\mathbf{y}}y_1 + \hat{\mathbf{z}}z_1$ and $d\mathbf{l}_2$ at $\mathbf{r}_2 = \hat{\mathbf{x}}x_2 + \hat{\mathbf{y}}y_2 + \hat{\mathbf{z}}z_2$, we have $R = [(x_2 - x_1)^2 + (y_2 - y_1)^2 + (z_2 - z_1)^2]^{1/2}$, and $\hat{\mathbf{R}} = (\mathbf{r}_2 - \mathbf{r}_1)/|\mathbf{r}_2 - \mathbf{r}_1| = \mathbf{R}/R$.

In his extensive experiments, Ampère found that the total vector force $\mathbf{F}_{12}$ exerted on C_2 by C_1 (both of which are located in free space) due to the mutual interaction of the currents I_1 and I_2 can be expressed[15] as

$$\boxed{\mathbf{F}_{12} = \frac{\mu_0}{4\pi}\oint_{C_2}\oint_{C_1}\frac{I_2 d\mathbf{l}_2 \times (I_1 d\mathbf{l}_1 \times \hat{\mathbf{R}})}{R^2}} \qquad [6.1]$$

Equation [6.1] is referred to as *Ampère's law of force* and constitutes the foundation of magnetostatics. In MKS units, $\mathbf{F}_{12}$ is measured in newtons, the currents I_1 and I_2 in amperes, and the lengths $d\mathbf{l}_1$, $d\mathbf{l}_2$ and R in meters. The proportionality constant is $\mu_0/(4\pi)$ because of our use of MKS units and includes the 4π "rationalization" factor so that a 4π factor does not appear in Maxwell's equations. In the MKS system of

---

[13]The circuits are idealized in the sense that the batteries that would have to be sustaining the currents are not shown. It is assumed that such batteries are some distance away and that the conducting leads from them are twisted closely together; one of Ampère's first experiments showed that two oppositely directed currents close together produced no effect on another current.

[14]That is, from the "source" point to the "field" point or "observation" point, since we are aiming to write an expression for the force experienced by the circuit C_2 due to the presence of C_1. A comparison between Figure 6.2 and Figure 4.3 of Chapter 4 illustrates the similarity between the fundamental force laws of magnetostatics and electrostatics.

[15]It may appear incredible that a formula with the generality implied in [6.1] could have been established from a few experiments on circuits of special and simple shapes as was done by Ampère. Indeed, [6.1] represents a generalization from results found in special arrangements of current loops; however, it should be noted that [6.1] continues to be valid (for steady currents), providing consistent results for every experiment involving a certain arrangement of loops that has been carried out since the time of Ampère.

units, μ_0 is *defined* to have precisely the value $\mu_0 = 4\pi \times 10^{-7}$ henrys per meter $\simeq$ 1.26 μH-m^{-1}. This constant is called the *permeability of free space;* for practical purposes μ_0 is also the permeability of air.[16] From [6.1], we see that the dimensions of μ_0 are force-(current)$^{-2}$, or N-A^{-2}. Thus, we have 1 H = 1 N-m-A^{-2} = 1 J-A^{-2}. Since the newton and the meter are determined independently, the above choice of the value of μ_0 constitutes a definition of the unit of electric current, or ampere (and hence also the unit electric charge, or coulomb).

Ampère, André-Marie *(French mathematician and physicist, b. January 22, 1775, Lyon; d. June 10, 1836, Marseille) Young Ampère was privately tutored and proved to be quite a phenomenon, devouring the encyclopedic works of Buffon [1707–1788] and Diderot [1713–1784] and mastering advanced mathematics by the age of twelve. He even learned Latin in order to read the works of those like Euler [1707–1783] who wrote in that language. The even tenor of his youth was, however, interrupted by the French Revolution.*

In 1793 Lyon revolted against the revolutionaries and was taken by the republican army. Ampère's father, who was a well-to-do merchant and one of the city's officials, was guillotined. Ampère went into a profound depression as a result, out of which, with the encouragement of the sympathetic Lalande [1732–1807], he struggled with difficulty. In 1803 his beloved wife of but four years died and this again hit him hard. Indeed, he never recovered from that blow. (In 1818 he married a second time, and this time the marriage was unhappy.)

Under Napoleon Ampère continued a fruitful career as a professor of physics and chemistry at Bourg, and then as a professor of mathematics in Paris. In 1808, Napoleon appointed him inspector general of the national university system.

Ampère and Arago [1786–1853] were in the forefront of the flurry of scientific activity that followed Oersted's [1777–1851] announcement to the French Academy of Sciences in 1820. Within one week, Ampère showed that the deflection of the needle could be expressed by what is now known as the "right-hand rule" to be the direction of the curling fingers as if a magnetic force circled the wire. This was the beginning of the concept of lines of force that Faraday [1791–1867] was later to generalize.

In setting up this right-hand screw rule, Ampère took the direction of current flow to be from positive to negative, using the earlier concept of Franklin [1706–1790] that the positive pole had the excess of "electrical fluid" and the negative pole the deficiency. This convention has been used ever since, but Franklin had guessed wrong and Ampère had gone wrong with him. We now know that the electric current is a movement of electrons flowing from the negative pole to the positive.

---

[16]The physical meaning of permeability and its dimensions will become clearer in Section 6.8, when we discuss magnetic materials and inductance. At this point we may note that permeability has the same significance for magnetostatics as permittivity has for electrostatics.

Ampère set up two parallel wires, one of which was freely movable back and forth. When both wires carried current in the same direction, the two wires clearly attracted each other. If the currents flowed in opposite directions, they repelled each other. Ampère also recognized, as did Arago, that from a theoretical standpoint a helix of wires would behave as though it were a bar magnet. He called such a helix a solenoid. This notion was put into practice by Sturgeon [1783–1850] and was then refined to a startling degree by Henry [1797–1878].

It was Ampère's experiments that founded the science of electric currents in motion, which Ampère named electrodynamics. He also introduced the term electrostatics *for the older study of stationary electric charges, in which Franklin's work had been so important.*

In 1823 Ampère advanced a theory that the magnet's properties arose from tiny electrical currents circling eternally within it. In this he was ahead of his time, for the existence of tiny electrically charged particles circling eternally was not to be known for three quarters of a century. Ampère's contemporaries received his theories with great skepticism.

Ampère died of pneumonia, and his judgment of his own life is indicated by the sorrow-laden epitaph he chose for his own gravestone: Tandem felix *(Happy, at last). In his honor, the quantity of electric current passing through a conductor is universally measured in amperes, a usage originated by Kelvin [1824–1907] in 1883. [Adapted with permission from I. Asimov,* Biographical Encyclopedia of Science and Technology, *Doubleday, 1982.]*

Example 6-1: An extra-high-voltage dc transmission line. An extra-high-voltage direct-current (dc) overhead transmission line consists of two very long parallel wires $a = 10$ m apart and each located $h = 35$ m above ground as shown in Figure 6.3.

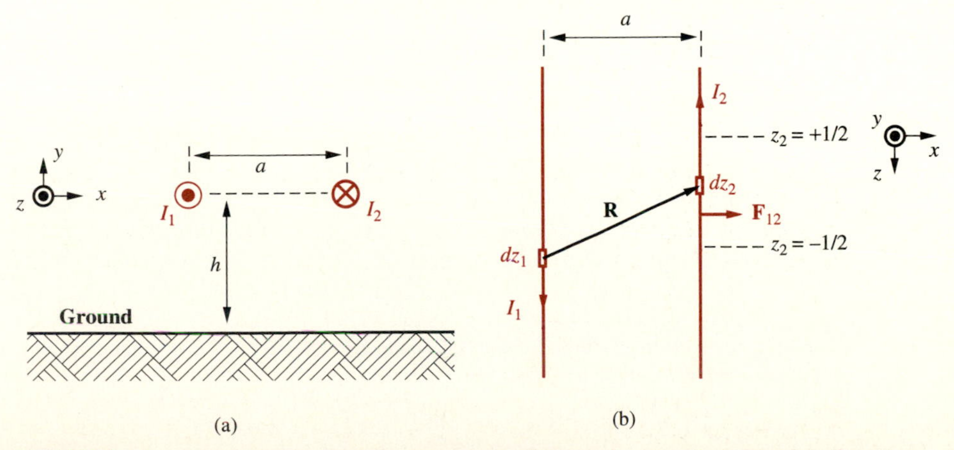

FIGURE 6.3. Extra-high-voltage dc transmission line. (a) Cross-sectional view. (b) Top view.

If the wires carry an equal current of $I = 3000$ A each flowing in opposite directions, find the repulsion force per unit length on each line. Neglect the effects of the ground plane.

Solution: The wires are assumed to be in the z direction, with $I_1\,d\mathbf{l}_1 = \hat{\mathbf{z}}I\,dz_1$ and $I_2\,d\mathbf{l}_2 = -\hat{\mathbf{z}}I\,dz_2$ as shown. Using Ampère's law of force we can calculate the net force exerted by one of the wires on a unit length section of the other wire. We can write

$$\mathbf{F}_{12} = \frac{\mu_0 I^2}{4\pi}\int_{z_2=-1/2}^{1/2}\int_{z_1=-\infty}^{\infty}\frac{(-dz_2)\hat{\mathbf{z}}\times(dz_1\hat{\mathbf{z}}\times\hat{\mathbf{R}})}{R^2}$$

where $R = \sqrt{(z_2-z_1)^2+a^2}$ is the magnitude of $\mathbf{R}$ shown in Figure 6.3b and $\hat{\mathbf{R}} = [\hat{\mathbf{z}}(z_2-z_1)+\hat{\mathbf{x}}a]/R$ and z_1 and z_2 are the integration variables over wires 1 and 2, respectively. Note that $\mathbf{F}_{12}$ is the force exerted on a 1-m long segment of wire 2 (thus the limits of the z_2 integral are from $-\frac{1}{2}$ to $+\frac{1}{2}$) by the entire wire 1 (thus the limits of $\pm\infty$ for the z_1 integral). Substituting for R and $\hat{\mathbf{R}}$, we have

$$\mathbf{F}_{12} = \frac{\mu_0 I^2}{4\pi}\int_{z_2=-1/2}^{1/2}(-dz_2)\hat{\mathbf{z}}\times\hat{\mathbf{y}}\int_{z_1=-\infty}^{\infty}\frac{a\,dz_1}{[(z_2-z_1)^2+a^2]^{3/2}}$$

where we have used $\hat{\mathbf{z}}\times\hat{\mathbf{R}} = \hat{\mathbf{y}}a[(z_2-z_1)^2+a^2]^{-1/2}$. Performing the integration[17] over z_1 we find

$$\mathbf{F}_{12} = \frac{\mu_0 I^2 a}{4\pi}\int_{z_2=-1/2}^{1/2}(-dz_2)\hat{\mathbf{z}}\times\hat{\mathbf{y}}\left[\frac{-(z_2-z_1)}{a^2\sqrt{a^2+(z_2-z_1)^2}}\right]_{z_1=-\infty}^{\infty}$$

$$= \hat{\mathbf{x}}\frac{\mu_0 I^2 a}{2\pi a^2}\int_{z_2=-1/2}^{1/2}dz_2 = \hat{\mathbf{x}}\frac{\mu_0 I^2}{2\pi a}$$

Using the numerical values given as $I = 3000$ A and $a = 10$ m, we find the repulsion force to be

$$\mathbf{F}_{12} = \hat{\mathbf{x}}\frac{(4\pi\times10^{-7}\ \text{H-m}^{-1})(3000\ \text{A})^2}{2\pi(10\ \text{m})} = 0.18\ \text{N-m}^{-1}$$

This lateral repulsion force can be compared with the weight of a line segment of 1 m length. Assuming the wires to have outside diameters of a few centimeters, their mass would be a few kg, and their weight a few tens of newtons. Thus, the per-unit-length magnetostatic repulsion force is quite small compared to the per-unit-length weight of the wires. However, it should be noted that the force

---

[17]This is a rather common integral already encountered in Example 4-5. Using integral tables, or simply a change of variables, it can be shown that

$$\int\frac{d\zeta}{[\zeta^2+b^2]^{3/2}} = \frac{\zeta}{b^2\sqrt{\zeta^2+b^2}} + \text{const.}$$

between the two wires is proportional to the square of the current, so it can reach substantial values when much larger currents (tens to hundreds of kA) may flow due to accidental shorts (see Problem 6-3).

Ampère's force law is another example of a force law describing "action at a distance," analogous in this sense to Coulomb's law for electrostatics. Just as it was useful to divide Coulomb's law into two parts by using the concept of an electric field as an intermediary to describe the interaction between charges, we can use Ampère's force law to define an appropriate field that may be regarded as the means by which currents exert forces on one another. The so-called[18] *magnetostatic induction field* or *magnetic flux density* **B** can be defined by rewriting [6.1] as follows:

$$\mathbf{F}_{12} = \oint_{C_2} I_2\, d\mathbf{l}_2 \times \underbrace{\frac{\mu_0}{4\pi}\oint_{C_1} \frac{I_1\, d\mathbf{l}_1 \times \hat{\mathbf{R}}}{R^2}}_{\mathbf{B}_{12}} \qquad [6.2]$$

where

$$\mathbf{B}_{12} = \frac{\mu_0}{4\pi}\oint_{C_1} \frac{I_1\, d\mathbf{l}_1 \times \hat{\mathbf{R}}}{R^2} \qquad [6.3]$$

where $\mathbf{B}_{12}$ is the **B** field at point $\mathbf{r}_2$ (i.e., at the location of $d\mathbf{l}_2$ at a distance R from $d\mathbf{l}_1$) due to the current I_1 in circuit C_1. Equation [6.2] evaluates the force on circuit C_2 in terms of the interaction of its current I_2 with the field $\mathbf{B}_{12}$, which is set up by the current I_1 in circuit C_1. The current–field interaction takes place over the entire circuit C_2, while the field $\mathbf{B}_{12}$ depends only on the current and configuration of the circuit C_1, which sets up the field. In the MKS system of units, the **B** field has units of[19] tesla (T) or weber per meter^2. Since one tesla is relatively large as a practical value, **B** is often also given in the CGS units of gauss (G), where $1\ T = 10^4$ G.

For reference purposes, the earth's magnetic field on the surface is ~0.5 G, the **B** fields of permanent magnets range from a few to thousands of gauss, while the **B** field at the surface of neutron stars is believed to be 10^{12} Gauss. The most powerful permanent magnets,[20] such as samarium-cobalt or neodymium-iron-boron magnets, have fields of 3000 to 4000 gauss, and several of them could easily lift an entire

---

[18]For historical reasons, the term "magnetic field" is generally used for a different vector, which will be defined and discussed in Section 6.8. In retrospect, it would have been more proper to refer to **B** as the magnetic field. The terms "magnetostatic induction field" or "magnetic flux density" are in fact quite inappropriate descriptions of the nature of the **B** field (also see Section 6.2.3). For this reason, and whenever possible, we shall try to refer to **B** simply as the "**B** field."

[19]Nikola Tesla (1856–1943) was a brilliant electrical engineer who invented the induction motor and many other useful electromagnetic devices and made alternating current practical. See M. Cheney, *Tesla: Man Out of Time,* Prentice Hall, Englewood Cliffs, New Jersey, 1981. Wilhelm E. Weber was a professor of physics at Göttingen and a close colleague of J. K. F. Gauss (see Section 4.5.2). For most of his professional life, Weber worked in collaboration with Gauss on the study of magnetic phenomena.

[20]P. Campbell, *Permanent Magnet Materials and Their Design,* Cambridge University Press, 1994.

TABLE 6.1. Typical **B** field values in selected applications

Application	B field (Gauss)
Sensitivity of a scanning SQUID microscope*	10^{-11}
Human brain	10^{-8}
Intergalactic and interstellar magnetic fields	10^{-6}
Human heart	10^{-4}–10^{-3}
Earth's magnetic field	0.5
Refrigerator memo magnets	10–800
Electron beam of CRT (computer or TV)	50–100
Magnetic read switch	100–200
1-horsepower electric motor	1000–2000
Cow magnets*	2000
Powerful permanent magnets*	3000–4000
Magnetic resonance imaging (MRI)	10^{3}–10^{5}
High-energy particle accelerators	10^{5}
Pulsed electromagnets*	10^{5}–10^{6}
White dwarf stars	10^{8}
Neutron stars	10^{12}

*See references in text.

refrigerator. World-record magnetic fields of nearly a million gauss have been achieved with pulsed electromagnets of novel designs.[21] Fields of such intensity represent a burst of energy comparable to an exploding stick of dynamite; the resultant forces imposed on the current-carrying wire surpass the tensile strength of copper and the resistive losses generate enough heat to melt the copper wire. Table 6-1 provides additional examples of **B** field values in different applications, ranging from high-technology applications such as superconducting quantum interference devices[22] to relatively low-technology (but no less important) applications such as cow magnets.[23]

[21]G. Boebinger, A. Passner, and J. Bevk, Building world-record magnets, *Scientific American,* pp. 58–66, June 1995.

[22]Superconducting quantum interference devices (SQUIDs) have recently become the basis for scanning SQUID microscopes, which can image magnetic fields at the surface of samples under study with an unprecedented sensitivity. SQUID imaging of fields as low as 10^{-11} gauss in a 1 cm^2 area was reported by J. Clarke, SQUIDs, *Scientific American,* pp. 46–53, August 1994. At the level of sensitivity of these devices, nearly everything is magnetic. Applications of SQUIDs range from diagnosis of brain tumors to tests of relativity. For further information, see J. Kirtley, Imaging magnetic fields, *IEEE Spectrum,* pp. 40–48, December 1996.

[23]The cow magnet is an alnico (Al-Ni-Co-Fe) cylinder three inches long and a half-inch in diameter; it has rounded ends and is usually coated with plastic for protection against corrosion and breakage. Grazing cows are encouraged to swallow the cow magnet as a protection against the so-called "hardware disease," which occurs when a grazing cow indiscriminately swallows sharp steel objects, such as bits of wire used to bale hay, which can cause damage to the walls of her intestines. The cow magnet remains in the cow's stomach, attracting to it any steel objects she later swallows, and can be retrieved and reused when the cow is slaughtered. For more on cow magnets and other interesting applications, see J. D. Livingston, *Driving Force: The Natural Magic of Magnets,* Harvard University Press, Cambridge, Massachusetts, 1996.

The vector field **B** can be calculated at any field point **r** using [6.3] even if there is no current element there to experience a magnetostatic force. When a current-carrying wire is placed in a region permeated by a **B** field, it experiences a magnetostatic force given by $\mathbf{F} = \int I d\mathbf{l} \times \mathbf{B}$, as is evident from [6.2], which can be calculated as illustrated in Example 6-2. As noted in Section 4.3, the notion of a "field," however abstract, not only is useful and convenient in thinking and working with action-at-a-distance phenomena but may also be regarded as an actual physical entity in its own right.[24] For our purposes in the rest of this chapter, it is more useful[25] to continue our discussions of the implications of [6.1] using the so-called *Biot–Savart law,* which quantifies the **B** field produced at a point by currents in its vicinity.

Example 6-2: A semicircular loop in a B field. A semicircular loop of wire of 1 m diameter carrying a current of $I = 10$ A lies in a uniform **B** field of magnitude $B_0 = 1.5$ T, as shown in Figure 6.4. Determine the total magnetostatic force experienced by the loop.

Solution: The total magnetostatic force on the straight portion of the current loop is

$$\mathbf{F}_{\text{str}} = \int_{-a}^{a} (-I\,dx)\hat{\mathbf{x}} \times \hat{\mathbf{z}}B_0 = \hat{\mathbf{y}}IB_0 \int_{-a}^{a} dx$$

$$= \hat{\mathbf{y}}2IB_0a = \hat{\mathbf{y}}2(10\text{ A})(1.5\text{ T})(0.5\text{ m}) = \hat{\mathbf{y}}15\text{ N}$$

The total force experienced by the semicircular arc portion of the current loop is

$$\mathbf{F}_{\text{arc}} = \int_{\phi=0}^{\pi} (-Ia\,d\phi)\hat{\boldsymbol{\phi}} \times \hat{\mathbf{z}}B_0 = -IB_0a\int_0^{\pi} \hat{\mathbf{r}}\,d\phi = -IB_0a\int_0^{\pi} (\hat{\mathbf{x}}\cos\phi + \hat{\mathbf{y}}\sin\phi)\,d\phi$$

$$= -IB_0a[\hat{\mathbf{x}}\sin\phi - \hat{\mathbf{y}}\cos\phi]_0^{\pi} = -\hat{\mathbf{y}}2IB_0a = -\hat{\mathbf{y}}15\text{ N}$$

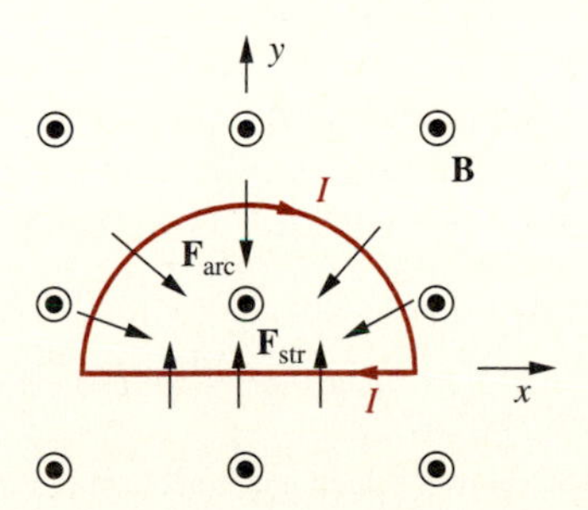

FIGURE 6.4. A semicircular current loop in a B field. The **B** field is directed out of the page, as represented with a circle having a central dot.

---

[24]Which it certainly was for M. Faraday; see footnote 42 in Section 4.3.

[25]Direct use of [6.1] to determine the force on a circuit is feasible only in relatively simple geometries. It is usually easier to determine the force on a particular circuit (C_2) through the concept of the **B** field produced by the current of another circuit (C_1).

The fact that the force is in the y direction makes sense, since the net effect of the x components of the force on the semicircular arc portion cancel out due to symmetry, as is evident in Figure 6.4. Thus, the net total magnetostatic force exerted on the semicircular loop by the **B** field is

$$\mathbf{F} = \mathbf{F}_{\text{str}} + \mathbf{F}_{\text{arc}} = \hat{\mathbf{y}}2IB_0a - \hat{\mathbf{y}}2IB_0a = 0$$

6.2 THE BIOT–SAVART LAW AND ITS APPLICATIONS

The Biot–Savart law is the most basic law of magnetostatics; it describes how the **B** field at a given point is produced by the moving charges (i.e., currents) in the neighborhood of that point. This basic law has been experimentally verified by innumerable measurements of the fields produced by many types of current distributions. A highly accurate assessment of the inverse square dependence is provided by geophysical measurements of the earth's magnetic field, which indicate[26] that not only is the $r^{-2+\zeta}$ dependence valid at distances of $\sim 10^4$ km but that $|\zeta| < 10^{-16}$.

We consider electric currents that are either steady or slowly varying.[27] Using superposition,[28] we can consider any current to consist of infinitesimal current elements. More complex circuits and current distributions can always be expressed as a sum of such elemental currents.

The **B** field vector at any point P identified by the position vector **r**, as shown in Figure 6.5a, due to a differential current element $I\,d\mathbf{l}'$ located at position $\mathbf{r}'$, follows from [6.3] as

$$d\mathbf{B}_{\text{p}} = \frac{\mu_0 I\,d\mathbf{l}' \times \hat{\mathbf{R}}}{4\pi R^2} \qquad [6.4]$$

where $\hat{\mathbf{R}}$ is the unit vector pointing *from* the location of the current element *to* the field point P, and $R = |\mathbf{r} - \mathbf{r}'|$ is the distance between them. Note that we resort to our practice of using primed quantities to represent the position vector or coordinates of source points, while unprimed quantities are used to represent the position vector or coordinates of points at which the **B** field is evaluated. The cross product in [6.4]

---

[26]For an excellent review, see A. S. Goldbaher and M. M. Nieto, Terrestrial and extraterrestrial limits on the photon mass, *Rev. Mod. Phys.*, 43, pp. 277–295, 1971.

[27]The currents may vary with time if the time interval over which appreciable variation occurs is much longer compared to r/c, where r is the maximum distance from any of the elementary source currents to the observation point where the magnetic field is to be computed, and c is the velocity of light in free space. Any given circuit that the current flows through undoubtedly has its internal time constant (due, for example, to its distributed capacitance and/or inductance); the time intervals over which the currents vary appreciably must also be much longer than this time constant.

[28]As in the case of electric fields due to charges, we are able to use superposition only because it is verified by experiment.

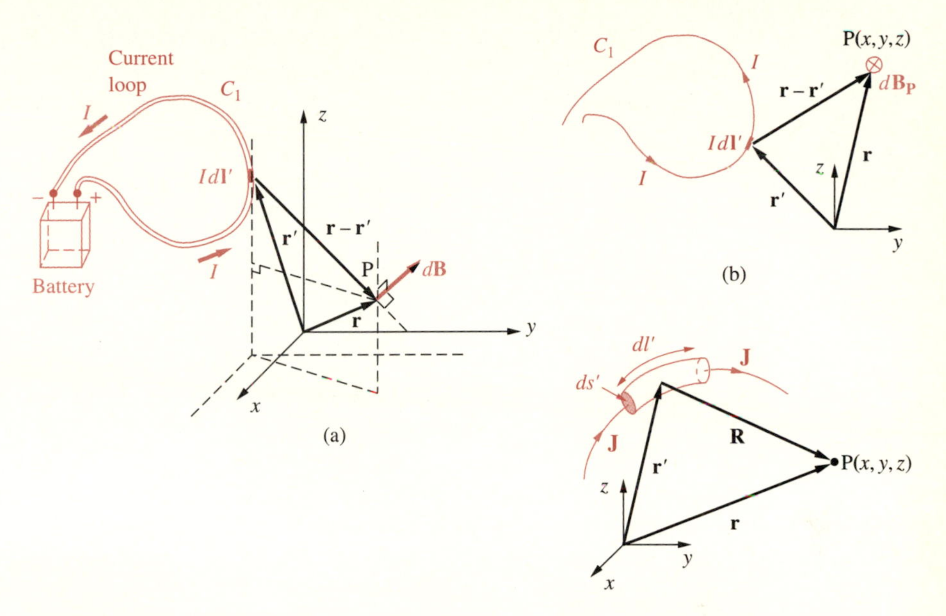

FIGURE 6.5. Magnetic field of a current element. (a) Note that the differential current element is necessarily part of a circuit or loop of current, driven by a battery or other source as shown. (b) A two-dimensional view with $d\mathbf{l}'$ and $\mathbf{R} = (\mathbf{r} - \mathbf{r}')$ on the plane of the paper, in which case $d\mathbf{B}_\text{P}$ is into or out of the paper. (c) An arbitrary volume distribution of current can be analyzed in terms of differential tubular current density elements.

indicates that $d\mathbf{B}$ is perpendicular to both $d\mathbf{l}'$ and $\hat{\mathbf{R}}$; the orientation of $d\mathbf{B}$ is determined by the right-hand rule. As indicated in Figure 6.6b, the *right-hand rule* states that if the fingers close up as they move from $I\,d\mathbf{l}'$ to $\hat{\mathbf{R}}$, then the thumb points in the direction of their cross product, namely in the direction of $d\mathbf{B}$. As indicated in Figure 6.6a, the right-hand rule also states that if the thumb of the right hand is pointed in the direction of the current, the curled right-hand fingers encircling the current element point in the direction of the **B** field. The latter form of the right-hand rule is more useful in interpreting the direction of the **B** field near current-carrying conductors. In relation to Figure 6.5b, where the current loop C_1 is assumed to be entirely confined to the yz plane, note that we have adopted the customary convention of representing the direction of vectors that are perpendicular to the page (in this case the **B** field) by a small circle with a cross at the center where the vector points into the page. Vectors pointing out of the page are represented with circles having a central dot, as was done in Figure 6.4.

Note that, in general, the current element $I d\mathbf{l}'$ is part of a thin filamentary[29] closed current loop C_1, with more complex circuits and current distributions represented as a superposition of many such closed filamentary loops of current. The total magnetic flux density at point P due to the filamentary current loop C_1 is given

[29]The diameter of the conductor is much smaller than any other dimensions in the system, so we can assume that all the current flows along the axis of the conductor.

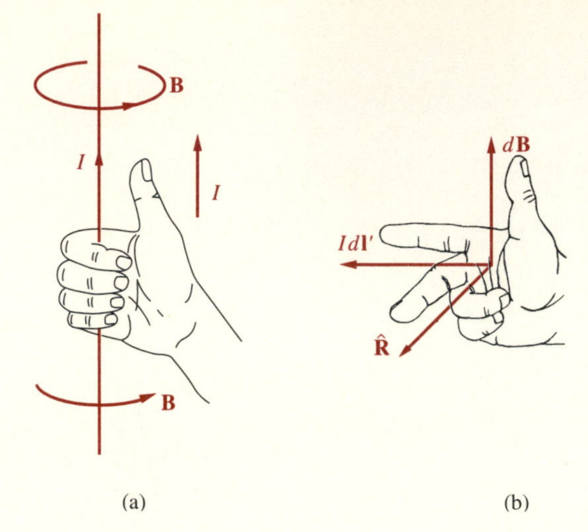

FIGURE 6.6. The right-hand rule. (a) Determination of the direction of the **B** field from the direction of current. (b) Determination of the direction of the cross product of two vectors, in this case $I\,d\mathbf{l}' \times \hat{\mathbf{R}}$.

by the sum (superposition) of the contributions from its individual differential current elements $I\,d\mathbf{l}'$, namely

$$\mathbf{B} = \frac{\mu_0}{4\pi}\oint_{C_1} \frac{I\,d\mathbf{l}' \times \hat{\mathbf{R}}}{R^2} \qquad [6.5]$$

Equations [6.4] and [6.5] are statements of the *Biot–Savart law.*

In practice, we do not always have currents flowing in thin conductors, and it is thus necessary to generalize the definition of **B** so that it applies for any arbitrary volume distribution of current. Since a steady electric current is divergence-free ($\nabla \cdot \mathbf{J} = 0$, see Section 5.4), all current flow lines form closed loops. Consider a differential length dl' of a single current-flow tube of differential cross-sectional area ds' as shown in Figure 6.5c, carrying a current with a current density $\mathbf{J}(\mathbf{r}')$. Since we can associate the direction with the current density vector $\mathbf{J}(\mathbf{r}')$ (rather than with an arc length dl'), a given current-flow tube element of length dl' and area ds' located at $\mathbf{r}'$ produces a field $d\mathbf{B}$ at point P given by

$$d\mathbf{B}_{\mathrm{P}} = \frac{\mu_0}{4\pi}\frac{\mathbf{J}(\mathbf{r}')\,ds'\,dl' \times \hat{\mathbf{R}}}{R^2}$$

since the total current flowing in this differential tube is $\mathbf{J}(\mathbf{r}')ds'$. The total current contained in a volume V' thus produces a **B** field given by

$$\mathbf{B}(\mathbf{r}) = \frac{\mu_0}{4\pi}\int_{V'} \frac{\mathbf{J}(\mathbf{r}') \times \hat{\mathbf{R}}}{R^2}\,dv' \qquad [6.6]$$

where the integration is to be carried out over the source coordinates x', y', and z', and dv' is an element of volume V' given by $dv' = ds'\,dl'$.

Biot, Jean Baptiste *(French physicist, b. April 21, 1774, Paris; d. February 3, 1862, Paris) Biot, the son of a treasury official, served a year in the artillery in 1792, fighting the British. Then he entered the École Polytechnique, studying under Lagrange [1736–1813] and Berthollet [1748–1822]. He and Malus [1775–1812] took part in a street riot in 1795 that was easily put down by Napoleon, marking the end of the French Revolution. Biot suffered imprisonment for a while as a result, and remained consistently anti-Napoleon in later life, being awarded the Legion of Honor by Louis XVIII, who succeeded the fallen emperor.*

Biot obtained an appointment as professor of mathematics at the University of Beauvais, and in 1800 moved on to the Collège de France through the sponsorship of Laplace [1749–1827], whose self-centered soul he had pleased by offering to read proof on the colossal Mécanique Céleste *and then actually doing it. He proved himself to be well deserving of the post.*

In 1803 he investigated a reported sighting of material falling from heaven, and his findings finally convinced a skeptical scientific world that meteorites existed. In 1804 Biot and Gay-Lussac [1778–1850] used a balloon left over from Napoleon's Egyptian campaign to study the earth's atmosphere. Loaded down with instruments, plus an assortment of small animals, Biot and Gay-Lussac flew the balloon on August 23, 1804, ran a number of experiments, and showed that terrestrial magnetism remained undiminished at heights of one to three miles.

Biot's most important work was in connection with what we now call polarized light, which could only be explained properly by a wave theory of light put forth by Fresnel [1788–1827]. Biot did not accept Fresnel's theories but worked fruitfully with polarized light. In 1815 he showed that organic substances might, in fact, rotate polarized light either clockwise or counterclockwise when the organic compounds were liquid or in solution. He suggested that this was due to an asymmetry that might exist in the molecules themselves. In 1840 he was awarded the Rumford medal.

He lived long enough to see Pasteur [1822–1895] prove asymmetry in organic crystals that had this twisting effect on polarized light, but (despite his longevity) not long enough to see the molecular asymmetry he had predicted become an important facet of organic chemistry through the work of Van't Hoff [1852–1911] and Le Bel [1847–1930]. Though an atheist most of his adult life, Biot returned to Catholicism in 1846. [Adapted with permission from I. Asimov, Biographical Encyclopedia of Science and Technology, *Doubleday, 1982.]*

1700 1774 1862 2000

Savart, Félix *(French physicist, b. June 30, 1791, Mézières, France; d. March 16, 1841, Paris, France) Savart made experimental studies of many phenomena involving vibration. With Biot he showed that the magnetic field produced by the current in a long, straight wire is inversely proportional to the distance from the wire.*

The son of Gérard Savart, an engineer at the military school of Metz, Savart studied medicine, first at the military hospital at Metz and then at the University of Strasbourg, where he received his medical degree in 1816. But his real interest was in the physics of the violin. In 1827, Savart replaced Fresnel [1788–1827] as a member of the Paris Academy, and in 1828 he became a professor of experimental physics at the Collège de France, where he taught acoustics.

In 1820, a few months after Oersted's [1777–1851] discovery of the magnetic field produced by a current, Biot and Savart determined the relative strength of the field by observing the rate of oscillation of a magnetic dipole suspended at various distances from a long, straight wire.

In his earliest work Savart gave the first explanation of the function of certain parts of the violin. Over a period of twenty years, Savart performed numerous experiments in acoustics and vibration. He generalized his work on the violin to analyze the vibrations of coupled systems. On the basis of his experience observing vibrational modes, Savart introduced a new way to learn about the structure of materials. The vibration of nodal patterns for laminae cut along different planes of a nonisotropic material indicated the orientational dependence of the material's elasticity. Savart sought the axes of elasticity of various substances, including certain crystalline ones. His papers on this subject were translated and reprinted.

Savart also studied aspects of the voice and of hearing. Savart was highly regarded as an experimenter; in connection with determining the lower frequency limit of hearing, he devised and used the rotating toothed wheel for producing sound of any frequency. [Adapted with permission from Dictionary of Scientific Biography, *Charles Scribner's Sons, New York, 1975.]*

1700 *1791* *1841* *2000*

6.2.1 Examples of the Use of the Biot–Savart Law

We now consider various examples of the application of the Biot–Savart law. In Example 6-3, we consider the **B** field at a point in the vicinity of a straight wire segment—which is largely a hypothetical problem, since current must flow in continuous closed paths. In Example 6-4, we take advantage of the fact that many actual closed circuits can be considered as a superposition of straight wire segments, using the result obtained in Example 6-3. Example 6-5 concerns the **B** field at different points in the vicinity of two infinitely long current-carrying wires, while in Example 6-6 we consider the **B** field along the axis of a circular loop of current. Example 6-7 is a generalization of the case considered in Example 6-3, while Example 6-8 illustrates a case in which the source current is not filamentary.

Example 6-3: Finite-length straight wire. Consider a straight, current-carrying filamentary conductor of length $2a$, as shown in Figure 6.7a. Find the **B** field at a point $P(r, \phi, 0)$ equidistant from the end points of the conductor.

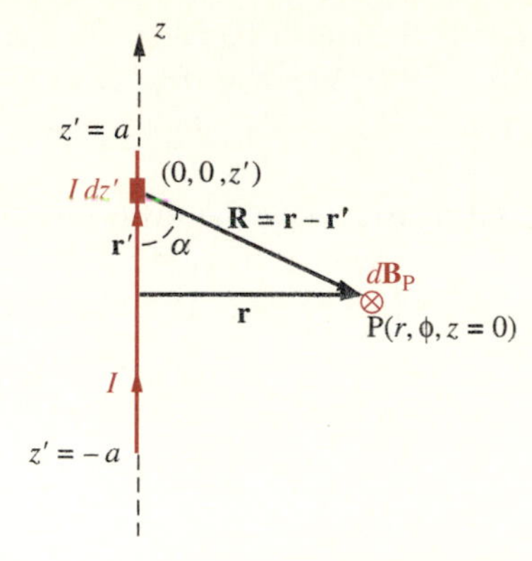

FIGURE 6.7. Finite-length current-carrying wire. Coordinate system and the source and observation points.

Solution: Since we have azimuthal symmetry (i.e., no dependence on ϕ), a cylindrical coordinate system with the wire oriented along the z-axis and centered at the origin is the most appropriate. The field $d\mathbf{B}_{\mathrm{P}}$ at point P (note that $\mathbf{r} = \hat{\mathbf{r}}r$) due to current element $I\,d\mathbf{l}'$, located at $\mathbf{r}' = \hat{\mathbf{z}}z'$ at a distance $R = |\mathbf{r} - \mathbf{r}'| = \sqrt{r^2 + (z')^2}$ from point P is

$$d\mathbf{B}_{\mathrm{P}} = \frac{\mu_0}{4\pi}\frac{I\,dz'\hat{\mathbf{z}} \times \hat{\mathbf{R}}}{R^2} = \hat{\boldsymbol{\phi}}\frac{\mu_0 I}{4\pi}\frac{\sin\alpha\, dz'}{[r^2 + (z')^2]} = \hat{\boldsymbol{\phi}}\frac{\mu_0 I}{4\pi}\frac{r\,dz'}{[r^2 + (z')^2]^{3/2}}$$

where we have used the fact that the magnitude of the cross product of two vectors is equal to the product of the magnitudes of the two vectors times the sine of the angle between them; that is, $\hat{\mathbf{z}} \times \hat{\mathbf{R}} = \hat{\boldsymbol{\phi}}|\hat{\mathbf{z}}||\hat{\mathbf{R}}|\sin\alpha = \hat{\boldsymbol{\phi}}\sin\alpha$. Note that

$$\sin\alpha = r[r^2 + (z')^2]^{-1/2}$$

with α as shown in Figure 6.7. We can find the total $\mathbf{B}_{\mathrm{P}}$ field at point P by integrating[30] $d\mathbf{B}_{\mathrm{P}}$ over the length of the wire,

$$\mathbf{B}_{\mathrm{P}} = \hat{\boldsymbol{\phi}}\frac{\mu_0 I r}{4\pi}\int_{z'=-a}^{+a}\frac{dz'}{[r^2 + (z')^2]^{3/2}}$$

$$= \hat{\boldsymbol{\phi}}\frac{\mu_0 I r}{4\pi}\left[\frac{z'}{r^2\sqrt{r^2 + (z')^2}}\right]_{z'=-a}^{+a} = \hat{\boldsymbol{\phi}}\frac{\mu_0 I a}{2\pi r\sqrt{r^2 + a^2}}$$

For an infinitely long conductor, or at small distances from a finite-length conductor (i.e., $r \ll a$), we have

$$\mathbf{B}_{\mathrm{P}} \simeq \hat{\boldsymbol{\phi}}\frac{\mu_0 I}{2\pi r}$$

analogous to $\mathbf{E} \simeq \hat{\mathbf{r}}\rho_l/(2\pi\epsilon_0 r)$ for an infinite line charge (see Example 4-5).

---

[30]Either by substitution of variables or by looking up integral tables, just as in the case of Example 4-5.

As a numerical example, the **B** field at a distance of 1 cm from a long straight wire carrying a current of $I = 5$ A is $B \simeq \mu_0 I/(2\pi r) = (4\pi \times 10^{-7})(5)/[2\pi(0.01)] = 10^{-4}$ T or 1 gauss.

Example 6-4: Square loop of current. Use the result derived in Example 6-3 to find the **B** field at the center of a square loop of current. Consider the square loop of side length $2a$ carrying a steady current I as shown in Figure 6.8.

Solution: First consider one side of the loop, say DE. Using the result of Example 6-3, the **B** field due to the current flowing along side DE is given by

$$\mathbf{B}_1 = -\hat{\mathbf{z}}\frac{\mu_0 I}{2\sqrt{2}\pi a}$$

From symmetry, the contributions by the other sides of the square loop are the same, since by the right-hand rule they all produce a magnetic field in the same direction. Thus, the total **B** field at the center is

$$\mathbf{B}_{\text{ctr}} = 4\mathbf{B}_1 = -\hat{\mathbf{z}}\frac{\sqrt{2}\mu_0 I}{\pi a}$$

Example 6-5: Two infinitely long parallel wires. Consider two infinitely long parallel wires, each carrying a current I in the z direction, one passing through the point $(x = 0, y = -a)$ and the other through $(x = 0, y = a)$, as shown in Figure 6.9. Find **B** at $P_1(x = 0, y = 0)$, $P_2(x = b, y = 0)$, and $P_3(x = 0, y = b)$.

Solution: Using the result of Example 6-3, namely that the **B** field at a distance r from an infinitely long straight wire carrying a current I in the z direction is $\mathbf{B} = \hat{\boldsymbol{\phi}}\mu_0 I/(2\pi r)$, we have

$$\mathbf{B}_{P_1} = \mathbf{B}_1 + \mathbf{B}_2 = \hat{\mathbf{x}}\frac{\mu_0 I}{2\pi a} - \hat{\mathbf{x}}\frac{\mu_0 I}{2\pi a} = 0$$

At P_2, we have

$$\begin{aligned}\mathbf{B}_{P_2} &= \mathbf{B}_1 + \mathbf{B}_2 \\ &= \frac{\mu_0 I}{2\pi\sqrt{a^2 + b^2}}(\hat{\mathbf{x}}\sin\psi + \hat{\mathbf{y}}\cos\psi) + \frac{\mu_0 I}{2\pi\sqrt{a^2 + b^2}}(-\hat{\mathbf{x}}\sin\psi + \hat{\mathbf{y}}\cos\psi) \\ &= \hat{\mathbf{y}}\frac{\mu_0 I b}{\pi(a^2 + b^2)}\end{aligned}$$

where we have used $\cos\psi = b(a^2 + b^2)^{-1/2}$. At P_3, we have

$$\begin{aligned}\mathbf{B}_{P_3} = \mathbf{B}_1 + \mathbf{B}_2 &= (-\hat{\mathbf{x}})\frac{\mu_0 I}{2\pi(b - a)} + (-\hat{\mathbf{x}})\frac{\mu_0 I}{2\pi(b + a)} \\ &= (-\hat{\mathbf{x}})\frac{\mu_0 I b}{\pi(b^2 - a^2)}\end{aligned}$$

Note that this result is valid regardless of whether $b > a$ or $a > b$.

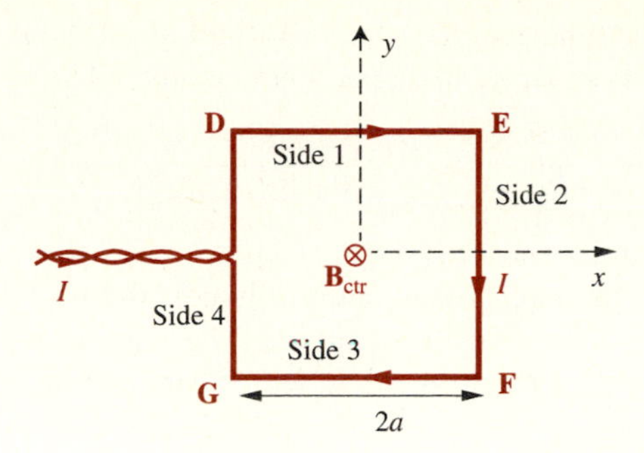

FIGURE 6.8. Square loop of current. A current I flows in a square wire loop.

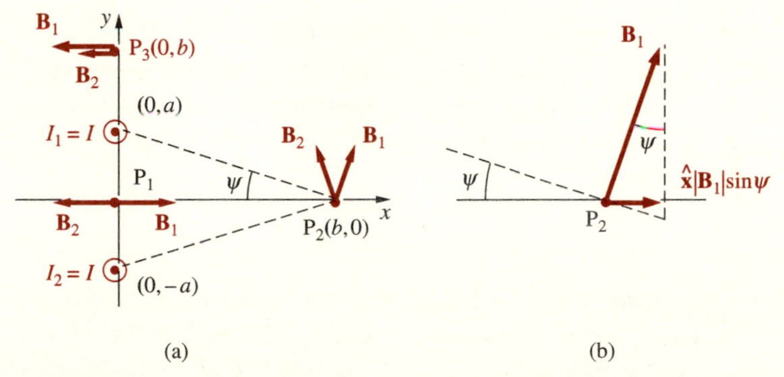

FIGURE 6.9. Two parallel wires. (a) Each wire is infinitely long carrying a current I in the z direction. (b) Expanded geometry of the vicinity of point P_2.

Example 6-6: Circular loop of current. Consider a circular loop of radius a, carrying a current I, and situated as shown in Figure 6.10. Find $\mathbf{B}$ at a point P on the axis of the loop.

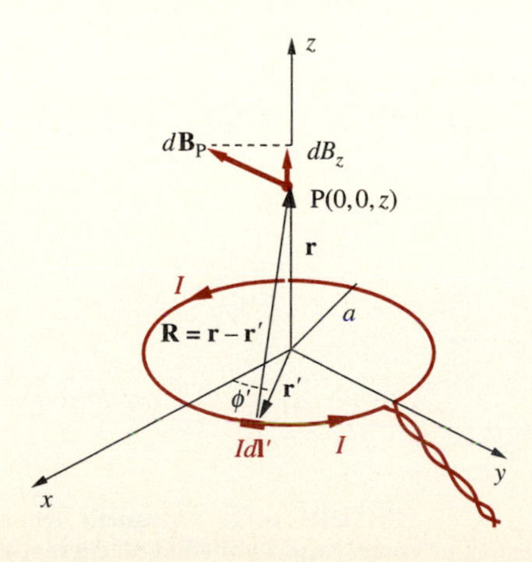

FIGURE 6.10. Circular loop of current. A circular loop of wire of radius a carrying a current I producing a $\mathbf{B}$ field along its axis with only a z component.

Solution: Using the cylindrical coordinate system, the **B** field at point P due to a current element $I\,d\mathbf{l}' = Ia\,d\phi'\hat{\boldsymbol{\phi}}$ located at $\mathbf{r}'$ is given by

$$d\mathbf{B}_{\mathrm{P}} = \frac{\mu_0 I\,d\mathbf{l}' \times \hat{\mathbf{R}}}{4\pi R^2} = \frac{\mu_0 Ia\,d\phi'\hat{\boldsymbol{\phi}} \times (\hat{\mathbf{z}}z - \hat{\mathbf{r}}a)}{4\pi(a^2+z^2)^{3/2}}$$

since in cylindrical coordinates we have $\mathbf{r} = \hat{\mathbf{z}}z$, $\mathbf{r}' = \hat{\mathbf{r}}a$, $R = \sqrt{a^2+z^2}$, and $\hat{\mathbf{R}} = (\mathbf{r}-\mathbf{r}')/R$. (Note that in rectangular coordinates, the vectors $\mathbf{r}$ and $\mathbf{r}'$ are given as $\mathbf{r} = z\hat{\mathbf{z}}$ and $\mathbf{r}' = \hat{\mathbf{x}}a\cos\phi' + \hat{\mathbf{y}}a\sin\phi'$). The cross product $\hat{\boldsymbol{\phi}} \times \hat{\mathbf{z}} = \hat{\mathbf{r}}$ gives the component of $d\mathbf{B}_{\mathrm{P}}$ in the $\hat{\mathbf{r}}$ direction, which cancels out upon integration over the whole loop, due to symmetry. The magnitude of the z component (dB_z), which results from the cross product $\hat{\boldsymbol{\phi}} \times (-\hat{\mathbf{r}}) = \hat{\mathbf{r}} \times \hat{\boldsymbol{\phi}} = \hat{\mathbf{z}}$, is

$$dB_z = \frac{\mu_0 Ia^2\,d\phi'}{4\pi(a^2+z^2)^{3/2}}$$

so that the total **B** at point P due to the whole loop is

$$\mathbf{B}_{\mathrm{P}} = \hat{\mathbf{z}}\int_{\phi'=0}^{2\pi} \frac{\mu_0 Ia^2}{4\pi(a^2+z^2)^{3/2}}\,d\phi' = \hat{\mathbf{z}}\frac{\mu_0 Ia^2}{4\pi(a^2+z^2)^{3/2}}\int_{\phi'=0}^{2\pi} d\phi' = \hat{\mathbf{z}}\frac{\mu_0 Ia^2}{2(a^2+z^2)^{3/2}}$$

Furthermore, at the center of the loop ($z = 0$), the **B** field simplifies to

$$\mathbf{B}_{\mathrm{ctr}} = \hat{\mathbf{z}}\frac{\mu_0 I}{2a}$$

Example 6-7: Finite-length wire B at an off-axis point. The problem arrangement is identical to that for Example 6-3, except that we now evaluate the **B** field at an arbitrary point P(r, ϕ, z), as shown in Figure 6.11.

Solution: From the Biot–Savart law, we have

$$d\mathbf{B}_{\mathrm{P}} = \frac{\mu_0}{4\pi}\frac{I\,dz'\,\hat{\mathbf{z}} \times \hat{\mathbf{R}}}{R^2} = \hat{\boldsymbol{\phi}}\frac{\mu_0}{4\pi}\frac{I\sin\alpha\,dz'}{[r^2+(z-z')^2]}$$

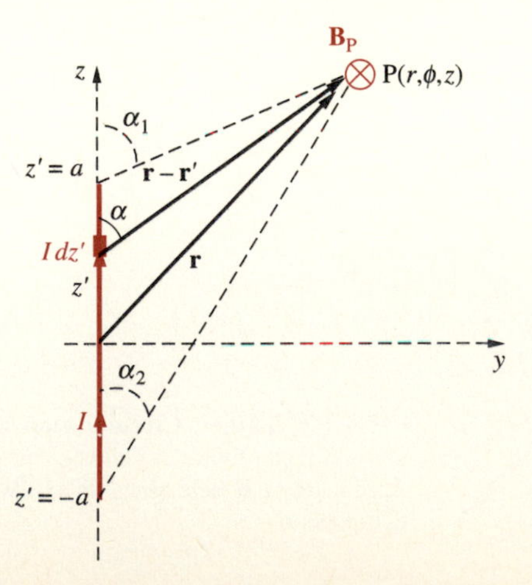

FIGURE 6.11. Magnetic field at an arbitrary point near a finite-length current-carrying wire.

where we note that $\mathbf{r} = \hat{\mathbf{r}}r + \hat{\mathbf{z}}z$ and $\mathbf{r}' = \hat{\mathbf{z}}z'$ so that $R = \sqrt{r^2 + (z - z')^2}$, and $\hat{\mathbf{R}} = (\mathbf{r} - \mathbf{r}')/R = [\hat{\mathbf{r}}r + \hat{\mathbf{z}}(z - z')]/R$. From Figure 6.11, we have

$$\sin\alpha = \frac{r}{\sqrt{r^2 + (z - z')^2}}$$

Substituting into the expression for $d\mathbf{B}_\text{P}$, we find

$$d\mathbf{B}_\text{P} = \hat{\boldsymbol{\phi}}\frac{\mu_0}{4\pi}\frac{Ir\,dz'}{[r^2 + (z - z')^2]^{3/2}}$$

Integrating over the full length of the wire, we find the total **B** field at point P as

$$\begin{aligned}\mathbf{B}_\text{P} &= \hat{\boldsymbol{\phi}}\frac{\mu_0 Ir}{4\pi}\int_{z'=-a}^{+a}\frac{dz'}{[r^2 + (z - z')^2]^{3/2}} \\ &= \hat{\boldsymbol{\phi}}\frac{\mu_0 Ir}{4\pi}\left[\frac{z - z'}{r^2\sqrt{r^2 + (z - z')^2}}\right]_{z'=+a}^{-a} \\ &= \hat{\boldsymbol{\phi}}\frac{\mu_0 I}{4\pi r}\left[\frac{z + a}{\sqrt{r^2 + (z + a)^2}} - \frac{z - a}{\sqrt{r^2 + (z - a)^2}}\right]\end{aligned}$$

where the integration can be carried out using either integral tables or substitution of variables, just as in the case of Example 4-5. We now note that

$$\frac{z + a}{\sqrt{r^2 + (z + a)^2}} = \cos\alpha_2 \qquad \text{and} \qquad \frac{z - a}{\sqrt{r^2 + (z - a)^2}} = \cos\alpha_1$$

so that we have

$$\mathbf{B}_\text{P} = \hat{\boldsymbol{\phi}}\frac{\mu_0 I}{4\pi r}(\cos\alpha_2 - \cos\alpha_1)$$

Note that for a point on the bisecting axis of the wire (i.e., at $z = 0$), we have

$$\cos\alpha_2 = -\cos\alpha_1 = \frac{a}{\sqrt{r^2 + a^2}}$$

and the result in Example 6-3 is obtained, namely

$$\mathbf{B}_\text{P} = \hat{\boldsymbol{\phi}}\frac{\mu_0 Ia}{2\pi r\sqrt{r^2 + a^2}}$$

Example 6-8: B field of a strip of sheet current. Consider an infinitely long perfectly conducting sheet of width d with a surface current density of $\mathbf{J}_s = \hat{\mathbf{z}}J_{s0}$ amperes per meter, as shown in Figure 6.12. Determine the **B** field produced by this strip at an arbitrary point P(x, y, z).

Solution: First, we can decompose the sheet into incremental filamentary currents, each located at $y = 0$, $x = x'$, carrying a current of $J_{s0}\,dx'$ and

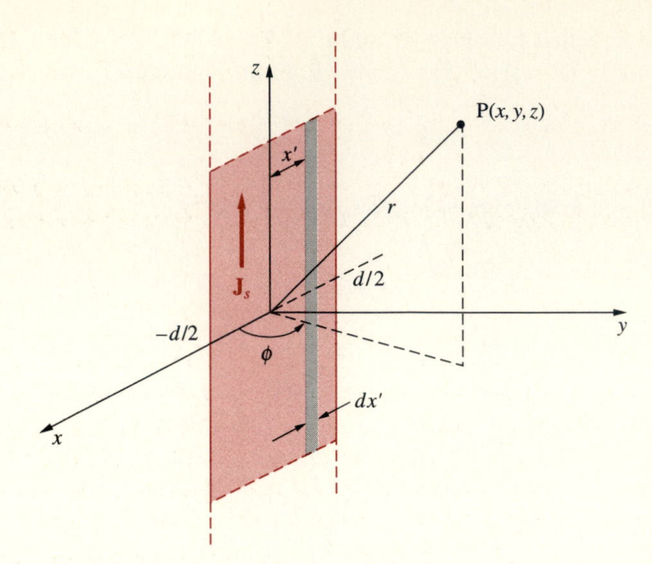

FIGURE 6.12. Surface current J_{s0} on an infinitely long sheet of width d. Note that the ribbonlike sheet lies in the xz (i.e., $y = 0$) plane.

producing a field $d\mathbf{B}$ as found in Example 6.3. This field would have components dB_x and dB_y given by

$$dB_x = \frac{-\mu_0 J_{s0}}{2\pi} \frac{y\,dx'}{y^2 + (x - x')^2} \qquad \text{and} \qquad dB_y = \frac{\mu_0 J_{s0}}{2\pi} \frac{(x - x')\,dx'}{y^2 + (x - x')^2}$$

Note that we used the basic result from Example 6-3 that for a strip of width dx' along the z axis, carrying a total current of $J_{s0}dx'$, we have

$$\begin{aligned} d\mathbf{B} &= \hat{\boldsymbol{\phi}} \frac{\mu_0 J_{s0} dx'}{2\pi\sqrt{x^2 + y^2}} = [-\hat{\mathbf{x}} \sin\phi + \hat{\mathbf{y}} \cos\phi] \frac{\mu_0 J_{s0} dx'}{2\pi\sqrt{x^2 + y^2}} \\ &= -\hat{\mathbf{x}} \frac{y\mu_0 J_{s0} dx'}{2\pi(x^2 + y^2)} + \hat{\mathbf{y}} \frac{x\mu_0 J_{s0} dx'}{2\pi(x^2 + y^2)} = \hat{\mathbf{x}} dB_x + \hat{\mathbf{y}} dB_y \end{aligned}$$

since $\cos\phi = x(x^2+y^2)^{-1/2}$ and $\sin\phi = y(x^2+y^2)^{-1/2}$ as shown in Appendix A. Replacing x with $(x-x')$ gives the dB_x and dB_y components due to a strip located at $x = x'$ instead of along the z axis. The total $\mathbf{B}$ field is found by integrating over x' from $-d/2$ to $d/2$. After manipulation of integrals[31] the components of the field are found to be

$$B_x(x, y, z) = \frac{-\mu_0 J_{s0}}{2\pi} \left[\tan^{-1}\left(\frac{x + d/2}{y}\right) - \tan^{-1}\left(\frac{x - d/2}{y}\right)\right]$$

---

[31] The essential integrals are

$$\int \frac{dx'}{y^2 + (x - x')^2} = -\frac{1}{y} \tan^{-1}\left(\frac{x - x'}{y}\right) \quad \text{and} \quad \int \frac{(x - x')\,dx'}{y^2 + (x - x')^2} = -\frac{1}{2} \ln[y^2 + (x - x')^2]$$

$$B_y(x, y, z) = \frac{\mu_0 J_{s0}}{4\pi} \ln\left[\frac{y^2 + (x + d/2)^2}{y^2 + (x - d/2)^2}\right]$$

Note that if the sheet current is infinitely wide (i.e., $d \to \infty$) in addition to being infinitely long (i.e., surface current flowing on the whole $y = 0$ plane), then the y components of the **B** field would cancel because of symmetry. Thus, the total **B** field in the x direction is given as

$$B_x = \begin{cases} -\dfrac{\mu_0 J_{s0}}{2} & y > 0 \\ \dfrac{\mu_0 J_{s0}}{2} & y < 0 \end{cases}$$

6.2.2 Magnetic Field Inside a Solenoid

The result found in Example 6-6 can be used directly to evaluate the **B** field for a commonly used practical circuit element, namely a solenoid. A solenoid is a coil consisting of many turns of an insulated wire wound on a cylindrical support with a circular cross section, as shown in Figure 6.13a. The cross section of the solenoid is shown in Figure 6.13b. Fairly strong fields can be produced by winding wires in the form of a solenoid so that the **B** fields due to the successive turns are additive.

Because of the coil being very tightly wound in a spiral, and since the wire diameter is typically much smaller than the length l of the solenoid, it is possible to treat the problem in terms of an equivalent surface current of density $J_s = (NI)/l$ A-m^{-1}, where N is the total number of turns and I is the current flowing in the wire, as shown in Figure 6.13c. A three-dimensional view of the equivalent sheet current $\mathbf{J}_s$ is also shown in Figure 6.13d. The current in an elemental length dz' of the solenoid is then given by

$$J_s\, dz' = \frac{NI}{l}\, dz'$$

We can view each dz' length of the solenoid as a thin circular current loop as in Example 6-6 and use the result derived there to find the **B** field at any point z along the axis of the solenoid. The differential field $d\mathbf{B}$ at any point z along the axis of the solenoid due to such a loop of differential thickness dz' located at a distance z' from the center ($z = 0$) of the solenoid is given by

$$d\mathbf{B} = \hat{\mathbf{z}} \frac{\mu_0 \left(\dfrac{NI}{l}\, dz'\right) a^2}{2[a^2 + (z - z')^2]^{3/2}}$$

To find the total field **B** of the solenoid at any point z along its axis, we integrate $d\mathbf{B}$ to sum the contributions of all such loops. We find

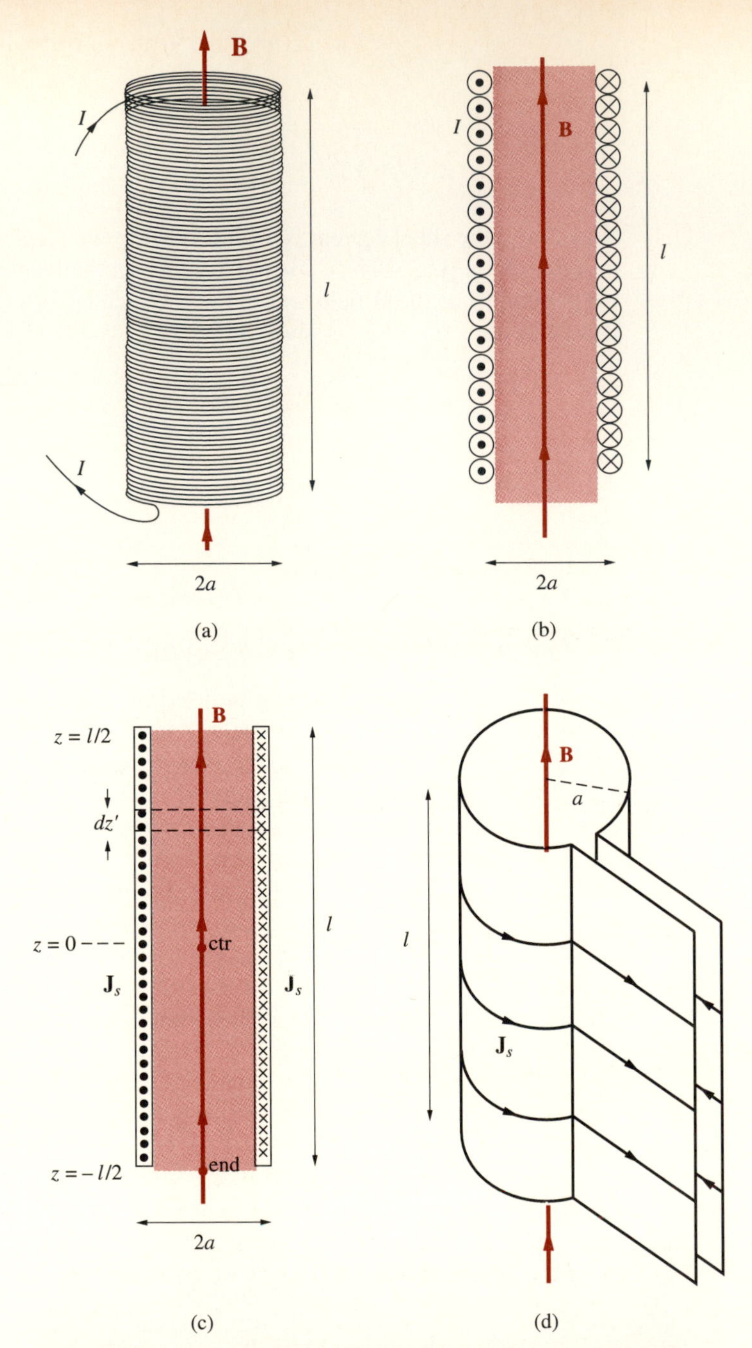

FIGURE 6.13. **Solenoid.** (a) Closely wound solenoid coil. (b) Cross section of the solenoid showing many fewer than the actual number of turns. (c) Equivalent surface current sheet of density $\mathbf{J}_s$ in A-m^{-1} representing the current flowing through the coil of the solenoid. (d) Three-dimensional view of equivalent surface current sheet of the solenoid.

$$\begin{aligned}
\mathbf{B}(z) &= \hat{\mathbf{z}}\frac{\mu_0 N I a^2}{2l}\int_{-l/2}^{+l/2}\frac{dz'}{[a^2+(z-z')^2]^{3/2}} \\
&= \hat{\mathbf{z}}\frac{\mu_0 N I a^2}{2l}\int_{z+l/2}^{z-l/2}\frac{-du}{(a^2+u^2)^{3/2}} \\
&= \hat{\mathbf{z}}\frac{\mu_0 N I a^2}{2la^2}\left[\frac{u}{\sqrt{a^2+u^2}}\right]_{z-l/2}^{z+l/2} \\
\mathbf{B}(z) &= \hat{\mathbf{z}}\frac{\mu_0 N I}{2l}\left[\frac{(z+l/2)}{\sqrt{a^2+(z+l/2)^2}}-\frac{(z-l/2)}{\sqrt{a^2+(z-l/2)^2}}\right]
\end{aligned}$$

where we have made the substitution $u = z - z'$, so that $du = -dz'$. To find the total field $\mathbf{B}$ at the center of the solenoid (the point marked "ctr" in Figure 6.13c), we substitute $z = 0$, yielding

$$\mathbf{B}_{\text{ctr}} = \hat{\mathbf{z}}\frac{\mu_0 N I}{\sqrt{(2a)^2+l^2}} = \hat{\mathbf{z}}\frac{\mu_0 N I}{l\sqrt{1+(2a/l)^2}}$$

For a very long solenoid or one with very small radius (i.e., $l \gg a$), the $\mathbf{B}$ field at the center of the solenoid is given by

$$\mathbf{B}_{\text{ctr}} \simeq \hat{\mathbf{z}}\frac{\mu_0 N I}{l}$$

and at the ends of its axis (i.e., $z = \pm l/2$) is

$$\mathbf{B}_{\text{end}} \simeq \hat{\mathbf{z}}\frac{\mu_0 N I}{2l}$$

We utilize these expressions for $\mathbf{B}_{\text{ctr}}$ and $\mathbf{B}_{\text{end}}$ for a specific case in Example 6-9.

Example 6-9: A solenoid-type magnetic filter. A magnetic filter involves a solenoid-type magnet used to separate very fine metallic particles from food products such as flours and corn starch. The solenoid has a 50 cm diameter and 1 m length. If the solenoid has 1000 turns of uniformly wound wire with a current of 100 A flowing through it, calculate the magnetic field (a) at its center and (b) at the end of its axis. Assume an air-core solenoid.

Solution: Using $a = 25$ cm and $l = 1$ m, we have

(a)

$$\mathbf{B}_{\text{ctr}} = \hat{\mathbf{z}}\frac{(4\pi\times10^{-7}\text{ H-m}^{-1})(1000\text{ turns})(100\text{ A})}{\sqrt{(0.5)^2+1^2}} \simeq 0.112\text{ T} = 1120\text{ gauss}$$

and

(b)

$$\mathbf{B}_{\text{end}} = \hat{\mathbf{z}}\frac{(4\pi \times 10^{-7}\ \text{H-m}^{-1})(1000\ \text{turns})(100\ \text{A})}{2\sqrt{(0.25)^2 + 1^2}} \simeq 0.061\ \text{T} = 610\ \text{gauss}$$

That $\mathbf{B}_{\text{end}}$ is exactly one-half of $\mathbf{B}_{\text{ctr}}$ can be seen by considering a very long solenoid of length $2l$ and with $2N$ turns. The field at the center of this solenoid is $\mu_0(2N)I/(2l)$, half of which is equal to $\mathbf{B}_{\text{end}}$ (by superposition, we expect the field to be halved when the contributions of the elemental loops from one half of the solenoid are removed). Thus, for a long solenoid, the **B** field at the ends of its axis is one-half that of the field at its center. The variation of the **B** field along the axis of the solenoid between its two ends is shown in Figure 6.14, for two different values of the ratio of the solenoid radius to its length. We see that for a long solenoid ($a/l = 0.1$), the magnetic field is effectively confined to the region inside the solenoid, and is nearly uniform except close to the edges. However, for a shorter solenoid ($a/l = 0.5$), there is more leakage of the **B** field outside the solenoid, and also the field is not as uniform within the solenoid. It is interesting to note, however, that the **B** field at the end (i.e., at $z = l/2$) of the solenoid is equal for both values of a/l.

In the foregoing discussion, we confined ourselves to the evaluation of the **B** field at points along the axis of the solenoid. For a long and slim solenoid (i.e., $l \gg a$), the field is nearly uniform within the solenoid even at points off its axis, as we shall show in the next section using Ampère's law.

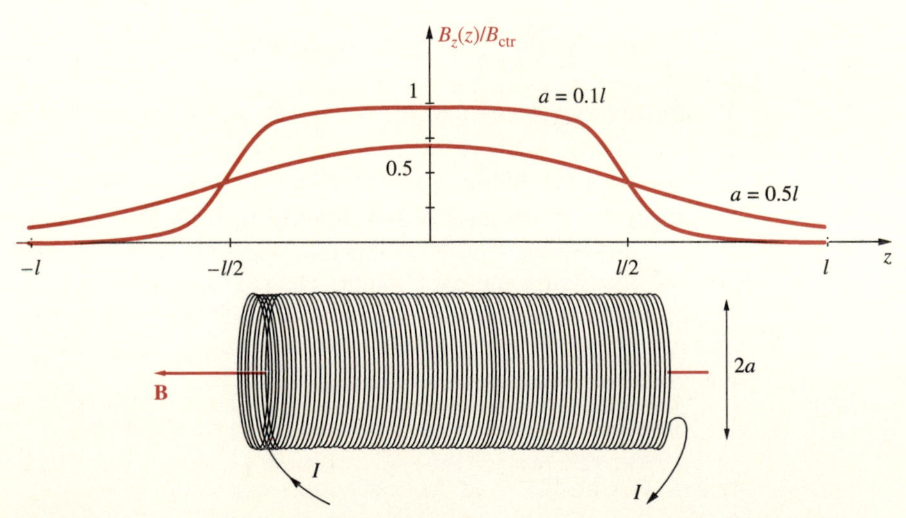

FIGURE 6.14. The variation of the B field along the axis of a solenoid. Normalized magnitude of B_z (i.e., $B(z)/B_{\text{ctr}}$) is plotted versus z for two different ratios of radius versus length, namely $a/l = 0.1$ and 0.5.

6.2.3 The Duality of the B and E Fields

The **B** field is in many ways analogous to the electric field **E** that we worked with in Chapter 4, rather than to the electric flux density **D**. In this section we comment on this duality. The basic experimental fact of magnetostatics, as elucidated[32] by Ampère, is that current-carrying wires exert forces on one another. The Biot–Savart law is one statement of this experimental fact, as discussed at the beginning of Section 6.2. In terms of Ampère's law of force, the magnetic force $\mathbf{F}_{\mathrm{m}}$ felt by a current element $I d\mathbf{l}$ in the presence of a magnetic field **B** is

$$\mathbf{F}_{\mathrm{m}} = I\, d\mathbf{l} \times \mathbf{B}$$

Noting that current is charge in motion, a charge q moving at a velocity **v** is equivalent to an element of current $I\, d\mathbf{l} = q\mathbf{v}$, and hence, in the presence of a magnetic field **B**, would experience a force $\mathbf{F}_{\mathrm{m}}$ given by

$$\mathbf{F}_{\mathrm{m}} = q\mathbf{v} \times \mathbf{B} \qquad [6.7]$$

This force is called the *Lorentz magnetic force,* and [6.7] is sometimes[33] taken as the defining equation for **B**.

In many ways, the quantity **B** is analogous to the electric field **E**, as can be seen by comparing the electric and magnetic force expressions given below:

$$\mathbf{F}_{\mathrm{e}} = q\mathbf{E} = q\int_{V'} \frac{\rho(\mathbf{r}')\hat{\mathbf{R}}\, dv'}{4\pi\epsilon_0 R^2}$$

$$\mathbf{F}_{\mathrm{m}} = q\mathbf{v} \times \mathbf{B} = q\mathbf{v} \times \int_{V'} \frac{\mathbf{J}(\mathbf{r}') \times \hat{\mathbf{R}}\, dv'}{4\pi\mu_0^{-1} R^2}$$

We can see that both **E** and **B** act on a charge q to produce force, both are related to their respective sources (ρ for **E** and **J** for **B**) in a similar manner, and both are medium-dependent (through ϵ_0^{-1} and μ_0) quantities. In view of this similarity, it is unfortunate that **B** is usually referred to as the *magnetic flux density,* which conveys the incorrect impression[34] of a similarity between **B** and the electric flux density **D**. In this text, we may succumb to this convention and on occasion refer to **B** as the *magnetic flux density,* although in most cases we simply refer to it as the "**B** field."

---

[32] A.-M. Ampère, On the mathematical theory of electrodynamic phenomena uniquely deduced from experiment, *Mem. Acad.*, pp. 175–388, 1825.

[33] The Lorentz force acting on a particle with charge q is the vector sum of the magnetic and electric Lorentz forces; that is, $\mathbf{F} = q\mathbf{E} + q\mathbf{v} \times \mathbf{B}$. Also see Section 6.10.1.

[34] The magnetostatic quantity that is the correct analogue of the electric flux density vector **D** is in fact a different vector **H,** which we introduce later in Section 6.8. Here, it suffices to mention that **H** is a medium-independent quantity defined by the relation $\mathbf{H} \equiv \mu^{-1}\mathbf{B}$, where μ is the magnetic permeability of the medium also defined and introduced in Section 6.8. Also see footnote 71.

6.3 AMPÈRE'S CIRCUITAL LAW

In electrostatics, problems involving symmetries of one form or another can often be solved more easily using Gauss's law than by direct application of Coulomb's law. For static magnetic fields, the Biot–Savart law and Ampère's force law are the point relationships that are analogous to Coulomb's law, whereas *Ampère's circuital law* serves the same purpose as Gauss's law. Ampère's circuital law, hereafter referred to simply as *Ampère's law,* is a mathematical consequence[35] of the Biot–Savart law, much as Gauss's law was that of Coulomb's law. Ampère's law can simply be stated as

$$\oint_C \mathbf{B} \cdot d\mathbf{l} = \int_S \mu_0 \mathbf{J} \cdot d\mathbf{s} = \mu_0 I \qquad [6.8]$$

where C is the contour that encloses the surface S. In words, [6.8] states that the line integral of $\mathbf{B}$ around any closed contour is equal to μ_0 times the total net current I passing through the surface S enclosed by the contour C. This law is particularly useful in solving magnetostatic problems having some degree of symmetry (usually cylindrical symmetry), when a contour C over which the $\mathbf{B}$ field is constant can be identified, in which case the integration on the left-hand side of [6.8] can be readily carried out.

Examples 6-10 through 6-15 illustrate the use of Ampère's law for some simple but useful current configurations.

Example 6-10: Infinitely long cylindrical conductor. Consider an infinitely long, straight cylindrical conductor of radius a carrying a steady current I, as shown in Figure 6.15a. Assume[36] the current density to be uniform over the cross section of the conductor. Find the magnetic field both inside and outside the conductor.

Solution: We choose to use the cylindrical coordinate system with the current I flowing in the z direction. In view of the cylindrical symmetry, it is convenient to use Ampère's law to find $\mathbf{B}$. According to the Biot–Savart law and the right-hand rule, $\mathbf{B}$ is in the $\hat{\boldsymbol{\phi}}$ direction, so that we have $\mathbf{B}(r) = B(r)\hat{\boldsymbol{\phi}}$. We take two circular contours, C_1 and C_2, respectively with radii $r_1 < a$ and $r_2 > a$, as shown in Figure 6.15a. Applying Ampère's law along path C_1 lying inside the conductor, we have

[35]A formal derivation of Ampère's law from the Biot–Savart law is provided in Appendix B.

[36]We shall see in Chapter 8 that the current can be uniformly distributed across the cross section only for steady currents. In most applications involving time-varying fields, the current is confined mostly to a narrow region at the outer edge of the conductor.

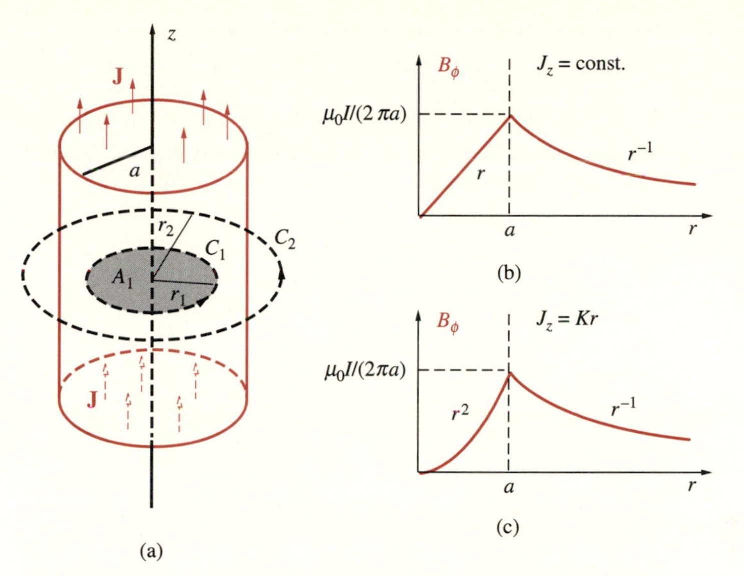

FIGURE 6.15. Infinitely long cylindrical conductor. (a) The cylindrical conductor carrying a current with current density **J**. (b) Magnetic flux density B_ϕ versus distance r for the case of uniform current distribution. (c) B_ϕ versus r when $J_z = Kr$.

$$\oint_{C_1} \mathbf{B} \cdot d\mathbf{l} = \mu_0 \int_{A_1} \mathbf{J} \cdot d\mathbf{s}$$

$$\int_{\phi=0}^{2\pi} (B_\phi \hat{\boldsymbol{\phi}}) \cdot (\hat{\boldsymbol{\phi}} r_1 \, d\phi) = \mu_0 \int_0^{2\pi} \int_0^{r_1} \frac{I}{\pi a^2} \hat{\mathbf{z}} \cdot \hat{\mathbf{z}} r \, dr \, d\phi = \mu_0 I_1$$

where I_1 is the portion of the total current I through the area $A_1 = \pi r_1^2$ enclosed by contour C_1. From the above we have

$$B_\phi r_1 \int_0^{2\pi} d\phi = \mu_0 \frac{I}{\pi a^2} A_1 = \mu_0 \frac{I}{\pi a^2} \pi r_1^2 \quad \rightarrow \quad B_\phi = \frac{\mu_0 I r_1}{2\pi a^2} \qquad r_1 \leq a$$

Similarly, by applying Ampère's law along contour C_2 lying outside the conductor (i.e., $r_2 > a$), we find

$$B_\phi(2\pi r_2) = \mu_0 I \quad \rightarrow \quad B_\phi = \frac{\mu_0 I}{2\pi r_2} \qquad r_2 > a$$

In summary, the **B** field at any position r is

$$\mathbf{B} = \begin{cases} \hat{\boldsymbol{\phi}} \dfrac{\mu_0 I r}{2\pi a^2} & r \leq a \\[2ex] \hat{\boldsymbol{\phi}} \dfrac{\mu_0 I}{2\pi r} & r > a \end{cases}$$

which is plotted in Figure 6.15b.

Example 6-11: Nonuniform distribution of current. Consider an infinitely long, straight cylindrical conductor of radius a carrying a steady current I, as in the previous example, except that the current is distributed in such a way that the current density J is proportional to r, the distance from the axis (assume $J_z = Kr$, where K is a constant). Find the magnetic field both inside and outside the conductor.

Solution: Since the total current flowing through the conductor is I, we have

$$I = \int_S \mathbf{J} \cdot d\mathbf{s} = \int_0^{2\pi} \int_0^a Kr\hat{\mathbf{z}} \cdot \hat{\mathbf{z}} r\, dr\, d\phi \quad \rightarrow \quad K = \frac{3I}{2\pi a^3}$$

So, the current density is given by $\mathbf{J} = \hat{\mathbf{z}}3Ir/(2\pi a^3)$. Applying Ampère's law along path C_1 inside the conductor, we can write

$$\oint_{C_1} \mathbf{B} \cdot d\mathbf{l} = \mu_0 \int_{A_1} \mathbf{J} \cdot d\mathbf{s}$$

$$\int_0^{2\pi} B_\phi \hat{\boldsymbol{\phi}} \cdot (\hat{\boldsymbol{\phi}} r_1\, d\phi) = \left(\frac{3\mu_0 I}{2\pi a^3}\right)\left(\int_0^{r_1} 2\pi r^2\, dr\right)$$

$$B_\phi(2\pi r_1) = \left(\frac{3\mu_0 I}{2\pi a^3}\right)\left(\frac{2\pi r_1^3}{3}\right)$$

from which we have

$$B_\phi = \frac{\mu_0 I r_1^2}{2\pi a^3} \qquad r_1 \leq a$$

Similarly, by applying Ampère's law along contour C_2 lying outside the conductor, we find

$$B_\phi(2\pi r_2) = \mu_0 I \quad \rightarrow \quad B_\phi = \frac{\mu_0 I}{2\pi r_2} \qquad r_2 > a$$

In summary, the **B** field at any position r is

$$\mathbf{B} = \begin{cases} \hat{\boldsymbol{\phi}} \dfrac{\mu_0 I r^2}{2\pi a^3} & r \leq a \\[2ex] \hat{\boldsymbol{\phi}} \dfrac{\mu_0 I}{2\pi r} & r > a \end{cases}$$

Figure 6.15c shows the variation of B_ϕ as a function of r.

Example 6-12: Coaxial line. Consider an infinitely long coaxial transmission line carrying a uniformly distributed current I in the inner conductor of radius a and $-I$ in the outer conductor of inner and outer radii b and c such that $c > b > a$, whose cross-sectional view is shown in Figure 6.16a. Find the **B** field everywhere.

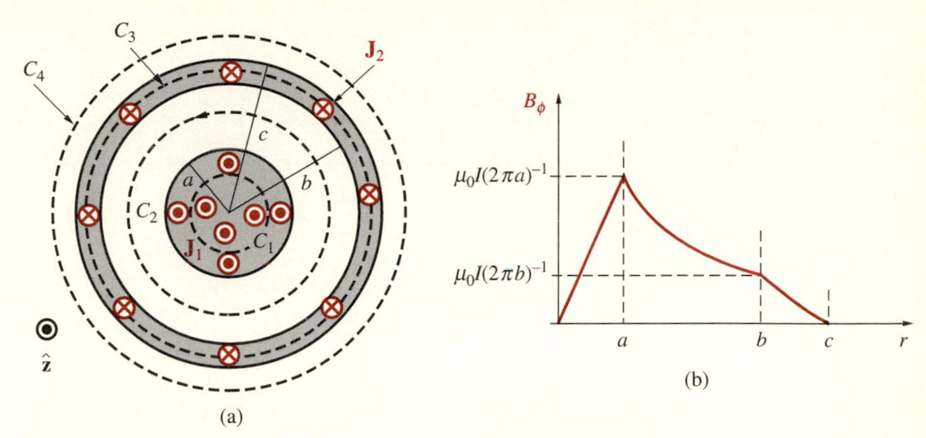

FIGURE 6.16. **Coaxial line.** (a) Cross-sectional view. (b) Variation of the **B** field as a function of r.

Solution: Following a similar approach as in Example 6-10, we can apply Ampère's law for $r_1 \leq a$ to write

$$\oint_{C_1} \mathbf{B} \cdot d\mathbf{l} = \oint_{C_1} (B_\phi \hat{\boldsymbol{\phi}}) \cdot (\hat{\boldsymbol{\phi}} r_1 \, d\phi) = \mu_0 \int_{A_1} \mathbf{J}_1 \cdot d\mathbf{s} = \mu_0 I_1$$

where I_1 is the portion of the current I passing through the area A_1. Noting that $\mathbf{J}_1 = \hat{\mathbf{z}} I / (\pi a^2)$, we have

$$B_\phi (2\pi r_1) = \mu_0 \frac{I}{\pi a^2} \pi r_1^2 \quad \rightarrow \quad B_\phi = \frac{\mu_0 I r_1}{2\pi a^2} \qquad r_1 \leq a$$

which is identical to that found in Example 6-10 for $r_1 < a$. For $a < r_2 < b$, we have

$$B_\phi = \frac{\mu_0 I}{2\pi r_2} \qquad a < r_2 \leq b$$

also identical to that in Example 6-10. For $b < r_3 \leq c$, using $\mathbf{J}_2 = -\hat{\mathbf{z}}(I/[\pi(c^2 - b^2)])$ we have

$$\begin{aligned} B_\phi (2\pi r_3) &= \mu_0 \left[\int_0^{2\pi} \int_0^a \frac{I}{\pi a^2} r \, dr \, d\phi + \int_0^{2\pi} \int_b^{r_3} \frac{-I}{\pi (c^2 - b^2)} r \, dr \, d\phi \right] \\ &= \mu_0 \left[I - \frac{r_3^2 - b^2}{c^2 - b^2} I \right] \\ &= \mu_0 I \frac{c^2 - r_3^2}{c^2 - b^2} \\ \rightarrow \quad B_\phi &= \frac{\mu_0 I}{2\pi r_3} \frac{c^2 - r_3^2}{c^2 - b^2} \qquad b < r_3 \leq c \end{aligned}$$

For $r_4 > c$, we have

$$B_\phi(2\pi r_4) = \mu_0\left[\underbrace{\int_0^{2\pi}\int_0^a \frac{I}{\pi a^2} r\,dr\,d\phi}_{I} + \underbrace{\int_0^{2\pi}\int_b^c \frac{-I}{\pi(c^2-b^2)} r\,dr\,d\phi}_{-I}\right] = 0$$

$$\rightarrow \quad B_\phi = 0 \qquad r_4 > c$$

since the net current linking the area enclosed by contour C_4 is zero. In summary, the **B** field at any position r is given by

$$\mathbf{B} = \begin{cases} \hat{\boldsymbol{\phi}}\dfrac{\mu_0 I r}{2\pi a^2} & r \le a \\ \hat{\boldsymbol{\phi}}\dfrac{\mu_0 I}{2\pi r} & a < r \le b \\ \hat{\boldsymbol{\phi}}\dfrac{\mu_0 I}{2\pi r}\dfrac{c^2 - r^2}{c^2 - b^2} & b < r \le c \\ 0 & r > c \end{cases}$$

Figure 6.16b shows the variation of B_ϕ as a function of r.

Example 6-13: Conductor with hole. A long cylindrical conductor of radius b and carrying a uniform current I in the z direction has a cylindrical hole of radius a along its entire length, as shown in Figure 6.17. The center of the hole is offset from the center of the conductor by a distance d. Find **B** inside the hole.

Solution: We can consider the region of the hole, within which the current density is zero, to result from two equal currents flowing in opposite directions.

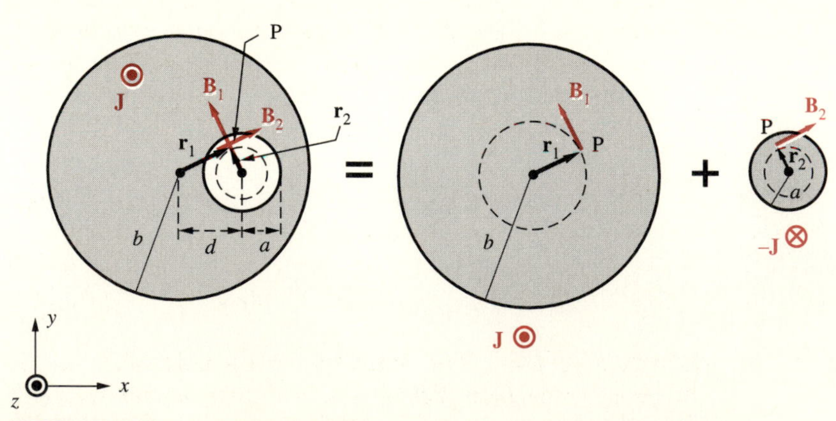

FIGURE 6.17. Cylindrical conductor with a hole. The conductor with a hole is viewed as a superposition of two solid cylindrical conductors of radii a and b, carrying equal but opposite current density **J**.

Accordingly, we can represent the problem in hand as a superposition of a cylindrical conductor of radius b carrying a current with density **J** without a hole and another of radius a with a current density of the same magnitude, but flowing in the opposite direction (i.e., $-\mathbf{J}$). This representation is indicated in Figure 6.17. Note that the current density **J** is

$$\mathbf{J} = \hat{\mathbf{z}}\frac{I}{\pi(b^2 - a^2)}$$

Applying Ampère's law, the **B** field at point $\mathrm{P}_1(r_1, \phi_1, z_1 = 0)$ in the hole region due to the current in the larger conductor (radius b) with no hole is

$$\mathbf{B}_1 = \hat{\boldsymbol{\phi}}\frac{\mu_0 I r_1}{2\pi(b^2 - a^2)}$$

based on the result derived in Example 6-10. This result can be rewritten as

$$\mathbf{B}_1 = \frac{\mu_0 I}{2\pi(b^2 - a^2)}(\hat{\mathbf{z}} \times \mathbf{r}_1) = \frac{\mu_0}{2}(\mathbf{J} \times \mathbf{r}_1)$$

where $\mathbf{r}_1 = \hat{\mathbf{r}}r_1$. Similarly, the **B** field at P_1 due to the current in the smaller conductor (radius a) is

$$\mathbf{B}_2 = \frac{\mu_0 I}{2\pi(b^2 - a^2)}[(-\hat{\mathbf{z}}) \times \mathbf{r}_2] = \frac{\mu_0}{2}[(-\mathbf{J}) \times \mathbf{r}_2]$$

where $\mathbf{r}_2 = \hat{\mathbf{r}}r_2$. Therefore, the total **B** field at P_1 is

$$\mathbf{B} = \mathbf{B}_1 + \mathbf{B}_2 = \frac{\mu_0}{2}[\mathbf{J} \times (\mathbf{r}_1 - \mathbf{r}_2)] = \frac{\mu_0}{2}[\mathbf{J} \times (\hat{\mathbf{x}}d)]$$

where we have noted from Figure 6.17 that $(\mathbf{r}_1 - \mathbf{r}_2) = \hat{\mathbf{x}}d$. This result shows that the magnetic field anywhere inside the hole is constant, and its direction is at right angles to the line joining the centers of the two circles. In the case shown, with the center of the hole on the x axis, the **B** field everywhere in the hole is constant and is in the y direction. Rewriting in terms of I, we then have

$$\mathbf{B} = \hat{\mathbf{y}}\frac{\mu_0 I d}{2\pi(b^2 - a^2)}$$

We can check the above result in two limiting cases. First, if the hole is directly at the center of the conductor (i.e., $d = 0$), we have $\mathbf{B} = 0$, as expected based on Ampère's law (i.e., no current enclosed). Secondly, if the radius a of the hole approaches zero ($a \to 0$), then **B** reduces to the form obtained in Example 6-10.

Example 6-14: The infinitely long solenoid. In Section 6.2.2 we considered the magnetic field inside a tightly wound solenoid of finite length and derived an expression for the **B** field along its axis using the Biot–Savart law. In this example, we consider a very long solenoid and find the **B** field inside it using Ampère's law.

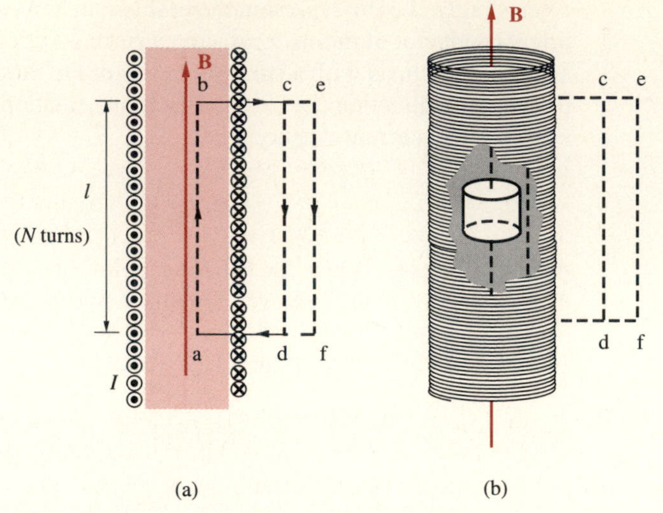

FIGURE 6.18. **Cross section of a solenoid.**

Solution: Consider the cross section of a closely wound solenoid as shown in Figure 6.18. For an infinitely long solenoid the **B** field everywhere inside must be parallel to the axis (i.e., $\mathbf{B} = \hat{\mathbf{z}}B_z$, with $B_r = 0$ and $B_\phi = 0$) of the solenoid,[37] and the field outside the solenoid must be zero.[38] Applying Ampère's law to a closed contour abcd, selected to be of length l and containing N turns of the solenoid coil, gives

$$\oint_{\text{abcd}} \mathbf{B} \cdot d\mathbf{l} = \int_{\text{ab}} + \underbrace{\int_{\text{bc}} + \int_{\text{cd}} + \int_{\text{da}}}_{=0}$$

Note that the integral along cd is zero because $\mathbf{B} = 0$, as argued above, and those along bc and da are zero since $\mathbf{B} = \hat{\mathbf{z}}B_z$ is perpendicular to $d\mathbf{l}$. Thus

$$\int_{\text{ab}} \mathbf{B} \cdot d\mathbf{l} = \mu_0 NI$$

$$Bl = \mu_0 NI$$

$$B = \frac{\mu_0 NI}{l}$$

[37]The reason for this is the fact that the integral of the **B** field over any closed surface is identically zero, as will be shown later in Section 6.7. Considering an axial cylinder of unit length inside the solenoid as shown in Figure 6.18b, the integral $\oint \mathbf{B} \cdot d\mathbf{s}$ over its surface is simply $2\pi r B_r$, since the integrals over the two end faces cancel. Thus, B_r must be zero in order for $\oint \mathbf{B} \cdot d\mathbf{s}$ to be zero. Note that $B_\phi = 0$ because all of the current flow is in the $\hat{\boldsymbol{\phi}}$ direction and we know from the Biot–Savart law that **B** must be perpendicular to the direction of current flow.

[38]If the field outside was not zero, then the line integral of **B** around the contour cdef would not be zero (as it should be, since it does not enclose any current) unless the field were the same along cd and ef (no contribution along the ce and df paths since $B_r = 0$), which would in turn imply that a constant field exists everywhere outside the solenoid, contrary to experimental observation.

which is identical to the result found in Section 6.2.2 for $\mathbf{B}_{\text{ctr}}$ for a very long solenoid. This result indicates also that the **B** field is constant across the cross section of the solenoid; note that we did not in any way restrict the location of the segment ab in Figure 6.18 to be on the axis of the solenoid.

In the context of the above formulation, it is easy to show that the z component of the **B** field at either end of a long solenoid would be half of that at its center. To see this, we can cut a very long solenoid into two. If the same current I is maintained in the two parts, the magnetic flux density in the newly created ends must drop to half of its original values; otherwise the field would not combine to its original value when the two ends were reconnected. Note, however, that the radial component B_r is not zero at the ends of a finite length solenoid (see Figure 6.40a in Section 6.7.3).

Example 6-15: The toroid. If the long solenoid of Example 6-14 is bent into a circle and closed on itself, we obtain a toroidal (doughnut-shaped) coil. A toroidal coil with a rectangular cross section is shown in Figure 6.19a. Find the **B** field inside the toroid.

Solution: Let the toroid consist of N turns of wire carrying a current I, uniformly and tightly distributed around its circumference as shown in Figure 6.19a. Because of cylindrical symmetry, the magnetic field **B** has only a ϕ component. We apply Ampère's law to any circular contour C_1 of radius r such that $a < r < b$ lying inside the toroid, as shown. By symmetry the field **B** must be the same everywhere along such a contour. Noting that the total current crossing through the area enclosed by the contour C_1 is NI, we have

$$\oint_{C_1} \mathbf{B} \cdot d\mathbf{l} = \oint_{C_1} (B_\phi \hat{\boldsymbol{\phi}}) \cdot (\hat{\boldsymbol{\phi}} r\, d\phi) = B_\phi(2\pi r) = \mu_0 NI \quad \rightarrow \quad B_\phi = \frac{\mu_0 NI}{2\pi r}$$

For any circular path C_2 that lies outside the toroid ($r > b$), the total net current crossing through the enclosed area of C_2 is $N(I - I) = 0$, so that $\oint_{C_2} \mathbf{B} \cdot d\mathbf{l} = 0$,

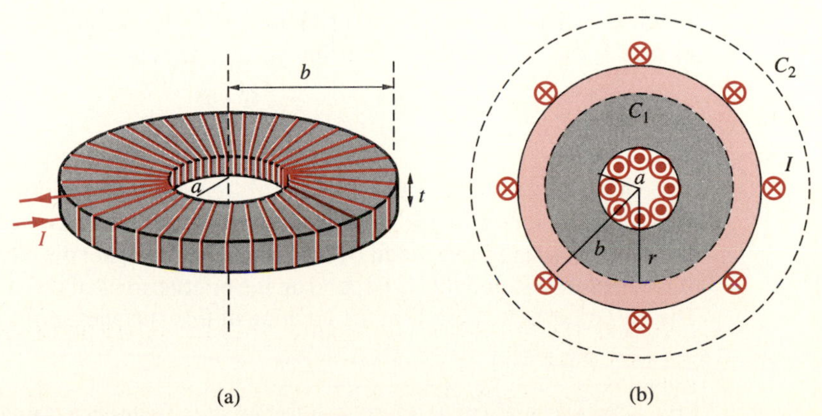

FIGURE 6.19. A toroid. (a) Three-dimensional view showing the windings. (b) Cross-sectional view showing two Amperian contours C_1 and C_2.

yielding $\mathbf{B} = 0$. Note that if we consider a contour with radius $r < a$, we can conclude by the same arguments that the magnetic field is also zero in the hole of the toroid. Thus, the magnetic field of a closely wound toroid is confined to the interior of its windings.

For a numerical example, consider the air-core toroid reactor vessel of the Tokamak Fusion Test Reactor[39] used for high-power plasma fusion experiments at Princeton University. This toroid has a mean radius of 2.2 m and consists of 20 coils, each with 44 turns per coil, uniformly wound around it. With each coil carrying a current of 73.3 kA, the **B** field at its mean radius can be found using the expression derived above. Namely,

$$B_\phi = \frac{(4\pi \times 10^{-7}\ \text{H-m}^{-1})(20 \times 44\ \text{turns})(73.3 \times 10^3\ \text{A})}{2\pi(2.2\ \text{m})} \simeq 5.86\ \text{T}$$

6.4 CURL OF THE MAGNETIC FIELD: DIFFERENTIAL FORM OF AMPÈRE'S LAW

In electrostatics, we discussed the concept of the divergence of a vector field, and used it to express the differential form of Gauss's law as $\nabla \cdot \mathbf{D} = \rho$. A second very important property of a vector field is its circulation. We can observe circulation as we watch water drain out of a bathtub or sink. For a perfectly symmetrical circulating fluid with angular velocity ω, a measure of the circulation at any radius r may be defined as the product of angular velocity and circumference [i.e., $\omega(2\pi r)$]. We can extend this concept of circulation to any vector field **G** by defining total or net circulation about any arbitrary closed path C as

$$\text{circulation} \equiv \oint_C \mathbf{G} \cdot d\mathbf{l}$$

which is called the *circulation* of **G** around C. Since any closed loop C encloses a surface S, the net circulation per unit surface area is

$$\frac{\oint_C \mathbf{G} \cdot d\mathbf{l}}{S}$$

Since a given vector field **G** may have different amounts of circulation at different points, we now consider a differential surface element $\Delta\mathbf{s} = \hat{\mathbf{s}}\Delta s$ located at some point in space, and surrounded by a contour C. We expect the circulation per unit area of the vector field about C to depend on the orientation $\hat{\mathbf{s}}$ of the surface enclosed by C. The *curl* of a vector field **G** at the location of this surface element is defined as an ax-

[39] For a description of fusion reactors and the Tokamak facility referred to here, see Chapter 9 of F. F. Chen, *Introduction to Plasma Physics,* Plenum Press, 1974.

ial vector (i.e., perpendicular to Δs) whose magnitude is the maximum (as the direction of the surface element $\Delta\mathbf{s} = \hat{\mathbf{s}}\Delta s$ is varied) net circulation of vector $\mathbf{G}$ per unit area as the area of the surface element approaches zero. In other words,

$$\text{curl } \mathbf{G} \equiv \left[\lim_{\Delta s \to 0} \frac{\oint_C \mathbf{G} \cdot d\mathbf{l}}{\Delta s} \right]_{\max} \hat{\mathbf{s}} \qquad [6.9]$$

Note that curl is clearly a directional quantity; by convention, the direction of curl is chosen to be the direction of the surface element $\Delta\mathbf{s} = \hat{\mathbf{s}}\Delta s$ that gives the maximum value for the magnitude of the net circulation per unit area in the defining expression [6.9]. Curl $\mathbf{G}$ is thus a vector; for example, the x component of the vector (curl $\mathbf{G}$) represents the line integral of $\mathbf{G}$ per unit area along an infinitesimally small (i.e., $\Delta s \to 0$) closed path lying in the yz plane.

For the $\mathbf{B}$ field, we know from Ampère's law that the circulation of $\mathbf{B}$ along any closed contour C (i.e., line integral of $\mathbf{B}$ along C) is equal to the total current passing through the surface area S enclosed by C times μ_0. For a differential surface element, we can assume that the current density is constant at all points over the surface, so the total current is $|\mathbf{J}|\Delta s$, if the surface is chosen to be orthogonal to the direction of the current flow so that the net circulation is maximized. Accordingly, we have

$$\text{curl } \mathbf{B} \equiv \left[\lim_{\Delta s \to 0} \frac{\oint_C \mathbf{B} \cdot d\mathbf{l}}{\Delta s} \right]_{\max} \hat{\mathbf{s}} = \mu_0 \mathbf{J}$$

6.4.1 Curl in the Rectangular Coordinate System

We now utilize the definition of curl to obtain a convenient expression in terms of partial derivatives. For this, we simply need to evaluate curl $\mathbf{B}$ at a general point in a given coordinate system. With reference to Figure 6.20, we can evaluate the component of curl $\mathbf{B}$ in the x direction by conducting a line integral along the path abcda, which, as shown, lies entirely in the yz plane. Note that this is an infinitesimally small path with dimensions Δy and Δz, so the magnitude of $\mathbf{B}$ at different sides can be simply related to the value of the field at the center point x_0, y_0, z_0. From Ampère's law, we have

$$\oint_{\text{abcda}} \mathbf{B} \cdot d\mathbf{l} = B_{1y}\Delta y + B_{2z}\Delta z - B_{3y}\Delta y - B_{4z}\Delta z = \mu_0 J_x \Delta y \Delta z$$

$$\underbrace{(B_{1y} - B_{3y})}_{-(\partial B_y/\partial z)\Delta z}\Delta y + \underbrace{(B_{2z} - B_{4z})}_{(\partial B_z/\partial y)\Delta y}\Delta z = \mu_0 J_x \Delta y \Delta z$$

$$\left[\frac{\partial B_z}{\partial y} - \frac{\partial B_y}{\partial z} \right] \Delta y \Delta z = \mu_0 J_x \Delta y \Delta z$$

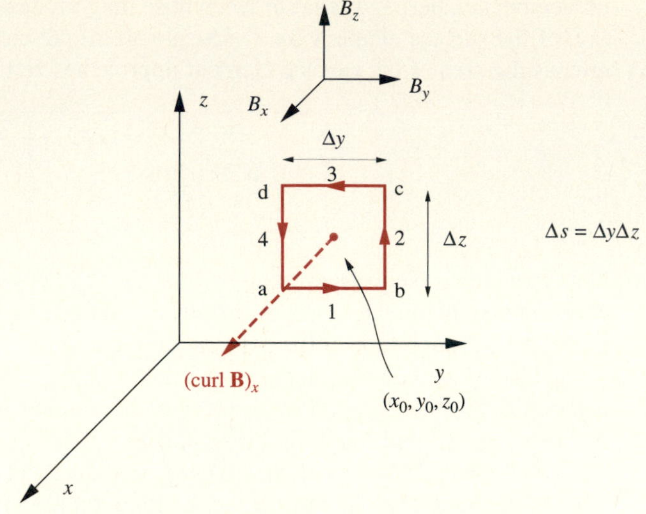

FIGURE 6.20. **Evaluation of the x component of curl B.**

from which

$$(\text{curl } \mathbf{B})_x = \lim_{\Delta s \to 0} \frac{\oint_{\text{abcda}} \mathbf{B} \cdot d\mathbf{l}}{\Delta s} = \frac{\partial B_z}{\partial y} - \frac{\partial B_y}{\partial z} = \mu_0 J_x$$

where the partial derivatives are to be evaluated at the center point (x_0, y_0, z_0). By taking line integrals along other contours lying in the xz and xy planes, we can show that

$$\text{curl } \mathbf{B} = \hat{\mathbf{x}}\left(\frac{\partial B_z}{\partial y} - \frac{\partial B_y}{\partial z}\right) + \hat{\mathbf{y}}\left(\frac{\partial B_x}{\partial z} - \frac{\partial B_z}{\partial x}\right) + \hat{\mathbf{z}}\left(\frac{\partial B_y}{\partial x} - \frac{\partial B_x}{\partial y}\right)$$

$$= \mu_0(\hat{\mathbf{x}}J_x + \hat{\mathbf{y}}J_y + \hat{\mathbf{z}}J_z) = \mu_0 \mathbf{J}$$

in rectangular coordinates. Using the *del* operator we can express the curl of **B** as

$$\boxed{\text{curl } \mathbf{B} = \nabla \times \mathbf{B} = \mu_0 \mathbf{J}} \qquad [6.10]$$

which is the differential form of Ampère's law. A convenient method for remembering the expression for $\nabla \times \mathbf{B}$ in a rectangular coordinate system is to use the determinant form

$$\nabla \times \mathbf{B} = \begin{vmatrix} \hat{\mathbf{x}} & \hat{\mathbf{y}} & \hat{\mathbf{z}} \\ \dfrac{\partial}{\partial x} & \dfrac{\partial}{\partial y} & \dfrac{\partial}{\partial z} \\ B_x & B_y & B_z \end{vmatrix}$$

Example 6-16 illustrates the concept of curl and circulation for the case of a fluid.

Example 6-16: Circulation in a fluid. Further physical insight into the nature of the curl operation can be gained by considering a simple example. Consider the vector field representing the velocity on a water surface. Suppose this velocity is given by

$$\mathbf{v} = \hat{\mathbf{y}}(v_0 x)$$

as shown in Figure 6.21. Find the curl of this vector field.

Solution: Consider the closed contour marked abcda in Figure 6.21. We can easily see that there is a net circulation $\oint_{\text{abcda}} \mathbf{v} \cdot d\mathbf{l}$ around this path, since the velocity field is stronger on side ab than it is on side cd, and there is no contribution to the line integral from the sides bc or ad. The line integral of the velocity vector around the contour abcda is thus given by

$$\oint_{\text{abcda}} \mathbf{v} \cdot d\mathbf{l} = v_0\left(x + \frac{\Delta x}{2}\right)\Delta y - v_0\left(x - \frac{\Delta x}{2}\right)\Delta y = v_0 \Delta x \Delta y$$

since the velocity $\mathbf{v}$ does not vary with y. Since $\Delta s = \Delta x \Delta y$, and $\hat{\mathbf{s}} = \hat{\mathbf{z}}$, we have

$$\text{curl}\,\mathbf{v} = \left[\lim_{\Delta s \to 0} \frac{\oint_C \mathbf{v} \cdot d\mathbf{l}}{\Delta s}\right]\hat{\mathbf{s}} = \hat{\mathbf{z}} v_0 \qquad \text{or} \qquad \nabla \times \mathbf{v} = \hat{\mathbf{z}} v_0$$

which is the same result that one would obtain by using the differential expression, namely,

$$\nabla \times \mathbf{v} = \hat{\mathbf{z}}\left(\frac{\partial v_y}{\partial x} - \frac{\partial v_x}{\partial y}\right) = \hat{\mathbf{z}}(v_0 + 0) = \hat{\mathbf{z}} v_0$$

We note that the velocity flow itself is entirely in the y direction and does not bend or curve around in any way. However, if we placed an infinitesimal wooden chip with its length in the x direction at the point on the surface where we have

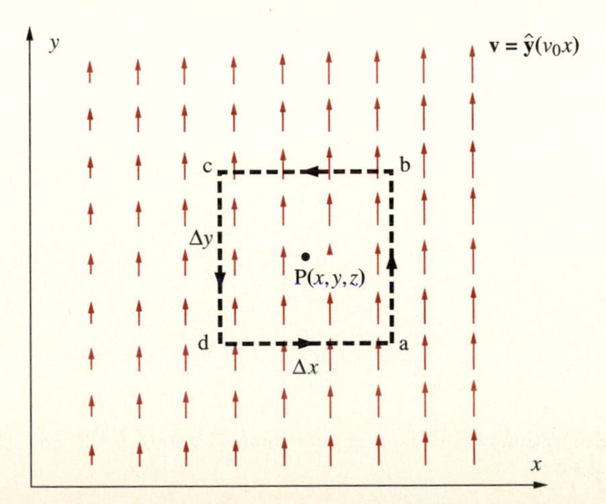

FIGURE 6.21. Velocity field on a water surface.

evaluated the curl, the chip would tend to rotate, since the speed of the flow is greater on side ab than on side cd. As viewed from above, the wooden chip rotates in the counterclockwise direction, which, according to our definitions is perpendicular to the plane of circulation and is in the direction defined by the fingers of the right hand when the thumb points in the direction of the curl vector (i.e., the right-hand rule).

6.4.2 Curl in Other Coordinate Systems

Although we use the notation $\nabla \times \mathbf{A}$ to indicate the curl of a vector $\mathbf{A}$, expression of curl as a cross product with a vector *del* operator (having the form $\nabla = \hat{\mathbf{x}}(\partial/\partial x) + \hat{\mathbf{y}}(\partial/\partial y) + \hat{\mathbf{z}}(\partial/\partial z)$ for rectangular coordinates) is useful only in the rectangular coordinate system. In other coordinate systems, we still denote the curl of $\mathbf{A}$ by $\nabla \times \mathbf{A}$, but note that the specific derivative expressions for each component of curl will need to be derived from the physical definition of curl as given in [6.9], using a differential surface element appropriate for that particular coordinate system.

To derive the differential expression for curl in the spherical coordinate system, we consider the cuboid volume element shown in Figure 6.22b. To determine the r component of the curl, we consider the curvilinear contour ABCDA. Note that this contour is perpendicular to $\hat{\mathbf{r}}$, and the sense of rotation is related to $\hat{\mathbf{r}}$ by the right-hand rule if we go around the contour in the A $\rightarrow$ B $\rightarrow$ C $\rightarrow$ D $\rightarrow$ A direction. The lengths of the sides are AB $= r\,d\theta$, CD $= r\,d\theta$, BC $= r\sin(\theta + d\theta)\,d\phi$, and DA $= r\sin\theta\,d\phi$. The line integral is given by

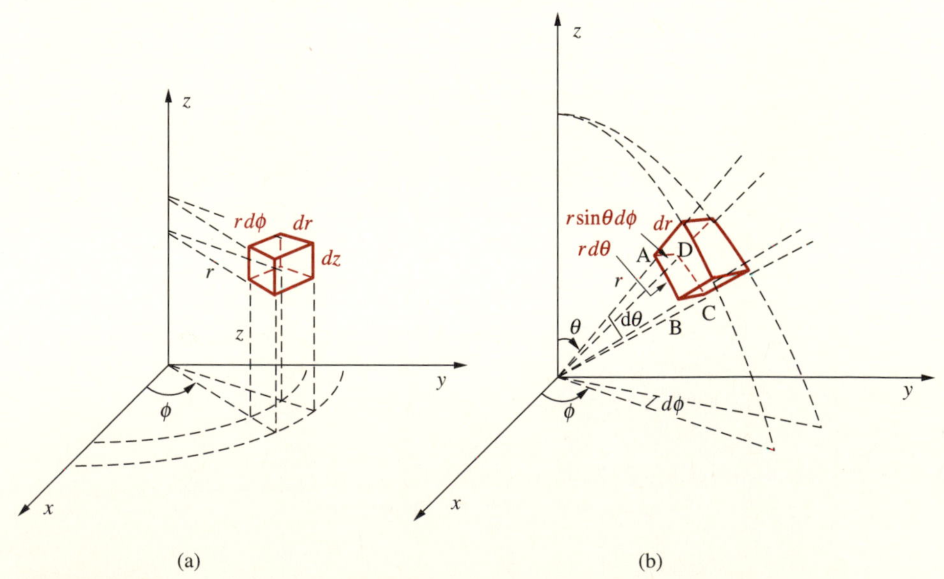

FIGURE 6.22. Derivation of curl in other coordinate systems. (a) Cylindrical coordinates. (b) Spherical coordinates.

$$\oint_{\text{ABCDA}} \mathbf{A} \cdot d\mathbf{l} = A_\theta r\, d\theta + \left(A_\phi + \frac{\partial A_\phi}{\partial \theta} d\theta\right) r \sin(\theta + d\theta)\, d\phi$$

$$- \left(A_\theta + \frac{\partial A_\theta}{\partial \phi} d\phi\right) r\, d\theta - A_\phi r \sin\theta\, d\phi$$

$$= -\frac{\partial A_\theta}{\partial \phi} r\, d\theta\, d\phi + \frac{\partial}{\partial \theta}(A_\phi \sin\theta) r\, d\theta\, d\phi$$

where we have retained small quantities up to second order. The area of the contour ABCDA is $\Delta s = r^2 \sin\theta\, d\theta\, d\phi$, so

$$[\nabla \times \mathbf{A}]_r = \frac{\oint_{\text{ABCDA}} \mathbf{A} \cdot d\mathbf{l}}{\Delta s} = \frac{1}{r \sin\theta}\left[\frac{\partial}{\partial \theta}(A_\theta \sin\theta) - \frac{\partial A_\phi}{\partial \phi}\right]$$

The other two components can be obtained by considering the other sides of the cuboid in Figure 6.22b, namely, the contour ABFEA for the ϕ component and the contour BCGFB for the θ component. The complete expression for the curl in spherical coordinates is

$$\text{curl}\,\mathbf{A} = \nabla \times \mathbf{A} = \hat{\mathbf{r}}\frac{1}{r \sin\theta}\left[\frac{\partial}{\partial \theta}(A_\phi \sin\theta) - \frac{\partial A_\theta}{\partial \phi}\right]$$
$$+ \hat{\boldsymbol{\theta}}\frac{1}{r}\left[\frac{1}{\sin\theta}\frac{\partial A_r}{\partial \phi} - \frac{\partial}{\partial r}(r A_\phi)\right] + \hat{\boldsymbol{\phi}}\frac{1}{r}\left[\frac{\partial}{\partial r}(r A_\theta) - \frac{\partial A_r}{\partial \theta}\right]$$

Similar reasoning in a cylindrical coordinate system using a volume element such as that shown in Figure 6.22a leads to the curl expression in cylindrical coordinates:

$$\text{curl}\,\mathbf{A} = \nabla \times \mathbf{A} = \hat{\mathbf{r}}\left[\frac{1}{r}\frac{\partial A_z}{\partial \phi} - \frac{\partial A_\phi}{\partial z}\right] + \hat{\boldsymbol{\phi}}\left[\frac{\partial A_r}{\partial z} - \frac{\partial A_z}{\partial r}\right] + \hat{\mathbf{z}}\frac{1}{r}\left[\frac{\partial}{\partial r}(r A_\phi) - \frac{\partial A_r}{\partial \phi}\right]$$

Examples 6-17 and 6-18 illustrate the evaluation of curl in cylindrical and spherical coordinate systems, respectively. Example 6-19 shows fields that have zero circulation or curl.

Example 6-17: Curl of the B field due to an infinitely long cylindrical conductor. In Example 6-10, the **B** field of an infinitely long straight cylindrical current-carrying conductor of radius a was found to be

$$\mathbf{B} = \begin{cases} \hat{\boldsymbol{\phi}}\dfrac{\mu_0 I r}{2\pi a^2} & r \le a \\[2ex] \hat{\boldsymbol{\phi}}\dfrac{\mu_0 I}{2\pi r} & r > a \end{cases}$$

Find $\nabla \times \mathbf{B}$.

Solution: Since $\mathbf{B}$ is expressed in cylindrical coordinates, we use the cylindrical coordinate expression for curl. Note that the only nonzero component of $\mathbf{B}$ is B_ϕ, which varies only with r. Thus, the curl of $\mathbf{B}$ is

$$\nabla \times \mathbf{B} = \hat{\mathbf{z}}\frac{1}{r}\frac{\partial}{\partial r}(rB_\phi) = \hat{\mathbf{z}}\frac{1}{r}\left[B_\phi + r\frac{\partial B_\phi}{\partial r}\right]$$

so that for $r \leq a$, we have

$$\nabla \times \mathbf{B} = \hat{\mathbf{z}}\left[\frac{\mu_0 I r}{2\pi a^2} + r\frac{\mu_0 I}{2\pi a^2}\right] = \hat{\mathbf{z}}\frac{\mu_0 I}{\pi a^2} = \mu_0 \mathbf{J}$$

where $\mathbf{J}$ is the current density vector in the conductor given by $\mathbf{J} = \hat{\mathbf{z}}I/(\pi a^2)$. Similarly, for $r > a$, we have

$$\nabla \times \mathbf{B} = \hat{\mathbf{z}}\left[\frac{\mu_0 I}{2\pi r} - r\frac{\mu_0 I}{2\pi r^2}\right] = 0$$

as expected, since $\mathbf{J} = 0$ for $r > a$.

Example 6-18: Curl of the electrostatic dipole field. As an example of a field that has both r and θ components, consider the electric field of the electric dipole as derived in Section 4.4.3. For a dipole consisting of two opposite charges $\pm Q$ separated by a distance d, the electric field at faraway points ($r \gg d$) was shown to be

$$\mathbf{E} = \frac{Qd}{4\pi\epsilon_0 r^3}[\hat{\mathbf{r}}2\cos\theta + \hat{\boldsymbol{\theta}}\sin\theta]$$

Find the curl of the electric field at faraway points.

Solution: The curl of $\mathbf{E}$ in spherical coordinates is given by

$$\nabla \times \mathbf{E} = \hat{\boldsymbol{\phi}}\frac{1}{r}\left[\frac{\partial}{\partial r}(rE_\theta) - \frac{\partial E_r}{\partial \theta}\right] = \hat{\boldsymbol{\phi}}\frac{1}{r}\left[-\frac{2\sin\theta}{r^3} + \frac{2\sin\theta}{r^3}\right] = 0$$

Example 6-19: Curl-free fields. Show that the following vector fields, shown in Figure 6.23, are curl-free:

$$\mathbf{A}_1 = -\hat{\mathbf{x}}x + \hat{\mathbf{y}}y$$

$$\mathbf{A}_2 = \hat{\mathbf{x}}\sin x - \hat{\mathbf{y}}\cos y$$

$$\mathbf{A}_3 = \hat{\mathbf{x}}\sin x\cos y + \hat{\mathbf{y}}\cos x\sin y$$

$$\mathbf{A}_4 = 1\hat{\mathbf{r}} = \hat{\mathbf{x}}\frac{x}{\sqrt{x^2+y^2}} + \hat{\mathbf{y}}\frac{y}{\sqrt{x^2+y^2}}$$

$$\mathbf{A}_5 = \hat{\mathbf{r}}2\phi r + \hat{\boldsymbol{\phi}}r$$

$$\mathbf{A}_6 = \hat{\boldsymbol{\phi}}\frac{1}{r}$$

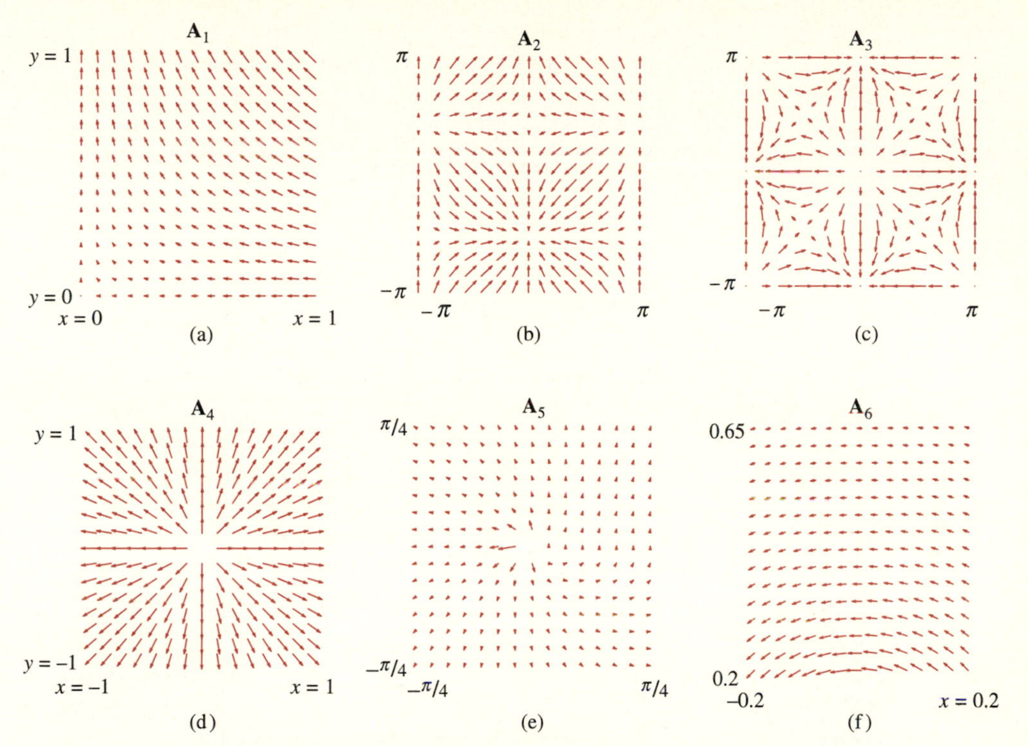

FIGURE 6.23. **Examples of curl-free vector fields.** For simplicity of presentation, two-dimensional vector fields are considered, with $A_z = 0$ and with no variation of any component with z. (a) $\mathbf{A}_1 = -\hat{\mathbf{x}}x + \hat{\mathbf{y}}y$. (b) $\mathbf{A}_2 = \hat{\mathbf{x}}\sin x - \hat{\mathbf{y}}\cos y$. (c) $\mathbf{A}_3 = \hat{\mathbf{x}}\sin x\cos y + \hat{\mathbf{y}}\cos x\sin y$. (d) $\mathbf{A}_4 = 1\hat{\mathbf{r}} = \hat{\mathbf{x}}(x/\sqrt{x^2+y^2}) + \hat{\mathbf{y}}(y/\sqrt{x^2+y^2})$. (e) $\mathbf{A}_5 = \hat{\mathbf{r}}2\phi r + \hat{\boldsymbol{\phi}}r$. (f) $\mathbf{A}_6 = \hat{\boldsymbol{\phi}}r^{-1}$.

Solution: Since these vector fields are all two-dimensional, they can have curl only in the z direction, given by

$$[\nabla\times\mathbf{A}]_z = \left(\frac{\partial A_y}{\partial x} - \frac{\partial A_x}{\partial y}\right)$$

in rectangular coordinates, and

$$[\nabla\times\mathbf{A}]_z = \frac{1}{r}\left[\frac{\partial(rA_\phi)}{\partial r} - \frac{\partial A_r}{\partial\phi}\right]$$

in cylindrical coordinates. We thus have

$$[\nabla\times\mathbf{A}_1]_z = \frac{\partial(y)}{\partial x} = \frac{\partial(-x)}{\partial y} = 0+0=0$$

$$[\nabla\times\mathbf{A}_2]_z = \frac{\partial(-\cos y)}{\partial x} - \frac{\partial(\sin x)}{\partial y} = 0+0=0$$

$$[\nabla\times\mathbf{A}_3]_z = -\sin x\sin y - (-\sin x\sin y) = 0$$

$$[\nabla \times \mathbf{A}_4]_z = -\frac{xy}{(x^2+y^2)^{3/2}} - \left(-\frac{xy}{(x^2+y^2)^{3/2}}\right) = 0$$

$$[\nabla \times \mathbf{A}_5]_z = \frac{1}{r}\left[\frac{\partial(r^2)}{\partial r} - \frac{\partial(2\phi r)}{\partial \phi}\right] = \frac{1}{r}[2r - 2r] = 0$$

$$[\nabla \times \mathbf{A}_6]_z = \frac{1}{r}\left[\frac{\partial(rr^{-1})}{\partial r} - 0\right] = 0$$

6.4.3 Skilling's Paddle Wheel

The term "curl" suggests an association with motion in curved lines. That such motion is not necessary in order to have nonzero curl was illustrated well in Example 6-16, where we found that straight-line motion may also have nonzero curl. If the velocity field considered there represents the water flow in a canal, every particle of water may indeed move in a straight line, but nevertheless there is curl. One of the best ways of thinking about the curl of a vector field is to visualize (see Figure 6.24)

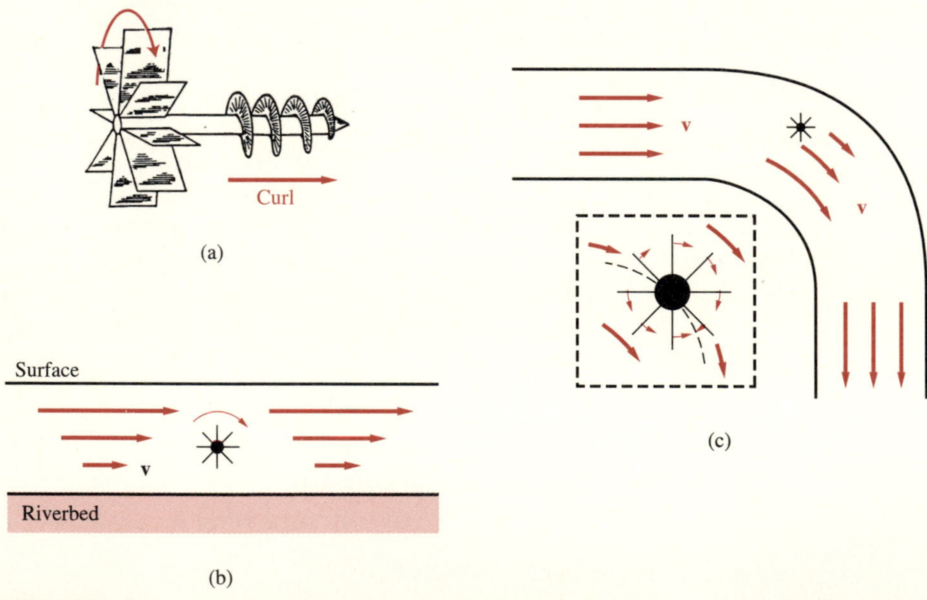

FIGURE 6.24. Skilling's paddle wheel. (a) If the paddle wheel turned a screw, its rotation would drive it in the direction of the curl. (b) Straight-line flow is not necessarily curl-free. (c) Flow fields that bend and go around corners can nevertheless be curl-free. (Figure adapted from H. H. Skilling, *Fundamentals of Electric Waves,* 2nd ed., John Wiley & Sons, Inc., New York, 1948.) H. H. Skilling was a pioneering electrical engineering educator and a prolific writer of many outstanding textbooks on electrical circuits and networks, electromechanics, transmission lines, and electromagnetics. He was the Chairman of the Electrical Engineering Department at Stanford University during 1941–1964.

a small paddle wheel[40] to be dipped into the flow field. The paddle wheel rotates only if the curl of the vector field is nonzero; the speed with which it would rotate is a measure of the magnitude of curl, while the direction it rotates as viewed from above (clockwise or counterclockwise) is an indication of the direction (down or up, respectively) of the curl vector. Another way to relate the direction of rotation to the curl vector is to note that if the paddle wheel turned a right-hand screw, it would drive the screw in the direction of the curl vector, as shown in Figure 6.24a.

Applying the paddle wheel concept to a case of straight-line flow as illustrated in Figure 6.24b (similar to that in Example 6-16), we clearly see that the flow has indeed nonzero curl. The paddle wheel turns in the clockwise direction, since the stream is more rapid on its upper blades than on its lower ones. Figure 6.24c shows water flow in another canal, in which the flow bends around a corner but is nevertheless curl-free. It is possible for water flow to turn around the corner with zero curl, provided that it flows faster along the inner margin of the channel by just the right amount. An enlarged view of the paddle wheel at the corner is shown in Figure 6.24c, with arrows indicating the force of water on each of its blades. Because of the curvature of the flow lines, more than half of its blades are driven clockwise. However, the velocity of water is greatest on the inner side, and, although fewer blades are driven counterclockwise, they are acted upon more forcefully. It is thus conceivable that the curvature and the variation of velocity be so related that the wheel does not rotate. Curved motion with zero curl is in fact characteristic of a truly frictionless fluid. In practice, it is often desirable to have a surface so that air or water can flow with minimum curl, since motion with nonzero curl develops eddies that waste energy.

Figure 6.23 shows some actual vector fields that exhibit flow lines that bend but are nevertheless curl-free. Two clear examples are $\mathbf{A}_1$ and $\mathbf{A}_6$. Since all of the vector fields shown in Figure 6.23 are curl-free, the reader is encouraged to visualize Skilling's paddle wheel placed in these fields and to consider whether it rotates.

Example 6-20 illustrates some vector fields with nonzero curl.

Example 6-20: Vector fields with nonzero curl. Find the curl of the following vector fields:

$$\begin{aligned}
\mathbf{A}_7 &= \hat{\mathbf{x}}\frac{x^2}{2} - \hat{\mathbf{y}}xy \\
\mathbf{A}_8 &= -\hat{\mathbf{x}}\sin y + \hat{\mathbf{y}}\cos x \\
\mathbf{A}_9 &= \hat{\mathbf{x}}\sin x\cos y - \hat{\mathbf{y}}\cos x\sin y \\
\mathbf{A}_{10} &= \hat{\mathbf{x}}\cos(xy) + \hat{\mathbf{y}}\sin(xy)
\end{aligned}$$

[40]This extremely useful concept is now widely used in many textbooks but was first introduced by H. H. Skilling, *Fundamentals of Electric Waves*, John Wiley & Sons, Inc., New York, 1942. Skilling's paddle wheel is mounted on frictionless bearings so that it can freely turn when dipped into the "flow"; it is also "small" so that it does not interfere with the flow.

Solution:

$$[\nabla \times \mathbf{A}_7]_z = \hat{\mathbf{z}}\left[\frac{\partial(-xy)}{\partial x} - \frac{\partial(x^2/2)}{\partial y}\right] = -\hat{\mathbf{z}}y$$

$$[\nabla \times \mathbf{A}_8]_z = \hat{\mathbf{z}}\left[\frac{\partial(\cos x)}{\partial x} - \frac{\partial(-\sin y)}{\partial y}\right] = \hat{\mathbf{z}}(-\sin x + \cos y)$$

$$[\nabla \times \mathbf{A}_9]_z = \hat{\mathbf{z}}\left[\frac{\partial(-\cos x \sin y)}{\partial x} - \frac{\partial(\sin x \cos y)}{\partial y}\right] = \hat{\mathbf{z}}(2 \sin x \sin y)$$

$$[\nabla \times \mathbf{A}_{10}]_z = \hat{\mathbf{z}}\left[\frac{\partial[\sin(xy)]}{\partial x} - \frac{\partial[\cos(xy)]}{\partial y}\right] = \hat{\mathbf{z}}[y\cos(xy) + x\sin(xy)]$$

Both the vector fields and their curls are plotted in Figures 6.25 and 6.26. Note that the curl of all of the vectors is exclusively in the z direction. For the curl

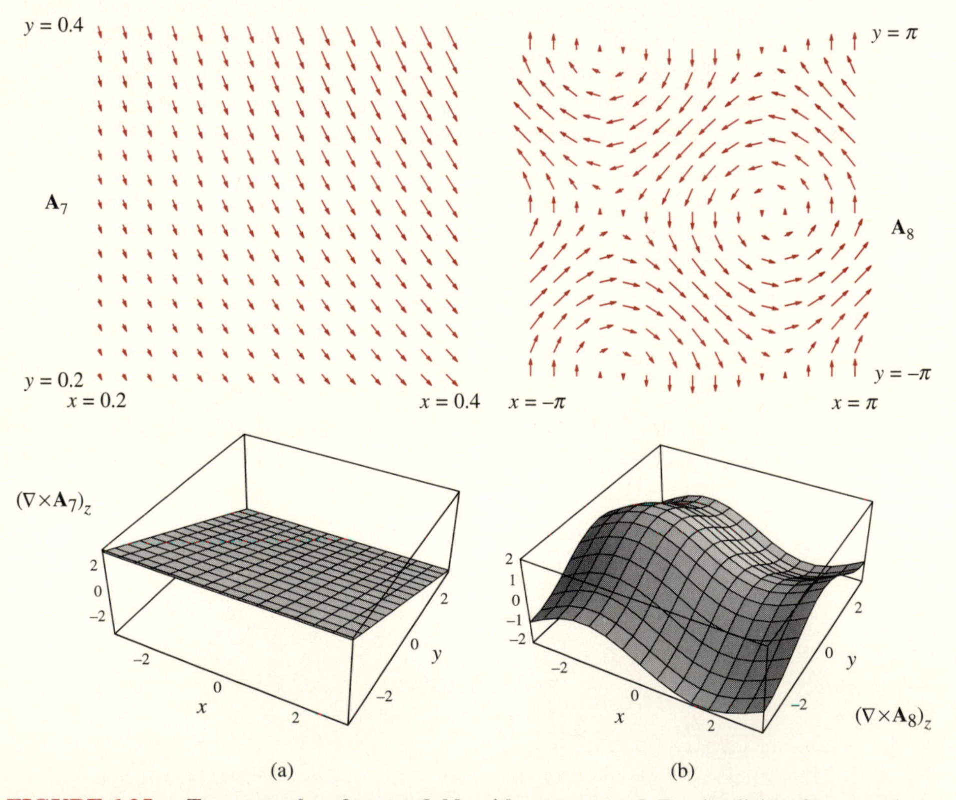

FIGURE 6.25. Two examples of vector fields with nonzero curl. For simplicity of presentation, two-dimensional vector fields are considered, with $A_z = 0$ and with no variation of any component with z. (a) $\mathbf{A}_7 = \hat{\mathbf{x}}(x^2/2) - \hat{\mathbf{y}}xy$. (b) $\mathbf{A}_8 = -\hat{\mathbf{x}}\sin y + \hat{\mathbf{y}}\cos x$.

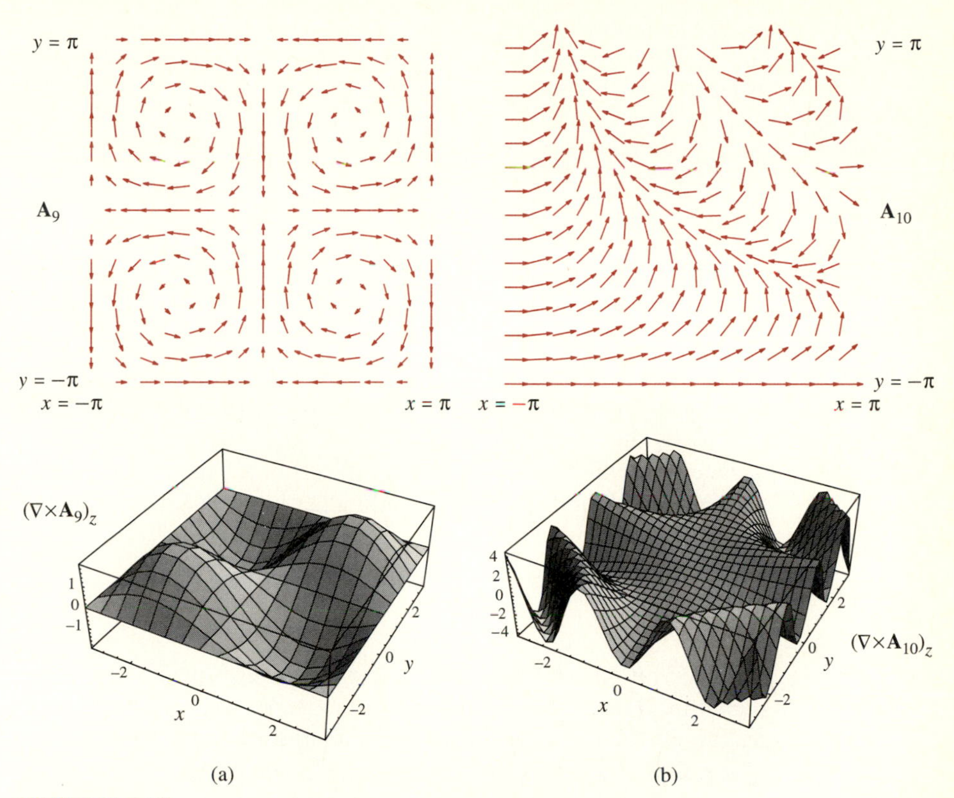

FIGURE 6.26. Two examples of vector fields with nonzero curl. For simplicity of presentation, two-dimensional vector fields are considered, with $A_z = 0$ and with no variation of any component with z. (a) $\mathbf{A}_9 = \hat{\mathbf{x}} \sin x \cos y - \hat{\mathbf{y}} \cos x \sin y$. (b) $\mathbf{A}_{10} = \hat{\mathbf{x}} \cos(xy) + \hat{\mathbf{y}} \sin(xy)$.

of these vectors to have an x or y component, either their z component must vary with y or x, respectively, or the y or x components must vary with z. Since $A_z = 0$ for all vectors, and since A_x and A_y are independent of z, their curl does not have x or y components. The reader is encouraged to visualize Skilling's paddle wheel placed in these fields. The places where the magnitude of the curl peaks are where the paddle wheel would rotate the fastest. Note that $\mathbf{A}_7$ is similar in appearance to the curl-free $\mathbf{A}_1$ and represents a flow that turns around a corner much like that shown in Figure 6.24c. However, the water velocity (i.e., $\mathbf{A}_7$) is apparently not larger on the inner side of the curve by the right amount, leading to nonzero curl.

6.4.4 Stokes's Theorem

Now that we have discussed circulation and curl of a vector field, we are ready to cite an important theorem widely used in electromagnetics and other fields. This theorem, known as Stokes's theorem, relates the line integral of a vector around a closed contour to the integral of its curl over the surface enclosed by the contour. For the magnetostatic field, we can combine the integral and differential forms of Ampère's law to find

$$\left.\begin{aligned}\oint_C \mathbf{B}\cdot d\mathbf{l} &= \int_S \mu_0\mathbf{J}\cdot d\mathbf{s}\\ \nabla\times\mathbf{B} &= \mu_0\mathbf{J}\end{aligned}\right\} \longrightarrow \boxed{\oint_C \mathbf{B}\cdot d\mathbf{l} = \int_S \nabla\times\mathbf{B}\cdot d\mathbf{s}}$$

where C is the contour that encloses the surface S. This integral equation is entirely consistent with the interpretation of the curl of a vector field as the circulation per unit surface area at a given point. Since the curl is the net circulation per unit surface area, it is to be expected that the total circulation around the boundary of the surface (i.e., along the contour C) can be obtained by integrating the curl over the enclosed surface. This relationship between the curl of a vector field and its line integral is true in general for *any* vector field. In other words, for any vector field $\mathbf{G}$ we have

$$\boxed{\oint_C \mathbf{G}\cdot d\mathbf{l} = \int_S (\nabla\times\mathbf{G})\cdot d\mathbf{s}} \qquad [6.11]$$

It is quite easy to see that [6.11] is true by considering a general surface as shown in Figure 6.27. Taking the surface to be planar for simplicity of discussion, we can envision it to be subdivided into differential surface elements Δs_i bounded by infinitesimal contours C_i as shown. We have

$$\oint_C \mathbf{G}\cdot d\mathbf{l} = \sum_i \int_{C_i} \mathbf{G}\cdot d\mathbf{l}$$

since the terms arising from the common boundaries between any two elements cancel out under the summation. The component of curl $\mathbf{G}$ normal to one of the surface elements is (curl $\mathbf{G}$) $\cdot\, \hat{\mathbf{s}}_i$, where $\hat{\mathbf{s}}_i$ is the unit vector in the direction of the surface element Δs_i. From the definition of curl as given in [6.9] we have

$$(\nabla\times\mathbf{G})\cdot\hat{\mathbf{s}}_i = \left[\lim_{\Delta s_i\to 0}\frac{\oint_{C_i}\mathbf{G}\cdot d\mathbf{l}}{\Delta s_i}\right]_{\max}$$

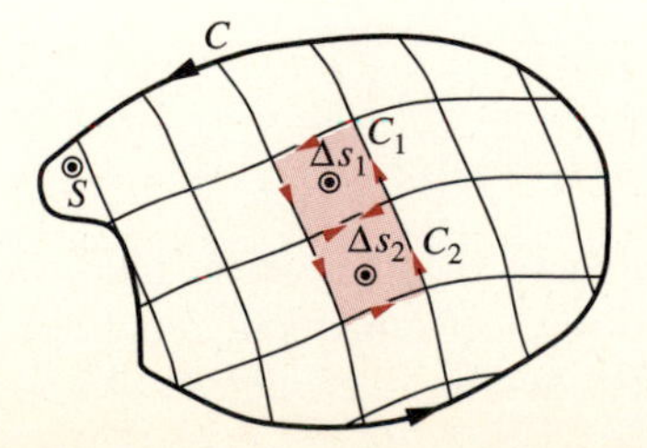

FIGURE 6.27. Stokes's theorem. The sum of the closed line integrals (i.e., circulations) about the perimeter of every small area element is the same as the closed line integral along the perimeter of S enclosed by C due to the cancellations at the edges of the interior paths.

where $\hat{\mathbf{s}}_i$ is the unit vector in the direction of $\Delta\mathbf{s}_i$. Thus we may write

$$\oint_C \mathbf{G} \cdot d\mathbf{l} = \lim_{\Delta s_i \to 0} \sum_i [(\nabla \times \mathbf{G}) \cdot \hat{\mathbf{s}}_i] \Delta s_i$$

In the limit, the summation becomes an integral, $\Delta\mathbf{s}_i$ becomes $d\mathbf{s}$, and we have

$$\oint_C \mathbf{G} \cdot d\mathbf{l} = \int_S (\nabla \times \mathbf{G}) \cdot d\mathbf{s}$$

thus proving the theorem.

One of the important consequences of Stokes's theorem is that the curl of a conservative vector field is zero. As was discussed in Chapter 4, a vector field is said to be *conservative* when $\oint_C \mathbf{G} \cdot d\mathbf{l} = 0$, a terminology that stems from an analogy with force fields. Let $\mathbf{G}$ be the force exerted on a particle at a given point. If the vector field does not include any friction or dissipation, then no net energy is required to move the particle around a closed path, so the energy is conserved. When $\oint_C \mathbf{G} \cdot d\mathbf{l} = 0$ for any arbitrary contour C, Stokes's theorem requires that $\nabla \times \mathbf{G} = 0$. For the electrostatic field, we then have

$$\oint_C \mathbf{E} \cdot d\mathbf{l} = 0 \quad \text{is equivalent to} \quad \nabla \times \mathbf{E} = 0$$

In other words, the electrostatic field is curl-free, as was indicated in passing in Section 4.11.

Stokes, Sir George Gabriel *(British mathematician and physicist, b. Skreen, Co. Sligo, Ireland, August 13, 1819; d. Cambridge, England, February 1, 1903)* *Stokes was the youngest child of a clergyman. He graduated from Cambridge in 1841 at the head of his class in mathematics, and his early promise was to be a true indication of a bright future. In 1849 he was appointed Lucasian professor of mathematics at Cambridge; in 1854, secretary of the Royal Society; and in 1885, president of the Royal Society. No one had held all three offices since Isaac Newton [1642–1727], a century and a half before. Stokes's vision is indicated by the fact that he was one of the first scientists to see the value of Joule's [1818–1889] work.*

Between 1845 and 1850 Stokes worked on the theory of viscous fluids. He also worked on fluorescence (a word he introduced in 1852), on sound, and on light. He studied ultraviolet radiation by means of the fluorescence it produced. He was the first to show that quartz was transparent to ultraviolet radiation, whereas ordinary glass was not.

In his lectures at Cambridge, Stokes announced interpretations of the significance of the Fraunhofer [1787–1826] lines, which were in effect anticipations of the later theories of Kirchhoff [1824–1887]. Although Stokes never published his views, others tried to award him the credit. Stokes himself (whose character was warm with generosity and modesty) always insisted that he had not seen certain key points that were involved and that he could lay no claim to priority.

In 1896, toward the end of his long life, Stokes was among the first to suggest that X rays, newly discovered by Roentgen [1845–1923], were electromagnetic radiation akin to light.

Stokes received the Rumford medal of the Royal Society in 1852 and its Copley medal in 1893. He served as a Conservative member in Parliament, sitting for Cambridge University (as once Newton had done, from 1887 to 1892) and was made a baronet in 1889. [Adapted with permission from I. Asimov, Biographical Encyclopedia of Science and Technology, *Doubleday, 1982.]*

1700 — *1819* — *1903* — *2000*

Example 6-21: Water drain in a sink. Consider water flow going down the drain in a sink, represented by a simplified vector field given by $\mathbf{v} = \hat{\boldsymbol{\phi}}\omega_0 r$, where ω_0 is a constant and r is the distance from the z axis (the axis of rotation) as shown in Figure 6.28. Verify Stokes's theorem using a flat circular (radius a) surface, shown as shaded in Figure 6.28, bounded by the contour marked C.

Solution: First, we take the line integral of the velocity vector $\mathbf{v}$ around the $r = a$ contour as

$$\oint_C \mathbf{v} \cdot d\mathbf{l} = \int_0^{2\pi} \omega_0 a \hat{\boldsymbol{\phi}} \cdot \hat{\boldsymbol{\phi}} a \, d\phi = \omega_0 a^2 \int_0^{2\pi} d\phi = 2\omega_0 a^2 \pi$$

Second, we take the curl of the velocity field as

$$\nabla \times \mathbf{v} = \hat{\mathbf{z}} \frac{1}{r} \frac{\partial}{\partial r}(r v_\phi) = \hat{\mathbf{z}} \frac{1}{r} \frac{\partial}{\partial r}(\omega_0 r^2) = \hat{\mathbf{z}} 2\omega_0$$

which we use to take the surface integral on the circular surface enclosed by $r = a$ as

$$\int_S (\nabla \times \mathbf{v}) \cdot d\mathbf{s} = \int_0^a \int_0^{2\pi} (2\omega_0 \hat{\mathbf{z}}) \cdot (\hat{\mathbf{z}} r \, dr \, d\phi) = 4\omega_0 \pi \int_0^a r \, dr = 2\omega_0 a^2 \pi$$

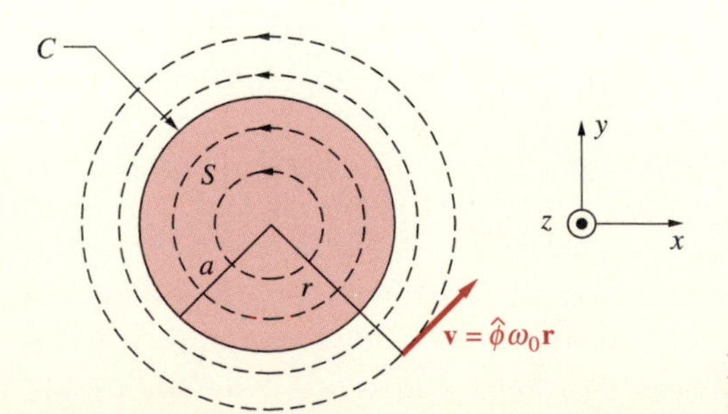

FIGURE 6.28. Velocity field of water drain.

which is the same result obtained from the line integral, as expected on the basis of Stokes's theorem.

6.5 VECTOR MAGNETIC POTENTIAL

We have seen the general form of the Biot–Savart law to be

$$\mathbf{B}(\mathbf{r}) = \frac{\mu_0}{4\pi}\int_{V'} \frac{\mathbf{J}(\mathbf{r}') \times \hat{\mathbf{R}}}{R^2}\,dv' \qquad [6.6]$$

where $\mathbf{R} = (\mathbf{r} - \mathbf{r}') = \hat{\mathbf{R}}R$, and $R = |\mathbf{r} - \mathbf{r}'|$. Note that $\hat{\mathbf{R}}$ is the unit vector directed *from* the source point *to* the observation point. We can manipulate [6.6] by noting[41] that $\hat{\mathbf{R}}/R^2 = -\nabla(1/R)$, in which case the integrand becomes $-\mathbf{J} \times \nabla(1/R)$. We then note that the *del* operator ∇ operates on the variables x, y, z and thus does not affect quantities such as $\mathbf{J}(\mathbf{r}')$ that are functions only of the source coordinates x', y', z'. Thus, we can take $\mathbf{J}(\mathbf{r}')$ inside the ∇ operator without affecting the integral as

$$\mathbf{B}(\mathbf{r}) = \frac{\mu_0}{4\pi}\int_{V'} \nabla \times \frac{\mathbf{J}}{R}\,dv' = \nabla \times \frac{\mu_0}{4\pi}\int_{V'} \frac{\mathbf{J}(\mathbf{r}')}{R}\,dv' \qquad [6.12]$$

where we have interchanged the $\nabla\times$ and the integration operations, since the former operates on the coordinates x, y, z, whereas the integration is carried out over the source coordinates x', y', z'.

It appears from [6.12] that we can express the magnetic field $\mathbf{B}(\mathbf{r})$ as the curl of another vector $\mathbf{A}$ given by

$$\mathbf{A}(\mathbf{r}) = \frac{\mu_0}{4\pi}\int_{V'} \frac{\mathbf{J}(\mathbf{r}')}{R}\,dv' \qquad [6.13]$$

---

[41] To see this we can expand $(1/R)$ into its components and manipulate as follows:

$$\frac{1}{R} = \frac{1}{[(x-x')^2 + (y-y')^2 + (z-z')^2]^{1/2}}$$

so that

$$\nabla\left(\frac{1}{R}\right) = \hat{\mathbf{x}}\frac{\partial}{\partial x}\left(\frac{1}{R}\right) + \hat{\mathbf{y}}\frac{\partial}{\partial y}\left(\frac{1}{R}\right) + \hat{\mathbf{z}}\frac{\partial}{\partial z}\left(\frac{1}{R}\right)$$

$$= -\frac{\hat{\mathbf{x}}(x-x') + \hat{\mathbf{y}}(y-y') + \hat{\mathbf{z}}(z-z')}{[(x-x')^2 + (y-y')^2 + (z-z')^2]^{3/2}}$$

$$= -\frac{\mathbf{R}}{R^3} = -\hat{\mathbf{R}}\frac{1}{R^2}$$

The same result was obtained using spherical coordinates in footnote 106 in Chapter 4.

The vector $\mathbf{A}$ is referred to as the *vector magnetic potential* and is in some ways analogous to the scalar electric potential Φ. In some cases, it is easier to evaluate $\mathbf{A}$ from [6.13] and then find $\mathbf{B}$ from $\mathbf{B} = \nabla \times \mathbf{A}$, rather than to evaluate $\mathbf{B}$ directly from [6.6]. The analogy between the scalar potential Φ and $\mathbf{A}$ can be seen by noting from [4.28] that Φ is related to the source charge distribution ρ in a similar manner. In other words, we have

$$\Phi(\mathbf{r}) = \frac{1}{4\pi\epsilon_o}\int_{V'} \frac{\rho(\mathbf{r}')}{R}\,dv'$$

which is a scalar integral similar to that for each component of [6.13]. In other words,

$$A_{x,y,\text{ or }z} = \frac{\mu_0}{4\pi}\int_{V'} \frac{J_{x,y,\text{ or }z}}{R}\,dv'$$

indicating that, for example, A_x is related to its source J_x in the same manner as Φ is related to ρ.

Note that for currents flowing in thin filamentary wires we have $\mathbf{J}\,dv' = JS'\,d\mathbf{l}' = I\,d\mathbf{l}'$, so [6.13] becomes

$$\mathbf{A} = \frac{\mu_0 I}{4\pi}\oint_C \frac{d\mathbf{l}'}{R}$$

where C is the contour over which the filamentary current I flows.

Examples 6-22 and 6-23, and the discussion of the magnetic dipole in the next section, demonstrate the use of $\mathbf{A}$ to determine $\mathbf{B}$ indirectly. For simple cases, such as for a finite-length straight wire, using vector potential to find the $\mathbf{B}$ field might actually lead to a more complicated solution. However, in more complicated cases of practical significance, such as the magnetic dipole, the use of the vector magnetic potential results in significant simplification. The concept of vector potential is commonly used in most antenna problems, where the common task is to relate source current distributions (on antenna wires) to radiated fields.

An interesting difference between Φ and $\mathbf{A}$ should be noted here. Electrostatic problems more commonly involve the determination of electric fields from specified potentials rather than from charges. Thus, solving for the potential Φ (usually by solving Poisson's equation) is actually more direct than first finding the charge distributions. In magnetostatics, on the other hand, we usually have a known distribution of currents from which the $\mathbf{B}$ field can be directly determined. The vector potential $\mathbf{A}$ is thus an auxiliary quantity that is used primarily for simplicity. Although $\mathbf{A}$ plays a similar role in magnetostatics as Φ does in electrostatics, it has not been given any simple physical meaning. In classical terms, it is possible to view $\mathbf{A}$ as an entity that is not a "real" field,[42] since in any region where $\mathbf{B}$ is zero a moving

[42]R. P. Feynman defines a "real" field as a mathematical function that is introduced to avoid the idea of action at a distance. In other words, a "real" field is a set of numbers we specify in such a way that what happens *at a point* depends only on the number *at that point*.

charged particle experiences no force (since $\mathbf{F}_m = q\mathbf{v}\times\mathbf{B} = 0$) even if $\mathbf{A} \neq 0$. However, there are phenomena in quantum mechanics that show that **A** is in fact a "real" field.[43] In terms of examples studied in this chapter, it can be shown[44] that although the **B** fields outside an infinitely long solenoid and an ideal toroid are identically zero, the **A** fields are not. For an infinitely long solenoid of radius a, the vector potentials inside and outside are given by

$$A_\phi = \begin{cases} \dfrac{\mu_0 N I a^2}{2r} & r > a \\[2ex] \frac{1}{2}\mu_0 N I r & r \le a \end{cases}$$

where r is the radial distance from the solenoid axis. Note that inside the solenoid $(r \le a)$, $\nabla \times \mathbf{A}$ yields the **B** field for an infinitely long solenoid, as found in Example 6-14. For $r > a$, with **A** having only a $\hat{\boldsymbol{\phi}}$ component proportional to r^{-1}, the curl of **A** is zero, since in cylindrical coordinates we have $\nabla\times\mathbf{A} = \hat{\mathbf{z}}(1/r)[\partial(rA_\phi)/\partial r] = 0$. This nonvanishing feature of **A** in regions where $\mathbf{B} = 0$ has important physical significance.[45]

Examples 6-22 and 6-23 illustrate the use of the vector potential to determine the **B** fields near a finite-length, filamentary, current-carrying wire and at the center of a spherical shell of current.

Example 6-22: Finite-length straight wire. Calculate the **B** field on the bisecting plane (i.e., the $z = 0$ plane) due to a current I flowing in a straight wire of length $2a$ oriented along the z axis and centered at $z = 0$ as shown in Figure 6.29. This problem constitutes a repeat of Example 6-3, except that we shall first find the vector potential **A** and then determine **B** using $\mathbf{B} = \nabla \times \mathbf{A}$.

Solution: Taking the current I to flow in the z direction and noting that $\int_S J\,ds' = I$, where S is the cross-sectional area of the wire, the vector potential **A** on the $z = 0$ plane is given by

$$\mathbf{A} = \frac{\mu_0}{4\pi}\int_{V'} \frac{\mathbf{J}}{R}\,dv' = \frac{\mu_0}{4\pi}\int_{-a}^{a} \frac{\hat{\mathbf{z}} I\,dz'}{R}$$

---

[43] For an interesting discussion, see Chapter II-15 of R. P. Feynman, R. B. Leighton, and M. Sands, *The Feynman Lectures in Physics*, Addison Wesley, 1964. Also see M. D. Semon and J. R. Taylor, Thoughts on the magnetic vector potential, *Amer. J. Phys.*, 64(11), pp. 1361–9, November 1996, and references therein.

[44] See, for example, Section 7.11 of W. T. Scott, *The Physics of Electricity and Magnetism*, 2nd ed., John Wiley & Sons, Inc., New York, 1966.

[45] For example, if a wire is looped around an infinitely long solenoid, Faraday's law (to be discussed in Chapter 7) requires that an electromotive force be induced on the loop if the current in the solenoid varies with time. We might then ask: How can a **B** field that never touches a wire have an effect on the charges in it? The apparent conceptual difficulty is resolved when we realize that the induced emf can be directly determined using **A**, which is nonzero.

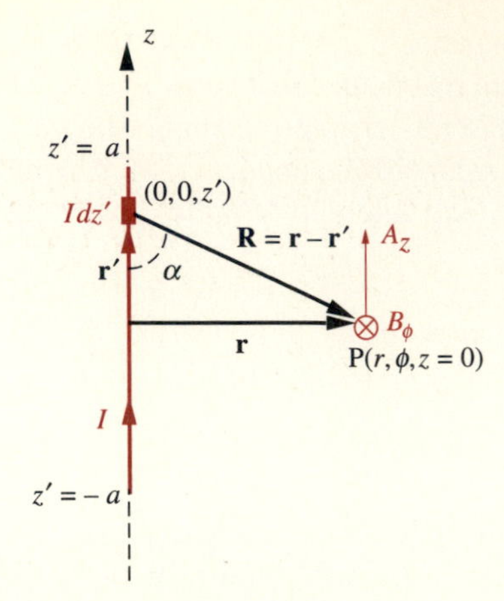

FIGURE 6.29. Vector potential A around a wire of length 2*a*.

From the geometry of the problem (Figure 6.29), and since $\mathbf{r}' = \hat{\mathbf{z}}z'$, $\mathbf{r} = \hat{\mathbf{r}}r$, we have $R = \sqrt{r^2 + (z')^2}$, so that the $\mathbf{A}$ field at a distance r from the wire on the $z = 0$ plane is given as

$$\mathbf{A} = \hat{\mathbf{z}}A_z = \frac{\mu_0 I}{4\pi}\int_{-a}^{a}\frac{dz'}{\sqrt{r^2+(z')^2}} = \hat{\mathbf{z}}\frac{\mu_0 I}{2\pi}\left[\ln\left(\frac{z'}{r}+\sqrt{\left(\frac{z'}{r}\right)^2+1}\right)\right]_0^a$$

$$= \hat{\mathbf{z}}\frac{\mu_0 I}{2\pi}\ln\left(\frac{a}{r}+\sqrt{\frac{a^2}{r^2}+1}\right)$$

where the integration is carried out either by change of variables or by using integral tables. By using $\mathbf{B} = \nabla \times \mathbf{A}$ in cylindrical coordinates, and noting that there is no dependence on ϕ, we have

$$\mathbf{B} = \nabla \times \mathbf{A} = \hat{\boldsymbol{\phi}}\left(-\frac{\partial A_z}{\partial r}\right)$$

so that on the bisecting plane, where $z = 0$, we have

$$B_\phi = -\frac{\mu_0 I}{2\pi}\frac{\partial}{\partial r}\ln\left(\frac{a}{r}+\sqrt{\frac{a^2}{r^2}+1}\right)$$

$$= \frac{\mu_0 I}{2\pi r}\frac{a/r}{\sqrt{1+(a/r)^2}} = \frac{\mu_0 I}{2\pi r}\frac{a}{\sqrt{r^2+a^2}}$$

Note that this result is identical to that obtained in Example 6-3. We can observe that for $r \ll a$, B_ϕ varies as $1/r$, consistent with what we obtained for an infinitely long wire. For $r \gg a$, on the other hand, we have B_ϕ varying as $1/r^2$, consistent with the field of a very short current element, as is expected from the Biot–Savart law.

Example 6-23: Field and potential at the center of a spherical shell. Consider a spherical shell of inner and outer radius a and b, respectively, and conducting a uniform current density $\mathbf{J} = \hat{\boldsymbol{\phi}} J_0$ (where J_0 is a constant), which is always tangent to circles of constant latitude, as shown in Figure 6.30. Find the vector potential $\mathbf{A}$ and the $\mathbf{B}$ field at the center of the sphere (i.e., at $z = 0$).

Solution: From [6.13] we have

$$\mathbf{A}(\mathbf{r} = 0) = \frac{\mu_0}{4\pi} \int_{V'} \frac{\mathbf{J}(\mathbf{r}')}{R} dv' = \frac{\mu_0}{4\pi} \int_{V'} \frac{\mathbf{J}(\mathbf{r}')}{|\mathbf{r}'|} dv'$$

where V' is the volume of the shell, and the origin of the coordinate system is located at the center of the sphere—that is, at the field point (in other words, $\mathbf{r} = 0$). Note that the magnitude of $\mathbf{J}$ is uniform throughout the volume V' of the spherical shell. Each current density element $\mathbf{J}\, dv'$ at a given point on the shell produces a differential vector potential $d\mathbf{A}$ at the center of the sphere, which is in the same direction as the current element. However, in view of the spherical symmetry, for each such element there exists another one located diametrically opposite and having oppositely directed current flow. Thus, the vector potential at the center of the sphere is $\mathbf{A}(\mathbf{r} = 0) = 0$. On the other hand, $\mathbf{B}(\mathbf{r} = 0)$ does not vanish. From [6.6] we have

$$\mathbf{B}(\mathbf{r} = 0) = \frac{\mu_0}{4\pi} \int_{V'} \frac{\mathbf{J}(\mathbf{r}') \times \hat{\mathbf{R}}}{R^2} dv'$$

where $\mathbf{J}(\mathbf{r}') = \hat{\boldsymbol{\phi}} J_0$ is the current flowing between $r = a$ and $r = b$, and $\hat{\mathbf{R}}$ is the unit vector pointing from a source point $\mathbf{r}'$ in the spherical shell to the origin (i.e., the observation point in this case) and is the inward normal to the spherical surface. From symmetry considerations, only the component of $\mathbf{J} \times \hat{\mathbf{R}}$ that is parallel to the polar axis of the sphere contributes to the $\mathbf{B}$ field at $\mathbf{r} = 0$.

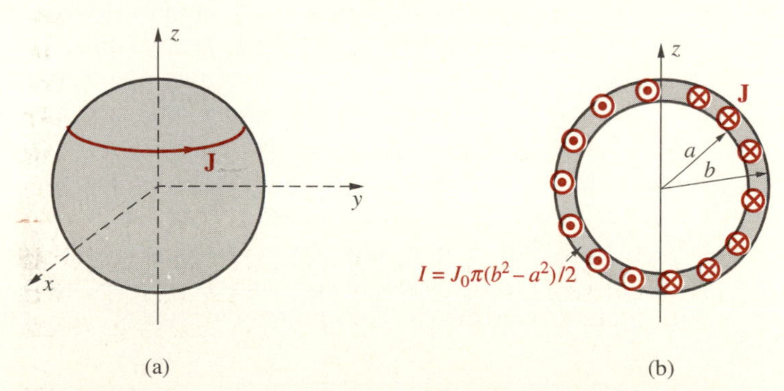

FIGURE 6.30. A spherical current-conducting shell. (a) The density $\mathbf{J}$ is everywhere tangent to circles of constant latitude. The polar axis of the sphere is along the z axis. (b) Vertical cross-sectional view along the polar axis.

Noting that this component is proportional to $\sin\theta'$, where θ' is the spherical coordinate of the source point, we have

$$\mathbf{B}(\mathbf{r}=0)=\frac{\mu_0}{4\pi}J_0\int_{V'}\hat{\mathbf{z}}\frac{\sin\theta'}{R^2}\,dv'$$

where $R^2=(r')^2$. Note also that because the spherical coordinate volume element is given as $dv'=(r')^2\sin\theta'\,d\theta'\,d\phi' dr'$, the volume integral can be easily carried out as

$$B_z(\mathbf{r}=0)=\frac{\mu_0 J_0}{4\pi}\int_a^b dr'\int_0^\pi \sin^2(\theta')\,d\theta'\int_0^{2\pi} d\phi'=\frac{\mu_0 J_0\pi^2}{4\pi}(b-a)$$

If the shell carries a total current I given by $I=(\pi/2)J_0(b^2-a^2)$ (see Figure 4.42b), we can write

$$B_z(\mathbf{r}=0)=\frac{\mu_0 I}{2(b+a)}$$

In the limit $b\to a$, with I fixed, we approach an infinitely thin shell conducting a finite total current I, and the **B** field at the center is

$$\mathbf{B}(\mathbf{r}=0)=\hat{\mathbf{z}}\frac{\mu_0 I}{4a}$$

The spherical shell carrying a uniform current provides an interesting example of a configuration for which $\mathbf{A}=0$ but $\mathbf{B}\neq 0$. This circumstance is not at all surprising when we realize that curl is essentially a vector differentiation; thus, the case in hand is analogous to a one-dimensional function that crosses zero with a nonzero slope.

Example 6-24: Saddle-shaped gradient coils. Magnetic resonance imaging (MRI) scanners use gradient coils to produce transverse gradients (nonuniformities) in the main field of the scanner in order for imaging to be possible. The magnetic fields of the gradient coils strengthen the main field (typically a constant field of a required strength and uniformity produced by superconducting or other types of coils or permanent magnets) in one region while weakening it in another, resulting in a total field whose strength varies in a continuous fashion in a desired direction. Two types[46] of gradient head coils are shown in Figure 6.31a and b. Typically, the gradient fields are much weaker than the main field, so much less powerful electromagnets can be used. Saddle-shaped coils are used in MRI scanners to produce transverse gradients. The optimum geometry[47] of saddle coils consists of two

[46]For more information, see W. H. Oldendorf, M.D. and W. Oldendorf, Jr., *Basics of Magnetic Resonance Imaging,* Martinus Nijhoff Publishing, 1988: also see W. H. Oldendorf, *MRI Primer,* Raven Press, New York, 1991.

[47]D. M. Ginsberg and M. J. Melchner, Optimum geometry of saddle shaped coils for generating a uniform magnetic field, *Rev. Sci. Instrum.*, 41(1), pp. 122–123, January 1970.

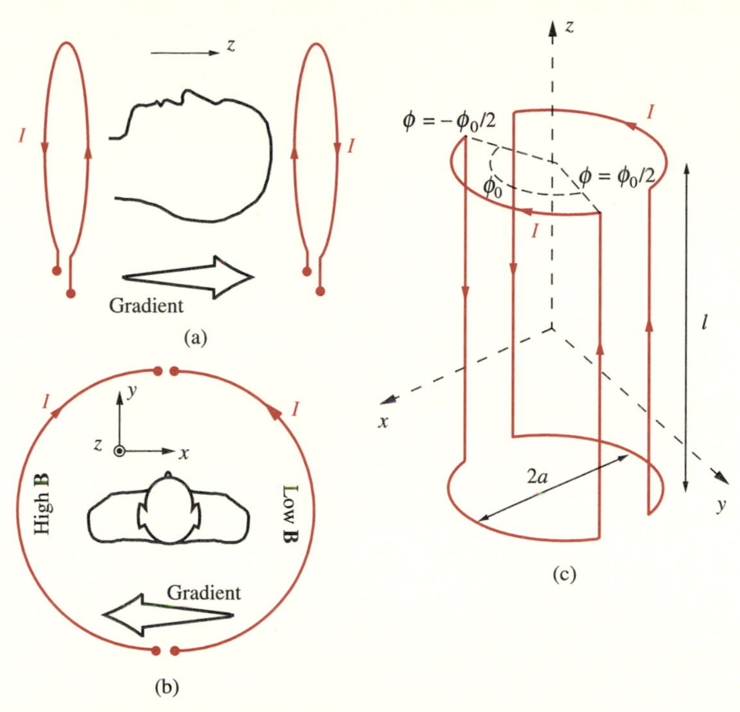

FIGURE 6.31. Saddle-shaped gradient coils. (a) A longitudinal z axis gradient is created by two circular coils. Current is passed in opposite directions through the two coils. The main **B** field (in the $+z$ direction) is reinforced near the coil on the right and is diminished near the coil on the left, resulting in a magnetic gradient between the two coils. (b) A transverse gradient is created using opposed semicircular loops in which current flows in opposite directions. The field from one semicircular loop adds to the main field (in the $+z$ direction), while the field from the other subtracts from it, forming a gradient. (c) The geometry of the saddle coils and the coordinate system.

semicircular loops of wire connected by straight wires. The semicircular portions are active in creating the gradient; the straight portions serve only to connect the loops and do not contribute. A pair of identical saddle-shaped coils wound on the surface of a cylinder with a circular cross section oriented in the z direction is shown in Figure 6.31c. Find the **B** field at the center of the imaging system (i.e., at the origin of the coordinate system in Figure 6.31c).

Solution: First we consider the **B** field due to the arcs of the two coils. The magnetic vector potential at a point along the axis due to an element of current on the circular arc is

$$d\mathbf{A} = \frac{\mu_0 I}{4\pi}\frac{d\mathbf{l}'}{R}$$

where $d\mathbf{l}' = \hat{\boldsymbol{\phi}} a\, d\phi' = (-a\sin\phi')\, d\phi'\hat{\mathbf{x}} + (a\cos\phi')\, d\phi'\hat{\mathbf{y}}$ and $R = |\mathbf{r} - \mathbf{r}'| = |\hat{\mathbf{z}}(z - z') - \hat{\mathbf{r}}a| = \sqrt{a^2 + (z - z')^2}$. Since the contributions from all four circular arcs add together, the total **B** field at the center (i.e., $|z - z'| = l/2$) is

$$\mathbf{B}_{\text{arcs}} = \nabla\times\mathbf{A} = \left\{\nabla\times\left[\frac{\mu_0 I}{\pi}\int_{-\phi_0/2}^{\phi_0/2}\frac{(-a\sin\phi')\hat{\mathbf{x}} + (a\cos\phi')\hat{\mathbf{y}}}{\sqrt{a^2+(z-z')^2}}\,d\phi'\right]\right\}_{z-z'=l/2}$$

$$= \left\{\nabla\times\left[\hat{\mathbf{y}}\frac{\mu_0 I}{\pi}\frac{2a\sin(\phi_0/2)}{\sqrt{a^2+(z-z')^2}}\right]\right\}_{z-z'=l/2}$$

$$= \hat{\mathbf{x}}\frac{\mu_0 I}{\pi}\frac{al\sin(\phi_0/2)}{[a^2+(l/2)^2]^{3/2}}$$

We next consider the **B** field due to the vertical portions of the coils. Using the results of Example 6-3, and realizing that the x components of the vertical wires add together while the y components cancel, the **B** field due to four vertical parts of the coils is

$$\mathbf{B}_{\text{vert}} = \hat{\mathbf{x}}4\frac{\mu_0 I(l/2)\sin(\phi_0/2)}{2\pi a\sqrt{a^2+(l/2)^2}}$$

$$= \hat{\mathbf{x}}\frac{\mu_0 Il}{\pi a}\frac{\sin(\phi_0/2)}{\sqrt{a^2+(l/2)^2}}$$

which is directed in the same direction as $\mathbf{B}_{\text{arcs}}$. Thus, the total **B** field at the center of the pair of saddle coils shown in Figure 6.31c is

$$\mathbf{B} = \mathbf{B}_{\text{arcs}} + \mathbf{B}_{\text{vert}} = \hat{\mathbf{x}}\frac{\mu_0 Il\sin(\phi_0/2)}{\pi\sqrt{a^2+(l/2)^2}}\left[\frac{a}{a^2+(l/2)^2}+\frac{1}{a}\right]$$

Determination of the **B** field at points off the center point is also straightforward but requires numerical evaluation of the resulting integrals. By examining the variation of the **B** field with position in the vicinity of the center point, it can be shown[48] that optimum homogeneity is obtained for $l = 2a$ and $\phi_0 = 120°$.

6.6 THE MAGNETIC DIPOLE

A small current-carrying loop constitutes the magnetic equivalent of the electric dipole. In this section we derive an expression for **B** at a large distance from such a small loop, which is referred to as the *magnetic dipole.* The concept of a magnetic dipole is useful in understanding the behavior of magnetic materials (Section 6.8), being analogous to the electric dipole, which was used in Section 4.10 to determine the response to an external electric field of dielectric materials.

[48]D. I. Hoult and R. E. Richards, The signal-to-noise ratio of the nuclear magnetic resonance experiment, *J. Magn. Reson.*, 24, pp. 71–85, 1976.

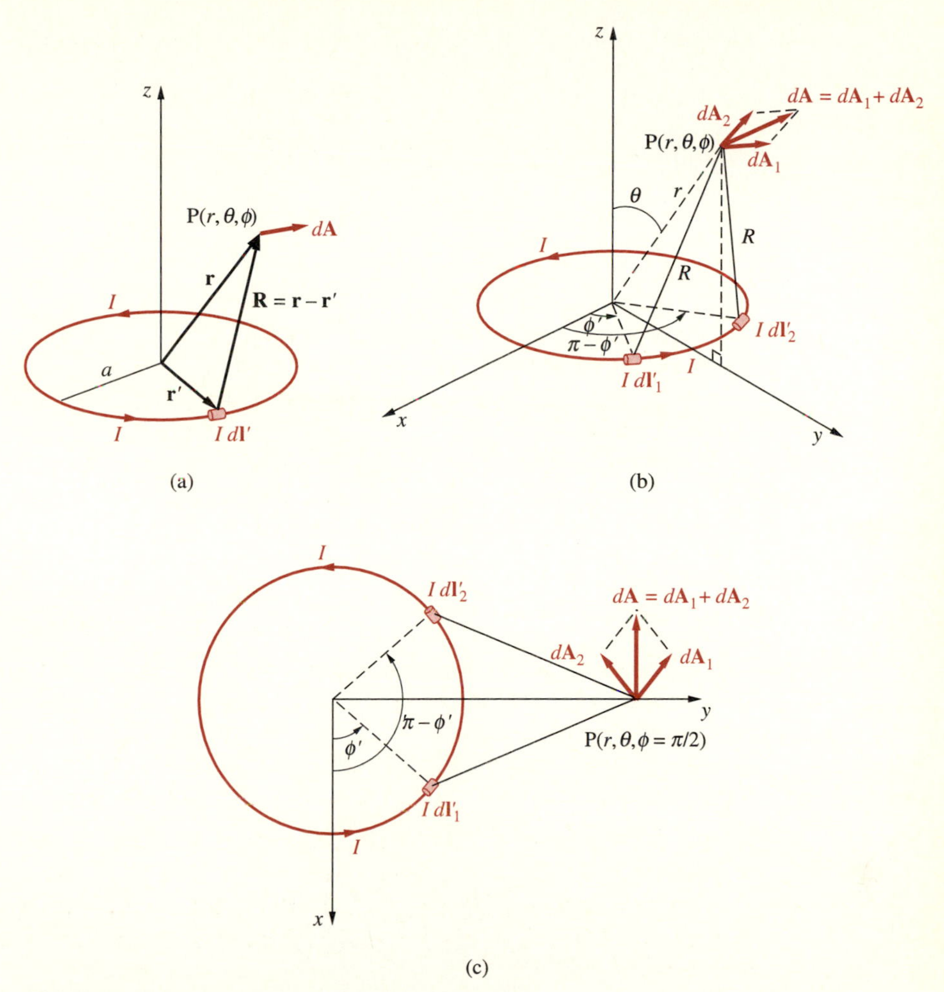

FIGURE 6.32. **A circular current loop in free space.** (a) A circular loop carrying a steady current I. (b) We lose no generality by taking the observation point to be in the yz plane ($\phi' = \pi/2$). (c) Top view, showing the vector potentials produced by two current elements symmetrical with respect to the y axis.

We consider a small circular loop of radius a carrying a steady current I with its axis coincident with the z axis and its center located at the origin, as shown in Figure 6.32a. The radius a of the loop is assumed to be much smaller than r (i.e., $a \ll r$), where r is the distance from the origin of the point P at which we wish to determine the **B** field. Under these conditions, the loop is considered to be a magnetic dipole. We would like to find the magnetic vector potential **A** at point P and use it to obtain the **B** field.

The exact expression for the vector **A** at point P (r, θ, ϕ) is given by

$$\mathbf{A} = \frac{\mu_0 I}{4\pi} \oint_C \frac{d\mathbf{l}'}{R}$$

where $d\mathbf{l}' = \hat{\boldsymbol{\phi}} a\, d\phi'$; noting that $\mathbf{r}' = \hat{\mathbf{x}} a \cos\phi' + \hat{\mathbf{y}} a \sin\phi'$, we have

$$R = |\mathbf{R}| = |\mathbf{r} - \mathbf{r}'| = \sqrt{(x - a\cos\phi')^2 + (y - a\sin\phi')^2 + z^2}$$

Substituting $x = r\sin\theta\cos\phi$, $y = r\sin\theta\sin\phi$, $z = r\cos\theta$, we can write

$$\begin{aligned} R = |\mathbf{R}| &= \sqrt{(r\sin\theta\cos\phi - a\cos\phi')^2 + (r\sin\theta\sin\phi - a\sin\phi')^2 + r^2\cos^2\theta} \\ &= \sqrt{a^2 + r^2 - 2ar\sin\theta(\cos\phi\cos\phi' + \sin\phi\sin\phi')} \\ &= \sqrt{a^2 + r^2 - 2ar\sin\theta\cos(\phi - \phi')} \\ &= r\sqrt{1 + \frac{a^2}{r^2} - \frac{2a}{r}\sin\theta\cos(\phi - \phi')} \end{aligned}$$

It is clear from symmetry that the potential is independent of the azimuth angle ϕ. We can consider straightforward evaluation of $\mathbf{A}(r, \theta)$ by substituting the expression for $\mathbf{R}$ and $d\mathbf{l}$ in the integral expression for $\mathbf{A}$. Thus, we have

$$\mathbf{A}(r, \theta) = \frac{\mu_0 I}{4\pi} \int_0^{2\pi} \frac{\hat{\boldsymbol{\phi}} a\, d\phi'}{\sqrt{a^2 + r^2 - 2ar\sin\theta\cos(\phi - \phi')}}$$

This integral does not have a closed-form solution and is not particularly revealing; however, direct numerical evaluation of $\mathbf{A}$ using the integral expression is straightforward, since the integrand is a well-behaved function.

In our case, we are interested in the fields at points far away from the loop, for which case useful approximate expressions can be derived for $\mathbf{A}$. Since $r \gg a$, we can use the Taylor series expansion[49] of R^{-1} in r to write

$$\frac{1}{R} \simeq \frac{1}{r}\left[1 + \frac{a}{r}\sin\theta\cos(\phi - \phi')\right]$$

Hence, it follows that

$$\mathbf{A} \simeq \frac{\mu_0 I a}{4\pi} \int_0^{2\pi} \hat{\boldsymbol{\phi}} \frac{1}{r}\left[1 + \frac{a}{r}\sin\theta\cos(\phi - \phi')\right] d\phi'$$

Without any loss of generality, the point P can be located directly above the y axis as shown in Figure 6.32b (i.e., $\phi = \pi/2$). The variable direction of the unit vector $\hat{\boldsymbol{\phi}}$ in the integrand is handled by considering two current elements, $I\, d\mathbf{l}_1$ and $I\, d\mathbf{l}_2$, which are symmetrically located with respect to the y axis. From the geometry, we have

$$I\, d\mathbf{l}_1 = (-\hat{\mathbf{x}}\sin\phi' + \hat{\mathbf{y}}\cos\phi') I a\, d\phi'$$

and

$$I\, d\mathbf{l}_2 = (-\hat{\mathbf{x}}\sin\phi' - \hat{\mathbf{y}}\cos\phi') I a\, d\phi'$$

---

[49] See Section 4.4.3 for the electric dipole, where a similar expansion was shown in detail.

the sum of which results in a net $d\mathbf{A}$ at point P that is $-x$-directed (or ϕ-directed; note that at point P with $\phi = \pi/2$ we have $-\hat{\mathbf{x}} = \hat{\boldsymbol{\phi}}$). This is more clearly seen from the top view shown in Figure 6.32c. Thus, the vector $\mathbf{A}$ for $r \gg a$ becomes

$$\mathbf{A} = \hat{\boldsymbol{\phi}}\frac{\mu_0 I a}{4\pi}2\int_{-\pi/2}^{\pi/2} \sin\phi' \frac{1}{r}\left[1 + \frac{a}{r}\sin\theta\sin\phi'\right]d\phi'$$

$$= \hat{\boldsymbol{\phi}}\frac{\mu_0 I a}{2\pi}\left[\frac{1}{r}\int_{-\pi/2}^{\pi/2} \sin\phi'\, d\phi' + \frac{a}{r^2}\sin\theta\int_{-\pi/2}^{\pi/2} \sin^2\phi'\, d\phi'\right]$$

where we have also noted that with $\phi = \pi/2$, $\cos(\phi - \phi') = \sin\phi'$. Note that the first integral is zero, so evaluating the second integral yields

$$\mathbf{A} = \hat{\boldsymbol{\phi}}\frac{\mu_0 I a^2 \sin\theta}{4r^2}$$

We can now find $\mathbf{B}$ from

$$\mathbf{B} = \nabla \times \mathbf{A}$$

Using spherical coordinates, since $A_r = A_\theta = 0$ and $\partial A_\phi/\partial\phi = 0$, it follows that

$$\mathbf{B} = \hat{\mathbf{r}}B_r + \hat{\boldsymbol{\theta}}B_\theta$$

where

$$B_r = \frac{1}{r\sin\theta}\frac{\partial}{\partial\theta}(A_\phi \sin\theta) = \frac{\mu_0 I a^2 \cos\theta}{2r^3}$$

and

$$B_\theta = -\frac{1}{r}\frac{\partial}{\partial r}(rA_\phi) = \frac{\mu_0 I a^2 \sin\theta}{4r^3}$$

It is interesting to compare these expressions with those obtained for the electric field due to an electric dipole. For the electric dipole (from Section 4.4.3), we have

$$\mathbf{E} = \hat{\mathbf{r}}\frac{p\cos\theta}{2\pi\epsilon_0 r^3} + \hat{\boldsymbol{\theta}}\frac{p\sin\theta}{4\pi\epsilon_0 r^3}$$

where $p = Qd$ is the electric dipole moment. For the magnetic dipole we have

$$\mathbf{B} = \hat{\mathbf{r}}\frac{\mu_0 I a^2 \cos\theta}{2r^3} + \hat{\boldsymbol{\theta}}\frac{\mu_0 I a^2 \sin\theta}{4r^3}$$

As these expressions are identical in form, we can associate the magnetic dipole with a dipole moment $\mathbf{m}$ such that $\mathbf{m} = \hat{\mathbf{z}}|\mathbf{m}| = \hat{\mathbf{z}}I(\pi a^2)$, where πa^2 is the total area of the circular loop. For a magnetic dipole with N turns, the magnetic dipole moment is $\mathbf{m} = \hat{\mathbf{z}}NI\pi a^2$. The $\mathbf{B}$ field for the magnetic dipole can be rewritten as

$$\mathbf{B} = \hat{\mathbf{r}}\frac{\mu_0|\mathbf{m}|\cos\theta}{2\pi r^3} + \hat{\boldsymbol{\theta}}\frac{\mu_0|\mathbf{m}|\sin\theta}{4\pi r^3} \quad [6.14]$$

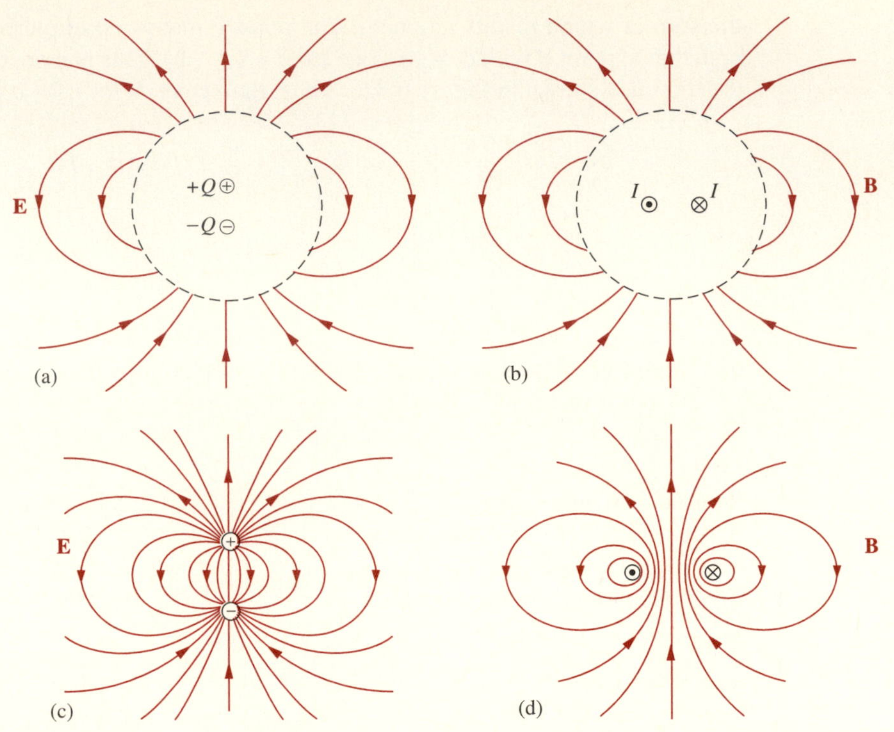

FIGURE 6.33. The field lines of electric and magnetic dipoles. (a) Field lines as defined by [4.27] for the **E** field at large distances from an electric dipole. (b) **B** field lines at large distances as given by [6.14] for a magnetic dipole. (c) Close-up view showing the termination of the **E** field on the charges. (d) Close-up view showing the self-closure of the **B** field lines.

Although [6.14] was derived for a circular loop, it is also valid for small, current-carrying loops of other symmetric shapes (e.g., a square loop), with $\mathbf{m} = \hat{\mathbf{z}}NIA$, where A is the area of the loop.

The similarity of the fields for the magnetic and electric dipoles is illustrated in Figure 6.33. The top panels show the normalized electric and magnetic dipole field lines as given, respectively, by [4.27] and [6.14], which of course have identical shapes but are valid only at large distances from the dipoles. The bottom panels show the close-up views, illustrating that the electric field lines terminate on the two charges, whereas the magnetic field lines close on themselves.

Note that magnetic dipole moment $\mathbf{m}$ is a vector, which has a magnitude $|\mathbf{m}|$ and a direction along the axis of the magnetic dipole loop determined by the direction of the current flow using the right-hand rule. The vector $\mathbf{A}$ due to a magnetic dipole can be written in terms of its dipole moment $\mathbf{m}$ as

$$\mathbf{A} = \frac{\mu_0 \mathbf{m} \times \hat{\mathbf{r}}}{4\pi r^2} = \hat{\boldsymbol{\phi}} \frac{\mu_0 |\mathbf{m}| \sin\theta}{4\pi r^2}$$

Noting that $\nabla r^{-1} = -\hat{\mathbf{r}}/r^2$ (see footnote 41 in Section 6.5), this result can also be written as

$$\mathbf{A} = \frac{\mu_0}{4\pi}\left(\mathbf{m} \times \frac{\hat{\mathbf{r}}}{r^2}\right) = \frac{\mu_0}{4\pi}\left(\nabla \times \frac{\mathbf{m}}{r}\right)$$

Using the above, we can also write an expression directly relating **B** and **m**. We have

$$\mathbf{B} = \nabla \times \mathbf{A} = \frac{\mu_0}{4\pi}\nabla \times \left(\nabla \times \frac{\mathbf{m}}{r}\right)$$

which can be simplified[50] to the form

$$\mathbf{B} = \frac{\mu_0}{4\pi}\nabla\left[\mathbf{m} \cdot \nabla\left(\frac{1}{r}\right)\right]$$

Example 6-25 illustrates the use of the magnetic dipole expressions for **B** and **A** to determine the **B** field of a long uniform solenoid.

Example 6-25: Field of a long uniform solenoid. As an application of the magnetic dipole results derived above, consider the **B** field of a long uniform solenoid as shown in Figure 6.34, having N turns per meter and carrying a current I. The length of the solenoid is l, and its radius is a. Determine the **B** field at point P(x, y, z) such that $r \gg a$, $R_1 \gg a$, and $R_2 \gg a$.

Solution: We can consider the solenoid as consisting of many magnetic dipoles stacked on top of one another, and since $r \gg a$ and $R_1, R_2 \gg a$, we can use the magnetic dipole expressions derived above for distant observation points. For a segment of length dz', we have $N dz'$ turns, each with current I, so that the magnetic dipole moment of that differential segment is $\mathbf{m} = \hat{\mathbf{z}} N I dz' \pi a^2$. Although we can first find the vector potential **A** and then use $\mathbf{B} = \nabla \times \mathbf{A}$ to find

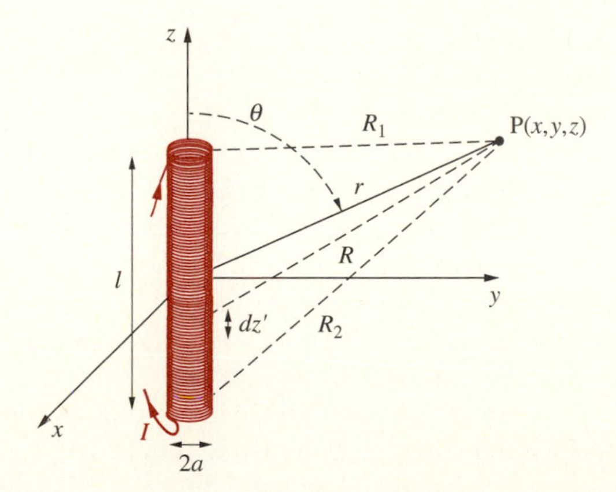

FIGURE 6.34. **B field of a solenoid.**

---

[50]Using the fact that $\nabla^2(1/r) = 0$ for $r \neq 0$ and the identity

$$\nabla \times \nabla \times \mathbf{G} \equiv \nabla(\nabla \cdot \mathbf{G}) - \nabla^2\mathbf{G}$$

the **B** field, we can also use the direct relationship between **B** and **m** as given above. The differential contribution to the field at P by the differential segment of the solenoid is

$$d\mathbf{B} = \frac{\mu_0}{4\pi}\nabla\left[\mathbf{m}\cdot\nabla\left(\frac{1}{R}\right)\right] = \frac{\mu_0}{4\pi}\nabla\left[NIdz'\pi a^2\hat{\mathbf{z}}\cdot\nabla\left(\frac{1}{R}\right)\right]$$

where R is the distance between the elemental magnetic dipole and point P as shown in Figure 6.34. Note that $\nabla(R)^{-1} = -\hat{\mathbf{R}}/R^2 = -\mathbf{R}/R^3$, as derived in footnote 41 in Section 6.5. To find the total **B** field at P we simply need to integrate over the length of the solenoid. In other words,

$$\mathbf{B} = \frac{\mu_0}{4\pi}\nabla\int_{-l/2}^{l/2}\nabla\left(\frac{1}{R}\right)\cdot\hat{\mathbf{z}}NI\pi a^2\,dz'$$

Noting that $R = |\mathbf{r}-\mathbf{r}'| = \hat{\mathbf{x}}x + \hat{\mathbf{y}}y + \hat{\mathbf{z}}(z-z')$, we have

$$\begin{aligned}
\mathbf{B} &= \frac{-NIa^2\mu_0}{4}\nabla\int_{-l/2}^{l/2} -\frac{\hat{\mathbf{R}}}{R^2}\cdot\hat{\mathbf{z}}\,dz' \\
&= \frac{-NIa^2\mu_0}{4}\nabla\int_{-l/2}^{l/2} -\frac{(z-z')\,dz'}{[x^2+y^2+(z-z')^2]^{3/2}} \\
&= \frac{-NIa^2\mu_0}{4}\nabla[x^2+y^2+(z-z')^2]^{-1/2}\Big|_{-l/2}^{l/2} \\
&= \frac{-NIa^2\mu_0}{4}\left[\nabla\left(\frac{1}{R}\right)\right]_{-l/2}^{l/2} = \frac{-NIa^2\mu_0}{4}\nabla\left[\frac{1}{R_1}-\frac{1}{R_2}\right]
\end{aligned}$$

where R_1 and R_2 are respectively the distances to the observation point from the top and bottom of the solenoid.

The evaluation of the gradient in the above expression for **B** is difficult in the general case, because of the complicated dependence of R_1 and R_2 on the spatial coordinates. However, when the point P is very far away from the solenoid (i.e., $r \gg l$), it can be shown that $R_1^{-1} \simeq r^{-1}[1-(l/2r)\cos\theta]$ and $R_2^{-1} \simeq r^{-1}[1+(l/2r)\cos\theta]$, so we have

$$\mathbf{B} = \frac{-NIa^2\mu_0}{4}\nabla\left(\frac{l\cos\theta}{r^2}\right) = \frac{\mu_0}{4\pi}\underbrace{(NI\pi a^2 l)}_{|\mathbf{m}_{\text{slnd}}|}\left(\hat{\mathbf{r}}\frac{2\cos\theta}{r^3}+\hat{\boldsymbol{\theta}}\frac{\sin\theta}{r^3}\right) \quad [6.15]$$

We note that the above expression for **B** is identical to [6.14], if we replace $|\mathbf{m}|$ with the dipole moment of the solenoid, namely $|\mathbf{m}_{\text{slnd}}| = NI\pi a^2 l$. Thus, as far as the field at large distances is concerned, the solenoid produces the same type of field structure as the elementary magnetic dipole, as if it had a total dipole magnetic moment of $NI\pi a^2 l$.

6.7 DIVERGENCE OF B, MAGNETIC FLUX, AND INDUCTANCE

As we discussed in Section 4.5, it is often advantageous to think of a vector field as representing the flow of something. Faraday's experiments on electric displacement were the basis for the concepts of electric flux and electric flux density $\mathbf{D}$, which was shown (Section 4.6) to originate from or terminate at sources (or electric charges), as described by the relation $\nabla \cdot \mathbf{D} = \rho$. In this section, we first discuss the concept of magnetic flux and the divergence of the $\mathbf{B}$ field. In fact, in engineering practice we more often think in terms of magnetic flux than electric flux. Much of our present view of magnetic flux has its origins in Faraday's visualizations of the magnetic force as stretching out in all directions from the electric current that produces it, filling all space as a magnetic field. Through this magnetic flux, which permeates all space, seemingly isolated circuits are inductively coupled; hence, we next consider the important concept of inductance.

6.7.1 Divergence of B and Magnetic Flux

We saw in Section 6.5 that the magnetic field $\mathbf{B}$ can be derived from the curl of an auxiliary vector potential function $\mathbf{A}$. This result leads at once to an important physical property for the $\mathbf{B}$ field. The divergence of the curl of any vector is identically equal to zero, and hence $\nabla \cdot \nabla \times \mathbf{A} = 0$, from which it follows that the divergence of $\mathbf{B}$ is identically zero; i.e.,

$$\boxed{\nabla \cdot \mathbf{B} = 0} \qquad [6.16]$$

This property of $\mathbf{B}$ is a mathematical consequence of the formulation of Ampère's experimentally based force law or its alternative statement in the Biot–Savart law. In Section 6.8, we show that the effects of material bodies on the distribution of the magnetic field can be accounted for by their equivalent volume and surface magnetization currents, which also produce magnetic fields, as given by [6.6]. Thus, even in the presence of material media it is possible to derive $\mathbf{B}$ from the curl of a vector potential $\mathbf{A}$, and $\nabla \cdot \mathbf{B} = 0$ is valid in general.

As was discussed in Chapter 5 in connection with steady currents for which $\nabla \cdot \mathbf{J} = 0$, the fact that $\nabla \cdot \mathbf{B} = 0$ requires that the flux lines of $\mathbf{B}$ are always continuous and form closed loops. Thus, [6.16] implies that there are no sources of magnetic fields. (In this connection, "source" is used in a mathematical sense; literally, electric current is the source of magnetic fields.) In a mathematical sense, [6.16] means that there exist no *magnetic charges* (or free magnetic poles), corresponding to electric charges, from which $\mathbf{B}$ field lines can emerge or onto which they can terminate. If the divergence of $\mathbf{B}$ were not zero, magnetic field lines could originate at sources (at which $\nabla \cdot \mathbf{B} \neq 0$), just as electric field lines originate at charges ($\nabla \cdot \mathbf{D} = \rho$). However, at the present time, there is no experimental evidence whatsoever for the existence of free magnetic

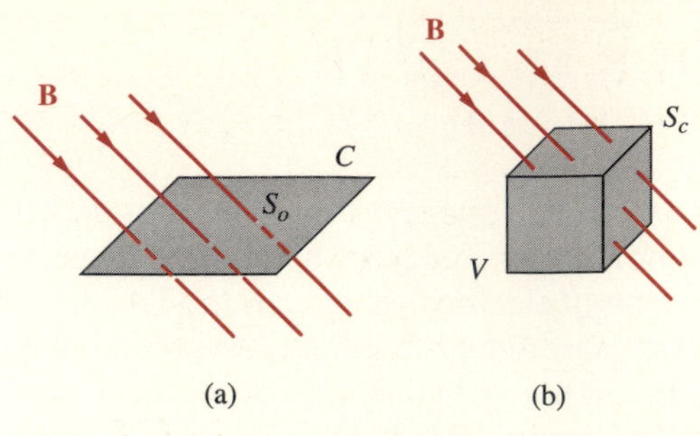

FIGURE 6.35. Magnetic flux through a surface S**.** (a) An open surface S_o enclosed by a contour C; (b) a closed surface S_c enclosing a volume V.

poles[51] or "magnetic charge." On the contrary, all available experimental data can be interpreted on the basis of $\nabla\cdot\mathbf{B} = 0$. All magnets have both a North and a South pole, and the field $\mathbf{B}$ is continuous through the magnet. For this reason the magnetostatic field $\mathbf{B}$ is fundamentally different from the electrostatic field $\mathbf{E}$.

Our discussion of the concepts of flux, flux tubes, and flux lines, given in Section 4.5 for electric fields, also applies to the $\mathbf{B}$ field. In thinking about the $\mathbf{B}$ field, it is often useful to consider it as representing the flow of something. It is in this context that the $\mathbf{B}$ field is often thought of as representing a magnetic flux density, although the implied analogy between $\mathbf{B}$ and the electric flux density $\mathbf{D}$ is somewhat misleading, as discussed in Section 6.2.3. The magnetic flux through a differential area ds is given by the dot product of $\mathbf{B}$ with $d\mathbf{s}$, namely $\mathbf{B}\cdot d\mathbf{s}$. The dot product selects the normal component of $\mathbf{B}$ through the surface ds, as shown in Figure 6.35a, so the total magnetic flux Ψ passing through the surface S is given by

$$\Psi = \int_S \mathbf{B}\cdot d\mathbf{s}$$

in units of webers (Wb). The magnetic flux passing through the surface S bounded by the contour C is said to *link* the contour C and is commonly referred to as the flux linkage.

For a closed surface S, however, as in Figure 6.35b, just as much flux leaves the surface as enters because of the continuous nature of the flux lines. Hence the integral of $\mathbf{B}\cdot d\mathbf{s}$ over a closed surface is always equal to zero. Mathematically, this result follows from [6.16] using the divergence theorem. Since $\nabla\cdot\mathbf{B} = 0$, we have

$$\int_V \nabla\cdot\mathbf{B}\, dv = \oint_S \mathbf{B}\cdot d\mathbf{s} = 0$$

[51]The lack of magnetic poles or charges is the one outstanding asymmetry between electrostatics and magnetostatics. If magnetic poles did exist, say with a volume density ρ_{m}, the divergence of the $\mathbf{B}$ field would then be $\nabla\cdot\mathbf{B} = \mu_0\rho_{\mathrm{m}}$. If the magnetic charges moved with a velocity $\mathbf{v}$, the quantity $\rho_{\mathrm{m}}\mathbf{v}$ would constitute a magnetic conduction current density $\mathbf{J}_{\mathrm{m}}$, which in turn would produce an electric field, just as electric conduction current density $\mathbf{J}$ produces a $\mathbf{B}$ field. P. A. M. Dirac [1902–1984] carried out an extensive theoretical investigation of magnetic poles and found no fundamental reason for isolated magnetic charges (*magnetic monopoles*) not to exist. However, he found that a quantum of magnetic charge would be so great that if pairs of opposite sign combined together, it would take highly energetic cosmic rays to separate them. See P. A. M. Dirac, The theory of magnetic poles, *Phys. Rev.,* 74, p. 817, 1948. For a brief discussion of experimental attempts to detect magnetic monopoles, see pp. 136–137 of M. Schwartz, *Principles of Electrodynamics,* Dover, 1972. Also see Sections 6.12 and 6.13 of J. D. Jackson, *Classical Electrodynamics,* Wiley, 2nd ed., 1975.

where the volume V is enclosed by the surface S. We have thus established the fact that the integral of the **B** field on any closed surface is zero, a fact that was utilized in Example 6-14 to deduce that the **B** field everywhere inside an infinitely long solenoid is directed along the axis of the solenoid.

The magnetic flux that links a contour C may also be expressed in terms of the vector potential **A**. Since $\mathbf{B} = \nabla \times \mathbf{A}$, we have

$$\Psi = \int_S \mathbf{B} \cdot d\mathbf{s} = \int_S \nabla \times \mathbf{A} \cdot d\mathbf{s}$$

The latter integral may be transformed to a contour integral by using Stokes's theorem:

$$\Psi = \int_S \nabla \times \mathbf{A} \cdot d\mathbf{s} = \oint_C \mathbf{A} \cdot d\mathbf{l}$$

This latter integral is sometimes more convenient to evaluate than $\int_S \mathbf{B} \cdot d\mathbf{s}$ when determining the magnetic flux linked by a contour C.

6.7.2 Inductance

In Chapter 4 we pointed out that electrostatic problems are often difficult because the distribution of the electric charges is not known and is determined by the configuration of metallic conductors, thus requiring the solution of Poisson's equation. The concept of capacitance was introduced as a measure of the distribution of the electric field in the vicinity of conductor configurations. In practice, calculation of magnetostatic fields is relatively simpler than that of electrostatic fields, because the magnetostatic problem usually involves a known distribution of currents, from which the **B** field can be found using the Biot–Savart law. However, practical problems are nevertheless complicated by the magnetic properties of the surrounding medium. Fortunately, a detailed description of the **B** field is rarely desirable; more often than not, some overall measure of the field is sufficient. *Inductance* is this single measure of the distribution of the magnetic field near a current-carrying conductor.

Capacitance was introduced as a measure of the ability of a conductor configuration to hold charge per unit applied voltage, or store electrical energy, and was shown to be a property of the physical arrangement of the conductors. In a similar vein, inductance is another property of the physical layout of conductors and is a measure of the ability of a conductor configuration to *link magnetic flux,* or store magnetic energy. For our purposes, we define *flux linkage* as the integral of the magnetic field over the area enclosed by a closed circuit.

Before defining inductance, we must introduce the concept of flux linkage. Consider two neighboring closed loops C_1 and C_2 as shown in Figure 6.36. If a current I_1 flows around the closed loop C_1, a magnetic field $\mathbf{B}_1$ is produced, and some of this magnetic field links C_2 (i.e., will pass through the area S_2 enclosed by C_2).[52] This

[52]We choose to not give a precise definition of the term "links," because intuition gives a better feel for this particular concept than any formal definition. Linking of one circuit by another can be thought of as the magnetic flux produced by one circuit "passing through" or connecting with another circuit.

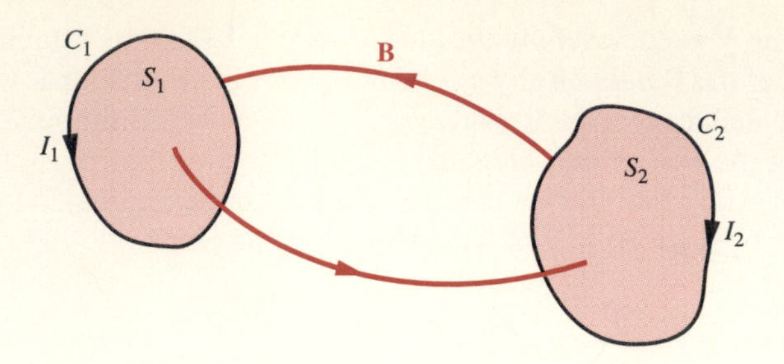

FIGURE 6.36. **Two magnetically coupled circuits.**

magnetic flux produced by the current I_1 flowing around C_1 but linked by the area S_2 enclosed by C_2 can be designated as

$$\Psi_{12} = \int_{S_2} \mathbf{B}_1 \cdot d\mathbf{s}_2$$

using the same subscript indexing as in previous cases, where the first subscript indicates the source of the quantity while the second indicates the location at which it is observed. In practice, inductances typically consist of coils with multiple turns. If C_1 consists of multiple turns N_1, then the total flux produced is N_1 times larger (see Section 6.2.2); namely, $\Lambda_{12} = N_1\Psi_{12}$.

The *mutual inductance* L_{12} between the two coils is thus defined as

$$L_{12} \equiv \frac{N_2\Lambda_{12}}{I_1} = \frac{N_2N_1\Psi_{12}}{I_1}$$

Noting that in general C_2 has N_2 turns, the magnetic flux produced by C_1 is thus linked N_2 times by C_2.

Some of the magnetic flux produced by I_1 links the area S_1 enclosed by the closed contour C_1, so that we can also define the *self-inductance* of loop C_1 as

$$L_{11} \equiv \frac{N_1\Lambda_{11}}{I_1} = \frac{N_1^2\Psi_{11}}{I_1}$$

where $\Lambda_{11} = N_1\Psi_{11}$ is the the total flux linked by a single turn of C_1, and Ψ_{11} is the magnetic flux produced by a single turn of C_1 and linked by a single turn of C_1.

Note that if a current I_2 flows around the closed loop C_2, it generates a magnetic flux that links C_1, so that the mutual inductance between the two loops is given by

$$L_{21} = \frac{N_1N_2\Psi_{21}}{I_2}$$

where Ψ_{21} is the flux generated by the current I_2 flowing around a single turn of C_2 and linking the area enclosed by a single turn of C_1.

To show that $L_{12} = L_{21}$, we can rely on the representation of the **B** field by a vector potential. Consider the expression for the mutual inductance L_{12}:

$$L_{12} = \frac{N_2}{I_1}\int_{S_2} \mathbf{B}_1 \cdot d\mathbf{s}_2$$

where $\mathbf{B}_1$ is the total field produced by C_1 (with N_1 turns) at the surface S_2 enclosed by C_2 (with N_2 turns). Noting that $\mathbf{B}_1 = \nabla \times \mathbf{A}_1$ we can rewrite L_{12} as

$$L_{12} = \frac{N_2}{I_1}\int_{S_2} \mathbf{B}_1 \cdot d\mathbf{s}_2 = \frac{N_2}{I_1}\int_{S_2} (\nabla \times \mathbf{A}_1) \cdot d\mathbf{s}_2 = \frac{N_2}{I_1}\oint_{C_2} \mathbf{A}_1 \cdot d\mathbf{l}_2$$

where we used Stokes's theorem. The vector potential $\mathbf{A}_1$ is related to its source current I_1 through [6.13], so we have

$$\mathbf{A}_1 = \frac{\mu_0 N_1 I_1}{4\pi}\oint_{C_1} \frac{d\mathbf{l}_1}{R}$$

where $R = |\mathbf{r} - \mathbf{r}'|$, with $\mathbf{r}$ and $\mathbf{r}'$ being the positions of the observation and source points respectively. We can thus write

$$L_{12} = \frac{\mu_0 N_1 N_2}{4\pi}\oint_{C_1}\oint_{C_2} \frac{d\mathbf{l}_1 \cdot d\mathbf{l}_2}{R} \qquad [6.17]$$

which indicates that $L_{12} = L_{21}$, since the dot product is commutative and the line integrals can be interchanged.

Equation [6.17] is known as the *Neumann formula* for mutual inductance. It underscores the fact that the mutual inductance is only a function of the geometrical arrangement of the conductors. We shall see in Section 6.8 that if the medium the circuits are located in is a magnetic material, the constant μ_0 needs to be replaced by another constant to be defined as magnetic permeability μ. Thus, inductance also depends on the magnetic properties of the medium that the circuits are located in, as represented by its permeability μ. In general, carrying out the double line integral is quite involved. In most cases, we take advantage of the symmetries inherent in the problem to determine the flux linkage or stored magnetic energy so that we can find the mutual inductance without resorting to [6.17]. In some cases, however, it actually is easier to use [6.17]; we shall use this formula in Example 6-31 to determine the self-inductance of a current-carrying loop.

For any given conductor configuration, the self-inductance of the closed loop C can be evaluated in a manner very similar to the procedure used to find the capacitance in electrostatics. We can first assume a current I to flow in the closed loop, from which we can determine $\mathbf{B}$ using Ampère's law or the Biot–Savart law or the vector potential $\mathbf{A}$. This magnetic field is proportional to the current I. We can then find the flux linked by the circuit by conducting an integral, namely

$$\Psi = \int_S \mathbf{B} \cdot d\mathbf{s}$$

where S is the area enclosed by the closed loop C. Given that the number of turns in the loop is N, and that all the turns produce and link the same flux, the self-inductance is then given by $L = (N^2\Psi)/I$. Note that I will cancel out so that L depends only on the geometrical arrangement (i.e., shape and dimensions) of the circuit C and the magnetic properties of the medium as mentioned in the preceding paragraph.

Example 6-26 illustrates the evaluation of the inductance of one of the most commonly encountered configurations, namely a long solenoid. Example 6-27 concerns the inductance of another common coil known as the toroid, whereas Examples 6-28 and 6-29 respectively examine the inductance of the coaxial line and the two-wire line, two of the commonly used transmission lines that were discussed in Section 2.7, and the inductances of which were tabulated in Table 2.2. Example 6-30 concerns the mutual inductance between two different multiturn coils.

Example 6-26: Self-inductance of a long solenoid. Find the self-inductance of a solenoidal coil of length l, radius a, and total number of turns equal to N.

Solution: In Section 6.2.2, it was shown that for a long solenoid the intensity of the **B** field at the end points of the axis is half that at the center because of the flux leakage near the ends. However, this leakage is mainly confined to the ends of the solenoid and can be neglected. So, a good approximation is to assume that the **B** field is constant over the entire interior of the solenoid, being equal to its value at the center given by the result derived in Example 6-14, namely

$$\mathbf{B} \simeq \mathbf{B}_{\text{ctr}} \simeq \hat{\mathbf{z}}\frac{\mu_0 N I}{l}$$

As a result, the total flux linkage of every individual turn of the solenoidal coil is

$$\Psi = B_z A = \frac{\mu_0 N I}{l} A$$

where A is the cross-sectional area of the solenoid (equal to $A = \pi a^2$ for a solenoid with a circular cross section of radius a). Since there are N turns, the total flux linkage of all N turns is

$$\Lambda = N\Psi = \frac{\mu_0 N^2 I A}{l}$$

Thus, the self-inductance of a long solenoid is

$$L = \frac{\Lambda}{I} = \frac{\mu_0 N^2 A}{l}$$

To get a feel about the orders of magnitude involved, let us calculate the self-inductance of a solenoid of 100 turns wound uniformly over a cylindrical wooden core ($\mu = \mu_0$) of length 10 cm and diameter 1 cm. Using the above equation, we have

$$L = \frac{(4\pi \times 10^{-7}\ \text{H-m}^{-1}) \times (100)^2 \times \pi(0.5 \times 10^{-2})^2\ \text{m}^2}{10 \times 10^{-2}\ \text{m}} \simeq 9.87\ \mu\text{H}$$

Using a linear ferromagnetic core material (Section 6.8) with $\mu = 1000\mu_0$ instead of the wooden core increases the self-inductance of the 100-turn solenoid by a factor of 1000, yielding $L \simeq 9.87$ mH.

Example 6-27: Inductance of a toroid. Find the inductance of a toroidal coil having N turns, similar to that shown in Figure 6.19 but having a circular cross section.

Solution: For a toroidal coil, we assume[53] that the mean radius of the toroid is much greater than the diameter of the coil; that is, $r_{\rm m} = (a + b)/2 \gg b - a$. With this assumption, the magnetic field **B** is approximately uniform throughout the inner part of the toroid (also referred to as the core) and is given by

$$\mathbf{B} \simeq \hat{\boldsymbol{\phi}} \frac{\mu_0 NI}{2\pi r_{\rm m}} \qquad a \le r \le b$$

Since the same flux links all the turns, the total flux linkage is

$$\Lambda = N \int_S \mathbf{B} \cdot d\mathbf{s} = NB_\phi \int_S ds = NB_\phi A = \frac{\mu_0 N^2 I A}{2\pi r_{\rm m}}$$

where $A = \pi[(b-a)/2]^2$ is the cross-sectional area of the toroid. Therefore, the inductance of the toroid is

$$L = \frac{\Lambda}{I} = \frac{\mu_0 N^2 A}{2\pi r_{\rm m}}$$

For example, the inductance of a 1000-turn air-core toroid having a mean radius of $r_{\rm m} = 5$ cm and a core diameter of $(b - a) = 1$ cm is approximately given by

$$L = \frac{(4\pi \times 10^{-7}) \times (1000)^2 \times \pi(0.5 \times 10^{-2})^2}{2\pi(5 \times 10^{-2})} \simeq 0.314 \text{ mH}$$

Again, the inductance of the toroid could be increased significantly by using a ferromagnetic type of core material.

Example 6-28: Coaxial inductor. In Example 4-27, we found the capacitance per unit length of an infinitely long coaxial line. Now we find the inductance per unit length of a coaxial line as shown in Figure 6.37. For simplicity, assume that the current flows only in thin layers[54] at $r = a$ and $r = b$ respectively, where $r = b$ is the inner radius of the outer conductor.

Solution: The current in the inner conductor is I and the outer conductor is of the same magnitude with opposite direction. Using the results found in Example 6-12, the **B** field is nonzero only between the two conductors, given by

$$B_\phi = \frac{\mu_0 I}{2\pi r} \qquad a \le r \le b$$

---

[53]This assumption is by no means necessary in order to allow a tractable solution. See Problems 6-38 and 6-39.

[54]This is effectively the case when the walls of the conductors are thin. In addition, at high frequencies the current is effectively confined to a thin layer (skin depth) at $r = a$ for the inner conductor and at $r = b$ for the outer, even if the conductors are not thin (the inner conductor is typically not hollow).

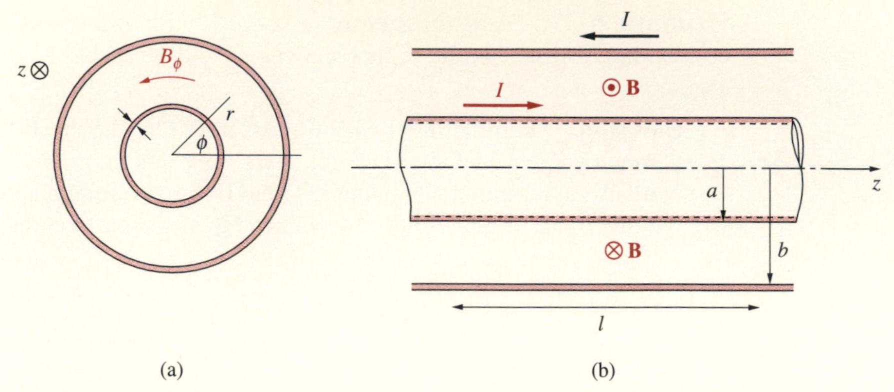

FIGURE 6.37. **Coaxial transmission line with a thin-walled inner conductor.** (a) Cross-sectional view. (b) Side view.

The total flux linkage of the coaxial line of length l can be found as

$$\Psi = \int_S \mathbf{B} \cdot d\mathbf{s} = \int_0^l \int_a^b B_\phi \hat{\boldsymbol{\phi}} \cdot (\hat{\boldsymbol{\phi}}\, dr\, dz) = \frac{\mu_0 I l}{2\pi} \int_a^b \frac{dr}{r} = \frac{\mu_0 I l}{2\pi} \ln\left(\frac{b}{a}\right)$$

Hence, the inductance of the coaxial line of length l is

$$L = \frac{\Psi}{I} = \frac{\mu_0 l}{2\pi} \ln\left(\frac{b}{a}\right) \text{ H}$$

or, the inductance *per unit length* of the coaxial line is

$$\mathcal{L} = \frac{L}{l} = \frac{\mu_0}{2\pi} \ln\left(\frac{b}{a}\right) = 0.2 \ln\left(\frac{b}{a}\right) \quad \mu\text{H-m}^{-1}$$

which is identical to the expression given in Table 2.2 for the inductance per unit length of a coaxial line.

Consider a practical coaxial line such as RG-214 (typically used for high-frequency (MHz range), high-power transmission) having an inner conductor diameter of ~2.256 mm and the inner diameter of its outer conductor of ~7.24 mm, filled with solid polyethylene (which has the same magnetic properties as air). Assuming the currents to be flowing in thin layers (excellent assumption at MHz frequencies, as we shall see in Chapter 8), the inductance per meter of this line is

$$\mathcal{L} \simeq 0.2 \ln \frac{7.24}{2.256} \simeq 0.233 \ \mu\text{H-m}^{-1}$$

Example 6-29: Inductance of the two-wire line. In Example 4-29, the capacitance per unit length of a two-wire line was calculated. Find the inductance per unit length of the two-wire line, as shown in Figure 6.38.

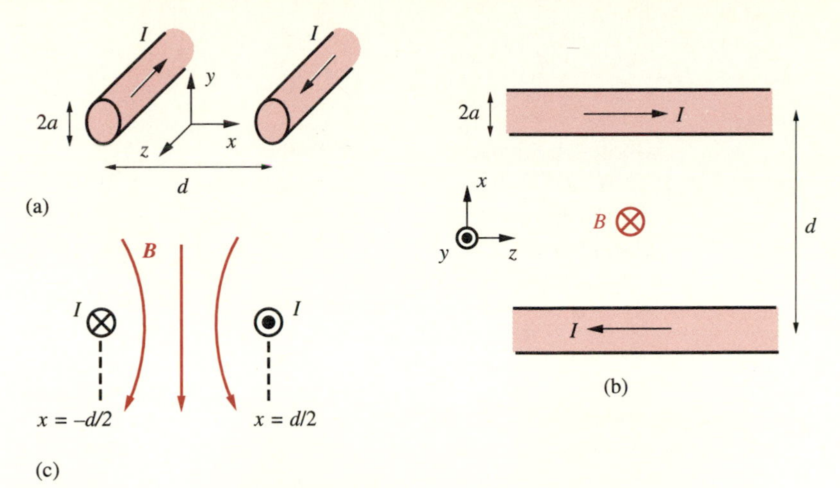

FIGURE 6.38. **Two-wire transmission line.** (a) Two parallel conductors, each carrying a current I, in opposite senses with radii a and separation distance d. (b) Horizontal cross-sectional view. (c) Vertical cross-sectional view.

Solution: For $d \gg a$, the steady currents in each conductor are assumed to be confined along their axes. Hence, the **B** field in the region of the $y = 0$ plane that extends from $x = -d/2 + a$ to $x = d/2 - a$ is approximately given by the sum of the fields of two infinitely long parallel wires as

$$B_y \simeq -\frac{\mu_0 I}{2\pi}\left(\frac{1}{x + d/2} - \frac{1}{x - d/2}\right)$$

The total flux linkage through the area on the $y = 0$ plane between the conductors each of length l can be found from

$$\Psi = \int_S \mathbf{B}\cdot d\mathbf{s} = \frac{\mu_0 I l}{2\pi}\int_{-d/2+a}^{d/2-a}\left[\frac{1}{x + d/2} - \frac{1}{x - d/2}\right]dx$$

$$= \frac{\mu_0 I l}{2\pi}\Big[\ln(x + d/2) - \ln(x - d/2)\Big]_{-d/2+a}^{d/2-a} = \frac{\mu_0 I l}{\pi}\ln\left(\frac{d-a}{a}\right)$$

Since $d \gg a$, we can write

$$\Psi \simeq \frac{\mu_0 I l}{\pi}\ln\frac{d}{a}$$

Therefore, the inductance per unit length of two-wire line is

$$\mathcal{L} = \frac{L}{l} = \frac{\Psi}{Il} = \frac{\mu_0}{\pi}\ln\frac{d}{a} = 0.4\ln\frac{d}{a} \quad \mu\text{H-m}^{-1}$$

The above result is for the case with $d \gg a$. A more accurate analysis for the general case shows that

$$L = \frac{\mu_0}{\pi}\ln\left[\frac{d}{2a} + \sqrt{\left(\frac{d}{2a}\right)^2 - 1}\right] \text{ H-m}^{-1} = 0.4\ln\left[\frac{d}{2a} + \sqrt{\left(\frac{d}{2a}\right)^2 - 1}\right] \quad \mu\text{H-m}^{-1}$$

which is the same as given in Table 2.2.

As a numerical example, consider a $Z_0 = 300\Omega$ air-spaced two-wire transmission line, which is commonly used in connecting a TV to an antenna. For this line, the ratio $d/(2a)$ is about 6, and the corresponding line inductance per unit length is approximately given as

$$L = \frac{4\pi \times 10^{-7}}{\pi}\ln[6 + \sqrt{6^2 - 1}] \simeq 0.991 \ \mu\text{H-m}^{-1}$$

Example 6-30: Two circular coils. Two circular coils with centers on a common axis as shown in Figure 6.39 have N_1 and N_2 turns, each of which is closely wound, and radii a and b, respectively. The coils are separated by a distance d, which is assumed to be much larger than both radii (i.e., $d \gg a, b$). Find the mutual inductance between the coils.

Solution: Assume a current I_1 flowing in coil 1. Using the results obtained for a magnetic dipole, the magnetic vector potential $\mathbf{A}_1$ due to coil 1 at an arbitrary point P on coil 2 can be written as

$$\mathbf{A}_1 = \hat{\boldsymbol{\phi}}\frac{\mu_0 N_1 I_1(\pi a^2)}{4\pi R^2}\sin\theta = \hat{\boldsymbol{\phi}}\frac{\mu_0 N_1 I_1 a^2}{4R^2}\frac{b}{R}$$

$$= \hat{\boldsymbol{\phi}}\frac{\mu_0 N_1 I_1 a^2 b}{4(d^2 + b^2)^{3/2}} \simeq \hat{\boldsymbol{\phi}}\frac{\mu_0 N_1 I_1 a^2 b}{4d^3}$$

since $d \gg b$. The total flux linkage through the area S_2 of coil 2 due to the field $\mathbf{B}_1$ produced by coil 1 can be written as

$$\Lambda_{12} = N_2\int_{S_2}\mathbf{B}_1 \cdot d\mathbf{s}_2 = N_2\int_{S_2}(\nabla \times \mathbf{A}_1)\cdot d\mathbf{s}$$

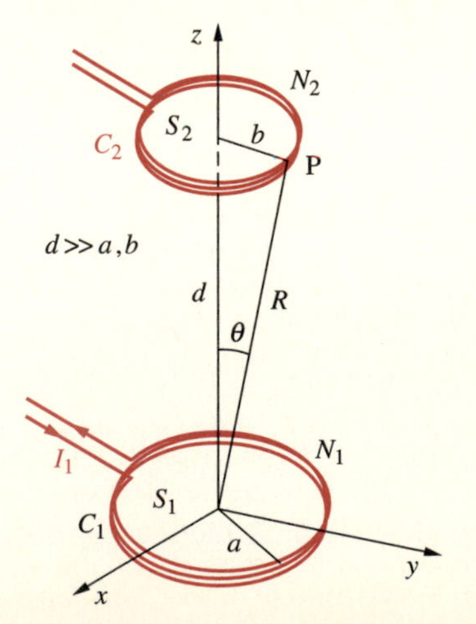

FIGURE 6.39. Two circular coils of turns N_1 and N_2.

Using Stokes's theorem, we can rewrite Λ_{12} as

$$\Lambda_{12} = N_2 \oint_{C_2} \mathbf{A}_1 \cdot d\mathbf{l} = N_2 \oint_{C_2} \frac{\mu_0 N_1 I_1 a^2 b}{4d^3} \hat{\boldsymbol{\phi}} \cdot (\hat{\boldsymbol{\phi}} b\, d\phi) = \frac{\mu_0 N_1 N_2 I_1 a^2 b}{4d^3}(2\pi b)$$

Therefore, the mutual inductance L_{12} is

$$L_{12} = \frac{\Lambda_{12}}{I_1} = \frac{\mu_0 \pi N_1 N_2 a^2 b^2}{2d^3}$$

6.7.3 Inductances of Some Practical Coils

The determination of the inductance of coils involving geometries more complicated than those considered in Examples 6-26 through 6-30 is a straightforward matter in principle and requires no new concepts but often involves difficult mathematical analyses. Much early work on electromagnetics involved the design of coils with optimum geometries for various applications.[55] In this section, we briefly discuss two types of practical coils: the finite-length solenoid and a single loop of wire.

Inductance of Solenoids of Finite Length The expression for the self-inductance of a solenoid as derived in Example 6-26 is valid only for an infinitely long solenoid and is a good approximation for solenoids with large length-to-radius ratio (i.e., $l \gg a$). For solenoids of finite length, it is clear that the inductance should be different, since the **B** field is obviously not constant throughout the solenoid as was assumed in Example 6-26. In fact, based on the discussion in Section 6.2.2, the axial **B** field at the ends of the solenoid is equal to one-half of that at the center, because of the leakage of the **B** field outside the solenoid. As a result of leakage, the **B** field also varies over the cross section of the solenoid, whereas in the infinitely long solenoid the **B** field is constant over the cross section as was shown in Example 6-14. The exact distribution of the **B** field for a finite-length solenoid is difficult to evaluate, because of the nature of the integrals involved; however, the shape of the **B** field lines is expected to be as shown in Figure 6.40a.

[55] J. C. Maxwell himself spent considerable effort on the choice of the geometrical arrangement that would produce the maximum inductance for a fixed length of wire. See Article 706 in Vol. 2 of J. C. Maxwell, *A Treatise on Electricity and Magnetism,* Dover, 1954 (reprinted from 1873 original). The answer, by the way, turns out to be the Brooks inductor; see P. N. Murgatroyd, The Brooks inductor: A study of optimal solenoid cross-sections, *IEEE Proc.,* B 133, pp. 309–314, 1986; P. N. Murgatroyd and A. D. Hinley, The well tempered coil winder, *IEEE Trans. Educ.*, 37(4), pp. 329–331, 1994. An excellent summary of inductance formulas complete enough to satisfy the ordinary needs of engineers and physicists and accurate to better than 0.5% is given in Section 2 of F. E. Terman, *Radio Engineers' Handbook,* McGraw-Hill, New York, 1943. More extensive collections of formulas are provided in E. B. Rosa and F. W. Grover, Formulas and tables for the calculation of mutual and self inductance, *Bur. Standards Bull.,* 8, pp. 1–237, January 1, 1912, and in F. W. Grover, *Inductance Calculations, Working Formulas and Tables,* Dover, 1946.

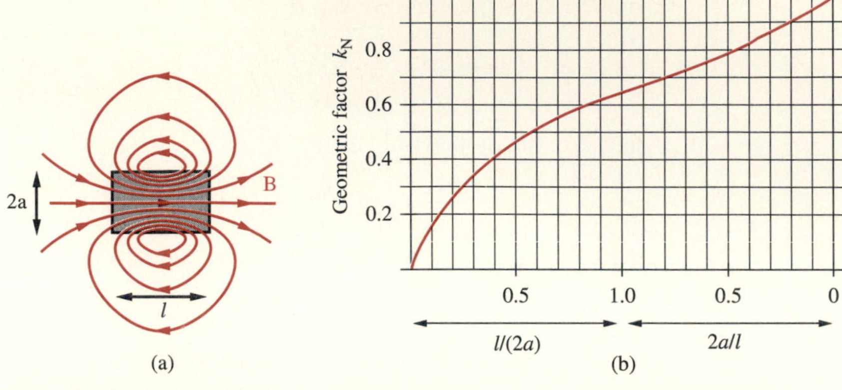

FIGURE 6.40. Inductance of a finite solenoid. (a) Lines of **B** field for a solenoid of finite length. Adapted from Figure 565 in E. Durand, *Électrostatique et Magnétostatique,* Masson et Cie, Paris, 1953. (b) The Nagaoka correction factor for the inductance of a solenoid of intermediate length.

For solenoids of intermediate length, empirical or semiempirical formulas for inductance are available. The most famous of these is the Nagaoka formula,[56] which applies a correction factor to the formula for an infinitely long solenoid. According to this, the self-inductance of a short solenoid is less than that of a long one by a factor of k_N,

$$L = k_N \frac{\mu_0 N^2 A}{l}$$

where k_N is a dimensionless number less than unity. Numerical values of k_N are plotted in Figure 6.40 for values of $l/(2a)$ from zero to unity and for values of $2a/l$ from unity to zero (where a is the radius of the cross section of the solenoid, so $A = \pi a^2$), thus covering the complete range from no solenoid to an infinitely long solenoid. A simple approximate formula valid to within less than a few percent for solenoidal coils with $l > 0.8a$ is[57]

$$L \simeq \frac{10\mu_0 \pi a^2 N^2}{9a + 10l}$$

[56]Prof. H. Nagaoka, Professor of Physics at the Imperial University, Tokyo, wrote a series of papers on self and mutual inductance of concentric coaxial solenoids and other coils between 1903 and 1912, starting with H. Nagaoka, On the potential and lines of force of a circular current, *Phil. Mag.,* 6, p. 19, 1903. Other papers included: *J. Coll. Sci., Tokyo,* 27, art. 6, 1909, and *Tokyo Math. Phys. Soc.,* 6, p. 10, 1911. Also see E. B. Rosa and F. W. Grover, Formulas and tables for the calculation of mutual and self inductance, *Bur. Standards Bull.,* 8, pp. 1–237, January 1, 1912. Nagaoka was also interested in atomic structure and proposed in 1904 that electrons encircled positively charged atoms much like the planets encircling the sun, rather than the atom being a sphere of positively charged matter with electrons placed on its surface, as was suggested by J. J. Thomson, who discovered the electron. Within two years, E. Rutherford showed that there was indeed a central positively charged nucleus.

[57]H. A. Wheeler, Simple inductance formulas for radio coils, *Proc. I. R. E.,* 16, pp. 1398–1400, 1928.

Inductance of a Single Loop of Wire The extreme case of a short solenoid is a single-conductor loop of arbitrary shape. In general, in evaluating the self-inductance of such a loop, we have to account for the internal and external inductances of the wire separately. For most circuits the total magnetic flux generated by a current can be partitioned into the portion lying outside the conductor plus the flux that lies wholly inside the conductor. The storage of magnetic energy and flux linkages associated with the internal flux lead to an *internal inductance*, while those associated with the fluxes outside the conductor are represented by the *external inductance*. Up to now, we have implicitly assumed circuits consisting of filamentary currents, thus neglecting internal inductance. For most high-frequency applications, internal inductance is negligible, since the magnetic fields (and thus magnetic fluxes) are confined to a very thin region on the surface of the current-carrying conductors. Determination of internal inductance using the linking of magnetic flux requires the introduction of the concept of partial flux linkages and unnecessarily complicated analysis.[58] It is generally much easier to evaluate the internal inductance by considering the magnetic energy stored inside the conductor, as will be done in Example 6-32, after we briefly introduce the concept of magnetic energy. We show in Example 6-32 that the dc internal inductance per unit length of an infinitely long wire (or the inner conductor of a coaxial line) is $\mu/(8\pi)$, where μ is the magnetic permeability (defined in Section 6.8) of the material the wire is made of. It is shown in Section 6.8 that $\mu \simeq \mu_0$ for most materials, including highly conducting metals such as copper and aluminum, and that μ is substantially different from μ_0 only for magnetic materials such as iron. Thus, the internal self-inductance per unit length of a wire is in most cases simply equal to $\mu_0/(8\pi)$.

For a thin wire of total length l bent into an arbitrary loop as shown in Figure 6.41, the magnetic field near the surface is very nearly the same as that for an infinitely long wire, provided the radius of curvature of the wire is much greater than the radius of the wire at all points. In other words, we may treat the wire locally as though it were part of an infinitely long wire. Thus, the internal self-inductance of

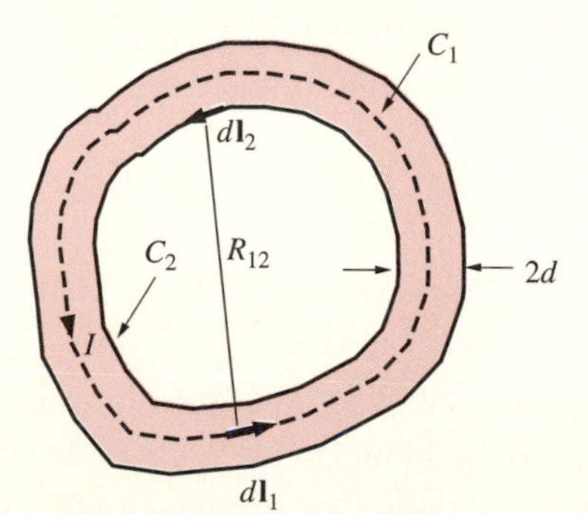

FIGURE 6.41. **A conductor of finite radius d bent into a closed loop.**

[58]For examples of partial flux linkage analyses, see Section 5-11 of C. T. A. Johnk, *Engineering Electromagnetic Fields and Waves*, John Wiley & Sons, New York, 1988, and Section 8.5 of R. Plonsey and R. E. Collin, *Principles and Applications of Electromagnetic Fields*, McGraw-Hill, New York, 1961.

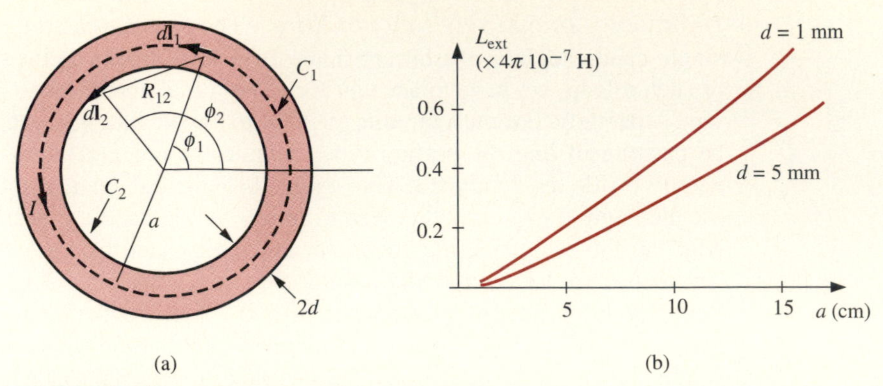

FIGURE 6.42. **A circular conducting loop.** (a) The circular loop showing various quantities referred to in the text. (b) Plots of the external self-inductance L_{ext} versus loop radius a for two different values of d, namely $d = 1$ mm and 5 mm.

any loop of mean length l is[59] $\mu_0 l/(8\pi)$. Thus, the internal self-inductance of a thin wire ($d \ll a$) circular loop (Figure 6.42) of radius a is $\mu_0(2\pi a)/(8\pi) = \mu_0 a/4$.

To evaluate the external self-inductance of an arbitrarily shaped loop, we can apply the Neumann formula [6.17] for the mutual inductance between two loops. With respect to Figure 6.41, consider the contour C_2 to coincide with the interior edge of the conducting loop. The external self-inductance is due to the flux produced by the current in the conductor that links the contour C_2. To evaluate the magnetic flux linking the interior contour C_2, we may assume with negligible error that the total current I flowing in the conductor is concentrated in an infinitely thin filamentary contour C_1 along the center of the conductor. The external self-inductance can now be evaluated by finding the mutual inductance between contours C_1 and C_2. We have

$$L_{ext} = L_{12} = \frac{\mu_0}{4\pi} \oint_{C_1} \oint_{C_2} \frac{d\mathbf{l}_1 \cdot d\mathbf{l}_2}{R_{12}} \qquad [6.18]$$

where R_{12} is the distance between the differential elements $d\mathbf{l}_1$ and $d\mathbf{l}_2$. In Example 6-31, we apply this method to determine the external self-inductance of a circular loop of conductor.

Example 6-31: Self-inductance of a circular loop. Consider a conductor of radius d bent into a circular shape of mean radius a, as shown in Figure 6.42a. Find the external self-inductance of this circular loop.

Solution: To determine the external inductance, we use the expression [6.18] for L_{ext}. From Figure 6.42a, the magnitudes of $d\mathbf{l}_1$ and $d\mathbf{l}_2$ are given by

[59]Note, however, that at frequencies in the radio range or higher the internal self-inductance is much smaller than the external one, since the magnetic flux is confined only to a small region on the surface, resulting in much smaller magnetic energy storage within the conductor.

$$|d\mathbf{l}_1| = a\,d\phi_1 \qquad |d\mathbf{l}_2| = (a-d)\,d\phi_2 \simeq a\,d\phi_2$$

where we have assumed $a \gg d$ (i.e., a thin wire loop). The angle between $d\mathbf{l}_1$ and $d\mathbf{l}_2$ is $(\phi_2 - \phi_1)$, and we have

$$d\mathbf{l}_1 \cdot d\mathbf{l}_2 = a^2 \cos(\phi_2 - \phi_1)\,d\phi_1\,d\phi_2$$

Using the law of cosines, the distance R_{12} between the two differential elements is given by

$$R_{12}^2 = a^2 + (a-d)^2 - 2a(a-d)\cos(\phi_2 - \phi_1)$$

Thus, the external self-inductance is given by

$$L_{\text{ext}} = \frac{\mu_0 a^2}{4\pi} \int_0^{2\pi}\int_0^{2\pi} \frac{\cos(\phi_2 - \phi_1)\,d\phi_1\,d\phi_2}{\sqrt{a^2 + (a-d)^2 - 2a(a-d)\cos(\phi_2 - \phi_1)}}$$

We can now integrate over ϕ_2 first, after changing variables to replace $(\phi_2 - \phi_1)$ by ϕ and $d\phi_2$ by $d\phi$; thus,

$$L_{\text{ext}} = \frac{\mu_0 a^2}{4\pi} \int_0^{2\pi}\int_0^{2\pi} \frac{\cos\phi\,d\phi\,d\phi_1}{\sqrt{d^2 + 2a(a-d)(1-\cos\phi)}}$$

where we have also rewritten the denominator in a more compact form. Note that the limits of integration for ϕ are kept as 0 to 2π, since the origin for ϕ_1 is arbitrary because of the circular symmetry. The integration over ϕ_1 simply brings in a multiplication with 2π, so we have

$$L_{\text{ext}} = \frac{\mu_0 a^2}{2} \int_0^{2\pi} \frac{\cos\phi\,d\phi}{\sqrt{d^2 + 2a(a-d)(1-\cos\phi)}}$$

The above integral does not have a compact analytical form, but it can be expressed in terms of elliptic integrals. However, it is even simpler to evaluate L_{ext} numerically for any set of parameters, since the integrand is a well-behaved function. Plots of L_{ext} for selected parameters are shown in Figure 6.42b. For d in the range of a few mm, and radius a being a few to tens of cm, the external self-inductance is of order μH or less. For the same range of parameters, the internal self-inductance is typically much smaller.

For a thin wire circular loop ($d \ll a$), the result of the above integral reduces to

$$L_{\text{ext}} \simeq \mu_0 a \left(\ln \frac{8a}{d} - 2 \right)$$

On the scale of Figure 6.42b, the values of L_{ext} given by this approximate expression are indistinguishable from those evaluated directly from the integral.

As a matter of practical interest, it is useful to consider the inductance of single loops of shapes other than a circle as in Example 6-31. Fundamentally, we expect the inductance to depend more on the area of the loop than on the particular shape; however, for a given length of conductor, the circular shape represents the best utilization for the purpose of producing and linking maximum magnetic flux. A simple formula[60] for a single-conductor thin-wire loop of fixed length l is

$$L \simeq 129l\left(2.303\log_{10}\frac{4l}{d} - \theta_s\right)$$

where θ_s is a shape factor. The values of θ_s for different shapes are 2.451, 2.636, 2.853, and 3.197, respectively for a circle, a regular hexagon, a square, and an equilateral triangle.

6.7.4 Energy in a Magnetic Field

Just as configurations of charges store electrostatic energy, configurations of stationary currents store magnetic energy. In electrostatics we were able to find expressions for energy stored in electric fields by considering the work necessary to bring charges from infinity to their locations. Such an approach is not appropriate in magnetostatics, since the creation of steady-state current configurations and associated magnetic fields involves an initial transient period during which currents are increased from zero to their final values. According to Faraday's law (Chapter 7), such time-varying fields generate electromotive forces against which work must be done. The work to establish the current distribution is thus done during this initial transient period, so we must consider time-varying fields in order to evaluate the total energy stored in a configuration of stationary currents.

As we are not yet equipped with Faraday's law and the concept of induced electric fields, we delay a complete consideration of energy in magnetic fields to Chapter 7. However, having just studied the concept of inductance, we provide below a preliminary discussion of stored magnetic energy, with expressions drawn from Chapter 7.

Analogous to the energy density $w_e = \frac{1}{2}\epsilon_0 E^2$ (J-m^{-3}) stored in electric fields, the magnetic energy density stored in a magnetic field configuration in free space is given by

$$w_m = \frac{1}{2\mu_0}B^2$$

in units of J-m^{-3}. The total magnetic energy W_m stored in a given steady-current configuration can be found by integrating the associated **B** field *over the entire volume V of its existence*, namely

$$W_m = \frac{1}{2}\int_V \frac{B^2}{\mu_0}dv' \qquad [6.19]$$

---

[60] See Section 2 of F. E. Terman, *Radio Engineers' Handbook*, McGraw-Hill, Inc., New York, 1943.

Given any configuration of current-carrying conductors, we know from circuit theory that we can alternatively express the stored magnetic energy using the inductance[61] L of the configuration. Namely,

$$W_{\mathrm{m}} = \tfrac{1}{2}LI^2$$

where I is the current in the circuit. This relation between W_{m} and inductance is formally derived in Section 7.3. We have thus arrived at an alternative method of calculating the inductance of a circuit; namely, we can determine W_{m} from [6.19] and then find L from

$$\boxed{L = \frac{2W_{\mathrm{m}}}{I^2}} \qquad [6.20]$$

We illustrate this procedure in Example 6-32.

Example 6-32: Inductance of a coaxial line. Consider a coaxial line with a solid inner conductor of radius a and an outer conductor of thickness $(c - b)$ as was shown in Figure 6.16, and discussed in Example 6-12. Find the inductance per unit length of this coaxial line using [6.20].

Solution: The magnetic field distribution for this case was found in Example 6-12 to be

$$B_\phi = \begin{cases} \dfrac{\mu_0 I r}{2\pi a^2} & r \leq a \\[2ex] \dfrac{\mu_0 I}{2\pi r} & a < r \leq b \\[2ex] \dfrac{\mu_0 I}{2\pi r}\dfrac{c^2 - r^2}{c^2 - b^2} & b < r \leq c \\[2ex] 0 & r > c \end{cases}$$

The total stored magnetic energy can be found by separately evaluating the energy stored in all three regions in which the field is nonzero. Considering a unit-length coaxial line, for $r \leq a$ we have

$$W_{\mathrm{m1}} = \frac{1}{2\mu_0}\int_0^a B_{1\phi}^2 2\pi r\, dr = \frac{\mu_0 I^2}{4\pi a^4}\int_0^a r^3\, dr = \frac{\mu_0 I^2}{16\pi}$$

---

[61]Note that the inductance we refer to in this context is the self-inductance L_{11} of a circuit C_1, in which a current I_1 is assumed to flow. In this context, we drop the subscript 1 and simply use L and I.

in units of J-m^{-1}. Similarly, for $a < r \le b$ we have

$$W_{\text{m2}} = \frac{1}{2\mu_0}\int_a^b B_{2\phi}^2 2\pi r\,dr = \frac{\mu_0 I^2}{4\pi}\int_a^b \frac{1}{r}dr = \frac{\mu_0 I^2}{4\pi}\ln\frac{b}{a}$$

and for $b < r \le c$

$$W_{\text{m3}} = \frac{1}{2\mu_0}\int_b^c B_{3\phi}^2 2\pi r\,dr = \frac{\mu_0 I^2}{4\pi(c^2-b^2)^2}\int_b^c \frac{(c^2-r^2)^2}{r}dr$$

$$= \frac{\mu_0 I^2}{4\pi(c^2-b^2)^2}\int_b^c\left(\frac{c^4}{r} - 2c^2 r + r^3\right)dr = \frac{\mu_0 I^2}{4\pi(c^2-b^2)^2}\left[c^4\ln r - c^2 r^2 + \frac{r^4}{4}\right]_b^c$$

$$= \frac{\mu_0 I^2}{4\pi}\left[\left(\frac{c^2}{c^2-b^2}\right)^2\ln\frac{c}{b} - \frac{c^2}{c^2-b^2} + \frac{1}{4}\left(\frac{c^2+b^2}{c^2-b^2}\right)\right]$$

The self-inductance of the coaxial line per unit length is then found to be

$$L = \frac{2(W_{\text{m1}} + W_{\text{m2}} + W_{\text{m3}})}{I^2}$$

$$= \underbrace{\frac{\mu_0}{2\pi}\ln\frac{b}{a}}_{\text{External inductance}} + \underbrace{\frac{\mu_0}{8\pi} + \frac{\mu_0}{2\pi}\left[\left(\frac{c^2}{c^2-b^2}\right)^2\ln\frac{c}{b} - \frac{c^2}{c^2-b^2} + \frac{1}{4}\left(\frac{c^2+b^2}{c^2-b^2}\right)\right]}_{\text{Internal inductance}}$$

The first term (due to W_{m2}) is identical to the self-inductance of the thin-walled coaxial line calculated in Example 6-28. The last two terms (due to W_{m1} and W_{m3}), represent the inductance due to the magnetic energy stored *inside* the inner and outer conductors, respectively. These inductances can be important only at near-dc frequencies, and if the wires are not completely nonmagnetic[62] (i.e., $\mu > \mu_0$). At any appreciable frequency, it will be shown in Chapter 8 that the current is effectively confined to a thin layer at $r = a$ on the inner and at $r = b$ on the outer conductor, resulting in negligible internal stored magnetic energy and, thus, negligible internal self-inductance. Therefore, the per unit length self-inductance of the coaxial line is for all practical purposes given by

$$L \simeq L_{\text{ext}} = \frac{\mu_0}{2\pi}\ln\left(\frac{b}{a}\right)$$

as was found using considerations of flux linkage in Example 6-28.

---

[62]For a quantitative discussion of cases in which internal inductance is important, see Chapter 7 of H. H. Skilling, *Electric Transmission Lines,* McGraw-Hill, 1951.

6.8 MAGNETIC FIELDS IN MATERIAL MEDIA

Our discussion of magnetostatics has so far been concerned with the physical effects produced by steady-state currents in vacuum or free space. In this section, we introduce and discuss the additional effects associated with the presence of material media. Our goals in this respect, as well as the level of discussion provided, are very similar to those in Section 4.10 on dielectric materials. We must necessarily start with a description of the material behavior at a microscopic level, but then perform suitable space and time averages to represent the microscopic effects in terms of macroscopic field quantities.

Materials are typically classified according to their most important characteristics. Glass is a *dielectric* because its dielectric constant is appreciably different from that of free space (i.e., $\epsilon \simeq 4$ to 9). The conductivity of glass is extremely small ($\sim 10^{-12}$ S-m^{-1} at 20°C), and its behavior in the presence of magnetic fields is not measurably different than that of free space. Copper, on the other hand, is a *conductor* because its conductivity is very large (5.7×10^8 S-m^{-1}), while its electric permittivity and magnetic properties are essentially the same as those of free space. The class of materials known as *magnetic* are those that exhibit magnetic properties. The most common example of such materials is iron, which is also a conductor of electricity ($\sigma_{\text{iron}} \simeq 10^7$ S-m^{-1}) but is better known for its striking magnetic behavior. Some other materials, such as nickel, cobalt, and a few special alloys, also exhibit similarly strong and clearly evident magnetic properties. Materials with such striking magnetic behavior are known as *ferromagnetic* materials; they exhibit permanent magnetic dipole moments and retain their magnetization after the external field is removed or when no macroscopic current flows in them or in their vicinity. The physics of ferromagnetic materials is inherently complex and is beyond the scope of this book. However, all substances exhibit some magnetic effects, although at levels substantially lower (typically 10^3 to 10^6 times less) than in ferromagnetic materials. In the following we confine our attention to a description of ordinary magnetism, which provides a basis for our formulation of magnetostatics in the presence of ordinary (i.e., nonferromagnetic) material media.

In Section 4.10 we stated that in isotropic materials electric polarization $\mathbf{P}$ was in the same direction as the electric field $\mathbf{E}$, and in Section 4.13 we showed that dielectrics are always attracted toward regions of higher electric field. Unlike this electrical effect in matter, some materials are attracted toward regions of higher magnetic fields, whereas others are repelled from such regions. Although the magnetic effect in ordinary materials is quite small, these two different tendencies of some materials (i.e., attraction or repulsion) can be easily demonstrated[63] using a strong electromagnet with a sharply pointed pole to create a gradient in the magnetic field. Substances such as bismuth are repelled away from the high-field region; they are said to acquire magnetization in a direction *opposite* (or antiparallel) to

[63]For further discussion, see R. P. Feynman, R. B. Leighton, and M. Sands, *The Feynman Lectures on Physics,* Addison Wesley, Reading, Massachusetts, 1964.

the external field and are called *diamagnetic* materials. Other materials, such as aluminum, are attracted toward the high-field region; they are said to be magnetized in the same direction (i.e., parallel) as the external field and are thus referred to as *paramagnetic* materials.

The magnetic behavior of materials on a macroscopic scale is inherently due to the fact that tiny currents exist in all materials on an atomic (microscopic) scale. Any medium is composed of atoms, and these atoms may be considered as consisting of electrons moving in orbits about fixed nuclei. Both these orbital motions and the inherent spins of the electrons about their axes constitute microscopic electric currents. On a macroscopic scale, these current loops are infinitesimal in size, so we can treat them as small magnetic dipoles. In ordinary materials the microscopic magnetic dipole moments are randomly oriented and cancel each other out when averaged over a finite volume. However, an external magnetic field leads to a net alignment of these magnetic dipoles, causing the medium to be magnetically polarized.

6.8.1 Microscopic Basis of Diamagnetism and Paramagnetism

Both diamagnetic and paramagnetic behavior of materials are inherently quantum mechanical phenomena that cannot be properly described in classical terms.[64] However, especially for diamagnetic materials, we can provide some semiquantitative classical arguments to illustrate the basic processes that lead to the formation of induced magnetic dipole moments in a direction opposite to the applied field. Paramagnetism is due almost entirely to the spin magnetic dipole of the electrons, so it is more difficult to discuss its bases in quantitative classical terms.

Diamagnetism can be qualitatively understood by considering the effects of an external magnetic field on electrons orbiting about their nuclei. For simplicity, we consider the case for which the applied magnetic field is aligned with the axis of revolution of an electron as shown in Figure 6.43. In the classical model, the electron orbits around the positively charged nuclei in such a way that the outward centrifugal force is balanced by the Coulomb attraction force. If the radius of the orbit is a and the angular velocity is ω_0 in the absence of the external field, we must have

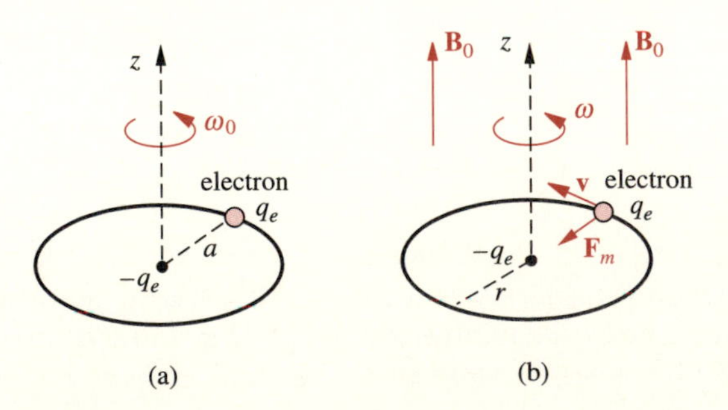

FIGURE 6.43. Perturbation of electron orbit by applied B-field. The nuclei are positively charged by an amount $-q_e \simeq 1.6 \times 10^{-19}$ C. (a) No external **B** field. (b) The presence of the external **B** field produces a magnetic force $\mathbf{F}_m$.

[64]See Section 34-5 of *The Feynman Lectures on Physics,* Addison Wesley, Reading, Massachusetts, 1964.

$$m_e\omega_0^2 a = \frac{q_e^2}{4\pi\epsilon_0 a^2} \quad \rightarrow \quad \omega_0^2 = \frac{q_e^2}{4\pi\epsilon_0 m_e a^3}$$

The magnetic dipole moment of this unperturbed orbit is the current times the area. The current is the charge per unit time that passes any point on the orbit, or simply the charge q_e divided by the time taken for one revolution, or $2\pi/\omega_0$. The unperturbed magnetic dipole moment of the electron orbit is thus

$$\mathbf{m}_\mathrm{u} = \hat{\mathbf{z}}I(\pi a^2) = \hat{\mathbf{z}}q_e\frac{\omega_0}{2\pi}\pi a^2 = \hat{\mathbf{z}}\frac{q_e\omega_0 a^2}{2}$$

Since the charge of the electron $q_e = -1.6 \times 10^{19}$ C, the current flow is in the direction opposite to that of the electron, so the dipole magnetic moment is oriented in the $-\hat{\mathbf{z}}$ direction.

In the presence of the external field $\mathbf{B} = \hat{\mathbf{z}}B_0$, the electron motion is influenced by an additional force given by $\mathbf{F}_\mathrm{m} = q_e\mathbf{v} \times \mathbf{B}$, where $\mathbf{v}$ is the linear velocity of the electron, whose magnitude is given by $|\mathbf{v}| = \omega r$, where ω is the new angular velocity and r is the new orbital radius, both of which may be different from their values in the absence of the external field. The force balance equation that is now required for the electron to orbit is

$$m_e\omega^2 r = \frac{q_e^2}{4\pi\epsilon_0 r^2} - q_e r\omega B_0$$

Assuming[65] that the orbital radius remains constant, so that we have $r \simeq a$, we have

$$\omega^2 = \omega_0^2 - \frac{q_e B_0}{m_e}\omega$$

Since we expect the applied field $\mathbf{B} = \hat{\mathbf{z}}B_0$ to produce only a small perturbation (i.e., $(\omega - \omega_0) \ll \omega_0$), we can write

$$\omega^2 - \omega_0^2 = (\omega - \omega_0)(\omega + \omega_0) \simeq 2\omega_0(\omega - \omega_0)$$

so that the net effect of the applied field is to accelerate the electron to a new angular speed given by

$$\omega \simeq \omega_0 - \frac{q_e B_0}{2m_e}$$

thus increasing the magnetic dipole moment of the electron orbit by an amount

[65]It is difficult to justify this assumption in the context of the classical model discussed here. However, an alternative way to study the effect of the external magnetic field on the electron orbit is to assume the field **B** to be slowly turned on. The associated time rate of change of the magnetic field would then induce an electric field via Faraday's law (see Chapter 7), which would accelerate the electron. It can be shown that the associated increase in the kinetic energy of the electron is exactly equal to that required to sustain circular motion at the same radius. For further discussion, see Chapter 9 of W. B. Cheston, *Elementary Theory of Electric and Magnetic Fields,* John Wiley & Sons, New York, 1964.

$$\Delta\mathbf{m} = -\hat{\mathbf{z}}\frac{q_e^2 a^2 B_0}{4m_e}$$

We note that this induced differential magnetic moment $\Delta\mathbf{m}$ is in the direction opposite, or *antiparallel,* to $\mathbf{B}$, hence the term *dia*magnetism. Note that if the applied field were in the $-\hat{\mathbf{z}}$ direction, the orbiting electron would then be slowed down by the $q_e\mathbf{v} \times \mathbf{B}$ force, so that the induced magnetic dipole moment would now be in the $+\hat{\mathbf{z}}$ direction—still opposite to the applied field.

In the absence of an applied field, the electron orbits are randomly oriented with respect to one another, and the orbital dipole moments cancel out when averaged over any finite volume. With the applied field, however, each atom acquires[66] a small additional dipole moment, and all these incremental moments are antiparallel (i.e., in the opposite direction) to the applied field. If N is the effective number of electron orbits per unit volume, the induced magnetic dipole moment per unit volume, also called *magnetization* and denoted by $\mathbf{M}$, is given by

$$\mathbf{M} = -\frac{Nq_e^2 a^2 \mathbf{B}}{4m_e} \quad \text{A-m}^{-1}$$

This quantity is entirely analogous to the polarization vector $\mathbf{P}$ for dielectric materials, as defined in [4.61]. In practice, it is often useful to lump together the various factors contributing to the magnetization in the following expression

$$\boxed{\mathbf{M} = \frac{\chi_{\mathrm{m}}\mathbf{B}}{\mu_0}} \qquad [6.21]$$

where

$$\chi_{\mathrm{m}} = \frac{-Nq_e^2 a^2 \mu_0}{4m_e}$$

is a dimensionless quantity called the *magnetic susceptibility* and $\mathbf{B}$ is the total applied macroscopic field. Substitution in this expression of typical values for N of 10^{28} to 10^{29} atoms-m^{-3} in a solid and a of order one angstrom shows that $|\chi_{\mathrm{m}}|$ is of order $\sim 10^{-6}$ to -10^{-5}, indicating that the diamagnetic effect is very small and negligible in most practical situations.

Example 6-33: Magnetic susceptibility of sodium. Find the magnetic susceptibility of sodium (Na) assuming that the atomic density is $N = 10^{29}$ m^{-3} and the electron radius is $a = 1$ angstrom.

---

[66]We should note that the geometry considered in Figure 6.43 is rather idealized, with **B** being along the axis of revolution. The interaction with the applied field of other electron orbits is more complicated. Nevertheless, the net effect is that while the electrons in an atom move nearly in the same orbits in the presence of an applied field as they do in zero field, superimposed upon their rapid orbital motion of the electron is a slow uniform precessional motion of the entire atom about the field direction.

Solution: Using the expression derived above for χ_m, we have

$$\chi_m = \frac{-(10^{29}\ \text{m}^{-3})(-1.6\times 10^{-19}\ \text{C})^2(10^{-10}\ \text{m})^2(4\pi\times 10^{-7}\ \text{H-m}^{-1})}{4\times 9.11\times 10^{-31}\ \text{kg}}$$

$$\simeq -8.83\times 10^{-6}$$

Based on the foregoing discussion, all materials should exhibit diamagnetic behavior. However, many materials exist whose measured susceptibilities are positive, a good example being aluminum. Materials with positive χ_m values are called *paramagnetic* materials. Paramagnetism is due almost entirely to the spin magnetic dipole moment of the electron. It is not really possible to understand paramagnetism in classical terms; however, we can note that the primary cause of the effect is that the atoms of some materials have a permanent magnetic dipole moment associated with them. In the absence of an external field, the atomic and molecular dipoles are randomly oriented, and the net resultant magnetic dipole moment, averaged over a finite volume, vanishes. However, in the presence of an applied magnetic field the individual electron spin dipoles are aligned with the field. The macroscopic moment per unit volume of magnetization **M** produced by this orientation is parallel to **B** for linear materials, thus corresponding to positive susceptibility χ_m. Typical values of χ_m for paramagnetic materials is of order 10^{-5} to 10^{-4} at room temperature. For example, oxygen gas is paramagnetic, with $\chi_m = 2.1\times 10^{-6}$ at atmospheric pressure and room temperature. Many metals are also paramagnetic, such as aluminum, with $\chi_m = 2.1\times 10^{-5}$ at room temperature. The magnetic susceptibility χ_m is typically inversely proportional to absolute temperature.[67] In materials that exhibit paramagnetic behavior, diamagnetic effects are also in general present. However, the diamagnetic reaction is completely masked by the paramagnetic effect, for reasons that cannot be described without constructing a microscopic theory of the material—a task well beyond the scope of this book.

6.8.2 Magnetic Field Intensity and the Permeability Concept

We saw in the previous section that the net effect of an applied magnetic field on a magnetic material is to create a macroscopic magnetization field **M**. We now proceed to consider the magnetic fields generated by this induced macroscopic magnetic dipole moment. In other words, we now set out to replace the material medium possessing an **M**-field in its interior by an equivalent distribution of currents residing in vacuum. The procedure we follow for this purpose is quite analogous to that used in Section 4.10.2 for the consideration of fields generated by induced electric dipole moments in dielectric materials.

[67]See Chapter 15 of C. Kittel, *Introduction to Solid State Physics,* Wiley, New York, 1971.

We start by recalling a result derived in Section 6.6, namely that the vector potential $\mathbf{A}$ at a point $\mathbf{r}$ due to an infinitesimal magnetic dipole $\mathbf{m}$ located at point $\mathbf{r}'$ is given by

$$\mathbf{A}(\mathbf{r}) = \frac{\mu_0}{4\pi}\frac{\mathbf{m}\times\hat{\mathbf{R}}}{R^2}$$

where $\hat{\mathbf{R}}$ is the unit vector pointing from the source point $\mathbf{r}'$ to the observation point $\mathbf{r}$ and is given by $\hat{\mathbf{R}} = \mathbf{R}/R$ where $\mathbf{R} = (\mathbf{r}-\mathbf{r}')$ and $R = |\mathbf{r}-\mathbf{r}'|$. If the material medium under consideration is subdivided into differential volumes dv', each possessing a dipole moment $\mathbf{M}(\mathbf{r}')dv'$, the vector potential due to all elemental magnetic dipoles in a given volume V' of the material can be written as

$$\mathbf{A}(\mathbf{r}) = \frac{\mu_0}{4\pi}\int_{V'}\frac{\mathbf{M}(\mathbf{r}')\times\hat{\mathbf{R}}}{R^2}dv' = \frac{\mu_0}{4\pi}\int_{V'}\left(\mathbf{M}(\mathbf{r}')\times\nabla'\frac{1}{R}\right)dv'$$

where we have used the identity $\nabla' R^{-1} = -\nabla R^{-1} = (\hat{\mathbf{R}}/R^2)$, as established in footnote 41 in Section 6.5, and where ∇' represents the del operator applied only to the source coordinates ($\mathbf{r}'$). We now use a vector identity, namely,

$$\nabla'\times\left[\frac{\mathbf{M}}{R}\right] \equiv \frac{1}{R}\nabla'\times\mathbf{M} - \mathbf{M}\times\nabla'\frac{1}{R}$$

Substituting this identity in the expression for $\mathbf{A}(\mathbf{r})$, we find

$$\mathbf{A}(\mathbf{r}) = \frac{\mu_0}{4\pi}\int_{V'}\frac{\nabla'\times\mathbf{M}(\mathbf{r}')}{R}dv' - \frac{\mu_0}{4\pi}\int_{V'}\nabla'\times\left[\frac{\mathbf{M}(\mathbf{r}')}{R}\right]dv'$$

The second integral in this expression can be transformed into an integral over a closed surface using an identity reminiscent of Stokes's theorem:

$$\int_V \nabla\times\mathbf{G}\,dv \equiv -\oint_S \mathbf{G}\times\hat{\mathbf{n}}\,ds$$

where $\hat{\mathbf{n}}$ is the unit outward normal to the surface S enclosing the volume V. We thus have

$$\mathbf{A}(\mathbf{r}) = \frac{\mu_0}{4\pi}\int_{V'}\frac{\nabla'\times\mathbf{M}(\mathbf{r}')}{R}dv' + \frac{\mu_0}{4\pi}\oint_{S'}\frac{\mathbf{M}(\mathbf{r}')\times\hat{\mathbf{n}}}{R}ds' \qquad [6.22]$$

We can now compare [6.22] with expression [6.13], which gives the vector potential from a distribution of true currents. This comparison suggests that the term $\nabla'\times\mathbf{M}$ in [6.22] can be interpreted as an equivalent *volume magnetization current* density $\mathbf{J}_{\text{m}}$ and that, similarly, the term $\mathbf{M}\times\hat{\mathbf{n}}$ can be interpreted as an equivalent *surface magnetization current* density $\mathbf{J}_{s\text{m}}$. In other words, if we define $\mathbf{J}_{\text{m}} = \nabla'\times\mathbf{M}$ (in A-m^{-2}) and $\mathbf{J}_{s\text{m}} = \mathbf{M}\times\hat{\mathbf{n}}$ (in A-m^{-1}), we can rewrite [6.22] as

$$\mathbf{A} = \frac{\mu_0}{4\pi}\left(\int_{V'}\frac{\mathbf{J}_m}{R}dv' + \oint_{S'}\frac{\mathbf{J}_{sm}}{R}ds'\right)$$

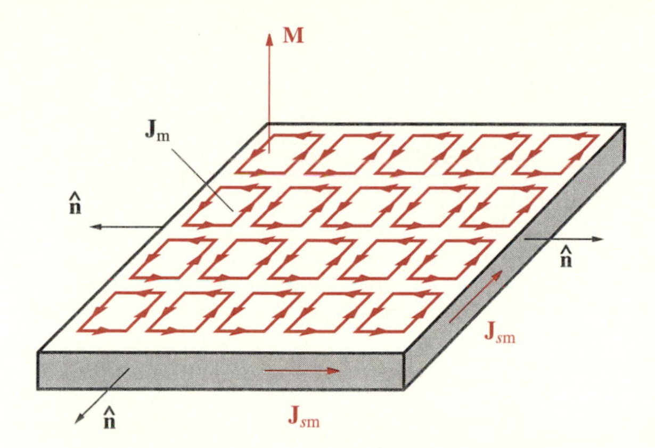

FIGURE 6.44. Equivalent currents in a magnetized material. Even if the magnetization was uniform throughout the material so that the volume density of the polarization current ($\mathbf{J}_m = \nabla' \times \mathbf{M}$) is zero, elementary current loops result in a surface polarization current $\mathbf{J}_{sm} = \mathbf{M} \times \hat{\mathbf{n}}$, which exists at the surface of the material.

The quantities $\mathbf{J}_m$ and $\mathbf{J}_{sm}$ are more than mathematical artifacts; they represent net effective currents, albeit inaccessible. A volume current $\mathbf{J}_m$ arises when adjacent molecular current loops do not completely cancel one another because the magnetization is inhomogeneous (i.e., $\mathbf{M}$ is not constant throughout the material). That a surface current $\mathbf{J}_{sm}$ exists at the outer boundary of a magnetic material can be seen by thinking of a uniformly magnetized medium in terms of tiny adjacent molecular currents, as shown in Figure 6.44. Although all internal currents are canceled by a contiguous current in the opposite direction, a net current is left uncanceled at the exterior edge of the material. This surface polarization current vanishes only when the magnetization $\mathbf{M}$ is perpendicular to the outer surface such that $\mathbf{M} \times \hat{\mathbf{n}}$ is zero. The quantities $\mathbf{J}_m$ and $\mathbf{J}_{sm}$ are analogous to the equivalent volume (ρ_p) and surface (ρ_{sp}) polarization charge densities for a dielectric as discussed in Section 4.10.

Example 6-34 applies the equivalent-current concepts discussed here to a cylindrical bar magnet.

Example 6-34: Field of a cylindrical bar magnet. Determine the $\mathbf{B}$ field produced by a cylindrical bar magnet at a point P(x, y, z) as shown in Figure 6.45. The length of the magnetic material is l and its radius is a. The magnet is assumed to be permanently magnetized with a uniform magnetization per unit volume of $\mathbf{M} = \hat{\mathbf{z}}M_0$.

Solution: Since the magnetization is in the z direction, the magnetization currents must flow in the $\hat{\boldsymbol{\phi}}$ direction. Since $\mathbf{M}$ is uniform, the volume density of the magnetization current $\mathbf{J}_m$ is zero. However, an uncanceled surface magnetization current flows circumferentially along the outer boundary of the cylindrical bar as shown. This current is given by $\mathbf{J}_{sm} = \mathbf{M} \times \hat{\mathbf{n}} = M_0\hat{\mathbf{z}} \times \hat{\mathbf{r}} = M_0\hat{\boldsymbol{\phi}}$. By comparing the geometry of Figure 6.45 to that of Figure 6.34 in Section 6.6, it is clear that the field produced at point P by the bar magnet will be the same as that from an equivalent solenoid having an effective surface current of M_0 amperes per meter. In other words, a closely wound solenoid of N turns per meter carrying a current of I is equivalent to a bar magnet of magnetization M_0 if $NI = M_0$.

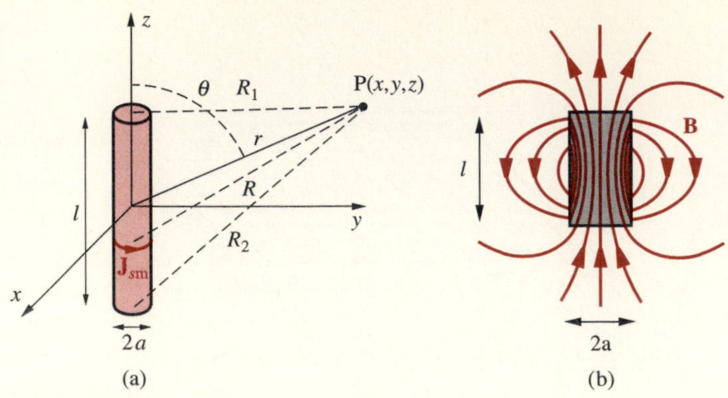

FIGURE 6.45. **B field of a cylindrical bar magnet.** (a) Coordinate system. Note the similarity with the solenoid geometry shown in Figure 6.34. (b) Approximate sketch of magnetic field lines in the yz plane. Note the similarity with Figure 6.40a.

The total magnetic dipole moment of the bar magnet can be found by integrating its magnetization density $\mathbf{M}$ over its volume; we find $|\mathbf{m}_{\text{bar}}| = \pi a^2 l M_0$ to be compared with the total moment of the solenoid of $|\mathbf{m}_{\text{slnd}}| = NI\pi a^2 l$. Substitution of $\mathbf{m}_{\text{bar}}$ instead of $\mathbf{m}_{\text{slnd}}$ in [6.15] gives the expression for the magnetic field of the bar magnet at distant ($r \gg l$) points.

While [6.15] is based on the dipole approximation (i.e., a distant observation point, or $r \gg a$ and $r \gg l$), it is possible to use the magnetization current sheet model of the permanent magnet to evaluate the **B** field everywhere, both inside and outside the magnet. Methods for such evaluations are described elsewhere.[68] However, an approximate sketch of the magnetic field lines around a cylindrical bar magnet is shown in Figure 6.45b. Note the similarity between the **B** field distribution for a bar magnet and that for a finite length solenoid, shown in Figure 6.40a.

We now consider the **B** field inside (i.e., $\mathbf{J}_{sm} = 0$) a magnetic material body in which a true current (i.e., not a magnetization current) distribution **J** exists. This current produces a macroscopic **B** field, which magnetizes the material, resulting in the magnetization current density $\mathbf{J}_{\text{m}}$. Since the total effective current density is then the vector sum of **J** and $\mathbf{J}_{\text{m}}$, we can write the differential form of Ampère's law as

$$\nabla \times \mathbf{B} = \mu_0(\mathbf{J} + \mathbf{J}_{\text{m}})$$

Noting that we have $\mathbf{J}_{\text{m}} = \nabla \times \mathbf{M}$, we can rewrite this as

$$\nabla \times \left(\frac{\mathbf{B}}{\mu_0} - \mathbf{M}\right) = \mathbf{J}$$

[68]For a comparative discussion of the **B** fields for a cylindrical rod magnet and a finite-length solenoid, see Section 9.4 of M. A. Plonus, *Applied Electromagnetics,* McGraw-Hill, 1978. For a discussion of the formal procedure used to determine the fields around a permanent magnet, see Section 6.6 of J. Van Bladel, *Electromagnetic Fields,* Revised Publishing, Hemisphere Publishing Corp., New York, 1985.

Therefore, the vector $[(\mathbf{B}/\mu_0) - \mathbf{M}]$ has as its source only the true current $\mathbf{J}$. Thus, to eliminate the necessity of dealing with the magnetization vector $\mathbf{M}$, we introduce a new vector $\mathbf{H}$, defined as

$$\mathbf{H} \equiv \frac{\mathbf{B}}{\mu_0} - \mathbf{M} \qquad [6.23]$$

The vector $\mathbf{H}$ is called the *magnetic field intensity* vector, and it is directly analogous to the electric flux density vector $\mathbf{D}$ in electrostatics in that both $\mathbf{D}$ and $\mathbf{H}$ are medium-independent and are directly related to their sources. Note that we can also write the following integral expressions relating $\mathbf{H}$ to the source current $\mathbf{J}$:

$$\mathbf{H} = \int_{V'} \frac{\mathbf{J} \times \hat{\mathbf{R}}}{4\pi R^2} dv'$$

The dimensions of $\mathbf{H}$ are amperes per meter. We can also express Ampère's law in terms of $\mathbf{H}$ as follows:

$$\oint_C \mathbf{H} \cdot d\mathbf{l} = \int_S \mathbf{J} \cdot d\mathbf{s} = I \qquad \text{or} \qquad \nabla \times \mathbf{H} = \mathbf{J}$$

where I is the total net current passing through the surface S enclosed by C. Note that the medium-independent nature of $\mathbf{H}$ is once again apparent. Taking the divergence of [6.23], we obtain

$$\nabla \cdot \mathbf{H} = \frac{1}{\mu_0} \nabla \cdot \mathbf{B} - \nabla \cdot \mathbf{M} = -\nabla \cdot \mathbf{M}$$

since $\nabla \cdot \mathbf{B} = 0$. Note that $\nabla \cdot \mathbf{H} = -\nabla \cdot \mathbf{M}$ is in general nonzero and can be thought to correspond mathematically to an equivalent magnetic charge.

For most materials other than those which are ferromagnetic, the magnetization $\mathbf{M}$ is directly proportional to the applied external field $\mathbf{B}$, so that, based on [6.23], $\mathbf{M}$ is also directly proportional to $\mathbf{H}$. It is thus customary to write

$$\mathbf{M} = \chi_{\mathrm{m}} \mathbf{H} \qquad [6.24]$$

where χ_{m} is the same quantity that was defined in equation [6.21] and is a dimensionless proportionality constant called the magnetic susceptibility of the material. The value of χ_{m} provides a measure of how susceptible a material is to becoming magnetized by an applied field. As was discussed in the previous subsection, χ_{m} is negative for diamagnetic materials and positive for paramagnetic materials. For ferromagnetic materials, χ_{m} in general depends on the applied magnetic field and the past history of magnetization of the material. Substituting [6.24] into [6.23], we find

$$\mathbf{B} = \mu_0(\mathbf{H} + \mathbf{M}) = \mu_0(1 + \chi_{\mathrm{m}})\mathbf{H} = \mu_0 \mu_r \mathbf{H} = \mu \mathbf{H} \qquad [6.25]$$

where the quantity $\mu = \mu_0(1+\chi_{\mathrm{m}}) = \mu_0 \mu_r$ is called the *magnetic permeability* and is entirely analogous to the electric permittivity ϵ in electrostatics. Note that μ_r is the relative permeability, analogous to ϵ_r in electrostatics. Just like electric permittivity ϵ, the permeability μ is a macroscopic parameter[69] that is used to account for the microscopic effects of material bodies in the presence of external magnetic fields. In practice, the permeability μ for a given material can be determined experimentally. When its value is known, μ can be used in [6.25] to relate **B** and **H** directly, thus eliminating the necessity of taking the magnetization **M** into account explicitly.

It is important to note that [6.25] only holds for a class of materials and under certain conditions. In particular, since [6.25] originates in [6.21], it depends primarily on the linear relationship between **M** and **B**. Magnetic materials, especially ferromagnetic materials discussed in the next section, often exhibit permanent magnetization and highly nonlinear behavior, in which case [6.21] (and thus [6.25]) is not valid. Equation [6.25] also requires that the material be isotropic—that is, that χ_m be independent of the direction of **B**. In anisotropic materials, such as ferrites (see Section 6.8.3), a **B** field in one direction can produce magnetization (and thus **M**) in another direction. Accordingly, the relation between **B** and **H** must be expressed[70] as a type of matrix, more commonly called a *tensor.*

In retrospect, it is quite unfortunate that the duality of **E** and **B** (see Section 6.2.3) and that of **D** and **H** were not appreciated during the early development of the theory of electromagnetics, since the use of the reciprocal of μ_0 in relating **H** to **B** would have been more appropriate. Also, the common references to **H** as the "magnetic field strength" and to **B** as the "magnetic flux density" are rather misleading.[71]

For paramagnetic and diamagnetic materials the value of μ differs only insignificantly from μ_0. For one of the strongest diamagnetic materials, bismuth, we have $\mu = 0.999833\mu_0$, whereas for the paramagnetic tungsten $\mu = 1.000078\mu_0$. The permeability of air is $\mu = 1.00000037\mu_0$. For copper and water, we have

---

[69] We should note here that in the interior of any material the values of **M** and **B** may vary rapidly with both time and space. However, in any practical problem that involves the measurement of fields, the field is sampled over regions of space much larger than atomic dimensions. Thus, the quantities **B**, **M**, and **H** in the above expressions are all macroscopic quantities.

[70] See Section 13.12 of S. Ramo, J. R. Whinnery, and T. Van Duzer, *Fields and Waves in Communication Electronics,* 3rd ed., John Wiley & Sons, Inc., 1994.

[71] The inappropriateness of the term "magnetic field strength" for **H** is discussed in Section 2 of A. Sommerfeld, *Electrodynamics,* Academic Press, 1952, reproduced here with permission of Academic Press, Inc. In Sommerfeld's words (p. 11): "We may indicate finally a subdivision of physical entities into entities of intensity and entities of quantity. **E** and **B** belong to the first class, **D** and **H**, to the second. The entities of the first class are answers to the question 'how strong,' those of the second class, to the question 'how much.' In the theory of elasticity, for example, the stress is an entity of intensity, the corresponding strain, one of quantity; in the theory of gases, pressure and volume form a corresponding pair of entities. In **D** the quantity character is clearly evident as the quantity of electricity (i.e., flux) that has passed through; in **H** the situation is slightly obscured by the fact there are no isolated magnetic poles. We are in general inclined to regard the entities of intensity as cause, the corresponding entities of quantity as their effect." On p. 45 of his text, Sommerfeld further states, "The unhappy term 'magnetic field' for **H** should be avoided as far as possible. It seems to us that this term has led into error none less than Maxwell himself." For an interesting discussion of the question of the relative physical merits of **B** versus **H** in connection with permanent magnets, see pp. 136–139 of M. Abraham and R. Becker, *Electricity and Magnetism,* Blackie & Son Limited, Glasgow, 1944.

TABLE 6.2. Relative permeability of selected materials

Material	Relative permeability (μ_r)
Air	1.00000037
Aluminum	1.000021
Bismuth	0.999833
Cobalt	250
Copper	0.9999906
Iron (Purified: 99.96% Fe)	280,000
Iron (Motor grade: 99.6%)	5,000
Lead	0.9999831
Manganese	1.001
Manganese-zinc ferrite	1,200
Mercury	0.999968
Nickel	600
Nickel-zinc ferrite	650
Oxygen	1.000002
Palladium	1.0008
Permalloy: 78.5% Ni, 21.5% Fe	70,000
Platinum	1.0003
Silver	0.9999736
Supermalloy: 79% Ni, 15% Fe, 5% Mo, 0.5% Mn	1,000,000
Tungsten	1.00008
Water	0.9999912

$\mu = 0.999991\mu_0$. Thus, it is very common to assume the relative permeability $\mu_r = 1$ for diamagnetic and paramagnetic materials. However, for ferromagnetic materials, discussed in the next section, μ_r is in general much larger than unity and in some special cases can be as large as 10^6! The relative permeability (μ_r) values for selected materials are given in Table 6.2.

6.8.3 Ferromagnetic Materials

The value of μ is generally much larger than μ_0 for *ferromagnetic* materials, which exhibit large magnetizations even in the presence of very weak magnetic fields. Only the three elements iron, nickel, and cobalt are ferromagnetic at room temperature and above. Almost all ferromagnetic alloys and compounds contain one or more of these three elements or manganese, which belongs to the same group of transition elements in the periodic table. The ease of magnetization of these materials results from quantum mechanical effects that tend to align neighboring atomic spins parallel to each other even in the absence of an applied magnetic field. In ferromagnetic materials the diamagnetic and paramagnetic effects also occur, but the contributions of these effects to the total magnetization is negligible even in relatively large fields. The permanent magnetization of cobalt, nickel, and iron leads to the magnetic permeabilities of $\mu_{\text{Co}} = 250\mu_0$, $\mu_{\text{Ni}} = 600\mu_0$, and $\mu_{\text{Fe}} = 6000\mu_0$. For purified iron, μ ranges from $10{,}000\mu_0$ to $200{,}000\mu_0$; for supermalloy (79% Ni, 15% Fe, 5% Mo, 0.5% Mn) μ can be up to $\sim 10^6\mu_0$. Typically, the value of μ for a ferromagnetic material is not unique, because of strong nonlinearities. In practice, the relation between **B** and **H** for ferromagnetic materials is represented graphically in the form of a curve known as the hysteresis curve. The magnetic properties of ferromagnetic materials are strongly temperature-dependent. In the absence of an applied field, they exhibit spontaneous magnetization below a temperature T_C, known as the Curie temperature, and are strongly paramagnetic above that temperature, susceptibilities (χ_m) decreasing with increasing temperature. The Curie temperatures for iron, nickel, and cobalt are, respectively, 1043 K, 631 K, and 1393 K.

Some elements that are neighbors of the ferromagnetic elements on the periodic table, such as chromium and manganese, also have strong quantum mechanical coupling forces between the atomic dipole moments. However, for these elements the coupling leads to antiparallel alignment of electron spins between adjacent atoms, so the net magnetic moment is zero. Elements with this property are called *antiferromagnetic*. Antiferromagnetism is also strongly temperature-dependent. At temperatures above the Curie temperature, the spin directions suddenly become random, and the material exhibits paramagnetic properties.

A number of oxides containing iron, nickel, or cobalt exhibit a magnetic behavior between ferromagnetism and antiferromagnetism and are called *ferrimagnetic* materials. The magnetic moments of these molecules alternate from atom to atom in an unequal manner, resulting in a net magnetic moment, but one that is much smaller than those of ferromagnetic materials. The most common ferrimagnetic materials are Fe_3O_4 and the family of ferrites described by the chemical formula $KO{\cdot}Fe_2O_3$, where K is any divalent metal such as Fe, Co, Ni, Mn, Mg, Cu, Zn, Cd, or a mixture of these. These are ceramiclike compounds with very low conductivities (for ferrites $\sigma = 10^{-4}$ to 1 S-m^{-1} as opposed to 10^7 S-m^{-1} for ferromagnetic Fe) and thus exhibit very low eddy current losses at high frequencies. Thus, ferrites are commonly used in high-frequency and microwave applications as cores for FM antennas, transformers, and phase shifters. An important property of ferrites is that they are *anisotropic* in the presence of magnetic fields, meaning that the **H** and **B** vectors are in different directions. Microwave devices that utilize the anisotropy of ferrites include isolators, gyrators, and some directional couplers.

Further discussion of ferromagnetic materials is beyond the scope of this book.[72]

6.8.4 The Meissner Effect in Superconductors

It was mentioned in Section 5.1 that some materials become superconducting below a critical temperature, with their conductivity becoming nearly infinite ($> 10^{20}$). This phenomenon is accompanied by a type of *repulsion* or *exclusion* of the **B** field from the interior of the material. In effect, the material becomes perfectly diamagnetic, with $\mathbf{B} = 0$ within the metal independent of the existence and nature of current carriers in the neighborhood of the metal, as if its permeability $\mu = 0$. This phenomenon was discovered in 1933 by W. Meissner and R. Ochsenfeld and is known

[72]For an excellent and relatively concise discussion at an appropriate level, see Chapters 9 and 10 of M. A. Plonus, *Applied Electromagnetics*, McGraw-Hill, New York, 1978.

as the *Meissner effect.* The description of the Meissner effect as a complete repulsion of the **B** field is an oversimplification. In fact, for a metal below the critical temperature the **B** field penetrates into the surfaces to optical distances, 50–200 nm. This result is true for dc fields as well as time-varying fields and may be understood by modeling the superconductor as a dense collisionless plasma.[73] The Meissner effect completes our understanding of a "perfect" conductor as one in which no static or time-varying electric or magnetic fields exist. A perfect conductor is usually understood to be one in which no electric field (at least none that varies more slowly than the relaxation time τ_r, which is typically of order $\sim 10^{-19}$ s for metallic conductors) can exist, since otherwise the abundantly available free electrons would be in perpetual motion. Maxwell's equations imply that there must then be no time-varying magnetic fields inside a perfect conductor either. However, a truly static magnetic field should in principle be completely uncoupled from the electric field and should not be affected by the conductivity of the material, even if $\sigma = \infty$. To the degree that superconductors are the physical materials that can be viewed most nearly as "perfect" conductors, the Meissner effect ensures that the interiors of perfect conductors are completely free of electric or magnetic fields.[74]

6.9 BOUNDARY CONDITIONS FOR MAGNETOSTATIC FIELDS

In some magnetostatic problems we deal with interfaces between two or more different types of materials. The manner in which **B** or **H** behaves across such interfaces is described by the boundary conditions. The boundary conditions are derived from fundamental laws of magnetostatics, which we reiterate below.

Starting with the experimental fact that electric currents create magnetic fields, as expressed by the Biot–Savart law, we have derived the following fundamental differential and integral laws of magnetostatics:

$$\nabla \times \mathbf{H} = \mathbf{J} \qquad \oint_C \mathbf{H} \cdot d\mathbf{l} = \int_S \mathbf{J} \cdot d\mathbf{s} \qquad \text{Ampère's law}$$

$$\nabla \cdot \mathbf{B} = 0 \qquad \oint_S \mathbf{B} \cdot d\mathbf{s} = 0 \qquad \text{No magnetic charges}$$

[73] T. Van Duzer and C. W. Turner, *Principles of Superconductive Devices and Circuits,* Elsevier, New York, 1981. Plasma is the fourth state of matter, consisting of an ionized collection of positive and negatively charged particles. Plasmas are inherently diamagnetic, since motions of both positively charged and negatively charged particles tend to reduce the magnetic field, as illustrated for a single electron in Figure 6.43.

[74] For further reading on this interesting subject, see Section 13.4 of S. Ramo, J. R. Whinnery, and T. Van Duzer, *Fields and Waves in Communication Electronics,* 3rd ed., John Wiley & Sons, Inc., New York, 1994; also see Section 9.10 of W. B. Cheston, *Elementary Theory of Electric and Magnetic Fields,* John Wiley & Sons, Inc., New York, 1964.

Note that Ampère's law was first derived in terms of **B** and used as such during most of this chapter. However, on the basis of our discussions in Section 6.8, we now understand that the more general version of Ampère's law relates the **H** field directly to the sources (i.e., **J**), independent of the material media. The magnetic properties of different physical media are accounted for via the *permeability* μ of the material. For linear (magnetization **M** proportional to **B**) and isotropic (magnetization **M** in the same direction as **B**) media, we have

$$\mathbf{H} = \mu^{-1}\mathbf{B} \qquad \text{or} \qquad \mathbf{B} = \mu\mathbf{H} \tag{6.26}$$

Note that [6.26] is valid even for inhomogeneous materials as long as we allow μ to be a function of position, i.e., $\mu(x, y, z)$.

A very common type of inhomogeneity encountered in practice occurs at the interface between two magnetically different materials. To establish a basis for solving such problems, we now study the behavior of **H** and **B** in crossing the boundary between two different materials and derive the conditions that **B** and **H** must satisfy at such interfaces. These conditions are referred to as the *boundary conditions;* they are similar in nature to the boundary conditions for electrostatic fields studied in Section 4.11. The boundary conditions must necessarily be derived using the integral forms of the fundamental magnetostatic laws, since the differential forms apply only at a point.

The methodology we use for derivation is identical to that used in Section 4.11 for electrostatics. For the normal component of the magnetic field, we consider the

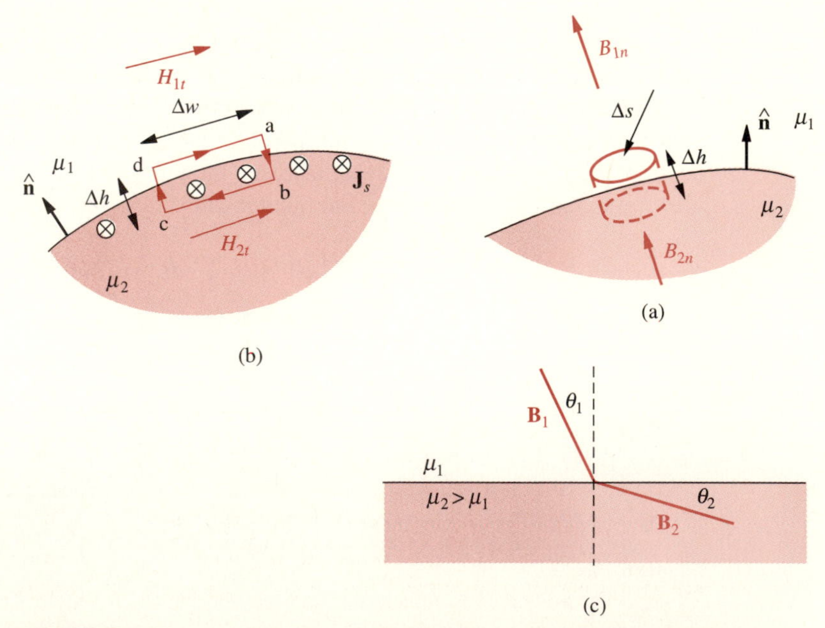

FIGURE 6.46. The boundary between two magnetic media. (a) A differential pillbox-shaped closed surface. (b) A differential closed contour. (c) **B** vectors at the boundary for the case $\mu_2 > \mu_1$.

surface of a differential pillbox, as shown in Figure 6.46a. Since the contributions from the side surfaces can be made infinitesimally small by taking $\Delta h \rightarrow 0$, we have

$$\oint_S \mathbf{B} \cdot d\mathbf{s} = \mathbf{B}_1 \cdot \hat{\mathbf{n}} - \mathbf{B}_2 \cdot \hat{\mathbf{n}} = 0 \quad \rightarrow \quad \boxed{B_{1n} = B_{2n}} \qquad [6.27]$$

Hence the normal components of the magnetic field **B** are always continuous across an interface between two different materials. Since $\mathbf{B} = \mu\mathbf{H}$, the magnetic field intensity **H** is not continuous across a boundary:

$$\mu_1 H_{1n} = \mu_2 H_{2n}$$

For the parallel components of the magnetic field, we consider the line integral of **H** around a closed contour abcda such as that shown in Figure 6.46b. In the absence of any currents at the interface (i.e., $\mathbf{J} = 0$), and since the contributions from the sides can be made infinitesimally small by taking $\Delta h \rightarrow 0$, we have

$$\int_{\mathrm{d}}^{\mathrm{a}} \mathbf{H}_1 \cdot d\mathbf{l} - \int_{\mathrm{b}}^{\mathrm{c}} \mathbf{H}_2 \cdot d\mathbf{l} = (H_{1t} - H_{2t})\Delta w = 0 \quad \rightarrow \quad \boxed{H_{1t} = H_{2t}} \qquad [6.28]$$

Thus the tangential component of the magnetic field is continuous across the interface between two materials, in the absence of any surface currents at the interface. Since $\mathbf{B} = \mu\mathbf{H}$, the tangential component of **B** is not continuous across a boundary. Instead we have

$$\frac{B_{1t}}{\mu_1} = \frac{B_{2t}}{\mu_2}$$

A consequence of [6.27] and [6.28] is the bending of magnetic field lines across material interfaces, as illustrated in Figure 6.46c. Consider a magnetic field **B** that is oriented at an angle θ_1 from the normal in medium 1. From [6.27] and [6.28] we have

$$B_1 \cos\theta_1 = B_2 \cos\theta_2$$

$$\mu_2 B_1 \sin\theta_1 = \mu_1 B_2 \sin\theta_2$$

where $B_{1,2}$ are the magnitudes of $\mathbf{B}_{1,2}$. From these results we can find

$$\tan\theta_2 = \frac{\mu_2}{\mu_1}\tan\theta_1$$

and

$$B_2 = B_1\sqrt{\left(\frac{\mu_2}{\mu_1}\sin\theta_1\right)^2 + \cos^2\theta_1}$$

The above relationships indicate that magnetic flux lines are *bent further away from the normal* in the medium with the *higher* magnetic permeability. Most materials have magnetic permeabilities of $\mu \simeq 1$, except for ferromagnetic materials, for which μ can be very large. If medium 2 is a ferromagnetic material and medium 1 is vacuum or air (or any other nonmagnetic material), $\mu_2 \gg \mu_1$, so $\tan\theta_1 \ll \tan\theta_2$, and the magnetic flux lines are for all practical purposes normal to the surface on the air side, provided θ_2 is not near 90°. This sharp bending of magnetic field lines is evident in Figure 6.45b, noting that the bar magnet must be a high-permeability material.

Discontinuity of* H *at Conductor Surfaces The relation [6.28] does not hold true if a current in a layer of vanishing thickness flows along the boundary between two materials, as occurs with time-varying fields and when one of the materials is an excellent conductor.[75]

From Figure 6.46b, and accounting for the possibility of a surface current in a microscopically small region near the surface, we have

$$\int_{\mathrm{d}}^{\mathrm{a}} \mathbf{H}_1 \cdot d\mathbf{l} - \int_{\mathrm{b}}^{\mathrm{c}} \mathbf{H}_2 \cdot d\mathbf{l} = \int_0^{\Delta w} \int_0^{\Delta h/2} J\, dh\, dw$$

$$\mathbf{H}_{1t}\Delta w - \mathbf{H}_{2t}\Delta w = J\Delta w \Delta h/2$$

$$H_{1t} - H_{2t} = J\Delta h/2 = J_s$$

Since the relationship between the surface current density $\mathbf{J}$ and $\mathbf{H}$ should be determined by the right-hand rule, the boundary condition can be compactly expressed as

$$\hat{\mathbf{n}} \times (\mathbf{H}_1 - \mathbf{H}_2) = \mathbf{J}_s$$

where $\hat{\mathbf{n}}$ is the *outward unit normal from medium 2* at the interface and $\mathbf{J}_s$ is normal to the page pointing inward. When the conductivities of both media are finite, currents are defined by volume current densities instead of surface currents, so $J\Delta h \rightarrow 0$ as $\Delta h \rightarrow 0$, and thus the tangential component of $\mathbf{H}$ is continuous across the interface. When one of the media (say, medium 2) is a perfect conductor, the magnetic field in its interior is zero (see the discussion in Section 6.8.4), so we have $\mathbf{H}_2 = 0$ and thus

$$\boxed{\hat{\mathbf{n}} \times \mathbf{H}_1 = \mathbf{J}_s} \qquad [6.29]$$

as the basic boundary condition on the surface of a perfect conductor. Note that another consequence of the fact that $\mathbf{B} = 0$ inside a perfect conductor is that $\mathbf{B}_{2n} = 0$, which implies that $\mathbf{B}_{1n} = 0$. In other words, magnetic fields cannot terminate

[75]The concept of a surface current presents somewhat of a difficulty and potential source of confusion. In electrostatics we saw a somewhat analogous concept of a surface charge, which exists on the surfaces of good conductors. However, we also observed that the surface charge exists in a microscopically infinitesimal (but nevertheless a few atomic dimensions thick, so that we can still think in terms of "density" of charge) layer. In practice, the depth to which a surface current on a good conductor is confined depends on the conductivity σ and the frequency of operation f. Quantitatively, the thickness of the surface current layer is given by the *skin depth* $\delta = (\pi f \mu \sigma)^{-1/2}$, as discussed in Chapter 8. For copper, for example, we have $\delta \simeq 0.07 f^{-1/2}$, so that at 1 MHz, $\delta \simeq 0.07$ mm, which is small but not infinitesimal on a macroscopic scale. Nevertheless, in most applications, a current $\mathbf{J}$ (A-m^{-2}) confined to a region of thickness $\delta \simeq 0.07$ mm can be considered to be a surface current of $\mathbf{J}_s = \mathbf{J}\delta$ (A-m^{-1}), for all practical purposes. The difficulty arises when one considers extremely low frequencies or purely static (dc) fields. In such cases, $\delta \rightarrow \infty$, and the current $\mathbf{J}$, and thus the associated magnetic field $\nabla \times \mathbf{H} = \mathbf{J}$, exists throughout the conductor. The only exception would be the case of perfect conductors, or superconductors, for which $\sigma = \infty$ and the skin depth $\delta \rightarrow 0$ even for static fields.

normally on a perfect conductor. The magnetic field immediately outside a perfect conductor must necessarily be tangential to the surface.

6.10 MAGNETIC FORCES AND TORQUES

We started this chapter with Ampère's force law, describing the manner in which current-carrying wires exert forces on one another, as the experimental basis of magnetostatics. The discovery that current-carrying wires experience physical forces in the presence of magnetic fields (created by other current-carrying wires or permanent magnets) immediately raised the possibility of using these forces to do work and indeed led to significant enhancements in industrial technology, including the development of direct- (i.e., steady-) current electric motors and sensitive instruments for electrical measurements (e.g., voltmeters and ammeters). Much of the machinery of our present-day industrial and technological environment is actually based on alternating currents and fields, which operate on the basis of Faraday's law of electromagnetic induction, covered in Chapter 7. However, forces and torques produced by static magnetic fields and direct-current motors and generators also account for a range of applications involving critical speed or position control requirements, including automobile power windows, food blenders, hand power tools, fans, electric cars, elevators, and hoists. In this section we briefly discuss magnetostatic forces and torques, which form the basic principles of direct-current rotating machines.

6.10.1 Magnetic Force on Moving Charges and the Hall Effect

Magnetic fields exert forces on matter via their influence on charged particles. In any region of space where a magnetic field $\mathbf{B}$ is present, the magnetic force $\mathbf{F}_{\rm m}$ on a charge q moving with a velocity $\mathbf{v}$ is given by

$$\mathbf{F}_{\rm m} = q\mathbf{v} \times \mathbf{B} \qquad [6.30]$$

which is equivalent to $\mathbf{F}_{\rm m} = I\,d\mathbf{l} \times \mathbf{B}$ when we note that a charge q moving with velocity $\mathbf{v}$ constitutes an element of current $I d\mathbf{l}$, as previously noted in Section 6.2.3.

Equation [6.30] describes the Lorentz magnetic force, which is the magnetic component of the total Lorentz force acting on a charged particle in the presence of electric and magnetic fields, namely,

$$\mathbf{F} = q[\mathbf{E} + \mathbf{v} \times \mathbf{B}] \qquad [6.31]$$

Equation [6.31] is the well-known Lorentz[76] force equation. The solution of the Lorentz force equation is required when determining the motion of electrons in

[76]H. A. Lorentz was the first to study the motion of electrons in the presence of electric and magnetic fields. In particular, he applied equation [6.31] to put forth an explanation for the Zeeman effect, involving the splitting of spectral lines in the presence of a magnetic field [P. Zeeman, Doublets and triplets in the spectrum produced by external magnetic forces, *Phil. Mag.* 43 (5), p. 226, 1897]. Zeeman was a student of Lorentz, and the two later shared the Nobel Prize in physics in 1902. Lorentz's famous theory of electrons is described in the extensive compilation of lectures that he gave at Columbia University in the Spring of 1906: H. A. Lorentz, *The Theory of Electrons and Its Applications to the Phenomena of Light and Radiant Heat*, Dover Publications, Inc., 1952.

electrostatic and magnetic deflection systems (such as in cathode-ray tubes, in microwave devices such as the klystron and magnetron, and in particle accelerators such as the cyclotron) and in other applications involving ionized gases or plasmas.

While the Lorentz force equation [6.31] is commonly used to study the motion of freely mobile electrons and positively charged particles such as those in an ionized gas, it is equally applicable to the dynamics of current carriers in solids, such as electrons and holes in semiconductors.[77] An interesting and important manifestation of the motion of electrons and holes under the influence of the Lorentz force is the Hall effect, which is described briefly below.

The Hall Effect In 1879, Edwin H. Hall, at that time conducting an experiment as a student in Baltimore, discovered[78] a new effect of a magnetic field on electric currents. Hall placed a strip of gold leaf mounted on glass, forming part of an electric circuit through which a current passed, between the poles of an electromagnet, the plane of the strip being perpendicular to the magnetic flux lines. The two ends of a sensitive voltmeter were then placed in contact with different parts of the strip, until two points at the same potential were found, so that the voltmeter reading was zero. When the magnetic field (i.e., the current of the electromagnet) was turned on or off, a deflection of the galvanometer needle was observed, indicating a potential difference between the voltmeter leads. In this way, Hall showed that the magnetic field produces a new electromotive force in the strip of gold, at right angles to the primary electromotive force and to the magnetic force, and proportional to the product of these two forces.[79] Physically, we can regard this effect as an additional electromotive force generated by the action of a magnetic field on the current. This phenomenon, called the Hall effect, is extremely useful in practice in the determination of charge densities in materials, especially in semiconductors.

Consider a rectangular semiconductor bar as shown in Figure 6.47. Assume that this is a p-type semiconductor, so that the majority charge carriers are holes, each with charge $|q_e|$ where $q_e \simeq -1.6 \times 10^{-19}$ C is the charge of an electron. A potential difference of V_0 applied between the two side faces sets up an applied electric field E_0 in the x direction, which causes a current $I_0 = J_x A = w d \sigma E_0$ to flow in the x direction, where σ is the conductivity of the semiconductor. In general, the conductivity for a semiconductor is given by

$$\sigma = |q_e|(\mu_e N_e + \mu_p N_p)$$

---

[77] A. S. Grove, *Physics and Technology of Semiconductor Devices*, John Wiley & Sons, Inc., New York, 1967.

[78] E. H. Hall, *Phil. Mag.* ix, p. 225, and x, p. 301, 1880.

[79] In his excellent historical account, *A History of the Theories of Aether and Electricity*, Thomas Nelson and Sons Ltd, 1951, E. T. Whittaker notes that in the early 1870s Oliver Lodge had done experiments on the flow of electricity in a metallic sheet and had come very close to discovering the Hall effect. However, he was deterred from making the crucial test upon reading a passage in Article 501 in Vol. ii of Maxwell's *Electricity and Magnetism:* "It must be carefully remembered, that the mechanical force which urges a conductor carrying a current across the lines of magnetic force acts, not on the electric current, but on the conductor which carries it."

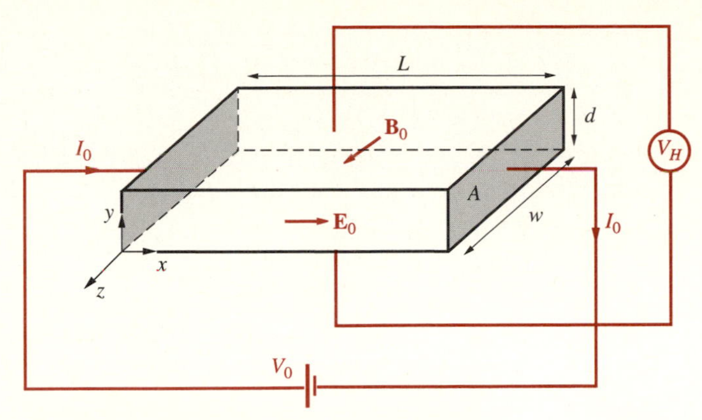

FIGURE 6.47. **Illustration of the Hall effect.**

where N_e and N_p are, respectively, the densities of electrons and holes, and μ_e and μ_p are, respectively, the mobilities of the electrons and holes. As was noted in Section 5.1 in connection with [5.1], mobility is the ratio between the drift velocity of the charge carriers and the applied electric field. For a p-type semiconductor, we have $N_p \gg N_e$, so that $\sigma \simeq |q_e|\mu_p N_p$, and the drift velocity is $v_d = \mu_p E_0$, where $\mu_p \simeq \sigma/(|q_e|N_p)$. If a magnetic field B_0 is applied in the z direction, the holes in the p-type semiconductor deflect in the $-y$ direction. Using vector notation, the total force on a single hole due to the electric and magnetic fields is given by [6.31], namely,

$$\begin{aligned}\mathbf{F} &= |q_e|(\mathbf{E} + \mathbf{v}_d \times \mathbf{B}) = |q_e|(\hat{\mathbf{x}}E_0 + \hat{\mathbf{x}}v_d \times \hat{\mathbf{z}}B_0)\\ &= |q_e|(\hat{\mathbf{x}}E_0 - \hat{\mathbf{y}}v_d B_0)\end{aligned}$$

According to this result, the holes in the semiconductor experience a magnetic force of $-\hat{\mathbf{y}}(|q_e|v_d B_0)$ and thus an acceleration in the $-y$ direction. However, the displacement of the holes in the $-y$ direction leads to the creation of an electric field E_y between the positively charged region into which the holes move and the negatively charged atoms that are left behind. Thus, this restoring electric field points in the y direction and at steady state must have a magnitude $v_d B_0$ to balance the magnetic force $|q_e|v_d B_0$. The formation of this electric field E_y sets up a potential difference V_H between the top and bottom ends of the bar, given by

$$V_H = E_y d = v_d B_0 d$$

and is called the Hall voltage.

Substituting for $v_d = \mu_p E_0$ and $\mu_p = r/(|q_e|N_p)$, and noting that $\mathbf{J} = \sigma\mathbf{E}$, the electric field E_y can be written in terms of the current density as

$$E_y = v_d B_0 \simeq \frac{\sigma E_0}{|q_e|N_p} B_0 = R_H J_x B_0$$

where $R_H = (|q_e|N_p)^{-1}$ is called the Hall coefficient and N_p is the hole concentration in the p-type semiconductor. The hole concentration N_p is given as

$$N_p = \frac{1}{|q_e|R_H} = \frac{J_x B_0}{|q_e|E_y} = \frac{[I_o/(wd)]B_0}{|q_e|(V_H/d)} = \frac{I_o B_0}{|q_e|wV_H}$$

Since I_0, B_0, w, $|q_e|$, and V_H are either known or can be measured, the Hall effect can be used to determine (quite accurately) the hole concentration in the p-type semiconductor. Note that the same is true for an n-type semiconductor, with the only difference being that the charge carriers are electrons.

The Hall effect can also be used to determine the conductivity of the semiconductor material. If the resistance of the semiconductor bar is measured ($R = V_0/I_0$), the conductivity is given by

$$\sigma = \frac{L}{Rwd} = \frac{LI_0}{wdV_0}$$

However, since the conductivity of the p-type bar is $\sigma \simeq |q_e|\mu_p N_p$, we can write

$$\mu_p \simeq \frac{\sigma}{|q_e|N_p}$$

so that by measuring the hole concentration N_p and the conductivity of the bar (inferred from the resistance of the bar), the mobility of the holes in the p-type material can be determined. Such measurements are very important in the analysis of semiconductor materials.

6.10.2 Magnetic Force on Current-Carrying Conductors

It is evident from [6.2] and [6.3], and the discussion in Section 6.2.3, that when a filamentary conductor carrying current I is placed in a region permeated by a magnetic field, it experiences a total net magnetic force given by

$$\mathbf{F}_\mathrm{m} = \int_C d\mathbf{F}_\mathrm{m} = \int_C I\, d\mathbf{l} \times \mathbf{B} \qquad [6.32]$$

where C is the contour defined by the filamentary conductor, which in many engineering applications is the length of a wire. Note that $d\mathbf{F}_\mathrm{m} = I\, d\mathbf{l} \times \mathbf{B}$ is the incremental magnetic force felt by an infinitesimal wire element $I\, d\mathbf{l}$.

In Example 6-1, we used Ampère's force law [6.1] to determine the repulsive magnetic force between two infinitely long wires carrying oppositely directed current. In the following Example 6-35, we use [6.32] to determine the magnetic attraction force between two infinitely long wires carrying equally directed current. Example 6-36 considers the magnetic force exerted on a rigid square loop by an infinitely long current-carrying wire.

Example 6-35: Magnetic force between two long wires. Two long wires in free space at a distance a from one another carry currents I_1 and I_2, both flowing in

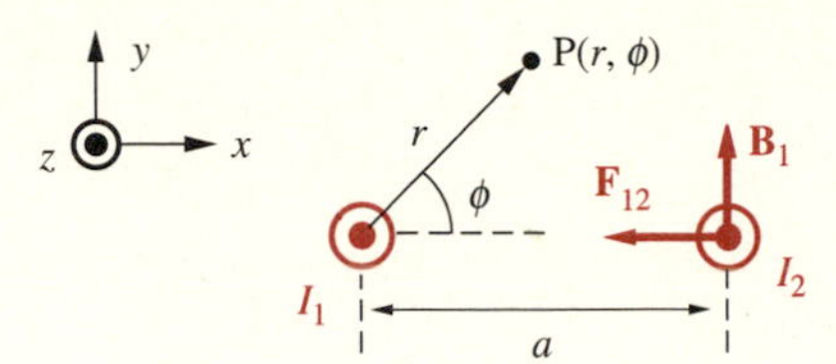

FIGURE 6.48. **Force between two long wires.**

the z direction, as shown in Figure 6.48. Find an expression for the magnetic force per unit length between the wires in terms of I_1, I_2, and a.

Solution: Based on the result of Example 6-3, the magnetic field $\mathbf{B}_1$ due to the current in wire 1 at any point P at a distance r outside the wire is $\mathbf{B}_1 = \hat{\boldsymbol{\phi}}\mu_0 I_1/(2\pi r)$, where r and ϕ are the cylindrical coordinate quantities, with wire 1 assumed to be located at the origin and oriented in the z direction. The $\mathbf{B}$ field at the location of wire 2 is then simply

$$[\mathbf{B}_1]_{\text{at wire 2}} = \hat{\mathbf{y}}\frac{\mu_0 I_1}{2\pi a}$$

since $r = a$ and $\hat{\boldsymbol{\phi}} = \hat{\mathbf{y}}$ for $\phi = 0$. The differential force on an elemental length of wire 2 is then given by

$$d\mathbf{F}_{12} = \hat{\mathbf{z}}(I_2\,dl_2) \times \hat{\mathbf{y}}\left(\frac{\mu_0 I_1}{2\pi a}\right)$$

which, when integrated over unit length, gives the magnetic force per unit length as

$$\mathbf{F}_{12} = -\hat{\mathbf{x}}\frac{\mu_0 I_1 I_2}{2\pi a} \quad \text{N-m}^{-1}$$

The force is attractive as expected. The reader is encouraged to evaluate the magnetic force per unit length $\mathbf{F}_{21}$ exerted on wire 1 by wire 2 and show that it is given by $\mathbf{F}_{21} = +\hat{\mathbf{x}}\mu_0 I_1 I_2/(2\pi a)$.

Example 6-36: Magnetic force on a current-carrying loop. A rigid rectangular loop carrying current I_2 is located near an infinitely long wire carrying current I_1 as shown in Figure 6.49. Find the magnetic force exerted by the long wire on the loop.

Solution: Note that the lead wires of the loop are twisted as in Example 6-4, so that the currents on the two wires (and thus the magnetic forces that they experience) effectively cancel. Thus, we need to consider only the force on the square loop. Once again using the result of Example 6-3, the $\mathbf{B}$ field produced by the long wire at any point in the plane of the loop and at a distance r from the long wire is

$$\mathbf{B}_1 = -\hat{\mathbf{x}}\frac{\mu_0 I_1}{2\pi r}$$

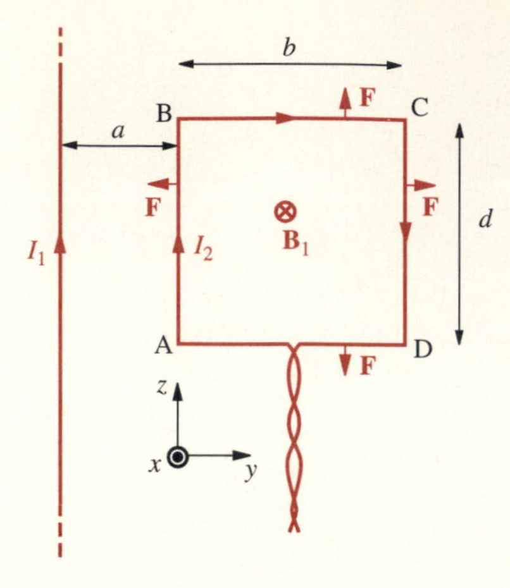

FIGURE 6.49. Force on a rectangular loop near a long wire. Coordinate system, dimensions, and loop orientation. The directions of the magnetic forces on each side of the loop are indicated.

Assuming that the loop is rigid (i.e., it maintains its shape even under the influence of the magnetic force; note that this can be achieved by simply winding the loop out of strong enough wire or just winding it on a nonmagnetic frame such as a wooden frame), the total magnetic force exerted on it is given by

$$\begin{aligned}\mathbf{F}_{\text{loop}} &= I_2 \oint_C d\mathbf{l}_2 \times \mathbf{B}_1 \\ &= I_2\left[\int_{\text{AB}} d\mathbf{l}_2 \times \mathbf{B}_1 + \int_{\text{BC}} d\mathbf{l}_2 \times \mathbf{B}_1 + \int_{\text{CD}} d\mathbf{l}_2 \times \mathbf{B}_1 + \int_{\text{DA}} d\mathbf{l}_2 \times \mathbf{B}_1\right] \\ &= \frac{\mu_0 I_1 I_2}{2\pi}\left\{\int_{\text{A}}^{\text{B}}\left[\hat{\mathbf{z}}\,dz \times \frac{-\hat{\mathbf{x}}}{a}\right] + \int_{\text{B}}^{\text{C}}\left[\hat{\mathbf{y}}\,dy \times \frac{-\hat{\mathbf{x}}}{y}\right]\right. \\ &\qquad \left. + \int_{\text{C}}^{\text{D}}\left[\hat{\mathbf{z}}\,dz \times \frac{-\hat{\mathbf{x}}}{a+b}\right] + \int_{\text{D}}^{\text{A}}\left[\hat{\mathbf{y}}\,dy \times \frac{-\hat{\mathbf{x}}}{y}\right]\right\}\end{aligned}$$

The directions of the forces on each side of the loop are shown in Figure 6.49. Note that the forces on the BC and DA sides are equal and opposite and thus cancel out. The forces on the AB and CD sides are also in opposing directions but their magnitudes are different, in view of their different distances from the long wire. The net total force on the loop is thus given by

$$\mathbf{F}_{\text{loop}} = -\hat{\mathbf{y}}\frac{\mu_0 I_1 I_2 d}{2\pi}\left(\frac{1}{a} - \frac{1}{a+b}\right)$$

Thus the loop is pulled toward the long wire. Note that if the polarity of either (but not both) of the two currents I_1 or I_2 were reversed, the loop would be pushed away from the long wire.

6.10.3 Torque on a Current-Carrying Loop

The orientation of the loop in Example 6-36 is such that the plane of the loop and thus all of its wires are entirely perpendicular to the **B** field generated by the long wire. If the loop is instead immersed in a region with a **B** field in the y direction, as shown in Figure 6.50a, its two sides AB and CD experience magnetic forces respectively into and out of the page, as shown, while the forces on the BC and DA sides are zero, since $d\mathbf{l}$ is parallel to **B**. Assuming again that the loop is rigid, it thus experiences a tendency to twist or turn around the z–z' axis shown in dashed lines. For given magnitudes of the forces $\mathbf{F}_{AB}$ and $\mathbf{F}_{CD}$, the loop has a greater tendency to rotate if the AB and CD sides are farther away from the z–z' axis. The quantitative measure of the tendency of a force to cause or change rotational motion is torque. Mathematically, when a force **F** acts on a body at a distance **r** from a reference axis, the torque **T** acting on a body with respect to that reference axis is defined as

$$\mathbf{T} = \mathbf{r} \times \mathbf{F}$$

The SI units for torque are newton-meters (N-m). The direction of torque is perpendicular to both **r** and **F** and defines the direction of rotation in the right-hand-rule sense. In other words, with the thumb of the right hand pointed in the direction of torque, rotation is in the direction of the fingers.

Let us return to the loop in Figure 6.50a and assume that it is mounted such that it is free to rotate about the z–z' axis and that the **B** field is uniform and is given by $\mathbf{B} = \hat{\mathbf{y}}B_0$. The total magnetic force on the loop can be calculated in the same manner as in Example 6-36. Noting once again that the forces on the BC and DA sides are zero since $d\mathbf{l}$ is parallel to **B**, we have

$$\begin{aligned}\mathbf{F}_{\text{loop}} &= \mathbf{F}_{AB} + \mathbf{F}_{CD} \\ &= I\left[\int_A^B d\mathbf{l} \times \mathbf{B} + \int_C^D d\mathbf{l} \times \mathbf{B}\right] \\ &= I\left[\int_A^B \hat{\mathbf{z}}\,dz \times \hat{\mathbf{y}}B_0 + \int_C^D \hat{\mathbf{z}}\,dz \times \hat{\mathbf{y}}B_0\right] = 0\end{aligned}$$

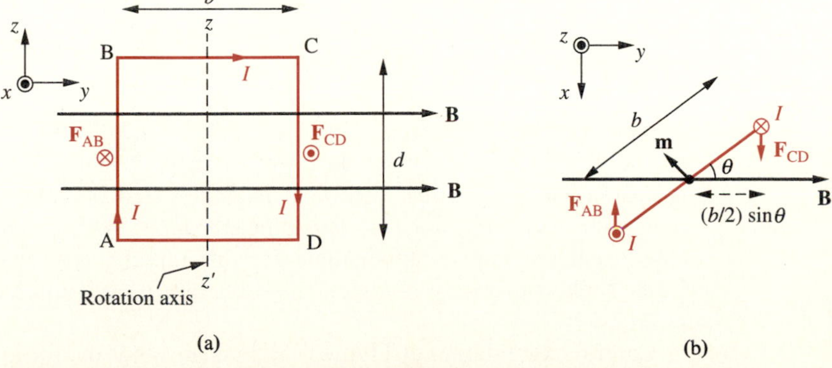

FIGURE 6.50. Torque on a loop in a B field. (a) Loop immersed in a uniform magnetic field **B**. The rotation axis is z–z'. (b) View from the top.

Note that the net total force acting on the loop is zero. However, since $\mathbf{F}_{AB}$ and $\mathbf{F}_{CD}$ act at different points, they produce a torque, as illustrated in Figure 6.50b. The magnitudes of $\mathbf{F}_{AB}$ and $\mathbf{F}_{CD}$ are equal, namely, $|\mathbf{F}_{AB}| = |\mathbf{F}_{CD}| = B_0 Id$, and each of these two forces produces torque in the same direction, at distances of $r = (b/2)\sin\theta$ with respect to the rotation axis. The resultant torque is thus given as

$$\begin{aligned}
\mathbf{T} &= \mathbf{r}\times\mathbf{F}_{AB} + \mathbf{r}\times\mathbf{F}_{CD} \\
&= \left(-\hat{\mathbf{y}}\frac{b}{2}\sin\theta\right)\times(-\hat{\mathbf{x}}B_0 Id) + \left(\hat{\mathbf{y}}\frac{b}{2}\sin\theta\right)\times(\hat{\mathbf{x}}B_0 Id) \\
&= -\hat{\mathbf{z}}B_0 Ibd\sin\theta
\end{aligned}$$

The sense of rotation of the coil as shown in Figure 6.50b is clockwise, consistent with the torque being in the $-z$ direction (i.e., if the right-hand thumb is pointed in the $-z$ direction, the fingers rotate clockwise).

The total torque acting on the loop is thus found to be proportional to the product (bd), which is the area A of the loop. Thus we can rewrite $\mathbf{T}$ as

$$\mathbf{T} = -\hat{\mathbf{z}}B_0 IA\sin\theta$$

where A is the area of the loop. As long as the loop is symmetrically positioned with respect to the z–z' axis of rotation and has the same area A, the torque acting on it can be shown to be independent of its shape; that is, a circular, hexagonal, or rectangular loop experiences the same torque as a square loop.

Using the definition of the magnetic dipole from Section 6.6, we can write $|\mathbf{m}| = IA$ and note that the direction of the dipole moment $\mathbf{m}$ is related to the current flow via the right-hand rule and is as shown in Figure 6.50b. Considering the relationship between $\mathbf{T}$, $\mathbf{m}$, and $\mathbf{B}$, we can see that the torque can be compactly expressed as

$$\boxed{\mathbf{T} = \mathbf{m}\times\mathbf{B}} \qquad [6.33]$$

Equation [6.33] is quite general and describes the torque acting on a loop when it is in a region permeated by a uniform magnetic field $\mathbf{B}$. The torque described by [6.33] is what aligns microscopic magnetic dipoles in magnetic materials and causes them to be magnetized as discussed in Section 6.8.1.

Direct-Current Machines Equation [6.33] indicates that rotational motion can be brought about from an arrangement of magnetostatic fields and steady electric current-carrying wires. Electrical machines that operate on this principle are called *direct-current* (or dc) *motors* and *generators*. Noting that the subject of electrical machinery and its applications is wide-ranging and well beyond the scope of this book, we briefly comment here on the basic principles of direct-current machines.

Careful examination of Figure 6.50 indicates that the torque acting on the loop tends to move it toward an equilibrium position at which the torque is zero. It is easy to see from Figure 6.50 that the torque is zero when **m** is parallel to **B**—that is, when the loop is perpendicular to the magnetic field. At this position there is no net force or torque acting on the loop. It turns out, however, that because of the finite mass of any coil, it approaches the equilibrium position with inertia and rotates through or overshoots it. As soon as the coil moves past the equilibrium position, the direction of torque is reversed (since $\sin\theta$ in Figure 6.50b is now negative) and it tends to rotate back toward the equilibrium position. Because of the damping effect of the surrounding air and mechanical friction, the coil eventually stops at the equilibrium position.

Even without continuous rotational motion, the fact that the torque produced is proportional to the current passing through the loop facilitates sensitive electrical measurements. The operation of voltmeters and ammeters is based on an arrangement such as that shown in Figure 6.51a, although an electromagnet may more often be used instead of a permanent magnet.

In order to obtain continuous rotation of the loop, it is necessary to prevent the reversal of the direction of torque as the loop moves past the equilibrium position. One possible way of achieving this is sketched in Figure 6.51b. Noting that the direction of the torque is dependent on the direction of current flowing in the loop, the direction of current is changed as soon as the loop reaches its equilibrium position, using a so-called *commutator* arrangement. The commutator in a dc machine changes the direction of current once per rotation so that the torque always remains in the same direction.

The simplistic dc motor depicted in Figure 6.51 is not useful, since the magnitude of the torque varies periodically with time, although its sense remains constant. Smoother operation can be achieved by designing the magnetic field structure to provide a nearly constant **B** field at all points during one rotational cycle.

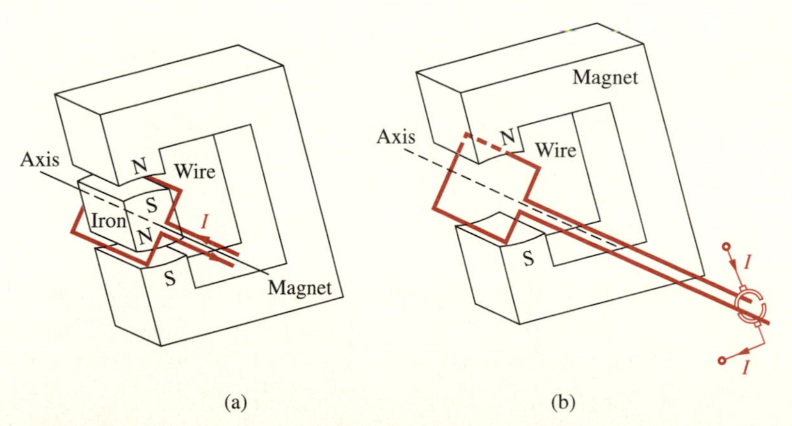

FIGURE 6.51. **Simple dc motors.** (a) Using a permanent magnet and soft iron material. In the case shown, there is no continuous motion, since the torque changes direction as the loop goes past the equilibrium position. (b) Using a permanent magnet and a commutator arrangement. By reversing the current every half cycle of rotation, continuous motion can be achieved.

6.11 SUMMARY

This chapter discussed the following topics:

- **Ampère's force law.** The magnetostatic force F_{12} exerted by a circuit C_1 carrying a current I_1 on another circuit C_2 carrying current I_2 is given by

$$\mathbf{F}_{12} = \frac{\mu_0}{4\pi}\oint_{C_2}\oint_{C_1}\frac{I_2\,d\mathbf{l}_2 \times (I_1\,d\mathbf{l}_1 \times \hat{\mathbf{R}})}{R^2}$$

  where $\hat{\mathbf{R}}$ is the unit vector pointing from current element $I_1\,d\mathbf{l}_1$ to $I_2\,d\mathbf{l}_2$ and R is the distance between them. Ampère's force law is experimentally based and constitutes the foundation of magnetostatics.

- **The magnetic field.** Any physical region of space in the vicinity of steady electric currents is said to be permeated by a magnetic field. The relationship between the **B** field and its source current is provided by the Biot–Savart law, which is given by

$$\mathbf{B} = \frac{\mu_0}{4\pi}\oint_{C'}\frac{I\,d\mathbf{l}' \times \hat{\mathbf{R}}}{R^2} \qquad \mathbf{B}(\mathbf{r}) = \frac{\mu_0}{4\pi}\int_{V'}\frac{\mathbf{J}(\mathbf{r}') \times \hat{\mathbf{R}}}{R^2}\,dv'$$

  depending on whether we have current I flowing in a filamentary contour C or a volume current $\mathbf{J}$, where $\hat{\mathbf{R}}$ is the unit vector pointing *from* the location of the current element *to* the observation point P and $R = |\mathbf{r} - \mathbf{r}'|$ is the distance between the current element and P. The Biot–Savart law is a direct consequence of Ampère's force law and is thus experimentally based.

- **Ampère's circuital law.** Ampère's circuital law states that the line integral of **B** over any closed contour C is a constant and is equal to μ_0 times the total net current I passing through the surface S enclosed by the contour. In other words,

$$\oint_C \mathbf{B}\cdot d\mathbf{l} = \int_S \mu_0\mathbf{J}\cdot d\mathbf{s} = \mu_0 I$$

  where C is the contour that encloses the surface S and I is the total current passing through the surface S. Ampère's circuital law is a direct consequence of the Biot–Savart law and is particularly useful for the analysis of problems involving a high degree of symmetry, such that contours over which the **B** field is constant can be identified.

- **Curl.** Curl is a measure of the circulation per unit area of a vector field. It is defined as an axial vector whose magnitude is the maximum net circulation of the vector field and whose direction is normal to the surface element that results in maximum circulation. For the magnetostatic field we have

$$\nabla \times \mathbf{B} = \mathbf{J}$$

which is the differential version of Ampère's law. Thus, the magnetostatic field has nonzero curl only at points where currents exist. Stokes's theorem is valid in general for any vector field,

$$\oint_C \mathbf{G} \cdot d\mathbf{l} = \int_S (\nabla \times \mathbf{G}) \cdot d\mathbf{s}$$

where S is the surface enclosed by the contour C.

- **Vector magnetic potential.** A direct consequence of the experimentally based Biot–Savart law is that the **B** field can be written as the curl of another vector **A** related to the source current **J** as

$$\mathbf{A}(\mathbf{r}) = \frac{\mu_0}{4\pi} \int_{V'} \frac{\mathbf{J}(\mathbf{r}')}{R} dv'$$

In some cases it is easier to evaluate **A** directly from the source currents using the above expression and subsequently find the **B** field using $\mathbf{B} = \nabla \times \mathbf{A}$.

- **The magnetic dipole.** A small current-carrying loop of radius a constitutes a magnetic dipole. The **B** field at large distances ($r \gg a$) produced by a small loop of radius a lying in the xy plane and carrying current I is given by

$$\mathbf{B} = \hat{\mathbf{r}} \frac{\mu_0 |\mathbf{m}| \cos\theta}{2\pi r^3} + \hat{\boldsymbol{\theta}} \frac{\mu_0 |\mathbf{m}| \sin\theta}{4\pi r^3}$$

where $\mathbf{m} = \hat{\mathbf{z}} I(\pi a^2)$ is the magnetic dipole moment.

- **Divergence of B and magnetic flux.** A direct consequence of the fact that we can write $\mathbf{B} = \nabla \times \mathbf{A}$ is that the divergence of the **B** field is identically zero. In other words,

$$\nabla \cdot \mathbf{B} = 0 \quad \rightarrow \quad \oint_S \mathbf{B} \cdot d\mathbf{s} = 0$$

which in turn means that the flux lines of the **B** field are always continuous and form closed loops.

- **Inductance.** The inductance of a circuit is a measure of its ability to link magnetic flux per unit current. The mutual inductance L_{12} between two coils C_1 and C_2 having, respectively, N_1 and N_2 turns is given by

$$L_{12} = \frac{N_1 N_2 \int_{S_2} \mathbf{B}_1 \cdot d\mathbf{s}_2}{I_1}$$

where S_2 is the surface enclosed by circuit C_2, I_1 is the current in loop C_1, and $\mathbf{B}_1$ is the field produced by a single turn of coil 1. In general, $L_{12} = L_{21}$. The self-inductance of a circuit C having N turns is given by

$$L = \frac{N^2 \int_S \mathbf{B} \cdot d\mathbf{s}}{I}$$

where S is the surface enclosed by the circuit C and $\mathbf{B}$ is the field produced by a single turn of C. To find the inductance of a circuit, we typically assume current I to flow and find the $\mathbf{B}$ field using the Biot–Savart law or Ampère's law and integrate it over the surface S to determine the flux linked by the circuit. A more general formula for the mutual inductance between two circuits C_1 and C_2 is given by the Neumann formula, namely,

$$L_{12} = \frac{\mu_0 N_1 N_2}{4\pi} \oint_{C_1} \oint_{C_2} \frac{d\mathbf{l}_1 \cdot d\mathbf{l}_2}{R}$$

where N_1 and N_2 are the number of turns in circuits C_1 and C_2, respectively.

- **Magnetic materials.** When a material is placed in an external $\mathbf{B}$ field, a magnetization distribution $\mathbf{M}$ is created, which is given by $\mathbf{M} = (\chi_{\rm m}/\mu_0)\mathbf{B}$, where $\chi_{\rm m}$ is the magnetic susceptibility, dependent on the particular microscopic properties of the material. The net effect of this magnetization within the material is that the total magnetic field is now given by $\mathbf{B} = \mu_0(1 + \chi_{\rm m})\mathbf{H}$, a fact that is typically accounted for by assigning to each material a magnetic permeability $\mu = \mu_0(1 + \chi_{\rm m})$ such that $\mathbf{B} = \mu\mathbf{H}$. For most materials, the value of $\chi_{\rm m}$ is quite small, so that $\mu \simeq \mu_0$. For ferromagnetic materials however, the values of μ can be very large, ranging from a few hundred up to a million.
- **Magnetostatic boundary conditions.** Experimentally established laws of magnetostatics dictate that the normal component of the $\mathbf{B}$ field is continuous across the interface between two materials with different permeability. The tangential component of the magnetic field $\mathbf{H}$ is always continuous across any interface except when there is surface current density $\mathbf{J}_s$ present, which typically occurs at the surfaces of metallic conductors. In summary we have

$$B_{1n} = B_{2n} \quad \text{and} \quad H_{1t} = H_{2t}$$

except at a perfect conductor surface, when we have

$$\hat{\mathbf{n}} \times \mathbf{H}_1 = \mathbf{J}_s$$

where $\hat{\mathbf{n}}$ is the outward normal unit vector from the perfect conductor surface.

- **Magnetic forces and torques.** Our starting point for magnetostatics was the experimental fact that current-carrying wires exert forces on one another. An alternative way of expressing this result is to state that a circuit C carrying current I placed in a magnetic field experiences a magnetic force given by

$$\mathbf{F}_{\rm m} = \int_C I\, d\mathbf{l} \times \mathbf{B}$$

When a loop of wire of area A, carrying current I and having a magnetic moment $\mathbf{m}$ such that $|\mathbf{m}| = IA$ is placed in a uniform magnetic field $\mathbf{B}$, it experiences a torque given by

$$\mathbf{T} = \mathbf{m} \times \mathbf{B}$$

6.12 PROBLEMS

6-1. Force between two infinitely long wires. A dc transmission line consists of two infinitely long, parallel wires separated by a distance of 50 cm. The two wires carry the same current I in opposite directions. Find I if the force per unit length experienced by each wire is 0.5 N-m^{-1}.

6-2. Forces between two wires. Consider a two-wire transmission line consisting of two parallel conductors of 1 mm diameter separated by 1 cm. If a potential difference of 10 V is applied between the conductors, equal charges of opposite sign would be induced on the two conductors, resulting in an electrostatic force of attraction between them. Is there a value of line current I for which this electric attraction force might be balanced by the magnetic force acting on the wires?

6-3. Bundle clash in a transmission line. Each phase of a three-phase alternating current transmission line consists of a bundle of two parallel wires each 4 cm in diameter and 50 cm apart, carrying current in the same direction. Under normal operation, the currents flowing in the wires are of the order of hundreds of amperes, and the magnetic force of attraction between the wires in the bundle is relatively small. An out-of-control airplane accidentally strikes the power transmission line, causing short-circuit loading, as a result of which the peak current in each wire in the bundle reaches a value of 100 kA. Calculate the peak magnetic force of attraction per unit length on each wire and comment on the possibility of a bundle clash. Note that similar short-circuit loading of power lines can also occur due, for example, to a tree falling on the line or a snake climbing up a power-line pole.

6-4. Magnetic force between current elements. Two elements of current are oriented in air in the same plane and positioned with respect to each other as illustrated in Figure 6.52a. (a) Find the magnitude and direction of the force exerted upon each element by the other. (b) Repeat (a) for the configuration of the current elements shown in Figure 6.52b.

6-5. Square loop. Does a square loop carrying current have a tendency to expand or contract?

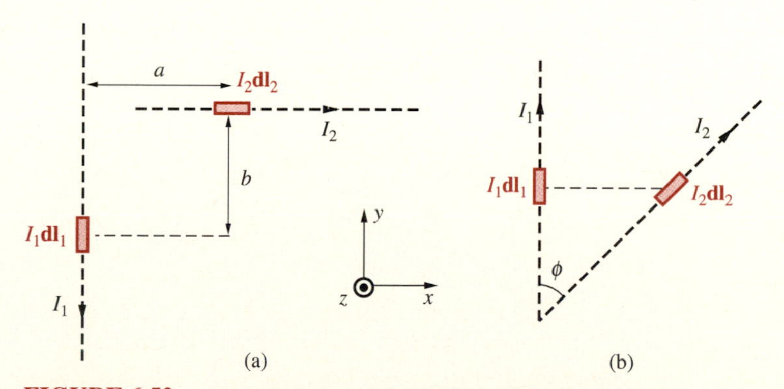

FIGURE 6.52. **Force between current elements.** Problem 6-4.

6-6. Force on a current-carrying wire. A single conductor of a transmission line extends in the east-west direction and carries a current of 1 kA. The earth's **B** field is directed essentially straight north at the point in question and has a value of ~0.5 T. What is the magnitude of the force per meter on the current-carrying conductor?

6-7. Rigid rectangular loop. A rigid rectangular loop carrying a current of 5 A is located in the xy plane with its four corners at (0,0), (1,0), (1,2), and (0,2). Determine the magnetic force exerted on each side of the loop if the region in which the loop is located is permeated with a **B** field given by (a) $\mathbf{B} = \hat{\mathbf{x}}1.5$ T, (b) $\mathbf{B} = \hat{\mathbf{z}}1.5$ T.

6-8. Two parallel wires. Consider two infinitely long parallel wires, each carrying a steady current of I in the z direction, one passing through the point (2,0,0) and the other through (0,2,0), as shown in Figure 6.53. (a) Find **B** at the origin. (b) Find **B** at point (1,1,0). (c) Find **B** at (2,2,0). (d) Repeat parts (a), (b), and (c) if the direction of the current in the wire on the x axis is switched.

6-9. An infinitely long L-shaped wire. Consider a single wire extending from infinity to the origin along the y axis and back to infinity along the x axis and carrying a current I. Find **B** at the following points: (a) $(-a,0,0)$; (b) $(0,-a,0)$; and (c) $(0,0,a)$.

6-10. Irregular loop. Find **B** at point P due to the wire carrying a current I in free space, as shown in Figure 6.54.

6-11. An N-sided regular loop. A regular polygon–shaped loop, with N sides, carries a current I. (a) Show that the magnetic field at the center of the loop has a magnitude given by $B = [\mu_0 NI/(\pi d)]\tan(\pi/N)$, where d is the diameter of the circle passing through the corners of the polygon. (b) Also show that as $N \to \infty$, B approaches that found in Example 6-6.

6-12. Square loop of current. A current I flows in a square loop of side length a as shown in Figure 6.55. Find the **B** field at the point P(a/4, a/4).

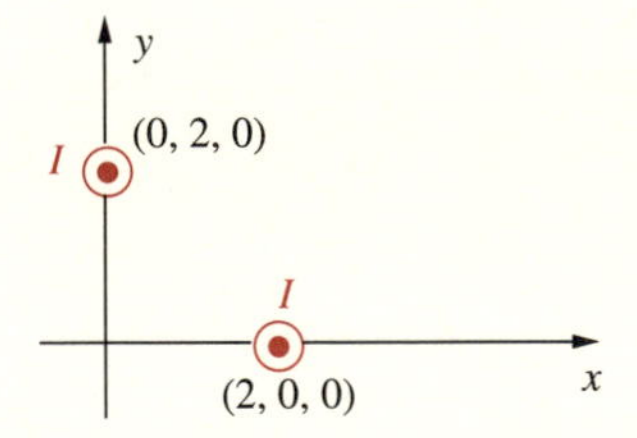

FIGURE 6.53. **Two parallel wires.** Problem 6-8.

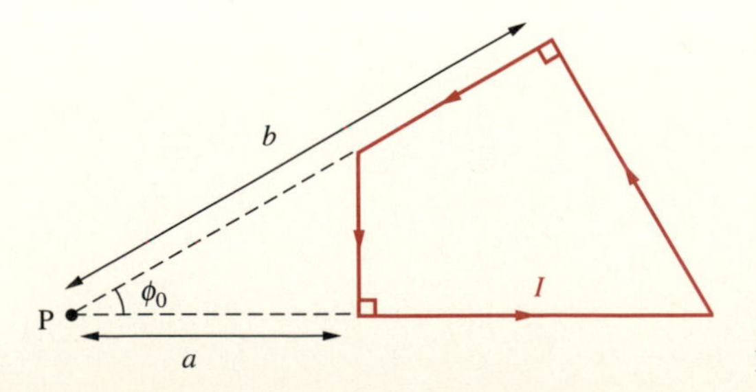

FIGURE 6.54. **Irregular loop.** Problem 6-10.

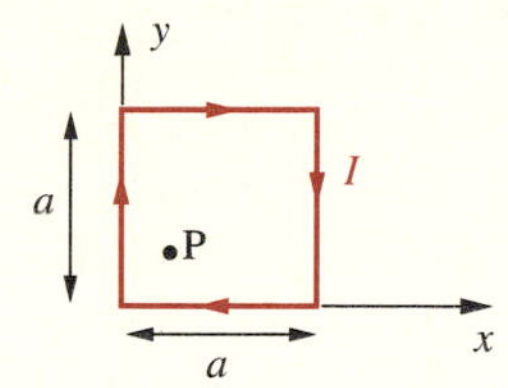

FIGURE 6.55. **Square loop of current.** Problem 6-12.

6-13. A wire with two circular arcs. Consider a loop of wire consisting of two circular and two straight segments carrying a current I, as shown in Figure 6.56. Find **B** at the center P of the circular arcs.

6-14. Wire with four 90° bends. An infinitely long wire carrying current $I = 1$ A has four sharp 90° bends 1 m apart as shown in Figure 6.57. Find the numerical value (in Wb-m^{-2}) and direction (i.e., the vector expression) of the **B** field at point P(1, 0, 0).

6-15. Infinitely long copper cylindrical wire carrying uniform current. An infinitely long solid copper cylindrical wire 2 cm in diameter carries a total current of 1000 A, distributed uniformly throughout its cross section. Find **B** at points inside and outside the wire.

6-16. Helmholtz coils. Two thin circular coaxial coils each of radius a, having N turns, carrying current I, and separated by a distance d, as shown in Figure 6.58, are referred to as *Helmholtz coils* for the case when $d = a$. This setup is well known for producing an approximately uniform magnetic field in the vicinity of its center of symmetry. (a) Find **B** on the axis of symmetry—the z axis—of the Helmholtz coils.

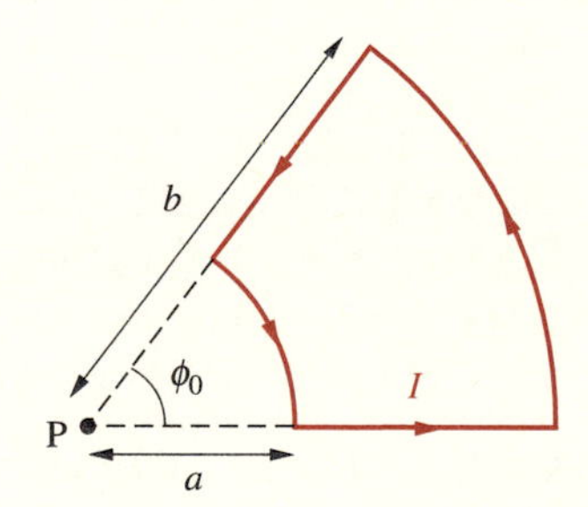

FIGURE 6.56. **Wire with two circular arcs.** Problem 6-13.

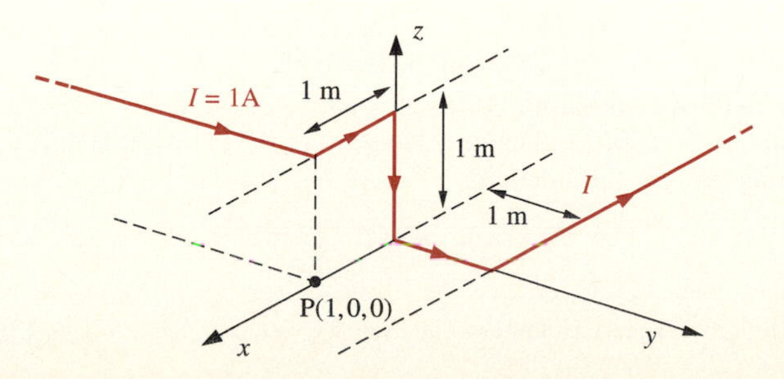

FIGURE 6.57. **Wire with four 90° bends.** Problem 6-14.

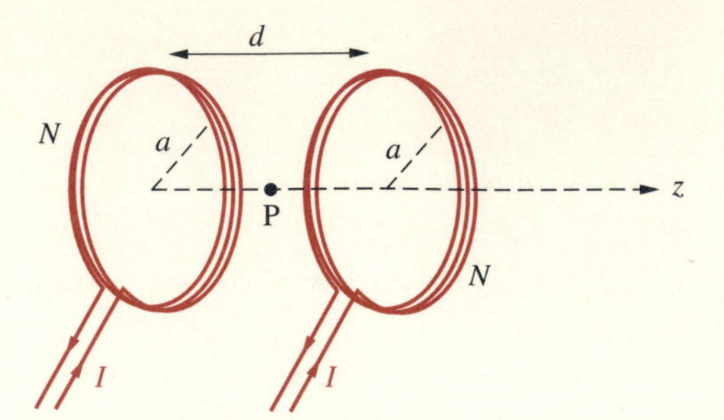

FIGURE 6.58. **Helmholtz coils.** Problem 6-16.

(b) Show that $dB_z/dz = 0$ at the point P midway between the two coils. (c) Show that both $d^2B_z/dz^2 = 0$ and $d^3B_z/dz^3 = 0$ at the midway when $d = a$. (Note that d is called the Helmholtz spacing, which corresponds to the coil separation for which the second derivative of B_z vanishes at the center.) (d) Show that B_z at the midpoint P between the Helmholtz coils is

$$B_z \simeq 0.8992 \times 10^{-6}\frac{NI}{a} \quad \text{T}$$

(e) Find B_z at the center of each loop and compare it with the value at the midpoint between the coils.

6-17. Helmholtz coils. Design a pair of Helmholtz coils, separated by 0.5 m, to produce a magnetic field of 100 μT midway between the two coils. Take $I = 1$ A.

6-18. Helmholtz coils. Consider the Helmholtz coils discussed in Problem 6-16. Calculate and sketch B_z along the axis of symmetry between the two coils if one of the coils is connected backward (i.e., its current is switched).

6-19. Square Helmholtz coils. Consider a pair of square coils, similar to Helmholtz coils, each of same side a, having N turns, and carrying current I, separated by a distance d. Show that the Helmholtz spacing d (i.e., the coil separation for which $d^2B_z/dz^2 = 0$ at the center) is equal[80] to $d \simeq 0.5445a$. (Note that the Helmholtz spacing for circular coils is $d = a$, where a is the radius of each coil.) To simplify your solution, take each side to be of unity length and use a simple iterative procedure.

6-20. A circular and a square coil. Consider a pair of coils, similar to Helmholtz coils, separated by a distance d, where one of the coils is circular in shape with radius a and the other square with side $2a$, each having N turns and carrying current I, and $d = a$. For $a = 25$ cm, $I = 1$ A, and $N = 20$, calculate and plot the magnitude of the **B** field along the axis of symmetry between the two coils.

6-21. Magnetic field of a solenoid. For an air-core solenoid of $N = 350$ turns, length $l = 40$ cm, radius $a = 2$ cm, and having a current of 1 A, find the **B** field (a) at the center; and (b) at the ends of the solenoid.

---

[80] A. H. Firester, Designs of square Helmholtz coil systems, *Rev. Sci. Instr.*, 37, pp. 1264–1265, 1966.

6-22. Magnetic field of a surface current distribution. A circular disk of radius a centered at the origin with its axis along the z axis carries a surface current flowing in a circular direction around its axis given by

$$\mathbf{J}_s = \hat{\boldsymbol{\phi}} Kr \text{ A-m}^{-2}$$

where K is a constant. Find $\mathbf{B}$ at a point P on the z axis.

6-23. B inside the solenoid. An air-core solenoid of 750 turns, 20 cm length, and 5 cm^2 cross-sectional area is carrying a current of 1 A. (a) Find $\mathbf{B}$ along the axis of the solenoid. (b) Find $\mathbf{B}_{\text{ctr}}$ at the center of the solenoid. (c) Sketch $|\mathbf{B}|$ along the axis of the solenoid.

6-24. B inside the solenoid. A long solenoid consists of a tightly wound coil around a magnetic core ($\mu_r = 300$) carrying a current of 10 A. (a) If the $\mathbf{B}$ field magnitude at the center of the core is $B_{\text{ctr}} \simeq 1.5$ T, find the number of turns of wire wound around the core per centimeter. (b) Repeat part (a) if this was an air-core solenoid.

6-25. Split washer. The thin flat washer shown in Figure 6.59 has an inner radius of $a =$ 10 mm and an outer radius of $b = 50$ mm. A uniform voltage V_0 is applied between the edges of the radial slot, resulting in an angular current of $I = 100$ A. Find the $\mathbf{B}$ field at the center of the washer. Note that the current density is a function of the radius from the center. Neglect the width of the slot. Note that the application of the voltage between the edges is as shown in Figure 5.6c.

6-26. Long wire encircled by small loop. A long straight wire oriented along the z axis carries a current of 1 A. A loop of 5 cm radius carrying a current of 100 A is located in the xy (i.e., $z = 0$) plane, with its center at the origin, concentric with the wire, so that the axis of the loop coincides with the z axis. Find the $\mathbf{B}$ field at a point on the xy plane at a distance of 1 m from the origin.

6-27. Flux through a rectangular loop. A long, straight wire carrying a current I and a rectangular loop of wire are separated by a distance a as shown in Figure 6.60. (a) If the sides of the rectangular loop parallel and perpendicular to the straight wire are a and $2a$, respectively, find the magnetic flux Ψ that links the rectangular loop due to the straight current-carrying wire. (b) The rectangular loop is rotated by 90° around its perpendicular symmetry axis. Find the percentage change in Ψ linking the loop.

6-28. Flux through a triangular loop. A long, straight wire carrying a current I and a triangular loop of wire are as shown in Figure 6.61. (a) Find the magnetic flux Ψ that links the triangular loop in terms of a, b, and ϕ_0. (b) Find Ψ if $I = 100$ A, $b = 4a = 40$ cm, and $\phi_0 = 45°$, respectively.

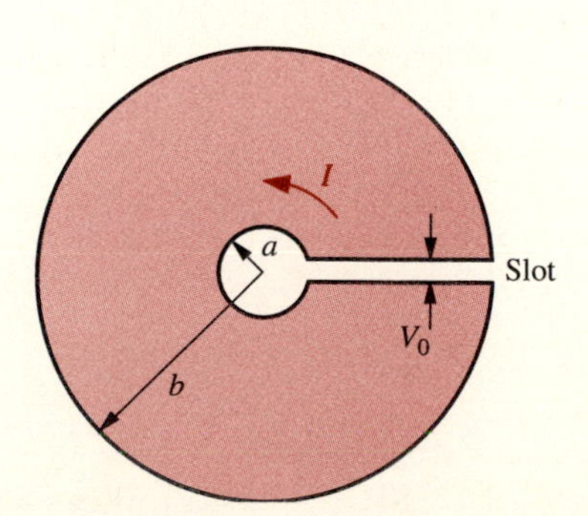

FIGURE 6.59. Split washer. Problem 6-25.

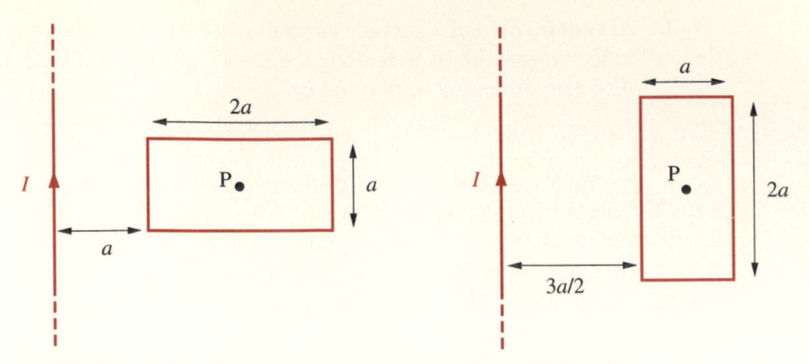

FIGURE 6.60. **Flux through rectangular loop.** Problem 6-27.

FIGURE 6.61. **Flux through a triangular loop.** Problem 6-28.

6-29. A toroidal coil around a long, straight wire. A long, straight wire carrying a current of 100 A coincides with the principal axis of symmetry of a 200-turn rectangular toroid of inner and outer radii $a = 4$ cm and $b = 6$ cm, thickness $t = 3$ cm, core material with $\mu_r = 250$, respectively. No current flows in the toroid. Find the total magnetic flux Ψ linking the toroid due to the current in the long, straight wire.

6-30. Infinitely long wire with a cylindrical hole. Consider an infinitely long cylindrical conductor wire of radius b, the cross section of which is as shown in Figure 6.62. The wire contains an infinitely long cylindrical hole of radius a parallel to the axis of the conductor. The axes of the two cylinders are apart by a distance d such that $d + a < b$. If the wire carries a total current I, (a) find the vector magnetic potential **A** and (b) use **A** to find the **B** field in the hole. Compare your result with that found in Example 6-13.

6-31. Finite-length straight wire: A at an off-axis point. Consider the straight, current-carrying filamentary conductor of length $2a$, as shown in Figure 6.11. (a) Find the magnetic vector potential **A** at an arbitrary point P(r, ϕ, z). (b) Find the **B** field using $\mathbf{B} = \nabla \times \mathbf{A}$ to verify the result of Example 6-7.

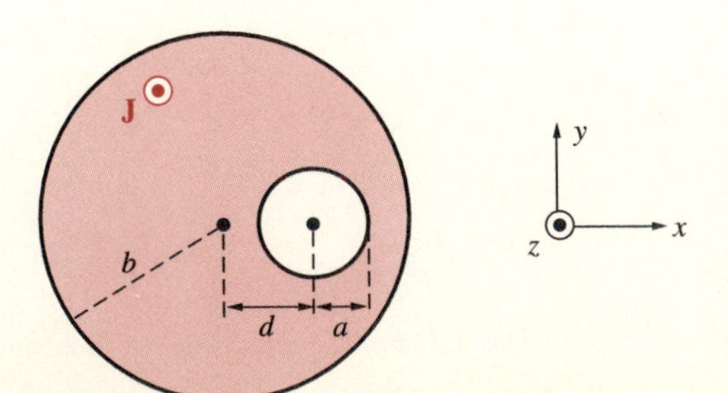

FIGURE 6.62. **Wire with hole.** Problem 6-30.

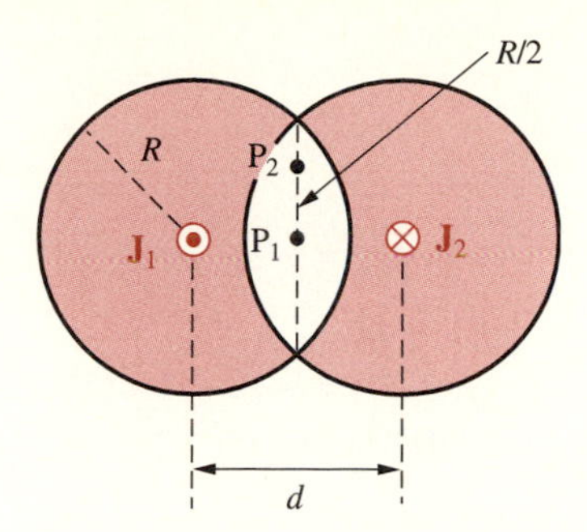

FIGURE 6.63. **Wire with oval-shaped hole.** Problem 6-33.

6-32. Two infinitely long wires. Consider two infinitely long parallel wires oriented in the z direction each carrying a current I in opposite directions respectively. Show that the magnetic vector potential $\mathbf{A}$ is give by $\mathbf{A} = \hat{\mathbf{z}}[\mu_0 I/(2\pi)]\ln(b/a)$, where a and b are the distances from the observation point to the wires.

6-33. Wire with oval-shaped hole. Consider a wire with the cross-sectional shape shown in Figure 6.63, having an oval-shaped axial hole. The wire segments on both sides of the hole carry uniform current densities $\mathbf{J}_1$ and $\mathbf{J}_2$ of equal magnitude J_0 and opposite sign. Find the magnetic field $\mathbf{B}$ at points P_1 and P_2 in the hole.

6-34. Inductance of a short solenoid. Find the inductance of a short air-core solenoid of 500 turns, 10 cm length, and 5.5 cm diameter. Use the Nagaoka formula given in Section 6.7.3.

6-35. Inductance of a finite-length solenoid. The secondary coil of a high-voltage electric generator is an air-core solenoid designed with the following parameters: $l = 1.8$ m, $a = 17.7$ cm, and $N = 780$ turns. Calculate the self-inductance of the coil by treating it (a) as an infinitely long solenoid, (b) as a finite-length solenoid. Compare your results.

6-36. Inductance of a rectangular toroid. Consider the toroidal coil of rectangular cross section shown in Figure 6.19. (a) Show that the inductance of this coil on an air core is in general given by

$$L = 2 \times 10^{-5} t N^2 \ln\left(\frac{b}{a}\right) \quad \text{H}$$

where N is the number of turns and t is the vertical thickness of the core in cm. (Do not assume $r_{\text{m}} \gg b - a$.) (b) If the dimensions of this toroid are $a = 1.2$ cm, $b = 2$ cm, and $t = 1.5$ cm, and $N = 1000$, find the inductance using the expression in part (a). (c) Find the inductance using the approximate expression derived in Example 6-27, assuming $r_{\text{m}} \gg b - a$ and compare it with the result of part (b).

6-37. Inductance of a rectangular toroid. Consider an air-core rectangular toroid as shown in Figure 6.19 with the values of its dimensions given by $a = 3.8$ mm, $b = 6.4$ mm, and $t = 4.8$ mm, respectively. (a) Find the total number of turns to be wound on this core such that its total inductance is around 1 mH. (b) Repeat part (a) for a powdered nickel-iron core (assume $\mu_r = 200$) having the same geometric dimensions.

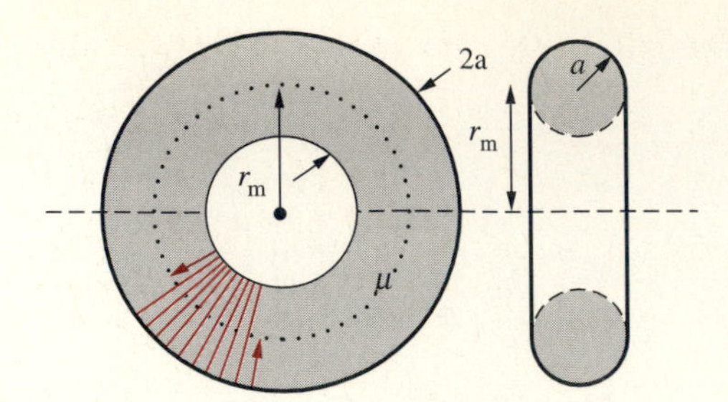

FIGURE 6.64. Inductance of a circular toroid. Problem 6-39.

6-38. Inductance of a rectangular toroid. A 50-mH toroid inductor is to be designed using a molypermalloy powder core with $\mu_r = 125$, $a = 7.37$ mm, and $b = 13.5$ mm, and $t = 11.2$ mm. Find the approximate number of turns N required.

6-39. Inductance of a circular toroid. Consider a toroid of circular cross section of radius a and mean radius r_m as shown in Figure 6.64. Show that the inductance of this coil is given by

$$L = \mu_r \mu_0 N^2 [r_m - (r_m^2 - a^2)^{1/2}]$$

where μ_r is the relative permeability of the core material.

6-40. Inductance of a two-wire line. Determine the inductance per unit length of a two-wire transmission line in air as shown in Figure 6.38, designed for an amateur radio transmitter, with conductor radius $a = 1$ mm and spacing $d = 6$ cm. (b) Repeat part (a) if the conductor spacing is doubled (i.e., $d = 12$ cm).

6-41. Inductance of a thin circular loop of wire. (a) Find the inductance of a single circular loop of wire with $d = 2$ mm and $a = 5$ cm. (b) Repeat part (a) with $d = 0.5$ mm and $a = 10$ cm. Refer to Figure 6.42.

6-42. Mutual inductance between two solenoidal coils. An air-core solenoid 25 cm long and 2.5 cm diameter is wound of 1000 turns of closely spaced, insulated wire. A separate coil of 100 turns, 3 cm long, with about the same diameter is located at the center of the longer coil (where **B** is approximately constant). Find the mutual inductance between the two coils.

6-43. Mutual inductance between a wire and a circular loop. Find the mutual inductance between an infinitely long, straight wire and a circular wire loop, as shown in Figure 6.65.

6-44. Long wire and loop. An infinitely long wire carrying current I passes just under a circular loop of radius a, also carrying the same current I as shown in Figure 6.66. (a) Find

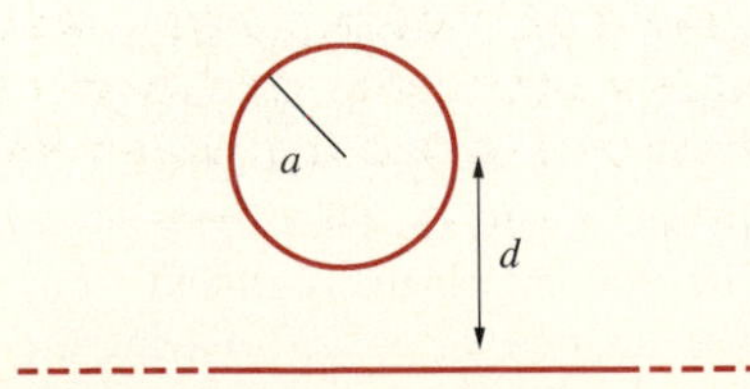

FIGURE 6.65. Wire and circular loop. Problem 6-43.

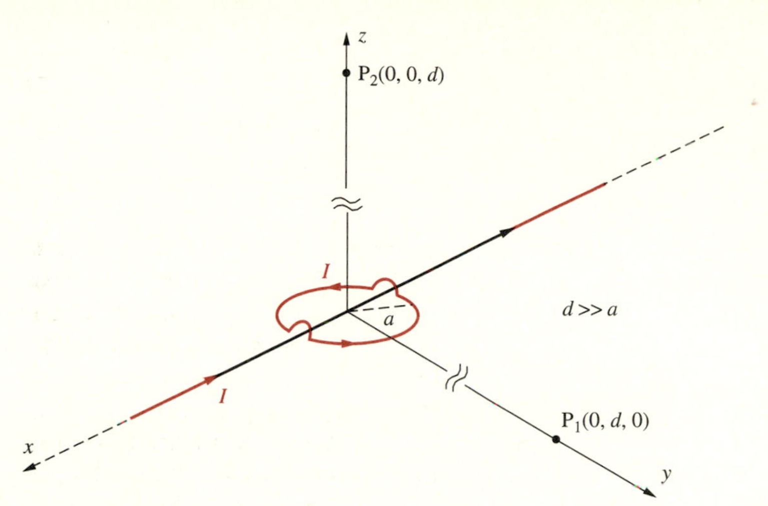

FIGURE 6.66. **Long wire and loop.** Problem 6-44.

the magnitude and direction of the **B** field at point P_1. Note that $d \gg a$. (b) Repeat (a) for point P_2. (c) What is the mutual inductance between the wire and the loop?

6-45. Inductance and energy of a solenoid. An air-core solenoid 30 cm long and 1 cm diameter is wound of 1000 turns of closely spaced, insulated wire. (a) Find the inductance of the solenoid. (b) Find the energy stored in the solenoid if it is carrying a current of 10 A.

6-46. Two square coils. Two square coils each of side length a are both located on the xy plane such that the distance between their centers is d. (a) Find a simple expression for the mutual inductance between the two loops for the case $d \gg a$. (b) Find the mutual inductance between the two coils for the case $d = 4a$. (c) For $d = 4a$, assume that coil #1 has 10 turns and that coil #2 has 100 turns. If an alternating current of 1 A of frequency 10 kHz is passed through coil #1, find the electromotive force induced across the terminals of coil #2.

6-47. Fixed-length piece of wire. You have a fixed length of copper wire of 1 mm diameter and 20 cm length. You can bend this wire in any shape or form to obtain as large an external self-inductance as you can. (a) What is the value of the maximum inductance, and what arrangement would work best? You cannot use any magnetic materials. (b) If you had the tools necessary to melt the copper wire and form it into a longer wire of diameter 0.1 mm instead, how would your answer to (a) change? Also compare and comment on the resistance of the two different wires.

6-48. Magnetic energy. A conductor consists of a cylinder with radius b with a hole of radius a ($a < b$) drilled coaxially through its axis. The current density is uniform and corresponds to a total current of I. Find the magnetic energy stored inside per unit length of the wire.

6-49. Explosion in a power transformer. A current transformer used for 500 kV transmission lines has a single primary coil which is connected to the high voltage line via two parallel wires that are 20 cm apart and carrying the same current in opposite

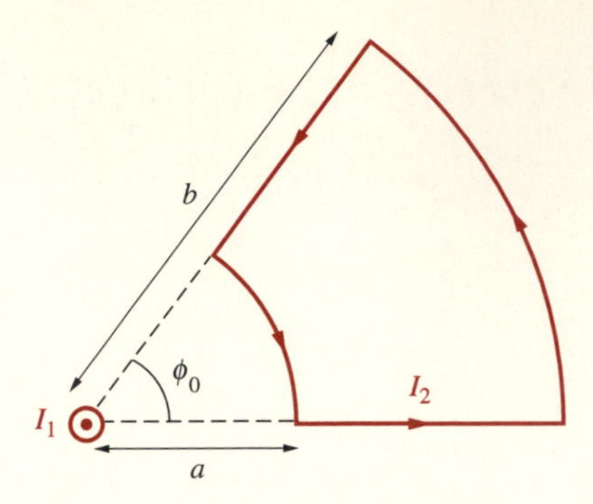

FIGURE 6.67. Force between a long wire and a loop. Problem 6-50.

directions. A fault occurs on the line and causes the current of each wire of the transformer to reach a peak value of 100 kA which results in an explosion in the oil/paper insulation structure of the wires. Calculate the per-unit-force on each wire and explain what happened.

6-50. Force between a long wire and a loop. A long wire extending along the z axis is situated near a rigid loop as shown in Figure 6.67. Find an expression for the force and torque (about the origin) experienced by the loop.

CHAPTER 7

Time-Varying Fields and Maxwell's Equations

Oersted's accidental discovery in 1820 of the close connection between electricity and magnetism led to almost immediate advances in technology. The fact that wires carrying current experience forces and torques when in the presence of a magnetic field was put to good use in direct-current electric motors (Section 6.10) and in the design of galvanometers, sensitive instruments for measuring electrical current. These instruments were crucial in facilitating scientific experimentation. However, a much broader range of industrial and technological applications of electricity and magnetism was yet to come, with Faraday's discovery in 1831 of electromagnetic induction—the fact that magnetic fields that change with time produce electric fields. This experimentally based fact constitutes the last of our three experimental pillars of electromagnetics[1] and is the subject of this chapter.

Electrostatics (Chapter 4) deals with the effects of stationary charges, the spatial distributions of which are determined by the presence of conducting bodies and applied potentials. The governing laws of electrostatics, derived from Coulomb's force law, can be summarized as follows:

$$\nabla \times \mathbf{E} = 0 \qquad \oint_C \mathbf{E} \cdot d\mathbf{l} = 0$$

$$\nabla \cdot \mathbf{D} = \rho \qquad \oint_S \mathbf{D} \cdot d\mathbf{s} = \int_V \rho \, dv$$

[1]The first pillar of electromagnetics is the fact that electric charges attract or repel one another in a manner inversely proportional to the distance between them (Coulomb's law). The second pillar is the fact that current-carrying wires create magnetic fields and exert forces on one another (Ampère's law of force or alternatively the Biot–Savart law).

where the fields **E** and **D** are related via $\mathbf{D} = \epsilon\mathbf{E}$, with ϵ being a scalar constant for linear, isotropic, homogenous and time-invariant media.

Magnetostatics (Chapter 6) deals with the physical effects produced by charges in motion (i.e., steady currents). The governing laws of magnetostatics, derived from Ampère's force law (expressed also in the Biot–Savart Law), can be summarized as follows:

$$\nabla \times \mathbf{H} = \mathbf{J} \qquad \oint_C \mathbf{H} \cdot d\mathbf{l} = \int_S \mathbf{J} \cdot d\mathbf{s}$$

$$\nabla \cdot \mathbf{B} = 0 \qquad \oint_S \mathbf{B} \cdot d\mathbf{s} = 0$$

where the fields **B** and **H** are related via $\mathbf{H} = \mu^{-1}\mathbf{B}$ where μ is a constant scalar for linear, isotropic, homogeneous, and time-invariant media.

Up to now, we have exclusively considered static cases and paid little attention to electric and magnetic effects produced by charges, currents, or circuits whose positions or intensities vary with time.[2] Under the static assumption, the electric and magnetic fields are unrelated, and the field vectors **E**, **D** and **B**, **H** form independent pairs.[3] However, following Oersted's findings that electric currents produce magnetic fields, M. Faraday and many others started thinking about the possibility that magnetic fields may in turn produce electric fields, and that the two sets of preceding equations may be coupled.

After many years of various trials,[4] trials, Faraday carried out the classic experiment[5] of induction on August 29, 1831, and showed that **E** and **B** were indeed related, but only for the case in which the fields change in time, that is, under nonstatic conditions. At about the same time, J. Henry of Albany Academy in New York studied similar effects and arrived at similar conclusions. However, Faraday's results were published earlier and had the greatest impact in the scientific world.[6] As a result, the law of electromagnetic induction is commonly referred to as Faraday's law.

---

[2]Occasionally, it was necessary to apply Gauss's law to time-dependent phenomena with little or no justification, for example, in the derivation of the continuity equation (Section 5.5).

[3]In a conducting medium ($\sigma \neq 0$), a static electric field causes a steady current to flow ($\mathbf{J} = \sigma\mathbf{E}$), which in turn gives rise to a static magnetic field. However, at steady state, the electric field can be completely determined from the static charges or potential distributions. The magnetic field is simply a consequence of the current and does not affect the electric field.

[4]For a brief historical account, see Section 5.1, or see R. Elliott, *Electromagnetics,* IEEE Press, New Jersey, 1993. Another excellent review of the history and development of electromagnetics, with Chapter VI devoted to Faraday, is given by E. Whittaker, *History of the Theories of the Aether and Electricity,* T. Nelson and Sons, Ltd., London, 1951. It is interesting, for example, that a number of scientists, including Ampère, only narrowly missed discovering the law of induction.

[5]M. Faraday, *Experimental Researches in Electricity,* Taylor, London, vol. I., 1839, pp. 1–109.

[6]Henry's lack of promptness in announcing his results, combined with the fact that the New World was remote from European centers, caused his results to be viewed as merely confirmation of Faraday's findings.

Faraday's law of electromagnetic induction, namely that magnetic fields that change with time induce[7] electromotive force, is the third and final experimental fact[8] that forms the basis for Maxwell's equations and is arguably the most important of our three experimental pillars of electromagnetics. Without it, electromagnetic waves would not propagate through free space, and wireless communication would not be possible. Faraday's discoveries also form the basic underpinnings of modern electrical technology; without electromagnetic induction, the workhorses of our industrial world—electric motors and generators—would not be possible.

Our coverage of Faraday's law in this chapter completes our understanding of the underlying physical basis of the transmission line behavior discussed in Chapters 2 and 3. The fact that temporal variation of a voltage applied between the two conductors of a transmission line induces electromotive force, which in turn affects the current flow, is the very basis of the distributed inductance of the line. Completing our formulation of Maxwell's equations later in the chapter also allows us to exhibit the similarity between the fundamental equations that relate electric and magnetic fields for electromagnetic waves and the transmission line equations that relate line voltages and currents. Natural solutions of both the electric/magnetic field equations and the voltage/current equations are in the form of waves that propagate with a finite speed, leading to the distributed circuit effects studied in Chapters 2 and 3, and the propagation, reflection, and guiding of electromagnetic waves covered in Chapter 8.

As we discuss time-varying magnetic fields and Faraday's law in the first three sections of this chapter, we are concerned with variations that are slow enough so that radiation effects are negligible. This is the so-called *quasi-static approximation,* which implies that the system of conductors carrying currents has dimensions much smaller than a wavelength. A wide range of problems in laboratory physics and engineering lie within the domain of the quasi-static approximation. The criteria for the validity of the quasi-static approximation are the same as those discussed in

---

[7]The word "induce" is used quite literally in this context; to induce means to cause the formation of. The word "induction," on the other hand, has several meanings in nontechnical parlance, including the initiation of a person into military service. In our context, the word *induction* refers to the process by which the electromotive force is brought into existence by the time-varying magnetic field.

[8]We note once again that the complete set of Maxwell's equations can be derived via transformation of Coulomb's law, using special theory of relativity. This was first shown in 1912 [L. Page, A derivation of the fundamental relations of electrodynamics from those of electrostatics, *Am. J. Sci.,* 34, pp. 57–68, 1912]; a clear treatment is given in Chapter 5 of R. S. Elliott, *Electromagnetics,* IEEE Press, New Jersey, 1993. However, such derivations must necessarily make other implicit assumptions; see Section 12.2 of J. D. Jackson, *Classical Electrodynamics,* Wiley, 2nd ed., 1975. Even before the introduction of the special theory of relativity, and about ten years after Faraday's discovery, H. von Helmholtz and Lord Kelvin recognized and suggested [Helmholtz, *Über die Erhaltung der Kraft,* 1847; Kelvin, *Trans. Brit. Assoc.,* 1848 and *Phil. Mag.,* Dec. 1851] that it is possible to deduce the existence of induced currents from energy considerations. For a brief discussion, see Article 481 of G. H. Livens, *The Theory of Electricity,* Cambridge Univ. Press, 1918. Nevertheless, and in view of the profound importance of the law of electromagnetic induction in forming the basis of Maxwell's equations, in this text we consider Faraday's law to be based on experimental fact.

Chapter 1 in connection with determining the applicability of the lumped analysis and the necessity of a distributed circuit treatment. Basically, the quasi-static approximation restricts us to considering circuits of such sizes and rates of change of current that the electromagnetic disturbance propagates over much of the useful parts of the circuit before the current has changed significantly. Under such conditions, we implicitly assume that at any given instant, the magnetic fields everywhere in the circuit are strictly proportional to the current at that instant. In general, the quasi-static approximation amounts to calculating all fields as if they were stationary, without having to account for the travel time of fields from one point of the circuit to another. Formally, the quasi-static approximation is valid so far as $\partial\rho/\partial t$ is negligible compared to $\nabla\cdot\mathbf{J}$. When free-charge density varies rapidly enough that $\partial\rho/\partial t$ is comparable to $\nabla\cdot\mathbf{J}$, time variations of electric fields produce magnetic fields, leading to propagation of electromagnetic waves, as is discussed in Section 7.4.

We start this Chapter by introducing Faraday's law and electromagnetic induction in Section 7.1, followed by a separate discussion of induction due to motion in Section 7.2. Section 7.3 covers the topic of magnetic energy stored by configurations of current-carrying loops and magnetic energy expressed in terms of fields. The new concept of displacement current and the formulation of Maxwell's equations in their final form is discussed in Section 7.4. Finally, Section 7.5 provides an overview of Maxwell's equations and electromagnetic boundary conditions, the development of which has been undertaken in Chapters 4 through 7.

Faraday, Michael *(English physicist and chemist, b. Newington, Surrey, September 22, 1791; d. Hampton Court, Middlesex, now part of greater London, August 25, 1867) Faraday was one of ten children born to a blacksmith and had no official education beyond reading and writing; he was apprenticed to a bookbinder in 1805. Being exposed to books, he worked his way through the electrical articles in the* Encyclopaedia Britannica, *for instance, and read Lavoisier's [1743–1794] great textbook of chemistry.*

In 1812 a customer gave Faraday tickets to attend the lectures of Davy [1778–1829] at the Royal Institution. Faraday took careful notes, which he further elaborated with colored diagrams. He ended up with 386 pages, which he bound in leather and sent to Banks [1743–1820], president of the Royal Society, and to Davy along with an application for a job as his assistant. Davy was enormously impressed, but did not oblige the young man until he later fired his assistant for brawling. Faraday accepted the offer at a salary less than his bookbinder's salary and was first assigned to wash bottles. He virtually lived in and for the laboratory, then and later, never using a collaborator or assistant. He became director of the laboratory in 1825 and professor of chemistry at the Royal Institution in 1833. He concentrated on his lone researches, refusing an ample income for services as an expert witness in court and other higher-paying positions.

Following Oersted's [1777–1851] announcement in 1820 that electric currents produced magnetic fields, Faraday worked hard for almost a decade to prove the reverse—that an electric current could be produced by magnetic attraction until 1831, when he observed electromagnetic induction in an experiment (see Figure 7.1). To Faraday, this observation required explanation. Because he was uneducated, and completely innocent of mathematics, he used his intuitive ability to pictorialize, an ability perhaps unequaled in scientific history. He dropped iron filings on a paper under which a magnet was located and noticed the regular patterns they took up when the paper was tapped. (So had Peter Peregrinus [1240–?] six centuries before.) He began to visualize the magnetic force, then, as stretching out in all directions from the electric current that produced it and filling space as a magnetic field. The iron filings aligned themselves along lines—Faraday called them lines of force—drawn through that field, thus making them "visible." It was possible to work out the form of the lines of force for wires, even for giant magnets such as the earth. This was the beginning of a picture of the universe consisting of fields of various types, a universe more subtle, flexible, and useful than the purely mechanical picture Galileo [1564–1642] and Newton [1642–1727] had described.

Faraday's pictorial and nonmathematical imagination visualized these lines of force as real lines. When a circuit was closed and electricity was set to flowing, the lines sprang outward into space. When the circuit was broken they collapsed inward again. Faraday decided then that an electric current was induced in a wire only when lines of force cut across it. In his transformer (Figure 7.1), when the current started in the first coil of wire, the expanding lines of force cut across the wire of the second coil and accounted for the short burst of electric current. Once the original current was established, the lines of force no longer moved, and there was no current in the second coil. When the circuit was broken, the collapsing lines of force cut across the second coil in the opposite direction, and a burst of electric current resulted again, but in the direction opposite to that in the first coil.

At the time, Faraday was giving enormously popular lectures in science for the general public; he was such an excellent lecturer that the novelist Charles Dickens, no mean lecturer himself, was among Faraday's admirers, along with Prince Albert, the husband of Queen Victoria, and Prince Edward, her son (and later Edward VII). It was during one of these lectures that Faraday demonstrated the theory involving the lines of force by inserting a magnet into a coil of wire attached to a galvanometer (Figure 7.2).

Once Faraday had demonstrated that electricity could be induced by magnetism, the next step was to do so continuously, not in short spurts. Faraday turned a copper wheel in such a way that its edge passed between the poles of a permanent magnet. An electric current was set up in the copper disc, and it continued to flow as long as the wheel continued to turn—Faraday had thus invented the first electric generator in 1831.

Faraday was an extremely religious man and strongly favored a more important role for science in education. His beliefs enabled him to solve decisively a problem that agonizes many scientists of our day—the conflict between the demands of one's nation and personal ethics. During the Crimean War of the 1850s, Faraday was asked by the British government whether it would be possible to prepare quantities of poison gas for use on the battlefield and whether he would head a project to perform the task. Faraday answered at once and with finality that the project was certainly feasible, but that he himself would have absolutely nothing to do with it.

Faraday kept a meticulous day-by-day record of his forty-two years of scientific labors (1820–1862). This was published in 1932 in seven volumes. He requested that he be buried under "a gravestone of the most ordinary kind" and that only a few relatives and friends attend his funeral, and this was done. His true memorial, of course, is our electrified world of today. The unit of capacitance, the farad, is named after him. [Adapted with permission from Biographical Encyclopedia of Science and Technology, *I. Asimov, Doubleday, 1982].*

7.1 FARADAY'S LAW

Michael Faraday was completely innocent of all mathematics, but at the same time he was one of the most imaginative and visual scientific thinkers of all time. Since Oersted demonstrated that an electric current could deflect a compass needle (i.e., generate a magnetic field), Faraday was convinced that a magnetic field or a magnet should be capable of producing a current. In 1831, Faraday set up an apparatus[9] consisting of an iron ring in the shape of a toroid on which were wound two coils of wire as sketched in Figure 7.1. The primary coil was connected through a switch to a voltaic cell; the ends of the secondary coil were joined with a wire that ran above a compass. Any current induced in the secondary coil would thus deflect the compass

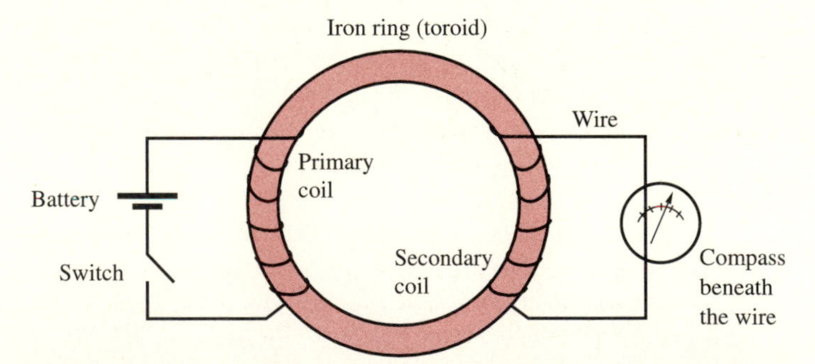

FIGURE 7.1. Faraday's iron-ring experiment. A sketch showing the elements of the apparatus used by Faraday that led to the discovery of the law of electromagnetic induction (also called Faraday's law). The iron ring provides a path for the magnetic flux coupling the two circuits.

[9] See H. Kondo, Michael Faraday, *Scientific American,* 10, pp. 91–98, 1953 and G. L. Verschuur, *Hidden Attraction, The Mystery and History of Magnetism,* Oxford University Press, New York, Chapter 6, 1993.

needle to the left or right of its normal position, depending on its flow direction. Upon closing the switch, current flowed in the primary coil, producing a magnetic field in the iron ring that passed through the secondary coil. Faraday observed a momentary deflection of the compass needle detecting a brief surge of current induced in the secondary coil. However, the compass needle quickly settled back to zero, indicating that the induced current existed only during the initial transient. In other words, no current flowed in the secondary coil once the current in the primary coil became steady and the field in the toroid reached its final value. When the switch was opened, terminating the current in the primary coil and thus eliminating the magnetic field in the toroid, another momentary deflection was observed, opposite in polarity with respect to the one observed when the switch was closed. Based on these extraordinary observations, Faraday deduced that a magnetic field could be generated by a steady current, but that a current could be induced only by a *changing* magnetic field.[10] This meant that the mathematical description of this phenomenon, which came to be known as electromagnetic induction, or simply induction, had to depend on the time variation of the magnetic flux.

Later, Faraday investigated the possibility of generating a steady current in the secondary coil and realized that a *continuously moving* magnetic field would be necessary. For this purpose he made a copper disk rotate with its edge between the poles of a magnet and found that current flowed from the center toward the edge of the disk (or vice versa). This was the world's first dynamo, or electric generator. Faraday also used the motion of a magnet in and out of a helical coil to create a current, as sketched in Figure 7.2. He also determined that a more efficient way to produce currents was to move coils of wire in a magnetic field, the principle on which electric generators are built.

These experiments enabled Faraday to formulate his famous law of electromagnetic induction, or *Faraday's law,* which states that when the magnetic flux enclosed by a loop of wire changes with time, a current is produced in the loop, indicating that a voltage, or an *electromotive force* (emf), is induced. The variation of the magnetic flux can result from a time-varying magnetic field linking a stationary and fixed loop of wire or from a loop of wire moving in a static magnetic field or from both, i.e., a loop of wire moving in a time-varying magnetic field.

---

[10]The transitory nature of the electromotive force produced, due to its dependence on the *change* in magnetic flux, contributed to the elusive nature of the effect. An outstanding example of this was the experiments of Jean Daniel Colladon in Geneva in 1825. Colladon made a large helix of insulated copper wire 8–10 cm long and 4–5 cm in diameter, and he connected the ends to a sensitive galvanometer. He then brought a permanent magnet up to one end of the helix, expecting to produce a permanent current flow to be registered by the galvanometer. However, being the careful scientist he was, he wanted to place the galvanometer far from the magnet so that it would not be affected by the magnetic field of the magnet itself. His solution was to connect the galvanometer to the helix with 50 m of wire and to place it under a glass jar in another room well away from the magnet. There is little doubt that Colladon's galvanometer registered the temporary current caused by his movement of the magnet; however, by the time he walked to the other room, the galvanometer needle had returned to its zero reading and Colladon missed his chance of discovering electromagnetic induction. [From W. A. Atherton, The history of electromagnetic induction, *Am. J. Phys.,* 48(9), pp. 781–782, September 1980.]

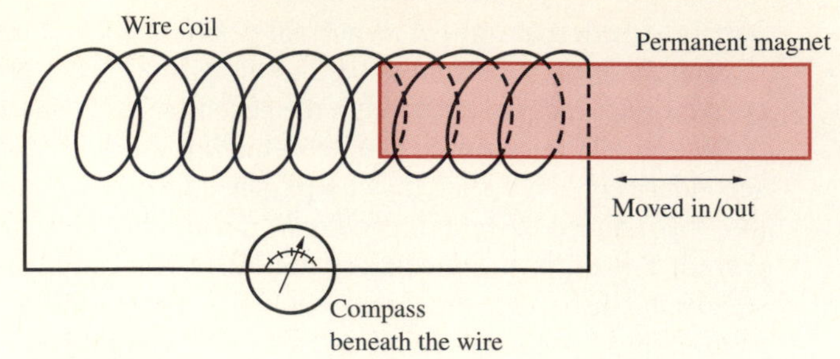

FIGURE 7.2. A magnet moving in a coil. A sketch of one of Faraday's experiments used to induce a current in a helical coil of wire by moving a magnet into and out of the coil—the principle of the electric generator. The induced current is detected by the deflection of the compass needle located under the wire combining the ends of the coil.

The mathematical form of Faraday's law is

$$\mathcal{V}_{\text{ind}} = -\frac{d\Psi}{dt} \qquad [7.1]$$

where $\mathcal{V}_{\text{ind}}$ is the induced voltage across the terminals of the loop C, and Ψ is the total magnetic flux linking[11] the closed loop C, as shown in Figure 7.3. Faraday's law states that the induced voltage $\mathcal{V}_{\text{ind}}$ around a closed loop C is equal to the negative of the time rate of change of the magnetic flux linking C. Note that Ψ is the total magnetic flux linking the contour C, given by

$$\Psi = \int_S \mathbf{B} \cdot d\mathbf{s}$$

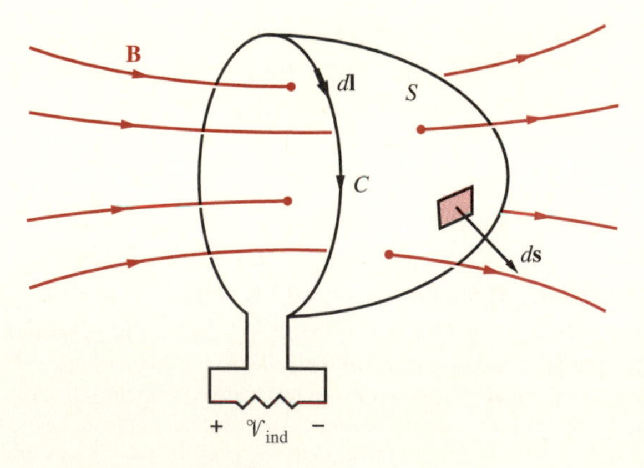

FIGURE 7.3. Illustration of Faraday's law. Voltage $\mathcal{V}_{\text{ind}}$ is induced between the terminals of loop C due to a time-varying magnetic field **B** that links the area S enclosed by C, where the direction of the differential length $d\mathbf{l}$ on the contour and the direction of the differential area $d\mathbf{s}$ on the surface S are related with the right-hand rule.

[11] With loops of N turns, where each turn links the same amount of flux, we can also write $\Lambda = N\Psi$, where Λ is the total flux linkage of N turns and Ψ is the flux linked by each turn. We then have $\mathcal{V}_{\text{ind}} = -N\,d\Psi/dt = -d\Lambda/dt$.

where S is the area of the surface enclosed by contour C, as shown in Figure 7.3. Having an induced voltage $\mathcal{V}_{\text{ind}}$ appear across the terminals of the loop means that current will flow if a resistance is connected across the terminals. In this sense, this induced voltage is similar to the electromotive force as defined and discussed in Chapter 5. The electromotive force is the agent that "pushes" the electrons that constitute the current. Faraday thus discovered that changing magnetic fields produce an electromotive force or emf, which acts just like the emf generated in other voltage sources, such as a chemical battery, a piezoelectric crystal (emf produced in response to mechanical pressure), or a thermocouple (emf resulting from a temperature gradient). The term "electromotive force" is somewhat misleading because an emf is actually not a force, but rather a line integral of a force per unit charge (i.e., an electric field). More precisely, *emf is defined as the tangential force per unit charge along the wire integrated over its length, around the complete circuit.*

Having a nonzero emf across the terminals of the loop in turn means that the line integral of the electric field around the loop is not zero, since the force that moves the current carriers is ultimately due to an electric field. Note that this electric field is certainly not an electrostatic field, since otherwise its integral around the closed circuit must be zero. Nor can the source of the force be a magnetostatic field, since the stationary charges cannot experience magnetic forces. The field that provides the force to drive the current is an entirely new kind of electric field produced by the new physical effect embodied in Faraday's law.[12] In words, Faraday's law states that *a changing magnetic field induces an electric field.* The induced voltage $\mathcal{V}_{\text{ind}}$ can thus be expressed as

$$\mathcal{V}_{\text{ind}} = \oint_C \mathbf{E} \cdot d\mathbf{l}$$

where $\mathbf{E}$ is the *induced* electric field. In other words, the mathematical statement of Faraday's law is

$$\boxed{\oint_C \mathbf{E} \cdot d\mathbf{l} = -\frac{d}{dt}\int_S \mathbf{B} \cdot d\mathbf{s}} \qquad [7.2]$$

where S is the surface enclosed by the contour C. According to [7.2], an electric field induced by a changing magnetic field exists in space regardless of whether or not there are conducting wires present. If conducting wires are present, an *induced* current flows through these wires. Faraday's law is the basic principle on which electric generators operate; mechanical energy is supplied to change the magnetic flux Ψ that links the coil C (for example by rotating coil C and thus the surface S) and thus to produce induced voltage across its terminals through [7.2]. Faraday himself invented the first direct-current generator in 1831 and the prototype of the modern electric generator in 1851.

[12]This new kind of physical force that drives electric current is also called an electric field to recognize the fact that it moves charges in the same manner as an electrostatic field. Keep in mind, however, that this new field is quite different from an electrostatic field; whereas the line integral of an electrostatic field around a closed path is identically zero, that of an induced field is obviously not zero.

The minus sign in [7.2] indicates that the induced emf is in a direction that opposes the change in the magnetic flux that caused the emf in the first place. In other words, the induced voltage leads to current flow[13] in a direction that produces an opposing magnetic flux Ψ. This statement of experimental fact[14] is known as *Lenz's law.* In terms of the correct polarities of the two integrals in [7.2], it is important to note that the orientation of the two integrals cannot be independently chosen. In other words, if $d\mathbf{l}$ is chosen such that one finds the first integral by going counterclockwise around the contour C, then the direction of $d\mathbf{s}$ in the second integral must be outward from the page assuming that the area S enclosed by loop C lies in the same plane as the loop. The right-hand rule applies here; with the fingers of the right hand pointing in the direction of $d\mathbf{l}$ along C, the thumb points in the direction of $d\mathbf{s}$.

The relationships between the polarity of the induced voltage $\mathcal{V}_{\text{ind}}$, the relative orientations of the induced electric field $\mathbf{E}$, and the other quantities (e.g., $d\mathbf{l}$) in [7.2]

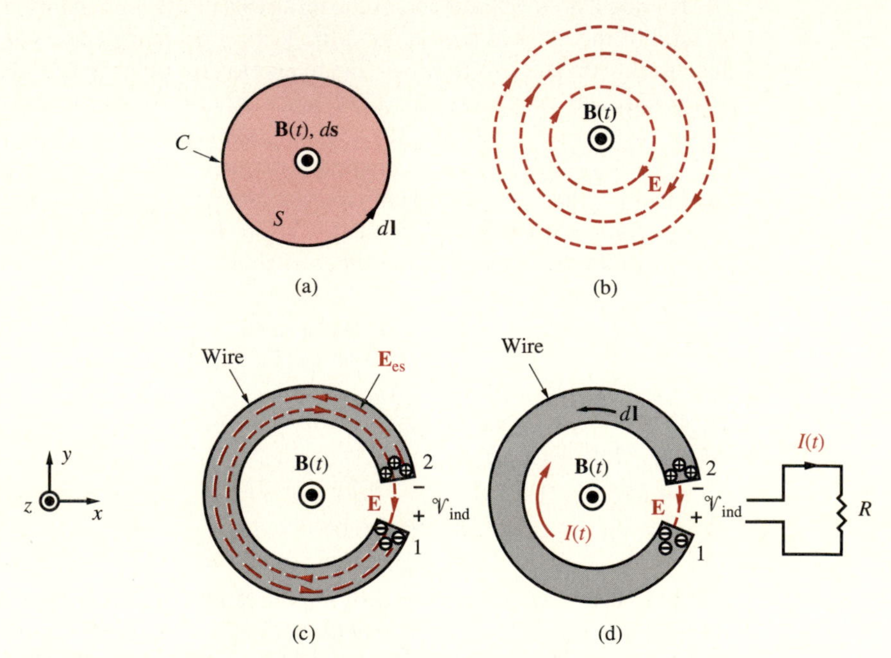

FIGURE 7.4. The polarity of the induced voltage and the electric field. The case illustrated is for $\partial B/\partial t > 0$.

[13]In typical circuits, which have nonzero resistance, the induced currents may be tiny fractions of the original currents that produce the magnetic fluxes linking the circuit. In other words, the magnetic fields produced by the induced currents are usually negligible, so that although the circuit opposes the changing flux by establishing a current flow, the resulting currents typically fall far short of fully compensating for the flux change.

[14]We note here that although Lenz's law is taken to be an experimental fact, it is in fact a consequence of the conservation of energy. In other words, if the polarity of the induced current were opposite to that dictated in Lenz's law, magnetic energy would not be conserved, as is discussed later in Section 7.3. We expect this result on a purely heuristic basis; if the polarity of the induced currents were such as to enhance the effect that produced the currents, the magnetic flux would increase without limit.

are depicted in Figure 7.4. We take the contour C to be enclosing the surface S, as shown in Figure 7.4a, and we take the orientation of the surface element $d\mathbf{s}$ to be in the z direction (out of the page), in which case $d\mathbf{l}$ must be counterclockwise, as shown in Figure 7.4a. For the purposes of this discussion, we take the magnetic field $\mathbf{B}$ to be pointing out of the page, uniformly present everywhere, and steadily increasing in time (i.e., $\partial B/\partial t > 0$). The orientation of the induced electric field $\mathbf{E}$ is shown in Figure 7.4b. Note that this field is induced regardless of whether or not conducting wires are present. Since $\partial B/\partial t > 0$ and $\mathbf{B}$ and $d\mathbf{s}$ are both out of the page, the right-hand side of [7.2] is negative. The fact that $d\mathbf{s}$ is out of the page requires that $d\mathbf{l}$ must be in the counterclockwise direction, so the induced electric field $\mathbf{E}$ must encircle the $\mathbf{B}$ field in the clockwise direction in order for the left-hand side of [7.2] to also be negative.

Now consider what happens when a conducting wire loop is present, with a small gap across which the induced voltage appears as shown in Figure 7.4c. Note that the thickness of the wire is exaggerated for convenience. At the instant this wire loop is introduced into the system, the induced electric field causes the free electrons inside the conductor to move to one end and leave the other end positively charged, as shown in Figure 7.4c. Assuming that the wire is a good conductor such as copper, this rearrangement of charge within the wire occurs in $\sim 10^{-19}$ s and produces an electrostatic field $\mathbf{E}_{\text{es}}$ due to the separation of charge, which cancels the induced field, so that the net electric field inside the conductor is zero. The potential difference between the ends 2 and 1 is

$$\Phi_{12} = \Phi_2 - \Phi_1 = -\int_1^2 \mathbf{E} \cdot d\mathbf{l}$$

This relation is consistent with the definition of electric potential in Section 4.4, where Φ_{12} is the work done in moving a unit positive test charge from point 1 to point 2 (e.g., see [4.18]) where $[W/q]_{a\to b} = \Phi_{ab} = \Phi(b) - \Phi(a) = -\int_a^b \mathbf{E} \cdot d\mathbf{l}$. Note that if we integrate the total electric field around the contour C in the counterclockwise direction, the only contribution to the integral comes from across the gap (since the total electric field is zero within the wire), where $\mathbf{E} \cdot d\mathbf{l}$ is negative, which in turn results in a positive value of Φ_{12}. If a load resistance R is connected across the gap (Figure 7.4d), current flows in the clockwise direction, tending to produce magnetic flux opposing the increase in $\mathbf{B}$; this is consistent with the orientation of the induced electric field, which drives the current carriers. If the loop did not have a gap, then current would still flow in the clockwise direction, in accordance with the wire resistance and self-inductance of the loop. In effect, the external resistor R can then be thought of as the internal resistance of the wire.

Turning now to the choice of the polarity of $\mathcal{V}_{\text{ind}}$, we note from [7.2] and the definition of $\mathcal{V}_{\text{ind}} = \oint \mathbf{E} \cdot d\mathbf{l}$ that for $\partial B/\partial t > 0$, and for $d\mathbf{s}$ chosen to be in the same direction as $\mathbf{B}$ as is the case in Figure 7.4, $\mathcal{V}_{\text{ind}}$ as determined by [7.2] is negative. Thus, the polarity of $\mathcal{V}_{\text{ind}}$ in Figure 7.4 must be chosen to be positive at end 1 and negative at end 2, so that we have $\mathcal{V}_{\text{ind}} = -\Phi_{12}$, i.e., a negative value, consistent with the result obtained from the equation [7.2]. More generally, in applying [7.2] to any loop with a gap, the polarity of $\mathcal{V}_{\text{ind}}$ must be defined to be positive on the terminal

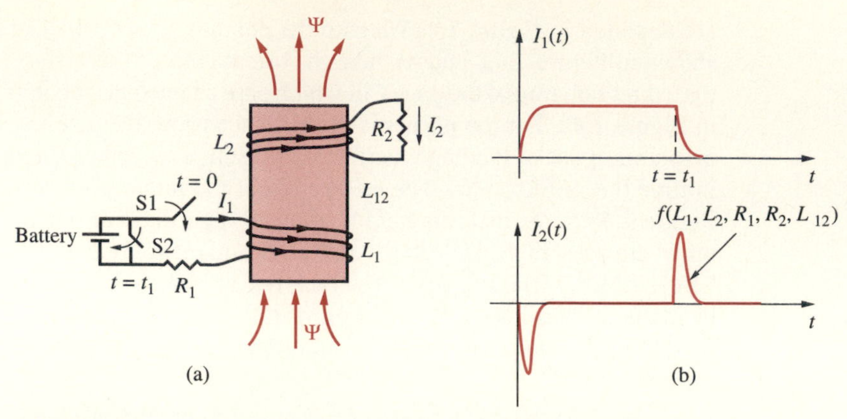

FIGURE 7.5. **Illustration of Lenz's law.** (a) A solenoidal core with two separately wound coils of wire. A battery is connected to or disconnected from the lower coil through switches S1 and S2. (b) The variation of the currents I_1 and I_2 as a function of time for the case when the switch S1 is closed at $t = 0$ and S2 is closed at $t = t_1$.

at which the $d\mathbf{l}$ element, the polarity of which is determined by the right hand rule once $d\mathbf{s}$ is chosen, points outward from that terminal.

For further elaboration of Lenz's law, consider the circuit shown in Figure 7.5a, which consists of a magnetic core material[15] having two separately wound coils of wire (similar to Faraday's experiment shown in Figure 7.1, except the circuit in Figure 7.5 is a solenoid instead of a toroid). The lower coil is connected to a battery through the switch S1. When the switch S1 is closed (say at time $t = 0$), a current I_1 starts to flow in the lower coil. As a result of the increase in the magnetic flux linking the upper coil, a current I_2 is induced in the direction opposite to I_1, opposing the change in the magnetic flux linking the upper coil. Note that, in general, the current I_1 does not begin to flow immediately after switch S1 is closed, but is instead governed by a time constant determined by the inductance and resistance of the lower coil, as well as on the mutual inductance between the two coupled coils. Similarly, the establishment of the current I_2 is governed by a time constant dependent on R_2, L_2, as well as on the mutual inductance between the coils. If the wire were a perfect conductor, the induced current I_2 would flow in the upper coil as long as current I_1 flows in the lower coil. However, in a practical coil, the resistive losses of the wire cause the current I_2 to decay exponentially with a time constant determined by the self-inductance and resistance (L_1, R_1, L_2, and R_2) of each coil, as well as the mutual inductance L_{12} between the two coupled coils. Thus, at steady state, the current I_2 becomes zero and a constant magnetic flux passes through the upper coil due to the current I_1 flowing in the lower coil. Later, when the switch S2 is closed (say at time $t = t_1$), the upper coil tries to keep the flux constant by inducing a current I_2 in the upper coil that is in the same direction as the current I_1 in the lower coil in order to maintain constant flux, until I_1 gradually decreases to zero. Both I_1 and the induced current I_2 die off

[15]The magnetic material enhances the mutual inductance, and thus the inductive coupling, between the two coils.

with time (in a manner defined by the solution of the differential equation describing the coupled circuit consisting of L_1, L_2, R_1, R_2, and L_{12}, i.e., not necessarily a simple exponential decay) due to the resistive losses in their respective loops. Figure 7.5b shows the variation of the two currents I_1 and I_2 as a function of time as just discussed.

Lenz, Heinrich Friedrich Emil *(Russian physicist, b. Dorpat (now Tartu), Estonia, February 24, 1804; d. Rome, Italy, February 10, 1865) Lenz, the son of a magistrate, had originally studied theology but grew interested in science. Between 1823 and 1826, he accompanied a scientific ocean voyage around the globe as Darwin [1809–1882] was to do a few years later. He then spent most of his life as professor of physics at the University of St. Petersburg.*

Lenz was investigating electrical induction at about the same time Faraday was [1791–1867], and Henry [1797–1878] was third in the field. In 1834 Lenz generalized that a current induced by electromagnetic forces always produces effects that oppose those forces.

In 1833, Lenz reported his discovery that the resistance of a metallic conductor changes with temperature, increasing with a rise in temperature and decreasing with a fall in temperature. [Adapted with permission from Biographical Encyclopedia of Science and Technology, *I. Asimov, Doubleday, 1982].*

1700 *1804* *1865* *2000*

Examples 7-1 through 7-4 illustrate Faraday's law and Lenz's law. In most of these examples, which consist typically of single-turn loops or circuits with relatively small inductance, we implicitly neglect the wire resistance and the self-inductance of the coils that link the changing magnetic flux. Thus, the induced current is implicitly assumed to respond instantaneously to the change in the magnetic flux linked.

Example 7-1: Emf induced in a coil. Consider a single-turn coil of wire of radius a, as shown in Figure 7.6a. The region the coil lies in is permeated by a magnetic field $\mathbf{B}(t) = \hat{\mathbf{z}}B_z(t) = \hat{\mathbf{z}}B_0\sin(\omega t)$. Find the induced voltage between the two open end terminals of the coil.

Solution: The induced voltage is given by

$$\mathcal{V}_{\text{ind}}(t) = -\frac{d}{dt}\int_S \mathbf{B}\cdot d\mathbf{s} = -\frac{d}{dt}\int_S [\hat{\mathbf{z}}B_0\sin(\omega t)]\cdot(\hat{\mathbf{z}}\,ds)$$

$$= -\frac{d}{dt}[B_0\sin(\omega t)](\pi a^2) = -\omega\pi a^2 B_0\cos(\omega t)$$

To appreciate the polarity of $\mathcal{V}_{\text{ind}}$, it is best to plot the variation of $\mathcal{V}_{\text{ind}}(t)$ over one cycle together with $B_z(t)$. In fact, it is more illuminating to think in terms of the current $I(t)$ that would flow in the coil if an external load were connected to the loop, as shown in Figure 7.6b.

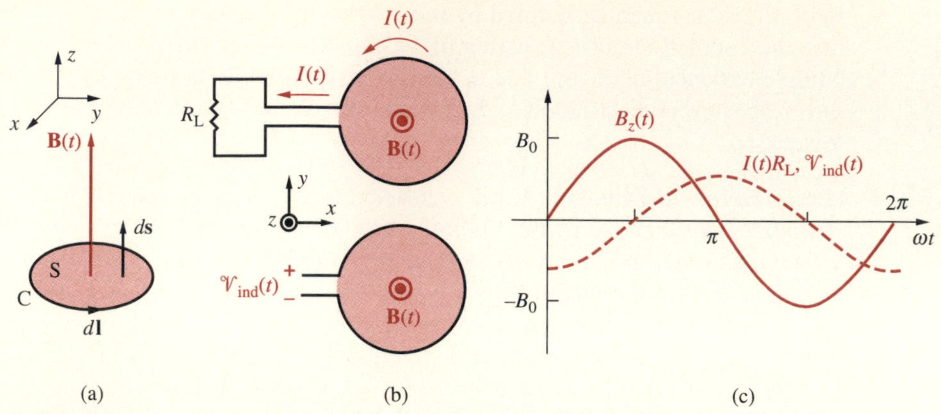

FIGURE 7.6. Emf induced in a coil. (a) A stationary coil located in a time-varying magnetic field. (b) The best way to think about the direction of the induced current and the polarity of the induced voltage $\mathcal{V}_{\text{ind}}$ is to view the coil as a voltage source. (c) The variation of the magnetic field B_z, the induced emf $\mathcal{V}_{\text{ind}} = IR_{\text{L}}$ as a function of time. The direction of induced current flow is determined by Lenz's law.

The plots of $B_z(t)$, $I(t)$, and $\mathcal{V}_{\text{ind}}(t)$ in Figure 7.6b clearly illustrate Lenz's law in effect; when the magnetic field $B_z(t)$ exhibits its largest rate of increase (at $t = 0$), the current $I(t)$ is maximally negative (i.e., flows clockwise so that its associated magnetic field is in the $-z$ direction and *opposes* the increase in $B_z(t)$). Negative (i.e., clockwise) current flows throughout the interval $0 < \omega t < \pi/2$, but with the magnitude of the current decreasing as the rate of change of $B_z(t)$ decreases, with $I(t)$ eventually equal to zero at $\omega t = \pi/2$ when $B_z(t)$ has reached its peak value B_0, so that $\partial B_z/\partial t = 0$. Similar manifestations of Lenz's law can be observed during the rest of the cycle. Note that, in this context, the coil is best viewed as a voltage source (or battery) for a load R_{L} connected to it.

If the coil consists of N turns, the surface S is enclosed N times, where each turn induces a voltage. These voltages all add in series, so that the total voltage induced across the terminals of the coil is N times greater and is given by

$$\mathcal{V}_{\text{ind}}(t) = -N\omega\pi a^2 B_0 \cos(\omega t)$$

The induced voltage is thus proportional to the rate of change of the **B** field (ω), the number of turns (N), and the intensity of the field (B_0) linking each turn with the area of the coil.

Example 7-2: Pair of lines and a rectangular loop. Consider a rectangular loop in the vicinity of a pair of infinitely long wires, as shown in Figure 7.7. The two wires carry a current I flowing in opposite directions, the magnitude of which increases at a rate $dI/dt > 0$. Find the electromotive force induced in the loop (i.e., the $\mathcal{V}_{\text{ind}}$ that would be measured across the small opening if the loop were broken) for the two different placements of the rectangular loop, as shown in Figure 7.7.

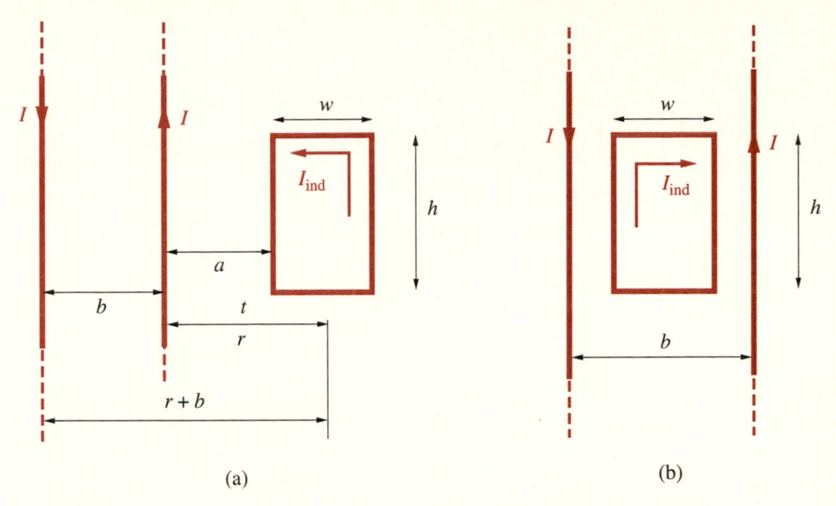

FIGURE 7.7. Pair of parallel wires and a loop. (a) The loop outside the region between the wires. (b) The loop in between the two wires equidistant from each wire.

Solution: We know from Chapter 6 that a current I in an infinitely long straight wire produces a field $\mathbf{B} = \hat{\boldsymbol{\phi}} B_\phi$ at a distance r of

$$B_\phi = \frac{\mu_0 I}{2\pi r}$$

(a) Since the **B** field due to both infinitely long straight wires is given by the same expression, the total flux linking the loop can be written as

$$\Psi = \frac{\mu_0 I}{2\pi}\left(\int_a^{a+w} \frac{h\,dr}{r} - \int_{b+a}^{b+a+w} \frac{h\,dr}{r}\right)$$

$$= \frac{\mu_0 h I}{2\pi} \ln\left[\frac{(b+a)(a+w)}{a(b+a+w)}\right]$$

The induced emf in the rectangular loop is then given by

$$\mathcal{V}_{\text{ind}} = -\frac{\mu_0 h}{2\pi} \ln\left[\frac{(b+a)(a+w)}{a(b+a+w)}\right]\frac{dI}{dt}$$

$$= \frac{\mu_0 h}{2\pi} \ln\left[\frac{a(b+a+w)}{(b+a)(a+w)}\right]\frac{dI}{dt}$$

Note that the polarity of the emf is such that the induced current I_{ind} in the loop flows in the *counterclockwise* direction when dI/dt is positive.

(b) When the loop is between the wires (Figure 7.7b), placed equidistant from both straight wires, the magnetic fluxes linking the loop due to the current flowing in each wire are equal in magnitude and in the same direction. Thus, the total flux linking the rectangular loop is

$$\Psi = 2\frac{\mu_0 I h}{2\pi}\int_{(b-w)/2}^{(b+w)/2}\frac{dr}{r} = \frac{\mu_0 I h}{\pi}\ln\left[\frac{b+w}{b-w}\right]$$

The induced emf is

$$\mathcal{V}_{\text{ind}} = -\frac{d\Psi}{dt} = -\frac{\mu_0 h}{\pi}\ln\left[\frac{b+w}{b-w}\right]\frac{dI}{dt}$$

Note that this time, the polarity of the induced voltage is such that the induced current I_{ind} in the loop flows in the *clockwise* direction when dI/dt is positive.

Example 7-3: Jumping ring. A conducting ring of radius a is placed on top of a solenoidal coil wound around an iron core, as shown in Figure 7.8. Describe what happens to the ring when the switch is closed.

Solution: Before the switch is closed, the conducting ring has zero flux linkage. When the current is first turned on in the solenoid by closing the switch, the flux produced by the solenoid links the ring in the upward direction, inducing a current in the ring opposite in direction to that in the coil (i.e., Lenz's law). Since currents in opposite directions repel, the repulsion force lifts the conducting ring off the top of the solenoid. This force is given by

$$\mathbf{F} = \int_C I\, d\mathbf{l}\times\mathbf{B} = \int_0^{2\pi} I_\phi(\hat{\boldsymbol{\phi}} a\, d\phi)\times(\hat{\mathbf{r}}B_r + \hat{\mathbf{z}}B_z)$$

$$= -\hat{\boldsymbol{\phi}}2\pi a I_\phi \times \hat{\mathbf{r}}B_r = 2\pi a I_\phi B_r\hat{\mathbf{z}}$$

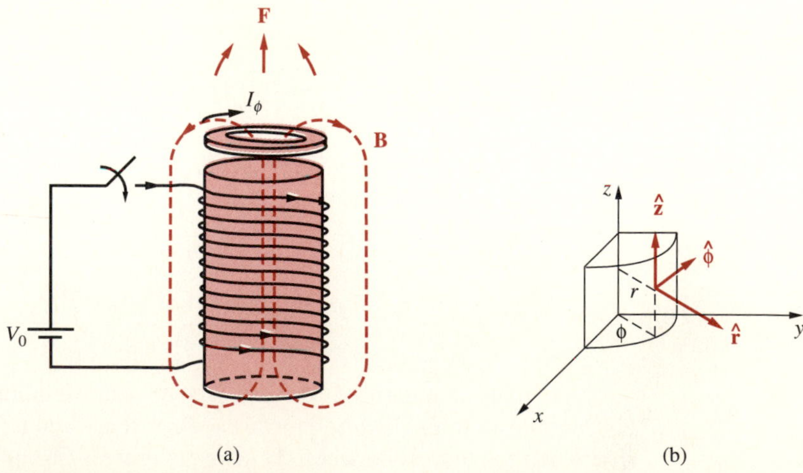

FIGURE 7.8. Jumping ring. (a) A conducting ring placed on top of a solenoidal coil. (b) A cylindrical coordinate system.

where I_ϕ is the magnitude of the induced current I (note that, by Lenz's law, this current must be in the direction $-\hat{\boldsymbol{\phi}}$, as shown), a is the radius of the ring, and B_r and B_z are the radial and vertical components of the **B** field. Note that the radial component of the force due to the B_z component cancels out because of the circular symmetry, so the resultant force is in the z direction. Note also that the actual dynamics of the motion of the ring are quite complicated indeed. The magnitude of the induced current I_ϕ depends on the mutual inductance between the solenoid and the ring; once the ring lifts off and begins to move away, the magnitude of the **B** field (and thus the magnetic force) experienced by the ring changes; the time rate of change of the magnetic flux (and hence the duration of the induced current I) depends on the inductance of the solenoidal coil, its internal resistance, and so on.

Example 7-4: Two concentric coils. Consider two circular coils of radii $a = 50$ cm and $b = 5$ cm, number of turns $N_1 = 25$ and $N_2 = 100$, and both centered at the origin and located on the same plane, as shown in Figure 7.9. (a) Find the mutual inductance of the two coils. (b) If the larger coil is carrying an alternating current of $I(t) = 10\sin(377t)$ A, find the induced emf across the terminals of the smaller coil.

Solution:

(a) In Example 6-6, we found out that the **B** field produced at the center of a circular current-carrying wire is given by

$$\mathbf{B}_{\text{ctr}} = \hat{\mathbf{z}}\frac{\mu_0 I}{2a}$$

The **B** field at the center of a circular wire of N_1 turns is

$$\mathbf{B}_{\text{ctr}} = \hat{\mathbf{z}}\frac{\mu_0 N_1 I}{2a}$$

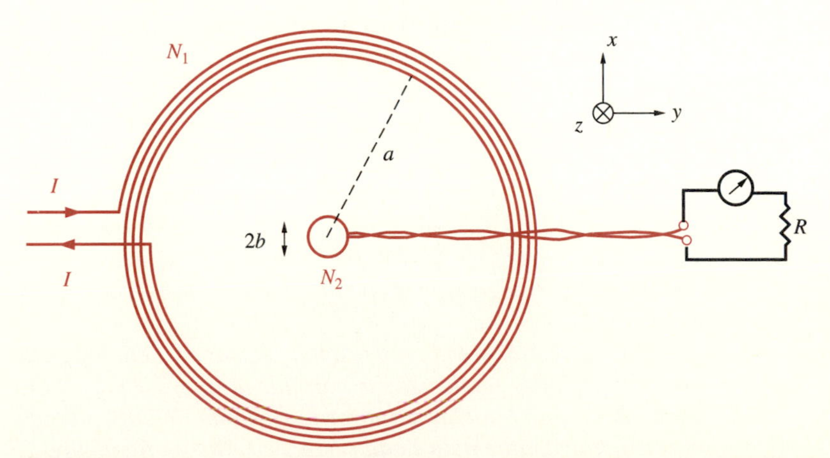

FIGURE 7.9. Two concentric coils. The circuit is arranged such that emf is to be measured across the terminals of the smaller coil.

Since the radius of the smaller loop $b \ll a$, we can assume the **B** field to be constant over its flat surface. The total magnetic flux linking the small loop due to the current flowing in the large loop is thus $\Lambda_{12} = N_2 B_{\text{ctr}} \pi b^2$. The mutual inductance is then

$$L_{12} = \frac{\Lambda_{12}}{I} = \frac{N_2 N_1 \pi b^2 \mu_0}{2a}$$

(b) Substituting the values given, we find

$$\mathbf{B}_{\text{ctr}} = \hat{\mathbf{z}}\frac{(4\pi \times 10^{-7}\ \text{H-m}^{-1})(25)[10\sin(377t)\ \text{A}]}{2 \times (0.5\ \text{m})} \simeq \hat{\mathbf{z}}\, 3.14 \times 10^{-4} \sin(377t)\ \text{T}$$

The total flux linked by a single turn of the small loop can be found as

$$\Psi_{12} = \int_S \mathbf{B}_{\text{ctr}} \cdot d\mathbf{s} \simeq |\mathbf{B}_{\text{ctr}}| \int_S ds = \frac{\mu_0 N_1 I}{2a}(\pi b^2)$$

$$= [\pi \times 10^{-4} \sin(377t)\ \text{T}][\pi \times (5 \times 10^{-2})^2]$$

$$\simeq 2.47 \times 10^{-6} \sin(377t)\ \text{Wb}$$

Therefore, the induced emf across the terminals of the smaller loop of $N_2 =$ 100 turns is found as

$$\mathcal{V}_{\text{ind}}(t) = -N_2 \frac{d\Psi_{12}}{dt}$$

$$\simeq -100 \times 2.47 \times 10^{-6} \times 377 \cos(377t) \simeq -93 \cos(377t)\ \text{mV}$$

Note that we are not concerned with the polarity of $\mathcal{V}_{\text{ind}}$ since the terminals of the small loop are not identified. However, Lenz's law tells us that when the current $I(t)$ increases during part of its alternating cycle, the induced current in the small loop must flow in the counterclockwise direction in order to oppose the increase of the **B** field, which is in the z direction.

Henry, Joseph *(American physicist, b. December 17, 1797, Albany, New York; d. May 13, 1878, Washington, D.C.)* *Henry, like Faraday [1791–1867], came from a poor family. He was the son of a day laborer, had little schooling, and was forced to go to work while young. At thirteen, he was apprenticed to a watchmaker.*

Henry was schooled at the Albany Academy, teaching at country schools and tutoring privately on the side to earn his tuition. He was set to study medicine when an offer of a job as surveyor turned him toward engineering. By 1826 he was teaching mathematics and science at Albany Academy. Like Faraday, he grew interested in the experiment of Oersted [1777–1851], and he became the first American to experiment with electricity in any important way since Franklin's [1706–1790] pioneering work.

Sturgeon [1783–1850] had put Oersted's work to use by creating an electromagnet. In 1829, Henry heard of this and thought he could improve a strong magnet by wrapping more coils of conducting wire around an iron core. To do this, it was necessary to insulate the wires so that they did not touch and cause a short circuit. Henry tore up one of his wife's silk petticoats for this purpose. In the years to come, he spent a great deal of his time in the boring task of slowly wrapping insulation about wires. However, the electromagnets he created were far more powerful than Sturgeon's. By 1831, he had developed a magnet that could lift 750 pounds, compared to Sturgeon's, which could lift only 9 pounds. The same year, in a demonstration at Yale University, another of his electromagnets, using the current from an ordinary battery, lifted more than a ton of iron. In 1832 he reaped his reward in the form of a professorial appointment at Princeton.

Henry also built small, delicate electromagnets that could be used for control. Imagine a small electromagnet at one end of a mile of wire, with a battery at the other end. Suppose you could send a current through the wire by pressing a key and closing the circuit. With the current flowing, the electromagnet, a mile away, could be made to attract a small iron bar. If the key were then released, the current would be broken, the electromagnet would lose its force, and the small iron bar would be pulled away by a spring attached to it. By opening and closing the key in a particular pattern, the iron bar a mile away could be made to open and close, clicking away in that same particular pattern. By 1831 Henry was already using telegraphy.

However, the longer the wire, the greater its resistance; this set a practical limit to the distance over which such a pattern can be sent. To circumvent this problem, Henry invented the electrical relay in 1835. A current just strong enough to activate an electromagnet would lift a small iron key. This key when lifted would close a second circuit with a current (from a nearby battery) flowing through it, in turn activating another relay. In this way the current would travel from relay to relay and could cover huge distances without weakening. The opening and closing of a key could then impress its peculiar pattern through any distance.

Henry had in effect invented the telegraph. However, he did not patent any of his devices for he believed that the discoveries of science were for the benefit of all humanity. As a result, it was Morse [1791–1872] who worked out the first telegraph put to practical use (between Baltimore and Washington, D.C., in 1844) and it is Morse who is usually credited as the inventor. In tackling the technical end of the problem, Morse, who was completely ignorant of science, was helped freely by Henry. In England, Wheatstone [1802–1875], after a long conference with Henry, worked up a telegraph in 1837. Henry, an idealist, didn't mind that he did not share in the financial rewards of the telegraph. It bothered him, however, that neither inventor ever publicly acknowledged Henry's help.

Henry missed the credit for another important discovery in a more heartbreaking way. At the Albany Academy, Henry's teaching duties were so heavy that he could do research only in August. In August 1830 he discovered the principle of electromagnetic induction, but did not quite finish his work at the end of the month and put it aside for next August. Well before next August, he read Faraday's preliminary note concerning his discovery of induction. Henry rushed back to his experiments and published his own work, but by then it was too late. Henry had done the key experiments ahead of Faraday, but Faraday had published first. Henry was not one to feel bitter and always freely admitted Faraday's priority.

In Henry's paper, however, he explained that the electric current in a coil can induce another current not only in another coil, but also in itself. The actual current observed in the coil is, then, the combination of the original current and the induced current. This discovery of self-induction is credited to Henry. Faraday discovered it independently by 1834, but this time he was second.

In 1831 Henry published a paper describing the electric motor. The importance of the motor cannot be overemphasized. The supply of cheap, abundant electricity made possible by Faraday's discovery of the generator would have been useless without some means of putting it conveniently to work. It is Henry's motor in vacuum cleaners, refrigerators, shavers, typewriters, and more than a hundred other electrical appliances that harnesses electricity.

In 1846 Henry was elected first director of the newly formed Smithsonian Institution. He made the Smithsonian a clearinghouse of scientific knowledge and encouraged scientific communication on a worldwide scale. He was one of the founders of the National Academy of Sciences of the United States and one of its early presidents. He also encouraged the growth of new sciences within the United States. He was interested in meteorology and set up a system of obtaining weather reports from all over the nation (by telegraph); this led to the founding of the United States Weather Bureau. When Henry died, his funeral was attended by high government officials, including Rutherford B. Hayes, then President of the United States. The unit of inductance, the henry, is named after him. [Adapted with permission from I. Asimov, Biographical Encyclopedia of Science and Technology, *Doubleday, 1982].*

1700 — 1797 — 1878 — 2000

7.2 INDUCTION DUE TO MOTION

When conductors move in the presence of magnetic fields, an induced voltage is produced, in addition to the induced voltage due to the variation of magnetic fields with time. Induction due to motion is the basis for most of the engines of our industrial society, including the giant generators of electricity in hydroelectric dams, large motors used in steel mills, and tiny motors in dentist's drills or children's toys.[16] In general, when a loop travels through space, its motion through a magnetic field may alter the amount of flux through the loop even though the magnetic field may be constant at all points in space. If, in addition to motion, the magnetic field itself varies with time, then electromotive force is induced due to both motion and the time variation of the magnetic field.

7.2.1 Motion in a Constant Magnetic Field

The magnitude of the voltage induced due to motion of a conductor in a constant magnetic field can be found from the Lorentz force, which was defined in Chapter 6 with [6.7]. The Lorentz force due to a magnetic field is the force $\mathbf{F}_{\rm m}$ that a $\mathbf{B}$ field exerts on a charged particle q moving with a velocity $\mathbf{v}$. In other words,

$$\mathbf{F}_{\rm m} = q\mathbf{v} \times \mathbf{B}$$

Note that this force acts in a direction perpendicular to both $\mathbf{v}$ and $\mathbf{B}$, and is in fact just another form of the Ampère's force law in expression [6.2], or $\mathbf{F}_{\rm m} = I d\mathbf{l} \times \mathbf{B}$, since $q\mathbf{v}$ can be viewed as a current element (see Section 6.2.3).

The fact that the charges in the wire experience a force of $q\mathbf{v} \times \mathbf{B}$ indicates that an electromotive force (emf) is induced across the ends of the wire. To see this, note that in Section 7.1, we defined emf as the tangential force per unit charge integrated over its length. Thus, in the case shown in Figure 7.10, the induced emf is given by the integral of $\mathbf{F}_{\rm m}$ over the length of the bar. If a resistance were connected across the ends of the moving wire, current flows, with the current carriers "pushed" by

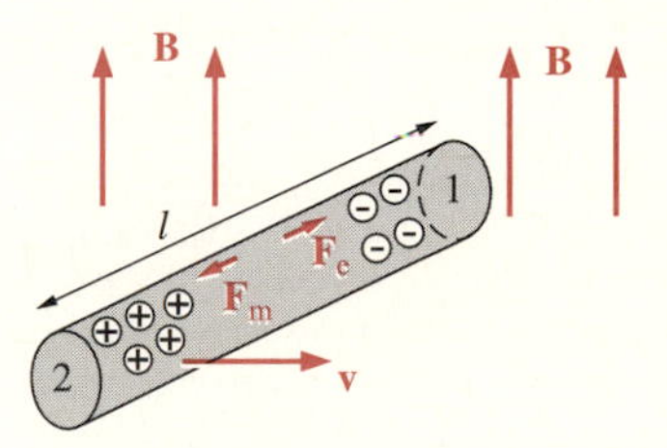

FIGURE 7.10. A conducting bar moving in a static magnetic field. The charges in the conducting bar redistribute themselves in such a way that an induced emf is generated across the bar.

[16]For an excellent qualitative and enlightening discussion of motors and generators and their applications, see Chapter 16 of R. P. Feynman, R. B. Leighton, and M. Sands, *The Feynman Lectures on Physics, Volume II,* Addison Wesley, 1964. As reproduced here with permission, Feynman discusses the engineering miracle of hot lights (electricity) produced from cold water (a hydroelectric dam) hundreds of miles away, all done with specially arranged pieces of copper and iron: "Hundreds of little wheels, turning in response to the turning of the big wheel at Boulder Dam. Stop the big wheel, and all the little wheels stop; the lights go out. They are really connected. Yet there is more. The same phenomena that take the tremendous power of the river and spread it through the countryside, until a few drops of the river are running the dentist's drill, come again into the building of extremely fine instruments . . . for the detection of incredibly small amounts of current . . . for the transmission of voices, music, and pictures . . . for computers . . . for automatic machines of fantastic precision."

the magnetic force, namely ($q\mathbf{v}\times\mathbf{B}$). Note that we do not need to invoke any *induced* electric field, since, due to the motion of the bar, the current carriers can be driven by a magnetic force.[17] The electromotive force between points 1 and 2 (Figure 7.10) is thus given by

$$\mathcal{V}_{\text{ind}} = \int_2^1 (\mathbf{v}\times\mathbf{B})\cdot d\mathbf{l} \qquad [7.3]$$

since tangential force per unit charge in the wire is ($\mathbf{v}\times\mathbf{B}$).

This expression is the mathematical formulation of Faraday's observation that moving conductors in the vicinity of magnets (or moving magnets in the vicinity of conductors) induce electromotive force. The induced voltage as given by [7.3] is sometimes referred to as the *motional emf.*

Example 7-5 illustrates a practical application of motional emf.

Example 7-5: Moving metal bar in uniform static magnetic field. Consider the arrangement shown in Figure 7.11, in which a metal bar can slide over conducting rails. The magnetic field is static (i.e., constant with respect to time) and uniform (i.e., the same everywhere) and is out of the page as shown and given by $\mathbf{B} = \hat{\mathbf{z}}B_0$. The bar and the rails constitute a conducting wire loop, the area

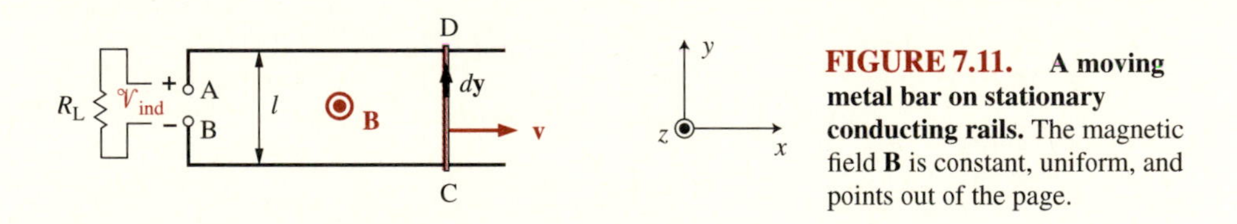

FIGURE 7.11. A moving metal bar on stationary conducting rails. The magnetic field **B** is constant, uniform, and points out of the page.

[17]An electrostatic field is in fact set up within the conductor as a result of separation of charge. When a conductor moves with velocity **v** in a static magnetic field **B**, as shown in Figure 7.10, the $q\mathbf{v}\times\mathbf{B}$ force causes the electrons inside the conductor to flow to one end of the conductor and leave the other end positively charged. This brief current flows until equilibrium is reached (very quickly, over a time of $\sim 10^{-19}$ s for most metals), when the coulomb force of attraction $\mathbf{F}_{\text{es}}$ due to charge separation cancels the force due to the magnetic field. In other words, we have $\mathbf{F} = \mathbf{F}_{\text{es}} + \mathbf{F}_{\text{m}} = q\mathbf{E}_{\text{es}} + q\mathbf{v}\times\mathbf{B} = 0$, or $\mathbf{E}_{\text{es}} = -(\mathbf{v}\times\mathbf{B})$, where $\mathbf{E}_{\text{es}}$ is the electrostatic field within the conductor. Note that the induced electric field **E**, which gives $\mathcal{V}_{\text{ind}}$ when integrated along any external path between the two ends of the bar, is related to ($\mathbf{v}\times\mathbf{B}$) via [7.3], or

$$\mathcal{V}_{\text{ind}} = \int_2^1 \mathbf{E}\cdot d\mathbf{l} = \int_2^1 (\mathbf{v}\times\mathbf{B})\cdot d\mathbf{l}$$

so that $\mathbf{E} = \mathbf{v}\times\mathbf{B}$, or $\mathbf{E} = -\mathbf{E}_{\text{es}}$. Note that unlike $\mathbf{E}_{\text{es}}$, the induced field **E** is nonconservative.

To an observer moving with the bar, the charges in the bar appears to be stationary, so the fact that they experience a force (equal to $q\mathbf{v}\times\mathbf{B}$) can be interpreted as being due to an electric field given by ($\mathbf{v}\times\mathbf{B}$), as long as $|v| \ll c$. The electrostatic field set up by the displaced charges may be observed in both a stationary frame of reference and a moving frame attached to the conductor.

of which expands as the bar moves to the right with a constant velocity $\mathbf{v} = \hat{\mathbf{x}}v_0$. Find the total emf induced in the loop.

Solution: The emf induced between terminals A and B can be calculated using Faraday's law. Since the magnetic field does not vary with time, it is appropriate to use [7.3] to find $\mathcal{V}_{\text{ind}}$. In applying [7.3], we need to integrate around the entire loop, part of which is the moving bar. In other words, we have

$$\mathcal{V}_{\text{ind}} = \int_C (\mathbf{v} \times \mathbf{B}) \cdot d\mathbf{l}$$

where C is the contour BCDA. However, since all other parts of the loop are stationary ($\mathbf{v} = 0$), the only contribution to the right-hand side is from the moving bar (i.e., the CD segment of the loop). Thus we have

$$\mathcal{V}_{\text{ind}} = \int_{\text{CD}} (\mathbf{v} \times \mathbf{B}) \cdot d\mathbf{l} = \int_{\text{CD}} (\hat{\mathbf{x}}v_0 \times \hat{\mathbf{z}}B_0) \cdot (\hat{\mathbf{y}}\, dy) = -v_0 B_0 l$$

Note that the polarity of the induced emf is such that terminal B is positive with respect to terminal A. In other words, if a load resistance is connected between terminals A and B, a current of $I = v_0 B_0 l / R_{\text{L}}$ flows in the *clockwise* direction through the loop to oppose the increasing magnetic flux linking the loop (i.e., Lenz's law). The electrical power dissipated in the load resistor R_{L} is

$$P_{\text{elec}} = I^2 R_{\text{L}} = \frac{(v_0 B_0 l)^2}{R_{\text{L}}}$$

Based on energy conservation, this electrical power must originate from the mechanical energy used to move the metal bar at a velocity $\mathbf{v}$. Energy is essentially transferred from the mechanical source to the load resistor. To quantitatively evaluate the energy transfer, we can consider the magnetic force experienced by the moving bar. In the presence of a magnetic field, a current-carrying wire element experiences a magnetic force given by

$$d\mathbf{F}_{\text{m}} = I\, d\mathbf{l} \times \mathbf{B}$$

so that the total magnetic force experienced by the metal bar is

$$\mathbf{F}_{\text{m}} = I \int_{\text{C}}^{\text{D}} -\hat{\mathbf{y}}\, dy \times \hat{\mathbf{z}}B_0 = -\hat{\mathbf{x}} I B_0 l$$

where the induced current in the loop flows in the clockwise direction, that is, $I\, d\mathbf{l} = -I\, d\hat{\mathbf{y}}\, dy$. The magnetic force experienced by the bar is thus in the $-x$ direction, *opposing* the motion. In order to sustain the motion of the bar at a constant velocity $\mathbf{v} = \hat{\mathbf{x}}v_0$, the mechanical force must be exactly equal and opposite to this magnetic force. The mechanical power is given by

$$P_{\text{mech}} = -\mathbf{F}_{\text{m}} \cdot \mathbf{v} = I v_0 B_0 l = \left(\frac{v_0 B_0 l}{R_{\text{L}}}\right) v_0 B_0 l = \frac{(v_0 B_0 l)^2}{R_{\text{L}}}$$

since $\mathbf{F}_m$ and $\mathbf{v}$ are parallel. Thus we see that, as expected, $P_{\text{mech}} = P_{\text{elec}}$. The system shown in Figure 7.11 is thus a simple example of an electric generator, where the applied mechanical energy is converted into electrical energy.

7.2.2 Moving Conductor in a Time-Varying Magnetic Field

We have seen that an electromotive force is induced through [7.2], when the magnetic field linking a stationary loop varies with time, or through [7.3], when a conductor moves through a static magnetic field. The general case involves motion of a loop C in addition to a time variation of the magnetic field. In such a case, the induced voltage is given by

$$\mathcal{V}_{\text{ind}} = -\int_S \frac{\partial \mathbf{B}}{\partial t} \cdot d\mathbf{s} + \oint_C (\mathbf{v} \times \mathbf{B}) \cdot d\mathbf{l} \quad [7.4]$$

where the contour C encloses the surface S. Care must be taken so that $d\mathbf{l}$ and $d\mathbf{s}$ are related via the right-hand rule, as discussed in connection with [7.2]. Note that the velocity of the different parts of the loop need not be the same, since C may be changing shape as well as undergoing translation (linear motion) and/or rotation.

The motional emf term (i.e., the second term) of [7.4] can in many cases be thought of in terms of Faraday's original changing-flux concept, as expressed in [7.1]. To see this, consider Figure 7.12. In a time interval dt, an element $d\mathbf{l}$ of C sweeps out an area $d\mathbf{s} = (\mathbf{v} \times d\mathbf{l})\,dt$. The change in magnetic flux $d\Psi_1$ caused by the motion of the element $d\mathbf{l}$ is equal to the integral of $\mathbf{B}$ through the swept-out area, or $d\Psi_1 = \mathbf{B} \cdot d\mathbf{s} = \mathbf{B} \cdot (\mathbf{v} \times d\mathbf{l})dt$. The total change in magnetic flux due to the displacement of the entire contour C is then given as $d\Psi = \oint_C d\Psi_1$ or

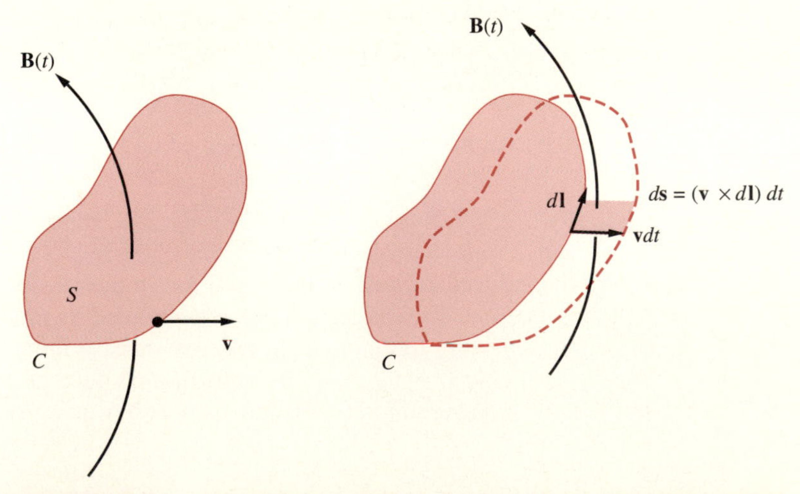

FIGURE 7.12. A conducting loop C moving in a time-varying magnetic field.

$$d\Psi = \oint_C \mathbf{B} \cdot (\mathbf{v} \times d\mathbf{l})\, dt \quad \rightarrow \quad -\frac{d\Psi}{dt} = -\oint_C \mathbf{B} \cdot (\mathbf{v} \times d\mathbf{l}) = \oint_C (\mathbf{v} \times \mathbf{B}) \cdot d\mathbf{l}$$

where $\oint_C (\mathbf{v} \times \mathbf{B}) \cdot d\mathbf{l}$ is the motional emf part of [7.4].

Based on the preceding expression, it appears that the most general form of Faraday's law, including the cases involving moving circuits, can be expressed as $\mathcal{V}_{\text{ind}} = -d\Psi/dt$, or

$$\boxed{\mathcal{V}_{\text{ind}} = \oint_C \mathbf{E} \cdot d\mathbf{l} = -\frac{d}{dt}\int_S \mathbf{B} \cdot d\mathbf{s}} \qquad [7.5]$$

where $\mathbf{E}$ is the induced electric field.

Although [7.5] applies in general, there are situations in which a closed contour cannot be identified.[18] An example in which it is not clear how the flux linkage can be evaluated is the single moving conductor of Figure 7.10. In such cases, the correct physics is always revealed by [7.4].

Examples 7-6 and 7-7 illustrate the uses of [7.4] and [7.5] to determine the induced emf in cases involving moving conductors.

Example 7-6: Elementary alternating current generator: A rotating conductor loop in a static magnetic field. This example illustrates the basic operating principle of an elementary alternating current generator, which typically involves an armature coil mechanically rotated in a steady magnetic field at a constant rate. Consider the single-turn rectangular coil shown in Figure 7.13, rotating with an angular velocity ω_0 about its axis (z axis) in the presence of a static uniform magnetic field $\mathbf{B} = \hat{\mathbf{x}}B_0$. Determine the voltage induced between terminals 1 and 2.

Solution: Since $\partial\mathbf{B}/\partial t = 0$, the induced emf is due solely to the motion of the conductor (i.e., motional emf), and we can use either [7.3] or [7.5] to determine $\mathcal{V}_{\text{ind}}$. We first select the polarity of $\mathcal{V}_{\text{ind}}$ to be such that the terminal marked 2 is positive. In other words, $\mathcal{V}_{\text{ind}} \equiv \Phi_{12} = \Phi_2 - \Phi_1$.

We proceed by using the generalized Faraday's law, namely [7.5]. At any instant of time t, the magnetic flux Ψ through the coil is

$$\Psi = \int_S \mathbf{B} \cdot d\mathbf{s} = \int_S B_0 \hat{\mathbf{x}} \cdot \hat{\mathbf{r}}\, ds = B_0 A \cos\phi = abB_0 \cos(\omega_0 t)$$

where $A = ab$ is the area of the loop. The induced voltage $\mathcal{V}_{\text{ind}}$ is then

$$\mathcal{V}_{\text{ind}} = -\frac{d\Psi}{dt} = \omega_0 abB_0 \sin(\omega_0 t)$$

---

[18]For examples and a discussion, see Chapter II-17 of R. P. Feynman, R. B. Leighton, and M. Sands, *The Feynman Lectures on Physics,* Addison Wesley, 1964.

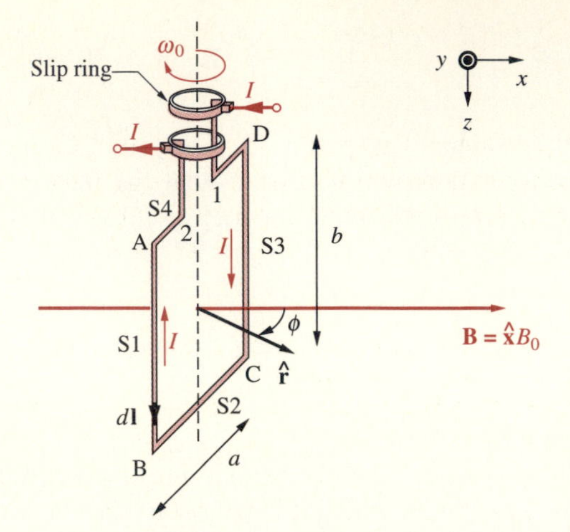

FIGURE 7.13. Elementary alternating current generator. A rectangular coil is rotated at a constant angular velocity in a uniform magnetic field. The slip rings maintain electrical contact with the terminals of the coil as it rotates.

The same result can also be obtained by using [7.3]. Note that the velocity of an electron along the vertical sides (S1 and S3) of the coil is $\mathbf{v} = \hat{\boldsymbol{\phi}}\omega_0(a/2)$ and the velocity $\mathbf{v}$ and the field $\mathbf{B} = \hat{\mathbf{x}}B_0$ are at an angle $\phi = \omega_0 t$, so that we have $|\mathbf{v} \times \mathbf{B}| = |v|B_0 \sin(\omega_0 t) = (a\omega_0/2)B_0 \sin(\omega_0 t)$. Note that the directions of $\mathbf{v} \times \mathbf{B}$ are opposite to each other on sides AB and CD. The sides BC and DA do not contribute to the line integral since $\mathbf{v} \times \mathbf{B}$ and $d\mathbf{l}$ are orthogonal to each other, and thus $(\mathbf{v} \times \mathbf{B}) \cdot d\mathbf{l} = 0$. Assuming that the loop wire is perfectly conducting, the only voltage drop is across terminals 1 and 2. Thus, we have

$$\begin{aligned}
\mathcal{V}_{\text{ind}} = \Phi_{12} &= \int_2^1 \mathbf{E} \cdot d\mathbf{l} = \int_2^1 (\mathbf{v} \times \mathbf{B}) \cdot d\mathbf{l} \\
&= \int_A^B \hat{\mathbf{z}}\frac{a\omega_0}{2}[B_0 \sin(\omega_0 t)] \cdot (\hat{\mathbf{z}}\,dz) + \int_C^D -\hat{\mathbf{z}}\frac{a\omega_0}{2}[B_0 \sin(\omega_0 t)] \cdot (\hat{\mathbf{z}}\,dz) \\
&= b\frac{a\omega_0}{2}B_0 \sin(\omega_0 t) + b\frac{a\omega_0}{2}B_0 \sin(\omega_0 t) \\
\mathcal{V}_{\text{ind}} &= ba\omega_0 B_0 \sin(\omega_0 t)
\end{aligned}$$

which is the same as the previous result obtained from [7.5].

In general, depending on the impedance of the external load connected across the generator terminals 1 and 2, the current I can be out of phase with respect to $\mathcal{V}_{\text{ind}}$. Assuming the current that flows is $I(t) = I_0 \cos(\omega_0 t + \theta)$, the instantaneous electric power delivered to the external load by the generator is

$$P_{\text{elec}} = \mathcal{V}_{\text{ind}} I(t) = [\omega_0 abB_0 \sin(\omega_0 t)][I_0 \cos(\omega_0 t + \theta)]$$

The source of this electrical power is the mechanical power that maintains the rotation of the coil at ω_0. To determine the mechanical power needed, we note that as it rotates in a constant magnetic field, the coil experiences a torque given by [6.33], namely $\mathbf{T} = \mathbf{m} \times \mathbf{B}$, where $\mathbf{m}$ is the magnetic dipole moment such that $\mathbf{m} = \hat{\mathbf{r}}abI(t) = \hat{\mathbf{r}}abI_0 \cos(\omega_0 t + \theta)$. Thus, the torque acting on the coil is

$$\mathbf{T} = [\hat{\mathbf{r}}abI_0\cos(\omega_0 t + \theta)] \times \hat{\mathbf{x}}B_0 = \hat{\mathbf{z}}[abI_0\cos(\omega_0 t + \theta)](B_0)\sin\phi$$
$$= \hat{\mathbf{z}}[abI_0\cos(\omega_0 t + \theta)](B_0)\sin(\omega_0 t)$$

Note that the sign of the torque is such that it opposes the rotation of the coil. In order to maintain the rotation of the coil at a fixed rate ω_0, the externally applied mechanical force must overcome this electromagnetic opposition. Thus, noting that mechanical power in rotational motion is the product of torque and angular velocity, we have

$$P_{\text{mech}} = \omega_0 T = \omega_0 abB_0 I_0 \sin(\omega_0 t)\cos(\omega_0 t + \theta)$$

Note that $P_{\text{elec}} = P_{\text{mech}}$, as was the case for the translational motion electrical generator of Example 7-5.

Example 7-7: Expanding circular loop. A circular conducting rubber loop expands at a constant radial velocity $\mathbf{v} = \hat{\mathbf{r}}v_0$ in a region (see Figure 7.14) permeated by a uniform magnetic field that is perpendicular to the loop and varies in time as $\mathbf{B} = \hat{\mathbf{z}}B_0 t$. The radius of the loop at any time t is $r_t = v_0 t$. Neglect the thickness of the conducting rubber wire. Find the induced emf $\mathcal{V}_{\text{ind}}$. How does the result change if the magnetic field is oriented in the $-z$ direction (i.e., $\mathbf{B} = -\hat{\mathbf{z}}B_0 t$)?

Solution: We can solve the problem using [7.5] or [7.4]. We first use equation [7.5]. Since the magnetic field is in the $+z$ direction, it is convenient to choose $d\mathbf{s}$ also in the $+z$ direction so that the integrand in [7.5] is positive. This in turn means that the contour integral in [7.5] should be taken in the counterclockwise direction, since the contour C and the surface S in [7.5] are related via the right-hand rule. Therefore, the polarity of $\mathcal{V}_{\text{ind}}$ is defined to be positive at terminal 1, at which $d\mathbf{l}$ is outward (see Figure 7.4 and accompanying discussion).

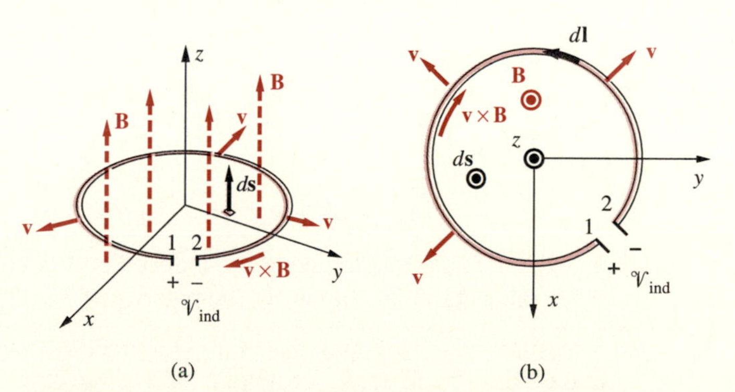

FIGURE 7.14. An expanding circular loop. (a) Coordinate system. (b) Top view with the magnetic field pointing out of the page ($+z$ direction).

Noting that we then have $d\mathbf{s} = r\,dr\,d\phi\,\hat{\mathbf{z}}$ and $\mathbf{B} = \hat{\mathbf{z}}B_0 t$, from [7.5] we have

$$\mathcal{V}_{\text{ind}} = -\frac{d}{dt}\int_{S(t)} \mathbf{B}\cdot d\mathbf{s} = -\frac{d}{dt}\int_0^{r_t = v_0 t}\int_0^{2\pi} B_0 t r\,dr\,d\phi$$

$$= -\frac{d}{dt}\left[2\pi B_0 t\left(\frac{1}{2}r^2\right)\right]_0^{v_0 t} = -\frac{d}{dt}[B_0 \pi v_0^2 t^3]$$

$$= -3B_0\pi v_0^2 t^2 = -3B_0\pi\left(\frac{r_t^2}{t^2}\right)t^2 = -3B_0\pi r_t^2$$

We now use [7.4] to determine $\mathcal{V}_{\text{ind}}$. We integrate around the loop in the counterclockwise direction, with $d\mathbf{l}$ in [7.5] chosen by convention to be in the $+\hat{\boldsymbol{\phi}}$ direction. We have

$$\mathbf{v}\times\mathbf{B} = v_0\hat{\mathbf{r}}\times\hat{\mathbf{z}}B_0 t = -B_0 v_0 t\hat{\boldsymbol{\phi}} = -B_0 r_t\hat{\boldsymbol{\phi}} \qquad \text{and} \qquad d\mathbf{l} = r_t\,d\phi\hat{\boldsymbol{\phi}}$$

so that

$$\oint_{C(t)} (\mathbf{v}\times\mathbf{B})\cdot d\mathbf{l} = \int_0^{2\pi} -B_0 r_t^2\,d\phi = -2\pi B_0 r_t^2$$

and

$$-\int_{S(t)} \frac{\partial\mathbf{B}}{\partial t}\cdot d\mathbf{s} = -\int_{S(t)} \frac{\partial}{\partial t}[B_0 t\hat{\mathbf{z}}]\cdot\hat{\mathbf{z}}r\,dr\,d\phi = -\int_0^{v_0 t}\int_0^{2\pi} B_0 r\,dr\,d\phi$$

$$= \left[-2\pi B_0\left(\frac{1}{2}r^2\right)\right]_0^{v_0 t} = -\pi B_0 v_0^2 t^2 = -\pi B_0 r_t^2$$

or

$$\mathcal{V}_{\text{ind}} = -\pi B_0 r_t^2 - 2\pi B_0 r_t^2 = -3\pi B_0 r_t^2$$

which is the same as the result we found using [7.5].

If the magnetic field points in the $-z$ direction and we maintain the same orientation for $d\mathbf{l}$ (and thus $d\mathbf{s}$), we have $d\mathbf{s}$ and $\mathbf{B}$ in the opposite directions, so that the contribution of the first term in [7.4] is the negative of that found in the preceding expression, or $+\pi B_0 r_t^2$. However, since $\mathbf{B} = -\hat{\mathbf{z}}B_0 t$ and $\mathbf{v} = \hat{\mathbf{r}}v_0$, we have $(\mathbf{v}\times\mathbf{B}) = +B_0 v_0 t\hat{\boldsymbol{\phi}}$, so that $(\mathbf{v}\times\mathbf{B})$ and $d\mathbf{l}$ are now also in opposite directions. Thus, the contribution from the second term in [7.4] also changes sign and is now $+2\pi B_0 r_t^2$. Thus, the induced emf is

$$\mathcal{V}_{\text{ind}} = \left(+\pi B_0 r_t^2\right) + \left(+2\pi B_0 r_t^2\right) = 3\pi B_0 r_t^2$$

As expected, the magnitude of the induced emf is not affected by the direction of the magnetic field. However, the polarity of $\mathcal{V}_{\text{ind}}$ is of course reversed.

7.2.3 Differential Form of Faraday's Law

The most general form of Faraday's law, namely [7.5], can be expressed in differential form by using Stokes's theorem. We have

$$\oint_C \mathbf{E} \cdot d\mathbf{l} = \int_S (\nabla \times \mathbf{E}) \cdot d\mathbf{s} = -\frac{d}{dt}\int_S \mathbf{B} \cdot d\mathbf{s}$$

which immediately yields a differential equation involving **E** and **B** *if* it is legitimate to carry out the following operation:

$$\frac{d}{dt}\int_S \mathbf{B} \cdot d\mathbf{s} = \int_S \frac{\partial \mathbf{B}}{\partial t} \cdot d\mathbf{s}$$

The interchange of integration over the variables in S and the differentiation with respect to time is legitimate if C is an arbitrary contour fixed in space, that is, if C is fixed and stationary. With this interpretation of C, the integrands in the preceding equation can be equated since S is arbitrary. Thus, the differential form of Faraday's law in a stationary frame of reference[19] is

$$\boxed{\nabla \times \mathbf{E} = -\frac{\partial \mathbf{B}}{\partial t}} \qquad [7.6]$$

Equation [7.6] is one of the fundamental laws of electromagnetics, namely Maxwell's equations. The fields **E** and **B** in [7.6] must be measured in the *same* coordinate system since the contour C must be fixed when converting from the integral form of Faraday's law to the differential form. In other words, the spatial derivatives of **E** are related to the time derivative of **B** at a point according to [7.6] only if measurements of these derivatives are made by the same observer.

7.2.4 Faraday's Law in Terms of the Vector Potential A

Faraday's law can be alternately stated in terms of the magnetic vector potential **A**. Since $\mathbf{B} = \nabla \times \mathbf{A}$, we have from [7.5]

$$\mathcal{V}_{\text{ind}} = -\frac{d\Psi}{dt}$$

$$\oint_C \mathbf{E} \cdot d\mathbf{l} = -\frac{d}{dt}\int_S (\nabla \times \mathbf{A}) \cdot d\mathbf{s}$$

$$\oint_C \mathbf{E} \cdot d\mathbf{l} = -\frac{d}{dt}\oint_C \mathbf{A} \cdot d\mathbf{l} \qquad [7.7]$$

where the last step follows from Stokes's theorem. Equation [7.7] is generally valid, even if **A** varies with time while the contour C is in motion. When contour C and surface S are stationary, we can rewrite [7.7] as

[19] It can be shown that Faraday's law for a moving medium is

$$\nabla \times \mathbf{E}' = -\frac{d\mathbf{B}}{dt} = -\left[\frac{\partial \mathbf{B}}{\partial t} + (\mathbf{v} \cdot \nabla)\mathbf{B}\right]$$

where $\mathbf{E}'$ is measured in the moving frame of reference and the velocity **v** is assumed to be small relative to the velocity of light in free space. See Appendix A for an expanded form of $(\mathbf{v} \cdot \nabla)\mathbf{B}$. However, a stationary observer sees an electric field $\mathbf{E} = \mathbf{E}' - \mathbf{v} \times \mathbf{B}$ for which we have $\nabla \times \mathbf{E} = -\partial \mathbf{B}/\partial t$. For further discussion, see Chapter 9 of W. K. H. Panofsky and M. Phillips, *Classical Electricity and Magnetism* 2nd. ed., Addison Wesley, 1962.

$$\oint_C \mathbf{E} \cdot d\mathbf{l} = -\oint_C \frac{\partial \mathbf{A}}{\partial t} \cdot d\mathbf{l} \qquad [7.8]$$

Since [7.8] must be valid for *any* path C, we must have

$$\mathbf{E} = -\frac{\partial \mathbf{A}}{\partial t}$$

at each point, as long as $\mathbf{E}$ represents the induced electric field. It is important here again to distinguish between induced electric field and the electric field associated with stationary electric charges. For electrostatic fields $\mathbf{E}_{\text{es}}$ associated with electric charges, we have $\oint \mathbf{E}_{\text{es}} \cdot d\mathbf{l} = 0$; for such fields, we can define an electrostatic potential Φ such that $\mathbf{E}_{\text{es}} = -\nabla\Phi$. Induced electric fields, on the other hand, are generated precisely because $\oint \mathbf{E} \cdot d\mathbf{l} \neq 0$. If stationary electrical charges are present in addition to electromagnetic induction, their effects can be described by a scalar potential Φ, producing an additional electric field $\mathbf{E}_{\text{es}} = -\nabla\Phi$. Since $\mathbf{E}_{\text{es}}$ is conservative and its curl is thus zero, the total electric field $\mathbf{E}_{\text{t}} = \mathbf{E}_{\text{es}} + \mathbf{E}$ still obeys [7.6]. In other words, we have

$$\nabla \times \mathbf{E}_{\text{t}} = -\frac{\partial \mathbf{B}}{\partial t}$$

$$\nabla \times \mathbf{E}_{\text{t}} = -\frac{\partial}{\partial t}(\nabla \times \mathbf{A})$$

$$\nabla \times \left(\mathbf{E}_{\text{t}} + \frac{\partial \mathbf{A}}{\partial t}\right) = 0$$

Since any vector with zero curl can be expressed as the gradient of a scalar (see Appendix A), we can write

$$\mathbf{E}_{\text{t}} + \frac{\partial \mathbf{A}}{\partial t} = -\nabla\Phi \quad \rightarrow \quad \boxed{\mathbf{E}_{\text{t}} = -\nabla\Phi - \frac{\partial \mathbf{A}}{\partial t}} \qquad [7.9]$$

showing that the total electric field $\mathbf{E}_{\text{t}}$ is produced either by electromagnetic induction $(-\partial\mathbf{A}/\partial t)$ or stationary electric charges $(-\nabla\Phi)$, or both. Note that for the static case, with $\partial\mathbf{A}/\partial t = 0$, [7.9] reduces to $\mathbf{E}_{\text{t}} = -\nabla\Phi$, so that $\mathbf{E}_{\text{t}}$ can be determined from Φ alone, and $\mathbf{B}$ can be determined from $\mathbf{A}$ alone. For time-varying fields, the total electric field $\mathbf{E}_{\text{t}}$ depends on both Φ and $\mathbf{A}$.

7.3 ENERGY IN A MAGNETIC FIELD

Since current-carrying loops exert forces and torques on each other, a finite amount of work must be expended in establishing or modifying a given configuration of current loops. It is thus to be expected that a system of conductors carrying steady-state currents stores magnetic energy (W_{m}) in a manner analogous to a configuration of electric charges storing electrostatic energy (W_{e}). In Chapter 6, we deferred the discussion of methods of evaluation of W_{m} until after we were equipped with Faraday's law, because the establishment of the current configurations necessarily involves a transient period during which currents and associated fields are increased from zero to their final values. We now consider the magnetic energy of current-carrying loops and express the stored energy solely in terms of the fields produced by the currents. Once this is done, we can ascribe energy to the magnetic field in the same manner that energy was associated with the electrostatic field in Section 4.12.

In electrostatics, when we considered the establishment of charge configurations, the calculation of the work expended depended on the existence of two well-defined physical quantities, namely the electric charge q and the electrostatic potential Φ. In magnetostatics, magnetic charges do not exist, and further, it is not possible, in general, to define a magnetostatic potential that is a single-valued function.[20] Thus the development of an expression for W_{m} must differ markedly from that used to calculate W_{e}.

We first consider placing all current-carrying loops at infinite distances apart from each other so that they do not initially exert forces or torques on one another. Then, one by one, the loops can be brought to their final positions and orientations while we calculate the work expended. Although such a procedure appears straightforward, we need to make two observations: (1) Each current-carrying loop or circuit contains finite energy; that is, the loop would change its shape if it were not for the mechanical stresses in the wires and so on that hold it together (this is the *self-energy* of a current-carrying loop); (2) a certain amount of work is necessary to keep the currents flowing in the loops as they are moved into their positions.[21] Alternatively, we can think of current-carrying loops as being initially at their fixed positions and orientations but carrying zero current. As we establish current in the loops one by one, we have to exert electrical energy to oppose the induced electromotive forces. Consider for a moment that we have only a single circuit with a constant current I flowing in it. Faraday's law states that an electromotive force $\mathcal{V}_{\text{ind}}$ is induced around it if the magnetic flux Ψ through the circuit changes (for example, if we bring the loop closer to other current-carrying loops). In order to keep the current I constant, the sources of the current must do work. The amount of work[22] that must be done in a time interval Δt is given by $\Delta W = -I\mathcal{V}_{\text{ind}}\Delta t = I\Delta\Psi$, where we have used $\mathcal{V}_{\text{ind}} = -\Delta\Psi/\Delta t$ (Faraday's law). The sources can supply this work by providing an applied voltage of $-\mathcal{V}_{\text{ind}}$ for the time interval Δt, while the current I flows.

---

[20]Although it is true that we can always find a vector $\mathbf{A}$ such that $\mathbf{B} = \nabla\times\mathbf{A}$, the vector $\mathbf{A}$ is not uniquely specified until we also define its divergence $\nabla\cdot\mathbf{A}$. In electromagnetic problems, it is customary to use the Lorentz gauge, or Lorentz condition, where the divergence of $\mathbf{A}$ is taken to be $\nabla\cdot\mathbf{A} = -(\mu\epsilon)\,\partial\Phi/\partial t$. This particular choice uncouples the equations for the magnetic vector potential $\mathbf{A}$ and the scalar electric potential Φ, making the formulation and solution of antenna radiation problems relatively easier.

[21]The work referred to here includes only the *reversible* work that must be expended because of the induced electromotive force (via Faraday's law); it does not include the Joule heating (i.e., ohmic losses, or I^2R losses). We are also not concerned here with the mechanical work required to move the coils physically.

[22]The time rate of change in energy of a particle with velocity $\mathbf{v}$ acted on by a force $\mathbf{F}$ is $\mathbf{F}\cdot\mathbf{v}$. When changing flux induces an additional electric field $\mathbf{E}$, the energy of each conduction electron of charge q_e and drift velocity $\mathbf{v}_d$ increases by $|q_e\mathbf{v}_d\cdot\mathbf{E}|$ per unit time. The power required to sustain the current (i.e., $-I\mathcal{V}_{\text{ind}}$) corresponds to a summation over all of the electrons in the circuit.

We now proceed to use Faraday's law to determine the self-energy of a single current loop, the magnetic energy of a system of current-carrying loops, and, finally, the magnetic energy in terms of the **B** and **H** fields.

7.3.1 Self-Energy of a Current-Carrying Loop

We consider the self-energy of a single filamentary loop carrying a current I. To determine the self-energy of such a loop, consider the magnetic flux Ψ threading through a loop carrying a current I. The magnetic flux Ψ can be written in terms of the **B** field or the vector potential **A** as

$$\Psi = \int_S \mathbf{B} \cdot d\mathbf{s} = \int_S (\nabla \times \mathbf{A}) \cdot d\mathbf{s} = \oint_C \mathbf{A} \cdot d\mathbf{l}$$

where S is the area enclosed by the loop defined by the contour C, and where we have invoked Stokes's theorem. From Section 6.5, we know that the vector potential is related to the current I as

$$\mathbf{A}(\mathbf{r}) = \frac{\mu_0 I}{4\pi} \oint_C \frac{d\mathbf{l}'}{R}$$

where $R = |\mathbf{r} - \mathbf{r}'|$. Thus, we have

$$\Psi = \oint_C \mathbf{A} \cdot d\mathbf{l} = I \left[\frac{\mu_0}{4\pi} \oint_C \oint_C \frac{d\mathbf{l} \cdot d\mathbf{l}'}{R} \right] = IL$$

In view of the definition of inductance as magnetic flux linked per unit current, the preceding term in square brackets is the *self-inductance* L of the loop. Note that this quantity depends solely on the geometry of the loop and is thus a constant for a loop with fixed geometry. Note also that the term in square brackets is a special case of the Neumann formula for inductance that was derived in Section 6.7 for the mutual inductance of two circuits.

The self-inductance of a circuit acts as a sort of electric inertia. Since $\Psi = IL$ and L is a constant determined by geometry, whenever the current in the circuit varies, there is a corresponding variation in magnetic flux, which by Faraday's law gives rise to an electromotive force tending to prevent the variation. The self-energy of a loop carrying current can thus be determined by considering the fact that the establishment of a current I requires energy in order to oppose this induced electromotive force. If the current in the loop is increased by an amount ΔI, the change in the magnetic flux threading through the surface S is $\Delta\Psi = L\Delta I$. The rate at which magnetic energy is added to the circuit when the current is increased by ΔI is

$$\frac{dW_{\text{m}}}{dt} = -I\mathcal{V}_{\text{ind}} = I\frac{\Delta\Psi}{\Delta t} = IL\frac{dI}{dt}$$

when we consider a differential change in current such that $\Delta I \to 0$. Consequently, the self-energy W_{ms} of the loop, defined as the total energy added to the loop to increase its current from zero to a final value of I, is

$$W_{\text{ms}} = \int_0^I \left(\frac{dW_{\text{ms}}}{dt}\right) dt = L\int_0^I I'\,dI' = \frac{1}{2}LI^2$$

where I' is simply used to denote the dummy variable of integration. We have thus arrived at the general result that we know from circuit theory, namely that the magnetic energy stored in an inductance L carrying current I is given by $W_m = \frac{1}{2}LI^2$.

7.3.2 Magnetic Energy of a System of Current-Carrying Loops

Consider two conducting loops C_1 and C_2, as shown in Figure 7.15, in which we would like to establish currents I_1 and I_2, respectively. Work must be done to increase the currents from zero to their final values. This work will result in energy being stored in the magnetic field surrounding the conductors. To evaluate this energy, we can start by assuming that both of the loops initially have zero current. We can then increase the current in C_1 from zero to I_1, while keeping the current in C_2 zero. This action will have two consequences: (1) An emf will be induced in C_1 due to the self-inductance of C_1; (2) an emf will be induced in C_2 due to the mutual inductance between C_1 and C_2. The work that must be supplied to overcome the first emf is the self-energy of C_1. An additional amount of energy must be supplied to C_2 in order to maintain zero current in it while the current in C_1 is being changed. Once the current I_1 is established in C_1, the next step is to increase the current in C_2 from zero to I_2. Once again, this action will have two consequences: (1) An emf will be induced in C_2 due to the self-inductance of C_2, requiring that energy be supplied that will eventually be stored as the self-energy of C_2; (2) an emf will be induced in C_2 due to the mutual inductance between C_2 and C_1, so that additional energy must be supplied in order to maintain the current in C_1 as I_1.

We now analyze the preceding sequence of events more quantitatively. Let us first maintain the current i_2 in C_2 at zero while we increase the current i_1 in C_1 from zero to I_1. When we change the current in C_1 by an amount di_1 in a time interval dt, the magnetic field $\mathbf{B}_1$ due to the current in C_1 changes at a rate of $d\mathbf{B}_1/dt$ and will thus produce an induced voltage $\mathcal{V}_{\text{ind}_1} = -d\Psi_{11}/dt$ in C_1 and an induced voltage $\mathcal{V}_{\text{ind}_2} = -d\Psi_{12}/dt$ in C_2. Thus, in order to increase the current in C_1 by an amount di_1 in a time interval dt, we must apply a voltage $-\mathcal{V}_{\text{ind}_1}$ in C_1. At the same time, we must apply a voltage $-\mathcal{V}_{\text{ind}_2}$ in C_2 in order to maintain[23] its current at zero. The work done in a time interval dt by an applied voltage of $-\mathcal{V}_{\text{ind}_1}$ is

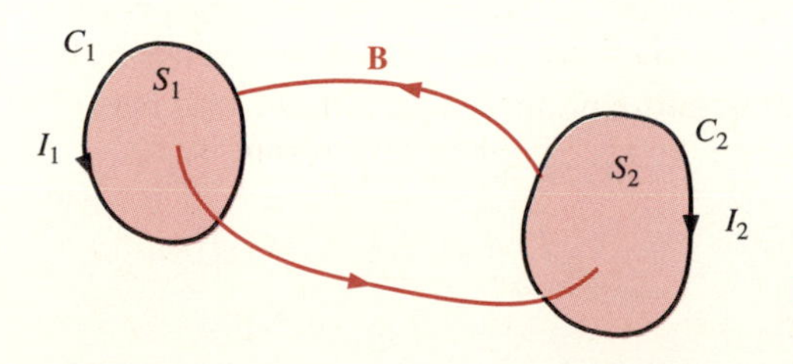

FIGURE 7.15. Two magnetically coupled current-carrying loops.

[23]Note that having a voltage of $\mathcal{V}_{\text{ind}_2} = -d\Psi_{12}/dt = -L_{12}\,di_2/dt$ induced in circuit C_2 means that the time derivative of i_2 is nonzero and that i_2 will thus change from its value of zero, unless we keep the net total voltage in circuit C_2 zero by applying $-\mathcal{V}_{\text{ind}_2}$.

$$dW_1 = -\mathcal{V}_{\text{ind}_1} i_1\, dt = i_1\, d\Psi_{11} = L_{11} i_1\, di_1$$

where we have used the definition of the self-inductance of loop C_1 as $L_{11} i_1 = \Psi_{11}$ and noted that $L_{11}\, di_1 = d\Psi_{11}$, because L_{11} is constant. The voltage $-\mathcal{V}_{\text{ind}_2}$ applied in C_2 does no work, because its current i_2 is kept at zero. The total work done in increasing i_1 from zero to I_1 is thus

$$W_1 = \int_0^{I_1} L_{11} i_1\, di_1 = \tfrac{1}{2} L_{11} I_1^2$$

which is the self-energy of the single loop C_1. Since the current i_2 in C_2 is kept at zero, the physical presence of the loop C_2 has no effect on loop C_1.

The next step is to keep $i_1 = I_1$ and increase i_2 by an amount di_2 in a time interval dt. This results in the induced voltages

$$\mathcal{V}_{\text{ind}_2} = -\frac{d\Psi_{22}}{dt} = -L_{22}\frac{di_2}{dt} \quad \text{in } C_2$$

$$\mathcal{V}_{\text{ind}_1} = -\frac{d\Psi_{21}}{dt} = -L_{21}\frac{di_2}{dt} \quad \text{in } C_1$$

To maintain i_1 at a constant value I_1, we must apply a voltage $-\mathcal{V}_{\text{ind}_1}$, which in time dt does work of an amount

$$dW_{21} = -\mathcal{V}_{\text{ind}_1} I_1\, dt = I_1 L_{21}\, di_2$$

Note that dW_{21} can be negative (work taken out of the system), depending on whether $\mathcal{V}_{\text{ind}_1}$ tends to increase or decrease i_1. Similarly, the work done by the voltage $-\mathcal{V}_{\text{ind}_2}$ that must be applied to change i_2 by an amount di_2 is

$$dW_2 = -\mathcal{V}_{\text{ind}_2} i_2\, dt = L_{22} i_2\, di_2$$

The total work done in raising i_2 from zero to I_2 while keeping i_1 at I_1 is then

$$\begin{aligned} W_{21} + W_{22} &= I_1 L_{21} \int_0^{I_2} di_2 + L_{22} \int_0^{I_2} i_2\, di_2 \\ &= I_1 I_2 L_{21} + \tfrac{1}{2} I_2^2 L_{22} \end{aligned}$$

The total work done on the system to establish the currents I_1 and I_2 in loops C_1 and C_2, respectively, is the sum of W_1 and $(W_{21} + W_{22})$, which represents the energy W_{m} stored in the magnetic field surrounding the two loops. This energy is thus given by

$$\begin{aligned} W_{\text{m}} &= W_1 + W_{21} + W_{22} \\ &= \tfrac{1}{2} L_{11} I_1^2 + L_{21} I_1 I_2 + \tfrac{1}{2} L_{22} I_2^2 \\ &= \tfrac{1}{2} \sum_{i=1}^{2} \sum_{j=1}^{2} L_{ij} I_i I_j \end{aligned}$$

The final result written in terms of a double summation can easily be generalized to a system of N loops by simply extending the summations for both i and j up to N. This equation for W_{m} provides an interpretation of the coefficients of inductance L_{ij} as the coefficients in the quadratic expression for the energy stored in a magnetic field. Note that L_{ij} $(i \neq j)$ may be either positive or negative, depending on the direction in which the mutual magnetic flux links the respective circuits. The sign of L_{ij} depends on the choices of positive directions for currents I_i and I_j. If the directions of I_i and I_j are such that they produce magnetic flux in the same direction, then L_{ij} is positive, since the stored magnetic energy is increased. If the magnetic fluxes produced by I_i and I_j tend to cancel one another, L_{ij} is negative.

Examples 7-8 and 7-9 illustrate the determination of the magnetic energy stored by coupled circuits.

Example 7-8: Magnetic energy stored by parallel coaxial coils. Consider the parallel coaxial coils considered in Example 6-30 and shown in Figure 6.39. If $a = 2$ cm, $b = 1$ cm, $d = 10$ cm, and $N_1 = N_2 = 10$, calculate the total energy stored in this system if currents of $I_1 = I_2 = 1$ mA flow in both coils. Assume that the thickness of the wire that the coils are made of is 0.4 mm.

Solution: The self-inductance of a single circular loop of wire was determined in Example 6-31. For wire thickness $2t$ much less than the loop radius a, the self-inductance is given by

$$L \simeq \mu_0 a \left(\ln\frac{8a}{t} - 2\right)$$

The self-inductance of a loop with N turns is simply N^2 times larger, assuming that each turn produces and links the same magnetic flux (such is the case if the coils are wound tightly enough and if the number of turns is not large, as shown in Figure 6.39). Using this formula, the self-inductances of coils 1 and 2, having radii of $a = 2$ cm and $b = 1$ cm, respectively, are

$$L_{11} \simeq N^2\mu_0 a \left(\ln\frac{8a}{t} - 2\right)$$

$$= 10^2(4\pi \times 10^{-7})(2 \times 10^{-2}\text{ m})\left[\ln\left(\frac{8 \times 2 \times 10^{-2}\text{ m}}{0.2 \times 10^{-3}\text{ m}}\right) - 2\right] \simeq 12\ \mu\text{H}$$

and $L_{22} \simeq 5\ \mu$H.

The mutual inductance between two parallel coaxial coils as shown in Figure 6.39 were evaluated in Example 6-30 to be

$$L_{12} \simeq \frac{\mu_0 \pi N_1 N_2 a^2 b^2}{2d^3}$$

for the case when $d \gg a, b$, which is the case in hand with $d = 0.1$ m. Using this expression for L_{12}, we find

$$L_{12} \simeq \frac{(4\pi \times 10^{-7})\pi(10)(10)(0.02)^2(0.01)^2}{2(0.1)^3} \simeq 0.008\ \mu\text{H}$$

Thus we see that the mutual inductance between the two coils is much less than their respective self-inductances. The main reason for this is that the coils are relatively far apart, with $d = 10$ cm. Note that L_{12} is inversely proportional to d^3, so that it is larger by approximately a factor of 1000 if the coils are 1 cm (instead of 10 cm) apart. However, care should be exercised here since the expression for L_{12} derived in Example 6-30 is only precisely valid for $d \gg a, b$.

The sign of the mutual inductance L_{12} depends on the definition of the polarity of the currents I_1 and I_2 in the two circuits. If these are defined such that positive values of I_1 and I_2 produce magnetic flux in the same direction, then L_{12} is positive. The total magnetic energy stored in the system of two parallel coaxial coils is thus

$$\begin{aligned} W_\text{m} &= \tfrac{1}{2}L_{11}I_1^2 \pm L_{12}I_1I_2 + \tfrac{1}{2}L_{22}I_2^2 \\ &\simeq \tfrac{1}{2}(12 \times 10^{-6}\ \text{H})(10^{-3}\ \text{A})^2 \pm (8 \times 10^{-9}\ \text{H})(10^{-3}\ \text{A})(10^{-3}\ \text{A}) \\ &\quad + \frac{1}{2}(5 \times 10^{-6}\ \text{H})(10^{-3}\ \text{A})^2 \\ &\simeq (8.5 \pm 0.008) \quad \text{pJ} \end{aligned}$$

Note that the mutual coupling, numerically small in this case, can either slightly decrease or slightly increase the stored energy, depending on the relative directions of the two currents I_1 and I_2.

Example 7-9: A solenoid with a secondary winding. Find the total magnetic energy stored by the solenoid shown in Figure 7.16, which has a secondary winding wound concentrically over the main winding. Neglecting leakage flux from the ends, determine the total magnetic energy stored in this system if currents of $I_1 = I_2 = 1$ mA are flowing through the two coils and other parameters are $N_1 = 1000$, $N_2 = 100$, $l_1 = 10$ cm, $l_2 = 5$ cm, and $a = 1$ cm. Assume a nonmagnetic core ($\mu = \mu_0$).

Solution: From Example 6-26, we know that the self-inductance of a solenoid of length l and having N turns with a cross-sectional area of A is $L = \mu_0 N^2 A/l$. Thus, we have

$$L_{11} = \frac{\mu_0 N_1^2 \pi a^2}{l_1} = \frac{(4\pi \times 10^{-7})(1000)^2[\pi(0.01)^2]}{0.10} \simeq 3.95 \text{ mH}$$

and similarly, we find $L_{22} \simeq 0.079$ mH. The mutual inductance between the two coils is given by $L_{12} = N_2\Lambda_{12}/I_1$. Once again invoking the result of Example 6-26, the flux produced by the primary coil is

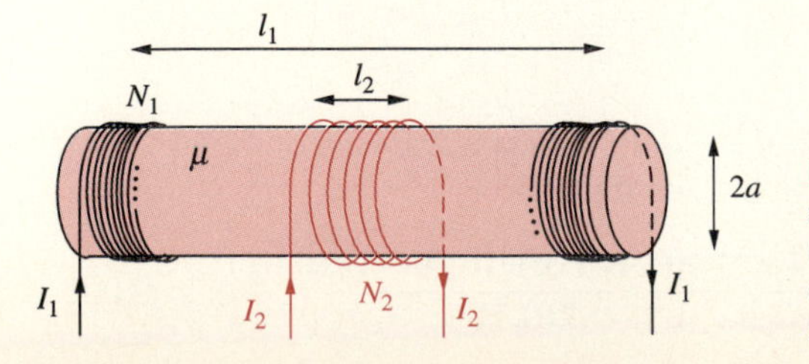

FIGURE 7.16. A solenoid with a secondary winding. The primary winding (N_1 turns) in fact covers the full length (l_1) of the core, but the middle portion is not shown to avoid clutter in the diagram.

$$\Lambda_{12} = \frac{\mu_0 N_1 I_1 \pi a^2}{l_1}$$

Since this flux is linked N_2 times by the secondary coil, we have

$$L_{12} = \frac{N_2 \Lambda_{12}}{I_1} = \frac{\mu_0 N_1 N_2 \pi a^2}{l_1}$$

Upon substitution of the parameter values, this yields $L_{12} \simeq 0.395$ mH. Thus we see that the mutual inductance between the two coils is larger than the self-inductance of the secondary coil.

The total magnetic energy stored in this system is given by

$$\begin{aligned} W_\mathrm{m} &= \tfrac{1}{2}L_{11}I_1^2 \pm L_{12}I_1I_2 + \tfrac{1}{2}L_{22}I_2^2 \\ &\simeq \tfrac{1}{2}[(3.95 \times 10^{-3})(10^{-3})^2 + (0.079 \times 10^{-3})(10^{-3})^2] \pm (0.395 \times 10^{-3})(10^{-3})(10^{-3}) \\ &\simeq (2.01 \pm 0.395) \quad \text{nJ} \end{aligned}$$

If the currents I_1 and I_2 are positive with directions as shown in Figure 7.16, they reinforce one another so that L_{12} is positive and $W_\mathrm{m} \simeq 2.41$ nJ. Otherwise, the magnetic flux produced by I_1 is reduced as a result of that produced by I_2, so that the total energy stored is $W_\mathrm{m} \simeq 1.62$ nJ.

7.3.3 Magnetic Energy in Terms of the Fields

As we did for electrostatics in Section 4.12, it is possible to express the stored magnetic energy of a configuration of current-carrying loops solely in terms of the **B** and **H** fields. To this end, we start with the general expression derived in Section 7.3.2 for the energy of N current-carrying loops, namely

$$W_\mathrm{m} = \tfrac{1}{2}\sum_{i=1}^{N}\sum_{j=1}^{N} L_{ij}I_iI_j$$

From Section 6.7, we recall the Neumann formula for the mutual inductance between two circuits C_i and C_j as

$$L_{ij} = \frac{\mu_0}{4\pi}\oint_{C_i}\oint_{C_j}\frac{d\mathbf{l}_i \cdot d\mathbf{l}_j}{R_{ij}}$$

where R_{ij} is the distance between the differential lengths $d\mathbf{l}_i$ and $d\mathbf{l}_j$. Using this, we have

$$\begin{aligned} W_\mathrm{m} &= \frac{1}{2}\sum_{i=1}^{N} I_i\left[\sum_{j=1}^{N} I_j L_{ij}\right] = \frac{1}{2}\sum_{i=1}^{N} I_i \sum_{j=1}^{N} I_j \left[\frac{\mu_0}{4\pi}\oint_{C_i}\oint_{C_j}\frac{d\mathbf{l}_i \cdot d\mathbf{l}_j}{R_{ij}}\right] \\ &= \frac{1}{2}\sum_{i=1}^{N} I_i \oint_{C_i}\left[\frac{\mu_0}{4\pi}\sum_{j=1}^{N} I_j \oint_{C_j}\frac{d\mathbf{l}_j}{R_{ij}}\right]\cdot d\mathbf{l}_i \end{aligned}$$

Recalling from Section 6.5 the definition of the vector potential **A**, we recognize that the expression in the square brackets is the vector potential $\mathbf{A}_i$ at a point on the loop identified by subscript i, due to (or produced by) currents flowing in all the loops. Therefore we can write

$$W_{\mathrm{m}} = \frac{1}{2}\sum_{i=1}^{N}\oint_{C_i} I_i d\mathbf{l}_i \cdot \mathbf{A}$$

In terms of the more general volume current density $\mathbf{J}$, and noting that $\oint_C I\, d\mathbf{l}$ is equivalent to $\int_V \mathbf{J}\, dv$, the preceding expression can be further generalized as

$$W_{\mathrm{m}} = \frac{1}{2}\int_V \mathbf{J}\cdot\mathbf{A}\, dv \qquad [7.10]$$

where the volume V is the entire region of space within which $\mathbf{J}\cdot\mathbf{A} \neq 0$.

Equation [7.10] is analogous to the electrostatic version, namely [4.73]. This expression identifies the quantity $\frac{1}{2}\mathbf{J}\cdot\mathbf{A}$ as the volume density of magnetostatic energy. Using the relationships between $\mathbf{A}$, $\mathbf{B}$ and $\mathbf{J}$, $\mathbf{H}$, we can write W_{m} entirely in terms of the $\mathbf{B}$ or $\mathbf{H}$ fields. To do this, we proceed in a manner identical to that used in electrostatics. Noting that in free space we have $\nabla\times\mathbf{H} = \mathbf{J}$, and using the vector identity (see Appendix A)

$$\nabla\cdot(\mathbf{H}\times\mathbf{A}) \equiv (\nabla\times\mathbf{H})\cdot\mathbf{A} - (\nabla\times\mathbf{A})\cdot\mathbf{H}$$

we have

$$\begin{aligned}
W_{\mathrm{m}} &= \frac{1}{2}\int_V \mathbf{J}\cdot\mathbf{A}\, dv = \frac{1}{2}\int_V (\nabla\times\mathbf{H})\cdot\mathbf{A}\, dv \\
&= \frac{1}{2}\int_V (\nabla\times\mathbf{A})\cdot\mathbf{H}\, dv + \frac{1}{2}\int_V \nabla\cdot(\mathbf{H}\times\mathbf{A})\, dv \\
&= \frac{1}{2}\int_V \mathbf{H}\cdot\mathbf{B}\, dv + \frac{1}{2}\oint_S (\mathbf{H}\times\mathbf{A})\cdot d\mathbf{s}
\end{aligned}$$

where S is the surface that encloses the volume V and where we have used Stokes's theorem to rewrite the second term as a surface integral. Note that the surface S and the volume V must be large enough to enclose all currents in the region of interest. Since there are no other restrictions on the choice of the surface S, we can extend it to infinity. In this case, the surface integral reduces to zero because dependencies of the various quantities at large distances from the currents (i.e., at points from which the entire distribution of currents looks just like an elemental current) are

$$|\mathbf{A}| \sim \frac{1}{r}, \qquad |\mathbf{H}| \sim \frac{1}{r^2}, \quad \text{and} \quad |d\mathbf{s}| \sim r^2$$

so that

$$\lim_{r\to\infty}(\mathbf{H}\times\mathbf{A})\cdot d\mathbf{s} \quad \to \lim_{r\to\infty}\frac{1}{r}\frac{1}{r^2}r^2 \to 0$$

In other words, the magnetostatic energy W_{m} reduces to simply the volume integral term, or

$$\boxed{W_{\rm m} = \tfrac{1}{2}\int_V \mathbf{H}\cdot\mathbf{B}\,dv} \qquad [7.11]$$

Note that since we let S extend to infinity, we must have $V \to \infty$, meaning that the integral in [7.11] has to be carried out over *all* space in which the magnetic field is nonzero. Based on [7.11], we can define the *volume energy density* for the magnetostatic field as

$$\boxed{w_{\rm m} = \tfrac{1}{2}\mathbf{H}\cdot\mathbf{B}} \qquad [7.12]$$

For linear[24] and isotropic materials, where the permeability μ is a simple constant, we have $\mathbf{H}\cdot\mathbf{B} = \mu H^2$, so that the magnetic energy density is $w_{\rm m} = \frac{1}{2}\mu H^2$ in units of J-m^{-3}. Alternatively, magnetic energy density can be expressed in terms of the $\mathbf{B}$ field as $w_{\rm m} = \frac{1}{2}(B^2/\mu)$. The relationship between magnetic energy density and the magnetic field $\mathbf{H}$ is entirely analogous to the corresponding expression for the electrostatic field, namely $w_{\rm e} = \frac{1}{2}\epsilon E^2$, found in Section 4.12.

As in the electrostatic case, we note that the two alternative expressions for magnetic energy density, namely $\frac{1}{2}\mathbf{J}\cdot\mathbf{A}$ and $\frac{1}{2}\mathbf{H}\cdot\mathbf{B}$, are quite different. The former implies that magnetic energy exists at places where currents exist and is zero where $\mathbf{J} = 0$. However, the latter indicates that magnetic energy exists wherever the fields exist. As before, both points of view have merit, and it is neither necessary nor possible to determine which one is more "correct." It is not possible to localize energy or to decide whether it is associated with the current or the field. Thus, the quantities $\frac{1}{2}\mathbf{J}\cdot\mathbf{A}$ and $\frac{1}{2}\mathbf{H}\cdot\mathbf{B}$ represent magnetic energy density only to the extent that their volume integral over space is the total stored magnetic energy.

Example 7-10 illustrates the use of [7.12] to determine the magnetic energy stored in the vicinity of infinitely long wires of circular cross section.

Example 7-10: Magnetic energy near long, straight, current-carrying wires. Compare the magnetic energy stored per unit length by two separate infinitely long, straight wires of different radii a and b such that $b > a$; each wire has a circular cross section and carries a uniformly distributed current I.

---

[24]Even Equation [7.11] is not the most general form of magnetic energy applicable for nonlinear materials. The work done by external sources in establishing a magnetic field of $\mathbf{B}$ in material media, which by definition is the magnetic energy stored in that field, is given by

$$W_{\rm m} = \int_V \left[\int_0^{\mathbf{B}} \mathbf{H}\cdot d\mathbf{B}\right] dv$$

where the integral is to be carried out over all space. For further discussion, see Section 2.15 of J. A. Stratton, *Electromagnetic Theory,* McGraw-Hill, New York, 1941. On the other hand, equations [7.11] and [7.12] apply to linear but anisotropic materials, in which the permeability is a tensor, such as in the case of ferrites. See §5 of A. Sommerfeld, *Electrodynamics,* Academic Press, New York, 1952.

Solution: The **B** field at any position r in the vicinity of a wire carrying uniformly distributed current and having a circular cross section of radius a was determined in Example 6-10 to be

$$\mathbf{B} = \begin{cases} \hat{\boldsymbol{\phi}} \dfrac{\mu_0 I r}{2\pi a^2} & r \leq a \\ \hat{\boldsymbol{\phi}} \dfrac{\mu_0 I}{2\pi r} & r > a \end{cases}$$

To determine the magnetic energy stored per unit length, we can integrate $\frac{1}{2}(B_\phi^2/\mu_0)$ over a cylindrical volume of 1-m axial length. In other words,

$$\begin{aligned} W_{\mathrm{m}} &= \frac{1}{2\mu_0} \int_V B_\phi^2 \, dv = \frac{1}{2\mu_0} \int_0^\infty B_\phi^2 2\pi r \, dr \\ &= \frac{1}{2\mu_0} \frac{\mu_0^2 I^2}{(2\pi a^2)^2} \int_0^a r^2 2\pi r \, dr + \frac{1}{2\mu_0} \frac{\mu_0^2 I^2}{(2\pi)^2} \int_a^\infty \frac{1}{r^2} 2\pi r \, dr \\ &= \frac{\mu_0 I^2}{4\pi a^4} \left[\frac{r^4}{4} \right]_{r=0}^{r=a} + \frac{\mu_0 I^2}{4\pi} [\ln r]_{r=a}^{r=\infty} \\ &= \frac{\mu_0 I^2}{16\pi} + \infty \end{aligned}$$

which indicates that the energy residing in the magnetic field surrounding a 1-m length of such a wire is infinite. This result is physically not surprising, since the very premise of sustaining a current I in an infinitely long wire is nonphysical. In reality, any such current carried in one direction by a long straight wire has to return via another wire or set of wires. This return current would generate **B** fields in the opposite direction, thus reducing the net total **B** field so that the stored magnetic energy is not divergent.

An interesting aspect of the preceding expression for W_{m} is the fact that the first term, which represents the magnetic energy stored within the wire in the region $r \leq a$, is independent of the radius a of the wire. This internal magnetic energy given by $(W_{\mathrm{m}})_{\mathrm{int}} = \mu_0 I^2/(16\pi)$ is the quantity represented by the internal inductance of a wire discussed in Section 6.7. In other words, we have

$$L_{\mathrm{int}} = \frac{2(W_{\mathrm{m}})_{\mathrm{int}}}{I^2} = \frac{\mu_0}{8\pi}$$

Although the total magnetic energy stored per unit length for a straight wire is divergent, the difference between the energy stored by two different wires of radii a and b is finite. Since the internal stored energy is the same for both radii, the difference between the total magnetic energies stored per unit length for the two different wires is given by

$$W_{\mathrm{m}}^a - W_{\mathrm{m}}^b = \frac{1}{2\mu_0} \frac{\mu_0^2 I^2}{(2\pi)^2} \int_a^b \frac{1}{r^2} 2\pi r \, dr = \frac{\mu_0 I^2}{4\pi} \ln \frac{b}{a}$$

7.3.4 Magnetic Forces and Torques in Terms of Stored Magnetic Energy

In Section 4.13, we saw that electrostatic forces can be expressed in terms of spatial derivatives of stored electrostatic energy. The underlying principle used was called the principle of virtual displacement, which involves consideration of how the energy of a system changes for a small virtual change in geometry. In an analogous manner, magnetic forces and torques can also be derived from stored magnetic energy using the principle of virtual displacement. In this manner, one can usually calculate the lifting force of magnets.[25] In this section, we briefly discuss this topic with analogy to Section 4.13.

In Section 4.13, we noted that the relationship between electrostatic force and stored energy depended on the premise under which we allowed the virtual displacement to occur. Specifically, it was possible to consider virtual displacement under the constraint of constant charge or constant voltage. For magnetostatics, we can consider a system of current-carrying loops with constant currents or constant magnetic flux linkages. In the following, we consider only the case of constant current.

Consider a solenoid with a movable, permeable core, where $\mu > \mu_0$, as shown in Figure 7.17. A constant-current generator is attached to the coil, and the core of the solenoid is allowed to move a distance Δx, as shown. Assume that the movement of the core by an amount Δx changes the magnetic flux linked by the coil from Ψ_1 to Ψ_2. Note that the magnetic energy stored in this system is given by $W_m = \frac{1}{2}LI^2$, or alternatively $W_m = \frac{1}{2}I\Psi$, since $L = \Psi/I$. Thus, the change in flux corresponds to a change in stored magnetic energy of

$$\Delta W_m = \tfrac{1}{2}I\Psi_2 - \tfrac{1}{2}I\Psi_1 = \tfrac{1}{2}I\Delta\Psi$$

This additional energy must be provided by the generator in order to maintain the current I constant. The generator can do this by canceling out the induced emf by supplying a voltage of $-\mathcal{V}_{ind}$, in which case the amount of work done by the generator is

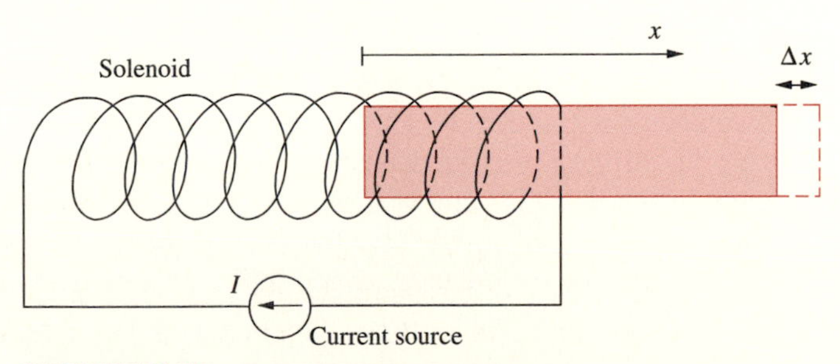

FIGURE 7.17. Solenoid with a movable core.

[25]For a discussion at an appropriate level, see Chapter 10 of M. A. Plonus, *Applied Electromagnetics*, McGraw-Hill, 1978.

$$\Delta W_{\text{g}} = I(-\mathcal{V}_{\text{ind}})\,\Delta t = I\frac{\Delta\Psi}{\Delta t}\,\Delta t = I\,\Delta\Psi = 2\,\Delta W_{\text{m}}$$

where Δt is the time duration over which the virtual displacement is assumed to occur. In order for energy to be conserved, we must have

$$F_x\,\Delta x + \Delta W_{\text{m}} = \Delta W_{\text{g}} \quad\rightarrow\quad F_x\,\Delta x + \Delta W_{\text{m}} = 2\,\Delta W_{\text{m}} \quad\rightarrow\quad F_x = \frac{\Delta W_{\text{m}}}{\Delta x}$$

where $F_x\,\Delta x$ is the virtual work done to move the core a distance Δx. In other words, F_x is the magnetic force exerted on the core by the coil, and F_x must be overcome in order to move the core. We have thus arrived at a result analogous to that found for electrostatic forces in Section 4.13.

For the case shown in Figure 7.17, under the constant-current constraint, the **H** field everywhere inside the coil remains constant. With the magnetic energy density being $\frac{1}{2}\mu H^2$, and for $\mu > \mu_0$ (i.e., paramagnetic or ferromagnetic material), the total stored magnetic energy decreases when the core moves to the right, so that less of it is within the coil. Thus, $F_x = dW_{\text{m}}/dx$ is negative, and the core experiences a force that tends to pull it within the coil, in a manner analogous to the case of the dielectric slab between parallel plates analyzed in Example 4-35. Note that if the material were diamagnetic, that is, if $\mu < \mu_0$, then the direction of the force would be reversed, tending to push the core out of the coil. In practice, however, μ is significantly different from μ_0 only for ferromagnetic materials, so that magnetic forces are significant only when the core is made out of such materials; in this case, the magnetic force tends to pull the core in, under constant-current conditions.

More generally, it can be shown that, under the constraint of constant currents, the magnetic force acting on a circuit is given by

$$[\mathbf{F}]_{I=\text{const.}} = \nabla W_{\text{m}}$$

If the circuit is constrained to rotate about the z axis, then the z component of the torque acting on it is given by

$$[T_z]_{I=\text{const.}} = \frac{\partial W_{\text{m}}}{\partial\phi}$$

Magnetic Forces in Terms of Mutual Inductance The principle of virtual displacement can also be used to determine magnetic forces and torques between current-carrying circuits. Consider, for example, two parallel coaxial loops carrying currents I_1 and I_2 and with the z axis passing through their centers, as shown in Figure 6.39. Based on what we found in Section 7.3.2, and also discussed in Example 7-8, the magnetic energy stored in this system is given by

$$W_{\text{m}} = \tfrac{1}{2}L_{11}I_1^2 \pm L_{12}I_1I_2 + \tfrac{1}{2}L_{22}I_2^2$$

If the currents I_1 and I_2 are maintained (i.e., kept constant) in both circuits (presumably by driving them with current sources), then the mutual inductance between the two coils is the only quantity that varies as a result of any relative motion of the two coils. Thus, an infinitesimal displacement by an amount Δz of either circuit with respect to the other results in a change of energy given by

$$\Delta W_m = I_1 I_2 \frac{\Delta L_{12}}{\Delta z} \Delta z = I_1 I_2 \, \Delta L_{12}$$

Note that ΔW_m is a positive quantity when I_1 and I_2 have the same sign and when ΔL_{12} is positive. Thus, the force between the two circuits, given by the gradient of W_m, tends to pull the circuits in the direction in which L_{12} increases most rapidly. In other words, the two circuits attract each other whenever their currents are in the same direction, since their mutual inductance obviously increases when they move closer. When I_1 and I_2 are in opposite directions, ΔW_m is negative and the circuits repel one another.

In general, it can be shown that under constant-current conditions, the magnetic force between two circuits is proportional to the gradient of mutual inductance, or

$$\mathbf{F} = I_1 I_2 \, \nabla L_{12}$$

and if the circuit is constrained to rotate about the z axis, then the z component of the torque acting on it is given by

$$T_z = I_1 I_2 \frac{\partial L_{12}}{\partial \phi}$$

7.4 DISPLACEMENT CURRENT AND MAXWELL'S EQUATIONS

In a series of brilliant papers between 1856 and 1865, culminating in his classic paper,[26] James Clerk Maxwell formulated the complete classical theory of electromagnetics.[27] He provided a mathematical framework for Faraday's results, clearly elucidated the different behavior of conductors and insulators under the influence of fields, imagined and introduced the concept of displacement current, and inferred the electromagnetic nature of light. His theoretical framework predicted the existence of electromagnetic waves in the absence of any experimental evidence. His hypotheses were to be confirmed 23 years later (in 1887) in the experiments of Heinrich Hertz.[28]

Hertz, Heinrich Rudolf *(German physicist, b. Hamburg, February 22, 1857; d. Bonn, January 1, 1894)* *Hertz studied physics at the University of Berlin under Helmholtz [1821–1894] and Kirchhoff [1824–1887]. He obtained his Ph.D.*

---

[26]For an excellent account with passages quoted from Maxwell's papers, see Chapter 5 of R. S. Elliott, *Electromagnetics,* IEEE Press, New Jersey, 1993.

[27]J. C. Maxwell, A dynamical theory of the electromagnetic field, *Phil. Trans. Royal Soc.* (London), 155, p. 450, 1865.

[28]H. Hertz, On the finite velocity of propagation of electromagnetic actions, *Sitzb. d. Berl. Akad. d. Wiss.,* Feb 2, 1888; for a collected English translation of this and other papers by H. Hertz, see H. Hertz, *Electric Waves,* MacMillan, London, 1893.

magna cum laude in 1880 and stayed on for two years more as an assistant to Helmholtz, his lifelong friend and mentor. Working at the University of Kiel in 1883, Hertz grew interested in Maxwell's equations.

In 1888, Hertz set up an electrical circuit that oscillated, providing current surging into first one, then another, of two metal balls separated by an air gap. Each time the potential reached a peak in one direction or the other, it sent a spark across the gap. With such an oscillating spark, Maxwell's equations predicted that electromagnetic radiation would be generated. Hertz used a simple loop of wire with a small air gap at one point to detect the possible radiation. Just as current gave rise to radiation in the first coil, so the radiation ought to give rise to a current in the second coil. Sure enough, Hertz was able to detect small sparks jumping across the gap in his detector coil. By moving his detector coil to various points in the room, Hertz could tell the shape of the waves by the intensity of spark formation, and he also showed that the waves involved both an electric and a magnetic field and were therefore electromagnetic in nature.

Hertz's experiments were quickly confirmed in England by Lodge [1851–1940], while Righi [1850–1920] in Italy demonstrated the relationship of these "Hertzian waves" to light. In 1889 Hertz became a professor of physics at the University of Bonn. Hertz died tragically before his thirty-seventh birthday, after a long illness due to chronic blood poisoning. [Adapted with permission from I. Asimov, Biographical Encyclopedia of Science and Technology, *Doubleday, 1982].*

1700 1857 1894 2000

The crucial step in Maxwell's development of the theory of electromagnetics was his introduction of the notion of a displacement current. In the following section, we discuss this concept.

7.4.1 Displacement Current and the General Form of Ampère's Law

Before Maxwell developed his theory of electromagnetics, Ampère's circuital law was known as the relationship between a static magnetic field $\mathbf{H}$ and the conduction-current density $\mathbf{J}$. From Chapter 6, the differential form of this relationship is

$$\nabla \times \mathbf{H} = \mathbf{J}$$

Another fundamental relation concerning conduction-current density was derived in Chapter 5 on the basis of conservation of charge. This relation, known as the continuity equation and given in [5.5], is repeated here:

$$\nabla \cdot \mathbf{J} + \frac{\partial \rho}{\partial t} = 0 \qquad [7.13]$$

In magnetostatics, we simply have $\nabla \cdot \mathbf{J} = 0$, since $\partial \rho / \partial t = 0$ by definition. We can then relate Ampère's law to the continuity equation by taking the divergence of the former,

$$\nabla \cdot (\nabla \times \mathbf{H}) = \nabla \cdot \mathbf{J} \quad \rightarrow \quad 0 = \nabla \cdot \mathbf{J}$$

where we have observed (see Appendix A) that the divergence of the curl of any vector is identically zero. We see that Ampère's law is indeed consistent with the principle of conservation of charge under static conditions. However, in the more general case when the field and source quantities are allowed to vary with time, Ampère's law, in its form used in Chapter 6, is clearly inconsistent with the fundamental principle of conservation of charge.

Faced with this inconsistency, Maxwell developed the time-varying form of Ampère's circuital law by postulating an additional current term $\mathbf{J}_d$ such that

$$\nabla \times \mathbf{H} = \mathbf{J}_c + \frac{\partial \mathbf{D}}{\partial t} = \mathbf{J}_c + \mathbf{J}_d \qquad [7.14]$$

where $\mathbf{J}_c$ is the conduction-current density and $\mathbf{J}_d = \partial\mathbf{D}/\partial t$ is the *displacement-current density*. Note that the divergence of [7.14] gives

$$\nabla \cdot (\nabla \times \mathbf{H}) = 0 = \nabla \cdot \mathbf{J}_c + \nabla \cdot \frac{\partial \mathbf{D}}{\partial t} = \nabla \cdot \mathbf{J}_c + \frac{\partial}{\partial t}(\nabla \cdot \mathbf{D})$$

Substituting for $\nabla \cdot \mathbf{D} = \rho$ from Gauss' law, we have

$$0 = \nabla \cdot \mathbf{J}_c + \frac{\partial \rho}{\partial t}$$

Thus we see that [7.14] is consistent with the continuity equation [7.13] and hence with the principle of conservation of charge.

Equation [7.14] states that the net circulation (or curl) of $\mathbf{H}$ is the sum of the conduction-current density $\mathbf{J}_c$ plus the displacement-current density $\mathbf{J}_d = \partial\mathbf{D}/\partial t$. Note that this new term has the dimensions of a current density and indicates that $\nabla \times \mathbf{H}$ may be generated (or sustained) at a point in space (even in the absence of $\mathbf{J}_c$) when a time-varying electric field is present. This relationship between $\mathbf{D}$ and $\mathbf{H}$ is analogous to Faraday's law [7.6], which states that $\nabla \times \mathbf{E}$ may be generated at a point in space when a time-varying magnetic field is present. The combined action of the $\nabla \times \mathbf{H}$ and $\nabla \times \mathbf{E}$ equations is precisely what leads to the propagation of electromagnetic waves, as discussed in Section 7.4.2 in connection with Figure 7.20. Without the displacement current, electromagnetic waves do not exist. It is not surprising that this term was not discovered earlier (e.g., in Ampère's experiments); in current-carrying wires, and especially at low frequencies, the magnitude of $\mathbf{J}_d$ is much smaller than that of $\mathbf{J}_c$ (see the discussion that follows). On the other hand, for electromagnetic waves propagating in free space, where there cannot be any conduction current, the displacement current $\mathbf{J}_d = \partial\mathbf{D}/\partial t$ is the only current term.

It is not surprising that the time rate of change of electric field $\mathbf{E}$ behaves like a current. In any given configuration of conductors, any increase (decrease) in $\mathbf{E}$ implies a buildup (decrease) of charge, which requires current flow. Consider a parallel-plate capacitor, as shown in Figure 7.18a. A current I_c charges up the left-hand plate to a surface charge density ρ_s, which means that current moves away from the right-hand plate to create an opposing charge density $-\rho_s$. The resulting electric field between the capacitor plates is $E = \rho_s/\epsilon_0$. If the current flow is continuous, ρ_s (and

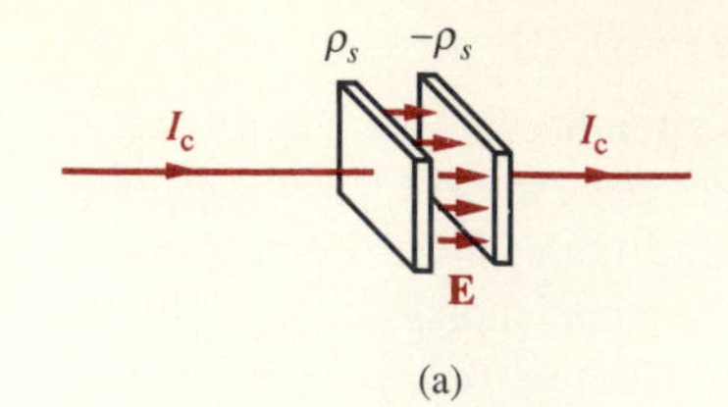

(a)

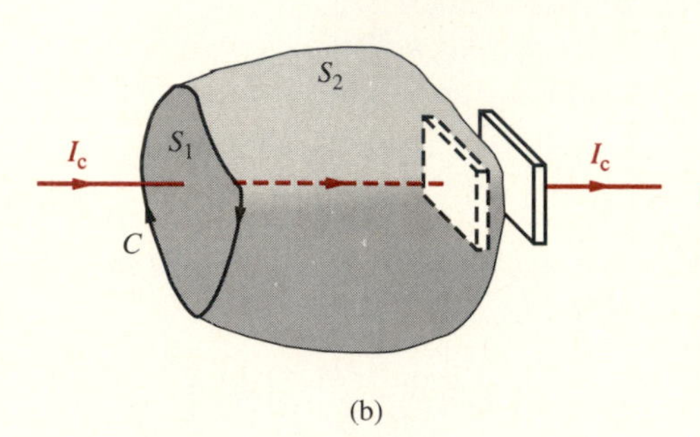

(b)

FIGURE 7.18. Displacement current concept. (a) Flow of current I_c charges up the left-hand plate, thus increasing the electric field. (b) Two different surfaces S_1 and S_2, enclosed by the same contour C. Note that S_2 is shaped like a sack, with C being the circumference of its opening.

thus E) increases with time. Thus, current flow into the capacitor is associated with a time rate of change of **E** (i.e., $\partial\mathbf{E}/\partial t$) between the capacitor plates.

The physical nature of the conduction and displacement currents is of course quite different. Conduction occurs as a result of the slow drift of charge carriers under the influence of an applied electric field. Conduction can occur only through material media since electrons or ions must be present to carry the current. Displacement current in material media can also be thought of in terms of movement or flow of charges. As **D** changes with time, the electrons in a dielectric are displaced with respect to their parent nuclei and move to and fro in accordance with $\partial\mathbf{D}/\partial t$. Although the charge never leaves the parent molecule, the back-and-forth motion constitutes an alternating current. Maxwell called the $\partial\mathbf{D}/\partial t$ term 'displacement current density' to indicate that it is based on the alternating displacement of charges from their parent molecules and to distinguish it from conduction current density.

The most brilliant achievement of Maxwell, however, was to imagine that the displacement current must also occur in vacuum, in the complete absence of material media. Thus, it is not necessary for charges to move to and fro for displacement current to exist. The real physical bases of the displacement current are our three experimental pillars of electromagnetics, namely Coulomb's law, the Biot–Savart law, and Faraday's law. After all, Maxwell introduced the $\partial\mathbf{D}/\partial t$ term in order to reconcile these physical experimental facts with the principle of conservation of charge (i.e., the continuity equation [5.5]). The conclusive experimental proof of the presence of displacement current was provided by Hertz's experiments in 1888, when he demonstrated that electromagnetic waves can travel through free space. Displacement current flows through free space, or time-varying electric fields produce magnetic fields, simply because physical experiments tell us this is so.

To further illustrate the need for a displacement current, consider an arbitrary contour C_1 encircling both the wire and two different surfaces S_1 and S_2, as shown in Figure 7.18b. Considering surface S_1, we have from Ampère's law

$$\oint_{C_1} \mathbf{H} \cdot d\mathbf{l} = \int_{S_1} \mathbf{J}_c \cdot d\mathbf{s} = I_c$$

where I_c is the conduction current in the wire, which passes through S_1. When we consider surface S_2, however, we have a problem. Without the displacement current term, we have no conduction current through the surface S_2 so that we would have to conclude $\oint_{C_1} \mathbf{H} \cdot d\mathbf{l} = 0$, which contradicts what we found for surface S_1 enclosed by the same contour C_1! This difficulty is resolved by including the displacement-current term. Let A be the area of the parallel-plate capacitor and a be the separation between its plates. The capacitance is then given by $C = \epsilon A/a$. When a displacement current I_d flows through the capacitor, the potential Φ across its plates is determined by the well-known voltage-current relationship for a capacitor, namely

$$I_d = C\frac{d\Phi}{dt}$$

assuming that the material between the capacitor plates is a perfect dielectric with $\sigma = 0$ (i.e., $I_c = 0$). But the potential difference across the plates is $\Phi = Ea$, where E is the electric field between the plates. Recalling from Chapter 4 that $C = \epsilon A/a$, we have

$$I_d = Ca\frac{dE}{dt} = \frac{Ca}{\epsilon A}\frac{d(\epsilon AE)}{dt} = \frac{d(\epsilon EA)}{dt} = \frac{d}{dt}\left(\int_{S_2} \mathbf{D} \cdot d\mathbf{s}\right) = \int_{S_2} \mathbf{J}_d \cdot d\mathbf{s}$$

Note that if we neglect any fringing effects so that the electric field between the capacitor plates is uniform over the area A and zero elsewhere, the integral of $\mathbf{D} \cdot d\mathbf{s}$ over any arbitrary surface is simply equal to ϵEA. Therefore, we see that the choice of either surface S_1 or S_2 gives the same result for $\oint_{C_1} \mathbf{H} \cdot d\mathbf{l}$, illustrating that $I_c = I_d$ and that consistent results are obtained by including the displacement-current term in Ampère's law.

Example 7-11 provides a quantitative comparison between displacement and conduction currents for a porcelain-filled capacitor.

Example 7-11: A parallel-plate capacitor. Consider a parallel-plate capacitor consisting of two metal plates of 50-cm^2 area each, separated by a porcelain layer of thickness $a = 1$ cm (for porcelain, $\epsilon_r = 5.5$ and $\sigma = 10^{-14}$ S-m^{-1}). If a voltage $\Phi(t) = 110\sqrt{2}\cos(120\pi t)$ V is applied across the capacitor plates, find (a) J_c, (b) J_d, and (c) the total current I through the capacitor.

Solution: The electric field in the porcelain layer is

$$E(t) = \frac{\Phi(t)}{a} = \frac{110\sqrt{2}\cos(120\pi t) \text{ V}}{0.01 \text{ m}} = (1.1\sqrt{2}) \times 10^4 \cos(120\pi t) \text{ V-m}^{-1}$$

The conduction (J_c) and the displacement-current (J_d) densities are then

$$\text{(a)} \quad J_c = \sigma E(t) = (10^{-14} \text{ S-m}^{-1})[(1.1\sqrt{2}) \times 10^4 \cos(120\pi t) \text{ V-m}^{-1}]$$
$$= 1.1\sqrt{2} \times 10^{-10} \cos(120\pi t) \text{ A-m}^{-2}$$

(b)

$$J_d = \frac{dD}{dt} = \epsilon\frac{dE(t)}{dt} = \epsilon_r\epsilon_0\frac{dE(t)}{dt}$$

$$= (5.5 \times 8.85 \times 10^{-12}\ \text{F-m}^{-1})[-1.1\sqrt{2} \times 10^4 \times 120\pi \sin(120\pi t)\ \text{V-(ms)}^{-1}]$$

$$\simeq -(2.86 \times 10^{-4})\sin(120\pi t)\ \ \text{A-m}^{-2}$$

Note that the conduction current is in phase with the electric field, whereas the displacement current is 90° out of phase. Also note that the amplitude of the conduction current is $\sim 5 \times 10^5$ times smaller than the displacement current, as expected since porcelain is an excellent insulator.

(c) The total current I can be found by integrating the total current density over the cross-sectional area of the capacitor plates as

$$I = \int_A (\mathbf{J}_c + \mathbf{J}_d) \cdot d\mathbf{s} = (J_c + J_d)A$$

since $\mathbf{J}_c$ and $\mathbf{J}_d$ are both approximately uniform and perpendicular to the plates throughout the dielectric region. Note also that $|\mathbf{J}_c|_{max} \ll |\mathbf{J}_d|_{max}$, so that for all practical purposes we can write

$$I \simeq J_d A \simeq [-2.86 \times 10^{-4}\sin(120\pi t)\ \text{A-m}^{-2}](50 \times 10^{-4}\ \text{m}^2)$$

$$\simeq -1.43 \times 10^{-6}\sin(120\pi t)\ \text{A} = -1.43\sin(120\pi t)\quad \mu\text{A}$$

Note that we could have easily obtained this result by using the familiar current-voltage relationship of an ideal capacitor, that is, $I(t) = C\,d\Phi(t)/dt$, where $C = \epsilon A/a \simeq 24.3$ pF.

It is instructive to estimate the ratio of conduction to displacement-current density in a metallic conductor. For an alternating electric field $E_0\cos(\omega t)$ within a metallic conductor of conductivity σ, the conduction-current density is $J_c = \sigma E_0\cos(\omega t)$, whereas the displacement-current density is $J_d = \partial D/\partial t = -\omega\epsilon E_0\sin(\omega t)$, where ϵ is the dielectric constant of the material. Thus we have

$$\frac{|J_c|_{max}}{|J_d|_{max}} = \frac{\sigma}{\omega\epsilon} \tag{7.15}$$

The dielectric constant ϵ for a metallic conductor is not easily measurable, since any polarization effect is completely overshadowed by conduction. Nevertheless, based on measurements of reflectivity of metals and the fact that atomic resonances for metals lie in the ultraviolet and x-ray ranges, metallic conductors can be treated as if their dielectric constant is ϵ_0 at frequencies up to and including the visible range (i.e., $\sim 10^{15}$ Hz). Taking $\sigma \simeq 10^7$ S-m^{-1} for a metallic conductor, we then have $|J_c|_{max}/|J_d|_{max} \simeq 10^{17}/f$, where $f = \omega/(2\pi)$ is the frequency of operation. Thus, the displacement current within a metallic conductor is completely negligible compared to the conduction current at frequencies below the optical range of up to $\sim 10^{15}$ Hz. Note that this calculation concerns the displacement current *inside* the wire. For propagation of electromagnetic waves in free space, however, where $\sigma = 0$, the only current term is the displacement current.

Example 7-12: Conduction to displacement current ratio for humid soil. Consider a certain type of humid soil with the following properties: $\sigma \simeq 10^{-2}$ S-m^{-1}, $\epsilon_r = 30$, and $\mu_r = 1$. Find the ratio of the amplitudes of the conduction and displacement currents at 1 kHz, 1 MHz, and 1 GHz.

Solution: Using $|J_c|_{\max}/|J_d|_{\max} = \sigma/(\omega\epsilon)$, we have

$$\frac{\sigma}{\omega\epsilon} = \frac{1}{2\pi f}\frac{10^{-2}\text{ S-m}^{-1}}{30 \times 8.85 \times 10^{-12}\text{ F-m}^{-1})} \simeq \frac{6 \times 10^6}{f}$$

Thus we have $\sigma/(\omega\epsilon) \simeq 6000$, 6, and 0.006, respectively, at 1 kHz, 1MHz, and 1 GHz.

7.4.2 Maxwell's Equations

We are now ready to examine the complete set of four Maxwell's equations:

$$\oint_C \overline{\mathscr{E}} \cdot d\mathbf{l} = \int_{S_C} \frac{\partial \overline{\mathscr{B}}}{\partial t} \cdot d\mathbf{s} \qquad \nabla \times \overline{\mathscr{E}} = -\frac{\partial \overline{\mathscr{B}}}{\partial t} \qquad [7.16a]$$

$$\oint_{S_V} \overline{\mathscr{D}} \cdot d\mathbf{s} = \int_V \tilde{\rho}\, dv \qquad \nabla \cdot \overline{\mathscr{D}} = \tilde{\rho} \qquad [7.16b]$$

$$\oint_C \overline{\mathscr{H}} \cdot d\mathbf{l} = \int_{S_C} \overline{\mathscr{J}} \cdot d\mathbf{s} + \int_{S_C} \frac{\partial \overline{\mathscr{D}}}{\partial t} \cdot d\mathbf{s} \qquad \nabla \times \overline{\mathscr{H}} = \overline{\mathscr{J}} + \frac{\partial \overline{\mathscr{D}}}{\partial t} \qquad [7.16c]$$

$$\oint_{S_V} \overline{\mathscr{B}} \cdot d\mathbf{s} = 0 \qquad \nabla \cdot \overline{\mathscr{B}} = 0 \qquad [7.16d]$$

where C is a closed contour that encloses the surface S_C, and S_V is a closed surface that encloses the volume V, as depicted in Figure 7.19. At this point, we introduce the notation $\overline{\mathscr{E}}$ (rather than **E**) for the various vector quantities in order to distinguish[29] real field quantities as functions of time and space $\overline{\mathscr{E}}(x, y, z, t)$ from complex phasors representing time-harmonic quantities $\mathbf{E}(x, y, z)$ (see Section 7.4.4). Similarly, the charge density $\tilde{\rho}(x, y, z, t)$ is distinguished from its phasor $\rho(x, y, z)$. Equations are written for a coordinate system fixed in space, which is why the integral equations

[29]Note that until now **E** was used to represent the real electric field vector as a function of time and space, that is, $\mathbf{E}(x, y, z, t)$. From here on, however, $\mathbf{E}(x, y, z)$ is a phasor, and is related to the real field quantity $\overline{\mathscr{E}}$ as $\overline{\mathscr{E}}(x, y, z, t) = \Re e\{\mathbf{E}(x, y, z)e^{j\omega t}\}$, as discussed in Section 7.4.4.

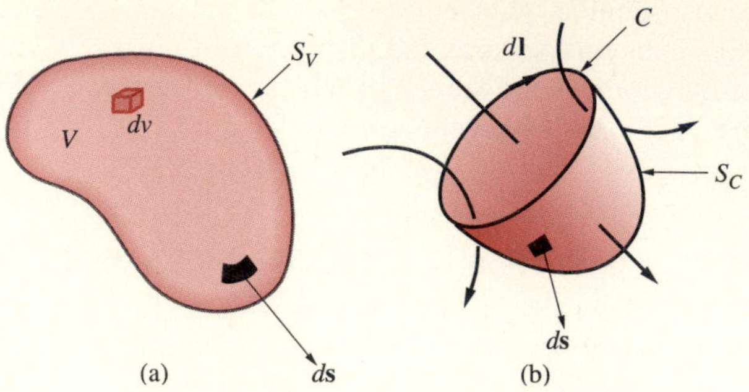

FIGURE 7.19. Contour, surface, and volume. (a) A closed surface S_V enclosing a volume V. (b) A closed contour C enclosing a surface S_C.

have $\partial/\partial t$ instead of d/dt. The current $\overline{\mathscr{J}}$ in [7.16c] can be a source current or a conduction current that flows due to the finite conductivity of the medium (i.e., $\overline{\mathscr{J}} = \sigma\overline{\mathscr{E}}$); note that, for the sake of ease of notation, we drop the subscript "c" in the designation of the conduction current and simply use $\overline{\mathscr{J}}$ rather than $\overline{\mathscr{J}}_c$. The complete specification of the field quantities must also involve the constitutive relations, which in simple media (i.e., linear, isotropic, homogeneous media) are $\overline{\mathscr{B}} = \mu\overline{\mathscr{H}}$ and $\overline{\mathscr{D}} = \epsilon\overline{\mathscr{E}}$.

It is possible to eliminate the vectors $\overline{\mathscr{D}}$ and $\overline{\mathscr{H}}$ from Maxwell's equations by substituting for them as follows:

$$\overline{\mathscr{D}} = \epsilon_0\overline{\mathscr{E}} + \overline{\mathscr{P}} \quad \text{and} \quad \overline{\mathscr{H}} = \frac{\overline{\mathscr{B}}}{\mu_0} - \overline{\mathscr{M}}$$

where $\overline{\mathscr{P}}$ is the polarization vector in a dielectric in units of coulombs-m^{-2}, and $\overline{\mathscr{M}}$ is the magnetization vector in a magnetic medium in units of ampere-m^{-1}. These two quantities account for the presence of matter at the points considered. Maxwell's equations then take the following form:

$$\nabla \times \overline{\mathscr{E}} = -\frac{\partial \overline{\mathscr{B}}}{\partial t}$$

$$\nabla \cdot \overline{\mathscr{E}} = \frac{1}{\epsilon_0}(\tilde{\rho} - \nabla \cdot \overline{\mathscr{P}})$$

$$\nabla \times \overline{\mathscr{B}} = \epsilon_0\mu_0 \frac{\partial \overline{\mathscr{E}}}{\partial t} + \mu_0\left(\overline{\mathscr{J}} + \frac{\partial \overline{\mathscr{P}}}{\partial t} + \nabla \times \overline{\mathscr{M}}\right)$$

$$\nabla \cdot \overline{\mathscr{B}} = 0$$

The preceding equations are completely general but are expressed in a way that stresses the contributions of the medium. Note that the presence of matter has the effect of adding the bound volume charge density $-\nabla \cdot \overline{\mathscr{P}}$ (Section 4.10), the polarization charge density $\partial\overline{\mathscr{P}}/\partial t$, and the equivalent volume magnetization current density $\nabla \times \overline{\mathscr{M}}$ (Section 6.8). Note that $\tilde{\rho}$ is free charge density, while $\overline{\mathscr{J}}$ could be a source current or a conduction current given by $\overline{\mathscr{J}} = \sigma\overline{\mathscr{E}}$, or a superposition thereof.

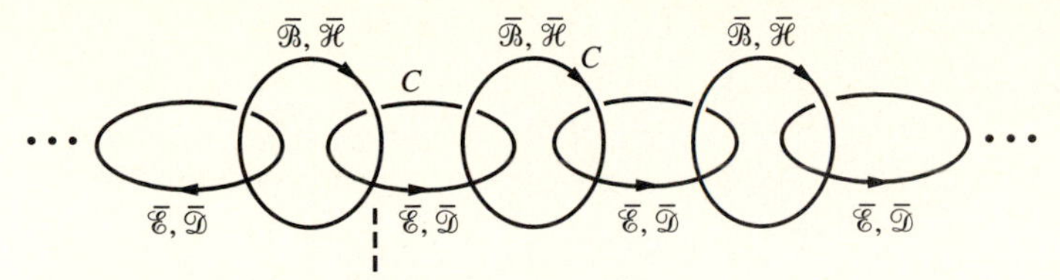

FIGURE 7.20. Plausibility of electromagnetic wave propagation as dictated by equations [7.16*a*] and [7.16*c*]. Starting with a time-varying magnetic field at the location of the dashed line, electric and magnetic fields are successively generated at surrounding regions.

Careful examination of [7.16] provides a qualitative understanding of the plausibility of electromagnetic wave propagation in space. Consider, for example, [7.16*a*] and [7.16*c*] in a nonconducting medium (i.e., $\sigma = 0$) and in the absence of sources (i.e., $\overline{\mathcal{J}}, \tilde{\rho} = 0$). According to [7.16*a*], any magnetic field $\overline{\mathcal{B}}$ that varies with time generates an electric field along a contour C surrounding it, as shown in Figure 7.20. On the other hand, according to [7.16*c*], this electric field $\overline{\mathcal{E}}$, which would typically vary in time because $\overline{\mathcal{B}}$ was taken to be varying with time, in turn generates a magnetic field along a contour C surrounding itself. This process indefinitely continues, as shown in Figure 7.20. It thus appears that if we start with a magnetic field at one point in space and vary it with time, Maxwell's equations dictate that magnetic fields and electric fields are created at surrounding points—that is, that the disturbance initiated by the changing magnetic field propagates away from its point of origin, indicated by the dashed line in Figure 7.20. Note that although Figure 7.20 shows propagation toward the left and right, the same sequence of events takes place in all directions.

7.4.3 Transverse Electromagnetic (TEM) Waves

Maxwell's equations are listed in Section 7.4.2 in their most general forms, when the various field quantities may be full vectors with all three components finite (i.e., $\mathcal{E}_x$, $\mathcal{E}_y$, and $\mathcal{E}_z$ all nonzero) and may vary with all three spatial coordinates as well as with time (i.e., $\overline{\mathcal{E}} = \overline{\mathcal{E}}(x, y, z, t)$). However, in a variety of useful practical cases, Maxwell's equations reduce to simpler forms. Consider, for example, an electric field that has only one nonzero component (say, $\mathcal{E}_x$) and that varies only with one spatial coordinate (say, z). In other words, $\overline{\mathcal{E}} = \hat{\mathbf{x}}\mathcal{E}_x(z, t)$. In this case, the integral form of [7.16*a*] reduces to

$$\oint_C \hat{\mathbf{x}}\mathcal{E}_x \cdot d\mathbf{l} = -\int_{S_C} \frac{\partial \overline{\mathcal{B}}}{\partial t} \cdot d\mathbf{s} \qquad [7.17]$$

Consider, for example, the contour C to be the square-shaped contour abcda in the xz plane that encloses the surface S_1, as shown in Figure 7.21. Noting that the side lengths are $\Delta x = \Delta z$, we evaluate the left-hand side of [7.17] as

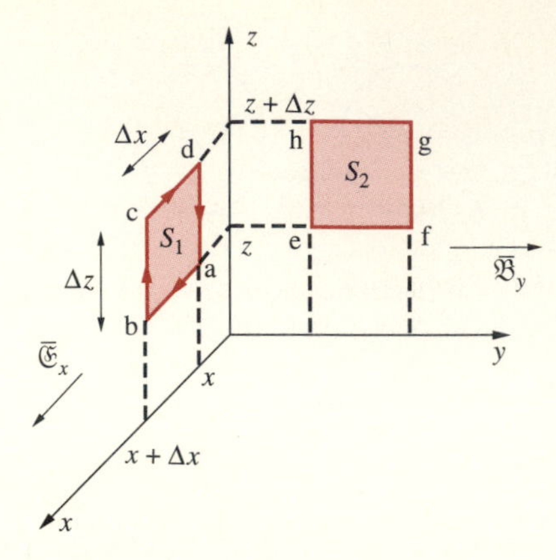

FIGURE 7.21. Transverse electromagnetic wave fields. Contours abcda and efghe used in the discussion of the one-dimensional version of Maxwell's equations for $\overline{\mathscr{E}} = \hat{\mathbf{x}}\mathscr{E}_x(z, t)$.

$$\int_{\text{abcda}} \hat{\mathbf{x}}\mathscr{E}_x \cdot d\mathbf{l} = \mathscr{E}_x(z, t)\,\Delta x + 0 - \mathscr{E}_x(z + \Delta z, t)\,\Delta x + 0 = -\frac{\partial \mathscr{E}_x}{\partial z}\,\Delta z\,\Delta x$$

assuming Δz and Δx are infinitesimally small. Since $\overline{\mathscr{B}}$ can be taken to be constant over such a differential surface element, the right-hand side of [7.17] reduces to

$$-\frac{\partial}{\partial t}\int_{S_1} \overline{\mathscr{B}} \cdot d\mathbf{s} = -\frac{\partial}{\partial t}\int_{S_1} \overline{\mathscr{B}} \cdot (-\hat{\mathbf{y}}\,ds) = \frac{\partial \mathscr{B}_y}{\partial t}\,\Delta x\,\Delta z$$

Note that only the $\mathscr{B}_y$ component contributes to the integral since the surface S_1 enclosed by contour abcda lies in the xz plane. Also note that the surface element $d\mathbf{s}$ points in the $-y$ direction according to the right-hand rule; with the fingers of the right hand describing the sense of line integration around contour C, the thumb points in the direction of the differential surface element. Equating both sides once again, we have

$$-\frac{\partial \mathscr{E}_x}{\partial z}\,\Delta z\,\Delta x = \frac{\partial \mathscr{B}_y}{\partial t}\,\Delta x\,\Delta z \quad \rightarrow \quad \frac{\partial \mathscr{E}_x}{\partial z} = -\frac{\partial \mathscr{B}_y}{\partial t}$$

which is actually the reduced form of the differential form of equation [7.16*a*]. Note that our choice of an electric field of the form $\overline{\mathscr{E}} = \hat{\mathbf{x}}\mathscr{E}_x(z, t)$ requires that the magnetic field $\overline{\mathscr{B}}$ have only a nonzero y component, so that $\overline{\mathscr{B}} = \hat{\mathbf{y}}\mathscr{B}_y(z, t)$.

Assuming that the physical regions in consideration are far from the sources, so that $\overline{\mathscr{J}}, \tilde{\rho} = 0$, and assuming free space conditions so that $\overline{\mathscr{B}} = \mu_0\overline{\mathscr{H}} = \hat{\mathbf{y}}\mu_0\mathscr{H}_y(z, t)$ and $\overline{\mathscr{D}} = \epsilon_0\overline{\mathscr{E}} = \hat{\mathbf{x}}\epsilon_0\mathscr{E}_x(z, t)$, we now consider, in a similar fashion, the integral form of equation [7.16*c*] for the contour efghe in the yz plane. We derive the corresponding differential form as

$$-\frac{\partial \mathscr{H}_y}{\partial z} = \epsilon_0\frac{\partial \mathscr{E}_x}{\partial t}$$

Electromagnetic waves that have single, orthogonal vector electric and magnetic field components (e.g., $\mathscr{E}_x$ and $\mathscr{H}_y$), both varying with a single coordinate of space (e.g., z), are known as *uniform plane waves* or *transverse electromagnetic (TEM) waves*. These types of waves will be considered explicitly and extensively in Chapter 8. At this point, we note that the voltage and current waves on lossless transmission lines as discussed in Chapters 2 and 3 were also TEM waves. To see this, consider the similarity between the fundamental transmission line equations and the preceding derived expressions for $\mathscr{E}_x$ and $\mathscr{H}_y$:

$$\frac{\partial \mathscr{E}_x}{\partial z} = -\mu_0 \frac{\partial \mathscr{H}_y}{\partial t} \qquad \longleftrightarrow \qquad \frac{\partial \mathscr{V}}{\partial z} = -L \frac{\partial \mathscr{I}}{\partial t}$$

$$\frac{\partial \mathscr{H}_y}{\partial z} = -\epsilon_0 \frac{\partial \mathscr{E}_x}{\partial t} \qquad \longleftrightarrow \qquad \frac{\partial \mathscr{I}}{\partial z} = -C \frac{\partial \mathscr{V}}{\partial t}$$

This correspondence between transmission line equations and the equations relating the electric and magnetic fields of transverse electromagnetic waves is more than a simple analogy. As we shall see in Chapter 8, the transmission line voltage and current waves that we worked with in Chapters 2 and 3 are indeed transverse electromagnetic waves and can be analyzed solely in terms of electric and magnetic fields. Their treatment in terms of distributed circuits concepts and in terms of voltage and current waves is in fact a simplification of their more general behavior, which is best expressed in terms of field quantities. By the same token, the fact that the equations relating $\mathscr{E}_x$ to $\mathscr{H}_y$ and $\mathscr{V}$ to $\mathscr{I}$ are virtually identical indicates that we can expect transverse electromagnetic waves to exhibit the full range of behavior of transmission lines in terms of both transient and steady-state waves as analyzed, respectively, in Chapters 2 and 3. The association between transmission line behavior and transverse electromagnetic wave behavior will be further discussed in Chapter 8.

7.4.4 Time-Harmonic Maxwell's Equations

Numerous practical applications (e.g., broadcast radio and TV, radar, optical, and microwave applications) involve transmitting sources that operate in such a narrow band of frequencies that the behavior of all the field components is very similar to that of the central single-frequency sinusoid (i.e., the carrier). Most generators also produce sinusoidal voltages and currents, and hence electric and magnetic fields that vary sinusoidally with time. In many applications, the transients involved at the time the signal is switched on (or off) are not of concern, so the steady-state sinusoidal approximation is most suitable. For example, for an AM broadcast station operating at a carrier frequency of 1 MHz, any turn-on transients would last only a few μs and are of little consequence to the practical application. For all practical purposes, the signal propagating from the transmitting antenna to the receivers can be treated as a sinusoid, with its amplitude modulated within a narrow bandwidth (e.g., ± 5 kHz) around the carrier frequency. Since the characteristics of the propagation medium do not vary significantly over this bandwidth, we can describe the propagation behavior of the AM broadcast signal by studying a single sinusoidal carrier at a frequency of 1 MHz.

The time-harmonic (sinusoidal steady-state) forms of Maxwell's equations[30] are listed here with their more general versions:

$$\nabla \times \overline{\mathscr{E}} = -\frac{\partial \overline{\mathscr{B}}}{\partial t} \qquad \nabla \times \mathbf{E} = -j\omega\mathbf{B} \qquad [7.18a]$$

$$\nabla \cdot \overline{\mathscr{D}} = \tilde{\rho} \qquad \nabla \cdot \mathbf{D} = \rho \qquad [7.18b]$$

$$\nabla \times \overline{\mathscr{H}} = \overline{\mathscr{J}} + \frac{\partial \overline{\mathscr{D}}}{\partial t} \qquad \nabla \times \mathbf{H} = \mathbf{J} + j\omega\mathbf{D} \qquad [7.18c]$$

$$\nabla \cdot \overline{\mathscr{B}} = 0 \qquad \nabla \cdot \mathbf{B} = 0 \qquad [7.18d]$$

Note that in [17.8*a*–*d*], $\overline{\mathscr{E}}$, $\overline{\mathscr{D}}$, $\overline{\mathscr{H}}$, and $\overline{\mathscr{B}}$ are real (measurable) quantities that can vary with time, whereas the vectors **E**, **D**, **H**, and **B** are complex phasors that do not vary with time. In general, we can obtain the former from the latter by multiplying by $e^{j\omega t}$ and taking the real part. For example,

$$\overline{\mathscr{E}}(x, y, z, t) = \mathscr{Re}\{\mathbf{E}(x, y, z)e^{j\omega t}\}$$

Note that the same is true for all of the quantities. For example,

$$\tilde{\rho}(x, y, z, t) = \mathscr{Re}\{\rho(x, y, z)e^{j\omega t}\}$$

Example 7-13 illustrates the use of equations [7.18].

Example 7-13: Aircraft VHF communication signal. The electric field component of an electromagnetic wave in air used by an aircraft to communicate with the air traffic control tower can be represented by

$$\overline{\mathscr{E}}(z, t) = \hat{\mathbf{y}}0.02\cos(7.5 \times 10^8 t - \beta z) \quad \text{V-m}^{-1}$$

Find the corresponding wave magnetic field $\overline{\mathscr{H}}(z, t)$ and the constant β.

Solution: In view of the sinusoidal nature of the electric field signal, it is appropriate to work with phasors. The phasor form of the electric field is

$$\mathbf{E}(z) = \hat{\mathbf{y}}E_y(z) = \hat{\mathbf{y}}0.02e^{-j\beta z} \quad \text{V-m}^{-1}$$

Using [7.18*a*], we can write

---

[30]The actual derivation of the time-harmonic form, for example, for [7.16*a*] is as follows:

$$\nabla \times \overline{\mathscr{E}} = -\frac{\partial \overline{\mathscr{B}}}{\partial t} \quad \rightarrow \quad \nabla \times \underbrace{[\mathscr{Re}\{\mathbf{E}(x, y, z)e^{j\omega t}\}]}_{\overline{\mathscr{E}}} = -\frac{\partial}{\partial t}\underbrace{[\mathscr{Re}\{\mathbf{B}(x, y, z)e^{j\omega t}\}]}_{\overline{\mathscr{B}}}$$

$$\rightarrow \quad \mathscr{Re}\{e^{j\omega t}\nabla \times \mathbf{E}\} = \mathscr{Re}\{-j\omega e^{j\omega t}\mathbf{B}\} \quad \rightarrow \quad \nabla \times \mathbf{E} = -j\omega\mathbf{B}$$

$$\mathbf{H}(z) = -\frac{1}{j\omega\mu_0}\nabla \times \mathbf{E}(z) = \hat{\mathbf{x}}\frac{1}{j\omega\mu_0}\frac{\partial E_y}{\partial z}$$
$$= -\hat{\mathbf{x}}\frac{\beta}{\omega\mu_0}E_y = -\hat{\mathbf{x}}\frac{0.02\beta}{\omega\mu_0}e^{-j\beta z}$$

where $\omega = 7.5 \times 10^8$ rad-s^{-1} and $\mu_0 = 4\pi \times 10^{-7}$ H-m^{-1}. Substituting the expression for $\mathbf{H}(z)$ into [7.18c], we find

$$\mathbf{E}(z) = \frac{1}{j\omega\epsilon_0}\nabla \times \mathbf{H}(z) = \hat{\mathbf{y}}\frac{1}{j\omega\epsilon_0}\frac{\partial H_x}{\partial z} = \hat{\mathbf{y}}\frac{0.02\beta^2}{\omega^2\mu_0\epsilon_0}e^{-j\beta z}$$

where $\epsilon_0 = 8.85 \times 10^{-12}$ F-m^{-1}. But this expression for $\mathbf{E}(z)$ must be the same as the electric field phasor expression we started with. Thus, we must have

$$\frac{0.02\beta^2}{\omega^2\mu_0\epsilon_0} = 0.02 \quad \rightarrow \quad \beta = \omega\sqrt{\mu_0\epsilon_0} = \frac{7.5 \times 10^8 \text{ rad-s}^{-1}}{3 \times 10^8 \text{ m-s}^{-1}} = 2.5 \text{ rad-m}^{-1}$$

where we have used the fact that $(\mu_0\epsilon_0)^{-1/2} = c$, the speed of light in free space.

The corresponding magnetic field $\mathbf{H}$ is then

$$\mathbf{H}(z) = -\hat{\mathbf{x}}\frac{(0.02)(2.5)}{(7.5 \times 10^8)(4\pi \times 10^{-7})}e^{-j2.5z} \simeq -\hat{\mathbf{x}}53.1 \times e^{-j2.5z} \quad \mu\text{A-m}^{-1}$$

and the instantaneous magnetic field $\overline{\mathcal{H}}(z, t)$ is

$$\overline{\mathcal{H}}(z, t) = \mathcal{R}e\{\mathbf{H}(z)e^{j\omega t}\} = \mathcal{R}e\{-\hat{\mathbf{x}}53.1e^{-j2.5z}e^{j\omega t}\}$$
$$= -\hat{\mathbf{x}}53.1\cos(7.5 \times 10^8 t - 2.5z) \quad \mu\text{A-m}^{-1}$$

Complex Permittivity: Dielectric Heating In most dielectrics that are good insulators, the direct conduction current (which is due to finite σ) is usually negligible. However, at high frequencies, an alternating current that is in phase with the applied field is present because the rapidly varying applied electric field has to do work against molecular forces[31] in alternately polarizing the bound electrons. As a

[31] For electronic polarization (see Section 4.10), the electrostatic force produced by the displacement of the electron cloud with respect to the nucleus acts as the restoring force proportional to displacement x and is given by $-kx$, where k is the constant evaluated in Section 4.10 to be $k = q_e^2/(4\pi\epsilon_0 a^3)$, with a being the radius of the electron cloud. The energy transferred by the electron to its surroundings during its rapid motion can be represented by a viscous (or frictional) damping force proportional to the mass and to the velocity of the electron; that is, $-m_e\kappa v_x$, where κ is a positive constant. The time-harmonic (i.e., phasor) form of the equation of motion for the electron is then given by

$$-m_e\omega^2 x = -kx - j\omega m_e\kappa x + q_e E_x$$

where E_x is the applied electric field. The solution of the above equation is $x = (q_e/m_e)E_x/(\omega_0^2 - \omega^2 + j\omega\kappa)$, where $\omega_0 = \sqrt{k/m_e}$ is the characteristic angular frequency of the bound electron. Assuming that there are N electrons per unit volume, the polarization is then a complex number given by $P_x = N(q_e^2/m_e)E_x/(\omega_0^2 - \omega^2 + j\omega\kappa)$, leading to complex susceptibility χ_e (since $P_x = \epsilon_0\chi_e E_x$) and thus complex permittivity $\epsilon_c \equiv \epsilon_0(1 + \chi_e)$.

result, materials that are good insulators at low frequencies can consume considerable energy when they are subjected to high-frequency fields. The heat generated as a result of such radio-frequency heating is used in molding plastics; in microwave cooking; and in microwave drying of paper, printing ink, glued products, leather, textile fibers, wood, foundry materials, rubbers, and plastics.[32]

The microphysical bases of such effects are different for solids, liquids, and gases and are too complex to be summarized here.[33] When an external time-varying field is applied to a material, displacement of bound charges occurs, giving rise to volume polarization density **P**, as discussed in Section 4.10. At sinusoidal steady state, the polarization **P** varies at the same frequency as the applied field **E**. At low frequencies, **P** is also in phase with **E**, both quantities reaching their maxima and minima at the same points in the radio-frequency cycle. As the frequency is increased, however, the inertia of the charged particles (not just because of their mass but also because of the elastic and frictional forces that keep them attached to their molecules) tends to prevent the polarization **P** from keeping in phase with the applied field. The work that must be done against the frictional damping forces causes the applied field to lose power, and this power is deposited in the medium as heat. This condition of out-of-phase polarization that occurs at higher frequencies can be characterized by a complex electric susceptibility χ_e, and hence a complex permittivity ϵ_c. In such cases, both the frictional damping and any other ohmic losses (due to nonzero conductivity σ) are included in the imaginary part of the complex dielectric constant ϵ_c:

$$\epsilon_c = \epsilon' - j\epsilon''$$

We can analyze the resultant effects by substituting the preceding equation into [7.18*c*]:

$$\nabla \times \mathbf{H} = +j\omega\epsilon_c\mathbf{E} = (\omega\epsilon'')\mathbf{E} + j\omega\epsilon'\mathbf{E}$$

Note that we have assumed $\mathbf{J} = \sigma\mathbf{E} = 0$ since the effects of any nonzero σ are included in ϵ''. It thus appears that the imaginary part of ϵ_c (namely, ϵ'') leads to a volume current density term that is in phase with the electric field, as if the material had an effective conductivity $\sigma_{\text{eff}} = \omega\epsilon''$. At low frequencies, $\omega\epsilon''$ is small due to ω being small and due to the fact that ϵ'' is itself small so that the losses are largely negligible. However, at high frequencies, $\omega\epsilon''$ increases and produces the same macroscopic effect as if the dielectric had effective conductivity $\sigma_{\text{eff}} = \omega\epsilon''$. In Section 5.8, we found that when a steady current $\mathbf{J} = \sigma\mathbf{E}$ flows in a conducting material in response to a constant applied field **E**, electrical power per unit volume dissipated in the material is given by $\mathbf{E} \cdot \mathbf{J} = \sigma E^2$ in units of W-m^{-3}. Similarly, when a dielectric is excited at frequencies high enough for

[32] J. Thuery, *Microwaves: Industrial, Scientific and Medical Applications*, Artech House, 1992.

[33] For further discussion at an appropriate level, see C. Kittel, *Introduction to Solid State Physics*, 5th ed., Wiley, New York, 1976.

$\omega\epsilon''$ to be appreciable, an alternating current density of $\omega\epsilon''\mathbf{E}$ flows (in addition to the displacement current density given by $j\omega\epsilon'\mathbf{E}$) in response to an applied alternating field $\mathbf{E}$, leading to an instantaneous power dissipation of $[(\omega\epsilon''\mathbf{E})\cdot\mathbf{E}] = \omega\epsilon''|\mathbf{E}|^2$ in units of W-m^{-3}. When the electric field is time-harmonic, i.e., $\mathscr{E}(t) = |\mathbf{E}|\cos(\omega t)$, the time-average power dissipated in the material is $T_p^{-1}\int_0^{T_p}\omega\epsilon''|\mathscr{E}(t)|^2 dt = T_p^{-1}\int_0^{T_p}\omega\epsilon''|\mathbf{E}|^2\cos^2(\omega t)dt = \frac{1}{2}\omega\epsilon''|\mathbf{E}|^2$. This dissipated power is the basis for microwave heating of dielectric materials. Since the values of ϵ'' are often determined by measurement, it is not necessary to distinguish between losses due to nonzero conductivity σ and the dielectric losses discussed here.

In general, both ϵ' and ϵ'' can depend on frequency in complicated ways, exhibiting several resonances over wide frequency ranges. The typical behavior in the vicinity of resonances is enhanced losses (i.e., large values of ϵ'') and a reduction of ϵ' to a new level. The frequency response of a hypothetical dielectric is shown in Figure 7.22, with contributions from the different types of polarizations (see Section 4.10) identified in terms of the frequency ranges in which they are significant. The various resonances illustrated have important practical implications. To cite a few examples, the broad orientational resonance in the microwave region is responsible for microwave heating applications. A resonance in water vapor molecules at 1.25-cm wavelength (24 GHz) limits the frequency of long-range radar. Another strong resonance at 0.5-cm wavelength (60 GHz) from oxygen, together with other higher-frequency absorptions, severely limits propagation of signals above 0.5-cm wavelength. Very strong resonances, due mostly to ozone and nitrogen at high altitudes, keep most of the ultraviolet energy of the sun from penetrating lower altitudes.

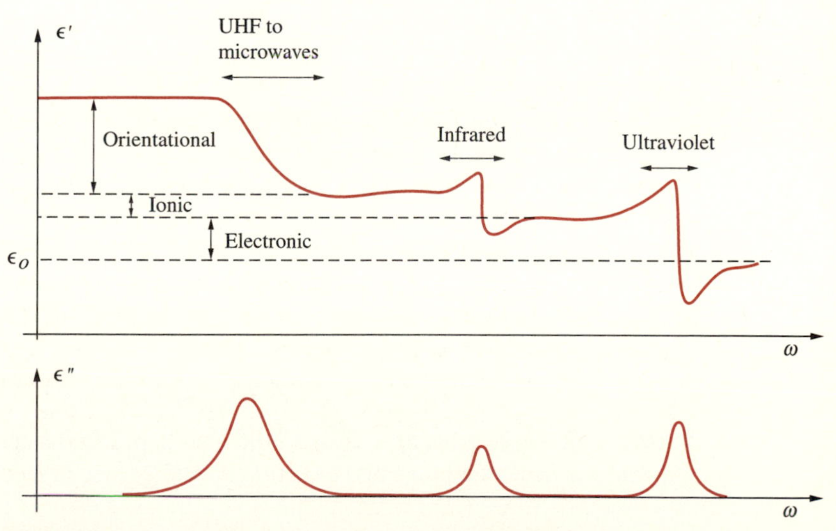

FIGURE 7.22. Dielectric constant as a function of frequency. At low frequencies, the permittivity differs from ϵ_0 by a constant multiplier. In the vicinity of the resonance, ϵ'' goes through a pronounced peak, while ϵ' generally decreases to a new level.

Different materials exhibit a large variety of resonances, which originate in the basic energy-level structure of the atoms and molecules.

It is clear from the preceding discussion that the degree to which losses in a dielectric are important depends on the ratio ϵ''/ϵ', and that the electrical properties of materials are functions of frequency. In tabulating the dielectric constants of different materials, it is often customary to give the values of the real part of the permittivity ϵ' and the so-called *loss-tangent,* $\tan\delta_c$, defined as

$$\tan\delta_c \equiv \frac{\epsilon''}{\epsilon'}$$

typically specified at given frequencies, in view of the fact that ϵ'' and ϵ' are both frequency-dependent, as shown in Figure 7.22.

Example 7-14 illustrates the use of complex permittivity to determine the power dissipated in microwave heating of milk.

Example 7-14: Microwave heating of milk. The dielectric properties of milk with 7.3% moisture content at 20°C and 3 GHz are $\epsilon_r' = 51$ and $\tan\delta_c = 0.59$. Calculate (a) ϵ_r'' and (b) the average dissipated power per unit volume if the peak electric field inside the dielectric is 30 kV-m^{-1}.

Solution: (a) From $\tan\delta_c = 0.59 = \epsilon_r''/\epsilon_r' = \epsilon_r''/51$, we find $\epsilon_r'' \simeq 30.1$. (b) The average power dissipated per unit volume is

$$\tfrac{1}{2}\omega\epsilon'' E_{\text{peak}}^2 \simeq \tfrac{1}{2}(2\pi \times 3 \times 10^9 \text{ rad-s}^{-1})(30.1 \times 8.85 \times 10^{-12} \text{ F-m}^{-1})(30 \text{ kV-m}^{-1})^2$$

$$\simeq 2.26 \times 10^9 \text{ W-m}^{-3} = 2.26 \text{ W-mm}^{-3}$$

Complex Permeability: Magnetic Relaxation If a magnetic field is suddenly applied to a paramagnetic material, the magnetization density **M** exhibits some inertia and does not immediately reach its static value, but instead approaches it gradually. Similar inertial delay also occurs when the field is turned off. The inertia exhibited is attributable to the energy exchanges between the spinning electrons and the lattice vibrations as well as energy exchanges between neighboring spins.[34] In analogy with dielectric relaxation, such effects can be described by introducing a complex permeability μ such that

$$\mu_c = \mu' - j\mu''$$

To represent the fact that this effect would lead to power dissipation in the medium, consider the time rate of change of magnetic energy density, or

---

[34]For further details, see Section 7.16 of R. S. Elliott, *Electromagnetics,* IEEE Press, 1993, and Chapter 16 of C. Kittel, *Introduction to Solid State Physics,* 5th ed., Wiley, New York, 1976.

$$\frac{\partial w_{\mathrm{m}}}{\partial t} = \frac{\partial}{\partial t}\left(\frac{1}{2}\overline{\mathcal{H}} \cdot \overline{\mathcal{B}}\right) \quad \rightarrow \quad \tfrac{1}{2} j\omega(\mathbf{H} \cdot \mathbf{B})$$

With $\mathbf{B} = (\mu' - j\mu'')\mathbf{H}$, we have

$$\tfrac{1}{2} j\omega(\mathbf{H} \cdot \mathbf{B}) = \tfrac{1}{2}(j\omega\mu')H^2 + \tfrac{1}{2}\omega\mu'' H^2$$

where $H = |\mathbf{H}|$. For a time-harmonic magnetic field, the real part of the preceding quantity represents the time-average power dissipated in a medium. We note that the real part of the first term is zero; that is, this term simply represents the volume density of energy stored in the magnetic field per unit time. The second term, on the other hand, is purely real and represents the time-average power density of the magnetization loss. The additional losses due to magnetization damping can be represented by an effective conductivity $\sigma_{\mathrm{eff}} = \epsilon'\omega\mu''/\mu'$.

In principle, diamagnetic materials also exhibit time-varying relaxation behavior. Although the resultant effects are so small that one might think such phenomena are of little practical interest, resonant absorption in diamagnetic liquids is the cause of the phenomenon of nuclear magnetic resonance,[35] which in turn is the basis for magnetic resonance imaging (MRI) technology.[36]

7.5 REVIEW OF MAXWELL'S EQUATIONS

A brief review of Maxwell's equations[37] and their underlying foundations is now in order, since their development was spread over Chapters 4 through 7. We have established these four fundamental equations of electromagnetics on the basis of three separate experimentally established facts, namely, Coulomb's law,[38] Ampère's law[39] (or the Biot–Savart law), Faraday's law,[40] and the principle of conservation of electric charge. The validity of Maxwell's equations is based on their consistency with all of our experimental knowledge to date concerning electromagnetic

---

[35]Resonant absorption occurs because of the highly frequency-dependent nature of the permeability μ, analogous to that shown in Figure 7.22 for ϵ_c. For further discussion and relevant references, see Chapter 16 of C. Kittel, *Introduction to Solid State Physics,* 5th ed., Wiley, New York, 1976.

[36]See M. A. Brown and R. C. Semelka, *MRI Basic Principles and Applications,* Wiley-Liss, 1995.

[37]J. C. Maxwell, *A Treatise in Electricity and Magnetism,* Clarendon Press, Oxford, 1892, Vol. 2, pp. 247–262.

[38]Electric charges attract or repel one another in a manner inversely proportional to the square of the distance between them; C. A. de Coulomb, Première mémoire sur l'électricité et magnétisme (First memoir on electricity and magnetism), *Histoire de l'Académie Royale des Sciences,* 1785, p. 569.

[39]Current-carrying wires create magnetic fields and exert forces on one another, with the intensity of the magnetic field (and thus the force) depending on the inverse square of the distance; A.-M. Ampère, *Recueil d'observations électrodynamiques,* Crochard, Paris, 1820–1833.

[40]Magnetic fields that change with time induce electromotive force; M. Faraday, *Experimental Researches in Electricity,* Taylor and Francis, London, Vol. I, 1839, pp. 1–109.

phenomena. The physical meaning of the equations is better perceived in the context of their integral forms, which are listed below together with their differential counterparts:

(1) Faraday's law is based on the experimental fact that time-changing magnetic flux induces electromotive force:

$$\oint_C \overline{\mathscr{E}} \cdot d\mathbf{l} = -\int_S \frac{\partial \overline{\mathscr{B}}}{\partial t} \cdot d\mathbf{s} \qquad \nabla \times \overline{\mathscr{E}} = -\frac{\partial \overline{\mathscr{B}}}{\partial t} \qquad [7.16a]$$

where the contour C is that which encloses the surface S, and where the sense of the line integration over the contour C (i.e., $d\mathbf{l}$) must be consistent with the direction of the surface vector $d\mathbf{s}$ in accordance with the right-hand rule.

(2) Gauss's law is a mathematical expression of the experimental fact that electric charges attract or repel one another with a force inversely proportional to the square of the distance between them (i.e., Coulomb's law):

$$\oint_S \overline{\mathscr{D}} \cdot d\mathbf{s} = \int_V \tilde{\rho}\, dv \qquad \nabla \cdot \overline{\mathscr{D}} = \tilde{\rho} \qquad [7.16b]$$

where the surface S encloses the volume V. The volume charge density is represented with $\tilde{\rho}$ to distinguish it from the phasor form ρ used in the time-harmonic form of Maxwell's equations.

(3) Maxwell's third equation is a generalization of Ampère's law, which states that the line integral of the magnetic field over any closed contour must equal the total current enclosed by that contour. Maxwell's third equation expresses the fact that time-varying electric fields produce magnetic fields. The first term of this equation (also referred to as the conduction-current term) is Ampère's law, which is a mathematical statement of the experimental findings of Oersted, whereas the second term, known as the displacement-current term, was introduced theoretically by Maxwell in 1862 and verified experimentally many years later in Hertz's experiments.[41]

$$\oint_C \overline{\mathscr{H}} \cdot d\mathbf{l} = \int_S \overline{\mathscr{J}} \cdot d\mathbf{s} + \int_S \frac{\partial \overline{\mathscr{D}}}{\partial t} \cdot d\mathbf{s} \qquad \nabla \times \overline{\mathscr{H}} = \overline{\mathscr{J}} + \frac{\partial \overline{\mathscr{D}}}{\partial t} \qquad [7.16c]$$

where the contour C is that which encloses the surface S.

(4) Maxwell's fourth equation is based on the fact that there are no magnetic charges (i.e., magnetic monopoles) and that, therefore, magnetic field lines always close on themselves:

[41] H. Hertz, On the finite velocity of propagation of electromagnetic actions, *Sitzb. d. Berl. Akad. d. Wiss.*, Feb 2, 1888; for a collected English translation of this and other papers by H. Hertz, see H. Hertz, *Electric Waves,* MacMillan, London, 1893.

$$\oint_S \overline{\mathscr{B}} \cdot d\mathbf{s} = 0 \qquad \nabla \cdot \overline{\mathscr{B}} = 0 \qquad [7.16d]$$

where the surface S encloses the volume V. As we have seen in Section 6.7, this equation can actually be derived from the Biot–Savart law, so it is not completely independent.[42]

The two constitutive relations $\overline{\mathscr{D}} = \epsilon\overline{\mathscr{E}}$ and $\overline{\mathscr{B}} = \mu\overline{\mathscr{H}}$ (more properly expressed as $\overline{\mathscr{H}} = \mu^{-1}\overline{\mathscr{B}}$ since, in the magnetostatic case, $\overline{\mathscr{H}}$ is the medium-independent quantity, as discussed in Section 6.2.3 and in footnote 71 in Section 6.8) govern the manner by which the electric and magnetic fields, $\overline{\mathscr{E}}$ and $\overline{\mathscr{B}}$, are related to the medium-independent quantities, $\overline{\mathscr{D}}$ and $\overline{\mathscr{H}}$, in material media, where, in general, $\epsilon \neq \epsilon_0$ and $\mu \neq \mu_0$. The volume charge density $\tilde{\rho}$ and conduction-current density $\overline{\mathscr{J}}$ are typically sources from which electric fields or magnetic fields originate, respectively. In conducting media ($\sigma \neq 0$), any present electric field $\overline{\mathscr{E}}$ must be accompanied by a volume current density of $\overline{\mathscr{J}} = \sigma\overline{\mathscr{E}}$.

Note that ϵ, μ, and σ are macroscopic parameters that describe the relationships among macroscopic field quantities, but they are based on the microscopic behavior of the atoms and molecules in response to the fields. These parameters are simple constants only for *simple* material media, which are linear, homogeneous, time-invariant, and isotropic. Otherwise, for complex material media that are nonlinear, inhomogeneous, time-variant, and anisotropic, ϵ, μ, and σ may depend on the magnitudes of $\overline{\mathscr{E}}$ and $\overline{\mathscr{B}}$ (nonlinear), on spatial coordinates (x, y, z) (inhomogeneous), on time (time-variant), or on the orientations of $\overline{\mathscr{E}}$ and $\overline{\mathscr{H}}$ (anisotropic). For anisotropic media, ϵ, μ, or σ is generally expressed as a matrix (i.e., as a *tensor*) whose entries relate each component (e.g., x, y, or z components) of $\overline{\mathscr{E}}$ (or $\overline{\mathscr{H}}$) to the other three components (e.g., x, y, and z components) of $\overline{\mathscr{D}}$ or $\overline{\mathscr{J}}$ (or $\overline{\mathscr{B}}$). In ferromagnetic materials, the magnetic field $\overline{\mathscr{B}}$ is determined by the past history of the field $\overline{\mathscr{H}}$ rather than by its instantaneous value. Such substances are said to exhibit *hysteresis*. Some hysteresis effects can also be seen in certain dielectric materials.

The continuity equation, which expresses the principle of conservation of charge in differential form, is contained in Maxwell's equations and in fact can be readily

[42] It is also interesting that [7.16d] can be derived from [7.16a] by taking the divergence of the latter and using the vector identity of $\nabla \cdot (\nabla \times \mathbf{G}) \equiv 0$, which is true for any vector $\mathbf{G}$. We find

$$\nabla \cdot (\nabla \times \overline{\mathscr{E}}) = -\nabla \cdot \left(\frac{\partial \overline{\mathscr{B}}}{\partial t}\right) \quad \rightarrow \quad 0 = -\frac{\partial(\nabla \cdot \overline{\mathscr{B}})}{\partial t} \quad \rightarrow \quad \text{const.} = \nabla \cdot \overline{\mathscr{B}}$$

The constant can then be shown to be zero by the following argument. If we suppose that the $\overline{\mathscr{B}}$ field was produced a finite time ago, that is, it has not always existed, then, if we go back far enough in time, we have $\overline{\mathscr{B}} = 0$ and therefore $\nabla \cdot \overline{\mathscr{B}} = 0$. Hence it would appear that

$$\nabla \cdot \overline{\mathscr{B}} = 0 \quad \text{and} \quad \int_S \overline{\mathscr{B}} \cdot d\mathbf{s} = 0$$

derived[43] by taking the divergence of [7.16c] and using [7.16b]. For the sake of completeness, we give the integral and differential forms of the continuity equation:

$$-\oint_S \overline{\mathcal{J}} \cdot d\mathbf{s} = \frac{\partial}{\partial t}\int_V \tilde{\rho}\, dv \qquad \nabla \cdot \overline{\mathcal{J}} = -\frac{\partial \tilde{\rho}}{\partial t} \qquad [7.19]$$

where the surface S encloses the volume V.

Note that for all of the equations [7.16] through [7.19], the differential forms can be derived from the integral forms (or vice versa) by using either Stokes's or the divergence theorem, both of which are valid for any arbitrary vector field $\mathcal{G}$. These theorems are

$$\oint_C \mathcal{G} \cdot d\mathbf{l} = \int_S (\nabla \times \mathcal{G}) \cdot d\mathbf{s} \qquad \text{(Stokes's theorem)}$$

where the contour C encloses the surface S, and

$$\oint_S \mathcal{G} \cdot d\mathbf{s} = \int_V (\nabla \cdot \mathcal{G})\, dv \qquad \text{(Divergence theorem)}$$

where the surface S encloses volume V.

Electromagnetic Boundary Conditions The integral forms of equations [7.16] can be used to derive the relationships between electric- and magnetic-field components on both sides of interfaces between two different materials (i.e., different μ, ϵ, or σ).

The electromagnetic boundary conditions can be summarized as follows:

(1) It follows from applying [7.16a] to a contour C, as shown in Figure 7.23b, that the tangential component of the electric field $\overline{\mathscr{E}}$ is continuous across any interface:

$$\hat{\mathbf{n}} \times [\overline{\mathscr{E}}_1 - \overline{\mathscr{E}}_2] = 0 \qquad \rightarrow \qquad \mathscr{E}_{1t} = \mathscr{E}_{2t} \qquad [7.20]$$

where $\hat{\mathbf{n}}$ is the unit vector perpendicular to the interface and outward from medium 2 (shown as $\hat{\mathbf{n}}$ in Figure 7.23b).

(2) It can be shown by applying [7.16c] to the same contour C in Figure 7.23b that the tangential component of the magnetic field $\overline{\mathscr{H}}$ is continuous across any interface:

$$\hat{\mathbf{n}} \times [\overline{\mathscr{H}}_1 - \overline{\mathscr{H}}_2] = 0 \qquad \rightarrow \qquad \mathscr{H}_{1t} = \mathscr{H}_{2t} \qquad [7.21]$$

except where surface currents ($\mathbf{J}_s$) may exist, such as at the surface of a perfect conductor (i.e., $\sigma = \infty$):

[43] That the continuity equation can be derived from equations [7.16b] and [7.16c] indicates that Maxwell's equations [7.16b] and [7.16c] are not entirely independent, if we accept conservation of electric charge as a fact.

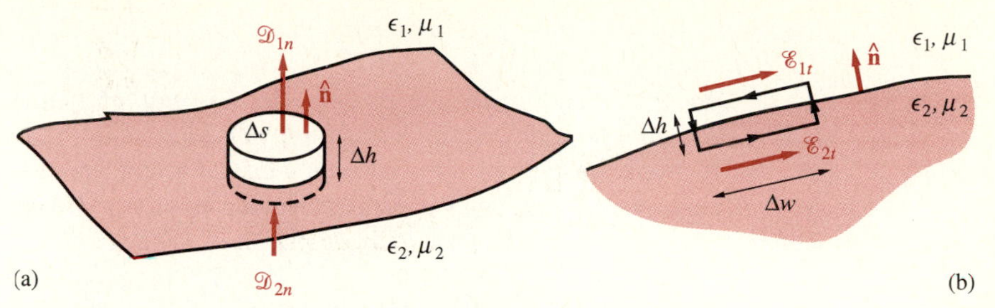

FIGURE 7.23. Interfaces between two different materials. The boundary conditions for the electromagnetic fields are derived by applying the surface integrals to the cylindrical surface as shown in (a) and the line integrals to the rectangular contour in (b).

$$\hat{\mathbf{n}} \times \overline{\mathscr{H}}_1 = \overline{\mathscr{J}}_s$$

Noting that the field $\overline{\mathscr{H}}_2$ inside the perfect conductor is zero, as discussed in Section 6.8.4.

(3) It can be shown by applying [7.16*b*] to the surface of the cylinder shown in Figure 7.23a that the normal component of electric flux density $\overline{\mathscr{D}}$ is continuous across interfaces, except where surface charge ($\tilde{\rho}_s$) may exist, such as at the surface of a metallic conductor or at the interface between two lossy dielectrics ($\sigma \neq 0$):

$$\hat{\mathbf{n}} \cdot (\overline{\mathscr{D}}_1 - \overline{\mathscr{D}}_2) = \tilde{\rho}_s \qquad \rightarrow \qquad \mathscr{D}_{1n} - \mathscr{D}_{2n} = \tilde{\rho}_s \qquad [7.22]$$

(4) A consequence of applying [7.16*d*] to the surface of the cylinder in Figure 7.23a is that the normal component of the magnetic field $\overline{\mathscr{B}}$ is continuous across interfaces:

$$\hat{\mathbf{n}} \cdot [\overline{\mathscr{B}}_1 - \overline{\mathscr{B}}_2] = 0 \qquad \rightarrow \qquad \mathscr{B}_{1n} = \mathscr{B}_{2n} \qquad [7.23]$$

(5) It follows from applying [7.19] to the surface of the cylinder in Figure 7.23a that, at the interface between two lossy media (i.e., $\sigma_1 \neq 0$, $\sigma_2 \neq 0$), the normal component of the current density $\overline{\mathscr{J}}$ is continuous, except where time-varying surface charge may exist, such as at the surface of a perfect conductor or at the interface between lossy dielectrics ($\epsilon_1 \neq \epsilon_2$ and $\sigma_1 \neq \sigma_2$):

$$\hat{\mathbf{n}} \cdot (\overline{\mathscr{J}}_1 - \overline{\mathscr{J}}_2) = -\frac{\partial \tilde{\rho}_s}{\partial t} \qquad \rightarrow \qquad J_{1n} - J_{2n} = -\frac{\partial \tilde{\rho}_s}{\partial t} \qquad [7.24]$$

Note that [7.24] is not completely independent of [7.20] through [7.23], since [7.19] is contained in [7.16*a*] through [7.16*d*], as mentioned previously. For the stationary case ($\partial/\partial t = 0$), [7.24] implies $\mathscr{J}_{1n} = \mathscr{J}_{2n}$, or $\sigma_1 \mathscr{E}_{1n} = \sigma_2 \mathscr{E}_{2n}$, which means that at the interface between lossy dielectrics (i.e., $\epsilon_1 \neq \epsilon_2$ and $\sigma_1 \neq \sigma_2$) there must in general be finite surface charge (i.e., $\tilde{\rho}_s \neq 0$) because otherwise [7.22] demands that $\epsilon_1 \mathscr{E}_{1n} = \epsilon_2 \mathscr{E}_{2n}$.

Maxwell, James Clerk *(Scottish mathematician and physicist, b. November 13, 1831, Edinburgh; d. November 5, 1879, Cambridge, England)* *Maxwell, born of a well-known Scottish family, was an only son. His mother died of cancer when he was eight, but except for that, he had a happy childhood.*

Early in life, he showed signs of mathematical talent. He entered Cambridge in 1850 and graduated second in his class in mathematics. Maxwell was appointed to his first professorship at Aberdeen in 1856. In 1857 he made his major contribution to astronomy by showing mathematically that Saturn's rings must consist of numerous small solid particles in order to be dynamically stable. In 1860 Maxwell brought his mathematics to bear upon the problem of particles making up gases. He considered the molecules to move not only in all directions but at all velocities, and bouncing off each other and off the walls of the container with perfect elasticity. Along with Boltzmann [1844–1906], he worked out the Maxwell–Boltzmann kinetic theory of gases, bringing about an entirely new view of heat and temperature.

The crowning work of Maxwell's life took place between 1864 and 1873, when he placed into mathematical form the speculations of Faraday [1791–1867] concerning magnetic lines of force. (Maxwell resembled Faraday, by the way, in possessing deep religious convictions and in having a childless, but very happy, marriage.)

In developing the concept of lines of force, Maxwell was able to work out a few simple equations that expressed all the varied phenomena of electricity and magnetism and bound them indissolubly together. Maxwell's theory showed that electricity and magnetism could not exist in isolation—where one exists, so does the other—so his work is usually referred to as the electromagnetic theory.

Maxwell showed that the oscillation of an electric charge produced an electromagnetic field that radiated outward from its source at a constant speed. This speed could be calculated by taking the ratio of certain units expressing magnetic phenomena to units expressing electrical phenomena. This ratio worked out to be just about 300,000 kilometers per second, or 186,300 miles per second, which is approximately the speed of light (for which the best available figure at present is 299,792.5 kilometers per second, or 186,282 miles per second).

To Maxwell, this seemed to be more than coincidence, and he suggested that light itself arose through an oscillating electric charge and was therefore an electromagnetic radiation. Furthermore, since charges could oscillate at any speed, it seemed to Maxwell that a whole family of electromagnetic radiations existed, of which visible light was only a small part.

Maxwell believed that not only were the waves of electromagnetic radiation carried by the ether, but that the magnetic lines of force were actually disturbances of the ether. In this way, he conceived he had abolished the notion of action at a distance. It had seemed to some experimenters in electricity and magnetism (Ampère [1775–1836], for instance), that a magnet attracted iron without actually making contact with the iron. To Maxwell it seemed that the disturbances in the ether set up by the magnet touched the iron and that everything could be worked out as action on contact. Maxwell also rejected the notion that electricity was particulate in nature, even though that was so strongly suggested by Faraday's laws of electrolysis.

Maxwell died of cancer before the age of fifty. Had he lived out what would today be considered a normal life expectancy, he would have seen his prediction of a broad spectrum of electromagnetic radiation verified by Hertz [1857–1894]. However, he would also have seen the ether, which his theory had seemed to establish firmly, brought into serious question by the epoch-making experiment of Michelson [1852–1931] and Morley [1838–1923], and he would have seen electricity proved to consist of particles after all. His electromagnetic equations did not depend on his own interpretations of the ether, however, and he had wrought better than he knew. When Einstein's [1879–1955] theories, a generation after Maxwell's death, upset almost all of classical physics, Maxwell's equations remained untouched—as valid as ever. [Adapted with permission from I. Asimov, Biographical Encyclopedia of Science and Technology, *Doubleday, 1982].*

1700 1831 1879 2000

7.6 SUMMARY

This chapter discussed the following topics:

- **Faraday's law.** When the magnetic flux enclosed by a loop of wire changes with time, an electromotive force is induced, which is given by

$$\mathcal{V}_{\text{ind}} = \oint_C \mathbf{E} \cdot d\mathbf{l} = -\frac{d}{dt}\int_S \mathbf{B} \cdot d\mathbf{s}$$

  where S is the surface enclosed by the contour C and where the orientations of the two integrals are related by the right-hand rule. The polarity of the induced emf, determined by Lenz's law, opposes the change in magnetic flux that causes it. More generally, Faraday's law states that an electric field is induced by a time-varying magnetic field in space, regardless of whether or not wires are present. The direction of the induced electric field is also determined by Lenz's law.

- **Induction due to motion.** When conductors move in the presence of magnetic fields, an induced voltage is produced even if the magnetic fields do not vary in time. The general expression for the induced electromotive force is given by

$$\mathcal{V}_{\text{ind}} = -\int_S \frac{\partial \mathbf{B}}{\partial t} \cdot d\mathbf{s} + \oint_C (\mathbf{v} \times \mathbf{B}) \cdot d\mathbf{l}$$

  where the contour C encloses the surface S and where $d\mathbf{l}$ and $d\mathbf{s}$ are related via the right-hand rule.

- **Differential form of Faraday's law.** The differential form of Faraday's law in a stationary frame of reference is

$$\nabla \times \mathbf{E} = -\frac{\partial \mathbf{B}}{\partial t}$$

which describes the fact that any time-varying magnetic field in space is encircled by an electric field **E**.

- **Energy in a magnetic field.** The self-energy of a current-carrying loop is given by

$$W_{\text{ms}} = \frac{1}{2}\int_V \mathbf{J}\cdot\mathbf{A}\,dv$$

where **A** is the magnetic vector potential. This expression reduces to the familiar expression for the energy stored in circuit with inductance L, namely, $(\frac{1}{2})LI^2$.

The magnetic energy stored by two current-carrying loops is given by

$$W_{\text{m}} = \tfrac{1}{2}L_{11}I_1^2 \pm L_{12}I_1I_2 + \tfrac{1}{2}L_{22}I_2^2$$

where L_{12} is the mutual inductance between the two circuits.

The magnetostatic energy stored in a distribution of magnetic fields is given by

$$W_{\text{m}} = \frac{1}{2}\int_V \mathbf{H}\cdot\mathbf{B}\,dv$$

where V is the entire volume in which the magnetic field is nonzero. The volume energy density of the magnetostatic field for a linear and isotropic medium is $w_{\text{m}} = \frac{1}{2}\mu H^2$.

- **Displacement current.** Maxwell developed the time-varying form of Ampère's law by postulating the displacement-current term such that

$$\nabla \times \mathbf{H} = \mathbf{J} + \frac{\partial \mathbf{D}}{\partial t}$$

The displacement current term makes Ampère's law consistent with the law of conservation of charge and indicates that time-varying electric fields produce magnetic fields. Together with Faraday's law, the $\partial\mathbf{D}/\partial t$ term makes it possible for electromagnetic waves to propagate through free space, a prediction confirmed in 1888 by Hertz's experiments.

- **Maxwell's equations.** Summarized and reviewed in Section 7.5.

7.7 PROBLEMS

7-1. Stationary rectangular loop. Consider a fixed single-turn rectangular loop of area A with its plane perpendicular to a uniform magnetic field. Find the voltage induced across the terminals of the loop if the magnetic flux density is given by (a) $B(t) = B_0 t e^{-\alpha t}$, and (b) $B(t) = B_0 e^{-\alpha t} \sin(\omega t)$.

7-2. Stationary circular loop. Consider a 20-turn circular loop of wire of 15 cm diameter with its plane perpendicular to a uniform magnetic field, as shown in Figure 7.6. If the magnetic flux density B is given by

$$B = 10\cos(120\pi t) \quad \text{G}$$

find the rms value of the induced current through the 10Ω resistor at the following time instants: (a) $t = 0$, (b) $t = 10$ ms, (c) $t = 100$ ms, and (d) $t = 1$ s.

7-3. Two circular coils. Two circular coils of 5 cm radius each have the same axis of symmetry and are 1 m apart from each other. Coil 1 has 10 turns and coil 2 has 100 turns. (a) If an alternating current of 100 A amplitude and 1 kHz frequency is passed through coil 1, find the amplitude of the induced voltage across the open terminals of coil 2. (b) Repeat part (a) for a frequency of 10 kHz.

7-4. Two concentric coils. Consider the two concentric coils shown in Figure 7.9. The magnetic flux density produced at the center of the two coils due to the current I flowing in the larger coil is given by

$$\mathbf{B} = \hat{\mathbf{z}}2.5 \times 10^{-5} I \text{ T}$$

(a) If the larger coil has 30 turns, find its radius. (b) If the smaller coil has 75 turns and its radius is 1 cm, find the total flux linking the smaller coil. (c) If the current I in the larger coil is given by $I(t) = 10\cos(120\pi t)$ A, find the induced current through the resistor R assuming $R = 10\Omega$.

7-5. Triangular loop and long wire. An equilateral triangular loop is situated near a long current-carrying wire, as shown in Figure 7.24. The wire is part of a power line carrying 60-Hz sinusoidal current. An ac ammeter inserted in the loop reads a current of amplitude 1 mA. Assume the total resistive impedance of the loop to be 0.01Ω. Find the amplitude of the current I in the long wire.

7-6. Toroidal coil around a long, straight wire. A long, straight wire carrying an alternating current of $I(t) = t00\cos(377t)$ A coincides with the principal axis of symmetry of a 200-turn coil wrapped uniformly around a rectangular, toroidal-shaped iron core of inner and outer radii $a = 6$ cm and $b = 8$ cm, thickness $t = 3$ cm, and relative permeability $\mu_r = 1000$, as shown in Figure 7.25. Calculate the induced voltage $\mathcal{V}_{\text{ind}}$ between the terminals of the coil. (This type of coil, called a *current transformer,* is used to measure the current in a conductor wire that passes through it.)

7-7. Current transformer. A current transformer is used to measure the current in a high-voltage transmission line, as shown in Figure 7.26. The circular toroidal core has a mean diameter of 6 cm, circular cross section of 1 cm diameter, and relative permeability of $\mu_r = 200$. The winding consists of $N = 300$ turns. If the 60-Hz current of

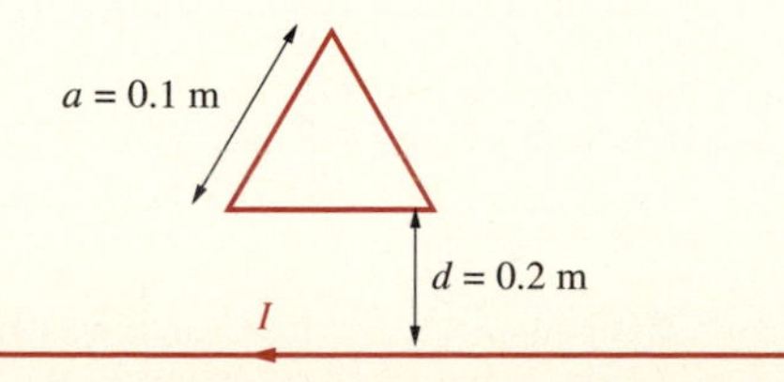

FIGURE 7.24. Triangular loop and long wire. Problem 7-5.

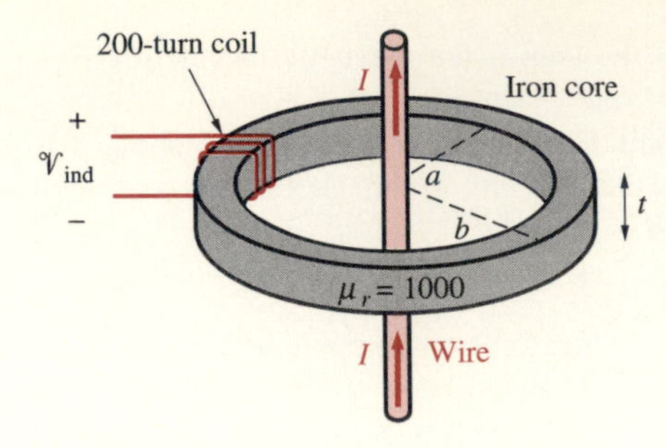

FIGURE 7.25. Toroidal coil around long wire. Problem 7-6.

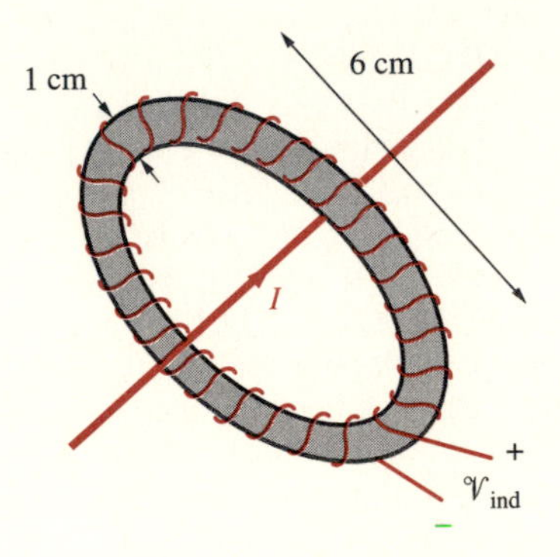

FIGURE 7.26. A current transformer. Problem 7-7.

amplitude 1000 A flows through the high-voltage line, find the rms value of the open circuit voltage induced across the terminals of the toroid.

7-8. Sliding bar in a constant magnetic field. A metal bar able to move on fixed rails is initially at rest on two stationary conducting rails that are $l = 50$ cm apart and connected to each other via a $V_0 = 10$ V voltage source in series with a $R = 10\Omega$ resistor, as shown in Figure 7.27. A constant magnetic field of $\mathbf{B}_0 = \hat{\mathbf{z}}1.5$ T is turned on at $t = 0$. Assume the rails and the metal bar to be perfectly conducting, and neglect the self-inductance of the loop formed by the rails and the bar. The initial resting position of the bar is at $y_0 = 75$ cm. (a) With no friction, and assuming that the mass of the bar is 0.5 kg, calculate the magnitudes and directions of its initial acceleration and its final velocity. Explain your reasoning. (b) Describe what happens if the voltage source is

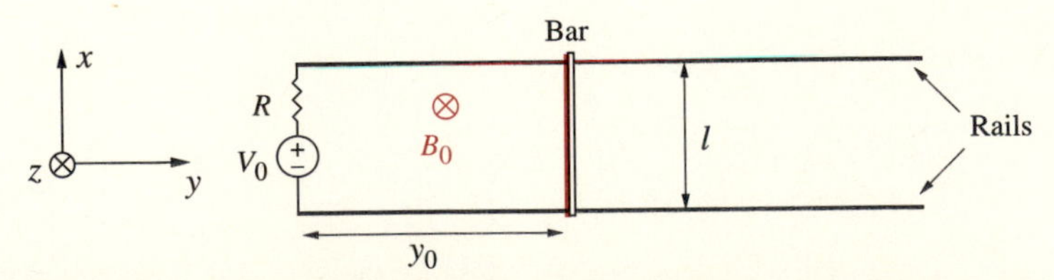

FIGURE 7.27. Sliding bar in a constant magnetic field. Problem 7-8.

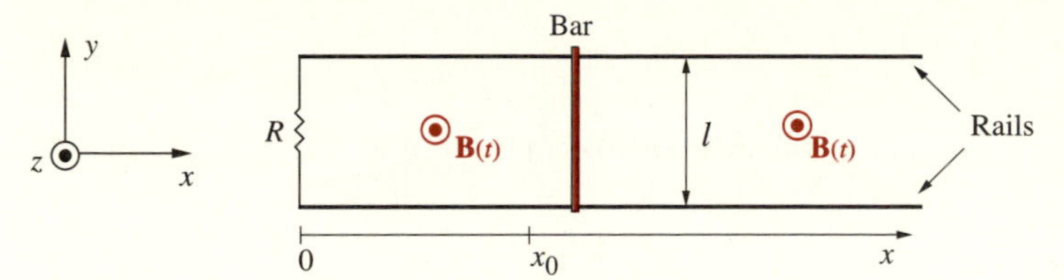

FIGURE 7.28. **Oscillating bar in a time-varying magnetic field.** Problem 7-9.

turned off (i.e., $V_0 = 0$) at $t = t_0$, when the bar has reached a velocity v_0. (c) What happens if the voltage source remains at V_0 but the magnetic field is turned off (i.e., $\mathbf{B} = 0$) at $t = t_0$?

7-9. Oscillating bar in a time-varying magnetic field. A metal wire is in contact with two long, parallel, stationary conducting rails connected with a resistance R at one end, as shown in Figure 7.28. The wire oscillates symmetrically around $x = x_0$ with a velocity $\mathbf{v}(t) = \hat{\mathbf{x}} v_0 \cos(\omega_0 t)$ in the presence of a time-varying magnetic field that is perpendicular to the plane of the rails. The magnetic field is given by $\mathbf{B}(t) = \hat{\mathbf{z}} B_0 \sin(\omega_0 t)$. Find the induced current through the resistance R.

7-10. Moving square loop. A uniform magnetic field of $B_0 = 1$ T is confined to a square-shaped area 10 cm to a side as shown in Figure 7.29, with zero magnetic field everywhere else. A square loop of side length 5 cm moves through the loop with a velocity of $v = 10$ cm-s^{-1}. Find the electromotive force induced in the square loop and plot its value as a function of distance x, for $0 \leq x \leq 15$ cm. Note that the plane of the loop is orthogonal to the magnetic field.

7-11. Rectangular loop and long wire. A rectangular loop of wire lies in the plane of a long, straight wire carrying a steady current I, as shown in Figure 7.30. If the loop is

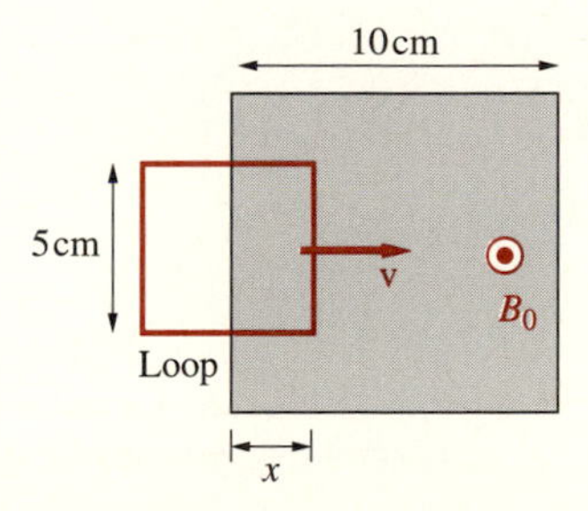

FIGURE 7.29. **Moving square loop.** Problem 7-10.

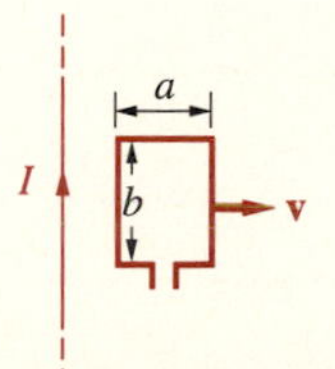

FIGURE 7.30. **Rectangular loop and long wire.** Problem 7-11.

moving radially away from the long wire with a velocity $\mathbf{v} = \hat{\mathbf{r}}v_0$ without changing its shape, find the voltage induced across the open terminals of the loop with its polarity indicated.

7-12. Rotating wire in a constant magnetic field. A semicircular-shaped wire of radius a is rotated in a constant magnetic field B_0 at a constant angular frequency ω, as shown in Figure 7.31. If the wire forms a closed loop with a resistance R, find the induced current in R. Neglect the resistance of the wires.

7-13. Rotating rectangular loop in a constant magnetic field. A single turn rectangular coil of dimensions 5 cm × 10 cm is rotated at 1500 rpm in a uniform magnetic field of 50 mT normal to the axis of rotation, as shown in Figure 7.13. Find the induced voltage across the open terminals of the loop.

7-14. Rotating rectangular loop in a nonuniform magnetic field. Consider the rectangular coil of 10 turns with dimensions 20 cm × 10 cm, rotating at an angular velocity of 2000 rpm about its axis, as shown in Figure 7.32, in the presence of a magnetic field. If the **B** field vector is in the cylindrical radial direction, which only varies with the azimuthal angle ϕ as $\mathbf{B} = \hat{\mathbf{r}}3\cos\phi$ mT, find the voltage $\mathcal{V}_{ab}$ induced across the open terminals of the coil.

7-15. Extracting power from a power line. Consider the possibility of putting your electromagnetics knowledge to practical use. You live out in the country and suddenly realize that a large power line passes by your farm. The power-line easement extends to 20 m on either side, so that the border of your property is at a ground distance of 20 m from the wire. You have at your disposal 200 m of number 6 gauge (4.1-mm diameter) copper wire, which can be deployed in the form of a 1-turn rectangular loop of side lengths a and b, as shown in Figure 7.33. (a) Assuming the power line to carry a sinusoidal (60-Hz) current of $I = 4000$ A, find the maximum amount of power (in

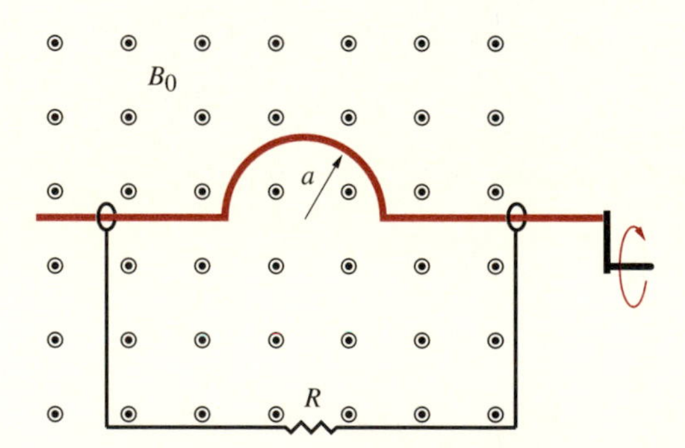

FIGURE 7.31. Rotating wire in constant magnetic field. Problem 7-12.

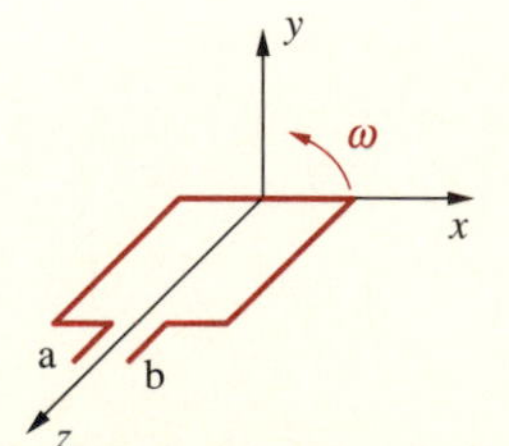

FIGURE 7.32. Rotating rectangular loop. Problem 7-14.

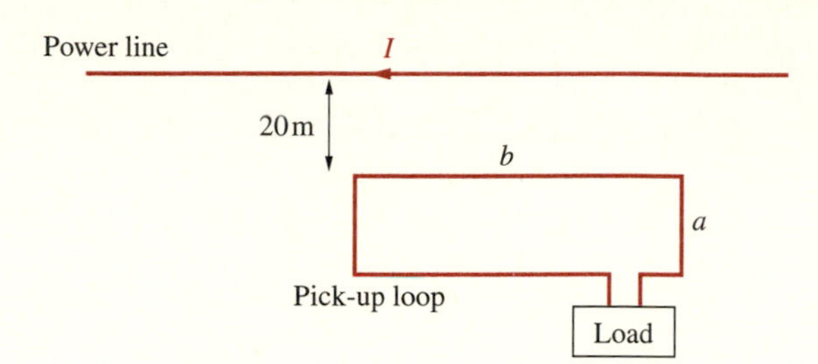

FIGURE 7.33. Extracting power from a power line. Problem 7-15.

watts) that you can extract from the power line. State all assumptions. (b) Could you extract more power by using your 200-m wire in the form of a multiturn loop? Note that this is a hypothetical problem, which, like many engineering problems, may have legal ramifications. The power company may claim that it owns the spillover energy; on the other hand, you can claim that the spillover energy is an unauthorized intrusion into your private property and that the least you can do is to take advantage of it.

7-16. Induction. Two infinitely long wires carrying currents I_1 and I_2 cross (without electrical contact) at the origin, as shown in Figure 7.34. A small rectangular loop is placed next to the wires. (a) If $I_1 = \cos(\omega t)$ and $I_2 = \sin(\omega t)$, determine the polarity and magnitude of the induced voltage $\mathcal{V}_{\text{ind}}(t)$. Sketch $\mathcal{V}_{\text{ind}}(t)$ together with $I_1(t)$ and $I_2(t)$. (b) If I_1 and I_2 are both constant so that $I_1 = I_2 = I$, and if we move the loop away from the wires at a constant velocity **v**, which direction should it be moved in order to produce the largest $|\mathcal{V}_{\text{ind}}|$? Find this value of $|\mathcal{V}_{\text{ind}}|$.

7-17. Induction. Consider the square loop located between two infinitely long parallel wires carrying currents I_1 and I_2, where the sides of the loop are equidistant from each wire, as shown in Figure 7.35. (a) Starting at $t = 0$, the current I_1 decreases exponentially over time according to $I_1 = I_0 e^{-t}$, while the current I_2 stays constant at $I_2 = I_0$. Find an expression for the induced voltage $\mathcal{V}_{\text{ind}}(t)$, and sketch it as a function of time, paying particular attention to its polarity. (b) Repeat part (a) for $I_1 = I_0 \sin(\omega t)$. Your sketch of $\mathcal{V}_{\text{ind}}(t)$ should cover the range $0 < t < 2\pi/\omega$.

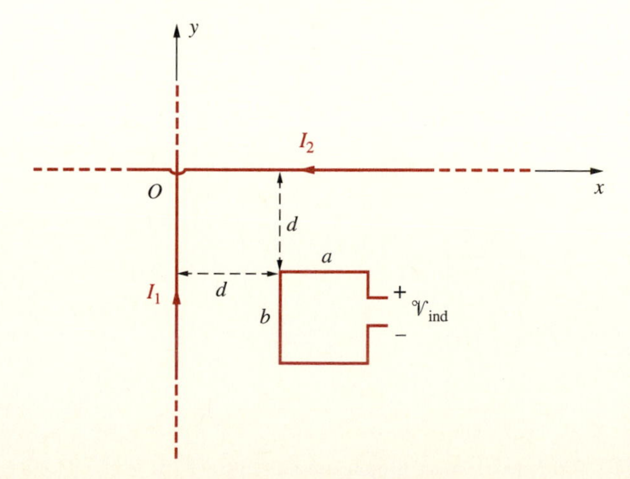

FIGURE 7.34. Induction. Problem 7-16.

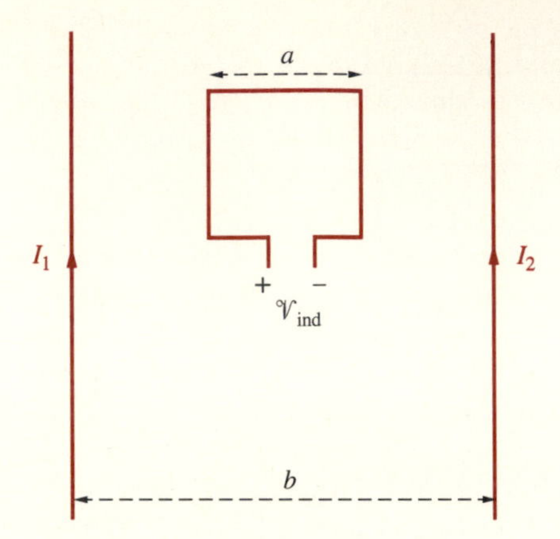

FIGURE 7.35. **Induction.** Problem 7-17.

7-18. Wave propagation. An electromagnetic pulse of length l traveling in the $+x$ direction with a constant velocity v_p has an electric field in the z direction, which is shown at $t = 0$ as a function of x. (See Figure 7.36.) An observer capable of measuring $\mathscr{E}_z$ is located at position $x = d$ (where $d > l$). (a) Sketch $\mathscr{E}_z$ as a function of x at $t = d - l/v_p$, $(d - l/2)/v_p$, $(d + l/2)/v_p$, and $(d + 3l)/v_p$. (b) Find $\mathscr{E}_z$ measured at $t = (d - l/3)/v_p$, $(d - 2l/3)/v_p$, and $(d + l)/v_p$.

7-19. Wave propagation. An observer located at $y = y_1$ measures the magnetic field of an electromagnetic pulse propagating with a constant velocity v_p in the $+y$ direction as

$$\overline{\mathscr{H}}(y_1, t) = \begin{cases} 0 & t < 0 \\ \hat{\mathbf{z}} H_0 t/t_1 & 0 < t < t_1 \\ \hat{\mathbf{z}} H_0 & t_1 < t < 3t_1 \\ 0 & t > 3t_1 \end{cases}$$

Sketch $\mathscr{H}_z$ versus y at the following times: (a) $t = 2t_1$ and (b) $t = 4t_1$. Indicate the position of the observer in each sketch.

7-20. Interference of two waves. Two transverse electromagnetic pulses A and B propagate in the $+z$ and $-z$ directions, respectively, each with a velocity v_p. At $t = 0$, the fronts of the two pulses are at a distance d from one another. The electric field waveform for pulse A is known, whereas the temporal shape of pulse B is unknown. If an observer

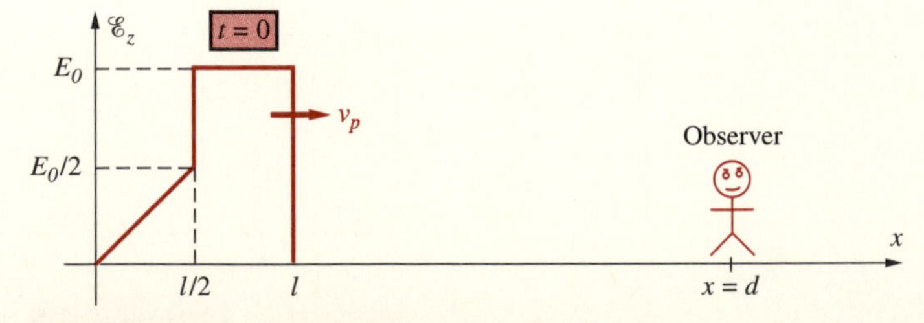

FIGURE 7.36. **Wave propagation.** Problem 7-18.

located midway between the two waves measures a total electric field as shown in Figure 7.37, sketch the electric field waveform of pulse B. Assume that the electric fields of both pulses are in the x direction.

7-21. Wave propagation. The electric field of an electromagnetic wave propagating in air has the shape of a Gaussian pulse, given by

$$\overline{\mathscr{E}}(z,t) = \hat{\mathbf{y}}75e^{-\pi(z-v_pt)^2}$$

in V-m^{-1}, where $v_p = 3 \times 10^8$ m-s^{-1}. An observer located at $z = 1$ km has a receiver that is capable of measuring a minimum electric field of 0.1 μV-m^{-1}. (a) How long does the observer continue to detect a signal? (b) At what time does the field measured by the observer reach a maximum? What is the maximum $\overline{\mathscr{E}}$ measured? (c) Sketch $\mathscr{E}_y(z,t)$ as a function of z at $t = 5$ μs with the position of the observer indicated. Does the observer detect a measurable signal at this time?

7-22. Phasors. Write the following real-time expressions in phasor form: (a) $\overline{\mathscr{E}}(z,t) = \hat{\mathbf{y}}\cos(\omega t - z)$ V-m^{-1}. (b) $\overline{\mathscr{H}}(x,t) = 0.1[\hat{\mathbf{y}}\cos(\omega t - 0.3x) + \hat{\mathbf{z}}0.5\sin(\omega t + 0.3x)]$ mA-m^{-1}. (c) $\overline{\mathscr{B}}(y,z,t) = \hat{\mathbf{x}}40\sin(3 \times 10^8 t + 0.8y - 0.6z + \pi/4)$ μT. (d) $\overline{\mathscr{E}}(x,y,t) = \hat{\mathbf{z}}E_0\sin(ax)\cos(\omega t + by)$.

7-23. Phasors. Express the following phasors as real-time quantities: (a) $\mathbf{E}(y) = \hat{\mathbf{z}}5e^{-j40\pi y}$ V-m^{-1}. (b) $\mathbf{B}(z) = \hat{\mathbf{x}}0.1e^{-j2\pi z} - \hat{\mathbf{y}}0.3je^{-j2\pi z}$ μT. (c) $\mathbf{E}(x) = \hat{\mathbf{z}}0.1(e^{-j18x} - 0.5e^{j18x})$ V-m^{-1}. (d) $\mathbf{H}(x,z) = \hat{\mathbf{y}}e^{-j48\pi x}e^{j64\pi z}$ mA-m^{-1}. (e) $\mathbf{J}(y) = \hat{\mathbf{x}}40e^{-0.1y(1+j)+j\pi/3}$ μA-m^{-2}.

7-24. Displacement current in a capacitor. Consider a parallel-plate capacitor with metal plates of 1-cm^2 area, each separated by mica ($\epsilon_r = 6$) 1 mm thick. If an alternating voltage of $V(t) = 10\cos(2\pi ft)$ V is applied across the capacitor plates, find the displacement current $\mathbf{J}_d$ through the capacitor at (a) $f = 10$ kHz, and (b) $f = 1$ MHz.

7-25. Propagation through lake water. The electric field of an electromagnetic wave propagating in a deep lake (for lake water, assume $\sigma = 4 \times 10^{-3}$ S-m^{-1}, $\epsilon_r = 81$, and

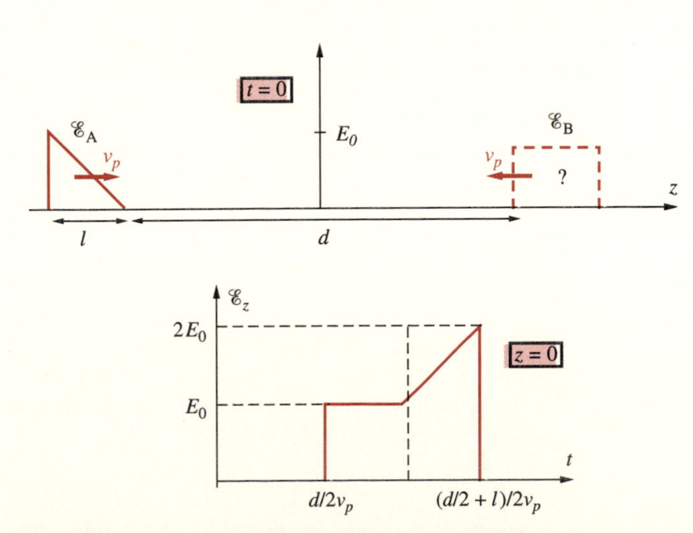

FIGURE 7.37. Interference of two waves. Problem 7-21.

$\mu_r = 1$) transmitted by an electromagnetic probe submerged in the lake is approximately given by

$$\overline{\mathscr{E}} \simeq \hat{\mathbf{x}}10e^{-0.08z}\cos(2.7 \times 10^6 \pi t - 0.27z) \text{ mV-m}^{-1}$$

Find the peak values of the vectors of conduction-current density $\overline{\mathscr{J}}_c$ and displacement-current density $\overline{\mathscr{J}}_d$ at (a) $z = 0$, (b) $z = 10$ m, and (c) $z = 100$ m.

7-26. Sea water. For sea water, $\sigma = 4$ S-m^{-1}, $\epsilon_r = 81$, and $\mu_r = 1$. (a) Find the ratio of the magnitudes of the conduction-current density and the displacement-current density at 10 kHz, 1 MHz, 100 MHz, and 10 GHz. (b) Find the frequency at which the ratio is 1.

7-27. Dry soil. For dry soil, assume $\sigma \simeq 10^{-4}$ S-m^{-1}, $\epsilon_r \simeq 3$, and $\mu_r = 1$. Find the frequency at which the ratio of the magnitudes of the conduction-current and displacement-current densities is unity.

7-28. Maxwell's equations. Consider an electromagnetic wave propagating in a source-free nonconducting medium represented by $\overline{\mathscr{E}}$ and having components given by

$$\mathscr{E}_x = p_1(z - v_p t) + p_2(z + v_p t) \qquad \mathscr{E}_y = \mathscr{E}_z = 0$$

where p_1 and p_2 are any two arbitrary functions, and $v_p = 1/\sqrt{(\mu\epsilon)}$. Show that $\overline{\mathscr{E}}$ satisfies all of Maxwell's equations, and find the components of the corresponding $\overline{\mathscr{H}}$.

7-29. AM radio waves. The electric-field and magnetic-field components of an AM radio signal propagating in air are given by

$$\overline{\mathscr{E}} = \hat{\mathbf{x}}E_0\cos(7.5 \times 10^6 t - \beta z)$$

$$\overline{\mathscr{H}} = \hat{\mathbf{y}}\frac{E_0}{\eta}\cos(7.5 \times 10^6 t - \beta z)$$

Find the values of β and η such that these expressions satisfy all of Maxwell's equations.

7-30. Maxwell's equations. The magnetic field phasor of an electromagnetic wave in air is given by

$$\mathbf{H}(y) = \hat{\mathbf{z}}1.83 \times 10^{-4}e^{-j4y} \text{ A-m}^{-1}$$

(a) Find the angular frequency ω of the wave such that $\mathbf{H}$ satisfies all of Maxwell's equations. (b) Find the corresponding time-harmonic electric field $\mathbf{E}$. (c) Find the electric flux density $\mathbf{D}$ and the displacement-current density $\mathbf{J}_d$.

7-31. Superposition of two waves. The sum of the electric fields of two time-harmonic (sinusoidal) electromagnetic waves propagating in opposite directions in air is given as

$$\overline{\mathscr{E}}(z, t) = \hat{\mathbf{x}}95\sin(\beta z)\sin(21 \times 10^9 \pi t) \text{ mV-m}^{-1}$$

(a) Find the constant β. (b) Find the corresponding $\overline{\mathscr{H}}$.

7-32. Electromagnetic wave in free space. An electromagnetic wave propagating in free space has an electric field given by

$$\overline{\mathscr{E}}(x, z, t) = \hat{\mathbf{y}}4.9\cos(1.8 \times 10^9 \pi t - ax - 2.5az) \text{ V-m}^{-1}$$

where a is a constant. Find the value of a and the corresponding expression for the magnetic field $\overline{\mathscr{H}}$.

7-33. Coaxial lines. The electric field of a transverse electromagnetic wave guided within a lossless coaxial transmission line along the z axis is expressed in cylindrical coordinates as

$$\mathbf{E}(r, z) = \hat{\mathbf{r}}\frac{E_0}{r}e^{-j\beta z} \qquad a \leq r \leq b$$

where $\beta = \omega\sqrt{\mu\epsilon}$; μ and ϵ are, respectively, the permeability and permittivity of the dielectric material, and a and b are the inner and the outer radii of the coaxial line. (a) Find the corresponding $\mathbf{H}$. (b) Write the time-domain expressions for $\overline{\mathscr{E}}$ and $\overline{\mathscr{H}}$. (c) Sketch both $\mathscr{E}_r$ and $\mathscr{H}_\phi$ as functions of r over the range $a \leq r \leq b$ at position $z = 0$ and time $t = 0$. (d) Sketch both $\mathscr{E}_r$ and $\mathscr{H}_\phi$ as functions of z at position $r = a$ and time $t = 0$.

7-34. Specific absorption rate (SAR). Consider a plane wave in an electrically conducting homogeneous biological tissue. As the wave attenuates, the power absorbed by the tissue produces local heating at a rate measured by the specific absorption rate (SAR), defined as the time-average power dissipated per unit mass of the tissue. The SAR in watts per kilogram is thus given by

$$\text{SAR} = \frac{\frac{1}{2}\sigma|\mathbf{E}|^2}{\rho_m} \text{ W-kg}^{-1}$$

where ρ_m is the mass density (or the specific gravity) of the tissue in units of kg-m^{-3}. Table 7-1 lists the specific gravity, relative permittivity, and the conductivity of various tissues at two mobile telephone frequencies.[44] For each tissue, calculate (a) the loss tangent, and (b) the SAR at both frequencies, assuming the internal electric field in each tissue to be 1 V-m^{-1}.

7-35. Dipole antenna next to a lossy sphere. Consider a center-fed half-wave dipole antenna radiating 1 W resonant at 835 MHz placed 25 mm from the surface of a 10.7 cm radius spherical glass phantom filled with liquid representing[45] human brain tissue.

TABLE 7.1. Dielectric properties and specific gravities of tissues

		835 MHz		1900 MHz	
Tissue	Specific gravity (10^3 kg-m^{-3})	ϵ_r	σ (S-m^{-1})	ϵ_r	σ (S-m^{-1})
Muscle	1.04	51.76	1.11	49.41	1.64
Fat	0.92	9.99	0.17	9.38	0.26
Bone (skull)	1.81	17.40	0.25	16.40	0.45
Cartilage	1.10	40.69	0.82	38.10	1.28
Skin	1.01	35.40	0.63	37.21	1.25
Brain	1.04	45.26	0.92	43.22	1.29

[44]Gandhi, O. P., G. Lazzi, and C. M. Furse, Electromagnetic absorption in the human head and neck for mobile telephones at 835 and 1900 MHz, *IEEE Transactions on Microwave Theory and Techniques,* 44(10), pp. 1884–1897, October 1996.

[45]Faraone, A. D. Simunic, and O. Balzano, Experimental dosimetry in a sphere of simulated brain tissue near a half-wave dipole antenna. *Proc. IEEE Intl. Symp. on Electromagnetic Compatibility,* Denver, August 23–28, 1998.

$\epsilon_r = 44, \sigma = 0.9$ S-m^{-1}. The peak value of the electric field magnitude at the point in the tissue nearest to the antenna is measured as 115 V-m^{-1}. (a) Calculate the peak specific absorption rate (SAR) at that point, assuming that the specific gravity of the brain tissue is 1 g-cm^{-3}. (b) Assuming the SAR decays exponentially inside the tissue with an attenuation constant $\alpha = 0.5$ np-(cm)$^{-1}$, compute the 1-g SAR average assuming a uniform SAR transverse distribution across 1 cm^2, and compare it with the ANSI-IEEE C95.1-1992 SAR standard, which is 1.6 W-(kg)$^{-1}$ for any 1 g of biological tissue as an upper limit for uncontrolled environments.[46]

7-36. Microwave heating of beef products. When microwaves are intercepted by dielectric materials such as food, they dissipate energy, which results in an increase in the temperature of the material. The dielectric properties of beef products at 2.45 GHz and 25°C are $\epsilon_r' = \epsilon'/\epsilon_0 = 52.4$ and $\tan\delta_c = 0.33$.[47] Calculate (a) ϵ_r'' and (b) the dissipated power density if the electric field inside the beef is 25 kV-m^{-1}.

7-37. Lossy capacitor. A parallel-plate capacitor with 1-cm^2 area and 2-mm plate separation is filled with polyethylene ($\epsilon_r' = 2.26$, $\epsilon_r'' = 0.0007$ at 3 GHz) and is operated at

TABLE 7.2. Dielectric properties of rice weevils and wheat at 24°C

	Rice weevils		Wheat	
Frequency	ϵ_r'	ϵ_r''	ϵ_r'	ϵ_r''
50 kHz	12.16	0.34	4.68	0.21
200	11.17	0.59	4.53	0.26
1 MHz	11.56	1.36	4.33	0.26
5	9.68	2.16	4.14	0.34
20	7.63	2.03	3.90	0.41
50	6.79	2.19	3.86	0.45
100	6.13	1.96	3.38	0.36
200	5.13	1.20	3.08	0.30
300	5.15	1.07	3.09	0.30
440	5.38	0.92	3.09	0.40
1.0 GHz	4.18	0.42	2.76	0.32
2.4	3.82	0.30	2.63	0.30
3.0	4.16	0.38	2.61	0.25
5.5	3.68	0.46	2.46	0.28
8.4	4.39	0.47	2.60	0.22
10.1	4.19	0.44	2.55	0.24
11.0	4.17	0.41	2.58	0.28
12.2	4.05	0.34	2.56	0.23

[46]American National Standards Institute—Safety levels with respect to exposure to radio frequency electromagnetic fields, 3 kHz to 300 GHz, ANSI/IEEE C95.1, 1992.

[47]D. I. C. Wang and S. A. Goldblith, Dielectric properties of foods, Technical Report TR-76-27-FEL, Massachusetts Institute of Technology, Dept. of Nutrition and Food Science, Cambridge, Massachusetts, November, 1976.

3 GHz. If the peak voltage of the applied signal is 5 V, find (a) the total current and (b) the total power lost in the form of heat. Neglect fringing fields.

7-38. Radio-frequency insect control. Experiments conducted over the past 50 years have shown that exposing grain infested by stored-grain insects to radio-frequency and microwave energy can control the insect infestations via selective dielectric heating of the insects and grain.[48] Table 7.2 shows the dielectric properties of adult insects of the rice weevil *Sitophilus oryzae* (L.) and bulk samples of hard red winter wheat at 24°C as a function of frequency ranging from 50 kHz to 12.2 GHz. Modeling[49] indicates that when the grain and insect mixture is subjected to an external electric field, the ratio of the electric field within the insects to that in the wheat is approximately given by

$$\frac{E_{\text{insect}}}{E_{\text{wheat}}} \simeq \frac{3}{2 + (\epsilon'_{\text{insect}}/\epsilon'_{\text{wheat}})}$$

(a) Calculate the loss tangent of adult rice weevils and hard red winter wheat at 1 MHz, 100 MHz, and 2.4 GHz. (b) Find the approximate frequency at which the ratio of the power dissipated per unit volume in adult rice weevils versus that in hard red winter wheat is maximized. Assume the electric field amplitude in the wheat to be 120 kV-m^{-1}. (c) Based on your results, comment on the controllability of stored-grain insects using radio-frequency dielectric heating. Which of the two frequencies, 50 MHz or 2.4 GHz, works the best for selective heating of insects in relation to the grain they infest? Why?

---

[48] S. O. Nelson, Review and assessment of radio-frequency and microwave energy for stored-grain insect control, *Trans. American Soc. Agricultural Engineers (ASAE)*, 39(4), pp. 1475–1484, 1996.

[49] S. O. Nelson and L. F. Charity, Frequency dependence of energy absorption by insects and grain in electric fields, *Trans. ASAE*, 15(6), pp. 1099–1102, 1972.

CHAPTER 8

Electromagnetic Waves

Having established the physical basis of the complete set of Maxwell's equations, we now proceed to discuss what is arguably their most important consequence: electromagnetic waves. Before we consider the space-time structure of electromagnetic waves and their propagation, reflection, transmission, and guiding, a few comments are in order concerning the general nature of waves.

We recognize a wave as some pattern of values in space that appears to move in time. In the introductions to Chapters 1, 2, and 3 we mentioned that the idea of waves is one of the great unifying concepts of physics, and we discussed examples of different kinds of waves. Now that we are ready to embark on a quantitative study of electromagnetic waves, we briefly revisit some of these examples of other waves in order to place the characteristics of electromagnetic waves in context. The reader has no doubt observed the activity on a line of cars stopping or starting at a traffic light. The cars in the line do not all start at the same time when the light turns green; rather, the act of starting travels backwards through the line with a certain speed. This "start-up" wave is initiated by the first car in the line, and the speed of this wave depends on the reaction time of the drivers and the response characteristics of the cars (including the inertial mass and engine power). An important point to note is that no mass is transported by the wave; what travels is merely the "act" of start-up. In fact, in this particular case, the transport of mass (i.e., the motion of cars) is in the opposite direction to that of the start-up wave.

Similarly, when the end of a stretched rope is suddenly moved sideways, the event of sideways motion travels along the rope as a wave with a certain speed, which depends on the tension in the rope and its mass. The initial sideways displacement propagates away from the source, appearing at farther distances from the end at proportionally later times. Once again, no mass is transported by the wave; in fact, the mass elements move in the sideways direction. In a compressional seismic wave, the elastic displacement is in the same direction as that of the wave; however, clearly what travels is the "state of being compressed" rather than mass.

Consider water waves as perhaps the most familiar example. If we somehow (e.g., by dropping a pebble or pouring a bucketful of water) initiate the vertical motion of a volume of water in the middle of a lake, the surface of the water rises and falls in an oscillatory manner. However, it is not possible for the bucketful of water to oscillate independently of the water surrounding it. Its periodic excesses and deficiencies of pressure are transmitted to the surrounding water, which thereby receives energy and is, in turn, put into motion. In its resulting motion this body of water also transfers energy to its surrounding outer regions. By this process, energy is transmitted to the adjacent bodies of water and, as a result, a water wave propagates across the surface of the lake. The fundamental basis of the water wave is the fact that the motion of and the pressure exerted by a given volume of water are not independent of the motion and pressure of the water in its surrounding volume.

The fundamental reason for the existence of electromagnetic waves is very similar. Since a changing electric field produces a magnetic field, and that in turn produces an electric field, and so on, a series of energy transfers is initiated whenever any electric or magnetic disturbance takes place. A changing magnetic field induces an electric field, both in the region in which the magnetic field changes and also in the surrounding region; similarly, a changing electric field produces a magnetic field in the region in which the change occurs and also in the surrounding regions. This process was qualitatively discussed in Section 7.4 when we considered the plausibility of electromagnetic wave propagation. Consequently, when a disturbance of either the electric or the magnetic field takes place in a given region of space, it cannot be confined to that space. The changing fields within the region induce fields in surrounding regions also; these induced fields induce fields in further surrounding regions, and so forth; as a result, the electromagnetic energy travels in the surrounding region. When there is an excess of electromagnetic energy anywhere in unbounded space, it cannot stand still, any more than a mound of displaced water can be stationary on the surface of a lake! It cannot simply subside in a limited region of unbounded space. It can only travel as a wave until its energy is dissipated.

Although other types of waves travel only through material media (such as sound waves, which travel through air or water, or seismic waves through earth), electromagnetic waves can also travel through free space (vacuum). Nineteenth-century physicists could not bring themselves to believe this fact and thought that vacuum of space was permeated by an elusive substance, referred to as the *ether,* that allowed the transmission of light waves.[1]

It was mentioned above that waves do not necessarily involve transport of any mass. Like moving objects, however, waves carry *energy* from one point to another. For example, the electromagnetic energy that arrives at the earth from the sun can be converted into electrical energy via solar cells and into chemical energy by plants, which is subsequently released as we burn wood or coal. A radio or television transmitter broadcasts its programs in the form of electromagnetic signals at power levels

[1]E. Whittaker, *A History of the Theories of Aether and Electricity,* Thomas Nelson and Sons Ltd., London, 1951.

of some tens of kilowatts. Each radio or TV receiver picks up a minute fraction of this power to reproduce the transmitted signal. Sound waves produced by thunder can shake a nearby house, while those produced by human voices dissipate to negligible levels within a few tens of feet. Although most waves carry energy, the amount of energy differs from case to case.

Like moving matter, waves also have *momentum.* When water waves are absorbed or reflected by an object, they push the object in their direction of travel. For example, a steady sound wave imparts momentum on a membrane, in addition to causing it to vibrate. The momentum of a sound wave must be taken into account in understanding the behavior of solids. Ordinarily, however, the momentum of waves is less noticeable than their energy. Light and other types of electromagnetic waves also have momentum, although extremely high intensities are necessary to produce perceptible effects.

Another important property of waves is that they each have a *velocity.* The speed of wave propagation differs widely between different types of waves and the types of materials in which they propagate. As discussed in Chapter 1, for example, sound waves propagate at a velocity of $\sim$340 m-s^{-1} in air but at $\sim$1500 m-s^{-1} in water; heat in a concrete dam propagates at a few meters per second, whereas electromagnetic waves in air or in free space propagate at 300,000 kilometers per second.

We now proceed with the formal treatment of the propagation of electromagnetic waves. In Section 8.1 we consider the propagation in an *unbounded, simple,* and *source-free* medium of a special type of electromagnetic waves known as *uniform plane waves.* Uniform plane waves are waves in which the amplitude and phase of the electric field (and the magnetic field) at any instant of time are constants over infinite planes orthogonal to the direction of propagation. The characteristics of uniform plane waves are particularly simple, so their study constitutes an excellent starting point in understanding more complicated electromagnetic waves. Furthermore, many practically important electromagnetic waves can be approximated as uniform plane waves, so our study of such waves is also of significant practical importance. Uniform plane waves are often also referred to as transverse electromagnetic (or TEM) waves, since both the electric and magnetic fields of the wave are transverse to the propagation direction. In Section 8.2 we consider the reflection and transmission of electromagnetic waves at planar boundaries between different material media. Many practical problems encountered in electromagnetics involve reflection of waves from interfaces between dielectrics and perfect conductors (e.g., air and copper) or between two different dielectrics (e.g., glass and air), and the treatment of such problems requires that we take into account the complicating effects of the boundary surfaces. In Section 8.3, we consider the guiding of electromagnetic waves by metallic structures. Guiding of electromagnetic energy by confining it in two dimensions and allowing it to propagate in the third dimension forms the basis of a wide range of applications, including the two-conductor transmission lines covered in Chapters 2 and 3; perhaps the most common example being the coaxial line.

Many of the electromagnetic wave concepts that we study in this chapter directly correspond to voltage and current wave concepts that we studied in Chapters 2 and 3. The basic wave equations and their general solutions are identical, as are the concepts of propagation constant, wavelength, phase velocity, and attenuation constant. We take note of this correspondence at various points during the chapter. The similarity of voltage and current waves and uniform plane waves is, of course, not just a coincidence. The uniform two-conductor transmission lines considered in Chapters 2 and 3 are merely special cases of more general guiding structures that can efficiently transmit electromagnetic energy from one point to another. Thus, the voltage and current waves of Chapters 2 and 3 are in fact uniform plane electromagnetic waves and can be studied entirely in terms of electric and magnetic fields, rather than voltages and currents. The relatively simple physical configuration of two-conductor transmission lines allows us to analyze the propagation of electromagnetic waves on them by means of voltages and currents, which is why we were able to study transmission lines before introducing Maxwell's equations.

8.1 PLANE ELECTROMAGNETIC WAVES IN AN UNBOUNDED MEDIUM

8.1.1 Plane Waves in a Simple, Source-Free, and Lossless Medium

We saw in Chapter 7 that two of Maxwell's equations, namely the two curl equations [7.16*a*] and [7.16*c*], which represent the fact that changing magnetic fields produce electric fields and vice versa, necessarily lead to propagation of electromagnetic waves. In this section, we study the characteristics of such electromagnetic waves, restricting our attention to propagation in unbounded, simple, source-free, and lossless media. Our starting point consists of Maxwell's equations, repeated below for convenience:

$$\nabla \times \overline{\mathscr{E}} = -\frac{\partial \overline{\mathscr{B}}}{\partial t} \qquad [8.1a]$$

$$\nabla \cdot \overline{\mathscr{D}} = \tilde{\rho} \qquad [8.1b]$$

$$\nabla \times \overline{\mathscr{H}} = \overline{\mathscr{J}} + \frac{\partial \overline{\mathscr{D}}}{\partial t} \qquad [8.1c]$$

$$\nabla \cdot \overline{\mathscr{B}} = 0 \qquad [8.1d]$$

The $\overline{\mathscr{J}}$ term in [8.1*c*] can in general be nonzero, either because of the presence of source currents $\overline{\mathscr{J}}_{\text{source}}$ (e.g., wires carrying current) or because of conduction current $\overline{\mathscr{J}}_c = \sigma\overline{\mathscr{E}}$, which flows in media with nonzero conductivity ($\sigma \neq 0$). The latter leads to loss of electromagnetic power, with volume density of power dissipation represented by $\overline{\mathscr{E}} \cdot \overline{\mathscr{J}}$, as discussed in Section 5.8. In this section, we consider electromagnetic wave propagation in source-free (i.e., $\overline{\mathcal{J}}_{\text{source}}, \tilde{\rho} = 0$), simple,[2] and lossless (i.e., $\sigma = 0$, and thus $\overline{\mathcal{J}}_c = 0$) media, so that $\overline{\mathcal{J}} = \overline{\mathcal{J}}_{\text{source}} + \overline{\mathcal{J}}_c = 0$. Our goal is to describe the properties of different types of electromagnetic waves that can exist (i.e., can satisfy Maxwell's equations) in regions without any source currents and charges, regardless of where and when those fields may have originated—undoubtedly at faraway sources $\overline{\mathcal{J}}_{\text{source}}$ and/or $\tilde{\rho}$.

In general, the two coupled equations [8.1*a*] and [8.1*c*] can be combined to obtain equations in terms of only $\overline{\mathscr{E}}$ (or $\overline{\mathscr{H}}$). Taking the curl of [8.1*a*] and substituting into [8.1*c*] gives

$$\nabla \times \nabla \times \overline{\mathscr{E}} = -\mu \frac{\partial}{\partial t} \nabla \times \overline{\mathscr{H}} = -\mu \frac{\partial^2 \overline{\mathscr{D}}}{\partial t^2}$$

where we have taken into account the fact that $\overline{\mathcal{J}} = 0$ and assumed that μ is a constant (i.e., the medium is magnetically linear, isotropic, homogeneous, and time-invariant) and that $\overline{\mathscr{H}}$ is a continuous function of time and space, so that spatial and temporal derivatives can be interchanged.

We now use the vector identity

$$\nabla \times \nabla \times \overline{\mathscr{E}} \equiv \nabla(\nabla \cdot \overline{\mathscr{E}}) - \nabla^2 \overline{\mathscr{E}}$$

and the fact that under source-free conditions ($\tilde{\rho} = 0$) we have

$$\nabla \cdot \overline{\mathscr{D}} = 0 \quad \rightarrow \quad \nabla \cdot (\epsilon \overline{\mathscr{E}}) = 0 \quad \rightarrow \quad \nabla \cdot \overline{\mathscr{E}} = 0$$

assuming that ϵ is constant (i.e., that the medium is electrically linear, isotropic, homogeneous, and time-invariant) to obtain

$$\boxed{\nabla^2 \overline{\mathscr{E}} - \mu\epsilon \frac{\partial^2 \overline{\mathscr{E}}}{\partial t^2} = 0} \qquad [8.2]$$

We can obtain a similar equation for the magnetic field $\overline{\mathscr{H}}$ by taking the curl of [8.1*c*] and making similar assumptions (i.e., a simple medium, being linear, isotropic, homogeneous, and time-invariant). We find

$$\boxed{\nabla^2 \overline{\mathscr{H}} - \mu\epsilon \frac{\partial^2 \overline{\mathscr{H}}}{\partial t^2} = 0} \qquad [8.3]$$

Equations of the type [8.2] and [8.3] are encountered in many branches of science and engineering and have natural solutions in the form of propagating waves; they are thus referred to as *wave equations*. The solutions of these equations describe the characteristics of the electromagnetic waves as dictated by Maxwell's equations and the properties (μ, ϵ) of lossless material media in regions without any source currents and charges. It is important to note that the equations [8.2] and [8.3] are not independent (since they were both obtained from [8.1*a*] and [8.1*c*]), and that either

---

[2]In this context, a "simple" medium is a material medium that is linear, time-invariant, isotropic, and homogeneous, so that ϵ, μ are simple constants.

[8.2] or [8.3] (together with [8.1a] or [8.1c]) can be used to solve for both $\overline{\mathscr{E}}$ and $\overline{\mathscr{H}}$. We follow the usual convention of solving for $\overline{\mathscr{E}}$ from the electric field equation [8.2] and then determining $\overline{\mathscr{H}}$ from [8.1*a*].

We note that in the general case when $\overline{\mathscr{E}}$ has three nonzero components ($\mathscr{E}_x$, $\mathscr{E}_y$, and $\mathscr{E}_z$), which may vary with the three Cartesian spatial coordinates (x, y, and z), [8.2] is actually a set of three scalar equations:

$$\left[\frac{\partial^2}{\partial x^2} + \frac{\partial^2}{\partial y^2} + \frac{\partial^2}{\partial z^2}\right]\mathscr{E}_x - \mu\epsilon\frac{\partial^2\mathscr{E}_x}{\partial t^2} = 0 \qquad [8.2a]$$

$$\left[\frac{\partial^2}{\partial x^2} + \frac{\partial^2}{\partial y^2} + \frac{\partial^2}{\partial z^2}\right]\mathscr{E}_y - \mu\epsilon\frac{\partial^2\mathscr{E}_y}{\partial t^2} = 0 \qquad [8.2b]$$

$$\left[\frac{\partial^2}{\partial x^2} + \frac{\partial^2}{\partial y^2} + \frac{\partial^2}{\partial z^2}\right]\mathscr{E}_z - \mu\epsilon\frac{\partial^2\mathscr{E}_z}{\partial t^2} = 0 \qquad [8.2c]$$

We do not need to consider the most general case in order to obtain the characteristics of propagating electromagnetic waves contained in the solution of [8.2]. For simplicity, we limit ourselves here to the special case in which the electric field $\overline{\mathscr{E}}$ is independent of two dimensions (say x and y) . Equation [8.2] then becomes

$$\frac{\partial^2\overline{\mathscr{E}}}{\partial z^2} - \mu\epsilon\frac{\partial^2\overline{\mathscr{E}}}{\partial t^2} = 0 \qquad [8.4]$$

which is equivalent to three scalar equations, one for each component of $\overline{\mathscr{E}}$. With no loss of generality, we further restrict our attention to one of the components, say $\mathscr{E}_x$, the equation for which is then

$$\boxed{\frac{\partial^2\mathscr{E}_x}{\partial z^2} - \mu\epsilon\frac{\partial^2\mathscr{E}_x}{\partial t^2} = 0} \qquad [8.5]$$

Second-order partial differential equations of the type [8.5] commonly occur in many branches of engineering and science. For example, replacing $\mathscr{E}_x(z, t)$ with $\mathscr{V}(z, t)$ results in the wave equation describing the voltage wave $\mathscr{V}(z, t)$ on a lossless transmission line (see equations [2.5] and [2.6]), whereas replacing $\mathscr{E}_x(z, t)$ with $u(z, t)$ becomes the wave equation describing the variation of velocity $u(z, t)$ for acoustic waves in a fluid.

In a similar manner to the solution of equation [2.7] provided by [2.8] as discussed in Chapter 2, the general solution of [8.5] is of the form

$$\mathscr{E}_x(z, t) = p_1(z - v_p t) + p_2(z + v_p t) \qquad [8.6]$$

where $v_p = 1/\sqrt{(\mu\epsilon)}$ and p_1 and p_2 are arbitrary functions representing the shape (e.g., square pulse, sinusoid with a Gaussian envelope, exponentially decaying pulse) of the electric field excited by a remote source. Examples of functions of $(z - v_p t)$ include $Ae^{-(z-v_p t)^2}$, $A\sqrt{(z - v_p t)}$, and $Ae^{-(z-v_p t)}\cos(z - v_p t)$. Note also that the functions p_1 and p_2 are not necessarily the same. The fact that [8.6] is a solution of [8.5] can be seen by simple substitution.

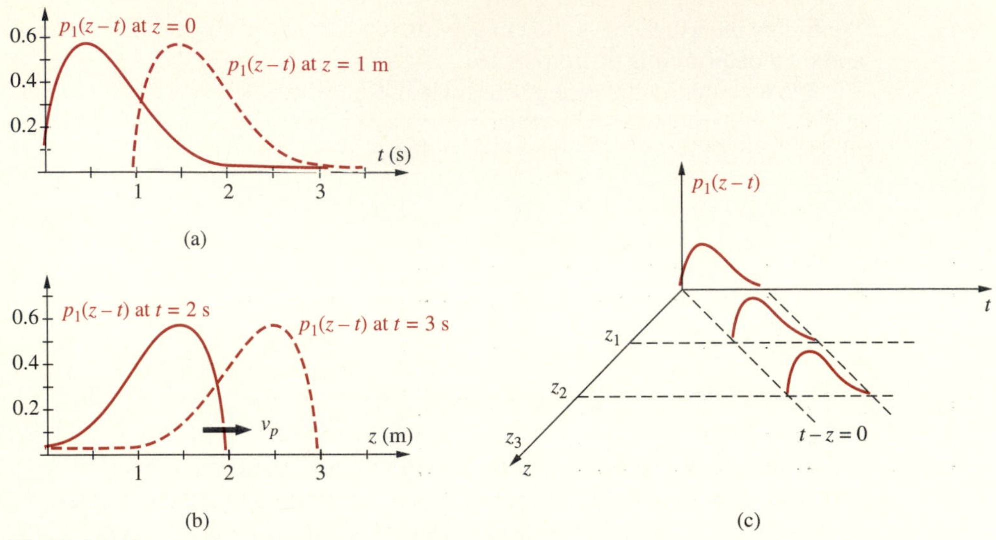

FIGURE 8.1. Variation in space and time of an arbitrary function $p_1(z - v_pt)$. The function $p_1(\zeta)$ shown above is $p_1(\zeta) = \zeta^{1/2}(\sin\zeta/\zeta)^6 u(\zeta)$, where $u(\zeta)$ is the unit step function (i.e., $u(\zeta) = 0$ for $\zeta < 0$ and $u(\zeta) = 1$ for $\zeta > 0$). For the purpose of the plots on the left-hand side, the speed of propagation is taken to be $v_p = 1$ m-s^{-1}. (a) $p_1(z - t)$ as a function of t at two positions $z = 0$ and $z = 1$ m. (b) $p_1(z - t)$ as a function of z at two time instants, showing how the pulse travels in the z directions as time progresses. (c) $p_1(z - t)$ as a function of z and t.

That the solutions $p_1(z - v_pt)$ and $p_2(z + v_pt)$ represent waves propagating respectively in the $+z$ and $-z$ directions can be seen by observing their variation with z at different times t_i, as shown for $p_1(z - v_pt)$ in Figure 8.1. In this context, a wave is to be understood as a physical phenomenon or a disturbance (e.g., an electric field variation) that occurs at one place at a given time and at other places at other times, with time delays proportional to the spatial separations from the first location. Since the wave travels with a velocity v_p, a time z/v_p elapses as the wave propagates from $z = 0$ to the position z. Thus, an observer at point z sees events that have actually occurred (at $z = 0$) at an earlier time. For example, light waves from a supernova explosion may arrive at earth millions of years after their source has been extinguished. A wave is not necessarily a repetitive or oscillatory disturbance in time; a single pulse moving in space also constitutes a wave, as happens when a transmission line is connected to a battery for a short period of time. A tsunami generated in an undersea earthquake is another example of a wave that is not repetitive in time.

Note that at any time, say t_1, the function $p_1(z - v_pt_1)$ is simply a function of z since v_pt_1 is a constant. At another time t_2, the function is $p_1(z - v_pt_2)$, which has exactly the same type of dependence on z, but displaced to the right by an amount $v_p(t_2 - t_1)$. In other words, the disturbance (in this case the electric field) represented by $p_1(z - v_pt)$ has traveled in the positive z direction with a velocity v_p. Note that in free space, $v_p = 1/\sqrt{(\mu_0\epsilon_0)} \simeq 3 \times 10^8$ m-s^{-1}, or the speed of light, a fact that led Maxwell to suggest that light is a form of electromagnetic radiation. The space-time dependence of a wave pulse such as $p_1(z - v_pt)$ is illustrated in Figure 8.1.

Note that the second term in [8.6], $p_2(z + v_p t)$, represents a wave traveling in the negative z direction. The general solution of [8.5] is thus a superposition of two waves: a forward wave traveling in the $+z$ direction (away from the source, if the source is assumed to be far away in the $-z$ direction) and a reverse wave traveling in the $-z$ direction (back toward the source). The reverse wave term is nonzero only if there are discontinuities (surfaces with different ϵ and/or μ) that "reflect" some of the forward-traveling wave energy back to the source. The reflection and refraction of electromagnetic waves at planar interfaces are studied in Section 8.2.

Electromagnetic waves for which the field components are functions of only one spatial coordinate (in this case z) are known as *plane waves*. This term originates from the fact that the surfaces over which the argument of the function is a constant (e.g., $(z - v_p t) =$ constant) are planes.[3] Waves for which the field amplitudes do not vary with position over the planes of constant phase are known as *uniform plane waves*. Note that although we studied the solution of [8.4] only for the x component of $\overline{\mathscr{E}}$, identical behavior would be expected for the y component $\mathscr{E}_y$, since the differential equation governing it is the y component of [8.4], which is identical in form to [8.5]. In general, both components may exist, in which case the total field would simply be a linear superposition of the two separate solutions of the corresponding differential equations. Although it would seem that the $\mathscr{E}_z$ would also behave in a similar manner, a uniform plane wave propagating in the z direction in a simple medium does not have a z component, as can be seen by examining equation $\nabla \cdot \overline{\mathscr{D}} = \rho$. We have

$$\nabla \cdot \overline{\mathscr{D}} = 0 \longrightarrow \nabla \cdot \overline{\mathscr{E}} = 0 \longrightarrow \frac{\partial \mathscr{E}_x}{\partial x} + \frac{\partial \mathscr{E}_y}{\partial y} + \frac{\partial \mathscr{E}_z}{\partial z} = 0$$

The first two terms ($\partial \mathscr{E}_x/\partial x$ and $\partial \mathscr{E}_y/\partial y$) are zero (since the field components do not vary with x or y), which requires that $\partial \mathscr{E}_z/\partial z = 0$, which in turn means that $\mathscr{E}_z$ can at most be a constant in space. Substituting $\partial \mathscr{E}_z/\partial z = 0$ into the z component of [8.4], we find $\partial^2 \mathscr{E}_z/\partial t^2 = 0$, which means that $\mathscr{E}_z$ can at most be a linearly increasing function of time. Since a quantity that is constant in space and linearly increasing in time cannot contribute to wave motion, a uniform plane wave propagating in the z direction has no z component.

Relation between $\overline{\mathscr{E}}$ and $\overline{\mathscr{H}}$ Although we have just discussed the solution of the wave equation for the electric field, we could have just as easily solved [8.3] for the magnetic field $\overline{\mathscr{H}}$ by making the same assumption (i.e., uniform plane wave) and restricting our attention to a single component of the magnetic field. However, once we have described a solution for the electric field of a uniform plane wave (say the component $\mathscr{E}_x(z)$), we are no longer free to choose arbitrarily one of the two components of its magnetic field (i.e., $\mathscr{H}_x$ or $\mathscr{H}_y$) to be nonzero. Because the wave equation [8.4] for the electric field was obtained using the entire set of Maxwell's equations, the

[3]Such surfaces are known as *surfaces of constant phase, phase surfaces,* or *phase fronts*. Although the concept of phase is more commonly associated with sinusoidal steady-state or time-harmonic waves, in which case the function $p_1(z - v_p t)$ has the form $\cos[\omega(t - z/v_p)]$, it is nevertheless equally valid for nonsinusoidal functions. Time-harmonic waves are discussed in Section 8.1.2.

magnetic field is already determined once we specify the electric field. The proper means for finding the magnetic field of a uniform plane wave is thus to derive it from the electric field using Maxwell's equations, for example [8.1*a*].

Let us again assume that we have a uniform plane wave having an electric field with only an x component propagating in a simple medium in the $+z$ direction:

$$\mathscr{E}_x(z, t) = p_1(z - v_p t)$$

Since $\mathscr{E}_y, \mathscr{E}_z = 0$, $\mathscr{E}_x$ varies only with z, and any components of the magnetic field $\overline{\mathscr{H}}$ also vary only with z, we have from [8.1*a*]

$$\nabla \times \overline{\mathscr{E}} = -\frac{\partial \overline{\mathscr{B}}}{\partial t} \quad \rightarrow \quad \hat{\mathbf{y}}\frac{\partial \mathscr{E}_x}{\partial z} = -\mu\hat{\mathbf{x}}\frac{\partial \mathscr{H}_x}{\partial t} - \mu\hat{\mathbf{y}}\frac{\partial \mathscr{H}_y}{\partial t}$$

whereas from [8.1*c*] we have

$$\nabla \times \overline{\mathscr{H}} = \frac{\partial \overline{\mathscr{D}}}{\partial t} \quad \rightarrow \quad -\hat{\mathbf{x}}\frac{\partial \mathscr{H}_y}{\partial z} + \hat{\mathbf{y}}\frac{\partial \mathscr{H}_x}{\partial z} = \epsilon\hat{\mathbf{x}}\frac{\partial \mathscr{E}_x}{\partial t}$$

Equating the different components of the above two equations we find

$$\frac{\partial \mathscr{E}_x}{\partial z} = -\mu\frac{\partial \mathscr{H}_y}{\partial t}; \qquad -\frac{\partial \mathscr{H}_y}{\partial z} = \epsilon\frac{\partial \mathscr{E}_x}{\partial t} \qquad [8.7]$$

It thus appears that the magnetic field of a uniform plane wave with an electric field of $\overline{\mathscr{E}} = \hat{\mathbf{x}}p_1(z - v_p t)$ has only a y component, which is related to $\mathscr{E}_x$ as described by [8.7]. For any given function $p_1(z - v_p t)$, $\mathscr{H}_y$ can be found using [8.7], as we will do later for sinusoidal uniform plane waves, for which case $p_1(z - v_p t) = A\cos(z - v_p t)$. By straightforward application of the chain rule and some algebraic manipulation, it can also be shown[4] that, in general, for $\mathscr{E}_x = p_1(z - v_p t)$, [8.7] leads to

$$\mathscr{H}_y = \left(\sqrt{\frac{\epsilon}{\mu}}\right)p_1(z - v_p t) = \frac{1}{\eta}p_1(z - v_p t)$$

where $\eta \equiv \sqrt{\mu/\epsilon}$ is a quantity that has units of impedance (ohms) and is defined as the *intrinsic impedance* of the medium. Thus, for uniform plane waves in a simple lossless medium, the ratio of the electric and magnetic fields is η and is determined only by the material properties of the medium (i.e., μ, ϵ). The intrinsic impedance of free space[5] is $\eta = \sqrt{\mu_0/\epsilon_0} \simeq 120\pi \simeq 377\Omega$.

An important characteristic of uniform plane waves is that their electric and magnetic fields are perpendicular to one another. In simple and lossless media, the variation of $\overline{\mathscr{E}}$ and $\overline{\mathscr{H}}$ in space and time are identical [i.e., they are both proportional to $p_1(z - v_p t)$]; in other words, $\overline{\mathscr{E}}$ and $\overline{\mathscr{H}}$ propagate in unison along z, reaching their maxima and minima at the same points in space and at the same times. The orientation of $\overline{\mathscr{E}}$ and $\overline{\mathscr{H}}$ is such that the vector $\overline{\mathscr{E}} \times \overline{\mathscr{H}}$ is in the $+z$ direction, which is the

---

[4] The same derivation was carried out in Section 2.2.2 for $\mathscr{V}(z, t)$ and $\mathscr{I}(z, t)$.

[5] Using $\mu_0 = 4\pi \times 10^{-7}$ H-m^{-1} and $\epsilon_0 = 1/(\mu_0 c^2) \simeq 10^{-9}/(36\pi)$ F-m^{-1}.

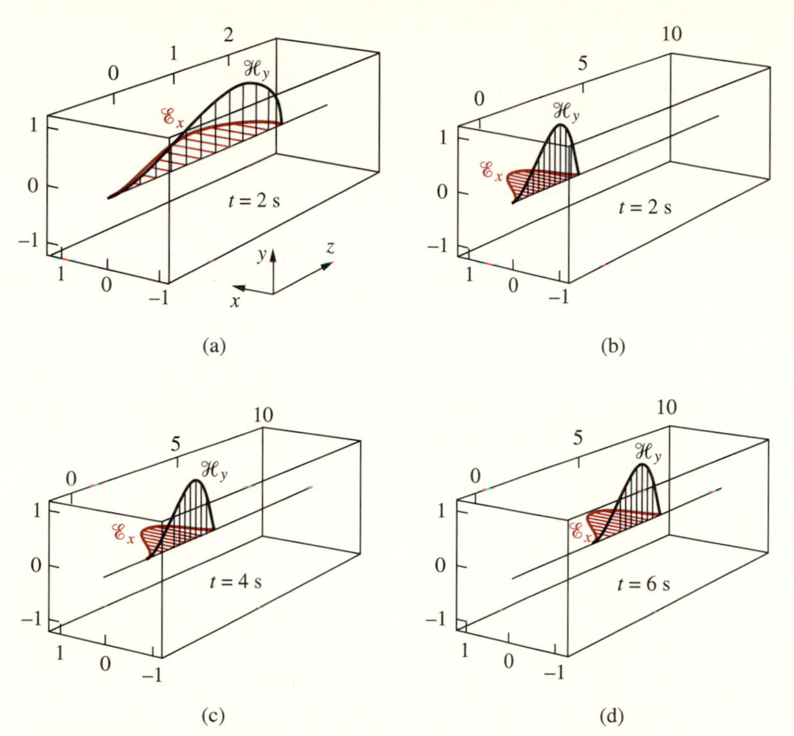

FIGURE 8.2. The propagation of a uniform plane electromagnetic disturbance in the z direction. The electric and magnetic fields are orthogonal at all times. For the purpose of this figure, the speed of propagation is taken to be $v_p = 1$ m-s^{-1}. (a) $\mathscr{E}_x(z, t)$ and $\eta\mathscr{H}_y(z, t)$ as a function of z at $t = 2$ s. (b) Same as (a) but shown on a compressed distance scale. (c) $\mathscr{E}_x(z, t)$ and $\eta\mathscr{H}_y(z, t)$ as a function of z at $t = 4$ s. (d) $\mathscr{E}_x(z, t)$ and $\eta\mathscr{H}_y(z, t)$ as a function of z at $t = 6$ s. The propagation of the pulse in the $+z$ direction is clearly evident. Note that the pulse shapes for the electric and magnetic fields are identical.

direction of propagation of the wave. The orthogonality of the electric and magnetic fields and the propagation of a disturbance are illustrated in Figure 8.2.

Note that since the choice of the coordinate system cannot affect the physical relationship between $\overline{\mathscr{E}}$ and $\overline{\mathscr{H}}$, if we start with $\overline{\mathscr{E}}$ having only a y component (i.e., $\overline{\mathscr{E}} = \hat{\mathbf{y}}\mathscr{E}_y$), $\overline{\mathscr{H}}$ then comes out to be $\overline{\mathscr{H}} = -\hat{\mathbf{x}}\mathscr{H}_x$. The relationship between $\overline{\mathscr{E}} = \hat{\mathbf{y}}\mathscr{E}_y$ and $\overline{\mathscr{H}} = -\hat{\mathbf{x}}\mathscr{H}_x$ for a uniform plane wave propagating in the $+z$ direction is depicted in Figure 8.3.

The concept of uniform plane waves makes the solution of the wave equation [8.2] tractable, as we have seen. In a strict sense, uniform plane waves can be excited (created) only by sources infinite in extent (e.g., uniformly distributed over the entire xy plane to produce a uniform plane wave propagating in the $+z$ direction). However, uniform plane waves are in fact often excellent approximations in practice, especially when we observe (or receive) electromagnetic waves at large distances (compared to wavelength) from their sources (e.g., in broadcast radio and television

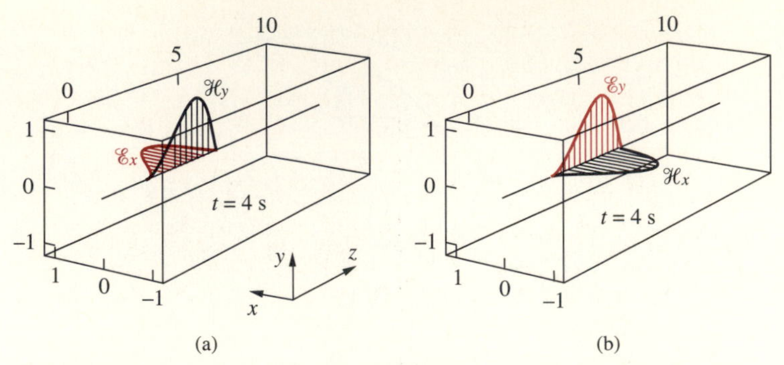

FIGURE 8.3. Electric and magnetic fields of a uniform plane wave. The relationship between $\overline{\mathcal{E}}$ and $\overline{\mathcal{H}}$ for a uniform plane wave is independent of the choice of a particular coordinate system. (a) $\mathcal{E}_x(z, t)$ and $\eta\mathcal{H}_y(z, t)$. (b) $\mathcal{E}_y(z, t)$ and $\eta\mathcal{H}_x(z, t)$.

applications). Electromagnetic fields emanating from a point source spread spherically, and since a very small portion of the surface of a large sphere is approximately planar, they can be considered plane waves at large distances from their sources. Uniform plane waves are also important because the electromagnetic fields of a nonuniform or nonplanar wave[6] can be expressed[7] as a superposition of component plane waves, much like a Fourier decomposition.

8.1.2 Time-Harmonic Uniform Plane Waves in a Lossless Medium

We now study uniform plane waves for which the temporal behavior is harmonic, or sinusoidal. It was pointed out in Section 7.4.4 that a large number of practical applications (e.g., broadcast radio and TV, radar, optical and microwave applications) involve sources (transmitters) that operate in such a narrow band of frequencies that the behavior of all of the field components is very similar to that of the central single frequency sinusoid (i.e., the carrier).

---

[6]An example of the electric field of a nonuniform plane wave is

$$\overline{\mathcal{E}}(z, t) = \hat{\mathbf{x}}C\cos(\pi y)\cos(\omega t - \beta z)$$

which is a plane wave because the constant phase surfaces are planes defined by $z = \text{const.}$, but is nonuniform because the amplitude of the field is a function of y and hence is not uniform over the planes of constant phase. In general, the electric field of a nonplanar time-harmonic wave can be expressed as

$$\overline{\mathcal{E}}(\mathbf{r},t) = \overline{\mathcal{E}}_0\cos[\omega t - g(\mathbf{r})]$$

where $g(\cdot)$ is some arbitrary function. For example, if $g(\mathbf{r}) = g(x, y, z) = xy$, then the surfaces of constant phase would be hyperbolic cylinders with infinite extent in z, defined by by $xy = \text{const.}$ A uniform cylindrical wave is one for which $g(\mathbf{r}) = \beta r$, with r being the radial cylindrical coordinate, in which case the surfaces of constant phase are cylinders with infinite extent in z.

[7]P. C. Clemmow, *The Plane Wave Spectrum Representation of Electromagnetic Fields,* Pergamon Press, Oxford, 1966.

The AM radio broadcast signal was mentioned in Section 7.4.4 as an example of a signal with a narrow bandwidth centered around a carrier frequency. Other examples include FM radio broadcasts, which utilize carrier frequencies of 88 to 108 MHz with a bandwidth of ~200 kHz, or UHF television broadcasts, where the signal bandwidth is ~6 MHz for carrier frequencies in the range 470 to 890 MHz. For all practical purposes, the propagation of the signals from the transmitters to receivers can be described by studying single sinusoids at the carrier frequency, since the characteristics of the propagation medium do not vary significantly over the signal bandwidth. In other words, while at the transmitter the actual signal is constituted by a superposition of Fourier components at different frequencies within the signal bandwidth, each of the frequency components propagates to the receiver in a manner identical to the propagation of the carrier.

In other applications, such as in the case of guided propagation of electromagnetic waves, the propagation characteristics may vary significantly over the bandwidth of the signal (especially when terminated transmission lines or waveguides are concerned). However, sinusoidal (or time-harmonic) analysis is still useful, because any arbitrary waveform can be represented as a linear superposition of its Fourier components whose behavior is well represented by the time-harmonic (or sinusoidal steady-state) analysis.

To study the propagation of time-harmonic electromagnetic waves, we use the time-harmonic form of Maxwell's equations as given in [7.18]. Note that these are written in terms of the vector phasor quantities represented with the boldface symbols (**E**, **H**, etc.,) which are related to the real space-time fields in the usual manner previously described in Section 7.4.4; for example,

$$\overline{\mathscr{E}}(x, y, z, t) = \mathscr{R}e\{\mathbf{E}e^{j\omega t}\}$$

For lossless ($\sigma = 0$), source-free ($\mathbf{J}_{\text{source}}, \rho = 0$), linear [i.e., $\epsilon, \mu \neq f(\mathbf{E},\mathbf{H})$], homogeneous [i.e., $\epsilon, \mu \neq f(x, y, z)$], isotropic [i.e., $\epsilon, \mu \neq f(\text{direction})$], and time-invariant [i.e., $\epsilon, \mu \neq f(t)$] medium, we can derive the wave equation for the electric field from the preceding equation by first taking the curl of [7.18*a*]:

$$\nabla \times \nabla \times \mathbf{E} = -j\omega\mu(\nabla \times \mathbf{H})$$

Note that μ must be independent of direction and spatial coordinates in order to write $\nabla \times \mathbf{B} = \mu\nabla \times \mathbf{H}$. We now use a vector identity to replace the left-hand side of the preceding equation:

$$\nabla \times \nabla \times \mathbf{E} \equiv \nabla(\nabla \cdot \mathbf{E}) - \nabla^2\mathbf{E}$$

Since ϵ is independent of direction and spatial coordinates, we have from [7.18*b*]

$$\nabla \cdot \mathbf{D} = 0 \longrightarrow \nabla \cdot (\epsilon\mathbf{E}) = \epsilon\nabla \cdot \mathbf{E} = 0 \longrightarrow \nabla \cdot \mathbf{E} = 0$$

Using [7.18*c*] further reduces the preceding equation to

$$-\nabla^2\mathbf{E} = -j\omega\mu(j\omega\epsilon\mathbf{E})$$

or

$$\boxed{\nabla^2\mathbf{E} + \beta^2\mathbf{E} = 0} \qquad [8.8]$$

where $\beta = \omega\sqrt{\mu\epsilon}$ is the phase constant, also often called the wave number, or propagation constant, in units of radians-m^{-1}. Note that the propagation constant $\beta = \omega\sqrt{\mu\epsilon}$ for uniform plane waves in free space is in fact identical[8] to the phase constant for voltage waves on uniform lossless transmission lines studied in Chapter 3, namely $\beta = \omega\sqrt{LC}$. Equation [8.8] is known as the *vector wave equation* or *Helmholtz equation* and is actually a set of three equations:

$$\left[\frac{\partial^2}{\partial x^2} + \frac{\partial^2}{\partial y^2} + \frac{\partial^2}{\partial z^2}\right]E_x + \beta^2 E_x = 0 \qquad [8.8a]$$

$$\left[\frac{\partial^2}{\partial x^2} + \frac{\partial^2}{\partial y^2} + \frac{\partial^2}{\partial z^2}\right]E_y + \beta^2 E_y = 0 \qquad [8.8b]$$

$$\left[\frac{\partial^2}{\partial x^2} + \frac{\partial^2}{\partial y^2} + \frac{\partial^2}{\partial z^2}\right]E_z + \beta^2 E_z = 0 \qquad [8.8c]$$

We now limit our attention to uniform plane waves, by considering solutions of [8.8] for which $\mathbf{E}$ has only an x component and is only a function of z. In other words, we let

$$\mathbf{E}(x, y, z) = \hat{\mathbf{x}}E_x(z)$$

The wave equation [8.8] then reduces to

$$\boxed{\frac{d^2E_x}{dz^2} + \beta^2 E_x = 0} \qquad [8.9]$$

which is a second-order ordinary differential equation, encountered in sinusoidal–steady state applications in many branches of physics and engineering. For example, replacing E_x with V gives the wave equation [3.6] for the voltage phasor on a lossless transmission line. The general solution of [8.9] is

$$E_x(z) = \underbrace{C_1 e^{-j\beta z}}_{E_x^+(z)} + \underbrace{C_2 e^{+j\beta z}}_{E_x^-(z)} \qquad [8.10]$$

where C_1 and C_2 are constants to be determined by boundary conditions. Note that the quantity $E_x(z)$ is a complex phasor. The real (or instantaneous) electric field $\overline{\mathscr{E}}(z, t)$ can be found from $E_x(z)$ as

$$\mathscr{E}_x(z, t) = \mathscr{R}e\{(C_1 e^{-j\beta z} + C_2 e^{+j\beta z})e^{j\omega t}\} \qquad [8.11]$$

or

$$\mathscr{E}_x(z, t) = \underbrace{C_1 \cos(\omega t - \beta z)}_{\mathscr{E}_x^+(z,t)} + \underbrace{C_2 \cos(\omega t + \beta z)}_{\mathscr{E}_x^-(z,t)} \qquad [8.12]$$

---

[8]The reader is encouraged to use the various capacitance and inductance expressions tabulated in Table 2.2 and derived in Chapters 4 and 6 for lossless transmission lines to show that $LC = \mu\epsilon$.

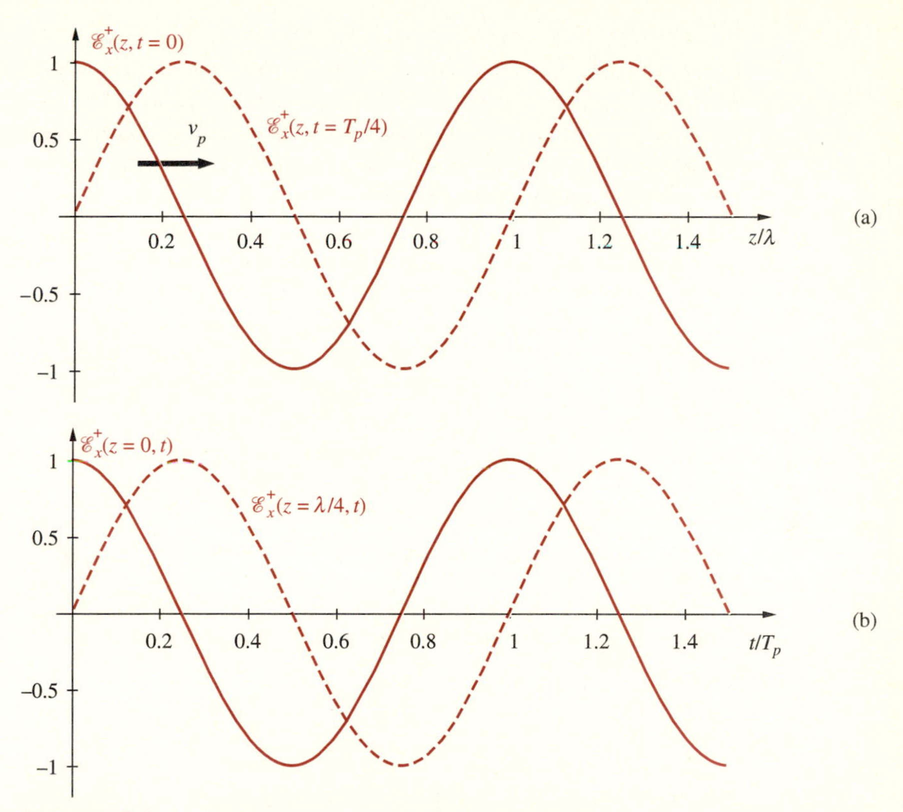

FIGURE 8.4 **Wave behavior in space and time.** (a) $\mathscr{E}_x^+(z, t) = \cos[(2\pi/T_p)t - (2\pi/\lambda)z]$ versus z/λ and for $t = 0$ and $t = T_p/4$. (b) $\mathscr{E}_x^+(z, t)$ versus t/T_p for $z = 0$ and $z = \lambda/4$.

assuming that C_1 and C_2 are real constants. The first term $\mathscr{E}_x^+(z, t)$ represents a wave traveling in the $+z$ direction, as can be seen from the successive snapshots in time plotted as a function of z in Figure 8.4a. The z axis is normalized to wavelength, which, as in Chapter 3, is defined as $\lambda = 2\pi/\beta$. Note that the electric field at any given point in space varies sinusoidally in time, as illustrated in Figure 8.4b.

By the same token, the second term $\mathscr{E}_x^-(z, t)$ in [8.12] represents a wave traveling in the $-z$ direction. Both terms represent waves traveling at a speed given by the phase velocity $v_p = \omega/\beta$, which is the velocity of travel of any point identified by a certain phase (i.e., argument) of the sinusoid, as defined in Chapter 3. In other words, if we were to observe a fixed phase point on the wave, we have

$$\omega t - \beta z = \text{const.} \quad \longrightarrow \quad \frac{dz}{dt} = v_p = \frac{\omega}{\beta}$$

Note that $v_p = \omega/(\omega\sqrt{\mu\epsilon}) = 1/\sqrt{\mu\epsilon}$, which for free space ($\mu = \mu_0$, $\epsilon = \epsilon_0$) is the speed of light in free space, or $v_p = c$. Note also that substituting $\beta = 2\pi/\lambda$ in $v_p = \omega/\beta$ gives the familiar expression $v_p = f\lambda$, where $f = \omega/2\pi$. It is interesting to note that, for a simple lossless medium, the phase velocity is a function only of the medium parameters and is independent of frequency. This is an important property of uniform plane waves in a simple lossless medium, which does not hold true for waves propagating in a lossy medium or those that propagate in guiding structures, as we shall see later. However, it should be noted that ϵ and μ are in general functions of frequency even in a lossless medium and can only be considered to be constant over limited frequency ranges.

Although we have discussed the solution of the wave equation for the electric field, we could have just as easily derived a wave equation identical to [8.8] for the magnetic field **H**, and we could have solved it by making the same assumption (i.e., uniform plane wave) and restricting our attention to a single component of the magnetic field. However, once we have described a solution for the electric field of a uniform plane wave (say the component E_x), we are no longer free to choose one of the two components of its magnetic field (i.e., H_x or H_y) arbitrarily to be nonzero, because the wave equation for the electric field (i.e., [8.8]) was obtained using the entire set of equations [7.18] and the associated magnetic field is already determined once we specify its electric field. The proper means for finding the magnetic field of a uniform plane wave is thus to derive it from the electric field using [7.18*a*]. We have

$$\nabla \times \mathbf{E} = \hat{\mathbf{y}}\frac{dE_x(z)}{dz} = -j\omega\mu[\hat{\mathbf{x}}H_x + \hat{\mathbf{y}}H_y + \hat{\mathbf{z}}H_z]$$

It thus appears that the corresponding **H** has only a y component, which for $E_x(z)$ as given in [8.10] can be found as

$$\begin{aligned} H_y(z) &= \frac{1}{-j\omega\mu}\frac{d}{dz}(C_1 e^{-j\beta z} + C_2 e^{j\beta z}) \\ &= \frac{\beta}{\omega\mu}(C_1 e^{-j\beta z} - C_2 e^{j\beta z}) = \frac{1}{\eta}(C_1 e^{-j\beta z} - C_2 e^{j\beta z}) \end{aligned}$$

where we have used $\eta = \sqrt{\mu/\epsilon} = \omega\mu/\beta$. Note that the intrinsic impedance η is a real number for a simple lossless medium, but, as we will see in Section 8.1.3, it is a complex number for a lossy medium.

The instantaneous magnetic field $\mathcal{H}(x, y, z, t)$ can be obtained from the phasor $\mathbf{H}(x, y, z)$ in the same way as in [8.11]. Considering only the first term in [8.12] and the associated magnetic field, the electric and magnetic fields of a time-harmonic uniform plane wave propagating in the $+z$ direction are

$$\begin{aligned} \mathscr{E}_x(z, t) &= C_1 \cos(\omega t - \beta z) \\ \mathscr{H}_y(z, t) &= \frac{1}{\eta} C_1 \cos(\omega t - \beta z) \end{aligned} \qquad [8.13]$$

We find, as before (Section 8.1.1), that the electric and magnetic fields of a uniform plane wave are perpendicular to one another. The $\mathscr{E}$ and $\mathscr{H}$ fields are also both perpendicular to the direction of propagation, which is why uniform plane waves are also often referred to as *transverse electromagnetic,* or *TEM,* waves. In a simple

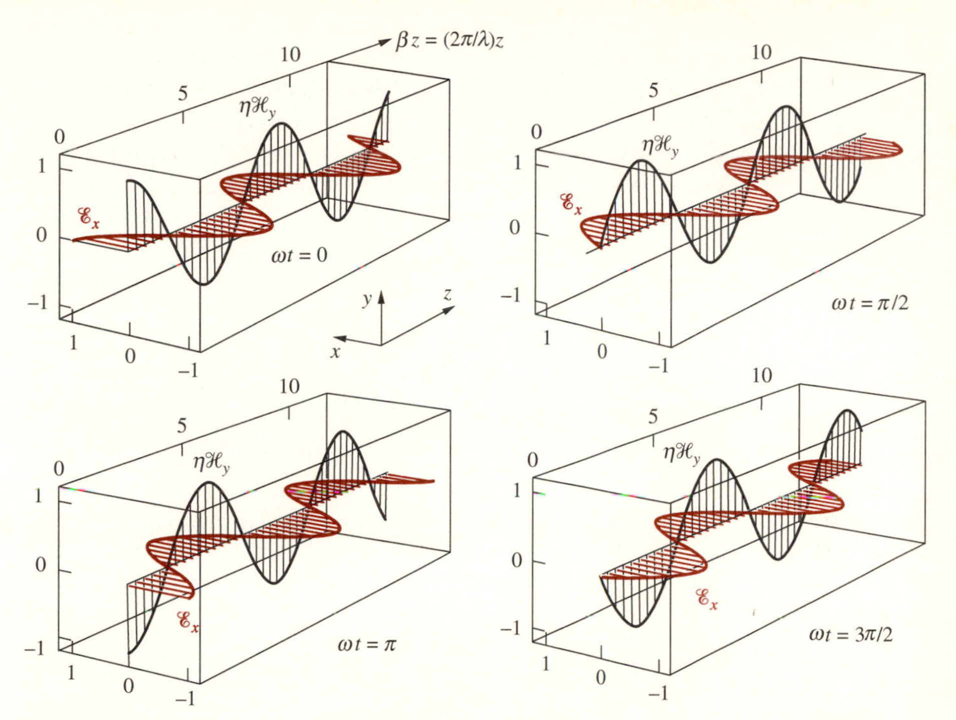

FIGURE 8.5. Electric and magnetic fields of a uniform plane wave in a lossless medium. Snapshots of $\mathcal{E}_x(z, t)$ and $\eta\mathcal{H}_y(z, t)$ for a sinusoidal uniform plane wave shown as a function of βz at $\omega t = 0, \pi/2, \pi$, and $3\pi/2$. The plots shown are for $C_1 = 1$.

lossless medium, the variation of $\overline{\mathcal{E}}$ and $\overline{\mathcal{H}}$ in space and time are identical (i.e., they are in phase); in other words, the $\overline{\mathcal{E}}$ and $\overline{\mathcal{H}}$ propagate in unison along z, reaching their maxima and minima at the same points in space and at the same times. The orientation of $\overline{\mathcal{E}}$ and $\overline{\mathcal{H}}$ is such that the vector $\overline{\mathcal{E}} \times \overline{\mathcal{H}}$ is in the $+z$ direction, which is the direction of propagation for the wave. Note that since the choice of the coordinate system cannot change the physical picture, if we start with $\overline{\mathcal{E}}$ having only a y component (i.e., $\overline{\mathcal{E}} = \hat{\mathbf{y}}\mathcal{E}_y$), then $\overline{\mathcal{H}}$ comes out to be $\overline{\mathcal{H}} = -\hat{\mathbf{x}}\mathcal{H}_x$, i.e., $\overline{\mathcal{E}} \times \overline{\mathcal{H}}$ would still be in the direction of propagation (i.e., the z direction). The relationship between $\overline{\mathcal{E}}$ and $\overline{\mathcal{H}}$ for a time-harmonic uniform plane wave is depicted in Figure 8.5. As time progresses, the electric and magnetic fields propagate in the $+z$ direction, staying in phase at all points and at all times.

Equations [8.13] and Figure 8.5 completely describe the properties of uniform plane waves in a simple lossless unbounded medium. As long as μ and ϵ are simple constants, an electromagnetic wave in free space differs from that in the material medium (with $\epsilon \neq \epsilon_0$ and/or $\mu \neq \mu_0$) primarily in terms of its wavelength, given by $\lambda = (f\sqrt{\mu\epsilon})^{-1}$. We shall see in the next section that the characteristics of uniform plane waves in a lossy medium are substantially more influenced by the material properties.

Examples 8-1 through 8-4 illustrate the use of [8.13] for different applications.

Example 8-1: AM broadcast signal. The instantaneous expression for the electric field component of an AM broadcast signal propagating in air is given by

$$\overline{\mathscr{E}}(x,t) = \hat{\mathbf{z}}10\cos(1.5\pi\times 10^6 t + \beta x) \quad \text{V-m}^{-1}$$

(a) Determine the direction of propagation and frequency f. (b) Determine the phase constant β and the wavelength λ. (c) Write the instantaneous expression for the corresponding magnetic field $\overline{\mathscr{H}}(x,t)$.

Solution:

(a) The wave propagates in the $-x$ direction. From $\omega = 2\pi f = 1.5\pi\times 10^6$ rad-s^{-1}, $f = 750$ kHz.

(b) The phase constant is given by

$$\beta = \omega\sqrt{\mu_0\epsilon_0} \simeq 1.5\pi\times 10^6/(3\times 10^8) = 0.005\pi \text{ rad-m}^{-1}$$

The wavelength in air is equal to that in free space, or

$$\lambda = 2\pi/\beta = 2\pi/0.005\pi = 400 \text{ m}$$

(c) The phasor electric field is given by

$$\mathbf{E}(x) = \hat{\mathbf{z}}10e^{+j\beta x} = \hat{\mathbf{z}}10e^{j0.005\pi x} \quad \text{V-m}^{-1}$$

To find the corresponding $\mathbf{H}$, we use [7.18c] to find

$$-j\omega\mu_0\mathbf{H} = \nabla\times\mathbf{E} \quad\rightarrow\quad H_y(x) = -\frac{1}{j\omega\mu_0}\left[-\frac{\partial E_z(x)}{\partial x}\right] = \frac{1}{\eta}E_z(x)$$

$$\mathbf{H}(x) = \hat{\mathbf{y}}\frac{1}{\eta}E_z(x) = \hat{\mathbf{y}}\frac{10\text{ V-m}^{-1}}{377\Omega}e^{j0.005\pi x}$$

Therefore,

$$\overline{\mathscr{H}}(x,t) \simeq \hat{\mathbf{y}}26.5\times 10^{-3}\cos(1.5\pi\times 10^6 t + 0.005\pi x) \quad \text{A-m}^{-1}$$

Example 8-2: FM broadcast signal. An FM radio broadcast signal traveling in the y direction in air has a magnetic field given by the phasor

$$\mathbf{H}(y) = 2.92\times 10^{-3}e^{-j0.68\pi y}(-\hat{\mathbf{x}} + \hat{\mathbf{z}}j) \quad \text{A-m}^{-1}$$

(a) Determine the frequency (in MHz) and wavelength (in m). (b) Find the corresponding $\mathbf{E}(y)$. (c) Write the instantaneous expression for $\overline{\mathscr{E}}(y,t)$ and $\overline{\mathscr{H}}(y,t)$.

Solution:

(a) We have

$$\beta = \omega\sqrt{\mu_0\epsilon_0} = 0.68\pi \text{ rad-m}^{-1}$$

from which we find

$$f = \frac{\omega}{2\pi} \simeq \frac{0.68\pi\text{ rad-m}^{-1}\times 3\times 10^8\text{ m-s}^{-1}}{2\pi} = 102 \text{ MHz}$$

(b) Using [7.18c]

$$\nabla \times \mathbf{H} = \hat{\mathbf{x}}\frac{\partial H_z}{\partial y} - \hat{\mathbf{z}}\frac{\partial H_x}{\partial y} = j\omega\epsilon_0\mathbf{E}$$

and performing the partial differentiation yields

$$\mathbf{E}(y) \simeq 1.1e^{-j0.68\pi y}(-\hat{\mathbf{x}}j - \hat{\mathbf{z}}) \quad \text{V-m}^{-1}$$

(c) The instantaneous expressions for $\overline{\mathscr{E}}(y,t)$ and $\overline{\mathscr{H}}(y,t)$ are given by

$$\overline{\mathscr{E}}(y,t) = -\hat{\mathbf{x}}1.1\cos(2.04\pi \times 10^8 t - 0.68\pi y + \pi/2) - \hat{\mathbf{z}}1.1\cos(2.04\pi \times 10^8 t - 0.68\pi y) \text{ V-m}^{-1}$$

$$\overline{\mathscr{H}}(y,t) = -\hat{\mathbf{x}}2.92 \times 10^{-3}\cos(2.04\pi \times 10^8 t - 0.68\pi y) + \hat{\mathbf{z}}2.92 \times 10^{-3}\cos(2.04\pi \times 10^8 t - 0.68\pi y + \pi/2) \quad \text{A-m}^{-1}$$

Example 8-3: Uniform plane wave. Consider a uniform plane wave traveling in the z direction in a simple lossless nonmagnetic medium (i.e., $\mu = \mu_0$) with a y-directed electric field of maximum amplitude of 60 V-m^{-1}. If the wavelength is 20 cm and the velocity of propagation is 10^8 m-s^{-1}, (a) determine the frequency of the wave and the relative permittivity of the medium; (b) write complete time-domain expressions for both the electric and magnetic field components of the wave.

Solution:

(a) We know that $f = v_p/\lambda = 10^8/0.2 = 500$ MHz. We also know that $v_p = 1/\sqrt{\mu\epsilon} = 1/\sqrt{\mu_0\epsilon_r\epsilon_0} \simeq 3 \times 10^8/\sqrt{\epsilon_r} = 10^8$, from which we find the relative permittivity to be $\epsilon_r = 9$. The phase constant is $\beta = 2\pi/\lambda = 2\pi/0.2 = 10\pi$ rad-m^{-1}, and the intrinsic impedance is $\eta = \sqrt{\mu_0/\epsilon} = \sqrt{\mu_0/(\epsilon_r\epsilon_0)} = 120\pi/3 \simeq 126\Omega$.

(b) Therefore, the instantaneous electric and magnetic fields are given by

$$\overline{\mathscr{E}}(z,t) = \hat{\mathbf{y}}60\cos(10^9\pi t - 10\pi z) \quad \text{V-m}^{-1}$$

$$\overline{\mathscr{H}}(z,t) \simeq -\hat{\mathbf{x}}0.477\cos(10^9\pi t - 10\pi z) \quad \text{A-m}^{-1}$$

Example 8-4: UHF cellular phone signal. The magnetic field component of a UHF electromagnetic signal transmitted by a cellular phone base station is given by

$$\mathbf{H}(y) = \hat{\mathbf{x}}50e^{-j(17.3y-\pi/3)} \quad \mu\text{A-m}^{-1}$$

where the coordinate system is defined such that the z axis is in the vertical direction above a horizontal ground (xy plane or the $z = 0$ plane). (a) Determine the frequency f and the wavelength λ. (b) Write the corresponding expression for the electric field, $\mathbf{E}(y)$. (c) An observer located at $y = 0$ is using a vertical electric dipole antenna to measure the electric field as a function of time. Assuming that the observer can measure the vertical component of the electric field without any loss, what is the electric field at the time instants corresponding to $\omega t_1 = 0$, $\omega t_2 = \pi/2$, $\omega t_3 = \pi$, $\omega t_4 = 3\pi/2$, $\omega t_5 = 2\pi$ radians?

Solution:

(a) From $\beta = \omega/v_p = 2\pi f/c = 17.3$ rad-m^{-1}, we find $f \simeq 826$ MHz. The corresponding wavelength is $\lambda = 2\pi/\beta \simeq 36.3$ cm.

(b) From $\nabla \times \mathbf{H} = j\omega\epsilon\mathbf{E}$, we can solve for $\mathbf{E}$ as

$$\mathbf{E}(y) = \hat{\mathbf{z}}\eta H_x(y) \simeq \hat{\mathbf{z}}377 \times 50 \times 10^{-6} e^{-j(17.3y-\pi/3)} = \hat{\mathbf{z}}18.85 e^{-j(17.3y-\pi/3)} \text{ mV-m}^{-1}$$

(c) The instantaneous electric field is given by

$$\overline{\mathscr{E}}(y,t) = \mathscr{Re}\{\mathbf{E}(y)e^{j\omega t}\} \simeq \hat{\mathbf{z}}18.85\cos(5.19 \times 10^9 t - 17.3y + \pi/3) \quad \text{mV-m}^{-1}$$

Substituting $y = 0$ and $t_1 = 0$ yields $\overline{\mathscr{E}}(0, t_1) = \hat{\mathbf{z}}9.425$ mV-m^{-1}. Similarly, $\overline{\mathscr{E}}(0, t_2) \simeq -\hat{\mathbf{z}}16.3$ mV-m^{-1}, $\overline{\mathscr{E}}(0, t_3) = -\hat{\mathbf{z}}9.425$ mV-m^{-1}, $\overline{\mathscr{E}}(0, t_4) \simeq \hat{\mathbf{z}}16.3$ mV-m^{-1}, and $\overline{\mathscr{E}}(0, t_5) = \hat{\mathbf{z}}9.425$ mV-m^{-1}.

The Electromagnetic Spectrum The properties of uniform plane waves, as defined by the electric and magnetic field expressions given in [8.13], are identical over all of the electromagnetic spectrum that has been investigated experimentally, frequencies ranging from millihertz to 10^{24} Hz. Regardless of their frequency, all uniform plane electromagnetic waves propagate in unbounded free space with the same velocity, namely $v_p = c$, but with different wavelengths, as determined by $\lambda = c/f$. The speed of propagation of electromagnetic waves is also independent of frequency in a simple lossless material medium. (However, the material properties themselves are often functions of frequency, so the assumption of a simple lossless material medium does not hold true over the entire electromagnetic spectrum for any material medium.) Table 8.1 lists the various designated frequency and wavelength ranges of the electromagnetic spectrum and selected applications for each range. Maxwell's equations and the results derived from them thus encompass a truly amazing range of physical phenomena and applications that affect nearly every aspect of human life and our physical environment. In this section, we briefly comment on a few of the many applications listed in Table 8.1.

At frequencies in the ultraviolet range and higher, physicists are more accustomed to thinking in terms of the associated energy level of the photon (a quantum of radiation), which is given by hf, where $h = 6.63 \times 10^{-34}$ J-s is *Planck's constant.* Cosmic rays, consisting typically of photons at energies 10 MeV or greater, are constantly present in our universe; they ionize the highest reaches of the earth's atmosphere and help maintain the ionosphere at night, in the absence of solar radiation. Short-duration bursts of γ-rays, which bathe our solar system (and our galaxy) about three times a day, are believed to be produced in the most powerful explosions in the universe, releasing (within a few seconds or minutes) energies of 10^{51} ergs—more energy than our sun will produce in its entire ten billion years of existence.[9]

---

[9]G. J. Fishman and D. H. Hartmann, Gamma-ray bursts, *Scientific American,* pp. 46–51, July 1997.

TABLE 8.1. The electromagnetic spectrum and related applications

Frequency	Designation	Selected Applications	Wavelength (in free space)
$> 10^{22}$ Hz	Cosmic rays	Astrophysics	
10^{18}–10^{22} Hz	γ-rays	Cancer therapy, astrophysics	
10^{16}–10^{21} Hz	X-rays	Medical diagnosis	
10^{15}–10^{18} Hz	Ultraviolet	Sterilization	0.3–300 nm
3.95×10^{14}–7.7×10^{14}Hz	Visible Light	Vision, astronomy, optical communications	390–760 nm
		Violet	390–455
		Blue	455–492
		Green	492–577
		Yellow	577–600
		Orange	600–625
		Red	625–760
10^{12}–10^{14} Hz	Infrared	Heating, night vision, optical communications	3–300 μm
0.3–1 THz	Millimeter	Astronomy, meteorology	0.3–1 mm
30–300 GHz	EHF	Radar, remote sensing	0.1–1 cm
80–100		W-band	
60–80		V-band	
40–60		U-band	
27–40		K_a-band	
3–30 GHz	SHF	Radar, satellite comm.	1–10 cm
18–27		K-band	
12–18		K_u-band	
8–12		X-band	
4–8		C-band	
0.3–3 GHz	UHF	Radar, TV, GPS, cellular phone	10–100 cm
2–4		S-band	
2.45		Microwave ovens	
1–2		L-band, GPS system	
470–890 MHz		TV Channels 14–83	
30–300 MHz	VHF	TV, FM, police,	1–10 m
174–216		TV Channels 7–13	
88–108		FM radio	
76–88		TV Channels 5–6	
54–72		TV Channels 2–4	
3–30 MHz	HF	Short-wave, Citizens' band	10–100 m
0.3–3 MHz	MF	AM broadcasting	0.1–1 km
30–300 kHz	LF	Navigation, radio beacons	1–10 km
3–30 kHz	VLF	Navigation, positioning, naval communications	10–100 km
0.3–3 kHz	ULF	Telephone, audio	0.1–1 Mm
30–300 Hz	SLF	Power transmission, submarine communications	1–10 Mm
3–30 Hz	ELF	Detection of buried metals	10–100 Mm
< 3 Hz		Geophysical prospecting	> 100 Mm

Recently, brief flashes of γ-rays have been observed to be originating from the planet earth, believed to be generated at high altitudes above thunderstorms.[10]

It is also interesting to note that only a very narrow portion of the electromagnetic spectrum is perceptible to human vision, namely the visible range. We can also "feel" infrared as heat, and our bodies can be damaged by excessive amounts of microwave radiation, X rays, and γ-rays. Applications of X rays, ultraviolet, visible, and infrared light are far too numerous to be commented on here and include vision, lasers, optical fiber communications, and astronomy.

At present, a relatively unexplored part of the electromagnetic spectrum is the transition region between the low-frequency end of infrared and the high-frequency end of the millimeter range. The frequency in this region is measured in terahertz (THz), or 10^{12} Hz, with the corresponding wavelengths being in the submillimeter range. Fabrication of antennas, transmission lines, and other receiver components usable in these frequencies requires the use of novel quasi-optical techniques.[11] At present, the terahertz region is primarily used in astronomy and remote sensing; however, industrial, medical, and other scientific applications are currently under investigation.

Each decade of the electromagnetic spectrum below the millimeter range of frequencies is divided[12] into designated ranges, with the acronyms indicated in Table 8.1: extremely high frequency (EHF); super high frequency (SHF); ultra high frequency (UHF); very high frequency (VHF); high frequency (HF); medium frequency (MF); low frequency (LF); very low frequency (VLF); ultra low frequency (ULF); super low frequency (SLF); extremely low frequency (ELF).

The microwave band of frequencies is vaguely defined as the range from 300 MHz up to 1 THz, including the millimeter range, and is extensively utilized for radar, remote sensing, and a host of other applications too numerous to cite here.[13] In radar work, the microwave band is further subdivided into bands with alphabetical designations, which are listed in Table 8.1.

The VLF range of frequencies is generally used for global navigation and naval communications, using transmitters that utilize huge radiating structures.[14] Although VLF transmissions are used for global communications with surface ships and submarines near the water surface, even lower frequencies are required for communica-

---

[10]G. J. Fishman, P. N. Bhat, R. Mallozzi, J. M. Horack, T. Koshut, C. Kouveliotou, G. N. Pendleton, C. A. Meegan, R. B. Wilson, W. S. Paciesas, S. J. Goodman, and H. J. Christian, Discovery of intense gamma-ray flashes of atmospheric origin, *Science,* 264, p. 1313, 1994.

[11]See Special Issue on Terahertz Technology of the *Proceedings of the IEEE,* 80 (11), November, 1992.

[12]There is a certain arbitrariness in the designations of these frequency ranges. In geophysics and solar terrestrial physics, the designation ELF is used for the range 3 Hz to 3 kHz, while ULF is used to describe frequencies typically below 3 Hz.

[13]See J. Thuery, *Microwaves: Industrial, Scientific and Medical Applications,* Artech House, 1992. Also see O. P. Gandhi, editor, *Biological Effects and Medical Applications of Electromagnetic Energy,* Prentice Hall, 1990.

[14]See A. D. Watt, *VLF Radio Engineering,* Pergamon Press, New York, 1967; J. C. Kim and E. I. Muehldorf, *Naval Shipboard Communications Systems,* Prentice Hall, 1995.

tion with deeply submerged submarines. The Sanguine system operated by the U.S. Navy utilizes two sets of orthogonal large (22.5 km length) horizontal antennas, located in Wisconsin and Michigan at a distance of 240 km apart, and operates[15] at 72 to 80 Hz.

The lowest frequencies of the experimentally investigated electromagnetic spectrum are commonly used to observe so-called micropulsations, which are electromagnetic waves at frequencies 0.001 to 10 Hz generated as a result of large-scale currents flowing in the earth's auroral regions and by the interaction between the earth's magnetic field and the energetic particles that stream out of the sun in the form of the solar wind.[16] These natural signals are also often used for geophysical prospecting and magnetotelluric studies.[17] Another potentially very important practical use of this band may yet emerge, based on experimental evidence of detectable electromagnetic precursors produced many hours prior to major earthquakes.[18]

8.1.3 Plane Waves in Lossy Media

Many of the more interesting electromagnetic applications involve the interactions between electric and magnetic fields and matter. We represent the microscopic interactions of electromagnetic waves with matter in terms of the macroscopic parameters ϵ, μ, and σ. In general, most dielectric media exhibit small but nonzero conductivity or complex permittivity and can absorb electromagnetic energy, resulting in the attenuation of an electromagnetic wave as it propagates through the medium. The performance of most practical transmission lines and other devices that are used to convey electromagnetic energy from one point to another is limited by small losses in conductors. The inherently lossy nature of some media (e.g., seawater, animal tissue) determines the range of important applications (e.g., submarine communications and medical diagnostic implants). All media exhibit losses in some frequency ranges; for example, although air is largely transparent (lossless) over the radio and microwave ranges, it is a highly lossy medium at optical frequencies. The fact that upper atmospheric ozone absorbs ultraviolet light (i.e., it is lossy at ultraviolet frequencies) protects life on planet earth from this harmful radiation.

When a material exhibits a nonzero conductivity σ, the electric field of a propagating wave causes a conduction current of $\mathbf{J}_c = \sigma\mathbf{E}$ to flow. This current, which

---

[15] C. H. Richard, *Sub vs. Sub,* Orion Books, New York, 1988; M. F. Genge and R. D. Carlson, Project ELF Electromagnetic Compatibility Assurance Program, *IEEE Journal of Oceanic Engineering,* 9(3), pp. 143–153, July 1984.

[16] J. A. Jacobs, *Geomagnetic Micropulsations,* Springer-Verlag, New York, Heidelberg, and Berlin, 1970; J. K. Hargreaves, *The Solar-Terrestrial Environment,* Cambridge University Press, 1992.

[17] M. N. Nabighian, editor, *Electromagnetic Methods in Applied Geophysics, Vol. 2: Application,* Parts A and B, Society of Exploration Geophysics, 1991.

[18] A. C. Fraser-Smith, A. Bernardi, P. R. McGill, M. E. Ladd, R. A. Helliwell, and O. G. Villard, Jr., Low-frequency magnetic field measurements near the epicenter of the M_s 7.1 Loma Prieta earthquake, *Geophys. Res. Lett.,* 17, pp. 1465–1468, 1990; also see B. Holmes, Radio hum may herald quakes, *New Scientist,* p. 15, 23/30 December 1995.

is in phase with the wave electric field, leads to dissipation of some of the wave energy as heat within the material, with the power dissipated per unit volume being given by $\mathbf{E} \cdot \mathbf{J}$, as discussed in Section 5.8. This dissipation requires that the wave electric and magnetic fields attenuate with distance as they propagate in the lossy material, much like the attenuation of waves propagating on a lossy transmission line as discussed in Section 3.8. To determine the characteristics of uniform plane waves in lossy media, we start with Maxwell's equations and follow a procedure quite similar to that used in the previous section.

In a conducting (or lossy) medium the $\mathbf{J}$ term in [7.18*c*] is nonzero even in the absence of sources, since a current of $\mathbf{J}_c = \sigma\mathbf{E}$ flows in response to the wave's electric field. We then have

$$\nabla \times \mathbf{H} = \sigma\mathbf{E} + j\omega\epsilon\mathbf{E} \qquad [8.14]$$

and, taking the curl of [7.18*a*], we have

$$\nabla \times \nabla \times \mathbf{E} \equiv \nabla(\nabla \cdot \mathbf{E}) - \nabla^2\mathbf{E} = -j\omega\nabla \times \mathbf{B}$$

Using the fact that $\nabla \cdot \mathbf{E} = 0$ for a simple, source-free medium, and substituting from [7.18*c*], we find

$$\nabla^2\mathbf{E} - j\omega\mu(\sigma + j\omega\epsilon)\mathbf{E} = 0$$

or

$$\boxed{\nabla^2\mathbf{E} - \gamma^2\mathbf{E} = 0} \qquad [8.15]$$

where $\gamma = \sqrt{j\omega\mu(\sigma + j\omega\epsilon)}$ is known as the propagation constant (or wave number) and is in general a complex quantity, which can be expressed in terms of its real and imaginary parts as $\gamma = \alpha + j\beta$. Note that the complex propagation constant γ for uniform plane waves in an unbounded lossy medium is equivalent to the complex propagation constant γ given in [3.61] and derived in Section 3.8 for propagation of voltage waves on a lossy transmission line.

Realizing that [8.15] represents three scalar equations similar to [8.8], we limit our attention once again to uniform plane waves propagating in the z direction (i.e., all field quantities varying only in z), with the electric field having only an x component, for which [8.15] becomes

$$\boxed{\frac{d^2E_x}{dz^2} - \gamma^2E_x = 0} \qquad [8.16]$$

which is identical to equation [3.60], except that the dependent variable is $E_x(z)$ rather than $V(z)$ in [3.60]. The general solution of [8.16] is

$$E_x(z) = C_1e^{-\gamma z} + C_2e^{+\gamma z} = \underbrace{C_1e^{-\alpha z}e^{-j\beta z}}_{E_x^+(z)} + \underbrace{C_2e^{+\alpha z}e^{+j\beta z}}_{E_x^-(z)} \qquad [8.17]$$

For $\alpha, \beta > 0$, the two terms $E_x^+(z)$ and $E_x^-(z)$ represent waves propagating in the $+z$ and $-z$ directions, respectively. Note that the constants C_1 and C_2 are in general complex numbers to be determined by boundary conditions at material interfaces.

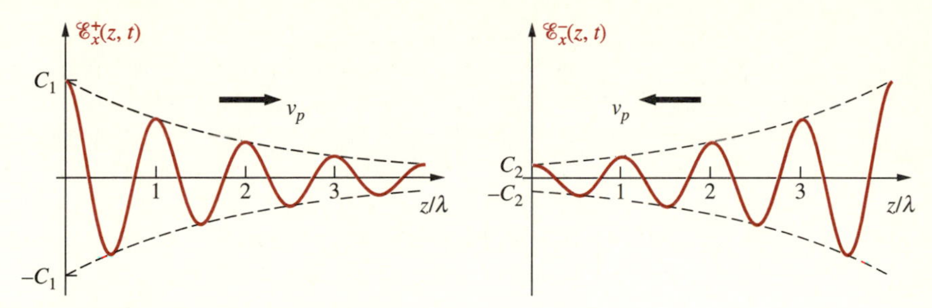

FIGURE 8.6. Snapshots of waves in a lossy medium. The two terms of [8.18] plotted as a function of z at $\omega t = 0$.

The instantaneous electric field can be found from [8.17] in the same way[19] as described in [8.11]. We find

$$\mathscr{E}_x(z, t) = \underbrace{C_1 e^{-\alpha z} \cos(\omega t - \beta z)}_{\mathscr{E}_x^+(z,t)} + \underbrace{C_2 e^{+\alpha z} \cos(\omega t + \beta z)}_{\mathscr{E}_x^-(z,t)} \qquad [8.18]$$

assuming C_1 and C_2 to be real. The nature of the waves described by the two terms of [8.18] is shown in Figure 8.6. We see, for example, that the wave propagating in the $+z$ direction has a decreasing amplitude with increasing distance at a fixed instant of time. In other words, the wave is attenuated as it propagates in the medium. The rate of this attenuation is given by the attenuation constant α, in units of np-m^{-1}; see Section 3.8.1 for a discussion of the unit nepers per meter (np-m^{-1}). The quantity β (in rad-m^{-1}) determines the phase velocity and wavelength of the wave and is referred to as the phase constant. We shall see below that the wavelength of a uniform plane wave in lossy materials can be substantially different from that in free space. We note once again that all aspects of the solution given in [8.18] are entirely analogous to that given in [3.66] for voltage waves on a lossy transmission line. In this connection, the similarity between Figure 8.6 and Figure 3.60a should also be noted.

Explicit expressions for α and β can be found by equating the real and imaginary parts of

$$\boxed{\gamma = \alpha + j\beta = \sqrt{j\omega\mu(\sigma + j\omega\epsilon)}}$$

We find

$$\boxed{\alpha = \omega\sqrt{\frac{\mu\epsilon}{2}}\left[\sqrt{1 + \left(\frac{\sigma}{\omega\epsilon}\right)^2} - 1\right]^{1/2} \text{ np-m}^{-1}} \qquad [8.19]$$

[19] $\mathcal{R}e\{C_1 e^{-\alpha z} e^{-j\beta z} e^{j\omega t}\} = C_1 e^{-\alpha z} \mathcal{R}e\{e^{j(\omega t - \beta z)}\} = C_1 e^{-\alpha z} \cos(\omega t - \beta z)$, where C_1 is a real constant.

and

$$\beta = \omega\sqrt{\frac{\mu\epsilon}{2}}\left[\sqrt{1+\left(\frac{\sigma}{\omega\epsilon}\right)^2}+1\right]^{1/2} \text{ rad-m}^{-1} \qquad [8.20]$$

Consider now only the wave propagating in the $+z$ direction, the first term of [8.17]:

$$E_x^+(z) = C_1 e^{-\alpha z} e^{-j\beta z} \qquad [8.21]$$

The wave magnetic field that accompanies this electric field can be found by substituting [8.21] into [7.18*a*], as we did for the lossless case, or by making an analogy with the lossless-media solution obtained in the previous section. Following the latter approach, we can rewrite [8.14] as

$$\nabla \times \mathbf{H} = \sigma\mathbf{E} + j\omega\epsilon\mathbf{E} = j\omega\epsilon_{\text{eff}}\mathbf{E}$$

where $\epsilon_{\text{eff}} = \epsilon - j\sigma/\omega$. It thus appears that the solutions we obtained in the previous section for a lossless medium can be used as long as we make the substitution $\epsilon \to \epsilon_{\text{eff}}$. On this basis, the intrinsic impedance of a conducting medium can be found as

$$\eta_c = |\eta_c|e^{j\phi_\eta} = \sqrt{\frac{\mu}{\epsilon_{\text{eff}}}} = \sqrt{\frac{\mu}{\epsilon - j\dfrac{\sigma}{\omega}}} = \frac{\sqrt{\dfrac{\mu}{\epsilon}}}{\left[1+\left(\dfrac{\sigma}{\omega\epsilon}\right)^2\right]^{1/4}} e^{j(1/2)\tan^{-1}(\sigma/\omega\epsilon)} \qquad [8.22]$$

and the associated magnetic field $\mathbf{H}$ is

$$\mathbf{H} = \hat{\mathbf{y}}H_y^+(z) = \hat{\mathbf{y}}\frac{1}{\eta_c}C_1 e^{-\alpha z}e^{-j\beta z}$$

The instantaneous magnetic field $\overline{\mathcal{H}}(z, t)$ can be found as

$$\mathcal{H}_y^+(z,t) = \Re e\left\{H_y^+(z)e^{j\omega t}\right\} = \Re e\left\{\frac{1}{|\eta_c|e^{j\phi_\eta}}C_1 e^{-\alpha z}e^{-j\beta z}e^{j\omega t}\right\}$$

$$= \frac{1}{|\eta_c|}C_1 e^{-\alpha z}\Re e\{e^{j(\omega t - \beta z)}e^{-j\phi_\eta}\} \qquad [8.23]$$

$$\mathcal{H}_y^+(z,t) = \frac{C_1}{|\eta_c|}e^{-\alpha z}\cos(\omega t - \beta z - \phi_\eta)$$

Thus we see that the electric and magnetic fields of a uniform plane wave in a conducting medium do not reach their maxima at the same time (at a fixed point) or at the same point (at a fixed time); in other words, $\overline{\mathcal{E}}(z, t)$ and $\overline{\mathcal{H}}(z, t)$ are *not in phase.* The magnetic field lags the electric field by a phase difference equal to the phase of the complex intrinsic impedance, ϕ_η. This relationship is depicted in Figure 8.7.

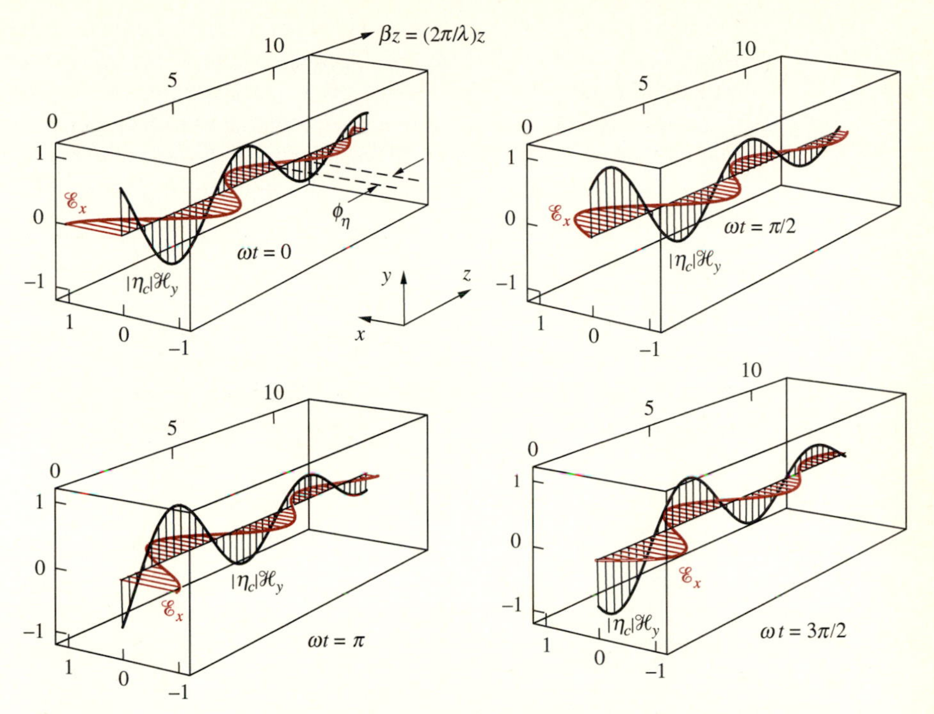

FIGURE 8.7. Time snapshots of the electric and magnetic fields in a conducting medium. The curves shown are for $C_1 = 1$, $\phi_\eta = 45°$ and $\alpha = 0.1\beta$, where $\beta = 2\pi/\lambda$.

We can see from [8.19], [8.20], and [8.22] that the rate of attenuation α, the phase constant β, and the intrinsic impedance η_c depend sensitively on $\sigma/(\omega\epsilon)$. For $\sigma \ll (\omega\epsilon)$, the attenuation rate is small, and the propagation constant and intrinsic impedance are only slightly different from that for a lossless medium; namely, $\beta \simeq \omega\sqrt{\mu\epsilon}$ and $\eta_c \simeq \sqrt{\mu/\epsilon}$. For $\sigma \gg (\omega\epsilon)$, the attenuation rate α is large, causing the uniform plane wave to decay rapidly with distance, and the intrinsic impedance is very small, approaching zero as $\sigma \to \infty$, as it does for a perfect conductor. A perfect conductor, in other words, is a medium with zero intrinsic impedance,[20] so the presence of an electric field of nonzero magnitude $C_1 \neq 0$ requires a magnetic field of an infinite magnitude. Thus, no time-harmonic electric or magnetic fields can exist in a perfect conductor.[21]

[20]A perfect conductor is thus the ultimate lossy material. Although this may at first appear counterintuitive, the behavior of a perfect conductor as a medium with zero impedance is analogous to a short circuit. The voltage drop across a short circuit is always zero, while the current through it is determined by the external circuits it is connected to. Similarly, we shall see in Section 8.2 that when a uniform plane wave is incident on a perfect conductor, the wave is perfectly reflected, since no electromagnetic fields can exist inside the conductor, while the current that flows on the surface of the perfect conductor is determined by the magnetic field intensity of the incident uniform plane wave.

[21]Note, from Section 4.7, that static electric fields inside metallic conductors must also be zero and, from Section 6.8.4, that static magnetic fields cannot exist in a perfect conductor because of the Meissner effect.

The quantity $\tan\delta_c = \sigma/(\omega\epsilon)$ is called the *loss tangent* of the medium and is a measure of the degree to which the medium conducts. A medium is considered to be a good conductor if $\tan\delta_c \gg 1$; most metals are good conductors at frequencies of up to 100 GHz or so. A medium is considered to be a poor conductor, or a good insulator, if $\tan\delta_c \ll 1$. Example 8-5 illustrates the use of expressions [8.19], [8.20], and [8.22] to determine the parameter values for uniform plane waves in muscle tissue.

Example 8-5: Microwave exposure of muscle tissue. Find the complex propagation constant γ and the intrinsic impedance η_c of a microwave signal in muscle tissue[22] at 915 MHz ($\sigma = 1.6$ S-m^{-1}, $\epsilon_r = 51$).

Solution: The complex propagation constant is given by

$$\gamma = \alpha + j\beta$$

where α and β are given by [8.19] and [8.20]. The loss tangent of the muscle tissue at 915 MHz is

$$\tan\delta_c = \frac{\sigma}{\omega\epsilon} = \frac{1.6\ \text{S-m}^{-1}}{2\pi \times 915 \times 10^6\ \text{r-s}^{-1} \times 51 \times 8.85 \times 10^{-12}\ \text{F-m}^{-1}} \simeq 0.617$$

So α and β are given by

$$\begin{Bmatrix}\alpha \\ \beta\end{Bmatrix} \simeq (2\pi \times 915 \times 10^6\ \text{rad-s}^{-1})\frac{\sqrt{51}}{\sqrt{2} \times 3 \times 10^8\ \text{m-s}^{-1}}\left[\sqrt{1 + (0.617)^2} \mp 1\right]^{1/2}$$

$$\simeq \begin{Bmatrix} 40.7\ \text{np-m}^{-1} \\ 142.7\ \text{rad-m}^{-1}\end{Bmatrix}$$

So the complex propagation constant is given by

$$\gamma \simeq 40.7 + j142.7$$

The complex intrinsic impedance can be found using [8.22] as

$$\eta_c \simeq \frac{377/\sqrt{51}}{[1 + (0.617)^2]^{1/4}}\exp\left[j\frac{1}{2}\tan^{-1}(0.617)\right] \simeq 48.7e^{j15.8^\circ}\ \Omega$$

At this point, it is important to remember that the material properties themselves (i.e., σ and ϵ) may well be functions of frequency (see Figure 7.22). For typical good conductors, both σ and ϵ are nearly independent of frequency, at frequencies below

[22] C. C. Johnson and A. W. Guy, Nonionizing electromagnetic wave effects in biological materials and systems, *Proceedings of the IEEE*, 60(6), pp. 692–718, June 1972.

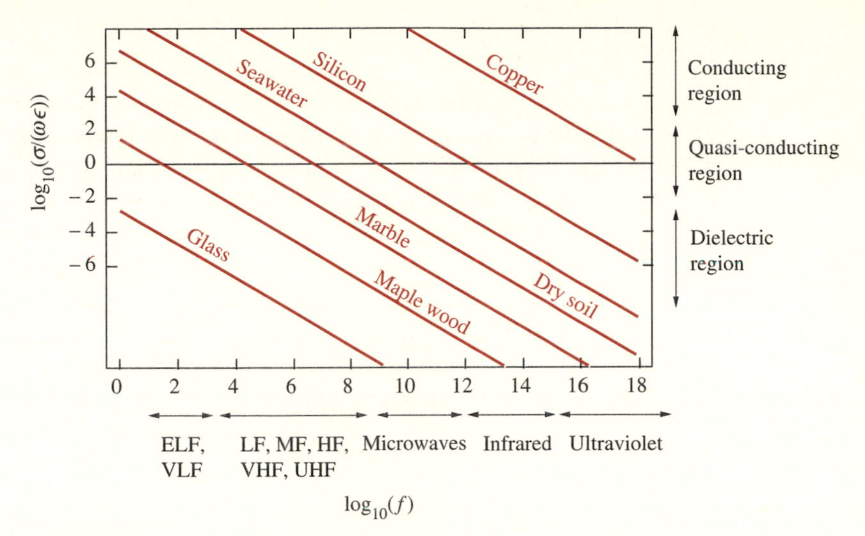

FIGURE 8.8. **Loss tangent versus frequency.** The loss tangent $\tan\delta_c = \sigma/(\omega\epsilon)$ for selected materials plotted as a function of frequency, assuming that the material constants σ and ϵ are as given in Table 8.2 and that they do not vary with frequency.

the optical range, but for lossy dielectrics the material constants σ and ϵ tend to be functions of frequency. For some materials, the loss tangent $\tan\delta_c = \sigma/(\omega\epsilon)$ tends to vary less over the frequency range of interest. Properties of dielectrics are usually given in terms of ϵ and $\tan\delta_c$. The ratio $\sigma/(\omega\epsilon)$ is plotted as a function of frequency in Figure 8.8 for some common materials, assuming the parameters ϵ and σ to be constants. The values above the microwave region (>10 GHz) are not likely to be accurate since σ and ϵ may vary significantly with frequency. Table 8.2 gives the values of σ and ϵ used in Figure 8.8.

The most common cases of practical interest involving the propagation of electromagnetic waves in lossy media concern lossy dielectrics that exhibit complex

TABLE 8.2. Relative permittivity and conductivity of selected materials

Medium	Relative permittivity ϵ_r (dimensionless)	Conductivity σ, (S-m^{-1})
Copper	1	5.8×10^7
Seawater	81	4
Doped silicon	12	10^3
Marble	8	10^{-5}
Maple wood	2.1	3.3×10^{-9}
Dry soil	3.4	10^{-4} to 10^{-2}
Fresh water	81	$\sim 10^{-2}$
Mica	6	10^{-15}
Flint glass	10	10^{-12}

permittivity,[23] and good conductors. We consider these two cases separately in the next two sections.

Uniform Plane Wave Propagation in Lossy Dielectrics Most dielectrics behave like lossy materials at high frequencies, not because they exhibit finite conductivity σ but because their permittivity is complex: $\epsilon_c = \epsilon' - j\epsilon''$. As was discussed in Section 7.4, the imaginary part of the permittivity represents the effect on a macroscopic scale of frictional and tensional forces that the wave electric field has to overcome in polarizing the material at high frequencies. The ability of a molecule to reorient itself (i.e., form a polarization dipole) in response to a rapidly varying applied electric field depends on its molecular shape. When the frequency of the applied field is higher than the characteristic relaxation (or resonance) frequency for the material, the system does not follow or respond to the field without losses. For example, for water at room temperature, the resonance frequency is $\sim$24 GHz, whereas for ice at $-20°$C the relaxation frequency is $\sim$1 kHz. Table 8.3 provides values of real and imaginary parts of the relative permittivity for selected materials.[24] Since these properties (especially ϵ'') are strongly dependent on frequency, and also to some degree on temperature, the frequency and temperature to which the given values correspond to are specified. In some cases, the moisture content of the material is also noted, since the permittivity values also depend on this parameter.

Considering [8.14] with $\sigma = 0$ but with $\epsilon_c = \epsilon' - j\epsilon''$, we find

$$\nabla \times \mathbf{H} = j\omega(\epsilon' - j\epsilon'')\mathbf{E} = \omega\epsilon''\mathbf{E} + j\omega\epsilon'\mathbf{E} \qquad [8.24]$$

---

[23]As discussed in Section 7.4, losses can also occur in magnetic materials due to relaxation effects that lead to complex permeability $\mu = \mu' - j\mu''$. The losses due to this effect can be represented by an effective conductivity $\sigma_{\text{eff}} = \omega\epsilon'\mu''/\mu'$. However, such effects have limited practical application and are beyond the scope of this book.

[24]The values in Table 8.3 were taken from the following sources: *Reference Data for Radio Engineers,* 8th ed., Sams Prentice Hall Computer Publishing, Carmel, Indiana, 1993; J. Thuery, *Microwaves: Industrial, Scientific and Medical Applications,* Artech House, Inc., 1992; D. I. C. Wang and A. Goldblith, Dielectric properties of food, *Tech. Rep. TR-76-27-FEL,* MIT, Dept. of Nutrition and Food Science, Cambridge, Massachusetts, November, 1976; S. O. Nelson, Comments on "Permittivity measurements of granular food products suitable for computer simulations of microwave cooking processes," *IEEE Trans. Instrum. Meas.,* 40(6), pp. 1048–1049, December 1991; A. D. Green, Measurement of the dielectric properties of cheddar cheese, *J. Microwave Power Electromagn. Energy,* 32(1), pp. 16–27, 1997; S. Puranik, A. Kumbharkhane, and S. Mehrotra, Dielectric properties of honey-water mixtures between 10 MHz to 10 GHz using time domain technique, *J. Microwave Power Electromagn. Energy,* 26(4), pp. 196–201,1991; H. C. Rhim and O. Buyukozturk, Electromagnetic properties of concrete at microwave frequency range, *ACI Materials Journal,* Vol. 95, No. 3, pp. 262–271, May–June 1998; J. M. Osepchuk, Sources and basic characteristics of microwave/RF radiation, *Bull. N. Y. Acad. Med.,* 55(11), pp. 976–998, December 1979; S. O. Nelson, Microwave dielectric properties of fresh onions, *Trans. Amer. Soc. Agr. Eng.,* 35(3), pp. 963–966, May–June 1992; R. F. Shiffmann, Understanding microwave reactions and interactions, *Food Product Design,* pp. 72–88, April 1993.

TABLE 8.3. Dielectric properties of selected materials

Material	f (GHz)	ϵ_r'	ϵ_r''	T (°C)
Aluminum oxide (Al_2O_3)	3.0	8.79	8.79×10^{-3}	25
Barium titanate ($BaTiO_3$)	3.0	600	180	26
Bread	4.6	1.20		
Bread dough	2.45	22.0	9.00	
Butter (salted)	2.45	4.6	0.60	20
Cheddar cheese	2.45	16.0	8.7	20
Concrete (dry)	2.45	4.5	0.05	25
Concrete (wet)	2.45	14.5	1.73	25
Corn (8% moisture)	2.45	2.2	0.2	24
Corn oil	2.45	2.5	0.14	25
Distilled water	2.45	78	12.5	20
Dry sandy soil	3.0	2.55	1.58×10^{-2}	25
Egg white	3.0	35.0	17.5	25
Frozen beef	2.45	4.4	0.528	−20
Honey (100% pure)	2.45	10.0	3.9	25
Ice (pure distilled)	3.0	3.2	2.88×10^{-3}	−12
Milk	3.0	51.0	30.1	20
Most plastics	2.45	2 to 4.5	0.002 to 0.09	20
Papers	2.45	2 to 3	0.1 to 0.3	20
Potato (78.9% moisture)	3.0	81.0	30.8	25
Polyethylene	3.0	2.26	7.01×10^{-4}	25
Polystyrene	3.0	2.55	8.42×10^{-4}	25
Polytetrafluoroethylene (Teflon)	3.0	2.1	3.15×10^{-4}	22
Raw beef	2.45	52.4	17.3	25
Snow (fresh fallen)	3.0	1.20	3.48×10^{-4}	−20
Snow (hard packed)	3.0	1.50	1.35×10^{-3}	−6
Some glasses (Pyrex)	2.45	~4.0	0.004 to 0.02	20
Smoked bacon	3.0	2.50	0.125	25
Soybean oil	3.0	2.51	0.151	25
Steak	3.0	40.0	12.0	25
White onion (78.7% moisture)	2.45	53.8	13.5	22
White rice (16% moisture)	2.45	3.8	0.8	24
Wood	2.45	1.2 to 5	0.01 to 0.5	25

By comparing [8.24] with [8.14], we see that the net effect of ϵ'' is to introduce a loss term, which can be represented in all the foregoing expressions by replacing σ with an "effective" conductivity $\sigma_{\text{eff}} = \omega\epsilon''$.

In practice, the electrical properties of a lossy dielectric are identified by specifying the real part of its dielectric constant (i.e., ϵ') and its loss tangent $\tan\delta_c$, given as

$$\tan\delta_c = \frac{\sigma_{\text{eff}}}{\omega\epsilon'} = \frac{\epsilon''}{\epsilon'}$$

If a given material has both nonzero conductivity ($\sigma \neq 0$) and complex permittivity, it can be treated as if it has an effective conductivity of $\sigma_{\text{eff}} = \sigma + \omega\epsilon''$ and a loss tangent $\tan\delta_c = (\sigma + \omega\epsilon'')/(\omega\epsilon')$. In practice, it is generally not necessary to distinguish between losses due to σ or $\omega\epsilon''$, since ϵ' and σ_{eff} (or $\tan\delta_c$) are often determined by measurement. The attenuation in a lossy dielectric is frequently expressed as the attenuation distance or penetration depth d over which the field intensity decreases by a factor of e^{-1}. Using [8.19] we can write

$$d \equiv \frac{1}{\alpha} = \frac{\sqrt{2}}{\omega\sqrt{\mu\epsilon'}[\sqrt{1+\tan^2\delta_c}-1]^{1/2}} = \frac{c\sqrt{2}}{\omega\sqrt{\mu_r\epsilon_r'}[\sqrt{1+\tan^2\delta_c}-1]^{1/2}} \quad [8.25]$$

where c is the speed of light in free space.

Examples 8-6 and 8-7 provide numerical values of penetration depth for two different lossy dielectric materials.

Example 8-6: Complex permittivity of distilled water. Consider distilled water at 25 GHz ($\epsilon_{cr} \simeq 34 - j9.01$). Calculate the attenuation constant α, propagation constant β, penetration depth d, and wavelength λ.

Solution: The loss tangent of distilled water at 25 GHz is $\tan\delta_c = \epsilon''/\epsilon' = 0.265$. From [8.19], the attenuation constant is then

$$\alpha \simeq \frac{(2\pi \times 25 \times 10^9 \text{ rad-m}^{-1})\sqrt{34}}{\sqrt{2}\times(3\times 10^8 \text{ m-s}^{-1})}[\sqrt{1+(0.265)^2}-1]^{1/2} \simeq 401 \text{ np-m}^{-1}$$

In a similar manner, the phase constant β can be found from [8.20] to be $\beta \simeq 3079$ rad-m^{-1}. The penetration depth is $d = \alpha^{-1} \simeq 2.49$ mm, and the wavelength is $\lambda = 2\pi/\beta \simeq 2.04$ mm.

In other words, the depth of penetration into distilled water is of the order of one wavelength. Although the condition $\tan\delta_c \ll 1$ is not satisfied for this case, the value of β is nevertheless within a few percent of that for a lossless dielectric with $\epsilon_r = 34$.

Example 8-7: Complex permittivity of bread dough. The dielectric properties of commercially available bread dough are investigated in order to develop more efficient microwave systems and sensors for household and industrial baking.[25] The material properties depend not only on the microwave frequencies used, but also on the time of the baking process. Table 8.4 provides the dielectric

TABLE 8.4. Relative permittivity $\epsilon_{cr} = \epsilon_r' - \epsilon_r''$ of bread dough versus frequency and baking time

	10 min	**20 min**	**30 min**
600 MHz	$23.1 - j11.85$	$16.78 - j6.66$	$8.64 - j2.51$
2.4 GHz	$12.17 - j4.54$	$9.53 - j3.12$	$4.54 - j1.22$

[25] J. Zuercher, L. Hoppie, R. Lade, S. Srinivasan, and D. Misra, Measurement of the complex permittivity of bread dough by an open-ended coaxial line method at ultra high frequencies, *J. Microwave Power Electromagn. Energy,* 25(3), pp. 161–167, 1990.

TABLE 8.5. Loss tangent $\tan\delta_c$ and depth of penetration $\boldsymbol{d}$ versus frequency and baking time

	10 min		20 min		30 min	
	$\tan\delta_c$	d	$\tan\delta_c$	d	$\tan\delta_c$	d
600 MHz	0.513	6.65 cm	0.397	9.97 cm	0.291	18.8 cm
2.4 GHz	0.373	3.11 cm	0.327	3.99 cm	0.269	7.01 cm

constants of the bread baked for 10, 20 (not done yet), and 30 (baking complete) minutes, respectively, at two microwave frequencies, 600 MHz and 2.4 GHz. Calculate the depth of penetration for each case and compare the results.

Solution: At $f = 600$ MHz, after baking for 10 minutes, the loss tangent of bread dough is

$$\tan\delta_c = \frac{11.85}{23.1} \simeq 0.513$$

from which the depth of penetration can be calculated as

$$d \simeq \frac{3\times 10^8 \text{ m-s}^{-1}}{2\pi\times 6\times 10^8 \text{ rad-m}^{-1}}\left[\frac{2}{23.1(\sqrt{1+(0.513)^2}-1)}\right]^{1/2} \simeq 6.65 \text{ cm}$$

Similarly, we can calculate the other values. The results are tabulated in Table 8.5.

As we see from Table 8.5, the loss tangent decreases with both increasing baking time and increasing frequency. The depth of penetration increases with baking time but decreases with frequency.

In most applications, we do not need to use the exact expressions for α, β, and η_c, given respectively by [8.19], [8.20], and [8.22], since materials can be classified either as good conductors or low-loss dielectrics, depending respectively on whether their loss tangents are much larger or much smaller than unity.

A low-loss dielectric (or a good dielectric) is one for which $\tan\delta_c \ll 1$. In many applications, dielectric materials that are used are very nearly perfect but nevertheless do cause some nonzero amount of loss. A good example is the dielectric fillings of a coaxial line; the materials used for this purpose are typically nearly perfect insulators, and an approximate evaluation of the small amount of losses is entirely adequate for most applications.

When $\tan\delta_c \ll 1$ the expressions [8.19] and [8.20] for α and β can be simplified using the binomial approximation[26]

[26] $(1\pm x)^k = 1 \pm kx + \dfrac{k(k-1)}{2}x^2 \pm \cdots$

$$\gamma = \alpha + j\beta = j\omega\sqrt{\mu\epsilon'}\left(1 - j\frac{\sigma_{\text{eff}}}{\omega\epsilon'}\right)^{-1/2} \simeq j\omega\sqrt{\mu\epsilon'}\left[1 - j\frac{\sigma_{\text{eff}}}{2\omega\epsilon'} + \frac{1}{8}\left(\frac{\sigma_{\text{eff}}}{\omega\epsilon'}\right)^2 + \cdots\right]$$

from which we can write

$$\alpha \simeq \frac{\sigma_{\text{eff}}}{2}\sqrt{\frac{\mu}{\epsilon'}} = \frac{\omega\epsilon''}{2}\sqrt{\frac{\mu}{\epsilon'}}\ \text{np-m}^{-1}$$

as the attenuation constant and

$$\beta \simeq \omega\sqrt{\mu\epsilon'}\left[1 + \frac{1}{8}\left(\frac{\sigma_{\text{eff}}}{\omega\epsilon'}\right)^2\right] = \omega\sqrt{\mu\epsilon'}\left[1 + \frac{1}{8}\left(\frac{\epsilon''}{\epsilon'}\right)^2\right]\ \text{rad-m}^{-1}$$

as the phase constant. Note that since $\tan\delta_c \ll 1$, the phase constant β (and therefore the wavelength $\lambda = 2\pi/\beta$) in a low-loss dielectric are very nearly equal to those in a lossless medium. The attenuation constant imposed on the wave by the medium is usually very small. For example, for dry earth (take $\sigma = 10^{-5}$ S-m^{-1} and $\epsilon = 3\epsilon_0$) at 10 MHz, $\tan\delta_c \simeq 0.006$, and $\alpha \simeq 10^{-3}$ np/m, so that the signal is attenuated down to $1/e$ of its value in ~920 m. The wavelength $(2\pi/\beta)$ on the other hand is ~17.3 m, so that the depth of penetration for a low-loss dielectric is typically many wavelengths. Such a picture is illustrated in Figure 8.9.

Note that a simplified expression for the phase velocity ($v_p = \omega/\beta$) can be obtained by using [8.20] and the binomial expansion for $(1 + x)^k$ and noting that the exponent k is negative:

$$v_p \simeq \frac{1}{\sqrt{\mu\epsilon'}}\left[1 - \frac{1}{8}\left(\frac{\sigma_{\text{eff}}}{\omega\epsilon'}\right)^2\right] = \frac{1}{\sqrt{\mu\epsilon'}}\left[1 - \frac{1}{8}\left(\frac{\epsilon''}{\epsilon'}\right)^2\right]\ \text{m-s}^{-1}$$

Since $1/\sqrt{\mu\epsilon'}$ is the phase velocity in the lossless medium, we see that the small loss introduced by the low-loss dielectric also slightly reduces the velocity of the wave.

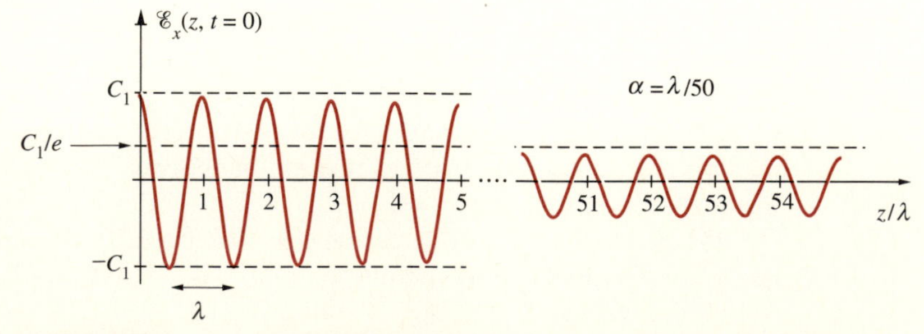

FIGURE 8.9. Low-loss dielectric. The electric field component of a uniform plane wave propagating in a low-loss (i.e., good) dielectric, in which the depth of penetration is typically much larger than a wavelength.

An approximate expression for the intrinsic impedance of a good dielectric can also be obtained, again using the binomial expansion. We find:

$$\eta_c = \sqrt{\frac{\mu}{\epsilon'}}\left(1 - j\frac{\sigma_{\text{eff}}}{\omega\epsilon'}\right)^{-1/2} \simeq \sqrt{\frac{\mu}{\epsilon'}}\left(1 + j\frac{\sigma_{\text{eff}}}{2\omega\epsilon'}\right) = \sqrt{\frac{\mu}{\epsilon'}}\left(1 + j\frac{\epsilon''}{2\epsilon'}\right) \quad \Omega$$

We see that the small amount of loss also introduces a small reactive (in this case inductive) component to the intrinsic impedance of the medium. This will in turn result in a slight phase shift between the electric and magnetic fields, since ϕ_η in [8.22] is nonzero. We note, however, that the phase shift is usually quite small; for dry earth, for example, at 10 MHz, $\phi_\eta \simeq 0.172°$.

The use of the approximate expressions for α, β, and η_c for a low-loss dielectric is illustrated in Example 8-8.

Example 8-8: Uniform plane wave in gallium arsenide. A uniform plane wave of frequency 10 GHz propagates in a sufficiently large sample of gallium arsenide (GaAs, $\epsilon_r \simeq 12.9$, $\mu_r = 1$, $\tan\delta_c \simeq 5 \times 10^{-4}$), which is a commonly used substrate material for high-speed solid-state circuits. Find (a) the attenuation constant α in np-m^{-1}, (b) phase velocity v_p in m-s^{-1}, and (c) intrinsic impedance η_c in Ω.

Solution: Since $\tan\delta_c = 5 \times 10^{-4} \ll 1$, we can use the approximate expressions for a good dielectric.

(a) We have

$$\alpha = \frac{\sigma}{2}\sqrt{\frac{\mu}{\epsilon}} = \frac{\omega\epsilon\tan\delta_c}{2}\sqrt{\frac{\mu}{\epsilon}}$$

$$= \frac{2\pi \times 10^{10} \times 5 \times 10^{-4}}{2}\sqrt{\mu\epsilon}$$

$$= \frac{2\pi \times 10^{10} \times 5 \times 10^{-4}\sqrt{\mu_r\epsilon_r}\sqrt{\mu_0\epsilon_0}}{2}$$

$$= \frac{2\pi \times 10^{10} \times 5 \times 10^{-4}}{2 \times 3 \times 10^8}\sqrt{12.9} \simeq 0.188 \text{ np-m}^{-1}$$

(b) Since phase velocity $v_p = \omega/\beta$ where $\beta \simeq \omega\sqrt{\mu\epsilon}$, we have $v_p \simeq 1/\sqrt{\mu\epsilon} = 3 \times 10^8/\sqrt{12.9} \simeq 8.35 \times 10^7$ m-s^{-1}. Note that the phase velocity is ~3.59 times slower than that in air.

(c) The intrinsic impedance $\eta_c \simeq \sqrt{\mu/\epsilon} = 377/\sqrt{12.9} \simeq 105\Omega$. Note that the intrinsic impedance is ~3.59 times smaller than that in air.

Uniform Plane Wave Propagation in a Good Conductor Another special case of wave propagation in lossy media that is of practical importance involves propagation in nearly perfect conductors. Many transmission lines and guiding systems are constructed using metallic conductors, such as copper, aluminum, or silver. In most cases these conductors are nearly perfect, so that the losses are relatively small. Nevertheless, these small losses often determine the range of applicability of the systems utilized and need to be quantitatively determined. Fortunately, approximate determination of the losses are perfectly adequate in most cases, using simplified formulas derived on the basis that the materials are good conductors; that is, that $\sigma \gg \omega\epsilon$.

For $\tan\delta_c \gg 1$, the propagation constant γ can be simplified as follows:

$$\gamma = \alpha + j\beta = \sqrt{(j\omega\mu\sigma)\left(1 + j\frac{\omega\epsilon}{\sigma}\right)} \simeq \sqrt{j\omega\mu\sigma} = \sqrt{\omega\mu\sigma}e^{j45^\circ}$$

where we have used the fact that $\sqrt{j} = e^{j\pi/4}$. The real and imaginary parts (i.e., the attenuation and phase constants) of γ for a good conductor are thus equal:

$$\boxed{\alpha = \beta \simeq \sqrt{\frac{\omega\mu\sigma}{2}}}$$

The phase velocity and wavelength can be obtained from β, as $v_p = \sqrt{2\omega/\mu\sigma)}$ and $\lambda = 2\sqrt{\pi/(f\mu\sigma)}$. Since both α and β are proportional to $\sigma^{1/2}$ and σ is large, it appears that uniform plane waves not only are attenuated heavily but also undergo a significant phase shift per unit length as they propagate in a good conductor. The phase velocity of the wave and the wavelength are both proportional to $\sigma^{-1/2}$ and are therefore significantly smaller than the corresponding values in free space.

For example, for copper ($\sigma = 5.8 \times 10^7$ S-m^{-1}) at 300 MHz, we have $v_p \simeq$ 7192 m-s^{-1}, and $\lambda \simeq 0.024$ mm, which are much smaller than the free-space values of $c \simeq 3 \times 10^8$ m-s^{-1} and $\lambda \simeq 1$ m at 300 MHz. At 60 Hz, the values for copper are even more dramatic, being $v_p \simeq 3.22$ m-s^{-1} and $\lambda \simeq 53.6$ mm, compared to the free-space wavelength of ~5000 km. As an example of a nonmetallic conductor, for seawater ($\epsilon = 81\epsilon_0$, $\sigma = 4$ S-m^{-1}), at 10 kHz we have $v_p \simeq 1.58 \times 10^5$ m-s^{-1} and $\lambda \simeq 15.8$ m, compared to a free-space wavelength of 30 km. It is interesting to note here that in the context of undersea (submarine) communications, a half-wavelength antenna (i.e., a reasonably efficient radiator) operating in seawater at 10 kHz has to be only ~7.91 m long, whereas a half-wavelength antenna for a ground-based very low frequency (VLF) transmitter radiating at 10 kHz in air would have to be ~15 km long. In practice, VLF communication and navigation transmitters use electrically short (a few hundred meters high) vertical monopole antennas above ground planes and/or in top-loaded fashion.[27]

[27] A. D. Watt, *VLF Radio Engineering,* Pergamon Press, New York, 1967.

Using [8.22] we can express the intrinsic impedance for $\sigma \gg \omega\epsilon$ as

$$\eta_c = \sqrt{\frac{\mu}{\epsilon}} \frac{1}{\left(1 - j\dfrac{\sigma}{\omega\epsilon}\right)^{1/2}} \simeq \sqrt{\frac{j\omega\mu}{\sigma}} = \sqrt{\frac{\mu\omega}{\sigma}} e^{j45^\circ}$$

The magnitude of η_c for good conductors is typically quite small (proportional to $\sigma^{-1/2}$), and the medium presents an inductive impedance (i.e., $\overline{\mathcal{H}}$ lags $\overline{\mathcal{E}}$). The phase of the intrinsic impedance (ϕ_η) is very nearly 45° for all good conductors. For copper at 300 MHz, the magnitude of the intrinsic impedance is $\sim 6.39 \times 10^{-3}\Omega$, compared to 377Ω in free space. Thus, for a given uniform plane-wave electric field intensity, the corresponding wave magnetic field in copper is $\sim 6 \times 10^4$ times as large as that in free space. For seawater at 10 kHz, $|\eta_c| \simeq 0.14\Omega$, which is a factor of ~2700 times smaller than the free-space value of 377Ω.

A useful parameter used in assessing the degree to which a good conductor is lossy and the degree to which electromagnetic waves can penetrate into a good conductor is the so-called *skin depth* δ, defined as the depth in which the wave is attenuated to $1/e$ (or ~36.8%) of its original intensity (i.e., its value immediately below the surface if a wave is incident on the material from above). The skin depth[28] is given by

$$\delta \equiv \frac{1}{\alpha} \simeq \sqrt{\frac{2}{\omega\mu\sigma}} = \frac{1}{\sqrt{\pi f \mu\sigma}} \qquad [8.26]$$

For example, for copper at 300 MHz, $\delta \simeq 0.00382$ mm $= 3.82$ μm, which is a very small distance both in absolute terms and also as compared to the wavelength ($\lambda \simeq 0.024$ mm $= 24$ μm, see above). The nature of the wave propagation in a good conductor, as shown in Figure 8.10, is quite different from the case of that in a good dielectric shown in Figure 8.9. In a good conductor, since $\lambda = 2\pi\delta$, the wave attenuates rapidly with distance and reaches negligible amplitudes after traveling only a fraction of a wavelength. Note that the skin depth in good conductors such as copper is very small even at audio or low radio frequencies (tens to hundreds of kHz). For example, at 30 kHz, we have $\delta \simeq 0.382$ mm, which means that any time-varying fields and current densities only exist in a thin layer on the surface. We shall see in Section 8.3 that in typical transmission lines (e.g., a coaxial line), electromagnetic power propagates in the dielectric region between the conductors, typically at the velocity of light in free space (or at the phase velocity appropriate for the dielectric medium). The electromagnetic wave is guided by the conductor structure

[28]Note that the definition of the skin depth δ as being equal to $1/\alpha$ is identical to that of the penetration depth d in [8.25]. However, the term "skin depth" is exclusively used for good conductors and emphasizes that the penetration is confined to a thin region on the skin of the conductor.

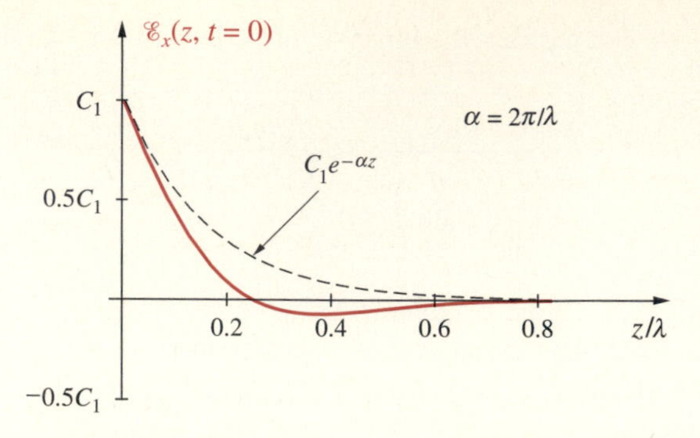

FIGURE 8.10. Good conductor. The electric field component of a uniform plane wave in a good conductor.

and penetrates into the conductor only to the degree that the conductor is imperfect. For example, in the case of a coaxial line, the propagation characteristics of the wave are not different in the case of a solid inner metallic conductor versus one that consists of a thin cylindrical shell of metallic conductor that is hollow or filled with any other material (conductor or insulator).

Examples 8-9, 8-10, and 8-11 quantitatively illustrate skin depth values in different cases.

Example 8-9: Comparison of skin depths. Calculate the skin depth in each of the following nonmagnetic ($\mu = \mu_0$) media at 10 kHz: Seawater ($\epsilon_r = 81$, $\sigma = 4$ S-m^{-1}); wet earth ($\epsilon_r = 10$, $\sigma = 10^{-2}$ S-m^{-1}); and dry earth ($\epsilon_r = 3$, $\sigma = 10^{-4}$ S-m^{-1}).

Solution: Note that for any medium, the loss tangent is $\tan\delta_c = \sigma/(\omega\epsilon)$. For seawater we have

$$\tan\delta_c = \frac{4 \text{ S-m}^{-1}}{(2\pi \times 10^4 \text{ rad-s}^{-1})(81 \times 8.85 \times 10^{-12} \text{ F-m}^{-1})} \simeq 8.88 \times 10^4$$

Similarly, for wet earth we find $\tan\delta_c \simeq 1.8 \times 10^3$, and for dry earth $\tan\delta_c \simeq 59.9$. Thus, at 10 kHz, all three media can be considered to be good conductors, and we can use the skin depth formula given by [8.26]. For seawater at 10 kHz we then have

$$\delta = \frac{1}{\sqrt{(\pi \times 10^4 \text{ rad-s}^{-1})(4\pi \times 10^{-7} \text{ H-m}^{-1})(4 \text{ S-m}^{-1})}} \simeq 2.52 \text{ m}$$

Similarly, the skin depths for wet and dry earth can be found to be, respectively, ~50.3 m and ~503 m. The relatively large skin depths facilitate the use of VLF frequencies (3–30 kHz) for various applications such as undersea communication, search for buried objects, and geophysical prospecting. Note that the skin depths are even larger for the ELF, SLF, and ULF ranges of frequencies (see Table 8.1), in the range 3 Hz to 3 kHz. These relatively large penetration distances facilitate the use of these frequencies for geophysical prospecting.

Example 8-10: Propagation of radio waves through lake water versus seawater. Consider the propagation of a 10 MHz radio wave in lake water (assume $\sigma = 4 \times 10^{-3}$ S-m^{-1} and $\epsilon_r = 81$) versus seawater ($\sigma = 4$ S-m^{-1} and $\epsilon_r = 81$). Calculate the attenuation constant and the penetration depth in each medium and comment on the difference. Note that both media are nonmagnetic.

Solution: For lake water, the loss tangent at 10 MHz is given by

$$\tan\delta_c = \frac{\sigma}{\omega\epsilon_r\epsilon_0} = \frac{4 \times 10^{-3}\text{ S-m}^{-1}}{2\pi \times 10^7\text{ rad-s}^{-1} \times 81 \times 8.85 \times 10^{-12}\text{ F-m}^{-1}} \simeq 8.88 \times 10^{-2} \ll 1$$

Therefore, at 10 MHz, lake water behaves like a low-loss dielectric. The penetration depth for lake water is

$$d = \frac{1}{\alpha} \simeq \frac{2}{\sigma}\sqrt{\frac{\epsilon_r\epsilon_0}{\mu_0}} = \left[\frac{2}{4 \times 10^{-3}\text{ S-m}^{-1}}\right]\left[\frac{\sqrt{81}}{377\Omega}\right] \simeq 11.9\text{ m} \simeq 12\text{ m}$$

For seawater, the loss tangent at 10 MHz is given by

$$\tan\delta_c = \frac{4\text{ S-m}^{-1}}{2\pi \times 10^7\text{ rad-s}^{-1} \times 81 \times 8.85 \times 10^{-12}\text{ F-m}^{-1}} \simeq 88.8 \gg 1$$

Therefore, at 10 MHz, seawater behaves like a good conductor. The skin depth for seawater is

$$\delta = \frac{1}{\alpha} \simeq \sqrt{\frac{2}{\omega\mu\sigma}}$$

$$= \frac{1}{\sqrt{\pi \times 10^7\text{ rad-s}^{-1} \times 4\pi \times 10^{-7}\text{ H-m}^{-1} \times 4\text{ S-m}^{-1}}}$$

$$= \frac{1}{4\pi}\text{ m} \simeq 8\text{ cm}$$

The results clearly indicate that communication through seawater at 10 MHz is not feasible, since the skin depth is only 8 cm. However, 10 MHz electromagnetic waves do penetrate lake water, making communication possible at relatively shallow depths of tens of meters.

Example 8-11: VLF waves in the ocean. The electric field component of a uniform plane VLF electromagnetic field propagating vertically down in the z direction in the ocean ($\sigma = 4$ S-m^{-1}, $\epsilon_r = 81$, $\mu_r = 1$) is approximately given by

$$\overline{\mathscr{E}}(z, t) = \hat{\mathbf{x}}E_0 e^{-\alpha z}\cos(6\pi \times 10^3 t - \beta z) \quad \text{V-m}^{-1}$$

where E_0 is the electric field amplitude at $z = 0^+$ immediately below the air-ocean interface, at $z = 0$. Note that the frequency of the wave is 3 kHz. (a) Find the attenuation constant α and phase constant β. (b) Find the wavelength λ, phase velocity v_p, skin depth δ, and intrinsic impedance η_c, and compare them to their values in air.

(c) Write the instantaneous expression for the corresponding magnetic field, $\overline{\mathcal{H}}(z, t)$. (d) A submarine located at a depth of 100 m has a receiver antenna capable of measuring electric fields with amplitudes of 1 μV-m^{-1} or greater. What is the minimum required electric field amplitude immediately below the ocean surface (i.e., E_0) in order to communicate with the submarine? What is the corresponding value for the amplitude of the magnetic field?

Solution: At $\omega = 6\pi \times 10^3$ rad-s^{-1}, the loss tangent of the ocean is given by

$$\tan\delta_c = \frac{\sigma}{\omega\epsilon_r\epsilon_0} = \frac{4}{6\pi \times 10^3 \text{ rad/s} \times 81 \times (8.85 \times 10^{-12} \text{ F/m})} \simeq 2.96 \times 10^5 \gg 1$$

Therefore, the ocean acts as a good conductor at $f = 3$ kHz.

(a) Using the approximate expressions we have

$$\alpha = \beta = \sqrt{\frac{\omega\mu\sigma}{2}} = \sqrt{\frac{(6\pi \times 10^3 \text{ rad-s}^{-1})(4\pi \times 10^{-7} \text{ H-m}^{-1})(4 \text{ S-m}^{-1})}{2}}$$

$$\simeq 0.218 \text{ np-m}^{-1} \text{or rad-m}^{-1}$$

(b) The wavelength in the ocean is

$$\lambda = \frac{2\pi}{\beta} \simeq \frac{2\pi}{0.218} \simeq 28.9 \text{ m}$$

Since in air $\lambda \simeq 100$ km at 3 kHz, the wavelength in the ocean is approximately 3464 times smaller than that in air. The phase velocity is

$$v_p = f\lambda \simeq 3 \times 10^3 \times 28.9 \simeq 8.66 \times 10^4 \text{ m-s}^{-1}$$

Compared to $c \simeq 3 \times 10^8$ m/s in air, the phase velocity in the ocean at 3 kHz is approximately ~3464 times smaller. The skin depth and the intrinsic impedance in the ocean are

$$\delta = 1/\alpha \simeq 4.59 \text{ m}$$

$$\eta_c = |\eta_c|e^{j\phi_\eta} \simeq \sqrt{\frac{\mu\omega}{\sigma}}e^{j45^\circ}$$

$$= \sqrt{\frac{(4\pi \times 10^{-7} \text{ H-m}^{-1})(6\pi \times 10^3 \text{ rad-s}^{-1})}{4 \text{ S-m}^{-1}}}e^{j45^\circ} \simeq 7.70 \times 10^{-2}e^{j45^\circ}\Omega$$

Compared to $\eta \simeq 377\Omega$ in air, the magnitude of η_c in the ocean is approximately ~4900 times smaller. In addition, η_c in the ocean is a complex number with a phase angle of ~45°.

(c) The corresponding magnetic field is given by

$$\overline{\mathcal{H}}(z, t) = \hat{\mathbf{y}}\frac{E_0}{|\eta_c|}e^{-\alpha z}\cos(6\pi \times 10^3 t - \beta z - \phi_\eta)$$

$$\simeq 13E_0 e^{-0.218z}\cos(6\pi \times 10^3 t - 0.218z - \pi/4) \text{ A-m}^{-1}$$

(d) At $z_1 = 100$ m, the amplitude of the electric field is given by

$$E_0 e^{-\alpha z_1} \simeq E_0 e^{-(0.218)(100)} \geq E_{\min} = 1\ \mu\text{V-m}^{-1}$$

Therefore, the minimum value for E_0 to establish communication with the submarine located at a depth of 100 m is

$$E_0 \cong 2.84\ \text{kV-m}^{-1}$$

The corresponding value for H_0 is

$$H_0 = \frac{E_0}{|\eta_c|} \simeq 13E_0 \simeq 36.9\ \text{kA-m}^{-1}$$

8.1.4 Electromagnetic Energy Flow and the Poynting Vector

Electromagnetic waves carry power through space, transferring energy and momentum from one set of charges and currents (i.e., the sources that generated them) to another (those at the receiving points). Our goal in this section is to derive (directly from Maxwell's equations) a simple relation between the rate of this energy transfer and the electric and magnetic fields $\overline{\mathscr{E}}$ and $\overline{\mathscr{H}}$. We can in fact attribute definite amounts of energy and momentum to each elementary volume of space occupied by the electromagnetic fields. The energy and momentum exchange between waves and particles is described by the Lorentz force equation, discussed in Section 6.10.1. When fields exert forces on the charges, the charges respond to the force (i.e., they move) and gain energy, which must be at the expense of the energy lost by the fields. It is thus clear that electromagnetic fields store energy that is available to charged particles when the fields exert forces on them. Since energy is transmitted to the particles at a specific rate, we can think of the fields as transmitting power.

Flow of Electromagnetic Energy and Poynting's Theorem We know from previous chapters that the densities of energy stored in static electric and magnetic fields in linear and isotropic media are, respectively, $\frac{1}{2}\epsilon E^2$ and $\frac{1}{2}\mu H^2$. When these fields vary with time, the associated stored energy densities can also be assumed to vary with time. If we consider a given volume of space, electromagnetic energy can be transported into or out of it by electromagnetic waves depending on whether the source of waves is inside or outside the volume. In addition, electromagnetic energy can be stored in the volume in the form of electric and magnetic fields, and electromagnetic power can be dissipated in it in the form of Joule heating (i.e., as represented by $\overline{\mathscr{E}} \cdot \overline{\mathscr{J}}$). Poynting's theorem concerns the balance between power flow into or out of a given volume and the rate of change of stored energy and power dissipation within it. This theorem provides the physical framework by which we can express the flow of electromagnetic power in terms of the fields $\overline{\mathscr{H}}$ and $\overline{\mathscr{E}}$.

We now proceed to find an expression for electromagnetic power in terms of the field quantities $\overline{\mathscr{E}}$ and $\overline{\mathscr{H}}$. Since we expect the power flow to be related to the volume density of power dissipation represented by $\overline{\mathscr{E}} \cdot \overline{\mathscr{J}}$, we can start with [8.1$c$]:

$$\overline{\mathcal{J}} = \nabla \times \overline{\mathcal{H}} - \frac{\partial \overline{\mathcal{D}}}{\partial t}$$

and take the dot product of both sides with $\overline{\mathcal{E}}$ to obtain $\overline{\mathcal{E}} \cdot \overline{\mathcal{J}}$ in units of W-m^{-3}:

$$\overline{\mathcal{E}} \cdot \overline{\mathcal{J}} = \overline{\mathcal{E}} \cdot (\nabla \times \overline{\mathcal{H}}) - \overline{\mathcal{E}} \cdot \frac{\partial \overline{\mathcal{D}}}{\partial t}$$

and use the vector identity

$$\nabla \cdot (\overline{\mathcal{E}} \times \overline{\mathcal{H}}) \equiv \overline{\mathcal{H}} \cdot (\nabla \times \overline{\mathcal{E}}) - \overline{\mathcal{E}} \cdot (\nabla \times \overline{\mathcal{H}})$$

to find

$$\overline{\mathcal{E}} \cdot \overline{\mathcal{J}} = \overline{\mathcal{H}} \cdot (\nabla \times \overline{\mathcal{E}}) - \nabla \cdot (\overline{\mathcal{E}} \times \overline{\mathcal{H}}) - \overline{\mathcal{E}} \cdot \frac{\partial \overline{\mathcal{D}}}{\partial t} \qquad [8.27]$$

We now use [8.1*a*],

$$\nabla \times \overline{\mathcal{E}} = -\frac{\partial \overline{\mathcal{B}}}{\partial t}$$

and substitute in [8.27] to obtain

$$\overline{\mathcal{E}} \cdot \overline{\mathcal{J}} = -\overline{\mathcal{H}} \cdot \frac{\partial \overline{\mathcal{B}}}{\partial t} - \overline{\mathcal{E}} \cdot \frac{\partial \overline{\mathcal{D}}}{\partial t} - \nabla \cdot (\overline{\mathcal{E}} \times \overline{\mathcal{H}}) \qquad [8.28]$$

We now consider the two terms with the time derivatives. For a simple medium[29] (i.e., ϵ and μ simple constants) we can write

$$\overline{\mathcal{H}} \cdot \frac{\partial \overline{\mathcal{B}}}{\partial t} = \overline{\mathcal{H}} \cdot \frac{\partial (\mu \overline{\mathcal{H}})}{\partial t} = \frac{1}{2} \frac{\partial (\mu \overline{\mathcal{H}} \cdot \overline{\mathcal{H}})}{\partial t} = \frac{\partial}{\partial t}\left(\frac{1}{2}\mu |\overline{\mathcal{H}}|^2\right)$$

and

$$\overline{\mathcal{E}} \cdot \frac{\partial \overline{\mathcal{D}}}{\partial t} = \overline{\mathcal{E}} \cdot \frac{\partial (\epsilon \overline{\mathcal{E}})}{\partial t} = \frac{1}{2} \frac{\partial (\epsilon \overline{\mathcal{E}} \cdot \overline{\mathcal{E}})}{\partial t} = \frac{\partial}{\partial t}\left(\frac{1}{2}\epsilon |\overline{\mathcal{E}}|^2\right)$$

respectively. With these substitutions, [8.28] can be rewritten as

$$\overline{\mathcal{E}} \cdot \overline{\mathcal{J}} = -\frac{\partial}{\partial t}\left(\frac{1}{2}\mu |\overline{\mathcal{H}}|^2\right) - \frac{\partial}{\partial t}\left(\frac{1}{2}\epsilon |\overline{\mathcal{E}}|^2\right) - \nabla \cdot (\overline{\mathcal{E}} \times \overline{\mathcal{H}}) \qquad [8.29]$$

---

[29]The more general form of Poynting's theorem, without these assumptions about the medium, is

$$\int_V \overline{\mathcal{E}} \cdot \overline{\mathcal{J}}\, dv = -\frac{\partial}{\partial t} \int_V \left(\overline{\mathcal{H}} \cdot \frac{\partial \overline{\mathcal{B}}}{\partial t} + \overline{\mathcal{E}} \cdot \frac{\partial \overline{\mathcal{D}}}{\partial t}\right) dv - \oint_S (\overline{\mathcal{E}} \times \overline{\mathcal{H}}) \cdot d\mathbf{s}$$

which is valid even for nonlinear media, as long as there are no hysteresis effects. For ferromagnetic materials, for which the relation between $\overline{\mathcal{B}}$ and $\overline{\mathcal{H}}$ is often multivalued due to hysteresis, an additional amount of energy is deposited within the material.

Integrating [8.29] over an arbitrary volume V we have

$$\int_V \overline{\mathscr{E}} \cdot \overline{\mathscr{J}}\, dv = -\frac{\partial}{\partial t}\int_V \left(\frac{1}{2}\mu|\overline{\mathscr{H}}|^2 + \frac{1}{2}\epsilon|\overline{\mathscr{E}}|^2\right) dv - \int_V \nabla \cdot (\overline{\mathscr{E}} \times \overline{\mathscr{H}})\, dv \qquad [8.30]$$

Using the divergence theorem on the last term in [8.30] we find

$$\boxed{\int_V \overline{\mathscr{E}} \cdot \overline{\mathscr{J}}\, dv = -\frac{\partial}{\partial t}\int_V \left(\frac{1}{2}\mu|\overline{\mathscr{H}}|^2 + \frac{1}{2}\epsilon|\overline{\mathscr{E}}|^2\right) dv - \oint_S (\overline{\mathscr{E}} \times \overline{\mathscr{H}}) \cdot d\mathbf{s}} \qquad [8.31]$$

where the surface S encloses the volume V.

We can now interpret the various terms in [8.31] physically. The left-hand term is the generalization of Joule's law and represents the instantaneous power dissipated in the volume V. If $\overline{\mathscr{E}}$ is the electric field that produces $\overline{\mathscr{J}}$ in a lossy medium, this term represents the ohmic (I^2R) power loss in the medium. Note that in a simple isotropic medium (i.e., σ a simple constant), $\overline{\mathscr{E}}$ and $\overline{\mathscr{J}}$ are in the same direction. However, in general this is not true. In Earth's ionosphere (which is an anisotropic medium because of the presence of Earth's magnetic field), for example, an applied electric field in one direction can cause current flow in that direction as well as in other directions. Even in such cases, however, $\overline{\mathscr{E}} \cdot \overline{\mathscr{J}}$ still represents the power dissipated per unit volume, although $\overline{\mathscr{J}}$ and $\overline{\mathscr{E}}$ are not parallel. Alternatively, there could be an energy source within the volume V, such as an antenna carrying current, in which case $\overline{\mathscr{E}} \cdot \overline{\mathscr{J}}$ is negative and represents power flow out of that region.

The first term on the right-hand side represents the rate at which the electromagnetic energy stored in volume V decreases (negative sign), with the terms $\frac{1}{2}\mu|\overline{\mathscr{H}}|^2$ and $\frac{1}{2}\epsilon|\overline{\mathscr{E}}|^2$ representing, respectively, the magnetic and electric energy densities. Note that, strictly speaking, the quantities $W_e = \frac{1}{2}\epsilon|\overline{\mathscr{E}}|^2$ and $W_m = \frac{1}{2}\mu|\overline{\mathscr{H}}|^2$ are known to represent electric and magnetic energy densities for static fields. However, it is generally assumed that these quantities also represent stored energy densities in the case of time-varying fields.[30]

From conservation of energy, the last term in [8.31] must represent the flow of energy inward or outward through the surface S enclosing the volume V. Thus, the vector $\overline{\mathscr{S}} = \overline{\mathscr{E}} \times \overline{\mathscr{H}}$, which has dimensions of watts per square meter (W-m^{-2}), appears to be a measure of the rate of energy flow per unit area at any point on the surface S. The direction of power flow is perpendicular to both $\overline{\mathscr{E}}$ and $\overline{\mathscr{H}}$. In other words, the power density in an electromagnetic wave is given by

[30]Such an assumption is entirely reasonable since the energy density is defined at a given point. From another point of view, we can consider Poynting's theorem or [8.31] to be the definition of energy density for time-varying fields. The correct amount of total electromagnetic energy is always obtained by assigning an amount $\frac{1}{2}(\overline{\mathscr{B}} \cdot \overline{\mathscr{H}} + \overline{\mathscr{D}} \cdot \overline{\mathscr{E}}) = \frac{1}{2}(\epsilon|\overline{\mathscr{E}}|^2 + \mu|\overline{\mathscr{H}}|^2)$ to each unit of volume. Other previously cited expressions for static energy densities, such as $\frac{1}{2}\rho\Phi$ for electrostatic fields, are not applicable for time-varying fields. See Chapter II-27 of R. P. Feynman, R. B. Leighton, and M. Sands, *The Feynman Lectures on Physics,* Addison-Wesley, 1964, and Section 2.19 of J. A. Stratton, *Electromagnetic Theory,* McGraw-Hill, 1941.

$$\boxed{\overline{\mathcal{S}} = \overline{\mathcal{E}} \times \overline{\mathcal{H}}}$$

Equation [8.31] is known as *Poynting's theorem,* and the vector $\overline{\mathcal{S}}$ is known as the *Poynting vector.* In words, we have, "Electromagnetic power flow into a closed surface at any instant equals the sum of the time rates of increase of the stored electric and magnetic energies plus the ohmic power dissipated (or electric power generated, if the surface encloses a source) within the enclosed volume."

Poynting, John Henry *(b. Sept 9, 1852; d. March 10, 1914) Poynting was a professor of physics at Mason Science College, later the University of Birmingham, from 1880 until his death. In papers published in 1884 and in 1885, he showed that the flow of energy at a point can be expressed by a simple formula in terms of the electric and magnetic forces at that point. This is Poynting's theorem. The value assigned to the rate of power flow of electrical energy is known as Poynting's vector. He also wrote papers on radiation and the pressure of light. After 12 years of experiments, he determined the mean density of the Earth in 1891, and in 1893 he determined the gravitational constant, a measure of the effect of gravity. He published his results in* The Mean Density of the Earth *(1894) and* The Earth: Its Shape, Size, Weight and Spin *(1893). [Adapted in part from* Encyclopaedia Britannica, *14th ed., 1929, vol. 18, p. 396.]*

Note that the interpretation of $\overline{\mathcal{S}}$ as the local power flux vector does not follow from [8.31] with mathematical rigor. In principle, any vector $\overline{\mathcal{Y}}$ for which the integral over the closed surface S is zero (as is true for any vector that is the curl of another vector)[31] can be added to $\overline{\mathcal{E}} \times \overline{\mathcal{H}}$ without changing the result. Similar concerns may be raised about the interpretation of the terms $\frac{1}{2}\epsilon|\overline{\mathcal{E}}|^2$ and $\frac{1}{2}\mu|\overline{\mathcal{H}}|^2$ as energy densities. In general, Poynting's theorem makes physical sense in terms of its integral form, enabling us to describe the net flow of electromagnetic power through a closed surface. However, we run into difficulty when we try to describe where the energy resides.[32] This problem is reminiscent of the potential energy of a raised object, which is given by the weight times the height above the ground, but for which we also find it difficult to determine *where* the energy actually resides. The fact that it is neither necessary nor possible to "determine" the location of energy definitively was also discussed in Sections 4.12 and 7.3.

The work done per unit time per unit volume by the fields (i.e., $\overline{\mathcal{J}} \cdot \overline{\mathcal{E}}$) is a conversion of electromagnetic energy into mechanical energy or heat. Since matter ultimately consists of charged particles (electrons and nuclei), we can think of this rate of conversion as a rate of increase of energy of the charged particles per unit volume.

[31]Let $\overline{\mathcal{Y}} = \nabla \times \overline{\mathcal{X}}$. Then $\oint_S \overline{\mathcal{Y}} \cdot d\mathbf{s} = \int_V \nabla \cdot (\nabla \times \overline{\mathcal{X}})\, dv \equiv 0$, since $\nabla \cdot \nabla \times \overline{\mathcal{X}} \equiv 0$ for any vector $\overline{\mathcal{X}}$.

[32]For a qualitative discussion see Section II-27-4 of *The Feynman Lectures on Physics,* Addison Wesley, 1964.

In this sense, Poynting's theorem for the microscopic fields can be interpreted as a statement of conservation of energy of the combined system of particles and fields.

Examples 8-12, 8-13, and 8-14 illustrate the application of Poynting's theorem to determine electromagnetic power flow in a wire carrying direct current, in a coaxial transmission line, and in a parallel-plate capacitor.

Example 8-12: Wire carrying direct current. Consider a long cylindrical conductor of conductivity σ and radius a, carrying a direct current I as shown in Figure 8.11. Find the power dissipated in a portion of the wire of length l, using (a) the left-hand side of [8.31]; (b) the right-hand side of [8.31]. Note that the cross-sectional area of the wire is $A = \pi a^2$.

Solution:

(a) The current density and electric field within the conductor are respectively $\overline{\mathcal{J}} = \hat{\mathbf{z}}(I/A)$ and $\overline{\mathcal{E}} = \hat{\mathbf{z}}I(\sigma A)$. We can evaluate the left-hand side of [8.31] for a cylindrical volume V of arbitrary radius greater than a as shown in Figure 8.11. Noting that the current density is zero outside the wire, the limits of integration for r need extend only up to $r = a$, and we have

$$\int_V \overline{\mathcal{E}} \cdot \overline{\mathcal{J}}\, dv = \int_0^l \int_0^{2\pi} \int_0^a \left(\frac{I}{\sigma A}\hat{\mathbf{z}}\right) \cdot \left(\hat{\mathbf{z}}\frac{I}{A}\right) r\, dr\, d\phi\, dz$$

$$= \frac{I^2}{\sigma A^2}(\pi a^2 l) = \frac{I^2 l}{\sigma A} = I^2 R$$

where $R = l/(\sigma A)$ is the resistance of the cylindrical conductor. Thus, the left-hand side of [8.31] represents the total ohmic losses in the volume V.

(b) Since the current I and thus the associated electric and magnetic fields are static ($\partial/\partial t = 0$), the rate of increase of the electric and magnetic energies within the conductor is zero. Thus, the right-hand side of [8.31] reduces to the surface integral of the $\overline{\mathcal{E}} \times \overline{\mathcal{H}}$ term. Using Ampère's law with a contour of radius $r \geq a$, the magnetic field surrounding the conductor is

$$\overline{\mathcal{H}} = \hat{\phi}\frac{I}{2\pi r}$$

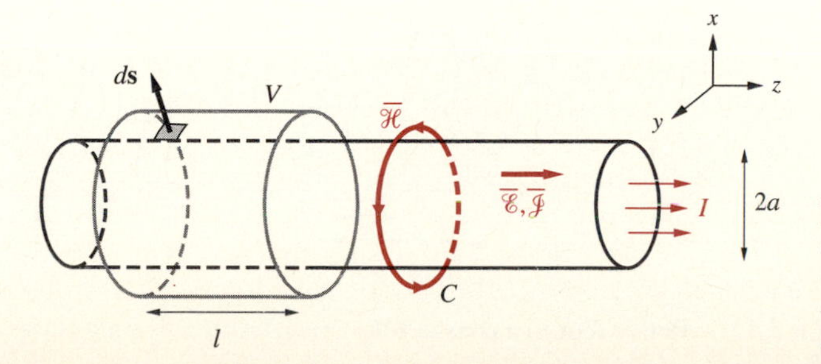

FIGURE 8.11. Long straight wire. The power dissipated in a cylindrical conductor carrying a direct current I is given by I^2R.

At the surface of the conductor ($r = a$), we thus have $\overline{\mathcal{H}} = \hat{\boldsymbol{\phi}} I/(2\pi a)$, so that the Poynting vector is

$$\overline{\mathcal{S}} = \overline{\mathcal{E}} \times \overline{\mathcal{H}} = \frac{I}{\sigma A}\hat{\mathbf{z}} \times \hat{\boldsymbol{\phi}}\frac{I}{2\pi a} = -\hat{\mathbf{r}}\frac{I^2}{2\pi a \sigma A}$$

which is directed radially inwards toward the surface of the wire. The right-hand side of [8.31] is simply the negative of the integral of $\overline{\mathcal{S}}$ over the closed surface of the cylindrical wire (note that there is no contribution from the ends of the cylinder, since $\overline{\mathcal{S}}$ is parallel to those surfaces):

$$-\oint_S \overline{\mathcal{S}} \cdot d\mathbf{s} = -\int_0^l \int_0^{2\pi} \frac{I^2}{2\pi a \sigma A}(-\hat{\mathbf{r}}) \cdot \hat{\mathbf{r}} a \, d\phi \, dz = \frac{I^2 l}{\sigma A} = I^2 R$$

which is equal to the left-hand side of [8.31] evaluated in part (a), thus verifying the Poynting theorem. Note here that if the wire were a perfect conductor ($\sigma = \infty$), then the electric field inside the wire (and thus just outside the wire) must be zero. In such a case, $\overline{\mathcal{E}} \times \overline{\mathcal{H}} = 0$ and there is no power flow into the wire, consistent with the fact that there is no power dissipation in the wire.

Example 8-13: Power flow in a coaxial line. Consider a coaxial line delivering power to a resistor as shown in Figure 8.12. Assume the wires to be perfect conductors, so that there is no power dissipation in the wires, and the electric field inside them is zero. The configurations of electric and magnetic field lines in the coaxial line are as shown, the electric field being due to the applied potential difference between the inner and outer conductors. Find the power delivered to the resistor R.

Solution: From Ampère's law we have

$$\overline{\mathcal{H}} = \hat{\boldsymbol{\phi}}\frac{I}{2\pi r} \qquad a \le r \le b$$

To find the electric field, we recall that the capacitance (per unit length) of a coaxial line is $C = (2\pi\epsilon)/\ln(b/a)$, so the applied voltage V induces a charge per unit length of $Q = CV$, from which we can find the E field using Gauss's law as

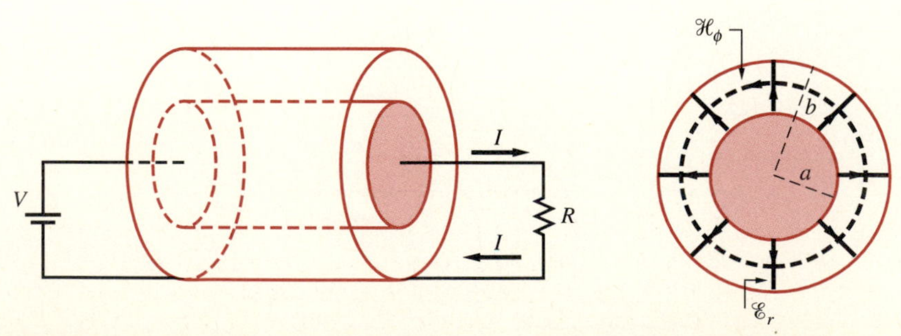

FIGURE 8.12. Power flow in a coaxial line. Coaxial wire delivering power to a resistor R.

$$\overline{\mathscr{E}} = \hat{\mathbf{r}}\frac{Q}{2\pi\epsilon r} = \hat{\mathbf{r}}\frac{V}{r\ln(b/a)}$$

Therefore, the power delivered through the cross-sectional area of the coaxial line can be found by integrating the Poynting vector $\overline{\mathscr{S}} = \overline{\mathscr{E}} \times \overline{\mathscr{H}}$ over the area

$$\oint_S \overline{\mathscr{E}} \times \overline{\mathscr{H}} \cdot d\mathbf{s} = \int_a^b \int_0^{2\pi} \frac{V}{r\ln(b/a)}\left(\frac{I}{2\pi r}\right)\hat{\mathbf{z}} \cdot (\hat{\mathbf{z}} r\, dr\, d\phi)$$

$$= \int_a^b \frac{V}{r\ln(b/a)}\left(\frac{I}{2\pi r}\right) 2\pi r\, dr = \frac{VI}{\ln(b/a)}\int_a^b \frac{dr}{r} = VI$$

as expected on a circuit theory basis. We note that the electromagnetic power is carried entirely outside the conductors. Even if the inner and outer conductors of the coaxial line were imperfect conductors, this simply leads to a component of the Poynting vector radially inward, as we saw in the previous example. No part of the axially directed Poynting flux (i.e., the power flux flowing along the coaxial line in the z direction) is carried within the conductors.

Example 8-14: Energy flow in a capacitor. Figure 8.13 shows a parallel-plate capacitor of capacitance $C = \epsilon A/d$, being charged by a current I flowing in

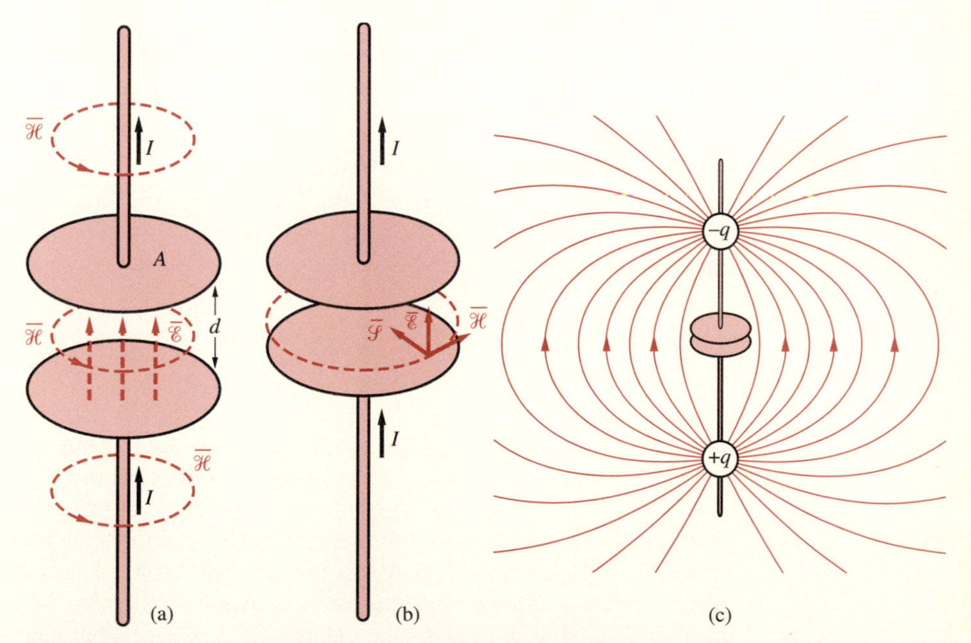

FIGURE 8.13. A parallel-plate capacitor being charged. (a) Magnetic fields encircling the conduction (outside the region between plates) or displacement (inside the region between the plates) currents. (b) Power evidently enters the storage region (i.e., the region between plates) through the sides via $\overline{\mathscr{E}} \times \overline{\mathscr{H}}$. (c) The intensification of the electric fields near the capacitor as two point charges come closer to the plates.

the connecting wires. When it is charged, an electric field $\mathscr{E}_z$ exists (taking the z axis to be along the current-carrying wire) between the capacitor plates, with $\frac{1}{2}\epsilon\mathscr{E}_z^2$ representing the stored energy. Investigate the manner in which the energy enters into the region between the capacitor plates.

Solution: Neglecting fringing fields, the electric field between the capacitor plates is uniform, so the total electrical energy stored between the plates is

$$W_e = \int_V \tfrac{1}{2}\epsilon\mathscr{E}_z^2 dv = (\tfrac{1}{2}\epsilon\mathscr{E}_z^2)Ad$$

The time rate of change of this energy is given by

$$\frac{dW_e}{dt} = Ad\frac{d}{dt}\left(\frac{1}{2}\epsilon\mathscr{E}_z^2\right) = \epsilon A d\mathscr{E}_z\frac{d\mathscr{E}_z}{dt}$$

which means that there must be a flow of energy into the volume between the plates from somewhere. Can it be arriving through the wire? Apparently not, according to Poynting's theorem, since we note that $\overline{\mathscr{E}}$ is perpendicular to the plates, and thus $\overline{\mathscr{E}} \times \overline{\mathscr{H}}$ must necessarily be parallel to the plates.

A magnetic field $\overline{\mathscr{H}}$ encircles the z axis, supported by the conduction current along the wires and the displacement current in the region between the plates, as shown in Figure 8.13a. In the region between the plates, where $\mathbf{J} = 0$, the magnetic field at any radial distance r is given by [7.16*c*] as

$$\oint_C \overline{\mathscr{H}}\cdot d\mathbf{l} = \int_S \frac{\partial(\epsilon\overline{\mathscr{E}})}{\partial t}\cdot d\mathbf{s} \quad\rightarrow\quad \mathscr{H}_\phi(2\pi r) = \epsilon(\pi r^2)\frac{\partial\mathscr{E}_z}{\partial t} \quad\rightarrow\quad \mathscr{H}_\phi = \frac{\epsilon r}{2}\frac{\partial\mathscr{E}_z}{\partial t}$$

Thus, Poynting's vector $\overline{\mathscr{S}} = \overline{\mathscr{E}} \times \overline{\mathscr{H}} = \mathscr{E}_z\hat{\mathbf{z}} \times \hat{\boldsymbol{\phi}}\mathscr{H}_\phi = \mathscr{E}_z\mathscr{H}_\phi(-\hat{\mathbf{r}})$ has only an r component, pointed inward, as shown in Figure 8.13b. The energy stored in the capacitor is apparently being supplied through the sides of the cylindrical region between the plates. The total power flow into the region can be found by integrating $\overline{\mathscr{S}}$ over the cylindrical side surface of the capacitor at $r = a$:

$$\oint_S \overline{\mathscr{S}}\cdot d\mathbf{s} = \int_0^{2\pi}\int_{z=0}^{d} \mathscr{E}_z\left(\frac{\epsilon a}{2}\frac{\partial\mathscr{E}_z}{\partial t}\right)a\,dz\,d\phi = \pi a^2 d\frac{\partial}{\partial t}\left(\frac{1}{2}\epsilon\mathscr{E}_z^2\right)$$

which is identical to the dW_e/dt expression we found above, where we note that $A = \pi a^2$. Thus, we see that the power supplied into the region from the sides is precisely equal to the time rate of change of stored energy.

This result is quite contrary to our intuitive expectations of the power being supplied to the capacitor via the wires. One way to think about this result[33] is illustrated in Figure 8.13c. Consider two opposite charges above and below the capacitor plates, approaching the capacitor plates via the wire. When the charges are far away from the plates, the electric field between the plates is weak but

[33]Taken from Section II-27-5 of R. P. Feynman, R. B. Leighton, and M. Sands, *The Feynman Lectures on Physics,* Addison Wesley, 1964.

spread out in the region surrounding the capacitor. As the charges approach the plates, the electric field between the capacitor plates becomes stronger (i.e., the field energy moves toward the capacitor) and in the limit is eventually completely confined to the region between the plates. Thus, as shown, the electromagnetic energy flows into the region between the capacitor plates through the openings on the sides.

Electromagnetic Power Carried by a Uniform Plane Wave in a Lossless Medium Consider the uniform plane wave propagating in a lossless medium that was studied in Section 8.1.2:

$$\mathscr{E}_x = E_0 \cos(\omega t - \beta z)$$

$$\mathscr{H}_y = \frac{1}{\eta} E_0 \cos(\omega t - \beta z)$$

where $\beta = \omega\sqrt{\mu\epsilon}$ and $\eta = \sqrt{\mu/\epsilon}$. The electric and magnetic fields of the wave are as shown in Figure 8.14a. The Poynting vector for this wave is given by

$$\overline{\mathscr{S}} = \overline{\mathscr{E}} \times \overline{\mathscr{H}} = \hat{\mathbf{z}} E_0 \left(\frac{E_0}{\eta}\right) \cos^2(\omega t - \beta z) = \hat{\mathbf{z}} \frac{E_0^2}{2\eta}[1 + \cos 2(\omega t - \beta z)]$$

We see from Figure 8.14b that $\overline{\mathscr{S}}$ everywhere is directed in the $+z$ direction. The power carried by the wave has a constant component and a component varying at a frequency twice that of the fields, exhibiting two positive peaks per wavelength of the wave, as shown in Figure 8.14b. In most cases, the constant component, called the time-average Poynting vector $\mathbf{S}_{\text{av}}$, is the quantity of interest. The quantity $\mathbf{S}_{\text{av}}$ is defined as the average over one period ($T_p = 2\pi/\omega$) of the wave of $\overline{\mathscr{S}}$. For a uniform plane wave in a lossless medium, we have

$$\mathbf{S}_{\text{av}} = \frac{1}{T_p}\int_0^{T_p} \overline{\mathscr{S}}(z,t)\,dt = \frac{1}{T_p}\int_0^{T_p} \hat{\mathbf{z}}\frac{E_0^2}{2\eta}[1 + \cos 2(\omega t - \beta z)]\,dt \quad \rightarrow \quad \boxed{\mathbf{S}_{\text{av}} = \hat{\mathbf{z}}\frac{E_0^2}{2\eta}} \qquad [8.32]$$

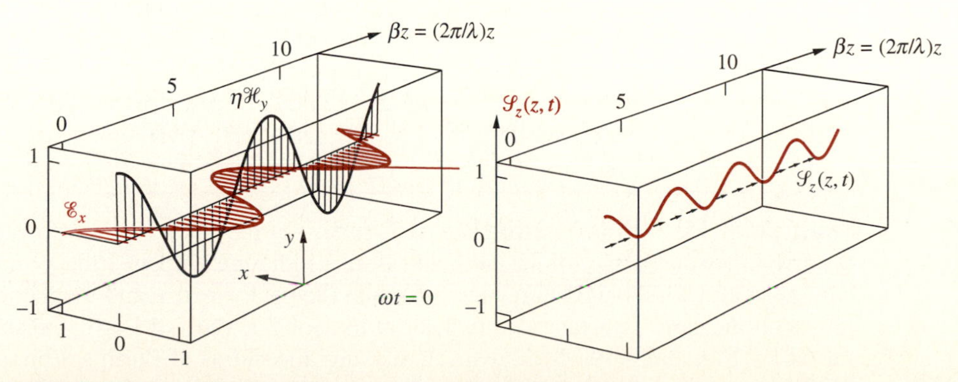

FIGURE 8.14. The Poynting vector of a uniform plane wave.

According to Poynting's theorem as expressed in [8.31], and since the medium is assumed to be lossless and source-free, so $\overline{\mathcal{J}} = 0$, the integral of the Poynting vector $\overline{\mathcal{S}}$ over any closed surface should be balanced by the time rate of change of the sum of stored electric and magnetic energies in the volume enclosed by that surface. The electric and magnetic energy densities for this uniform plane wave are

$$W_{\mathrm{e}} = \tfrac{1}{2}\epsilon|\mathscr{E}_x|^2 = \tfrac{1}{2}\epsilon E_0^2\cos^2(\omega t - \beta z) \qquad [8.33a]$$

$$W_{\mathrm{m}} = \frac{1}{2}\mu|\mathscr{H}_y|^2 = \frac{1}{2}\frac{\mu}{(\sqrt{\mu/\epsilon})^2}E_0^2\cos^2(\omega t - \beta z) = \frac{1}{2}\epsilon E_0^2\cos^2(\omega t - \beta z) \qquad [8.33b]$$

We thus note that $W_{\mathrm{e}} = W_{\mathrm{m}}$, indicating that the instantaneous values of the electric and magnetic stored energies are equal at all points in space and at all times. The reader is encouraged to choose a suitable closed surface and show that the volume integral of the partial derivative with respect to time of $W_{\mathrm{m}} + W_{\mathrm{e}}$ is indeed equal to the surface integral of $\overline{\mathcal{S}}$. Alternatively, one can consider the differential form of Poynting's theorem, as given in [8.29], which for $\overline{\mathscr{E}} \cdot \overline{\mathcal{J}} = 0$ reduces to

$$\frac{\partial}{\partial t}\left(\frac{1}{2}\mu|\overline{\mathscr{H}}|^2\right) + \frac{\partial}{\partial t}\left(\frac{1}{2}\epsilon|\overline{\mathscr{E}}|^2\right) = -\nabla\cdot(\overline{\mathscr{E}}\times\overline{\mathscr{H}})$$

and show its validity by simply substituting [8.33*a*] and [8.33*b*] and carrying out the differentiation with respect to time and the curl operation. The manipulations are left as an exercise for the reader.

The instantaneous variations of the stored energies are often not of practical interest. Instead, one is often interested in the time averages of these quantities over one period ($T_p = 2\pi/\omega$). Denoting these time averages as $\overline{W}_{\mathrm{e}}$ and $\overline{W}_{\mathrm{m}}$, we have

$$\begin{aligned}\overline{W}_{\mathrm{e}} = \overline{W}_{\mathrm{m}} &= \frac{1}{T_p}\int_0^{T_p}\frac{1}{2}\epsilon E_0^2\cos^2(\omega t - \beta z)\,dt \\ &= \frac{1}{2}\epsilon E_0^2\frac{1}{T_p}\int_0^{T_p}\left\{\frac{1}{2} + \frac{1}{2}\cos[2(\omega t - \beta z)]\right\}dt = \frac{1}{4}\epsilon E_0^2\end{aligned} \qquad [8.34]$$

We undertake a more general discussion of instantaneous and time-average electromagnetic power in the next section, applicable to uniform plane waves in lossy media. Examples 8-15 and 8-16 illustrate the use of [8.32] to determine the solar radiation flux on the earth and to determine the intensity at earth of a signal transmitted from a spaceship near the moon. Examples 8-17 and 8-18 concern the application of [8.32] to determine the power density of an FM and a VHF/UHF broadcast signal.

Example 8-15: Solar radiation at Earth. If the time-average power density arriving at the surface of the earth due to solar radiation is about 1400 W-m^{-2} (an average over a wide band of frequencies), find the peak electric and magnetic field values on the earth's surface due to solar radiation. The radii of Earth and the Sun are ~6.37 Mm and ~700 Mm, respectively, and the radius of Earth's orbit around the Sun is ~1.5×10^8 km.

Solution: Using [8.32] we can write

$$|\mathbf{S}_{\text{av}}| = \frac{1}{2}\frac{E_0^2}{\eta} = 1400 \text{ W-m}^{-2}$$

where $\eta \simeq 377\Omega$. Solving for E_0 we find $E_0 \simeq \sqrt{2 \times 377 \times 1400} \simeq 1027 \text{ V-m}^{-1}$. The corresponding magnetic field H_0 is $H_0 = E_0/\eta \simeq 1027/377 \simeq 2.73 \text{ A-m}^{-1}$.

Example 8-16: A spaceship in lunar orbit. A spaceship in lunar orbit (the Earth-Moon distance is ~380 Mm) transmits plane waves with an antenna operating at 1 GHz and radiating a total power of 1 MW isotropically (i.e., equally in all directions). Find (a) the time-average power density and (b) the peak electric field value on the earth's surface, and (c) the time it takes for these waves to travel from the spaceship to the earth.

Solution:

(a) The time-average power density on the earth's surface a distance $R \simeq 380$ Mm away from the spaceship radiating a total power of $P_{\text{tot}} = 1$ MW isotropically is

$$|\mathbf{S}_{\text{av}}| = \frac{P_{\text{tot}}}{4\pi R^2} = \frac{10^6 \text{ W}}{4\pi(380 \times 10^6)^2 \text{ m}^2} \simeq 5.51 \times 10^{-13} \text{ W-m}^{-2}$$

(b) The corresponding peak electric field value E_0 can be found from [8.32]:

$$|\mathbf{S}_{\text{av}}| = \frac{1}{2}\frac{E_0^2}{\eta} \quad \rightarrow \quad E_0 \simeq \sqrt{2 \times 377 \times 5.51 \times 10^{-13}} \simeq 2.04 \times 10^{-5} \text{ V-m}^{-1}$$

(c) The time it takes for the waves to travel from the spaceship to the earth is given by

$$t = \frac{R}{c} \simeq \frac{380 \times 10^6 \text{ m}}{3 \times 10^8 \text{ m-s}^{-1}} \simeq 1.27 \text{ s}$$

Example 8-17: FM broadcasting. A transmitter antenna for an FM radio broadcasting station operates around 92.3 MHz (KGON station in Portland, Oregon) with an average radiated power of 100 kW as limited by Federal Communications Commission (FCC) regulations to minimize interference problems with other broadcast stations outside its coverage area. For simplicity, assume isotropic radiation, i.e., radiation equally distributed in all directions, such as from a point source. (a) For a person standing 50 m away from the antenna, calculate the time-average power density of the radiated wave and compare it with the IEEE safety limit[34]

---

[34]IEEE c95.1-1991, *IEEE Standard for Safety Levels with Respect to Human Exposure to Radio Frequency Electromagnetic Fields,* 3 kHz to 300 GHz. For a related discussion, see M. Fischetti, The cellular phone scare, *IEEE Spectrum,* pp. 43–47, June, 1993.

for an uncontrolled environment at that frequency, namely 2 W-m^{-2} or 0.2 mW-cm^{-2}. (b) Find the minimum distance such that the time-average power density is equal to the IEEE safety standard value at that frequency. (c) If the antenna is mounted on a tower that is 200 m above the ground, repeat part (a) for a person standing 50 m away from the foot of the tower.

Solution: Assume the antenna to be a point source and neglect all effects from boundaries.

(a) The wavelength at 92.3 MHz is

$$\lambda = \frac{3 \times 10^8}{92.3 \times 10^6} \cong 3.25 \text{ m}$$

Since 50 m $\gg$ 3.25 m, we can assume far field, as if the wave were a uniform plane wave. The total power of $P_{\text{tot}} = 100$ kW is radiated equally in all directions, so if we take a sphere of radius 50 m centered at the antenna, the average power density on the surface of the sphere is given by

$$|\mathbf{S}_{\text{av}}| = \frac{P_{\text{tot}}}{4\pi R^2} = \frac{100 \times 10^3}{4\pi(50)^2}$$

$$\simeq 3.18 \text{ W-m}^{-2} = 0.318 \text{ mW-cm}^{-2} > 0.2 \text{ mW-cm}^{-2}$$

Therefore, it is not safe to stand at a distance of 50 m from this FM broadcast antenna.

(b) From

$$\frac{P_{\text{tot}}}{4\pi R_{\text{min}}^2} = \frac{100 \times 10^3}{4\pi R_{\text{min}}^2} = 2 \text{ W-m}^{-2}$$

we find $R_{\text{min}} \simeq 63.1$ m.

(c) Considering a 200 m tower height and a person standing on the ground at a distance 50 m from the base of the tower, the distance between the antenna and the person is $\sqrt{(200)^2 + (50)^2}$, so that we have

$$|\mathbf{S}_{\text{av}}| = \frac{100 \times 10^3}{4\pi[(200)^2 + (50)^2]} \simeq 0.187 \text{ W-m}^{-2} < 2 \text{ W-m}^{-2}$$

Therefore, the radiation level on the ground is well below the IEEE safety standard.

Example 8-18: VHF/UHF broadcast radiation. A survey[35] conducted in the United States indicates that ~50% of the population is exposed to average power densities of approximately 0.005 μW-cm^{-2} due to VHF and UHF broadcast radiation. Find the corresponding amplitudes of the electric and magnetic fields.

---

[35] R. A. Tell and E. D. Mantiply, Population exposure to VHF and UHF broadcast radiation in the United States, *Proc. IEEE,* 68(1), pp. 6–12, January 1980.

Solution: By direct application of [8.32] we have

$$|\mathbf{S}_{\text{av}}| = \frac{1}{2}\frac{E_0^2}{\eta} = 0.005\ \mu\text{W-cm}^{-2}$$

so

$$E_0 \simeq \sqrt{2 \times 377 \times 5 \times 10^{-9}/10^{-4}} \simeq 194\ \text{mV-m}^{-1}$$

$$H_0 = \frac{E_0}{\eta} \simeq \frac{194\ \text{mV-m}^{-1}}{377\Omega} \simeq 515\ \mu\text{A-m}^{-1}$$

where we have noted that $\eta \simeq 377\Omega$ in air.

Instantaneous and Time-Average Power and the Complex Poynting Theorem In most applications we are interested in the Poynting vector averaged over time rather than in its instantaneous value. Consider the expressions for the instantaneous electric and magnetic fields of a time-harmonic uniform plane wave in the general case of a lossy medium. From [8.18] and [8.23] we have

$$\overline{\mathscr{E}}(z, t) = \hat{\mathbf{x}} C_1 e^{-\alpha z} \cos(\omega t - \beta z)$$

and

$$\overline{\mathscr{H}}(z, t) = \hat{\mathbf{y}} \frac{C_1}{|\eta_c|} e^{-\alpha z} \cos(\omega t - \beta z - \phi_\eta)$$

Noting that $\hat{\mathbf{x}} \times \hat{\mathbf{y}} = \hat{\mathbf{z}}$, the instantaneous Poynting vector given by the cross product $\overline{\mathscr{E}} \times \overline{\mathscr{H}}$ is in the $+z$ direction:

$$\overline{\mathscr{S}}(z, t) = \overline{\mathscr{E}}(z, t) \times \overline{\mathscr{H}}(z, t) = \hat{\mathbf{z}} \frac{C_1^2}{|\eta_c|} e^{-2\alpha z} \cos(\omega t - \beta z) \cos(\omega t - \beta z - \phi_\eta) \quad [8.35]$$

As expected, the electromagnetic power carried by a uniform plane wave propagating in the $+z$ direction flows in the $+z$ direction. We can simplify [8.35] to

$$\overline{\mathscr{S}}(z, t) = \overline{\mathscr{E}}(z, t) \times \overline{\mathscr{H}}(z, t) = \hat{\mathbf{z}} \frac{C_1^2}{2|\eta_c|} e^{-2\alpha z} [\cos(\phi_\eta) + \cos(2\omega t - 2\beta z - \phi_\eta)] \quad [8.36]$$

We see that the instantaneous Poynting vector has a component that does not change with time and a component that oscillates at twice the rate at which electric and magnetic fields change in time. In most applications, the time-average value of the power transmitted by a wave is more significant than the fluctuating component. The relationship between the electric and magnetic fields and the Poynting vector $\overline{\mathscr{S}}(z, t)$ is illustrated in Figure 8.15. Note that the instantaneous Poynting vector itself can be negative in certain regions; however, the time-average value is always positive for a uniform plane wave and represents real power flow in the $+z$ direction.

The time-average value can be obtained from [8.36] by integrating over one period T_p of the sinusoidal variation. We find

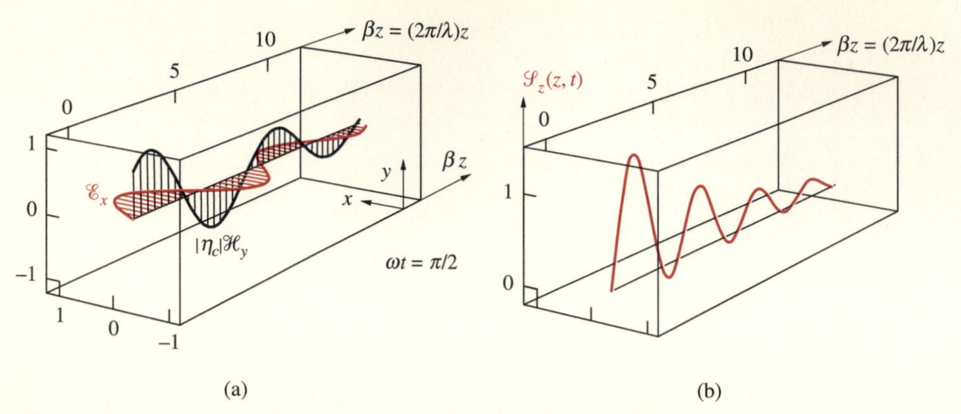

FIGURE 8.15. The Poynting vector and fields for a uniform plane wave in a lossy medium. (a) The wave electric and magnetic fields at $\omega t = \pi/2$; (b) the Poynting vector $\overline{\mathscr{S}}(z, \omega t = \pi/2)$.

$$\mathbf{S}_{\text{av}}(z) = \frac{1}{T_p}\int_0^{T_p} \overline{\mathscr{S}}(z, t)\, dt = \hat{\mathbf{z}}\frac{C_1^2}{2|\eta_c|}e^{-2\alpha z}\cos(\phi_\eta) \qquad [8.37]$$

As an example, consider a uniform plane wave at 300 MHz in copper, for which $\eta_c \simeq 6.4 \times 10^{-3}e^{j45°}\Omega$. The time-average electromagnetic power carried by a wave with 1 V-m^{-1} field amplitude at $z = 0$ (i.e., $C_1 = 1$ V-m^{-1}) is ~55.3 W-m^{-2}, compared to ~0.0013 W-m^{-2} in free space for the same C_1.

Finding the time-average Poynting flux $\mathbf{S}_{\text{av}}$ by an integration over one period as in [8.37] is in general not necessary for sinusoidal signals, since $\mathbf{S}_{\text{av}}$ can be directly obtained from the phasors $\mathbf{E}$ and $\mathbf{H}$, as shown in the next section.

The Poynting theorem is expressed mathematically in [8.31] in terms of the instantaneous field quantities $\overline{\mathscr{E}}$ and $\overline{\mathscr{H}}$. The complex version of the theorem in terms of the phasors $\mathbf{E}$ and $\mathbf{H}$ cannot be obtained from [8.31] by substituting $\partial/\partial t \to j\omega$, since the terms in [8.31] involve vector products of field quantities. Thus, the complex Poynting theorem has to be derived from the phasor form of Maxwell's equations [7.18*a*] and [7.18*c*]:

$$\nabla \times \mathbf{E} = -j\omega\mathbf{B} \qquad \nabla \times \mathbf{H} = \mathbf{J} + j\omega\mathbf{D}$$

We start from the complex form of the vector identity

$$\nabla \cdot (\mathbf{E} \times \mathbf{H}^*) = \mathbf{H}^* \cdot (\nabla \times \mathbf{E}) - \mathbf{E} \cdot (\nabla \times \mathbf{H}^*)$$

Substituting the two curl equations into this identity, we have

$$\nabla \cdot (\mathbf{E} \times \mathbf{H}^*) = \mathbf{H}^* \cdot (-j\omega\mathbf{B}) - \mathbf{E} \cdot (\mathbf{J}^* - j\omega\mathbf{D}^*) \qquad [8.38]$$

Integrating [8.38] over a volume V and using the divergence theorem,

$$\int_V \nabla \cdot (\mathbf{E} \times \mathbf{H}^*)\, dv = \oint_S (\mathbf{E} \times \mathbf{H}^*) \cdot d\mathbf{s} = \int_V -[\mathbf{E} \cdot \mathbf{J}^* + j\omega(\mathbf{H}^* \cdot \mathbf{B} - \mathbf{E} \cdot \mathbf{D}^*)]\, dv \qquad [8.39]$$

If the medium is isotropic and if all the losses occur through conduction currents $\mathbf{J} = \sigma\mathbf{E}$, [8.39] becomes

$$\oint_S (\mathbf{E} \times \mathbf{H}^*) \cdot d\mathbf{s} = -\int_V \sigma\mathbf{E} \cdot \mathbf{E}^*\, dv - j\omega \int_V (\mu\mathbf{H} \cdot \mathbf{H}^* - \epsilon\mathbf{E} \cdot \mathbf{E}^*)\, dv \qquad [8.40]$$

In the case of lossy dielectrics, for which losses occur through effective conduction currents $\overline{\mathbf{J}} = \sigma_{\text{eff}}\overline{\mathbf{E}} = \omega\epsilon''\overline{\mathbf{E}}$, [8.39] is valid when we replace σ with $\sigma_{\text{eff}} = \omega\epsilon''$ and ϵ with ϵ'. Recognizing from [8.34] that $\overline{W}_{\text{e}} = \frac{1}{4}\epsilon\mathbf{E}\cdot\mathbf{E}^*$ and $\overline{W}_{\text{m}} = \frac{1}{4}\mu\mathbf{H}\cdot\mathbf{H}^*$ are the time averages of the energy densities $W_{\text{e}} = \frac{1}{2}\epsilon\overline{\mathscr{E}}\cdot\overline{\mathscr{E}}$ and $W_{\text{m}} = \frac{1}{2}\mu\overline{\mathscr{H}}\cdot\overline{\mathscr{H}}$ respectively, and noting that the volume density of time-average power dissipation in conduction currents is $P_c = \frac{1}{2}\sigma\mathbf{E}\cdot\mathbf{E}^*$, we can equate the real and imaginary parts of [8.40] as

$$\mathscr{R}e\left\{\oint_S (\mathbf{E}\times\mathbf{H}^*)\cdot d\mathbf{s}\right\} = -2\int_V P_c\, dv \qquad [8.41]$$

$$\mathscr{I}m\left\{\oint_S (\mathbf{E}\times\mathbf{H}^*)\cdot d\mathbf{s}\right\} = -4\omega\int_V (\overline{W}_{\text{m}} - \overline{W}_{\text{e}})\, dv \qquad [8.42]$$

The instantaneous Poynting vector $\overline{\mathscr{S}}(z, t)$ can be written in terms of the phasors $\mathbf{E}(z)$ and $\mathbf{H}(z)$, as follows:

$$\overline{\mathscr{S}}(z, t) = \mathscr{R}e\{\mathbf{E}(z)e^{j\omega t}\}\times\mathscr{R}e\{\mathbf{H}(z)e^{j\omega t}\} = \tfrac{1}{2}\mathscr{R}e\{\mathbf{E}(z)\times\mathbf{H}^*(z) + \mathbf{E}(z)\times\mathbf{H}(z)e^{j2\omega t}\}$$

where we have used the fact that in general, for two different complex vectors $\mathbf{G}$ and $\mathbf{F}$, and noting that $(\mathbf{G}\times\mathbf{F}^*)^* = \mathbf{G}^*\times\mathbf{F}$, we have

$$\begin{aligned}\mathscr{R}e\{\mathbf{G}\}\times\mathscr{R}e\{\mathbf{F}\} &= \tfrac{1}{2}(\mathbf{G}+\mathbf{G}^*)\times\tfrac{1}{2}(\mathbf{F}+\mathbf{F}^*)\\ &= \tfrac{1}{4}[(\mathbf{G}\times\mathbf{F}^* + \mathbf{G}^*\times\mathbf{F}) + (\mathbf{G}\times\mathbf{F} + \mathbf{G}^*\times\mathbf{F}^*)]\\ &= \tfrac{1}{2}\mathscr{R}e\{\mathbf{G}\times\mathbf{F}^* + \mathbf{G}\times\mathbf{F}\}\end{aligned}$$

since in general we have $\mathscr{R}e\{\mathbf{G}\} = \frac{1}{2}(\mathbf{G}+\mathbf{G}^*)$ for any complex vector $\mathbf{G}$. The time-average power density, which was denoted $\mathbf{S}_{\text{av}}$ in [8.37], can then be obtained by integrating $\overline{\mathscr{S}}(z, t)$ over one period $T_p = 2\pi/\omega$. Since the time-average of the $\mathbf{E}(z)\times\mathbf{H}(z)e^{j2\omega t}$ term vanishes, we find that the time-average Poynting vector is given by

$$\boxed{\mathbf{S}_{\text{av}} = \tfrac{1}{2}\mathscr{R}e[\mathbf{E}\times\mathbf{H}^*]} \qquad [8.43]$$

This result allows us to evaluate $\mathbf{S}_{\text{av}}$ conveniently from the phasor field quantities. Note that $\mathbf{S} = \mathbf{E}\times\mathbf{H}^*$ is referred to as the *complex Poynting vector.* The imaginary part of the integral of $\mathbf{S}$ as given in [8.42] is equal to the difference between the average energies stored in the magnetic and the electric fields and represents reactive power flowing back and forth to supply the instantaneous changes in the net stored energy in the volume.

Example 8-19 illustrates the application of the complex Poynting theorem to determine electromagnetic power flow in selected practical cases.

Example 8-19: VLF waves in the ocean. Consider VLF wave propagation in the ocean as discussed in Example 8-11. Find the time-average Poynting flux at the sea surface ($z = 0$) and at the depth of the submarine ($z = 100$ m).

Solution: The phasor expressions for the minimum electric and magnetic fields to communicate with the submarine at 100 m depth are given by

$$\mathbf{E}(z) \simeq \hat{\mathbf{x}}2.84e^{-0.218z}e^{-j0.218z} \text{ kV-m}^{-1}$$

$$\mathbf{H}(z) \simeq \hat{\mathbf{y}}36.9e^{-0.218z}e^{-j0.218z}e^{-j\pi/4} \text{ kA-m}^{-1}$$

Using [8.43], the time-average Poynting vector is given by

$$\begin{aligned} \mathbf{S}_{\text{av}}(z) &= \tfrac{1}{2}\mathcal{R}e[\mathbf{E} \times \mathbf{H}^*] \\ &\simeq \hat{\mathbf{z}}\tfrac{1}{2}(2.84)(36.9)e^{-2(0.218z)}\cos(\pi/4) \\ &\simeq \hat{\mathbf{z}}36.9e^{-0.435z} \text{ MW-m}^{-2} \end{aligned}$$

At the sea surface (i.e., $z = 0^+$), $\mathbf{S}_{\text{av}} \simeq \hat{\mathbf{z}}36.9$ MW-m^{-2}. At the location of the submarine (i.e., $z = 100$ m), $\mathbf{S}_{\text{av}} \simeq \hat{\mathbf{z}}4.59 \times 10^{-12}$ W-m^{-2}. The power density required at the sea surface is extremely high, indicating that it is not feasible to use 3-kHz signals to communicate with a submarine at 100 m depth.

8.2 REFLECTION AND TRANSMISSION OF WAVES AT PLANAR INTERFACES

Up to now, we have considered the relatively simple propagation of uniform plane waves in homogeneous and unbounded media. However, practical problems usually involve waves propagating in bounded regions in which different media may be present, and require that we take account of the complicating effects at boundary surfaces. Typical boundary surfaces may lie between regions of different permittivity (e.g., glass and air) or between regions of different conductivity (e.g., air and copper). Boundary surfaces between regions of different permeability (e.g., air and iron) are also interesting but of less practical importance for electromagnetic wave phenomena.

When a wave encounters the boundary between two different homogeneous media, it is split into two waves: a *reflected* wave, propagating back to the first medium, and a *transmitted* (or *refracted*) wave, which proceeds into the second medium. Reflection of waves is very much a part of our everyday experience. When a guitar string is plucked, a wave is generated that runs back and forth between the ends of the string, reflecting repeatedly at each end. When we shout toward a cliff, we hear the reflection of the sound waves we generated as an echo. The basis for operation of radars, sonars, and lidars is the reflection of radio, acoustic, or light waves from various objects. When sound waves of sufficient intensity (e.g., thunder) hit an object, they transmit energy into it by making it shake for a while. Seismic waves transfer their energy quite efficiently through boundaries between different layers of the earth, and earthquake waves generated deep under the ocean floors transmit wave energy into the ocean, generating tsunamis.

When we look at the way ocean waves hit the beach or go around a ship standing still, we realize that the reflection of waves from boundaries can be very complicated indeed. The shape of a wave can be greatly altered on reflection, and a wave striking a rough (nonsmooth) surface can be reflected in all directions. We also know from our experiences with physical optics (e.g., refraction of light by prisms and rain drops, the latter leading to the formation of rainbows) that the direction and magnitude of the waves reflected and refracted at a boundary depend on the angle of arrival of the waves at the boundary. In this book, we confine our attention to planar waves of infinite extent vertically (or normally) incident on planar boundaries. A number of important applications involving reflection and refraction of obliquely incident light waves (most importantly the guiding of light waves in optical fibers) are thus not covered. However, many other applications (as mentioned below) can be effectively modeled in terms of normal incidence of plane waves, and are covered in full detail. Furthermore, most of the fundamental principles of reflection and transmission of electromagnetic waves at boundaries are well illustrated in the context of planar waves of infinite extent normally incident on planar boundaries.

When uniform plane waves encounter planar interfaces, the reflected and transmitted waves are also planar, and expressions for their directions of propagation and their amplitudes and phases can be derived with relative ease. In fact, the reflection and transmission of uniform plane waves from planar interfaces is in many ways analogous to the reflection and transmission of waves at junctions between different transmission lines or at the terminations of transmission lines at a load, topics which were studied extensively in Chapters 2 and 3. Accordingly, we make references to this analogy as we study uniform plane wave reflection and transmission, and in some cases we will utilize it for more efficient solution of interface problems.

In general, when a uniform plane wave propagating in one medium is *incident* on a boundary, part of its energy reflects back into the first medium and is carried away from the boundary by a reflected wave, and the other part propagates into the second medium, carried by a transmitted wave. The existence of these two waves (reflected and transmitted) is a direct result of the boundary conditions imposed by the properties of the media, which in general cannot be satisfied without postulating the presence of both of these waves. Using Maxwell's equations with the appropriate boundary conditions, we can find mathematical expressions for the reflected and transmitted waves in terms of the properties of the incident wave and the two media. We limit our coverage to interfaces that are infinitely sharp, with the electromagnetic properties (e.g., dielectric constant) changing suddenly (within a distance very small compared to the wavelength) from that of one medium to that of the other. For example, for optical applications we require the medium properties near the interface to change from one medium to another over distances much smaller than the free-space wavelength for visible light, ranging from 390 to 760 nm. This is why the surfaces of optical coatings have to be very smooth, since any thin layer of impurities will change the reflection properties.

We now proceed with our coverage of reflection and transmission of electromagnetic waves by considering cases of increasing complexity. We start in Section 8.2.1 with normal incidence on perfect conductors, followed in Section 8.2.2 with normal incidence on single lossless dielectric interfaces. In Section 8.2.3 we consider normal incidence on multiple dielectric interfaces, and the important applications of coating, and in Section 8.2.4 the reflection and transmission of waves at interfaces between lossy media.

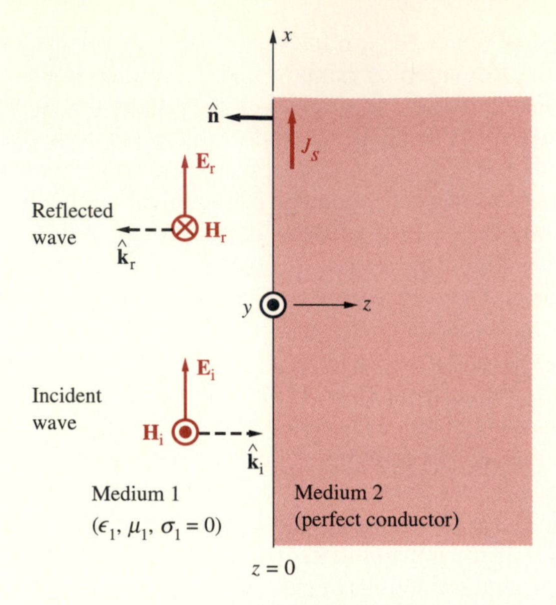

FIGURE 8.16. Uniform plane wave incident normally on a plane, perfectly conducting boundary. The unit vectors $\hat{\mathbf{k}}_i$ and $\hat{\mathbf{k}}_r$ represent the propagation directions of the incident and reflected waves, respectively.

8.2.1 Normal Incidence on a Perfect Conductor

We first consider the case of a uniform plane wave propagating in the $+z$ direction in a simple lossless ($\sigma_1 = 0$) medium occupying the half-space $z < 0$ and normally incident from left on a perfectly conducting ($\sigma_2 = \infty$) medium which occupies the $z > 0$ half-space as shown in Figure 8.16. The interface between the two media is the entire xy plane (i.e., infinite in transverse extent). We arbitrarily[36] assume that the electric field of the incident wave is oriented in the x direction and that therefore its magnetic field is in the y direction. The phasor field expressions for the incident wave (presumably originating at $z = -\infty$) and the reflected wave are given as

Incident wave:	Reflected wave:
$\mathbf{E}_i(z) = \hat{\mathbf{x}}E_{i0}e^{-j\beta_1 z}$	$\mathbf{E}_r(z) = \hat{\mathbf{x}}E_{r0}e^{+j\beta_1 z}$
$\mathbf{H}_i(z) = \hat{\mathbf{y}}\dfrac{E_{i0}}{\eta_1}e^{-j\beta_1 z}$	$\mathbf{H}_r(z) = -\hat{\mathbf{y}}\dfrac{E_{r0}}{\eta_1}e^{+j\beta_1 z}$

where $\beta_1 = \omega\sqrt{\mu_1\epsilon_1}$ and $\eta_1 = \sqrt{\mu_1/\epsilon_1}$ are, respectively, the propagation constant and the intrinsic impedance for medium 1. The presence of the boundary requires that a reflected wave propagating in the $-z$ direction exist in medium 1, with phasor field expressions as given above. Note that the reflected wave propagates in the $-z$ direction, so that its magnetic field is oriented in the $-y$ direction, for an x-directed electric field.

[36]Note that for normal incidence on a planar boundary of infinite extent in two dimensions, we lose no generality by taking the direction in which the electric field of the incident wave vibrates to be one of the principal axes.

The boundary condition on the surface of the conductor requires the total tangential electric field to vanish for $z = 0$ for *all x and y,* since the electric field inside medium 2 (perfect conductor) must be zero. This condition requires that the electric field of the reflected wave is also confined to the x direction. The total electric and magnetic fields in medium 1 are

$$\mathbf{E}_1(z) = \mathbf{E}_\mathrm{i}(z) + \mathbf{E}_\mathrm{r}(z) = \hat{\mathbf{x}}[E_{\mathrm{i}0}e^{-j\beta_1 z} + E_{\mathrm{r}0}e^{+j\beta_1 z}]$$

$$\mathbf{H}_1(z) = \mathbf{H}_\mathrm{i}(z) + \mathbf{H}_\mathrm{r}(z) = \hat{\mathbf{y}}\left[\frac{E_{\mathrm{i}0}}{\eta_1}e^{-j\beta_1 z} - \frac{E_{\mathrm{r}0}}{\eta_1}e^{+j\beta_1 z}\right]$$

Since $\mathbf{E}_1(z)$ is tangential to the boundary, application of the boundary condition at $z = 0$, namely that $\mathbf{E}_1(z = 0) = 0$, gives

$$\mathbf{E}_1(z = 0) = \hat{\mathbf{x}}[E_{\mathrm{i}0}e^{-j\beta_1 z} + E_{\mathrm{r}0}e^{+j\beta_1 z}]_{z=0} = \hat{\mathbf{x}}[E_{\mathrm{i}0} + E_{\mathrm{r}0}] = 0$$

from which

$$\boxed{E_{\mathrm{r}0} = -E_{\mathrm{i}0}}$$

so that

$$\mathbf{E}_1(z) = \hat{\mathbf{x}}E_{\mathrm{i}0}\underbrace{[e^{-j\beta_1 z} - e^{+j\beta_1 z}]}_{-j2\sin(\beta_1 z)} = -\hat{\mathbf{x}}E_{\mathrm{i}0}j2\sin(\beta_1 z)$$

With the constant $E_{\mathrm{r}0}$ determined in terms of $E_{\mathrm{i}0}$ (i.e., $E_{\mathrm{r}0} = -E_{\mathrm{i}0}$) we can also write the total magnetic field in medium 1 as

$$\mathbf{H}_1(z) = \mathbf{H}_\mathrm{i}(z) + \mathbf{H}_\mathrm{r}(z) = \hat{\mathbf{y}}\frac{1}{\eta_1}[E_{\mathrm{i}0}e^{-j\beta_1 z} + E_{\mathrm{i}0}e^{+j\beta_1 z}] = \hat{\mathbf{y}}\frac{2E_{\mathrm{i}0}}{\eta_1}\cos(\beta_1 z)$$

The corresponding space-time functions $\overline{\mathscr{E}}_1(z, t)$ and $\overline{\mathscr{H}}_1(z, t)$ can be found from the phasors $\mathbf{E}_1(z)$ and $\mathbf{H}_1(z)$ as

$$\overline{\mathscr{E}}_1(z, t) = \mathscr{R}e\{\mathbf{E}_1(z)e^{j\omega t}\} = \mathscr{R}e\{\hat{\mathbf{x}}2E_{\mathrm{i}0}\sin(\beta_1 z)e^{j\omega t}\underbrace{e^{-j\pi/2}}_{-j}\}$$

$$\overline{\mathscr{H}}_1(z, t) = \mathscr{R}e\{\mathbf{H}_1(z)e^{j\omega t}\} = \mathscr{R}e\left\{\hat{\mathbf{y}}\frac{2E_{\mathrm{i}0}}{\eta_1}\cos(\beta_1 z)e^{j\omega t}\right\}$$

which gives

$$\boxed{\begin{aligned}\overline{\mathscr{E}}_1(z, t) &= \hat{\mathbf{x}}2E_{\mathrm{i}0}\sin(\beta_1 z)\sin(\omega t)\\ \overline{\mathscr{H}}_1(z, t) &= \hat{\mathbf{y}}2\frac{E_{\mathrm{i}0}}{\eta_1}\cos(\beta_1 z)\cos(\omega t)\end{aligned}} \qquad [8.44]$$

assuming that $E_{\mathrm{i}0}$ is a real number.[37]

[37] If, instead, $E_{\mathrm{i}0}$ was a complex number, $E_{\mathrm{i}0} = |E_{\mathrm{i}0}|e^{j\zeta}$, the expressions for $\overline{\mathscr{E}}_1$ and $\overline{\mathscr{H}}_1$ would be

$$\overline{\mathscr{E}}_1(z, t) = \hat{\mathbf{x}}2|E_{\mathrm{i}0}|\sin(\beta_1 z)\sin(\omega t + \zeta)$$

$$\overline{\mathscr{H}}_1(z, t) = \hat{\mathbf{y}}2\frac{|E_{\mathrm{i}0}|}{\eta_1}\cos(\beta_1 z)\cos(\omega t + \zeta)$$

amounting to a simple shift of the time origin.

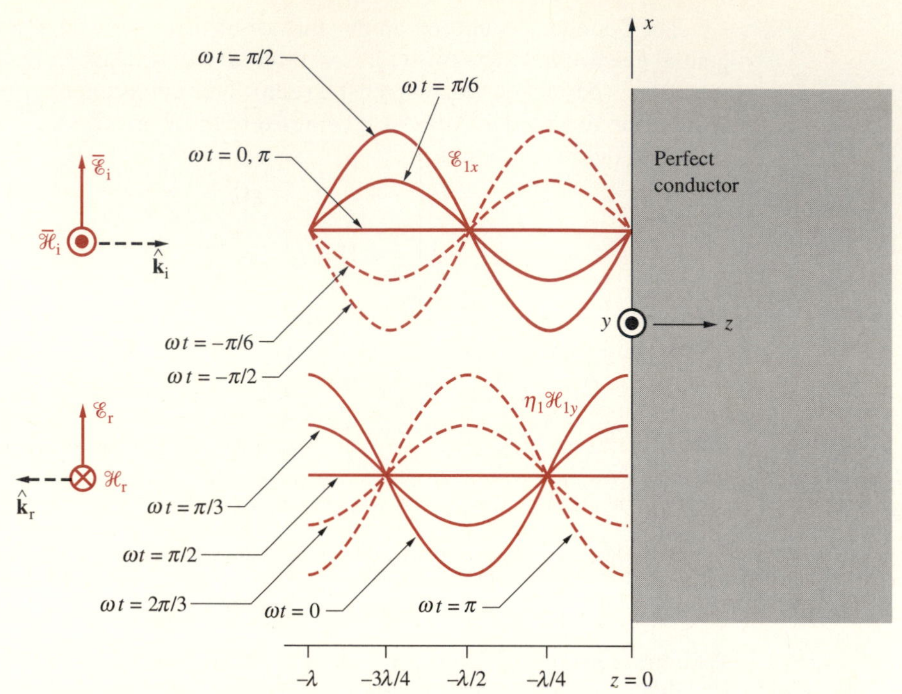

FIGURE 8.17. Instantaneous electric and magnetic field waveforms. Total electric and magnetic field waveforms in medium 1 are shown at selected instants of time. The "standing" nature of the waves is apparent as the peaks and nulls remain at the same point in space as time progresses. Note that medium 1 is a perfect dielectric, while medium 2 is a perfect conductor.

The fields described by [8.44] are plotted in Figure 8.17 as a function of distance z at different time instants. Note that the $\overline{\mathscr{E}}$ and $\overline{\mathscr{H}}$ fields do not represent a propagating wave, since as time advances, the peaks (or nulls) always occur at the same points in space. Such a wave is termed to be a *pure standing wave;* it consists of a superposition of two waves traveling in opposite directions. The maxima and minima *stand* at the same location as time advances. This standing wave, which is established when a uniform plane wave is incident on a perfect conductor, is analogous in all respects to the one that occurs in the case of sinusoidal excitation of a lossless transmission line terminated in a short circuit (Chapter 3). This analogy is illustrated in Figure 8.18.

A pure standing wave such as the one just described does not carry electromagnetic energy, as expected on physical grounds, since the perfect conductor boundary reflects all of the incident energy. To verify this notion, we can consider the Poynting vector for the standing wave just described. Using the phasor forms of the fields and noting that E_{i0} is in general complex, we have

$$(\mathbf{S}_{\text{av}})_1 = \frac{1}{2}\mathcal{R}e[\mathbf{E}_1 \times \mathbf{H}_1^*] = \frac{1}{2}\mathcal{R}e\left\{\hat{\mathbf{x}}[-2jE_{i0}\sin(\beta_1 z)] \times \hat{\mathbf{y}}\left[\frac{2E_{i0}^*}{\eta_1}\cos(\beta_1 z)\right]\right\}$$

$$= \frac{1}{2}\mathcal{R}e\left\{\hat{\mathbf{z}}\left[-j\frac{2|E_{i0}|^2}{\eta_1}\sin(2\beta_1 z)\right]\right\} = 0$$

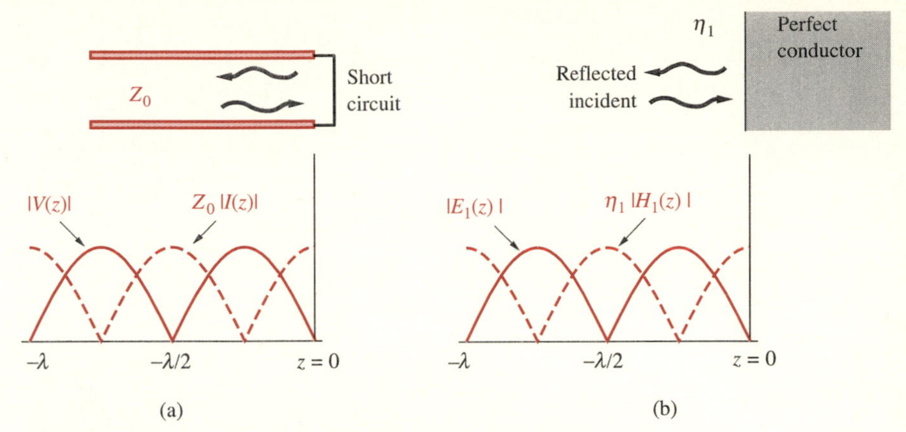

FIGURE 8.18. **Transmission line analogy.** (a) The voltage/current standing-wave patterns on a short-circuited lossless transmission line. (b) The electric and magnetic fields for normal incidence of a uniform plane wave from a lossless medium on a perfect conductor.

The instantaneous Poynting vector can also be obtained from $\overline{\mathscr{E}}_1$ and $\overline{\mathscr{H}}_1$ as

$$\overline{\mathscr{S}}_1(z,t) = \overline{\mathscr{E}}_1(z,t) \times \overline{\mathscr{H}}_1(z,t) = \hat{\mathbf{z}}\frac{4|E_{i0}|^2}{\eta_1}\sin(\beta_1 z)\cos(\beta_1 z)\sin(\omega t + \zeta)\cos(\omega t + \zeta)$$
$$= \hat{\mathbf{z}}\frac{|E_{i0}|^2}{\eta_1}\sin(2\beta_1 z)\sin[2(\omega t + \zeta)]$$

where ζ is the arbitrary phase angle of E_{i0}; that is, we assume that $E_{i0} = |E_{i0}|e^{j\zeta}$. The time average value of $\overline{\mathscr{S}}$ is clearly zero. Although electromagnetic energy does not flow, it surges back and forth. In other words, at any given point z, the Poynting vector fluctuates between positive and negative values at a rate of 2ω.

With respect to the discussion of electromagnetic power flow in Section 8.1.4, we note from [8.29] that since there is no dissipation term (i.e., $\overline{\mathscr{J}}_1 = 0$ and thus $\overline{\mathscr{E}}_1 \cdot \overline{\mathscr{J}}_1 = 0$), the fluctuating Poynting vector $\overline{\mathscr{S}}_1(z,t)$ represents the time variation of the stored electric and magnetic energy densities, w_e and w_m, which are given by

$$w_e(z,t) = \tfrac{1}{2}\epsilon_1|\overline{\mathscr{E}}_1(z,t)|^2 = 2\epsilon_1|E_{i0}|^2\sin^2(\beta_1 z)\sin^2(\omega t + \zeta) \quad \text{J-m}^{-3}$$
$$w_m(z,t) = \tfrac{1}{2}\mu_1|\overline{\mathscr{H}}_1(z,t)|^2 = 2\epsilon_1|E_{i0}|^2\cos^2(\beta_1 z)\cos^2(\omega t + \zeta) \quad \text{J-m}^{-3}$$

The variations of w_m and w_e with z at different times are shown in Figure 8.19. We can see that the stored energy at point z alternates between fully magnetic energy $(\omega t + \zeta = 0)$ and fully electric energy $(\omega t + \zeta = \pi/2)$, much like the fluctuation of stored energy between the inductance and capacitance in a resonant LC circuit.

Note that the total wave magnetic field phasor exhibits a maximum at the conductor surface $(z = 0)$. This magnetic field is supported by a surface current that flows on the surface of the conductor, the density of which can be obtained from the boundary condition

$$\mathbf{J}_s = \hat{\mathbf{n}} \times \mathbf{H}_1(0) = (-\hat{\mathbf{z}} \times \hat{\mathbf{y}})\frac{2E_{i0}}{\eta_1} = \hat{\mathbf{x}}\frac{2E_{i0}}{\eta_1} \quad \text{A-m}^{-1}$$

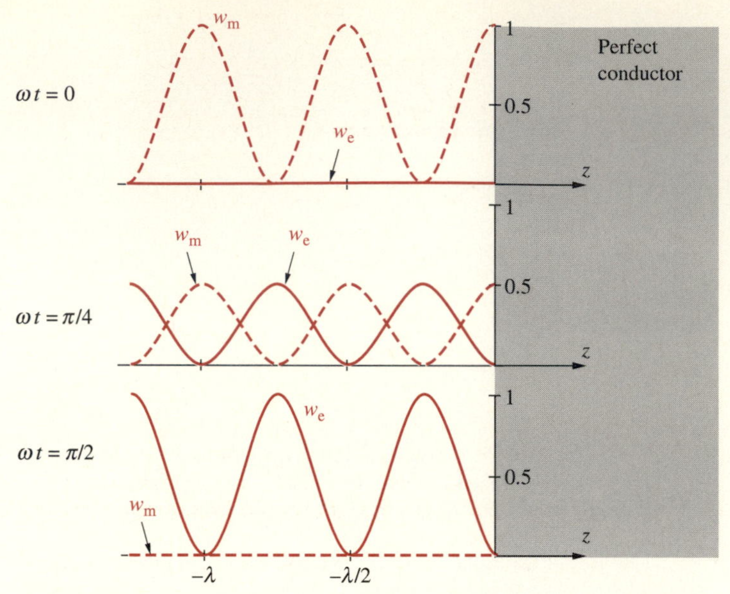

FIGURE 8.19. Instantaneous electric and magnetic energy densities. Oscillation of the total wave energy between electric and magnetic fields in a standing wave. For the purpose of these plots, we have taken E_{i0} to be real and equal to $1/\sqrt{2\epsilon_1}$, so that $\zeta = 0$.

where $\hat{\mathbf{n}}$ is the outward normal to the first medium as shown in Figure 8.16. Note that the direction of the surface current $\mathbf{J}_s$ is related to that of $\mathbf{H}_1$ by Ampère's law, with its sense determined by the right-hand rule. If the conductor is not perfect, this current flow leads to dissipation of electromagnetic power, as discussed in Section 8.2.4.

Example 8-20 quantitatively illustrates the reflection of a uniform plane wave normally incident on a perfect conductor.

Example 8-20: Uniform plane wave incident on an air–conductor interface. A uniform plane wave having a z directed electric field with peak value 10 V-m^{-1} and operating at 1.5 GHz is normally incident from air on a perfectly conducting surface located at $y = 0$ as shown in Figure 8.20. (a) Write the phasor expressions for the total electric and magnetic fields in air. (b) Determine the nearest location to the reflecting surface where the total electric field is zero at all times. (c) Determine the nearest location to the reflecting surface where the total magnetic field is zero at all times.

Solution:

(a) The electric and magnetic fields of the incident wave are

$$\mathbf{E}_i(y) = \hat{\mathbf{z}}10e^{-j\beta_1 y} \quad \text{V-m}^{-1}$$

$$\mathbf{H}_i(y) = \hat{\mathbf{x}}\frac{10}{\eta_1}e^{-j\beta_1 y} \quad \text{A-m}^{-1}$$

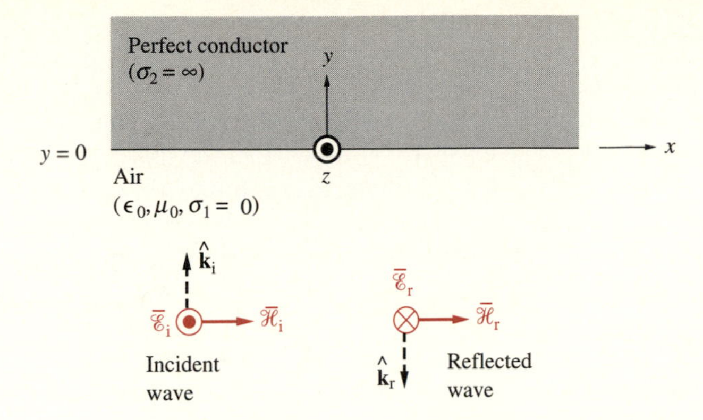

FIGURE 8.20. Air–perfect conductor interface. A uniform plane wave incident from air on a perfect conductor.

where $\beta_1 \simeq [2\pi(1.5 \times 10^9)\text{ rad-s}^{-1}]/(3 \times 10^8\text{ m-s}^{-1}) = 10\pi\text{ rad-m}^{-1}$ and $\eta_1 \simeq 377\Omega$. The electric and magnetic fields of the reflected wave are

$$\mathbf{E}_r(y) = -\hat{\mathbf{z}}10e^{+j\beta_1 y}\text{ V-m}^{-1}$$

$$\mathbf{H}_r(y) = \hat{\mathbf{x}}\frac{10}{\eta_1}e^{+j\beta_1 y}\text{ A-m}^{-1}$$

The total fields in air are

$$\mathbf{E}_1(y) = \mathbf{E}_i(y) + \mathbf{E}_r(y) \simeq -\hat{\mathbf{z}}j20\sin(10\pi y) \quad \text{V-m}^{-1}$$

$$\mathbf{H}_1(y) = \mathbf{H}_i(y) + \mathbf{H}_r(y) \simeq \hat{\mathbf{x}}\frac{20}{377}\cos(10\pi y) \simeq \hat{\mathbf{x}}(0.0531)\cos(10\pi y) \quad \text{A-m}^{-1}$$

The corresponding time-domain expressions for the total fields in air are

$$\overline{\mathscr{E}}_1(y, t) \simeq \hat{\mathbf{z}}20\sin(10\pi y)\sin(3\pi \times 10^9 t) \quad \text{V-m}^{-1}$$

$$\overline{\mathscr{H}}_1(y, t) \simeq \hat{\mathbf{x}}53.1\cos(10\pi y)\cos(3\pi \times 10^9 t) \quad \text{mA-m}^{-1}$$

(b) Therefore, the nearest location to $y = 0$ boundary where the total electric field is zero at all times can be found from

$$10\pi y = -\pi \quad \longrightarrow \quad y = -0.1\text{ m} = -10\text{ cm}$$

(c) Similarly, the nearest location to $y = 0$ where the total magnetic field is zero at all times can be found from

$$10\pi y = -\pi/2 \quad \longrightarrow \quad y = -5\text{ cm}$$

Note that the two answers correspond to 10 cm $= \lambda/2$ and 5 cm $= \lambda/4$, as expected on the basis of Figure 8.17.

8.2.2 Normal Incidence on a Lossless Dielectric

In the previous section we considered the special case of normal incidence on a plane boundary in which the second medium was a perfect conductor, within which no electromagnetic field can exist. We now consider the more general case in which the second medium is also a lossless dielectric, within which a nonzero electromagnetic field can propagate. When a uniform plane wave propagating in medium 1 is normally incident on an interface with a second medium with a different dielectric constant as shown in Figure 8.21, some of the incident wave energy is transmitted into medium 2 and continues to propagate to the right ($+z$ direction). In the following discussion, we assume both media to be lossless dielectrics (i.e., $\sigma_1, \sigma_2 = 0$). Once again we assume, with no loss of generality under conditions of normal incidence on a planar boundary, that the incident electric field is oriented in the x direction. We also assume that the amplitude E_{i0} of the incident wave is real—once again, with no loss of generality, since this basically amounts to the choice of the time origin, as shown in the previous section. The phasor fields for the incident, reflected, and transmitted waves are given as

$$\left.\begin{aligned}\mathbf{E}_i(z) &= \hat{\mathbf{x}}E_{i0}e^{-j\beta_1 z}\\ \mathbf{H}_i(z) &= \hat{\mathbf{y}}\frac{E_{i0}}{\eta_1}e^{-j\beta_1 z}\end{aligned}\right\}\quad \text{Incident wave}$$

$$\left.\begin{aligned}\mathbf{E}_r(z) &= \hat{\mathbf{x}}E_{r0}e^{+j\beta_1 z}\\ \mathbf{H}_r(z) &= -\hat{\mathbf{y}}\frac{E_{r0}}{\eta_1}e^{+j\beta_1 z}\end{aligned}\right\}\quad \text{Reflected wave}$$

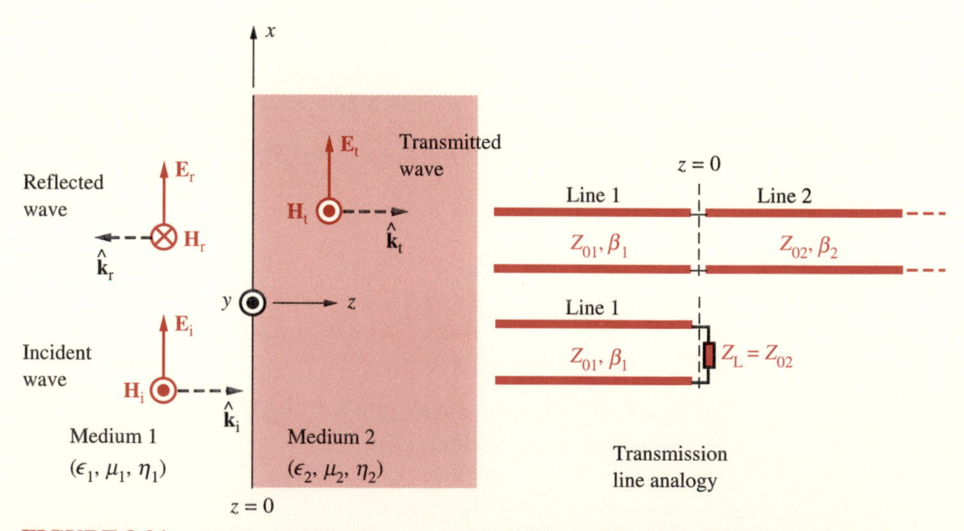

FIGURE 8.21. Uniform plane wave normally incident on a lossless dielectric boundary. Also shown on the right is the transmission line equivalent. Note that the second medium can either be thought of as an infinitely long line with characteristic impedance Z_{02} or as a load with load impedance $Z_L = Z_{02}$.

$$\left.\begin{aligned}\mathbf{E}_t(z) &= \hat{\mathbf{x}}E_{t0}e^{-j\beta_2 z}\\ \mathbf{H}_t(z) &= \hat{\mathbf{y}}\frac{E_{t0}}{\eta_2}e^{-j\beta_2 z}\end{aligned}\right\}\quad \text{Transmitted wave}$$

where $\beta_2 = \omega\sqrt{\mu_2\epsilon_2}$ and $\eta_2 = \sqrt{\mu_2/\epsilon_2}$ are, respectively, the phase constant and the intrinsic impedance for medium 2. Note that E_{t0} is the amplitude (yet to be determined) of the transmitted wave at $z = 0$. In Figure 8.21 we have defined the polarities of $\mathbf{E}_i$ and $\mathbf{E}_r$ to be the same, and taken (as before in Figure 8.16) $\mathbf{H}_r$ to be in the $-y$ direction so that $\mathbf{E}_r \times \mathbf{H}_r$ is in the $-z$ direction. Note that, at this point, the selected orientations of $\mathbf{E}$ and $\mathbf{H}$ for the different waves (incident, reflected, transmitted) are simply convenient choices.[38] The boundary conditions will determine whether the phasor fields at the boundary are positive or negative according to these assumed conventions.

Reflection and Transmission Coefficients We now proceed by taking the incident wave as given and determine the properties of the reflected and transmitted waves so that the fundamental boundary conditions for electromagnetic fields are satisfied at the interface, where all three waves can be related to one another. We have two unknown quantities E_{r0} and E_{t0} to be determined in terms of the incident field amplitude E_{i0}. We will use two boundary conditions to determine them. These two conditions are the continuity of the tangential components of both the electric and magnetic fields across the interface. We thus have

$$\mathbf{E}_i(z=0) + \mathbf{E}_r(z=0) = \mathbf{E}_t(z=0) \quad \longrightarrow \quad E_{i0} + E_{r0} = E_{t0}$$

$$\mathbf{H}_i(z=0) + \mathbf{H}_r(z=0) = \mathbf{H}_t(z=0) \quad \longrightarrow \quad \left(\frac{E_{i0}}{\eta_1} - \frac{E_{r0}}{\eta_1}\right) = \frac{E_{t0}}{\eta_2}$$

The solution of these two equations yields

$$E_{r0} = \frac{\eta_2 - \eta_1}{\eta_2 + \eta_1}E_{i0} \qquad E_{t0} = \frac{2\eta_2}{\eta_2 + \eta_1}E_{i0}$$

We now define[39] the reflection and transmission coefficients as follows:

$$\Gamma = \frac{E_{r0}}{E_{i0}} = \frac{\eta_2 - \eta_1}{\eta_2 + \eta_1} \quad \text{Reflection coefficient} \qquad [8.45]$$

$$\mathcal{T} = \frac{E_{t0}}{E_{i0}} = \frac{2\eta_2}{\eta_2 + \eta_1} \quad \text{Transmission coefficient} \qquad [8.46]$$

---

[38]The fact that selected orientations are simply convenient choices is important, especially since a different convention may be adopted in other texts.

[39]Note that definition of Γ as the ratio of the reflected to the incident *electric* fields (rather than magnetic field) is simply a matter of convention.

where we note that $(1+\Gamma) = \mathcal{T}$. Although we limited our discussions here to lossless dielectrics, the above relations are fully applicable when the media are dissipative, as long as the proper complex values of η_1 and η_2 are used. Note that physically, the above coefficients are derived from application of the boundary conditions, which are valid for all media[40] in general. Complex reflection and transmission coefficients may result when η_2 and/or η_1 are complex (i.e., one or both of the media are lossy), meaning that in addition to the differences in amplitudes, phase shifts are also introduced between the incident, reflected, and transmitted fields at the interface.

The reflection coefficient expression also could have been obtained on the basis of the transmission line analogy depicted in Figure 8.21. We know from Chapter 3 that for a transmission line of characteristic impedance Z_{01} terminated in a load impedance Z_L, the load reflection coefficient is given by

$$\Gamma_L = \frac{Z_L - Z_{01}}{Z_L + Z_{01}}$$

which, upon substitution of $Z_L \rightarrow \eta_2$ and $Z_{01} \rightarrow \eta_1$, is identical to [8.45]. Note that the load could just as well be an infinitely long transmission line with characteristic impedance $Z_{02} \neq Z_{01}$, presenting an impedance $Z_L = Z_{02}$ at the junction, as depicted in Figure 8.21. The analogy between the junction between two transmission lines of infinite extent and the interface between two different dielectric media of infinite extent is thus quite clear.

For most dielectrics and insulators, the magnetic permeability does not differ appreciably from its free-space value, so the expressions derived for Γ and $\mathcal{T}$ can also be simply rewritten in terms of ϵ_1 and ϵ_2. So, when $\mu_1 = \mu_2 = \mu_0$, we have

$$\Gamma = \frac{\sqrt{\epsilon_1} - \sqrt{\epsilon_2}}{\sqrt{\epsilon_1} + \sqrt{\epsilon_2}} \qquad \mathcal{T} = \frac{2\sqrt{\epsilon_1}}{\sqrt{\epsilon_1} + \sqrt{\epsilon_2}}$$

In most optical applications, the electromagnetic properties of the dielectric materials are expressed in terms of the *refractive index,* defined as $n \equiv c/v_p = \beta c/\omega = \sqrt{\mu_r \epsilon_r}$, where v_p is the wave phase velocity. For the nonmagnetic case $(\mu_1 = \mu_2 = \mu_0)$, the reflection and transmission coefficients can also be expressed in terms of the refractive indices as

$$\Gamma = \frac{n_1 - n_2}{n_1 + n_2} \qquad \mathcal{T} = \frac{2n_1}{n_1 + n_2}$$

Examples 8-21 and 8-22 illustrate the application of [8.45] and [8.46] to simple dielectric interfaces.

Example 8-21: Air–water interface. Calculate the reflection and transmission coefficients when a radio frequency uniform plane wave traveling in air is

---

[40]Note that when medium 2 is a perfect conductor, for which we have $\eta_2 = 0$, the expressions reduce to $\Gamma = -1$ and $\mathcal{T} = 0$, as expected.

incident normally upon a calm lake. Assume the water in the lake to be lossless with a relative dielectric constant of $\epsilon_r = 81$. Note that for water, we have $\mu_r = 1$.

Solution: The reflection coefficient is given by

$$\Gamma = \frac{\sqrt{\epsilon_{1r}} - \sqrt{\epsilon_{2r}}}{\sqrt{\epsilon_{1r}} + \sqrt{\epsilon_{2r}}} = \frac{1 - \sqrt{81}}{1 + \sqrt{81}} = -0.8$$

and the transmission coefficient is

$$\mathcal{T} = \frac{2\sqrt{\epsilon_{1r}}}{\sqrt{\epsilon_{1r}} + \sqrt{\epsilon_{2r}}} = \frac{2 \times 1}{1 + \sqrt{81}} = 0.2$$

both of which satisfy the relationship $1 + \Gamma = \mathcal{T}$.

Example 8-22: Air–germanium interface. Germanium (Ge) is a popular material used in infrared optical system designs in either the 3 to 5 μm or 8 to 12 μm spectral bands. Germanium has an index of refraction of approximately 4.0 at these wavelengths. Calculate the reflection and transmission coefficients for an uncoated air–germanium surface.

Solution: The reflection coefficient is given by

$$\Gamma = \frac{n_1 - n_2}{n_1 + n_2} = \frac{1 - 4}{1 + 4} = -0.6$$

and the transmission coefficient is

$$\mathcal{T} = \frac{2n_1}{n_1 + n_2} = \frac{2 \times 1}{1 + 4} = 0.4$$

Once again, the relationship $1 + \Gamma = \mathcal{T}$ is satisfied.

Propagating and Standing Waves In Section 8.2.1, we determined that for a uniform plane wave normally incident on a perfect conductor, the total electric field in medium 1 consisted of a purely standing wave. In the case of normal incidence on a lossless dielectric, we expect at least a portion of the total wave in medium 1 to be propagating in the z direction in order to supply the electromagnetic power taken away from the interface by the transmitted wave in medium 2. To determine the nature of the wave in medium 1, we now examine the total electric field in medium 1. We have

$$\begin{aligned}\mathbf{E}_1(z) &= \mathbf{E}_i(z) + \mathbf{E}_r(z) = \hat{\mathbf{x}}E_{i0}(e^{-j\beta_1 z} + \Gamma e^{+j\beta_1 z}) \\ &= \hat{\mathbf{x}}E_{i0}[(1 + \Gamma)e^{-j\beta_1 z} - \Gamma e^{-j\beta_1 z} + \Gamma e^{+j\beta_1 z}] \\ \mathbf{E}_1(z) &= \hat{\mathbf{x}}E_{i0}[(1 + \Gamma)e^{-j\beta_1 z} + \Gamma j2\sin(\beta_1 z)]\end{aligned}$$

The corresponding space-time field is

$$\boxed{\vec{\mathscr{E}}_1(z, t) = \hat{\mathbf{x}}E_{i0}[\underbrace{\mathcal{T}\cos(\omega t - \beta_1 z)}_{\text{Propagating wave}} + \underbrace{(-2\Gamma)\sin(\beta_1 z)\sin(\omega t)}_{\text{Standing wave}}]} \qquad [8.47]$$

where we note once again that E_{i0} was assumed to be real. Note that the propagating wave in medium 1 sustains the transmitted wave in medium 2, whereas the standing wave is produced by the sum of the reflected wave and a portion of the incident wave. We can also express the total electric field phasor in medium 1 as

$$\boxed{\mathbf{E}_1(z) = \hat{\mathbf{x}}E_{i0}e^{-j\beta_1 z}(1 + \Gamma e^{j2\beta_1 z})}$$

The associated total magnetic field phasor in medium 1 is

$$\begin{aligned}
\mathbf{H}_1(z) &= \hat{\mathbf{y}}\frac{E_{i0}}{\eta_1}(e^{-j\beta_1 z} - \Gamma e^{j\beta_1 z}) \\
&= \hat{\mathbf{y}}\frac{E_{i0}}{\eta_1}[(1 + \Gamma)e^{-j\beta_1 z} - \Gamma e^{-j\beta_1 z} - \Gamma e^{+j\beta_1 z}] \\
&= \hat{\mathbf{y}}\frac{E_{i0}}{\eta_1}[(1 + \Gamma)e^{-j\beta_1 z} - 2\Gamma\cos(\beta_1 z)]
\end{aligned}$$

The corresponding space-time function is

$$\boxed{\overline{\mathscr{H}}_1(z, t) = \hat{\mathbf{y}}\frac{E_{i0}}{\eta_1}[\underbrace{\mathcal{T}\cos(\omega t - \beta_1 z)}_{\text{Propagating wave}} + \underbrace{(-2\Gamma)\cos(\beta_1 z)\cos(\omega t)}_{\text{Standing wave}}]}$$

The total magnetic field phasor in medium 1 can also be expressed in a compact form as

$$\boxed{\mathbf{H}_1(z) = \hat{\mathbf{y}}\frac{E_{i0}}{\eta_1}e^{-j\beta_1 z}(1 - \Gamma e^{j2\beta_1 z})}$$

In medium 2, we only have one wave propagating in the $+z$ direction, represented as

$$\mathbf{E}_2(z) = \mathbf{E}_t(z) = \hat{\mathbf{x}}\mathcal{T}E_{i0}e^{-j\beta_2 z}$$

$$\mathbf{H}_2(z) = \mathbf{H}_t(z) = \hat{\mathbf{y}}\frac{\mathcal{T}}{\eta_2}E_{i0}e^{-j\beta_2 z}$$

with the corresponding space-time field expressions of

$$\boxed{\begin{aligned}
\overline{\mathscr{E}}_t(z, t) &= \hat{\mathbf{z}}\mathcal{T}E_{i0}\cos(\omega t - \beta_2 z) \\
\overline{\mathscr{H}}_t(z, t) &= \hat{\mathbf{y}}\frac{\mathcal{T}}{\eta_2}E_{i0}\cos(\omega t - \beta_2 z)
\end{aligned}}$$

To better understand the relationships between the incident, reflected, and transmitted waves, and the matching of the fields at the boundary, we can examine the instantaneous electric field waveforms shown in Figure 8.22a. These displays are

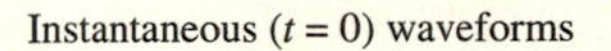

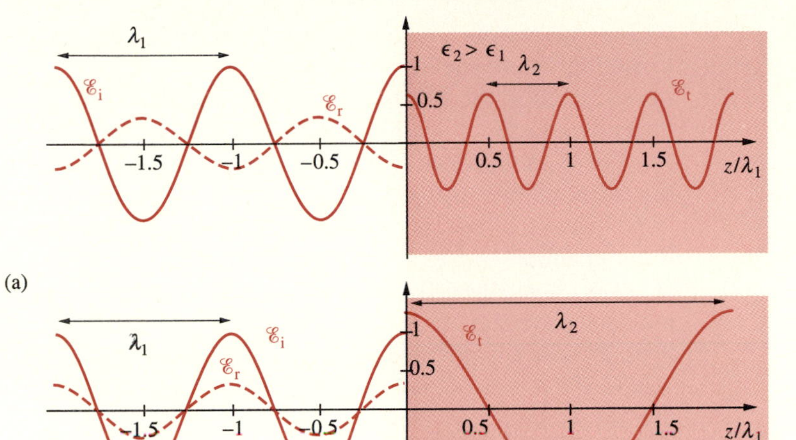

Standing-wave patterns

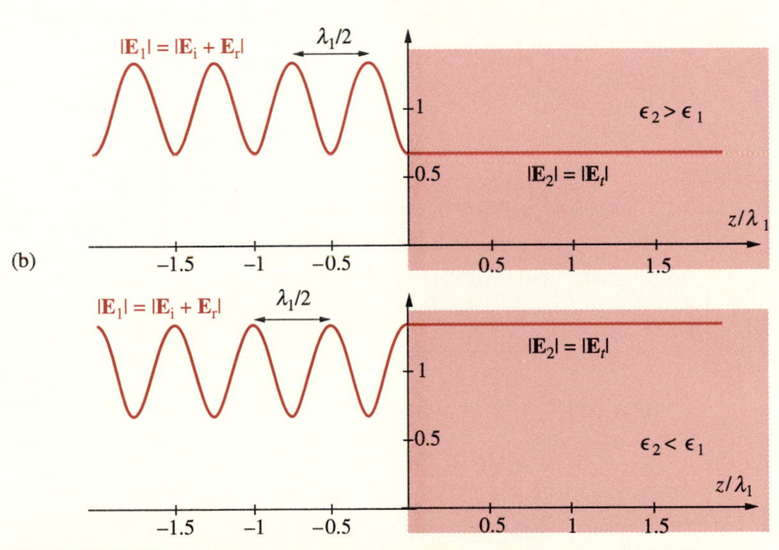

FIGURE 8.22. Electric field waveforms and standing-wave patterns on both sides of the interface. (a) Electric field waveforms are given separately for $\epsilon_1 < \epsilon_2$ (in the example shown, $\epsilon_{2r} = 4\epsilon_{1r}$) and for $\epsilon_1 > \epsilon_2$ ($\epsilon_{1r} = 4\epsilon_{2r}$). The electric field waveforms are shown as a function of position (z) at time instant $\omega t = 0$, with E_{i0} taken to be unity. (b) Also shown are the standing-wave patterns for the same two cases, $\epsilon_{2r} = 4\epsilon_{1r}$ and $\epsilon_{1r} = 4\epsilon_{2r}$. Note that the standing-wave patterns are identical in form to those given in Figure 3.15 for voltage waves on a transmission line.

similar to the waveforms in Figure 8.17, except that the incident and reflected waves in medium 1 are shown separately and only at one instant of time, namely, at $\omega t = 0$. The waveforms are shown for the two different cases of $\epsilon_1 < \epsilon_2$ and $\epsilon_1 > \epsilon_2$ assuming nonmagnetic media, or $\mu_1 = \mu_2 = \mu_0$. The difference in wavelength ($\lambda = 2\pi/\beta$) in the two media is readily apparent; for the cases shown, ϵ_1 and ϵ_2 differ by a factor of 4, which means that wavelengths differ by a factor of 2.

For $\epsilon_1 < \epsilon_2$, once again assuming $\mu_1 = \mu_2$, we note that the amplitude of the transmitted wave is smaller than that of the incident wave (i.e., $E_{t0} < E_{i0}$, or $\mathcal{T} < 1$) and the reflection coefficient Γ is negative. On the other hand, for $\epsilon_1 > \epsilon_2$, the transmitted wave amplitude is actually larger than that of the incident wave ($E_{t0} > E_{i0}$), and the reflected wave electric field is in the same direction as that of the incident wave (i.e., $\Gamma > 0$) as expected from [8.45]. Note, however, that $E_{t0} > E_{i0}$ does not pose any problems with the conservation of wave energy, since the Poynting vector of the transmitted wave is given by $E_{t0}^2/(2\eta_2)$, and power conservation holds true, as we shall see below.

Also shown, in Figure 8.22b, are the standing-wave patterns: the magnitude of $\mathbf{E}_1(z)$ in both media for the two different cases of $\epsilon_1 < \epsilon_2$ and $\epsilon_1 > \epsilon_2$. These are entirely analogous to standing-wave patterns for transmission lines, studied in Chapter 3. In medium 2, there is only one wave, and the standing-wave ratio is thus unity. In medium 1, however, the interference between the reflected and incident waves produces a standing-wave ratio of 3, since we have chosen $\epsilon_{2r} = 4\epsilon_{1r}$ or $\epsilon_{1r} = 4\epsilon_{2r}$ for the two cases shown. Note that in this context, S is given as

$$S = \frac{|\mathbf{E}_1(z)|_{\text{max}}}{|\mathbf{E}_1(z)|_{\text{min}}} = \frac{1 + |\Gamma|}{1 - |\Gamma|} = \frac{1 + 1/3}{1 - 1/3} = 2$$

Electromagnetic Power Flow We can now examine the electromagnetic power flow in both medium 1 and medium 2. Noting that using the complex Poynting vector and the phasor expressions for the fields is most convenient, we have in medium 2

$$(\mathbf{S}_{\text{av}})_2 = \frac{1}{2}\mathcal{R}e[\mathbf{E}_t(z) \times \mathbf{H}_t^*(z)] = \hat{\mathbf{z}}\frac{E_{i0}^2}{2\eta_2}\mathcal{T}^2$$

which follows from Section 8.1.4 assuming E_{i0}^2 is real. In medium 1 we have

$$(\mathbf{S}_{\text{av}})_1 = \tfrac{1}{2}\mathcal{R}e\{\mathbf{E}_1 \times \mathbf{H}_1^*\}$$

$$(\mathbf{S}_{\text{av}})_1 = \hat{\mathbf{z}}\frac{{E_{i0}}^2}{2\eta_1}\mathcal{R}e\{e^{-j\beta_1 z}(1 + \Gamma e^{j2\beta_1 z})e^{+j\beta_1 z}(1 - \Gamma e^{-j2\beta_1 z})\}$$

$$= \hat{\mathbf{z}}\frac{{E_{i0}}^2}{2\eta_1}\mathcal{R}e\{(1 - \Gamma^2) + \Gamma(e^{j2\beta_1 z} - e^{-j2\beta_1 z})\}$$

$$= \hat{\mathbf{z}}\frac{{E_{i0}}^2}{2\eta_1}\mathcal{R}e\{(1 - \Gamma^2) + j2\Gamma\sin(2\beta_1 z)\}$$

$$(\mathbf{S}_{\text{av}})_1 = \hat{\mathbf{z}}\frac{{E_{i0}}^2}{2\eta_1}(1 - \Gamma^2)$$

Note that the time-average power in medium 1 propagates in the $+z$ direction and basically supplies the power carried by the transmitted wave in medium 2. The average power flows in both media must be equal, since no energy is stored or dissipated at the interface between the two media. We thus have

$$(\mathbf{S}_{\text{av}})_1 = (\mathbf{S}_{\text{av}})_2 \quad \longrightarrow \quad \boxed{1 - \Gamma^2 = \frac{\eta_1}{\eta_2}\mathcal{T}^2}$$

which can also be independently verified using the expressions [8.45] and [8.46] for Γ and $\mathcal{T}$.

Example 8-23 applies the above expressions for electromagnetic power flow to an air–glass interface.

Example 8-23: Reflectance of glass. A beam of light is incident normally from air on a BK-7 glass interface. Calculate the reflection coefficient and percent of incident energy reflected if the BK-7 glass has an index of refraction of $n = 1.52$.

Solution: Assuming $\mu_1 = \mu_2$, the reflection coefficient is given by

$$\Gamma = \frac{n_1 - n_2}{n_1 + n_2} = \frac{1 - 1.52}{1 + 1.52} \simeq -0.206$$

The power density of the incident wave is given by $|(\mathbf{S}_{\text{av}})_{\text{i}}| = \frac{1}{2}C_1^2/\eta_1$, where C_1 is the magnitude of the electric field of the incident wave. The power density of the reflected wave is $|(\mathbf{S}_{\text{av}})_{\text{r}}| = \frac{1}{2}(\Gamma C_1)^2/\eta_1 = \Gamma^2|(\mathbf{S}_{\text{av}})_{\text{i}}|$. Therefore, the percent of incident energy reflected back is given by

$$\frac{|(\mathbf{S}_{\text{av}})_{\text{r}}|}{|(\mathbf{S}_{\text{av}})_{\text{i}}|} \times 100 \simeq \Gamma^2 \times 100 \simeq (0.206)^2 \times 100 \simeq 4.26\%$$

Thus, only $\sim$4% of the incident power is reflected by the glass interface. In some applications, this loss may be considered as a significant loss. For example, a camera lens often consists of three or more separate lenses, representing six or more air–glass interfaces; if $\sim$4% of the incoming energy reflects back every time light passes through one of these interfaces, up to $\sim$22% of the original energy is lost during each traverse of light through the lens. It is possible to reduce these losses significantly by introducing antireflection coating on the glass surface, as will be discussed in the next section.

8.2.3 Multiple Dielectric Interfaces

Many practical applications involve the reflection and refraction of electromagnetic waves from dielectric or metallic surfaces that are coated with another dielectric material to reduce reflections and to improve the coupling of the wave energy. The underlying principle in such cases is identical to that in the case of half-wave or quarter-wave transformer matching of transmission lines, studied in

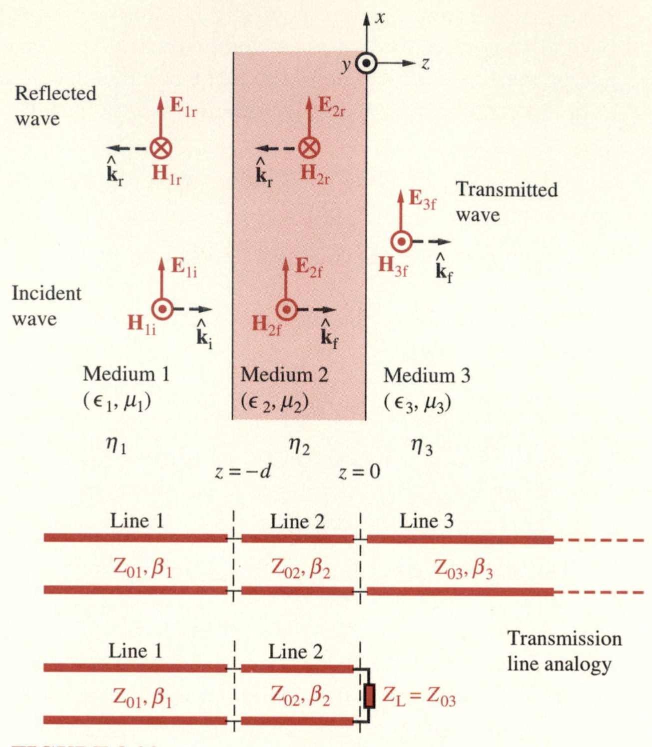

FIGURE 8.23. Normal incidence at multiple dielectric interfaces. Normal incidence of a uniform plane on a multiple dielectric interface is directly analogous to that involving a transmission line with characteristic impedance Z_{01} (medium 1) coupled into a load (medium 3) via another transmission line (characteristic impedance Z_{02}) segment of length d, as shown above.

Chapter 3. Some of the applications include antireflection coatings to improve light transmission of a lens, thin-film coatings on optical components to reduce losses selectively over narrow wavelength ranges, metal mirror coatings, flexible metal foils to provide EMI shielding, radar domes (radomes), and stealth technology involving specially coated aircraft to reduce radar reflectivity.[41]

We formulate the multiple dielectric interface problem as shown in Figure 8.23, in terms of three different dielectric media characterized by intrinsic impedances η_1, η_2, and η_3 and separated by infinite planar boundaries located at $z = -d$ and $z = 0$, so that medium 2 is a layer of thickness d.

---

[41] An example of a coating material that reduces reflections is ferrite-titanate, for which $\mu_r \simeq \epsilon_r \simeq 60(2 - j1)$ at 100 MHz, so that its characteristic impedance, $\eta = \sqrt{\mu/\epsilon}$, is approximately equal to that of free space (377Ω). See Sections 12–16 of J. D. Kraus, *Electromagnetics,* 4th ed., McGraw-Hill, 1992. To provide radar invisibility over a range of frequencies, multilayered coatings are used. See J. A. Adam, How to design an "invisible" aircraft, *IEEE Spectrum,* p. 30, April 1988.

We assume a uniform plane wave propagating in medium 1 to be normally incident at the boundary located at $z = -d$. In attempting to determine the amount of wave energy reflected back, we might at first be tempted to treat the problem in terms of a series of reflections, with a portion of the energy of the incident wave to be transmitted into medium 2; which then propagates to the right and encounters the boundary at $z = 0$, at which time a portion of this energy is reflected; this reflected energy propagates to the left in medium 2, encountering the boundary at $z = -d$ from the right, at which some of it is reflected back and some transmitted into medium 1, and so on. Even though it might lead to the "correct" answer in some cases, such thinking, which implies a "sequence of events" type of a scenario, is inconsistent with our initial steady-state (or time-harmonic) assumption.

Application of Boundary Conditions A fundamentally based determination of the reflected and transmitted waves requires the use of the general solutions of the wave equation with two undetermined constants (amplitudes of component waves propagating, respectively, in the $+z$ and $-z$ directions) in each medium, and application of the boundary conditions to determine these unknown constants. An alternative method is to rely on a transmission line analogy as implied in Figure 8.23, which reduces the problem to one of impedance transformation. Note that the latter approach is valid because *within* each medium the wave propagation obeys equations identical to the transmission line equations, and because the boundary conditions of the continuity of the tangential electric and magnetic fields are precisely equivalent to the boundary conditions for the continuity of voltage and current at a line termination. We proceed with the general solution of the wave equation but also take advantage of the transmission line analogy in interpreting and generalizing our results. The impedance transformation method is discussed in the next subsection.

The expressions for the total wave fields in the three media shown in Figure 8.23 can be written in their most general form as

$$\left.\begin{aligned}\mathbf{E}_1(z) &= \mathbf{E}_{1\mathrm{i}} + \mathbf{E}_{1\mathrm{r}} = \hat{\mathbf{x}}E_{\mathrm{i}0}[e^{-j\beta_1(z+d)} + \Gamma_{\mathrm{eff}}e^{+j\beta_1(z+d)}] \\ \mathbf{H}_1(z) &= \mathbf{H}_{1\mathrm{i}} + \mathbf{H}_{1\mathrm{r}} = \hat{\mathbf{y}}\frac{E_{\mathrm{i}0}}{\eta_1}[e^{-j\beta_1(z+d)} - \Gamma_{\mathrm{eff}}e^{+j\beta_1(z+d)}]\end{aligned}\right\} \quad z < -d \quad \text{Medium 1}$$

$$\left.\begin{aligned}\mathbf{E}_2(z) &= \mathbf{E}_{2\mathrm{f}} + \mathbf{E}_{2\mathrm{r}} = \hat{\mathbf{x}}E_{20}[e^{-j\beta_2 z} + \Gamma_{23}e^{+j\beta_2 z}] \\ \mathbf{H}_2(z) &= \mathbf{H}_{2\mathrm{f}} + \mathbf{H}_{2\mathrm{r}} = \hat{\mathbf{y}}\frac{E_{20}}{\eta_2}[e^{-j\beta_2 z} - \Gamma_{23}e^{+j\beta_2 z}]\end{aligned}\right\} \quad -d < z < 0 \quad \text{Medium 2}$$

$$\left.\begin{aligned}\mathbf{E}_3(z) &= \mathbf{E}_{3\mathrm{f}} = \hat{\mathbf{x}}\mathcal{T}_{\mathrm{eff}}E_{\mathrm{i}0}e^{-j\beta_3 z} \\ \mathbf{H}_3(z) &= \mathbf{H}_{3\mathrm{f}} = \hat{\mathbf{y}}\frac{\mathcal{T}_{\mathrm{eff}}E_{\mathrm{i}0}}{\eta_3}e^{-j\beta_3 z}\end{aligned}\right\} \quad z > 0 \quad \text{Medium 3}$$

where the wave components $\mathbf{E}_{2\mathrm{f}}$ and $\mathbf{E}_{2\mathrm{r}}$ in medium 2 are, respectively, the forward- and reverse-propagating waves that constitute the general solution of the wave equation, and where we recognize that there is no reverse-propagating wave in medium 3. Note that Γ_{23} is the reflection coefficient at the interface $z = 0$, whereas Γ_{eff} is an "effective" reflection coefficient[42] at the interface $z = -d$ that accounts for the effects of the presence of the third medium (and the second interface at $z = 0$). Similarly, $\mathcal{T}_{\text{eff}}$ is an "effective" transmission coefficient. However, due to the presence of the second boundary, Γ_{eff} is *not* simply the single-layer reflection coefficient of $(\eta_2 - \eta_1)/(\eta_2 + \eta_1)$. The solution written above for medium 1 simply recognizes that the amplitudes of the wave components traveling in the $+z$ and $-z$ directions are represented by E_{i0} and $\Gamma_{\text{eff}} E_{i0}$, respectively. However, viewing Γ_{eff} as a reflection coefficient is useful, since in practice (e.g., dielectric coating of glass) the purpose of the second layer is typically to make $\Gamma_{\text{eff}} = 0$ in order to eliminate reflections from the entire multiple-dielectric interface system. Note that although the single-boundary reflection coefficient Γ was necessarily real (i.e., $\Gamma = \rho e^{j(0 \text{ or } \pi)}$) for lossless media, the effective reflection coefficient $\Gamma_{\text{eff}} = \rho_{\text{eff}} e^{j\phi_\Gamma}$ is in general complex even in the lossless case. That this should be so can be easily understood by considering the transmission line analogy shown in Figure 8.23 and referring to [3.31]; the dielectric layer of thickness d simply acts as an impedance transformer, which takes the intrinsic impedance η_3 of medium 3 and presents it at the $z = -d$ interface as another impedance, which in general is complex. Similarly, the effective transmission coefficient $\mathcal{T}_{\text{eff}} = |\mathcal{T}_{\text{eff}}| e^{j\phi_\mathcal{T}}$ is in general also complex.

We can now apply the electromagnetic boundary conditions, namely, the requirement that the tangential components of the electric and magnetic fields be continuous at the two interfaces. At $z = -d$, we find

$$E_{i0}(1 + \Gamma_{\text{eff}}) = E_{20}(e^{+j\beta_2 d} + \Gamma_{23} e^{-j\beta_2 d}) \qquad [8.48a]$$

$$\frac{E_{i0}}{\eta_1}(1 - \Gamma_{\text{eff}}) = \frac{E_{20}}{\eta_2}(e^{+j\beta_2 d} - \Gamma_{23} e^{-j\beta_2 d}) \qquad [8.48b]$$

Similarly, at $z = 0$ we find

$$E_{20}(1 + \Gamma_{23}) = \mathcal{T}_{\text{eff}} E_{i0} \qquad [8.49a]$$

$$\frac{E_{20}}{\eta_2}(1 - \Gamma_{23}) = \frac{\mathcal{T}_{\text{eff}} E_{i0}}{\eta_3} \qquad [8.49b]$$

Note that we have four simultaneous equations in terms of the four unknowns E_{20}, $\mathcal{T}_{\text{eff}}$, Γ_{eff}, and Γ_{23}, which can be determined in terms of the incident field amplitude E_{i0} and the parameters of the media, η_1, η_2, and η_3.

To solve, we start by multiplying [8.49*b*] by η_3 and subtracting from [8.49*a*] to find

$$E_{20}\left(1 - \frac{\eta_3}{\eta_2}\right) + \Gamma_{23} E_{20}\left(1 + \frac{\eta_3}{\eta_2}\right) = 0$$

---

[42]In this context, by an "effective" reflection coefficient we mean one that is defined as the ratio of the reflected to the incident electric field phasors at the boundary at $z = -d$, so that Γ_{eff} determines the amplitude and phase of the reflected wave in medium 1 in terms of that of the incident wave. Note, however, that $\Gamma_{\text{eff}} \neq (\eta_2 - \eta_1)/(\eta_2 + \eta_1)$, since Γ_{eff} must also include the effect of the second boundary at $z = 0$.

or

$$\Gamma_{23} = \frac{\dfrac{\eta_3}{\eta_2} - 1}{\dfrac{\eta_3}{\eta_2} + 1} = \frac{\eta_3 - \eta_2}{\eta_3 + \eta_2}$$

which is as expected, since there is only one wave in medium 3, and therefore the matching of the boundary conditions at the $z = 0$ interface is identical to that of a two-medium, single-interface problem as studied in the previous section. Accordingly, the expression for Γ_{23} is simply [8.45] applied to the interface between media 2 and 3.

We can now rewrite equations [8.48*a*] and [8.48*b*] using the above expression for Γ_{23}, divide them by one another, and manipulate further to find

$$\Gamma_{\text{eff}} = \frac{(\eta_2 - \eta_1)(\eta_3 + \eta_2) + (\eta_2 + \eta_1)(\eta_3 - \eta_2)e^{-2j\beta_2 d}}{(\eta_2 + \eta_1)(\eta_3 + \eta_2) + (\eta_2 - \eta_1)(\eta_3 - \eta_2)e^{-2j\beta_2 d}} \qquad [8.50]$$

The effective transmission coefficient $\mathcal{T}_{\text{eff}}$ can be also determined by similar manipulations of equations [8.48*a*] and [8.48*b*]; we find

$$\mathcal{T}_{\text{eff}} = \frac{4\eta_2\eta_3 e^{-j\beta_2 d}}{(\eta_2 + \eta_1)(\eta_3 + \eta_2) + (\eta_2 - \eta_1)(\eta_3 - \eta_2)e^{-j2\beta_2 d}} \qquad [8.51]$$

Note that the foregoing expressions for Γ_{eff} and $\mathcal{T}_{\text{eff}}$ are completely general, since they were derived by applying the fundamental electromagnetic boundary conditions. Although our treatment in this section is limited to the cases of interfaces involving lossless dielectrics, [8.50] and [8.51] are also fully applicable when one or more of the three media is lossy, in which case η_1, η_2, and η_3 may be complex and $j\beta_2$ must be replaced by $\gamma_2 = \alpha_2 + j\beta_2$. We discuss multiple interfaces involving lossy media in Section 8.2.4, using the generalized forms of [8.50] and [8.51].

Based on power arguments similar to those used in Section 8.2.2 for single interfaces, the net time-average electromagnetic power propagating in the $+z$ direction in medium 1 must be equal to that carried in the same direction by the transmitted wave in medium 3, because medium 2 is lossless and therefore cannot dissipate power. Accordingly, the following relation between Γ_{eff} and $\mathcal{T}_{\text{eff}}$ must hold true:

$$1 - |\Gamma_{\text{eff}}|^2 = \frac{\eta_1}{\eta_3}|\mathcal{T}_{\text{eff}}|^2$$

To illustrate the dependence of the reflection coefficient on the thickness d of the dielectric slab, we study the practically useful case of nonmagnetic media with $\epsilon_1 < \epsilon_2 < \epsilon_3$ with $\epsilon_1 = \epsilon_0$, and for two different example media 3, namely, water ($\epsilon_3 = 81\epsilon_0$) and glass ($\epsilon_3 = 2.25\epsilon_0$). Plots of the magnitudes and phases of Γ_{eff} for the two cases are shown in Figure 8.24. For each case, we show results for three different values of ϵ_2.

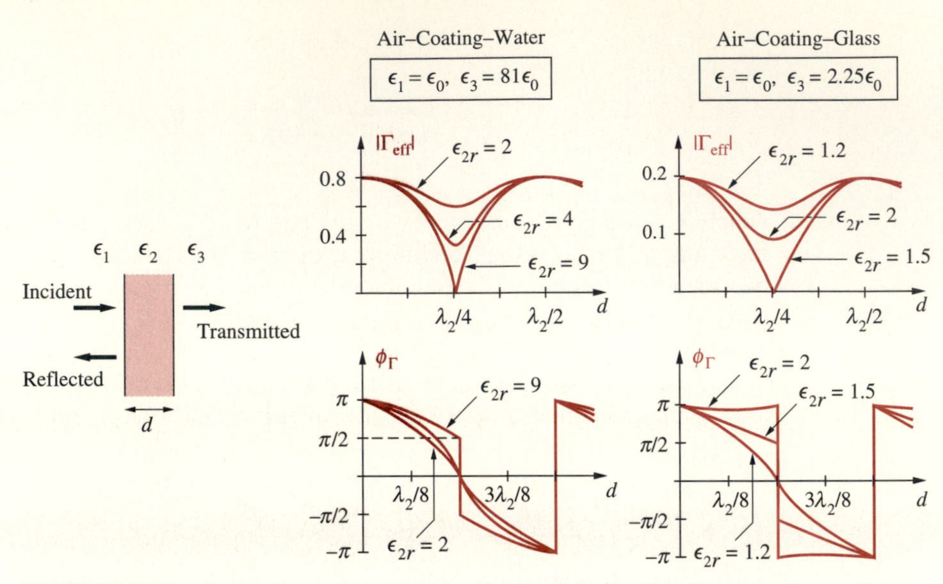

FIGURE 8.24. Reflection and transmission from a dielectric slab. The reflection coefficients of dielectric slabs are shown as functions of the slab thickness d (given in terms of the wavelength λ_2 in the slab) and dielectric constant ϵ_{2r}. Results are given for two examples, namely, the air–slab–water and air–slab–glass interfaces.

We note from Figure 8.24 that Γ_{eff} varies significantly as a function of d. In each case, the minimum value of $|\Gamma_{\text{eff}}|$ (and hence the maximum value of $|\mathcal{T}_{\text{eff}}|$) occurs at $d = \lambda_2/4$, and further, $|\Gamma_{\text{eff}}| = 0$ when $\epsilon_2 = \sqrt{\epsilon_1\epsilon_3}$. Under these conditions, all the energy is transmitted from medium 1 into medium 3. We note that with $\epsilon_2 = \sqrt{\epsilon_1\epsilon_3}$ (or $\eta_2^2 = \eta_1\eta_3$) and $d = \lambda_2/4$ we have from [8.51]

$$\mathcal{T}_{\text{eff}} = \sqrt{\frac{\eta_3}{\eta_1}}e^{-j\pi/2}$$

so that the condition

$$1 - |\Gamma_{\text{eff}}|^2 = \frac{\eta_1}{\eta_3}|\mathcal{T}_{\text{eff}}|^2$$

is satisfied with $\Gamma_{\text{eff}} = 0$ for $d = \lambda_2/4$. The $e^{-j\pi/2}$ factor simply represents the phase difference over a distance $d = \lambda_2/4$, or the fact that, for example at $\omega t = 0$, the peak of the incident wave is at $z = -d = -\lambda_2/4$, whereas the peak of the transmitted wave at $z = 0$ as measured occurs at $\omega t = \pi/2$ rather that at $\omega t = 0$.

When the slab thickness is $d = \lambda_2/2$, we have from [8.50]

$$\Gamma_{\text{eff}} = \frac{\eta_3 - \eta_1}{\eta_3 + \eta_1}$$

so that the system behaves as if the slab were not present. In the air-slab-water example we have $\Gamma_{13} = (1 - \sqrt{81})/(1 + \sqrt{81}) = -0.8 = 0.8e^{j\pi}$, as can be also seen from Figure 8.24. These results are reminiscent of the quarter-wavelength or half-wavelength transmission line impedance transformers studied in Chapter 3 and in fact can be arrived at by using the transmission line analogy implied in Figure 8.23, as we show in the next subsection.

Impedance Transformation and Transmission Line Analogy The transmission line analogy implied in Figure 8.23 provides a useful and systematic method of solution of the problem of multiple dielectric interfaces, especially those involving more than three dielectric regions. Such a solution method, based on successive impedance transformations on transmission lines, can follow along lines very similar to methods used in Chapter 3, for example, in determining the input impedance of cascaded transmission lines, in Examples 3-14 and 3-19. For this purpose, we define the concept of *wave impedance* of the total electromagnetic field at any position z in a given medium as

$$Z(z) \equiv \frac{[E_x(z)]_{\text{total}}}{[H_y(z)]_{\text{total}}}$$

where we implicitly assume a wave with an x-directed electric field propagating in the z direction. Note that, in this context, $E_x(z)$ and $H_y(z)$ are scalar quantities entirely analogous to $V(z)$ and $I(z)$ on a transmission line.

Referring back to the three-media problem depicted in Figure 8.23, the wave impedance of the total field in medium 3 is

$$Z_3(z) = \frac{E_{3x}(z)}{H_{3y}(z)} = \eta_3$$

Note that since the field in medium 3 consists only of a single wave traveling in the $+z$ direction, $Z_3(z) = \eta_3$ is independent of position z. This result is similar to the line impedance on an infinitely long or match-terminated transmission line, and η_3 is entirely analogous to the characteristic impedance of this line.

In media 2 and 1, on the other hand, the total fields are constituted by two waves traveling in the $+z$ and $-z$ directions, so the wave impedances are not simply η_2 or η_1. In medium 2, for example, the wave impedance has to take into account medium 3, which lies beyond the $z = 0$ interface and acts as a "load" (of impedance $Z_L = \eta_3$) on the second "transmission line." To find the ratio of E_{2x} and H_{2y}, we need to know the relative amplitudes of the two traveling waves within the layer; that is, Γ_{23}. We already know that the reverse-propagating wave in medium 2 (i.e., $E_{20}\Gamma_{23}e^{+j\beta_2 z}$) is related to the forward-propagating wave in medium 3. In other words, the "reflection coefficient" that we "see" looking toward the load from a line with characteristic impedance Z_0 at the $z = 0$ interface is

$$\Gamma_{23} = \frac{Z_L - Z_0}{Z_L + Z_0} \quad \rightarrow \quad \Gamma_{23} = \frac{\eta_3 - \eta_2}{\eta_3 + \eta_2}$$

This result is also expected on the basis of the fact that the field structure at the last interface is exactly the three-wave problem we studied in the case of a single dielectric interface, with an incident wave within the layer (here of unknown amplitude

E_{20} rather than E_{i0}) and a reflected wave (of amplitude $\Gamma_{23}E_{20}$ rather than ΓE_{i0}). The above result for Γ_{23} then follows directly from expression [8.45] for Γ.

Using [3.31], the wave impedance in medium 2 is then given by

$$Z_2(z) = \frac{E_{2x}(z)}{H_{2y}(z)} = \eta_2 \frac{e^{-j\beta_2 z} + \Gamma_{23}e^{+j\beta_2 z}}{e^{-j\beta_2 z} - \Gamma_{23}e^{+j\beta_2 z}}$$

which in general results in a complex impedance. At $z = 0$ the above reduces to

$$Z_2(z = 0) = \eta_2 \frac{1 + \Gamma_{23}}{1 - \Gamma_{23}} = \eta_2 \frac{1 + \dfrac{\eta_3 - \eta_2}{\eta_3 + \eta_2}}{1 - \dfrac{\eta_3 - \eta_2}{\eta_3 + \eta_2}} = \eta_2 \frac{2\eta_3}{2\eta_2} = \eta_3$$

which is as expected. The wave impedance in medium 2 at $z = -d$ is given by

$$Z_2(-d) = \eta_2 \frac{e^{+j\beta_2 d} + \Gamma_{23}e^{-j\beta_2 d}}{e^{+j\beta_2 d} - \Gamma_{23}e^{-j\beta_2 d}} = \eta_2 \frac{\eta_3 + j\eta_2 \tan(\beta_2 d)}{\eta_2 + j\eta_3 \tan(\beta_2 d)} \qquad [8.52]$$

Using the expressions for the total electric and magnetic fields in medium 1, we can write the wave impedance in medium 1 as

$$Z_1(z) = \frac{E_{1x}(z)}{H_{1y}(z)} = \eta_1 \frac{e^{-j\beta_1(z+d)} + \Gamma_{\text{eff}}e^{+j\beta_1(z+d)}}{e^{-j\beta_1(z+d)} - \Gamma_{\text{eff}}e^{+j\beta_1(z+d)}}$$

which at $z = -d$ reduces to

$$Z_1(z = -d) = \eta_1 \frac{1 + \Gamma_{\text{eff}}}{1 - \Gamma_{\text{eff}}}$$

Since the electromagnetic boundary conditions require the continuity of the tangential electric (E_x) and magnetic (H_y) field components, and since wave impedance is defined as the ratio of these two quantities, the wave impedance on both sides of any of the interfaces must be equal. In other words, we must have

$$Z_1(z = -d) = \eta_1 \frac{1 + \Gamma_{\text{eff}}}{1 - \Gamma_{\text{eff}}} = Z_2(z = -d) \quad \rightarrow \quad \Gamma_{\text{eff}} = \frac{Z_2(-d) - \eta_1}{Z_2(-d) + \eta_1}$$

where $Z_2(-d)$ is given by [8.52].

In summary, a dielectric slab of thickness d (i.e., medium 2) inserted between two media (i.e., media 1 and 3) essentially works like a transmission line impedance transformer, transforming the impedance of medium 3 from its intrinsic value of η_3 to $Z_2(-d)$. The impedance $Z_2(-d)$ is what is "seen" looking toward the "load" at the $z = -d$ interface. Without the slab, the reflection coefficient at the interface of media 1 and 3 is given by

$$\Gamma_{13} = \frac{\eta_3 - \eta_1}{\eta_3 + \eta_1}$$

With the slab present, the effective reflection coefficient Γ_{eff} can be simply found by substituting $Z_2(-d)$ for η_3, namely,

$$\Gamma_{\text{eff}} = \rho_{\text{eff}} e^{j\phi_\Gamma} = \frac{Z_2(-d) - \eta_1}{Z_2(-d) + \eta_1} \qquad [8.53]$$

As an example, consider the case when medium 1 and medium 3 are the same (i.e., $\eta_1 = \eta_3$) and the thickness of the medium 2 is an integer multiple of a half-wavelength in that medium. Since $\beta_2 d = (2\pi d)/\lambda_2$, where $d = (n\lambda_2)/2$, $n = 0, 1, 2, \ldots$, we have $\tan(\beta_2 d) = 0$, yielding

$$Z_2(-d) = \eta_3$$

In addition, since $\eta_1 = \eta_3$, we find $\Gamma_{\text{eff}} = 0$, resulting in total transmission from medium 1 into medium 3. Note that this is true for any dielectric slab (i.e., a dielectric material with any ϵ_2) as long as its thickness is an integer multiple of a half-wavelength in that dielectric at the frequency of operation.

Another case of interest is that in which the intrinsic impedance of medium 2 is the geometric mean of the intrinsic impedances of media 1 and 3 (i.e., $\eta_2 = \sqrt{\eta_1\eta_3}$), and the thickness of medium 2 is an odd integer multiple of a quarter-wavelength in that medium. Since $\beta_2 d = (2\pi d)/\lambda_2$, where $d = [(2n+1)\lambda_2]/4$, $n = 0, 1, 2, \ldots$, we have $\tan(\beta_2 d) = \pm\infty$, yielding

$$\lim_{\tan(\beta_2 d)\to\pm\infty} [Z_2(-d)] = \lim_{\tan(\beta_2 d)\to\pm\infty} \left[\eta_2 \frac{\eta_3 + j\eta_2\tan(\beta_2 d)}{\eta_2 + j\eta_3\tan(\beta_2 d)}\right] = \frac{\eta_2^2}{\eta_3}$$

But in addition, since $\eta_2^2 = \eta_1\eta_3$, we have $Z_2(-d) = \eta_1$, resulting in total transmission from medium 1 into medium 3.

The results displayed in Figure 8.24 can now be better understood as simply a manifestation of the quarter-wave type of transmission line matching. The variation of $|\Gamma_{\text{eff}}|$ with normalized slab thickness d/λ_2 provides a measure of the frequency bandwidth over which the matching is effective, since for a given value of d the curves show the variation of $|\Gamma_{\text{eff}}|$ with wavelength or frequency.

The transmission line analogy can easily be extended to problems involving more than two dielectric boundaries, which can be modeled in the same manner by using a series of cascaded transmission lines. Multiple-layer coatings to reduce reflections may be utilized because single-layer coating materials with the desired refractive index (i.e., n_2) are not convenient to use (e.g., they are not structurally self-supporting at the thicknesses desired) or are not available. Also, multiple-layer coatings are generally more desirable in order to achieve the "matching" (i.e., no reflection) condition over a broader range of wavelengths,[43] in a manner quite analogous to the multiple-stage quarter-wavelength transmission line matching discussed in Example 3-19.

Examples 8-24, 8-25, and 8-26 illustrate selected applications of reflection from multiple dielectric interfaces.

---

[43]For extensive discussion of multiple-layer coatings for optical applications see J. D. Rancourt, *Optical Thin Films User's Handbook,* McGraw-Hill, 1987.

Example 8-24: Radome design. A radar dome, or *radome,* is a protective dielectric enclosure for a microwave antenna. A ground-based C-band microwave landing system used to help airplanes to land is to be protected from weather by enclosing it in a radome. The center frequency of the operating frequency band is 5 GHz. Thermoplastic PEI material (assume lossless, nonmagnetic, with $\epsilon_{2r} \simeq 3$) is chosen for the design. (a) Assuming a flat planar radome as shown in Figure 8.25a, determine the minimum thickness of the radome that will give no reflections at 5 GHz. (b) If the frequency is changed to 4 GHz and the thickness of the radome remains as in part (a), what percentage of the incident power is reflected? (c) Repeat part (b) for 6 GHz.

Solution:

(a) At 5 GHz, the wavelength in thermoplastic PEI material is

$$\lambda_2 = \frac{\lambda_1}{\sqrt{\epsilon_{2r}}} = \frac{c}{f\sqrt{\epsilon_{2r}}} \simeq \frac{3 \times 10^8 \text{ m-s}^{-1}}{(5 \times 10^9 \text{ Hz})\sqrt{3}} \simeq 3.46 \times 10^{-2} \text{ m} = 3.46 \text{ cm}$$

In order not to affect the operation of the microwave landing system, the thickness d of the radome layer should be an integer multiple of $\lambda_2/2$, where $\lambda_2/2 \simeq 1.73$ cm, which is also the minimum thickness required.

(b) Using the transmission line method (see Figure 8.25b), we start by evaluating $Z_2(z = -d)$. Assuming that the radome is designed with the minimum thickness required (i.e., $d = d_{\text{min}} \simeq 1.73$ cm) and noting that at $f = 4$ GHz, $\lambda_1 = (3 \times 10^8)/(4 \times 10^9) = 0.075$ m or 7.5 cm, $\lambda_2 = \lambda_1/\sqrt{\epsilon_{2r}} = 7.5/\sqrt{3} \simeq 4.33$ cm, and $\tan(\beta_2 d) = \tan[(2\pi/\lambda_2)d] \simeq \tan[(2\pi/4.33)1.73] \simeq -0.727$, and $\eta_2 = \eta_1/\sqrt{\epsilon_{2r}} \simeq 377/\sqrt{3} \simeq 218\Omega$, we have from [8.52]

$$\begin{aligned} Z_2(z = -d) &= \eta_2 \frac{\eta_3 + j\eta_2 \tan(\beta_2 d)}{\eta_2 + j\eta_3 \tan(\beta_2 d)} \\ &\simeq 218 \frac{377 + j218(-0.727)}{218 + j377(-0.727)} \\ &\simeq 218 \frac{409e^{-j22.8°}}{350e^{-j51.5°}} \simeq 254e^{j28.8°}\,\Omega \end{aligned}$$

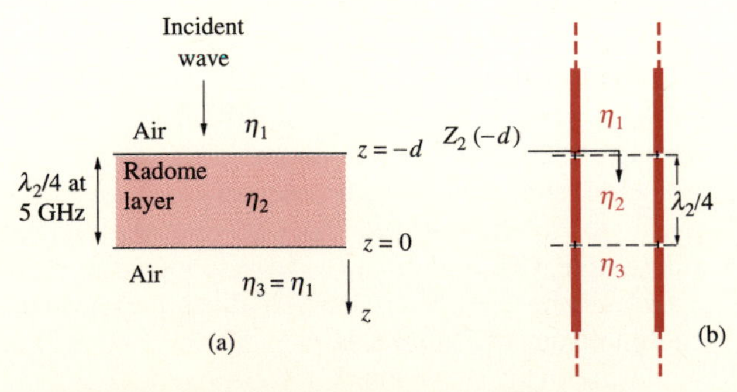

FIGURE 8.25. Radome design. (a) Geometry of the radome layer. (b) Transmission line analog.

Therefore, using [8.53], the effective reflection coefficient is

$$\Gamma_{\text{eff}} = \frac{Z_2(-d) - \eta_1}{Z_2(-d) + \eta_1} \simeq \frac{254e^{j28.8^\circ} - 377}{254e^{j28.8^\circ} + 377}$$

$$\simeq \frac{-154 + j122}{600 + j122} \simeq \frac{197e^{j142^\circ}}{612e^{j11.5^\circ}} \simeq 0.321e^{j130^\circ}$$

Hence, the percentage of the incident power that reflects back can be calculated as

$$\frac{|(\mathbf{S}_{\text{av}})_{\text{r}}|}{|(\mathbf{S}_{\text{av}})_{\text{i}}|} \times 100 = |\Gamma_{\text{eff}}|^2 \times 100 \simeq (0.321)^2 \times 100 \simeq 10.3\%$$

(c) Following a similar procedure, at $f = 6$ GHz, we have $\lambda_1 = 5$ cm, $\lambda_2 \simeq 2.89$ cm, $\tan(\beta_2 d) = \tan(216^\circ) \simeq 0.727$, $\eta_2 \simeq 218\Omega$, and [8.52] yields

$$Z_2(z = -d) \simeq 218\frac{377 + j218(0.727)}{218 + j377(0.727)} \simeq 218\frac{409e^{j22.8^\circ}}{350e^{j51.5^\circ}} \simeq 254e^{-j28.8^\circ}\Omega$$

Using [8.53], the effective reflection coefficient is

$$\Gamma_{\text{eff}} = \frac{254e^{-j28.8^\circ} - 377}{254e^{-j28.8^\circ} + 377} \simeq \frac{-154 - j122}{600 - j122} \simeq 0.321e^{-130^\circ}$$

Hence, the percentage of the incident power that reflects back is $(0.321)^2 \times 100 \simeq 10.3\%$. Thus we see that the effective reflection coefficient varies quite symmetrically around the design frequency of 5 GHz, being down by the same amount in magnitude at 4 GHz as at 6 GHz.

Example 8-25: Coated glass surface. Consider Example 8-23 on the reflectance of glass. Determine the refractive index and minimum thickness of a film to be deposited on the glass surface ($n_3 = 1.52$) such that no normally incident visible light of free-space wavelength 550 nm (i.e., ~545 THz) is reflected.

Solution: Since the permittivity of air is equal to ϵ_0, the wavelength in air (medium 1) is $\lambda_1 = 550$ nm. Since we have nonmagnetic media the requirement of $\eta_2 = \sqrt{\eta_1\eta_3}$ is equivalent to $n_2 = \sqrt{n_1 n_3}$, and we can find the refractive index of the film as $n_2 = \sqrt{1 \times 1.52} \simeq 1.23$. The minimum thickness of the film is

$$d = d_{\text{min}} = \frac{\lambda_2}{4} = \frac{\lambda_1}{4n_2} = \frac{550 \times 10^{-9}}{4 \times 1.23} \simeq 0.112\ \mu\text{m}$$

Note that once this film is deposited on glass, it will eliminate reflections completely only at 550 nm. Although there would still be reflections at other wavelengths in the vicinity of 550 nm, the percentage of the reflected power will be less than the ~4% found in Example 8-23.

In practice, it is typically not possible to manufacture antireflection coating materials that have precisely the desired refractive index. A practical coating

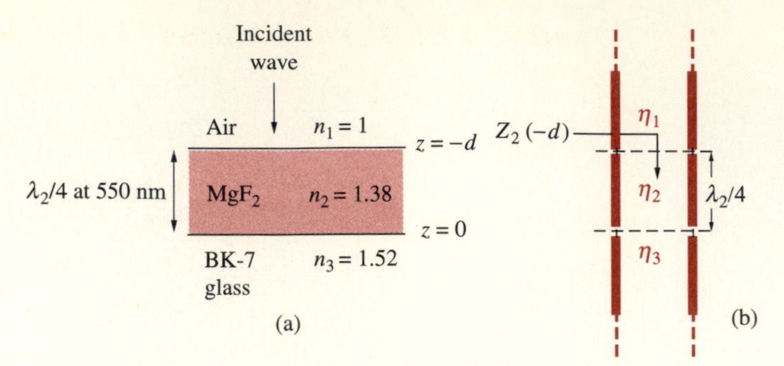

FIGURE 8.26. Air–glass interface with single MgF_2 coating layer. (a) Geometry of the antireflection coating layer. (b) Transmission line analog.

material that is commonly used is magnesium fluoride (MgF_2, $n_2 = 1.38$; see Figure 8.26a). For visible light with free-space wavelength of 550 nm, the wavelength in MgF_2 is $\lambda_2 = 550/1.38 \simeq 399$ nm, so the thickness of the quarter-wavelength MgF_2 layer is $d = \lambda_2/4 \simeq 99.6$ nm.

Since we know the matching is not perfect (i.e., $n_{MgF_2} \neq 1.23$) we can calculate the reflection coefficient for $\lambda_1 = 550$ nm. For this purpose, we can plug in values of η_1, η_2, η_3, and d in [8.50] or use the transmission line analogy described earlier. We use the latter approach here for the purposes of illustration and also because the transmission line method can be applied to multiple coating layers, as discussed in the next example. To use the transmission line method, we start by evaluating $Z_2(z = -d)$. Noting that $d = \lambda_2/4$, so that we have $\tan(\beta_2 d) = \tan[(2\pi/\lambda_2)(\lambda_2/4)] = \infty$, and that $\eta_1 = 377\Omega$, $\eta_2 = \eta_1/1.38$, and $\eta_3 = \eta_1/1.52$, we have [8.52]

$$Z_2(z = -d) = \eta_2 \frac{\eta_3 + j\eta_2 \tan(\beta_2 d)}{\eta_2 + j\eta_3 \tan(\beta_2 d)} = \frac{\eta_2^2}{\eta_3} = \frac{(377)(1.52)}{(1.38)^2} \simeq 301\Omega$$

Therefore, using [8.53] we have

$$\Gamma_{\text{eff}} = \frac{Z_2(-d) - \eta_1}{Z_2(-d) + \eta_1} \simeq \frac{301 - 377}{301 + 377} \simeq -0.112$$

so the fraction of the incident power reflected at 550 nm is $\rho_{\text{eff}}^2 \simeq (0.112)^2 \simeq 0.0126$, or 1.26%, substantially less than the ~4% reflection that occurs without any coating. To assess the effectiveness of the MgF_2 coating layer in reducing reflections over the visible spectrum, we can calculate the reflection coefficient at other wavelengths. For example, at 400 nm (violet light) we have

$$\tan(\beta_2 d) = \tan\left(\frac{2\pi}{\lambda_2} \cdot \frac{\lambda_2}{4} \cdot \frac{550}{400}\right) = \tan(123.75°) \simeq -1.50$$

so we have

$$Z_2(z = -d) \simeq \left(\frac{377}{1.38}\right)\frac{(377/1.52) + j(377/1.38)(-1.50)}{(377/1.38) + j(377/1.52)(-1.50)}$$

$$= \left(\frac{377}{1.38}\right)\frac{1.38 - j(1.50)(1.52)}{1.52 - j(1.50)(1.38)} \simeq 283.45e^{-j5.11^\circ}$$

and therefore from [8.53]

$$\Gamma_{\text{eff}} = \frac{Z_2(-d) - \eta_1}{Z_2(-d) + \eta_1} \simeq \frac{283e^{-j5.11^\circ} - 377}{283e^{-j5.11^\circ} + 377} \simeq \frac{-94.67 - j25.24}{659.3 - j25.24} \simeq 0.1485e^{-j163^\circ}$$

so the percentage of reflected power in terms of the incident power is

$$\Gamma_{\text{eff}}^2 \times 100 = (0.149)^2 \times 100 \simeq 2.21\%$$

At the other end of the visible light spectrum, for red light at 750 nm, we have

$$\tan(\beta_2 d) = \tan\left(\frac{2\pi}{\lambda_2}\cdot\frac{\lambda_2}{4}\cdot\frac{550}{750}\right) \simeq 2.246$$

and using [8.52], we have

$$Z_2(-d) \simeq \left(\frac{377}{1.38}\right)\frac{(1.52)^{-1} + j(1.38)^{-1}(2.246)}{(1.38)^{-1} + j(1.52)^{-1}(2.246)} \simeq 290.65 + j20.90$$

which from [8.53] gives

$$\Gamma_{\text{eff}} = \frac{Z_2(-d) - \eta_1}{Z_2(-d) + \eta_1} \simeq 0.13e^{j164.6^\circ}$$

so the percentage of incident power that reflects back is $\sim$1.7%, still substantially lower than the $\sim$4% we have without any coating. The above results for MgF_2 coating are consistent with the plot of $|\Gamma_{\text{eff}}|$ versus d given in Figure 8.24 for $\epsilon_{3r} = 2.25$ (glass with $n_3 = 1.5$) and $\epsilon_{2r} = 2$ (close to $\epsilon_{MgF_2} \simeq 1.9$ or $n_{MgF_2} \simeq 1.38$).

Example 8-26: Multiple-layered coatings. Antireflection coatings are required in most optical applications in order to reduce unwanted reflections at the surfaces of optical elements. In most cases, it is desirable to reduce the surface reflectivity (or simply the reflection coefficient) to an extremely low value over an extended spectral region so as to maintain proper color balance and provide optimum efficiency. Single-layer coatings of magnesium fluoride generally do not satisfy these requirements because, as seen in Example 8-25, they can only reduce reflectivity from $\sim$4% down to $\sim$2.2%. A commonly used technique to achieve the desired requirements is triple-layer coatings.[44] To illustrate the improvements brought about by the use of multiple-layer coating, consider a wideband nonmagnetic antireflection

[44]First put forth by J. T. Cox, G. Hass, and A. Theelen, Triple-layer antireflection coatings on glass for the visible and near infrared, *J. Opt. Soc. Am.*, 52, p. 965, 1962; also see J. D. Rancourt, *Optical Thin Film User's Handbook*, McGraw-Hill, 1987.

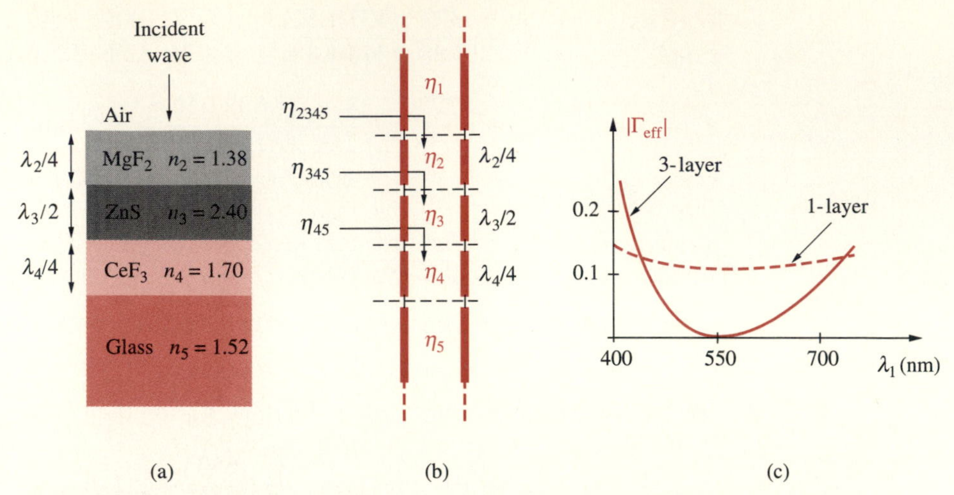

FIGURE 8.27. **Triple-layer coating.** (a) A modern quarter-half-quarter wideband antireflection triple-layer coating system for air-glass interface. The outermost layer is MgF_2, the middle layer is ZnS, and the bottom layer is CeF_3. (b) The transmission line analog is also shown for reference. (c) Plot of $|\Gamma_{\rm eff}|$ versus free space wavelength for single- and three-layer coatings.

coating system[45] for an air–glass interface, involving three layers of coating materials, as shown in Figure 8.27a. Analyze this system by extending the transmission line analogy method and plot $|\Gamma_{\rm eff}|$ over the wavelength range 400 to 750 nm.

Solution: We take the design wavelength for this system to be green light, with $\lambda_1 = 550$ nm. With the refractive indices of the layers being $n_2 = 1.38$, $n_3 = 2.4$, and $n_4 = 1.70$, the thicknesses of the coating layers are

$$d_2 = \frac{\lambda_2}{4} = \frac{\lambda_1}{4n_2} \simeq 99.6 \text{ nm}$$

$$d_3 = \frac{\lambda_3}{2} = \frac{\lambda_1}{2n_3} \simeq 115 \text{ nm}$$

$$d_4 = \frac{\lambda_4}{4} = \frac{\lambda_1}{4n_4} \simeq 80.9 \text{ nm}$$

The phase constants in the three coating materials are

$$\beta_2 = \frac{2\pi}{\lambda_2} = \frac{2\pi}{\lambda_1/n_2} = \frac{2\pi(1.38)}{\lambda_1}$$

$$\beta_3 = \frac{2\pi}{\lambda_3} = \frac{2\pi}{\lambda_1/n_3} = \frac{2\pi(2.4)}{\lambda_1}$$

$$\beta_4 = \frac{2\pi}{\lambda_4} = \frac{2\pi}{\lambda_1/n_4} = \frac{2\pi(1.70)}{\lambda_1}$$

[45] J. T. Cox, G. Hass, and A. Theelen, Triple-layer antireflection coatings on glass for the visible and near infrared, *J. Opt. Soc. Am.*, 529, pp. 965–969, 1962.

With reference to Figure 8.27b, we start at the glass (i.e., the "load") end and simply transform impedances toward air (i.e., "source" end). We have

$$\eta_{45} = \eta_4 \frac{\eta_5 + j\eta_4 \tan(\beta_4 d_4)}{\eta_4 + j\eta_5 \tan(\beta_4 d_4)}$$

where $\eta_4 = \eta_1/n_4 = 377/1.7 \simeq 222\Omega$ and $\eta_5 = \eta_1/n_5 = 377/1.52 = 248\Omega\text{m}$ and the subscript "45" indicates that η_{45} is the intrinsic impedance η_5 transformed by layer (or line) 4. The combined equivalent impedance η_{45} of materials 4 and 5 is now transformed over the next line segment into η_{345}:

$$\eta_{345} = \eta_3 \frac{\eta_{45} + j\eta_3 \tan(\beta_3 d_3)}{\eta_3 + j\eta_{45} \tan(\beta_3 d_3)}$$

where $\eta_3 = \eta_1/n_3 = 377/2.4 \simeq 157\Omega$. Finally we have

$$\eta_{2345} = \eta_2 \frac{\eta_{345} + j\eta_2 \tan(\beta_2 d_2)}{\eta_2 + j\eta_{345} \tan(\beta_2 d_2)}$$

where $\eta_2 = \eta_1/n_2 = 377/1.38 \simeq 273\Omega$. The effective reflection coefficient at the air interface is now given as

$$\Gamma_{\text{eff}} = \frac{\eta_{2345} - \eta_1}{\eta_{2345} + \eta_1}$$

where $\eta_1 = 377\Omega$. The magnitude of Γ_{eff} is plotted in Figure 8.27c over the wavelength range 400–750 nm of the visible spectrum. The wideband performance of the triple-layer system is evident, with extremely low ($<0.01\%$) value of $|\Gamma_{\text{eff}}|$ over a broad frequency range. For comparison, $|\Gamma_{\text{eff}}|$ for a single MgF_2 layer (Example 8-25) is also shown.

8.2.4 Normal Incidence on a Lossy Medium

Up to now, we have studied reflection and refraction of uniform plane waves at interfaces between different lossless media or between a lossless dielectric and a perfect conductor. In practice, all media have some loss, leading to absorption of a portion of the incident electromagnetic energy. In some cases such attenuation may be undesirable but unavoidable; in other cases the heat produced by the attenuation of the wave in the lossy material may constitute the primary application. In this section we consider two important cases involving reflection from an imperfect conducting plane and multiple lossy interfaces.

Reflection from an Imperfect Conducting Plane We now consider the special case of incidence of a uniform plane wave on a "good" conductor with finite conductivity σ. We will show that the total current flowing within the conducting material is essentially independent of the conductivity. As the conductivity approaches infinity, the total current is squeezed into a narrower and narrower layer, until in the limit a true surface current (as discussed in Section 8.2.1 for the case of a perfectly conducting boundary) is obtained. We will further show that the conductor can be

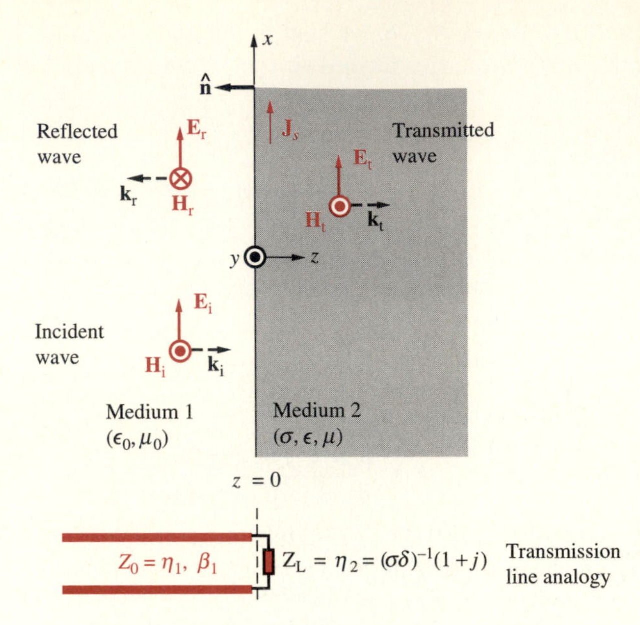

FIGURE 8.28. Uniform plane wave normally incident on the boundary between a dielectric and an imperfect conductor.

characterized as a boundary exhibiting a surface impedance of $Z_s = (\sigma\delta)^{-1}(1 + j)$, where δ is the skin depth for the conductor, being $\delta = \sqrt{2/(\omega\mu\sigma)}$, as given in [8.26].

Let a uniform plane wave be normally incident on a conducting interface located at $z = 0$ (i.e., the half-space $z \geq 0$ is filled with a conducting medium), as shown in Figure 8.28. The phasor fields for the incident, reflected, and transmitted waves are

$$\mathbf{E}_\mathrm{i}(z) = \hat{\mathbf{x}}E_{\mathrm{i}0}e^{-j\beta_1 z}$$

$$\mathbf{H}_\mathrm{i}(z) = \hat{\mathbf{y}}\frac{E_{\mathrm{i}0}}{\eta_1}e^{-j\beta_1 z}$$

$$\mathbf{E}_\mathrm{r}(z) = \hat{\mathbf{x}}E_{\mathrm{r}0}e^{+j\beta_1 z}$$

$$\mathbf{H}_\mathrm{r}(z) = -\hat{\mathbf{y}}\frac{E_{\mathrm{r}0}}{\eta_1}e^{+j\beta_1 z}$$

$$\mathbf{E}_\mathrm{t}(z) = \hat{\mathbf{x}}E_{\mathrm{t}0}e^{-\gamma z}$$

$$\mathbf{H}_\mathrm{t}(z) = \hat{\mathbf{y}}\frac{E_{\mathrm{t}0}}{\eta_c}e^{-\gamma z}$$

where

$$\beta_1 = \omega\sqrt{\mu_0\epsilon_0} \qquad \eta_1 = \sqrt{\mu_0/\epsilon_0} \simeq 377\Omega$$

$$\gamma = \sqrt{j\omega\mu(\sigma + j\omega\epsilon)} \qquad \eta_c = \sqrt{j\omega\mu/(\sigma + j\omega\epsilon)}$$

In a good conducting medium (i.e., $\sigma \gg \omega\epsilon$) we have $\nabla \times \mathbf{H} = (j\omega\epsilon + \sigma)\mathbf{E} \simeq \sigma\mathbf{E}$, since the conduction current is much greater than the displacement current. Rewriting this equation as $\nabla \times \mathbf{H} = j\omega(\sigma/j\omega)\mathbf{E}$, we recall from Section 8.1.3 that

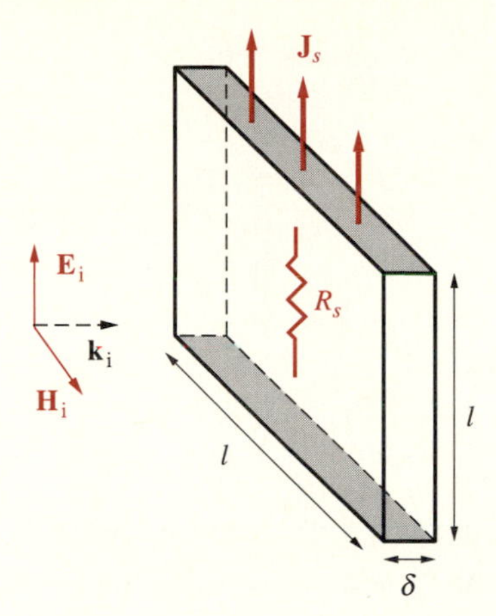

FIGURE 8.29. Surface resistance concept. The resistance between the shaded faces is independent of the linear dimension l and hence can be thought of as a surface resistance.

$\epsilon_{\text{eff}} = \sigma/j\omega$ may be considered as the permittivity ϵ in Maxwell's equations in a lossless medium. Thus, propagation constant and instrinsic impedance of the lossy medium can be obtained from that of a lossless medium by the substitution $\epsilon_{\text{eff}} \longrightarrow \epsilon$. We thus have the propagation constant given by

$$\gamma = j\omega\sqrt{\mu\epsilon_{\text{eff}}} = j\omega\left(\frac{\mu\sigma}{j\omega}\right)^{1/2} = (j\omega\mu\sigma)^{1/2} = \delta^{-1}(1 + j) \qquad [8.54]$$

and the complex intrinsic impedance given by

$$\eta_c \equiv Z_s = R_s + jX_s = \sqrt{\frac{\mu}{\epsilon_{\text{eff}}}} = \left(\frac{j\omega\mu}{\sigma}\right)^{1/2} = \frac{\gamma}{\sigma} = (\sigma\delta)^{-1}(1 + j) \qquad [8.55]$$

where $\delta = (\pi f \mu\sigma)^{-1/2}$ is the skin depth. Note that for medium 1 we have the propagation constant and intrinsic impedance of a lossless medium, namely, $\beta_1 = \omega\sqrt{\mu_1\epsilon_1}$ and $\eta_1 = \sqrt{\mu_1/\epsilon_1}$.

The conductor presents an impedance $Z_s = \eta_c$ to the electromagnetic wave with equal inductive and resistive parts, defined above as R_s and jX_s, respectively. The resistance part is simply the resistance of a sheet of metal of 1 meter square and of thickness δ, as illustrated in Figure 8.29; actually, the resistance is independent of the area (l^2) of the square plate. Thus, the resistance between the two shaded faces (perpendicular to the current flow) is given by

$$R_s = \frac{l}{l\delta\sigma} = \frac{1}{\delta\sigma} \qquad [8.56]$$

Since the resistance is independent of the linear dimension l, it is called a *surface resistance,* and the complex intrinsic impedance η_c can be thought of as the *surface impedance* of the conductor.

To find the reflection and transmission coefficients, we follow the same procedure as before and apply the boundary conditions:

$$E_{i0} + E_{r0} = E_{t0} \quad \text{and} \quad \frac{1}{\eta_1}(E_{i0} - E_{r0}) = \frac{E_{t0}}{Z_s}$$

Solving for E_{r0}/E_{i0} and E_{t0}/E_{i0}, we find

$$\Gamma = \frac{E_{r0}}{E_{i0}} = \frac{Z_s - \eta_1}{Z_s + \eta_1} = \rho e^{j\phi_\Gamma} \qquad \mathcal{T} = \frac{E_{t0}}{E_{i0}} = \frac{2Z_s}{Z_s + \eta_1} = \tau e^{j\phi_\mathcal{T}}$$

Note that we use the more general notation $\Gamma = \rho e^{j\phi_\Gamma}$ and $\mathcal{T} = \tau e^{j\phi_\mathcal{T}}$, since the reflection and transmission coefficients are in general complex.

For any reasonably good conductor, Z_s is very small compared to η_1 for free space (i.e., 377Ω), as was discussed in Section 8.1.3. For example, for copper ($\sigma = 5.8 \times 10^7$ S-m^{-1}) at 1 MHz, $\delta \simeq 66.1$ μm and $R_s = 2.61 \times 10^{-8}\Omega$. The reflection and transmission coefficients, respectively, are $\Gamma \simeq 0.9999986e^{j179.999921°}$ and $\mathcal{T} \simeq 2 \times 10^{-6}e^{45°}$. We thus note that for all practical purposes the field in front of the conductor ($z < 0$) is the same as exists for a perfect conductor, since $\rho = |\Gamma|$ is very close to unity. For the same reason, only a very small amount of power is transmitted into the conductor; that is, $\tau = |\mathcal{T}|$ is small. Nevertheless, this small amount of power that penetrates into the walls can cause significant attenuation of waves propagating in waveguides and coaxial lines, especially when these types of transmission lines are relatively long. Next we discuss a method for calculating the finite amount of power dissipated in the conductor.

In the case when medium 2 is a good conductor (i.e., $\sigma_2/\omega\epsilon_2 \gg 1$), we can find approximate expressions for Γ and $\mathcal{T}$ as follows:

$$\Gamma \simeq \frac{\delta - \lambda_1(1-j)}{\delta + \lambda_1(1-j)} \quad \text{and} \quad \mathcal{T} \simeq \frac{2\delta}{\lambda_1(1-j)}$$

where $\lambda_1 = 2\pi/\beta_1$ is the wavelength in the first medium. Note that since $\lambda_1 \gg \delta$, we have $\Gamma \simeq 1e^{j\pi}$ and $\tau = |\mathcal{T}| \ll 1$. Other useful approximate expressions can be obtained for Γ by using power series expansions. For example,

$$\Gamma \simeq \left(-1 + \sqrt{\frac{2\omega\epsilon_1}{\sigma_2}}\right) + j\sqrt{\frac{2\omega\epsilon_1}{\sigma_2}}$$

Based on these expressions, the fraction of the incident power reflected is approximately given by

$$|\Gamma|^2 \simeq 1 - 4\sqrt{\frac{\omega\epsilon_1}{2\sigma_2}}$$

and the fraction transmitted into the conductor is

$$1 - |\Gamma|^2 \simeq 4\sqrt{\frac{\omega\epsilon_1}{2\sigma_2}}$$

The current density $\mathbf{J}$ in the conductor is $\mathbf{J} = \sigma\mathbf{E}_t = \hat{\mathbf{x}}\sigma\mathcal{T}E_{i0}e^{-\gamma z}$. The total current per unit width of the conductor is

$$J_s = \sigma \mathcal{T} E_{i0} \int_0^\infty e^{-\gamma z} dz = \frac{\sigma \mathcal{T} E_{i0}}{\gamma} \qquad [8.57]$$

Note that while $\mathbf{J}$ is in units of A-m^{-2}, the current per unit width is a surface current in units of A-m^{-1}. We can relate this surface current to the magnetic field at the surface of the conductor, namely, the total magnetic field $\mathbf{H}_1$ in medium 1. Note that the continuity of the tangential magnetic field at the interface requires that $\mathbf{H}_1(z = 0) = \mathbf{H}_t(z = 0)$. Since the magnetic field of the transmitted wave is $\mathbf{H}_t(z) = \hat{\mathbf{y}}(E_{t0}/\eta_c)e^{-\gamma z}$, and substituting $\eta_c = Z_s$ and $\gamma = \delta^{-1}(1 + j)$ we have

$$\mathbf{H}_1(0) = \mathbf{H}_t(0) \quad \rightarrow \quad \mathbf{H}_1(0) = \hat{\mathbf{y}}\frac{\mathcal{T} E_{i0}}{Z_s} \quad \rightarrow \quad \mathcal{T} E_{i0} = Z_s H_{1y}(0)$$

Substituting in [8.57] and using [8.54] and [8.55] we have

$$J_s = \frac{\sigma Z_s H_{1y}(0)}{\gamma} = \frac{\sigma(\sigma\delta)^{-1}(1 + j)}{\delta^{-1}(1 + J)} H_{1y}(0) = H_{1y}(0)$$

It is thus apparent that if we let $\sigma \rightarrow \infty$, we have $\delta \rightarrow 0$, $\rho \rightarrow -1$, but the total current J_s does not vanish; it remains equal to H_{1y} or more generally, we have $\mathbf{J}_s = \hat{\mathbf{n}} \times \mathbf{H}_1(0)$, as was found in Section 8.2.1 for the case of normal incidence of a uniform plane wave on a perfect conductor. However, since $\delta \rightarrow 0$ as $\sigma \rightarrow \infty$, the current is squeezed into a narrower and narrower layer and in the limit becomes a true surface current. When σ is finite, the current density $\overline{\mathcal{J}}$ inside the conductor varies with z in the same manner as $\mathbf{E}_t(z)$ and is confined to a region of thickenss δ, with the total current per unit width of the conductor being $\overline{\mathcal{J}}_s$.

The time-average power loss per unit area in the xy plane may be evaluated from the complex Poynting vector at the surface. We have

$$|\mathbf{S}_{av}| = \left|\frac{1}{2}\mathcal{R}e\{\mathbf{E} \times \mathbf{H}^*\}\right| = \frac{1}{2}\mathcal{R}e\{E_{1x}H_{1y}^*\} = \frac{1}{2}\mathcal{R}e\left\{\mathcal{T} E_{i0}\left(\frac{\mathcal{T}^* E_{i0}^*}{Z_s^*}\right)\right\}$$

$$= \frac{1}{2}|\mathcal{T} E_{i0}|^2 \mathcal{R}e\left\{\frac{1}{Z_s^*}\right\} = \frac{1}{2}|\mathcal{T} E_{i0}|^2 \mathcal{R}e\left\{\frac{1}{R_s + jX_s}\left(\frac{R_s - jX_s}{R_s - jX_s}\right)\right\}$$

$$= \frac{1}{2}|\mathcal{T} E_{i0}|^2 \mathcal{R}e\left\{\frac{R_s - jX_s}{R_s^2 + X_s^2}\right\} = \frac{1}{2}|\mathcal{T} E_{i0}|^2 \frac{R_s}{R_s^2 + X_s^2} = \frac{1}{4}|\mathcal{T} E_{i0}|^2 \sigma\delta$$

where we have used the fact that $R_s = X_s = (\sigma\delta)^{-1}$. This Poynting flux represents the total average power per unit area entering the conductor. All of this power must be dissipated in the conductor due to $\mathbf{E} \cdot \mathbf{J}$ losses, which can be evaluated by integrating over a volume with 1 m^2 cross section in the xy plane. In other words,

$$P_{loss} = \frac{1}{2}\int_V \mathbf{E} \cdot \mathbf{J}^* dv = \frac{1}{2}\int_0^1 \int_0^1 \int_0^\infty (E_x)(\sigma E_x^*)\, dx\, dy\, dz$$

$$= \frac{\sigma}{2}|E_{i0}\tau|^2 \int_0^\infty e^{-2z/\delta} dz = \frac{1}{4}|\mathcal{T} E_{i0}|^2 \sigma\delta$$

Thus we find that, as expected, the total power entering the conductor through a unit area on its surface is equal to the total power dissipated in the volume of the conductor behind the unit area.

We can express P_{loss} in a more useful form by replacing $|E_{i0}\mathcal{T}|$ by $|J_s\gamma/\sigma|$ using [8.57]. We find

$$P_{\text{loss}} = \frac{1}{4}|J_s\gamma|^2\frac{\delta}{\sigma} = \frac{1}{2}|J_s|^2 R_s \qquad [8.58]$$

which underscores the term *surface resistance* for R_s.

In practice, an approximate method is generally used to evaluate power loss per unit area in conducting walls of waveguides and coaxial lines. The electric and magnetic field configurations are found using the assumption that the conductors are perfect (i.e., $\sigma = \infty$). The surface current density is then determined from the boundary condition

$$\mathbf{J}_s = \hat{\mathbf{n}} \times \mathbf{H} \quad \rightarrow \quad J_s = H_t$$

where $\hat{\mathbf{n}}$ is the outward normal to the conductor surface and H_t is the tangential field at the surface, evaluated for $\sigma = \infty$. Once $\mathbf{J}_s$ is found, we can use the surface resistance $R_s = (\sigma\delta)^{-1}$ to calculate the power loss per unit area using [8.58].

Note that the value of H_t calculated assuming $\sigma = \infty$ is a very good approximation of the actual field, since $\eta_1 \gg |Z_s|$. In the case of infinite conductivity the tangential electric field at the conductor surface is zero. However, for finite σ there has to be a finite value of tangential electric field to support a component of the Poynting vector directed into the conductor. It can be shown that this tangential electric field at the surface is $\mathbf{E}_t = \mathbf{J}_s Z_s$ and is thus normally quite small.

Example 8-27 illustrates the absorption of incident electromagnetic power at an air–copper interface, while Example 8-28 shows the effectiveness of a thin foil of aluminum in shielding electromagnetic fields.

Example 8-27: Air–copper interface. Consider a uniform plane wave propagating in air incident normally on a large copper block. Find the percentage time-average power absorbed by the copper block at 1, 10, and 100 MHz and 1 GHz.

Solution: The amount of power absorbed by the copper block is the power transmitted into medium 2, which is equal to the Poynting flux moving in the $+z$ direction in medium 1:

$$(\mathbf{S}_{\text{av}})_1 = \hat{\mathbf{z}}\frac{E_{i0}^2}{2\eta_1}(1 - |\Gamma|^2)$$

as derived in Section 8.2.2. Since the power of the incident wave is $E_{i0}^2/(2\eta_1)$, the fraction of the incident power transmitted into medium 2 is $(1 - |\Gamma|^2)$. Thus, we need to determine the reflection coefficient Γ. The intrinsic impedance of copper is given by

$$\eta_c = \sqrt{\frac{j\omega\mu_0}{\sigma}} = \sqrt{\frac{\omega\mu_0}{\sigma}}e^{j45^\circ} = \frac{2\pi \times 10^{-7}\sqrt{f}}{\sqrt{5.8}}(1 + j)$$
$$\simeq 2.61 \times 10^{-7}\sqrt{f}(1 + j)\Omega$$

where we have used $\sigma = 5.8 \times 10^7$ S-m^{-1} for copper and $\sqrt{j} = e^{j\pi/4}$. The reflection coefficient at the air–copper interface is

$$\Gamma = \frac{\eta_c - \eta_0}{\eta_c + \eta_0} \simeq \frac{2.61 \times 10^{-7}\sqrt{f}(1 + j) - 377}{2.61 \times 10^{-7}\sqrt{f}(1 + j) + 377}$$

At $f = 1$ MHz, we have

$$\Gamma \simeq \frac{2.61 \times 10^{-4}\sqrt{f}(1 + j) - 377}{2.61 \times 10^{-4}\sqrt{f}(1 + j) + 377} \simeq 0.9999986e^{179.99992^\circ}$$

so that the percentage of incident power absorbed by the copper block is

$$P_{\text{Cu}} \simeq 1 - |\Gamma|^2 \times 100 \simeq 0.000277\%$$

Similar calculations give the following results

10 MHz	$\Gamma \simeq 0.9999956e^{j179.99975^\circ}$	$P_{\text{Cu}} \simeq 0.000875\%$
100 MHz	$\Gamma \simeq 0.9999862e^{j179.99921^\circ}$	$P_{\text{Cu}} \simeq 0.00277\%$
1000 MHz	$\Gamma \simeq 0.9999562e^{j179.99749^\circ}$	$P_{\text{Cu}} \simeq 0.00875\%$

We see that the percentage of the incident power absorbed by the copper block increases with the frequency of the incident wave. However, note that as the frequency is increased, more of the total absorbed power is dissipated in a narrower region near the surface (i.e., the skin depth in copper, proportional to $f^{-1/2}$).

Example 8-28: Transmission through a metal foil: RF shielding. Consider an x-polarized uniform plane radio-frequency (RF) wave propagating in air, incident normally on a metal foil of thickness d at $z = 0$ as shown in Figure 8.30. (a) Find a relationship between the electric field of the transmitted wave (i.e., $E_{t0} = |\mathbf{E}_t|$ and that of the incident wave (i.e., $E_{i0} = |\mathbf{E}_i|$. Assume the foil to be thick enough that multiple reflections can be neglected. (b) Consider an ordinary aluminum foil, which is approximately 0.025 mm thick. If a 100 MHz plane wave is normally incident from one side of the foil, find the percentage of the incident power transmitted to the other side. For aluminum, take $\sigma = 3.54 \times 10^7$ S-m^{-1}, $\epsilon = \epsilon_0$, and $\mu = \mu_0$.

Solution:

(a) Neglecting multiple reflections allows us to treat each boundary of the metal foil as a separate interface. Thus, the amplitude of the wave transmitted into the metal foil due to the incident wave at the first boundary is given by

$$|\mathbf{E}_{\text{m}}(z = 0)| = \tau_1|\mathbf{E}_{\text{i}}(z = 0)|$$

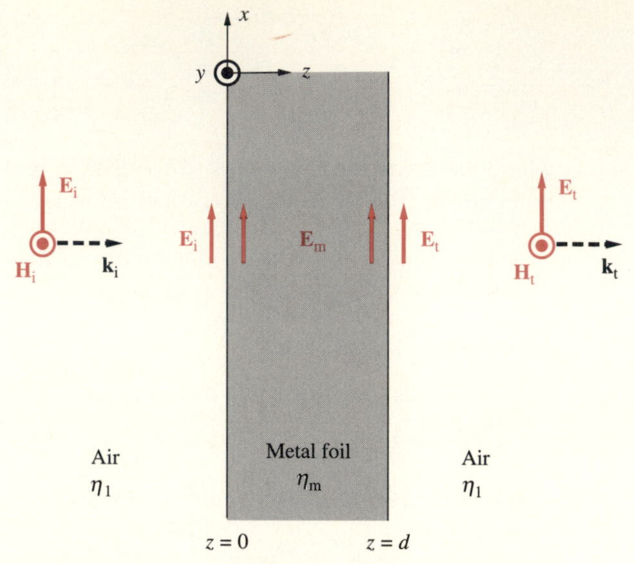

FIGURE 8.30. **RF shielding.** A metal foil of thickness d designed to shield RF energy. Note that typically we have $|\mathbf{E}_t| \ll |\mathbf{E}_i|$.

where $\tau_1 = |\mathcal{T}_1|$ is the magnitude of the transmission coefficient at the air–metal interface. Note that we have

$$\mathcal{T}_1 = \frac{2\eta_m}{\eta_1 + \eta_m}$$

where $\eta_1 = 377\Omega$ and $\eta_m \simeq \sqrt{j\omega\mu_0/\sigma} = \sqrt{\omega\mu_0/(2\sigma)}(1 + j)$. This transmitted wave exponentially attenuates as it propagates into the metal foil, so its amplitude at $z = d$ is

$$|\mathbf{E}_m(z = d)| = |\mathbf{E}_m(z = 0)|e^{-d/\delta}$$

where $\delta = (\pi f \mu_0 \sigma)^{-1/2}$ is the skin depth in the metal. The attenuated transmitted wave incident on the metal–air boundary at $z = d$ transmits wave energy into medium 3 (air). The amplitude of the wave transmitted into medium 3 can be written as

$$|\mathbf{E}_t(z = d)| = \tau_2 |\mathbf{E}_m(z = d)|$$

where τ_2 is the amplitude of the transmission coefficient $\mathcal{T}_2$ at the metal–air boundary, given by

$$\mathcal{T}_2 = \frac{2\eta_1}{\eta_c + \eta_1}$$

The incident and transmitted electric field magnitudes are related by

$$|\mathbf{E}_t| = \tau_1 e^{-d/\delta} \tau_2 |\mathbf{E}_i|$$

(b) We can first calculate the skin depth of aluminum at 100 MHz:

$$\delta = \frac{1}{\sqrt{\pi f \mu_0 \sigma}} = \frac{1}{\sqrt{\pi(10^8)(4\pi \times 10^7)(3.54 \times 10^7)}} \simeq 8.46 \times 10^{-6}\text{ m} = 8.46\ \mu\text{m}$$

Noting that the thickness of the aluminum foil is $d = 0.025$ mm $\simeq 2.96\delta$, our assumption of neglecting multiple reflections is easily justified. In this connection, note that the reflected wave will be attenuated by an additional exponential factor of e^{-3} in the course of its propagation back through the metal to the first boundary. Thus, using the result found in (a) we have

$$\frac{|\mathbf{E}_\text{t}|}{|\mathbf{E}_\text{i}|} = \tau_1 e^{-d/\delta} \tau_2$$

where

$$|\tau_1 \tau_2| = \left|\frac{2\eta_\text{m}}{\eta_\text{m} + \eta_1}\right| \left|\frac{2\eta_1}{\eta_1 + \eta_\text{m}}\right| = \left|\frac{4\eta_\text{m}\eta_1}{(\eta_1 + \eta_\text{m})^2}\right|$$

and $e^{-d/\delta} \simeq e^{-2.96} \simeq 0.0521$. Noting that $\eta_1 = 377\Omega$ and

$$\eta_\text{m} = \sqrt{\frac{j\omega\mu_0}{\sigma}} \simeq \frac{2\pi \times 10^{-7}\sqrt{f}}{\sqrt{3.54}}(1 + j) \simeq 3.34 \times 10^{-3}(1 + j)\Omega$$

Since $\eta_1 \gg \eta_\text{m}$, we have

$$\tau_1 \tau_2 \simeq \left|\frac{4\eta_1\eta_\text{m}}{\eta_1^2}\right| = \frac{4|\eta_\text{m}|}{\eta_1} \simeq \frac{4\sqrt{2}(3.34 \times 10^{-3})}{377} \simeq 5.01 \times 10^{-5}$$

so $|\mathbf{E}_\text{t}|/|\mathbf{E}_\text{i}| \simeq (0.0521)(5.01 \times 10^{-5}) \simeq 2.61 \times 10^{-6}$. Therefore, the percentage of the incident power that transmits to the other side of the foil is

$$\frac{|\mathbf{E}_\text{t}|^2/(2\eta_1)}{|\mathbf{E}_\text{i}|^2/(2\eta_1)} \times 100 = \frac{|\mathbf{E}_\text{t}|^2}{|\mathbf{E}_\text{i}|^2} \times 100 \simeq 6.80 \times 10^{-10}\ !$$

Thus, the thin aluminum foil works quite well indeed as a shield for RF fields at 100 MHz.

Reflection from Multiple Lossy Interfaces In the case of multiple interfaces involving conducting (lossy) media, the various expressions for the effective reflection and transmission coefficients that we derived in Section 8.2.3 apply if we substitute the complex propagation constant $\gamma_2 = \alpha_2 + j\beta_2$ instead of $j\beta_2$ and allow the various intrinsic impedances (η) to be complex. In other words, we have [8.50]

$$\Gamma_\text{eff} = \rho e^{j\phi_\Gamma} = \frac{(\eta_2 - \eta_1)(\eta_3 + \eta_2) + (\eta_2 + \eta_1)(\eta_3 - \eta_2)e^{-2\gamma_2 d}}{(\eta_2 + \eta_1)(\eta_3 + \eta_2) + (\eta_2 - \eta_1)(\eta_3 - \eta_2)e^{-2\gamma_2 d}}$$

TABLE 8.6. ϵ_r and σ for biological tissues

	Muscle, skin, and tissues with high water content		Fat, bone, and tissues with low water content	
f (MHz)	$(\epsilon_m)_r$	σ_m (S-m^{-1})	$(\epsilon_f)_r$	σ_f (mS-m^{-1})
100	71.7	0.889	7.45	19.1–75.9
300	54	1.37	5.7	31.6–107
750	52	1.54	5.6	49.8–138
915	51	1.60	5.6	55.6–147
1,500	49	1.77	5.6	70.8–171
2,450	47	2.21	5.5	96.4–213
5,000	44	3.92	5.5	162–309
10,000	39.9	10.3	4.5	324–549

*Note that σ_m is in S-m^{-1} whereas σ_f is in mS-m^{-1}.

The effective transmission coefficient $\mathcal{T}_{\text{eff}}$ can also be determined in a similar manner from [8.51] as

$$\mathcal{T}_{\text{eff}} = \tau e^{j\phi_{\mathcal{T}}} \frac{4\eta_2\eta_3 e^{-\gamma_2 d}}{(\eta_2 + \eta_1)(\eta_3 + \eta_2) + (\eta_2 - \eta_1)(\eta_3 - \eta_2)e^{-2\gamma_2 d}}$$

We note, however, that when the second medium is lossy, we have

$$1 - |\Gamma_{\text{eff}}|^2 \neq \frac{\eta_1}{|\eta_3|}|\mathcal{T}_{\text{eff}}|^2$$

since some of the incident power is absorbed in medium 2.

Reflection and penetration (transmission) of electromagnetic signals at biological tissue interfaces constitute an interesting and important application of multiple lossy interfaces and are briefly covered in Examples 8-29 through 8-31. In order to better appreciate these applications, it should be noted that the dielectric properties of different biological tissues are different and have been studied over a very broad frequency range. The dielectric constant of tissues decreases gradually over many orders of magnitude as the frequency is varied from a few Hz to tens of GHz. The effective conductivity, on the other hand, rises with frequency, initially very slowly, and then more rapidly above 1 GHz. The permittivity of tissues depends on the tissue type and widely varies between different tissues. Tissues of higher water content, such as muscle, brain, kidney, heart, liver, and pancreas, have larger dielectric constant and conductivity than low-water-content tissues such as bone, fat, and lung. Table 8.6 lists the relative dielectric constant and conductivity of biological tissues with high water content versus those with low water content at discrete source frequencies over the radio-frequency spectrum.[46]

[46]C. C. Johnson and A. W. Guy, Nonionizing electromagnetic wave effects in biological materials and systems, *Proc. IEEE,* 60(6), pp. 692–718, June 1972.

Examples 8-29 through 8-31 provide quantitative values for the reflection and transmission coefficients at planar biological interfaces.

Example 8-29: Air–muscle interface. Consider a planar interface between air and muscle tissue. If a plane wave is normally incident at this boundary, find the percentage of incident power absorbed by the muscle tissue at (a) 100 MHz; (b) 300 MHz; (c) 915 MHz; and (d) 2.45 GHz.

Solution:

(a) The reflection coefficient is given by

$$\Gamma = \rho e^{j\phi_\perp} = \frac{\eta_m - \eta_1}{\eta_m + \eta_1}$$

where $\eta_1 = 377\Omega$ and the intrinsic impedance of muscle tissue η_m is given by [8.22]. So, at 100 MHz, the loss tangent of muscle is given by

$$\tan\delta_m = \frac{\sigma_m}{\omega\epsilon_m} = \frac{0.889 \text{ S-m}^{-1}}{2\pi \times 10^8 \text{ rad-s}^{-1} \times 71.7 \times 8.85 \times 10^{-12} \text{ F-m}^{-1}} \simeq 2.23$$

and the intrinsic impedance is

$$\eta_m \simeq \frac{377/\sqrt{71.7}}{[1 + (2.23)^2]^{1/4}} e^{j(1/2)\tan^{-1}(2.23)} \simeq 28.5e^{j32.9^\circ}\,\Omega$$

Substituting, we find

$$\Gamma \simeq \frac{28.5e^{j32.9^\circ} - 377}{28.5e^{j32.9^\circ} + 377} \simeq \frac{23.9 + j15.5 - 377}{23.9 + j15.5 + 377} \simeq \frac{353e^{j177^\circ}}{401e^{j2.21^\circ}} \simeq 0.881e^{j175^\circ}$$

So the percentage of incident power absorbed by the muscle tissue can be calculated as

$$\frac{|(\mathbf{S}_{av})_t|}{|(\mathbf{S}_{av})_i|} \times 100\% = (1 - \rho^2) \times 100\% \simeq [1 - (0.881)^2] \times 100\% \simeq 22.4\%$$

(b) Similarly, at 300 MHz, we have $\tan\delta_m \simeq 1.52$, $\eta_m \simeq 38e^{j28.3^\circ}\,\Omega$, and the reflection coefficient is

$$\Gamma \simeq \frac{38e^{j28.3^\circ} - 377}{38e^{j28.3^\circ} + 377} \simeq 0.837e^{j174^\circ}$$

so that the percentage absorbed power is $\sim[1 - (0.837)^2] \times 100 \simeq 29.9\%$.

(c) At 915 MHz, $\tan\delta_m \simeq 0.617$, $\eta_m \simeq 48.7e^{j15.8^\circ}\,\Omega$, and

$$\Gamma \simeq \frac{48.7e^{j15.8^\circ} - 377}{48.7e^{j15.8^\circ} + 377} \simeq 0.779e^{j176^\circ}$$

so that the percentage absorbed power is $\sim[1 - (0.779)^2] \times 100 \simeq 39.3\%$.

(d) At 2.45 GHz, $\tan\delta_m \simeq 0.345$, $\eta_m \simeq 53.5e^{j9.52^\circ}\,\Omega$, and

$$\Gamma \simeq \frac{53.5e^{j9.52^\circ} - 377}{53.5e^{j9.52^\circ} + 377} \simeq 0.755e^{j177^\circ}$$

So the percentage of incident power absorbed is

$$(1-\rho^2)\times 100 \simeq [1-(0.755)^2]\times 100 \simeq 43\%$$

Example 8-30: Muscle–fat interface. Consider a planar interface between muscle and fat tissues. If a 1 mW-cm^{-2} plane wave in muscle is normally incident at this boundary at 2.45 GHz, find the power density of the wave transmitted into the fat tissue. For fat tissue, take $\sigma_f = 155$ mS-m^{-1}.

Solution: At 2.45 GHz, the intrinsic impedances of muscle and fat tissues, η_m and η_f, are given by

$$\eta_m \simeq 53.5e^{j9.52^\circ}\,\Omega \text{ (from previous example)}$$

$$\eta_f \simeq \frac{377/\sqrt{5.5}}{[1+(0.207)^2]^{1/4}}\exp\left[j\frac{1}{2}\tan^{-1}(0.207)\right] \simeq 159e^{j5.84^\circ}\,\Omega$$

The reflection coefficient can be calculated as

$$\Gamma = \frac{\eta_m-\eta_f}{\eta_m+\eta_f} \simeq \frac{53.5e^{j9.52^\circ}-159e^{j5.84^\circ}}{53.5e^{j9.52^\circ}+159e^{j5.84^\circ}}$$

$$\simeq \frac{52.7+j8.84-158-j16.2}{52.7+j8.84+158+j16.2} \simeq 0.498e^{j177^\circ}$$

Therefore the power transmitted into the fat tissue is given by

$$|(\mathbf{S}_{av})_t| = (1-\rho^2)|(\mathbf{S}_{av})_i| \simeq [1-(0.498)^2](1\text{ mW-cm}^{-2}) \simeq 0.752\text{ mW-cm}^{-2}$$

Example 8-31: Microwave treatment of hypothermia in newborn piglets. Newly born piglets are very vulnerable to cold temperatures, and many of them die because of hypothermia. At the moment, hypothermia is treated by placing the piglets under infrared lamps, which are not very effective and are very costly. It has been proposed[47] that microwaves can be used to treat hypothermia. Compared to the infrared lamp, a microwave heater is more expensive to build, but is more effective and consumes less power.

Consider a plane wave normally incident at the surface of the body of a pig (which to first order can be assumed to be a plane boundary). The body of the pig can be approximately modeled as a layer of fat tissue of a certain thickness followed by muscle tissue (which is assumed to be infinite in extent), as shown in Figure 8.31. For a newly born piglet, the fat layer is so thin (less than a mm) that it can be neglected, so the problem reduces to that of a single boundary. However, for a developed pig, the thickness of the fat layer can vary anywhere from 2 to 5 cm (taken below to be 4 cm for a mature pig), so it must be taken into account. Calculate the percentage of the incident microwave power reflected back into air, the percentage of power dissipated

[47] M. Allen, Pinky perks up when popped in the microwave, *New Scientist,* p. 19, January 23, 1993.

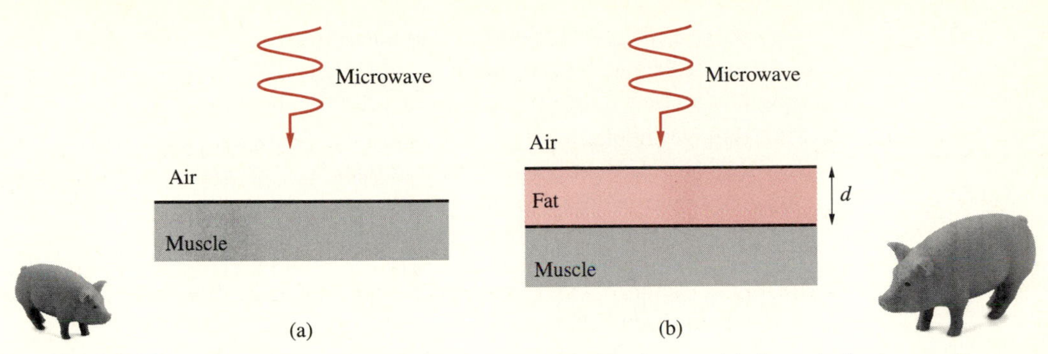

FIGURE 8.31. Microwave warming of pigs. (a) Air–piglet interface, modeled as a single air–muscle interface. (b) Air–adult pig interface, modeled as an air–fat–muscle interface.

in the fat layer, the percentage of power transmitted into the muscle tissue, and the depth of penetration into the muscle tissue of the newly born piglet at (a) 915 MHz; (b) 2.45 GHz. (c) Repeat (a) and (b) for a mature pig. Use the same parameters for tissues as given in Table 8.6.

Solution:

(a) For the newly born piglet, using the results of Example 8-30, for 915 MHz, the percentage of incident power reflected is

$$\frac{|(\mathbf{S}_{\text{av}})_{\text{r}}|}{|(\mathbf{S}_{\text{av}})_{\text{i}}|} \times 100 = \rho^2 \times 100 \simeq (0.779)^2 \times 100 \simeq 60.7\%$$

and the percentage of incident power transmitted is

$$\frac{|(\mathbf{S}_{\text{av}})_{\text{t}}|}{|(\mathbf{S}_{\text{av}})_{\text{i}}|} \times 100 = (1 - \rho^2) \times 100 \simeq 39.3\%$$

No significant power is dissipated in the fat layer, since its thickness is negligible. The depth of penetration into the muscle layer at 915 MHz can be calculated from [8.25] as

$$d = \frac{c}{\omega}\left[\frac{2}{\epsilon_r(\sqrt{1 + \tan^2\delta} - 1)}\right]^{1/2}$$

$$\simeq \frac{(3 \times 10^{10}\ \text{cm-s}^{-1})}{2\pi(915 \times 10^6\ \text{Hz})}\left[\frac{2}{51(\sqrt{1 + (0.617)^2} - 1)}\right]^{1/2} \simeq 2.47\ \text{cm}$$

(b) Similarly, for 2.45 GHz, the percentage of the incident power reflected is $(0.755)^2 \times 100 \simeq 57\%$, and the percentage of incident power transmitted is $(1 - 0.755^2) \times 100 \simeq 43\%$. The depth of penetration at 2.45 GHz is

$$d \simeq \frac{(3 \times 10^{10}\ \text{cm-s}^{-1})}{2\pi(2.45 \times 10^9\ \text{Hz})}\left[\frac{2}{47(\sqrt{1 + (0.345)^2} - 1)}\right]^{1/2} \simeq 1.67\ \text{cm}$$

As expected, the depth of penetration at 2.45 GHz is less than that at 915 MHz. Thus, it may be more desirable to use 915 MHz rather than the more typical commercial microwave oven frequency of 2.45 GHz. At the lower frequency the microwave energy can penetrate deeper into the piglet's body and thus provide more effective heating.

(c) For the mature pig, we have a two-boundary problem. To calculate the percentages of incident power reflected back into air and transmitted into the muscle tissue, we need to use the effective reflection (Γ_{eff}) and transmission ($\mathcal{T}_{\text{eff}}$) coefficient expressions. First, we need to find the intrinsic impedances of fat and muscle tissues at 915 MHz and 2.45 GHz. At 915 MHz, for fat tissue (assume $\sigma_{\text{f}} = 0.1$ S-m^{-1}) we have

$$\tan\delta_{\text{f}} = \frac{0.1 \text{ S-m}^{-1}}{(2\pi \times 915 \times 10^6 \text{ rad-s}^{-1}) \times (5.6 \times 8.85 \times 10^{-12} \text{ F-m}^{-1})} \simeq 0.351$$

so that from [8.31], the intrinsic impedance is

$$\eta_{\text{f}} \simeq \frac{377/\sqrt{5.6}}{[1 + (0.351)^2]^{1/4}} \exp\left[j\frac{1}{2}\tan^{-1}(0.351)\right] \simeq 155e^{j9.67^\circ}\,\Omega$$

and the propagation constant is

$$\gamma_{\text{f}} = \alpha_{\text{f}} + j\beta_{\text{f}}$$

where from [8.19] and [8.20], we have

$$\left\{\begin{matrix}\alpha_{\text{f}}\\ \beta_{\text{f}}\end{matrix}\right\} \simeq \frac{2\pi \times 915 \times 10^6}{3 \times 10^8}\sqrt{\frac{5.6}{2}}\left[\sqrt{1 + (0.351)^2} \mp 1\right]^{1/2} \simeq \left\{\begin{matrix}7.84 & \text{np-m}^{-1}\\ 46.0 & \text{rad-m}^{-1}\end{matrix}\right.$$

At 915 MHz, for muscle tissue, we have from the previous example,

$$\eta_{\text{m}} \simeq 48.7e^{j15.8^\circ}\,\Omega$$

so the effective reflection and transmission coefficients are given by

$$\Gamma_{\text{eff}} = \frac{(\eta_{\text{f}} - \eta_1)(\eta_{\text{m}} + \eta_{\text{f}}) + (\eta_{\text{f}} + \eta_1)(\eta_{\text{m}} - \eta_{\text{f}})e^{-2\gamma_{\text{f}}d}}{(\eta_{\text{f}} + \eta_1)(\eta_{\text{m}} + \eta_{\text{f}}) + (\eta_{\text{f}} - \eta_1)(\eta_{\text{m}} - \eta_{\text{f}})e^{-2\gamma_{\text{f}}d}} \simeq 0.414e^{j124^\circ}$$

and

$$\mathcal{T}_{\text{eff}} = \frac{4\eta_{\text{f}}\eta_{\text{m}}e^{-\gamma_{\text{f}}d}}{(\eta_{\text{f}} + \eta_1)(\eta_{\text{m}} + \eta_{\text{f}}) + (\eta_{\text{f}} - \eta_1)(\eta_{\text{m}} - \eta_{\text{f}})e^{-2\gamma_{\text{f}}d}} \simeq 0.261e^{-j47.8^\circ}$$

Therefore, the percentages of the power reflected back to air and power transmitted into the muscle tissue are $\rho_{\text{eff}}^2 \times 100 \simeq (0.414)^2 \times 100 \simeq 17.1\%$ and $[\eta_1\tau_{\text{eff}}^2/|\eta_{\text{m}}|] \simeq [377(0.261)^2/48.7] \times 100 \simeq 52.7\%$, respectively, from which the percentage of power absorbed by the fat later can be found as $\sim$30.2%. These results show that the mature pig absorbs, overall, a much higher percentage of the incident power than the newborn piglet ($\sim$83% versus $\sim$39%) does, because the fat layer acts as an impedance transformer and reduces the amount of reflections significantly. The calculations at 2.45 GHz are left as an exercise for the reader.

8.3 GUIDED WAVES

When an electromagnetic wave propagates out from a point source through empty space, it does so as a spherical wave, traveling with equal speed in all directions. At large distances ($r \gg \lambda$) from its source, a spherical electromagnetic wave is well approximated as a uniform plane wave; in Section 8.1, we considered the propagation of uniform plane waves in unbounded media, and in Section 8.2 their reflections from planar interfaces between different simple media. We now consider the guiding of electromagnetic waves from one point to another by means of metallic boundaries.

In a wide range of electromagnetic applications, it is necessary to convey electromagnetic waves efficiently from one point to another, to transmit either energy or information (signals). The transmission of energy and information can be achieved by means of *unguided* electromagnetic waves propagating in free space. Some applications, such as line-of-sight microwave (radio, television, and satellite) links, optical links (bar-code readers, infrared smoke detectors, infrared remote controllers), high-power lasers (used for materials processing, laser printers, and surgery), and radar, require directional transmission of waves, whereas others, such as navigation, radio communication, and radio and TV broadcast use omnidirectional transmissions. In unguided transmission of electromagnetic energy, the characteristics of the transmitting antennas determine the intensity of the waves radiated in different directions. In addition to the spherical spreading of the available energy, broadcast applications do not represent efficient use of the electromagnetic spectrum, since only one transmitter can operate at a given frequency in a given region.

In many other applications it is necessary to transmit electromagnetic energy or signals by using waves that are *guided* by metallic or dielectric structures, minimizing radiation losses and unnecessary spreading of the energy. Guided transmission of waves also facilitates the use of the same frequency band to convey multiple signals simultaneously, since most waveguides can be laid close to one another without significant coupling between them. Guiding of electromagnetic waves is generally achieved by confining electromagnetic waves in two dimensions and allowing them to propagate freely in the third dimension. The simplest example is a coaxial line, which completely confines the energy to the region between the inner and outer conductors. However, the principle of operation of the two-wire line is identical; although the electric and magnetic fields are not completely confined in an enclosure, their amplitudes decrease rapidly with transverse distance away from the conductors. Thus, most of the electromagnetic energy is confined to the region in the immediate vicinity of the conductors, with the surfaces of the conducting wires providing boundaries on which the electric field lines can terminate so that the wave can propagate as a plane wave following the conductor from end to end. A power transmission line and a telephone wire are examples of such two-wire lines.

Wave-guiding structures of all types can be called waveguides, although the term *waveguide* is sometimes used specifically for the particular guiding structure consisting of a metallic cylindrical tube. The two guiding systems explicitly considered in this section are the parallel-plate waveguide and the coaxial line. However, all of the methodologies used are directly applicable to other transmission systems utilizing two separate conductors symmetrically arranged across the transverse plane (e.g., the two-wire line, twisted-pair, striplines), which can support the propagation of transverse electromagnetic (TEM) waves. In most cases, TEM waves guided by two-conductor systems are not *uniform* plane waves, since the amplitudes of the wave field components vary over the planar constant-phase fronts, for example decreasing with distance away from the conductors for the two-wire line (hence the confinement of the energy to the vicinity of the conductors, which "guide" the wave).

Our coverage of guided waves in this chapter completes our discussions in Chapters 2 and 3 of the propagation of voltage and current waves on transmission lines. The TEM waves that propagate on these and other two-conductor lines are one and the same with the voltage and current waves that we studied in Chapters 2 and 3. In this section we are finally able to show the precise definition of voltage and current on such lines and their relationship with the propagating electric and magnetic fields. Our extensive transmission line analyses in Chapters 2 and 3 are thus directly applicable to any of the two-conductor structures that support TEM waves. The only differences between the various structures are in the values of their particular parameters, such as the per-unit-length inductance, capacitance, conductance, and resistance values, which are listed in Table 2.2.

The mathematical foundation of guided electromagnetic waves and modern transmission line theory was laid down by O. Heaviside, who wrote a series of 47 papers between 1885 and 1887, introducing the vector notation currently in use today and many other concepts that have since proven to be so useful in electrical engineering. Most of the early experimental work on guided waves was carried out by H. Hertz (see short biography in Section 7.4); both Hertz and Heaviside exclusively considered two-conductor transmission lines, with Heaviside even declaring that two conductors were definitely necessary for transmission of electromagnetic energy.[48] Heaviside was wrong in this prophecy, since electromagnetic waves propagate very effectively in hollow metallic cylindrical guides.[49] Nevertheless, the series of papers written by Heaviside and his three-volume work published in 1893 stand out as the mathematical foundations of guided electromagnetic waves.

Heaviside, Oliver *(b. May 18, 1850, London; d. February 3, 1925, Paignton, Devonshire) Heaviside was the son of an artist. Like Edison [1847–1931], he*

[48]On p. 399 of his three-volume work *Electromagnetic Theory,* Benn, London, 1893, or on p. 100 of the reprinted edition by Dover, 1950, Heaviside stated that guided waves needed "two leads as a pair of parallel wires; or if but one be used, there is the earth, or something equivalent, to make another." For an interesting discussion of the early history, see J. H. Bryant, Coaxial transmission lines, related two-conductor transmission lines, connectors, and components: A U.S. historical perspective, *IEEE Trans. Microwave Theory and Techniques,* 32(9), pp. 970–983, September 1984.

[49]Discussion of electromagnetic wave propagation in metallic waveguides in the form of hollow tubes is beyond the scope of this book. However, propagation of waves in such structures can be analyzed by straightforward extensions of the formulation used in Sections 8.3.1 and 8.3.2. For coverage of this subject at an appropriate level, see Chapter 10 of D. K. Cheng, *Field and Wave Electromagnetics,* 2nd ed., Addison Wesley Publishing Company, Inc., Reading, Massachusetts, 1989.

had no formal education past the elementary level, became a telegrapher, and was hampered by deafness. With the encouragement of his uncle (by marriage), Charles Wheatstone [1802–1875], he inaugurated a program of self-education that succeeded admirably. He could concentrate on it more since he never married and lived with his parents till they died. He did important work in applying mathematics to the study of electrical circuits and extended Maxwell's [1831–1879] work on electromagnetic theory. Perhaps because of his unorthodox education, he made use of mathematical notations and methods of his own that were greeted by the other (and lesser) physicists with disdain. For instance, he used vector notation where many physicists, notably Kelvin [1824–1907], did not. It was Kelvin, however, who first brought Heaviside's work to the notice of the scientific community. Heaviside was forced to publish his papers at his own expense because of their unorthodoxy. After the discovery of radio waves by Hertz [1857–1894], Heaviside applied his mathematics to wave motion and published a large three-volume work, Electromagnetic Theory. *In it he predicted the existence of an electrically charged layer in the upper atmosphere just months after Kennelly [1861–1939] had done so. This layer is now often referred to as the Kennelly–Heaviside layer. Heaviside spent his last years poor and alone and died in a nursing home. [Adapted with permission from I. Asimov,* Biographical Encyclopedia of Science and Technology, *Doubleday, 1982].*

1700 *1850* *1925* *2000*

Our purpose in this section is not necessarily to provide comprehensive coverage of electromagnetic wave-guiding systems that are in practical use at the present time but rather to provide adequate coverage of the fundamental principles of the guiding of waves, using parallel-plate and coaxial lines as two examples. The practical use of a wave-guiding system in any given application is determined by the frequency range of operation, available technology, the particular needs of the application in hand, and engineering feasibility and cost. For example, two-wire (power lines) and twisted-pair (telephone lines) transmission lines work well at relatively low frequencies. However, radiation effects become intolerable at higher frequencies, either because of power losses or because of unwanted coupling of energy between circuits. At VHF/UHF frequencies, coaxial lines or the encapsulated two-wire lines (i.e., the household TV antenna lead wire) are equally usable, although the latter is generally less desirable, since it is an unshielded guiding structure. Nevetheless, the encapsulated two-wire line is relatively inexpensive to manufacture and is widely used in the lower range of the microwave spectrum. Coaxial lines are more expensive to manufacture but provide excellent shielding and are thus used in applications where it is important to minimize interference.

Coaxial lines can in principle operate at any frequency, but they become increasingly lossy and difficult to construct for use at higher frequencies. At frequencies above about 3 GHz, hollow metallic tubes have practical dimensions, provide much lower loss, and have substantially larger power capacity. However, such waveguides are often bulky and expensive to manufacture because of the precision required, and their applicability at the higher end of the microwave spectrum is thus limited. In the 1970s, the stripline and the microstrip line (Figure 2.1d,e) emerged as a practical alternative to waveguides in some applications, primarily because these structures could be manufactured relatively easily (hence at low cost) using planar integrated circuit technologies. At optical frequencies ($>10^{12}$ Hz), guiding structures made of conductors become impractical, but dielectric waveguides and optical fibers provide efficient guiding of electromagnetic energy. Coverage of propagation of electromagnetic waves in an optical fiber or in dielectric slabs or rods is beyond the scope of this book.[50]

We now proceed to analyze electromagnetic wave propagation between two parallel metallic plates in Sections 8.3.1 and 8.3.2 and between two coaxial conductors in Section 8.3.3.

8.3.1 Waves between Parallel Metal Plates: Field Solutions for Propagating Modes

The simplest type of waveguide consists of two perfectly conducting plates between which electromagnetic waves of various types can be guided. Accordingly, we start our considerations of guided waves by analyzing the so-called *parallel-plate waveguide,* as shown in Figure 8.32. The plates have an infinite extent in the z direction in which the waves are to be guided. In practice the conductor plates would have a finite extent b in the y direction, typically much larger than the plate separation a, so that the effects of fringing fields at the edges are negligible. For our purposes here, we assume that the conductor plates have infinite extent ($b \gg a$) in the y direction, so that the electric and magnetic field do not vary with y.

Just as in the case of waves propagating in unbounded media, the configuration and propagation of the electromagnetic fields in any guiding structure are governed by Maxwell's equations and the wave equations derived from them. However, in the case of metallic waveguides, the solutions of the wave equations are now subject to the following boundary conditions at the conductor surfaces:

$$E_{\text{tangential}} = 0 \qquad H_{\text{normal}} = 0$$

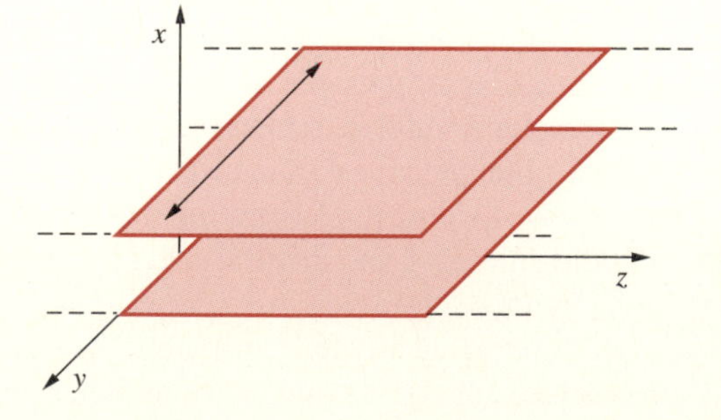

FIGURE 8.32. Parallel-plate waveguide. In an ideal parallel-plate waveguide, the plates are assumed to be infinite in extent in the y and z directions. In practice, the width b of the guide in the y direction is finite but is typically much larger than a. The material between the plates is assumed to be a lossless dielectric.

[50]For a recent textbook, and also a comparative discussion of optical fibers versus metallic waveguides, see Chapter 9 of W. Tomasi, *Advanced Electronic Communication Systems,* Prentice Hall, 1998.

Assuming that the medium between the perfectly conducting plates is source-free, simple (i.e., ϵ and μ are simple constants), and lossless[51] (i.e., $\sigma = 0$ and ϵ is real), the governing equations are the time-harmonic form of Maxwell's equations, namely

$$\nabla \times \mathbf{H} = j\omega\epsilon\mathbf{E} \qquad [8.59a]$$

$$\nabla \cdot \mathbf{E} = 0 \qquad [8.59b]$$

$$\nabla \times \mathbf{E} = -j\omega\mu\mathbf{H} \qquad [8.59c]$$

$$\nabla \cdot \mathbf{H} = 0 \qquad [8.59d]$$

from which the wave equations for **E** and **H** can be derived in a straightforward manner following the same procedure as in Section 8.1.1. We have

$$\nabla^2\mathbf{E} = -\beta^2\mathbf{E} \qquad [8.60a]$$

$$\nabla^2\mathbf{H} = -\beta^2\mathbf{H} \qquad [8.60b]$$

where $\beta = \omega\sqrt{\mu\epsilon}$.

Equations [8.59a] and [8.59c] can be written in component form for rectangular coordinates (note that the conductor configuration is rectangular; otherwise, for example for circularly shaped conductors, it is more appropriate to write the equations in terms of cylindrical coordinates r, ϕ, and z) for the region between plates (assumed to be a lossless dielectric, with both σ and ϵ'' equal to zero) as

$$\underbrace{\begin{aligned} \frac{\partial H_z}{\partial y} - \frac{\partial H_y}{\partial z} &= j\omega\epsilon E_x \\ \frac{\partial H_x}{\partial z} - \frac{\partial H_z}{\partial x} &= j\omega\epsilon E_y \\ \frac{\partial H_y}{\partial x} - \frac{\partial H_x}{\partial y} &= j\omega\epsilon E_z \end{aligned}}_{\nabla\times\mathbf{H}=j\omega\epsilon\mathbf{E}} \qquad \underbrace{\begin{aligned} \frac{\partial E_z}{\partial y} - \frac{\partial E_y}{\partial z} &= -j\omega\mu H_x \\ \frac{\partial E_x}{\partial z} - \frac{\partial E_z}{\partial x} &= -j\omega\mu H_y \\ \frac{\partial E_y}{\partial x} - \frac{\partial E_x}{\partial y} &= -j\omega\mu H_z \end{aligned}}_{\nabla\times\mathbf{E}=-j\omega\mu\mathbf{H}} \qquad [8.61]$$

Note that since the extent of the conductor plates in the y direction is infinite, none of the field quantities vary in the y direction, and as a result, the $\partial(\cdots)/\partial y$ terms in the above are all zero.

We assume propagation in the $+z$ direction, so that all field components vary as $e^{-\bar{\gamma}z}$, where $\bar{\gamma} = \bar{\alpha} + j\bar{\beta}$, with $\bar{\alpha}$ and $\bar{\beta}$ being real.[52] In the rest of this chapter,

---

[51] We shall later relax this requirement and calculate the losses in the dielectric medium using a perturbation solution.

[52] Note here that $\bar{\gamma}$ is different from the complex propagation constant of the medium $\gamma = \alpha + j\beta$ used earlier for uniform plane waves. Although γ, α, and β are dependent only on the wave frequency and the medium parameters (ϵ, μ, σ), the waveguide propagation constant $\bar{\gamma}$ depends in general also on the dimensions of the guiding structure. Even though the propagation medium may be nondissipative, $\bar{\gamma}$ may be either imaginary ($\bar{\gamma} = j\bar{\beta}$) or real ($\bar{\gamma} = \bar{\alpha}$), depending on the frequency of operation, as we shall see later. Note also that the quantities $\bar{\gamma}$, $\bar{\alpha}$, and $\bar{\beta}$ are not vectors; the bar in this case is simply used to distinguish them from corresponding quantities for waves propagating in simple, unbounded media. The $\bar{\gamma}$, $\bar{\alpha}$, and $\bar{\beta}$ notation used here is adapted from E. C. Jordan and K. G. Balmain, *Electromagnetic Waves and Radiating Systems,* Prentice Hall, 1968.

we consider the two special cases for which $\bar{\beta} = 0$ or for which $\bar{\alpha} = 0$. When the phasor fields vary in this manner, the real fields vary as

$$\Re e\{e^{-\bar{\gamma}z}e^{j\omega t}\} = \begin{cases} e^{-\bar{\alpha}z}\cos(\omega t) & \bar{\gamma} = \bar{\alpha} \quad \text{Evanescent wave} \\ \cos(\omega t - \bar{\beta}z) & \bar{\gamma} = j\bar{\beta} \quad \text{Propagating wave} \end{cases}$$

An evanescent[53] wave is one that does not propagate but instead exponentially decays with distance.

The wave equation(s) [8.60] can be rewritten as

$$\nabla^2\mathbf{E} = -\omega^2\mu\epsilon\mathbf{E} \quad \rightarrow \quad \frac{\partial^2\mathbf{E}}{\partial x^2} + \frac{\partial^2\mathbf{E}}{\partial y^2} + \frac{\partial^2\mathbf{E}}{\partial z^2} = -\omega^2\mu\epsilon\mathbf{E}$$

$$\nabla^2\mathbf{H} = -\omega^2\mu\epsilon\mathbf{H} \quad \rightarrow \quad \frac{\partial^2\mathbf{H}}{\partial x^2} + \frac{\partial^2\mathbf{H}}{\partial y^2} + \frac{\partial^2\mathbf{H}}{\partial z^2} = -\omega^2\mu\epsilon\mathbf{H}$$

Once again, we assume that none of the field quantities vary in the y direction. In other words, we have

$$\frac{\partial^2}{\partial y^2}(\cdots) = 0$$

The field variations in the x direction cannot be similarly limited, since the conductor imposes boundary conditions in the x direction, at $x = 0$ and $x = a$. Note that for any of the field components, for example for H_y, we can explicitly show the variations in the z direction by expressing $H_y(x, z)$ as a product of two functions having x and z, respectively, as their independent variables, namely,

$$H_y(x, z) = H_y^0(x)e^{-\bar{\gamma}z}$$

where $H_y^0(x)$ is a function only of x. We then have

$$\frac{\partial}{\partial z}H_y(x, z) = -\bar{\gamma}H_y^0(x)e^{-\bar{\gamma}z} = -\bar{\gamma}H_y(x, z) \quad \longrightarrow \quad \frac{\partial}{\partial z} \rightarrow -\bar{\gamma}$$

The wave equations [8.60] can thus be rewritten as

$$\frac{\partial^2\mathbf{E}}{\partial x^2} + \bar{\gamma}^2\mathbf{E} = -\omega^2\mu\epsilon\mathbf{E} \qquad [8.62]$$

$$\frac{\partial^2\mathbf{H}}{\partial x^2} + \bar{\gamma}^2\mathbf{H} = -\omega^2\mu\epsilon\mathbf{H} \qquad [8.63]$$

Substituting $\partial/\partial z \rightarrow -\bar{\gamma}$ and $\partial/\partial y = 0$ in [8.61], we find

$$\begin{aligned} \bar{\gamma}E_y &= -j\omega\mu H_x & \bar{\gamma}H_y &= j\omega\epsilon E_x \\ -\bar{\gamma}E_x - \frac{\partial E_z}{\partial x} &= -j\omega\mu H_y & -\bar{\gamma}H_x - \frac{\partial H_z}{\partial x} &= j\omega\epsilon E_y \\ \frac{\partial E_y}{\partial x} &= -j\omega\mu H_z & \frac{\partial H_y}{\partial x} &= j\omega\epsilon E_z \end{aligned} \qquad [8.64]$$

[53]The dictionary meaning of the word *evanescent* is "tending to vanish like vapor."

The foregoing relationships between the field components are valid in general for any time-harmonic electromagnetic wave solution for which (i) there are no variations in the y direction (i.e., $\partial(\cdot)/\partial y = 0$) and (ii) the field components vary in the z direction as $e^{-\bar{\gamma}z}$. In general, if a solution of the wave equation subject to the boundary conditions is obtained for any one of the field components, the other components can be found from [8.64].

In some cases, it is convenient to express the various field components explicitly in terms of the axial components (i.e., E_z and H_z). Introducing $h^2 = \bar{\gamma}^2 + \omega^2\mu\epsilon$, we can rewrite equations [8.64] in the form

$$H_x = -\frac{\bar{\gamma}}{h^2}\frac{\partial H_z}{\partial x} \quad [8.65a]$$

$$H_y = -\frac{j\omega\epsilon}{h^2}\frac{\partial E_z}{\partial x} \quad [8.65b]$$

$$E_x = -\frac{\bar{\gamma}}{h^2}\frac{\partial E_z}{\partial x} \quad [8.65c]$$

$$E_y = +\frac{j\omega\mu}{h^2}\frac{\partial H_z}{\partial x} \quad [8.65d]$$

These expressions are useful because the different possible solutions of the wave equation are categorized in terms of the axial components (i.e., z components), as discussed in the next subsection.

In general, all of the field components, including both E_z and H_z, may need to be nonzero to satisfy the boundary conditions imposed by a conductor or dielectric structure. However, it is convenient and customary to divide the solutions of guided wave equations into three categories:

$$\left.\begin{array}{l} E_z = 0 \\ H_z \neq 0 \end{array}\right\} \text{Transverse electric (TE) waves}$$

$$\left.\begin{array}{l} E_z \neq 0 \\ H_z = 0 \end{array}\right\} \text{Transverse magnetic (TM) waves}$$

$$E_z, H_z = 0 \quad \longrightarrow \quad \text{Transverse electromagnetic (TEM) waves}$$

where z is the axial direction of the guide along which the wave propagation takes place. In the following, we separately examine the field solutions for TE, TM, and TEM waves.

Transverse Electric (TE) Waves We first consider transverse electric (TE) waves, for which we have $E_z = 0$, $H_z \neq 0$. With $E_z = 0$, the electric field vector of a TE wave is always and everywhere transverse to the direction of propagation. We can solve the wave equation for any of the field components; for the parallel plate case, it is most convenient[54] to solve for E_y, from which all the other field components can be found using equations [8.64]. Rewriting the wave equation [8.62] for the y component, we have

$$\frac{\partial^2 E_y}{\partial x^2} + \bar{\gamma}^2 \mathrm{E}_y = -\omega^2 \mu \epsilon \mathrm{E}_y \qquad [8.66]$$

or

$$\frac{\partial^2 E_y}{\partial x^2} = -h^2 E_y \qquad [8.67]$$

where $h^2 = \bar{\gamma}^2 + \omega^2 \mu \epsilon$. We now express $E_y(x, z)$ as a product of two functions, one varying with x and another varying with z as

$$E_y(x, z) = E_y^0(x) e^{-\bar{\gamma} z}$$

Note that z variation of all the field components has the form $e^{-\bar{\gamma} z}$.

We can now rewrite [8.67] as an ordinary differential equation in terms of $E_y^0(x)$:

$$\frac{d^2 E_y^0(x)}{dx^2} = -h^2 E_y^0(x) \qquad [8.68]$$

Note that [8.68] is a second-order differential equation that has the general solution of

$$E_y^0(x) = C_1 \sin(hx) + C_2 \cos(hx) \qquad [8.69]$$

where C_1, C_2, and h are constants to be determined by the boundary conditions. Note that the complete phasor for the y component of the electric field is

$$E_y(x, z) = [C_1 \sin(hx) + C_2 \cos(hx)] e^{-\bar{\gamma} z}$$

We now consider the boundary condition requiring the tangential electric field to be zero at the perfectly conducting plates. In other words,

$$\left.\begin{aligned} E_y &= 0 \text{ at } x = 0 \\ E_y &= 0 \text{ at } x = a \end{aligned}\right\} \text{for all } y \text{ and } z$$

Thus $C_2 = 0$, and

$$E_y(x, z) = C_1 \sin(hx) e^{-\bar{\gamma} z}$$

results from the first boundary condition. The second condition (i.e., $E_y = 0$ at $x = a$) imposes a restriction on h; that is,

$$h = \frac{m\pi}{a} \qquad m = 1, 2, 3, \ldots$$

[54]Since $E_z \equiv 0$ for a TE wave, and since there is no variation with y, it is clear that there can be a y component of the electric field, as long as E_y, which is tangential to the conductors, varies with x such that it is zero at the perfectly conducting walls.

This result illustrates the particular importance of the constant h. Waveguide field solutions such as [8.69] can satisfy the necessary boundary conditions only for a discrete set of values of h. These values of h for which the field solutions of the wave equation can satisfy the boundary conditions are known as the *characteristic values* or *eigenvalues*.[55]

The other wave field components can all be obtained from E_y using [8.64]. We find $E_x = H_y = 0$, and

Parallel-plate TE_m, $m = 0, \pm 1, \pm 2, \ldots$

$$\begin{aligned} E_y &= C_1 \sin\left(\frac{m\pi}{a}x\right)e^{-\bar{\gamma}z} \\ H_z &= -\frac{1}{j\omega\mu}\frac{\partial E_y}{\partial x} = -\frac{m\pi}{j\omega\mu a}C_1\cos\left(\frac{m\pi}{a}x\right)e^{-\bar{\gamma}z} \\ H_x &= -\frac{\bar{\gamma}}{j\omega\mu}E_y = -\frac{\bar{\gamma}}{j\omega\mu}C_1\sin\left(\frac{m\pi}{a}x\right)e^{-\bar{\gamma}z} \end{aligned} \qquad [8.70]$$

Note that $m = 0$ makes all fields vanish. Each integer value of m specifies a given field configuration, or *mode*, referred to in this case as the TE_m mode. The configurations of fields for the cases of $m = 1$ and 2 are shown in Figure 8.33.

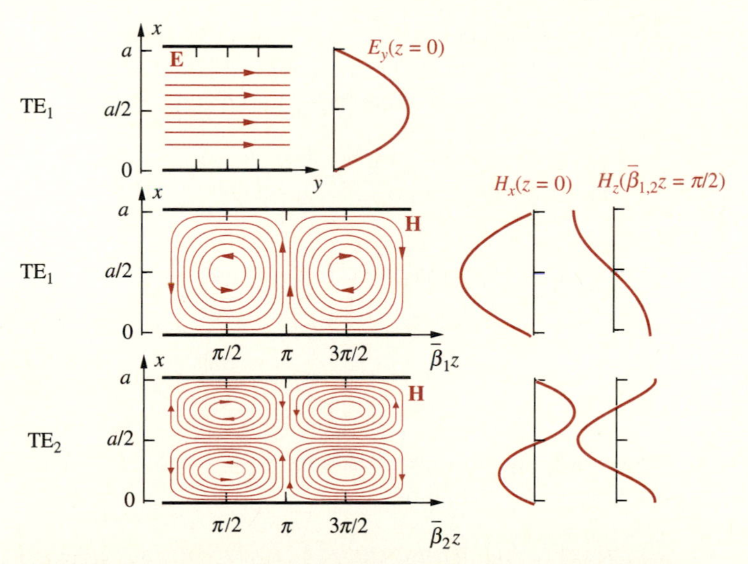

FIGURE 8.33. TE_1 and TE_2 modes. The electric and magnetic field distributions for the TE_1 and the magnetic field distribution for the TE_2 modes in a parallel-plate waveguide. Careful examination of the field structure for the TE_1 mode indicates that the magnetic field lines encircle the electric field lines (i.e., displacement current) in accordance with $\nabla \times \mathbf{H} = j\omega\epsilon\mathbf{E}$ and the right-hand rule. The same is true for the TE_2 mode, although the electric field distribution for this mode is not shown.

[55] *Eigen* is from the German word for "characteristic" or "one's own."

Transverse Magnetic (TM) Waves For TM waves the magnetic field vector is always and everywhere transverse to the direction of propagation (i.e., $H_z = 0$), while the axial component of the electric field is nonzero (i.e., $E_z \neq 0$). We proceed with the solution in a similar manner as was done for TE waves, except that it is more convenient to use the wave equation for **H**, namely [8.63]. The y component of this equation for H_y is in the form of [8.66] for E_y. Following the same procedure as was done for the TE case for E_y, the general solution for H_y is

$$H_y = [C_3 \sin(hx) + C_4 \cos(hx)]e^{-\bar{\gamma}z}$$

Since the boundary condition for the magnetic field is in terms of its component normal to the conductor boundary, it cannot be applied to H_y directly. However, using [8.64], we can write the tangential component of the electric field E_z in terms of H_y as

$$E_z = \frac{1}{j\omega\epsilon}\frac{\partial H_y}{\partial x} = \frac{h}{j\omega\epsilon}[C_3 \cos(hx) - C_4 \sin(hx)]e^{-\bar{\gamma}z}$$

We now note that

$$E_z = 0 \quad \text{at } x = 0 \quad \rightarrow C_3 = 0$$

and

$$E_z = 0 \quad \text{at } x = a \quad \rightarrow h = \frac{m\pi}{a} \qquad m = 0, 1, 2, 3, \ldots$$

The only other nonzero field component is E_x, which can be simply found from H_y or E_z. The nonzero field components for this mode, denoted as the TM_m mode, are then

Parallel-plate TM_m, $m = 0, \pm 1, \pm 2, \ldots$

$$\boxed{\begin{aligned} H_y &= C_4 \cos\left(\frac{m\pi}{a}x\right)e^{-\bar{\gamma}z} \\ E_x &= \frac{\bar{\gamma}}{j\omega\epsilon}H_y = \frac{\bar{\gamma}}{j\omega\epsilon}C_4 \cos\left(\frac{m\pi}{a}x\right)e^{-\bar{\gamma}z} \\ E_z &= \frac{jm\pi}{\omega\epsilon a}C_4 \sin\left(\frac{m\pi}{a}x\right)e^{-\bar{\gamma}z} \end{aligned}} \qquad [8.71]$$

The configurations of the fields for $m = 1$ and 2 (i.e., for the TM_1 and TM_2 modes) are shown in Figure 8.34.

Transverse Electromagnetic (TEM) Waves Note that contrary to the TE case, the TM solutions do not all vanish for $m = 0$. Since E_z is zero for $m = 0$, the TM_0 mode is actually a transverse electromagnetic (or TEM) wave. In this case, we have

Parallel-plate TEM

$$\boxed{\begin{aligned} H_y &= C_4 e^{-\bar{\gamma}z} \\ E_x &= \frac{\bar{\gamma}}{j\omega\epsilon}C_4 e^{-\bar{\gamma}z} \\ E_z &= 0 \end{aligned}} \qquad [8.72]$$

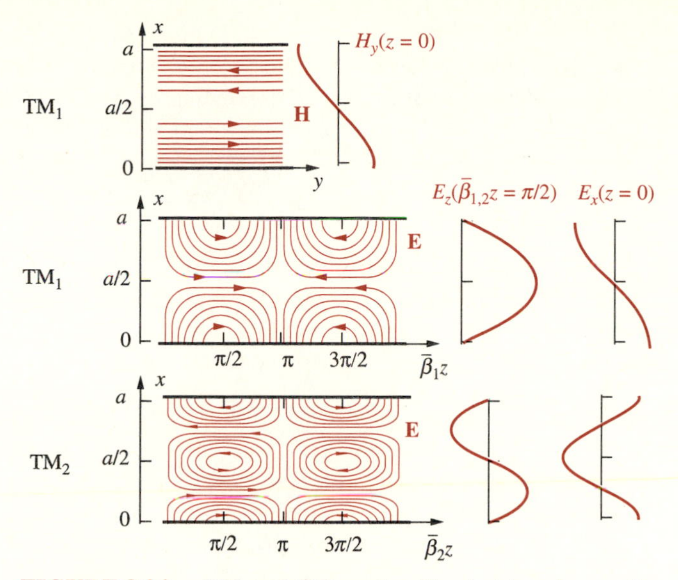

FIGURE 8.34. **TM$_1$ and TM$_2$ modes.** The electric and magnetic field distributions for the TM$_1$ and TM$_2$ modes in a parallel-plate waveguide. Note that for TM$_1$ the electric field lines encircle the magnetic field lines (Faraday's law) in accordance with $\nabla \times \mathbf{E} = -j\omega\mu\mathbf{H}$. Same is true for TM$_2$, although the magnetic field distribution for TM$_2$ is not shown.

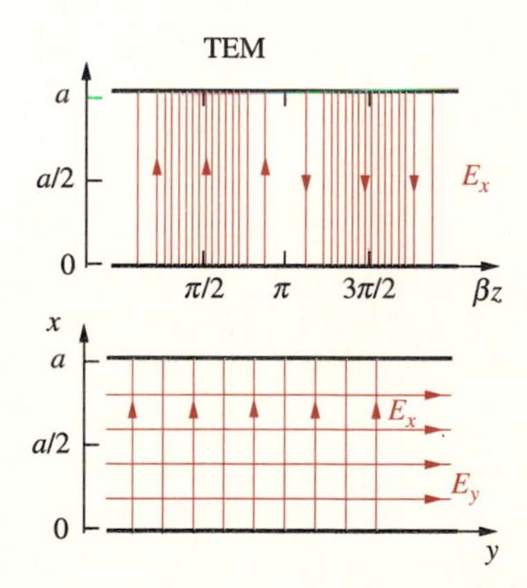

FIGURE 8.35. **TEM mode.** Electric and magnetic fields between parallel planes for the TEM (TM$_0$) mode. Only electric field lines are shown in the top panel; the magnetic field lines are out of (or into) the page.

For this case, we have $h = 0$, so $\bar{\gamma} = j\bar{\beta} = j\beta$, where $\beta = \omega\sqrt{\mu\epsilon}$ is the phase constant for a uniform plane wave in the unbounded lossless medium. The field configuration for the TM$_0$ or TEM wave is shown in Figure 8.35.

Cutoff Frequency, Phase Velocity, and Wavelength Close examination of the solutions for TE and TM waves reveal a number of common characteristics: (i) **E** and **H** have sinusoidal standing-wave distributions in the x direction; (ii) any xy plane is an equiphase plane, or, in other words, the TE and TM waves are plane waves with surfaces of constant phase given by z = const.; and (iii) these equiphase surfaces progress (propagate) along the waveguide with a velocity $\bar{v}_p = \omega/\bar{\beta}$. To see this, consider any of the field components—say, E_y, for TE waves. For a propagating wave, with $\bar{\gamma} = j\bar{\beta}$, and assuming C_1 to be real, we have

$$\mathscr{E}_y(x, z, t) = C_1 \sin\left(\frac{m\pi}{a}x\right)\cos(\omega t - \bar{\beta}z)$$

To determine the conditions under which we can have a propagating wave (i.e., $\bar{\gamma}$ purely imaginary, or $\bar{\gamma} = j\bar{\beta}$), consider the definition of the eigenvalue parameter h:

$$h^2 = \bar{\gamma}^2 + \omega^2\mu\epsilon \quad \rightarrow \quad \bar{\gamma} = \sqrt{h^2 - \omega^2\mu\epsilon} = \sqrt{\left(\frac{m\pi}{a}\right)^2 - \omega^2\mu\epsilon}$$

Note that this expression for h is valid for both TE and TM modes. Thus, the expressions given below for propagation constant, phase velocity, and wavelength are also equally valid for both TE and TM modes.

From the preceding definition of h it is apparent that for each mode (TE or TM) identified by mode index m, there exists a critical or *cutoff* frequency f_{c_m} (or wavelength λ_{c_m}) such that

$$\bar{\gamma} = 0 \quad \rightarrow \quad f_{c_m} = \frac{mv_p}{2a} = \frac{m}{2a\sqrt{\mu\epsilon}} \quad \text{or} \quad \lambda_{c_m} = \frac{v_p}{f_{c_m}} = \frac{2a}{m} \qquad [8.73]$$

where $v_p = 1/\sqrt{\mu\epsilon}$ is the *intrinsic* phase velocity for uniform plane waves in the unbounded lossless medium. Note that $v_p = (\mu\epsilon)^{-1/2}$ depends only on ϵ and μ and is thus indeed an intrinsic property of the medium. For $f > f_{c_m}$, the propagation constant $\bar{\gamma}$ is purely imaginary and is given by

$$\boxed{\bar{\gamma} = j\bar{\beta}_m = j\sqrt{\omega^2\mu\epsilon - \left(\frac{m\pi}{a}\right)^2} = j\beta\sqrt{1 - \left(\frac{f_{c_m}}{f}\right)^2}, \qquad f > f_{c_m}} \qquad [8.74]$$

where we now start to use $\bar{\beta}_m$ instead of $\bar{\beta}$ to underscore the dependence of the phase constant on the mode index m. Note that $\bar{\beta}_m$ is the phase constant (sometimes referred to as the longitudinal phase constant or the propagation constant) for mode m, and $\beta = \omega\sqrt{\mu\epsilon}$ is the intrinsic phase constant, being that for uniform plane waves in the unbounded lossless medium. For $f < f_{c_m}$, $\bar{\gamma}$ is a real number, given by

$$\boxed{\bar{\gamma} = \bar{\alpha}_m = \sqrt{\left(\frac{m\pi}{a}\right)^2 - \omega^2\mu\epsilon} = \beta\sqrt{\left(\frac{f_{c_m}}{f}\right)^2 - 1}, \qquad f < f_{c_m}} \qquad [8.75]$$

where $\bar{\alpha}_m$ is the attenuation constant, in which case the fields will attenuate exponentially in the z direction, without any wave motion. For example, the expression for the $\mathscr{E}_y$ component for TE waves with C_1 real for the case when $f < f_{c_m}$ is

$$\mathscr{E}_y(x, z, t) = C_1 e^{-\bar{\alpha}z} \sin\left(\frac{m\pi}{a}x\right)\cos(\omega t)$$

Such a nonpropagating mode is called an evanescent wave.[56] For $f > f_{c_m}$, the propagating wave has a wavelength $\bar{\lambda}_m$ (sometimes referred to as the guide wavelength) and phase velocity $\bar{v}_{p_m}$ given as

$$\bar{\lambda}_m = \frac{2\pi}{\bar{\beta}_m} = \frac{\lambda}{\sqrt{1 - (f_{c_m}/f)^2}}$$
$$\bar{v}_{p_m} = \frac{\omega}{\bar{\beta}_m} = \frac{v_p}{\sqrt{1 - (f_{c_m}/f)^2}} \quad [8.76]$$

where $\lambda = 2\pi/\beta = v_p/f$ is the intrinsic wavelength, being the wavelength for uniform plane waves in the unbounded lossless medium, and thus depending only on μ, ϵ, and frequency f. Note that for the TM_0 mode (which is in fact a TEM wave, as discussed above) $\bar{v}_p$ is equal to the intrinsic phase velocity $v_p = (\sqrt{\mu\epsilon})^{-1}$ in the unbounded lossless medium and is thus independent of frequency. Furthermore, the phase constant $\bar{\beta}_m$ and wavelength $\bar{\lambda}_m$ are also equal to their values for a uniform plane wave in an unbounded lossless medium. There is no cutoff frequency for this mode, since the propagating field solutions are valid at any frequency.

For TE and TM waves in general, it is clear from this discussion that the wavelength $\bar{\lambda}_m$, as observed along the guide, is longer than the corresponding wavelength in an unbounded lossless medium by the factor $[\sqrt{1 - (f_{c_m}/f)^2}]^{-1}$. Also, the velocity of phase progression inside the guide is likewise greater than the corresponding intrinsic phase velocity. Note, however, that $\bar{v}_p$ is not the velocity with which energy or information propagates,[57] so $\bar{v}_p > v_p$ does not pose any particular dilemma.

Examples 8-32 and 8-33 illustrate the application of the various parallel-plate waveguide relationships derived above.

Example 8-32: Parallel-plate waveguide modes. An air-filled parallel-plate waveguide has a plate separation of 1.25 cm. Find (a) the cutoff frequencies of the TE_0, TM_0, TE_1, TM_1, and TM_2 modes, (b) the phase velocities of those modes at

[56]It is important to note that the attenuation of the field in the z direction is not due to any energy losses; this condition results simply from the fact that the boundary conditions cannot be satisfied by a TE or TM wave at frequency ω. If a TE or TM field configuration is somehow excited (e.g., by an appropriate configuration of source currents or charges) at a given point z, the fields attenuate exponentially with distance away from the excitation point, with no energy being carried away.

[57]The velocity at which information (e.g., the envelope of a modulated signal) travels is the so-called *group velocity* v_g, given by $v_g = d\omega/d\beta$, which for metalllic waveguides is $v_g = c^2/\bar{v}_p$. See Section 5.15 of S. Ramo, J. R. Whinnery, and T. Van Duzer, *Fields and Waves in Communication Electronics,* 3rd ed., John Wiley & Sons, Inc., New York, 1994.

15 GHz, (c) the lowest-order TE and TM mode that cannot propagate in this waveguide at 25 GHz.

Solution:

(a) As we have seen in earlier sections, TE_0 mode does not exist in a parallel-plate waveguide. TM_0 is equivalent to the TEM mode, and this mode can propagate at all frequencies (i.e., $f_{c_0} = 0$). The cutoff frequencies of the other modes can be calculated using

$$f_{c_m} = \frac{m}{2a\sqrt{\mu_0\epsilon_0}} = \frac{m(3 \times 10^{10} \text{ cm-s}^{-1})}{2(1.25 \text{ cm})} = 1.2 \times 10^{10} m \text{ Hz} = 12m \text{ GHz}$$

The results are summarized in the following table:

Mode	TM_0	TE_1	TM_1	TE_2	TM_2
f_{c_m} (GHz)	0	12	12	24	24

(b) For the TM_0 mode, $\bar{v}_p = c$. For TE_1 and TM_1 modes, we have

$$\bar{v}_p = \frac{c}{\sqrt{1 - (f_{c_1}/f)^2}} = \frac{c}{\sqrt{1 - (12/15)^2}} = \frac{5c}{4}$$

Note that the TE_2 and TM_2 modes do not propagate at 15 GHz; that is, their cutoff frequencies are above 15 GHz.

(c) The lowest-order modes that cannot propagate in this waveguide are TE_3 and TM_3, since

$$f_{c_3} = 12 \times 3 = 36 \text{ GHz} > 25 \text{ GHz}$$

Example 8-33: ELF propagation in the earth–ionosphere waveguide. Extremely low frequencies (ELF) are ideal for communicating with deeply submerged submarines because, below 1 kHz, electromagnetic waves penetrate deeply into seawater.[58] Propagation at these frequencies takes place in the "waveguide" formed between the earth and the ionosphere; low propagation losses allow nearly worldwide communication from a single ELF transmitter.

In J. R. Wait's simple model,[59] the surface of the earth and the bottom of the ionosphere form the boundaries of a terrestrial "parallel"-plate waveguide with lossy walls. The ionosphere is approximated by an isotropic layer beginning at an altitude h and extending to infinity with no horizontal variations allowed. Energy is lost through the "walls" either into the finitely conducting ionosphere or into the ground, with the

[58] S. L. Bernstein, et al., Long-range communications at extremely low frequencies, *Proc. IEEE*, 62(3), pp. 292–312, March 1974.

[59] J. R. Wait, Earth-ionosphere cavity resonances and the propagation of ELF radio waves, *Radio Sci.*, 69D, pp. 1057–1070, August 1965.

former loss being dominant. The important feature of propagation below 1 kHz is that there is a single propagating mode, a so-called quasi-TEM mode. All the other modes are evanescent and are almost undetectable at distances in excess of 1000 km. In the far field, the wave consists of a vertical electric field and a horizontal magnetic field transverse to the direction of propagation (Figure 8.36). The leakage of energy from this wave into the ocean gives rise to a plane wave propagating downward, and it is this wave that the submarine receiver detects.

Consider an ideal earth–ionosphere waveguide where both the ionosphere and the earth are assumed to be perfect conductors. In addition, neglect the curvature of the waveguide and assume it to be flat. The height of the terrestrial waveguide can vary anywhere from 70 km to 90 km depending on conditions; for our purposes, assume it to be 80 km. Find all the propagating modes at an operating frequency of (a) 100 Hz (ELF); (b) 1 kHz (ELF); and (c) 10 kHz (VLF).

Solution: As we already know, the TEM mode (or TM_0 mode) exists in all cases. The cutoff frequencies of other modes can be found as

$$f_{c_m} = \frac{m(3 \times 10^8 \text{ m-s}^{-1})}{2 \times 80 \times 10^3 \text{ m}} = 1875m \text{ Hz}$$

So the cutoff frequencies of some of the lower-order modes are

Mode	TM_0	TE_1	TM_1	TE_2	TM_2	TE_3	TM_3
f_{c_m} (Hz)	0	1875	1875	3750	3750	5625	5625

Therefore, (a) at 100 Hz and (b) at 1 kHz, only one mode (the TEM mode) can propagate, whereas (c) at 10 kHz, eleven different modes (i.e., TEM, TE_1, TM_1,

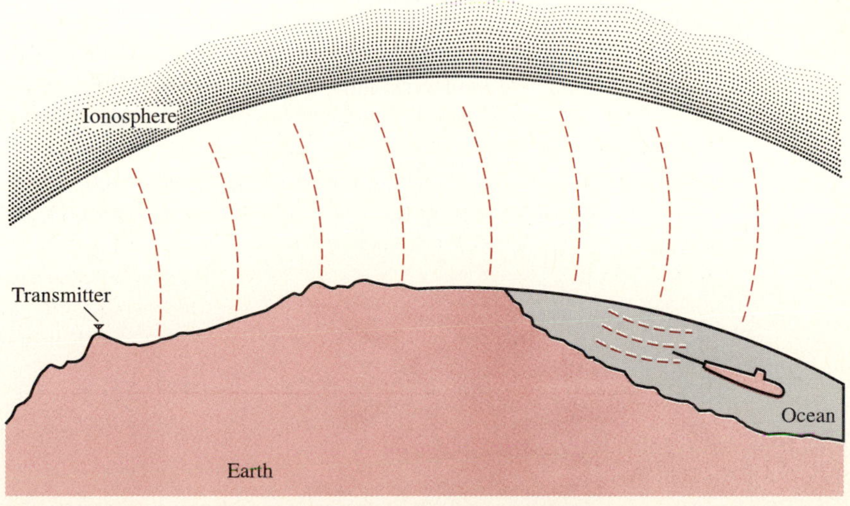

FIGURE 8.36. ELF propagation and submarine reception.

TE_2, TM_2, TE_3, TM_3, TE_4, TM_4, TE_5, and TM_5) can propagate in the earth–ionosphere waveguide.

Voltage, Current, and Impedance in a Parallel-Plate Waveguide The TEM wave on a parallel-plate line is the propagation mode that corresponds to the voltage and current waves on two-conductor transmission lines, which were discussed in detail in Chapters 2 and 3. The voltage between the two conductors and the total current flowing through a 1-meter wide segment of each of the plates are related to the electric and magnetic fields of the TEM wave as

$$V_{\text{TEM}}(z) = -\int_{x=0}^{a} \mathbf{E}\cdot d\mathbf{l} = \int_{a}^{0} E_x(z)\,dx = \frac{\beta}{\omega\epsilon}C_4 e^{-j\beta z}\int_{a}^{0} dx = V^+ e^{-j\beta z}$$

$$I_{\text{TEM}}(z) = |\mathbf{J}_s|(1\text{ m}) = |\hat{\mathbf{n}}\times\mathbf{H}| = |\pm\hat{\mathbf{x}}\times\hat{\mathbf{y}}[H_x]_{x=0}| = |\hat{\mathbf{x}}\times\hat{\mathbf{y}}C_4 e^{-j\beta z}|$$

$$= C_4 e^{-j\beta z} = \frac{V^+}{Z_{\text{TEM}}}e^{-j\beta z}$$

where we have noted that for the TEM mode $\bar{\beta} = \beta = \omega\sqrt{\mu\epsilon}$, defined $V^+ = -\beta C_4 a/(\omega\epsilon) = -\eta a C_4$ as the peak voltage, and introduced the quantity Z_{TEM}, called the wave impedance, also discussed in Section 8.2.3. The wave impedance is defined as the ratio of the transverse components of the electric and magnetic fields, namely

$$Z_{\text{TEM}} \equiv \frac{E_x}{H_y} = \frac{[\beta/(\omega\epsilon)]C_4 e^{-j\beta z}}{C_4 e^{-j\beta z}} = \sqrt{\frac{\mu}{\epsilon}} = \eta$$

By identifying $V(z) = V_{\text{TEM}}$, $I(z) = I_{\text{TEM}}$, and $Z_0 = Z_{\text{TEM}}$ respectively as the line voltage, line current, and characteristic line impedance, all of the transmission line analysis techniques that were studied in Chapters 2 and 3 are applicable to parallel-plate waveguides operating in the TEM mode. As an example, the time-average power flowing down the transmission line, given by $\frac{1}{2}|V^+|^2/Z_0$, is equal to the time-average transmitted power P_{av} evaluated in the next section using the phasor expressions for the electric and magnetic fields. We note that the phase velocity $\bar{v}_p = \omega/\bar{\beta} = 1/\sqrt{\mu\epsilon}$ and the wave impedance Z_{TEM} for TEM waves are both independent of frequency.

In general, for TE or TM modes, the definitions of equivalent voltage and current become ambiguous. To see this, consider the propagating TE_m mode, the fields for which are

$$E_y = C_1\sin\left(\frac{m\pi}{a}x\right)e^{-j\bar{\beta}z}$$

$$H_z = -\frac{m\pi}{j\omega\mu a}C_1\cos\left(\frac{m\pi}{a}x\right)e^{-j\bar{\beta}z}$$

$$H_x = -\frac{\bar{\beta}}{\omega\mu}C_1\sin\left(\frac{m\pi}{a}x\right)e^{-j\bar{\beta}z}$$

where $\bar{\beta} = \sqrt{\omega^2\mu\epsilon - (m\pi/a)^2}$. We note that the wave electric field is in the y direction, so the quantity that we would normally define as voltage—that is, the line integral of the electric field from one plate to another—is zero. In other words,

$$-\int_0^a \mathbf{E} \cdot d\mathbf{l} = -\int_0^a \hat{\mathbf{y}} E_y \cdot \hat{\mathbf{x}}\, dx = 0$$

Similarly, the line current, which we would normally think of as a z-directed current in the plates, actually flows in the y direction. To see this, note that the only nonzero magnetic field at the surfaces of the conducting plates is H_z, which requires a surface current of

$$\mathbf{J}_s = \hat{\mathbf{n}} \times \hat{\mathbf{z}}[H_z]_{x=0} = \hat{\mathbf{x}} \times \hat{\mathbf{z}}\left(\frac{-m\pi}{j\omega\mu a}\right)e^{-j\bar{\beta}z} = \hat{\mathbf{y}}\left(\frac{m\pi}{j\omega\mu a}\right)e^{-j\bar{\beta}z}$$

along the $x = 0$ plate. In practice, the lack of unique definitions of voltage and current for TE and TM modes does not preclude the use of transmission line techniques in analyzing the behavior of microwave networks. In most cases, useful definitions suitable for the particular problems in hand can be put forth.[60]

8.3.2 Waves between Parallel Plates: Attenuation Due to Conduction and Dielectric Losses

Up to now, we have considered the field structures and other characteristics of propagating modes, assuming the waveguide walls to be perfectly conducting and the dielectric between the plates to be lossless, and thus implicitly neglecting losses. In this section we consider attenuation of the fields due to conduction or dielectric losses.

Attenuation Due to Conduction Losses In all of the TE and TM mode solutions we have discussed, the wave magnetic field has a nonzero component parallel to the waveguide walls. As a result of the boundary condition $\hat{\mathbf{n}} \times \mathbf{H} = \mathbf{J}_s$, this parallel magnetic field component causes surface currents to flow in the metallic waveguide walls. If the walls are made of perfectly conducting materials, this current flow does not have any effect on the wave fields. However, in actual waveguides, the fact that the metallic walls have finite conductivity results in losses,[61] and the fields attenuate with distance.[62] In this section, we adopt the common practice of assuming that the

---

[60]See R. E. Collin, *Foundations of Microwave Engineering,* 2nd ed., McGraw-Hill, Inc., 1992.

[61]Attenuation in waveguides does not only occur because of wall losses; it can also come about if the dielectric material between the plates is lossy (i.e., $\epsilon'' \neq 0$ and/or $\sigma \neq 0$). We consider dielectric losses (represented by attenuation constant α_d) in the next section in connection with hollow cylindrical waveguides; expressions and methodologies used there are directly applicable to the case of the parallel-plate waveguide.

[62]Note that this attenuation is in general quite small, unlike the very rapid attenuation of an evanescent wave. In other words, typically the value of $\bar{\alpha}$ for a nonpropagating mode is much larger than the attenuation constant for conducting losses; that is, α_c.

attenuation is exponential in form (i.e., the fields vary as $e^{-\alpha_c z}$) and determine the attenuation constant α_c using a perturbation solution.[63]

Since the fields vary as $e^{-\alpha_c z}$, the average wave power P_{av} transmitted along the waveguide (i.e., in the z direction) varies as $e^{-2\alpha_c z}$. The rate of decrease of transmitted power along the waveguide is then

$$-\frac{\partial P_{\text{av}}}{\partial z} = +2\alpha_c P_{\text{av}}$$

The reduction in power per unit length must be equal to the power lost or dissipated per unit length. Therefore we have

$$\frac{\text{Power lost per unit length}}{\text{Power transmitted}} = \frac{2\alpha_c P_{\text{av}}}{P_{\text{av}}} = 2\alpha_c$$

or

$$\alpha_c = \frac{\text{Power lost per unit length}}{2 \times \text{Power transmitted}} = \frac{P_{\text{loss}}}{P_{\text{av}}} \qquad [8.77]$$

We now use this definition of α_c for the parallel-plate waveguide.

We first consider attenuation due to conductor losses for a TEM wave. Using the solutions obtained in the previous subsection, we have

$$H_y = C_4 e^{-j\bar{\beta}z} \qquad E_x = \frac{\bar{\beta}}{\omega\epsilon} C_4 e^{-j\bar{\beta}z}$$

The current density in each of the conducting plates is

$$\mathbf{J}_s = \hat{\mathbf{n}} \times \mathbf{H}$$

The current in each plate is $J_s = C_4 e^{-j\bar{\beta}z}$ so that $|J_s| = C_4$, assuming that C_4 is positive and real.

The total loss (sum of that in upper and lower plates) for a length of 1 meter and a width of b meters of the guide is

$$P_{\text{loss}} = 2\int_0^1 \int_0^b \tfrac{1}{2}|J_s|^2 R_s \, dy \, dz = C_4^2 R_s$$

where R_s is the resistive part of the surface impedance discussed in Section 8.2.4. From [8.55], the surface impedance of a conductor is

$$\eta_c = Z_s = R_s + jX_s = (\sigma\delta)^{-1}(1 + j)$$

where $\delta = \sqrt{2/(\omega\mu_0\sigma)}$ is the skin depth, assuming nonmagnetic, conducting walls with permeability $\mu = \mu_0$. Another way of expressing R_s is $R_s = \sqrt{\omega\mu_0/(2\sigma)}$.

[63] In actual conductors such as copper or brass, and at the typical frequencies utilized, the finite conductivity of the walls has negligible effect on the field configuration. Therefore, in most cases it is possible to find the surface currents that flow on the walls by using the fields determined under the assumption of perfectly conducting boundaries. The attenuation is then estimated from the currents using the surface resistance R_s of the conductor.

The time-average power density transmitted down the guide per unit cross-sectional area is

$$\mathbf{S}_{\text{av}} = \tfrac{1}{2}\Re e\{\mathbf{E}\times\mathbf{H}^*\}$$

Since $|E_x| = \eta|H_y|$, with C_4 assumed to be real, we have $|\mathbf{S}_{\text{av}}| = \frac{1}{2}\eta C_4^2$. For a spacing of a meters and a width of b meters, the cross-sectional area is ba; thus the total time-average power transmitted through such a cross-sectional area is

$$P_{\text{av}} = \tfrac{1}{2}\eta C_4^2 ba$$

Thus, using the definition of α_c in [8.77], we have

Parallel-plate TEM (TM_0)

$$\alpha_{c_{\text{TEM}}} = \frac{C_4^2 R_s b}{2(\frac{1}{2}\eta C_4^2 ba)} = \frac{R_s}{\eta a} = \frac{1}{\eta a}\sqrt{\frac{\omega\mu_0}{2\sigma}} \qquad [8.78]$$

Next we consider conductor losses for TE waves. Note that the nonzero field components for TE waves in a parallel-plate waveguide are as given in [8.70]. The amplitude of the current density is determined by the tangential $\mathbf{H}$ (i.e., H_z) at $x = 0$ and $x = a$ to be (for C_1 real and positive)

$$|J_{sy}| = |H_z| = \frac{m\pi C_1}{\omega\mu a}$$

It is interesting to note that J_{sy} is the only nonzero current component and does not flow in the propagation direction (i.e., z direction). The total power loss (sum of those in the two plates) for a length of 1 meter and a width of b meters of the guide is

$$P_{\text{loss}} = 2\int_0^1\int_0^b\left[\frac{1}{2}|J_{sy}|^2 R_s\right]dy\,dz = \frac{2bm^2\pi^2 C_1^2\sqrt{\omega\mu_0/(2\sigma)}}{2\omega^2\mu^2 a^2}$$

The time-average power density transmitted in the z direction per unit cross-sectional area is

$$|\mathbf{S}_{\text{av}}| = \frac{1}{2}\Re e\{\mathbf{E}\times\mathbf{H}^*\}\cdot\hat{\mathbf{z}} = -\frac{1}{2}E_y H_x^* = \frac{\bar{\beta}C_1^2}{2\omega\mu}\sin^2\left(\frac{m\pi}{a}x\right)$$

The total power through a guide of cross-sectional area of 1 meter width and a meters height can be determined by integrating $|\mathbf{S}_{\text{av}}|$ in x and y:

$$P_{\text{av}} = \int_0^a \frac{\bar{\beta}C_1^2}{2\omega\mu}\sin^2\left(\frac{m\pi}{a}x\right)dx = \frac{\bar{\beta}C_1^2 a}{4\omega\mu}$$

The attenuation constant α_c is then given by

Parallel-plate TE_m

$$\alpha_{c_{\text{TE}_m}} = \frac{2m^2\pi^2\sqrt{\omega\mu_0/(2\sigma)}}{\bar{\beta}\omega\mu a^3} = \frac{2R_s(f_{c_m}/f)^2}{\eta a\sqrt{1-(f_{c_m}/f)^2}} \qquad [8.79]$$

where we have used $\bar{\beta} = \beta\sqrt{1-(f_{c_m}/f)^2}$.

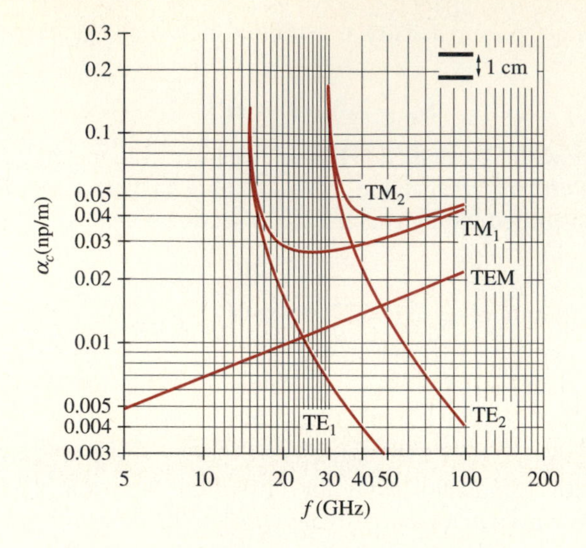

FIGURE 8.37. Attenuation versus frequency for parallel-plate waveguide. Attenuation-versus-frequency characteristics of waves guided by parallel plates.

The derivation of a corresponding expression for α_c for TM waves proceeds in the same manner as for TE waves. We find

$$\text{Parallel-plate TM}_m \qquad \boxed{\alpha_{c_{\text{TM}m}} = \frac{2R_s}{\eta a\sqrt{1-(f_{c_m}/f)^2}}} \qquad [8.80]$$

Attenuation due to conduction losses for TEM, TE, and TM waves is plotted as a function of frequency in Figure 8.37. It is interesting to note that the losses are generally higher for TM than for TE waves. This is because the current **J** flows in support of the tangential component of **H**, and for TM waves the transverse component of the magnetic field (H_y) is tangential. For TE waves, the currents are related to H_z, whose magnitude approaches zero as frequency increases.

Example 8-34 compares the numerical values of attenuation constant α_c for different waveguide modes.

Example 8-34: Parallel-copper-plate waveguide. Consider an air-filled parallel-copper-plate waveguide with an inner plate separation distance of 1 cm. Calculate the attenuation constant due to conductor losses, α_c, in decibels per meter (dB-m^{-1}), for TEM, TE$_1$, and TM$_1$ modes at operating frequencies of (a) 20 GHz and (b) 30 GHz.

Solution: For copper, the surface resistance R_s is given by

$$R_s = \sqrt{\frac{\omega\mu_0}{2\sigma}} = \sqrt{\frac{2\pi f \times 4\pi \times 10^{-7}}{2 \times 5.8 \times 10^7}} \simeq 2.61 \times 10^{-7}\sqrt{f}\ \Omega$$

Thus, at 20 GHz, we have $R_s \simeq 3.69 \times 10^{-2}\Omega$, whereas at 30 GHz, $R_s \simeq 4.52 \times 10^{-2}\Omega$. For the TEM mode, the attenuation constant α_c is given by

$$\alpha_{c_{\text{TEM}}} = \frac{R_s}{\eta a}$$

So, at 20 GHz, we have $\alpha_{c_{\text{TEM}}} \simeq 9.79 \times 10^{-3}$ nepers-m^{-1} (or[64] $\sim 8.50 \times 10^{-2}$ dB-m^{-1}). At 30 GHz, we have $\alpha_{c_{\text{TEM}}} \simeq 0.104$ dB-m^{-1}.

The cutoff frequencies for the TE_1 and TM_1 modes are $f_{c_1} = (3 \times 10^8)/0.02 = 15$ GHz < 20 GHz. Thus, both TE_1 and TM_1 modes propagate at 20 GHz and 30 GHz. At 20 GHz for the TE_1 mode, we have

$$\alpha_{c_{\text{TE}_1}} = \frac{2R_s(f_{c_1}/f)^2}{\eta a\sqrt{1-(f_{c_1}/f)^2}} \simeq \frac{2 \times 3.69 \times 10^{-2}(15/20)^2}{377 \times 10^{-2} \times \sqrt{1-(15/20)^2}} \simeq 1.66 \times 10^{-2} \text{ np-m}^{-1}$$

corresponding to an attenuation rate of ~ 0.145 dB-m^{-1}. Similarly at 30 GHz, the attenuation rate is $\sim 6.01 \times 10^{-2}$ dB-m^{-1}.

For the TM_1 mode at 20 GHz, we have

$$\alpha_{c_{\text{TM}_1}} = \frac{2R_s}{\eta a\sqrt{1-(f_{c_1}/f)^2}} \simeq \frac{2 \times 3.69 \times 10^{-2}}{377 \times 10^{-2} \times \sqrt{1-(15/20)^2}} \simeq 2.96 \times 10^{-2} \text{ np-m}^{-1}$$

or an attenuation rate of ~ 0.257 dB-m^{-1}. At 30 GHz, the attenuation rate is ~ 0.240 dB-m^{-1}.

Attenuation Due to Dielectric Losses We have so far considered losses due to imperfectly conducting parallel plate walls. In practice, losses also occur because of imperfections of the dielectric material between the conductors. Such losses can typically be accounted for by taking the permittivity of the dielectric to be complex, or

$$\epsilon_c = \epsilon' - j\epsilon''$$

For the TEM mode of propagation, simple substitution of $\epsilon_c = \epsilon' - j\epsilon''$ instead of ϵ in all field expressions gives an attenuation constant in the same manner as we have obtained the fields for a uniform TEM plane wave in a lossy dielectric by making the substitution $\sigma \rightarrow \omega\epsilon''$.

For TE or TM modes, the analysis of the field structures (i.e., mode configurations) was developed by assuming $\bar{\gamma}^2$ to be real ($\bar{\gamma}^2 = \bar{\alpha}^2$ or $-\bar{\beta}^2$). If the dielectric losses are substantial, the field configurations may well have to be different; however, in most cases, the losses are small enough that we can assume that the character of the fields remains the same, and that the modes and their cutoff frequencies can be calculated assuming no losses. Consider

$$\bar{\gamma} = j[\mu\epsilon(\omega^2 - \omega_{c_m}^2)]^{1/2}$$

which, upon substitution of $\epsilon \rightarrow \epsilon' - j\epsilon''$ and some manipulation, can be shown to be

---

[64]As discussed in Section 3.8, conversion from nepers-m^{-1} to dB-m^{-1} simply requires multiplication by $20 \log_{10} e \simeq 8.686$.

$$\bar{\gamma} = \alpha_d + j\bar{\beta}_m \simeq \frac{\omega\sqrt{\mu_0\epsilon'}\epsilon''/\epsilon'}{2\sqrt{1-(\omega_{c_m}/\omega)^2}} + j\sqrt{(\omega_{c_m}^2 - \omega^2)\mu\epsilon'} \qquad [8.81]$$

Thus, we see that the propagation constant (i.e., the imaginary part of $\bar{\gamma}$) is

$$\boxed{\bar{\beta}_m = \sqrt{(\omega^2 - \omega_{c_m}^2)\mu\epsilon'}} \qquad [8.82]$$

which is simply the waveguide propagation constant for a lossless dielectric with permittivity ϵ'. The attenuation constant due to dielectric losses is the real part of $\bar{\gamma}$ in [8.81]:

$$\boxed{\alpha_d \simeq \frac{\omega\sqrt{\mu_0\epsilon'}\epsilon''/\epsilon'}{2\sqrt{1-(\omega_{c_m}/\omega)^2}}} \qquad [8.83]$$

Note that in most cases the dielectric materials used are nonmagnetic, so $\mu = \mu_0$. Expressions [8.82] and [8.83] for α_d and $\bar{\beta}_m$ are valid for dielectrics with small losses such that $\epsilon'' \ll \epsilon'$. Note that equation [8.83] for α_d is valid for all TE or TM modes, and in fact also for all cross-sectional shapes of the waveguide, as long as f_{c_m} is the cutoff frequency for the particular mode under consideration. The attenuation rate due to dielectric losses for the TEM mode can be found from [8.83] by substituting $\omega_{c_m} = 0$, which gives $\alpha_d \simeq \omega\epsilon''\sqrt{\mu_0/\epsilon'}/2$.

8.3.3 Waves in Coaxial Lines

The coaxial line, illustrated in Figure 8.38, is one of the most common[65] two-conductor transmission systems and is widely used at frequencies mostly below about 5 GHz. However, specially constructed coaxial lines are used when possible at frequencies up to 20 GHz. The physical structure of a coaxial line consists of a braided outer conductor with many strands of thin copper wire wound in a

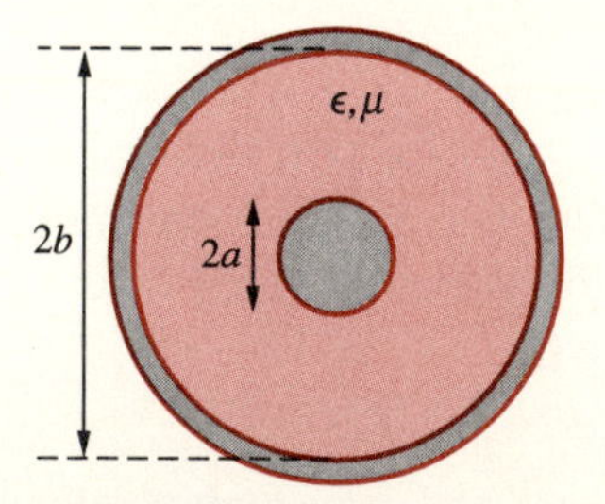

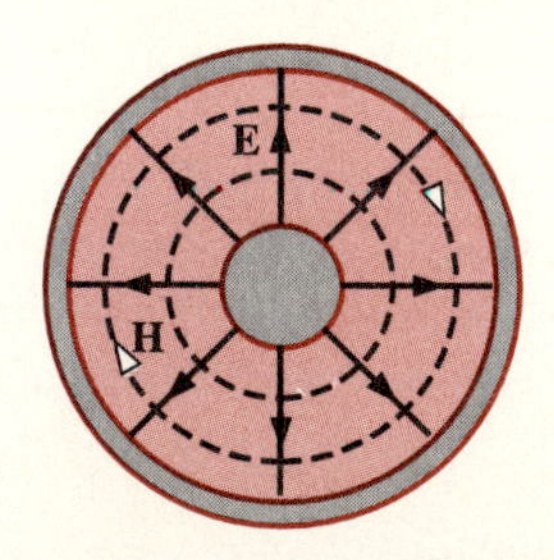

FIGURE 8.38. Coaxial transmission line. Coordinate system and dimensions for a coaxial line.

[65]For an excellent discussion of coaxial lines, including the history of their development, see J. H. Bryant, Coaxial transmission lines, related two-conductor transmission lines, connectors, and components: A U.S. historical perspective, *IEEE Trans. Microwave Theory and Techniques,* 32(9), pp. 970–983, 1984.

helix shape. The inner conductor is also made out of many strands of wire in order to provide flexibility. The dielectric filling used in most microwave applications is polyethylene or Teflon. A commonly used coaxial line, the RG58 (see Example 2-20), has a nominal attenuation rate at 1 GHz of $\sim$0.76 dB-m^{-1}. Coaxial lines for use in the 5 to 20 GHz range typically have smaller overall diameters and silver-coated conductors and may have attenuation rates of a few dB-m^{-1}.

From the point of view of the distribution of the electric and magnetic fields, the coaxial line can essentially be viewed as a cylindrical-coordinate version of the parallel-plate waveguide studied in Section 8.3.1. In this section, we discuss the dominant TEM mode in a coaxial transmission line and briefly comment on the higher-order TE or TM modes.

TEM Mode in a Coaxial Line As in the case of other two-conductor systems with symmetry in the plane transverse to the direction of propagation, the dominant mode of propagation for the coaxial line is the TEM mode. The fields for this mode can be derived by solving the wave equation [8.62] written in cylindrical coordinates. However, we can arrive at the TEM solution more simply by considering the boundary conditions on the inner and outer conductor surfaces. The fact that the tangential electric field and the normal magnetic field components must be zero on the conductor surfaces requires that $E_\phi = 0$ and $H_r = 0$ at $r = a, b$. Thus, nonzero E_ϕ and H_r components of a TEM solution could exist in the space between the two conductors only if these components varied with r. However, an r variation of a ϕ-directed electric field would necessitate the presence of axial components,[66] contradicting the premise of a TEM wave. Thus, a TEM solution can exist only with $\mathbf{E} = \hat{\mathbf{r}}E_r(r, z)$ and $\mathbf{H} = \hat{\phi}H_\phi(r, z)$. Note that there is no ϕ dependence, because of azimuthal symmetry. Under these conditions we can write [8.59a] in component form as

$$-\frac{\partial H_\phi}{\partial z} = j\omega\epsilon E_r \qquad \rightarrow \qquad j\beta H_\phi^0(r) = j\omega\epsilon E_r^0(r) \qquad [8.84a]$$

$$\frac{1}{r}H_\phi + \frac{\partial H_\phi}{\partial r} = 0 \qquad \rightarrow \qquad \frac{1}{r}H_\phi^0(r) + \frac{\partial H_\phi^0(r)}{\partial r} = 0 \qquad [8.84b]$$

where we have assumed solutions propagating in the z direction, with E_r and H_ϕ expressed as products of two functions respectively of r and z. Namely, $\mathbf{H} = \hat{\boldsymbol{\phi}}H_\phi(r, z) = \hat{\boldsymbol{\phi}}H_\phi^0(r)e^{-j\beta z}$, and $\mathbf{E} = \hat{\mathbf{r}}E_r^0(r)e^{-j\beta z}$, with $\beta = \omega\sqrt{\mu\epsilon}$. Equation [8.84$b$] can be directly solved for $H_\phi^0(r)$ to find

$$H_\phi^0(r) = \frac{H_0}{r} \qquad \rightarrow \qquad \mathbf{H} = \hat{\boldsymbol{\phi}}\frac{H_0}{r}e^{-j\beta z}$$

---

[66]To see this, consider that

$$[\nabla \times \mathbf{E}]_z = \frac{1}{r}\left[\frac{\partial}{\partial r}(rE_\phi) - \frac{\partial E_r}{\partial \phi}\right]$$

where H_0 is a constant. The associated wave electric field can be found from [8.84*a*] as

$$\mathbf{E} = \hat{\mathbf{r}}E_r(r, z) = \hat{\mathbf{r}}\frac{H_0\eta}{r}e^{-j\beta z}$$

Note that there are no cutoff conditions inherent in any of these discussions, since the phase constant β is that of a uniform plane wave in the unbounded dielectric medium. Thus the TEM mode can propagate at all frequencies.

Attenuation of the TEM Mode in a Coaxial Line The nonzero tangential magnetic fields on the surfaces of the inner and outer conductor necessitate the flow of surface currents on the conductors. We have

$$\mathbf{J}_s(z) = \hat{\mathbf{n}} \times \mathbf{H} = \begin{cases} \hat{\mathbf{z}}H_\phi(a, z) & \text{on the inner conductor } (\hat{\mathbf{n}} = \hat{\mathbf{r}}) \\ -\hat{\mathbf{z}}H_\phi(b, z) & \text{on the outer conductor } (\hat{\mathbf{n}} = -\hat{\mathbf{r}}) \end{cases}$$

These surface currents lead to power loss in the inner and outer conductors and hence attenuation of the TEM wave as it propagates along the line. The attenuation constant α_c due to conduction losses in a coaxial line can be found as usual from [8.77]. Following a procedure similar to that in previous sections, and assuming H_0 is real, we first find the power transmitted as

$$P_{\text{av}} = \frac{1}{2}\Re e\left[\int_a^b \int_0^{2\pi} E_r H_\phi^* r\, d\phi\, dr\right] = \pi\eta H_0^2[\ln r]_a^b = \pi\eta H_0^2 \ln\left(\frac{b}{a}\right)$$

and the power lost per unit length is given as

$$P_{\text{loss}}(z) = \frac{R_s}{2}\int_S |J_s|^2\, ds = \frac{R_s}{2}\int_0^1\left[\int_0^{2\pi} |H_\phi(a, z)|^2 a\, d\phi + \int_0^1\int_0^{2\pi} |H_\phi(b, z)|^2 b\, d\phi\right] dz$$

$$= \pi R_s H_0^2\left[\frac{1}{a} + \frac{1}{b}\right]$$

The attenuation constant α_c is then given by

$$\alpha_{c_{\text{TEM}}} = \frac{P_{\text{loss}}(z)}{2P_{\text{av}}} = \frac{\pi R_s H_0^2\left[\frac{1}{a} + \frac{1}{b}\right]}{2\pi\eta H_0^2 \ln(b/a)} = \frac{R_s}{2\eta \ln(b/a)}\left[\frac{1}{a} + \frac{1}{b}\right] \qquad [8.85]$$

Noting that $R_s = \sqrt{\omega\mu_0/(2\sigma)}$, we see that $\alpha_{c_{\text{TEM}}}$ for a coaxial line increases with frequency in the same manner as that for the parallel-plate line as given by [8.78].

Attenuation of TEM waves propagating in a coaxial line can also occur because of dielectric losses. To determine the attenuation constant α_d due to dielectric losses, we can use the general expression for α_d derived in the previous section. Noting that the cutoff frequency for the TEM is $f_c = 0$ (i.e., there is no cutoff), we have

$$\alpha_d \simeq \left[\frac{\omega\sqrt{\mu_0\epsilon'}\epsilon''/\epsilon'}{2\sqrt{1-(\omega_c/\omega)^2}}\right]_{\omega_c=0} = \frac{\omega\epsilon''\sqrt{\mu_0}}{2\sqrt{\epsilon'}} \simeq \frac{\omega\epsilon''\eta}{2} \qquad [8.86]$$

The intrinsic impedance of a low-loss nonmagnetic dielectric is $\eta \simeq \sqrt{\mu_0/\epsilon'}$.

Coaxial Transmission Line Voltage, Current, and Characteristic Impedance As was discussed previously in connection with the parallel-plate waveguide, the TEM wave on a two-conductor line is the propagation mode that corresponds to the voltage and current waves on two-conductor transmission lines, which were discussed in detail in Chapters 2 and 3. The voltage between the two conductors and the total current on the inner conductor of the coaxial line are related to the electric and magnetic fields of the TEM wave as

$$V(z) = -\int_a^b \mathbf{E}\cdot d\mathbf{l} = -\int_a^b E_r\hat{\mathbf{r}}\cdot\hat{\mathbf{r}}\,dr = -\int_a^b E_r(r,z)\,dr$$
$$= -\eta H_0 \ln(b/a)e^{-j\beta z} = V^+e^{-j\beta z}$$

and

$$I(z) = \oint_C \mathbf{H}\cdot d\mathbf{l} = \int_0^{2\pi} H_\phi(a,z)\hat{\boldsymbol{\phi}}\cdot\hat{\boldsymbol{\phi}}a\,d\phi = 2\pi H_0 e^{-j\beta z} = \frac{2\pi V^+e^{-j\beta z}}{\eta \ln b/a} = \frac{V^+}{Z_0}e^{-j\beta z}$$

where we have defined the peak voltage to be $V^+ = \eta H_0 \ln(b/a)$ and Z_0 is the characteristic impedance of the coaxial line given by

$$Z_0 = \frac{\eta \ln b/a}{2\pi} \simeq 60 \ln\frac{b}{a} \qquad [8.87]$$

since $\eta \simeq 120\pi$. Note that [8.87] is identical to the expression given in Table 2.2 for the characteristic impedance of a coaxial line. Upon substitution of $V^+ = \eta H_0 \ln(b/a)$ in the expressions for P_{av} and P_{loss}, the transmitted power flowing down the coaxial line and the power lost per unit line can be shown to be

$$P_{\text{av}}(z) = \frac{1}{2}\frac{|V^+|^2}{Z_0} \qquad P_{\text{loss}}(z) = \frac{R_s|V^+|^2}{4\pi Z_0^2}\left(\frac{1}{a}+\frac{1}{b}\right)$$

Optimum Coaxial Lines The (b/a) for a coaxial line can be varied to achieve different objectives, such as maximizing power- or voltage-handling capacity or minimizing α_c. Since most of these characteristics can be improved by using a larger coaxial line, the optimum dimensions are generally determined for a fixed value of the outer radius b. For this purpose, we assume the line to be terminated in its characteristic impedance, treat b as a constant, and find the ratio (b/a) for the optimum property desired. Note from [8.87] that the characteristic impedance of an air-filled coaxial line is completely determined by the value of (b/a).

As an example, we can determine the best Z_0 for handling maximum power in the TEM mode. The maximum power capacity is determined by the maximum value of the electric field beyond which dielectric breakdown and sparking occur in an air-filled line. Since the electric field varies as r^{-1}, the maximum field occurs at minimum radius, or at $r = a$. The maximum voltage V_{max} and the maximum electric field E_{max} are related by

$$E_{\text{max}} = \frac{V_{\text{max}}}{a\ln(b/a)} = \frac{V_{\text{max}}\zeta}{b\ln\zeta} \quad\rightarrow\quad V_{\text{max}} = \frac{E_{\text{max}} b \ln\zeta}{\zeta}$$

where $\zeta = b/a$. The time-average power transmitted by the coaxial transmission line at this maximum voltage is given by

$$P_{\text{av}} = \frac{V_{\text{max}}^2}{2Z_0} = \frac{(E_{\text{max}}b)^2(\ln\zeta)^2}{120\zeta^2\ln\zeta} = K\frac{\ln\zeta}{\zeta^2}$$

where K is a constant. To determine the value of ζ for maximum P_{av} we set $dP_{\text{av}}/d\zeta = 0$:

$$\frac{dP_{\text{av}}}{d\zeta} = K\left(\frac{1}{\zeta^2} - \frac{2\ln\zeta}{\zeta^2}\right) = 0 \quad\rightarrow\quad \ln\zeta = \frac{1}{2} \quad\rightarrow\quad \frac{b}{a} \simeq 1.65$$

which gives a characteristic impedance of $Z_0 \simeq 60\ln(1.65) \simeq 30\Omega$.

Similar analyses indicate that the maximum voltage-handling capacity (i.e., maximum $V(z) = \int_a^b \mathbf{E}\cdot d\mathbf{l}$) is achieved with a 60Ω line and that minimum conduction losses (minimum $\alpha_{c_{\text{TEM}}}$) occur for a 77Ω line.

TE and TM Modes in a Coaxial Line Transverse electric and transverse magnetic propagation modes are in principle possible on any transmission system with boundaries such that the modal field equations can be satisfied. As in the case of the parallel-plate waveguide, TE and TM modes may exist in a coaxial line, in addition to the principal (dominant) TEM mode already discussed. These modes are generally undesirable, and attempts are made in practice to avoid their excitation. The mathematical procedure for determination of the TE and TM modes on a coaxial cable is very similar to that used for finding the TE and TM modes in a parallel-plate waveguide, except that the wave equation [8.62] has to be written and solved in cylindrical coordinates. The existence of the inner and outer conductors requires that the tangential component of the electric field must vanish at $r = a$ and $r = b$. The mode with the lowest cutoff frequency has the smallest eigenvalue h, which is approximately[67] given by

$$h \simeq \frac{2}{a+b}$$

---

[67]See Section 8.10 of S. Ramo, J. R. Whinnery, and T. Van Duzer, *Fields and Waves in Communication Electronics,* John Wiley & Sons, Inc., New York, 1994.

provided that $b \leq 5a$. This eigenvalue leads to a cutoff wavelength and cutoff frequency for the lowest-order TE mode of

$$\lambda_c = \frac{2\pi}{h} \simeq \pi(a+b) \quad \text{and} \quad f_c \simeq \frac{1}{\pi(a+b)\sqrt{\mu\epsilon}}$$

respectively. In practice, it is usually desirable to have only the dominant TEM mode propagate and to have the TE or TM modes be evanescent. For a given frequency of operation, the dimensions of the coaxial line are usually chosen to be sufficiently small so that λ_c is less than $\lambda = (f\sqrt{\mu\epsilon})^{-1}$ or the frequency of operation f is less than f_c. This criterion restricts the value of b for fixed inner radius a and also determines the practical upper limit of frequency for which the coaxial line is useful.[68]

8.4 SUMMARY

This chapter discussed the following topics:

- **Uniform plane waves.** The characteristics of electromagnetic waves in source-free and simple media are governed by the wave equation derived from Maxwell's equations:

$$\nabla^2\overline{\mathscr{E}} - \mu\epsilon\frac{\partial^2\overline{\mathscr{E}}}{\partial t^2} = 0$$

  Uniform plane waves are the simplest type of solution of the wave equation, with the electric and magnetic fields both lying in the direction transverse to the direction of propagation (i.e., z direction):

$$\mathscr{E}_x(z,t) = p_1(z - v_p t)$$
$$\mathscr{H}_y(z,t) = \frac{1}{\eta}p_1(z - v_p t)$$

  where $v_p = 1/\sqrt{\mu\epsilon}$ is the phase velocity, $\eta = \sqrt{\mu/\epsilon}$ is the intrinsic impedance of the medium, and p_1 is an arbitrary function.

- **Time-harmonic waves in a lossless medium.** The electric and magnetic fields of a time-harmonic uniform plane wave propagating in the z direction in a simple lossless medium are

$$\mathbf{E}(z) = \hat{\mathbf{x}}C_1 e^{-j\beta z} \qquad \overline{\mathscr{E}}(z,t) = \hat{\mathbf{x}}C_1\cos(\omega t - \beta z)$$

  and

$$\mathbf{H}(z) = \hat{\mathbf{y}}\frac{C_1}{\eta}e^{-j\beta z} \qquad \overline{\mathscr{H}}(z,t) = \hat{\mathbf{y}}\frac{C_1}{\eta}\cos(\omega t - \beta z)$$

[68]For very similar reasons, it can be shown that a two-wire transmission line propagates only the TEM mode if the wavelength is smaller than twice the spacing between the wires. However, the practical upper limit of frequency for the two-wire line is usually much lower because of radiation losses.

where $\mathbf{E}$ and $\mathbf{H}$ are the phasor quantities, and $\beta = \omega\sqrt{\mu\epsilon}$ is the propagation constant.

- **Uniform plane waves in a lossy medium.** The electric and magnetic fields of a uniform plane wave propagating in the z direction in a lossy medium are

$$\mathbf{E}(z) = \hat{\mathbf{x}}C_1 e^{-\gamma z} \qquad \overline{\mathscr{E}}(z, t) = \hat{\mathbf{x}}C_1 e^{-\alpha z}\cos(\omega t - \beta z)$$

and

$$\mathbf{H}(z) = \hat{\mathbf{y}}\frac{C_1}{\eta_c}e^{-\gamma z} \qquad \overline{\mathscr{H}}(z, t) = \hat{\mathbf{y}}\frac{C_1 e^{-\alpha z}}{|\eta_c|}\cos(\omega t - \beta z - \phi_\eta)$$

where $\gamma = \sqrt{j\omega\mu(\sigma + j\omega\epsilon)} = \alpha + j\beta$ is the complex propagation constant and $\eta_c = \sqrt{\mu/[\epsilon - j(\sigma/\omega)]} = |\eta_c|e^{j\phi_\eta}$ is the complex intrinsic impedance of the medium.

The characteristics of uniform plane waves in a lossy medium are largely determined by the loss tangent $\tan\delta_c = \sigma/(\omega\epsilon)$. When $\tan\delta_c \ll 1$, the material is classified as a low-loss dielectric, and the propagation constant β is only negligibly different from that in a lossless medium. The attenuation and propagation constants α, β, and the intrinsic impedance for a low-loss dielectric are given by

$$\alpha \simeq \frac{\omega\epsilon''}{2}\sqrt{\frac{\mu}{\epsilon'}} \qquad \beta \simeq \omega\sqrt{\mu\epsilon'}\left[1 + \frac{1}{8}\left(\frac{\epsilon''}{\epsilon'}\right)^2\right] \qquad \eta_c \simeq \sqrt{\frac{\mu}{\epsilon'}}\left(1 + j\frac{\epsilon''}{2\epsilon'}\right)$$

Material media for which $\tan\delta_c \gg 1$ are classified as good conductors. The attenuation and propagation constants α and β and the intrinsic impedance for a good conductor are given by

$$\alpha = \beta \simeq \sqrt{\frac{\omega\mu\sigma}{2}} \qquad \eta_c \simeq \sqrt{\frac{\mu\omega}{\sigma}}e^{j45^\circ}$$

Another important parameter for good conductors is the skin depth δ, given by

$$\delta = \frac{1}{\alpha} \simeq \frac{1}{\sqrt{\pi f \mu\sigma}}$$

The skin depth for metallic conductors is typically extremely small, being $\sim$3.82 μm for copper at 300 MHz.

- **Electromagnetic power flow and the Poynting vector.** Poynting's theorem states that electromagnetic power flow entering into a given volume through the surface enclosing it equals the sum of the time rates of increase of the stored electric and magnetic energies and the ohmic power dissipated within the volume. The instantaneous power density of the electromagnetic wave is identified as

$$\overline{\mathscr{S}}(z, t) = \overline{\mathscr{E}}(z, t) \times \overline{\mathscr{H}}(z, t)$$

although in most cases the quantity of interest is the time-average power, which can be found either from $\overline{\mathscr{S}}(z, t)$ or directly from the phasor fields $\mathbf{E}$ and $\mathbf{H}$ as

$$\mathbf{S}_{\text{av}}(z) = \frac{1}{T_p}\int_0^{T_p} \overline{\mathcal{S}}(z,t)\,dt \qquad \mathbf{S}_{\text{av}}(z) = \frac{1}{2}\mathcal{R}e\{\mathbf{E}\times\mathbf{H}^*\}$$

The time-average Poynting flux for a uniform plane wave propagating in the z direction in an unbounded medium is

$$\mathbf{S}_{\text{av}}(z) = \hat{\mathbf{z}}\frac{C_1^2}{2|\eta_c|}e^{-2\alpha z}\cos(\phi_\eta)$$

- **Normal incidence on a perfect conductor.** When a uniform plane wave is normally incident from a dielectric (medium 1) onto a perfect conductor, a standing wave pattern is produced with the total electric and magnetic fields in medium 1 given by

$$\overline{\mathscr{E}}_1(z,t) = \hat{\mathbf{x}}2E_{i0}\sin(\beta_1 z)\sin(\omega t)$$

$$\overline{\mathscr{H}}_1(z,t) = \hat{\mathbf{y}}2\frac{E_{i0}}{\eta_1}\cos(\beta_1 z)\cos(\omega t)$$

The standing wave is entirely analogous to the standing wave that is produced on a short-circuited lossless uniform transmission line and represents purely reactive power, with the time-average power being identically zero.

- **Normal incidence on a lossless dielectric.** The reflection and transmission coefficients for the case of normal incidence of a uniform plane wave on a lossless dielectric are respectively given by

$$\Gamma = \frac{E_{r0}}{E_{i0}} = \frac{\eta_2 - \eta_1}{\eta_2 + \eta_1} \qquad \mathcal{T} = \frac{E_{t0}}{E_{i0}} = \frac{2\eta_2}{\eta_2 + \eta_1}$$

where η_1 and η_2 are the intrinsic impedances of the two dielectric media. Note that $(1+\Gamma) = \mathcal{T}$. The total electric and magnetic fields in medium 1 each consist of a component propagating in the z direction and another that is a standing wave, and can be expressed in phasor form as

$$\mathbf{E}_1(z) = \hat{\mathbf{x}}E_{i0}e^{-j\beta_1 z}(1 + \Gamma e^{j2\beta_1 z})$$

$$\mathbf{H}_1(z) = \hat{\mathbf{y}}\frac{E_{i0}}{\eta_1}e^{-j\beta_1 z}(1 - \Gamma e^{j2\beta_1 z})$$

The time-average Poynting flux in medium 1 is equal to that in medium 2 and is given as

$$(\mathbf{S}_{\text{av}})_1 = \hat{\mathbf{z}}\frac{E_{i0}^2}{2\eta_1}(1 - \Gamma^2)$$

so the fraction of the incident power transmitted into medium 1 is determined by $(1 - \Gamma^2)$.

- **Multiple dielectric interfaces.** The effective reflection and transmission coefficients for a uniform plane wave incident on an interface consisting of two dielectrics are given as

$$\Gamma_{\text{eff}} = \frac{(\eta_2 - \eta_1)(\eta_3 + \eta_2) + (\eta_2 + \eta_1)(\eta_3 - \eta_2)e^{-2j\beta_2 d}}{(\eta_2 + \eta_1)(\eta_3 + \eta_2) + (\eta_2 - \eta_1)(\eta_3 - \eta_2)e^{-2j\beta_2 d}}$$

$$\mathcal{T}_{\text{eff}} = \frac{4\eta_2\eta_3 e^{-j\beta_2 d}}{(\eta_2 + \eta_1)(\eta_3 + \eta_2) + (\eta_2 - \eta_1)(\eta_3 - \eta_2)e^{-j2\beta_2 d}}$$

These expressions are valid even for lossy media, as long as the proper complex intrinsic impedances are used and $j\beta_2$ is replaced by $\gamma_2 = \alpha_2 + j\beta_2$. In general, interfaces involving multiple dielectrics can be analyzed by relying on the transmission line analogy and treating the problem as one involving impedance transformation by a set of cascaded transmission lines.

- **Normal incidence on a lossy medium.** When a uniform plane wave is normally incident on an imperfect conductor, the surface current that flows within the conductor leads to power dissipation, although the magnitude of the reflection coefficient is very nearly unity. The amount of power lost in the conductor is given by

$$P_{\text{loss}} = \tfrac{1}{2}|J_s|^2 R_s$$

where the surface resistance $R_s = (\delta\sigma)^{-1}$, with δ being the skin depth of the imperfect conductor, and $|J_s| = |\mathbf{H}_1(0)|$ with $\mathbf{H}_1(0)$ being the total tangential magnetic field on the conductor surface.

- **Guided waves.** The configuration and propagation of electromagnetic waves guided by metallic conductors are governed by the solutions of the wave equations subject to the boundary conditions:

$$E_{\text{tangential}} = 0 \qquad H_{\text{normal}} = 0$$

The basic method of solution assumes propagation in the z direction, so all field components vary as $e^{-\bar{\gamma}z}$, where $\bar{\gamma}$ is equal to either $\bar{\alpha}$ or $j\bar{\beta}$. The z variations of real fields then have the form

$$\bar{\gamma} = \bar{\alpha} \qquad e^{-\bar{\alpha}z}\cos(\omega t) \qquad \text{Evanescent wave}$$

$$\bar{\gamma} = j\bar{\beta} \qquad \cos(\omega t - \bar{\beta}z) \qquad \text{Propagating wave}$$

The different type of possible solutions are classified as TE, TM, or TEM waves:

$$E_z = 0,\ H_z \neq 0 \qquad \text{Transverse electric (TE) waves}$$

$$E_z \neq 0,\ H_z = 0 \qquad \text{Transverse magnetic (TM) waves}$$

$$E_z,\ H_z = 0 \qquad \text{Transverse electromagnetic (TEM) waves}$$

Assuming that the wave-guiding structures extend in the z direction, the transverse field components can be derived from the axial (z direction) components, so that the z components act as a generating function for the transverse field components.

The form of the z components of the propagating TE_m and TM_m modes between parallel plates are

$$\text{TE}_m: \quad H_z = H_0 \cos\left(\frac{m\pi}{a}x\right)e^{-j\bar{\beta}_m z}$$

$$\text{TM}_m: \quad E_z = E_0 \sin\left(\frac{m\pi}{a}x\right)e^{-j\bar{\beta}_m z}$$

The cutoff frequencies f_c and cutoff wavelengths λ_c for the TE_m or TM_m modes are

$$f_{c_m} = \frac{m}{2a\sqrt{\mu\epsilon}} \qquad \lambda_{c_m} = \frac{2a}{m}$$

Note that f_{c_m} determines all the other quantities, such as $\bar{\beta}_m$, $\bar{\lambda}_m$, and $\bar{v}_{p_m}$, via [8.74] and [8.76].

- **TEM modes on two-conductor transmission lines.** Two-conductor structures such as the parallel-plate waveguide and the coaxial line allow the propagation of TEM modes. With the voltage and current properly related to the electric and magnetic fields, respectively, the TEM mode of propagation is identified as the mode that was studied extensively in Chapters 2 and 3 in terms of voltage and current waves. The TEM mode electric and magnetic field phasors for the parallel-plate and coaxial lines are given by

Parallel-plate:

$$\mathbf{E} = \hat{\mathbf{x}}E_0 e^{-j\beta z}$$

$$\mathbf{H} = \hat{\mathbf{y}}\frac{E_0}{\eta}e^{-j\beta z}$$

Coaxial:

$$\mathbf{E} = \hat{\mathbf{r}}\frac{E_0}{r}e^{-j\beta z}$$

$$\mathbf{H} = \hat{\boldsymbol{\phi}}\frac{E_0}{\eta r}e^{-j\beta z}$$

where $\beta = \omega\sqrt{\mu\epsilon}$. TEM modes do not exhibit any cutoff frequency and can in principle propagate at any frequency. Typical practice is to operate two-conductor lines at frequencies below the lowest cutoff frequency of the TE and TM modes so that TEM is the only propagating mode. Attenuation due to conductor losses typically increases with frequency for TEM modes, the attenuation constants due to conduction losses being given by

Parallel-plate:

$$\alpha_c = \frac{1}{\eta a}\sqrt{\frac{\omega\mu}{2\sigma}}$$

Coaxial:

$$\alpha_c = \frac{\sqrt{\omega\mu/(2\sigma)}}{2\eta\ln(b/a)}\left[\frac{1}{a} + \frac{1}{b}\right]$$

The attenuation coefficient for dielectric losses is $\alpha_d = \omega\epsilon''\sqrt{\mu/\epsilon'}/2$ for both parallel-plate and coaxial lines.

8.5 PROBLEMS

8-1. Uniform plane wave. A uniform plane wave propagates in the y direction in air with its electric field oriented in the z direction. At $t = 0$, the wave electric field has a maximum value of 100 mV-m^{-1} at position $y = 1$ m and an adjacent minimum value of -100 mV-m^{-1} at position $y = 2.25$ m. (a) Find the electric field $\overline{\mathscr{E}}(y, t)$. (b) Sketch $\overline{\mathscr{E}}(y)$ at $t = 0$ and $t = 12.5$ ns. (c) Sketch $\overline{\mathscr{E}}(t)$ at $y = 0$ and $y = 12.5$ m. (d) An observer measures $\overline{\mathscr{E}}(t)$ at $y = 25$ m, starting at $t = 0$. Find the first time instant at which the observer measures a maximum electric field.

8-2. TV broadcast signal. The magnetic field of a TV broadcast signal propagating in air is given as

$$\overline{\mathscr{H}}(z, t) = \hat{\mathbf{x}}0.1\sin(\omega t - 9.3z)\ \text{mA-m}^{-1}$$

(a) Find the wave frequency $f = \omega/(2\pi)$. (b) Find the corresponding $\overline{\mathscr{E}}(z, t)$.

8-3. Uniform plane wave. The electric field phasor of an 18-GHz uniform plane wave propagating in free space is given by

$$\mathbf{E}(y) = \hat{\mathbf{x}}5e^{j\beta y}\ \text{V-m}^{-1}$$

(a) Find the phase constant, β, and wavelength, λ. (b) Find the corresponding magnetic field phasor $\mathbf{H}(y)$.

8-4. Lossless nonmagnetic medium. The magnetic field component of a uniform plane wave propagating in a lossless simple nonmagnetic medium ($\mu = \mu_0$) is given by

$$\overline{\mathscr{B}}(x, t) = \hat{\mathbf{y}}0.25\sin[2\pi(10^8 t + 0.5x - 0.125)]\ \mu\text{T}$$

(a) Find the frequency, wavelength, and the phase velocity. (b) Find the relative permittivity, ϵ_r, and the intrinsic impedance, η, of the medium. (c) Find the corresponding $\overline{\mathscr{E}}$. (d) Find the time-average power density carried by this wave.

8-5. Short electric dipole antenna. Consider a short electric dipole antenna of length l, carrying a current $I_0\cos(\omega t)$, located at the origin and oriented along the z axis in free space. The electric field component of the wave at distances r very much greater than the wavelength is approximately given by

$$\mathbf{E}(r, \theta) = \hat{\theta}j\frac{30\beta I_0 l}{r}\sin\theta\, e^{-j\beta r}$$

where $\beta = \omega\sqrt{\mu_0\epsilon_0}$. (a) Find the corresponding magnetic field, $\mathbf{H}(r, \theta)$. (b) Find the time-averaged Poynting vector, $\mathbf{S}_{\text{av}}$. (c) Find the total power radiated by the dipole source, by integrating the radial Poynting vector over a spherical surface centered at the dipole.

8-6. Cellular phones. The electric field component of a uniform plane wave in air emitted by a mobile communication system is given by

$$\overline{\mathscr{E}}(x, y, t) = \hat{\mathbf{z}}100\cos(\omega t + 4.8\pi x - 3.6\pi y + \theta)\ \text{mV-m}^{-1}$$

(a) Find the frequency f and wavelength λ. (b) Find θ if $\overline{\mathscr{E}}(x, y, t) \simeq \hat{\mathbf{z}}80.9$ mV-m^{-1} at $t = 0$ and at $y = 2x = 0.2$ m. (c) Find the corresponding magnetic field $\overline{\mathscr{H}}(x, y, t)$. (d) Find the time-average power density carried by this wave.

8-7. VHF TV signal. The magnetic field component of a 10 μW-m^{-2}, 200 MHz TV signal in air is given by

$$\overline{\mathcal{H}}(x, y, t) = \hat{\mathbf{z}} H_0 \sin(\omega t - ax - ay + \pi/3)$$

(a) What are the values of H_0 and a? (b) find the corresponding electric field $\overline{\mathcal{E}}(x, y, t)$. (c) An observer at $z = 0$ is equipped with a wire antenna capable of detecting the component of the electric field along its length. Find the maximum value of the measured electric field if the antenna wire is oriented along the (i) x direction, (ii) y direction, (iii) 45° line between the x and y directions.

8-8. Unknown material. The intrinsic impedance and phase velocity of a uniform plane wave traveling in an unknown material at 2 GHz are measured to be 98Ω and 7.8×10^7 m-s^{-1} respectively. Determine the constitutive parameters (i.e., ϵ_r and μ_r) of the material.

8-9. Uniform plane wave. A 2.5 cm wavelength uniform plane wave with maximum electric field of 12 V-m^{-1} propagates in air in the direction of a unit vector given by

$$\hat{\mathbf{u}} = \frac{(\hat{\mathbf{y}}\sqrt{3} - \hat{\mathbf{z}})}{2}$$

The wave magnetic field has only an x component, which is approximately equal to 15.9 mA-m^{-1} at $y = z = t = 0$. Find $\overline{\mathcal{E}}$ and $\overline{\mathcal{H}}$.

8-10. Propagation through wet versus dry earth. Assume the conductivities of wet and dry earth to be $\sigma_{\text{wet}} = 10^{-2}$ S-m^{-1} and $\sigma_{\text{dry}} = 10^{-4}$ S-m^{-1}, respectively, and the corresponding permittivities to be $\epsilon_{\text{wet}} = 10\epsilon_0$ and $\epsilon_{\text{dry}} = 3\epsilon_0$. Both media are known to be nonmagnetic (i.e., $\mu_r = 1$). Determine the attenuation constant, phase constant, wavelength, phase velocity, penetration depth, and intrinsic impedance for a uniform plane wave of 20 MHz propagating in (a) wet earth, (b) dry earth. Use approximate expressions whenever possible.

8-11. Propagation in lossy media. (a) Show that the penetration depth (i.e., the depth at which the field amplitude drops to $1/e$ of its value at the surface) in a lossy medium with $\mu = \mu_0$ is approximately given by

$$d \simeq \frac{0.225(c/f)}{\sqrt{\epsilon_r'}\sqrt{\sqrt{1 + \tan^2\delta_c} - 1}}$$

where $\tan\delta_c$ is the loss tangent.[69] (b) For $\tan\delta_c \ll 1$, show that the above equation can be further approximated as $d \simeq [0.318(c/f)]/[\tan\delta_c\sqrt{\epsilon_r'}]$. (c) Assuming the properties of fat tissue at 2.45 GHz to be $\sigma = 0.12$ S-m^{-1}, $\epsilon_r = 5.5$, and $\mu_r = 1$, find the penetration depth of a 2.45 GHz plane wave in fat tissue using both expressions, and compare the results.

8-12. Ice in the polar regions. A cold polar glacier such as the Antarctic ice sheet has the parameters $\epsilon_r' = 3.17$ and $\tan\delta_c = 0.002$ in the frequency range of 100 to 600 MHz.[70] Find the penetration depth and the attenuation (in dB-m^{-1}) in the ice at (a) 100 MHz and (b) 600 MHz. Assume the ice to be an unbounded nonmagnetic medium.

8-13. Good dielectric. Alumina (Al_2O_3) is a low-loss ceramic material that is commonly used as a substrate for printed circuit boards. At 10 GHz, the relative permittivity and

---

[69] J. M. Osepchuk, Sources and basic characteristics of microwave radiation, *Bull. N. Y. Acad. Med.*, 55(11), pp. 976–998, December 1979.

[70] V. V. Bogorodsky, C. R. Bentley, and P. E. Gudmandsen, *Radioglaciology*, Reidel Publishers, 1985.

loss tangent of alumina are approximately equal to $\epsilon_r = 9.7$ and $\tan\delta_c = 2 \times 10^{-4}$. Assume $\mu_r = 1$. For a 10 GHz uniform plane wave propagating in a sufficiently large sample of alumina, determine the following: (a) attenuation constant, α, in np-m^{-1}; (b) penetration depth, d; (c) total attenuation in dB over thicknesses of 1 cm and 1 m.

8-14. Unknown medium. The electric field of a uniform plane wave in a nonmagnetic medium is given by

$$\overline{\mathscr{E}}(z,t) = \hat{\mathbf{y}}1000e^{-2.3z}\cos(75 \times 10^7 t - 14.328z + \pi/3) \text{ V-m}^{-1}$$

(a) Find the conductivity σ and relative permittivity ϵ_r of the medium. (b) Find the corresponding magnetic field $\overline{\mathscr{H}}(z,t)$.

8-15. Propagation in seawater. Transmission of electromagnetic energy through the ocean is practically impossible at high frequencies because of the high attenuation rates encountered. For seawater, take $\sigma = 4$ S-m^{-1}, $\epsilon_r = 81$, and $\mu_r = 1$, respectively. (a) Show that seawater is a good conductor for frequencies much less than ~890 MHz. (b) For frequencies less than 100 MHz, calculate, as a function of frequency (in Hz), the approximate distance over which the amplitude of the electric field is reduced by a factor of 10.

8-16. Wavelength in seawater. Find and sketch the wavelength in seawater as a function of frequency. Calculate λ_{sw} at the following frequencies: 1 Hz, 1 kHz, 1 MHz, and 1 GHz. Sketch $\log \lambda_{sw}$ vs. $\log f$. Use the following properties for seawater: $\sigma = 4$ S-m^{-1}, $\epsilon_r = 81$, and $\mu_r = 1$.

8-17. Communication in seawater. ELF communication signals ($f \leq 3$ kHz) can more effectively penetrate seawater than VLF signals (3 kHz $\leq f \leq$ 30 kHz). In practice, an ELF signal used for communication can penetrate and be received at a depth of up to 80 m below the ocean surface.[71] (a) Find the ELF frequency at which the skin depth in seawater is equal to 80 m. For seawater, use $\sigma = 4$ S-m^{-1}, $\epsilon_r = 81$, and $\mu_r = 1$. (b) Find the ELF frequency at which the skin depth is equal to half of 80 m. (c) At 100 Hz, find the depth at which the peak value of the electric field propagating vertically downward in seawater is 40 dB less than its value immediately below the surface of the sea. (d) A surface vehicle–based transmitter operating at 1 kHz generates an electromagnetic signal of peak value 1 V-m^{-1} immediately below the sea surface. If the antenna and the receiver system of the submerged vehicle can measure a signal with a peak value of as low as 1 μV-m^{-1}, calculate the maximum depth beyond which the two vehicles cannot communicate.

8-18. Submarine communication near a river delta. A submarine submerged in the sea ($\sigma = 4$ S-m^{-1}, $\epsilon_r = 81$, $\mu_r = 1$) wants to receive the signal from a VLF transmitter operating at 20 kHz. (a) How close must the submarine be to the surface in order to receive 0.1% of the signal amplitude immediately below the sea surface? (b) Repeat part (a) if the submarine is submerged near a river delta, where the average conductivity of seawater is ten times smaller.

8-19. Muscle and fat. Two novel materials have been developed to simulate the dielectric properties of human muscle and fat at 1 GHz.[72] The high-permittivity material, EWSG,

---

[71] J. C. Kim and E. I. Muehldorf, *Naval Shipboard Communications Systems*, Prentice-Hall, 1995, p. 111.

[72] M. P. Robinson, M. J. Richardson, J. L. Green, and A. W. Preece, New materials for dielectric simulation of tissues, *Phys. Med. Biol.*, 36(12), pp. 1565–1571, 1991.

TABLE 8.7. Measured Complex Permittivities.

Material	Frequency (GHz)	$\epsilon_r' \pm 0.6$	$\epsilon_r'' \pm 0.4$
EWSG (muscle)	0.5	53.6	33.5
	1.0	49.4	24.4
	2.45	39.5	22.5
EGP (fat)	0.5	11.4	3.5
	1.0	8.2	3.6
	2.45	6.0	3.6

simulates muscle. The low-permittivity material, EGP, simulates fat. Using the data given in Table 8.7, calculate and compare the penetration depth of each material at 1 GHz. Which material has a higher microwave absorption rate?

8-20. Unknown medium. A uniform plane wave propagates in the x direction in a certain type of material with unknown properties. At $t = 0$, the wave electric field is measured to vary with x as shown in Figure 8.39a. At $x = 40$ m, the temporal variation of the wave electric field is measured to be in the form shown in Figure 8.39b. Using the data in these two figures, find (a) σ and ϵ_r (assume nonmagnetic case), (b) the depth of penetration and the attenuation in dB-m^{-1}, and (c) the total dB attenuation and the phase shift over a distance of 100 m through this medium.

8-21. Thickness of beef products. Microwave heating is generally uniform over the entire body of the product being heated if the thickness of the product does not exceed about 1 to 1.5 times its penetration depth.[73] (a) Consider a beef product to be heated in a microwave oven operating at 2.45 GHz. The dielectric properties of raw beef at 2.45 GHz and 25°C are $\epsilon_r' = 52.4$ and $\tan\delta_c = 0.33$.[74] What is the maximum thickness of this beef product for it to be heated uniformly? (b) Microwave ovens operating at 915 MHz are evidently more appropriate for cooking products with large cross sections and high dielectric loss factors. The dielectric properties of raw beef at 915 MHz

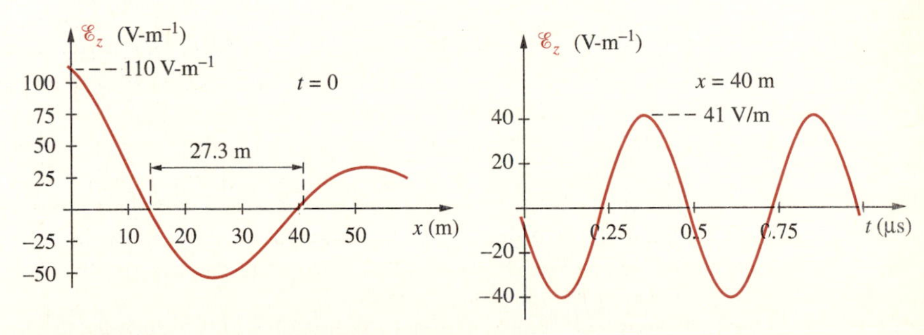

FIGURE 8.39. **Unknown medium.** Problem 8-20.

---

[73] J. Thuery, *Microwaves: Industrial, Scientific and Medical Applications*, Artech House, 1992.

[74] D. I. C. Wang and A. Goldblith, Dielectric properties of food, *Tech. Rep. TR-76-27-FEL*, MIT, Dept. of Nutrition and Food Science, Cambridge, Massachusetts, November, 1976.

and 25°C are $\epsilon_r' = 54.5$ and $\tan\delta_c = 0.411$. Find the maximum thickness of the beef product at 915 MHz and compare it with the results of part (a).

8-22. Beef versus bacon. The dielectric properties of cooked beef and smoked bacon at 25°C are given by $\epsilon_r \simeq 31.1 - j10.3$ at 2.45 GHz and $\epsilon_r \simeq 2.5 - j0.125$ at 3 GHz, respectively (see the references in the preceding problem). Calculate the loss tangent and the penetration depth for each and explain the differences.

8-23. Aluminum foil. A sheet of aluminum foil of thickness ~25 μm is used to shield an electronic instrument at 100 MHz. Find the dB attenuation of a plane wave that travels from one side to the other side of the aluminum foil. (Neglect the effects from the boundaries.) For aluminum, $\sigma = 3.54 \times 10^7$ S-m^{-1} and $\epsilon_r = \mu_r = 1$.

8-24. Soil with unknown properties. Using the results of a reflection measurement technique, the intrinsic impedance of a certain type of soil at 25 MHz was found to be approximately given by

$$\eta = 90e^{j12.1°}\Omega$$

Assuming that the soil is nonmagnetic, find the conductivity σ and the relative dielectric constant ϵ_r' of this soil.

8-25. Poynting flux. The electric and magnetic field expressions for a uniform plane wave propagating in a lossy medium are as follows:

$$\mathscr{E}_x(z,t) = 2e^{-4z}\cos(\omega t - \beta z) \text{ V-m}^{-1}$$

$$\mathscr{H}_y(z,t) = H_0 e^{-4z}\cos(\omega t - \beta z - \zeta) \text{ A-m}^{-1}$$

The frequency of operation is $f = 10^8$ Hz, and the electrical parameters of the medium are $\epsilon_r = 18.5\epsilon_0$, μ_0, and σ. (a) Find the time-average electromagnetic power density *entering* a rectangular box-shaped surface like that shown in Figure 8.40, assuming $a = d = 1$ m and $b = 0.5$ m. (b) Determine the power density *exiting* this region and compare with (a). (c) The difference between your results in (a) and (b) should represent electromagnetic power lost in the region enclosed by the square-box region. Can you calculate this dissipated power by any other method (i.e., without using the Poynting vector)? If yes, carry out this calculation. *Hint.* You may first need to find σ.

8-26. Laser beams. The electric field component of a laser beam propagating in the z direction is approximated by

$$\mathscr{E}_r = E_0 e^{-r^2/\omega^2}\cos(\omega t - \beta z)$$

where E_0 is the amplitude on the axis and ω is the effective beam radius, where the electric field amplitude is a factor of e^{-1} lower than E_0. (a) Find the corresponding

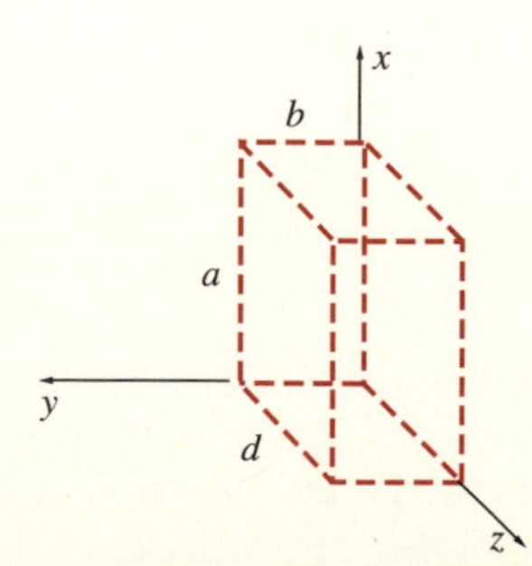

FIGURE 8.40. Poynting flux. Problem 8-25.

expression for the magnetic field $\overline{\mathcal{H}}$. (b) Show that the time-average power density at the center of the laser beam is given by

$$|\mathbf{S}_{\mathrm{av}}| = \frac{E_0^2}{2\eta} e^{-2r^2/\omega^2} \text{ W-m}^{-2}$$

where $\eta = 377\Omega$. (c) Find the total power of the laser beam. Consider a typical laboratory helium-neon laser with a total power of 5 mW and an effective radius of $\omega = 400$ μm. What is the power density at the center of the beam? (d) The power density of solar electromagnetic radiation at the Earth's surface is 1400 W-m^{-2}. At what distance from the Sun would its power density be equal to that for the helium-neon laser in part (c)? (e) One of the highest-power lasers built for fusion experiments operates at $\lambda = 1.6$ μm, produces 10.2 kJ for 0.2 ns, and is designed for focusing on targets of 0.5 mm diameter. Estimate the electric field strength at the center of the beam. Is the field large enough to break down air? What is the radiation pressure of the laser beam? How much weight can be lifted with the pressure of this beam?

8-27. Maxwell's equations. Consider the parallel-plate transmission line with perfectly conducting plates of large extent, separated by a distance of d meters. As shown in Figure 8.41, an alternating surface current density J_s in the z direction flows on the conductor surface:

$$\overline{\mathcal{J}}_s(z, t) = \hat{\mathbf{z}} J_0 \cos\left[\omega\left(t - \frac{z}{c}\right)\right] \text{ A-m}^{-1}$$

(a) Find an expression for the electric field, and determine the voltage between the plates, for $d = 0.1$ m and $J_0 = 1$ A-m^{-1}. (b) Use the continuity equation to find an expression for the surface charge density $\rho_s(z, t)$.

8-28. Uniform plane wave. A uniform plane electromagnetic wave propagates in free space with electric and magnetic field components as shown in Figure 8.42:

$$\overline{\mathscr{E}} = \hat{\mathbf{x}} E_0 \cos(\omega t - \beta z)$$

$$\overline{\mathcal{H}} = \hat{\mathbf{y}} \frac{E_0}{\eta} \cos(\omega t - \beta z)$$

The wave frequency is 300 MHz and the electric field amplitude $E_0 = 1$ V-m^{-1}. A square loop antenna with side length $a = 10$ cm is placed at $z = 2$ m as shown. (a) Find the voltage $\mathcal{V}_{\mathrm{ind}}(t)$ induced at the terminals of the loop. (b) Repeat (a) for the loop located at a distance of $d = 3$ m from the x axis instead of 2 m as shown. Compare your answers in (a) and (b).

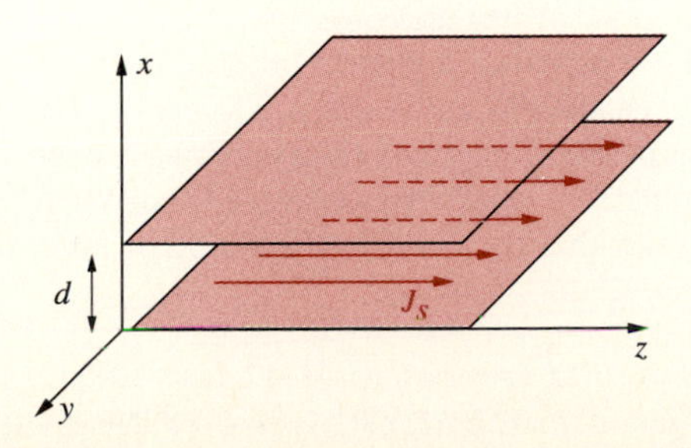

FIGURE 8.41. Surface current. Problem 8-27.

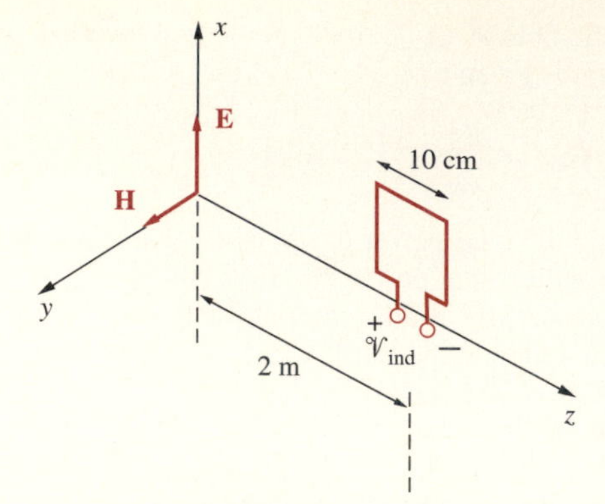

FIGURE 8.42. **Uniform plane wave.** Problem 8-28.

8-29. Mobile phones. Cellular phone antennas installed on cars have a maximum output power of 3 W, set by the Federal Communications Commission (FCC) standards. The passengers in the car are exposed to incident electromagnetic energy that doesn't pose any health threats, both because they are some distance away from the antenna and also because the body of the car and glass window shield them from much of the radiation.[75] (a) For a car with a synthetic roof, the maximum localized power density in the passenger seat is about 0.3 mW-cm^{-2}. For cars with metal roofs, this value reduces to 0.02 mW-cm^{-2} or less. Antennas mounted on the trunk or in the glass of the rear windshield deliver power densities of about 0.35 to 0.07 mW-cm^{-2} to passengers in the back seat. Compare these values with the IEEE safety limits (IEEE Standard C95.1-1991) in the cellular phone frequency range (which is typically 800 to 900 MHz) and comment on the safety of the passengers. Note that from 300 MHz to 15 GHz, IEEE safety limits specify a maximum allowable power density that increases linearly with frequency as $f/1500$ in mW-cm^{-2}, where the frequency f is in MHz. For reference, the maximum allowable power density is 1 mW-cm^{-2} at 1.5 GHz. (b) Calculate the maximum output power of a cellular phone antenna installed on the metal roof of a car such that the localized power density in the passenger compartment is equal to the IEEE safety limit at 850 MHz.

8-30. Microwave cataracts in humans. Over 50 cases of human cataract induction have been attributed to microwave exposures, primarily encountered in occupational situations involving acute exposure to presumably relatively high intensity fields.[76] The following are three reported incidents of cataracts caused by microwave radiation: (1) A 22-year-old technician exposed approximately five times to 3 GHz radiation at an estimated average power density of 300 mW-cm^{-2} for 3 min/exposure developed bilateral cataracts. (2) A person was exposed to microwaves for durations of approximately 50 hr/month over a 4-year period at average power densities of less than 10 mW-cm^{-2} in most instances, but with a period of 6 months or more during which the average power density was approximately 1 W-cm^{-2}. (3) A 50-year-old female was intermittently exposed to leakage radiation from a 2.45 GHz microwave oven of approximately 1 mW-cm^{-2} during oven operation with levels of up to 90 mW-cm^{-2} when the oven door was open, which presumably occurred over a period of approximately 6 years

[75] M. Fischetti, The cellular phone scare, *IEEE Spectrum,* pp. 43–47, June, 1993.

[76] S. F. Cleary, Microwave cataractogenesis, *Proc. IEEE,* 68(1), pp. 49–55, January 1980.

prior to developing cataract. For each of these above cases, compare the power densities with the IEEE standards (see Problem 8-29) and comment.

8-31. Air–perfect conductor interface. A uniform plane wave traveling in air with its electric field given by

$$\overline{\mathscr{E}}_i(y,t) = \hat{\mathbf{z}}E_0\cos(\omega t - \beta y)$$

is normally incident on a perfect conductor boundary located at $y = 0$. If the measured distance between any two successive zeros of the total electric field in air is 6 cm and the maximum value of the electric field measured at $y = -75$ cm is 3 V-m^{-1}, determine the following: (a) the frequency (in GHz) and the power density (in μW-cm^{-2}) of the incident wave; (b) the instantaneous expression for the total electric field, $\overline{\mathscr{E}}_1(y,t)$, in air; (c) the instantaneous expression for the total magnetic field, $\overline{\mathscr{H}}_1(y,t)$, in air; (d) the maximum value of the total magnetic field measured at $y = -75$ cm.

8-32. Air–perfect conductor interface. A uniform plane wave of time-average power density 10 mW-cm^{-2} in air is normally incident on the surface of a perfect conductor located at $z = 0$. The total magnetic field phasor in air is given by

$$\mathbf{H}_1(z) = \hat{\mathbf{y}}H_0\cos(2\pi z)$$

(a) What is H_0? (b) What is the frequency, f? (c) At $z = -3.5$ m, what is the total electric field $\overline{\mathscr{E}}(z,t)$?

8-33. Air–perfect conductor interface. A uniform plane wave propagating in air given by

$$\mathbf{E}_i(x) = 60e^{-j40\pi x}(\hat{\mathbf{y}} - j\hat{\mathbf{z}}) \text{ V-m}^{-1}$$

is normally incident on a perfectly conducting plane located at $x = 0$. (a) Find the frequency and the wavelength of this wave. (b) Find the corresponding magnetic field $\mathbf{H}_i(x)$. (c) Find the electric and magnetic field vectors of the reflected wave [i.e., $\mathbf{E}_r(x)$ and $\mathbf{H}_r(x)$]. (d) Find the total electric field in air [i.e., $\mathbf{E}_1(x)$] and plot the magnitude of each one of its components as a function of x.

8-34. Unknown material. When a uniform plane wave in air is normally incident onto a planar lossless material medium, the reflection coefficient is measured to be -0.25 and the phase velocity of the wave is reduced by a factor of 3. Find the relative permittivity and permeability of the unknown material.

8-35. Dielectric–dielectric interface. A uniform plane wave propagates from one dielectric into another at normal incidence. Find the ratio of the dielectric constants such that the magnitude of the reflection and transmission coefficients are both equal to 0.5. Assume lossless nonmagnetic materials.

8-36. Air–GaAs interface. A uniform plane wave having a magnetic field given by

$$\mathbf{H}_i(z) = \hat{\mathbf{x}}10e^{j210z} \text{ mA-m}^{-1}$$

is normally incident from air onto a plane air–gallium arsenide (GaAs) interface located at $z = 0$. Assume GaAs to be a perfect dielectric with $\epsilon_r \simeq 13$. (a) Find the reflected ($\mathbf{H}_r(z)$) and the transmitted ($\mathbf{H}_t(z)$) fields. (b) Calculate the power density of the incident, reflected, and transmitted waves independently and verify the conservation of energy principle. (c) Find the expression for the total magnetic field ($\mathbf{H}_1(z)$) in air and sketch its magnitude as a function of z between $z = 0$ and $z = 3$ cm.

8-37. Aircraft–submarine communication. A submarine submerged in the ocean is trying to communicate with a Navy airplane equipped with a VLF transmitter operating at 20 kHz, which is approximately 10 km immediately overhead from the location of the submarine. If the output power of the VLF transmitter is 200 kW and the receiver sensitivity of the submarine is 1 μV-m^{-1}, calculate the maximum depth of the submarine from the surface of the ocean for it to be able to communicate with the transmitter. Assume the transmitter is radiating its power isotropically, and assume normal incidence at the air–ocean boundary. Use $\sigma = 4$ S-m^{-1}, $\epsilon_r = 81$, and $\mu_r = 1$ for the properties of the ocean.

8-38. Air–fat interface. Consider a planar interface between air and fat tissue (assume it to be semi-infinite extent). If a plane wave is normally incident from air at this boundary, find the percentage of the power absorbed by the fat tissue at (a) 100 MHz; (b) 300 MHz; (c) 915 MHz; and (d) 2.45 GHz, and compare your results with the results of Example 8-29. Use Table 8-6 for the parameters of the fat tissue.

8-39. Shielding with a copper foil. A 1 GHz 1 kW-m^{-2} microwave beam is incident upon a sheet of copper foil of 10 μm thickness. Consider neglecting multiple reflections, if justified. (a) Find the power density of the reflected wave. (b) Find the power density transmitted into the foil. (c) Find the power density of the wave that emerges from the other side of the foil. Comment on the shielding effectiveness of this thin copper foil.

8-40. Absorbing material. Consider a commercial absorber slab made of EHP-48 material of 1 m thickness backed by a perfectly conducting metal plate. A 100 MHz uniform plane wave is normally incident from the air side at the air–absorber–metal interface. Find the percentage of incident power lost in the absorber material. For EHP-48, use $\epsilon_r = 6.93 - j8.29$ at 100 MHz.

8-41. Radome design. A radome is to be designed for the nose of an aircraft to protect an X-band weather radar operating between 8.5 to 10.3 GHz. A new type of foam material with $\epsilon_r = 2$ (assume lossless) is chosen for the design. (a) Assuming a flat planar radome, determine the minimum thickness of the foam that will give no reflections at the center frequency of the band. (b) Using the thickness found in part (a), what percentage of the incident power is reflected at each end of the operating frequency band? (c) A thin layer of a different material ($\epsilon_r = 4.1$, $\tan\delta_c = 0.04$, thickness 0.25 mm) is added on one side of the radome designed in part (a) to prevent the radome from rain erosion. What percent of the incident power is reflected at the center frequency?

8-42. Transmission through a multilayered dielectric. (a) Find the three lowest frequencies at which all of the incident power would be transmitted through the three-layer structure shown in Figure 8.43. The permeability of all three media is μ_0. (b) If complete transmission is required for *any* thickness of the center medium, what is the lowest usable frequency? (c) Find the bandwidth of the transmission, defined as the range between the two lowest percentage values adjacent to and on either side of the frequency found in (b). Also find the lowest percentage values of transmission. (d) Why does the reflection from multiply coated optical lenses tend to be purple in color?

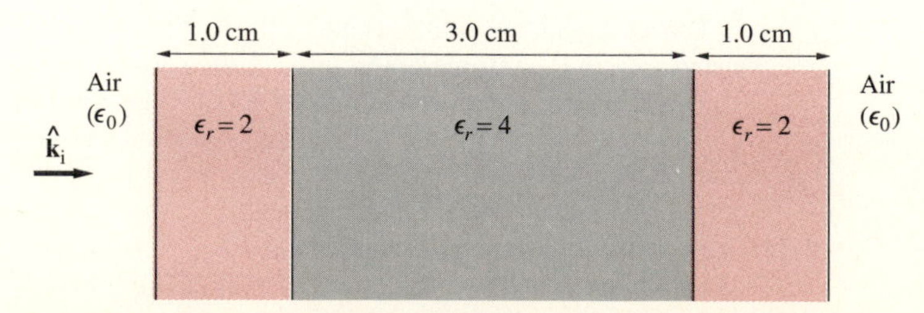

FIGURE 8.43. Multilayered dielectric. Problem 8-43.

8-43. Glass slab. Consider a 1-cm thick slab of crown glass, with index of refraction $n = 1.52$. (a) If a beam of visible light at 500 nm is normally incident from one side of the slab, what percentage of the incident power transmits to the other side? (b) Repeat for 400 and 600 nm.

8-44. Antireflection coating on a glass slab. A beam of light is normally incident on one side of a 1-cm thick slab of flint glass (assume $n = 1.86$) at 550 nm. (a) What percentage of the incident power reflects back? (b) To minimize reflections, the glass is coated with a thin layer of antireflection coating material on both sides. The material chosen is magnesium fluoride (MgF_2), which has a refractive index of around 1.38 at 550 nm. Find the approximate thickness of each coating layer of MgF_2 needed. (c) If a beam of light at 400 nm is normally incident on the coated glass with the thickness of the coating layers found in part (b), what percentage of the incident power reflects back? (d) Repeat part (c) for a light beam at 700 nm.

8-45. Infrared antireflection coating. To minimize reflections at the air–germanium interface in the infrared frequency spectrum, a coating material with an index of refraction of 2.04 is introduced as shown in Figure 8.44. (a) If the thickness of the coating material is adjusted to be a quarter-wavelength in the coating material for a free-space wavelength of 4 μm, find the effective reflection coefficient at free-space wavelengths of 4 and 8 μm. (b) Repeat part (a) if the thickness of the coating material is adjusted to be a quarter-wavelength in the coating material at 8 μm.

8-46. A snow-covered glacier. A glacier is a huge mass of ice that sits over land. Glaciers are formed in the cold polar regions and in high mountains. Most glaciers range in thickness from about 100 m to 3000 m. In Antarctica, the deepest ice on the polar plateau is 4.7 km. Consider a large glacier in Alaska covered with a layer of snow of 1 m thickness during late winter. A radar signal operating at 56 MHz is normally incident from air onto the air–snow interface. Assume both the snow and the ice to be lossless; for ice, $\epsilon_r = 3.2$, and for snow, ϵ_r can vary between 1.2 and 1.8. Assuming both the snow and the ice to be homogeneous and the ice to be semi-infinite in extent, calculate the reflection coefficient at the air–snow interface for three different permittivity values of snow: $\epsilon_r = 1.2$, 1.5, and 1.8. For which case is the snow layer most transparent (invisible) to the radar signal? Why?

8-47. Minimum ice thickness. Consider a 500-MHz uniform plane wave radiated by an aircraft radar normally incident on a fresh-water ($\epsilon_r = 88$) lake covered with a layer of ice ($\epsilon_r = 3.2$) as shown in Figure 8.45. (a) Find the minimum thickness of the ice such that the reflected wave has maximum strength. Assume the lake water to be very deep. (b) What is the ratio of the amplitudes of the reflected and incident electric fields?

8-48. A snow–ice covered lake. An interior lake in Alaska can be 30 to 100 m deep and is covered with ice and snow on the top, each of which can be about a meter deep in late winter.[77] Consider a 5-GHz C-band radar signal normally incident from air onto the

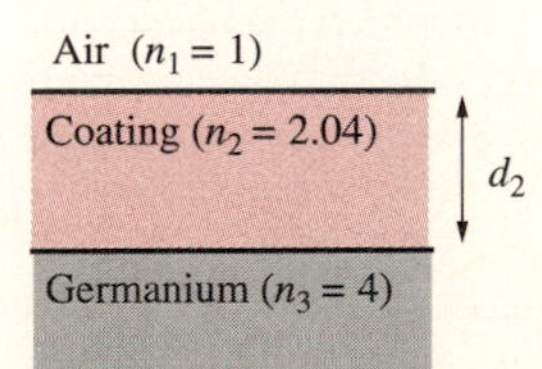

FIGURE 8.44. Infrared antireflection coating. Problem 8-45.

[77] S. A. Arcone, N. E. Yankielun, and E. F. Chacho, Jr., Reflection profiling of Arctic lake ice using microwave FM-CW radar, *IEEE Trans. Geosci. Remote Sensing,* 35(2), pp. 436–443, March 1997.

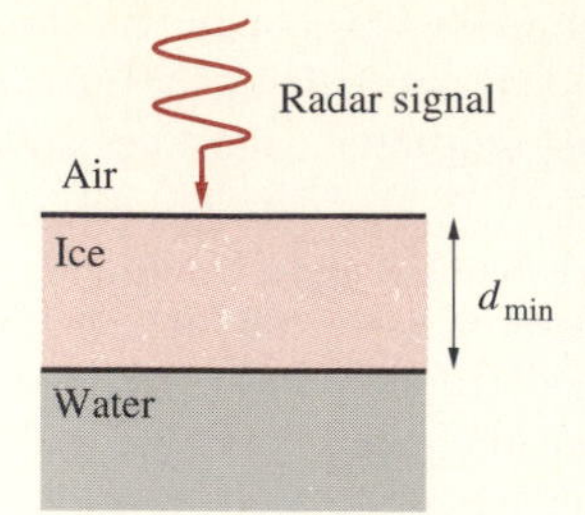

FIGURE 8.45. Aircraft radar signal incident on an icy lake surface. Problem 8-47.

surface of a lake that is 50 m deep, covered with a layer of snow (assume $\epsilon_r = 1.5$) of 60 cm thickness over a layer of ice ($\epsilon_r = 3.2$) of 1.35 m. Assume both the snow and the ice to be lossless. Also assume the lake water, with $\epsilon_{cr} = 68 - j35$ at 5 GHz at 0°C, to be slightly brackish (salty), with an approximate conductivity of $\sigma = 0.01$ S-m^{-1}, and the bottom of the lake to consist of thick silt with $\epsilon_r \simeq 50$. (a) Calculate the reflection coefficient at the air interface with and without the snow layer. (b) Repeat part (a) at an X-band radar frequency of 10 GHz. Assume all the other parameters to be the same except for the lake water, $\epsilon_{cr} = 42 - j41$ at 10 GHz and 0°C. Use any approximations possible, with the condition that sufficient justifications are provided.

8-49. A dielectric-coated reflector antenna. A dielectric-coated reflector antenna consists of a metal reflector (assume to be a perfect conductor) covered with a layer of dielectric material. Typical examples include some commercial K_u-band reflectors made of fiberglass layers bonded to a thin metal screen, as well as metal reflectors that are painted to protect against corrosion and solar heating. Consider a uniform plane wave operating at 12 GHz to be normally incident on a planar metal surface coated with a dielectric coating of thickness d and relative permittivity $\epsilon_r = 4.5 - j0.009$ (typical for both fiberglass and paint).[78] (a) If $d = 1.75$ mm, what percentage of the incident power is reflected? (b) When the surface of the dielectric coating becomes wet from condensation or rain, the reflector antenna performance can be seriously degraded. To study this effect, consider a thin lossy water layer of thickness 0.1 mm and relative permittivity $\epsilon_r = 55.4 - j35.45$ at 20°C on the top of the dielectric coating (use $d = 1.75$ mm). What percentage of the incident power is reflected in this case?

8-50. Parallel-plate waveguide modes. A parallel-plate air waveguide has a plate separation of 6 mm and width 10 cm. (a) List the cutoff frequencies of the seven lowest-order modes (TE_m and TM_m) that can propagate in this guide. (b) Find all the propagating modes (TE_m and TM_m) at 40 GHz. (c) Find all the propagating modes at 60 GHz. (d) Repeat part (c) if the waveguide is filled with polyethylene (assume lossless, with $\epsilon_r \simeq 2.25$).

8-51. Waveguide in the earth's crust. It has been proposed[79] that radio waves may propagate in a waveguide deep in the earth's crust, in which the basement rock has very low conductivity and is sandwiched between the conductive layers near the surface and the high-temperature conductive layer far below the surface. The upper boundary of this

[78]K. W. Lo and T. S. Bird, Effects of dielectric coating on parabolic reflector antenna performance, *Microwave and Optical Technology Letters,* 8(1), pp. 1–4, January 1995.

[79]H. A. Wheeler, Radio-wave propagation in the earth's crust, *J. Res. NBS,* 65D(2), pp. 189–191, March–April 1961. Also see H. Mott and A. W. Biggs, Very-low-frequency propagation below the bottom of the sea, *IEEE Trans. Antennas and Propagation,* AP-11, pp. 323–329, May 1963.

waveguide is on the order of 1 km to several kilometers below the earth's surface. The depth of the dielectric layer of the waveguide (the basement rock) can vary anywhere from 2 to 20 km, with a conductivity of 10^{-6} to 10^{-11} S-m^{-1} and a relative dielectric constant of $\sim$6. This waveguide may be used for communication from a shore sending station to an underwater receiving station. Consider such a waveguide and assume it to be an ideal parallel-plate waveguide. (a) If the depth of the dielectric layer of the guide is 2 km, find all the propagating modes of this guide below an operating frequency of 2 kHz. (b) Repeat part (a) (same depth) below 5 kHz. (c) Repeat parts (a) and (b) for a dielectric layer depth of 20 km.

8-52. Parallel-plate waveguide design. Design a parallel-plate air waveguide to operate at 5 GHz such that the cutoff frequency of the TE_1 mode is at least 25% less than 5 GHz, the cutoff frequency of the TE_2 mode is at least 25% greater than 5 GHz, and the power-carrying capability of the guide is maximized.

8-53. Parallel-plate waveguide modes. The electric field of a particular mode in a parallel-plate air waveguide with a plate separation of 2 cm is given by

$$E_x(y, z) = 10e^{-60\pi y}\sin(100\pi z)\text{ kV-m}^{-1}$$

(a) What is this mode? Is it a propagating or a nonpropagating mode? (b) What is the operating frequency? (c) What is the similar highest-order mode (TE_m or TM_m) that can propagate in this waveguide?

8-54. Power-handling capacity of a parallel-plate waveguide. A parallel-plate air waveguide with a plate separation of 1.5 cm is operated at a frequency of 15 GHz. Determine the maximum time-average power per unit guide width in units of kW-cm^{-1} that can be carried by the TE_1 mode in this guide, using breakdown strength of air taken as 15 kV-cm^{-1} (safety factor of approximately 2 to 1) at sea level.

8-55. Power capacity of a parallel-plate waveguide. Show that the maximum power-handling capability of a TM_m mode propagating in a parallel-plate waveguide without dielectric breakdown is determined only by the longitudinal component of the electric field for $f_{c_m} < f < \sqrt{2}f_{c_m}$ and by the transverse component of the electric field for $f > \sqrt{2}f_{c_m}$.

8-56. Power capacity of a parallel-plate waveguide. For a parallel-plate waveguide formed of two perfectly conducting plates separated by air at an operating frequency of $f = 1.5f_{c_m}$, find the maximum time-average power per unit area of the waveguide that can be carried without a dielectric breakdown (use 15 kV-cm^{-1} for maximum allowable electric field in air, which is half of the breakdown electric field in air at sea level) for the following modes: (a) TEM; (b) TE_1; and (c) TM_1.

8-57. Attenuation in a parallel-plate waveguide. Consider a parallel-plate waveguide of plate separation a having a lossless dielectric medium with properties ϵ and μ. (a) Find the frequency in terms of the cutoff frequency (i.e., find f/f_{c_m}) such that the attenuation constants α_c due to conductor losses of the TEM and the TE_m modes are equal. (b) Find the attenuation α_c for the TEM, TE_m, and TM_m modes at that frequency.

8-58. Attenuation in a parallel-plate waveguide. For a TM_m mode propagating in a parallel-plate waveguide, do the following: (a) Show that the attenuation constant α_c due to conductor losses for the propagating TM_m mode is given by

$$\alpha_c = \frac{2\omega\epsilon R_s}{\beta a}$$

(b) Find the frequency in terms of f_{c_m} (i.e., find f/f_{c_m}) such that the attenuation constant α_c found in part (a) is minimum. (c) Find the minimum for α_c. (d) For an air-filled waveguide made of copper plates 2.5 cm apart, find α_c for TEM, TE_1, and TM_1 modes at the frequency found in part (a).

8-59. Parallel-plate waveguide: phase velocity, wavelength, and attenuation. A parallel-plate waveguide is formed by two parallel brass plates ($\sigma = 2.56 \times 10^7$ S-m^{-1}) separated by a 1.6-cm thick polyethylene slab ($\epsilon_r' \simeq 2.25$, $\tan\delta \simeq 4 \times 10^{-4}$) to operate at a frequency of 10 GHz. For TEM, TE_1, and TM_1 modes, find (a) the phase velocity $\bar{v}_p$ and the waveguide wavelength $\bar{\lambda}$, (b) the attenuation constants α_c due to conductor losses and α_d due to dielectric losses.

8-60. Attenuation in a coaxial line. An air-filled coaxial line has thick walls made of copper. Its attenuation constant is $\alpha_c = 0.003$ np-m^{-1} at 300 MHz. (a) What is its attenuation constant at 600 MHz? (b) If all of its cross-sectional linear dimensions (i.e., inner and outer radii a, b, and thickness d of the outer conductor) are doubled, what is the new value of α_c at 300 MHz? (c) Suppose the original line is filled with a dielectric for which $\epsilon' = 2\epsilon_0$ and the loss tangent $\tan\delta_c = (15/\pi) \times 10^{-4}$ at 300 MHz. Find the total attenuation constant $(\alpha_c + \alpha_d)$ at 300 MHz. (d) Repeat (c) for 600 MHz, assuming that ϵ' and $\tan\delta_c$ remain the same.

8-61. Optimum coaxial lines. Consider a coaxial line with an outer conductor of fixed inside radius b. (a) Show that α_c is a minimum when $a = 3.6b$, where a is the outside radius of the inner conductor. (b) What is the characteristic impedance of such a coaxial line designed for minimum attenuation? (c) If you start instead with a coaxial line of fixed inner conductor radius a, can you find a value of outer conductor radius b for which α_c is minimized? If so, what is the minimizing value of b, and what is the characteristic impedance of such a coaxial line?

8-62. Power capacity of a coaxial line. A coaxial transmission line has an inner radius of 0.5 mm and an outer radius of 2.5 mm. What is the maximum power that may be carried in the TEM mode? Assume the breakdown voltage to be 10 kV-(mm)$^{-1}$, and the maximum allowed surface current density to be 10 A-(mm)$^{-1}$. Discuss design changes that could be made in order to increase the power-handling capacity.

APPENDIX A

Vector Analysis

Vector analysis is used extensively in electromagnetics and in other fields of engineering and physics in which the physical quantities involved have both direction and magnitude; that is, they are *vectors*. Examples of physical quantities that are vectors include velocity, momentum, force, displacement, electric field, current, and magnetic field. Vector quantities are inherently different from *scalars,* which are physical quantities that are specified entirely by one value, such as the number of coins in your pocket, time, mass, the temperature in a room, electric charge density, electric potential and energy. In this appendix, we review some basic rules and techniques of vector analysis that are particularly relevant to this book. More general topics of vector algebra or vector calculus are covered elsewhere[1] and are not discussed here.

Many of the essential aspects of vector manipulations and definitions needed to work with the material in this book were covered either in the body of the text or in footnotes in those sections where they first appeared. This appendix simply collects these vector rules and techniques in one handy place. The appendix also covers some of the most basic definitions that were taken for granted, such as the definition of vectors, position vectors, unit vectors, and coordinate systems. Some of the more important concepts of vector calculus, including gradient, divergence, curl, and the associated theorems (the divergence theorem and Stokes's theorem), are covered extensively in the text and are not included here.

Throughout this text, vector quantities are often written either using boldface symbols (e.g., **G**) or with a bar above the symbol (e.g., $\overline{\mathcal{G}}$). In Section 7.4 and Chapter 8, the latter notation is used to represent real physical quantities and the

[1]Simpler treatments are available in most textbooks; for a complete treatment of the subject, see Chapter 1 of G. B. Arfken and H. J. Weber, *Mathematical Methods for Physicists,* Academic Press, 1995.

former for their corresponding complex phasors, as needed for sinusoidal (or time-harmonic) applications. In this appendix, we represent vectors using boldface symbols.

To accurately describe vectors requires that we specify their direction as well as their magnitude. To do this, we need a framework to orient the vector in three-dimensional space and show its different projections or components. The most common means of doing this is to use the so-called rectangular or Cartesian coordinate system, which we use in Sections A.1 and A.2 to describe the basic aspects of vector addition and multiplication. We then briefly review, in Section A.3, two other commonly used coordinate systems, namely the cylindrical and spherical coordinate systems. Section A.4 presents some commonly encountered vector identities.

A.1 VECTOR COMPONENTS, UNIT VECTORS, AND VECTOR ADDITION

The Cartesian or rectangular coordinate system is illustrated in Figure A.1. To describe a vector **A** in the rectangular coordinate system, we can represent it as extending outward from the origin in a direction determined by the magnitudes of each of its three components, A_x, A_y, and A_z, as shown in Figure A.1a. It is customary to write the vector **A** as

$$\mathbf{A} = \hat{\mathbf{x}}A_x + \hat{\mathbf{y}}A_y + \hat{\mathbf{z}}A_z$$

where $\hat{\mathbf{x}}$, $\hat{\mathbf{y}}$, and $\hat{\mathbf{z}}$ are the rectangular coordinate unit vectors, as shown in Figure A.1b. In this text, we represent unit vectors with the "hat," or circumflex, notation.

The magnitude of vector **A** is denoted $|\mathbf{A}|$ (or sometimes simply A) and is given by

$$A = |\mathbf{A}| = \sqrt{A_x^2 + A_y^2 + A_z^2}$$

Sometimes it is convenient to define a unit vector that is in the direction of a given vector **A**. The unit vector in the direction of vector **A** is denoted $\hat{\mathbf{A}}$ and is defined as

$$\hat{\mathbf{A}} = \frac{\mathbf{A}}{|\mathbf{A}|} = \frac{\mathbf{A}}{\sqrt{A_x^2 + A_y^2 + A_z^2}}$$

Note that $|\hat{\mathbf{A}}| = 1$. An alternative way of expressing vector **A** is $\mathbf{A} = \hat{\mathbf{A}}|\mathbf{A}| = \hat{\mathbf{A}}A$.

Two vectors **A** and **B** can be added together to produce another vector $\mathbf{C} = \mathbf{A} + \mathbf{B}$. In rectangular coordinates, vector addition can be carried out component by component. In other words,

$$\mathbf{C} = \mathbf{A} + \mathbf{B} = \hat{\mathbf{x}}(A_x + B_x) + \hat{\mathbf{y}}(A_y + B_y) + \hat{\mathbf{z}}(A_z + B_z) = \hat{\mathbf{x}}C_x + \hat{\mathbf{y}}C_y + \hat{\mathbf{z}}C_z$$

Geometrically, the orientation and magnitude of the sum vector **C** can be determined using the parallelogram method depicted in Figure A.2a. Subtraction of two vectors

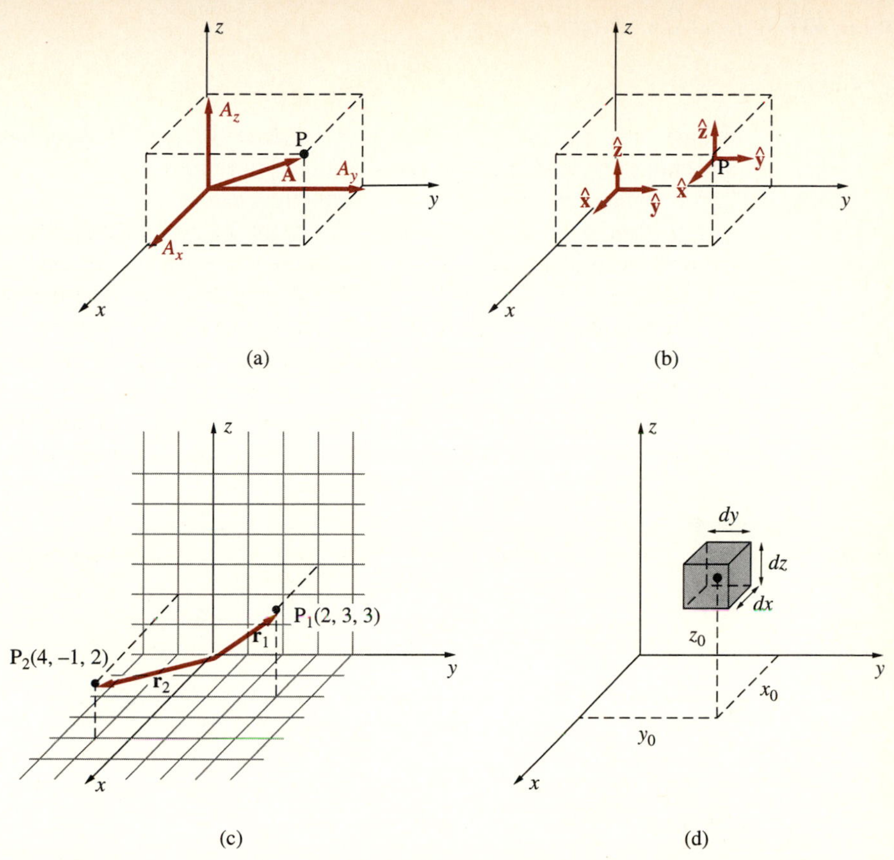

FIGURE A.1. **Rectangular coordinate system.** (a) Vector **A** pointing from the origin to point $P(A_x, A_y, A_z)$. (b) The three rectangular coordinate unit vectors. (c) Position vectors $\mathbf{r}_1 = \hat{\mathbf{x}}2 + \hat{\mathbf{y}}3 + \hat{\mathbf{z}}3$ and $\mathbf{r}_2 = \hat{\mathbf{x}}4 - \hat{\mathbf{y}} + \hat{\mathbf{z}}2$. (d) The differential volume element in a rectangular coordinate system.

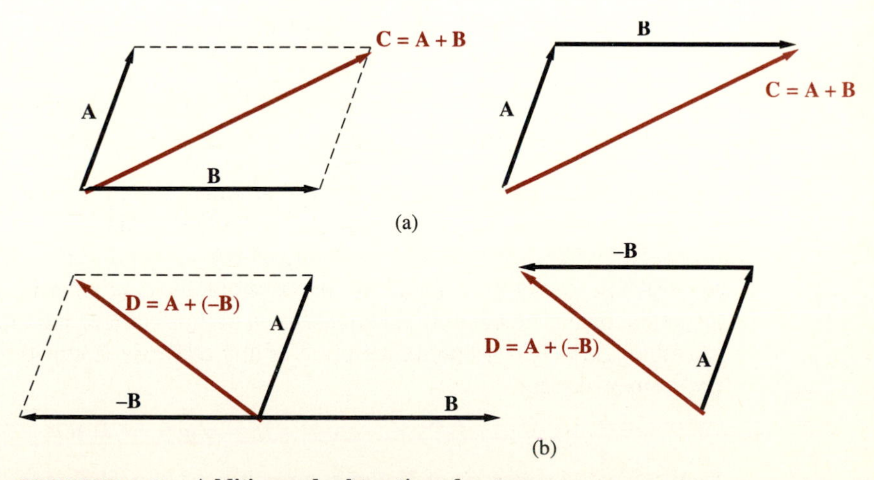

FIGURE A.2. **Addition and subtraction of vectors.**

is done similarly; subtraction of vector **B** from vector **A** gives the same result as the addition of vectors **A** and $-\mathbf{B}$, as depicted in Figure A.2b.

A point P in a rectangular coordinate system is represented by its coordinates $P(x, y, z)$. The position vector **r** of a point P is defined as the directed distance from the origin to point P. For example, the position vector of point P_1 in Figure A.1c is $\mathbf{r}_1 = \hat{\mathbf{x}}2 + \hat{\mathbf{y}}3 + \hat{\mathbf{z}}3$. The position vector of point P_2 in Figure A.1c is $\mathbf{r}_2 = \hat{\mathbf{x}}4 - \hat{\mathbf{y}} + \hat{\mathbf{z}}2$. The vector pointing from point P_1 to point P_2 is known as the distance vector (or separation vector) and is given by

$$\begin{aligned}\mathbf{R} = \mathbf{r}_2 - \mathbf{r}_1 &= \hat{\mathbf{x}}4 - \hat{\mathbf{y}} + \hat{\mathbf{z}}2 - (\hat{\mathbf{x}}2 + \hat{\mathbf{y}}3 + \hat{\mathbf{z}}3)\\ &= \hat{\mathbf{x}}2 - \hat{\mathbf{y}}4 - \hat{\mathbf{z}}\end{aligned}$$

Note that the magnitude of **R** is $R = |\mathbf{R}| = \sqrt{2^2 + 4^2 + 1^2} = \sqrt{21}$. The unit vector in the direction of **R** is thus

$$\hat{\mathbf{R}} = \frac{\mathbf{R}}{R} = \hat{\mathbf{x}}\frac{2}{\sqrt{21}} - \hat{\mathbf{y}}\frac{4}{\sqrt{21}} - \hat{\mathbf{z}}\frac{1}{\sqrt{21}}$$

A.2 VECTOR MULTIPLICATION

A.2.1 The Dot Product

The *dot product* of two vectors **A** and **B** is a scalar denoted by $\mathbf{A} \cdot \mathbf{B}$. It is equal to the product of the magnitudes $|\mathbf{A}|$ and $|\mathbf{B}|$ of vectors **A** and **B** and the cosine of the angle ψ_{AB} between vectors **A** and **B**. Namely,

$$\mathbf{A} \cdot \mathbf{B} \equiv |\mathbf{A}||\mathbf{B}| \cos \psi_{AB}$$

(The dot product is sometimes referred to as the *scalar product* since the result is a scalar quantity.) Noting that in rectangular coordinates we have $\mathbf{A} = \hat{\mathbf{x}}A_x + \hat{\mathbf{y}}A_y + \hat{\mathbf{z}}A_z$ and $\mathbf{B} = \hat{\mathbf{x}}B_x + \hat{\mathbf{y}}B_y + \hat{\mathbf{z}}B_z$, an alternative expression for the dot product is

$$\begin{aligned}\mathbf{A} \cdot \mathbf{B} &= (\hat{\mathbf{x}}A_x + \hat{\mathbf{y}}A_y + \hat{\mathbf{z}}A_z) \cdot (\hat{\mathbf{x}}B_x + \hat{\mathbf{y}}B_y + \hat{\mathbf{z}}B_z)\\ \mathbf{A} \cdot \mathbf{B} &= A_xB_x + A_yB_y + A_zB_z\end{aligned}$$

where we have used the fact that the dot product of one rectangular unit vector with a different rectangular unit vector is zero, whereas the dot product of any rectangular unit vector with itself is unity. For example, $\hat{\mathbf{x}} \cdot \hat{\mathbf{y}} = 0$ and $\hat{\mathbf{x}} \cdot \hat{\mathbf{z}} = 0$, but $\hat{\mathbf{x}} \cdot \hat{\mathbf{x}} = 1$. This feature of the dot product facilitates determining the component of a vector in a given direction, that is, determining the *projection* of the vector along a given direction. For example, the projection along the y axis of any arbitrary vector $\mathbf{B} = \hat{\mathbf{x}}B_x + \hat{\mathbf{y}}B_y + \hat{\mathbf{z}}B_z$ is simply given by

$$\mathbf{B} \cdot \hat{\mathbf{y}} = (\hat{\mathbf{x}}B_x + \hat{\mathbf{y}}B_y + \hat{\mathbf{z}}B_z) \cdot \hat{\mathbf{y}} = B_y$$

Similarly, the projection of **B** along any other arbitrary unit vector $\hat{\mathbf{a}}$ is represented by $\mathbf{B} \cdot \hat{\mathbf{a}}$, with the resulting scalar being equal to the length of the vector **B** projected

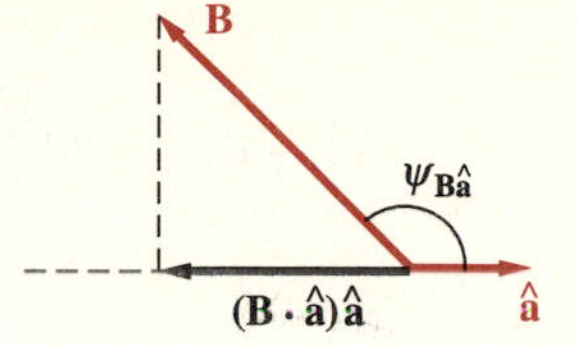

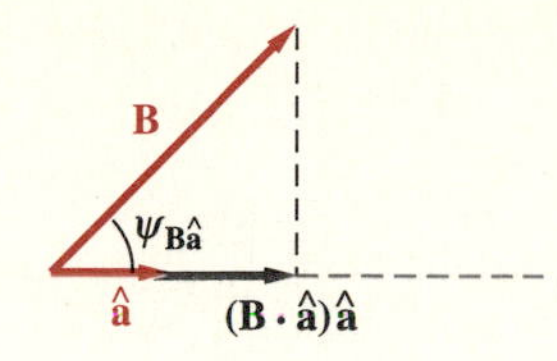

FIGURE A.3. The dot product. The dot product of **B** with any arbitrary unit vector $\hat{\mathbf{a}}$ gives the projection of **B** along $\hat{\mathbf{a}}$, which is negative if $|\psi_{\mathbf{B}\hat{\mathbf{a}}}| > \pi/2$ or positive if $|\psi_{\mathbf{B}\hat{\mathbf{a}}}| < \pi/2$.

on vector $\hat{\mathbf{a}}$. Note that, depending on whether the angle $\psi_{\mathbf{B}\hat{\mathbf{a}}}$ is acute or obtuse, the projection could be positive or negative, as illustrated in Figure A.3.

The dot product is both commutative and distributive; in other words,

$$\mathbf{A}\cdot\mathbf{B} = \mathbf{B}\cdot\mathbf{A}$$

$$\mathbf{A}\cdot(\mathbf{B}+\mathbf{C}) = \mathbf{A}\cdot\mathbf{B} + \mathbf{A}\cdot\mathbf{C}$$

A.2.2 The Cross Product

The *cross product* of two vectors **A** and **B** is a vector **C** denoted by $\mathbf{C} = (\mathbf{A}\times\mathbf{B})$. The magnitude of **C** is equal to the product of the magnitudes of vectors **A** and **B** and the sine of the angle ψ_{AB} between vectors **A** and **B**. The direction of **C**, according to the right-hand rule, follows that of the thumb of the right hand when the fingers rotate from **A** to **B** through the angle ψ_{AB}, as depicted in Figure A.4. In other words,

$$\mathbf{A}\times\mathbf{B} \equiv \hat{\mathbf{n}}|\mathbf{A}||\mathbf{B}|\sin\psi_{\mathrm{AB}}$$

where $\hat{\mathbf{n}}$ is the unit vector normal to both **A** and **B** and points in the direction in which a right-handed screw advances as **A** is turned toward **B**. (The cross product is sometimes referred to as the *vector* product since the result is a vector quantity.)

Noting that $\hat{\mathbf{x}}\times\hat{\mathbf{y}} = \hat{\mathbf{z}}$, $\hat{\mathbf{y}}\times\hat{\mathbf{z}} = \hat{\mathbf{x}}$, and $\hat{\mathbf{z}}\times\hat{\mathbf{x}} = \hat{\mathbf{y}}$, in rectangular coordinates we can use the distributive property of the cross product to write

$$\begin{aligned}\mathbf{C} = \mathbf{A}\times\mathbf{B} &= (\hat{\mathbf{x}}A_x + \hat{\mathbf{y}}A_y + \hat{\mathbf{z}}A_z)\times(\hat{\mathbf{x}}B_x + \hat{\mathbf{y}}B_y + \hat{\mathbf{z}}B_z)\\ &= \hat{\mathbf{x}}(A_yB_z - A_zB_y) + \hat{\mathbf{y}}(A_zB_x - A_xB_z) + \hat{\mathbf{z}}(A_xB_y - A_yB_x)\\ &= \hat{\mathbf{x}}C_x + \hat{\mathbf{y}}C_y + \hat{\mathbf{z}}C_z\end{aligned}$$

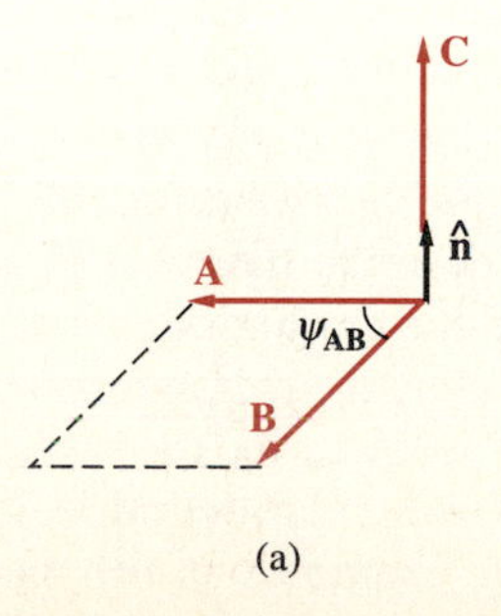

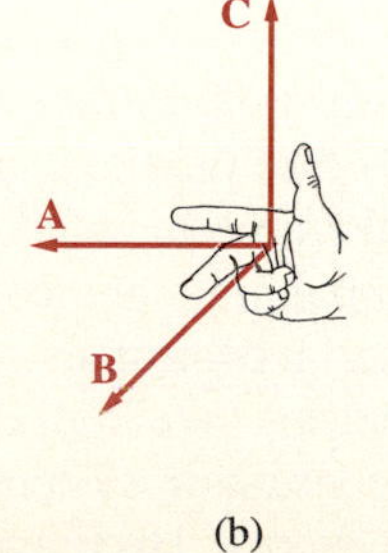

FIGURE A.4. The cross product. The cross product of two vectors **A** and **B** is another vector $\mathbf{C} = \mathbf{A}\times\mathbf{B}$, which is perpendicular to both **A** and **B** and is related to them via the right-hand rule.

It is often convenient to write the cross product in determinant form:

$$\mathbf{A}\times\mathbf{B} = \begin{vmatrix} \hat{\mathbf{x}} & \hat{\mathbf{y}} & \hat{\mathbf{z}} \\ A_x & A_y & A_z \\ B_x & B_y & B_z \end{vmatrix}$$

Using the definition of the cross product and following the right-hand rule, it is clear that

$$\mathbf{B}\times\mathbf{A} = -\mathbf{A}\times\mathbf{B}$$

so that the cross product is not commutative. The cross product is also not associative since we have

$$\mathbf{A}\times(\mathbf{B}\times\mathbf{C}) \neq (\mathbf{A}\times\mathbf{B})\times\mathbf{C}$$

It is clear, however that the cross product is distributive; in other words,

$$\mathbf{A}\times(\mathbf{B}+\mathbf{C}) = \mathbf{A}\times\mathbf{B}+\mathbf{A}\times\mathbf{C}$$

A.2.3 Triple Products

Given three vectors **A**, **B**, and **C**, there are in general two different ways we can form a triple product. The *scalar triple product* is defined as

$$\mathbf{A}\cdot(\mathbf{B}\times\mathbf{C}) = \mathbf{B}\cdot(\mathbf{C}\times\mathbf{A}) = \mathbf{C}\cdot(\mathbf{A}\times\mathbf{B})$$

Note that the result of the scalar triple product is a scalar.

The *vector triple product* is defined as $\mathbf{A}\times(\mathbf{B}\times\mathbf{C})$ and can be expressed as the difference of two simpler vectors as follows:

$$\mathbf{A}\times(\mathbf{B}\times\mathbf{C}) = \mathbf{B}(\mathbf{A}\cdot\mathbf{C}) - \mathbf{C}(\mathbf{A}\cdot\mathbf{B})$$

A.3 CYLINDRICAL AND SPHERICAL COORDINATE SYSTEMS

The physical properties of electromagnetic fields and waves as studied in this book do not depend on the particular coordinate system used to describe the vector quantities. An electric field exists, and may or may not have nonzero divergence or curl, depending on its physical properties with no regard to any system of coordinates. Scalar products, cross products, and other vector operations are all independent of the mathematical frame of reference used to describe them. In practice, however, description, manipulation, and computation of vector quantities in a given problem may well be easier in coordinate systems other than the rectangular system, depending on the symmetries involved. In this section we describe the two most commonly used additional coordinate systems: namely, the circular cylindrical and the spherical coordinate systems. These two systems are examples of orthogonal curvilinear coordinate systems; other examples not covered here include the elliptic cylindrical, parabolic cylindrical, conical, and prolate spheroidal systems.

A.3.1 Cylindrical Coordinates

The circular cylindrical system is an extension to three dimensions of the polar coordinate system of analytical geometry. In view of its common use, it is often referred to simply as the cylindrical coordinate system, although cylindrical systems with other cross-sectional shapes, such as elliptic or hyperbolic, are also used in special applications. The three cylindrical coordinates are r, ϕ, and z. The radial coordinate r of any point P is the closest distance from the z axis to that point. A given value of r specifies a circular cylindrical surface on which the point P resides, with the particular position of P on the sphere further specified by ϕ and z. The convention is to measure ϕ from the x axis, in the right-handed sense. In other words, with the thumb pointed in the z direction, the fingers trace the direction of increasing ϕ. The angle ϕ is commonly called the azimuthal angle, and it varies from 0 to 2π.

The cylindrical coordinate unit vectors $\hat{\mathbf{r}}$, $\hat{\boldsymbol{\phi}}$, and $\hat{\mathbf{z}}$ are shown in Figure A.5b. Note that the unit vectors are mutually orthogonal and that $\hat{\mathbf{r}} \times \hat{\boldsymbol{\phi}} = \hat{\mathbf{z}}$. A good way

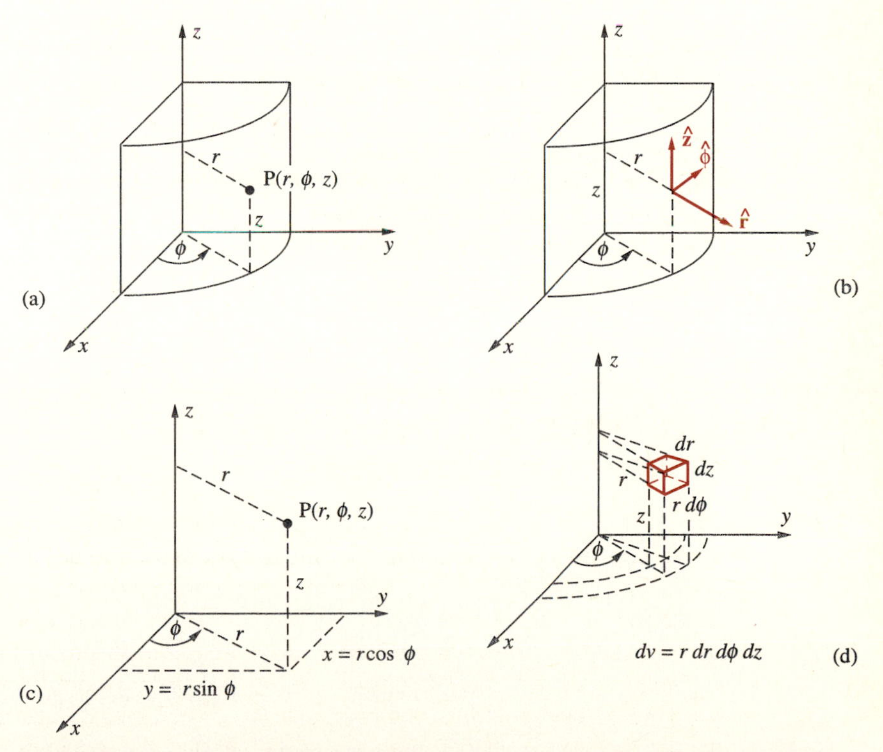

FIGURE A.5. The cylindrical coordinate system. (a) The three cylindrical coordinates. (b) The three cylindrical coordinate unit vectors. (c) The relationships between rectangular and cylindrical coordinates. (d) The differential volume element in a cylindrical coordinate system.

to remember the orientation of the unit vectors is to note that they obey the right-hand rule, with the thumb, forefinger, and the middle finger representing the directions of $\hat{\mathbf{r}}$, $\hat{\boldsymbol{\phi}}$, and $\hat{\mathbf{z}}$, respectively.

A vector **A** in cylindrical coordinates is specified by means of the values of its components A_r, A_ϕ, and A_z and is typically written as

$$\mathbf{A} = \hat{\mathbf{r}}A_r + \hat{\boldsymbol{\phi}}A_\phi + \hat{\mathbf{z}}A_z$$

The magnitude of **A** is given by $|\mathbf{A}| = \sqrt{\mathbf{A}\cdot\mathbf{A}} = \sqrt{A_r^2 + A_\phi^2 + A_z^2}$.

The expressions for the scalar and vector products of vectors for cylindrical coordinates are very similar to those for rectangular coordinates. We have

$$\mathbf{A}\cdot\mathbf{B} = A_rB_r + A_\phi B_\phi + A_zB_z$$

$$\mathbf{A}\times\mathbf{B} = \begin{vmatrix} \hat{\mathbf{r}} & \hat{\boldsymbol{\phi}} & \hat{\mathbf{z}} \\ A_r & A_\phi & A_z \\ B_r & B_\phi & B_z \end{vmatrix}$$

Cylindrical-to-Rectangular Transformations A vector specified in cylindrical coordinates can be transformed into rectangular coordinates and vice versa. The relationships between the (r, ϕ, z) and (x, y, z) are illustrated in Figure A.5c. Consider a vector

$$\mathbf{A} = \hat{\mathbf{r}}A_r + \hat{\boldsymbol{\phi}}A_\phi + \hat{\mathbf{z}}A_z$$

where in general each of the component values A_r, A_ϕ, and A_z may themselves be functions of spatial coordinates r, ϕ, and z. The rules for transforming from cylindrical to rectangular coordinates are given as follows, where the set of expressions on the left are for transforming scalar quantities (such as A_r), and those on the right are for transforming the vectors.

$$r = \sqrt{x^2 + y^2} \qquad \hat{\mathbf{r}} = \hat{\mathbf{x}}\cos\phi + \hat{\mathbf{y}}\sin\phi$$

$$\phi = \tan^{-1}\left[\frac{y}{x}\right] \qquad \hat{\boldsymbol{\phi}} = -\hat{\mathbf{x}}\sin\phi + \hat{\mathbf{y}}\cos\phi$$

$$z = z \qquad \hat{\mathbf{z}} = \hat{\mathbf{z}}$$

Note that the preceding relationships can be deduced from careful examination of Figure A.5b (unit vectors) and Figure A.5c (coordinates). A similar set of rules can be given for transformation from rectangular to cylindrical coordinates:

$$x = r\cos\phi \qquad \hat{\mathbf{x}} = \hat{\mathbf{r}}\cos\phi - \hat{\boldsymbol{\phi}}\sin\phi$$

$$y = r\sin\phi \qquad \hat{\mathbf{y}} = \hat{\mathbf{r}}\sin\phi + \hat{\boldsymbol{\phi}}\cos\phi$$

$$z = z \qquad \hat{\mathbf{z}} = \hat{\mathbf{z}}$$

The transformation of a vector given in the cylindrical (rectangular) coordinate system can be accomplished by direct substitution of the preceding unit vector and coordinate variable expressions. However, doing so in fact amounts to taking the dot

TABLE A.1. Dot products of unit vectors

	Cylindrical coordinates			Spherical coordinates		
	$\hat{\mathbf{r}}$	$\hat{\boldsymbol{\phi}}$	$\hat{\mathbf{z}}$	$\hat{\mathbf{r}}$	$\hat{\boldsymbol{\theta}}$	$\hat{\boldsymbol{\phi}}$
$\hat{\mathbf{x}}\cdot$	$\cos\phi$	$-\sin\phi$	0	$\sin\theta\cos\phi$	$\cos\theta\cos\phi$	$-\sin\phi$
$\hat{\mathbf{y}}\cdot$	$\sin\phi$	$\cos\phi$	0	$\sin\theta\sin\phi$	$\cos\theta\sin\phi$	$\cos\phi$
$\hat{\mathbf{z}}\cdot$	0	0	1	$\cos\theta$	$-\sin\theta$	0

product of the vector as written in cylindrical (rectangular) coordinates with the rectangular (cylindrical) coordinate unit vectors and substituting coordinate variables, as illustrated in Example A-1. The dot products of the cylindrical coordinate unit vectors with their rectangular coordinate counterparts are provided in Table A.1.

Example A-1: Cylindrical-to-rectangular and rectangular-to-cylindrical transformations. (a) Transform a vector

$$\mathbf{A} = \hat{\mathbf{r}}\left(\frac{1}{r}\right) + \hat{\boldsymbol{\phi}}\left(\frac{2}{r}\right)$$

given in cylindrical coordinates to rectangular coordinates. (b) Transform a vector $\mathbf{B} = \hat{\mathbf{x}}x^2 + \hat{\mathbf{y}}xy$ given in rectangular coordinates to cylindrical coordinates.

Solution:

(a) We set out to find A_x, A_y, and A_z. We have

$$A_x = \mathbf{A}\cdot\hat{\mathbf{x}} = \left[\hat{\mathbf{r}}\left(\frac{1}{r}\right) + \hat{\boldsymbol{\phi}}\left(\frac{2}{r}\right)\right]\cdot\hat{\mathbf{x}} = \frac{\cos\phi}{r} - \frac{2\sin\phi}{r} = \frac{x-2y}{x^2+y^2}$$

where we have used the unit vector dot products as given in Table A.1 and substituted $\cos\phi = x/r$, $\sin\phi = y/r$, and $r = \sqrt{x^2+y^2}$. Similarly, we have

$$A_y = \mathbf{A}\cdot\hat{\mathbf{y}} = \left[\hat{\mathbf{r}}\left(\frac{1}{r}\right) + \hat{\boldsymbol{\phi}}\left(\frac{2}{r}\right)\right]\cdot\hat{\mathbf{y}} = \frac{\sin\phi}{r} + \frac{2\cos\phi}{r} = \frac{y+2x}{x^2+y^2}$$

Note that $A_z = 0$, since $\mathbf{A}\cdot\hat{\mathbf{z}} = 0$. Thus we have

$$\mathbf{A} = \hat{\mathbf{x}}\frac{x-2y}{x^2+y^2} + \hat{\mathbf{y}}\frac{y+2x}{x^2+y^2}$$

(b) We set out to find B_r, B_ϕ, and B_z. We have

$$B_r = \mathbf{B}\cdot\hat{\mathbf{r}} = (\hat{\mathbf{x}}x^2 + \hat{\mathbf{y}}xy)\cdot\hat{\mathbf{r}} = x^2\cos\phi + xy\sin\phi$$

$$= [r^2\cos^2\phi\cos\phi + (r\cos\phi)(r\sin\phi)\sin\phi] = r^2\cos\phi[\cos^2\phi + \sin^2\phi] = r^2\cos\phi$$

where we have used the unit vector dot products as given in Table A.1 and substituted $x = r\cos\phi$ and $y = r\sin\phi$. Similarly, we have

$$B_\phi = \mathbf{B}\cdot\hat{\boldsymbol{\phi}} = (\hat{\mathbf{x}}x^2 + \hat{\mathbf{y}}xy)\cdot\hat{\boldsymbol{\phi}} = -x^2\sin\phi + xy\cos\phi$$
$$= [-r^2\cos^2\phi\sin\phi + (r\cos\phi)(r\sin\phi)\cos\phi] = 0$$

Note that $B_z = 0$ since $\mathbf{B}\cdot\hat{\mathbf{z}} = 0$. Thus we have

$$\mathbf{B} = \hat{\mathbf{r}}r^2\cos\phi$$

Length, Surface, and Volume Elements The general expression for differential length $d\mathbf{l}$ in cylindrical coordinates is

$$d\mathbf{l} = \hat{\mathbf{r}}\,dr + \hat{\boldsymbol{\phi}}r\,d\phi + \hat{\mathbf{z}}\,dz$$

as is evident from Figure A.5d. By inspection of the volume element sketched in Figure A.5d, we can see that the cylindrical coordinate surface elements in the three coordinate directions are

$$(ds)_r = r\,d\phi\,dz$$
$$(ds)_\phi = dr\,dz$$
$$(ds)_z = r\,dr\,d\phi$$

and the volume element in cylindrical coordinates is:

$$dv = r\,dr\,d\phi\,dz$$

A.3.2 Spherical Coordinates

The three spherical coordinates are r, θ, and ϕ. The radial coordinate r of any point P is simply the distance from the origin to that point. A given value of r specifies a sphere on which the point P resides, with the particular position of P on the sphere further specified by θ and ϕ. The best way to think of the spherical coordinate system and especially the coordinates θ and ϕ is in terms of the latitude/longitude system of identifying a point on the earth's surface. The earth is a sphere with radius $r \simeq$ 6370 km. With respect to Figure A.6a, latitude corresponds to θ while longitude corresponds to ϕ. The convention is that the positive z axis points toward north, and ϕ is measured from the x axis, in the right-handed sense. In other words, with the thumb pointed in the z direction, the fingers trace the direction of increasing ϕ. The Greenwich meridian, from which positive (east) longitude is measured, coincides with the positive x axis. The angle ϕ, commonly called the azimuthal angle, varies from 0 to 2π. Unlike geographic latitude, which is commonly measured with respect to the equatorial plane and which thus varies in the range from $-\pi/2$ to $+\pi/2$, the spherical coordinate θ is measured from the positive z axis (see Figure A.6) and thus varies between $\theta = 0$ (north pole) and $\theta = \pi$ (south pole).

The spherical coordinate unit vectors $\hat{\mathbf{r}}$, $\hat{\boldsymbol{\phi}}$, and $\hat{\boldsymbol{\theta}}$ are shown in Figure A.6b. Note that $\hat{\mathbf{r}}$ points radially outward, $\hat{\boldsymbol{\phi}}$ points "east," and $\hat{\boldsymbol{\theta}}$ points "south." Note that the unit vectors are mutually orthogonal and that $\hat{\mathbf{r}}\times\hat{\boldsymbol{\theta}} = \hat{\boldsymbol{\phi}}$. A good way to remember

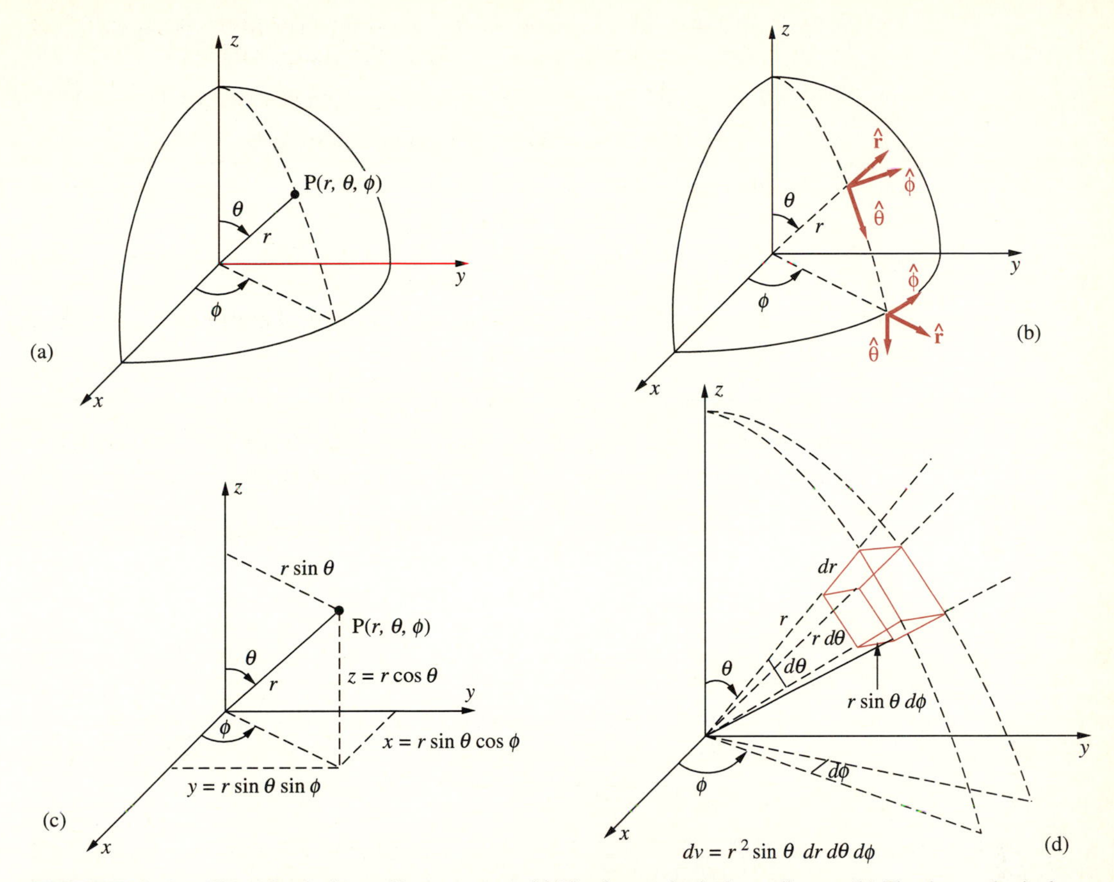

FIGURE A.6. **The spherical coordinate system.** (a) The three spherical coordinates. (b) The three spherical coordinate unit vectors. (c) The relationships between rectangular and spherical coordinates. (d) The differential volume element in a spherical coordinate system.

the orientation of the unit vectors is to note that they obey the right-hand rule, with the thumb, forefinger, and middle finger representing the directions of $\hat{\mathbf{r}}$, $\hat{\boldsymbol{\theta}}$, and $\hat{\boldsymbol{\phi}}$, respectively.

A vector $\mathbf{A}$ in spherical coordinates is specified by means of the values of its components A_r, A_θ, and A_ϕ and is typically written as

$$\mathbf{A} = \hat{\mathbf{r}}A_r + \hat{\boldsymbol{\theta}}A_\theta + \hat{\boldsymbol{\phi}}A_\phi$$

The magnitude of $\mathbf{A}$ is given by $|\mathbf{A}| = \mathbf{A}\cdot\mathbf{A} = \sqrt{A_r^2 + A_\theta^2 + A_\phi^2}$.

The expressions for the scalar and vector products of vectors for spherical coordinates are very similar to those for rectangular coordinates. We have

$$\mathbf{A}\cdot\mathbf{B} = A_rB_r + A_\theta B_\theta + A_\phi B_\phi$$

$$\mathbf{A}\times\mathbf{B} = \begin{vmatrix} \hat{\mathbf{r}} & \hat{\boldsymbol{\theta}} & \hat{\boldsymbol{\phi}} \\ A_r & A_\theta & A_\phi \\ B_r & B_\theta & B_\phi \end{vmatrix}$$

Spherical to Rectangular Transformations A vector specified in spherical coordinates can be transformed into rectangular coordinates and vice versa. The relationship between (r, θ, ϕ) and (x, y, z) is illustrated in Figure A.6c. Consider a vector

$$\mathbf{A} = \hat{\mathbf{r}}A_r + \hat{\boldsymbol{\theta}}A_\theta + \hat{\boldsymbol{\phi}}A_\phi$$

where in general each of the component values A_r, A_θ, and A_ϕ may themselves be functions of spatial coordinates r, θ, and ϕ. The rules for transformation from spherical coordinates to rectangular coordinates are given as follows, where the set of expressions on the left-hand side are for transforming scalar quantities (such as A_r), and those on the right are for transforming vectors.

Coordinate variables:	Unit vectors:
$r = \sqrt{x^2 + y^2 + z^2}$	$\hat{\mathbf{r}} = \hat{\mathbf{x}} \sin\theta\cos\phi + \hat{\mathbf{y}}\sin\theta\sin\phi + \hat{\mathbf{z}}\cos\theta$
$\theta = \cos^{-1}\left[\dfrac{z}{\sqrt{x^2 + y^2 + z^2}}\right]$	$\hat{\boldsymbol{\theta}} = \hat{\mathbf{x}}\cos\theta\cos\phi + \hat{\mathbf{y}}\cos\theta\sin\phi - \hat{\mathbf{z}}\sin\theta$
$\phi = \tan^{-1}\left[\dfrac{y}{x}\right]$	$\hat{\boldsymbol{\phi}} = -\hat{\mathbf{x}}\sin\phi + \hat{\mathbf{y}}\cos\phi$

Note that the preceding relationships can be deduced from careful examination of Figures A.6b (unit vectors) and A.6c (quantities). A similar set of rules can be given for transformation from rectangular to spherical coordinates:

Coordinate variables:	Unit vectors:
$x = r\sin\theta\cos\phi$	$\hat{\mathbf{x}} = \hat{\mathbf{r}}\sin\theta\cos\phi - \hat{\boldsymbol{\theta}}\cos\theta\cos\phi - \hat{\boldsymbol{\phi}}\sin\phi$
$y = r\sin\theta\sin\phi$	$\hat{\mathbf{y}} = \hat{\mathbf{r}}\sin\theta\sin\phi + \hat{\boldsymbol{\theta}}\cos\theta\sin\phi + \hat{\boldsymbol{\phi}}\cos\phi$
$z = r\cos\theta$	$\hat{\mathbf{z}} = \hat{\mathbf{r}}\cos\theta - \hat{\boldsymbol{\theta}}\sin\theta$

The transformation of a vector given in the spherical (rectangular) coordinate system can be accomplished by direct substitution of the unit vector and the preceding coordinate variable expressions. However, doing so in fact amounts to taking the dot products of the vector, as written in spherical (rectangular) coordinates, with the rectangular (spherical) coordinate unit vectors and substituting coordinate variables, as illustrated in Example A-1. The dot products of the spherical coordinate unit vectors with their rectangular coordinate counterparts are provided in Table A.1.

Example A-2: Spherical-to-rectangular and rectangular-to-spherical transformations. (a) Transform the vector

$$\mathbf{E} = \hat{\mathbf{r}}\frac{2\cos\theta}{r^3} + \hat{\boldsymbol{\theta}}\sin\theta$$

given in spherical coordinates to rectangular coordinates. (b) Transform a vector $\mathbf{B} = \hat{\mathbf{x}}x^2 + \hat{\mathbf{y}}xy$ given in rectangular coordinates to spherical coordinates.

Solution:

(a) Note that this vector $\mathbf{E}$ has the same form as the distant electric field of an electric dipole (Equation [4.27]) or the distant magnetic field of a magnetic dipole (Equation [6.14]). We set out to determine E_x, E_y, and E_z. We have

$$E_x = \mathbf{E}\cdot\hat{\mathbf{x}} = \frac{2\cos\theta}{r^3}\sin\theta\cos\phi + \sin\theta\cos\theta\cos\phi = \left(\frac{2}{r^3}+1\right)\sin\theta\cos\theta\cos\phi$$

$$= \left(\frac{2+r^3}{r^3}\right)\left(\frac{x}{r}\right)\left(\frac{z}{r}\right) = \frac{[2+(x^2+y^2+z^2)^{3/2}]xz}{(x^2+y^2+z^2)^{5/2}}$$

where we have used the unit vector dot products as given in Table A.1 and substituted $x = r\sin\theta\cos\phi$ and $z = r\cos\theta$. Similarly we have

$$E_y = \mathbf{E}\cdot\hat{\mathbf{y}} = \frac{2\cos\theta}{r^3}\sin\theta\sin\phi + \sin\theta\cos\theta\sin\phi = \left(\frac{2}{r^3}+1\right)\cos\theta\sin\theta\cos\phi$$

which is identical to E_x. The z component is given by

$$E_z = \mathbf{E}\cdot\hat{\mathbf{z}} = \frac{2\cos\theta}{r^3}\cos\theta - \sin\theta\sin\theta = \frac{2\cos^2\theta}{r^3} - \sin^2\theta = \frac{2\cos^2\theta}{r^3} - (1-\cos^2\theta)$$

$$= \frac{2z^2}{r^5} - 1 + \frac{z^2}{r^2} = \frac{2z^2}{(x^2+y^2+z^2)^{5/2}} + \frac{z^2}{x^2+y^2+z^2} - 1$$

Thus we have

$$\mathbf{E} = [\hat{\mathbf{x}}+\hat{\mathbf{y}}]\left\{\frac{[2+(x^2+y^2+z^2)^{3/2}]xz}{(x^2+y^2+z^2)^{3/2}}\right\} + \hat{\mathbf{z}}\left[\frac{2z^2}{(x^2+y^2+z^2)^{5/2}} + \frac{z^2}{x^2+y^2+z^2} - 1\right]$$

The rather complex nature of the rectangular form of vector $\mathbf{E}$ underscores the usefulness of spherical coordinates for problems involving spherical symmetry, such as the electric and magnetic dipoles.

(b) Note that the vector $\mathbf{B}$ is the same as that considered in Example A-1. We set out to find B_r, B_θ, and B_ϕ. We have

$$B_r = \mathbf{B}\cdot\hat{\mathbf{r}} = x^2\sin\theta\cos\phi + xy\sin\theta\sin\phi$$

$$= r^2\sin^2\theta\cos^2\phi\sin\theta\cos\phi + (r\sin\theta\cos\phi)(r\sin\theta\sin\phi)\sin\theta\sin\phi$$

$$= r^2\sin^3\theta\cos\phi(\cos^2\phi+\sin^2\phi) = r^2\sin^3\theta\cos\phi$$

where we have used the unit vector dot products as given in Table A.1 and substituted $x = r\sin\theta\cos\phi$ and $y = r\sin\theta\sin\phi$. Similarly, we have

$$B_\theta = \mathbf{B} \cdot \hat{\boldsymbol{\theta}} = x^2 \cos\theta \cos\phi + xy \cos\theta \sin\phi$$
$$= r^2 \sin^2\theta \cos^2\phi \cos\theta \cos\phi + (r \sin\theta \cos\phi)(r \sin\theta \sin\phi) \cos\theta \sin\phi$$
$$= r^2 \sin^2\theta \cos\theta \cos\phi(\cos^2\phi + \sin^2\phi)$$
$$= r^2 \sin^2\theta \cos\theta \cos\phi$$

And the ϕ component is given by

$$B_\phi = \mathbf{B} \cdot \hat{\boldsymbol{\phi}} = x^2(-\sin\phi) + xy \cos\phi$$
$$= -r^2 \sin^2\theta \cos^2\phi \sin\phi + (r \sin\theta \cos\phi)(r \sin\theta \sin\phi) \cos\phi = 0$$

Thus we have

$$\mathbf{B} = \hat{\mathbf{r}} r^2 \sin^3\theta \cos\phi + \hat{\boldsymbol{\theta}} r^2 \sin^2\theta \cos\theta \cos\phi$$

Length, Surface, and Volume Elements The general expression for differential length $d\mathbf{l}$ in spherical coordinates is

$$d\mathbf{l} = \hat{\mathbf{r}}\, dr + \hat{\boldsymbol{\theta}} r\, d\theta + \hat{\boldsymbol{\phi}} r \sin\theta\, d\phi$$

as is evident from Figure A.6d. By inspection of the volume element sketched in Figure A.6d, we can see that the spherical coordinate surface elements in the three coordinate directions are

$$(ds)_r = r^2 \sin\theta\, d\theta\, d\phi$$
$$(ds)_\theta = r \sin\theta\, dr\, d\phi$$
$$(ds)_\phi = r\, dr\, d\theta$$

and the volume element in spherical coordinates is:

$$dv = r^2 \sin\theta\, dr\, d\theta\, d\phi$$

A.4 VECTOR IDENTITIES

In Section 4.4 we introduced the nabla (del or grad) operator, which is defined in rectangular coordinates as

$$\nabla \equiv \hat{\mathbf{x}}\frac{\partial}{\partial x} + \hat{\mathbf{y}}\frac{\partial}{\partial y} + \hat{\mathbf{z}}\frac{\partial}{\partial z}$$

and we used it extensively in the rest of the book. A number of vector identities involving the nabla operator are quite useful and are given in this section. Each of these identities can be verified by direct reduction of both sides of the equation. Arbitrary vectors are represented as $\mathbf{A}$ and $\mathbf{B}$, whereas Φ and Ψ denote scalars.

Note that the preceding definition of the nabla operator is valid and useful only for a rectangular coordinate system. In cylindrical or spherical coordinate systems, the orientation of the unit vectors depends on their position, so that the quantity $\nabla \times \mathbf{A}$ is not simply the cross product of a corresponding nabla operator and the vector **A**. If any of the following vector identities were used in a cylindrical or spherical coordinate system, the quantities $\nabla \cdot \mathbf{A}$ and $\nabla \times \mathbf{A}$ should be taken to be symbolic representations of the cylindrical- or spherical-coordinate divergence and curl expressions, given in Sections 4.6 and 6.4. For example, in spherical coordinates, the quantity $\nabla \cdot \mathbf{A}$ is the quantity div **A** given by equation [4.37].

$$\nabla(\Phi + \Psi) \equiv \nabla\Phi + \nabla\Psi$$

$$\nabla(\Phi\Psi) \equiv \Phi\nabla\Psi + \Psi\nabla\Phi$$

$$\nabla \cdot (\Phi\mathbf{A}) \equiv \mathbf{A} \cdot \nabla\Phi + \Phi\nabla \cdot \mathbf{A}$$

$$\nabla \cdot (\mathbf{A} \times \mathbf{B}) \equiv \mathbf{B} \cdot (\nabla \times \mathbf{A}) - \mathbf{A} \cdot (\nabla \times \mathbf{B})$$

$$\nabla \times (\Phi\mathbf{A}) \equiv \nabla\Phi \times \mathbf{A} + \Phi\nabla \times \mathbf{A}$$

$$\nabla \times (\mathbf{A} \times \mathbf{B}) \equiv \mathbf{A}(\nabla \cdot \mathbf{B}) - \mathbf{B}(\nabla \cdot \mathbf{A}) + (\mathbf{B} \cdot \nabla)\mathbf{A} - (\mathbf{A} \cdot \nabla)\mathbf{B}$$

$$\nabla(\mathbf{A} \cdot \mathbf{B}) \equiv (\mathbf{A} \cdot \nabla)\mathbf{B} + (\mathbf{B} \cdot \nabla)\mathbf{A} + \mathbf{A} \times (\nabla \times \mathbf{B}) + \mathbf{B} \times (\nabla \times \mathbf{A})$$

$$\nabla \cdot \nabla\Phi \equiv \nabla^2\Phi$$

$$\nabla \cdot \nabla \times \mathbf{A} \equiv 0$$

$$\nabla \times \nabla\Phi \equiv 0$$

$$\nabla \times \nabla \times \mathbf{A} \equiv \nabla(\nabla \cdot \mathbf{A}) - \nabla^2\mathbf{A}$$

The operator formed by the dot product of a vector **A** and the del operator forms a new operator that is a scalar operation of the form

$$\mathbf{A} \cdot \nabla = A_x\frac{\partial}{\partial x} + A_y\frac{\partial}{\partial y} + A_z\frac{\partial}{\partial z}$$

which, when applied to a vector **B**, gives

$$(\mathbf{A} \cdot \nabla)\mathbf{B} = \hat{\mathbf{x}}\left[A_x\frac{\partial B_x}{\partial x} + A_y\frac{\partial B_x}{\partial y} + A_z\frac{\partial B_x}{\partial z}\right] + \hat{\mathbf{y}}\left[A_x\frac{\partial B_y}{\partial x} + A_y\frac{\partial B_y}{\partial y} + A_z\frac{\partial B_y}{\partial z}\right] + \hat{\mathbf{z}}\left[A_x\frac{\partial B_z}{\partial x} + A_y\frac{\partial B_z}{\partial y} + A_z\frac{\partial B_z}{\partial z}\right]$$

Special caution is needed in interpreting the Laplacian $\nabla^2\mathbf{A}$, which in rectangular coordinates can be simply expanded as $\nabla^2\mathbf{A} = \hat{\mathbf{x}}\nabla^2 A_x + \hat{\mathbf{y}}\nabla^2 A_y + \hat{\mathbf{z}}\nabla^2 A_z$. In spherical and cylindrical coordinates, the last identity given here, namely $\nabla^2\mathbf{A} \equiv \nabla(\nabla \cdot \mathbf{A}) - \nabla \times \nabla \times \mathbf{A} =$ grad(div **A**) − curl(curl **A**) is in fact the defining relation for the quantity $\nabla^2\mathbf{A}$.

Derivation of Ampère's Circuital Law from the Biot–Savart Law

In Section 6.3, we introduced Ampère's circuital law, expressed in [6.8], as a mathematical consequence of the Biot–Savart law, which is given by expression [6.6]. In this appendix, we mathematically derive [6.10] from [6.6]. Since [6.8] can be derived from [6.10] simply by using Stokes's theorem, the analysis that follows[1] constitutes a derivation of Ampère's circuital law.

Our starting point is thus the experimentally established Biot–Savart law:

$$\mathbf{B}(\mathbf{r}) = \frac{\mu_0}{4\pi}\int_{V'} \frac{\mathbf{J}(\mathbf{r}') \times \hat{\mathbf{R}}}{R^2}\, dv' \qquad [6.6]$$

which, as we recall from Section 6.5, can be rewritten as [6.12], or

$$\mathbf{B}(\mathbf{r}) = \nabla \times \frac{\mu_0}{4\pi}\int_{V'} \frac{\mathbf{J}(\mathbf{r}')}{R}\, dv' \qquad [6.12]$$

Since our goal is to show that [6.12] (and thus [6.6]) leads to [6.10], we start by taking the curl of both sides of [6.12]:

$$\nabla \times \mathbf{B}(\mathbf{r}) = \nabla \times \nabla \times \frac{\mu_0}{4\pi}\int_{V'} \frac{\mathbf{J}(\mathbf{r}')}{R}\, dv' \qquad [B.1]$$

We now use the same vector identity used in deriving the wave equation in Section 8.1, namely equation [8.2], to expand the curl-curl operator and to rewrite [B.1] as

$$\nabla \times \mathbf{B}(\mathbf{r}) = \frac{\mu_0}{4\pi}\int_{V'} \left\{ \nabla\left[\nabla \cdot \left(\frac{\mathbf{J}(\mathbf{r}')}{R}\right)\right] - \mathbf{J}(\mathbf{r}')\nabla^2\left(\frac{1}{R}\right)\right\} dv' \qquad [B.2]$$

First we consider the first term in [B.2], which can be rewritten as

$$\int_{V'} \nabla\left[\nabla \cdot \left(\frac{\mathbf{J}(\mathbf{r}')}{R}\right)\right] dv' = \nabla \int_{V'} \left[\nabla \cdot \left(\frac{\mathbf{J}(\mathbf{r}')}{R}\right)\right] dv' \qquad [B.3]$$

[1] The derivation presented here was adapted from that given in Section 6.6 of R. Plonsey and R. E. Collin, *Principles and Applications of Electromagnetic Fields,* McGraw-Hill, New York, 1961.

where one of the del operators is brought outside the integral. We can now rewrite the integrand as

$$\nabla\cdot\left(\frac{\mathbf{J}(\mathbf{r}')}{R}\right)=\mathbf{J}\cdot\nabla\left(\frac{1}{R}\right)=-\mathbf{J}\cdot\nabla'\left(\frac{1}{R}\right)=-\nabla'\cdot\left(\frac{\mathbf{J}}{R}\right)$$

where ∇' is the del operator applied to the source coordinates ($\mathbf{r}'$), and where we have used $\nabla'\cdot(\mathbf{J}/R) = (1/R)\nabla'\cdot\mathbf{J}+\mathbf{J}\cdot\nabla'(1/R)$ and noted that $\nabla'\cdot\mathbf{J} = 0$ for stationary currents, as required by the equation of continuity. Also note that since $R = |\mathbf{r}-\mathbf{r}'| = [(x-x')^2+(y-y')^2+(z-z')^2]^{1/2}$, we have $\nabla(1/R) = -\nabla'(1/R)$. Equation [B.3] can now be written as

$$\nabla\int_{V'}\left[\nabla\cdot\left(\frac{\mathbf{J}(\mathbf{r}')}{R}\right)\right]dv' = -\nabla\int_{V'}\left[\nabla'\cdot\left(\frac{\mathbf{J}(\mathbf{r}')}{R}\right)\right]dv' = -\nabla\oint_{S'}\left(\frac{\mathbf{J}}{R}\right)\cdot d\mathbf{s}' \quad \text{[B.4]}$$

where we have invoked the divergence theorem. Since $\mathbf{J}$ is a stationary current and is confined to a finite region of space, the surface S' can always be chosen to be large enough to include all currents within it, so that $\mathbf{J}\cdot d\mathbf{s}' = 0$ on it. Thus, [B.4] vanishes, and thus [B.3] is zero, and [B.2] is left with only the second term on the right-hand side. We have

$$\nabla\times\mathbf{B}(\mathbf{r}) = -\frac{\mu_0}{4\pi}\int_{V'}\mathbf{J}(\mathbf{r}')\nabla^2\left(\frac{1}{R}\right)dv' \quad \text{[B.5]}$$

Since our goal is to arrive at [6.10], we need to show that the right-hand side of [B.5] reduces to $\mu_0\mathbf{J}(\mathbf{r})$. Direct differentiation indicates that $\nabla^2(1/R) = 0$ for all finite values of R. Thus, if the observation point $\mathbf{r}$ is outside a source region of finite extent, then $R = |\mathbf{r}-\mathbf{r}'|$ is never zero, and we have $\nabla\times\mathbf{B} = 0$. However, if the observation point $\mathbf{r}$ is within the region where the sources are located, then in the course of the integration of [B.5], R can be zero at certain points. We must thus be able to show that the singularity of $\nabla^2(1/R)$ at $R = 0$ is integrable and yields a finite result.

Consider a point at which $R = 0$, where $\mathbf{r} = \mathbf{r}'$ or $x = x'$, $y = y'$, and $z = z'$. In the immediate neighborhood of this point, we can assume that the current density does not vary much from its value at $\mathbf{r} = \mathbf{r}'$. Since the integrand of [B.5] is zero everywhere except at $R = 0$, we can integrate [B.5] over a sphere centered at $\mathbf{r}$ that is small enough so that the current density everywhere within it is the same (i.e., $\mathbf{J}(\mathbf{r}') = \mathbf{J}(\mathbf{r})$). Thus, [B.5] becomes

$$\nabla\times\mathbf{B}(\mathbf{r}) = -\frac{\mu_0}{4\pi}\mathbf{J}(\mathbf{r})\int_{V'_s}\nabla^2\left(\frac{1}{R}\right)dv' \quad \text{[B.6]}$$

where V'_s is the small spherical volume surrounding the point at which $R = 0$. As noted previously in connection with [B.3], we have $\nabla(1/R) = -\nabla'(1/R)$ and $\nabla^2(1/R) = (\nabla')^2(1/R) = \nabla'\cdot\nabla'(1/R)$. Thus, we can write [B.6] as

$$\nabla\times\mathbf{B}(\mathbf{r}) = -\frac{\mu_0}{4\pi}\mathbf{J}(\mathbf{r})\int_{V'_s}\nabla'\cdot\nabla'\left(\frac{1}{R}\right)dv' = -\frac{\mu_0}{4\pi}\mathbf{J}(\mathbf{r})\oint_{S'_s}\nabla'\left(\frac{1}{R}\right)\cdot d\mathbf{s}' \quad \text{[B.7]}$$

We have invoked the divergence theorem, where S'_s is the spherical surface that encloses the spherical volume V'_s centered at $\mathbf{r}$. As was shown in footnote 41 in Section 6.5, we have $\nabla(1/R) = -\hat{\mathbf{R}}/R^2$, and thus $\nabla'(1/R) = -\nabla(1/R) = \hat{\mathbf{R}}/R^2$. Note that since we took the volume V'_s to be centered at $\mathbf{r}$, the vector $\mathbf{R} = \mathbf{r} - \mathbf{r}'$ points inward toward the center point $\mathbf{r}$. Thus, we have $\hat{\mathbf{R}} = -\hat{\mathbf{r}}$ so that $\nabla'(1/R) = -\hat{\mathbf{r}}/R^2$. Noting that the surface element for spherical coordinates is $\hat{\mathbf{r}}ds' = \hat{\mathbf{r}}R^2 \sin\theta' \, d\phi' \, d\theta'$, we can write [B.7] as

$$\nabla \times \mathbf{B}(\mathbf{r}) = -\frac{\mu_0}{4\pi}\mathbf{J}(\mathbf{r}) \int_0^{\pi} \int_0^{2\pi} \left(\frac{-\hat{\mathbf{r}}}{R^2}\right) \cdot \hat{\mathbf{r}}R^2 \sin\theta' \, d\phi' \, d\theta'$$

$$= \frac{\mu_0}{4\pi}\mathbf{J}(\mathbf{r}) \int_0^{\pi} \int_0^{2\pi} \sin\theta' \, d\phi' \, d\theta' = \mu_0 \mathbf{J}(\mathbf{r})$$

which is [6.10]. We have thus shown that [6.10] is a mathematical consequence of the Biot–Savart law, or [6.6]. Since Ampère's circuital law, or [6.8], can be obtained from [6.6] simply by using Stokes's theorem, we have also shown that Ampère's circuital law is a direct requirement of the Biot–Savart law, which is an expression of one of the three experimental pillars of our formulation of electromagnetics.

Symbols and Units for Basic Quantities

Symbol	Quantity	SI Unit	Comments
$\mathbf{A}$	Magnetic vector potential	webers (Wb)-m^{-1}	
B	Susceptance	siemens (S)	Ch. 3 only
$\mathbf{B}$	Magnetic field (**B** field)	Wb-m^{-2}	Phasor in Sec. 7.4 and Ch. 8
$\overline{\mathcal{B}}$	Magnetic field (**B** field)	Wb-m^{-2}	
C	Capacitance	farads (F)	Per unit length (F-m^{-1}) in Ch. 2 & 3
$\mathbf{D}$	Electric flux density	C-m^{-2}	Phasor in Sec. 7.4 and Ch. 8
$\overline{\mathcal{D}}$	Electric flux density	Wb-m^{-2}	
$\mathbf{E}$	Electric field intensity	V-m^{-1}	Phasor in Sec. 7.4 and Ch. 8
$\overline{\mathcal{E}}$	Electric field intensity	V-m^{-1}	
$\mathbf{F}$	Force	N	Newtons
f	Frequency	hertz (Hz)	
G	Conductance	S	Per unit length (S-m^{-1}) in Chs. 2 and 3
$\mathbf{H}$	Magnetic field intensity	V-m^{-1}	Phasor in Sec. 7.4 and Ch. 8
$\overline{\mathcal{H}}$	Magnetic field intensity	V-m^{-1}	
I	Current	amperes (A)	Phasor
$\mathcal{I}$	Current	A	
$\mathbf{J}$	Current density	A-m^{-2}	Phasor in Sec. 7.4 and Ch. 8
$\overline{\mathcal{J}}$	Current density	A-m^{-2}	
$\mathbf{J}_s$	Surface current density	A-m^{-1}	Phasor in Sec. 7.4 and Ch. 8
$\overline{\mathcal{J}}_s$	Surface current density	A-m^{-1}	
L	Inductance	henrys (H)	Per unit length (H-m^{-1}) in Ch. 2 & 3
l	Length	m	
$\mathbf{M}$	Magnetization vector	A-m^{-1}	
m	Mass	kg	
$\mathbf{m}$	Magnetic dipole moment	A-m^{-2}	
n	Index of refraction		
P	Power	watts (W)	
$\mathbf{P}$	Electric polarization vector	C-m^{-2}	
$\mathbf{p}$	Electric dipole moment	C-m	
Q, q	Electric charge	coulombs (C)	
Q	Quality factor		Sec. 3.9 only
R	Resistance	Ω	Per unit length (Ω-m^{-1}) in Ch. 2 & 3
R_s	Surface resistance	Ω	
S	Standing wave ratio		
$\overline{\mathcal{S}}$	Poynting vector	W-m^{-2}	
$\mathbf{S}_{\text{av}}$	Poynting vector	W-m^{-2}	Time-average quantity
$\mathbf{S}$	Complex Poynting vector	W-m^{-2}	
T	Temperature	°C	Ch. 5 only
$\mathbf{T}$	Torque	N-m	
$\mathcal{T}$	Transmission coefficient		$\mathcal{T} = \tau e^{j\phi_{\mathcal{T}}}$
t	Time	s	
t_d	One-way travel time	s	
t_r	Rise time	s	

Symbol	Quantity	SI Unit	Comments
$\mathbf{v}$	Velocity vector	m-s^{-1}	
v_p	Phase velocity	m-s^{-1}	
V	Voltage	volts (V)	Phasor
$\mathcal{V}$	Voltage	V	
$\mathcal{V}_{\text{emf}}$	Electromotive force	V	
$\mathcal{V}_{\text{ind}}$	Induced emf	V	
W	Work (energy)	joules (J)	
w	Energy density	J-m^{-3}	
X	Reactance	Ω	
Y	Admittance	S	
Z	Impedance	Ω	
Z_0	Characteristic impedance	Ω	$Z_0 = \sqrt{L/C}$ for lossless line
α	Attenuation constant	nepers (np)-m^{-1}	Real part of γ
$\bar{\alpha}$	Attenuation constant	np-m^{-1}	Real part of $\bar{\gamma}$
α_c	Attenuation constant	np-m^{-1}	Due to conduction losses
α_d	Attenuation constant	np-m^{-1}	Due to dielectric losses
$\alpha_e, \alpha_i, \alpha_T$	Polarizability	m^{-3}	Section 4.10 only
β	Propagation constant	rad-m^{-1}	$\beta = \omega\sqrt{\mu\epsilon}$
β	Phase constant	rad-m^{-1}	Imaginary part of γ
$\bar{\beta}$	Phase constant	rad-m^{-1}	Imaginary part of $\bar{\gamma}$
Γ	Reflection coefficient		$\Gamma = \rho e^{j\psi}$ or $\Gamma = \rho e^{j\phi_\Gamma}$
γ	Propagation constant	m^{-1}	$\gamma = \alpha + j\beta$
$\bar{\gamma}$	Propagation constant	m^{-1}	$\bar{\gamma} = \bar{\alpha}$ or $\bar{\gamma} = j\bar{\beta}$
δ	Skin depth	m	$\delta = (\pi f \mu\sigma)^{-1/2}$
$\tan\delta_c$	Loss tangent		$\tan\delta_c = \epsilon''/\epsilon'$
ϵ, ϵ_0	Permittivity	F-m^{-1}	
ϵ_r	Relative permittivity		
ϵ_c	Complex permittivity	F-m^{-1}	$\epsilon_c = \epsilon' - j\epsilon''$
η	Intrinsic impedance	Ω	$\eta = \sqrt{\mu/\epsilon}$
η_c	Impedance of a lossy medium	Ω	$\eta_c = \lvert\eta_c\rvert e^{j\phi_\eta}$
λ	Wavelength	m	$\lambda = 2\pi/\beta$
$\bar{\lambda}$	Wavelength in a waveguide	m	$\bar{\lambda} = 2\pi/\bar{\beta}$
μ, μ_0	Permeability	H-m^{-1}	
μ_r	Relative permeability		
ρ	Magnitude of Γ		In Ch. 3 and Sec. 8.2
ρ	Volume charge density	C-m^{-3}	Phasor in Sec. 7.4 and Ch. 8
$\tilde{\rho}$	Volume charge density	C-m^{-3}	
ρ_s	Surface charge density	C-m^{-2}	
ρ_l	Linear charge density	C-m^{-1}	
σ	Conductivity	S	
τ	Magnitude of $\mathcal{T}$		In Ch. 3 and Sec. 8.2
τ	Time constant	s	
τ_r	Relaxation time	s	$\tau_r = \epsilon/\sigma$
Φ	Electrostatic potential	volts (V)	
χ_e	Electrical susceptibility		
χ_m	Magnetic susceptibility		
ψ	Phase of reflection coefficient	rad	Ch. 3 only
Ψ	Magnetic flux	Wb	
ω	Angular frequency	radians-s^{-1}	

General Bibliography

In addition to the specific references to numerous books and articles provided in the footnotes, the following texts on electromagnetic fields and waves were found to be generally useful in developing this book.

Cheng, D. K., *Field and Wave Electromagnetics,* Addison Wesley, 2nd ed., 1989.

Cheston, W. B., *Elementary Theory of Electric and Magnetic Fields,* Wiley, 1964.

Elliott, R. S., *Electromagnetics: History, Theory, and Applications,* IEEE Press, 1992.

Feynman, R. P., R. O. Leighton, and M. Sands, *The Feynman Lectures on Physics,* Addison Wesley, 1964.

Jordan, E. C., and K. G. Balmain, *Electromagnetic Waves and Radiating Systems,* 2nd ed., Prentice-Hall, 1968.

Kraus, J. D., *Electromagnetics,* 4th ed., McGraw-Hill, 1992.

Moore, R. K., *Travelling-Wave Engineering,* McGraw-Hill, 1960.

Plonsey, R., and R. E. Collin, *Principles and Applications of Electromagnetic Fields,* 2nd ed., McGraw-Hill, 1982.

Plonus, M. A., *Applied Electromagnetics,* McGraw-Hill, 1978.

Ramo, S., J. R. Whinnery, and T. Van Duzer, *Fields and Waves in Communication Electronics,* 3rd ed., Wiley, 1994.

Scott, W. T., *The Physics of Electricity and Magnetism,* 2nd ed., Wiley, 1966.

Skilling, H. H., *Fundamentals of Electric Waves,* 2nd ed., Wiley, 1948; reprinted by Robert E. Krieger Publishing Co., 1974.

Answers to Odd-Numbered Problems

CHAPTER 2 **2-1.** $\mathcal{V}_{\rm L} = \{0, t < t_d; 2\ {\rm V}, t_d < t < 3t_d; 0, 3t_d < t < 4t_d\}$ and periodic with $T_p = 4t_d$. **2-3.** $\mathcal{V}_{\rm L} = \{0, t < 1.8\ {\rm ns}; 6\ {\rm V}, 1.8\ {\rm ns} < t < 5.4\ {\rm ns}; 2\ {\rm V}, 5.4\ {\rm ns} < t < 9\ {\rm ns}; \ldots, V_{\rm ss} = 3.6\ {\rm V}\}$. **2-5.** (a) $\mathcal{V}_{\rm s} = \{0.75\ {\rm V}, t < 1\ {\rm ns}; 0.375\ {\rm V}, 1\ {\rm ns} < t < 2\ {\rm ns}; 0.1875\ {\rm V}, 2\ {\rm ns} < t < 3\ {\rm ns}; \ldots\}$. (b) $\mathcal{V}_{\rm s} = \{0.75\ {\rm V}, t < 1\ {\rm ns}; -0.375\ {\rm V}, 1\ {\rm ns} < t < 2\ {\rm ns}; -0.1875\ {\rm V}, 2\ {\rm ns} < t < 2\ {\rm ns}; \ldots\}$. **2-7.** $Z_0 = 50\Omega, R_{\rm L} \simeq 42.9\Omega$. (b) $\mathcal{V}_{\rm ctr} = \{0, t < t_d/2; 0.5\ {\rm V}, t_d/2 < t < 3t_d/2; 0.3\ {\rm V}, 3t_d/2 < t < 2t_d; -0.2\ {\rm V}, 2t_d < t < 3t_d; 0, t > 3t_d\}$. **2-9.** $Z_{01} = 50\Omega, l_1 = 10\ {\rm cm}, R_1 = 50\Omega$. **2-11.** $Z_{01} = 75\Omega, l = 37.5\ {\rm cm}, Z_{02} = 125\Omega, R_{\rm L} = \infty$. **2-13.** (a) $\mathcal{V}_{\rm L1} = \mathcal{V}_{\rm L2} = \{0, t < 3\ {\rm ns}; 0.5\ {\rm V}, t > 3\ {\rm ns}\}$. (b) $\mathcal{V}_{\rm L1} = \{0, t < 3\ {\rm ns}; 0.5\ {\rm V}, 3\ {\rm ns} < t < 6\ {\rm ns}; 0.75\ {\rm V}, 6\ {\rm ns} < t < 9\ {\rm ns}; \ldots\}$, $\mathcal{V}_{\rm L2} = \{0, t < 3\ {\rm ns}; 1\ {\rm V}, 3\ {\rm ns} < t < 6\ {\rm ns}; 0.5\ {\rm V}, 6\ {\rm ns} < t < 9\ {\rm ns}; \ldots\}$, $V_{\rm ss} \simeq 0.667\ {\rm V}$. **2-15.** $\mathcal{V}_1 = \{4\ {\rm V}, t < 4\ {\rm ns}; 3.68\ {\rm V}, 4\ {\rm ns} < t < 8\ {\rm ns}; \sim 3.95\ {\rm V}, 8\ {\rm ns} < t < 12\ {\rm ns}; \ldots\}$, $\mathcal{V}_2 = \{0, t < 4\ {\rm ns}; 3.84\ {\rm V}, 4\ {\rm ns} < t < 8\ {\rm ns}; \sim 4.45\ {\rm V}, 8\ {\rm ns} < t < 12\ {\rm ns}; \ldots\}$. **2-17.** (a) $Z_0 = 50\Omega, v_p = 20\ {\rm cm\text{-}(ns)}^{-1}$. (b) Both $\mathcal{V}_{\rm L1}$ and $\mathcal{V}_{\rm L2}$ ring. (c) No ringing. **2-19.** $\mathcal{V}_{\rm L}$ reaches steady state at $t \simeq 750$ ps (reflected and transmitted signals cancel internally). **2-21.** $\mathcal{V}_{\rm L} \simeq \{-1.67\ {\rm V}, t < 1.2\ {\rm ns}; -0.231\ {\rm V}, 1.2\ {\rm ns} < t < 3.6\ {\rm ns}; -1.32\ {\rm V}, 3.6\ {\rm ns} < t < 6\ {\rm ns}; \ldots\}$. **2-23.** (a) Series termination: $R_{\rm T} = 36\Omega$, parallel: $R_{\rm T} = 50\Omega$. (b) No steady-state power dissipation for the series case. **2-25.** Yes, it causes significant reflections. **2-27.** $\mathcal{V}_{\rm L} = \{0, t < 2t_d; 4V_0/3, 2t_d < t < 4t_d; 8V_0/9, 4t_d < t < 6t_d; \ldots V_{\rm ss} = V_0\}$. **2-29.** $\mathcal{V}_{\rm L}(t) = -5e^{(t-3\ {\rm ns})/(0.5\ {\rm ns})}u(t - 3\ {\rm ns})$. **2-31.** (a) $L \simeq 6.67$ nH. **2-33.** $\mathcal{V}_{\rm L} = \{1.6\ {\rm V}, t < 8\ {\rm ns}; 0.4e^{(t-8\ {\rm ns})/(4\ {\rm ns})}\ {\rm V}, t > 8\ {\rm ns}\}$. **2-35.** (a) $\mathcal{V}_{\rm L} = \{0, t < 3t_d; 2.5\ {\rm V}, 3t_d < t < 5t_d; 0, 5t_d < t < 7t_d; -1.25\ {\rm V}, 7t_d < t < 11t_d; \ldots\}$. **2-37.** $\mathcal{V}_{s1} = \{V_0, t < 2t_d; V_0/2, t > 2t_d\}, \mathcal{V}_{s2} = \{0, t < 2t_d; V_0/2, t > 2t_d\}$. **2-39.** (b) Steady state reached in $\sim 13t_d$. **2-41.** (a) $v_p = 20\ {\rm cm\text{-}(ns)}^{-1}$. Since $t_r/t_d = 4$, lumped analysis is valid. (b) Since $t_r/t_d = 0.4$, severe transmission line effects (i.e., ringing) occur. **2-43.** $L \simeq 1.42\mu{\rm H\text{-}m}^{-1}, C \simeq 7.82\ {\rm pF\text{-}m}^{-1}, R \simeq 1.96\Omega{\rm \text{-}m}^{-1}$, $G = 0$, and $Z_0 \simeq 426.5\Omega$.

CHAPTER 3 **3-1.** (a) $l/\lambda \simeq 0.185$. (b) $\sim$ \$2.54 pF. (c) $\sim$ \$4.67 pF. **3-3.** (a) $l_{\rm sc} \simeq 2.4$ cm. (b) $l_{\rm oc} \simeq 5.4$ cm. **3-5.** (a) $Z_{\rm in} = 200(1+j)\Omega$. (b) $Z_{\rm in} = 100(1+j)\Omega$. **3-7.** (a) $S \simeq 3.73$. (b) $z_{\rm max} \simeq -0.417$ cm. (c) $z_{\rm min} \simeq -2.92$ cm. **3-9.** $Z(l \simeq 0.077\lambda) \simeq 252.6\Omega$. **3-11.** 60Ω. **3-13.** (a) $Z_0/2$. (b) 0. (c) $Z_0/4$. **3-15.** (b) $Z_{\rm L} = 25(\sqrt{3} - j)\Omega, \Gamma_{\rm L} \simeq 0.268e^{-j90°}$. **3-17.** (a) $P_{\rm L1} = P_{\rm L2} = 0.25$ W. (b) $P_{\rm L1} = P_{\rm L2} = 0.16$ W. **3-19.** (a) $\sim$0.444 W. (b) 0.5 W. **3-21.** $P_{\rm L} = 1.25$ W. **3-23.** (a) 0%, 50%. (b) 50%, 0%. (c) $\sim$3.92%, $\sim$22.3%. **3-25.** $l \simeq 2.63$ cm, $l_{\rm sc} \simeq 3.27$ cm. **3-27.** (a) $l_{\rm s} = 0.375\lambda$. (b) $S = \infty$. (c) $S \simeq 2.04$. **3-29.** (a) $Z_{\rm Q} \simeq 83.7\Omega$. (b) $S \simeq 1.374$. (c) $S \simeq 1.5$. **3-31.** (a) $Z_{\rm Q1} \simeq 84.1\Omega, Z_{\rm Q2} \simeq 237.8\Omega$. **3-33.** (a) A2 with $C_2 \simeq 2.44$ pF, $L_2 \simeq 15.4$ nH. (b) B1 with $L_1 \simeq 16.2$ nH, $L_2 \simeq 64.2$ nH. **3-35.** (a) $C \simeq 0.107$ pF, $L \simeq 4.27$ nH. (b) $l_{s1} \simeq 1.66$ cm, $l_{s2} \simeq 1.16$ cm. **3-37.** (a) $S = 1$. (b) $S \simeq 6.66$. (c) $S = \infty$. **3-39.** (a) $l = l_{\rm s} \simeq 0.152\lambda = 1.52$ cm. (b) $S = \infty$. **3-41.** $Z_{\rm L} \simeq (85 - j72.1)\Omega$. **3-43.** (a) $L \simeq 1.67\ \mu{\rm H\text{-}m}^{-1}, C \simeq 6.67\ {\rm pF\text{-}m}^{-1}$, and $R \simeq 1.95\Omega{\rm \text{-}m}^{-1}$. (b) $\gamma \simeq 1.945 \times 10^{-3}\ {\rm np\text{-}m}^{-1} + j3.016\ {\rm rad\text{-}m}^{-1}$ and $Z_0 \simeq (500 - j0.323)\Omega$. **3-45.** $L \simeq 126\ {\rm nH\text{-}m}^{-1}, C \simeq 88.5\ {\rm pF\text{-}m}^{-1}, R \simeq 0.33\ \Omega{\rm \text{-}m}^{-1}$, $Z_0 \simeq (37.7 - j0.00787)\Omega$, and $\alpha \simeq 0.00437\ {\rm np\text{-}m}^{-1}$. **3-47.** $R \simeq 241\ \Omega{\rm \text{-}m}^{-1}$, $L \simeq 9.10\ {\rm nH\text{-}(cm)}^{-1}, C \simeq 0.869\ {\rm pF\text{-}(cm)}^{-1}$, and $G \simeq 7.30 \times 10^{-5}\ {\rm S\text{-}(cm)}^{-1}$. **3-49.** (a) $Q \simeq 1839$. (b) $Q \simeq 1442$. (c) $Z_0 \simeq 83.1 - j0.0226\Omega$.

CHAPTER 4 **4-1.** $\sim 10.5\mu$C. **4-3.** $\sim$29.9°. **4-5.** $\sqrt{3}kQ^2/a^2$. **4-7.** $\mathbf{E}_1 \simeq (kQ/a^2)(\hat{\mathbf{y}}0.032 + \hat{\mathbf{z}}0.0871), \mathbf{E}_2 \simeq (kQ/a^2)(\hat{\mathbf{y}}0.032 - \hat{\mathbf{z}}0.0871)$. **4-9.** (0, $\sim$6.21) cm.

4-11. $\Phi(0,0,z) \simeq 1800(5\times10^{-3}+z^2)^{-1/2}$ V; $E_z(0,0,z) \simeq 1800(5\times10^{-3}+z^2)^{-3/2}$ V-m^{-1}. **4-13.** (a) $\Phi_P \simeq 2288$ V. (b) $\mathbf{E}_P \simeq \hat{\mathbf{x}}410 - \hat{\mathbf{y}}995$ V-m^{-1}. **4-15.** (a) $\Phi(1,0) \simeq 0.693$ V. (b) $\mathbf{E} = -\hat{\mathbf{x}}(11/12)$ V-m^{-1}. $\Phi = 0$ along the y axis. $\mathbf{E} = 0$ at infinity. **4-17.** (a) $\mathbf{E} = -\hat{\mathbf{y}}\rho_l/(4\epsilon_0 a)$. (b) $\mathbf{E} = -\hat{\mathbf{y}}\rho_0/(2\pi\epsilon_0 a)$. **4-19.** (a) $\Phi = \infty$. (b) $\mathbf{E}(0,0,z) = \hat{\mathbf{z}}\rho_s z/(2\epsilon_0\sqrt{a^2+z^2})$. **4-21.** (a) $E_r = [q_e/(4\pi\epsilon_0 r^2)]\{1-e^{-2r/a}[2(r/a)^2 + 2(r/a)+1]\}$. (b) $E_r = -q_e e^{-2r/a}[2(r/a)^2 + 2(r/a) + 1]/(4\pi\epsilon_0 r^2)$. **4-23.** (a) $Q = 8\pi\rho_0 a^3/15$. (b) $E_r = 2\rho_0 r/(15\epsilon_0)$, $r < a$; $E_r = 2\rho_0 a^3/(15\epsilon_0 r^2)$, $r > a$. (c) $\Phi = [\rho_0/(5\epsilon_0)][a^2 - r^2/3]$, $r < a$; $\Phi = 2\rho_0 a^3/(15\epsilon_0 r)$, $r > a$. **4-25.** $\mathbf{E}_c = 0$ and $\Phi_c \simeq 90$ kV. **4-27.** (a) $C_{ox} \simeq 1.73$ fF (μm)$^{-2}$. (b) ~431.6 fF. **4-29.** (a) $r_{min} \simeq 16.7$ cm. (b) $r_{min} \simeq 6.67$ cm. (c) $r_{min} \simeq 3.33$ cm. (d) $r_{min} = 2.5$ mm. **4-31.** (a) $C = (9\epsilon_0 A/d)\ln(10)$. (b) $C/C_{air} \simeq 3.91$. **4-33.** (a) $C = \pi(\epsilon_{1r} + \epsilon_{2r})\epsilon_0 l/[\ln(b/a)]$. (b) $C \simeq 9.75$ pF. (c) $C \simeq 5.82$ pF. **4-35.** (a) $C \simeq 709$ μF. (b) $Q \simeq 4.52 \times 10^5$ C and $W \simeq 160$ J. (c) $Q_{max} \simeq 1.35 \times 10^{10}$ C and $W_{max} \simeq 4.8$ MJ. **4-37.** Only $\Phi_2 = e^{-y}\cos x$ since $\nabla^2\Phi_2 = 0$ for $|y| > 0$ and corresponding $\rho_s(x,z) = 2\epsilon_0 \cos x$. **4-39.** (a) ~27.9 kV-m^{-1}. (b) ~51.1 kV. (c) Yes. **4-41.** (a) $W \simeq 2.54$ TJ. (b) $\Phi \simeq 16.9$ GV. (c) $E \simeq 3.39$ MV-m^{-1}.

CHAPTER 5 **5-1.** ~6.24×10^{18} electrons. **5-3.** (a) $E \simeq 26.2$ mV-m^{-1} (b) $\Phi \simeq 26.2$ mV. **5-5** (a) $\mu_e \simeq 42.8$ cm^2-V^{-1}s^{-1}. (b) $v_d \simeq 9.41 \times 10^{-5}$ m-s^{-1}. **5-7.** (a) $R \simeq 0.0185\Omega$-m^{-1}. (b) $R \simeq 0.022\Omega$-m^{-1}. (c) $R \simeq 0.0254\Omega$-m^{-1}. **5-9.** (a) $N_i \simeq 6.13 \times 10^9$ cm^{-3}, $\sigma \simeq 1.87 \times 10^{-4}$ S-m^{-1}. (b) $N_i \simeq 2.25 \times 10^{13}$ cm^{-3}, $\sigma \simeq 2.09$ S-m^{-1}. **5-11.** $R \simeq 85.5\Omega$, $R_{\text{pure Si}} \simeq 5.36 \times 10^8\Omega$. **5-13.** (c) $R = 1.5$ kΩ. **5-15.** $R_{sq} \simeq 347\Omega$ sq^{-1}, $l \simeq 17.3$μm. **5-17.** ~14.3 μm. **5-19.** (a) 4, ~6.67, ~8.33, 10, ~11.9, fS-(mm)$^{-1}$. **5-21.** $R \simeq 12.7$ kΩ. **5-23.** $R \simeq 11$ kΩ-(km)$^{-1}$. **5-25.** (a) $\Phi_{12} \simeq 133$ kV, (b) $\Phi_{12} \simeq 14.5$ kV. **5-27.** (a) $E_1 = [\sigma_2/(\sigma_1+\sigma_2)](V_0/d)$, $E_2 = [\sigma_1/(\sigma_1+\sigma_2)](V_0/d)$. (b) $\rho_s = (V_0/d)(\epsilon_2\sigma_1 - \epsilon_1\sigma_2)/(\sigma_1+\sigma_2)$. **5-29.** $R \simeq \ln(h/a)/(2\pi\sigma)$.

CHAPTER 6 **6-1.** $I \simeq 1118$ A. **6-3.** $F = 4000$ N-m^{-1}. **6-5.** Expand. **6-7.** (a) $\mathbf{F}_1 = -\mathbf{F}_3 = \hat{\mathbf{z}}15$ N, $\mathbf{F}_2 = \mathbf{F}_4 = 0$. (b) $\mathbf{F}_1 = -\mathbf{F}_3 = \hat{\mathbf{x}}15$ N, $\mathbf{F}_2 = -\mathbf{F}_4 = \hat{\mathbf{y}}15$ N. **6-9.** (a) $\mathbf{B} = \hat{\mathbf{z}}[\mu_0 I/(4\pi a)]$. (b) $\mathbf{B} = \hat{\mathbf{z}}[\mu_0 I/(4\pi a)]$. (c) $\mathbf{B} = (\hat{\mathbf{x}} + \hat{\mathbf{y}})[\mu_0 I/(4\pi a)]$. **6-13.** $B = [\mu_0\phi_0 I(a^{-1}-b^{-1})/(4\pi)]$, into the page. **6-15.** $B_\phi = \{2r$ T, $r \leq 10^{-2}$ m; $2\times10^{-4}r^{-1}$ T, $r > 10^{-2}$ m$\}$. **6-17.** $N \simeq 56$. **6-21.** (a) $\mathbf{B}_{ctr} \simeq \hat{\mathbf{z}}1.09$ mT. (b) $\mathbf{B}_{end} \simeq \hat{\mathbf{z}}0.55$ mT. **6-23.** $\mathbf{B}_{ctr} \simeq \hat{\mathbf{z}}46.8$ G. **6-25.** $\mathbf{B}_{ctr} \simeq \hat{\mathbf{z}}31.2$ G. **6-27.** (a) $\Psi = \mu_0 a l \ln 3/(2\pi)$. (b) ~7% reduction. **6-29.** $\Psi \simeq 1.22 \times 10^{-2}$ Wb. **6-31.** (a) $A_z = [\mu_0 I/(4\pi)]\{\ln[(z-a) + \sqrt{r^2+(z-a)^2}] - \ln[(z+a) + \sqrt{r^2+(z+a)^2}]\}$. **6-33.** $\mathbf{B}_1 = \mathbf{B}_2 = \hat{\mathbf{z}}\mu_0 J_0 d/2$. **6-35.** (a) $L \simeq 41.8$ mH. (b) $L \simeq 38$ mH. **6-37.** (a) $N \simeq 141$. (b) $N \simeq 10$. **6-41.** (a) $L \simeq 0.207$ μH. (b) $L \simeq 0.676$ μH. **6-43.** $L_{12} = \mu_0(d - \sqrt{d^2 - a^2})$. **6-45.** (a) $L \simeq 0.329$ mH. (b) $W \simeq 16.4$ mJ. **6-47.** (a) ~5 circular turns. (b) ~169 circular turns. **6-49.** 10^5 N-m^{-1}. This huge repulsion force causes the container of the wires to explode.

CHAPTER 7 **7-1.** (a) $\mathcal{V}_{ind} = B_0 A e^{-\alpha t}(1 - \alpha t)$. (b) $\mathcal{V}_{ind} = B_0 A e^{-\alpha t}[-\alpha\sin(\omega t) + \omega\cos(\omega t)]$. **7-3.** (a) $\mathcal{V}_{ind2} \simeq 7.72$ mV. (b) $\mathcal{V}_{ind2} \simeq 77.2$ mV. **7-5.** $I_0 \simeq 6.96$ A. **7-7.** $\mathcal{V}_{ind} \simeq 8.43$ V. **7-9.** $I = (B_0 l/R)\cos(\omega_0 t)[x_0\omega_0 + 2v_0\sin(\omega_0 t)]$. **7-11.** $\mathcal{V}_{ind} = \{\mu_0 I a b v_0/[2\pi(r_1 + v_0 t)(r_1 + v_0 t + a)]\}$, $t \geq 0$. **7-13.** $\mathcal{V}_{ind} \simeq 39.3\sin(50\pi t + \zeta)$ mV. **7-15.** (a) $P_{av} \simeq 190$ W. (b) No, using multiple turns does not increase power. **7-17.** (a) $\mathcal{V}_{ind} = \mu_0 b I_0 e^{-t}\ln[(a/d)+1]/(2\pi)$. (b) $\mathcal{V}_{ind} = -\mu_0 b\omega I_0\cos(\omega t)\ln[(a/d)+1]/(2\pi)$. **7-19.** (a) $\mathcal{H}_z(y, 2t_1) = \{0, y < y_1 - v_p t_1; H_0, y_1 - v_p t_1 < y < y_1 +$

$v_pt_1; H_0[1 - (y - y_1 - v_pt_1)/(v_pt_1)], y_1 + v_pt_1 < y < y_1 + 2v_pt_1; 0, y > y_1 + 2v_pt_1\}$. **7-21.** (a) $\Delta t \simeq 17$ ns. (b) $t_{\max} \simeq 3.33$ μs, $E_{\max} = 75$ V-m^{-1}. (c) No signal detected. **7-23.** (a) $\overline{\mathscr{E}}(y,t) = \hat{\mathbf{z}}5\cos(\omega t - 40\pi y)$ V-m^{-1}. (b) $\overline{\mathscr{B}}(z,t) = \hat{\mathbf{x}}0.1\cos(\omega t - 2\pi z) + \hat{\mathbf{y}}0.3\sin(\omega t - 2\pi z)$ μT. (c) $\overline{\mathscr{E}}(x,t) = \hat{\mathbf{z}}0.1[\cos(\omega t - 18x) - 0.5\cos(\omega t + 18x)]$ V-m^{-1}. (d) $\overline{\mathscr{H}}(x,z,t) = \hat{\mathbf{y}}\cos(\omega t - 48\pi x + 64\pi z)$ mA-m^{-1}. (e) $\overline{\mathscr{J}}(y,t) = \hat{\mathbf{x}}40e^{-0.1y}\cos(\omega t - 0.1y + \pi/3)$ μA-m^{-2}. **7-25.** (a) 40 μA-m^{-2}, ~60.8 μA-m^{-2}. (b) ~18 μA-m^{-2}, ~27.3 μA-m^{-2}. (c) ~1.34×10^{-2} μA-m^{-2}, ~2.04×10^{-2} μA-m^{-2}. **7-27.** $f \simeq 600$ kHz. **7-29.** $\beta = 0.025$ rad-m^{-1}, $\eta \simeq 377\Omega$. **7-31.** (a) $\beta = 70\pi$. (b) $\overline{\mathscr{H}} \simeq \hat{\mathbf{y}}252\cos(21\times10^9\pi t)\cos(70\pi z)$ μA-m^{-1}. **7-33.** (a) $\mathbf{H} = \hat{\boldsymbol{\phi}}E_0e^{-j\beta z}/(r\sqrt{\mu/\epsilon})$. (b) $\overline{\mathscr{E}} = \hat{\mathbf{r}}E_0\cos(\omega t - \beta z)/r$ and $\overline{\mathscr{H}} = \hat{\boldsymbol{\phi}}E_0\cos(\omega t - \beta z)/(r\sqrt{\mu/\epsilon})$. **7-35.** (a) $\text{SAR}_{\text{peak}} \simeq 5.95$ W-(kg)$^{-1}$. (b) $\text{SAR}_{\text{av}} \simeq 4.68$ W-(kg)$^{-1}$. **7-37.** (a) $I(t) \simeq 94.3\cos(6\pi\times10^9t + 89.98°)$ mA. (b) $P \simeq 73$ μW.

CHAPTER 8 **8-1.** (a) $\overline{\mathscr{E}}(y,t) = \hat{\mathbf{z}}0.1\cos(24\pi\times10^7t - 0.8\pi y + 144°)$ V-m^{-1}. (d) $t_{\max} \simeq 0.833$ ns. **8-3.** (a) $\lambda \simeq 1.67$ cm, $\beta = 120\pi$ rad-m^{-1}. (b) $\mathbf{H}(y) = \hat{\mathbf{z}}(24\pi)^{-1}e^{j120\pi y}$ A-m^{-1}. **8-5.** (a) $\mathbf{H}(r,\theta) = \hat{\boldsymbol{\phi}}[j\beta I_0 l\sin\theta e^{-j\beta r}/(4\pi r)]$ (b) $\mathbf{S}_{\text{av}} = \hat{\mathbf{r}}[30\beta^2I_0^2l^2\sin^2\theta/(4\pi r^2)]$ (c) $P_{\text{t}} = 20\beta^2I_0^2l^2$. **8-7.** (a) $H_0 \simeq 0.23$ mA-m^{-1}, $a \simeq 2.96$ rad-m^{-1}. (b) $\overline{\mathscr{E}}(x,y,t) \simeq 61.3[-\hat{\mathbf{x}} + \hat{\mathbf{y}}]\cos(4\pi\times10^8t - 2.96x - 2.96y - \pi/6)$ mV-m^{-1}. (c) (i) ~61.3 mV-m^{-1}, (ii) ~61.3 mV-m^{-1}, and (iii) 0. **8-9.** $\overline{\mathscr{H}}(y,z,t) = \hat{\mathbf{x}}(10\pi)^{-1}\cos(24\pi\times10^9t - 40\sqrt{3}\pi y + 40\pi z + \pi/3)$ A-m^{-1}, $\overline{\mathscr{E}}(y,z,t) = 6(\hat{\mathbf{y}} + \hat{\mathbf{z}}\sqrt{3})\cos(24\pi\times10^9t - 40\sqrt{3}\pi y + 40\pi z + \pi/3)$ V-m^{-1}. **8-11.** (c) $d \simeq 10.4$ cm. **8-13.** (a) $\alpha \simeq 0.0653$ np-m^{-1}. (b) $d \simeq 15.3$ m. (c) ~−0.00567 dB and ~−0.567 dB. **8-15.** (b) ~$(579.4/\sqrt{f})$ m. **8-17.** (a) $f \simeq 9.89$ Hz. (b) $f \simeq 39.6$ Hz. (c) $d \simeq 116$ m. (d) $d_{\max} \simeq 110$ m. **8-19.** $d_{\text{EWSG}} \simeq 2.83$ cm and $d_{\text{EGP}} \simeq 7.77$ cm, therefore EWSG has a higher absorption rate. **8-21.** (a) ~2.48 cm. (b) ~5.26 cm. **8-23.** ~−25.7 dB. **8-25.** (a) $|\mathbf{S}_{\text{av}}|_{\text{in}} \simeq 25$ mW-m^{-2}. (b) $|\mathbf{S}_{\text{av}}|_{\text{out}} \simeq 8.38$ μW-m^{-2}. **8-27.** (a) $\overline{\mathscr{E}}(z,t) = \hat{\mathbf{x}}120\pi\cos[\omega(t - z/c)]$ V-m^{-1}. $V(t) = 12\pi\cos[\omega(t - z/c)]$ V. (b) $\rho_s(z,t) = (J_0/c)\cos[\omega(t - z/c)]$ C-m^{-2}. **8-29.** (a) All less than 0.533 mW-(cm)$^{-2}$ at 800 MHz. (b) $P_{\max} = 85$ W. **8-31.** (a) $f = 2.5$ GHz, $|\mathbf{S}_{\text{av}}| = 3$ mW-m^{-2}. (b) $\overline{\mathscr{E}}_1(y,t) = \hat{\mathbf{z}}3\sin(50\pi y/3)\sin(5\pi\times10^9t)$ V-m^{-1}. (c) $\overline{\mathscr{H}}_1(y,t) = \hat{\mathbf{x}}(40\pi)^{-1}\cos(50\pi/3)\cos(5\pi\times10^9t)$ A-m^{-1}. (d) $|\mathbf{H}(y = -75\text{ cm})|_{\max} = 0$. **8-33.** (a) $f = 6$ GHz, $\lambda = 5$ cm. (b) $\mathbf{H}_{\text{i}}(x) = 0.5\pi^{-1}e^{-j40\pi x}(\hat{\mathbf{z}} + j\hat{\mathbf{y}})$ A-m^{-1}. (c) $\mathbf{E}_{\text{r}}(x) = 60e^{j40\pi x}(-\hat{\mathbf{y}} + j\hat{\mathbf{z}})$ V-m^{-1}, $\mathbf{H}_{\text{r}}(x) = 0.5\pi^{-1}e^{j40\pi x}(\hat{\mathbf{z}} + j\hat{\mathbf{y}})$ A-m^{-1}. **8-35.** $\epsilon_2/\epsilon_1 = 9$. **8-37.** $d_{\max} \simeq 7$ m. **8-39.** (a) ~999.9125 W-m^{-2}. (b) ~87.5 mW-m^{-2}. (c) ~1.46×10^{-11} W-m^{-2}. **8-41.** (a) $d_{\min} \simeq 1.13$ cm. (b) 1.08% at both 8.5 and 10.3 GHz. **8-43.** (a) 100%. (b) 100% at 400 nm; 87.8% at 600 nm **8-45.** (a) $\Gamma_{\text{eff}} \simeq -0.02$ at 4 μm and $\simeq 0.469e^{143°}$ at 8 μm. (b) $\Gamma_{\text{eff}} = -0.6$ at 4 μm and $\simeq -0.02$ at 8 μm. **8-47.** (a) $d_{\min} \simeq 16.8$ cm. (b) ~−0.807. **8-49.** (a) ~100%. (b) 62.2% **8-51.** (a) TEM. (b) TEM. (c) TEM for $f <$ 2 kHz and TEM, TE_1, TM_1 for $f < 5$ kHz. **8-53.** (a) Evanescent TE_2 mode. (b) $f = 12$ GHz. (c) TE_1. **8-57.** (a) $f/f_{c_m} \simeq 1.6$. (b) $\alpha_{c_{\text{TEM}}} = \alpha_{c_{\text{TE}_m}} = R_s/(\eta a)$, $\alpha_{c_{\text{TM}_m}} \simeq 2.56R_s/(\eta a)$. **8-59.** (a) For TEM mode, $\bar{v}_p = 2\times10^8$ m-s^{-1} and $\bar{\lambda} = 2$ cm. For TE_1 and TM_1 modes, $\bar{v}_p \simeq 2.56\times10^8$ m-s^{-1} and $\bar{\lambda} \simeq 2.56$ cm. (b) $\alpha_{c_{\text{TEM}}} \simeq 2.5\times10^{-6}$ np-m^{-1}; $\alpha_{c_{\text{TE}_1}} \simeq 2.5\times10^{-6}$ np-m^{-1}; $\alpha_{c_{\text{TM}_1}} \simeq 6.41\times10^{-6}$ np-m^{-1}; $\alpha_d \simeq 8.05\times10^{-2}$ np-m^{-1}. **8-61.** (b) $Z_0 \simeq 77\Omega$.

Index